PEARSON

ALWAYS LEARNING

Lisa A. Urry • Michael L. Cain • Steven A. Wasserman
Peter V. Minorsky • Robert B. Jackson • Jane B. Reece

Campbell Biology in Focus

Custom Edition for Anne Arundel Community College for BIO 101

Taken from:
Campbell Biology in Focus
by Lisa A. Urry, Michael L. Cain, Steven A. Wasserman,
Peter V. Minorsky, Robert B. Jackson, and Jane B. Reece

Cover Art: Courtesy of PhotoDisc/Getty Images.

Taken from:

Campbell Biology in Focus
by Lisa A. Urry, Michael L. Cain, Steven A. Wasserman, Peter V. Minorsky, Robert B. Jackson, and Jane B. Reece
Copyright © 2014 by Pearson Education, Inc.
New York, New York 10013

This special edition published in cooperation with Pearson Learning Solutions.

Pearson Learning Solutions, 330 Hudson Street, New York, New York 10013
A Pearson Education Company
www.pearsoned.com

Printed in the United States of America

3 4 5 6 7 8 9 10 V0UD 19 18 17 16 15

000200010271982770

AD

ISBN 10: 1-323-23910-3
ISBN 13: 978-1-323-23910-0

Preface

The short-toed snake eagle (*Circaetus gallicus*) that gazes from the cover of this book has an eye much like our own, yet evolutionary forces have honed its ability to spot a snake from a quarter mile up in the air. The eagle's keen eye is a metaphor for our goal in writing this text: to focus with high intensity on the core concepts that biology majors need to master in the introductory biology course. The current explosion of biological information, while exhilarating in its scope, poses a significant challenge—how best to teach a subject that is constantly expanding its own boundaries. In particular, instructors have become increasingly concerned that their students are overwhelmed by a growing volume of detail and are losing sight of the important ideas in biology.

In response to this challenge, groups of biologists have initiated efforts to refine and in some cases redesign the introductory biology course, summarizing their findings in reports that include *Bio 2010: Transforming Undergraduate Education for Future Research Biologists*[1] and *Vision and Change in Undergraduate Biology Education.*[2] Clear recommendations emerging from these initiatives are to focus course material and instruction on key ideas while transforming the classroom through active learning and scientific inquiry. Many instructors have embraced such approaches and changed how they teach. Cutting back on the amount of detail they present, they focus on core biological concepts, explore select examples, and engage in a rich variety of active learning exercises. We were inspired by the ongoing changes in biology education to develop this text, *CAMPBELL BIOLOGY IN FOCUS*. Based on the best-selling *CAMPBELL BIOLOGY*, this new, shorter textbook provides undergraduate biology majors and their instructors with a more focused exploration of the key questions, approaches, and ideas of modern biology.

Our Guiding Principles

Our objective in creating *CAMPBELL BIOLOGY IN FOCUS* was to produce a shorter text by streamlining selected material, while emphasizing conceptual understanding and maintaining clarity, proper pacing, and rigor. Here, briefly, are the four guiding principles for our approach.

Focus on Core Concepts and Skills

We developed this text to help students master the fundamental content and scientific skills they need as college biology majors. In structuring the text, we were guided by discussions with many biology professors, analysis of hundreds of syllabi, study of the debates in the literature of scientific pedagogy, and our experience as instructors at a range of institutions. The result is a **briefer book for biology majors** that is designed to inform, engage, and inspire.

Evolution as the Foundation of Biology

Evolution is the central theme of all biology, and it is the core theme of this text, as exemplified by the various ways that evolution is integrated into the text:

- Every chapter explicitly addresses the topic of evolution through an **Evolution section** that leads students to consider the material in the context of natural selection and adaptation.
- Each Chapter Review includes a **Focus on Evolution Question** that asks students to think critically about how an aspect of the chapter relates to evolution.
- Evolution is the unifying idea of **Chapter 1**, *Introduction: Evolution and the Foundations of Biology*, which outlines five key themes that students will encounter throughout the text and introduces the process of scientific inquiry.
- Following the in-depth coverage of evolutionary mechanisms in Unit 3, evolution also provides the storyline for the **novel approach to presenting biological diversity** in Unit 4, The Evolutionary History of Life. Focusing on landmark events in the history of life, the text highlights how

---

[1] The National Research Council of the National Academies, 2003

[2] The American Association for the Advancement of Science, supported by the National Science Foundation, the National Institutes of Health, and Howard Hughes Medical Institute, 2009

key adaptations arose within groups of organisms and how evolutionary events led to the diversity of life on Earth today.

Engaging Students in Scientific Thinking

Helping students learn to "think like a scientist" is a nearly universal goal of introductory biology courses. Students need to understand how to formulate and test hypotheses, design experiments, and interpret data. Scientific thinking and data interpretation skills top lists of learning outcomes and foundational skills desired for students entering higher-level courses. CAMPBELL BIOLOGY IN FOCUS meets this need in several ways:

- **Scientific Skills Exercises** in every chapter use real data to build skills in graphing, interpreting data, designing experiments, and working with math—skills essential for students to succeed in biology. These exercises can also be assigned and automatically graded in MasteringBiology.
- **Scientific Inquiry Questions** in the end-of-chapter material give students further practice in scientific thinking.
- **Inquiry Figures** and **Research Method Figures** reveal *how* we know *what* we know and model the process of scientific inquiry.

Outstanding Pedagogy

Since the publication of the first edition in 1987, CAMPBELL BIOLOGY has been praised for its clear and engaging narrative, superior pedagogy, and innovative use of art to promote student learning. These hallmark values are also at the core of CAMPBELL BIOLOGY IN FOCUS:

- In each chapter, a framework of carefully selected **Key Concepts** helps students distinguish the "forest" from the "trees."
- Questions throughout the text catalyze learning by encouraging students to **actively engage with and synthesize key material**:
  - To counter students' tendencies to compartmentalize information, **Make Connections Questions** ask students to connect what they are learning in a particular chapter to material covered in other chapters or units.
  - **Figure Legend Questions** foster student interaction with the figures.
  - Tiered **Concept Check Questions** test comprehension, require application, and prompt synthesis.
  - **Draw It Exercises** encourage students to test their understanding of biology through drawing.
  - **Summary of Key Concepts Questions** make reading the summary an active learning experience.

Our overall aim is to help students see biology as a whole, with each chapter adding to the network of knowledge they are building. To support this goal further, each unit in CAMPBELL

BIOLOGY IN FOCUS opens with a **visual preview** that tells the story of the chapters' contents, showing how the material in the unit fits into a larger context.

Organization of the Text

CAMPBELL BIOLOGY IN FOCUS is organized into an introductory chapter and seven units that cover thoughtfully paced core concepts. In the course of streamlining this material, we have worked diligently to maintain the finely tuned coverage of fundamental concepts found in CAMPBELL BIOLOGY. As we developed this alternative text, we carefully considered each chapter of CAMPBELL BIOLOGY. Based on surveys and discussions with instructors and analyses of hundreds of syllabi and reviews, we made informed choices about how to design each chapter of CAMPBELL BIOLOGY IN FOCUS to meet the needs of instructors and students. In some chapters, we retained most of the material; in other chapters, we pruned material; and in still others, we completely reconfigured the material. We summarize the highlights here.

Chapter 1: Introduction: Evolution and the Foundations of Biology

Chapter 1 introduces the **five biological themes** woven throughout this text: the core theme of **Evolution**, together with **Organization**, **Information**, **Energy and Matter**, and **Interactions**. Chapter 1 also explores the process of scientific inquiry through a case study describing experiments on the evolution of coat color in the beach mouse. The chapter concludes with a discussion of the importance of diversity within the scientific community.

Unit 1: Chemistry and Cells

A succinct, two-chapter treatment of basic chemistry provides the foundation for this unit focused on cell structure and function. The related topics of cell membranes and cell signaling are consolidated into one chapter. Due to the importance of the fundamental concepts in Units 1 and 2, much of the material in the rest of these two units has been retained from CAMPBELL BIOLOGY.

Unit 2: Genetics

Topics in this unit include meiosis and classical genetics as well as the chromosomal and molecular basis for genetics and gene expression. We also include a chapter on the regulation of gene expression and one on the role of gene regulation in development, stem cells, and cancer. Methods in biotechnology are integrated into appropriate chapters. The stand-alone chapter on viruses can be taught at any point in the course. The final chapter in the unit, on genome evolution, provides both a capstone for the study of genetics and a bridge to the evolution unit.

Unit 3: Evolution

This unit provides in-depth coverage of essential evolutionary topics, such as mechanisms of natural selection, population genetics, and speciation. Early in the unit, Chapter 20 introduces "tree thinking" to support students in interpreting phylogenetic trees and thinking about the big picture of evolution. Chapter 23 focuses on mechanisms that have influenced long-term patterns of evolutionary change. Throughout the unit, new discoveries in fields ranging from paleontology to phylogenomics highlight the interdisciplinary nature of modern biology.

Unit 4: The Evolutionary History of Life

This unit employs a novel approach to studying the evolutionary history of biodiversity. Each chapter focuses on one or more major steps in the history of life, such as the origin of cells or the colonization of land. Likewise, the coverage of natural history and biological diversity emphasizes the evolutionary process—how factors such as the origin of key adaptations have influenced the rise and fall of different groups of organisms over time.

Unit 5: Plant Form and Function

The form and function of higher plants are often treated as separate topics, thereby making it difficult for students to make connections between the two. In Unit 5, plant anatomy (Chapter 28) and the acquisition and transport of resources (Chapter 29) are bridged by a discussion of how plant architecture influences resource acquisition. Chapter 30 provides a solid introduction to plant reproduction. It also explores crop domestication, examining controversies surrounding the genetic engineering of crop plants. The final chapter explores how environmental sensing and the integration of information by plant hormones influence plant growth and reproduction.

Unit 6: Animal Form and Function

A focused exploration of animal physiology and anatomy applies a comparative approach to a limited set of examples to bring out fundamental principles and conserved mechanisms. Students are first introduced to the closely related topics of homeostasis and endocrine signaling in an integrative introductory chapter. Additional melding of interconnected material is reflected in chapters that combine treatment of circulation and gas exchange, reproduction and development, neurons and nervous systems, and motor mechanisms and behavior.

Unit 7: Ecology

This unit applies the key themes of the text, including evolution, interactions, and energy and matter, to help students learn ecological principles. Chapter 40 integrates material on population growth and Earth's environment, highlighting the importance of both biological and physical processes in determining where species are found. Chapter 43 ends the book with a focus on global ecology and conservation biology. This chapter illustrates the threats to all species from increased human population growth and resource use. It begins with local factors that threaten individual species and ends with global factors that alter ecosystems, landscapes, and biomes.

MasteringBiology®

MasteringBiology is the most widely used online assessment and tutorial program for biology, providing an extensive library of homework assignments that are graded automatically. Self-paced tutorials provide individualized coaching with specific hints and feedback on the most difficult topics in the course. For example:

- The **Scientific Skills Exercises** from the text can be assigned and automatically graded in MasteringBiology.
- **Make Connections Tutorials** help students connect what they are learning in one chapter with material they have learned in another chapter.
- **Data Analysis Tutorials** allow students to analyze real data from online databases.
- **BioFlix® Tutorials** use 3-D movie-quality animations to help students master tough topics.

In addition, Reading Quiz questions, Student Misconception questions, and approximately 3,000 Test Bank questions are available for assignment.

MasteringBiology and the text work together to provide an unparalleled learning experience.

* * *

Our overall goal in developing this text was to assist instructors and students in their exploration of biology by emphasizing essential content and skills while maintaining rigor. Although this first edition is now completed, we recognize that CAMPBELL BIOLOGY IN FOCUS, like its subject, will evolve. As its authors, we are eager to hear your thoughts, questions, comments, and suggestions for improvement. We are counting on you—our teaching colleagues and all students using this book—to provide us with this feedback, and we encourage you to contact us directly by e-mail:

Lisa Urry (Chapter 1, Units 1 and 2): lurry@mills.edu

Michael Cain (Chapter 1, Units 3 and 4): mcain@bowdoin.edu

Peter Minorsky (Unit 5): pminorsky@mercy.edu

Steven Wasserman (Chapter 1, Unit 6): stevenw@ucsd.edu

Rob Jackson (Unit 7): jackson@duke.edu

Jane Reece: janereece@cal.berkeley.edu

Michael L. Cain

Michael Cain (Chapter 1 and Units 3 and 4) is an ecologist and evolutionary biologist who is now writing full-time. Michael earned a joint degree in biology and math at Bowdoin College, an M.Sc. from Brown University, and a Ph.D. in ecology and evolutionary biology from Cornell University. As a faculty member at New Mexico State University and Rose-Hulman Institute of Technology, he taught a wide range of courses, including introductory biology, ecology, evolution, botany, and conservation biology. Michael is the author of dozens of scientific papers on topics that include foraging behavior in insects and plants, long-distance seed dispersal, and speciation in crickets. In addition to his work on CAMPBELL *BIOLOGY IN FOCUS*, Michael is also the lead author of an ecology textbook.

The author team's contributions reflect their biological expertise as researchers and their teaching sensibilities gained from years of experience as instructors at diverse institutions. They are also experienced textbook authors, having written CAMPBELL *BIOLOGY* in addition to CAMPBELL *BIOLOGY IN FOCUS*.

Steven A. Wasserman

Steve Wasserman (Chapter 1 and Unit 6) is Professor of Biology at the University of California, San Diego (UCSD). He earned his A.B. in biology from Harvard University and his Ph.D. in biological sciences from MIT. Through his research on regulatory pathway mechanisms in the fruit fly *Drosophila*, Steve has contributed to the fields of developmental biology, reproduction, and immunity. As a faculty member at the University of Texas Southwestern Medical Center and UCSD, he has taught genetics, development, and physiology to undergraduate, graduate, and medical students. He currently focuses on teaching introductory biology. He has also served as the research mentor for more than a dozen doctoral students and more than 50 aspiring scientists at the undergraduate and high school levels. Steve has been the recipient of distinguished scholar awards from both the Markey Charitable Trust and the David and Lucille Packard Foundation. In 2007, he received UCSD's Distinguished Teaching Award for undergraduate teaching.

Lisa A. Urry

Lisa Urry (Chapter 1 and Units 1 and 2) is Professor of Biology and Chair of the Biology Department at Mills College in Oakland, California, and a Visiting Scholar at the University of California, Berkeley. After graduating from Tufts University with a double major in biology and French, Lisa completed her Ph.D. in molecular and developmental biology at Massachusetts Institute of Technology (MIT) in the MIT/Woods Hole Oceanographic Institution Joint Program. She has published a number of research papers, most of them focused on gene expression during embryonic and larval development in sea urchins. Lisa has taught a variety of courses, from introductory biology to developmental biology and senior seminar. As a part of her mission to increase understanding of evolution, Lisa also teaches a nonmajors course called Evolution for Future Presidents and is on the Teacher Advisory Board for the Understanding Evolution website developed by the University of California Museum of Paleontology. Lisa is also deeply committed to promoting opportunities for women and underrepresented minorities in science.

Peter V. Minorsky

Peter Minorsky (Unit 5) is Professor of Biology at Mercy College in New York, where he teaches introductory biology, evolution, ecology, and botany. He received his A.B. in biology from Vassar College and his Ph.D. in plant physiology from Cornell University. He is also the science writer for the journal *Plant Physiology*. After a postdoctoral fellowship at the University of Wisconsin at Madison, Peter taught at Kenyon College, Union College, Western Connecticut State University, and Vassar College. His research interests concern how plants sense environmental change. Peter received the 2008 Award for Teaching Excellence at Mercy College.

Robert B. Jackson

Rob Jackson (Unit 7) is Professor of Biology and Nicholas Chair of Environmental Sciences at Duke University. Rob holds a B.S. in chemical engineering from Rice University, as well as M.S. degrees in ecology and statistics and a Ph.D. in ecology from Utah State University. Rob directed Duke's Program in Ecology for many years and just finished a term as the Vice President of Science for the Ecological Society of America. Rob has received numerous awards, including a Presidential Early Career Award in Science and Engineering from the National Science Foundation. He also enjoys popular writing, having published a trade book about the environment, *The Earth Remains Forever*, and two books of poetry for children, *Animal Mischief* and *Weekend Mischief*.

Jane B. Reece

The head of the author team for recent editions of *CAMPBELL BIOLOGY*, Jane Reece was Neil Campbell's longtime collaborator. Earlier, Jane taught biology at Middlesex County College and Queensborough Community College. She holds an A.B. in biology from Harvard University, an M.S. in microbiology from Rutgers University, and a Ph.D. in bacteriology from the University of California, Berkeley. Jane's research as a doctoral student and postdoctoral fellow focused on genetic recombination in bacteria. Besides her work on the Campbell textbooks for biology majors, she has been an author of *Campbell Biology: Concepts & Connections*, *Campbell Essential Biology*, and *The World of the Cell*.

Neil A. Campbell

Neil Campbell (1946–2004) combined the investigative nature of a research scientist with the soul of an experienced and caring teacher. He earned his M.A. in zoology from the University of California, Los Angeles, and his Ph.D. in plant biology from the University of California, Riverside, where he received the Distinguished Alumnus Award in 2001. Neil published numerous research articles on desert and coastal plants and how the sensitive plant (*Mimosa*) and other legumes move their leaves. His 30 years of teaching in diverse environments included introductory biology courses at Cornell University, Pomona College, and San Bernardino Valley College, where he received the college's first Outstanding Professor Award in 1986. Neil was a visiting scholar in the Department of Botany and Plant Sciences at the University of California, Riverside.

Focus on the Big Picture

See the Story of the Unit

Each unit begins with a **visual preview** of the chapters' contents, showing how the material in the unit fits into a larger context.

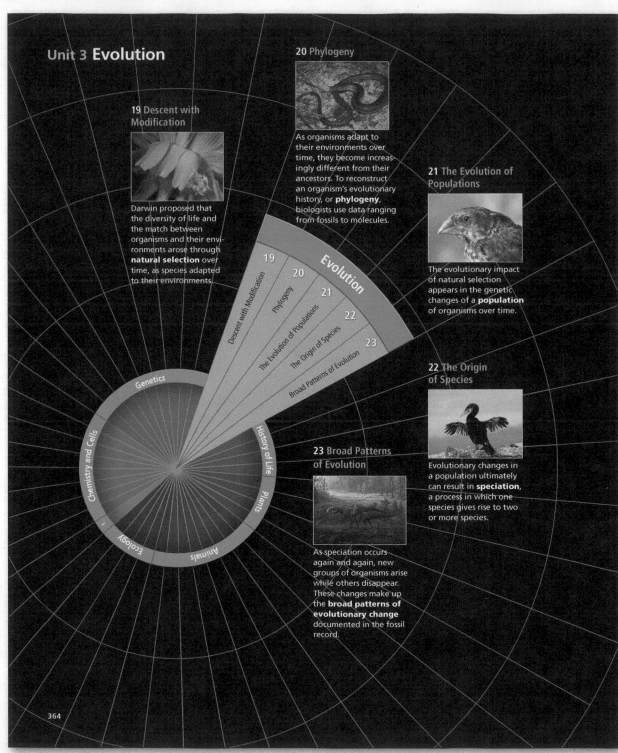

Unit 3 Evolution

19 Descent with Modification

Darwin proposed that the diversity of life and the match between organisms and their environments arose through **natural selection** over time, as species adapted to their environments.

20 Phylogeny

As organisms adapt to their environments over time, they become increasingly different from their ancestors. To reconstruct an organism's evolutionary history, or **phylogeny**, biologists use data ranging from fossils to molecules.

21 The Evolution of Populations

The evolutionary impact of natural selection appears in the genetic changes of a **population** of organisms over time.

22 The Origin of Species

Evolutionary changes in a population ultimately can result in **speciation**, a process in which one species gives rise to two or more species.

23 Broad Patterns of Evolution

As speciation occurs again and again, new groups of organisms arise while others disappear. These changes make up the **broad patterns of evolutionary change** documented in the fossil record.

364

Focus on the Key Concepts

Each chapter is organized around a framework of 3 to 6 **Key Concepts** that focus on the big picture and provide a context for the supporting details.

Students can get oriented by reading the **list of Key Concepts,** which introduces the big ideas covered in the chapter.

Each **Key Concept** serves as the heading for a major section of the chapter.

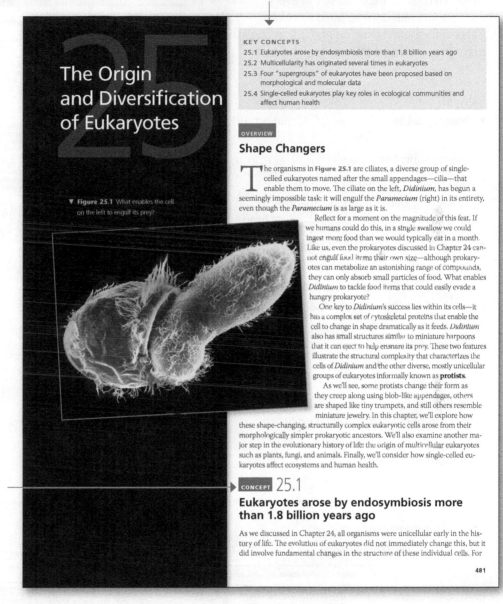

The Origin and Diversification of Eukaryotes

25

KEY CONCEPTS

25.1 Eukaryotes arose by endosymbiosis more than 1.8 billion years ago

25.2 Multicellularity has originated several times in eukaryotes

25.3 Four "supergroups" of eukaryotes have been proposed based on morphological and molecular data

25.4 Single-celled eukaryotes play key roles in ecological communities and affect human health

OVERVIEW

Shape Changers

The organisms in **Figure 25.1** are ciliates, a diverse group of single-celled eukaryotes named after the small appendages—cilia—that enable them to move. The ciliate on the left, *Didinium*, has begun a seemingly impossible task: it will engulf the *Paramecium* (right) in its entirety, even though the *Paramecium* is as large as it is.

Reflect for a moment on the magnitude of this feat. If we humans could do this, in a single swallow we could ingest more food than we would typically eat in a month. Like us, even the prokaryotes discussed in Chapter 24 cannot engulf food items their own size—although prokaryotes can metabolize an astonishing range of compounds, they can only absorb small particles of food. What enables *Didinium* to tackle food items that could easily evade a hungry prokaryote?

One key to *Didinium*'s success lies within its cells—it has a complex set of cytoskeletal proteins that enable the cell to change in shape dramatically as it feeds. *Didinium* also has small structures similar to miniature harpoons that it can eject to help ensnare its prey. These two features illustrate the structural complexity that characterizes the cells of *Didinium* and the other diverse, mostly unicellular groups of eukaryotes informally known as **protists**.

As we'll see, some protists change their form as they creep along using blob-like appendages, others are shaped like tiny trumpets, and still others resemble miniature jewelry. In this chapter, we'll explore how these shape-changing, structurally complex eukaryotic cells arose from their morphologically simpler prokaryotic ancestors. We'll also examine another major step in the evolutionary history of life: the origin of multicellular eukaryotes such as plants, fungi, and animals. Finally, we'll consider how single-celled eukaryotes affect ecosystems and human health.

▼ **Figure 25.1** What enables the cell on the left to engulf its prey?

CONCEPT 25.1

Eukaryotes arose by endosymbiosis more than 1.8 billion years ago

As we discussed in Chapter 24, all organisms were unicellular early in the history of life. The evolution of eukaryotes did not immediately change this, but it did involve fundamental changes in the structure of these individual cells. For

481

After reading a concept section, students can check their understanding using the **Concept Check questions** on their own or in a study group.

Make Connections questions ask students to relate content in the chapter to a concept presented earlier in the course.

What If? questions ask students to apply what they've learned.

CONCEPT CHECK 25.2

1. Summarize the evidence that choanoflagellates are the sister group of animals.

2. **MAKE CONNECTIONS** Describe how the origin of multicellularity in animals illustrates Darwin's concept of descent with modification (see Concept 19.2).

3. **WHAT IF?** Cells in *Volvox*, plants, and fungi are similar in being enclosed by a cell wall. Predict whether the cell-to-cell attachments of these organisms form using similar or different molecules. Explain.

For suggested answers, see Appendix A.

Focus on Scientific Skills

Practice Scientific Skills

Scientific Skills Exercises in every chapter use real data to build key skills needed for biology, including data analysis, graphing, experimental design, and math skills.

Selected Scientific Skills Exercises include:

- Making a Line Graph and Calculating a Slope
- Interpreting Histograms
- Using the Chi-Square (χ^2) Test
- Analyzing DNA Deletion Experiments
- Making and Testing Predictions
- Interpreting Data in a Phylogenetic Tree
- Using the Hardy-Weinberg Equation to Interpret Data and Make Predictions
- Understanding Experimental Design and Interpreting Data
- Interpreting Data Values Expressed in Scientific Notation
- Designing an Experiment Using Genetic Mutants
- Interpreting a Graph with Log Scales
- Using the Logistic Equation to Model Population Growth

Scientific Skills Exercise

Interpreting a Scatter Plot with a Regression Line

How Does the Carbonate Ion Concentration of Seawater Affect the Calcification Rate of a Coral Reef? Scientists predict that acidification of the ocean due to higher levels of atmospheric CO_2 will lower the concentration of dissolved carbonate ions, which living corals use to build calcium carbonate reef structures. In this exercise, you will analyze data from a controlled experiment that examined the effect of carbonate ion concentration ($[CO_3^{2-}]$) on calcium carbonate deposition, a process called calcification.

How the Experiment Was Done The Biosphere 2 aquarium in Arizona contains a large coral reef system that behaves like a natural reef. For several years, a group of researchers measured the rate of calcification by the reef organisms and examined how the calcification rate changed with differing amounts of dissolved carbonate ions in the seawater.

Data from the Experiment The black data points in the graph below form a scatter plot. The red line, known as a linear regression line, is the best-fitting straight line for these points. These data are from one set of experiments, in which the pH, temperature, and calcium ion concentration of the seawater were held constant.

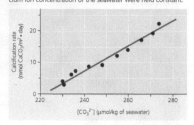

Interpret the Data
1. When presented with a graph of experimental data, the first step in analysis is to determine what each axis represents. (a) In words,

explain what is being shown on the x-axis. Be sure to include the units. (b) What is being shown on the y-axis (including units)? (c) Which variable is the independent variable—the variable that was *manipulated* by the researchers? (d) Which variable is the dependent variable—the variable that responded to or depended on the treatment, which was *measured* by the researchers? (For additional information about graphs, see the Scientific Skills Review in Appendix F and in the Study Area in MasteringBiology.)

2. Based on the data shown in the graph, describe in words the relationship between carbonate ion concentration and calcification rate.

3. (a) If the seawater carbonate ion concentration is 270 μmol/kg, what is the approximate rate of calcification, and approximately how many days would it take 1 square meter of reef to accumulate 30 mmol of calcium carbonate ($CaCO_3$)? To determine the rate of calcification, draw a vertical line up from the x-axis at the value of 270 μmol/kg until it intersects the red line. Then draw a horizontal line from the intersection over to the y-axis to see what the calcification rate is at that carbonate ion concentration. (b) If the seawater carbonate ion concentration is 250 μmol/kg, what is the approximate rate of calcification, and approximately how many days would it take 1 square meter of reef to accumulate 30 mmol of calcium carbonate? (c) If carbonate ion concentration decreases, how does the calcification rate change, and how does that affect the time it takes coral to grow?

4. (a) Referring to the equations in Figure 2.24, determine which step of the process is measured in this experiment. (b) Do the results of this experiment support the hypothesis that increased atmospheric $[CO_2]$ will slow the growth of coral reefs? Why or why not?

Data from C. Langdon et al., Effect of calcium carbonate saturation state on the calcification rate of an experimental coral reef, *Global Biogeochemical Cycles* 14:639–654 (2000).

A version of this Scientific Skills Exercise can be assigned in MasteringBiology.

MasteringBiology®
www.masteringbiology.com

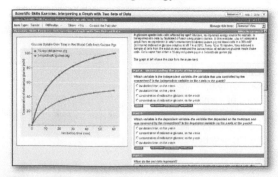

Scientific Skills Exercises from the text have assignable versions in MasteringBiology.

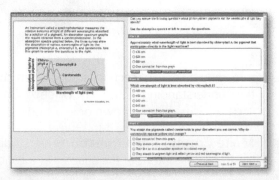

Interpreting Data Tutorials coach students on how to read and interpret data and graphs.

Analyzing Polypeptide Sequence Data

Are Rhesus Monkeys or Gibbons More Closely Related to Humans? As discussed in Concept 3.6, DNA and polypeptide sequences from closely related species are more similar to each other than are sequences from more distantly related species. In this exercise, you will look at amino acid sequence data for the β polypeptide chain of hemoglobin, often called β-globin. You will then interpret the data to hypothesize whether the monkey or the gibbon is more closely related to humans.

How Such Experiments Are Done Researchers can isolate the polypeptide of interest from an organism and then determine the amino acid sequence. More frequently, the DNA of the relevant gene is sequenced, and the amino acid sequence of the polypeptide is deduced from the DNA sequence of its gene.

Data from the Experiments In the data below, the letters give the sequence of the 146 amino acids in β-globin from humans, rhesus monkeys, and gibbons. Because a complete sequence would not fit on one line here, the sequences are broken into three segments. Note that the sequences for the three different species are aligned so that you can compare them easily. For example, you can see that for all three species, the first amino acid is V (valine; see Figure 3.17) and the 146th amino acid is H (histidine).

Interpret the Data

1. Scan along the monkey and gibbon sequences, letter by letter, circling any amino acids that do not match the human sequence. (a) How many amino acids differ between the monkey and the human sequences? (b) Between the gibbon and human?
2. For each nonhuman species, what percent of its amino acids are identical to the human sequence of β-globin?
3. Based on these data alone, state a hypothesis for which of these two species is more closely related to humans. What is your reasoning?
4. What other evidence could you use to support your hypothesis?

Data from Human: http://www.ncbi.nlm.nih.gov/protein/AAA21113.1; rhesus monkey: http://www.ncbi.nlm.nih.gov/protein/122634; gibbon: http://www.ncbi.nlm.nih.gov/protein/122616

🔎 A version of this Scientific Skills Exercise can be assigned in MasteringBiology.

Species	Alignment of Amino Acid Sequences of β-globin
Human	1 VHLTPEEKSA VTALWGKVNV DEVGGEALGR LLVVYPWTQR FFESFGDLST PDAVMGNPKV
Monkey	1 VHLTPEEKNA VTTLWCKVNV DEVGGEALGR LLLVYPWTQR FFEOFGDLSS PDAVMGNPKV
Gibbon	1 VHLTPEEKSA VTALWGKVNV DEVGGEALGR LLVVYPWTQR FFESFGDLST PDAVMGNPKV
Human	61 KAHGKKVLGA FSDGLAHLDN LKGTFATLSE LHCDKLHVDP ENFRLLGNVL VCVLAHHFGK
Monkey	61 KAHGKKVLGA FSDGLNHLDN LKGTFAQLSE LHCDKLHVDP ENFKLLGNVL VCVLAHHFGK
Gibbon	61 KAHGKKVLGA FSDGLAHLDN LKGTFAQLSE LHCDKLHVDP ENFRLLGNVL VCVLAHHFGK
Human	121 EFTPPVQAAY QKVVAGVANA LAHKYH
Monkey	121 EFTPQVQAAY QKVVAGVANA LAHKYH
Gibbon	121 EFTPQVQAAY QKVVAGVANA LAHKYH

▼ **Inquiry Figures** reveal "how we know what we know" by highlighting how researchers designed an experiment, interpreted their results, and drew conclusions.

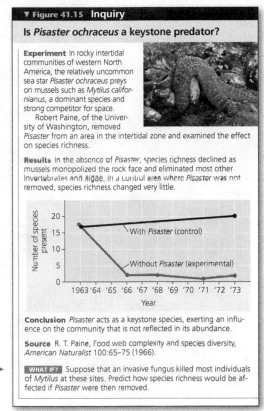

▼ Figure 41.15 **Inquiry**

Is *Pisaster ochraceus* a keystone predator?

Experiment In rocky intertidal communities of western North America, the relatively uncommon sea star *Pisaster ochraceus* preys on mussels such as *Mytilus californianus*, a dominant species and strong competitor for space.

Robert Paine, of the University of Washington, removed *Pisaster* from an area in the intertidal zone and examined the effect on species richness.

Results In the absence of *Pisaster*, species richness declined as mussels monopolized the rock face and eliminated most other invertebrates and algae. In a control area where *Pisaster* was not removed, species richness changed very little.

Conclusion *Pisaster* acts as a keystone species, exerting an influence on the community that is not reflected in its abundance.

Source R. T. Paine, Food web complexity and species diversity, *American Naturalist* 100:65–75 (1966).

WHAT IF? Suppose that an invasive fungus killed most individuals of *Mytilus* at these sites. Predict how species richness would be affected if *Pisaster* were then removed.

After exploring the featured experiment, ▶ students test their analytical skills by answering the **What If? question**.

Experimental Inquiry Tutorials, based on some of biology's most influential experiments, give students practice analyzing experimental design and data and help students understand how to reach conclusions based on collected data. Topics include:

- What Can You Learn About the Process of Science from Investigating a Cricket's Chirp?
- Which Wavelengths of Light Drive Photosynthesis?
- Does DNA Replication Follow the Conservative, Semiconservative, or Dispersive Model?
- Did Natural Selection of Ground Finches Occur When the Environment Changed?
- What Factors Influence the Loss of Nutrients from a Forest Ecosystem?

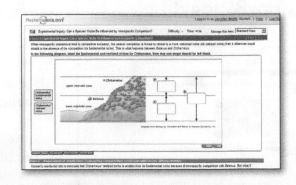

Synthesize and Assess

Make Connections Across Biology

By relating the content of a chapter to material presented earlier in the course, **Make Connections questions** help students develop a deeper understanding of biological principles.

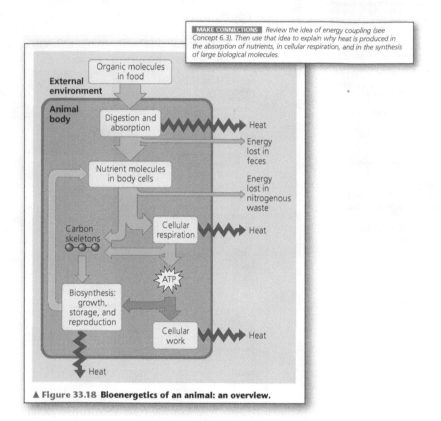

MAKE CONNECTIONS *Review the idea of energy coupling (see Concept 6.3). Then use that idea to explain why heat is produced in the absorption of nutrients, in cellular respiration, and in the synthesis of large biological molecules.*

▲ **Figure 33.18 Bioenergetics of an animal: an overview.**

CONCEPT CHECK 28.3
1. Contrast primary growth in roots and shoots.
2. **WHAT IF?** If a plant species has vertically oriented leaves, would you expect its mesophyll to be divided into spongy and palisade layers? Explain.
3. **MAKE CONNECTIONS** How are root hairs and microvilli analogous structures? (See Figure 4.7 and the discussion of analogy in Concept 20.2.)

▼ **Figure 4.7**   **Exploring Eukaryotic Cells**

Animal Cell (cutaway view of generalized cell)

ENDOPLASMIC RETICULUM (ER): network of membranous sacs and tubes; active in membrane synthesis and other synthetic and metabolic processes; has rough (ribosome-studded) and smooth regions

Flagellum: motility structure present in some animal cells, composed of a cluster of microtubules within an extension of the plasma membrane

Rough ER   Smooth ER

Centrosome: region where the cell's microtubules are initiated; contains a pair of centrioles

CYTOSKELETON: reinforces cell's shape; functions in cell movement; components are made of protein. Includes:

Microfilaments

Intermediate filaments

Microtubules

Microvilli: projections that increase the cell's surface area

Peroxisome: organelle with various specialized metabolic functions; produces hydrogen peroxide as a by-product, then converts it to water

Mitochondrion: organelle where cellular respiration occurs and most ATP is generated

Lysoso organel macrom hydroly

MasteringBiology®
www.masteringbiology.com

Make Connections Tutorials help students connect biological concepts across chapters in an interactive way.

Focus on Evolution

Every chapter has a section explicitly relating the chapter content to **evolution**, the fundamental theme of biology. Each section is highlighted by an Evolution banner.

The Evolutionary Origins of Mitochondria and Chloroplasts

EVOLUTION Mitochondria and chloroplasts display similarities with bacteria that led to the **endosymbiont theory**, illustrated in **Figure 4.16**. This theory states that an early ancestor of eukaryotic cells engulfed an oxygen-using non-photosynthetic prokaryotic cell. Eventually, the engulfed cell formed a relationship with the host cell in which it was enclosed, becoming an *endosymbiont* (a cell living within another cell). Indeed, over the course of evolution, the host cell and its endosymb eukaryotic cell with a cells may have then ta becoming the ancesto chloroplasts.

This theory is consis of mitochondria and ch bounded by a single m membrane system, mit two membranes surrou an internal system of m that the ancestral prop

Endoplasmic reticulum

Nucleus

Nuclear-envelope

Engulfing of oxygen-using nonphotosynthetic prokaryote, which becomes a mitochondrion

Mitochondrion

Ancestor of eukaryotic cells (host cell)

Nonphotosynthetic eukaryote

At least one cell

Engulfing of photosynthetic prokaryote

Chloroplast

Mitochondrion

Photosynthetic eukaryote

Review and Test Understanding

Chapter Reviews help students master the chapter content by focusing on the main points and offering opportunities to practice for exams.

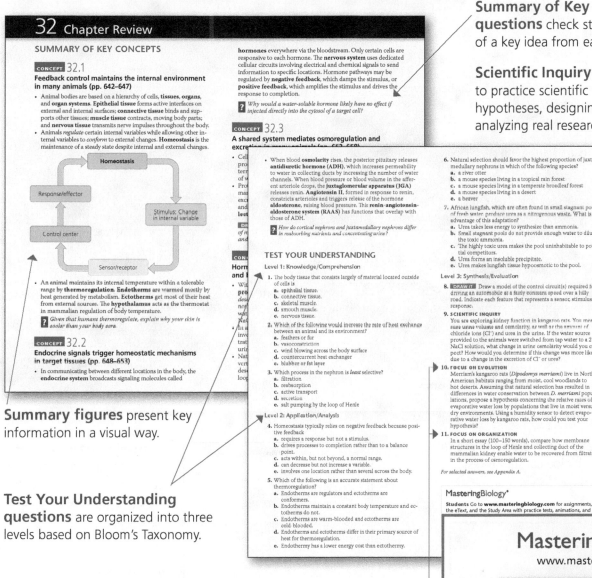

Summary of Key Concepts questions check students' understanding of a key idea from each concept.

Scientific Inquiry questions ask students to practice scientific thinking by developing hypotheses, designing experiments, and analyzing real research data.

Summary figures present key information in a visual way.

Test Your Understanding questions are organized into three levels based on Bloom's Taxonomy.

Focus on a Theme questions give students practice writing a short essay that connects the chapter's content to the five bookwide themes introduced in Chapter 1: **Evolution, Organization, Information, Energy and Matter,** and **Interactions.**

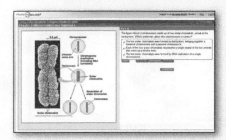

MasteringBiology®
www.masteringbiology.com

Student Misconception Questions provide assignable quizzes for each chapter to assess and remediate common student misconceptions.

Visualize Biology

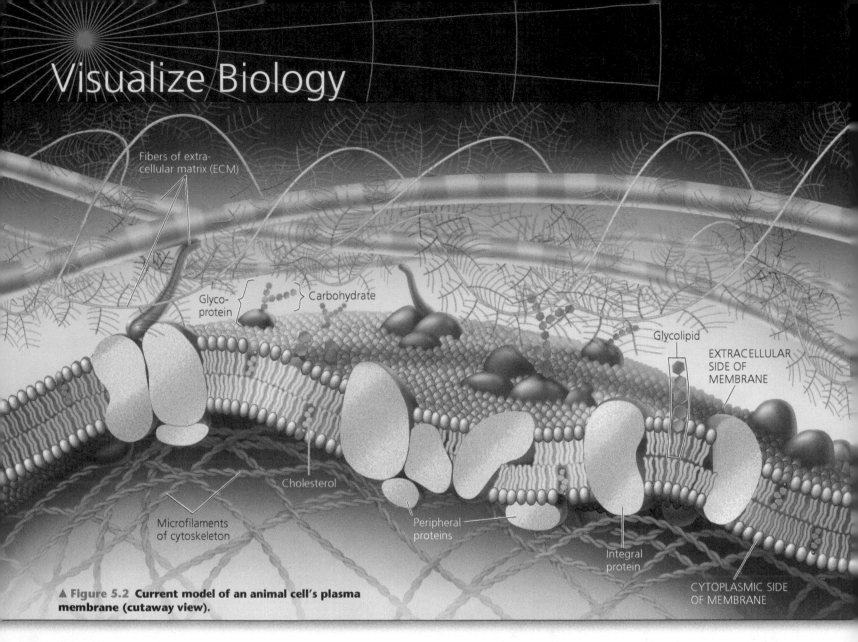

Fibers of extra-cellular matrix (ECM)

Glyco-protein

Carbohydrate

Glycolipid

EXTRACELLULAR SIDE OF MEMBRANE

Cholesterol

Microfilaments of cytoskeleton

Peripheral proteins

Integral protein

CYTOPLASMIC SIDE OF MEMBRANE

▲ **Figure 5.2 Current model of an animal cell's plasma membrane (cutaway view).**

▲ Selected figures are rendered in a **3-D style** to help students visualize biological structures.

MasteringBiology®

www.masteringbiology.com

Many **Tutorials** and **Activities** integrate art from the textbook, providing a unified learning experience.

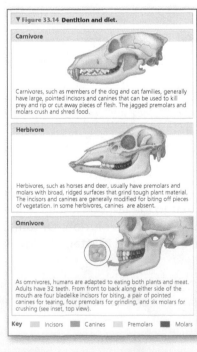

▼ Figure 33.14 **Dentition and diet.**

Carnivore

Carnivores, such as members of the dog and cat families, generally have large, pointed incisors and canines that can be used to kill prey and rip or cut away pieces of flesh. The jagged premolars and molars crush and shred food.

Herbivore

Herbivores, such as horses and deer, usually have premolars and molars with broad, ridged surfaces that grind tough plant material. The incisors and canines are generally modified for biting off pieces of vegetation. In some herbivores, canines are absent.

Omnivore

As omnivores, humans are adapted to eating both plants and meat. Adults have 32 teeth. From front to back along either side of the mouth are four bladelike incisors for biting, a pair of pointed canines for tearing, four premolars for grinding, and six molars for crushing (see inset, top view).

Key ☐ Incisors ☐ Canines ☐ Premolars ■ Molars

◄ **Visual Organizers** highlight the main parts of a figure, helping students see the key categories at a glance.

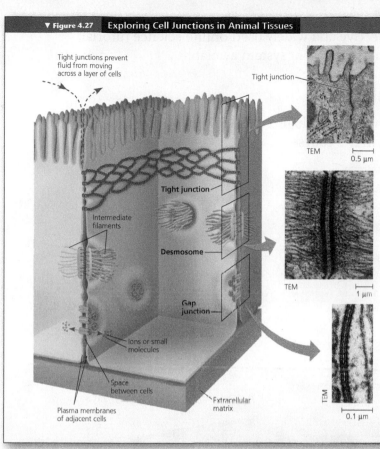

▼ Figure 4.27 Exploring Cell Junctions in Animal Tissues

Tight junctions prevent fluid from moving across a layer of cells

Tight junction

Intermediate filaments

Tight junction

Desmosome

Gap junction

Ions or small molecules

Space between cells

Plasma membranes of adjacent cells

Extracellular matrix

Tight Junctions

Tight junction

TEM — 0.5 μm

At **tight junctions**, the plasma membranes of neighboring cells are very tightly pressed against each other, bound together by specific proteins (purple). Forming continuous seals around the cells, tight junctions prevent leakage of extracellular fluid across a layer of epithelial cells. For example, tight junctions between skin cells make us watertight by preventing leakage between cells in our sweat glands.

Desmosomes

TEM — 1 μm

Desmosomes (also called anchoring junctions) function like rivets, fastening cells together into strong sheets. Intermediate filaments made of sturdy keratin proteins anchor desmosomes in the cytoplasm. Desmosomes attach muscle cells to each other in a muscle. Some "muscle tears" involve the rupture of desmosomes.

Gap Junctions

TEM — 0.1 μm

Gap junctions (also called communicating junctions) provide cytoplasmic channels from one cell to an adjacent cell and in this way are similar in their function to the plasmodesmata in plants. Gap junctions consist of membrane proteins that surround a pore through which ions, sugars, amino acids, and other small molecules may pass. Gap junctions are necessary for communication between cells in many types of tissues, such as heart muscle, and in animal embryos.

◀ By integrating text, art, and photos, **Exploring Figures** help students access information efficiently.

MasteringBiology®
www.masteringbiology.com

BioFlix

BioFlix® 3-D Animations help students visualize biology with movie-quality animations that can be presented in class, reviewed by students on their own in the Study Area, and assigned in MasteringBiology. **BioFlix Tutorials** use the animations as a jumping-off point for coaching exercises on tough topics in MasteringBiology. Tutorials and animations include:

- A Tour of the Animal Cell
- A Tour of the Plant Cell
- Membrane Transport
- Cellular Respiration
- Photosynthesis
- Mitosis
- Meiosis
- DNA Replication
- Protein Synthesis
- Mechanisms of Evolution
- Water Transport in Plants
- Homeostasis: Regulating Blood Sugar
- Gas Exchange
- How Neurons Work
- How Synapses Work
- Muscle Contraction
- Population Ecology
- The Carbon Cycle

MasteringBiology®

MasteringBiology® is the most effective and widely used online science tutorial, homework, and assessment system available.

www.masteringbiology.com

Personalized Coaching and Feedback

Assign self-paced **MasteringBiology tutorials** that provide individualized coaching with specific hints and feedback on the toughest topics in the course.

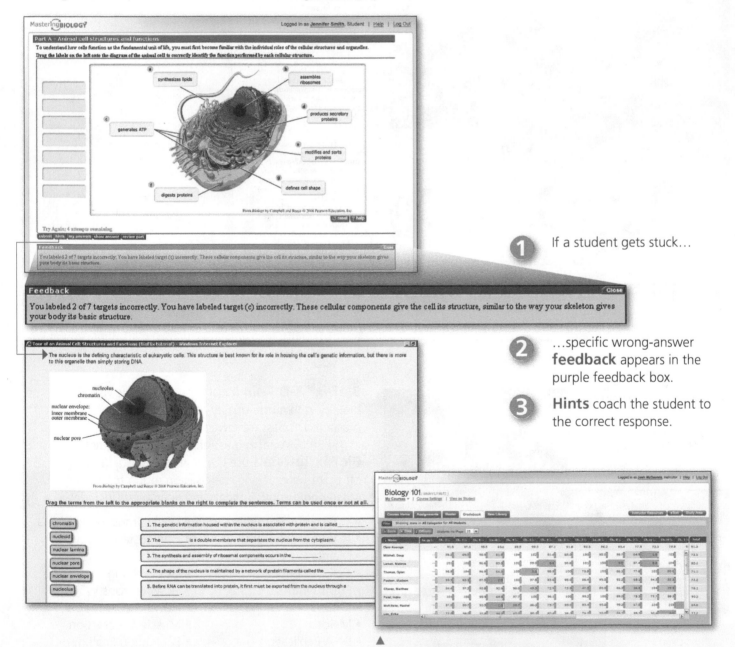

1 If a student gets stuck…

2 …specific wrong-answer **feedback** appears in the purple feedback box.

3 **Hints** coach the student to the correct response.

▲ The MasteringBiology **gradebook** provides instructors with quick results and easy-to-interpret insights into student performance. Every assignment is **automatically graded** and shades of red highlight vulnerable students and challenging assignments.

Students can use the Study Area on their own or in a study group.

BioFlix

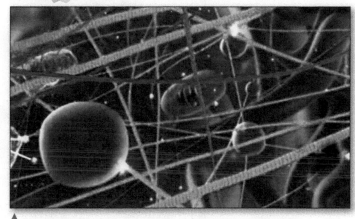

BioFlix 3-D Animations explore the most difficult biology topics, reinforced with tutorials, quizzes, and more.

The **Study Area** also includes:

- Scientific Skills Review
- Cumulative Test
- MP3 Tutor Sessions
- Videos
- Activities
- Investigations
- Lab Media
- Audio Glossary
- Word Roots
- Key Terms
- Flashcards
- Art

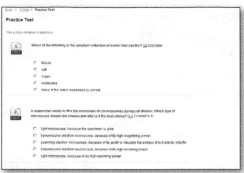

▲ **Practice Tests** help students assess their understanding of each chapter, providing feedback for right and wrong answers.

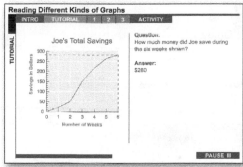

▲ **Get Ready for Biology** helps students get up to speed for their course by covering study skills, basic math review, terminology, biology basics, chemistry, and cell biology.

Access CAMPBELL BIOLOGY IN FOCUS Online

◄ The Pearson **eText** gives students access to the text whenever and wherever they can access the Internet. The eText can be viewed on PCs, Macs, and tablets, including iPad and Android. The eText includes powerful interactive and customization functions:

- Write notes
- Highlight text
- Bookmark pages
- Zoom
- Click hyperlinked words to view definitions
- Search
- Link to media activities and quizzes

Instructors can even write notes for the class and highlight important material using a tool that works like an electronic pen on a whiteboard.

Supplements

For Instructors

Instructor's Resource DVD Package
978-0-321-83322-8 • 0-321-83322-8

Assets for each chapter include:

- Editable figures (art and photos) and tables from the text in PowerPoint®
- Prepared PowerPoint Lecture Presentations for each chapter, with lecture notes, editable figures, tables, and links to animations and videos
- JPEG Images, including labeled and unlabeled art, photos from the text, and extra photos
- Clicker Questions in PowerPoint
- 250+ Instructor Animations and Videos, including BioFlix® 3-D Animations and *ABC News* Videos
- Test Bank questions in TestGen® software and Microsoft® Word
- Digital Transparencies
- Quick Reference Guide

BioFlix® Animations invigorate classroom lectures with 3-minute "movie-quality" 3-D graphics.

▼

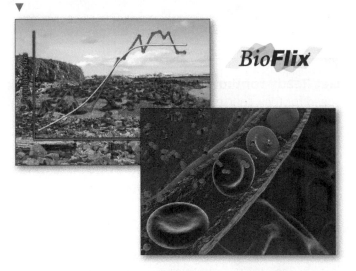

Customizable PowerPoints provide a jumpstart for each lecture.

▼

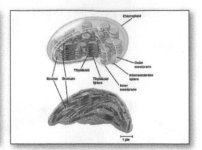

Clicker Questions can be used to stimulate effective classroom discussions (for use with or without clickers).

▼

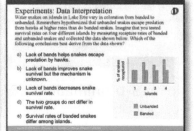

All of the art and photos from the book are provided with customizable labels.

▼

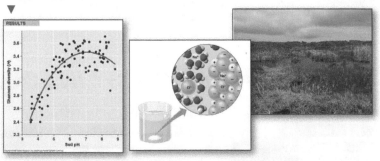

Test Bank
978-0-321-83153-8 • 0-321-83153-5

The Test Bank is available as part of the Instructor's Resource DVD Package or separately.

Course Management Systems

Content is available in Blackboard. Also, **MasteringBiology New Design** offers the usual Mastering features plus:

- Blackboard integration with single sign-on
- Temporary access (grace period)
- Discussion boards
- Email
- Chat and class live (synchronous whiteboard presentation)
- Submissions (dropbox)

Instructor Resources Area in MasteringBiology

This area includes:

- Figures and Tables in PowerPoint®
- PowerPoint Lecture Presentations
- JPEG Images
- Clicker Questions
- Animations
- Videos
- Test Bank Files
- Digital Transparencies
- Quick Reference Guide
- Instructor Guides for Supplements
- Rubric and Tips for Grading Short-Answer Essays
- Suggested Answers for Essay Questions
- Lab Media

For Students

Study Guide
by Martha R. Taylor, Cornell University
978-0-321-86499-4 • 0-321-86499-9

This study guide helps students extract key ideas from the textbook and organize their knowledge of biology. Exercises include concept maps, chapter summaries, word roots, chapter tests, and a variety of interactive questions in various formats.

Inquiry in Action: Interpreting Scientific Papers, Second Edition*
by Ruth Buskirk, University of Texas at Austin, and Christopher M. Gillen, Kenyon College
978-0-321-68336-6 • 0-321-68336-6

Nine research papers are summarized in Inquiry Figures. Each complete original research paper is also reprinted and accompanied by questions that help students analyze the paper.

Practicing Biology: A Student Workbook, Fourth Edition*
by Jean Heitz and Cynthia Giffen, University of Wisconsin, Madison
978-0-321-68328-1 • 0-321-68328-5

This workbook offers a variety of activities to suit different learning styles. Activities such as modeling and mapping allow students to visualize and understand biological processes. Other activities focus on basic skills, such as reading and drawing graphs.

Biological Inquiry: A Workbook of Investigative Cases, Third Edition*
by Margaret Waterman, Southeast Missouri State University, and Ethel Stanley, BioQUEST Curriculum Consortium and Beloit College
978-0-321-68320-5 • 0-321-68320-X

This workbook offers ten investigative cases. A link to a student website is in the Study Area in MasteringBiology.

Study Card
978-0-321-68322-9 • 0-321-68322-6

This quick-reference card provides an overview of the entire field of biology and helps students see the connections between topics.

Spanish Glossary
by Laura P. Zanello, University of California, Riverside
978-0-321-68321-2 • 0-321-68321-8

This resource provides definitions in Spanish for all the glossary terms.

Into the Jungle: Great Adventures in the Search for Evolution
by Sean B. Carroll, University of Wisconsin, Madison
978-0-321-55671-4 • 0-321-55671-2

These nine short tales vividly depict key discoveries in evolutionary biology and the excitement of the scientific process. Online resources available at www.aw-bc.com/carroll.

* An Instructor Guide is available for download in the Instructor Resources Area at www.masteringbiology.com.

Get Ready for Biology
by Lori K. Garrett, Parkland College
978-0-321-50057-1 • 0-321-50057-1

This engaging workbook helps students brush up on important math and study skills and get up to speed on biological terminology and the basics of chemistry and cell biology. Also available online through MasteringBiology.

A Short Guide to Writing About Biology, Seventh Edition
by Jan A. Pechenik, Tufts University
978-0-321-66838-7 • 0-321-66838-3

This best-selling writing guide teaches students to think as biologists and to express ideas clearly and concisely through their writing.

An Introduction to Chemistry for Biology Students, Ninth Edition
by George I. Sackheim, University of Illinois, Chicago
978-0-8053-9571-6 • 0-8053-9571-7

This text/workbook helps students review and master all the basic facts, concepts, and terminology of chemistry that they need for their life science course.

For Lab

Investigating Biology Laboratory Manual, Seventh Edition
by Judith Giles Morgan, Emory University, and M. Eloise Brown Carter, Oxford College of Emory University
978-0-321-66821-9 • 0-321-66821-9

The Seventh Edition emphasizes connections to recurring themes in biology, including structure and function, unity and diversity, and the overarching theme of evolution.

Annotated Instructor Edition for Investigating Biology Laboratory Manual, Seventh Edition
by Judith Giles Morgan, Emory University, and M. Eloise Brown Carter, Oxford College of Emory University
978-0-321-67668-9 • 0-321-67668-8

Preparation Guide for Investigating Biology Laboratory Manual, Seventh Edition
by Judith Giles Morgan, Emory University, and M. Eloise Brown Carter, Oxford College of Emory University
978-0-321-67669-6 • 0-321-67669-6

Symbiosis: The Pearson Custom Library for the Biological Sciences
www.pearsonlearningsolutions.com/custom-library/symbiosis
Professors can create a custom lab manual.

MasteringBiology® Virtual Biology Labs
www.masteringbiology.com

This online environment promotes critical thinking skills using virtual experiments and explorations that may be difficult to perform in a wet lab environment due to time, cost, or safety concerns. Designed to supplement or substitute for existing wet labs, this product offers students unique learning experiences and critical thinking exercises in the areas of microscopy, molecular biology, genetics, ecology, and systematics.

Featured Figures

Exploring Figures

1.3 Levels of Biological Organization 2
3.21 Levels of Protein Structure 56
4.3 Microscopy 68
4.7 Eukaryotic Cells 72
4.27 Cell Junctions in Animal Tissues 90
5.18 Endocytosis in Animal Cells 107
9.7 Mitosis in an Animal Cell 178
10.8 Meiosis in an Animal Cell 198
13.21 Chromatin Packing in a Eukaryotic Chromosome 260
22.3 Reproductive Barriers 420
23.4 The Origin of Mammals 441
24.19 Major Groups of Bacteria 472
25.2 The Early Evolution of Eukaryotes 482
25.9 Eukaryotic Diversity 490
26.6 Alternation of Generations 506
26.15 Fungal Diversity 512
26.25 Angiosperm Phylogeny 520
27.11 The Diversity of Invertebrate Bilaterians 535
27.14 Vertebrate Diversity 538
27.26 Reptilian Diversity 545
28.9 Examples of Differentiated Plant Cells 558
29.15 Unusual Nutritional Adaptations in Plants 586
30.6 Flower Pollination 602
30.12 Fruit and Seed Dispersal 607
32.2 Structure and Function in Animal Tissues 643
32.11 The Human Endocrine System 650
32.19 The Mammalian Excretory System 657
33.5 Four Main Feeding Mechanisms of Animals 669
36.10 Human Gametogenesis 736
38.6 The Organization of the Human Brain 772
38.20 The Structure of the Human Ear 783
38.25 The Structure of the Human Eye 786
39.6 The Regulation of Skeletal Muscle Contraction 796
40.2 The Scope of Ecological Research 819
40.3 Global Climate Patterns 820
40.9 Terrestrial Biomes 824
40.10 Aquatic Biomes 827
40.23 Mechanisms of Density-Dependent Regulation 841
42.13 Water and Nutrient Cycling 874
42.17 Restoration Ecology Worldwide 879

Inquiry Figures

1.19 Does camouflage affect predation rates on two populations of mice? 13
3.24 What can the 3-D shape of the enzyme RNA polymerase II tell us about its function? 59
5.4 Do membrane proteins move? 96
†8.9 Which wavelengths of light are most effective in driving photosynthesis? 161
9.9 At which end do kinetochore microtubules shorten during anaphase? 181
9.14 Do molecular signals in the cytoplasm regulate the cell cycle? 184
11.3 When F$_1$ hybrid pea plants self- or cross-pollinate, which traits appear in the F$_2$ generation? 208
11.8 Do the alleles for one character segregate into gametes dependently or independently of the alleles for a different character? 212
†12.4 In a cross between a wild-type female fruit fly and a mutant white-eyed male, what color eyes will the F$_1$ and F$_2$ offspring have? 231
12.9 How does linkage between two genes affect inheritance of characters? 235
13.2 Can a genetic trait be transferred between different bacterial strains? 246
13.4 Is protein or DNA the genetic material of phage T2? 247
*†13.11 Does DNA replication follow the conservative, semiconservative, or dispersive model? 253
16.10 Could Bicoid be a morphogen that determines the anterior end of a fruit fly? 319
16.11 Can the nucleus from a differentiated animal cell direct development of an organism? 321
19.14 Can a change in a population's food source result in evolution by natural selection? 373
20.6 What is the species identity of food being sold as whale meat? 384
*21.16 Do females select mates based on traits indicative of "good genes"? 413
22.8 Can divergence of allopatric populations lead to reproductive isolation? 424
22.10 Does sexual selection in cichlids result in reproductive isolation? 426
22.16 How does hybridization lead to speciation in sunflowers? 432
23.20 What caused the loss of spines in lake stickleback fish? 452
24.14 Can prokaryotes evolve rapidly in response to environmental change? 467
25.21 Where is the root of the eukaryotic tree? 497
26.28 Do endophytes benefit a woody plant? 523
27.12 Did the arthropod body plan result from new Hox genes? 536
*29.14 Does the invasive weed garlic mustard disrupt mutualistic associations between native tree seedlings and arbuscular mycorrhizal fungi? 585
31.2 What part of a grass coleoptile senses light, and how is the signal transmitted? 618
†31.3 Does asymmetric distribution of a growth-promoting chemical cause a coleoptile to grow toward the light? 619
31.4 What causes polar movement of auxin from shoot tip to base? 621
31.13 How does the order of red and far-red illumination affect seed germination? 628
34.21 What causes respiratory distress syndrome? 704
39.18 Does a digger wasp use landmarks to find her nest? 808
40.13 Does feeding by sea urchins limit seaweed distribution? 831
†41.3 Can a species' niche be influenced by interspecific competition? 847
41.15 Is Pisaster ochraceus a keystone predator? 854
41.22 How does species richness relate to area? 860
42.7 Which nutrient limits phytoplankton production along the coast of Long Island? 869
42.12 How does temperature affect litter decomposition in an ecosystem? 873
*43.12 What caused the drastic decline of the Illinois greater prairie chicken population? 889

Research Method Figures

8.8 Determining an Absorption Spectrum 161
10.3 Preparing a Karyotype 194
11.2 Crossing Pea Plants 207
11.7 The Testcross 211
12.11 Constructing a Linkage Map 239
13.25 The Polymerase Chain Reaction (PCR) 264
15.16 RT-PCR Analysis of the Expression of Single Genes 308
16.12 Reproductive Cloning of a Mammal by Nuclear Transplantation 321
20.14 Applying Parsimony to a Problem in Molecular Systematics 390
29.8 Hydroponic Culture 578
37.8 Intracellular Recording 756
41.11 Determining Microbial Diversity Using Molecular Tools 852
42.5 Determining Primary Production with Satellites 867

*The Inquiry Figure, original research paper, and a worksheet to guide you through the paper are provided in *Inquiry in Action: Interpreting Scientific Papers*, Second Edition.

†A related Experimental Inquiry Tutorial can be assigned in MasteringBiology.

1 **Interpreting a Pair of Bar Graphs**
How Much Does Camouflage Affect Predation on Mice by Owls with and without Moonlight? 15

2 **Interpreting a Scatter Plot with a Regression Line**
How Does the Carbonate Ion Concentration of Seawater Affect the Calcification Rate of a Coral Reef? 37

3 **Analyzing Polypeptide Sequence Data**
Are Rhesus Monkeys or Gibbons More Closely Related to Humans? 63

4 **Using a Scale Bar to Calculate Volume and Surface Area of a Cell**
How Much New Cytoplasm and Plasma Membrane Are Made by a Growing Yeast Cell? 74

5 **Interpreting a Graph with Two Sets of Data**
Is Glucose Uptake into Cells Affected by Age? 103

6 **Making a Line Graph and Calculating a Slope**
Does the Rate of Glucose 6-Phosphatase Activity Change over Time in Isolated Liver Cells? 128

7 **Making a Bar Graph and Evaluating a Hypothesis**
Does Thyroid Hormone Level Affect Oxygen Consumption in Cells? 149

8 **Making Scatter Plots with Regression Lines**
Does Atmospheric Carbon Dioxide Concentration Affect the Productivity of Agricultural Crops? 170

9 **Interpreting Histograms**
At What Phase Is the Cell Cycle Arrested by an Inhibitor? 188

10 **Making a Line Graph and Converting Between Units of Data**
How Does DNA Content Change as Budding Yeast Cells Proceed Through Meiosis? 202

11 **Making a Histogram and Analyzing a Distribution Pattern**
What Is the Distribution of Phenotypes Among Offspring of Two Parents Who Are Both Heterozygous for Three Additive Genes? 219

12 **Using the Chi-Square (χ^2) Test**
Are Two Genes Linked or Unlinked? 238

13 **Working with Data in a Table**
Given the Percentage Composition of One Nucleotide in a Genome, Can We Predict the Percentages of the Other Three Nucleotides? 249

14 **Interpreting a Sequence Logo**
How Can a Sequence Logo Be Used to Identify Ribosome-Binding Sites? 284

15 **Analyzing DNA Deletion Experiments**
What Control Elements Regulate Expression of the *mPGES-1* Gene? 303

16 **Analyzing Quantitative and Spatial Gene Expression Data**
How Is a Particular *Hox* Gene Regulated During Paw Development? 316

17 **Analyzing a DNA Sequence-Based Phylogenetic Tree to Understand Viral Evolution**
How Can DNA Sequence Data Be Used to Track Flu Virus Evolution During Pandemic Waves? 340

18 **Reading an Amino Acid Sequence Identity Table**
How Have Amino Acid Sequences of Human Globin Genes Diverged During Their Evolution? 356

19 **Making and Testing Predictions**
Can Predation Result in Natural Selection for Color Patterns in Guppies? 378

20 **Interpreting Data in a Phylogenetic Tree**
What Are the Evolutionary Relationships Among Bears? 394

21 **Using the Hardy-Weinberg Equation to Interpret Data and Make Predictions**
Is Evolution Occurring in a Soybean Population? 406

22 **Identifying Independent and Dependent Variables, Making a Scatter Plot, and Interpreting Data**
Does Distance Between Salamander Populations Increase Their Reproductive Isolation? 427

23 **Estimating Quantitative Data from a Graph and Developing Hypotheses**
Do Ecological Factors Affect Evolutionary Rates? 443

24 **Making a Bar Graph and Interpreting Data**
Do Soil Microorganisms Protect Against Crop Disease? 477

25 **Interpreting Comparisons of Genetic Sequences**
Which Prokaryotes Are Most Closely Related to Mitochondria? 485

26 **Synthesizing Information from Multiple Data Sets**
Can Mycorrhizae Help Plants Cope with High-Temperature Soils? 510

27 **Understanding Experimental Design and Interpreting Data**
Is There Evidence of Selection for Defensive Adaptations in Mollusc Populations Exposed to Predators? 549

28 **Using Bar Graphs to Interpret Data**
Nature versus Nurture: Why Are Leaves from Northern Red Maples "Toothier" Than Leaves from Southern Red Maples? 560

29 **Calculating and Interpreting Temperature Coefficients**
Does the Initial Uptake of Water by Seeds Depend on Temperature? 576

30 **Using Positive and Negative Correlations to Interpret Data**
Do Monkey Flower Species Differ in Allocating Energy to Sexual Versus Asexual Reproduction? 610

31 **Interpreting Experimental Results from a Bar Graph**
Do Drought-Stressed Plants Communicate Their Condition to Their Neighbors? 634

32 **Describing and Interpreting Quantitative Data**
How Do Desert Mice Maintain Osmotic Homeostasis? 655

33 **Interpreting Data from an Experiment with Genetic Mutants**
What Are the Roles of the *ob* and *db* Genes in Appetite Regulation? 681

34 **Interpreting Data in Histograms**
Does Inactivating the PCSK9 Enzyme Lower LDL Levels in Humans? 699

35 **Comparing Two Variables on a Common *x*-Axis**
How Does the Immune System Respond to a Changing Pathogen? 726

36 **Making Inferences and Designing an Experiment**
What Role Do Hormones Play in Making a Mammal Male or Female? 739

37 **Interpreting Data Values Expressed in Scientific Notation**
Does the Brain Have Specific Protein Receptors for Opiates? 765

38 **Designing an Experiment Using Genetic Mutants**
Does the SCN Control the Circadian Rhythm in Hamsters? 774

39 **Interpreting a Graph with Log Scales**
What Are the Energy Costs of Locomotion? 803

40 **Using the Logistic Equation to Model Population Growth**
What Happens to the Size of a Population When It Overshoots Its Carrying Capacity? 838

41 **Using Bar Graphs and Scatter Plots to Present and Interpret Data**
Can a Native Predator Species Adapt Rapidly to an Introduced Prey Species? 849

42 **Interpreting Quantitative Data in a Table**
How Efficient Is Energy Transfer in a Salt Marsh Ecosystem? 871

43 **Graphing Cyclic Data**
How Does the Atmospheric CO_2 Concentration Change During a Year and from Decade to Decade? 898

MB • Each Scientific Skills Exercise is also available as an automatically graded Tutorial in MasteringBiology.

The authors wish to express their gratitude to the global community of instructors, researchers, students, and publishing professionals who have contributed to the first edition of *CAMPBELL BIOLOGY IN FOCUS*.

As authors of this text, we are mindful of the daunting challenge of keeping up to date in all areas of our rapidly expanding subject. We are grateful to the many scientists who helped shape this text by discussing their research fields with us, answering specific questions in their areas of expertise, and sharing their ideas about biology education. We are especially grateful to the following, listed alphabetically: Monika Abedin, John Archibald, Daniel Boyce, Nick Butterfield, Jean DeSaix, Eileen Gregory, Hopi Hoekstra, Fred Holtzclaw, Theresa Holtzclaw, Azarias Karamanlidis, Patrick Keeling, David Lamb, Teri Liegler, Thomas Montavon, Joe Montoya, Kevin Peterson, Michael Pollock, Susannah Porter, Andrew Roger, Andrew Schaffner, Tom Schneider, Alastair Simpson, Doug Soltis, Pamela Soltis, and George Watts. In addition, the biologists listed on pages xxiii–xxiv provided detailed reviews, helping us ensure the text's scientific accuracy and improve its pedagogical effectiveness. We thank Marty Taylor, author of the Study Guide, for her many contributions to the accuracy, clarity, and consistency of the text; and we thank Carolyn Wetzel, Ruth Buskirk, Joan Sharp, Jennifer Yeh, and Charlene D'Avanzo for their contributions to the Scientific Skills Exercises.

Thanks also to the other professors and students, from all over the world, who contacted the authors directly with useful suggestions. We alone bear the responsibility for any errors that remain, but the dedication of our consultants, reviewers, and other correspondents makes us confident in the accuracy and effectiveness of this text.

The value of *CAMPBELL BIOLOGY IN FOCUS* as a learning tool is greatly enhanced by the supplementary materials that have been created for instructors and students. We recognize that the dedicated authors of these materials are essentially writing mini (and not so mini) books. We appreciate the hard work and creativity of all the authors listed, with their creations, on page xix. We are also grateful to Kathleen Fitzpatrick and Nicole Tunbridge (PowerPoint® Lecture Presentations) and Fleur Ferro, Brad Stith, and Loraine Washburn (Clicker Questions).

MasteringBiology® and the electronic media for this text are invaluable teaching and learning aids. We thank the hardworking, industrious instructors who worked on the revised and new media: Willy Cushwa, Tom Kennedy, Michael Pollock, and Heather Wilson-Ashworth. We are also grateful to the many other people—biology instructors, editors, and production experts—who are listed in the credits for these and other elements of the electronic media that accompany the text.

CAMPBELL BIOLOGY IN FOCUS results from an unusually strong synergy between a team of scientists and a team of publishing professionals. Our editorial team at Pearson Science again demonstrated unmatched talents, commitment, and pedagogical insights. Our Senior Acquisitions Editor, Josh Frost, brought publishing savvy, intelligence, and a much appreciated level head to leading the whole team. The clarity and effectiveness of every page owe much to our extraordinary Supervising Editors Pat Burner and Beth Winickoff, who worked with a top-notch team of Developmental Editors in Matt Lee and Mary Ann Murray. Our unsurpassed Senior Editorial Manager Ginnie Simione Jutson, Executive Director of Development Deborah Gale, Assistant Editor Katherine Harrison-Adcock, and Editor-in-Chief Beth Wilbur were indispensable in moving the project in the right direction. We also want to thank Robin Heyden for organizing the annual Biology Leadership Conferences and keeping us in touch with the world of AP Biology.

You would not have this beautiful text if not for the work of the production team: Director of Production Erin Gregg; Managing Editor Michael Early; Assistant Managing Editor Shannon Tozier; Senior Photo Editor Donna Kalal; Photo Researcher Maureen Spuhler; Copy Editor Janet Greenblatt; Proofreader Joanna Dinsmore; Permissions Editor Sue Ewing; Permissions Project Manager Joe Croscup; Permissions Manager Tim Nicholls; Senior Project Editor Emily Bush, Paging Specialists Jodi Gaherty and Donna Healy, and the rest of the staff at S4Carlisle; Art Production Manager Kristina Seymour, Developmental Artist Andrew Recher, and the rest of the staff at Precision Graphics; Design Manager Marilyn Perry; Text Designer Gary Hespenheide; Cover Designer Yvo Riezebos; and Manufacturing Buyer Michael Penne. We also thank those who worked on the text's supplements: Susan Berge, Brady Golden, Jane Brundage, David Chavez, Kris Langan, Pete Shanks, and John Hammett. And for creating the wonderful package of electronic media that accompanies the text, we are grateful to Tania Mlawer (Director of Content Development for MasteringBiology), Sarah Jensen, Jonathan Ballard, Brienn Buchanan, Katie Foley, and Caroline Ross, as well as Director of Media Development Lauren Fogel and Director of Media Strategy Stacy Treco.

For their important roles in marketing the text and media, we thank Christy Lesko, Lauren Harp, Scott Dustan, Chris Hess, Lillian Carr, Jane Campbell, and Jessica Perry. For their market development support, we thank Brooke Suchomel, Michelle Cadden, and Cassandra Cummings. We are grateful to Paul Corey, President of Pearson Science, for his enthusiasm, encouragement, and support.

The Pearson sales team, which represents *CAMPBELL BIOLOGY IN FOCUS* on campus, is an essential link to the users of the text. They tell us what you like and don't like about the text, communicate the features of the text, and provide prompt service. We thank them for their hard work and professionalism. For representing our text to our international audience, we thank our sales and marketing partners throughout the world. They are all strong allies in biology education.

Finally, we wish to thank our families and friends for their encouragement and patience throughout this long project. Our special thanks to Lily, Ross, Lily-too, and Alex (L.A.U.); Debra and Hannah (M.L.C.); Harry, Elga, Aaron, Sophie, Noah, and Gabriele (S.A.W.); Natalie (P.V.M.); Sally, Robert, David, and Will (R.B.J.); and Paul, Dan, Maria, Armelle, and Sean (J.B.R.). And, as always, thanks to Rochelle, Allison, Jason, McKay, and Gus.

Lisa A. Urry, Michael L. Cain, Steve A. Wasserman,
Peter V. Minorsky, Robert B. Jackson, and Jane B. Reece

Ann Aguanno, *Marymount Manhattan College*
Marc Albrecht, *University of Nebraska*
John Alcock, *Arizona State University*
Eric Alcorn, *Acadia University*
Rodney Allrich, *Purdue University*
John Archibald, *Dalhousie University*
Terry Austin, *Temple College*
Brian Bagatto, *University of Akron*
Virginia Baker, *Chipola College*
Teri Balser, *University of Wisconsin, Madison*
Bonnie Baxter, *Westminster College*
Marilee Benore, *University of Michigan, Dearborn*
Catherine Black, *Idaho State University*
William Blaker, *Furman University*
Edward Blumenthal, *Marquette University*
David Bos, *Purdue University*
Scott Bowling, *Auburn University*
Beverly Brown, *Nazareth College*
Beth Burch, *Huntington University*
Warren Burggren, *University of North Texas*
Dale Burnside, *Lenoir-Rhyne University*
Ragan Callaway, *The University of Montana*
Kenneth M. Cameron, *University of Wisconsin, Madison*
Patrick Canary, *Northland Pioneer College*
Cheryl Keller Capone, *Pennsylvania State University*
Mickael Cariveau, *Mount Olive College*
Karen I. Champ, *Central Florida Community College*
David Champlin, *University of Southern Maine*
Brad Chandler, *Palo Alto College*
Wei-Jen Chang, *Hamilton College*
Jung Choi, *Georgia Institute of Technology*
Steve Christenson, *Brigham Young University, Idaho*
Reggie Cobb, *Nashville Community College*
James T. Colbert, *Iowa State University*
Sean Coleman, *University of the Ozarks*
William Cushwa, *Clark College*
Deborah Dardis, *Southeastern Louisiana University*
Shannon Datwyler, *California State University, Sacramento*
Melissa Deadmond, *Truckee Meadows Community College*
Eugene Delay, *University of Vermont*
Daniel DerVartanian, *University of Georgia*
Jean DeSaix, *University of North Carolina, Chapel Hill*
Janet De Souza-Hart, *Massachusetts College of Pharmacy & Health Sciences*
Jason Douglas, *Angelina College*
Kathryn A. Durham, *Lorain Community College*
Anna Edlund, *Lafayette College*
Curt Elderkin, *College of New Jersey*
Mary Ellard-Ivey, *Pacific Lutheran University*
Kurt Elliott, *Northwest Vista College*
George Ellmore, *Tufts University*
Rob Erdman, *Florida Gulf Coast College*
Dale Erskine, *Lebanon Valley College*
Robert C. Evans, *Rutgers University, Camden*
Sam Fan, *Bradley University*
Paul Farnsworth, *University of New Mexico*
Myriam Alhadeff Feldman, *Cascadia Community College*
Teresa Fischer, *Indian River Community College*
David Fitch, *New York University*
T. Fleming, *Bradley University*
Robert Fowler, *San Jose State University*
Robert Franklin, *College of Charleston*
Art Fredeen, *University of Northern British Columbia*
Kim Fredericks, *Viterbo University*
Matt Friedman, *University of Chicago*
Cynthia M. Galloway, *Texas A&M University, Kingsville*

Kristen Genet, *Anoka Ramsey Community College*
Phil Gibson, *University of Oklahoma*
Eric Gillock, *Fort Hayes State University*
Simon Gilroy, *University of Wisconsin, Madison*
Edwin Ginés-Candelaria, *Miami Dade College*
Jim Goetze, *Laredo Community College*
Lynda Goff, *University of California, Santa Cruz*
Roy Golsteyn, *University of Lethbridge*
Barbara E. Goodman, *University of South Dakota*
Eileen Gregory, *Rollins College*
Bradley Griggs, *Piedmont Technical College*
David Grise, *Texas A&M University, Corpus Christi*
Edward Gruberg, *Temple University*
Karen Guzman, *Campbell University*
Carla Haas, *Pennsylvania State University*
Pryce "Pete" Haddix, *Auburn University*
Heather Hallen-Adams, *University of Nebraska, Lincoln*
Monica Hall-Woods, *St. Charles Community College*
Bill Hamilton, *Washington & Lee University*
Devney Hamilton, *Stanford University (student)*
Matthew B. Hamilton, *Georgetown University*
Dennis Haney, *Furman University*
Jean Hardwick, *Ithaca College*
Luke Harmon, *University of Idaho*
Jeanne M. Harris, *University of Vermont*
Stephanie Harvey, *Georgia Southwestern State University*
Bernard Hauser, *University of Florida*
Chris Haynes, *Shelton State Community College*
Andreas Hejnol, *Sars International Centre for Marine Molecular Biology*
Albert Herrera, *University of Southern California*
Chris Hess, *Butler University*
Kendra Hill, *San Diego State University*
Jason Hodin, *Stanford University*
Laura Houston, *Northeast Lakeview College*
Sara Huang, *Los Angeles Valley College*
Catherine Hurlbut, *Florida State College, Jacksonville*
Diane Husic, *Moravian College*
Thomas Jacobs, *University of Illinois*
Kathy Jacobson, *Grinnell College*
Mark Jaffe, *Nova Southeastern University*
Emmanuelle Javaux, *University of Liege, Belgium*
Douglas Jensen, *Converse College*
Lance Johnson, *Midland Lutheran College*
Roishene Johnson, *Bossier Parish Community College*
Cheryl Jorcyk, *Boise State University*
Caroline Kane, *University of California, Berkeley*
The-Hui Kao, *Pennsylvania State University*
Nicholas Kapp, *Skyline College*
Jennifer Katcher, *Pima Community College*
Judy Kaufman, *Monroe Community College*
Eric G. Keeling, *Cary Institute of Ecosystem Studies*
Chris Kennedy, *Simon Fraser University*
Hillar Klandorf, *West Virginia University*
Mark Knauss, *Georgia Highlands College*
Charles Knight, *California Polytechnic State University*
Roger Koeppe, *University of Arkansas*
Peter Kourtev, *Central Michigan University*
Jacob Krans, *Western New England University*
Eliot Krause, *Seton Hall University*
Steven Kristoff, *Ivy Tech Community College*
William Kroll, *Loyola University*
Barb Kuemerle, *Case Western Reserve University*
Rukmani Kuppuswami, *Laredo Community College*
Lee Kurtz, *Georgia Gwinnett College*
Michael P. Labare, *United States Military Academy, West Point*
Ellen Lamb, *University of North Carolina, Greensboro*

(Continued)

William Lamberts, *College of St. Benedict and St. John's University*
Tali D. Lee, *University of Wisconsin, Eau Claire*
Hugh Lefcort, *Gonzaga University*
Alcinda Lewis, *University of Colorado, Boulder*
Jani Lewis, *State University of New York*
Graeme Lindbeck, *Valencia Community College*
Hannah Lui, *University of California, Irvine*
Nancy Magill, *Indiana University*
Cindy Malone, *California State University, Northridge*
Mark Maloney, *University of South Mississippi*
Julia Marrs, *Barnard College (student)*
Kathleen Marrs, *Indiana University-Purdue University, Indianapolis*
Mike Mayfield, *Ball State University*
Kamau Mbuthia, *Bowling Green State University*
Tanya McGhee, *Craven Community College*
Darcy Medica, *Pennsylvania State University*
Susan Meiers, *Western Illinois University*
Mike Meighan, *University of California, Berkeley*
Jan Mikesell, *Gettysburg College*
Alex Mills, *University of Windsor*
Sarah Milton, *Florida Atlantic University*
Eli Minkoff, *Bates College*
Subhash Minocha, *University of New Hampshire*
Ivona Mladenovic, *Simon Fraser University*
Linda Moore, *Georgia Military College*
Courtney Murren, *College of Charleston*
Karen Neal, *Reynolds University*
Ross Nehm, *Ohio State University*
Kimberlyn Nelson, *Pennsylvania State University*
Jacalyn Newman, *University of Pittsburgh*
Kathleen Nolta, *University of Michigan*
Gretchen North, *Occidental College*
Margaret Olney, *St. Martin's University*
Aharon Oren, *The Hebrew University*
Rebecca Orr, *Spring Creek College*
Henry R. Owen, *Eastern Illinois University*
Matt Palmtag, *Florida Gulf Coast University*
Stephanie Pandolfi, *Michigan State University*
Nathalie Pardigon, *Institut Pasteur*
Cindy Paszkowski, *University of Alberta*
Andrew Pease, *Stevenson University*
Nancy Pelaez, *Purdue University*
Irene Perry, *University of Texas of the Permian Basin*
Roger Persell, *Hunter College*
Eric Peters, *Chicago State University*
Larry Peterson, *University of Guelph*
Mark Pilgrim, *College of Coastal Georgia*
Vera M. Piper, *Shenandoah University*
Deb Pires, *University of California, Los Angeles*
Crima Pogge, *City College of San Francisco*
Michael Pollock, *Mount Royal University*
Roberta Pollock, *Occidental College*
Therese M. Poole, *Georgia State University*
Angela R. Porta, *Kean University*
Jason Porter, *University of the Sciences, Philadelphia*
Robert Powell, *Avila University*
Elena Pravosudova, *University of Nevada, Reno*
Eileen Preston, *Tarrant Community College Northwest*
Terrell Pritts, *University of Arkansas, Little Rock*
Pushpa Ramakrishna, *Chandler-Gilbert Community College*
David Randall, *City University Hong Kong*
Monica Ranes-Goldberg, *University of California, Berkeley*
Robert S. Rawding, *Gannon University*
Robert Reavis, *Glendale Community College*
Sarah Richart, *Azusa Pacific University*

Todd Rimkus, *Marymount University*
John Rinehart, *Eastern Oregon University*
Kenneth Robinson, *Purdue University*
Deb Roess, *Colorado State University*
Heather Roffey, *Marianopolis College*
Suzanne Rogers, *Seton Hill University*
Patricia Rugaber, *College of Coastal Georgia*
Scott Russell, *University of Oklahoma*
Glenn-Peter Saetre, *University of Oslo*
Sanga Saha, *Harold Washington College*
Kathleen Sandman, *Ohio State University*
Louis Santiago, *University of California, Riverside*
Tom Sawicki, *Spartanburg Community College*
Andrew Schaffner, *California Polytechnic State University, San Luis Obispo*
Thomas W. Schoener, *University of California, Davis*
Patricia Schulte, *University of British Columbia*
Brenda Schumpert, *Valencia Community College*
David Schwartz, *Houston Community College*
Duane Sears, *University of California, Santa Barbara*
Brent Selinger, *University of Lethbridge*
Alison M. Shakarian, *Salve Regina University*
Joan Sharp, *Simon Fraser University*
Robin L. Sherman, *Nova Southeastern University*
Eric Shows, *Jones County Junior College*
Sedonia Sipes, *Southern Illinois University, Carbondale*
John Skillman, *California State University, San Bernardino*
Doug Soltis, *University of Florida, Gainesville*
Joel Stafstrom, *Northern Illinois University*
Alam Stam, *Capital University*
Judy Stone, *Colby College*
Cynthia Surmacz, *Bloomsburg University*
David Tam, *University of North Texas*
Yves Tan, *Cabrillo College*
Emily Taylor, *California Polytechnic State University*
Marty Taylor, *Cornell University*
Franklyn Tan Te, *Miami Dade College*
Kent Thomas, *Wichita State University*
Mike Toliver, *Eureka College*
Saba Valadkhan, *Center for RNA Molecular Biology*
Sarah VanVickle-Chavez, *Washington University, St. Louis*
William Velhagen, *New York University*
Amy Volmer, *Swarthmore College*
Janice Voltzow, *University of Scranton*
Margaret Voss, *Penn State Erie*
Charles Wade, *C.S. Mott Community College*
Claire Walczak, *Indiana University*
Jerry Waldvogel, *Clemson University*
Robert Lee Wallace, *Ripon College*
James Wandersee, *Louisiana State University*
Fred Wasserman, *Boston University*
James Wee, *Loyola University*
John Weishampel, *University of Central Florida*
Susan Whittemore, *Keene State College*
Murray Wiegand, *University of Winnipeg*
Kimberly Williams, *Kansas State University*
Janet Wolkenstein, *Hudson Valley Community College*
Grace Wyngaard, *James Madison University*
Shuhai Xiao, *Virginia Polytechnic Institute*
Paul Yancey, *Whitman College*
Anne D. Yoder, *Duke University*
Ed Zalisko, *Blackburn College*
Nina Zanetti, *Siena College*
Sam Zeveloff, *Weber State University*
Theresa Zucchero, *Methodist University*

Detailed Contents

1 Introduction: Evolution and the Foundations of Biology 1

OVERVIEW Inquiring About Life 1

CONCEPT 1.1 Studying the diverse forms of life reveals common themes 2

Theme: New Properties Emerge at Successive Levels of Biological Organization 2

Theme: Life's Processes Involve the Expression and Transmission of Genetic Information 5

Theme: Life Requires the Transfer and Transformation of Energy and Matter 6

Theme: Organisms Interact with Other Organisms and the Physical Environment 7

Evolution, the Core Theme of Biology 7

CONCEPT 1.2 The Core Theme: Evolution accounts for the unity and diversity of life 7

Classifying the Diversity of Life: The Three Domains of Life 7

Charles Darwin and the Theory of Natural Selection 9

The Tree of Life 10

CONCEPT 1.3 Biological inquiry entails forming and testing hypotheses based on observations of nature 11

Making Observations 11

Forming and Testing Hypotheses 12

A Case Study in Scientific Inquiry: Investigating Coat Coloration in Mouse Populations 12

Theories in Science 14

Science as a Social Process: Community and Diversity 14

UNIT 1 Chemistry and Cells 18

2 The Chemical Context of Life 19

OVERVIEW A Chemical Connection to Biology 19

CONCEPT 2.1 Matter consists of chemical elements in pure form and in combinations called compounds 19

Elements and Compounds 20

The Elements of Life 20

Evolution of Tolerance to Toxic Elements 20

CONCEPT 2.2 An element's properties depend on the structure of its atoms 20

Subatomic Particles 20

Atomic Number and Atomic Mass 21

Isotopes 21

The Energy Levels of Electrons 22

Electron Distribution and Chemical Properties 23

CONCEPT 2.3 The formation and function of molecules depend on chemical bonding between atoms 24

Covalent Bonds 24

Ionic Bonds 25

Weak Chemical Bonds 26

Molecular Shape and Function 27

CONCEPT 2.4 Chemical reactions make and break chemical bonds 28

CONCEPT 2.5 Hydrogen bonding gives water properties that help make life possible on Earth 29

Cohesion of Water Molecules 30

Moderation of Temperature by Water 30

Floating of Ice on Liquid Water 32

Water: The Solvent of Life 33

Acids and Bases 34

3 Carbon and the Molecular Diversity of Life 40

OVERVIEW Carbon Compounds and Life 40

CONCEPT 3.1 Carbon atoms can form diverse molecules by bonding to four other atoms 41

The Formation of Bonds with Carbon 41

Molecular Diversity Arising from Variation in Carbon Skeletons 42

The Chemical Groups Most Important to Life 42

ATP: An Important Source of Energy for Cellular Processes 44

CONCEPT 3.2 Macromolecules are polymers, built from monomers 44

The Synthesis and Breakdown of Polymers 44

The Diversity of Polymers 45

CONCEPT 3.3 Carbohydrates serve as fuel and building material 45

Sugars 45

Polysaccharides 47

CONCEPT 3.4 Lipids are a diverse group of hydrophobic molecules 49

Fats 49

Phospholipids 50

Steroids 50

CONCEPT 3.5 Proteins include a diversity of structures, resulting in a wide range of functions 51

Amino Acids 52

Polypeptides 54

Protein Structure and Function 54

CONCEPT 3.6 Nucleic acids store, transmit, and help express hereditary information 60

The Roles of Nucleic Acids 60

The Components of Nucleic Acids 60

Nucleotide Polymers 61

The Structures of DNA and RNA Molecules 62

DNA and Proteins as Tape Measures of Evolution 62

4 A Tour of the Cell 66

OVERVIEW The Fundamental Units of Life 66

CONCEPT 4.1 Biologists use microscopes and the tools of biochemistry to study cells 67
  Microscopy 67
  Cell Fractionation 69

CONCEPT 4.2 Eukaryotic cells have internal membranes that compartmentalize their functions 69
  Comparing Prokaryotic and Eukaryotic Cells 69
  A Panoramic View of the Eukaryotic Cell 71

CONCEPT 4.3 The eukaryotic cell's genetic instructions are housed in the nucleus and carried out by the ribosomes 74
  The Nucleus: Information Central 74
  Ribosomes: Protein Factories 76

CONCEPT 4.4 The endomembrane system regulates protein traffic and performs metabolic functions in the cell 76
  The Endoplasmic Reticulum: Biosynthetic Factory 77
  The Golgi Apparatus: Shipping and Receiving Center 78
  Lysosomes: Digestive Compartments 79
  Vacuoles: Diverse Maintenance Compartments 80
  The Endomembrane System: *A Review* 81

CONCEPT 4.5 Mitochondria and chloroplasts change energy from one form to another 81
  The Evolutionary Origins of Mitochondria and Chloroplasts 82
  Mitochondria: Chemical Energy Conversion 82
  Chloroplasts: Capture of Light Energy 83
  Peroxisomes: Oxidation 84

CONCEPT 4.6 The cytoskeleton is a network of fibers that organizes structures and activities in the cell 84
  Roles of the Cytoskeleton: Support and Motility 84
  Components of the Cytoskeleton 85

CONCEPT 4.7 Extracellular components and connections between cells help coordinate cellular activities 88
  Cell Walls of Plants 88
  The Extracellular Matrix (ECM) of Animal Cells 88
  Cell Junctions 90
  The Cell: A Living Unit Greater Than the Sum of Its Parts 91

5 Membrane Transport and Cell Signaling 94

OVERVIEW Life at the Edge 94

CONCEPT 5.1 Cellular membranes are fluid mosaics of lipids and proteins 94
  The Fluidity of Membranes 95
  Evolution of Differences in Membrane Lipid Composition 96
  Membrane Proteins and Their Functions 97
  The Role of Membrane Carbohydrates in Cell-Cell Recognition 98
  Synthesis and Sidedness of Membranes 98

CONCEPT 5.2 Membrane structure results in selective permeability 99
  The Permeability of the Lipid Bilayer 99
  Transport Proteins 99

CONCEPT 5.3 Passive transport is diffusion of a substance across a membrane with no energy investment 99
  Effects of Osmosis on Water Balance 100
  Facilitated Diffusion: Passive Transport Aided by Proteins 102

CONCEPT 5.4 Active transport uses energy to move solutes against their gradients 103
  The Need for Energy in Active Transport 103
  How Ion Pumps Maintain Membrane Potential 104
  Cotransport: Coupled Transport by a Membrane Protein 105

CONCEPT 5.5 Bulk transport across the plasma membrane occurs by exocytosis and endocytosis 106
  Exocytosis 106
  Endocytosis 106

CONCEPT 5.6 The plasma membrane plays a key role in most cell signaling 108
  Local and Long-Distance Signaling 108
  The Three Stages of Cell Signaling: *A Preview* 109
  Reception, the Binding of a Signaling Molecule to a Receptor Protein 109
  Transduction by Cascades of Molecular Interactions 111
  Response: Regulation of Transcription or Cytoplasmic Activities 113
  The Evolution of Cell Signaling 113

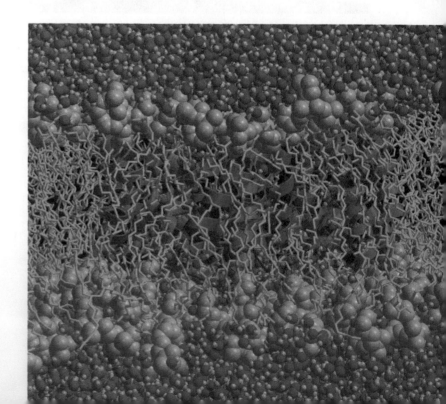

6 An Introduction to Metabolism 116

OVERVIEW The Energy of Life 116

CONCEPT 6.1 An organism's metabolism transforms matter and energy 116
  Metabolic Pathways 116
  Forms of Energy 117
  The Laws of Energy Transformation 118

CONCEPT 6.2 The free-energy change of a reaction tells us whether or not the reaction occurs spontaneously 119
  Free-Energy Change (ΔG), Stability, and Equilibrium 119
  Free Energy and Metabolism 120

CONCEPT 6.3 ATP powers cellular work by coupling exergonic reactions to endergonic reactions 122
  The Structure and Hydrolysis of ATP 122
  How the Hydrolysis of ATP Performs Work 123
  The Regeneration of ATP 124

CONCEPT 6.4 Enzymes speed up metabolic reactions by lowering energy barriers 125
  The Activation Energy Barrier 125
  How Enzymes Speed Up Reactions 126
  Substrate Specificity of Enzymes 126
  Catalysis in the Enzyme's Active Site 127
  Effects of Local Conditions on Enzyme Activity 129
  The Evolution of Enzymes 130

CONCEPT 6.5 Regulation of enzyme activity helps control metabolism 130
  Allosteric Regulation of Enzymes 130
  Specific Localization of Enzymes Within the Cell 132

7 Cellular Respiration and Fermentation 135

OVERVIEW Life Is Work 135

CONCEPT 7.1 Catabolic pathways yield energy by oxidizing organic fuels 136
  Catabolic Pathways and Production of ATP 136
  Redox Reactions: Oxidation and Reduction 136
  The Stages of Cellular Respiration: *A Preview* 139

CONCEPT 7.2 Glycolysis harvests chemical energy by oxidizing glucose to pyruvate 141

CONCEPT 7.3 After pyruvate is oxidized, the citric acid cycle completes the energy-yielding oxidation of organic molecules 142

CONCEPT 7.4 During oxidative phosphorylation, chemiosmosis couples electron transport to ATP synthesis 143
  The Pathway of Electron Transport 144
  Chemiosmosis: The Energy-Coupling Mechanism 145
  An Accounting of ATP Production by Cellular Respiration 147

CONCEPT 7.5 Fermentation and anaerobic respiration enable cells to produce ATP without the use of oxygen 148
  Types of Fermentation 150
  Comparing Fermentation with Anaerobic and Aerobic Respiration 150
  The Evolutionary Significance of Glycolysis 151

CONCEPT 7.6 Glycolysis and the citric acid cycle connect to many other metabolic pathways 151
  The Versatility of Catabolism 151
  Biosynthesis (Anabolic Pathways) 152

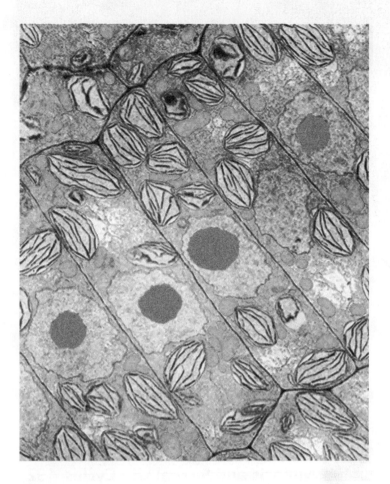

8 Photosynthesis 155

OVERVIEW The Process That Feeds the Biosphere 155

CONCEPT 8.1 Photosynthesis converts light energy to the chemical energy of food 156
  Chloroplasts: The Sites of Photosynthesis in Plants 156
  Tracking Atoms Through Photosynthesis: *Scientific Inquiry* 157
  The Two Stages of Photosynthesis: *A Preview* 158

CONCEPT 8.2 The light reactions convert solar energy to the chemical energy of ATP and NADPH 159
  The Nature of Sunlight 160
  Photosynthetic Pigments: The Light Receptors 160
  Excitation of Chlorophyll by Light 162
  A Photosystem: A Reaction-Center Complex Associated with Light-Harvesting Complexes 163
  Linear Electron Flow 164
  A Comparison of Chemiosmosis in Chloroplasts and Mitochondria 165

CONCEPT 8.3 The Calvin cycle uses the chemical energy of ATP and NADPH to reduce CO_2 to sugar 167
  Evolution of Alternative Mechanisms of Carbon Fixation in Hot, Arid Climates 169
  The Importance of Photosynthesis: *A Review* 171

9 The Cell Cycle 174

OVERVIEW The Key Roles of Cell Division 174

CONCEPT 9.1 Most cell division results in genetically identical daughter cells 175
Cellular Organization of the Genetic Material 175
Distribution of Chromosomes During Eukaryotic Cell Division 175

CONCEPT 9.2 The mitotic phase alternates with interphase in the cell cycle 177
Phases of the Cell Cycle 177
The Mitotic Spindle: *A Closer Look* 177
Cytokinesis: *A Closer Look* 180
Binary Fission in Bacteria 182
The Evolution of Mitosis 183

CONCEPT 9.3 The eukaryotic cell cycle is regulated by a molecular control system 183
Evidence for Cytoplasmic Signals 184
Checkpoints of the Cell Cycle Control System 184
Loss of Cell Cycle Controls in Cancer Cells 187

UNIT 2 Genetics 191

10 Meiosis and Sexual Life Cycles 192

OVERVIEW Variations on a Theme 192

CONCEPT 10.1 Offspring acquire genes from parents by inheriting chromosomes 193
Inheritance of Genes 193
Comparison of Asexual and Sexual Reproduction 193

CONCEPT 10.2 Fertilization and meiosis alternate in sexual life cycles 194
Sets of Chromosomes in Human Cells 194
Behavior of Chromosome Sets in the Human Life Cycle 195
The Variety of Sexual Life Cycles 196

CONCEPT 10.3 Meiosis reduces the number of chromosome sets from diploid to haploid 197
The Stages of Meiosis 197
A Comparison of Mitosis and Meiosis 201

CONCEPT 10.4 Genetic variation produced in sexual life cycles contributes to evolution 201
Origins of Genetic Variation Among Offspring 201
The Evolutionary Significance of Genetic Variation Within Populations 204

11 Mendel and the Gene Idea 206

OVERVIEW Drawing from the Deck of Genes 206

CONCEPT 11.1 Mendel used the scientific approach to identify two laws of inheritance 207
Mendel's Experimental, Quantitative Approach 207
The Law of Segregation 207
The Law of Independent Assortment 211

CONCEPT 11.2 The laws of probability govern Mendelian inheritance 213
The Multiplication and Addition Rules Applied to Monohybrid Crosses 213
Solving Complex Genetics Problems with the Rules of Probability 214

CONCEPT 11.3 Inheritance patterns are often more complex than predicted by simple Mendelian genetics 214
Extending Mendelian Genetics for a Single Gene 215
Multiple Alleles 216
Extending Mendelian Genetics for Two or More Genes 217
Nature and Nurture: The Environmental Impact on Phenotype 218
Integrating a Mendelian View of Heredity and Variation 218

CONCEPT 11.4 Many human traits follow Mendelian patterns of inheritance 219
Pedigree Analysis 220
Recessively Inherited Disorders 220
Dominantly Inherited Disorders 222
Multifactorial Disorders 223
Genetic Counseling Based on Mendelian Genetics 223

12 The Chromosomal Basis of Inheritance 228

OVERVIEW Locating Genes Along Chromosomes 228

CONCEPT 12.1 Mendelian inheritance has its physical basis in the behavior of chromosomes 228
Morgan's Experimental Evidence: *Scientific Inquiry* 230

CONCEPT 12.2 Sex-linked genes exhibit unique patterns of inheritance 231
The Chromosomal Basis of Sex 231
Inheritance of X-Linked Genes 232
X Inactivation in Female Mammals 233

CONCEPT 12.3 Linked genes tend to be inherited together because they are located near each other on the same chromosome 234
How Linkage Affects Inheritance 234
Genetic Recombination and Linkage 235
Mapping the Distance Between Genes Using Recombination Data: *Scientific Inquiry* 237

CONCEPT 12.4 Alterations of chromosome number or structure cause some genetic disorders 240
Abnormal Chromosome Number 240
Alterations of Chromosome Structure 241
Human Disorders Due to Chromosomal Alterations 241

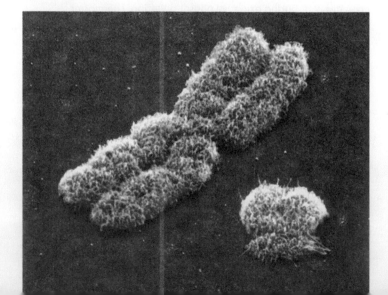

13 The Molecular Basis of Inheritance 245

OVERVIEW Life's Operating Instructions 245

CONCEPT 13.1 DNA is the genetic material 245
The Search for the Genetic Material: *Scientific Inquiry* 246
Building a Structural Model of DNA: *Scientific Inquiry* 248

CONCEPT 13.2 Many proteins work together in DNA replication and repair 251
The Basic Principle: Base Pairing to a Template Strand 252
DNA Replication: *A Closer Look* 252
Proofreading and Repairing DNA 257
Evolutionary Significance of Altered DNA Nucleotides 258
Replicating the Ends of DNA Molecules 258

CONCEPT 13.3 A chromosome consists of a DNA molecule packed together with proteins 259

CONCEPT 13.4 Understanding DNA structure and replication makes genetic engineering possible 261
DNA Cloning: Making Multiple Copies of a Gene or Other DNA Segment 262
Using Restriction Enzymes to Make Recombinant DNA 262
Amplifying DNA *in Vitro*: The Polymerase Chain Reaction (PCR) and Its Use in Cloning 264
DNA Sequencing 265

14 Gene Expression: From Gene to Protein 268

OVERVIEW The Flow of Genetic Information 268

CONCEPT 14.1 Genes specify proteins via transcription and translation 269
Evidence from the Study of Metabolic Defects 269
Basic Principles of Transcription and Translation 270
The Genetic Code 272

CONCEPT 14.2 Transcription is the DNA-directed synthesis of RNA: *a closer look* 274
Molecular Components of Transcription 274
Synthesis of an RNA Transcript 274

CONCEPT 14.3 Eukaryotic cells modify RNA after transcription 276
Alteration of mRNA Ends 276
Split Genes and RNA Splicing 277

CONCEPT 14.4 Translation is the RNA-directed synthesis of a polypeptide: *a closer look* 278
Molecular Components of Translation 278
Building a Polypeptide 281
Completing and Targeting the Functional Protein 282
Making Multiple Polypeptides in Bacteria and Eukaryotes 286

CONCEPT 14.5 Mutations of one or a few nucleotides can affect protein structure and function 288
Types of Small-Scale Mutations 288
Mutagens 290
What Is a Gene? *Revisiting the Question* 290

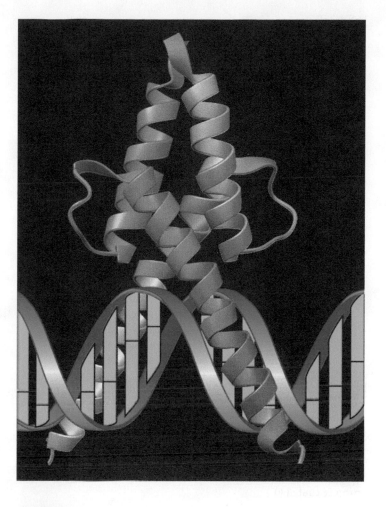

15 Regulation of Gene Expression 293

OVERVIEW Differential Expression of Genes 293

CONCEPT 15.1 Bacteria often respond to environmental change by regulating transcription 293
Operons: The Basic Concept 294
Repressible and Inducible Operons: Two Types of Negative Gene Regulation 295
Positive Gene Regulation 297

CONCEPT 15.2 Eukaryotic gene expression is regulated at many stages 298
Differential Gene Expression 298
Regulation of Chromatin Structure 299
Regulation of Transcription Initiation 299
Mechanisms of Post-Transcriptional Regulation 304

CONCEPT 15.3 Noncoding RNAs play multiple roles in controlling gene expression 305
Effects on mRNAs by MicroRNAs and Small Interfering RNAs 305
Chromatin Remodeling and Effects on Transcription by ncRNAs 306

CONCEPT 15.4 Researchers can monitor expression of specific genes 307
Studying the Expression of Single Genes 307
Studying the Expression of Groups of Genes 308

16 Development, Stem Cells, and Cancer 311

OVERVIEW Orchestrating Life's Processes 311

CONCEPT 16.1 A program of differential gene expression leads to the different cell types in a multicellular organism 312
    A Genetic Program for Embryonic Development 312
    Cytoplasmic Determinants and Inductive Signals 312
    Sequential Regulation of Gene Expression during Cellular Differentiation 313
    Pattern Formation: Setting Up the Body Plan 317

CONCEPT 16.2 Cloning of organisms showed that differentiated cells could be "reprogrammed" and ultimately led to the production of stem cells 320
    Cloning Plants and Animals 320
    Stem Cells of Animals 322

CONCEPT 16.3 Abnormal regulation of genes that affect the cell cycle can lead to cancer 324
    Types of Genes Associated with Cancer 324
    Interference with Cell-Signaling Pathways 325
    The Multistep Model of Cancer Development 326
    Inherited Predisposition and Other Factors Contributing to Cancer 327

17 Viruses 330

OVERVIEW A Borrowed Life 330

CONCEPT 17.1 A virus consists of a nucleic acid surrounded by a protein coat 330
    Viral Genomes 331
    Capsids and Envelopes 331

CONCEPT 17.2 Viruses replicate only in host cells 332
    General Features of Viral Replicative Cycles 332
    Replicative Cycles of Phages 333
    Replicative Cycles of Animal Viruses 335
    Evolution of Viruses 336

CONCEPT 17.3 Viruses are formidable pathogens in animals and plants 338
    Viral Diseases in Animals 338
    Emerging Viruses 338
    Viral Diseases in Plants 341

18 Genomes and Their Evolution 343

OVERVIEW Reading the Leaves from the Tree of Life 343

CONCEPT 18.1 The Human Genome Project fostered development of faster, less expensive sequencing techniques 344

CONCEPT 18.2 Scientists use bioinformatics to analyze genomes and their functions 345
    Centralized Resources for Analyzing Genome Sequences 345
    Understanding the Functions of Protein-Coding Genes 345
    Understanding Genes and Gene Expression at the Systems Level 346

CONCEPT 18.3 Genomes vary in size, number of genes, and gene density 347
    Genome Size 348
    Number of Genes 348
    Gene Density and Noncoding DNA 349

CONCEPT 18.4 Multicellular eukaryotes have much noncoding DNA and many multigene families 349
    Transposable Elements and Related Sequences 350
    Other Repetitive DNA, Including Simple Sequence DNA 351
    Genes and Multigene Families 352

CONCEPT 18.5 Duplication, rearrangement, and mutation of DNA contribute to genome evolution 353
    Duplication of Entire Chromosome Sets 353
    Alterations of Chromosome Structure 353
    Duplication and Divergence of Gene-Sized Regions of DNA 354
    Rearrangements of Parts of Genes: Exon Duplication and Exon Shuffling 355
    How Transposable Elements Contribute to Genome Evolution 357

CONCEPT 18.6 Comparing genome sequences provides clues to evolution and development 357
    Comparing Genomes 358
    Comparing Developmental Processes 360

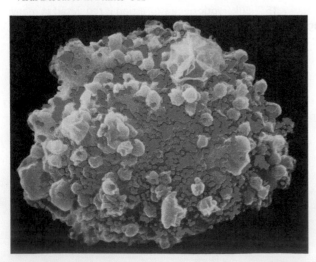

UNIT 3 Evolution 364

19 Descent with Modification 365

OVERVIEW Endless Forms Most Beautiful 365

CONCEPT 19.1 The Darwinian revolution challenged traditional views of a young Earth inhabited by unchanging species 366
- Scala Naturae and Classification of Species 366
- Ideas About Change over Time 366
- Lamarck's Hypothesis of Evolution 367

CONCEPT 19.2 Descent with modification by natural selection explains the adaptations of organisms and the unity and diversity of life 368
- Darwin's Research 368
- Ideas from The Origin of Species 370

CONCEPT 19.3 Evolution is supported by an overwhelming amount of scientific evidence 373
- Direct Observations of Evolutionary Change 373
- Homology 375
- The Fossil Record 376
- Biogeography 377
- What Is Theoretical About Darwin's View of Life? 379

21 The Evolution of Populations 399

OVERVIEW The Smallest Unit of Evolution 399

CONCEPT 21.1 Genetic variation makes evolution possible 400
- Genetic Variation 400
- Sources of Genetic Variation 401

CONCEPT 21.2 The Hardy-Weinberg equation can be used to test whether a population is evolving 402
- Gene Pools and Allele Frequencies 402
- The Hardy-Weinberg Principle 403

CONCEPT 21.3 Natural selection, genetic drift, and gene flow can alter allele frequencies in a population 406
- Natural Selection 407
- Genetic Drift 407
- Gene Flow 409

CONCEPT 21.4 Natural selection is the only mechanism that consistently causes adaptive evolution 410
- Natural Selection: A Closer Look 410
- The Key Role of Natural Selection in Adaptive Evolution 412
- Sexual Selection 412
- The Preservation of Genetic Variation 413
- Why Natural Selection Cannot Fashion Perfect Organisms 414

22 The Origin of Species 418

OVERVIEW That "Mystery of Mysteries" 418

CONCEPT 22.1 The biological species concept emphasizes reproductive isolation 418
- The Biological Species Concept 419
- Other Definitions of Species 422

CONCEPT 22.2 Speciation can take place with or without geographic separation 423
- Allopatric ("Other Country") Speciation 423
- Sympatric ("Same Country") Speciation 425
- Allopatric and Sympatric Speciation: A Review 427

CONCEPT 22.3 Hybrid zones reveal factors that cause reproductive isolation 428
- Patterns Within Hybrid Zones 428
- Hybrid Zones over Time 429

CONCEPT 22.4 Speciation can occur rapidly or slowly and can result from changes in few or many genes 430
- The Time Course of Speciation 430
- Studying the Genetics of Speciation 432
- From Speciation to Macroevolution 433

34 Circulation and Gas Exchange 684

OVERVIEW Trading Places 684

CONCEPT 34.1 Circulatory systems link exchange surfaces with cells throughout the body 685
Gastrovascular Cavities 685
Open and Closed Circulatory Systems 685
Organization of Vertebrate Circulatory Systems 686

CONCEPT 34.2 Coordinated cycles of heart contraction drive double circulation in mammals 688
Mammalian Circulation 688
The Mammalian Heart: *A Closer Look* 688
Maintaining the Heart's Rhythmic Beat 690

CONCEPT 34.3 Patterns of blood pressure and flow reflect the structure and arrangement of blood vessels 690
Blood Vessel Structure and Function 691
Blood Flow Velocity 691
Blood Pressure 692
Capillary Function 693
Fluid Return by the Lymphatic System 693

CONCEPT 34.4 Blood components function in exchange, transport, and defense 695
Blood Composition and Function 695
Cardiovascular Disease 697

CONCEPT 34.5 Gas exchange occurs across specialized respiratory surfaces 698
Partial Pressure Gradients in Gas Exchange 699
Respiratory Media 700
Respiratory Surfaces 700
Gills in Aquatic Animals 701
Tracheal Systems in Insects 702
Lungs 702

CONCEPT 34.6 Breathing ventilates the lungs 704
How a Mammal Breathes 704
Control of Breathing in Humans 705

CONCEPT 34.7 Adaptations for gas exchange include pigments that bind and transport gases 706
Coordination of Circulation and Gas Exchange 706
Respiratory Pigments 706
Carbon Dioxide Transport 708
Respiratory Adaptations of Diving Mammals 708

35 The Immune System 711

OVERVIEW Recognition and Response 711

CONCEPT 35.1 In innate immunity, recognition and response rely on traits common to groups of pathogens 712
Innate Immunity of Invertebrates 712
Innate Immunity of Vertebrates 713
Evasion of Innate Immunity by Pathogens 715

CONCEPT 35.2 In adaptive immunity, receptors provide pathogen-specific recognition 715
Antigen Recognition by B Cells and Antibodies 716
Antigen Recognition by T Cells 716
B Cell and T Cell Development 717

CONCEPT 35.3 Adaptive immunity defends against infection of body fluids and body cells 720
Helper T Cells: A Response to Nearly All Antigens 720
Cytotoxic T Cells: A Response to Infected Cells 721
B Cells and Antibodies: A Response to Extracellular Pathogens 722
Summary of the Humoral and Cell-Mediated Immune Responses 722
Active and Passive Immunization 723
Antibodies as Tools 724
Immune Rejection 724
Disruptions in Immune System Function 724
Cancer and Immunity 726

36 Reproduction and Development 729

OVERVIEW Pairing Up for Sexual Reproduction 729

CONCEPT 36.1 Both asexual and sexual reproduction occur in the animal kingdom 729
Mechanisms of Asexual Reproduction 730
Sexual Reproduction: An Evolutionary Enigma 730
Reproductive Cycles 731
Variation in Patterns of Sexual Reproduction 731
External and Internal Fertilization 732
Ensuring the Survival of Offspring 732

CONCEPT 36.2 Reproductive organs produce and transport gametes 733
Variation in Reproductive Systems 733
Human Male Reproductive Anatomy 734
Human Female Reproductive Anatomy 735
Gametogenesis 735

CONCEPT 36.3 The interplay of tropic and sex hormones regulates reproduction in mammals 738
Hormonal Control of the Male Reproductive System 739
Hormonal Control of Female Reproductive Cycles 739
Human Sexual Response 741

CONCEPT 36.4 Fertilization, cleavage, and gastrulation initiate embryonic development 742
Fertilization 743
Cleavage and Gastrulation 744
Human Conception, Embryonic Development, and Birth 745
Contraception 747
Infertility and *In Vitro* Fertilization 748

UNIT 7 Ecology 817

40 Population Ecology and the Distribution of Organisms 818

OVERVIEW Discovering Ecology 818

CONCEPT 40.1 Earth's climate influences the structure and distribution of terrestrial biomes 821
Global Climate Patterns 821
Regional Effects on Climate 821
Climate and Terrestrial Biomes 822
General Features of Terrestrial Biomes 823

CONCEPT 40.2 Aquatic biomes are diverse and dynamic systems that cover most of Earth 827
Zonation in Aquatic Biomes 827

CONCEPT 40.3 Interactions between organisms and the environment limit the distribution of species 830
Dispersal and Distribution 830
Biotic Factors 831
Abiotic Factors 831

CONCEPT 40.4 Dynamic biological processes influence population density, dispersion, and demographics 832
Density and Dispersion 832
Demographics 834

CONCEPT 40.5 The exponential and logistic models describe the growth of populations 835
Per Capita Rate of Increase 835
Exponential Growth 836
Carrying Capacity 836
The Logistic Growth Model 837
The Logistic Model and Real Populations 838

CONCEPT 40.6 Population dynamics are influenced strongly by life history traits and population density 839
"Trade-offs" and Life Histories 839
Population Change and Population Density 840
Mechanisms of Density-Dependent Population Regulation 840
Population Dynamics 840

41 Species Interactions 845

OVERVIEW Communities in Motion 845

CONCEPT 41.1 Interactions within a community may help, harm, or have no effect on the species involved 846
Competition 846
Predation 848
Herbivory 849
Symbiosis 849
Facilitation 851

CONCEPT 41.2 Diversity and trophic structure characterize biological communities 851
Species Diversity 851
Diversity and Community Stability 853
Trophic Structure 853
Species with a Large Impact 854
Bottom-Up and Top-Down Controls 855

CONCEPT 41.3 Disturbance influences species diversity and composition 856
Characterizing Disturbance 856
Ecological Succession 857
Human Disturbance 858

CONCEPT 41.4 Biogeographic factors affect community diversity 859
Latitudinal Gradients 859
Area Effects 860

CONCEPT 41.5 Pathogens alter community structure locally and globally 861
Effects on Community Structure 861
Community Ecology and Zoonotic Diseases 861

42 Ecosystems and Energy 864

OVERVIEW Cool Ecosystem 864

CONCEPT 42.1 Physical laws govern energy flow and chemical cycling in ecosystems 865
Conservation of Energy 865
Conservation of Mass 865
Energy, Mass, and Trophic Levels 866

CONCEPT 42.2 Energy and other limiting factors control primary production in ecosystems 866
Ecosystem Energy Budgets 867
Primary Production in Aquatic Ecosystems 868
Primary Production in Terrestrial Ecosystems 869

CONCEPT 42.3 Energy transfer between trophic levels is typically only 10% efficient 870
Production Efficiency 870
Trophic Efficiency and Ecological Pyramids 871

CONCEPT 42.4 Biological and geochemical processes cycle nutrients and water in ecosystems 872
Decomposition and Nutrient Cycling Rates 873
Biogeochemical Cycles 873
Case Study: Nutrient Cycling in the Hubbard Brook Experimental Forest 876

CONCEPT 42.5 Restoration ecologists help return degraded ecosystems to a more natural state 877
Bioremediation 877
Biological Augmentation 878
Restoration Projects Worldwide 878

43 Global Ecology and Conservation Biology 882

OVERVIEW Psychedelic Treasure 882

CONCEPT 43.1 Human activities threaten Earth's biodiversity 883
Three Levels of Biodiversity 883
Biodiversity and Human Welfare 884
Threats to Biodiversity 885

CONCEPT 43.2 Population conservation focuses on population size, genetic diversity, and critical habitat 888
Small-Population Approach 888
Declining-Population Approach 890
Weighing Conflicting Demands 891

CONCEPT 43.3 Landscape and regional conservation help sustain biodiversity 891
Landscape Structure and Biodiversity 892
Establishing Protected Areas 893

CONCEPT 43.4 Earth is changing rapidly as a result of human actions 895
Nutrient Enrichment 895
Toxins in the Environment 896
Greenhouse Gases and Climate Change 897

CONCEPT 43.5 The human population is no longer growing exponentially but is still increasing rapidly 900
The Global Human Population 900
Global Carrying Capacity 901

CONCEPT 43.6 Sustainable development can improve human lives while conserving biodiversity 902
Sustainable Development 902
The Future of the Biosphere 903

Appendix A   Answers A-1
Appendix B   Periodic Table of the Elements B-1
Appendix C   The Metric System C-1
Appendix D   A Comparison of the Light Microscope and the Electron Microscope D-1
Appendix E   Classification of Life E-1
Appendix F   Scientific Skills Review F-1
Glossary   G-1
Index   I-1

Introduction: Evolution and the Foundations of Biology

KEY CONCEPTS

1.1 Studying the diverse forms of life reveals common themes

1.2 The Core Theme: Evolution accounts for the unity and diversity of life

1.3 Biological inquiry entails forming and testing hypotheses based on observations of nature

OVERVIEW

Inquiring About Life

The brilliant white sand dunes and sparse clumps of beach grass along the Florida seashore afford little cover for the beach mice that live there. However, a beach mouse's light, dappled fur acts as camouflage, allowing the mouse to blend into its surroundings **(Figure 1.1)**. Although mice of the same species (oldfield mice, *Peromyscus polionotus*) also inhabit nearby inland areas, the inland mice are much darker in color, matching the darker soil and vegetation where they live **(Figure 1.2)**. This close match of each mouse to its environment is vital for survival, since hawks, herons, and other sharp-eyed predators periodically scan the landscape for food. How has the color of each mouse come to be so well matched, or *adapted*, to the local background?

An organism's adaptations to its environment, such as camouflage that helps protect it from predators, are the result of **evolution**, the process of change that has transformed life from its beginnings to the astounding array of organisms today. Evolution is the fundamental principle of biology and the core theme of this book.

Although biologists know a great deal about life on Earth, many mysteries remain. The question of how the mice's coats have come to match the colors of their habitats is just one example. Posing questions about the living world and seeking answers through scientific inquiry are the central activities of **biology**, the scientific study of life. Biologists' questions can be ambitious. They may ask how a single tiny cell becomes a tree or a dog, how the human mind works, or how the different forms of life in a forest interact. When questions occur to you as you observe the living world, you are already thinking like a biologist.

How do biologists make sense of life's diversity and complexity? This opening chapter sets up a framework for answering this question. The first part of the chapter provides a panoramic view of the biological "landscape," organized around a set of unifying themes. We'll then focus on biology's core theme, evolution. Finally, we'll examine the process of scientific inquiry—how scientists ask and attempt to answer questions about the natural world.

▼ **Figure 1.1** What can this beach mouse teach us about biology?

▶ **Figure 1.2 An "inland" oldfield mouse (*Peromyscus polionotus*).** This mouse has a much darker back, side, and face than mice of the same species that inhabit sand dunes.

Studying the diverse forms of life reveals common themes

Biology is a subject of enormous scope, and exciting new biological discoveries are being made every day. How can you organize and make sense of all the information you'll encounter as you study biology? Focusing on a few big ideas—ways of thinking about life that will still hold true decades from now—will help. Here, we'll describe five unifying themes to serve as touchstones as you proceed through this book.

Theme: New Properties Emerge at Successive Levels of Biological Organization

ORGANIZATION The study of life extends from the microscopic scale of the molecules and cells that make up organisms to the global scale of the entire living planet. As biologists, we can divide this enormous range into different levels of biological organization.

Imagine zooming in from space to take a closer and closer look at life on Earth. It is spring in Ontario, Canada, and our destination is a local forest, where we will eventually narrow our focus down to the molecules that make up a maple leaf. **Figure 1.3** narrates this journey into life, as the numbers guide

▼ Figure 1.3 Exploring Levels of Biological Organization

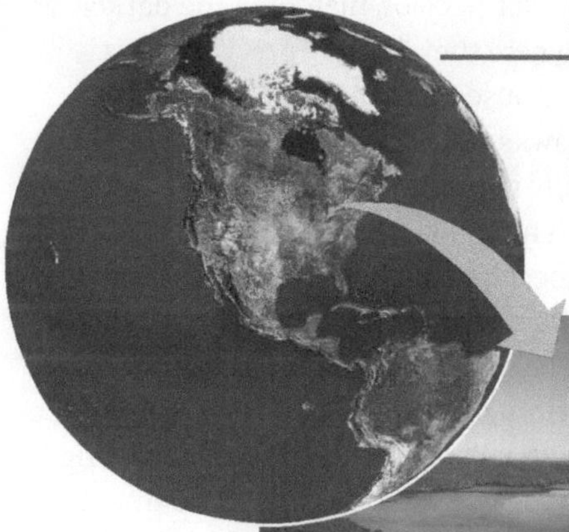

◄ 1 The Biosphere

Even from space, we can see signs of Earth's life—in the green mosaic of the forests, for example. We can also see the scale of the entire biosphere, which consists of all life on Earth and all the places where life exists: most regions of land, most bodies of water, the atmosphere to an altitude of several kilometers, and even sediments far below the ocean floor.

◄ 2 Ecosystems

Our first scale change brings us to a North American forest with many deciduous trees (trees that lose their leaves and grow new ones each year). A deciduous forest is an example of an ecosystem, as are grasslands, deserts, and coral reefs. An ecosystem consists of all the living things in a particular area, along with all the nonliving components of the environment with which life interacts, such as soil, water, atmospheric gases, and light.

► 3 Communities

The array of organisms inhabiting a particular ecosystem is called a biological community. The community in our forest ecosystem includes many kinds of trees and other plants, various animals, mushrooms and other fungi, and enormous numbers of diverse microorganisms, which are living forms, such as bacteria, that are too small to see without a microscope. Each of these forms of life is called a *species*.

► 4 Populations

A population consists of all the individuals of a species living within the bounds of a specified area. For example, our forest includes a population of sugar maple trees and a population of white-tailed deer. A community is therefore the set of populations that inhabit a particular area.

▲ 5 Organisms

Individual living things are called organisms. Each of the maple trees and other plants in the forest is an organism, and so is each deer, frog, beetle, and other forest animals. The soil teems with microorganisms such as bacteria.

you through photographs illustrating the hierarchy of biological organization.

Zooming in at ever-finer resolution illustrates the principle of *reductionism*—the approach of reducing complex systems to simpler components that are more manageable to study. Reductionism is a powerful strategy in biology. For example, by studying the molecular structure of DNA that had been extracted from cells, James Watson and Francis Crick inferred the chemical basis of biological inheritance. However, although it has propelled many major discoveries, reductionism provides a necessarily incomplete view of life on Earth, as we'll discuss next.

Emergent Properties

Let's reexamine Figure 1.3, beginning this time at the molecular level and then zooming out. Viewed this way, we see that at each level, novel properties emerge that are absent from the preceding one. These **emergent properties** are due to the arrangement and interactions of parts as complexity increases. For example, although photosynthesis occurs in an intact chloroplast, it will not take place in a disorganized test-tube mixture of chlorophyll and other chloroplast molecules. The coordinated processes of photosynthesis require a specific organization of these molecules in the chloroplast. Isolated components of living systems, acting as the objects of study in

▼ 6 Organs and Organ Systems

The structural hierarchy of life continues to unfold as we explore the architecture of more complex organisms. A maple leaf is an example of an organ, a body part that carries out a particular function in the body. Stems and roots are the other major organs of plants. The organs of complex animals and plants are organized into organ systems, each a team of organs that cooperate in a larger function. Organs consist of multiple tissues.

▶ 10 Molecules

Our last scale change drops us into a chloroplast for a view of life at the molecular level. A molecule is a chemical structure consisting of two or more units called atoms, represented as balls in this computer graphic of a chlorophyll molecule. Chlorophyll is the pigment molecule that makes a maple leaf green, and it absorbs sunlight during photosynthesis. Within each chloroplast, millions of chlorophyll molecules are organized into systems that convert light energy to the chemical energy of food.

▶ 9 Organelles

Chloroplasts are examples of organelles, the various functional components present in cells. This image, taken by a powerful microscope, shows a single chloroplast.

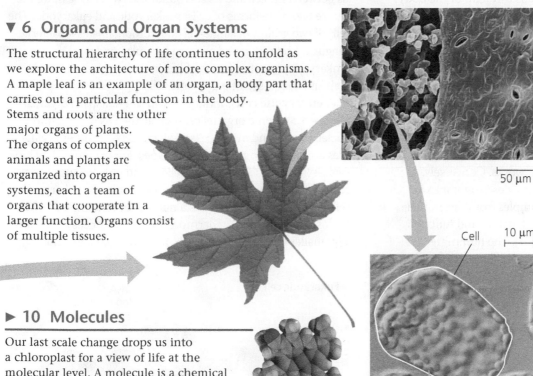

Atoms

Chlorophyll molecule

Cell   10 μm

50 μm

Chloroplast

1 μm

◀ 7 Tissues

To see the tissues of a leaf requires a microscope. Each tissue is a group of cells that work together, performing a specialized function. The leaf shown here has been cut on an angle. The honeycombed tissue in the interior of the leaf (left side of photo) is the main location of photosynthesis, the process that converts light energy to the chemical energy of sugar. The jigsaw puzzle–like "skin" on the surface of the leaf is a tissue called epidermis (right side of photo). The pores through the epidermis allow entry of the gas CO_2, a raw material for sugar production.

▲ 8 Cells

The cell is life's fundamental unit of structure and function. Some organisms are single cells, while others are multicellular. A single cell performs all the functions of life, while a multicellular organism has a division of labor among specialized cells. Here we see a magnified view of cells in a leaf tissue. One cell is about 40 micrometers (μm) across—about 500 of them would reach across a small coin. As tiny as these cells are, you can see that each contains numerous green structures called chloroplasts, which are responsible for photosynthesis.

a reductionist approach to biology, typically lack some of the properties that emerge at higher levels of organization.

Emergent properties are not unique to life. A box of bicycle parts won't transport you anywhere, but if they are arranged in a certain way, you can pedal to your chosen destination. Compared to such nonliving examples, however, the unrivaled complexity of biological systems makes the emergent properties of life especially challenging to study.

To fully explore emergent properties, biologists today complement reductionism with **systems biology**, the exploration of a biological system by analyzing the interactions among its parts. A single leaf cell can be considered a system, as can a frog, an ant colony, or a desert ecosystem. By examining and modeling the dynamic behavior of an integrated network of components, systems biology enables us to pose new kinds of questions. For example, how does a drug that lowers blood pressure affect the functioning of organs throughout the body? At a larger scale, how does a gradual increase in atmospheric carbon dioxide alter ecosystems and the entire biosphere? Systems biology can be used to study life at all levels.

Structure and Function

At each level of the biological hierarchy, we find a correlation of structure and function. Consider the leaf in Figure 1.3: Its thin, flat shape maximizes the capture of sunlight by chloroplasts. More generally, analyzing a biological structure gives us clues about what it does and how it works. Conversely, knowing the function of something provides insight into its structure and organization. Many examples from the animal kingdom show a correlation between structure and function, including the hummingbird **(Figure 1.4)**. The hummingbird's anatomy allows the wings to rotate at the shoulder, so hummingbirds have the ability, unique among birds, to fly backward or hover in place. Hovering, the birds can extend their long slender beaks into flowers and feed on nectar. The

elegant match of form and function in the structures of life is explained by natural selection, as we'll explore shortly.

The Cell: An Organism's Basic Unit of Structure and Function

In life's structural hierarchy, the cell is the smallest unit of organization that can perform all required activities. In fact, the activities of organisms are all based on the activities of cells. For instance, the movement of your eyes as you read this sentence results from the activities of muscle and nerve cells. Even a process that occurs on a global scale, such as the recycling of carbon atoms, is the cumulative product of cellular functions, including the photosynthetic activity of chloroplasts in leaf cells.

All cells share certain characteristics. For instance, every cell is enclosed by a membrane that regulates the passage of materials between the cell and its surroundings. Nevertheless, we recognize two main forms of cells: prokaryotic and eukaryotic. The cells of two groups of single-celled microorganisms—bacteria (singular, *bacterium*) and archaea (singular, *archaean*)—are prokaryotic. All other forms of life, including plants and animals, are composed of eukaryotic cells.

A **eukaryotic cell** contains membrane-enclosed organelles **(Figure 1.5)**. Some organelles, such as the DNA-containing nucleus, are found in the cells of all eukaryotes; other organelles are specific to particular cell types. For example, the chloroplast in Figure 1.3 is an organelle found only in eukaryotic cells that carry out photosynthesis. In contrast to eukaryotic cells, a **prokaryotic cell** lacks a nucleus or other membrane-enclosed organelles. Furthermore, prokaryotic cells are generally smaller than eukaryotic cells, as shown in Figure 1.5.

▲ **Figure 1.4 Form fits function in a hummingbird's body.** The unusual bone structure of a hummingbird's wing allows the bird to rotate its wings in all directions, enabling it to fly backward and to hover while it feeds.

? *What other examples of form fitting function do you observe in this photograph?*

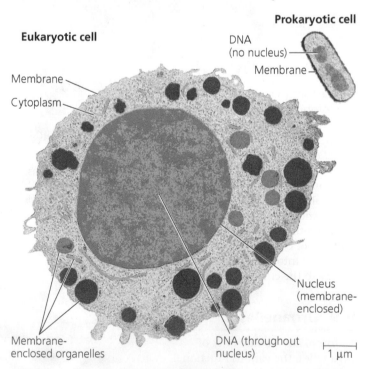

Eukaryotic cell

Prokaryotic cell

DNA (no nucleus)

Membrane

Membrane

Cytoplasm

Nucleus (membrane-enclosed)

Membrane-enclosed organelles

DNA (throughout nucleus)

1 μm

▲ **Figure 1.5 Contrasting eukaryotic and prokaryotic cells in size and complexity.**

Theme: Life's Processes Involve the Expression and Transmission of Genetic Information

INFORMATION Within cells, structures called chromosomes contain genetic material in the form of **DNA (deoxyribonucleic acid)**. In cells that are preparing to divide, the chromosomes may be made visible using a dye that appears blue when bound to the DNA **(Figure 1.6)**.

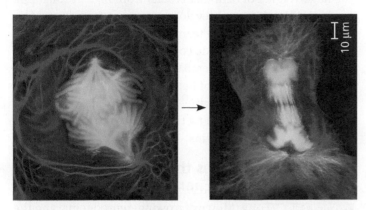

▲ **Figure 1.6 A lung cell from a newt divides into two smaller cells that will grow and divide again.**

DNA Structure and Function

Each time a cell divides, the DNA is first *replicated*, or copied, and each of the two cellular offspring inherits a complete set of chromosomes, identical to that of the parent cell. Each chromosome contains one very long DNA molecule with hundreds or thousands of **genes**, each a stretch of DNA arranged along the chromosome. Transmitted from parents to offspring, genes are the units of inheritance. They encode the information necessary to build all of the molecules synthesized within a cell, which in turn establish that cell's identity and function. Each of us began as a single cell stocked with DNA inherited from our parents. The replication of that DNA during each round of cell division transmitted copies of the DNA to what eventually became the trillions of cells of the human body. As the cells grew and divided, the genetic information encoded by the DNA directed our development **(Figure 1.7)**.

The molecular structure of DNA accounts for its ability to store information. A DNA molecule is made up of two long chains, called strands, arranged in a double helix. Each chain is made up of four kinds of chemical building blocks called nucleotides, abbreviated A, T, C, and G **(Figure 1.8)**. The way DNA encodes information is analogous to how we arrange the letters of the alphabet into words and phrases with specific meanings. The word *rat*, for example, evokes a rodent; the words *tar* and *art*, which contain the same letters, mean very different things. We can think of nucleotides as a four-letter alphabet. Specific sequences of these four nucleotides encode the information in genes.

DNA provides the blueprints for making proteins, which are the major players in building and maintaining the cell and

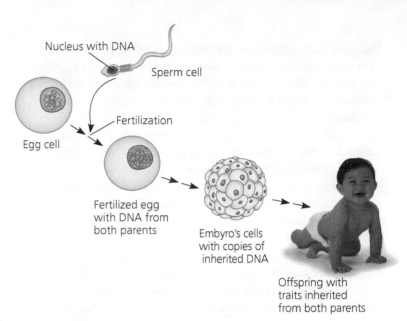

▲ **Figure 1.7 Inherited DNA directs development of an organism.**

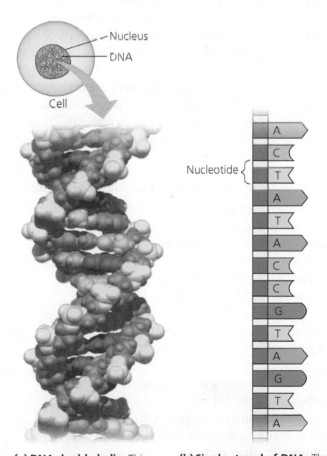

(a) DNA double helix. This model shows each atom in a segment of DNA. Made up of two long chains of building blocks called nucleotides, a DNA molecule takes the three-dimensional form of a double helix.

(b) Single strand of DNA. These geometric shapes and letters are simple symbols for the nucleotides in a small section of one chain of a DNA molecule. Genetic information is encoded in specific sequences of the four types of nucleotides. (Their names are abbreviated A, T, C, and G.)

▲ **Figure 1.8 DNA: The genetic material.**

carrying out its activities. For instance, a particular bacterial gene may specify a certain enzyme protein required to assemble the cell membrane, while a human gene may denote an antibody protein that helps fight off infection.

Genes control protein production indirectly, using a related molecule called RNA as an intermediary. The sequence of nucleotides along a gene is transcribed into RNA, which is then translated into a specific protein with a unique shape and function. This entire process, by which the information in a gene directs the manufacture of a cellular product, is called **gene expression**.

In translating genes into proteins, all forms of life employ essentially the same genetic code: A particular sequence of nucleotides says the same thing in one organism as it does in another. Differences between organisms reflect differences between their nucleotide sequences rather than between their genetic codes.

Not all RNA molecules in the cell are translated into protein; some RNAs carry out other important tasks. For example, we have known for decades that some types of RNA are actually components of the cellular machinery that manufactures proteins. Recently, scientists have discovered whole new classes of RNA that play other roles in the cell, such as regulating the functioning of protein-coding genes. All these RNAs are specified by genes, and the production of these RNAs is also referred to as gene expression. By carrying the instructions for making proteins and RNAs and by replicating with each cell division, DNA ensures faithful inheritance of genetic information from generation to generation.

Genomics: Large-Scale Analysis of DNA Sequences

The entire "library" of genetic instructions that an organism inherits is called its **genome**. A typical human cell has two similar sets of chromosomes, and each set has approximately 3 billion nucleotide pairs of DNA. If the one-letter abbreviations for the nucleotides of one strand in a set were written in letters the size of those you are now reading, the genetic text would fill about 800 introductory biology textbooks.

Since the early 1990s, the pace at which researchers can determine the sequence of a genome has accelerated at an almost unbelievable rate, enabled by a revolution in technology. The entire sequence of nucleotides in the human genome is now known, along with the genome sequences of many other organisms, including other animals and numerous plants, fungi, bacteria, and archaea. To make sense of the deluge of data from genome-sequencing projects and the growing catalog of known gene functions, scientists are applying a systems biology approach at the cellular and molecular levels. Rather than investigating a single gene at a time, researchers study whole sets of genes in one or more species—an approach called **genomics**.

Three important research developments have made the genomic approach possible. One is "high-throughput" technology, tools that can analyze biological materials very rapidly. The second major development is **bioinformatics**, the use of computational tools to store, organize, and analyze the huge volume of data that results from high-throughput methods. The third key development is the formation of interdisciplinary research teams—melting pots of diverse specialists that may include computer scientists, mathematicians, engineers, chemists, physicists, and, of course, biologists from a variety of fields. Researchers in such teams aim to learn how the activities of all the proteins and non-translated RNAs encoded by the DNA are coordinated in cells and in whole organisms.

Theme: Life Requires the Transfer and Transformation of Energy and Matter

ENERGY AND MATTER Moving, growing, reproducing, and the various cellular activities of life are work, and work requires energy. Input of energy, primarily from the sun, and transformation of energy from one form to another make life possible **(Figure 1.9)**. Chlorophyll molecules within plants' leaves convert the energy of sunlight to the chemical energy of food, the sugars produced during photosynthesis (see Figure 1.3). The chemical energy in sugar is then passed along by plants and other photosynthetic organisms (producers) to consumers. Consumers are organisms, such as animals, that feed on producers and other consumers.

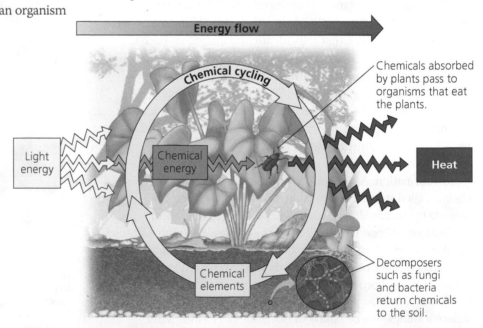

▲ **Figure 1.9 Energy flow and chemical cycling.** There is a one-way flow of energy in an ecosystem: During photosynthesis, plants convert energy from sunlight to chemical energy (stored in sugars), which is used by plants and other organisms to do work and is eventually lost from the ecosystem as heat. In contrast, chemicals cycle between organisms and the physical environment.

When an organism uses chemical energy to perform work, some of that energy is converted to thermal energy and is dissipated to the surroundings as heat. As a result, energy flows *through* an ecosystem, usually entering as light and exiting as heat. In contrast, chemical elements are recycled *within* an ecosystem (see Figure 1.9). Chemicals that a plant absorbs from the air or soil may be incorporated into the plant's body, then passed to an animal that eats the plant. Eventually, these chemicals will be returned to the environment by decomposers, such as bacteria and fungi, that break down waste products, organic debris, and the bodies of dead organisms. The chemicals are then available to be taken up by plants again, thereby completing the cycle.

Theme: Organisms Interact with Other Organisms and the Physical Environment

INTERACTIONS Turn again to Figure 1.3, this time focusing on the ecosystem, including the forest and its surroundings. Each organism interacts continuously with physical factors in its environment. The leaves of a tree, for example, absorb light from the sun, take in carbon dioxide from the air, and release oxygen to the air. The environment is also affected by the organisms living there. For example, a plant takes up water and minerals from the soil through its roots, and its roots break up rocks, thereby contributing to the formation of soil. On a global scale, plants and other photosynthetic organisms have generated all the oxygen in the atmosphere.

A tree also interacts with other organisms, such as soil microorganisms associated with its roots, insects that live in the tree, and animals that eat its leaves and fruit. Such interactions between organisms include those that are mutually beneficial **(Figure 1.10)**; those in which one species benefits and the other is harmed (as when a lion kills and eats a zebra); and those in which both species are harmed (as when two plants compete for a soil resource that is in short supply). As we'll see, interactions between organisms not only affect the participants; they also affect how populations evolve over time.

▼ **Figure 1.10 An interaction between species that benefits both participants.** These surgeonfish feed on small organisms living on the sea turtle's skin. The sea turtle benefits from the removal of parasites, and the surgeonfish gain a meal and protection from enemies.

Evolution, the Core Theme of Biology

Having considered four of the unifying themes that run through this text, let's now turn to biology's core theme—evolution. Evolution makes sense of everything we know about living organisms. Life has been evolving on Earth for billions of years, resulting in a vast diversity of past and present organisms. But along with the diversity are many shared features. For example, while sea horses, jackrabbits, hummingbirds, crocodiles, and giraffes all look very different, their skeletons are basically similar. The scientific explanation for this unity and diversity—as well as for the adaptation of organisms to their environments—is evolution: the idea that the organisms living on Earth today are the modified descendants of common ancestors. In other words, we can explain traits shared by two organisms with the idea that they have descended from a common ancestor, and we can account for differences with the idea that heritable changes have occurred along the way. Many kinds of evidence support the occurrence of evolution and the theory that describes how it takes place. In the next section, we'll consider the fundamental concept of evolution in greater detail.

CONCEPT CHECK 1.1

1. For each biological level in Figure 1.3, write a sentence that includes components from the previous (lower) level of biological organization; for example: "A community consists of populations of the various species inhabiting a certain area."
2. Identify the theme or themes exemplified by (a) the sharp spines of a porcupine, (b) the cloning of a plant from a single cell, and (c) a hummingbird using sugar to power its flight.
3. **WHAT IF?** For each theme discussed in this section, give an example not mentioned in the text.

For suggested answers, see Appendix A.

CONCEPT 1.2

The Core Theme: Evolution accounts for the unity and diversity of life

EVOLUTION Diversity is a hallmark of life. To date, biologists have identified and named about 1.8 million species of organisms, and estimates of the number of living species range from about 10 million to over 100 million. The remarkably diverse forms of life on this planet arose by evolutionary processes. Before exploring the core theme of evolution further, let's first consider how biologists make sense of the great variety of life-forms on this planet.

Classifying the Diversity of Life: The Three Domains of Life

Humans have a tendency to group diverse items according to their similarities and relationships to each other. Following this inclination, biologists have long used careful comparisons of form and function to classify life-forms into a hierarchy of increasingly inclusive groups. Consider, for example, the species known as the American black bear (*Ursus americanus*). Black bears belong to the same genus (*Ursus*) as the brown bear species and the polar bear species. Bringing together several similar genera forms a family, which in turn is a component of an order and then a class. For the black bear, this means being grouped with panda bears, raccoons, and others in the family Ursidae, with wolves in the order Carnivora, and with dolphins in the class Mammalia. These animals can be classified into still broader groupings: the phylum Chordata and the kingdom Animalia.

In the last few decades, new methods of assessing species relationships, especially comparisons of DNA sequences, have led to a reevaluation of the larger groupings. Although the reevaluation is ongoing, there is consensus among biologists that the kingdoms of life, whatever their number, can be further grouped into three so-called domains: Bacteria, Archaea, and Eukarya **(Figure 1.11)**.

As you read earlier, the organisms making up two of the three domains—**Bacteria** and **Archaea**—are prokaryotic. All the eukaryotes (organisms with eukaryotic cells) are grouped in domain **Eukarya**. This domain includes three kingdoms of multicellular eukaryotes: Plantae, Fungi, and Animalia. These three kingdoms are distinguished partly by their modes of nutrition. Plants produce their own sugars and other food molecules by photosynthesis; fungi absorb dissolved nutrients from their surroundings; and animals obtain food by eating and digesting other organisms. Animalia is, of course, our own kingdom.

Unity in the Diversity of Life

As diverse as life is, it also displays remarkable unity. Earlier we mentioned both the similar skeletons of different vertebrate

▼ **Figure 1.11 The three domains of life.**

(a) Domain Bacteria

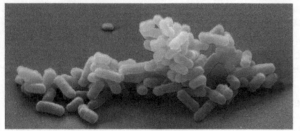

2 μm

Bacteria are the most diverse and widespread prokaryotes and are now classified into multiple kingdoms. Each rod-shaped structure in this photo is a bacterial cell.

(b) Domain Archaea

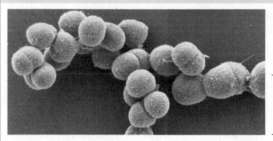

2 μm

Some of the prokaryotes known as **archaea** live in Earth's extreme environments, such as salty lakes and boiling hot springs. Domain Archaea includes multiple kingdoms. Each round structure in this photo is an archaeal cell.

(c) Domain Eukarya

◀ **Kingdom Animalia** consists of multicellular eukaryotes that ingest other organisms.

100 μm

▲ **Kingdom Plantae** consists of terrestrial multicellular eukaryotes (land plants) that carry out photosynthesis, the conversion of light energy to the chemical energy in food.

▶ **Kingdom Fungi** is defined in part by the nutritional mode of its members (such as this mushroom), which absorb nutrients from outside their bodies.

▶ **Protists** are mostly unicellular eukaryotes and some relatively simple multicellular relatives. Pictured here is an assortment of protists inhabiting pond water. Scientists are currently debating how to classify protists in a way that accurately reflects their evolutionary relationships.

animals and the universal genetic language of DNA (the genetic code). In fact, similarities between organisms are evident at all levels of the biological hierarchy.

How can we account for life's dual nature of unity and diversity? The process of evolution, explained next, illuminates both the similarities and differences in the world of life and introduces another dimension of biology: the passage of time.

Charles Darwin and the Theory of Natural Selection

The history of life, as documented by fossils and other evidence, is the saga of a changing Earth billions of years old, inhabited by an evolving cast of living forms **(Figure 1.12)**. This view of life came into sharp focus in November 1859, when Charles Robert Darwin published one of the most influential books ever written, *On the Origin of Species by Means of Natural Selection* **(Figure 1.13)**.

▶ **Figure 1.12 Digging into the past.** Paleontologists carefully excavate the hind leg of a long-necked dinosaur (*Rapetosaurus krausei*) from rocks in Madagascar.

▶ **Figure 1.13 Charles Darwin in 1840.** His revolutionary book *The Origin of Species* was published in 1859.

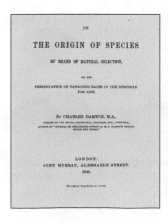

The Origin of Species articulated two main points. The first was that species have arisen from a succession of ancestors that differed from them. Darwin called this process "descent with modification." It was an insightful phrase, as it captured the duality of life's unity and diversity—unity in the kinship among species that descended from common ancestors, diversity in the modifications that evolved as species branched from their common ancestors **(Figure 1.14)**. Darwin's second main point was his proposal that "natural selection" is a mechanism for descent with modification.

Darwin developed his theory of natural selection from observations that by themselves were not revolutionary. Others had described the pieces of the puzzle, but Darwin saw how they fit together. He started with the following three observations from nature: First, individuals in a population vary in their traits, many of which seem to be heritable, passed on from parents to offspring. Second, a population can produce far more offspring than can survive to produce offspring of their own. Competition is thus inevitable. Third, species generally are suited to their environments—in other words, they are adapted to their environments. For instance, various birds that feed on hard seeds tend to have especially strong beaks.

Darwin inferred that individuals with inherited traits that are better suited to the local environment are more likely to survive and reproduce than are less well-suited individuals. As a result, over many generations, a higher and higher proportion of individuals in a population will have the advantageous traits. Darwin called this mechanism of evolutionary

▲ **Figure 1.14 Unity and diversity among birds.** These three birds are variations on a common body plan. For example, each has feathers, a beak, and wings, but these features are highly specialized for the birds' diverse lifestyles.

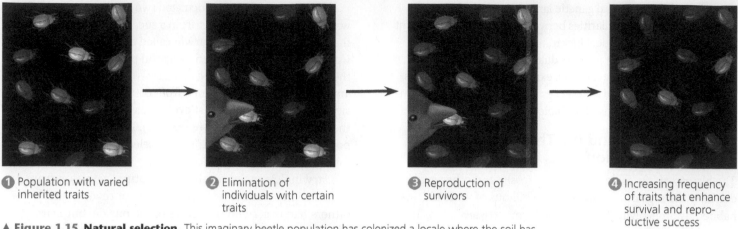

① Population with varied inherited traits

② Elimination of individuals with certain traits

③ Reproduction of survivors

④ Increasing frequency of traits that enhance survival and reproductive success

▲ **Figure 1.15 Natural selection.** This imaginary beetle population has colonized a locale where the soil has been blackened by a recent brush fire. Initially, the population varies extensively in the inherited coloration of the individuals, from very light gray to charcoal. For birds that prey on the beetles, it is easiest to spot the lighter ones.

adaptation **natural selection** because the natural environment "selects" for the propagation of certain traits among naturally occurring variant traits in the population **(Figure 1.15)**.

The Tree of Life

For another example of unity and diversity, consider the human arm. Your forelimb has the same bones, joints, nerves, and blood vessels found in other limbs as diverse as the foreleg of a horse, the flipper of a whale, and the wing of a bat. Indeed, all mammalian forelimbs are anatomical variations of a common architecture. According to the Darwinian concept of descent with modification, the shared anatomy of mammalian limbs reflects inheritance of the limb structure from a common ancestor—the "prototype" mammal from which all other mammals descended. The diversity of mammalian forelimbs results from modification by natural selection operating over millions of years in different environmental contexts.

Darwin proposed that natural selection, by its cumulative effects over time, could cause an ancestral species to give rise to two or more descendant species. This could occur, for example, if one population of organisms fragmented into several subpopulations isolated in different environments. In these separate arenas of natural selection, a species could gradually radiate into multiple species as the geographically isolated populations adapted over many generations to different environmental conditions.

The "family tree" of six finch species in **Figure 1.16** illustrates a famous example of this process of radiation.

Darwin collected specimens of these birds during his 1835 visit to the remote Galápagos Islands, 900 kilometers (km) off the Pacific coast of South America. The Galápagos finches are believed to have descended from an ancestral finch species that reached the archipelago from South America or the Caribbean. Over time, the Galápagos finches diversified from their ancestor as they adapted to different food sources on the various islands. Years after Darwin collected the finches, researchers began to sort out their evolutionary relationships, first from anatomical and geographic data and more recently using DNA sequence comparisons.

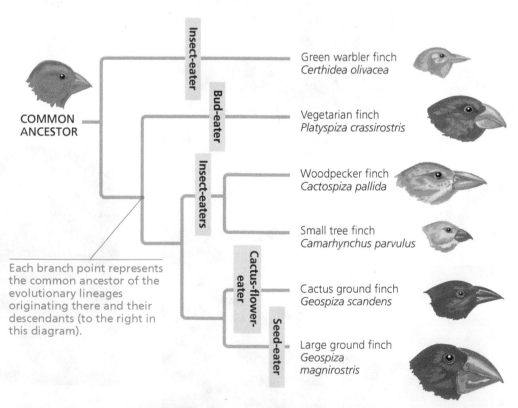

COMMON ANCESTOR

Insect-eater
Green warbler finch
Certhidea olivacea

Bud-eater
Vegetarian finch
Platyspiza crassirostris

Insect-eaters
Woodpecker finch
Cactospiza pallida

Small tree finch
Camarhynchus parvulus

Cactus-flower-eater
Cactus ground finch
Geospiza scandens

Seed-eater
Large ground finch
Geospiza magnirostris

Each branch point represents the common ancestor of the evolutionary lineages originating there and their descendants (to the right in this diagram).

▲ **Figure 1.16 Descent with modification: finches on the Galápagos Islands.** This "tree" diagram illustrates a current model for the evolutionary relationships among some of the finches on the Galápagos. Note the different beaks, which are adapted to food sources on the different islands.

Biologists' diagrams of such evolutionary relationships generally take treelike forms, though the trees are often turned sideways as in Figure 1.16. Tree diagrams make sense: Just as an individual has a genealogy that can be diagrammed as a family tree, each species is one twig of a branching tree of life extending back in time through ancestral species more and more remote. Species that are very similar, such as the Galápagos finches, share a relatively recent common ancestor. But through an ancestor that lived much farther back in time, finches are related to sparrows, hawks, penguins, and all other birds. And birds, mammals, and all other vertebrates share a common ancestor even more ancient. Trace life back far enough, and we reach the early prokaryotes that inhabited Earth 3.5 billion years ago. We can recognize their vestiges in our own cells—in the universal genetic code, for example. Indeed, all of life is connected through its long evolutionary history.

CONCEPT CHECK 1.2

1. How is a mailing address analogous to biology's hierarchical classification system?
2. Explain why "editing" is an appropriate metaphor for how natural selection acts on a population's heritable variation.
3. **WHAT IF?** Recent evidence indicates that fungi and animals are more closely related to each other than either of these kingdoms is to plants. Draw a simple branching pattern that symbolizes the proposed relationship between these three kingdoms of multicellular eukaryotes.

For suggested answers, see Appendix A.

CONCEPT 1.3

Biological inquiry entails forming and testing hypotheses based on observations of nature

The word *science* is derived from a Latin verb meaning "to know." **Science** is a way of knowing—an approach to understanding the natural world. It developed out of our human curiosity about ourselves, other life-forms, our planet, and the universe. Striving to make sense of our experiences seems to be one of our basic urges.

At the heart of science is **inquiry**, a search for information and explanations of natural phenomena. There is no formula for successful scientific inquiry, no single scientific method that researchers must rigidly follow. As in all quests, science includes elements of challenge, adventure, and luck, along with careful planning, reasoning, creativity, patience, and the persistence to overcome setbacks. Such diverse elements of inquiry make science far less structured than most people realize. That said, it is possible to distill certain characteristics that help to distinguish science from other ways of describing and explaining nature.

Scientists use a process of inquiry that includes making observations, forming logical hypotheses, and testing them. The process is necessarily repetitive: In testing a hypothesis, our observations may lead to conclusions that inspire revision of the original hypothesis or formation of a new one, thus leading to further testing. In this way, scientists circle closer and closer to their best estimation of the laws governing nature.

Making Observations

In the course of their work, scientists describe natural structures and processes as accurately as possible through careful observation and analysis of data. Observation is the use of the senses to gather information either directly or indirectly, such as with the help of microscopes or other tools that extend our senses. Recorded observations are called **data**. Put another way, data are items of information on which scientific inquiry is based.

The term *data* implies numbers to many people. But some data are *qualitative*, often in the form of recorded descriptions. For example, British primate researcher Jane Goodall spent decades recording her observations of chimpanzee behavior during field research in a Tanzanian jungle **(Figure 1.17)**. She also documented her observations with photographs and movies. Along with these qualitative data, Goodall also gathered and recorded volumes of *quantitative* data, a type of information generally expressed as numerical measurements and often organized into tables or graphs.

Collecting and analyzing observations can lead to important conclusions based on a type of logic called **inductive reasoning**. Through induction, we derive generalizations from a large number of specific observations. The generalization "All

▲ **Figure 1.17 Jane Goodall collecting qualitative data on chimpanzee behavior.** Goodall recorded her observations in field notebooks, often with sketches of the animals' behavior.

organisms are made of cells" was based on two centuries of microscopic observations made by biologists examining cells in diverse biological specimens. Careful observations and data analyses, along with the generalizations reached by induction, are fundamental to our understanding of nature.

Forming and Testing Hypotheses

Our innate curiosity often stimulates us to pose questions about the natural basis for the phenomena we observe in the world. What *caused* the diversification of finches on the Galápagos Islands? What *explains* the variation in coat color among mice of a single species, such as the beach and inland mice pictured in Figures 1.1 and 1.2? In science, answering such questions usually involves proposing and testing hypothetical explanations—that is, hypotheses.

In science, a **hypothesis** is a tentative answer to a well-framed question; it is an explanation on trial. The hypothesis is usually a rational accounting for a set of observations, based on the available data and guided by inductive reasoning. A scientific hypothesis leads to predictions that can be tested by making additional observations or by performing experiments.

We all use hypotheses in solving everyday problems. Let's say, for example, that your flashlight fails during a camp-out. That's an observation. The question is obvious: Why doesn't the flashlight work? Two reasonable hypotheses based on your experience are that (1) the batteries in the flashlight are dead or (2) the bulb is burnt out. Each of these alternative hypotheses leads to predictions you can test with experiments. For example, the dead-battery hypothesis predicts that replacing the batteries will fix the problem. Figuring things out in this way by trial and error is a hypothesis-based approach.

Deductive Reasoning

A type of logic called deduction is also built into the use of hypotheses in science. While induction entails reasoning from a set of specific observations to reach a general conclusion, **deductive reasoning** involves logic that flows in the opposite direction, from the general to the specific. From general premises, we extrapolate to the specific results we should expect if the premises are true. In the scientific process, deductions usually take the form of predictions of results that will be found if a particular hypothesis (premise) is correct. We then test the hypothesis by carrying out experiments or observations to see whether or not the results are as predicted. This deductive testing takes the form of "*If . . . then*" logic. In the case of the flashlight example: *If* the dead-battery hypothesis is correct, *then* the flashlight should work when you replace the batteries with new ones.

The flashlight inquiry demonstrates two other key points about the use of hypotheses in science. First, the initial observations may give rise to multiple hypotheses. The ideal is to design experiments to test all these candidate explanations. For instance, another of the many possible alternative hypotheses to explain our dead flashlight is that *both* the batteries *and*

the bulb are bad, and you could design an experiment to test this. Second, we can never *prove* that a hypothesis is true. The dead-battery hypothesis stands out as the most likely explanation, but testing supports that hypothesis *not* by proving that it is correct, but rather by not eliminating it through falsification (proving it false). Replacing the batteries might have fixed the flashlight, but perhaps the endpiece had simply not been screwed on tight enough in the first place. No amount of experimental testing can prove a hypothesis beyond a shadow of doubt, because it is impossible to test *all* alternative hypotheses. A hypothesis gains credibility by surviving multiple attempts to falsify it while alternative hypotheses are eliminated (falsified) by testing.

Questions That Can and Cannot Be Addressed by Science

Scientific inquiry is a powerful way to learn about nature, but there are limitations to the kinds of questions it can answer. A scientific hypothesis must be falsifiable; there must be some observation or experiment that could reveal if such an idea is actually *not* true. The hypothesis that dead batteries are the sole cause of the broken flashlight could be falsified by replacing the old batteries with new ones and finding that the flashlight still doesn't work.

Not all hypotheses meet the criteria of science: You wouldn't be able to falsify the hypothesis that invisible campground ghosts are fooling with your flashlight! Because science requires natural explanations for natural phenomena, it can neither support nor falsify hypotheses that angels, ghosts, or spirits, whether benevolent or evil, cause storms, rainbows, illnesses, and cures. Such supernatural explanations are simply outside the bounds of science, as are religious matters, which are issues of personal faith.

A Case Study in Scientific Inquiry: Investigating Coat Coloration in Mouse Populations

Now that we have highlighted the key features of scientific inquiry—making observations and forming and testing hypotheses—you should be able to recognize these features in a case study of actual scientific research.

The story begins with a set of observations and inductive generalizations. Color patterns of animals vary widely in nature, sometimes even between members of the same species. What accounts for such variation? As you may recall, the two mice depicted at the beginning of this chapter are members of the same species (*Peromyscus polionotus*), but they reside in very different habitats. Beach mice live along the ocean on white sand dunes, whereas "inland" mice live on darker, loamy soil away from the shore **(Figure 1.18)**. Even a brief glance at the photographs in Figure 1.18 reveals a striking match of mouse coloration to environment. The natural predators of these mice, including hawks, owls, foxes, and coyotes, are all visual hunters

► **Figure 1.18 Different coloration in two beach and inland populations of *Peromyscus polionotus*.**

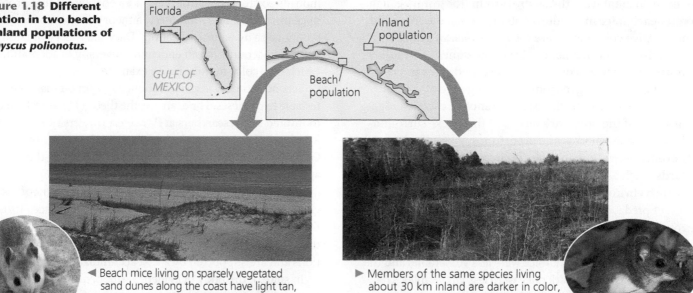

◄ Beach mice living on sparsely vegetated sand dunes along the coast have light tan, dappled coats that allow them to blend into their surroundings (camouflage).

► Members of the same species living about 30 km inland are darker in color, camouflaging them against the dark ground of their habitat.

(they use their eyes to look for prey). It was logical, therefore, for Francis Bertody Sumner, a naturalist studying populations of these mice in the 1920s, to hypothesize that their color patterns had evolved as adaptations that camouflage the mice in their native environments, protecting them from predation.

As obvious as the camouflage hypothesis may seem, it still required testing. In 2010, biologist Hopi Hoekstra of Harvard University and a group of her students headed to Florida to test the prediction that mice with coloration that did not match their habitat would be preyed on more heavily than the native, well-matched mice. **Figure 1.19** summarizes this field experiment, introducing a format we will use throughout the book to walk through other examples of biological inquiry.

The researchers built hundreds of silicone models of mice and spray-painted them to resemble either beach or inland mice, so that the models differed only in their color patterns. The researchers placed equal numbers of these model mice randomly in both habitats and left them overnight. The mouse models resembling the native

▼ **Figure 1.19 Inquiry**

Does camouflage affect predation rates on two populations of mice?

Experiment Hopi Hoekstra and colleagues tested the hypothesis that coloration of beach and inland popluations of oldfield mice (*Peromyscus polionotus*) provides camouflage that protects them from predation in their respective habitats. The researchers made mouse models with either light or dark color patterns that matched those of the beach and inland mice, then placed models with both patterns in each of the habitats. The next morning, they counted damaged or missing mice.

Results The researchers calculated the proportion of attacked mice that were camouflaged or non-camouflaged for each habitat. In both cases, the mice whose pattern did not match their surroundings suffered a much higher predation rate than did the camouflaged mice.

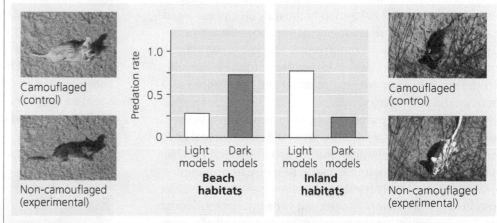

Camouflaged (control)

Non-camouflaged (experimental)

Camouflaged (control)

Non-camouflaged (experimental)

Conclusion The results do not falsify the researchers' prediction that mouse models with camouflage coloration would be preyed on less often than non-camouflaged mouse models. Thus, the experiment supports the camouflage hypothesis.

Source S. N. Vignieri, J. G. Larson, and H. E. Hoekstra, The selective advantage of crypsis in mice, *Evolution* 64:2153–2158 (2010).

WHAT IF? If you found a habitat with reddish, iron-rich soil, what would you predict with respect to the coat color of resident mice? What prediction would you make about the predation rate on beach mice and inland mice if you placed them in this new habitat?

mice in the habitat were the *control* group (for instance, light-colored beach mice in the dune habitat), while the mice with the non-native coloration were the *experimental* group (for example, the darker mainland mice in the dunes). The following morning, the team counted and recorded signs of predation events, which ranged from bites and gouge marks on some models to the outright disappearance of others. Judging by the shape of the predator's bites and the tracks surrounding the experimental sites, the predators appeared to be split fairly evenly between mammals (such as foxes and coyotes) and birds (such as owls, herons, and hawks).

For each environment, the researchers then calculated the fraction of predation events that targeted camouflaged mice. The results were clear-cut: Camouflaged mice showed much lower predation rates than those lacking camouflage in both the dune habitat (where light mice were less vulnerable) and the mainland habitat (where dark mice were less vulnerable). The data thus fit the key prediction of the camouflage hypothesis.

Experimental Controls

The mouse camouflage experiment described in Figure 1.19 is an example of a **controlled experiment**, one that is designed to compare an experimental group (the non-camouflaged mice, in this case) with a control group (the camouflaged mice normally resident in that area). Ideally, the experimental and control groups differ only in the one factor the experiment is designed to test—in our example, the effect of mouse coloration on the behavior of predators. Without the control group, the researchers would not have been able to rule out other factors as causes of the more frequent attacks on the non-camouflaged mice—such as different numbers of predators or different temperatures in the different test areas. The clever experimental design left coloration as the only factor that could account for the low predation rate on the camouflaged mice placed in their normal environment. It was not the absolute number of attacks on the non-camouflaged mice that counted, but the difference between that number and the number of attacks on the camouflaged mice.

A common misconception is that the term *controlled experiment* means that scientists control the experimental environment to keep everything constant except the one variable being tested. But that's impossible in field research and not realistic even in highly regulated laboratory environments. Researchers usually "control" unwanted variables not by *eliminating* them by regulating the environment, but by *canceling out* their effects using control groups.

Theories in Science

"It's just a theory!" Our everyday use of the term *theory* often implies an untested speculation. But the term *theory* has a different meaning in science. What is a scientific theory, and how is it different from a hypothesis or from mere speculation?

First, a scientific **theory** is much broader in scope than a hypothesis. *This* is a hypothesis: "A match of the coloration of a mouse's coat to its environment is an adaptation that protects mice from predators." But *this* is a theory: "Evolutionary adaptations arise by natural selection." Darwin's theory of natural selection accounts for an enormous diversity of adaptations, of which coat color in mice is one example.

Second, a theory is general enough to spin off many new, testable hypotheses. For example, the theory of natural selection motivated two researchers at Princeton University, Peter and Rosemary Grant, to test the specific hypothesis that the beaks of Galápagos finches evolve in response to changes in the types of available food. (For the results, see the Chapter 21 Overview.)

And third, compared to any one hypothesis, a theory is generally supported by a much greater body of evidence. Those theories that become widely adopted in science (such as the theory of natural selection) explain a great diversity of observations and are supported by a vast accumulation of evidence.

In spite of the body of evidence supporting a widely accepted theory, scientists must sometimes modify or even reject theories when new research methods produce results that don't fit. For example, biologists once lumped bacteria and archaea together as a kingdom of prokaryotes. When new methods for comparing cells and molecules could be used to test such relationships, the evidence led scientists to reject the theory that bacteria and archaea are members of the same kingdom. If there is "truth" in science, it is conditional, based on the weight of available evidence.

Science as a Social Process: Community and Diversity

The great scientist Sir Isaac Newton once said: "To explain all nature is too difficult a task for any one man or even for any one age. 'Tis much better to do a little with certainty, and leave the rest for others that come after you. . . .'" Anyone who becomes a scientist, driven by curiosity about nature, is sure to benefit from the rich storehouse of discoveries by others who have come before. In fact, while movies and cartoons sometimes portray scientists as loners working in isolated labs, science is an intensely social activity. Most scientists work in teams, which often include graduate and undergraduate students **(Figure 1.20)**.

▲ **Figure 1.20 Science as a social process.** Lab members help each other interpret data, troubleshoot experiments, and plan future research.

Science is rarely perfectly objective, but it is continuously vetted through the expectation that observations and experiments be repeatable and hypotheses be falsifiable. Scientists working in the same research field often check one another's claims by attempting to confirm observations or repeat experiments. In fact, Hopi Hoekstra's experiment benefited from the work of another researcher, D. W. Kaufman, 40 years earlier. You can study the design of Kaufman's experiment and interpret the results in the **Scientific Skills Exercise**.

If experimental results cannot be repeated by scientific colleagues, this failure may reflect some underlying weakness in the original claim, which will then have to be revised. In this sense, science polices itself. Integrity and adherence to high professional standards in reporting results are central to the scientific endeavor. After all, the validity of experimental data is key to designing further lines of inquiry.

Biologists may approach questions from different angles. Some biologists focus on ecosystems, while others study natural phenomena at the level of organisms or cells. This text is divided into units that focus on biology at different levels. Yet any given problem can be addressed from many perspectives, which in fact complement each other. For example, Hoekstra's work uncovered at least one genetic mutation that underlies the differences between beach and inland mouse coloration. Her lab includes biologists with different specialties, allowing discoveries on topics that range from evolutionary adaptations to their molecular basis in DNA.

The research community is part of society at large. The relationship of science to society becomes clearer when we add technology to the picture. The goal of **technology** is to *apply* scientific knowledge for some specific purpose. Because scientists put new technology to work in their research, science and technology are interdependent.

In centuries past, many major technological innovations originated along trade routes, where a rich mix of different cultures ignited new ideas. For example, the printing press was invented by Johannes Gutenberg around 1440, living in what is now Germany. This invention relied on several innovations from China,

---

Scientific Skills Exercise

Interpreting a Pair of Bar Graphs

How Much Does Camouflage Affect Predation on Mice by Owls with and without Moonlight? Nearly half a century ago, D. W. Kaufman investigated the effect of prey camouflage on predation. Kaufman tested the hypothesis that the amount of contrast between the coat color of a mouse and the color of its surroundings would affect the rate of nighttime predation by owls. He also hypothesized that the color contrast would be affected by the amount of moonlight. In this exercise, you will analyze data from his owl-mouse predation studies.

How the Experiment Was Done Pairs of mice (*Peromyscus polionotus*) with different coat colors, one light brown and one dark brown, were released simultaneously into an enclosure that contained a hungry owl. The researcher recorded the color of the mouse that was first caught by the owl. If the owl did not catch either mouse within 15 minutes, the test was recorded as a zero. The release trials were repeated multiple times in enclosures with either a dark-colored soil surface or a light-colored soil surface. The presence or absence of moonlight during each assay was recorded.

Data from the Experiment

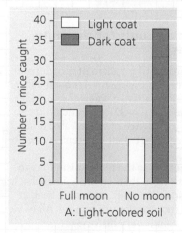
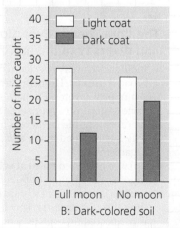

Interpret the Data

1. First, make sure you understand how the graphs are set up. Graph A shows data from the light-colored soil enclosure and Graph B from the dark-colored enclosure, but in all other respects the graphs are the same. (a) There is more than one independent variable in these graphs. What are the independent variables, the variables that were tested by the researcher? Which axis of the graphs has the independent variables? (b) What is the dependent variable, the response to the variables being tested? Which axis of the graphs has the dependent variable?

2. (a) How many dark brown mice were caught in the light-colored soil enclosure on a moonlit night? (b) How many dark brown mice were caught in the dark-colored soil enclosure on a moonlit night? (c) On a moonlit night, would a dark brown mouse be more likely to escape predation by owls on dark- or light-colored soil? Explain your answer.

3. (a) Is a dark brown mouse on dark-colored soil more likely to escape predation under a full moon or with no moon? (b) A light brown mouse on light-colored soil? Explain.

4. (a) Under which conditions would a dark brown mouse be most likely to escape predation at night? (b) A light brown mouse?

5. (a) What combination of independent variables led to the highest predation level in enclosures with light-colored soil? (b) What combination of independent variables led to the highest predation level in enclosures with dark-colored soil? (c) What relationship, if any, do you see in your answers to parts (a) and (b)?

6. What conditions are most deadly for both colors of mice?

7. Combining the data shown in both graphs, estimate the total number of mice caught in moonlight versus no-moonlight conditions. Which condition is optimal for predation by the owl on mice? Explain your answer.

Data from D. W. Kaufman, Adaptive coloration in *Peromyscus polionotus*: Experimental selection by owls, *Journal of Mammalogy* 55:271–283 (1974).

(MB) A version of this Scientific Skills Exercise can be assigned in MasteringBiology.

including paper and ink, and from Iraq, where technology was developed for the mass production of paper. Like technology, science stands to gain much from embracing a diversity of backgrounds and viewpoints among its practitioners.

The scientific community reflects the customs and behaviors of society at large. It is therefore not surprising that until recently, women and certain minorities have faced huge obstacles in their pursuit to become professional scientists. Over the past 50 years, changing attitudes about career choices have increased the proportion of women in biology and several other sciences, and now women constitute roughly half of undergraduate biology majors and biology Ph.D. students. The pace has been slow at higher levels in the profession, however, and women and many racial and ethnic groups are still significantly underrepresented in many branches of science. This lack of diversity hampers the progress of science. The more voices that are heard at the table, the more robust and productive the scientific conversation will be. The authors of this textbook welcome all students to the community of biologists, wishing you the joys and satisfactions of this exciting field of science.

CONCEPT CHECK 1.3

1. Contrast inductive reasoning with deductive reasoning.
2. What variable was tested in Hoekstra's mouse experiment?
3. Why is natural selection called a theory rather than a hypothesis?
4. How does science differ from technology?

For suggested answers, see Appendix A.

---

1 Chapter Review

SUMMARY OF KEY CONCEPTS

CONCEPT 1.1

Studying the diverse forms of life reveals common themes (pp. 2–7)

Theme: Organization

- The hierarchy of life unfolds as follows: biosphere > ecosystem > community > population > organism > organ system > organ > tissue > cell > organelle > molecule > atom. With each step up, new properties emerge (**emergent properties**) as a result of interactions among components at the lower levels.
- Structure and function are correlated at all levels of biological organization. The cell is the lowest level of organization that can perform all activities required for life. Cells are either prokaryotic or eukaryotic. **Eukaryotic cells** have a DNA-containing nucleus and other membrane-enclosed organelles. **Prokaryotic cells** lack such organelles.

Theme: Information

- Genetic information is encoded in the nucleotide sequences of **DNA**. It is DNA that transmits heritable information from parents to offspring. DNA sequences (called **genes**) program a cell's protein production by being transcribed into RNA and then translated into specific proteins, a process called **gene expression**. Gene expression also produces RNAs that are not translated into protein but serve other important functions.

Theme: Energy and Matter

- Energy flows through an ecosystem. All organisms must perform work, which requires energy. Producers convert energy from sunlight to chemical energy, some of which is then passed on to consumers (the rest is lost from the ecosystem as heat). Chemicals cycle between organisms and the environment.

Theme: Interactions

- Organisms interact continuously with physical factors. Plants take up nutrients from the soil and chemicals from the air and use energy from the sun. Interactions among plants, animals, and other organisms affect the participants in varying ways.

Core Theme: Evolution

- Evolution accounts for the unity and diversity of life and also for the match of organisms to their environments.

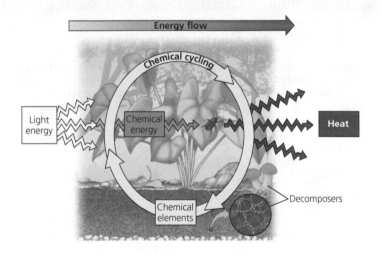

? *Thinking about the muscles and nerves in your hand, how does the activity of text messaging reflect the five unifying themes of biology described in this chapter?*

CONCEPT 1.2

The Core Theme: Evolution accounts for the unity and diversity of life (pp. 7–11)

- Biologists classify species according to a system of broader and broader groups. Domain **Bacteria** and domain **Archaea** consist of prokaryotes. Domain **Eukarya**, the eukaryotes, includes various groups of protists as well as plants, fungi, and animals. As diverse as life is, there is also evidence of remarkable unity, which is revealed in the similarities between different kinds of organisms.
- Darwin proposed **natural selection** as the mechanism for evolutionary adaptation of populations to their environments. Each species is one twig of a branching tree of life extending back in time through ancestral species more and more remote. All of life is connected through its long evolutionary history.

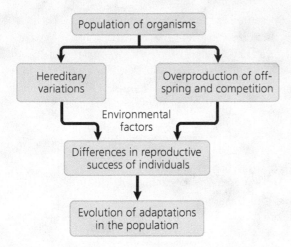

Population of organisms

Hereditary variations

Overproduction of offspring and competition

Environmental factors

Differences in reproductive success of individuals

Evolution of adaptations in the population

? *How could natural selection have led to the evolution of adaptations such as camouflaging coat color in beach mice?*

CONCEPT 1.3

Biological inquiry entails forming and testing hypotheses based on observations of nature (pp. 11–16)

- In scientific **inquiry**, scientists make observations (collect **data**) and use **inductive reasoning** to draw a general conclusion, which can be developed into a testable **hypothesis**. **Deductive reasoning** makes predictions that can be used to test hypotheses. Scientific hypotheses must be falsifiable.
- **Controlled experiments**, such as the study investigating coat color in mouse populations, are designed to demonstrate the effect of one variable by testing control groups and experimental groups that differ in only that one variable.
- A scientific **theory** is broad in scope, generates new hypotheses, and is supported by a large body of evidence.
- Scientists must be able to repeat each other's results, so integrity is key. Biologists approach questions at different levels; their approaches complement each other. **Technology** is a method or device that applies scientific knowledge for some specific purpose that affects society as well as for scientific research. Diversity among scientists promotes progress in science.

? *What are the roles of inductive and deductive reasoning in scientific inquiry?*

TEST YOUR UNDERSTANDING

Level 1: Knowledge/Comprehension

1. All the organisms on your campus make up
   a. an ecosystem.
   b. a community.
   c. a population.
   d. an experimental group.
   e. a domain.

2. Which of the following best demonstrates the unity among all organisms?
   a. identical DNA sequences
   b. descent with modification
   c. the structure and function of DNA
   d. natural selection
   e. emergent properties

3. A controlled experiment is one that
   a. proceeds slowly enough that a scientist can make careful records of the results.
   b. tests experimental and control groups in parallel.
   c. is repeated many times to make sure the results are accurate.
   d. keeps all variables constant.
   e. is supervised by an experienced scientist.

4. Which of the following statements best distinguishes hypotheses from theories in science?
   a. Theories are hypotheses that have been proved.
   b. Hypotheses are guesses; theories are correct answers.
   c. Hypotheses usually are relatively narrow in scope; theories have broad explanatory power.
   d. Hypotheses and theories are essentially the same thing.
   e. Theories are proved true; hypotheses are often falsified.

Level 2: Application/Analysis

5. Which of the following is an example of qualitative data?
   a. The temperature decreased from 20°C to 15°C.
   b. The plant's height is 25 centimeters (cm).
   c. The fish swam in a zigzag motion.
   d. The six pairs of robins hatched an average of three chicks.
   e. The contents of the stomach are mixed every 20 seconds.

6. Which of the following best describes the logic of scientific inquiry?
   a. If I generate a testable hypothesis, tests and observations will support it.
   b. If my prediction is correct, it will lead to a testable hypothesis.
   c. If my observations are accurate, they will support my hypothesis.
   d. If my hypothesis is correct, I can expect certain test results.
   e. If my experiments are set up right, they will lead to a testable hypothesis.

7. **DRAW IT** With rough sketches, draw a biological hierarchy similar to the one in Figure 1.3 but using a coral reef as the ecosystem, a fish as the organism, its stomach as the organ, and DNA as the molecule. Include all levels in the hierarchy.

Level 3: Synthesis/Evaluation

8. **SCIENTIFIC INQUIRY**
   Based on the results of the mouse coloration case study, suggest another hypothesis to extend the investigation.

9. **FOCUS ON EVOLUTION**
   In a short essay (100–150 words), discuss Darwin's view of how natural selection resulted in both unity and diversity of life on Earth. Include in your discussion some of his evidence. (A suggested grading rubric and tips for writing good essays can be found in the Study Area of MasteringBiology.)

10. **FOCUS ON INFORMATION**
    A typical prokaryotic cell has about 3,000 genes in its DNA, while a human cell has about 20,500 genes. About 1,000 of these genes are present in both types of cells. Based on your understanding of evolution, explain how such different organisms could have this same subset of genes. What sorts of functions might these shared genes have?

For selected answers, see Appendix A.

MasteringBiology®

Students Go to **MasteringBiology** for assignments, the eText, and the Study Area with practice tests, animations, and activities.

Instructors Go to **MasteringBiology** for automatically graded tutorials and questions that you can assign to your students, plus Instructor Resources.

Unit 1 Chemistry and Cells

2 The Chemical Context of Life

The structures and functions of living organisms are based on the **chemistry** of atoms and molecules.

3 Carbon and the Molecular Diversity of Life

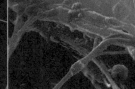

The **carbon** atom is the foundation of all organic molecules, and its versatility gives rise to the **molecular diversity of life**.

4 A Tour of the Cell

The basic structural and functional unit of life is the **cell**.

5 Membrane Transport and Cell Signaling

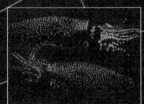

The **plasma membrane** regulates the passage of substances into and out of the cell and enables **signaling** between cells.

6 An Introduction to Metabolism

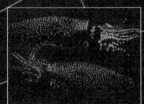

The cellular processes that transform matter and energy make up **cell metabolism**.

7 Cellular Respiration and Fermentation

Organisms obtain energy from food by breaking it down by means of **cellular respiration** or **fermentation**.

8 Photosynthesis

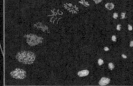

Photosynthesis is the basis of life on planet Earth: Photosynthetic organisms capture light energy and use it to make the food that all organisms depend on.

9 The Cell Cycle

A eukaryotic cell grows and then divides in two, passing along identical genetic information to its daughter cells via **mitosis**. The **cell cycle** describes this progression.

Chemistry and Cells

2 The Chemical Context of Life
3 Carbon and the Molecular Diversity of Life
4 A Tour of the Cell
5 Membrane Transport and Cell Signaling
6 An Introduction to Metabolism
7 Cellular Respiration and Fermentation
8 Photosynthesis
9 The Cell Cycle

Ecology
Animals
Plants
History of Life
Evolution
Genetics

The Chemical Context of Life

2

KEY CONCEPTS

2.1 Matter consists of chemical elements in pure form and in combinations called compounds

2.2 An element's properties depend on the structure of its atoms

2.3 The formation and function of molecules depend on chemical bonding between atoms

2.4 Chemical reactions make and break chemical bonds

2.5 Hydrogen bonding gives water properties that help make life possible on Earth

OVERVIEW

A Chemical Connection to Biology

Like other animals, beetles have structures and mechanisms that defend them from attack. The soil-dwelling bombardier beetle **(Figure 2.1)** has a particularly effective mechanism for dealing with the ants that plague it. Upon detecting an ant on its body, the beetle ejects a spray of boiling hot liquid from glands in its abdomen, aiming the spray directly at the ant. (In the photograph, the beetle aims its spray at a scientist's forceps.) The spray contains irritating chemicals that are generated at the moment of ejection by the explosive reaction of two sets of chemicals stored separately in the glands. The reaction produces heat and an audible pop.

Research on the bombardier beetle is only one example of the relevance of chemistry to the study of life. Unlike a list of college courses, nature is not neatly packaged into the individual natural sciences—biology, chemistry, physics, and so forth. Biologists specialize in the study of life, but organisms and their environments are natural systems to which the concepts of chemistry and physics apply. Biology is a multidisciplinary science.

This unit of chapters starts with some basic concepts of chemistry that apply to the study of life. In the unit, we will travel from atoms to molecules to cells and their main activities. Somewhere in the transition from molecules to cells, we will cross the blurry boundary between nonlife and life. This chapter introduces the chemical components that make up all matter, with a final section on the substance that supports all of life—water.

▼ **Figure 2.1** What is this bombardier beetle doing?

CONCEPT 2.1

Matter consists of chemical elements in pure form and in combinations called compounds

Organisms are composed of **matter**, which is defined as anything that takes up space and has mass. Matter exists in many diverse forms. Rocks, metals, oils, gases, and living organisms are just a few examples of what seems an endless assortment of matter.

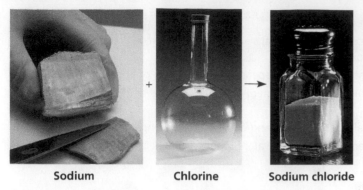

Sodium   Chlorine   Sodium chloride

▲ **Figure 2.2 The emergent properties of a compound.** The metal sodium combines with the poisonous gas chlorine, forming the edible compound sodium chloride, or table salt.

Elements and Compounds

Matter is made up of elements. An **element** is a substance that cannot be broken down to other substances by chemical reactions. Today, chemists recognize 92 elements occurring in nature; gold, copper, carbon, and oxygen are examples. Each element has a symbol, usually the first letter or two of its name. Some symbols are derived from Latin or German; for instance, the symbol for sodium is Na, from the Latin word *natrium*.

A **compound** is a substance consisting of two or more different elements combined in a fixed ratio. Table salt, for example, is sodium chloride (NaCl), a compound composed of the elements sodium (Na) and chlorine (Cl) in a 1:1 ratio. Pure sodium is a metal, and pure chlorine is a poisonous gas. When combined, however, sodium and chlorine form an edible compound. Water (H_2O), another compound, consists of the elements hydrogen (H) and oxygen (O) in a 2:1 ratio. These compounds provide simple examples of organized matter having *emergent properties*, ones not possessed by its constituents: A compound has chemical and physical characteristics different from those of its elements **(Figure 2.2)**.

The Elements of Life

Of the 92 natural elements, about 20–25% are **essential elements** that an organism needs to live a healthy life and reproduce. The essential elements are similar among organisms, but there is some variation—for example, humans need 25 elements, but plants need only 17.

Just four elements—oxygen (O), carbon (C), hydrogen (H), and nitrogen (N)—make up 96% of living matter. Calcium (Ca), phosphorus (P), potassium (K), sulfur (S), and a few other elements account for most of the remaining 4% of an organism's mass. **Trace elements** are required by an organism in only minute quantities. Some trace elements, such as iron (Fe), are needed by all forms of life; others are required only by certain species. For example, in vertebrates (animals with backbones), the element iodine (I) is an essential ingredient of a hormone produced by the thyroid gland. A daily intake of only 0.15 milligram (mg) of iodine is adequate for

normal activity of the human thyroid. An iodine deficiency in the diet causes the thyroid gland to grow to abnormal size, a condition called goiter. Consuming seafood or iodized salt reduces the incidence of goiter.

Evolution of Tolerance to Toxic Elements

EVOLUTION Some naturally occurring elements are toxic to organisms. In humans, for instance, the element arsenic has been linked to numerous diseases and can be lethal. Some species, however, have become adapted to environments containing elements that are usually toxic. For example, sunflower plants can take up lead, zinc, and other heavy metals in concentrations that would kill most organisms. (This capability enabled sunflowers to be used to detoxify contaminated soils after Hurricane Katrina.) Presumably, variants of ancestral sunflower species arose in heavy metal-laden soils, and subsequent natural selection resulted in their survival and reproduction.

CONCEPT CHECK 2.1

1. Is a trace element an essential element? Explain.
2. **WHAT IF?** In humans, iron is a trace element required for the proper functioning of hemoglobin, the molecule that carries oxygen in red blood cells. What might be the effects of an iron deficiency?

For suggested answers, see Appendix A.

CONCEPT 2.2

An element's properties depend on the structure of its atoms

Each element consists of a certain type of atom that is different from the atoms of any other element. An **atom** is the smallest unit of matter that still retains the properties of an element. Atoms are so small that it would take about a million of them to stretch across the period at the end of this sentence. We symbolize atoms with the same abbreviation used for the element that is made up of those atoms. For example, C stands for both the element carbon and a single carbon atom.

Subatomic Particles

Although the atom is the smallest unit having the properties of an element, these tiny bits of matter are composed of even smaller parts, called *subatomic particles*. Using high-energy collisions, physicists have produced more than a hundred types of particles from the atom, but only three kinds of particles are relevant here: **neutrons**, **protons**, and **electrons**. Protons and electrons are electrically charged. Each proton has one unit of positive charge, and each electron has one unit of negative charge. A neutron, as its name implies, is electrically neutral.

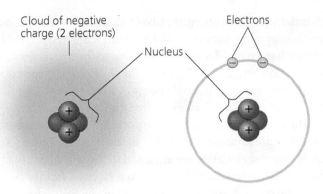

Cloud of negative
charge (2 electrons)

Nucleus

Electrons

(a) This model represents the two electrons as a cloud of negative charge.

(b) In this more simplified model, the electrons are shown as two small yellow spheres on a circle around the nucleus.

▲ **Figure 2.3 Simplified models of a helium (He) atom.** The helium nucleus consists of 2 neutrons (brown) and 2 protons (pink). Two electrons (yellow) exist outside the nucleus. These models are not to scale; they greatly overestimate the size of the nucleus in relation to the electron cloud.

Protons and neutrons are packed together in a dense core, or **atomic nucleus**, at the center of an atom. Protons give the nucleus a positive charge. The electrons form a cloud of negative charge around the nucleus, and it is the attraction between opposite charges that keeps the electrons in the vicinity of the nucleus. **Figure 2.3** shows two commonly used models for the structure of the helium atom as an example.

The neutron and proton are almost identical in mass, each about 1.7×10^{-24} gram (g). Grams and other conventional units are not very useful for describing the mass of objects so minuscule. Thus, for atoms and subatomic particles (and for molecules, too), we use a unit of measurement called the **dalton** (the same as the *atomic mass unit*, or *amu*). Neutrons and protons have masses close to 1 dalton. Because the mass of an electron is only about 1/2,000 that of a neutron or proton, we can ignore electrons when computing the total mass of an atom.

Atomic Number and Atomic Mass

Atoms of the various elements differ in their number of subatomic particles. All atoms of a particular element have the same number of protons in their nuclei. This number of protons, which is unique to that element, is called the **atomic number** and is written as a subscript to the left of the symbol for the element. The abbreviation $_2$He, for example, tells us that an atom of the element helium has 2 protons in its nucleus. Unless otherwise indicated, an atom is neutral in electrical charge, which means that its protons must be balanced by an equal number of electrons. Therefore, the atomic number tells us the number of protons and also the number of electrons in an electrically neutral atom.

We can deduce the number of neutrons from a second quantity, the **mass number**, which is the sum of protons plus neutrons in the nucleus of an atom. The mass number is written as a superscript to the left of an element's symbol. For example, we can use this shorthand to write an atom of helium as $_2^4$He. Because the atomic number indicates how many protons there are, we can determine the number of neutrons by subtracting the atomic number from the mass number: The helium atom $_2^4$He has 2 neutrons. For sodium (Na):

$$_{11}^{23}\text{Na}$$

Mass number = number of protons + neutrons
= 23 for sodium

Atomic number = number of protons
= 11 for sodium

Number of neutrons = mass number − atomic number
= 23 − 11 = 12 for sodium

The simplest atom is hydrogen $_1^1$H, which has no neutrons; it consists of a single proton with a single electron.

As we've seen, almost all of an atom's mass is concentrated in its nucleus. And because neutrons and protons each have a mass very close to 1 dalton, the mass number is an approximation of the total mass of an atom, called its **atomic mass**. So we might say that the atomic mass of sodium ($_{11}^{23}$Na) is 23 daltons, although more precisely it is 22.9898 daltons.

Isotopes

All atoms of a given element have the same number of protons, but some atoms have more neutrons than other atoms of the same element. These different atomic forms of the same element are called **isotopes** of the element. In nature, an element occurs as a mixture of its isotopes. For example, consider the three naturally occurring isotopes of the element carbon, which has the atomic number 6. The most common isotope is carbon-12, $_6^{12}$C, which accounts for about 99% of the carbon in nature. The isotope $_6^{12}$C has 6 neutrons. Most of the remaining 1% of carbon consists of atoms of the isotope $_6^{13}$C, with 7 neutrons. A third, even rarer isotope, $_6^{14}$C, has 8 neutrons. Notice that all three isotopes of carbon have 6 protons; otherwise, they would not be carbon. Although the isotopes of an element have slightly different masses, they behave identically in chemical reactions.

Both ^{12}C and ^{13}C are stable isotopes, meaning that their nuclei do not have a tendency to lose particles. The isotope ^{14}C, however, is unstable, or radioactive. A **radioactive isotope** is one in which the nucleus decays spontaneously, giving off particles and energy. When the decay leads to a change in the number of protons, it transforms the atom to an atom of a different element. For example, when an atom of ^{14}C decays, it becomes an atom of nitrogen.

Radioactive isotopes have many useful applications in biology. For example, researchers use measurements of radioactivity in fossils to date these relics of past life (see Chapter 23). Radioactive isotopes are also useful as tracers to follow atoms through metabolism, the chemical processes of an organism. Cells use the radioactive atoms as they would use

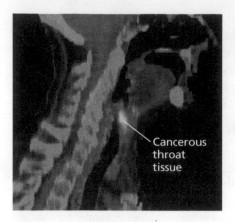

◀ Figure 2.4 A PET scan, a medical use for radioactive isotopes. PET, an acronym for positron-emission tomography, detects locations of intense chemical activity in the body. The bright yellow spot marks an area with an elevated level of radioactively labeled glucose, which in turn indicates the presence of cancerous tissue.

Cancerous throat tissue

nonradioactive isotopes of the same element, but the radioactive tracers can be readily detected.

Radioactive tracers are important diagnostic tools in medicine. For example, certain kidney disorders can be diagnosed by injecting small doses of substances containing radioactive isotopes into the blood and then measuring the amount of tracer excreted in the urine. Radioactive tracers are also used in combination with sophisticated imaging instruments. PET scanners, for instance, can monitor chemical processes, such as those involved in cancerous growth, as they actually occur in the body **(Figure 2.4)**.

Although radioactive isotopes are useful in research and medicine, radiation from decaying isotopes also poses a hazard to life by damaging cellular molecules. The severity of this damage depends on the type and amount of radiation an organism absorbs. One of the most serious environmental threats is radioactive fallout from nuclear accidents. The doses of isotopes used in medical diagnosis, however, are relatively safe.

The Energy Levels of Electrons

The simplified models of the atom in Figure 2.3 greatly exaggerate the size of the nucleus relative to the volume of the whole atom. If an atom of helium were the size of a typical football stadium, the nucleus would be the size of a pencil eraser in the center of the field. Moreover, the electrons would be like two tiny gnats buzzing around the stadium. Atoms are mostly empty space.

When two atoms approach each other during a chemical reaction, their nuclei do not come close enough to interact. Of the three kinds of subatomic particles we have discussed, only electrons are directly involved in the chemical reactions between atoms.

An atom's electrons vary in the amount of energy they possess. **Energy** is defined as the capacity to cause change—for instance, by doing work. **Potential energy** is the energy that matter possesses because of its location or structure. For example, water in a reservoir on a hill has potential energy because of its altitude. When the gates of the reservoir's dam are opened and the water runs downhill, the energy can be used to do work, such as moving the blades of turbines to generate

electricity. Because energy has been expended, the water has less energy at the bottom of the hill than it did in the reservoir. Matter has a natural tendency to move to the lowest possible state of potential energy; in this example, the water runs downhill. To restore the potential energy of a reservoir, work must be done to elevate the water against gravity.

The electrons of an atom have potential energy because of how they are arranged in relation to the nucleus. The negatively charged electrons are attracted to the positively charged nucleus. It takes work to move a given electron farther away from the nucleus, so the more distant an electron is from the nucleus, the greater its potential energy. Unlike the continuous flow of water downhill, changes in the potential energy of electrons can occur only in steps of fixed amounts. An electron having a certain amount of energy is something like a ball on a staircase **(Figure 2.5a)**. The ball can have different amounts of potential energy, depending on which step it is on, but it cannot spend much time between the steps. Similarly, an electron's potential energy is determined by its energy level. An electron cannot exist between energy levels.

An electron's energy level is correlated with its average distance from the nucleus. Electrons are found in different **electron shells**, each with a characteristic average distance and energy level. In diagrams, shells can be represented by concentric circles **(Figure 2.5b)**. The first shell is closest to the nucleus, and electrons in this shell have the lowest potential energy. Electrons in the second shell have more energy, and electrons in the third shell even more energy. An electron can change the shell it occupies, but only by absorbing or losing an

(a) A ball bouncing down a flight of stairs provides an analogy for energy levels of electrons, because the ball can come to rest only on each step, not between steps.

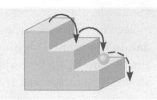

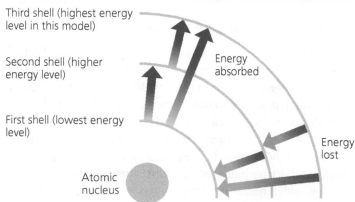

Third shell (highest energy level in this model)

Second shell (higher energy level)

First shell (lowest energy level)

Energy absorbed

Energy lost

Atomic nucleus

(b) An electron can move from one shell to another only if the energy it gains or loses is exactly equal to the difference in energy between the energy levels of the two shells. Arrows in this model indicate some of the stepwise changes in potential energy that are possible.

▲ Figure 2.5 Energy levels of an atom's electrons. Electrons exist only at fixed levels of potential energy called electron shells.

amount of energy equal to the difference in potential energy between its position in the old shell and that in the new shell. When an electron absorbs energy, it moves to a shell farther out from the nucleus. For example, light energy can excite an electron to a higher energy level. (Indeed, this is the first step taken when plants harness the energy of sunlight for photosynthesis, the process that produces food from carbon dioxide and water.) When an electron loses energy, it "falls back" to a shell closer to the nucleus, and the lost energy is usually released to the environment as heat.

Electron Distribution and Chemical Properties

The chemical behavior of an atom is determined by the distribution of electrons in the atom's electron shells. Beginning with hydrogen, the simplest atom, we can imagine building the atoms of the other elements by adding 1 proton and 1 electron at a time (along with an appropriate number of neutrons). **Figure 2.6**, an abbreviated version of what is called the *periodic table of the elements*, shows this distribution of electrons for the first 18 elements, from hydrogen ($_1$H) to argon ($_{18}$Ar). The elements are arranged in three rows, or periods, corresponding to the number of electron shells in their atoms. The left-to-right sequence of elements in each row corresponds to

the sequential addition of electrons and protons. (See Appendix B for the complete periodic table.)

Hydrogen's 1 electron and helium's 2 electrons are located in the first shell. Electrons, like all matter, tend to exist in the lowest available state of potential energy. In an atom, this state is in the first shell. However, the first shell can hold no more than 2 electrons; thus, hydrogen and helium are the only elements in the first row of the table. An atom with more than 2 electrons must use higher shells because the first shell is full. The next element, lithium, has 3 electrons. Two of these electrons fill the first shell, while the third electron occupies the second shell. The second shell holds a maximum of 8 electrons. Neon, at the end of the second row, has 8 electrons in the second shell, giving it a total of 10 electrons.

The chemical behavior of an atom depends mostly on the number of electrons in its *outermost* shell. We call those outer electrons **valence electrons** and the outermost electron shell the **valence shell**. In the case of lithium, there is only 1 valence electron, and the second shell is the valence shell. Atoms with the same number of electrons in their valence shells exhibit similar chemical behavior. For example, fluorine (F) and chlorine (Cl) both have 7 valence electrons, and both form compounds when combined with the element sodium (see

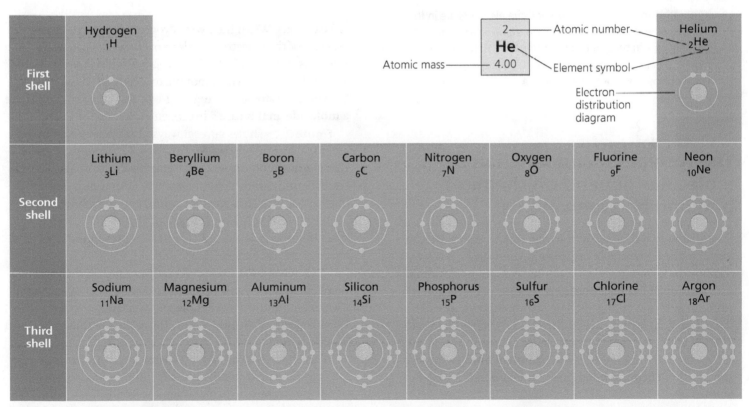

▲ **Figure 2.6 Electron distribution diagrams for the first 18 elements in the periodic table.** In a standard periodic table (see Appendix B), information for each element is presented as shown for helium in the inset. In the diagrams in this table, electrons are represented as yellow dots and electron shells as concentric circles. These diagrams are a convenient way to picture the distribution of an atom's electrons among its electron shells, but these simplified models do not accurately represent the shape of the atom or the location of its electrons. The elements are arranged in rows, each representing the filling of an electron shell. As electrons are added, they occupy the lowest available shell.

? *What is the atomic number of magnesium? How many protons and electrons does it have? How many electron shells? How many valence electrons?*

Figure 2.2). An atom with a completed valence shell is unreactive; that is, it will not interact readily with other atoms. At the far right of the periodic table are helium, neon, and argon, the only three elements shown in Figure 2.6 that have full valence shells. These elements are said to be *inert*, meaning chemically unreactive. All the other atoms in Figure 2.6 are chemically reactive because they have incomplete valence shells.

Notice that as we "build" the atoms in Figure 2.6, the first 4 electrons added to the second and third shells are not shown in pairs; only after 4 electrons are present do the next electrons complete pairs. The reactivity of an atom arises from the presence of one or more unpaired electrons in its valence shell. As you will see in the next section, atoms interact in a way that completes their valence shells. When they do so, it is the *unpaired* electrons that are involved.

CONCEPT CHECK 2.2

1. A nitrogen atom has 7 protons, and the most common isotope of nitrogen has 7 neutrons. A radioactive isotope of nitrogen has 8 neutrons. Write the atomic number and mass number of this radioactive nitrogen as a chemical symbol with a subscript and superscript.

2. How many electrons does fluorine have? How many electron shells? How many electrons are needed to fill the valence shell?

3. **WHAT IF?** In Figure 2.6, if two or more elements are in the same row, what do they have in common? If two or more elements are in the same column, what do they have in common?

For suggested answers, see Appendix A.

CONCEPT 2.3

The formation and function of molecules depend on chemical bonding between atoms

Now that we have looked at the structure of atoms, we can move up the hierarchy of organization and see how atoms combine to form molecules and ionic compounds. Atoms with incomplete valence shells can interact with certain other atoms in such a way that each partner completes its valence shell: The atoms either share or transfer valence electrons. These interactions usually result in atoms staying close together, held by attractions called **chemical bonds**. The strongest kinds of chemical bonds are covalent bonds and ionic bonds.

Covalent Bonds

A **covalent bond** is the sharing of a pair of valence electrons by two atoms. For example, let's consider what happens when two hydrogen atoms approach each other. Recall that hydrogen has 1 valence electron in the first shell, but the shell's capacity

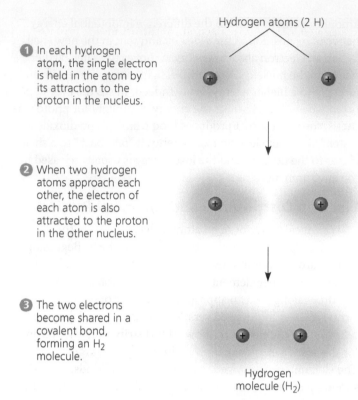

① In each hydrogen atom, the single electron is held in the atom by its attraction to the proton in the nucleus.

② When two hydrogen atoms approach each other, the electron of each atom is also attracted to the proton in the other nucleus.

③ The two electrons become shared in a covalent bond, forming an H_2 molecule.

Hydrogen molecule (H_2)

▲ **Figure 2.7 Formation of a covalent bond.**

is 2 electrons. When the two hydrogen atoms come close enough for their electron shells to overlap, they can share their electrons **(Figure 2.7)**. Each hydrogen atom is now associated with 2 electrons in what amounts to a completed valence shell. Two or more atoms held together by covalent bonds constitute a **molecule**, in this case a hydrogen molecule.

Figure 2.8a shows several ways of representing a hydrogen molecule. Its *molecular formula*, H_2, simply indicates that the molecule consists of two atoms of hydrogen. Electron sharing can be depicted by an electron distribution diagram or by a *structural formula*, H—H, where the line represents a **single bond**, a pair of shared electrons. A space-filling model comes closest to representing the actual shape of the molecule.

Oxygen has 6 electrons in its second electron shell and therefore needs 2 more electrons to complete its valence shell. Two oxygen atoms form a molecule by sharing *two* pairs of valence electrons **(Figure 2.8b)**. The atoms are thus joined by a **double bond** (O=O).

Each atom that can share valence electrons has a bonding capacity corresponding to the number of covalent bonds the atom can form. When the bonds form, they give the atom a full complement of electrons in the valence shell. The bonding capacity of oxygen, for example, is 2. This bonding capacity is called the atom's **valence** and usually equals the number of electrons required to complete the atom's outermost (valence) shell. See if you can determine the valences of hydrogen, oxygen, nitrogen, and carbon by

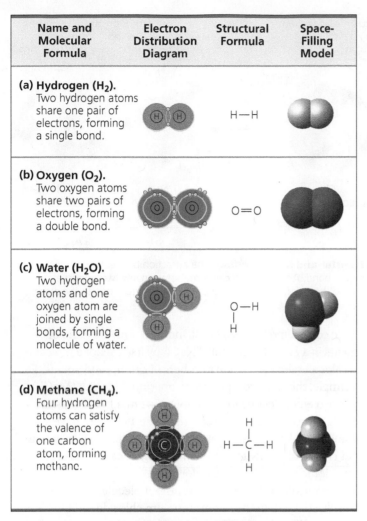

Name and Molecular Formula	Electron Distribution Diagram	Structural Formula	Space-Filling Model
(a) Hydrogen (H₂). Two hydrogen atoms share one pair of electrons, forming a single bond.		H—H	
(b) Oxygen (O₂). Two oxygen atoms share two pairs of electrons, forming a double bond.		O=O	
(c) Water (H₂O). Two hydrogen atoms and one oxygen atom are joined by single bonds, forming a molecule of water.		O—H $\mid$ H	
(d) Methane (CH₄). Four hydrogen atoms can satisfy the valence of one carbon atom, forming methane.		H $\mid$ H—C—H $\mid$ H	

▲ **Figure 2.8 Covalent bonding in four molecules.** The number of electrons required to complete an atom's valence shell generally determines how many covalent bonds that atom will form. This figure shows several ways of indicating covalent bonds.

studying the electron distribution diagrams in Figure 2.6. You can see that the valence of hydrogen is 1; oxygen, 2; nitrogen, 3; and carbon, 4. However, the situation is more complicated for elements in the third row of the periodic table. Phosphorus, for example, can have a valence of 3, as we would predict from the presence of 3 unpaired electrons in its valence shell. In some molecules that are biologically important, however, phosphorus can form three single bonds and one double bond. Therefore, it can also have a valence of 5.

The molecules H₂ and O₂ are pure elements rather than compounds because a compound is a combination of two or more *different* elements. Water, with the molecular formula H₂O, is a compound. Two atoms of hydrogen are needed to satisfy the valence of one oxygen atom. **Figure 2.8c** shows the structure of a water molecule. Water is so important to life that the last section of this chapter, Concept 2.5, is devoted to its structure and behavior.

Methane, the main component of natural gas, is a compound with the molecular formula CH₄. It takes four hydrogen atoms, each with a valence of 1, to complement one atom of carbon, with its valence of 4 **(Figure 2.8d)**. (We will look at many other compounds of carbon in Chapter 3.)

Atoms in a molecule attract shared electrons to varying degrees, depending on the element. The attraction of a particular atom for the electrons of a covalent bond is called its **electronegativity**. The more electronegative an atom is, the more strongly it pulls shared electrons toward itself. In a covalent bond between two atoms of the same element, the electrons are shared equally because the two atoms have the same electronegativity—the tug-of-war is at a standoff. Such a bond is called a **nonpolar covalent bond**. For example, the single bond of H₂ is nonpolar, as is the double bond of O₂. However, when an atom is bonded to a more electronegative atom, the electrons of the bond are not shared equally. This type of bond is called a **polar covalent bond**. Such bonds vary in their polarity, depending on the relative electronegativity of the two atoms. For example, the bonds between the oxygen and hydrogen atoms of a water molecule are quite polar **(Figure 2.9)**. Oxygen is one of the most electronegative of all the elements, attracting shared electrons much more strongly than hydrogen does. In a covalent bond between oxygen and hydrogen, the electrons spend more time near the oxygen nucleus than they do near the hydrogen nucleus. Because electrons have a negative charge and are pulled toward oxygen in a water molecule, the oxygen atom has a partial negative charge (indicated by the Greek letter δ with a minus sign, δ−, or "delta minus"), and each hydrogen atom has a partial positive charge (δ+, or "delta plus"). In contrast, the individual bonds of methane (CH₄) are much less polar because the electronegativities of carbon and hydrogen are similar.

Ionic Bonds

In some cases, two atoms are so unequal in their attraction for valence electrons that the more electronegative atom strips an electron completely away from its partner. This is what happens when an atom of sodium (₁₁Na) encounters an atom of

Because oxygen (O) is more electronegative than hydrogen (H), shared electrons are pulled more toward oxygen.

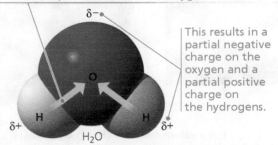

This results in a partial negative charge on the oxygen and a partial positive charge on the hydrogens.

▲ **Figure 2.9 Polar covalent bonds in a water molecule.**

chlorine ($_{17}$Cl) **(Figure 2.10)**. A sodium atom has a total of 11 electrons, with its single valence electron in the third electron shell. A chlorine atom has a total of 17 electrons, with 7 electrons in its valence shell. When these two atoms meet, the lone valence electron of sodium is transferred to the chlorine atom, and both atoms end up with their valence shells complete. (Because sodium no longer has an electron in the third shell, the second shell is now the valence shell.)

The electron transfer between the two atoms moves one unit of negative charge from sodium to chlorine. Sodium, now with 11 protons but only 10 electrons, has a net electrical charge of 1+. A charged atom (or molecule) is called an **ion**. When the charge is positive, the ion is specifically called a **cation**; the sodium atom has become a cation. Conversely, the chlorine atom, having gained an extra electron, now has 17 protons and 18 electrons, giving it a net electrical charge of 1−. It has become a chloride ion—an **anion**, or negatively charged ion. Because of their opposite charges, cations and anions attract each other; this attraction is called an **ionic bond**. The transfer of an electron is not the formation of a bond; rather, it allows a bond to form because it results in two ions of opposite charge. Any two ions of opposite charge can form an ionic bond. The ions do not need to have acquired their charge by an electron transfer with each other.

Compounds formed by ionic bonds are called **ionic compounds**, or **salts**. We know the ionic compound sodium chloride (NaCl) as table salt **(Figure 2.11)**. Salts are often found in nature as crystals of various sizes and shapes. Each salt crystal is an aggregate of vast numbers of cations and anions bonded by their electrical attraction and arranged in a three-dimensional lattice. Unlike a covalent compound, which consists of molecules having a definite size and number of atoms, an ionic compound does not consist of molecules. The formula for an

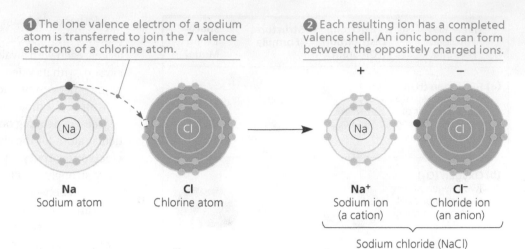

❶ The lone valence electron of a sodium atom is transferred to join the 7 valence electrons of a chlorine atom.

❷ Each resulting ion has a completed valence shell. An ionic bond can form between the oppositely charged ions.

Na
Sodium atom

Cl
Chlorine atom

Na⁺
Sodium ion
(a cation)

Cl⁻
Chloride ion
(an anion)

Sodium chloride (NaCl)

▲ **Figure 2.10 Electron transfer and ionic bonding.** The attraction between oppositely charged atoms, or ions, is an ionic bond. An ionic bond can form between any two oppositely charged ions, even if they have not been formed by transfer of an electron from one to the other.

ionic compound, such as NaCl, indicates only the ratio of elements in a crystal of the salt. "NaCl" by itself is not a molecule.

Not all salts have equal numbers of cations and anions. For example, the ionic compound magnesium chloride ($MgCl_2$) has two chloride ions for each magnesium ion. Magnesium ($_{12}$Mg) must lose 2 outer electrons if the atom is to have a complete valence shell, so it tends to become a cation with a net charge of 2+ (Mg^{2+}). One magnesium cation can therefore form ionic bonds with two chloride anions.

The term *ion* also applies to entire molecules that are electrically charged. In the salt ammonium chloride (NH_4Cl), for instance, the anion is a single chloride ion (Cl^-), but the cation is ammonium (NH_4^+), a nitrogen atom covalently bonded to four hydrogen atoms. The whole ammonium ion has an electrical charge of 1+ because it is 1 electron short.

Environment affects the strength of ionic bonds. In a dry salt crystal, the bonds are so strong that it takes a hammer and chisel to break enough of them to crack the crystal in two. If the same salt crystal is dissolved in water, however, the ionic bonds are much weaker because each ion is partially shielded by its interactions with water molecules. Most drugs are manufactured as salts because they are quite stable when dry but can dissociate (come apart) easily in water.

Weak Chemical Bonds

In organisms, most of the strongest chemical bonds are covalent bonds, which link atoms to form a cell's molecules. But weaker bonding within and between molecules is also indispensable in the cell, contributing greatly to the properties of life. Many large biological molecules are held in their functional form by weak bonds. In addition, when two molecules in the cell make contact, they may adhere temporarily by weak bonds. The reversibility of weak bonding can be an advantage: Two molecules can come together, respond to one another in some way, and then separate.

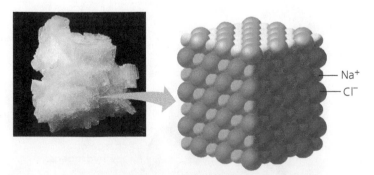

Na⁺

Cl⁻

▲ **Figure 2.11 A sodium chloride (NaCl) crystal.** The sodium ions (Na⁺) and chloride ions (Cl⁻) are held together by ionic bonds. The formula NaCl tells us that the ratio of Na⁺ to Cl⁻ is 1:1.

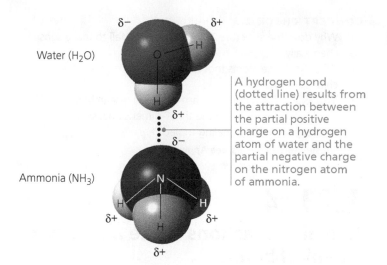

Water (H₂O)

Ammonia (NH₃)

A hydrogen bond (dotted line) results from the attraction between the partial positive charge on a hydrogen atom of water and the partial negative charge on the nitrogen atom of ammonia.

▲ **Figure 2.12 A hydrogen bond.**

Several types of weak chemical bonds are important in organisms. One is the ionic bond as it exists between ions dissociated in water, which we just discussed. Hydrogen bonds and van der Waals interactions are also crucial to life.

Hydrogen Bonds

Among the various kinds of weak chemical bonds, hydrogen bonds are so important in the chemistry of life that they deserve special attention. The partial positive charge on a hydrogen atom that is covalently bonded to an electronegative atom allows the hydrogen to be attracted to a different electronegative atom nearby. This noncovalent attraction between a hydrogen and an electronegative atom is called a **hydrogen bond**. In living cells, the electronegative partners are usually oxygen or nitrogen atoms. Refer to **Figure 2.12** to examine the simple case of hydrogen bonding between water (H_2O) and ammonia (NH_3).

Van der Waals Interactions

Even a molecule with nonpolar covalent bonds may have positively and negatively charged regions. Electrons are not always symmetrically distributed in such a molecule; at any instant, they may accumulate by chance in one part of the molecule or another. The results are ever-changing regions of positive and negative charge that enable all atoms and molecules to stick to one another. These **van der Waals interactions** are individually weak and occur only when atoms and molecules are very close together. When many such interactions occur simultaneously, however, they can be powerful: Van der Waals interactions are the reason a gecko lizard (right) can walk straight up a wall! Each gecko toe has hundreds of thousands of tiny hairs, with multiple

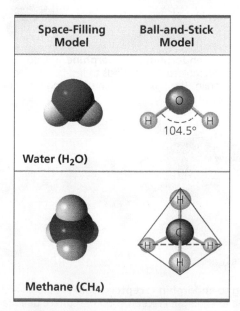

projections at each hair's tip that increase surface area. Apparently, the van der Waals interactions between the hair tip molecules and the molecules of the wall's surface are so numerous that despite their individual weakness, together they can support the gecko's body weight.

Van der Waals interactions, hydrogen bonds, ionic bonds in water, and other weak bonds may form not only between molecules but also between parts of a large molecule, such as a protein. The cumulative effect of weak bonds is to reinforce the three-dimensional shape of the molecule. (You will learn more about the very important biological roles of weak bonds in Chapter 3.)

Molecular Shape and Function

A molecule has a characteristic size and shape. The precise shape of a molecule is usually very important to its function in the living cell.

A molecule consisting of two atoms, such as H_2 or O_2, is always linear, but most molecules with more than two atoms have more complicated shapes. To take a very simple example, a water molecule (H_2O) is shaped roughly like a V, with its two covalent bonds spread apart at an angle of 104.5° (**Figure 2.13**). A methane molecule (CH_4) has a geometric shape called a tetrahedron, a pyramid with a triangular base. The carbon nucleus is inside, at the center, with its four covalent bonds radiating to hydrogen nuclei at the corners of the tetrahedron. Larger molecules containing multiple carbon atoms, including many of the molecules that make up living matter, have more complex overall shapes. However, the tetrahedral shape of a carbon atom bonded to four other atoms is often a repeating motif within such molecules.

Space-Filling Model	Ball-and-Stick Model
Water (H₂O)	104.5°
Methane (CH₄)	

▲ **Figure 2.13 Models showing the shapes of two small molecules.** Each of the molecules, water and methane, is represented in two different ways.

Molecular shape is crucial in biology because it determines how biological molecules recognize and respond to one another with specificity. Biological molecules often bind temporarily to each other by forming weak bonds, but this can happen only if their shapes are complementary. We can see this specificity in the effects of opiates, drugs derived from opium. Opiates, such as morphine and heroin, relieve pain and alter mood by weakly binding to specific receptor molecules on the surfaces of brain cells. Why would brain cells carry receptors for opiates, compounds that are not made by our bodies? The discovery of endorphins in 1975 answered this question. Endorphins are signaling molecules made by the pituitary gland that bind to the receptors, relieving pain and producing euphoria during times of stress, such as intense exercise. It turns out that opiates have shapes similar to endorphins and mimic them by binding to endorphin receptors in the brain. That is why opiates (such as morphine) and endorphins have similar effects (**Figure 2.14**).

Key

- ■ Carbon
- ■ Hydrogen
- ■ Nitrogen
- ■ Sulfur
- ■ Oxygen

Natural endorphin

Morphine

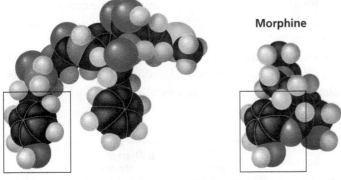

(a) Structures of endorphin and morphine. The boxed portion of the endorphin molecule (left) binds to receptor molecules on target cells in the brain. The boxed portion of the morphine molecule (right) is a close match.

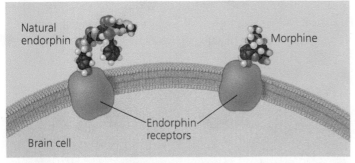

(b) Binding to endorphin receptors. Both endorphin and morphine can bind to endorphin receptors on the surface of a brain cell.

▲ **Figure 2.14 A molecular mimic.** Morphine affects pain perception and emotional state by mimicking the brain's natural endorphins.

CONCEPT CHECK 2.3

1. Why does the structure H—C≡C—H fail to make sense chemically?
2. What holds the atoms together in a crystal of magnesium chloride ($MgCl_2$)?
3. **WHAT IF?** If you were a pharmaceutical researcher, why would you want to learn the three-dimensional shapes of naturally occurring signaling molecules?

For suggested answers, see Appendix A.

CONCEPT 2.4

Chemical reactions make and break chemical bonds

The making and breaking of chemical bonds, leading to changes in the composition of matter, are called **chemical reactions**. An example is the reaction between hydrogen and oxygen molecules that forms water:

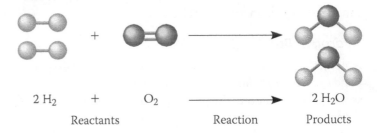

2 H_2	+	O_2		2 H_2O
Reactants			Reaction	Products

This reaction breaks the covalent bonds of H_2 and O_2 and forms the new bonds of H_2O. When we write a chemical reaction, we use an arrow to indicate the conversion of the starting materials, called the **reactants**, to the **products**. The coefficients indicate the number of molecules involved; for example, the coefficient 2 in front of H_2 means that the reaction starts with two molecules of hydrogen. Notice that all atoms of the reactants must be accounted for in the products. Matter is conserved in a chemical reaction: Reactions cannot create or destroy matter but can only rearrange it.

Photosynthesis, which takes place within the cells of green plant tissues, is an important biological example of how chemical reactions rearrange matter. Humans and other animals ultimately depend on photosynthesis for food and oxygen, and this process is at the foundation of almost all ecosystems. The following chemical shorthand summarizes the process of photosynthesis:

$$6\ CO_2 + 6\ H_2O \rightarrow C_6H_{12}O_6 + 6\ O_2$$

The raw materials of photosynthesis are carbon dioxide (CO_2), which is taken from the air, and water (H_2O), which is absorbed from the soil. Within the plant cells, sunlight powers the conversion of these ingredients to a sugar called glucose ($C_6H_{12}O_6$) and oxygen molecules (O_2), a by-product that the plant releases into the surroundings (**Figure 2.15**). Although

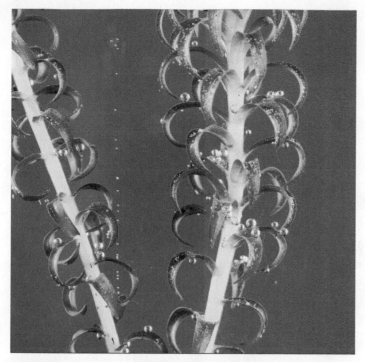

▲ Figure 2.15 Photosynthesis: a solar-powered rearrangement of matter. *Elodea*, a freshwater plant, produces sugar by rearranging the atoms of carbon dioxide and water in the chemical process known as photosynthesis, which is powered by sunlight. Much of the sugar is then converted to other food molecules. Oxygen gas (O_2) is a by-product of photosynthesis; notice the bubbles of O_2-containing gas escaping from the leaves in the photo.

? *Explain how this photo relates to the reactants and products in the equation for photosynthesis given in the text. (You will learn more about photosynthesis in Chapter 8.)*

photosynthesis is actually a sequence of many chemical reactions, we still end up with the same number and types of atoms that we had when we started. Matter has simply been rearranged, with an input of energy provided by sunlight.

All chemical reactions are reversible, with the products of the forward reaction becoming the reactants of the reverse reaction. For example, hydrogen and nitrogen molecules can combine to form ammonia, but ammonia can also decompose to regenerate hydrogen and nitrogen:

$$3\,H_2 + N_2 \rightleftharpoons 2\,NH_3$$

The two opposite-headed arrows indicate that the reaction is reversible.

One of the factors affecting the rate of a reaction is the concentration of reactants. The greater the concentration of reactant molecules, the more frequently they collide with one another and have an opportunity to react and form products. The same holds true for products. As products accumulate, collisions resulting in the reverse reaction become more frequent. Eventually, the forward and reverse reactions occur at the same rate, and the relative concentrations of products and reactants stop changing. The point at which the reactions offset one another exactly is called **chemical equilibrium**. This is a dynamic equilibrium; reactions are still going on, but with no net effect on the concentrations of reactants and products. Equilibrium does *not* mean that the reactants and products are equal in concentration, but only that their concentrations have stabilized at a particular ratio. The reaction involving ammonia reaches equilibrium when ammonia decomposes as rapidly as it forms. In some chemical reactions, the equilibrium point may lie so far to the right that these reactions go essentially to completion; that is, virtually all the reactants are converted to products.

To conclude this chapter, we focus on water, the substance in which all the chemical processes of organisms occur.

CONCEPT CHECK 2.4

1. Which type of chemical reaction occurs faster at equilibrium, the formation of products from reactants or reactants from products?

2. **WHAT IF?** Write an equation that uses the products of photosynthesis as reactants and the reactants of photosynthesis as products. Add energy as another product. This new equation describes a process that occurs in your cells. Describe this equation in words. How does this equation relate to breathing?

For suggested answers, see Appendix A.

CONCEPT 2.5

Hydrogen bonding gives water properties that help make life possible on Earth

All organisms are made mostly of water and live in an environment dominated by water. Most cells are surrounded by water, and cells themselves are about 70–95% water. Water is so common that it is easy to overlook the fact that it is an exceptional substance with many extraordinary qualities. We can trace water's unique behavior to the structure and interactions of its molecules. As you saw in Figure 2.9, the connections between the atoms of a water molecule are polar covalent bonds. The unequal sharing of electrons and water's V-like shape make it a **polar molecule**, meaning that its overall charge is unevenly distributed: The oxygen region of the molecule has a partial negative charge ($\delta-$), and each hydrogen has a partial positive charge ($\delta+$).

The properties of water arise from attractions between oppositely charged atoms of different water molecules: The slightly positive hydrogen of one molecule is attracted to the slightly negative oxygen of a nearby molecule. The two molecules are thus held together by a hydrogen bond. When water is in its liquid form, its hydrogen bonds are very fragile, each only about 1/20 as strong as a covalent bond. The hydrogen bonds form, break, and re-form with great frequency. Each lasts only a few trillionths of a second, but the molecules are constantly forming new hydrogen bonds with a succession

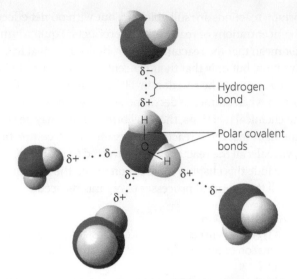

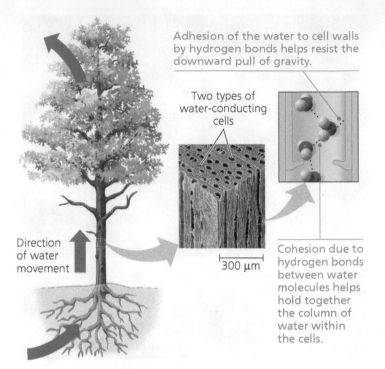

Adhesion of the water to cell walls by hydrogen bonds helps resist the downward pull of gravity.

Two types of water-conducting cells

300 μm

Cohesion due to hydrogen bonds between water molecules helps hold together the column of water within the cells.

Direction of water movement

▲ **Figure 2.16 Hydrogen bonds between water molecules.** The charged regions in a water molecule are due to its polar covalent bonds. Oppositely charged regions of neighboring water molecules are attracted to each other, forming hydrogen bonds. Each molecule can hydrogen-bond to multiple partners, and these associations are constantly changing.

DRAW IT *Draw partial charges on all the atoms of the water molecule on the far left above, and draw two more water molecules hydrogen-bonded to it.*

▲ **Figure 2.17 Water transport in plants.** Evaporation from leaves pulls water upward from the roots through water-conducting cells. Because of the properties of cohesion and adhesion, the tallest trees can transport water more than 100 m upward—approximately one-quarter the height of the Empire State Building in New York City.

 ANIMATION *BioFlix* Visit the Study Area in **MasteringBiology** for the BioFlix® 3-D Animation on Water Transport in Plants.

of partners. Therefore, at any instant, a substantial percentage of all the water molecules are hydrogen-bonded to their neighbors **(Figure 2.16)**. The extraordinary qualities of water emerge in large part from the hydrogen bonding that organizes water molecules into a higher level of structural order. We will examine four emergent properties of water that contribute to Earth's suitability as an environment for life: cohesive behavior, ability to moderate temperature, expansion upon freezing, and versatility as a solvent. After that, we'll discuss a critical aspect of water chemistry—acids and bases.

Cohesion of Water Molecules

Water molecules stay close to each other as a result of hydrogen bonding. At any given moment, many of the molecules in liquid water are linked by multiple hydrogen bonds. These linkages make water more structured than most other liquids. Collectively, the hydrogen bonds hold the substance together, a phenomenon called **cohesion**.

Cohesion due to hydrogen bonding contributes to the transport of water and dissolved nutrients against gravity in plants **(Figure 2.17)**. Water from the roots reaches the leaves through a network of water-conducting cells. As water evaporates from a leaf, hydrogen bonds cause water molecules leaving the veins to tug on molecules farther down, and the upward pull is transmitted through the water-conducting cells all the way to the roots. **Adhesion**, the clinging of one substance to another, also plays a role. Adhesion of water to cell walls by hydrogen bonds helps counter the downward pull of gravity.

Related to cohesion is **surface tension**, a measure of how difficult it is to stretch or break the surface of a liquid. The hydrogen bonds in water give it an unusually high surface tension, making it behave as though it were coated with an invisible film. You can observe the surface tension of water by slightly overfilling a drinking glass; the water will stand above the rim. The spider in **Figure 2.18** takes advantage of the surface tension of water to walk across a pond without breaking the surface.

Moderation of Temperature by Water

Water moderates air temperature by absorbing heat from air that is warmer and releasing the stored heat to air that is cooler. Water is effective as a heat bank because it can absorb or release a relatively large amount of heat with only a slight change in its own temperature. To understand this capability of water, we must first look briefly at temperature and heat.

Temperature and Heat

Anything that moves has **kinetic energy**, the energy of motion. Atoms and molecules have kinetic energy because they are always moving, although not necessarily in any particular direction. The faster a molecule moves, the greater its kinetic energy. The kinetic energy associated with the random movement of

▲ **Figure 2.18 Walking on water.** The high surface tension of water, resulting from the collective strength of its hydrogen bonds, allows this raft spider to walk on the surface of a pond.

atoms or molecules is called **thermal energy**. The *total* thermal energy of a body of matter depends in part on the matter's volume. Although thermal energy is related to temperature, they are not the same thing. **Temperature** represents the *average* kinetic energy of the molecules, regardless of volume. When water is heated in a coffeemaker, the average speed of the molecules increases, and the thermometer records this as a rise in temperature of the liquid. The amount of thermal energy also increases in this case. Note, however, that although the pot of coffee has a much higher temperature than, say, the water in a swimming pool, the swimming pool contains more thermal energy because of its much greater volume.

Whenever two objects of different temperature are brought together, thermal energy passes from the warmer to the cooler object until the two are the same temperature. Molecules in the cooler object speed up at the expense of the thermal energy of the warmer object. An ice cube cools a drink not by adding coldness to the liquid, but by absorbing thermal energy from the liquid as the ice itself melts. Thermal energy in transfer from one body of matter to another is defined as **heat**.

One convenient unit of heat used in this book is the **calorie (cal)**. A calorie is the amount of heat it takes to raise the temperature of 1 g of water by 1°C. Conversely, a calorie is also the amount of heat that 1 g of water releases when it cools by 1°C. A **kilocalorie (kcal)**, 1,000 cal, is the quantity of heat required to raise the temperature of 1 kilogram (kg) of water by 1°C. (The "calories" on food packages are actually kilocalories.) Another energy unit used in this book is the **joule (J)**. One joule equals 0.239 cal; one calorie equals 4.184 J.

Water's High Specific Heat

The ability of water to stabilize temperature stems from its relatively high specific heat. The **specific heat** of a substance is defined as the amount of heat that must be absorbed or lost for 1 g of that substance to change its temperature by 1°C. We already know water's specific heat because we have defined a calorie as the amount of heat that causes 1 g of water to change its temperature by 1°C. Therefore, the specific heat of water is 1 calorie per gram per degree Celsius, abbreviated as 1 cal/g·°C. Compared with most other substances, water has an unusually high specific heat. As a result, water will change its temperature less than other liquids when it absorbs or loses a given amount of heat. The reason you can burn your fingers by touching the side of an iron pot on the stove when the water in the pot is still lukewarm is that the specific heat of water is ten times greater than that of iron. In other words, the same amount of heat will raise the temperature of 1 g of the iron much faster than it will raise the temperature of 1 g of the water. Specific heat can be thought of as a measure of how well a substance resists changing its temperature when it absorbs or releases heat. Water resists changing its temperature; when it does change its temperature, it absorbs or loses a relatively large quantity of heat for each degree of change.

We can trace water's high specific heat, like many of its other properties, to hydrogen bonding. Heat must be absorbed in order to break hydrogen bonds; by the same token, heat is released when hydrogen bonds form. A calorie of heat causes a relatively small change in the temperature of water because much of the heat is used to disrupt hydrogen bonds before the water molecules can begin moving faster. And when the temperature of water drops slightly, many additional hydrogen bonds form, releasing a considerable amount of energy in the form of heat.

What is the relevance of water's high specific heat to life on Earth? A large body of water can absorb and store a huge amount of heat from the sun in the daytime and during summer while warming up only a few degrees. At night and during winter, the gradually cooling water can warm the air. This is the reason coastal areas generally have milder climates than inland regions **(Figure 2.19)**. The high specific heat of water also tends to stabilize ocean temperatures, creating a favorable environment for marine life. Thus, because of its high specific heat, the water that covers most of Earth keeps temperature fluctuations on land and in water within limits that permit life.

▲ **Figure 2.19 Effect of a large body of water on climate.** By absorbing or releasing heat, oceans moderate coastal climates. In this example from an August day in Southern California, the relatively cool ocean reduces coastal air temperatures by absorbing heat. (The temperatures are in degrees Fahrenheit.)

Also, because organisms are made primarily of water, they are better able to resist changes in their own temperature than if they were made of a liquid with a lower specific heat.

Evaporative Cooling

Molecules of any liquid stay close together because they are attracted to one another. Molecules moving fast enough to overcome these attractions can depart the liquid and enter the air as a gas. This transformation from a liquid to a gas is called vaporization, or *evaporation*. Recall that the speed of molecular movement varies and that temperature is the *average* kinetic energy of molecules. Even at low temperatures, the speediest molecules can escape into the air. Some evaporation occurs at any temperature; a glass of water at room temperature, for example, will eventually evaporate completely. If a liquid is heated, the average kinetic energy of molecules increases and the liquid evaporates more rapidly.

Heat of vaporization is the quantity of heat a liquid must absorb for 1 g of it to be converted from the liquid to the gaseous state. For the same reason that water has a high specific heat, it also has a high heat of vaporization relative to most other liquids. To evaporate 1 g of water at 25°C, about 580 cal of heat is needed—nearly double the amount needed to vaporize a gram of alcohol, for example. Water's high heat of vaporization is another property emerging from the strength of its hydrogen bonds, which must be broken before the molecules can make their exodus from the liquid.

The high amount of energy required to vaporize water has a wide range of effects. On a global scale, for example, it helps moderate Earth's climate. A considerable amount of solar heat absorbed by tropical seas is consumed during the evaporation of surface water. Then, as moist tropical air circulates poleward, it releases heat as it condenses and forms rain. On an organismal level, water's high heat of vaporization accounts for the severity of steam burns. These burns are caused by the heat energy released when steam condenses into liquid on the skin.

As a liquid evaporates, the surface of the liquid that remains behind cools down. This **evaporative cooling** occurs because the "hottest" molecules, those with the greatest kinetic energy, are the ones most likely to leave as gas. It is as if the hundred fastest runners at a college transferred to another school; the average speed of the remaining students would decline.

Evaporative cooling of water contributes to the stability of temperature in lakes and ponds and also provides a mechanism that prevents terrestrial organisms from overheating. For example, evaporation of water from the leaves of a plant helps keep the tissues in the leaves from becoming too warm in the sunlight. Evaporation of sweat from human skin dissipates body heat and helps prevent overheating on a hot day or when excess heat is generated by strenuous activity. High humidity on a hot day increases discomfort because the high concentration of water vapor in the air inhibits the evaporation of sweat from the body.

Floating of Ice on Liquid Water

Water is one of the few substances that are less dense as a solid than as a liquid. In other words, ice floats on liquid water. While other materials contract and become denser when they solidify, water expands. The cause of this exotic behavior is, once again, hydrogen bonding. At temperatures above 4°C, water behaves like other liquids, expanding as it warms and contracting as it cools. As the temperature falls from 4°C to 0°C, water begins to freeze because more and more of its molecules are moving too slowly to break hydrogen bonds. At 0°C, the molecules become locked into a crystalline lattice, each water molecule hydrogen-bonded to four partners **(Figure 2.20)**. The hydrogen bonds keep the molecules at "arm's length," far enough apart to make ice about 10% less dense than liquid water at 4°C. When ice absorbs enough heat for its temperature to rise above 0°C, hydrogen bonds between molecules are disrupted. As the crystal collapses, the ice melts, and molecules are free to slip closer together. Water reaches its greatest density at 4°C and then begins to expand as the molecules move faster.

The ability of ice to float due to its lower density is an important factor in the suitability of the environment for life.

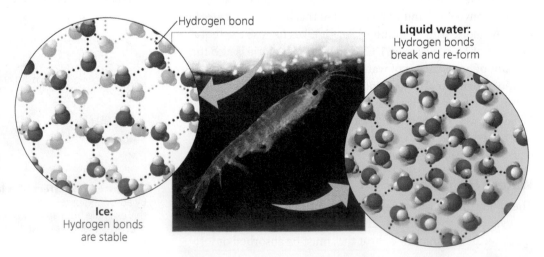

▶ **Figure 2.20 Ice: crystalline structure and floating barrier.** In ice, each molecule is hydrogen-bonded to four neighbors in a three-dimensional crystal. Because the crystal is spacious, ice has fewer molecules than an equal volume of liquid water. In other words, ice is less dense than liquid water. Floating ice becomes a barrier that protects the liquid water below from the colder air. The marine organism shown here is a type of shrimp called krill; it was photographed beneath floating ice in the Southern Ocean near Antarctica.

WHAT IF? *If water did not form hydrogen bonds, what would happen to the shrimp's environment?*

Hydrogen bond

Ice:
Hydrogen bonds are stable

Liquid water:
Hydrogen bonds break and re-form

If ice sank, then eventually all ponds, lakes, and even oceans would freeze solid, making life as we know it impossible on Earth. During summer, only the upper few inches of the ocean would thaw. Instead, when a deep body of water cools, the floating ice insulates the liquid water below, preventing it from freezing and allowing life to exist under the frozen surface, as shown in the photo in Figure 2.20.

Water: The Solvent of Life

A sugar cube placed in a glass of water will dissolve. Eventually, the glass will contain a uniform mixture of sugar and water; the concentration of dissolved sugar will be the same everywhere in the mixture. A liquid that is a completely homogeneous mixture of two or more substances is called a **solution**. The dissolving agent of a solution is the **solvent**, and the substance that is dissolved is the **solute**. In this case, water is the solvent and sugar is the solute. An **aqueous solution** is one in which water is the solvent.

Water is a very versatile solvent, a quality we can trace to the polarity of the water molecule. Suppose, for example, that a spoonful of table salt, the ionic compound sodium chloride (NaCl), is placed in water **(Figure 2.21)**. At the surface of each grain, or crystal, of salt, the sodium and chloride ions are exposed to the solvent. These ions and regions of the water molecules are attracted to each other owing to their opposite charges. The oxygen regions of the water molecules are negatively charged and are attracted to sodium cations. The hydrogen regions are positively charged and are attracted to chloride

anions. As a result, water molecules surround the individual sodium and chloride ions, separating and shielding them from one another. The sphere of water molecules around each dissolved ion is called a **hydration shell**. Working inward from the surface of each salt crystal, water eventually dissolves all the ions. The result is a solution of two solutes, sodium cations and chloride anions, homogeneously mixed with water, the solvent. Other ionic compounds also dissolve in water. Seawater, for instance, contains a great variety of dissolved ions, as do living cells.

A compound does not need to be ionic to dissolve in water; many compounds made up of nonionic polar molecules, such as sugars, are also water-soluble. Such compounds dissolve when water molecules surround each of the solute molecules, forming hydrogen bonds with them. Even molecules as large as proteins can dissolve in water if they have ionic and polar regions on their surface **(Figure 2.22)**. Many different kinds of polar compounds are dissolved (along with ions) in the water of such biological fluids as blood, the sap of plants, and the liquid within all cells. Water is the solvent of life.

Hydrophilic and Hydrophobic Substances

Any substance that has an affinity for water is said to be **hydrophilic** (from the Greek *hydro*, water, and *philos*, loving). In some cases, substances can be hydrophilic without actually dissolving. For example, some molecules in cells are so large that they do not dissolve. Another example of a hydrophilic substance that does not dissolve is cotton, a plant product. Cotton consists of giant molecules of cellulose, a compound with numerous regions of partial positive and partial negative charges that can form hydrogen bonds with water. Water adheres to the cellulose fibers. Thus, a cotton towel does a great job of drying the body, yet it does not dissolve in the washing

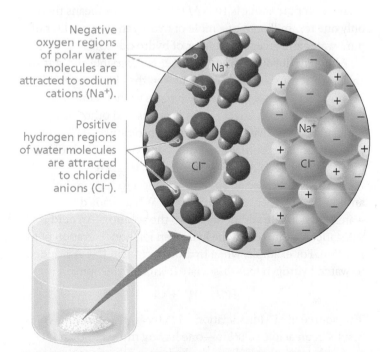

Negative oxygen regions of polar water molecules are attracted to sodium cations (Na⁺).

Positive hydrogen regions of water molecules are attracted to chloride anions (Cl⁻).

▲ **Figure 2.21 Table salt dissolving in water.** A sphere of water molecules, called a hydration shell, surrounds each solute ion.

WHAT IF? *What would happen if you heated this solution for a long time?*

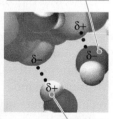

This oxygen is attracted to a slight positive charge on the lysozyme molecule.

This hydrogen is attracted to a slight negative charge on the lysozyme molecule.

▲ **Figure 2.22 A water-soluble protein.** Human lysozyme is a protein found in tears and saliva that has antibacterial action. This model shows the lysozyme molecule (purple) in an aqueous environment. Ionic and polar regions on the protein's surface attract water molecules.

machine. Cellulose is also present in the walls of plant cells that conduct water; you read earlier how the adhesion of water to these hydrophilic walls helps water move up the plant against gravity.

There are, of course, substances that do not have an affinity for water. Substances that are nonionic and nonpolar (or otherwise cannot form hydrogen bonds) actually seem to repel water; these substances are said to be **hydrophobic** (from the Greek *phobos*, fearing). An example from the kitchen is vegetable oil, which, as you know, does not mix stably with water-based substances such as vinegar. The hydrophobic behavior of the oil molecules results from a prevalence of relatively nonpolar covalent bonds, in this case bonds between carbon and hydrogen, which share electrons almost equally. Hydrophobic molecules related to oils are major ingredients of cell membranes. (Imagine what would happen to a cell if its membrane dissolved!)

Solute Concentration in Aqueous Solutions

Most of the chemical reactions in organisms involve solutes dissolved in water. To understand such reactions, we must know how many atoms and molecules are involved and be able to calculate the concentration of solutes in an aqueous solution (the number of solute molecules in a volume of solution).

When carrying out experiments, we use mass to calculate the number of molecules. We first calculate the **molecular mass**, which is simply the sum of the masses of all the atoms in a molecule. As an example, let's calculate the molecular mass of table sugar (sucrose), $C_{12}H_{22}O_{11}$. In round numbers, sucrose has a molecular mass of $(12 \times 12) + (22 \times 1) + (11 \times 16) = 342$ daltons. Because we can't measure out small numbers of molecules, we usually measure substances in units called moles. Just as a dozen always means 12 objects, a **mole (mol)** represents an exact number of objects: 6.02×10^{23}, which is called Avogadro's number. There are 6.02×10^{23} daltons in 1 g. Once we determine the molecular mass of a molecule such as sucrose, we can use the same number (342), but with the unit *gram*, to represent the mass of 6.02×10^{23} molecules of sucrose, or 1 mol of sucrose. To obtain 1 mol of sucrose in the lab, therefore, we weigh out 342 g.

The practical advantage of measuring a quantity of chemicals in moles is that a mole of one substance has exactly the same number of molecules as a mole of any other substance. Measuring in moles makes it convenient for scientists working in the laboratory to combine substances in fixed ratios of molecules.

How would we make a liter (L) of solution consisting of 1 mol of sucrose dissolved in water? We would measure out 342 g of sucrose and then add enough water to bring the total volume of the solution up to 1 L. At that point, we would have a 1-molar (1 M) solution of sucrose. **Molarity**—the number of moles of solute per liter of solution—is the unit of concentration most often used by biologists for aqueous solutions.

Acids and Bases

Occasionally, a hydrogen atom participating in a hydrogen bond between two water molecules shifts from one molecule to the other. When this happens, the hydrogen atom leaves its electron behind, and what is actually transferred is a **hydrogen ion** (H^+), a single proton with a charge of 1+. The water molecule that lost a proton is now a **hydroxide ion** (OH^-), which has a charge of 1−. The proton binds to the other water molecule, making that molecule a **hydronium ion** (H_3O^+).

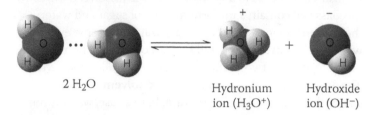

$2\ H_2O$ Hydronium ion (H_3O^+)   Hydroxide ion (OH^-)

By convention, H^+ (the hydrogen ion) is used to represent H_3O^+ (the hydronium ion), and we follow that practice here. Keep in mind, though, that H^+ does not exist on its own in an aqueous solution. It is always associated with another water molecule in the form of H_3O^+.

As indicated by the double arrows, this is a reversible reaction that reaches a state of dynamic equilibrium when water molecules dissociate at the same rate that they are being re-formed from H^+ and OH^-. At this equilibrium point, the concentration of water molecules greatly exceeds the concentrations of H^+ and OH^-. In pure water, only one water molecule in every 554 million is dissociated; the concentration of each ion in pure water is $10^{-7}\ M$ (at 25°C). This means there is only one ten-millionth of a mole of hydrogen ions per liter of pure water and an equal number of hydroxide ions.

Although the dissociation of water is reversible and statistically rare, it is exceedingly important in the chemistry of life. H^+ and OH^- are very reactive. Changes in their concentrations can drastically affect a cell's proteins and other complex molecules. As we have seen, the concentrations of H^+ and OH^- are equal in pure water, but adding certain kinds of solutes, called acids and bases, disrupts this balance.

What would cause an aqueous solution to have an imbalance in H^+ and OH^- concentrations? When acids dissolve in water, they donate additional H^+ to the solution. An **acid** is a substance that increases the hydrogen ion concentration of a solution. For example, when hydrochloric acid (HCl) is added to water, hydrogen ions dissociate from chloride ions:

$$HCl \rightarrow H^+ + Cl^-$$

This source of H^+ (dissociation of water is the other source) results in an acidic solution—one having more H^+ than OH^-.

A substance that *reduces* the hydrogen ion concentration of a solution is called a **base**. Some bases reduce the H^+ concentration directly by accepting hydrogen ions. Ammonia (NH_3), for instance, acts as a base when the unshared electron pair in nitrogen's valence shell attracts a hydrogen ion from the solution, resulting in an ammonium ion (NH_4^+):

$$NH_3 + H^+ \rightleftharpoons NH_4^+$$

Other bases reduce the H^+ concentration indirectly by dissociating to form hydroxide ions, which combine with hydrogen ions and form water. One such base is sodium hydroxide (NaOH), which in water dissociates into its ions:

$$NaOH \rightarrow Na^+ + OH^-$$

In either case, the base reduces the H^+ concentration. Solutions with a higher concentration of OH^- than H^+ are known as basic solutions. A solution in which the H^+ and OH^- concentrations are equal is said to be neutral.

Notice that single arrows were used in the reactions for HCl and NaOH. These compounds dissociate completely when mixed with water, so hydrochloric acid is called a strong acid and sodium hydroxide a strong base. In contrast, ammonia is a relatively weak base. The double arrows in the reaction for ammonia indicate that the binding and release of hydrogen ions are reversible reactions, although at equilibrium there will be a fixed ratio of NH_4^+ to NH_3.

There are also weak acids, which reversibly release and accept back hydrogen ions. An example is carbonic acid:

$$\underset{\substack{\text{Carbonic} \\ \text{acid}}}{H_2CO_3} \rightleftharpoons \underset{\substack{\text{Bicarbonate} \\ \text{ion}}}{HCO_3^-} + \underset{\substack{\text{Hydrogen} \\ \text{ion}}}{H^+}$$

Here the equilibrium so favors the reaction in the left direction that when carbonic acid is added to pure water, only 1% of the molecules are dissociated at any particular time. Still, that is enough to shift the balance of H^+ and OH^- from neutrality.

The pH Scale

In any aqueous solution at 25°C, the *product* of the H^+ and OH^- concentrations is constant at 10^{-14}. This can be written

$$[H^+][OH^-] = 10^{-14}$$

In such an equation, brackets indicate molar concentration. In a neutral solution at room temperature (25°C), $[H^+] = 10^{-7}$ and $[OH^-] = 10^{-7}$, so in this case, 10^{-14} is the product of $10^{-7} \times 10^{-7}$. If enough acid is added to a solution to increase $[H^+]$ to 10^{-5} M, then $[OH^-]$ will decline by an equivalent amount to 10^{-9} M (note that $10^{-5} \times 10^{-9} = 10^{-14}$). This constant relationship expresses the behavior of acids and bases in an aqueous solution. An acid not only adds hydrogen ions to a solution, but also removes hydroxide ions because of the tendency for H^+ to combine with OH^-, forming water. A base has the opposite effect, increasing OH^- concentration but also reducing H^+ concentration by the formation of water. If enough of a base is added to raise the OH^- concentration to 10^{-4} M, it will cause the H^+ concentration to drop to 10^{-10} M. Whenever we know the concentration of either H^+ or OH^- in an aqueous solution, we can deduce the concentration of the other ion.

Because the H^+ and OH^- concentrations of solutions can vary by a factor of 100 trillion or more, scientists have developed a way to express this variation more conveniently than in moles per liter. The pH scale (Figure 2.23) compresses the range of H^+ and OH^- concentrations by employing logarithms. The **pH** of a solution is defined as the negative logarithm (base 10) of the hydrogen ion concentration:

$$pH = -\log [H^+]$$

For a neutral aqueous solution, $[H^+]$ is 10^{-7} M, giving us

$$-\log 10^{-7} = -(-7) = 7$$

Notice that pH *declines* as H^+ concentration *increases*. Notice, too, that although the pH scale is based on H^+ concentration, it also implies OH^- concentration. A solution of pH 10 has a hydrogen ion concentration of 10^{-10} M and a hydroxide ion concentration of 10^{-4} M.

The pH of a neutral aqueous solution at 25°C is 7, the midpoint of the pH scale. A pH value less than 7 denotes an acidic solution; the lower the number, the more acidic the solution.

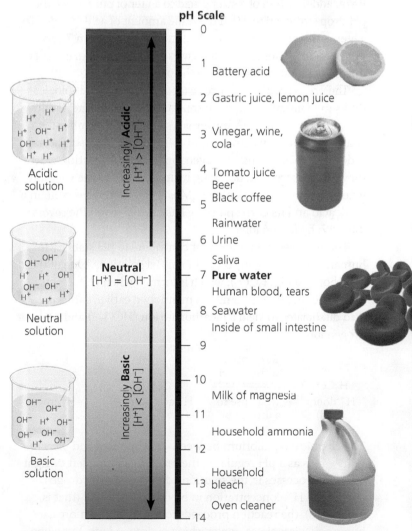

pH Scale

pH	
0	
1	Battery acid
2	Gastric juice, lemon juice
3	Vinegar, wine, cola
4	Tomato juice
	Beer
	Black coffee
5	
	Rainwater
6	Urine
	Saliva
7	**Pure water**
	Human blood, tears
8	Seawater
	Inside of small intestine
9	
10	
	Milk of magnesia
11	
	Household ammonia
12	
13	Household bleach
	Oven cleaner
14	

Acidic solution — Increasingly Acidic $[H^+] > [OH^-]$

Neutral solution — Neutral $[H^+] = [OH^-]$

Basic solution — Increasingly Basic $[H^+] < [OH^-]$

▲ **Figure 2.23 The pH scale and pH values of some aqueous solutions.**

The pH for basic solutions is above 7. Most biological fluids are within the range pH 6–8. There are a few exceptions, however, including the strongly acidic digestive juice of the human stomach, which has a pH of about 2.

Remember that each pH unit represents a tenfold difference in H^+ and OH^- concentrations. It is this mathematical feature that makes the pH scale so compact. A solution of pH 3 is not twice as acidic as a solution of pH 6, but a thousand times ($10 \times 10 \times 10$) more acidic. When the pH of a solution changes slightly, the actual concentrations of H^+ and OH^- in the solution change substantially.

Buffers

The internal pH of most living cells is close to 7. Even a slight change in pH can be harmful because the chemical processes of the cell are very sensitive to the concentrations of hydrogen and hydroxide ions. The pH of human blood is very close to 7.4, which is slightly basic. A person cannot survive for more than a few minutes if the blood pH drops to 7 or rises to 7.8, and a chemical system exists in the blood that maintains a stable pH. If you add 0.01 mol of a strong acid to a liter of pure water, the pH drops from 7.0 to 2.0. If the same amount of acid is added to a liter of blood, however, the pH decrease is only from 7.4 to 7.3. Why does the addition of acid have so much less of an effect on the pH of blood than it does on the pH of water?

The presence of substances called buffers allows biological fluids to maintain a relatively constant pH despite the addition of acids or bases. A **buffer** is a substance that minimizes changes in the concentrations of H^+ and OH^- in a solution. It does so by accepting hydrogen ions from the solution when they are in excess and donating hydrogen ions to the solution when they have been depleted. Most buffer solutions contain a weak acid and its corresponding base, which combine reversibly with hydrogen ions.

There are several buffers that contribute to pH stability in human blood and many other biological solutions. One of these is carbonic acid (H_2CO_3), which is formed when CO_2 reacts with water in blood plasma. As mentioned earlier, carbonic acid dissociates to yield a bicarbonate ion (HCO_3^-) and a hydrogen ion (H^+):

$$H_2CO_3 \underset{\text{Response to a drop in pH}}{\overset{\text{Response to a rise in pH}}{\rightleftharpoons}} HCO_3^- + H^+$$

H_2CO_3 — H^+ donor (acid); HCO_3^- — H^+ acceptor (base); H^+ — Hydrogen ion

The chemical equilibrium between carbonic acid and bicarbonate acts as a pH regulator, the reaction shifting left or right as other processes in the solution add or remove hydrogen ions. If the H^+ concentration in blood begins to fall (that is, if pH rises), the reaction proceeds to the right and more carbonic acid dissociates, replenishing hydrogen ions. But when H^+ concentration in blood begins to rise (when pH drops), the reaction proceeds to the left, with HCO_3^- (the base) removing the hydrogen ions from the solution and forming H_2CO_3. Thus, the carbonic acid–bicarbonate buffering system consists of an acid and a base in equilibrium with each other. Most other buffers are also acid-base pairs.

Acidification: A Threat to Our Oceans

Among the many threats to water quality posed by human activities is the burning of fossil fuels, which releases gaseous compounds into the atmosphere. When certain of these compounds react with water, the water becomes more acidic, altering the delicate balance of conditions for life on Earth.

Carbon dioxide is the main product of fossil fuel combustion. About 25% of human-generated CO_2 is absorbed by the oceans. In spite of the huge volume of water in the oceans, scientists worry that the absorption of so much CO_2 will harm marine ecosystems.

Recent data have shown that such fears are well founded. When CO_2 dissolves in seawater, it reacts with water to form carbonic acid, which lowers ocean pH, causing ocean acidification (see **Figure 2.24**). Based on measurements of CO_2 levels in air bubbles trapped in ice over thousands of years, scientists calculate that the pH of the oceans is 0.1 pH unit lower now than at any time in the past 420,000 years. Recent studies predict that it will drop another 0.3–0.5 pH unit by the end of this century.

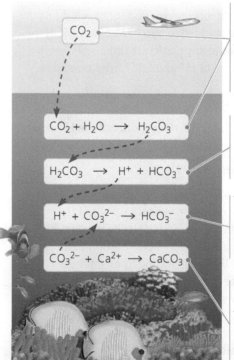

Some carbon dioxide (CO_2) in the atmosphere dissolves in the ocean, where it reacts with water to form carbonic acid (H_2CO_3).

$$CO_2 + H_2O \rightarrow H_2CO_3$$

Carbonic acid dissociates into hydrogen ions (H^+) and bicarbonate ions (HCO_3^-).

$$H_2CO_3 \rightarrow H^+ + HCO_3^-$$

The added H^+ combines with carbonate ions (CO_3^{2-}), forming more HCO_3^-.

$$H^+ + CO_3^{2-} \rightarrow HCO_3^-$$

Less CO_3^{2-} is available for calcification — the formation of calcium carbonate ($CaCO_3$)— by marine organisms such as corals.

$$CO_3^{2-} + Ca^{2+} \rightarrow CaCO_3$$

▲ **Figure 2.24 Atmospheric CO_2 from human activities and its fate in the ocean.**

WHAT IF? *Would lowering the ocean's carbonate concentration have any effect, even indirectly, on organisms that don't form $CaCO_3$? Explain.*

Interpreting a Scatter Plot with a Regression Line

How Does the Carbonate Ion Concentration of Seawater Affect the Calcification Rate of a Coral Reef? Scientists predict that acidification of the ocean due to higher levels of atmospheric CO_2 will lower the concentration of dissolved carbonate ions, which living corals use to build calcium carbonate reef structures. In this exercise, you will analyze data from a controlled experiment that examined the effect of carbonate ion concentration ($[CO_3{}^{2-}]$) on calcium carbonate deposition, a process called calcification.

How the Experiment Was Done The Biosphere 2 aquarium in Arizona contains a large coral reef system that behaves like a natural reef. For several years, a group of researchers measured the rate of calcification by the reef organisms and examined how the calcification rate changed with differing amounts of dissolved carbonate ions in the seawater.

Data from the Experiment The black data points in the graph below form a scatter plot. The red line, known as a linear regression line, is the best-fitting straight line for these points. These data are from one set of experiments, in which the pH, temperature, and calcium ion concentration of the seawater were held constant.

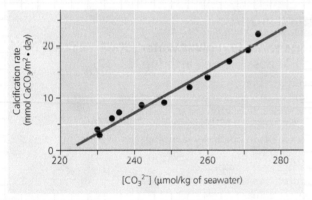

Interpret the Data

1. When presented with a graph of experimental data, the first step in analysis is to determine what each axis represents. (a) In words,

explain what is being shown on the *x*-axis. Be sure to include the units. (b) What is being shown on the *y*-axis (including units)? (c) Which variable is the independent variable—the variable that was *manipulated* by the researchers? (d) Which variable is the dependent variable—the variable that responded to or depended on the treatment, which was *measured* by the researchers? (For additional information about graphs, see the Scientific Skills Review in Appendix F and in the Study Area in MasteringBiology.)

2. Based on the data shown in the graph, describe in words the relationship between carbonate ion concentration and calcification rate.

3. (a) If the seawater carbonate ion concentration is 270 µmol/kg, what is the approximate rate of calcification, and approximately how many days would it take 1 square meter of reef to accumulate 30 mmol of calcium carbonate ($CaCO_3$)? To determine the rate of calcification, draw a vertical line up from the *x*-axis at the value of 270 µmol/kg until it intersects the red line. Then draw a horizontal line from the intersection over to the *y*-axis to see what the calcification rate is at that carbonate ion concentration. (b) If the seawater carbonate ion concentration is 250 µmol/kg, what is the approximate rate of calcification, and approximately how many days would it take 1 square meter of reef to accumulate 30 mmol of calcium carbonate? (c) If carbonate ion concentration decreases, how does the calcification rate change, and how does that affect the time it takes coral to grow?

4. (a) Referring to the equations in Figure 2.24, determine which step of the process is measured in this experiment. (b) Do the results of this experiment support the hypothesis that increased atmospheric $[CO_2]$ will slow the growth of coral reefs? Why or why not?

Data from C. Langdon et al., Effect of calcium carbonate saturation state on the calcification rate of an experimental coral reef, *Global Biogeochemical Cycles* 14:639–654 (2000).

(MB) A version of this Scientific Skills Exercise can be assigned in MasteringBiology.

As seawater acidifies, the extra hydrogen ions combine with carbonate ions ($CO_3{}^{2-}$) to form bicarbonate ions ($HCO_3{}^{-}$), thereby reducing the carbonate ion concentration (see Figure 2.24). Scientists predict that ocean acidification will cause the carbonate ion concentration to decrease by 40% by the year 2100. This is of great concern because carbonate ions are required for calcification, the production of calcium carbonate ($CaCO_3$), by many marine organisms, including reef-building corals and animals that build shells. The **Scientific Skills Exercise** gives you an opportunity to work with data from an experiment examining the effect of carbonate ion concentration on coral reefs. Coral reefs are sensitive ecosystems that act as havens for a great diversity of marine life. The disappearance of coral reef ecosystems would be a tragic loss of biological diversity.

CONCEPT CHECK 2.5

1. Describe how properties of water contribute to the upward movement of water in a tree.
2. How can the freezing of water crack boulders?
3. The concentration of the appetite-regulating hormone ghrelin is about 1.3×10^{-10} *M* in a fasting person. How many molecules of ghrelin are in 1 L of blood?
4. Compared with a basic solution at pH 9, the same volume of an acidic solution at pH 4 has ___ times as many hydrogen ions (H^+).
5. **WHAT IF?** What would be the effect on the properties of the water molecule if oxygen and hydrogen had equal electronegativity?

For suggested answers, see Appendix A.

SUMMARY OF KEY CONCEPTS

CONCEPT 2.1

Matter consists of chemical elements in pure form and in combinations called compounds (pp. 19–20)

- **Elements** cannot be broken down chemically to other substances. A **compound** contains two or more different elements in a fixed ratio. Oxygen, carbon, hydrogen, and nitrogen make up approximately 96% of living matter.

> **?** *In what way does the need for iodine or iron in your diet differ from your need for calcium or phosphorus?*

CONCEPT 2.2

An element's properties depend on the structure of its atoms (pp. 20–24)

- An **atom**, the smallest unit of an element, has the following components:

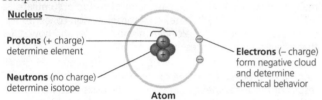

Nucleus

Protons (+ charge) determine element

Neutrons (no charge) determine isotope

Electrons (– charge) form negative cloud and determine chemical behavior

Atom

- An electrically neutral atom has equal numbers of electrons and protons; the number of protons determines the **atomic number**. **Isotopes** of an element differ from each other in neutron number and therefore mass. Unstable isotopes give off particles and energy as radioactivity.
- In an atom, electrons occupy specific **electron shells**; the electrons in a shell have a characteristic energy level. Electron distribution in shells determines the chemical behavior of an atom. An atom that has an incomplete outer shell, the **valence shell**, is reactive.

> **DRAW IT** *Draw the electron distribution diagrams for neon ($_{10}$Ne) and argon ($_{18}$Ar). Why are they chemically unreactive?*

CONCEPT 2.3

The formation and function of molecules depend on chemical bonding between atoms (pp. 24–28)

- **Chemical bonds** form when atoms interact and complete their valence shells. **Covalent bonds** form when pairs of electrons are shared. H_2 has a **single bond**: H — H. A **double bond** is the sharing of two pairs of electrons, as in O ═ O.
- **Molecules** consist of two or more covalently bonded atoms. The attraction of an atom for the electrons of a covalent bond is its **electronegativity**. Electrons of a **polar covalent bond** are pulled closer to the more electronegative atom.
- An **ion** forms when an atom or molecule gains or loses an electron and becomes charged. An **ionic bond** is the attraction between two oppositely charged ions, such as Na^+ and Cl^-.
- Weak bonds reinforce the shapes of large molecules and help molecules adhere to each other. A **hydrogen bond** is an attraction between a hydrogen atom carrying a partial positive charge ($\delta+$) and an electronegative atom ($\delta-$). **Van der Waals interactions** occur between transiently positive and negative regions of molecules.

- Molecular shape is usually the basis for the recognition of one biological molecule by another.

> **?** *In terms of electron sharing between atoms, compare nonpolar covalent bonds, polar covalent bonds, and the formation of ions.*

CONCEPT 2.4

Chemical reactions make and break chemical bonds (pp. 28–29)

- **Chemical reactions** change **reactants** into **products** while conserving matter. All chemical reactions are theoretically reversible. **Chemical equilibrium** is reached when the forward and reverse reaction rates are equal.

> **?** *What would happen to the concentration of products if more reactants were added to a reaction that was in chemical equilibrium? How would this addition affect the equilibrium?*

CONCEPT 2.5

Hydrogen bonding gives water properties that help make life possible on Earth (pp. 29–37)

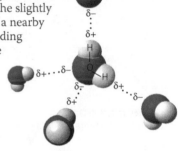

- A hydrogen bond forms when the slightly negatively charged oxygen of one water molecule is attracted to the slightly positively charged hydrogen of a nearby water molecule. Hydrogen bonding between water molecules is the basis for water's properties.
- Hydrogen bonding keeps water molecules close to each other, giving water **cohesion**. Hydrogen bonding is also responsible for water's **surface tension**.
- Water has a high **specific heat**: Heat is absorbed when hydrogen bonds break and is released when hydrogen bonds form. This helps keep temperatures relatively steady, within limits that permit life. **Evaporative cooling** is based on water's high **heat of vaporization**. The evaporative loss of the most energetic water molecules cools a surface.
- Ice floats because it is less dense than liquid water. This property allows life to exist under the frozen surfaces of lakes and seas.

Ice: stable hydrogen bonds

Liquid water: transient hydrogen bonds

- Water is an unusually versatile **solvent** because its polar molecules are attracted to ions and polar substances that can form hydrogen bonds. **Hydrophilic** substances have an affinity for water; **hydrophobic** substances do not. **Molarity**, the number of moles of **solute** per liter of **solution**, is used as a measure of solute concentration in solutions. A **mole** is a certain number of molecules of a substance. The mass of a mole of a substance in grams is the same as the **molecular mass** in daltons.
- A water molecule can transfer an H^+ to another water molecule to form H_3O^+ (represented simply by H^+) and OH^-.

- The concentration of H^+ is expressed as **pH**; pH = $-\log$ [H^+]. A **buffer** consists of an acid-base pair that combines reversibly with hydrogen ions, allowing it to resist pH changes.
- The burning of fossil fuels increases the amount of CO_2 in the atmosphere. Some CO_2 dissolves in the oceans, causing ocean acidification, which has potentially grave consequences for coral reefs.

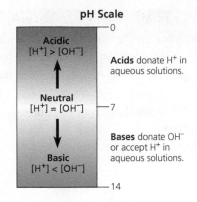

pH Scale

Acidic [H^+] > [OH^-] — 0

Acids donate H^+ in aqueous solutions.

Neutral [H^+] = [OH^-] — 7

Bases donate OH^- or accept H^+ in aqueous solutions.

Basic [H^+] < [OH^-] — 14

? *Describe how the properties of water result from the molecule's polar covalent bonds and how these properties contribute to Earth's suitability for life.*

TEST YOUR UNDERSTANDING

Level 1: Knowledge/Comprehension

1. The reactivity of an atom arises from
   a. the average distance of the outermost electron shell from the nucleus.
   b. the existence of unpaired electrons in the valence shell.
   c. the sum of the potential energies of all the electron shells.
   d. the potential energy of the valence shell.
   e. the energy differences between the electron shells.

2. Which of the following statements correctly describes any chemical reaction that has reached equilibrium?
   a. The concentrations of products and reactants are equal.
   b. The reaction is now irreversible.
   c. Both forward and reverse reactions have halted.
   d. The rates of the forward and reverse reactions are equal.
   e. No reactants remain.

3. Many mammals control their body temperature by sweating. Which property of water is most directly responsible for the ability of sweat to lower body temperature?
   a. water's change in density when it condenses
   b. water's ability to dissolve molecules in the air
   c. the release of heat by the formation of hydrogen bonds
   d. the absorption of heat by the breaking of hydrogen bonds
   e. water's high surface tension

4. We can be sure that a mole of table sugar and a mole of vitamin C are equal in their
   a. mass in daltons.   d. number of atoms.
   b. mass in grams.   e. number of molecules.
   c. volume.

5. Measurements show that the pH of a particular lake is 4.0. What is the hydrogen ion concentration of the lake?
   a. $4.0\ M$ b. $10^{-10}\ M$ c. $10^{-4}\ M$ d. $10^4\ M$ e. 4%

Level 2: Application/Analysis

6. The atomic number of sulfur is 16. Sulfur combines with hydrogen by covalent bonding to form a compound, hydrogen sulfide. Based on the number of valence electrons in a sulfur atom, predict the molecular formula of the compound.
   a. HS   b. HS_2 c. H_2S d. H_3S_2 e. H_4S

7. What coefficients must be placed in the following blanks so that all atoms are accounted for in the products?
$$C_6H_{12}O_6 \rightarrow \underline{\quad} C_2H_6O + \underline{\quad} CO_2$$
   a. 1; 2   b. 3; 1   c. 1; 3   d. 1; 1   e. 2; 2

8. A slice of pizza has 500 kcal. If we could burn the pizza and use all the heat to warm a 50-L container of cold water, what would be the approximate increase in the temperature of the water? (*Note*: A liter of cold water weighs about 1 kg.)
   a. 50°C   b. 5°C   c. 1°C   d. 100°C   e. 10°C

9. **DRAW IT** Draw the hydration shells that form around a potassium ion and a chloride ion when potassium chloride (KCl) dissolves in water. Label the positive, negative, and partial charges on the atoms.

Level 3: Synthesis/Evaluation

10. **SCIENTIFIC INQUIRY**
Female silkworm moths (*Bombyx mori*) attract males by emitting chemical signals that spread through the air. A male hundreds of meters away can detect these molecules and fly toward their source. The sensory organs responsible for this behavior are the comblike antennae visible in the photograph shown here. Each filament of an antenna is equipped with thousands of receptor cells

that detect the sex attractant. Based on what you learned in this chapter, propose a hypothesis to account for the ability of the male moth to detect a specific molecule in the presence of many other molecules in the air. What predictions does your hypothesis make? Design an experiment to test one of these predictions.

11. **FOCUS ON EVOLUTION**
The percentages of naturally occurring elements making up the human body are similar to the percentages of these elements found in other organisms. How could you account for this similarity among organisms?

12. **FOCUS ON ORGANIZATION**
Several emergent properties of water contribute to the suitability of the environment for life. In a short essay (100–150 words), describe how the ability of water to function as a versatile solvent arises from the structure of water molecules.

For selected answers, see Appendix A.

MasteringBiology®

Students Go to **MasteringBiology** for assignments, the eText, and the Study Area with practice tests, animations, and activities.

Instructors Go to **MasteringBiology** for automatically graded tutorials and questions that you can assign to your students, plus Instructor Resources.

Carbon and the Molecular Diversity of Life

3

KEY CONCEPTS

3.1 Carbon atoms can form diverse molecules by bonding to four other atoms

3.2 Macromolecules are polymers, built from monomers

3.3 Carbohydrates serve as fuel and building material

3.4 Lipids are a diverse group of hydrophobic molecules

3.5 Proteins include a diversity of structures, resulting in a wide range of functions

3.6 Nucleic acids store, transmit, and help express hereditary information

OVERVIEW

Carbon Compounds and Life

Water is the universal medium for life on Earth, but water aside, living organisms are made up of chemicals based mostly on the element carbon. Of all chemical elements, carbon is unparalleled in its ability to form molecules that are large, complex, and varied. Hydrogen (H), oxygen (O), nitrogen (N), sulfur (S), and phosphorus (P) are other common ingredients of these compounds, but it is the element carbon (C) that accounts for the enormous variety of biological molecules. For historical reasons, a compound containing carbon is said to be an **organic compound**; furthermore, almost all organic compounds associated with life contain hydrogen atoms in addition to carbon atoms. Different species of organisms and even different individuals within a species are distinguished by variations in their large organic compounds.

Given the rich complexity of life on Earth, it may surprise you to learn that the critically important large molecules of all living things—from bacteria to elephants—fall into just four main classes: carbohydrates, lipids, proteins, and nucleic acids. On the molecular scale, members of three of these classes—carbohydrates, proteins, and nucleic acids—are huge and are therefore called **macromolecules**. For example, a protein may consist of thousands of atoms that form a molecular colossus with a mass well over 100,000 daltons. Considering the size and complexity of macromolecules, it is noteworthy that biochemists have determined the detailed structure of so many of them. The scientist in the foreground of **Figure 3.1** is using 3-D glasses to help her visualize the structure of the protein displayed on her screen. The structures of macromolecules can provide important information about their functions.

In this chapter, we'll first investigate the properties of small organic molecules and then go on to discuss the larger biological molecules. After considering how macromolecules are built, we'll examine the structure and function of all four classes of large biological molecules. The architecture of a large biological molecule helps explain how that molecule works. Like small molecules, large biological molecules exhibit unique emergent properties arising from the orderly arrangement of their atoms.

▼ **Figure 3.1** Why do scientists study the structures of macromolecules?

Name and Comment	Molecular Formula	Structural Formula	Ball-and-Stick Model (molecular shape in pink)	Space-Filling Model
(a) Methane. When a carbon atom has four single bonds to other atoms, the molecule is tetrahedral.	CH_4			
(b) Ethane. A molecule may have more than one tetrahedral group of single-bonded atoms. (Ethane consists of two such groups.)	C_2H_6			
(c) Ethene (ethylene). When two carbon atoms are joined by a double bond, all atoms attached to those carbons are in the same plane; the molecule is flat.	C_2H_4			

3.1

Carbon atoms can form diverse molecules by bonding to four other atoms

The key to an atom's chemical characteristics is its electron configuration. This configuration determines the kinds and number of bonds an atom will form with other atoms, and it is the source of carbon's versatility.

The Formation of Bonds with Carbon

Carbon has 6 electrons, with 2 in the first electron shell and 4 in the second shell; thus, it has 4 valence electrons in a shell that holds 8 electrons. A carbon atom usually completes its valence shell by sharing its 4 electrons with other atoms so that 8 electrons are present. Each pair of shared electrons constitutes a covalent bond (see Figure 2.8d). In organic molecules, carbon usually forms single or double covalent bonds. Each carbon atom acts as an intersection point from which a molecule can branch off in as many as four directions. This ability is one facet of carbon's versatility that makes large, complex molecules possible.

When a carbon atom forms four single covalent bonds, the bonds angle toward the corners of an imaginary tetrahedron. The bond angles in methane (CH_4) are 109.5° **(Figure 3.2a)**, and they are roughly the same in any group of atoms where carbon has four single bonds. For example, ethane (C_2H_6) is

shaped like two overlapping tetrahedrons **(Figure 3.2b)**. In molecules with more carbons, every grouping of a carbon bonded to four other atoms has a tetrahedral shape. But when two carbon atoms are joined by a double bond, as in ethene (C_2H_4), the atoms joined to those carbons are in the same plane as the carbons **(Figure 3.2c)**. We find it convenient to write molecules as structural formulas, as if the molecules being represented are two-dimensional, but keep in mind that molecules are three-dimensional and that the shape of a molecule often determines its function.

The electron configuration of carbon gives it covalent compatibility with many different elements. **Figure 3.3** shows electron distribution diagrams for carbon and its most frequent partners—hydrogen, oxygen, and nitrogen. These are the four major atomic components of organic molecules. The number of unpaired electrons in the valence shell of an atom is generally equal to the atom's **valence**, the number of covalent bonds it can form. Let's consider how valence and

Hydrogen (valence = 1)	Oxygen (valence = 2)	Nitrogen (valence = 3)	Carbon (valence = 4)

▲ **Figure 3.3 Valences of the major elements of organic molecules.** Valence is the number of covalent bonds an atom can form. It is generally equal to the number of electrons required to complete the valence (outermost) shell (see Figure 2.6). Note that carbon can form four bonds.

the rules of covalent bonding apply to carbon atoms with partners other than hydrogen. We'll first look at the simple example of carbon dioxide.

In the carbon dioxide molecule (CO_2), a single carbon atom is joined to two atoms of oxygen by double covalent bonds. The structural formula for CO_2 is shown here:

$$O=C=O$$

Each line in a structural formula represents a pair of shared electrons. Thus, the two double bonds in CO_2 have the same number of shared electrons as four single bonds. The arrangement completes the valence shells of all atoms in the molecule. Because CO_2 is a very simple molecule and lacks hydrogen, it is often considered inorganic, even though it contains carbon. Whether we call CO_2 organic or inorganic, however, it is clearly important to the living world as the source of carbon for all organic molecules in organisms.

Carbon dioxide is a molecule with only one carbon atom. But as Figure 3.2 shows, a carbon atom can also use one or more valence electrons to form covalent bonds to other carbon atoms, linking the atoms into chains of seemingly infinite variety.

Molecular Diversity Arising from Variation in Carbon Skeletons

Carbon chains form the skeletons of most organic molecules. The skeletons vary in length and may be straight, branched, or arranged in closed rings (**Figure 3.4**). Some carbon skeletons have double bonds, which vary in number and location. Such variation in carbon skeletons is one important source of the molecular complexity and diversity that characterize living matter. In addition, atoms of other elements can be bonded to the skeletons at available sites.

All of the molecules shown in Figures 3.2 and 3.4 are **hydrocarbons**, organic molecules consisting of only carbon and hydrogen. Atoms of hydrogen are attached to the carbon skeleton wherever electrons are available for covalent bonding. Hydrocarbons are the major components of petroleum, which is called a fossil fuel because it consists of the partially decomposed remains of organisms that lived millions of years ago. Although hydrocarbons are not prevalent in most living organisms, many of a cell's organic molecules have regions consisting of only carbon and hydrogen. For example, the molecules known as fats have long hydrocarbon tails attached to a nonhydrocarbon component (as you will see in Figure 3.12). Neither petroleum nor fat dissolves in water; both are hydrophobic compounds because the great majority of their bonds are relatively nonpolar carbon-to-hydrogen linkages. Another characteristic of hydrocarbons is that they can undergo reactions that release a relatively large amount of energy. The gasoline that fuels a car consists of hydrocarbons, and the hydrocarbon tails of fats serve as stored fuel for animals.

▼ **Figure 3.4 Four ways that carbon skeletons can vary.**

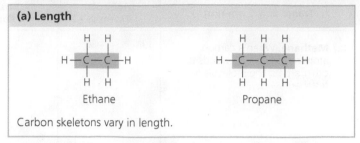

(a) Length

Ethane   Propane

Carbon skeletons vary in length.

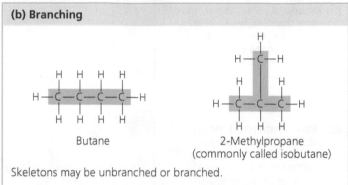

(b) Branching

Butane   2-Methylpropane
(commonly called isobutane)

Skeletons may be unbranched or branched.

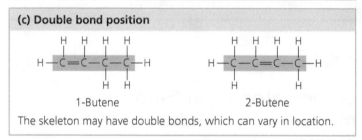

(c) Double bond position

1-Butene   2-Butene

The skeleton may have double bonds, which can vary in location.

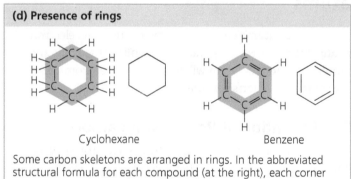

(d) Presence of rings

Cyclohexane   Benzene

Some carbon skeletons are arranged in rings. In the abbreviated structural formula for each compound (at the right), each corner represents a carbon and its attached hydrogens.

The Chemical Groups Most Important to Life

The distinctive properties of an organic molecule depend not only on the arrangement of its carbon skeleton but also on the chemical groups attached to that skeleton (**Figure 3.5**). We can think of hydrocarbons, the simplest organic molecules, as the underlying framework for more complex organic molecules. A number of chemical groups can replace one or more of the hydrogens bonded to the carbon skeleton of the hydrocarbon. The number and arrangement of chemical groups help give each organic molecule its unique properties.

▼ **Figure 3.5 Some biologically important chemical groups.**

Chemical Group	Compound Name	Examples
Hydroxyl group (—OH) —OH  (may be written HO—)	**Alcohol** (The specific name usually ends in -ol.)	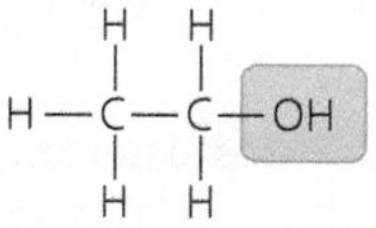 **Ethanol**, the alcohol present in alcoholic beverages
Carbonyl group ($>$C$=$O) 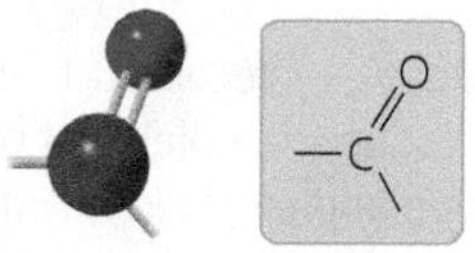 	**Ketone** if the carbonyl group is within a carbon skeleton **Aldehyde** if the carbonyl group is at the end of a carbon skeleton	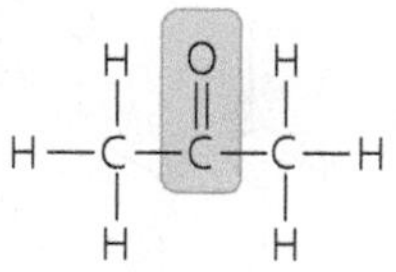 **Acetone**, the simplest ketone 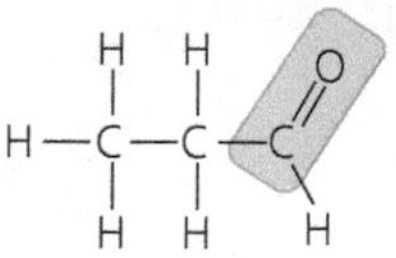 **Propanal**, an aldehyde
Carboxyl group (—COOH) 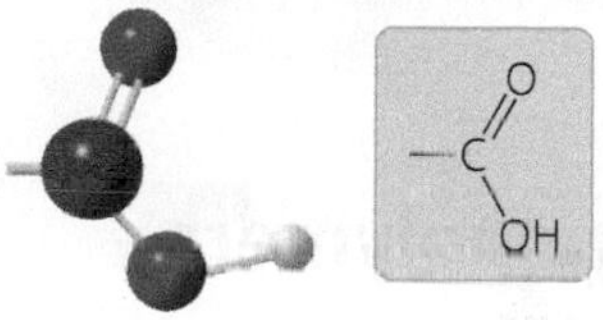 	**Carboxylic acid**, or **organic acid**	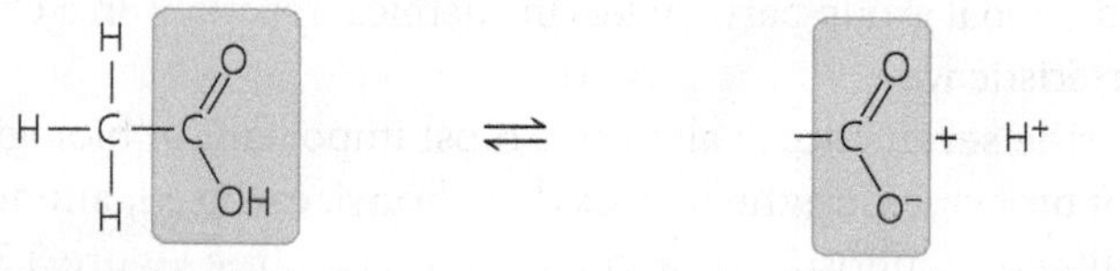 **Acetic acid**, which gives vinegar its sour taste Ionized form of —COOH (carboxylate ion), found in cells
Amino group (—NH$_2$) 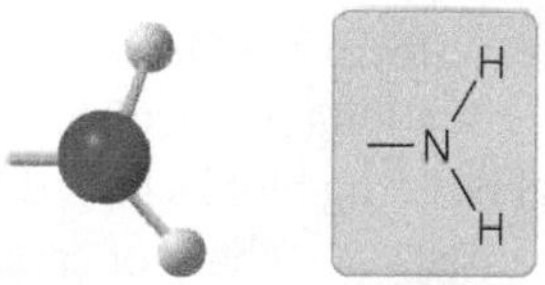 	**Amine**	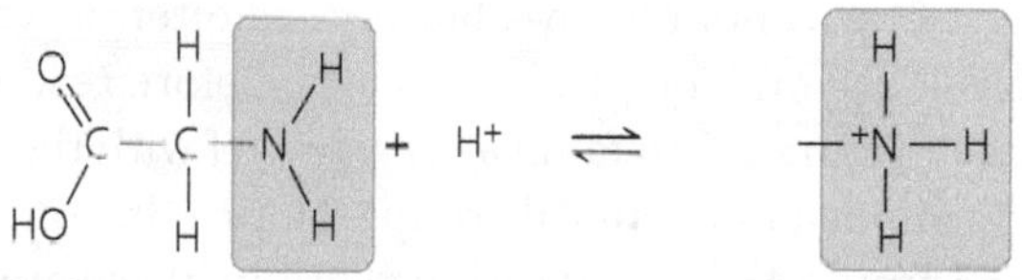 **Glycine**, an amino acid (note its carboxyl group) Ionized form of —NH$_2$, found in cells
Sulfhydryl group (—SH) 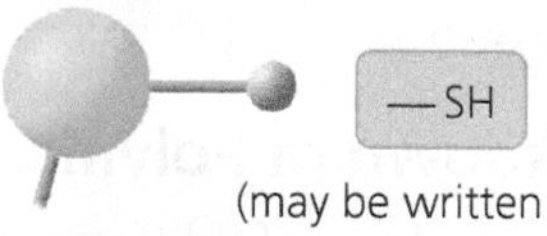 —SH (may be written HS —)	**Thiol**	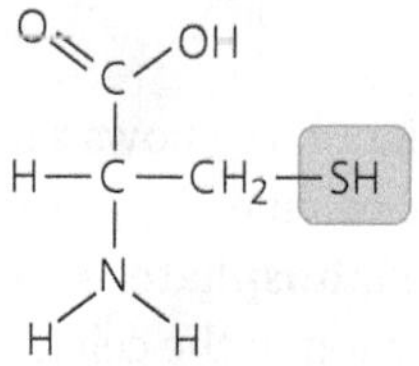 **Cysteine**, a sulfur-containing amino acid
Phosphate group (—OPO$_3^{2-}$)  	**Organic phosphate**	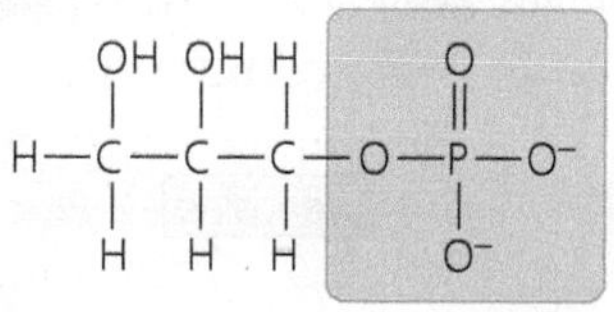 **Glycerol phosphate**, which takes part in many important chemical reactions in cells
Methyl group (—CH$_3$) 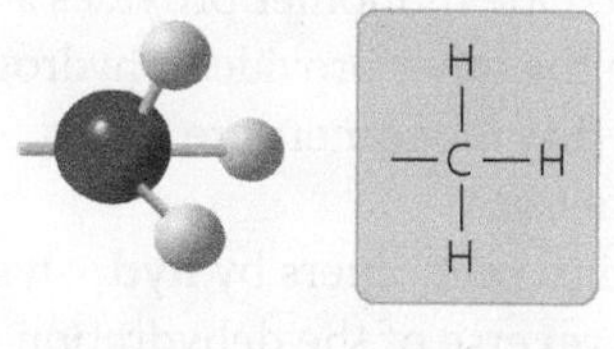 	**Methylated compound**	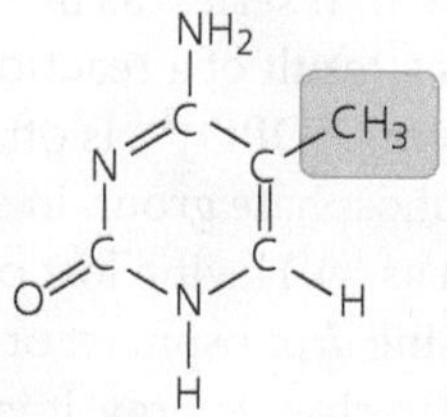 **5-Methyl cytosine**, a component of DNA that has been modified by addition of a methyl group

In some cases, chemical groups contribute to function primarily by affecting the molecule's shape. This is true for the steroid sex hormones estradiol (a type of estrogen) and testosterone, which differ in attached chemical groups.

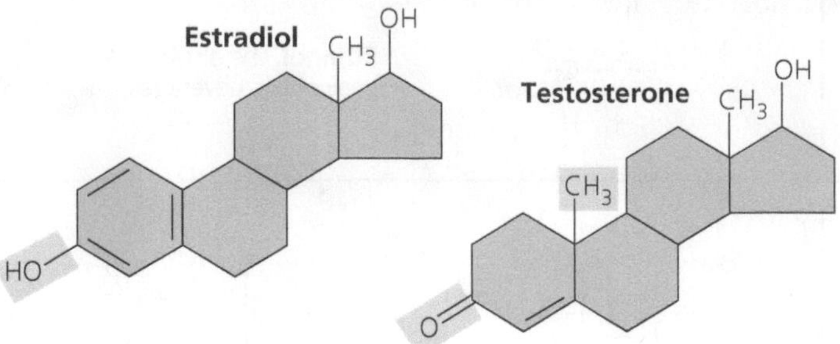

Estradiol

Testosterone

In other cases, the chemical groups affect molecular function by being directly involved in chemical reactions; these important chemical groups are known as **functional groups**. Each functional group participates in chemical reactions in a characteristic way.

The seven chemical groups most important in biological processes are the hydroxyl, carbonyl, carboxyl, amino, sulfhydryl, phosphate, and methyl groups (see Figure 3.5). The first six groups can act as functional groups; also, except for the sulfhydryl, they are hydrophilic and thus increase the solubility of organic compounds in water. The last group, the methyl group, is not reactive, but instead often serves as a recognizable tag on biological molecules. Before reading further, study Figure 3.5 to familiarize yourself with these biologically important chemical groups. Notice the ionized forms of the amino group and carboxyl group; these are the forms of these groups at normal cellular pH.

ATP: An Important Source of Energy for Cellular Processes

The "phosphate group" row in Figure 3.5 shows a simple example of an organic phosphate molecule. A more complicated organic phosphate, **adenosine triphosphate**, or **ATP**, is worth mentioning here because its function in the cell is so important. ATP consists of an organic molecule called adenosine attached to a string of three phosphate groups:

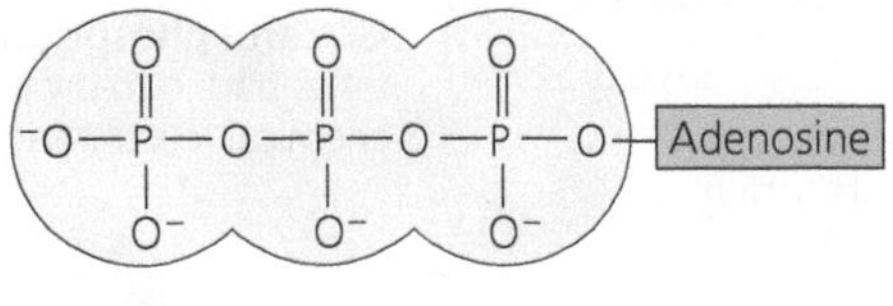

Where three phosphates are present in series, as in ATP, one phosphate may be split off as a result of a reaction with water. This inorganic phosphate ion, $HOPO_3^{2-}$, is often abbreviated Ⓟ$_i$ in this book, and a phosphate group in an organic molecule is often written as Ⓟ. Having lost one phosphate, ATP becomes adenosine *di*phosphate, or ADP. Although ATP is sometimes said to store energy, it is more accurate to think of it as storing the potential to react with

water. This reaction releases energy that can be used by the cell. (You will learn about this in more detail in Chapter 6.)

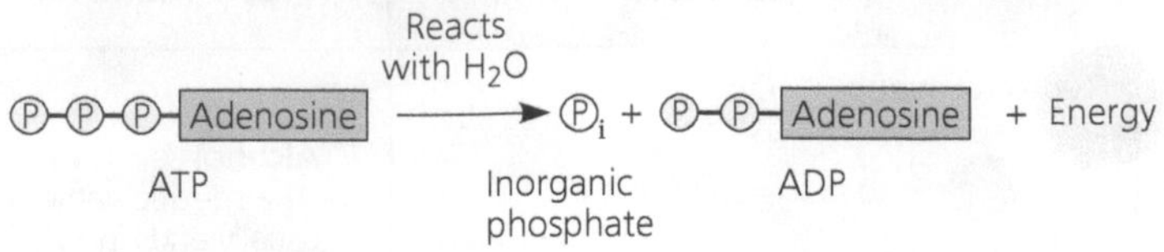

CONCEPT CHECK 3.1

1. How are gasoline and fat chemically similar?
2. What does the term *amino acid* signify about the structure of such a molecule?
3. **WHAT IF?** Suppose you had an organic molecule such as cysteine (see Figure 3.5, sulfhydryl group example), and you chemically removed the —NH₂ group and replaced it with —COOH. How would this change the chemical properties of the molecule?

For suggested answers, see Appendix A.

CONCEPT 3.2

Macromolecules are polymers, built from monomers

The macromolecules in three of the four classes of life's organic compounds—carbohydrates, proteins, and nucleic acids—are chain-like molecules called polymers (from the Greek *polys*, many, and *meros*, part). A **polymer** is a long molecule consisting of many similar or identical building blocks linked by covalent bonds, much as a train consists of a chain of cars. The repeating units that serve as the building blocks of a polymer are smaller molecules called **monomers** (from the Greek *monos*, single). Some of the molecules that serve as monomers also have other functions of their own.

The Synthesis and Breakdown of Polymers

Although each class of polymer is made up of a different type of monomer, the chemical mechanisms by which cells make and break down polymers are basically the same in all cases. In cells, these processes are facilitated by **enzymes**, specialized macromolecules (usually proteins) that speed up chemical reactions. Monomers are connected by a reaction in which two molecules are covalently bonded to each other, with the loss of a water molecule; this is known as a **dehydration reaction (Figure 3.6a)**. When a bond forms between two monomers, each monomer contributes part of the water molecule that is released during the reaction: One monomer provides a hydroxyl group (—OH), while the other provides a hydrogen (—H). This reaction is repeated as monomers are added to the chain one by one, making a polymer.

Polymers are disassembled to monomers by **hydrolysis**, a process that is essentially the reverse of the dehydration reaction **(Figure 3.6b)**. Hydrolysis means breakage using water

(a) Dehydration reaction: synthesizing a polymer

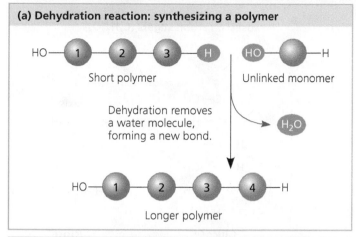

Short polymer

Unlinked monomer

Dehydration removes a water molecule, forming a new bond.

H_2O

Longer polymer

(b) Hydrolysis: breaking down a polymer

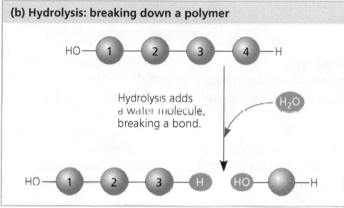

Hydrolysis adds a water molecule, breaking a bond.

H_2O

(from the Greek *hydro*, water, and *lysis*, break). The bond between the monomers is broken by the addition of a water molecule, with a hydrogen from the water attaching to one monomer and the hydroxyl group attaching to the adjacent monomer. An example of hydrolysis working within our bodies is the process of digestion. The bulk of the organic material in our food is in the form of polymers that are much too large to enter our cells. Within the digestive tract, various enzymes attack the polymers, speeding up hydrolysis. The released monomers are then absorbed into the bloodstream for distribution to all body cells. Those cells can then use dehydration reactions to assemble the monomers into new, different polymers that can perform specific functions required by the cell.

The Diversity of Polymers

Each cell has thousands of different macromolecules; the collection varies from one type of cell to another even in the same organism. The inherent differences between, for example, human siblings reflect small variations in polymers, particularly DNA and proteins. Molecular differences between unrelated individuals are more extensive and those between species greater still. The diversity of macromolecules in the living world is vast, and the possible variety is effectively limitless.

What is the basis for such diversity in life's polymers? These molecules are constructed from only 40 to 50 common monomers and some others that occur rarely. Building a huge variety of polymers from such a limited number of monomers is analogous to constructing hundreds of thousands of words from only 26 letters of the alphabet. The key is arrangement—the particular linear sequence that the units follow. However, this analogy falls far short of describing the great diversity of macromolecules because most biological polymers have many more monomers than the number of letters in the longest word. Proteins, for example, are built from 20 kinds of amino acids arranged in chains that are typically hundreds of amino acids long. The molecular logic of life is simple but elegant: Small molecules common to all organisms are ordered into unique macromolecules.

Despite this immense diversity, molecular structure and function can still be grouped roughly by class. Let's examine each of the four major classes of large biological molecules. For each class, the large molecules have emergent properties not found in their individual building blocks.

CONCEPT CHECK 3.2

1. How many molecules of water are needed to completely hydrolyze a polymer that is ten monomers long?
2. **WHAT IF?** Suppose you eat a serving of fish. What reactions must occur for the amino acid monomers in the protein of the fish to be converted to new proteins in your body?

For suggested answers, see Appendix A.

CONCEPT 3.3

Carbohydrates serve as fuel and building material

Carbohydrates include both sugars and polymers of sugars. The simplest carbohydrates are the monosaccharides, or simple sugars; these are the monomers from which more complex carbohydrates are constructed. Disaccharides are double sugars, consisting of two monosaccharides joined by a covalent bond. Carbohydrates also include macromolecules called polysaccharides, polymers composed of many sugar building blocks joined together by dehydration reactions.

Sugars

Monosaccharides (from the Greek *monos*, single, and *sacchar*, sugar) generally have molecular formulas that are some multiple of the unit CH_2O. Glucose ($C_6H_{12}O_6$), the most common monosaccharide, is of central importance in the chemistry of life. In the structure of glucose, we can see the trademarks of a sugar: The molecule has a carbonyl group

▼ Figure 3.7 Examples of monosaccharides. Sugars vary in the location of their carbonyl groups (orange) and the length of their carbon skeletons.

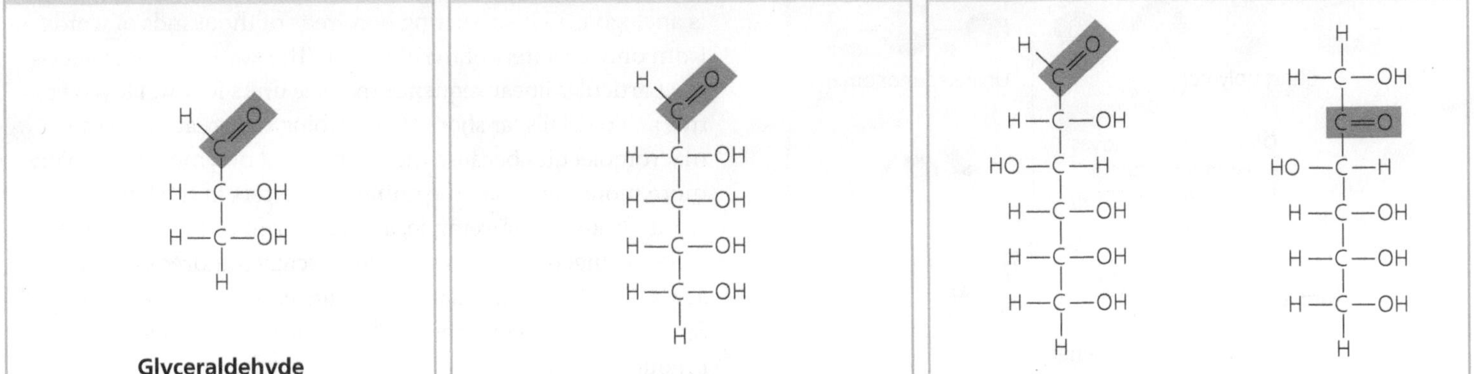

Triose: 3-carbon sugar ($C_3H_6O_3$)

Glyceraldehyde
An initial breakdown product of glucose in cells

Pentose: 5-carbon sugar ($C_5H_{10}O_5$)

Ribose
A component of RNA

Hexoses: 6-carbon sugars ($C_6H_{12}O_6$)

Glucose **Fructose**
Energy sources for organisms

(C$=$O) and multiple hydroxyl groups (—OH) **(Figure 3.7)**. The carbonyl group can be on the end of the linear sugar molecule, as in glucose, or attached to an interior carbon, as in fructose. (Thus, sugars are either aldehydes or ketones; see Figure 3.5.) The carbon skeleton of a sugar molecule ranges from three to seven carbons long. Glucose, fructose, and other sugars that have six carbons are called hexoses. Trioses (three-carbon sugars) and pentoses (five-carbon sugars) are also common. Note that most names for sugars end in *-ose*.

Although it is convenient to draw glucose with a linear carbon skeleton, this representation is not completely accurate. In aqueous solutions, glucose molecules, as well as most other five- and six-carbon sugars, form rings **(Figure 3.8)**.

Monosaccharides, particularly glucose, are major nutrients for cells. In the process known as cellular respiration, cells extract energy from glucose in a series of reactions that break down its molecules. Also, the carbon skeletons of sugars serve as raw material for the synthesis of other types of small organic molecules, such as amino acids. Sugar molecules that are not immediately used in these ways are generally incorporated as monomers into disaccharides or polysaccharides.

A **disaccharide** consists of two monosaccharides joined by a **glycosidic linkage**, a covalent bond formed between two monosaccharides by a dehydration reaction. The most prevalent disaccharide is sucrose, which is table sugar. Its two monomers are glucose and fructose **(Figure 3.9)**. Plants generally transport carbohydrates from leaves to roots and other nonphotosynthetic organs in the form of sucrose. Other disaccharides are lactose, the sugar present in milk, and maltose, an ingredient used in making beer.

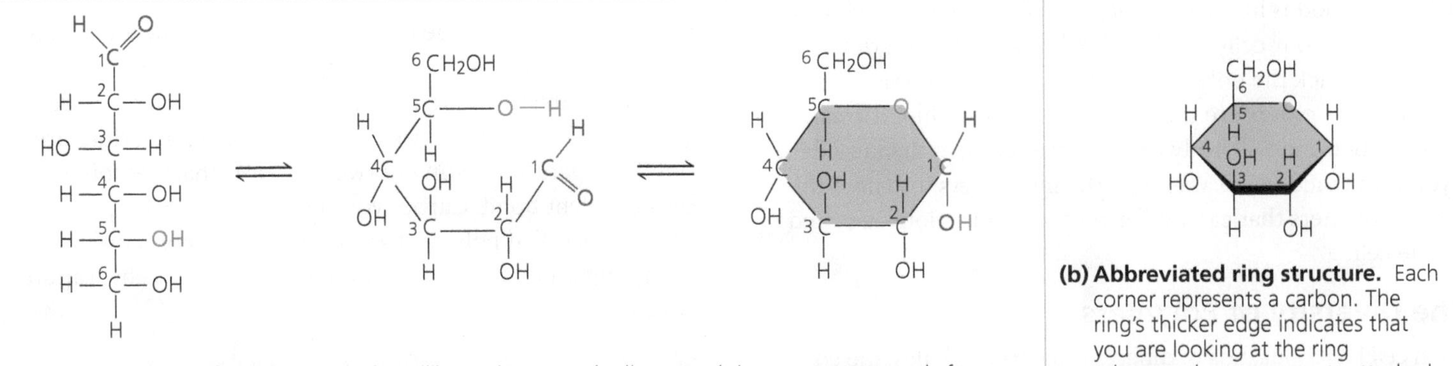

(a) Linear and ring forms. Chemical equilibrium between the linear and ring structures greatly favors the formation of rings. The carbons of the sugar are numbered 1 to 6, as shown. To form the glucose ring, carbon 1 bonds to the oxygen attached to carbon 5.

(b) Abbreviated ring structure. Each corner represents a carbon. The ring's thicker edge indicates that you are looking at the ring edge-on; the components attached to the ring lie above or below the plane of the ring.

▲ Figure 3.8 Linear and ring forms of glucose.

DRAW IT *Start with the linear form of fructose (see Figure 3.7) and draw the formation of the fructose ring in two steps. First, number the carbons starting at the top of the linear structure. Then attach carbon 5 via its oxygen to carbon 2. Compare the number of carbons in the fructose and glucose rings.*

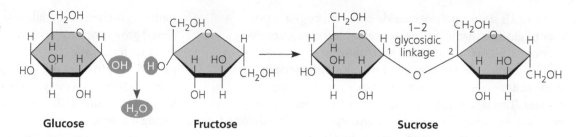

▶ **Figure 3.9 Disaccharide synthesis.** Sucrose is a disaccharide formed from glucose and fructose by a dehydration reaction. Notice that fructose, though a hexose like glucose, forms a five-sided ring.

DRAW IT *Referring to Figure 3.8, number the carbons in each sugar in this figure. Show how the numbering is consistent with the name of the glycosidic linkage.*

Glucose   **Fructose**   **Sucrose**

Polysaccharides

Polysaccharides are macromolecules, polymers with a few hundred to a few thousand monosaccharides joined by glycosidic linkages. Some polysaccharides serve as storage material, hydrolyzed as needed to provide sugar for cells. Other polysaccharides serve as building material for structures that protect the cell or the whole organism. The structure and function of a polysaccharide are determined by its sugar monomers and by the positions of its glycosidic linkages.

Storage Polysaccharides

Both plants and animals store sugars for later use in the form of storage polysaccharides **(Figure 3.10)**. Plants store **starch**, a polymer of glucose monomers, as granules within cells.

Synthesizing starch enables the plant to stockpile surplus glucose. Because glucose is a major cellular fuel, starch represents stored energy. The sugar can later be withdrawn from this carbohydrate "bank" by hydrolysis, which breaks the bonds between the glucose monomers. Most animals, including humans, also have enzymes that can hydrolyze plant starch, making glucose available as a nutrient for cells. Potato tubers and grains—the fruits of wheat, maize (corn), rice, and other grasses—are the major sources of starch in the human diet.

Most of the glucose monomers in starch are joined by 1–4 linkages (number 1 carbon to number 4 carbon). The simplest form of starch, amylose, is unbranched, as shown in Figure 3.10. Amylopectin, a more complex starch, is a branched polymer with 1–6 linkages at the branch points.

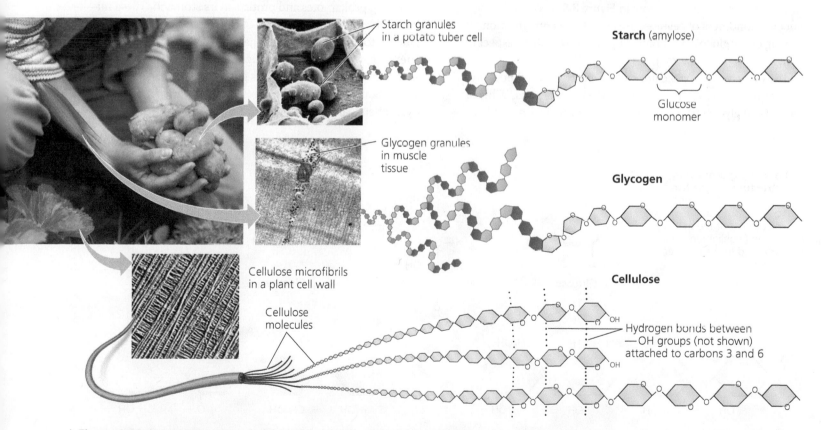

▲ **Figure 3.10 Polysaccharides of plants and animals.** The polysaccharides shown are composed entirely of glucose monomers, represented here by hexagons. In starch and glycogen, the polymer chains tend to form helices in unbranched regions because of the angle of the 1–4 linkage between the glucose monomers. Cellulose, with a different kind of 1–4 linkage, is always unbranched.

Animals store a polysaccharide called **glycogen**, a polymer of glucose that is like amylopectin but more extensively branched. Humans and other vertebrates store glycogen mainly in liver and muscle cells. Hydrolysis of glycogen in these cells releases glucose when the demand for sugar increases. This stored fuel cannot sustain an animal for long, however. In humans, for example, glycogen stores are depleted in about a day unless they are replenished by eating.

Structural Polysaccharides

Organisms build strong materials from structural polysaccharides. The polysaccharide called **cellulose** is a major component of the tough walls that enclose plant cells (see Figure 3.10). On a global scale, plants produce almost 10^{14} kg (100 billion tons) of cellulose per year; it is the most abundant organic compound on Earth. Like starch and glycogen, cellulose is a polymer of glucose with 1–4 glycosidic linkages, but the linkages in cellulose are different. The difference is based on the fact that there are actually two slightly different ring structures for glucose (**Figure 3.11a**). When glucose forms a ring, the hydroxyl group attached to the number 1 carbon is positioned either below or above the plane of the ring. These two ring forms for glucose are called alpha (α) and beta (β), respectively. In starch, all the glucose monomers are in the α configuration (**Figure 3.11b**), the arrangement we saw in Figure 3.8. In contrast, the glucose monomers of cellulose are all in the β configuration, making every glucose monomer "upside down" with respect to its neighbors (**Figure 3.11c**).

The differing glycosidic linkages in starch and cellulose give the two molecules distinct three-dimensional shapes. Whereas starch (and glycogen) molecules are largely helical, a cellulose molecule is straight. Cellulose is never branched, and some hydroxyl groups on its glucose monomers are free to hydrogen-bond with the hydroxyls of other cellulose molecules lying parallel to it. In plant cell walls, parallel cellulose molecules held together in this way are grouped into units called microfibrils (see Figure 3.10). These cable-like microfibrils are a strong building material for plants and an important substance for humans because cellulose is the major component of paper and the only constituent of cotton.

Enzymes that digest starch by hydrolyzing its α linkages are unable to hydrolyze the β linkages of cellulose because of the distinctly different shapes of these two molecules. In fact, few organisms possess enzymes that can digest cellulose. Animals, including humans, do not; the cellulose in our food passes through the digestive tract and is eliminated with the feces. Along the way, the cellulose abrades the wall of the digestive tract and stimulates the lining to secrete mucus, which aids in the smooth passage of food through the tract. Thus, although cellulose is not a nutrient for humans, it is an important part of a healthful diet. Most fresh fruits, vegetables, and whole grains are rich in cellulose. On food packages, "insoluble fiber" refers mainly to cellulose.

Some microorganisms can digest cellulose, breaking it down into glucose monomers. A cow harbors cellulose-digesting prokaryotes and protists in its stomach. These microbes hydrolyze the cellulose of hay and grass and convert the glucose to other compounds that nourish the cow. Similarly, a termite, which is unable to digest cellulose by itself, has prokaryotes or protists living in its gut that can make a meal of wood. Some fungi can also digest cellulose, thereby helping recycle chemical elements within Earth's ecosystems.

(a) α and β glucose ring structures. These two interconvertible forms of glucose differ in the placement of the hydroxyl group (highlighted in blue) attached to the number 1 carbon.

α Glucose

β Glucose

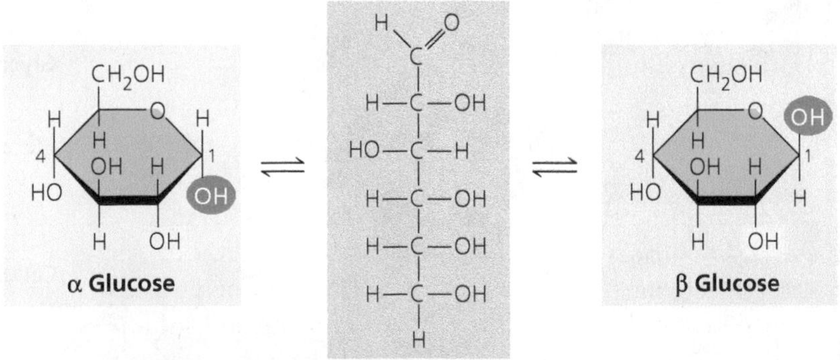

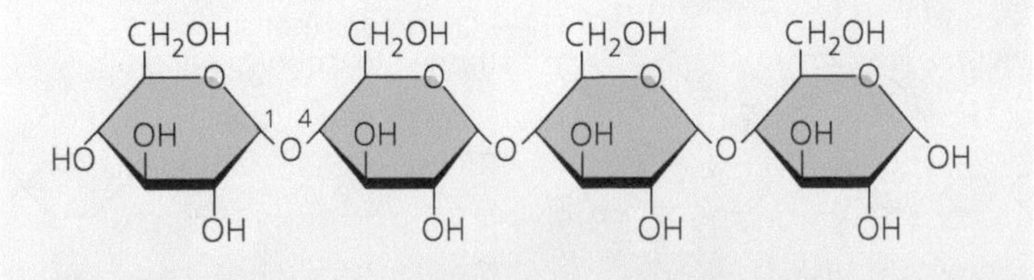

(b) Starch: 1–4 linkage of α glucose monomers. All monomers are in the same orientation. Compare the positions of the —OH groups highlighted in yellow with those in cellulose (c).

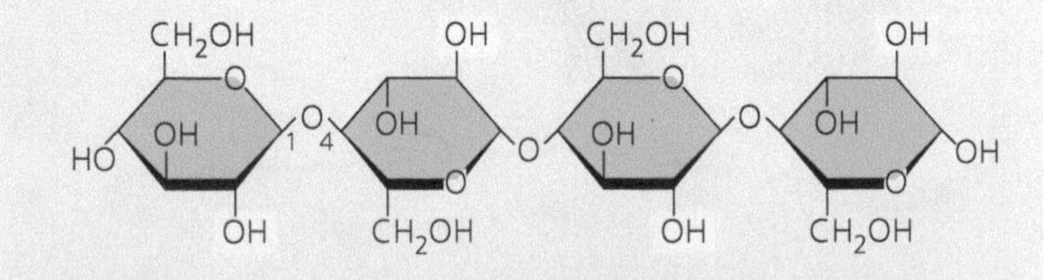

(c) Cellulose: 1–4 linkage of β glucose monomers. In cellulose, every β glucose monomer is upside down with respect to its neighbors.

▲ **Figure 3.11 Monomer structures of starch and cellulose.**

Another important structural polysaccharide is **chitin**, the carbohydrate used by arthropods (insects, spiders, crustaceans, and related animals) to build their exoskeletons—hard cases that surround the soft parts of these animals. Chitin is also found in many fungi, which use this polysaccharide as the building material for their cell walls. Chitin is similar to cellulose except that the glucose monomer of chitin has a nitrogen-containing appendage.

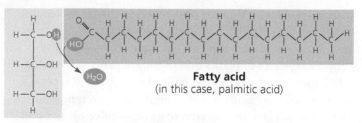

Fatty acid
(in this case, palmitic acid)

Glycerol

(a) One of three dehydration reactions in the synthesis of a fat

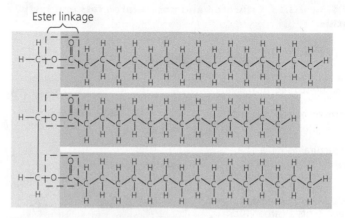

(b) Fat molecule (triacylglycerol)

▲ **Figure 3.12 The synthesis and structure of a fat, or triacylglycerol.** The molecular building blocks of a fat are one molecule of glycerol and three molecules of fatty acids. **(a)** One water molecule is removed for each fatty acid joined to the glycerol. **(b)** A fat molecule with three fatty acid units, two of them identical. The carbons of the fatty acids are arranged zigzag to suggest the actual orientations of the four single bonds extending from each carbon (see Figure 3.2a).

CONCEPT CHECK 3.3

1. Write the formula for a monosaccharide that has three carbons.
2. A dehydration reaction joins two glucose molecules to form maltose. The formula for glucose is $C_6H_{12}O_6$. What is the formula for maltose?
3. **WHAT IF?** After a cow is given antibiotics to treat an infection, a vet gives the animal a drink of "gut culture" containing various prokaryotes. Why is this necessary?

For suggested answers, see Appendix A.

 **3.4**

Lipids are a diverse group of hydrophobic molecules

Lipids are the one class of large biological molecules that does not include true polymers, and they are generally not big enough to be considered macromolecules. The compounds called **lipids** are grouped together because they share one important trait: They mix poorly, if at all, with water. The hydrophobic behavior of lipids is based on their molecular structure. Although they may have some polar bonds associated with oxygen, lipids consist mostly of hydrocarbon regions. Lipids are varied in form and function. They include waxes and certain pigments, but we will focus on the most biologically important types of lipids: fats, phospholipids, and steroids.

Fats

Although fats are not polymers, they are large molecules assembled from smaller molecules by dehydration reactions. A **fat** is constructed from two kinds of smaller molecules: glycerol and fatty acids **(Figure 3.12a)**. Glycerol is an alcohol; each of its three carbons bears a hydroxyl group. A **fatty acid** has a long carbon skeleton, usually 16 or 18 carbon atoms in length. The carbon at one end of the skeleton is part of a carboxyl group, the functional group that gives these molecules the name fatty *acid*. The rest of the skeleton consists of a hydrocarbon chain. The relatively nonpolar C—H bonds in the hydrocarbon chains of fatty acids are the reason fats are hydrophobic. Fats separate from water because the water molecules hydrogen-bond to one another and exclude the fats. This is the reason that vegetable oil (a liquid fat) separates from the aqueous vinegar solution in a bottle of salad dressing.

In making a fat, three fatty acid molecules are each joined to glycerol by an ester linkage, a bond between a hydroxyl group and a carboxyl group. The resulting fat, also called a **triacylglycerol**, thus consists of three fatty acids linked to one glycerol molecule. (Still another name for a fat is *triglyceride*, a word often found in the list of ingredients on packaged foods.) The fatty acids in a fat can be the same, or they can be of two or three different kinds, as in **Figure 3.12b**.

The terms *saturated fats* and *unsaturated fats* are commonly used in the context of nutrition. These terms refer to the structure of the hydrocarbon chains of the fatty acids. If there are no double bonds between carbon atoms composing a chain, then as many hydrogen atoms as possible are bonded to the carbon skeleton. Such a structure is said to be *saturated* with hydrogen, and the resulting fatty acid is called a **saturated fatty acid**. An **unsaturated fatty acid** has one or more double bonds, with one fewer hydrogen atom on each double-bonded carbon. Nearly every double bond in naturally occurring fatty acids has an orientation that creates a kink in the hydrocarbon chain.

A fat made from saturated fatty acids is called a saturated fat. Most animal fats are saturated: The hydrocarbon chains of their fatty acids—the "tails" of the fat molecules—lack double bonds, and their flexibility allows the fat molecules to pack together tightly. Saturated animal fats—such as lard and butter—are solid at room temperature (**Figure 3.13a**). In contrast, the fats of plants and fishes are generally unsaturated, meaning that they are built of one or more types of unsaturated fatty acids. Usually liquid at room temperature, plant and fish fats are referred to as oils—olive oil and cod liver oil are examples (**Figure 3.13b**). The kinks where the double bonds are located prevent the molecules from packing together closely enough to solidify at room temperature. The phrase "hydrogenated vegetable oils" on food labels means that unsaturated fats have been converted to saturated fats by adding hydrogen.

The major function of fats is energy storage. The hydrocarbon chains of fats are similar to gasoline molecules and just as rich in energy. A gram of fat stores more than twice as much energy as a gram of a polysaccharide, such as starch. Because plants are relatively immobile, they can function with bulky energy storage in the form of starch. (Vegetable oils are generally obtained from seeds, where more compact storage is an asset to the plant.) Animals, however, must carry their energy stores with them, so there is an advantage to having a more compact reservoir of fuel—fat.

Phospholipids

Cells could not exist without another type of lipid—**phospholipid**. Phospholipids are essential for cells because they are major constituents of cell membranes. Their structure provides a classic example of how form fits function at the molecular level. As shown in **Figure 3.14**, a phospholipid is similar to a fat molecule but has only two fatty acids attached to glycerol rather than three. The third hydroxyl group of glycerol is joined to a phosphate group, which has a negative electrical charge in the cell. Additional small molecules, which are usually charged or polar, can be linked to the phosphate group to form a variety of phospholipids.

The two ends of a phospholipid exhibit different behavior toward water. The hydrocarbon tails are hydrophobic and are excluded from water. However, the phosphate group and its attachments form a hydrophilic head that has an affinity for water. When phospholipids are added to water, they self-assemble into double-layered structures called "bilayers," shielding their hydrophobic portions from water (see Figure 3.14d).

At the surface of a cell, phospholipids are arranged in a similar bilayer. The hydrophilic heads of the molecules are on the outside of the bilayer, in contact with the aqueous solutions inside and outside of the cell. The hydrophobic tails point toward the interior of the bilayer, away from the water. The phospholipid bilayer forms a boundary between the cell and its external environment; the existence of cells depends on phospholipids.

Steroids

Steroids are lipids characterized by a carbon skeleton consisting of four fused rings. Different steroids are distinguished by the particular chemical groups attached to this ensemble of rings. Shown in **Figure 3.15**, **cholesterol** is a crucial steroid in animals. It is a common component of animal cell membranes and is also the precursor from which other steroids are synthesized, such as the vertebrate sex hormones estrogen and testosterone (see Concept 3.1).

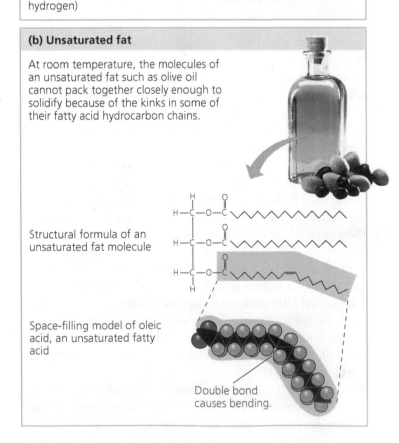

▼ **Figure 3.13 Saturated and unsaturated fats and fatty acids.**

(a) Saturated fat

At room temperature, the molecules of a saturated fat, such as the fat in butter, are packed closely together, forming a solid.

Structural formula of a saturated fat molecule (Each hydrocarbon chain is represented as a zigzag line, where each bend represents a carbon atom and hydrogens are not shown.)

Space-filling model of stearic acid, a saturated fatty acid (red = oxygen, black = carbon, gray = hydrogen)

(b) Unsaturated fat

At room temperature, the molecules of an unsaturated fat such as olive oil cannot pack together closely enough to solidify because of the kinks in some of their fatty acid hydrocarbon chains.

Structural formula of an unsaturated fat molecule

Space-filling model of oleic acid, an unsaturated fatty acid

Double bond causes bending.

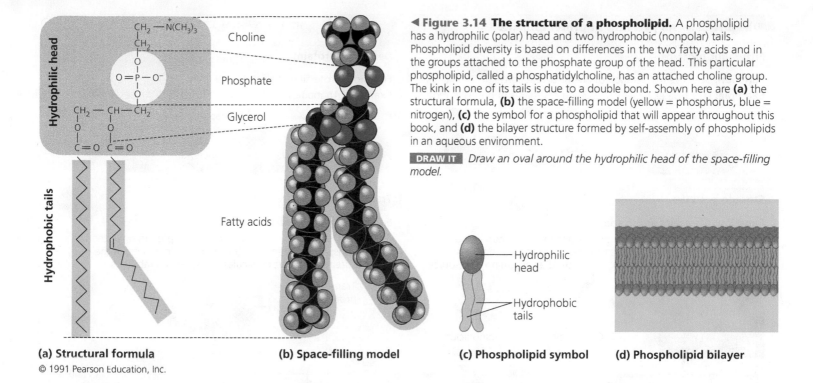

(a) Structural formula

(b) Space-filling model

© 1991 Pearson Education, Inc.

Hydrophilic head

CH₂—N(CH₃)₃ — Choline

O=P—O⁻ — Phosphate

CH₂—CH—CH₂ — Glycerol

Fatty acids

Hydrophobic tails

◄ **Figure 3.14 The structure of a phospholipid.** A phospholipid has a hydrophilic (polar) head and two hydrophobic (nonpolar) tails. Phospholipid diversity is based on differences in the two fatty acids and in the groups attached to the phosphate group of the head. This particular phospholipid, called a phosphatidylcholine, has an attached choline group. The kink in one of its tails is due to a double bond. Shown here are **(a)** the structural formula, **(b)** the space-filling model (yellow = phosphorus, blue = nitrogen), **(c)** the symbol for a phospholipid that will appear throughout this book, and **(d)** the bilayer structure formed by self-assembly of phospholipids in an aqueous environment.

DRAW IT *Draw an oval around the hydrophilic head of the space-filling model.*

Hydrophilic head

Hydrophobic tails

(c) Phospholipid symbol

(d) Phospholipid bilayer

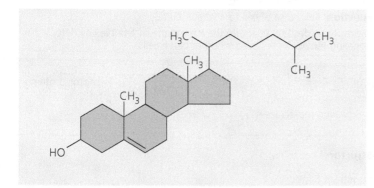

▲ **Figure 3.15 Cholesterol, a steroid.** Cholesterol is the molecule from which other steroids, including the sex hormones, are synthesized. Steroids vary in the chemical groups attached to their four interconnected rings (shown in gold).

In vertebrates, cholesterol is synthesized in the liver and is also obtained from the diet. A high level of cholesterol in the blood may contribute to atherosclerosis. In fact, saturated fats exert their negative impact on health by affecting cholesterol levels.

CONCEPT CHECK 3.4

1. Compare the structure of a fat (triacylglycerol) with that of a phospholipid.
2. Why are human sex hormones considered lipids?
3. **WHAT IF?** Suppose a membrane surrounded an oil droplet, as it does in the cells of plant seeds. Describe and explain the form it might take.

For suggested answers, see Appendix A.

CONCEPT 3.5

Proteins include a diversity of structures, resulting in a wide range of functions

Nearly every dynamic function of a living being depends on proteins. In fact, the importance of proteins is underscored by their name, which comes from the Greek word *proteios*, meaning "first," or "primary." Proteins account for more than 50% of the dry mass of most cells, and they are instrumental in almost everything organisms do. Some proteins speed up chemical reactions, while others play a role in defense, storage, transport, cellular communication, movement, or structural support. **Figure 3.16** shows examples of proteins with these functions (which you'll learn more about in later chapters).

Life would not be possible without enzymes, most of which are proteins. Enzymatic proteins regulate metabolism by acting as **catalysts**, chemical agents that selectively speed up chemical reactions without being consumed by the reaction. Because an enzyme can perform its function over and over again, these molecules can be thought of as workhorses that keep cells running by carrying out the processes of life.

A human has tens of thousands of different proteins, each with a specific structure and function; proteins, in fact, are the most structurally sophisticated molecules known. Consistent with their diverse functions, they vary extensively in structure, each type of protein having a unique three-dimensional shape.

▼ Figure 3.16 An overview of protein functions.

Enzymatic proteins

Function: Selective acceleration of chemical reactions

Example: Digestive enzymes catalyze the hydrolysis of bonds in food molecules.

Defensive proteins

Function: Protection against disease

Example: Antibodies inactivate and help destroy viruses and bacteria.

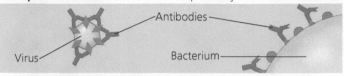

Storage proteins

Function: Storage of amino acids

Examples: Casein, the protein of milk, is the major source of amino acids for baby mammals. Plants have storage proteins in their seeds. Ovalbumin is the protein of egg white, used as an amino acid source for the developing embryo.

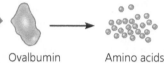

Ovalbumin → Amino acids for embryo

Transport proteins

Function: Transport of substances

Examples: Hemoglobin, the iron-containing protein of vertebrate blood, transports oxygen from the lungs to other parts of the body. Other proteins transport molecules across cell membranes.

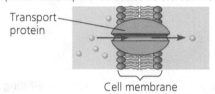

Transport protein

Cell membrane

Hormonal proteins

Function: Coordination of an organism's activities

Example: Insulin, a hormone secreted by the pancreas, causes other tissues to take up glucose, thus regulating blood sugar concentration.

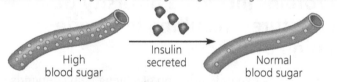

High blood sugar — Insulin secreted — Normal blood sugar

Receptor proteins

Function: Response of cell to chemical stimuli

Example: Receptors built into the membrane of a nerve cell detect signaling molecules released by other nerve cells.

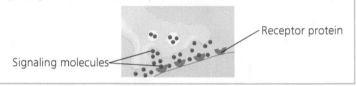

Receptor protein

Signaling molecules

Contractile and motor proteins

Function: Movement

Examples: Motor proteins are responsible for the undulations of cilia and flagella. Actin and myosin proteins are responsible for the contraction of muscles.

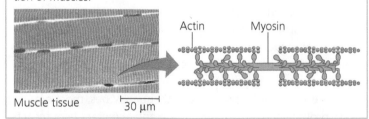

Actin   Myosin

Muscle tissue   30 μm

Structural proteins

Function: Support

Examples: Keratin is the protein of hair, horns, feathers, and other skin appendages. Insects and spiders use silk fibers to make their cocoons and webs, respectively. Collagen and elastin proteins provide a fibrous framework in animal connective tissues.

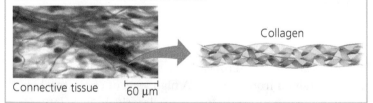

Collagen

Connective tissue   60 μm

Proteins are made up of polymers of amino acids called **polypeptides**. A **protein** is a biologically functional molecule that consists of one or more polypeptides folded and coiled into a specific three-dimensional structure.

Amino Acids

Polypeptides are all unbranched polymers constructed from the same set of 20 amino acids, and all amino acids share a common structure. An **amino acid** is an organic molecule with both an amino group and a carboxyl group. The figure at the right shows the general formula for an amino acid. At the center of the amino acid is a carbon atom called the *alpha* (α) *carbon*. Its four different partners are an amino group, a carboxyl group, a hydrogen atom, and a variable group symbolized by R. The R group, also called the side chain, differs with each amino acid **(Figure 3.17)**.

Side chain (R group)

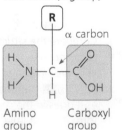

α carbon

Amino group   Carboxyl group

▼ **Figure 3.17 The 20 amino acids of proteins.** The amino acids are grouped here according to the properties of their side chains (R groups) and shown in their prevailing ionic forms at pH 7.2, the pH within a cell. The three-letter and one-letter abbreviations for the amino acids are in parentheses.

Nonpolar side chains; hydrophobic

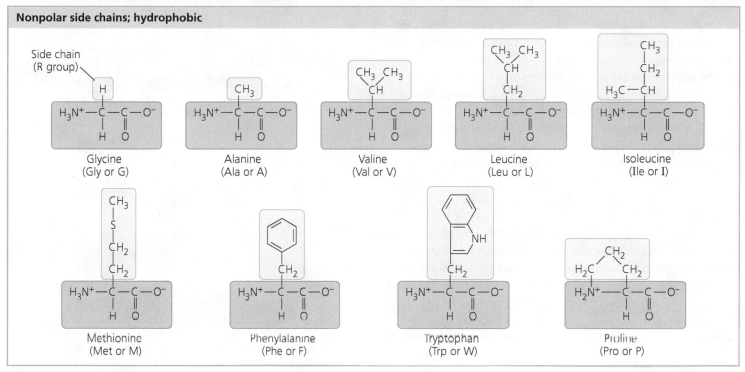

Glycine
(Gly or G)

Alanine
(Ala or A)

Valine
(Val or V)

Leucine
(Leu or L)

Isoleucine
(Ile or I)

Methionine
(Met or M)

Phenylalanine
(Phe or F)

Tryptophan
(Trp or W)

Proline
(Pro or P)

Polar side chains; hydrophilic

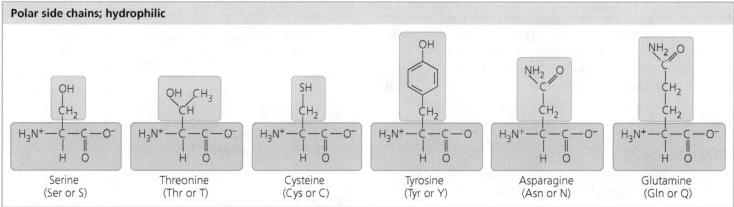

Serine
(Ser or S)

Threonine
(Thr or T)

Cysteine
(Cys or C)

Tyrosine
(Tyr or Y)

Asparagine
(Asn or N)

Glutamine
(Gln or Q)

Electrically charged side chains; hydrophilic

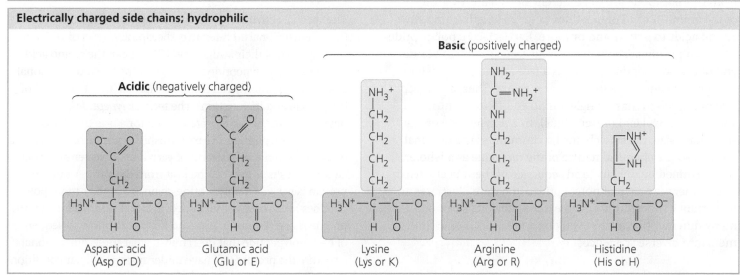

Aspartic acid
(Asp or D)

Glutamic acid
(Glu or E)

Lysine
(Lys or K)

Arginine
(Arg or R)

Histidine
(His or H)

The 20 amino acids in Figure 3.17 are the ones cells use to build their proteins. Here the amino groups and carboxyl groups are all depicted in ionized form, the way they usually exist at the pH found in a cell. The side chain (R group) may be as simple as a hydrogen atom, as in the amino acid glycine, or it may be a carbon skeleton with various functional groups attached, as in glutamine.

The physical and chemical properties of the side chain determine the unique characteristics of a particular amino acid, thus affecting its functional role in a polypeptide. In Figure 3.17, the amino acids are grouped according to the properties of their side chains. One group consists of amino acids with nonpolar side chains, which are hydrophobic. Another group consists of amino acids with polar side chains, which are hydrophilic. Acidic amino acids are those with side chains that are generally negative in charge owing to the presence of a carboxyl group, which is usually dissociated (ionized) at cellular pH. Basic amino acids have amino groups in their side chains that are generally positive in charge. (Notice that *all* amino acids have carboxyl groups and amino groups; the terms *acidic* and *basic* in this context refer only to groups on the side chains.) Because they are charged, acidic and basic side chains are also hydrophilic.

Polypeptides

Now that we have examined amino acids, let's see how they are linked to form polymers **(Figure 3.18)**. When two amino acids are positioned so that the carboxyl group of one is adjacent to the amino group of the other, they can become joined by a dehydration reaction, with the removal of a water molecule. The resulting covalent bond is called a **peptide bond**. Repeated over and over, this process yields a polypeptide, a polymer of many amino acids linked by peptide bonds.

The repeating sequence of atoms highlighted in purple in Figure 3.18 is called the polypeptide backbone. Extending from this backbone are the different side chains (R groups) of the amino acids. Polypeptides range in length from a few amino acids to a thousand or more. Each specific polypeptide has a unique linear sequence of amino acids. Note that one end of the polypeptide chain has a free amino group, while the opposite end has a free carboxyl group. Thus, a polypeptide of any length has a single amino end (N-terminus) and a single carboxyl end (C-terminus). In a polypeptide of any significant size, the side chains far outnumber the terminal groups, so the chemical nature of the molecule as a whole is determined by the kind and sequence of the side chains. The immense variety of polypeptides in nature illustrates an important concept introduced earlier—that cells can make many different polymers by linking a limited set of monomers into diverse sequences.

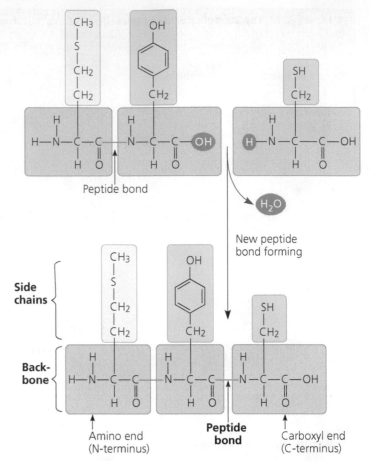

▲ **Figure 3.18 Making a polypeptide chain.** Peptide bonds are formed by dehydration reactions, which link the carboxyl group of one amino acid to the amino group of the next. The peptide bonds are formed one at a time, starting with the amino acid at the amino end (N-terminus). The polypeptide has a repetitive backbone (purple) from which the amino acid side chains (yellow and green) extend.

DRAW IT *At the top of the figure, circle and label the carboxyl and amino groups that will form the new peptide bond.*

Protein Structure and Function

The specific activities of proteins result from their intricate three-dimensional architecture, the simplest level of which is the sequence of their amino acids. What can the amino acid sequence of a polypeptide tell us about the three-dimensional structure (commonly referred to simply as "the structure") of the protein and its function? The term *polypeptide* is not synonymous with the term *protein*. Even for a protein consisting of a single polypeptide, the relationship is somewhat analogous to that between a long strand of yarn and a sweater of particular size and shape that can be knit from the yarn. A functional protein is not *just* a polypeptide chain, but one or more polypeptides precisely twisted, folded, and coiled into a molecule of unique shape **(Figure 3.19)**. And it is the amino acid sequence of each polypeptide that determines what three-dimensional structure the protein will have under normal cellular conditions.

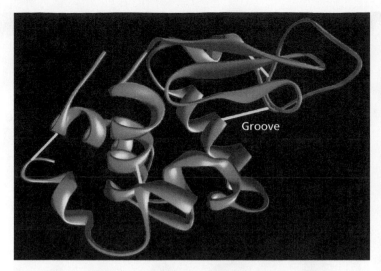

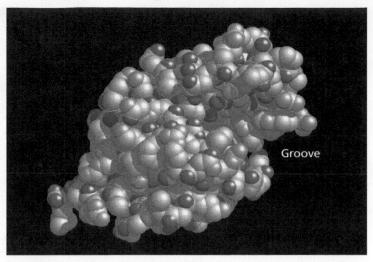

(a) A **ribbon model** shows how the single polypeptide chain folds and coils to form the functional protein. (The yellow lines represent disulfide bridges that stabilize the protein's shape.)

(b) A **space-filling model** shows more clearly the globular shape seen in many proteins, as well as the specific three-dimensional structure unique to lysozyme.

▲ **Figure 3.19 Structure of a protein, the enzyme lysozyme.** Present in our sweat, tears, and saliva, lysozyme is an enzyme that helps prevent infection by binding to and catalyzing the destruction of specific molecules on the surface of many kinds of bacteria. The groove is the part of the protein that recognizes and binds to the target molecules on bacterial walls.

When a cell synthesizes a polypeptide, the chain generally folds spontaneously, assuming the functional structure for that protein. This folding is driven and reinforced by the formation of various bonds between parts of the chain, which in turn depend on the sequence of amino acids. Many proteins are roughly spherical (*globular proteins*), while others are shaped like long fibers (*fibrous proteins*). Even within these broad categories, countless variations exist.

A protein's specific structure determines how it works. In almost every case, the function of a protein depends on its ability to recognize and bind to some other molecule. In an especially striking example of the marriage of form and function, **Figure 3.20** shows the exact match of shape between an antibody (a protein in the body) and the particular foreign substance on a flu virus that the antibody binds to and marks for destruction. (In Chapter 35, you'll learn more about how the immune system generates antibodies that match the shapes of specific foreign molecules so well.)

Another example of molecules with matching shapes is that of endorphin molecules—or morphine molecules—that fit into receptor molecules on the surface of brain cells in humans, producing euphoria and relieving pain. Morphine, heroin, and other opiate drugs are able to mimic endorphins because they all share a similar shape with endorphins and can thus fit into and bind to endorphin receptors in the brain. This fit is very specific, something like a lock and key (see Figure 2.14). The endorphin receptor, like other receptor molecules, is a protein. The function of a protein—for instance, the ability of a receptor protein to bind to a particular pain-relieving signaling molecule—is an emergent property resulting from exquisite molecular order.

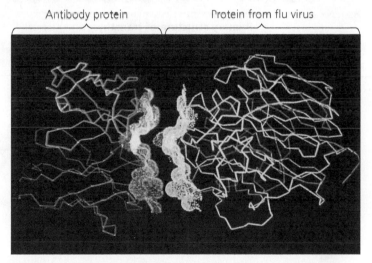

▲ **Figure 3.20 An antibody binding to a protein from a flu virus.** A technique called X-ray crystallography was used to generate a computer model of an antibody protein (blue and orange, left) bound to a flu virus protein (green and yellow, right). Computer software was then used to back the images away from each other, revealing the exact complementarity of shape between the two protein surfaces.

Four Levels of Protein Structure

In spite of their great diversity, all proteins share three superimposed levels of structure, known as primary, secondary, and tertiary structure. A fourth level, quaternary structure, arises when a protein consists of two or more polypeptide chains. **Figure 3.21** describes these four levels of protein structure. Be sure to study this figure thoroughly before going on to the next section.

Primary Structure
Linear chain of amino acids

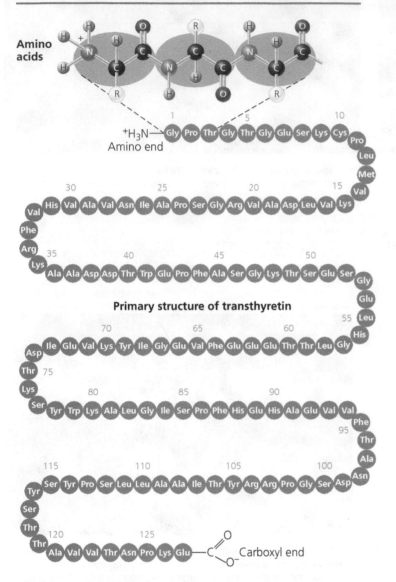

Amino acids

Primary structure of transthyretin

Amino end

Carboxyl end

Secondary Structure
Regions stabilized by hydrogen bonds between atoms of the polypeptide backbone

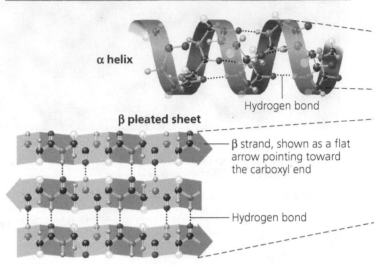

α helix

Hydrogen bond

β pleated sheet

β strand, shown as a flat arrow pointing toward the carboxyl end

Hydrogen bond

The **primary structure** of a protein is its sequence of amino acids. As an example, let's consider transthyretin, a globular blood protein that transports vitamin A and one of the thyroid hormones throughout the body. Transthyretin is made up of four identical polypeptide chains, each composed of 127 amino acids. Shown here is one of these chains unraveled for a closer look at its primary structure. Each of the 127 positions along the chain is occupied by one of the 20 amino acids, indicated here by its three-letter abbreviation.

The primary structure is like the order of letters in a very long word. If left to chance, there would be 20^{127} different ways of making a polypeptide chain 127 amino acids long. However, the precise primary structure of a protein is determined not by the random linking of amino acids, but by inherited genetic information. The primary structure in turn dictates secondary and tertiary structure, due to the chemical nature of the backbone and the side chains (R groups) of the amino acids along the polypeptide.

Most proteins have segments of their polypeptide chains repeatedly coiled or folded in patterns that contribute to the protein's overall shape. These coils and folds, collectively referred to as **secondary structure**, are the result of hydrogen bonds between the repeating constituents of the polypeptide backbone (not the amino acid side chains). Within the backbone, the oxygen atoms have a partial negative charge, and the hydrogen atoms attached to the nitrogens have a partial positive charge (see Figure 2.12); therefore, hydrogen bonds can form between these atoms. Individually, these hydrogen bonds are weak, but because they are repeated many times over a relatively long region of the polypeptide chain, they can support a particular shape for that part of the protein.

One such secondary structure is the **α helix**, a delicate coil held together by hydrogen bonding between every fourth amino acid, as shown above. Although each transthyretin polypeptide has only one α helix region (see tertiary structure), other globular proteins have multiple stretches of α helix separated by nonhelical regions (see hemoglobin). Some fibrous proteins, such as α-keratin, the structural protein of hair, have the α helix formation over most of their length.

The other main type of secondary structure is the **β pleated sheet**. As shown above, in this structure two or more segments of the polypeptide chain lying side by side (called β strands) are connected by hydrogen bonds between parts of the two parallel segments of polypeptide backbone. β pleated sheets make up the core of many globular proteins, as is the case for transthyretin (see tertiary structure), and dominate some fibrous proteins, including the silk protein of a spider's web. The teamwork of so many hydrogen bonds makes each spider silk fiber stronger than a steel strand of the same weight.

▼ Spiders secrete silk fibers made of a structural protein containing β pleated sheets, which allow the spider web to stretch and recoil.

Tertiary Structure

Three-dimensional shape stabilized by interactions between side chains

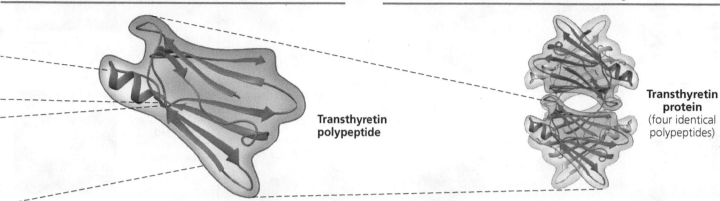

Transsthyretin polypeptide

Quaternary Structure

Association of two or more polypeptides (some proteins only)

Transsthyretin protein
(four identical polypeptides)

Superimposed on the patterns of secondary structure is a protein's tertiary structure, shown above in a ribbon model of the transthyretin polypeptide. While secondary structure involves interactions between backbone constituents, **tertiary structure** is the overall shape of a polypeptide resulting from interactions between the side chains (R groups) of the various amino acids. One type of interaction that contributes to tertiary structure is called—somewhat misleadingly—a **hydrophobic interaction**. As a polypeptide folds into its functional shape, amino acids with hydrophobic (nonpolar) side chains usually end up in clusters at the core of the protein, out of contact with water. Thus, a "hydrophobic interaction" is actually caused by the exclusion of nonpolar substances by water molecules. Once nonpolar amino acid side chains are close together, van der Waals interactions help hold them together. Meanwhile, hydrogen bonds between polar side chains and ionic bonds between positively and negatively charged side chains also help stabilize tertiary structure. These are all weak interactions in the aqueous cellular environment, but their cumulative effect helps give the protein a unique shape.

Covalent bonds called **disulfide bridges** may further reinforce the shape of a protein. Disulfide bridges form where two cysteine monomers, which have sulfhydryl groups (—SH) on their side chains (see Figure 3.5), are brought close together by the folding of the protein. The sulfur of one cysteine bonds to the sulfur of the second, and the disulfide bridge (—S—S—) rivets parts of the protein together (see yellow lines in Figure 3.19a). All of these different kinds of interactions can contribute to the tertiary structure of a protein, as shown here in a small part of a hypothetical protein:

Some proteins consist of two or more polypeptide chains aggregated into one functional macromolecule. **Quaternary structure** is the overall protein structure that results from the aggregation of these polypeptide subunits. For example, shown above is the complete globular transthyretin protein, made up of its four polypeptides.

Another example is collagen, shown below, which is a fibrous protein that has three identical helical polypeptides intertwined into a larger triple helix, giving the long fibers great strength. This suits collagen fibers to their function as the girders of connective tissue in skin, bone, tendons, ligaments, and other body parts. (Collagen accounts for 40% of the protein in a human body.)

Collagen

Hemoglobin, the oxygen-binding protein of red blood cells shown below, is another example of a globular protein with quaternary structure. It consists of four polypeptide subunits, two of one kind (α) and two of another kind (β). Both α and β subunits consist primarily of α-helical secondary structure. Each subunit has a nonpolypeptide component, called heme, with an iron atom that binds oxygen.

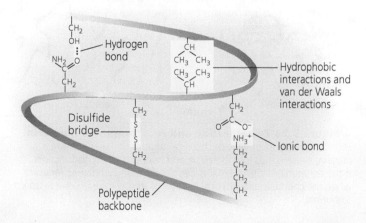

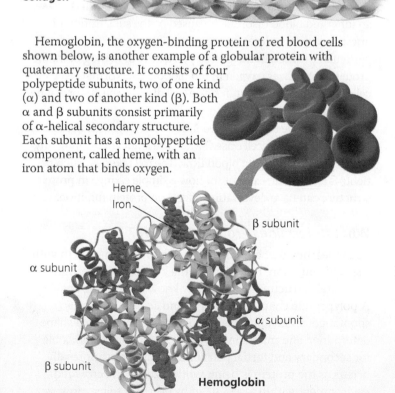

Heme
Iron
β subunit
α subunit
α subunit
β subunit
Hemoglobin

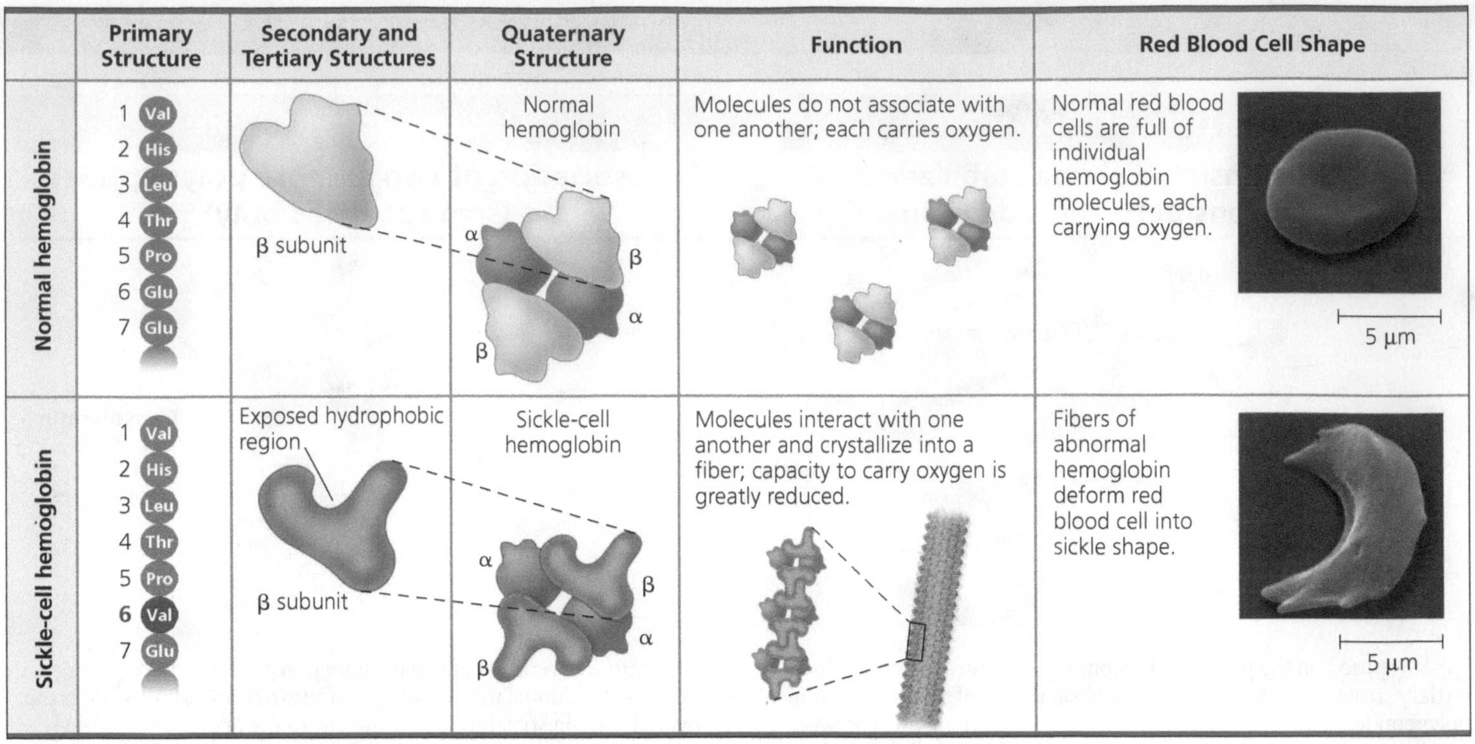

	Primary Structure	Secondary and Tertiary Structures	Quaternary Structure	Function	Red Blood Cell Shape
Normal hemoglobin	1 Val 2 His 3 Leu 4 Thr 5 Pro 6 Glu 7 Glu	β subunit	Normal hemoglobin α β α β	Molecules do not associate with one another; each carries oxygen.	Normal red blood cells are full of individual hemoglobin molecules, each carrying oxygen. 5 μm
Sickle-cell hemoglobin	1 Val 2 His 3 Leu 4 Thr 5 Pro 6 Val 7 Glu	Exposed hydrophobic region β subunit	Sickle-cell hemoglobin α β α β	Molecules interact with one another and crystallize into a fiber; capacity to carry oxygen is greatly reduced.	Fibers of abnormal hemoglobin deform red blood cell into sickle shape. 5 μm

▲ **Figure 3.22 A single amino acid substitution in a protein causes sickle-cell disease.**

MAKE CONNECTIONS *Considering the chemical characteristics of the amino acids valine and glutamic acid (see Figure 3.17), propose a possible explanation for the dramatic effect on protein function that occurs when valine is substituted for glutamic acid.*

Sickle-Cell Disease: A Change in Primary Structure

Even a slight change in primary structure can affect a protein's shape and ability to function. For instance, **sickle-cell disease**, an inherited blood disorder, is caused by the substitution of one amino acid (valine) for the normal one (glutamic acid) at a particular position in the primary structure of hemoglobin, the protein that carries oxygen in red blood cells. Normal red blood cells are disk-shaped, but in sickle-cell disease, the abnormal hemoglobin molecules tend to crystallize, deforming some of the cells into a sickle shape **(Figure 3.22)**. A person with the disease has periodic "sickle-cell crises" when the angular cells clog tiny blood vessels, impeding blood flow. The toll taken on such patients is a dramatic example of how a simple change in protein structure can have devastating effects on protein function.

What Determines Protein Structure?

You've learned that a unique shape endows each protein with a specific function. But what are the key factors determining protein structure? You already know most of the answer: A polypeptide chain of a given amino acid sequence can spontaneously arrange itself into a three-dimensional shape determined and maintained by the interactions responsible for secondary and tertiary structure. This folding normally occurs as the protein is being synthesized in the crowded environment within a cell, aided by other proteins. However,

protein structure also depends on the physical and chemical conditions of the protein's environment. If the pH, salt concentration, temperature, or other aspects of its environment are altered, the weak chemical bonds and interactions within a protein may be destroyed, causing the protein to unravel and lose its native shape, a change called **denaturation (Figure 3.23)**. Because it is misshapen, the denatured protein is biologically inactive.

Most proteins become denatured if they are transferred from an aqueous environment to a nonpolar solvent, such as

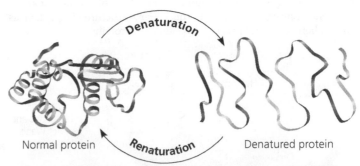

Denaturation

Renaturation

Normal protein   Denatured protein

▲ **Figure 3.23 Denaturation and renaturation of a protein.** High temperatures or various chemical treatments will denature a protein, causing it to lose its shape and hence its ability to function. If the denatured protein remains dissolved, it can sometimes renature when the chemical and physical aspects of its environment are restored to normal.

ether or chloroform; the polypeptide chain refolds so that its hydrophobic regions face outward toward the solvent. Other denaturation agents include chemicals that disrupt the hydrogen bonds, ionic bonds, and disulfide bridges that maintain a protein's shape. Denaturation can also result from excessive heat, which agitates the polypeptide chain enough to overpower the weak interactions that stabilize the structure. The white of an egg becomes opaque during cooking because the denatured proteins are insoluble and solidify. This also explains why excessively high fevers can be fatal: Proteins in the blood can denature at very high body temperatures.

When a protein in a test-tube solution has been denatured by heat or chemicals, it can sometimes return to its functional shape when the denaturing agent is removed. We can conclude that the information for building a specific shape is intrinsic to the protein's primary structure. The sequence of amino acids determines the protein's shape—where an α helix can form, where β pleated sheets can exist, where disulfide bridges are located, where ionic bonds can form, and so on. But how does protein folding occur in the cell?

Protein Folding in the Cell

Biochemists now know the amino acid sequence for more than 10 million proteins and the three-dimensional shape for more than 20,000. Researchers have tried to correlate the primary structure of many proteins with their three-dimensional structure to discover the rules of protein folding. Unfortunately, however, the protein-folding process is not that simple. Most proteins probably go through several intermediate structures on their way to a stable shape, and looking at the mature structure does not reveal the stages of folding required to achieve that form. However, biochemists have developed methods for tracking a protein through such stages. They are still working to develop computer programs that can predict the 3-D structure of a polypeptide from its primary structure alone.

Misfolding of polypeptides is a serious problem in cells. Many diseases, such as Alzheimer's, Parkinson's, and mad cow disease, are associated with an accumulation of misfolded proteins. In fact, misfolded versions of the transthyretin protein featured in Figure 3.21 have been implicated in several diseases, including one form of senile dementia.

Even when scientists have a correctly folded protein in hand, determining its exact three-dimensional structure is not simple, for a single protein molecule has thousands of atoms. The method most commonly used to determine the 3-D shape of a protein is **X-ray crystallography**, which depends on the diffraction of an X-ray beam by the atoms of a crystallized molecule. Using this technique, scientists can build a 3-D model that shows the exact position of every atom in a protein molecule **(Figure 3.24)**. Nuclear magnetic resonance (NMR) spectroscopy and bioinformatics (see Chapter 1) are complementary approaches to understanding protein structure and function.

▼ **Figure 3.24 Inquiry**

What can the 3-D shape of the enzyme RNA polymerase II tell us about its function?

Experiment In 2006, Roger Kornberg was awarded the Nobel Prize in Chemistry for using X-ray crystallography to determine the 3-D shape of RNA polymerase II, which binds to the DNA double helix and synthesizes RNA. After crystallizing a complex of all three components, Kornberg and his colleagues aimed an X-ray beam through the crystal. The atoms of the crystal diffracted (bent) the X-rays into an orderly array that a digital detector recorded as a pattern of spots called an X-ray diffraction pattern.

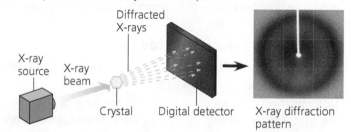

Results Using data from X-ray diffraction patterns, as well as the amino acid sequence determined by chemical methods, Kornberg and colleagues built a 3-D model of the complex with the help of computer software.

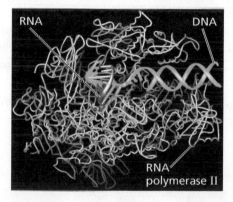

Conclusion By analyzing their model, the researchers developed a hypothesis about the functions of different regions of RNA polymerase II. For example, the region above the DNA may act as a clamp that holds the nucleic acids in place. (You'll learn more about RNA polymerase in Chapter 14.)

Further Reading A. L. Gnatt et al., Structural basis of transcription: an RNA polymerase II elongation complex at 3.3Å, *Science* 292:1876–1882 (2001).

WHAT IF? If you were one of the researchers and were describing the model, what type of protein structure would you call the small polypeptide spirals in RNA polymerase II?

CONCEPT CHECK 3.5

1. Why does a denatured protein no longer function normally?
2. What parts of a polypeptide participate in the bonds that hold together secondary structure? Tertiary structure?
3. **WHAT IF?** Where would you expect a polypeptide region that is rich in the amino acids valine, leucine, and isoleucine to be located in the folded polypeptide? Explain.

For suggested answers, see Appendix A.

Nucleic acids store, transmit, and help express hereditary information

If the primary structure of polypeptides determines a protein's shape, what determines primary structure? The amino acid sequence of a polypeptide is programmed by a discrete unit of inheritance known as a **gene**. Genes consist of DNA, which belongs to the class of compounds called nucleic acids. **Nucleic acids** are polymers made of monomers called nucleotides.

The Roles of Nucleic Acids

The two types of nucleic acids, **deoxyribonucleic acid (DNA)** and **ribonucleic acid (RNA)**, enable living organisms to reproduce their complex components from one generation to the next. Unique among molecules, DNA provides directions for its own replication. DNA also directs RNA synthesis and, through RNA, controls protein synthesis **(Figure 3.25)**.

DNA is the genetic material that organisms inherit from their parents. Each chromosome contains one long DNA molecule, usually carrying several hundred or more genes. When a cell reproduces itself by dividing, its DNA molecules are copied and passed along from one generation of cells to the next. Encoded in the structure of DNA is the information that

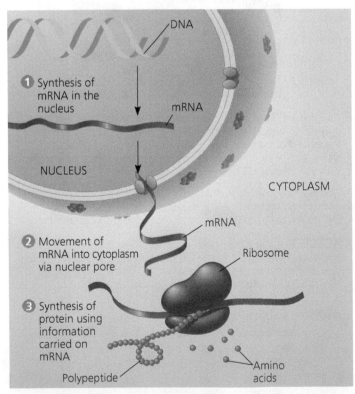

1 Synthesis of mRNA in the nucleus

DNA

mRNA

NUCLEUS

CYTOPLASM

mRNA

2 Movement of mRNA into cytoplasm via nuclear pore

Ribosome

3 Synthesis of protein using information carried on mRNA

Polypeptide

Amino acids

▲ **Figure 3.25 DNA → RNA → protein.** In a eukaryotic cell, DNA in the nucleus programs protein production in the cytoplasm by dictating synthesis of messenger RNA (mRNA). (The cell nucleus is actually much larger relative to the other elements of this figure.)

programs all the cell's activities. The DNA, however, is not directly involved in running the operations of the cell, any more than computer software by itself can print a bank statement or read the bar code on a box of cereal. Just as a printer is needed to print out a statement and a scanner is needed to read a bar code, proteins are required to implement genetic programs. The molecular hardware of the cell—the tools for biological functions—consists mostly of proteins. For example, the oxygen carrier in red blood cells is the protein hemoglobin, not the DNA that specifies its structure.

How does RNA, the other type of nucleic acid, fit into gene expression, the flow of genetic information from DNA to proteins? Each gene along a DNA molecule directs synthesis of a type of RNA called *messenger RNA (mRNA)*. The mRNA molecule interacts with the cell's protein-synthesizing machinery to direct production of a polypeptide, which folds into all or part of a protein. We can summarize the flow of genetic information as DNA → RNA → protein (see Figure 3.25). The sites of protein synthesis are tiny structures called ribosomes. In a eukaryotic cell, ribosomes are in the cytoplasm, the region between the nucleus and a cell's outer membrane, but DNA resides in the nucleus. Messenger RNA conveys genetic instructions for building proteins from the nucleus to the cytoplasm. Prokaryotic cells lack nuclei but still use mRNA to convey a message from the DNA to ribosomes and other cellular equipment that translate the coded information into amino acid sequences. In recent years, the spotlight has been turned on other, previously unknown types of RNA that play many other roles in the cell. As is so often true in biology, the story is still being written! (You'll hear more about the newly discovered functions of RNA molecules in Chapter 15.)

The Components of Nucleic Acids

Nucleic acids are macromolecules that exist as polymers called **polynucleotides (Figure 3.26a)**. As indicated by the name, each polynucleotide consists of monomers called **nucleotides**. A nucleotide, in general, is composed of three parts: a nitrogen-containing (nitrogenous) base, a five-carbon sugar (a pentose), and one or more phosphate groups **(Figure 3.26b)**. In a polynucleotide, each monomer has only one phosphate group. The portion of a nucleotide without any phosphate groups is called a *nucleoside*.

To build a nucleotide, let's first consider the nitrogenous bases **(Figure 3.26c)**. Each nitrogenous base has one or two rings that include nitrogen atoms. (They are called nitrogenous *bases* because the nitrogen atoms tend to take up H+ from solution, thus acting as bases.) There are two families of nitrogenous bases: pyrimidines and purines. A **pyrimidine** has one six-membered ring of carbon and nitrogen atoms. The members of the pyrimidine family are cytosine (C), thymine (T), and uracil (U). **Purines** are larger, with a six-membered ring fused to a five-membered ring. The purines are adenine (A) and guanine (G). The specific pyrimidines and purines differ in

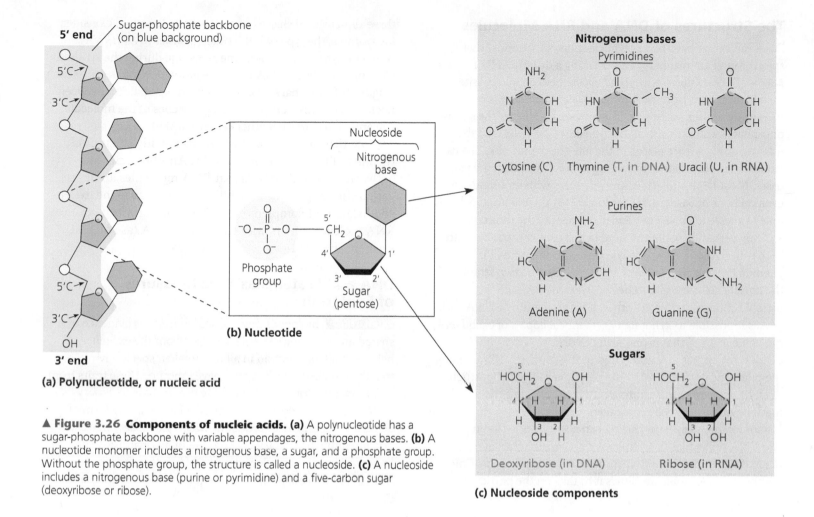

▲ **Figure 3.26 Components of nucleic acids. (a)** A polynucleotide has a sugar-phosphate backbone with variable appendages, the nitrogenous bases. **(b)** A nucleotide monomer includes a nitrogenous base, a sugar, and a phosphate group. Without the phosphate group, the structure is called a nucleoside. **(c)** A nucleoside includes a nitrogenous base (purine or pyrimidine) and a five-carbon sugar (deoxyribose or ribose).

the chemical groups attached to the rings. Adenine, guanine, and cytosine are found in both DNA and RNA; thymine is found only in DNA and uracil only in RNA.

Now let's add a sugar to the nitrogenous base. In DNA the sugar is **deoxyribose**; in RNA it is **ribose** (See Figure 3.26c). The only difference between these two sugars is that deoxyribose lacks an oxygen atom on the second carbon in the ring; hence the name *deoxy*ribose. To distinguish the numbers of the sugar carbons from those used for the ring atoms of the attached nitrogenous base, we add a prime (′) after the sugar carbon numbers of a nucleoside or nucleotide. Thus, the second carbon in the sugar ring is the 2′ ("2 prime") carbon, and the carbon that sticks up from the ring is called the 5′ carbon.

To complete the construction of a nucleotide, we attach a phosphate group to the 5′ carbon of the sugar (see Figure 3.26b). The molecule is now a nucleoside monophosphate, better known as a nucleotide.

Nucleotide Polymers

Now let's see how these nucleotides are linked together to build a polynucleotide. Adjacent nucleotides are joined by a phosphodiester linkage, which consists of a phosphate group that links the sugars of two nucleotides. This bonding results

in a backbone with a repeating pattern of sugar-phosphate units (see Figure 3.26a). (Note that the nitrogenous bases are not part of the backbone.) The two free ends of the polymer are distinctly different from each other. One end has a phosphate attached to a 5′ carbon, and the other end has a hydroxyl group on a 3′ carbon; we refer to these as the 5′ end and the 3′ end, respectively. We can say that a polynucleotide has a built-in directionality along its sugar-phosphate backbone, from 5′ to 3′, somewhat like a one-way street. All along this sugar-phosphate backbone are appendages consisting of the nitrogenous bases.

The sequence of bases along a DNA (or mRNA) polymer is unique for each gene and provides very specific information to the cell. Because genes are hundreds to thousands of nucleotides long, the number of possible base sequences is effectively limitless. A gene's meaning to the cell is encoded in its specific sequence of the four DNA bases. For example, the sequence 5′-AGGTAACTT-3′ means one thing, whereas the sequence 5′-CGCTTTAAC-3′ has a different meaning. (Entire genes, of course, are much longer.) The linear order of bases in a gene specifies the amino acid sequence—the primary structure—of a protein, which in turn specifies that protein's three-dimensional structure and its function in the cell.

The Structures of DNA and RNA Molecules

DNA molecules have two polynucleotides, or "strands," that spiral around an imaginary axis, forming a **double helix (Figure 3.27a)**. The two sugar-phosphate backbones run in opposite $5' \rightarrow 3'$ directions from each other; this arrangement is referred to as **antiparallel**, somewhat like a divided highway. The sugar-phosphate backbones are on the outside of the helix, and the nitrogenous bases are paired in the interior of the helix. The two strands are held together by hydrogen bonds between the paired bases. Most DNA molecules are very long, with thousands or even millions of base pairs. One long DNA double helix includes many genes, each one a particular segment of the molecule.

Only certain bases in the double helix are compatible with each other. Adenine (A) always pairs with thymine (T), and guanine (G) always pairs with cytosine (C). The two strands of the double helix are said to be *complementary*, each the predictable counterpart of the other. It is this feature of DNA that makes it possible to generate two identical copies of each DNA molecule in a cell that is preparing to divide. When the cell divides, the copies are distributed to the daughter cells, making them genetically identical to the parent cell. Thus, the structure of DNA accounts for its function of transmitting genetic information whenever a cell reproduces.

Complementary base pairing can also occur between two RNA molecules or even between two stretches of nucleotides in the *same* RNA molecule. In fact, base pairing within an RNA molecule allows it to take on the particular

three-dimensional shape necessary for its function. Consider, for example, the type of RNA called *transfer RNA* (*tRNA*), which brings amino acids to the ribosome during the synthesis of a polypeptide. A tRNA molecule is about 80 nucleotides in length. Its functional shape results from base pairing between nucleotides where complementary stretches of the molecule run antiparallel to each other **(Figure 3.27b)**.

Note that in RNA, adenine (A) pairs with uracil (U); thymine (T) is not present in RNA. Another difference between RNA and DNA is that RNA molecules are more variable in shape. This variability arises because the extent and location of complementary base pairing within an RNA molecule differs with the type of RNA (as you will learn in Chapter 14).

DNA and Proteins as Tape Measures of Evolution

EVOLUTION Biologists think of shared traits as evidence of shared ancestry. For example, we infer from the existence of hair and milk production in all mammalian species living today that the members of this group have inherited these traits from common ancestors that lived in the distant past.

Now we have additional kinds of evidence in the form of genes and their protein products, which like observable traits document the hereditary background of an organism. The linear sequences of nucleotides in DNA molecules are passed from parents to offspring; these sequences determine the

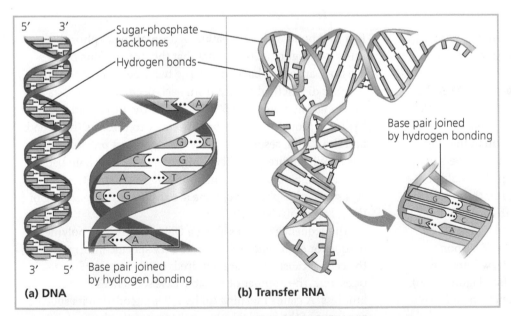

(a) DNA   **(b) Transfer RNA**

▲ **Figure 3.27 The structures of DNA and tRNA molecules. (a)** The DNA molecule is usually a double helix, with the sugar-phosphate backbones of the antiparallel polynucleotide strands (symbolized here by blue ribbons) on the outside of the helix. Holding the two strands together are pairs of nitrogenous bases attached to each other by hydrogen bonds. As illustrated here with symbolic shapes for the bases, adenine (A) can pair only with thymine (T), and guanine (G) can pair only with cytosine (C). Each DNA strand in this figure is the structural equivalent of the polynucleotide diagrammed in Figure 3.26a. **(b)** A tRNA molecule has a roughly L-shaped structure, with complementary base pairing of antiparallel stretches of RNA. In RNA, A pairs with U.

amino acid sequences of proteins. As a result, siblings have greater similarity in their DNA and proteins than do unrelated individuals of the same species. Given the validity of evolutionary theory, we can extend this concept of "molecular genealogy" to relationships between species: We would expect two species that appear to be closely related based on anatomical evidence (possibly including fossil evidence) to also share a greater proportion of their DNA and protein sequences than do more distantly related species. In fact, that is the case. An example is the comparison of the β polypeptide chain of human hemoglobin with the corresponding hemoglobin polypeptide in other vertebrates. In this chain of 146 amino acids, humans and gorillas differ in just 1 amino acid, while humans and frogs, more distantly related, differ in 67 amino acids. In the **Scientific Skills Exercise**, you can apply this sort of reasoning to additional species. Molecular biology has added a new tape measure to the toolkit biologists use to assess evolutionary kinship.

CONCEPT CHECK 3.6

1. **DRAW IT** Go to Figure 3.26a and, for the top three nucleotides, number all the carbons in the sugars, circle the nitrogenous bases, and star the phosphates.

2. **DRAW IT** In a DNA double helix, a region along one DNA strand has the following sequence of nitrogenous bases: 5'-TAGGCCT-3'. Copy this sequence, and write down its complementary strand, clearly indicating the 5' and 3' ends of the complementary strand.

3. **WHAT IF?** (a) Suppose a substitution occurred in one DNA strand of the double helix in question 2, resulting in

   5'-TAAGCCT-3'
   3'-ATCCGGA-5'

   Copy these two strands, and circle and label the mismatched bases. (b) If the modified top strand is used by the cell to construct a complementary strand, what would that matching strand be?

For suggested answers, see Appendix A.

Scientific Skills Exercise

Analyzing Polypeptide Sequence Data

Are Rhesus Monkeys or Gibbons More Closely Related to Humans? As discussed in Concept 3.6, DNA and polypeptide sequences from closely related species are more similar to each other than are sequences from more distantly related species. In this exercise, you will look at amino acid sequence data for the β polypeptide chain of hemoglobin, often called β-globin. You will then interpret the data to hypothesize whether the monkey or the gibbon is more closely related to humans.

How Such Experiments Are Done Researchers can isolate the polypeptide of interest from an organism and then determine the amino acid sequence. More frequently, the DNA of the relevant gene is sequenced, and the amino acid sequence of the polypeptide is deduced from the DNA sequence of its gene.

Data from the Experiments In the data below, the letters give the sequence of the 146 amino acids in β-globin from humans, rhesus monkeys, and gibbons. Because a complete sequence would not fit on one line here, the sequences are broken into three segments. Note that the sequences for the three different species are aligned so that you can compare them easily. For example, you can see that for all three species, the first amino acid is V (valine; see Figure 3.17) and the 146th amino acid is H (histidine).

Interpret the Data

1. Scan along the monkey and gibbon sequences, letter by letter, circling any amino acids that do not match the human sequence. (a) How many amino acids differ between the monkey and the human sequences? (b) Between the gibbon and human?

2. For each nonhuman species, what percent of its amino acids are identical to the human sequence of β-globin?

3. Based on these data alone, state a hypothesis for which of these two species is more closely related to humans. What is your reasoning?

4. What other evidence could you use to support your hypothesis?

Data from Human: http://www.ncbi.nlm.nih.gov/protein/AAA21113.1; rhesus monkey: http://www.ncbi.nlm.nih.gov/protein/122634; gibbon: http://www.ncbi.nlm.nih.gov/protein/122616

MB A version of this Scientific Skills Exercise can be assigned in MasteringBiology.

Species	Alignment of Amino Acid Sequences of β-globin
Human	1 VHLTPEEKSA VTALWGKVNV DEVGGEALGR LLVVYPWTQR FFESFGDLST PDAVMGNPKV
Monkey	1 VHLTPEEKNA VTTLWGKVNV DEVGGEALGR LLLVYPWTQR FFESFGDLSS PDAVMGNPKV
Gibbon	1 VHLTPEEKSA VTALWGKVNV DEVGGEALGR LLVVYPWTQR FFESFGDLST PDAVMGNPKV
Human	61 KAHGKKVLGA FSDGLAHLDN LKGTFATLSE LHCDKLHVDP ENFRLLGNVL VCVLAHHFGK
Monkey	61 KAHGKKVLGA FSDGLNHLDN LKGTFAQLSE LHCDKLHVDP ENFKLLGNVL VCVLAHHFGK
Gibbon	61 KAHGKKVLGA FSDGLAHLDN LKGTFAQLSE LHCDKLHVDP ENFRLLGNVL VCVLAHHFGK
Human	121 EFTPPVQAAY QKVVAGVANA LAHKYH
Monkey	121 EFTPQVQAAY QKVVAGVANA LAHKYH
Gibbon	121 EFTPQVQAAY QKVVAGVANA LAHKYH

3 Chapter Review

SUMMARY OF KEY CONCEPTS

CONCEPT 3.1

Carbon atoms can form diverse molecules by bonding to four other atoms (pp. 41–44)

- Carbon, with a **valence** of 4, can bond to various other atoms, including O, H, and N. Carbon can also bond to other carbon atoms, forming the carbon skeletons of **organic compounds**. These skeletons vary in length and shape.
- Chemical groups attached to the carbon skeletons of organic molecules participate in chemical reactions (**functional groups**) or contribute to function by affecting molecular shape.
- **ATP (adenosine triphosphate)** can react with water, releasing energy that can be used by the cell.

? *In what ways does a methyl group differ chemically from the other six important chemical groups shown in Figure 3.5?*

CONCEPT 3.2

Macromolecules are polymers, built from monomers (pp. 44–45)

- Proteins, nucleic acids, and large carbohydrates (polysaccharides) are **polymers**, which are chains of **monomers**. Monomers form larger molecules by **dehydration reactions**, in which water molecules are released. Polymers can disassemble by the reverse process, **hydrolysis**. In cells, dehydration reactions and hydrolysis are catalyzed by enzymes. An immense variety of polymers can be built from a small set of monomers.

? *What is the fundamental basis for the differences between carbohydrates, proteins, and nucleic acids?*

Large Biological Molecules	Components	Examples	Functions
CONCEPT 3.3 **Carbohydrates serve as fuel and building material (pp. 45–49)** **?** *Compare the composition, structure, and function of starch and cellulose. What roles do starch and cellulose play in the human body?*	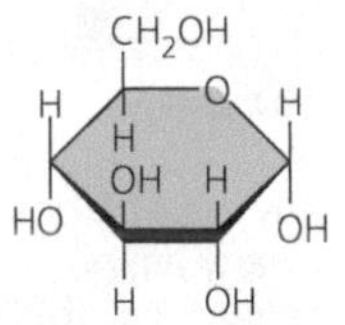 Monosaccharide monomer	**Monosaccharides:** glucose, fructose **Disaccharides:** lactose, sucrose	Fuel; carbon sources that can be converted to other molecules or combined into polymers
		Polysaccharides: • Cellulose (plants) • Starch (plants) • Glycogen (animals) • Chitin (animals and fungi)	• Strengthens plant cell walls • Stores glucose for energy in plants • Stores glucose for energy in animals • Strengthens exoskeletons and fungal cell walls
CONCEPT 3.4 **Lipids are a diverse group of hydrophobic molecules (pp. 49–51)** **?** *Why are lipids not considered to be macromolecules or polymers?*	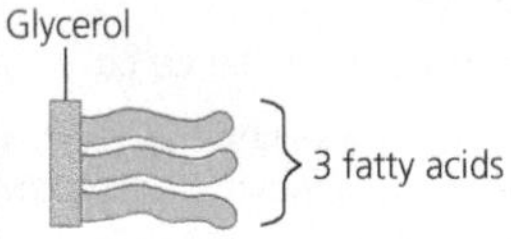 Glycerol / 3 fatty acids	**Triacylglycerols** (fats or oils): glycerol + 3 fatty acids	Important energy source
	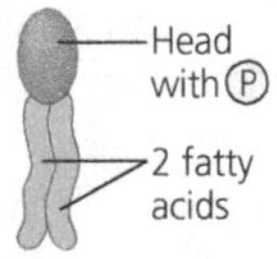 Head with (P) / 2 fatty acids	**Phospholipids:** phosphate group + glycerol + 2 fatty acids	Lipid bilayers of membranes 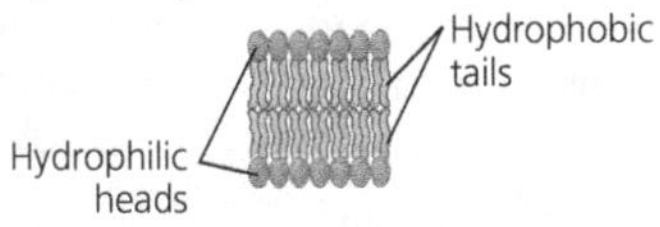 Hydrophilic heads / Hydrophobic tails
	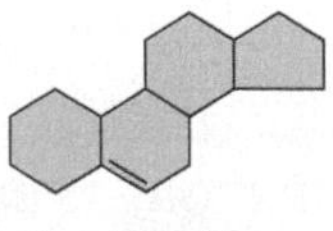 Steroid backbone	**Steroids:** four fused rings with attached chemical groups	• Component of cell membranes (cholesterol) • Signaling molecules that travel through the body (hormones)
CONCEPT 3.5 **Proteins include a diversity of structures, resulting in a wide range of functions (pp. 51–59)** **?** *Explain the basis for the great diversity of proteins.*	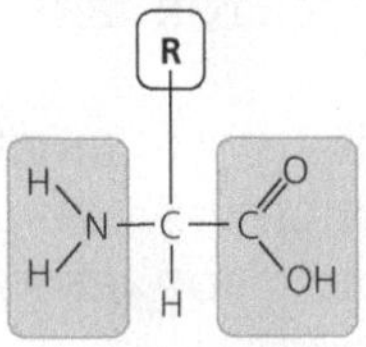 Amino acid monomer (20 types)	• Enzymes • Structural proteins • Storage proteins • Transport proteins • Hormones • Receptor proteins • Motor proteins • Defensive proteins	• Catalyze chemical reactions • Provide structural support • Store amino acids • Transport substances • Coordinate organismal responses • Receive signals from outside cell • Function in cell movement • Protect against disease

Large Biological Molecules	Components	Examples	Functions
CONCEPT 3.6 **Nucleic acids store, transmit, and help express hereditary information (pp. 60–63)** ❓ *What role does complementary base pairing play in the functions of nucleic acids?*	Nitrogenous base Phosphate group ℗—CH_2 O Sugar Nucleotide monomer	**DNA:** ∿∿∿∿∿ • Sugar = deoxyribose • Nitrogenous bases = C, G, A, T • Usually double-stranded	Stores hereditary information
		RNA: ∿ • Sugar = ribose • Nitrogenous bases = C, G, A, U • Usually single-stranded	Various functions in gene expression, including carrying instructions from DNA to ribosomes

TEST YOUR UNDERSTANDING

Level 1: Knowledge/Comprehension

1. Which functional group is *not* present in this molecule?
   a. carboxyl
   b. sulfhydryl
   c. hydroxyl
   d. amino

   HO—C=O / H
H—C—C—OH
N / H
H   H

2. **MAKE CONNECTIONS** Which chemical group is most likely to be responsible for an organic molecule behaving as a base (see Concept 2.5)?
   a. hydroxyl
   b. carbonyl
   c. carboxyl
   d. amino
   e. phosphate

3. Which of the following categories includes all others in the list?
   a. monosaccharide
   b. disaccharide
   c. starch
   d. carbohydrate
   e. polysaccharide

4. Which of the following statements concerning *unsaturated* fats is true?
   a. They are more common in animals than in plants.
   b. They have double bonds in the carbon chains of their fatty acids.
   c. They generally solidify at room temperature.
   d. They contain more hydrogen than do saturated fats having the same number of carbon atoms.
   e. They have fewer fatty acid molecules per fat molecule.

5. The structural level of a protein *least* affected by a disruption in hydrogen bonding is the
   a. primary level.
   b. secondary level.
   c. tertiary level.
   d. quaternary level.
   e. All structural levels are equally affected.

Level 2: Application/Analysis

6. Which of the following hydrocarbons has a double bond in its carbon skeleton?
   a. C_3H_8
   b. C_2H_6
   c. CH_4
   d. C_2H_4
   e. C_2H_2

7. The molecular formula for glucose is $C_6H_{12}O_6$. What would be the molecular formula for a polymer made by linking ten glucose molecules together by dehydration reactions?
   a. $C_{60}H_{120}O_{60}$
   b. $C_6H_{12}O_6$
   c. $C_{60}H_{102}O_{51}$
   d. $C_{60}H_{100}O_{50}$
   e. $C_{60}H_{111}O_{51}$

8. Rewrite the following table. Start with the left column, and then rearrange the terms in the second and third columns so they line up correctly. Label the columns and rows.

Monosaccharides	Polypeptides	Phosphodiester linkages
Fatty acids	Triacylglycerols	Peptide bonds
Amino acids	Polynucleotides	Glycosidic linkages
Nucleotides	Polysaccharides	Ester linkages

Level 3: Synthesis/Evaluation

9. **SCIENTIFIC INQUIRY**
   Suppose you are a research assistant in a lab studying DNA-binding proteins. You have been given the amino acid sequences of all the proteins encoded by the genome of a certain species and have been asked to find candidate proteins that could bind DNA. What type of amino acids would you expect to see in the DNA-binding regions of such proteins? Why?

10. **FOCUS ON EVOLUTION**
    Comparisons of amino acid sequences can shed light on the evolutionary divergence of related species. If you were comparing two living species, would you expect all proteins to show the same degree of divergence? Why or why not?

11. **FOCUS ON ORGANIZATION**
    Proteins, which have diverse functions in a cell, are all polymers of the same kinds of monomers—amino acids. Write a short essay (100–150 words) that discusses how the structure of amino acids allows this one type of polymer to perform so many functions.

For selected answers, see Appendix A.

MasteringBiology®

Students Go to **MasteringBiology** for assignments, the eText, and the Study Area with practice tests, animations, and activities.

Instructors Go to **MasteringBiology** for automatically graded tutorials and questions that you can assign to your students, plus Instructor Resources.

A Tour of the Cell

KEY CONCEPTS

4.1 Biologists use microscopes and the tools of biochemistry to study cells

4.2 Eukaryotic cells have internal membranes that compartmentalize their functions

4.3 The eukaryotic cell's genetic instructions are housed in the nucleus and carried out by the ribosomes

4.4 The endomembrane system regulates protein traffic and performs metabolic functions in the cell

4.5 Mitochondria and chloroplasts change energy from one form to another

4.6 The cytoskeleton is a network of fibers that organizes structures and activities in the cell

4.7 Extracellular components and connections between cells help coordinate cellular activities

OVERVIEW

The Fundamental Units of Life

Given the scope of biology, you may wonder sometimes how you will ever learn all the material in this course! The answer involves cells, which are as fundamental to the living systems of biology as the atom is to chemistry. The contraction of muscle cells moves your eyes as you read this sentence. The words on the page are translated into signals that nerve cells carry to your brain, where they are passed on to still other nerve cells. **Figure 4.1** shows extensions from one nerve cell (purple) making contact with another nerve cell (orange) in the brain. As you study, your goal is to make connections like these that solidify memories and permit learning to occur.

All organisms are made of cells. In the hierarchy of biological organization, the cell is the simplest collection of matter that can be alive. Indeed, many forms of life exist as single-celled organisms. (You may be familiar with single-celled eukaryotic organisms that live in pond water, such as paramecia.) Larger, more complex organisms, including plants and animals, are multicellular; their bodies are cooperatives of many kinds of specialized cells that could not survive for long on their own. Even when cells are arranged into higher levels of organization, such as tissues and organs, the cell remains the organism's basic unit of structure and function.

All cells are related by their descent from earlier cells. Furthermore, they have been modified in many different ways during the long evolutionary history of life on Earth. But although cells can differ substantially from one another, they share common features. It is these features that we focus on in most of this chapter.

We begin the chapter with a discussion of microscopy and some other techniques used by cell biologists. Next comes an overview of the cellular structures revealed by these methods. In the rest of the chapter, we explore cellular structures and their functions in more detail.

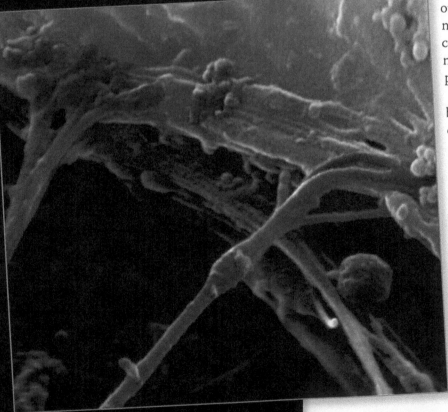

▼ **Figure 4.1** How do your brain cells help you learn about biology?

Biologists use microscopes and the tools of biochemistry to study cells

How can cell biologists investigate the inner workings of a cell, usually too small to be seen by the unaided eye? Before we tour the cell, it will be helpful to learn how cells are studied.

Microscopy

The development of instruments that extend the human senses has gone hand in hand with the advance of science. The discovery and early study of cells progressed with the invention of microscopes in 1590 and their refinement during the 1600s. Cell walls were first seen by Robert Hooke in 1665 as he looked through a microscope at dead cells from the bark of an oak tree. But it took the wonderfully crafted lenses of Antoni van Leeuwenhoek to visualize living cells. Imagine Hooke's awe when he visited van Leeuwenhoek in 1674 and the world of microorganisms—what his host called "very little animalcules"—was revealed to him.

The microscopes first used by Renaissance scientists, as well as the microscopes you are likely to use in the laboratory, are all light microscopes. In a **light microscope (LM)**, visible light is passed through the specimen and then through glass lenses. The lenses refract (bend) the light in such a way that the image of the specimen is magnified as it is projected into the eye or into a camera (see Appendix D).

Three important parameters in microscopy are magnification, resolution, and contrast. *Magnification* is the ratio of an object's image size to its real size. Light microscopes can magnify effectively to about 1,000 times the actual size of the specimen; at greater magnifications, additional details cannot be seen clearly. *Resolution* is a measure of the clarity of the image; it is the minimum distance two points can be separated and still be distinguished as separate points. For example, what appears to the unaided eye as one star in the sky may be resolved as twin stars with a telescope, which has a higher resolving ability than the eye. Similarly, using standard techniques, the light microscope cannot resolve detail finer than about 0.2 micrometer (μm), or 200 nanometers (nm), regardless of the magnification **(Figure 4.2)**. The third parameter, *contrast*, is the difference in brightness between the light and dark areas of an image. Methods for enhancing contrast in light microscopy include staining or labeling cell components to stand out visually. **Figure 4.3** shows some different types of microscopy; study this figure as you read the rest of this section.

Until recently, the resolution barrier prevented cell biologists from using standard light microscopy to study **organelles**, the membrane-enclosed structures within eukaryotic cells. To see these structures in any detail required the development of a new instrument. In the 1950s, the

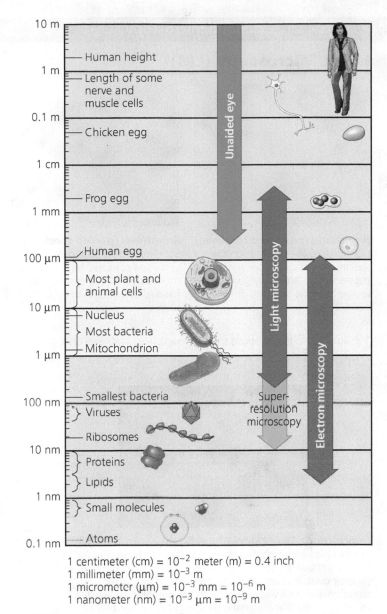

1 centimeter (cm) = 10^{-2} meter (m) = 0.4 inch
1 millimeter (mm) = 10^{-3} m
1 micrometer (μm) = 10^{-3} mm = 10^{-6} m
1 nanometer (nm) = 10^{-3} μm = 10^{-9} m

▲ **Figure 4.2 The size range of cells and how we view them.** Most cells are between 1 and 100 μm in diameter (yellow region of chart), and their components are even smaller, as are viruses. Notice that the scale along the left side is logarithmic to accommodate the range of sizes shown. Starting at the top of the scale with 10 m, each reference measurement marks a tenfold decrease in diameter or length. For a complete table of the metric system, see Appendix C.

electron microscope was introduced to biology. Rather than using light, an **electron microscope (EM)** focuses a beam of electrons through a specimen or onto its surface (see Appendix D). Resolution is inversely related to the wavelength of the radiation a microscope uses for imaging, and electron beams have much shorter wavelengths than visible light. Modern electron microscopes can theoretically achieve a resolution of about 0.002 nm, though in practice they usually cannot resolve structures smaller than about 2 nm across. Still, this is a hundredfold improvement over the standard light microscope.

Light Microscopy (LM)

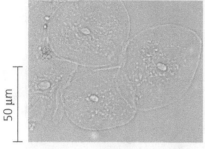

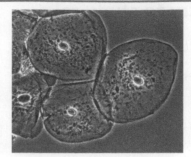

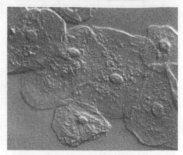

50 μm

Brightfield (unstained specimen). Light passes directly through the specimen. Unless the cell is naturally pigmented or artificially stained, the image has little contrast.

Brightfield (stained specimen). Staining with various dyes enhances contrast. Most staining procedures require that cells be fixed (preserved), thereby killing them.

Phase-contrast. Variations in density within the specimen are amplified to enhance contrast in unstained cells; this is especially useful for examining living, unpigmented cells.

Differential-interference contrast (Nomarski). As in phase-contrast microscopy, optical modifications are used to exaggerate differences in density; the image appears almost 3-D.

The light micrographs above show human cheek epithelial cells; the scale bar pertains to all four micrographs.

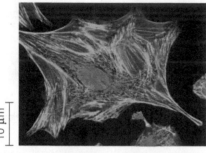

10 μm

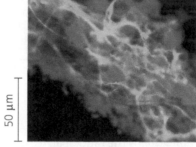

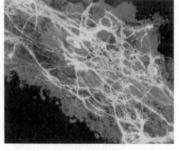

50 μm

Fluorescence. The locations of specific molecules in the cell can be revealed by labeling the molecules with fluorescent dyes or antibodies; some cells have molecules that fluoresce on their own. Fluorescent substances absorb ultraviolet radiation and emit visible light. In this fluorescently labeled uterine cell, nuclear material is blue, organelles called mitochondria are orange, and the cell's "skeleton" is green.

Confocal. The left image is a standard fluorescence micrograph of fluorescently labeled nervous tissue (nerve cells are green, support cells are orange, and regions of overlap are yellow); at right is a confocal image of the same tissue. Using a laser, this "optical sectioning" technique eliminates out-of-focus light from a thick sample, creating a single plane of fluorescence in the image. By capturing sharp images at many different planes, a 3-D reconstruction can be created. The standard image is blurry because out-of-focus light is not excluded.

Electron Microscopy (EM)

Scanning electron microscopy (SEM). Micrographs taken with a scanning electron microscope show a 3-D image of the surface of a specimen. This image shows the surface of a cell from a trachea (windpipe) covered with cilia. (Beating of the cilia helps move inhaled debris upward toward the throat.) The two micrographs shown here have been artificially colorized. Electron micrographs are black and white but are often artificially colorized to highlight particular structures.

Abbreviations used in figure legends throughout this book:
LM = Light Micrograph
SEM = Scanning Electron Micrograph
TEM = Transmission Electron Micrograph

Cilia

Longitudinal section of cilium

Cross section of cilium

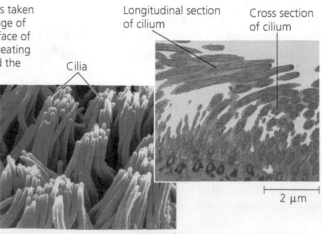

2 μm

2 μm

Transmission electron microscopy (TEM). A transmission electron microscope profiles a thin section of a specimen. Here we see a section through a tracheal cell, revealing its internal structure. In preparing the specimen, some cilia were cut along their lengths, creating longitudinal sections, while other cilia were cut straight across, creating cross sections.

The **transmission electron microscope (TEM)** is used to study the internal structure of cells (see Figure 4.3). The TEM aims an electron beam through a very thin section of a specimen, much as a light microscope aims light through a sample on a slide. For the TEM, the specimen has been stained with atoms of heavy metals, which attach to certain cellular structures, thus enhancing the electron density of some parts of the cell more than others. The electrons passing through the specimen are scattered more in the denser regions, so fewer are transmitted. The image displays the pattern of transmitted electrons. Instead of using glass lenses, the TEM uses electromagnets as lenses to bend the paths of the electrons, ultimately focusing the image onto a monitor for viewing.

The **scanning electron microscope (SEM)** is especially useful for detailed study of the topography of a specimen (see Figure 4.3). Controlled by electromagnetic "lenses" as in a TEM, an electron beam scans the surface of the sample, usually coated with a thin film of gold. The beam excites electrons on the surface, and these secondary electrons are detected by a device that translates the pattern of electrons into an electronic signal to a video screen. The result is an image of the specimen's surface that appears three-dimensional.

Electron microscopes have revealed many organelles and other subcellular structures that were impossible to resolve with the light microscope. But the light microscope offers advantages, especially in studying living cells. A disadvantage of electron microscopy is that the methods used to prepare the specimen kill the cells. Specimen preparation for any type of microscopy can introduce artifacts, structural features seen in micrographs that do not exist in the living cell.

In the past several decades, light microscopy has been revitalized by major technical advances. Labeling individual cellular molecules or structures with fluorescent markers has made it possible to see such structures with increasing detail. In addition, confocal and other newer types of fluorescent light microscopy have produced sharpened images of three-dimensional tissues and cells. Finally, new techniques and labeling molecules have in recent years allowed researchers to break the resolution barrier and distinguish subcellular structures as small as 10–20 nm across. As this "super-resolution microscopy" becomes more widespread, the images we'll see of living cells may well be as awe-inspiring to us as van Leeuwenhoek's were to Robert Hooke 350 years ago.

Microscopes are the most important tools of *cytology*, the study of cell structure. To understand the function of each structure, however, required the integration of cytology and *biochemistry*, the study of the chemical processes of cells.

Cell Fractionation

A useful technique for studying cell structure and function is **cell fractionation**. Broken-up cells are placed in a tube that is spun in a centrifuge. The resulting force causes the largest cell components to settle to the bottom of the tube, forming a pellet. The liquid above the pellet is poured into a new tube and centrifuged at a higher speed for a longer time. This process is repeated several times, resulting in a series of pellets that consist of nuclei, mitochondria (and chloroplasts if the cells are from a photosynthetic organism), pieces of membrane, and ribosomes, the smallest components.

Cell fractionation enables researchers to prepare specific cell components in bulk and identify their functions, a task not usually possible with intact cells. For example, in one of the cell fractions resulting from centrifugation, biochemical tests showed the presence of enzymes involved in cellular respiration, while electron microscopy revealed large numbers of the organelles called mitochondria. Together, these data helped biologists determine that mitochondria are the sites of cellular respiration. Biochemistry and cytology thus complement each other in correlating cell function with structure.

CONCEPT CHECK 4.1

1. How do stains used for light microscopy compare with those used for electron microscopy?
2. **WHAT IF?** Which type of microscope would you use to study (a) the changes in shape of a living white blood cell and (b) the details of surface texture of a hair?

For suggested answers, see Appendix A.

CONCEPT 4.2

Eukaryotic cells have internal membranes that compartmentalize their functions

Cells—the basic structural and functional units of every organism—are of two distinct types: prokaryotic and eukaryotic. Organisms of the domains Bacteria and Archaea consist of prokaryotic cells. Protists, fungi, animals, and plants all consist of eukaryotic cells.

Comparing Prokaryotic and Eukaryotic Cells

All cells share certain basic features: They are all bounded by a selective barrier, called the *plasma membrane*. Inside all cells is a semifluid, jellylike substance called **cytosol**, in which subcellular components are suspended. All cells contain *chromosomes*, which carry genes in the form of DNA. And all cells have *ribosomes*, tiny complexes that make proteins according to instructions from the genes.

A major difference between prokaryotic and eukaryotic cells is the location of their DNA. In a **eukaryotic cell**, most of the DNA is in an organelle called the *nucleus*, which is bounded by a double membrane (see Figure 4.7). In a **prokaryotic cell**, the DNA is concentrated in the **nucleoid**,

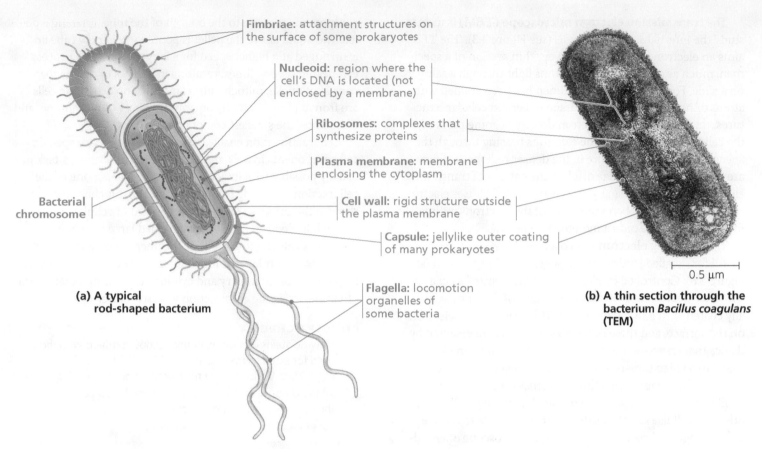

Fimbriae: attachment structures on the surface of some prokaryotes

Nucleoid: region where the cell's DNA is located (not enclosed by a membrane)

Ribosomes: complexes that synthesize proteins

Plasma membrane: membrane enclosing the cytoplasm

Cell wall: rigid structure outside the plasma membrane

Capsule: jellylike outer coating of many prokaryotes

Bacterial chromosome

Flagella: locomotion organelles of some bacteria

(a) A typical rod-shaped bacterium

0.5 μm

(b) A thin section through the bacterium *Bacillus coagulans* (TEM)

▲ **Figure 4.4 A prokaryotic cell.** Lacking a true nucleus and the other membrane-enclosed organelles of the eukaryotic cell, the prokaryotic cell is much simpler in structure. Prokaryotes include bacteria and archaea; the general cell structure of the two domains is essentially the same.

a region that is not bounded by a membrane **(Figure 4.4)**. The word *eukaryotic* means "true nucleus" (from the Greek *eu*, true, and *karyon*, kernel, here referring to the nucleus), and the word *prokaryotic* means "before nucleus" (from the Greek *pro*, before), reflecting the fact that prokaryotic cells evolved before eukaryotic cells.

The interior of either type of cell is called the **cytoplasm**; in eukaryotic cells, this term refers only to the region between the nucleus and the plasma membrane. Within the cytoplasm of a eukaryotic cell, suspended in cytosol, are a variety of organelles of specialized form and function. These membrane-bounded structures are absent in prokaryotic cells. Thus, the presence or absence of a true nucleus is just one aspect of the disparity in structural complexity between the two types of cells.

Eukaryotic cells are generally much larger than prokaryotic cells (see Figure 4.2). Size is a general feature of cell structure that relates to function. The logistics of carrying out cellular metabolism sets limits on cell size. At the lower limit, the smallest cells known are bacteria called mycoplasmas, which have diameters between 0.1 and 1.0 μm. These are perhaps the smallest packages with enough DNA to program metabolism and enough enzymes and other cellular equipment to carry out the activities necessary for a cell to sustain itself and

reproduce. Typical bacteria are 1–5 μm in diameter, about ten times the size of mycoplasmas. Eukaryotic cells are typically 10–100 μm in diameter.

Metabolic requirements also impose theoretical upper limits on the size that is practical for a single cell. At the boundary of every cell, the **plasma membrane** functions as a selective barrier that allows passage of enough oxygen, nutrients, and wastes to service the entire cell **(Figure 4.5)**. For each square micrometer of membrane, only a limited amount of a particular substance can cross per second, so the ratio of surface area to volume is critical. As a cell (or any other object) increases in size, its volume grows proportionately more than its surface area. (Area is proportional to a linear dimension squared, whereas volume is proportional to the linear dimension cubed.) Thus, a smaller object has a greater ratio of surface area to volume **(Figure 4.6)**. The **Scientific Skills Exercise** for this chapter (on p. 74) gives you a chance to calculate the volumes and surface areas of two actual cells—a mature yeast cell and a cell budding from it.

The need for a surface area sufficiently large to accommodate the volume helps explain the microscopic size of most cells and the narrow, elongated shapes of others, such as nerve cells. Larger organisms do not generally have *larger* cells than

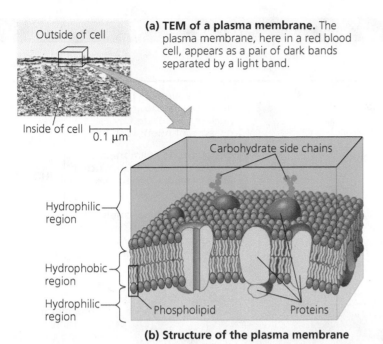

(a) TEM of a plasma membrane. The plasma membrane, here in a red blood cell, appears as a pair of dark bands separated by a light band.

Outside of cell

Inside of cell |———| 0.1 μm

Carbohydrate side chains

Hydrophilic region

Hydrophobic region

Hydrophilic region

Phospholipid

Proteins

(b) Structure of the plasma membrane

© 1996 Pearson Education, Inc.

▲ **Figure 4.5 The plasma membrane.** The plasma membrane and the membranes of organelles consist of a double layer (bilayer) of phospholipids with various proteins attached to or embedded in it. The hydrophobic parts, including phospholipid tails and interior portions of membrane proteins, are found in the interior of the membrane. The hydrophilic parts, including phospholipid heads, exterior portions of proteins, and channels of proteins, are in contact with aqueous solution. Carbohydrate side chains may be attached to proteins or lipids on the outer surface of the plasma membrane.

MAKE CONNECTIONS *Review Figure 3.14 and describe the characteristics of phospholipids that allow them to function as the major material of the plasma membrane.*

smaller organisms—they simply have *more* cells. A sufficiently high ratio of surface area to volume is especially important in cells that exchange a lot of material with their surroundings, such as intestinal cells. Such cells may have many thin projections from their surface called *microvilli*, which increase surface area without an appreciable increase in volume.

The evolutionary relationships between prokaryotic and eukaryotic cells will be discussed later in this chapter. Most of the discussion of cell structure that follows in this chapter applies to eukaryotic cells. (Prokaryotic cells will be described in detail in Chapter 24.)

A Panoramic View of the Eukaryotic Cell

In addition to the plasma membrane at its outer surface, a eukaryotic cell has extensive and elaborately arranged internal membranes. These membranes divide the cell into compartments—the organelles mentioned earlier. The cell's compartments provide different local environments that facilitate specific metabolic functions, so incompatible processes can go on simultaneously inside a single cell. The plasma membrane and organelle membranes also participate

Surface area increases while total volume remains constant

	1	5	1
Total surface area [sum of the surface areas (height × width) of all box sides × number of boxes]	6	150	750
Total volume [height × width × length × number of boxes]	1	125	125
Surface-to-volume ratio [surface area ÷ volume]	6	1.2	6

▲ **Figure 4.6 Geometric relationships between surface area and volume.** In this diagram, cells are represented as boxes. Using arbitrary units of length, we can calculate the cell's surface area (in square units, or units^2), volume (in cubic units, or units^3), and ratio of surface area to volume. A high surface-to-volume ratio facilitates the exchange of materials between a cell and its environment.

directly in the cell's metabolism, because many enzymes are built right into the membranes.

The basic fabric of most biological membranes is a double layer of phospholipids and other lipids. Embedded in this lipid bilayer or attached to its surface are diverse proteins (see Figure 4.5). However, each type of membrane has a unique composition of lipids and proteins suited to that membrane's specific functions. For example, enzymes embedded in the membranes of the organelles called mitochondria function in cellular respiration. (Because membranes are so fundamental to the organization of the cell, Chapter 5 will discuss them in more detail.)

Before continuing with this chapter, examine the eukaryotic cells in **Figure 4.7**. The generalized diagrams of an animal cell and a plant cell introduce the various organelles and highlight the key differences between animal and plant cells. The micrographs at the bottom of the figure give you a glimpse of cells from different types of eukaryotic organisms.

CONCEPT CHECK 4.2

1. After carefully reviewing Figure 4.7, briefly describe the structure and function of the nucleus, the mitochondrion, the chloroplast, and the endoplasmic reticulum.

2. **WHAT IF?** Imagine an elongated cell (such as a nerve cell) that measures 125 × 1 × 1 arbitrary units. Predict how its surface-to-volume ratio would compare with those in Figure 4.6. Then calculate the ratio and check your prediction.

For suggested answers, see Appendix A.

Animal Cell (cutaway view of generalized cell)

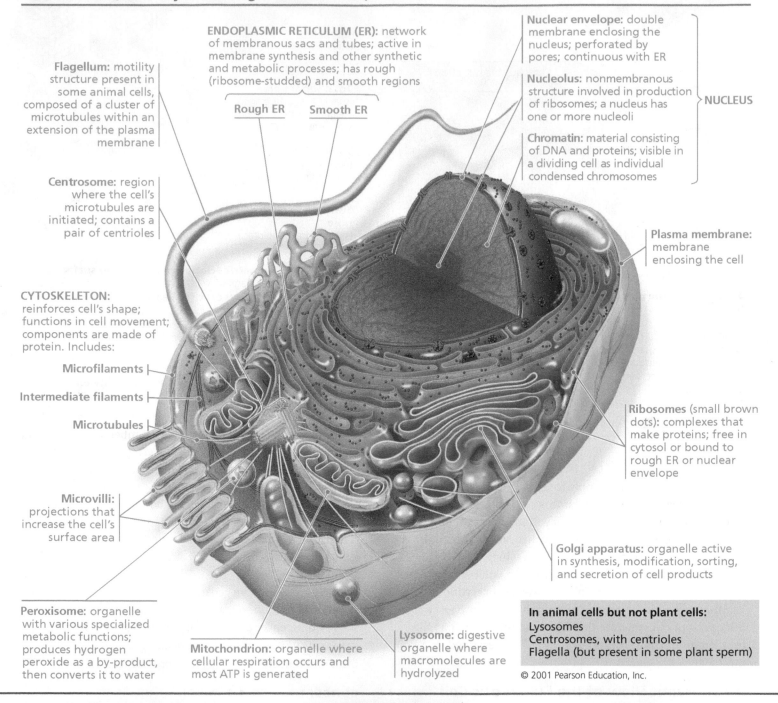

ENDOPLASMIC RETICULUM (ER): network of membranous sacs and tubes; active in membrane synthesis and other synthetic and metabolic processes; has rough (ribosome-studded) and smooth regions

Rough ER   Smooth ER

Flagellum: motility structure present in some animal cells, composed of a cluster of microtubules within an extension of the plasma membrane

Centrosome: region where the cell's microtubules are initiated; contains a pair of centrioles

Nuclear envelope: double membrane enclosing the nucleus; perforated by pores; continuous with ER

Nucleolus: nonmembranous structure involved in production of ribosomes; a nucleus has one or more nucleoli

NUCLEUS

Chromatin: material consisting of DNA and proteins; visible in a dividing cell as individual condensed chromosomes

Plasma membrane: membrane enclosing the cell

CYTOSKELETON: reinforces cell's shape; functions in cell movement; components are made of protein. Includes:

Microfilaments
Intermediate filaments
Microtubules

Ribosomes (small brown dots): complexes that make proteins; free in cytosol or bound to rough ER or nuclear envelope

Microvilli: projections that increase the cell's surface area

Golgi apparatus: organelle active in synthesis, modification, sorting, and secretion of cell products

Peroxisome: organelle with various specialized metabolic functions; produces hydrogen peroxide as a by-product, then converts it to water

Mitochondrion: organelle where cellular respiration occurs and most ATP is generated

Lysosome: digestive organelle where macromolecules are hydrolyzed

In animal cells but not plant cells:
Lysosomes
Centrosomes, with centrioles
Flagella (but present in some plant sperm)

© 2001 Pearson Education, Inc.

Animal Cells

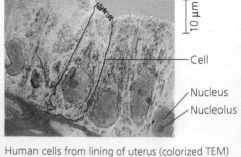

10 µm

Cell

Nucleus

Nucleolus

Human cells from lining of uterus (colorized TEM)

Fungal Cells

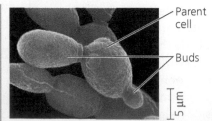

Parent cell

Buds

5 µm

Yeast cells: reproducing by budding (above, colorized SEM) and a single cell (right, colorized TEM)

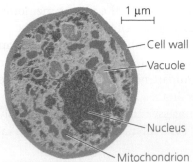

1 µm

Cell wall

Vacuole

Nucleus

Mitochondrion

Plant Cell (cutaway view of generalized cell)

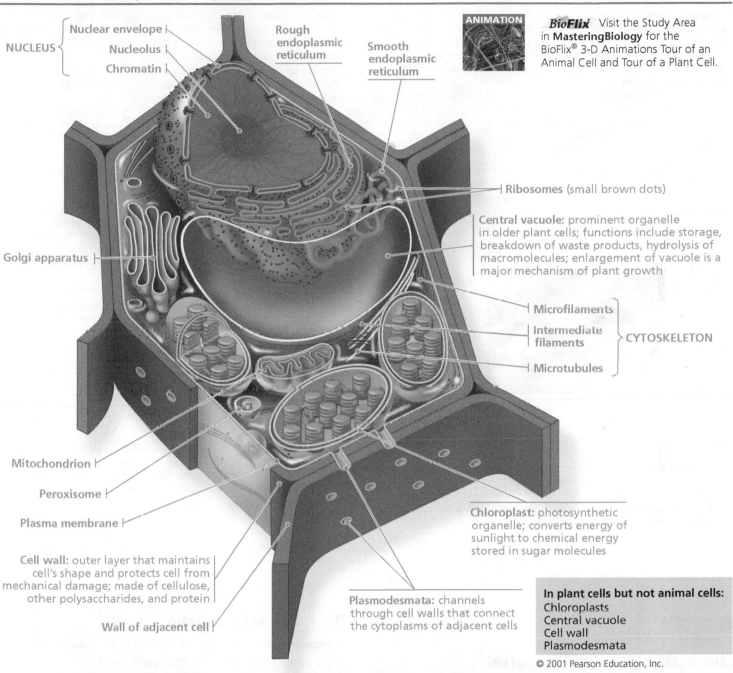

ANIMATION

BioFlix Visit the Study Area in **MasteringBiology** for the BioFlix® 3-D Animations Tour of an Animal Cell and Tour of a Plant Cell.

NUCLEUS
- Nuclear envelope
- Nucleolus
- Chromatin

Rough endoplasmic reticulum

Smooth endoplasmic reticulum

Ribosomes (small brown dots)

Central vacuole: prominent organelle in older plant cells; functions include storage, breakdown of waste products, hydrolysis of macromolecules; enlargement of vacuole is a major mechanism of plant growth

Golgi apparatus

Microfilaments

Intermediate filaments

Microtubules

CYTOSKELETON

Mitochondrion

Peroxisome

Plasma membrane

Chloroplast: photosynthetic organelle; converts energy of sunlight to chemical energy stored in sugar molecules

Cell wall: outer layer that maintains cell's shape and protects cell from mechanical damage; made of cellulose, other polysaccharides, and protein

Wall of adjacent cell

Plasmodesmata: channels through cell walls that connect the cytoplasms of adjacent cells

In plant cells but not animal cells:
Chloroplasts
Central vacuole
Cell wall
Plasmodesmata

© 2001 Pearson Education, Inc.

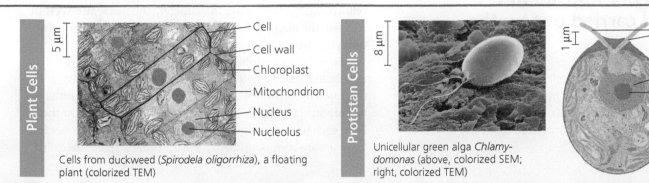

Plant Cells

5 µm

- Cell
- Cell wall
- Chloroplast
- Mitochondrion
- Nucleus
- Nucleolus

Cells from duckweed (*Spirodela oligorrhiza*), a floating plant (colorized TEM)

Protistan Cells

8 µm

Unicellular green alga *Chlamy-domonas* (above, colorized SEM; right, colorized TEM)

1 µm

- Flagella
- Nucleus
- Nucleolus
- Vacuole
- Chloroplast
- Cell wall

Using a Scale Bar to Calculate Volume and Surface Area of a Cell

How Much New Cytoplasm and Plasma Membrane Are Made by a Growing Yeast Cell? The unicellular yeast *Saccharomyces cerevisiae* divides by budding off a small new cell that then grows to full size. During its growth, the new cell synthesizes new cytoplasm, which increases its volume, and new plasma membrane, which increases its surface area.

In this exercise, you will use a scale bar to determine the sizes of a mature parent yeast cell and a cell budding from it. You will then calculate the volume and surface area of each cell. You will use your calculations to determine how much cytoplasm and plasma membrane the new cell needs to synthesize to grow to full size.

How the Experiment Was Done Yeast cells were grown under conditions that promoted division by budding. The cells were then viewed with a differential interference contrast light microscope and photographed.

Data from the Experiment This light micrograph shows a budding yeast cell about to be released from the mature parent cell:

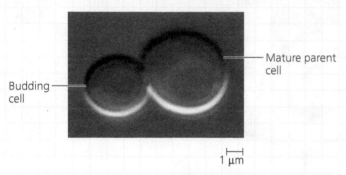

Budding cell

Mature parent cell

1 μm

Interpret the Data

1. Examine the micrograph of the yeast cells. The scale bar under the photo is labeled 1 μm. The scale bar works the same way as a scale on a map, where, for example, 1 inch equals 1 mile. In this case the bar represents a much smaller distance. Using the scale bar as a basic unit, determine the diameter of the mature parent cell and the new cell. Start by measuring the scale bar and then

each cell diameter. The units you use are irrelevant, but working in millimeters is convenient. Divide each diameter by the length of the scale bar and then multiply by the scale bar's label to give you the diameter in micrometers.

2. The shape of a yeast cell can be approximated by a sphere.
(a) Calculate the volume of each cell using the formula for the volume of a sphere:

$$V = \frac{4}{3} \pi r^3$$

Note that π (the Greek letter pi) is a constant with an approximate value of 3.14, *d* stands for diameter, and *r* stands for radius, which is half the diameter. (b) How much new cytoplasm will the new cell have to synthesize as it matures? To determine this, calculate the difference between the volume of the full-size cell and the volume of the new cell.

3. As the new cell grows, its plasma membrane needs to expand to contain the increased volume of the cell. (a) Calculate the surface area of each cell using the formula for the surface area of a sphere:

$$A = 4\pi r^2$$

(b) How much area of new plasma membrane will the new cell have to synthesize as it matures?

4. When the new cell matures, it will be approximately how many times greater in volume and how many times greater in surface area than its current size?

Micrograph from Kelly Tatchell, using yeast cells grown for experiments described in L. Kozubowski et al., Role of the septin ring in the asymmetric localization of proteins at the mother-bud neck in *Saccharomyces cerevisiae, Molecular Biology of the Cell* 16:3455–3466 (2005).

(MB) A version of this Scientific Skills Exercise can be assigned in MasteringBiology.

CONCEPT 4.3

The eukaryotic cell's genetic instructions are housed in the nucleus and carried out by the ribosomes

On the first stop of our detailed tour of the cell, let's look at two cellular components involved in the genetic control of the cell: the nucleus, which houses most of the cell's DNA, and the ribosomes, which use information from the DNA to make proteins.

The Nucleus: Information Central

The **nucleus** contains most of the genes in the eukaryotic cell. (Some genes are located in mitochondria and chloroplasts.) It is generally the most conspicuous organelle in a eukaryotic cell, averaging about 5 μm in diameter. The **nuclear envelope** encloses the nucleus **(Figure 4.8)**, separating its contents from the cytoplasm.

The nuclear envelope is a *double* membrane. The two membranes, each a lipid bilayer with associated proteins, are separated by a space of 20–40 nm. The envelope is perforated by pore structures that are about 100 nm in diameter. At the lip of each pore, the inner and outer membranes of the nuclear envelope are continuous. An intricate protein structure called a

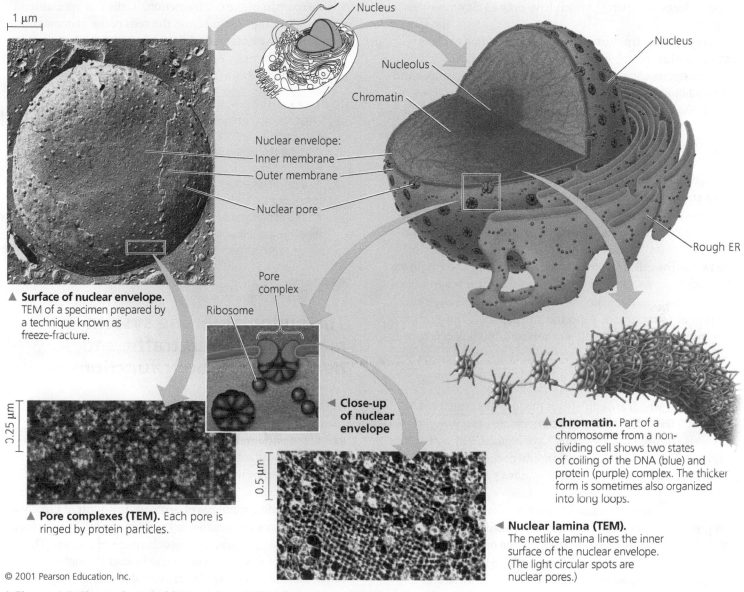

1 μm

Nucleus

Nucleus

Nucleolus

Nucleus

Chromatin

Nuclear envelope:
Inner membrane
Outer membrane

Nuclear pore

Rough ER

▲ **Surface of nuclear envelope.** TEM of a specimen prepared by a technique known as freeze-fracture.

Pore complex

Ribosome

0.25 μm

◄ **Close-up of nuclear envelope**

0.5 μm

▲ **Chromatin.** Part of a chromosome from a non-dividing cell shows two states of coiling of the DNA (blue) and protein (purple) complex. The thicker form is sometimes also organized into long loops.

▲ **Pore complexes (TEM).** Each pore is ringed by protein particles.

◄ **Nuclear lamina (TEM).** The netlike lamina lines the inner surface of the nuclear envelope. (The light circular spots are nuclear pores.)

▲ **Figure 4.8 The nucleus and its envelope.** Within the nucleus are the chromosomes, which appear as a mass of chromatin (DNA and associated proteins), and one or more nucleoli (singular, *nucleolus*), which function in ribosome synthesis. The nuclear envelope, which consists of two membranes separated by a narrow space, is perforated with pores and lined by the nuclear lamina.

MAKE CONNECTIONS *Since the chromosomes contain the genetic material and reside in the nucleus, how does the rest of the cell get access to the information they carry? See Figure 3.25.*

pore complex lines each pore and plays an important role in the cell by regulating the entry and exit of proteins and RNAs, as well as large complexes of macromolecules. Except at the pores, the nuclear side of the envelope is lined by the **nuclear lamina**, a netlike array of protein filaments that maintains the shape of the nucleus by mechanically supporting the nuclear envelope.

Within the nucleus, the DNA is organized into discrete units called **chromosomes**, structures that carry the genetic information. Each chromosome contains one long DNA molecule associated with proteins. Some of the proteins help coil the

DNA molecule of the chromosome, reducing its length and allowing it to fit into the nucleus. The complex of DNA and proteins making up chromosomes is called **chromatin**. When a cell is not dividing, stained chromatin appears as a diffuse mass in micrographs, and the chromosomes cannot be distinguished from one another, even though discrete chromosomes are present. As a cell prepares to divide, however, the chromosomes coil (condense) further, becoming thick enough to be distinguished as separate structures. Each eukaryotic species has a characteristic number of chromosomes. For example, a typical human cell has 46 chromosomes in its nucleus; the exceptions are the sex cells (eggs and sperm), which have only 23 chromosomes in humans.

A prominent structure within the nondividing nucleus is the **nucleolus** (plural, *nucleoli*), which appears through the electron microscope as a mass of densely stained granules and fibers adjoining part of the chromatin. Here a type of RNA called *ribosomal RNA* (*rRNA*) is synthesized from instructions in the DNA. Also in the nucleolus, proteins imported from the cytoplasm are assembled with rRNA into large and small subunits of ribosomes. These subunits then exit the nucleus through the nuclear pores to the cytoplasm, where a large and a small subunit can assemble into a ribosome. Sometimes there are two or more nucleoli.

The nucleus directs protein synthesis by synthesizing messenger RNA (mRNA) according to instructions provided by the DNA. The mRNA is then transported to the cytoplasm via the nuclear pores. Once an mRNA molecule reaches the cytoplasm, ribosomes translate the mRNA's genetic message into the primary structure of a specific polypeptide. (This process of transcribing and translating genetic information is outlined in Figure 3.25 and described in detail in Chapter 14.)

Ribosomes: Protein Factories

Ribosomes, which are complexes made of ribosomal RNA and protein, are the cellular components that carry out protein synthesis **(Figure 4.9)**. Cells that have high rates of protein synthesis have particularly large numbers of ribosomes. Not surprisingly, cells active in protein synthesis also have prominent nucleoli.

Ribosomes build proteins in two cytoplasmic locales. At any given time, *free ribosomes* are suspended in the cytosol, while *bound ribosomes* are attached to the outside of the endoplasmic reticulum or nuclear envelope (see Figure 4.9). Bound and free ribosomes are structurally identical, and ribosomes can alternate between the two roles. Most of the proteins made on free ribosomes function within the cytosol; examples are enzymes that catalyze the first steps of sugar breakdown. Bound ribosomes generally make proteins that are destined for insertion into membranes, for packaging within certain organelles such as lysosomes (see Figure 4.7), or for export from the cell (secretion). Cells that specialize in protein secretion—for instance, the cells of the pancreas that secrete digestive enzymes—frequently have a high proportion of bound ribosomes. (You will learn more about ribosome structure and function in Chapter 14.)

CONCEPT CHECK 4.3

1. What role do ribosomes play in carrying out genetic instructions?
2. Describe the molecular composition of nucleoli, and explain their function.
3. As a cell begins the process of dividing, its chromosomes become shorter, thicker, and individually visible in an LM. Explain what is happening at the molecular level.

For suggested answers, see Appendix A.

CONCEPT 4.4

The endomembrane system regulates protein traffic and performs metabolic functions in the cell

Many of the different membranes of the eukaryotic cell are part of the **endomembrane system**, which includes the nuclear envelope, the endoplasmic reticulum, the Golgi apparatus, lysosomes, various kinds of vesicles and vacuoles, and the plasma membrane. This system carries out a variety of tasks in the cell, including synthesis of proteins, transport of proteins into membranes and organelles or out of the cell, metabolism and movement of lipids, and detoxification of poisons. The membranes of this system are related either through direct physical continuity or by the transfer of membrane segments as tiny **vesicles** (sacs made of membrane). Despite these relationships, the various membranes are not identical in structure and function. Moreover, the thickness, molecular composition, and types of chemical reactions carried out in a given membrane are not fixed, but may be modified several times during

▶ **Figure 4.9 Ribosomes.** This electron micrograph of part of a pancreas cell shows many ribosomes, both free (in the cytosol) and bound (to the endoplasmic reticulum). The simplified diagram of a ribosome shows its two subunits.

DRAW IT *After you have read the section on ribosomes, circle a ribosome in the micrograph that might be making a protein that will be secreted.*

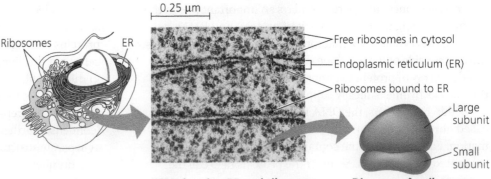

0.25 μm

Ribosomes   ER

Free ribosomes in cytosol
Endoplasmic reticulum (ER)
Ribosomes bound to ER
Large subunit
Small subunit

TEM showing ER and ribosomes   **Diagram of a ribosome**

the membrane's life. Having already discussed the nuclear envelope, we will now focus on the endoplasmic reticulum and the other endomembranes to which the endoplasmic reticulum gives rise.

The Endoplasmic Reticulum: Biosynthetic Factory

The **endoplasmic reticulum (ER)** is such an extensive network of membranes that it accounts for more than half the total membrane in many eukaryotic cells. (The word *endoplasmic* means "within the cytoplasm," and *reticulum* is Latin for "little net.") The ER consists of a network of membranous tubules and sacs called cisternae (from the Latin *cisterna*, a reservoir for a liquid). The ER membrane separates the internal compartment of the ER, called the ER lumen (cavity) or cisternal space, from the cytosol. And because the ER membrane is continuous with the nuclear envelope, the space between the two membranes of the envelope is continuous with the lumen of the ER **(Figure 4.10)**.

There are two distinct, though connected, regions of the ER that differ in structure and function: smooth ER and rough ER. **Smooth ER** is so named because its outer surface lacks ribosomes. **Rough ER** is studded with ribosomes on the outer surface of the membrane and thus appears rough through the electron microscope. As already mentioned, ribosomes are also attached to the cytoplasmic side of the nuclear envelope's outer membrane, which is continuous with rough ER.

Functions of Smooth ER

The smooth ER functions in diverse metabolic processes, which vary with cell type. These processes include synthesis of lipids, metabolism of carbohydrates, detoxification of drugs and poisons, and storage of calcium ions.

Enzymes of the smooth ER are important in the synthesis of lipids, including oils, phospholipids, and steroids. Among the steroids produced by the smooth ER in animal cells are the sex hormones of vertebrates and the various steroid hormones secreted by the adrenal glands. The cells that synthesize and secrete these hormones—in the testes and ovaries, for example—are rich in smooth ER, a structural feature that fits the function of these cells.

Other enzymes of the smooth ER help detoxify drugs and poisons, especially in liver cells. Detoxification usually involves adding hydroxyl groups to drug molecules, making them more soluble and easier to flush from the body. The sedative phenobarbital and other barbiturates are examples of drugs metabolized in this manner by smooth ER in liver cells. In fact, barbiturates, alcohol, and many other drugs induce the proliferation of smooth ER and its associated detoxification enzymes, thus increasing the rate of detoxification. This, in turn, increases tolerance to the drugs, meaning that higher doses are required to achieve a particular effect, such as sedation. Also, because some of the detoxification enzymes have relatively

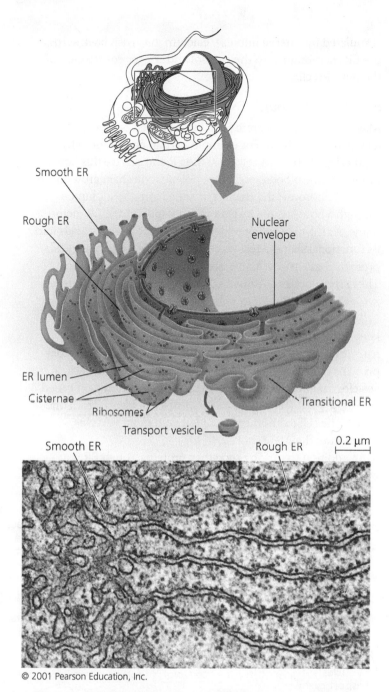

Smooth ER

Rough ER

Nuclear envelope

ER lumen

Cisternae

Ribosomes

Transitional ER

Transport vesicle

Smooth ER

Rough ER

0.2 μm

© 2001 Pearson Education, Inc.

▲ **Figure 4.10 Endoplasmic reticulum (ER).** Λ membranous system of interconnected tubules and flattened sacs called cisternae, the ER is also continuous with the nuclear envelope. (The drawing is a cutaway view.) The membrane of the ER encloses a continuous compartment called the ER lumen (or cisternal space). Rough ER, which is studded on its outer surface with ribosomes, can be distinguished from smooth ER in the electron micrograph (TEM). Transport vesicles bud off from a region of the rough ER called transitional ER and travel to the Golgi apparatus and other destinations.

broad action, the proliferation of smooth ER in response to one drug can increase tolerance to other drugs as well. Barbiturate abuse, for example, can decrease the effectiveness of certain antibiotics and other useful drugs.

The smooth ER also stores calcium ions. In muscle cells, for example, the smooth ER membrane pumps calcium ions from the cytosol into the ER lumen. When a muscle cell is stimulated by a nerve impulse, calcium ions rush back across the ER membrane into the cytosol and trigger contraction of the muscle cell.

Functions of Rough ER

Many types of cells secrete proteins produced by ribosomes attached to rough ER. For example, certain pancreatic cells synthesize the protein insulin in the ER and secrete this hormone into the bloodstream. As a polypeptide chain grows from a bound ribosome, the chain is threaded into the ER lumen through a pore formed by a protein complex in the ER membrane. As the new polypeptide enters the ER lumen, it folds into its functional shape. Most secretory proteins are **glycoproteins**, proteins that have carbohydrates covalently bonded to them. The carbohydrates are attached to the proteins in the ER by enzymes built into the ER membrane.

After secretory proteins are formed, the ER membrane keeps them separate from proteins that are produced by free ribosomes and that will remain in the cytosol. Secretory proteins depart from the ER wrapped in the membranes of vesicles that bud like bubbles from a specialized region called transitional ER (see Figure 4.10). Vesicles in transit from one part of the cell to another are called **transport vesicles**; we will discuss their fate shortly.

In addition to making secretory proteins, rough ER is a membrane factory for the cell; it grows in place by adding membrane proteins and phospholipids to its own membrane. As polypeptides destined to be membrane proteins grow from the ribosomes, they are inserted into the ER membrane itself and anchored there by their hydrophobic portions. Like the smooth ER, the rough ER also makes membrane phospholipids; enzymes built into the ER membrane assemble phospholipids from precursors in the cytosol. The ER membrane expands, and portions of it are transferred in the form of transport vesicles to other components of the endomembrane system.

The Golgi Apparatus: Shipping and Receiving Center

After leaving the ER, many transport vesicles travel to the **Golgi apparatus**. We can think of the Golgi primarily as a warehouse for receiving, sorting, and shipping, although some manufacturing also occurs there. In the Golgi, products of the ER, such as proteins, are modified and stored and then sent to other destinations. Not surprisingly, the Golgi apparatus is especially extensive in cells specialized for secretion.

The Golgi apparatus consists of flattened membranous sacs—cisternae—looking like a stack of pita bread (**Figure 4.11**). A cell may have many, even hundreds, of these stacks. The membrane of each cisterna in a stack separates its internal space from the cytosol. Vesicles concentrated in the vicinity of the Golgi apparatus are engaged in the transfer of material between parts of the Golgi and other structures.

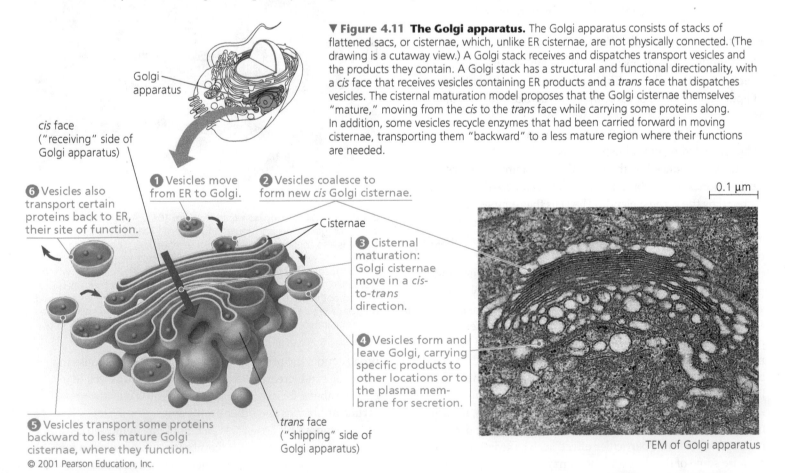

▼ **Figure 4.11 The Golgi apparatus.** The Golgi apparatus consists of stacks of flattened sacs, or cisternae, which, unlike ER cisternae, are not physically connected. (The drawing is a cutaway view.) A Golgi stack receives and dispatches transport vesicles and the products they contain. A Golgi stack has a structural and functional directionality, with a *cis* face that receives vesicles containing ER products and a *trans* face that dispatches vesicles. The cisternal maturation model proposes that the Golgi cisternae themselves "mature," moving from the *cis* to the *trans* face while carrying some proteins along. In addition, some vesicles recycle enzymes that had been carried forward in moving cisternae, transporting them "backward" to a less mature region where their functions are needed.

Golgi apparatus

cis face ("receiving" side of Golgi apparatus)

6 Vesicles also transport certain proteins back to ER, their site of function.

1 Vesicles move from ER to Golgi.

2 Vesicles coalesce to form new *cis* Golgi cisternae.

Cisternae

3 Cisternal maturation: Golgi cisternae move in a *cis*-to-*trans* direction.

4 Vesicles form and leave Golgi, carrying specific products to other locations or to the plasma membrane for secretion.

5 Vesicles transport some proteins backward to less mature Golgi cisternae, where they function.

trans face ("shipping" side of Golgi apparatus)

© 2001 Pearson Education, Inc.

0.1 μm

TEM of Golgi apparatus

A Golgi stack has a distinct structural directionality, with the membranes of cisternae on opposite sides of the stack differing in thickness and molecular composition. The two sides of a Golgi stack are referred to as the *cis* face and the *trans* face; these act, respectively, as the receiving and shipping departments of the Golgi apparatus. The *cis* face is usually located near the ER. Transport vesicles move material from the ER to the Golgi apparatus. A vesicle that buds from the ER can add its membrane and the contents of its lumen to the *cis* face by fusing with a Golgi membrane. The *trans* face gives rise to vesicles that pinch off and travel to other sites.

Products of the endoplasmic reticulum are usually modified during their transit from the *cis* region to the *trans* region of the Golgi apparatus. For example, glycoproteins formed in the ER have their carbohydrates modified, first in the ER itself, then as they pass through the Golgi. The Golgi removes some sugar monomers and substitutes others, producing a large variety of carbohydrates. Membrane phospholipids may also be altered in the Golgi.

In addition to its finishing work, the Golgi apparatus also manufactures some macromolecules. Many polysaccharides secreted by cells are Golgi products. For example, pectins and certain other noncellulose polysaccharides are made in the Golgi of plant cells and then incorporated along with cellulose into their cell walls. Like secretory proteins, nonprotein Golgi products that will be secreted depart from the *trans* face of the Golgi inside transport vesicles that eventually fuse with the plasma membrane.

The Golgi manufactures and refines its products in stages, with different cisternae containing unique teams of enzymes. Until recently, biologists viewed the Golgi as a static structure, with products in various stages of processing transferred from one cisterna to the next by vesicles. While this may occur, recent research has given rise to a new model of the Golgi as a more dynamic structure. According to the *cisternal maturation model*, the cisternae of the Golgi actually progress forward from the *cis* to the *trans* face, carrying and modifying their cargo as they move. Figure 4.11 shows the details of this model.

Before a Golgi stack dispatches its products by budding vesicles from the *trans* face, it sorts these products and targets them for various parts of the cell. Molecular identification tags, such as phosphate groups added to the Golgi products, aid in sorting by acting like ZIP codes on mailing labels. Finally, transport vesicles budded from the Golgi may have external molecules on their membranes that recognize "docking sites" on the surface of specific organelles or on the plasma membrane, thus targeting the vesicles appropriately.

Lysosomes: Digestive Compartments

A **lysosome** is a membranous sac of hydrolytic enzymes that an animal cell uses to digest (hydrolyze) macromolecules. Lysosomal enzymes work best in the acidic environment found in lysosomes. If a lysosome breaks open or leaks its contents,

the released enzymes are not very active because the cytosol has a neutral pH. However, excessive leakage from a large number of lysosomes can destroy a cell by self-digestion.

Hydrolytic enzymes and lysosomal membrane are made by rough ER and then transferred to the Golgi apparatus for further processing. At least some lysosomes probably arise by budding from the *trans* face of the Golgi apparatus (see Figure 4.11). How are the proteins of the inner surface of the lysosomal membrane and the digestive enzymes themselves spared from destruction? Apparently, the three-dimensional shapes of these lysosomal proteins protect vulnerable bonds from enzymatic attack.

Lysosomes carry out intracellular digestion in a variety of circumstances. Amoebas and many other protists eat by engulfing smaller organisms or food particles, a process called **phagocytosis** (from the Greek *phagein*, to eat, and *kytos*, vessel, referring here to the cell). The *food vacuole* formed in this way then fuses with a lysosome, whose enzymes digest the food (**Figure 4.12**, bottom). Digestion products, including simple sugars, amino acids, and other monomers, pass into the

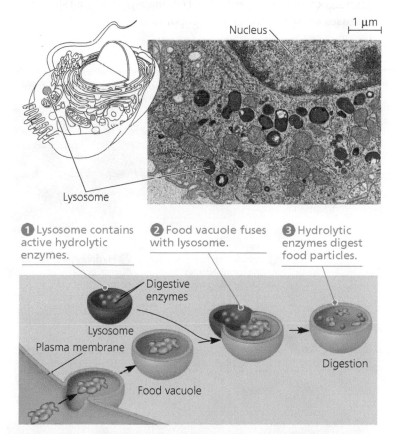

▲ Figure 4.12 Lysosomes: Phagocytosis. In phagocytosis, lysosomes digest (hydrolyze) materials taken into the cell. *Top*: In this macrophage (a type of white blood cell) from a rat, the lysosomes are very dark because of a stain that reacts with one of the products of digestion inside the lysosome (TEM). Macrophages ingest bacteria and viruses and destroy them using lysosomes. *Bottom*: This diagram shows a lysosome fusing with a food vacuole during the process of phagocytosis by a protist.

cytosol and become nutrients for the cell. Some human cells also carry out phagocytosis. Among them are macrophages, a type of white blood cell that helps defend the body by engulfing and destroying bacteria and other invaders (see Figure 4.12, top, and Figure 4.28).

Lysosomes also use their hydrolytic enzymes to recycle the cell's own organic material, a process called *autophagy*. During autophagy, a damaged organelle or small amount of cytosol becomes surrounded by a double membrane, and a lysosome fuses with the outer membrane of this vesicle **(Figure 4.13)**. The lysosomal enzymes dismantle the enclosed material, and the resulting small organic compounds are released to the cytosol for reuse. With the help of lysosomes, the cell continually renews itself. A human liver cell, for example, recycles half of its macromolecules each week.

The cells of people with inherited lysosomal storage diseases lack a functioning hydrolytic enzyme normally present in lysosomes. The lysosomes become engorged with indigestible material, which begins to interfere with other cellular activities. In Tay-Sachs disease, for example, a lipid-digesting enzyme is missing or inactive, and the brain becomes impaired by an accumulation of lipids in the cells. Fortunately, lysosomal storage diseases are rare in the general population.

Vacuoles: Diverse Maintenance Compartments

Vacuoles are large vesicles derived from the endoplasmic reticulum and Golgi apparatus. Thus, vacuoles are an integral part of a cell's endomembrane system. Like all cellular membranes, the vacuolar membrane is selective in transporting solutes; as a result, the solution inside a vacuole differs in composition from the cytosol.

Vacuoles perform a variety of functions in different kinds of cells. **Food vacuoles**, formed by phagocytosis, have already been mentioned (see Figure 4.12). Many freshwater protists have **contractile vacuoles** that pump excess water out of the cell, thereby maintaining a suitable concentration of ions and molecules inside the cell (see Figure 5.12). In plants and fungi, certain vacuoles carry out enzymatic hydrolysis, a function shared by lysosomes in animal cells. (In fact, some biologists consider these hydrolytic vacuoles to be a type of lysosome.) In plants, small vacuoles can hold reserves of important organic compounds, such as the proteins stockpiled in the storage cells in seeds. Vacuoles may also help protect the plant against herbivores by storing compounds that are poisonous or unpalatable to animals. Some plant vacuoles contain pigments, such as the red and blue pigments of petals that help attract pollinating insects to flowers.

Mature plant cells generally contain a large **central vacuole (Figure 4.14)**, which develops by the coalescence of smaller vacuoles. The solution inside the central vacuole, called cell sap, is the plant cell's main repository of inorganic ions, including potassium and chloride. The central vacuole plays a major role in the growth of plant cells, which enlarge as the vacuole absorbs water, enabling the cell to become larger with a minimal investment in new cytoplasm. The cytosol often occupies only a thin layer between the central vacuole and the plasma

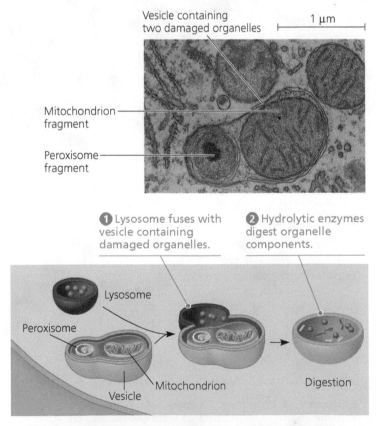

① Lysosome fuses with vesicle containing damaged organelles.

② Hydrolytic enzymes digest organelle components.

Vesicle containing two damaged organelles

1 μm

Mitochondrion fragment

Peroxisome fragment

Lysosome

Peroxisome

Vesicle

Mitochondrion

Digestion

▲ **Figure 4.13 Lysosomes: Autophagy.** In autophagy, lysosomes recycle intracellular materials. *Top*: In the cytoplasm of this rat liver cell is a vesicle containing two disabled organelles; the vesicle will fuse with a lysosome in the process of autophagy (TEM). *Bottom*: This diagram shows fusion of such a vesicle with a lysosome and the subsequent digestion of the damaged organelles.

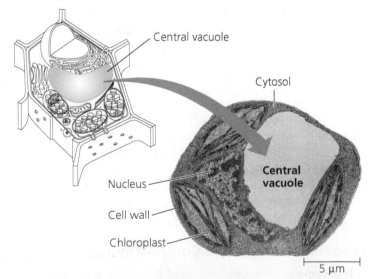

Central vacuole

Cytosol

Nucleus

Central vacuole

Cell wall

Chloroplast

5 μm

▲ **Figure 4.14 The plant cell vacuole.** The central vacuole is usually the largest compartment in a plant cell; the rest of the cytoplasm is often confined to a narrow zone between the vacuolar membrane and the plasma membrane (TEM).

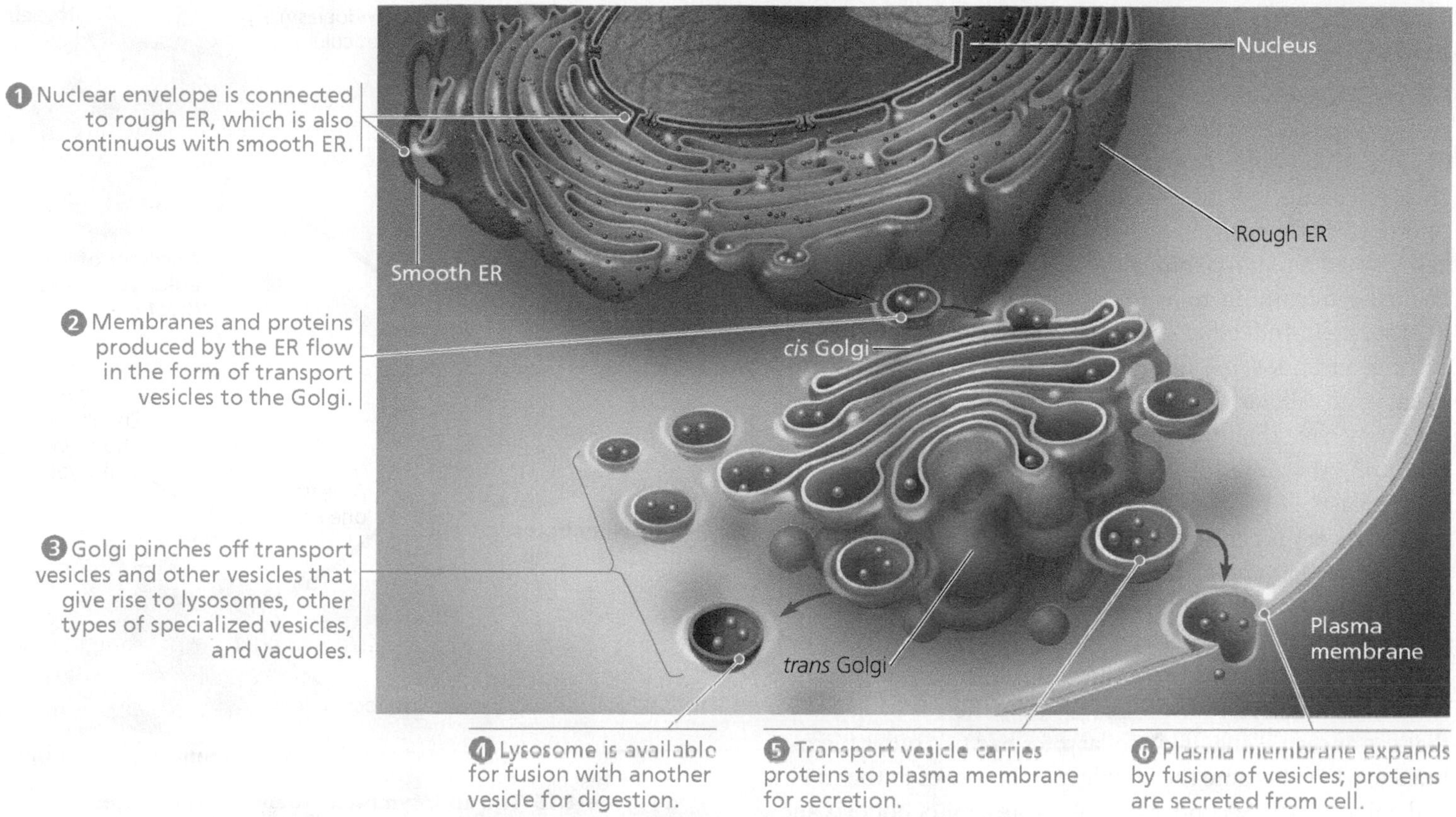

① Nuclear envelope is connected to rough ER, which is also continuous with smooth ER.

Smooth ER

② Membranes and proteins produced by the ER flow in the form of transport vesicles to the Golgi.

③ Golgi pinches off transport vesicles and other vesicles that give rise to lysosomes, other types of specialized vesicles, and vacuoles.

Nucleus

Rough ER

cis Golgi

trans Golgi

Plasma membrane

④ Lysosome is available for fusion with another vesicle for digestion.

⑤ Transport vesicle carries proteins to plasma membrane for secretion.

⑥ Plasma membrane expands by fusion of vesicles; proteins are secreted from cell.

▲ **Figure 4.15 Review: relationships among organelles of the endomembrane system.** The red arrows show some of the migration pathways for membranes and the materials they enclose.

membrane, so the ratio of plasma membrane surface to cytosolic volume is sufficient, even for a large plant cell.

The Endomembrane System: *A Review*

Figure 4.15 reviews the endomembrane system, showing the flow of membrane lipids and proteins through the various organelles. As the membrane moves from the ER to the Golgi and then elsewhere, its molecular composition and metabolic functions are modified, along with those of its contents. The endomembrane system is a complex and dynamic player in the cell's compartmental organization.

We'll continue our tour of the cell with some organelles that are not closely related to the endomembrane system but play crucial roles in the energy transformations carried out by cells.

CONCEPT CHECK 4.4

1. Describe the structural and functional distinctions between rough and smooth ER.
2. Describe how transport vesicles integrate the endomembrane system.
3. **WHAT IF?** Imagine a protein that functions in the ER but requires modification in the Golgi apparatus before it can achieve that function. Describe the protein's path through the cell, starting with the mRNA molecule that specifies the protein.

For suggested answers, see Appendix A.

CONCEPT 4.5

Mitochondria and chloroplasts change energy from one form to another

Organisms transform the energy they acquire from their surroundings. In eukaryotic cells, mitochondria and chloroplasts are the organelles that convert energy to forms that cells can use for work. **Mitochondria** (singular, *mitochondrion*) are the sites of cellular respiration, the metabolic process that uses oxygen to generate ATP by extracting energy from sugars, fats, and other fuels. **Chloroplasts**, found in plants and algae, are the sites of photosynthesis. These organelles convert solar energy to chemical energy by absorbing sunlight and using it to drive the synthesis of organic compounds such as sugars from carbon dioxide and water.

In addition to having related functions, mitochondria and chloroplasts share similar evolutionary origins, which we'll discuss briefly before describing their structures. In this section, we will also consider the peroxisome, an oxidative organelle. The evolutionary origin of the peroxisome, as well as its relation to other organelles, is still under debate.

The Evolutionary Origins of Mitochondria and Chloroplasts

EVOLUTION Mitochondria and chloroplasts display similarities with bacteria that led to the **endosymbiont theory**, illustrated in **Figure 4.16**. This theory states that an early ancestor of eukaryotic cells engulfed an oxygen-using nonphotosynthetic prokaryotic cell. Eventually, the engulfed cell formed a relationship with the host cell in which it was enclosed, becoming an *endosymbiont* (a cell living within another cell). Indeed, over the course of evolution, the host cell and its endosymbiont merged into a single organism, a eukaryotic cell with a mitochondrion. At least one of these cells may have then taken up a photosynthetic prokaryote, becoming the ancestor of eukaryotic cells that contain chloroplasts.

This theory is consistent with many structural features of mitochondria and chloroplasts. First, rather than being bounded by a single membrane like organelles of the endomembrane system, mitochondria and typical chloroplasts have two membranes surrounding them. (Chloroplasts also have an internal system of membranous sacs.) There is evidence that the ancestral engulfed prokaryotes had two outer membranes, which became the double membranes of mitochondria and chloroplasts. Second, like prokaryotes, mitochondria and chloroplasts contain ribosomes, as well as multiple circular DNA molecules attached to their inner membranes. The DNA in these organelles programs the synthesis of some of their own proteins, which are made on the ribosomes inside the organelles. Third, also consistent with their probable evolutionary origins as cells, mitochondria and chloroplasts are autonomous (somewhat independent) organelles that grow and reproduce within the cell. (We will discuss the endosymbiont theory in more detail in Chapter 25.)

Next we focus on the structure of mitochondria and chloroplasts, while providing an overview of their functions.

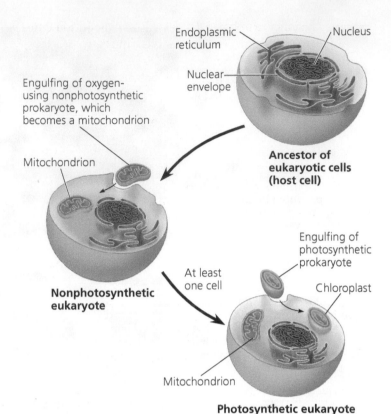

▲ Figure 4.16 The endosymbiont theory of the origin of mitochondria and chloroplasts in eukaryotic cells. According to this theory, the proposed ancestors of mitochondria were oxygen-using nonphotosynthetic prokaryotes, while the proposed ancestors of chloroplasts were photosynthetic prokaryotes. The large arrows represent change over evolutionary time; the small arrows inside the cells show the process of the endosymbiont becoming an organelle.

Mitochondria: Chemical Energy Conversion

Each of the two membranes enclosing a mitochondrion is a phospholipid bilayer with a unique collection of embedded proteins **(Figure 4.17)**. The outer membrane is smooth, but the inner membrane is convoluted, with infoldings called

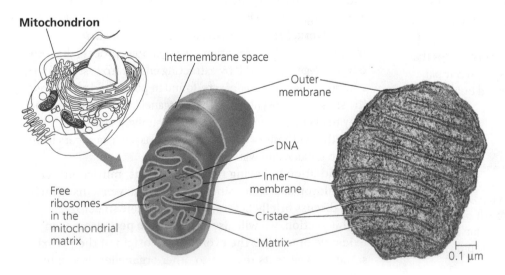

◀ Figure 4.17 The mitochondrion, site of cellular respiration. The inner and outer membranes of the mitochondrion are evident in this drawing and electron micrograph (TEM). The cristae are infoldings of the inner membrane, which increase its surface area. The cutaway drawing shows the two compartments bounded by the membranes: the intermembrane space and the mitochondrial matrix. Many respiratory enzymes are found in the inner membrane and the matrix. Free ribosomes are also present in the matrix. The circular DNA molecules are attached to the inner mitochondrial membrane.

cristae. The inner membrane divides the mitochondrion into two internal compartments. The first is the intermembrane space, the narrow region between the inner and outer membranes. The second compartment, the **mitochondrial matrix**, is enclosed by the inner membrane. The matrix contains many different enzymes as well as the mitochondrial DNA and ribosomes. Enzymes in the matrix catalyze some of the steps of cellular respiration. Other proteins that function in respiration, including the enzyme that makes ATP, are built into the inner membrane. As highly folded surfaces, the cristae give the inner mitochondrial membrane a large surface area, thus enhancing the productivity of cellular respiration. This is another example of structure fitting function. (Chapter 7 discusses cellular respiration in detail.)

Mitochondria are found in nearly all eukaryotic cells, including those of plants, animals, fungi, and most protists. Some cells have a single large mitochondrion, but more often a cell has hundreds or even thousands of mitochondria; the number correlates with the cell's level of metabolic activity. For example, cells that move or contract have proportionally more mitochondria per volume than less active cells.

Mitochondria are generally in the range of 1–10 μm long. Time-lapse films of living cells reveal mitochondria moving around, changing their shapes, and fusing or dividing in two, unlike the static structures seen in electron micrographs of dead cells.

Chloroplasts: Capture of Light Energy

Chloroplasts contain the green pigment chlorophyll, along with enzymes and other molecules that function in the photosynthetic production of sugar. These lens-shaped organelles, about 3–6 μm in length, are found in leaves and other green organs of plants and in algae (**Figure 4.18**).

The contents of a chloroplast are partitioned from the cytosol by an envelope consisting of two membranes separated by a very narrow intermembrane space. Inside the chloroplast is another membranous system in the form of flattened, interconnected sacs called **thylakoids**. In some regions, thylakoids are stacked like poker chips; each stack is called a **granum** (plural, *grana*). The fluid outside the thylakoids is the **stroma**, which contains the chloroplast DNA and ribosomes as well as many enzymes. The membranes of the chloroplast divide the chloroplast space into three compartments: the intermembrane space, the stroma, and the thylakoid space. This compartmental organization enables the chloroplast to convert light energy to chemical energy during photosynthesis. (You will learn more about photosynthesis in Chapter 8.)

As with mitochondria, the static and rigid appearance of chloroplasts in micrographs or schematic diagrams is not true to their dynamic behavior in the living cell. Their shape is changeable, and they grow and occasionally pinch in two, reproducing themselves. They are mobile and, along with mitochondria and other organelles, move around the cell along tracks of the cytoskeleton, a structural network we will consider later in this chapter.

The chloroplast is a specialized member of a family of closely related plant organelles called **plastids**. One type of plastid, the *amyloplast*, is a colorless organelle that stores starch (amylose), particularly in roots and tubers. Another is the *chromoplast*, which has pigments that give fruits and flowers their orange and yellow hues.

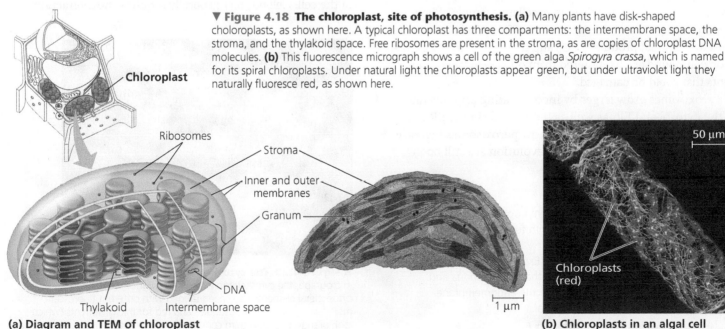

▼ **Figure 4.18 The chloroplast, site of photosynthesis. (a)** Many plants have disk-shaped choloroplasts, as shown here. A typical chloroplast has three compartments: the intermembrane space, the stroma, and the thylakoid space. Free ribosomes are present in the stroma, as are copies of chloroplast DNA molecules. **(b)** This fluorescence micrograph shows a cell of the green alga *Spirogyra crassa*, which is named for its spiral chloroplasts. Under natural light the chloroplasts appear green, but under ultraviolet light they naturally fluoresce red, as shown here.

Chloroplast

Ribosomes

Stroma

Inner and outer membranes

Granum

DNA

Thylakoid   Intermembrane space

(a) Diagram and TEM of chloroplast

1 μm

50 μm

Chloroplasts (red)

(b) Chloroplasts in an algal cell

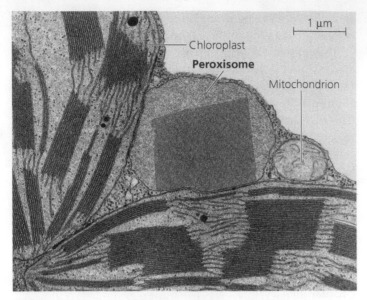

▲ **Figure 4.19 A peroxisome.** Peroxisomes are roughly spherical and often have a granular or crystalline core that is thought to be a dense collection of enzyme molecules. This peroxisome is in a leaf cell (TEM). Notice its proximity to two chloroplasts and a mitochondrion. These organelles cooperate with peroxisomes in certain metabolic functions.

Peroxisomes: Oxidation

The **peroxisome** is a specialized metabolic compartment bounded by a single membrane **(Figure 4.19)**. Peroxisomes contain enzymes that remove hydrogen atoms from certain molecules and transfer them to oxygen (O_2), producing hydrogen peroxide (H_2O_2). These reactions have many different functions. For example, peroxisomes in the liver detoxify alcohol and other harmful compounds by transferring hydrogen from the poisons to oxygen. The H_2O_2 formed by peroxisomes is itself toxic, but the organelle also contains an enzyme that converts H_2O_2 to water. This is an excellent example of how the cell's compartmental structure is crucial to its functions: The enzymes that produce H_2O_2 and those that dispose of this toxic compound are sequestered from other cellular components that could be damaged.

Peroxisomes grow larger by incorporating proteins made in the cytosol and ER, as well as lipids made in the ER and within the peroxisome itself. But how peroxisomes increase in number and how they arose in evolution are still open questions.

CONCEPT CHECK 4.5

1. Describe two characteristics that chloroplasts and mitochondria have in common. Consider both function and membrane structure.
2. Do plant cells have mitochondria? Explain.
3. **WHAT IF?** A classmate proposes that mitochondria and chloroplasts should be classified in the endomembrane system. Argue against the proposal.

For suggested answers, see Appendix A.

The cytoskeleton is a network of fibers that organizes structures and activities in the cell

In the early days of electron microscopy, biologists thought that the organelles of a eukaryotic cell floated freely in the cytosol. But improvements in both light microscopy and electron microscopy have revealed the **cytoskeleton**, a network of fibers extending throughout the cytoplasm **(Figure 4.20)**. The cytoskeleton plays a major role in organizing the structures and activities of the cell.

Roles of the Cytoskeleton: Support and Motility

The most obvious function of the cytoskeleton is to give mechanical support to the cell and maintain its shape. This is especially important for animal cells, which lack walls. The remarkable strength and resilience of the cytoskeleton as a whole is based on its architecture. Like a dome tent, the cytoskeleton is stabilized by a balance between opposing forces exerted by its elements. And just as the skeleton of an animal helps fix the positions of other body parts, the cytoskeleton provides anchorage for many organelles and even cytosolic enzyme molecules. The cytoskeleton is more dynamic than an animal skeleton, however. It can be quickly dismantled in one part of the cell and reassembled in a new location, changing the shape of the cell.

Several types of cell motility (movement) also involve the cytoskeleton. The term *cell motility* encompasses both changes in cell location and more limited movements of parts of the cell. Cell motility generally requires the interaction of

▲ **Figure 4.20 The cytoskeleton.** As shown in this fluorescence micrograph, the cytoskeleton extends throughout the cell. The cytoskeletal elements have been tagged with different fluorescent molecules: green for microtubules and red for microfilaments (which look orangish here). A third component of the cytoskeleton, intermediate filaments, is not evident. (The blue area is DNA in the nucleus.)

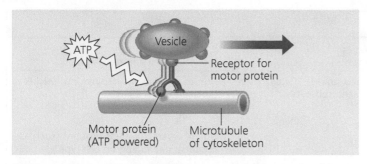

(a) Motor proteins that attach to receptors on vesicles can "walk" the vesicles along the cytoskeletal fibers called microtubules or, in some cases, along microfilaments. ATP powers the movement.

Microtubule   Vesicles   0.25 μm

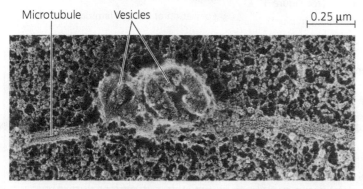

(b) In this SEM of a squid giant axon (a nerve cell extension), two vesicles containing neurotransmitters migrate toward the tip of the axon via the mechanism shown in (a).

▲ **Figure 4.21 Motor proteins and the cytoskeleton.**

the cytoskeleton with **motor proteins**. Examples of such cell motility abound. Cytoskeletal elements and motor proteins work together with plasma membrane molecules to allow whole cells to move along fibers outside the cell. Inside the cell, vesicles and other organelles often use motor protein "feet" to "walk" to their destinations along a track provided by the cytoskeleton. For example, this is how vesicles containing neurotransmitter molecules migrate to the tips of axons, the long extensions of nerve cells that release these molecules as chemical signals to adjacent nerve cells **(Figure 4.21)**. The vesicles that bud off from the ER also travel along cytoskeletal tracks as they make their way to the Golgi. And the cytoskeleton can manipulate the plasma membrane so that it bends inward to form food vacuoles or other phagocytic vesicles.

Components of the Cytoskeleton

Let's look more closely at the three main types of fibers that make up the cytoskeleton: *Microtubules* are the thickest, *microfilaments* (actin filaments) are the thinnest, and *intermediate filaments* are fibers with diameters in a middle range. **Table 4.1** (next page) summarizes the properties of these fibers.

Microtubules

All eukaryotic cells have **microtubules**, hollow rods constructed from a globular protein called tubulin. Each

tubulin protein is a *dimer*, a molecule made up of two subunits. A tubulin dimer consists of two slightly different polypeptides, α-tubulin and β-tubulin. Microtubules grow in length by adding tubulin dimers; they can also be disassembled and their tubulin used to build microtubules elsewhere in the cell.

Microtubules shape and support the cell and serve as tracks along which organelles equipped with motor proteins can move (see Figure 4.21). Microtubules are also involved in the separation of chromosomes during cell division.

Centrosomes and Centrioles In animal cells, microtubules grow out from a **centrosome**, a region that is often located near the nucleus and is considered a "microtubule-organizing center." These microtubules function as compression-resisting girders of the cytoskeleton. Within the centrosome is a pair of **centrioles**, each composed of nine sets of triplet microtubules arranged in a ring **(Figure 4.22)**. Before an animal cell divides, the centrioles replicate. Although centrosomes with centrioles may help organize microtubule assembly in animal cells, they are not essential for this function in all eukaryotes; fungi and almost all plant cells lack centrosomes with centrioles but have well-organized microtubules. Apparently, other microtubule-organizing centers play the role of centrosomes in these cells.

Cilia and Flagella In eukaryotes, a specialized arrangement of microtubules is responsible for the beating of **flagella** (singular, *flagellum*) and **cilia** (singular, *cilium*), microtubule-containing extensions that project from some cells. (The bacterial flagellum, shown in Figure 4.4, has a completely different

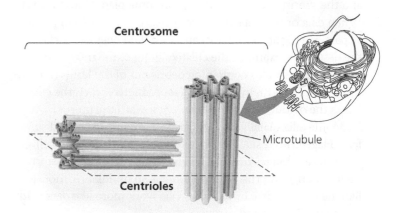

▲ **Figure 4.22 Centrosome containing a pair of centrioles.** Most animal cells have a centrosome, a region near the nucleus where the cell's microtubules are initiated. Within the centrosome is a pair of centrioles, each about 250 nm (0.25 μm) in diameter. The two centrioles are at right angles to each other, and each is made up of nine sets of three microtubules. The blue portions of the drawing represent nontubulin proteins that connect the microtubule triplets.

? *How many microtubules are in a centrosome? In the drawing, circle and label one microtubule and describe its structure. Circle and label a triplet.*

Table 4.1 The Structure and Function of the Cytoskeleton

Property	Microtubules (Tubulin Polymers)	Microfilaments (Actin Filaments)	Intermediate Filaments
Structure	Hollow tubes; wall consists of 13 columns of tubulin molecules	Two intertwined strands of actin, each a polymer of actin subunits	Fibrous proteins supercoiled into thicker cables
Diameter	25 nm with 15-nm lumen	7 nm	8–12 nm
Protein subunits	Tubulin, a dimer consisting of α-tubulin and β-tubulin	Actin	One of several different proteins (such as keratins), depending on cell type
Main functions	Maintenance of cell shape	Maintenance of cell shape	Maintenance of cell shape
	Cell motility (as in cilia or flagella)	Changes in cell shape	Anchorage of nucleus and certain other organelles
	Chromosome movements in cell division (see Figure 9.7)	Muscle contraction (see Figure 39.4)	Formation of nuclear lamina
	Organelle movements	Cytoplasmic streaming in plants	
		Cell motility (as in amoeboid movement)	
		Division of animal cells (see Figure 9.10)	

Column of tubulin dimers

25 nm

α β Tubulin dimer

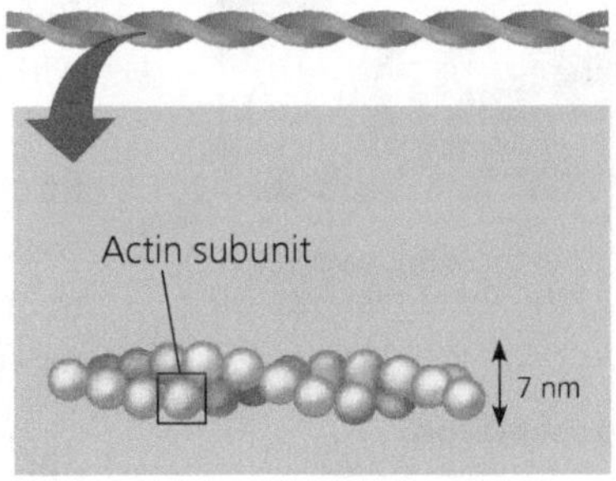

Actin subunit

7 nm

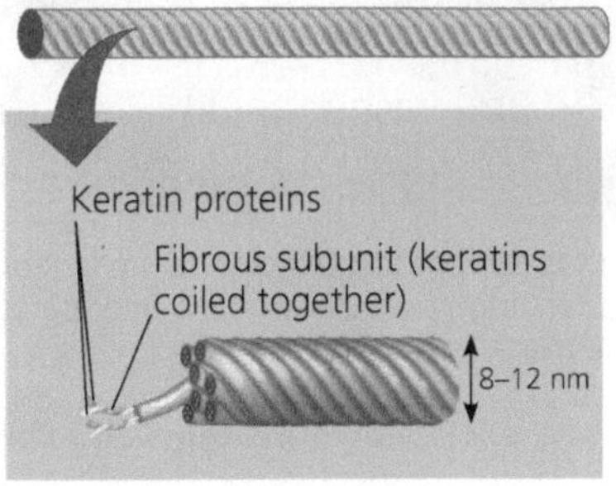

Keratin proteins

Fibrous subunit (keratins coiled together)

8–12 nm

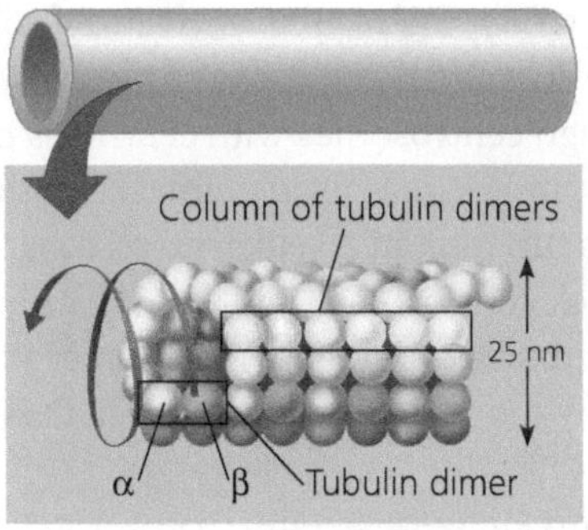

© 2000 Pearson Education, Inc.

structure.) Many unicellular eukaryotes are propelled through water by cilia or flagella that act as locomotor appendages, and the sperm of animals, algae, and some plants have flagella. When cilia or flagella extend from cells that are held in place as part of a tissue layer, they can move fluid over the surface of the tissue. For example, the ciliated lining of the trachea (windpipe) sweeps mucus containing debris out of the lungs (see the EMs in Figure 4.3). In a woman's reproductive tract, the cilia lining the oviducts help move an egg toward the uterus.

Motile cilia usually occur in large numbers on the cell surface. Flagella are usually limited to just one or a few per cell, and they are longer than cilia. Flagella and cilia also differ in their beating patterns. A flagellum has an undulating motion like the tail of a fish. In contrast, cilia work more like oars, with alternating power and recovery strokes.

A cilium may also act as a signal-receiving antenna for the cell. Cilia that have this function are generally nonmotile, and there is only one per cell. (In fact, in vertebrate animals, it appears that almost all cells have such a cilium, which is called a *primary cilium*.) Membrane proteins on this kind of cilium transmit molecular signals from the cell's environment to its interior, triggering signaling pathways that may lead to changes in the cell's activities. Cilium-based signaling appears to be crucial to brain function and to embryonic development.

Though different in length, number per cell, and beating pattern, motile cilia and flagella share a common structure. Each motile cilium or flagellum has a group of microtubules sheathed in an extension of the plasma membrane **(Figure 4.23)**. Nine doublets of microtubules are arranged in a ring; in the center of the ring are two single microtubules. This arrangement, referred to as the "9 + 2" pattern, is found in nearly all eukaryotic flagella and motile cilia. (Nonmotile primary cilia have a "9 + 0" pattern, lacking the central pair of microtubules.) The microtubule assembly of a cilium or flagellum is anchored in the cell by a **basal body**, which is structurally like a centriole, with microtubule triplets in a "9 + 0" pattern. In fact, in many animals (including humans), the basal body of the fertilizing sperm's flagellum enters the egg and becomes a centriole.

How does the microtubule assembly produce the bending movements of flagella and motile cilia? Bending involves large motor proteins called **dyneins** (red in the diagram) that are attached along each outer microtubule doublet. A typical dynein protein has two "feet" that "walk" along the microtubule of the adjacent doublet, using ATP for energy. One foot maintains contact while the other releases and reattaches farther along the microtubule (see Figure 4.21). The outer doublets and two central microtubules are held together by flexible cross-linking

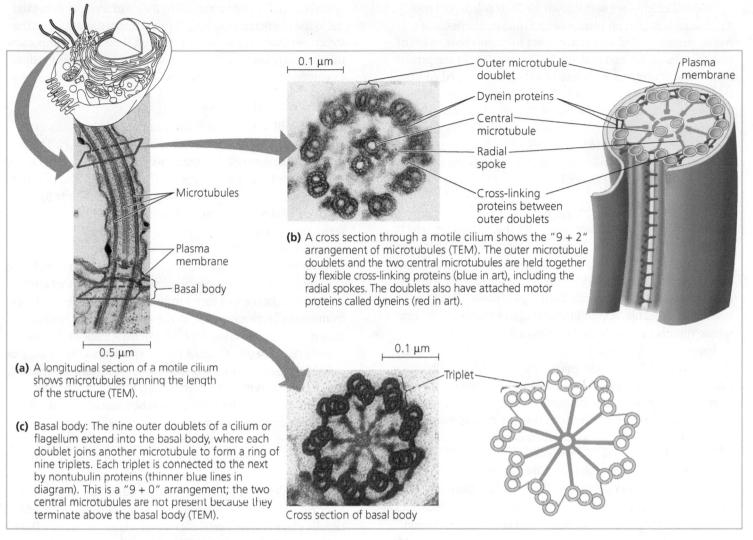

(a) A longitudinal section of a motile cilium shows microtubules running the length of the structure (TEM).

0.5 μm

Microtubules

Plasma membrane

Basal body

0.1 μm

Outer microtubule doublet

Dynein proteins

Central microtubule

Radial spoke

Cross-linking proteins between outer doublets

Plasma membrane

(b) A cross section through a motile cilium shows the "9 + 2" arrangement of microtubules (TEM). The outer microtubule doublets and the two central microtubules are held together by flexible cross-linking proteins (blue in art), including the radial spokes. The doublets also have attached motor proteins called dyneins (red in art).

0.1 μm

Triplet

Cross section of basal body

(c) Basal body: The nine outer doublets of a cilium or flagellum extend into the basal body, where each doublet joins another microtubule to form a ring of nine triplets. Each triplet is connected to the next by nontubulin proteins (thinner blue lines in diagram). This is a "9 + 0" arrangement; the two central microtubules are not present because they terminate above the basal body (TEM).

▲ **Figure 4.23 Structure of a flagellum or motile cilium.**

DRAW IT *In* **(a)**, *circle the central pair of microtubules. Show where they terminate, and explain why they aren't seen in the cross section of the basal body in* **(c)**.

proteins. If the doublets were not held in place, the walking action would make them slide past each other. Instead, the movements of the dynein feet cause the microtubules—and the organelle as a whole—to bend.

Microfilaments (Actin Filaments)

Microfilaments are thin solid rods. They are also called actin filaments because they are built from molecules of **actin**, a globular protein. A microfilament is a twisted double chain of actin subunits (see Table 4.1). Besides occurring as linear filaments, microfilaments can form structural networks when certain proteins bind along the side of such a filament and allow a new filament to extend as a branch.

The structural role of microfilaments in the cytoskeleton is to bear tension (pulling forces). A three-dimensional network formed by microfilaments just inside the plasma membrane helps support the cell's shape. In some kinds of animal cells, such as nutrient-absorbing intestinal cells, bundles of

microfilaments make up the core of microvilli, delicate projections that increase the cell's surface area (**Figure 4.24**).

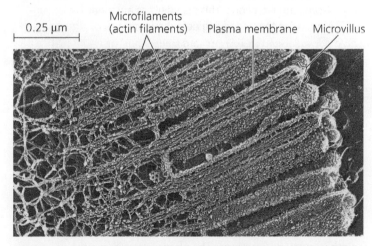

0.25 μm

Microfilaments (actin filaments)

Plasma membrane

Microvillus

▲ **Figure 4.24 A structural role of microfilaments.** The surface area of this intestinal cell is increased by its many microvilli (singular, *microvillus*), cellular extensions reinforced by bundles of microfilaments (TEM).

Microfilaments are well known for their role in cell motility. Thousands of actin filaments and thicker filaments of a motor protein called **myosin** interact to cause contraction of muscle cells (described in detail in Chapter 39). In the protist *Amoeba* and some of our white blood cells, localized contractions brought about by actin and myosin are involved in the amoeboid (crawling) movement of the cells. In plant cells, actin-myosin interaction contributes to *cytoplasmic streaming*, a circular flow of cytoplasm within cells. This movement, which is especially common in large plant cells, speeds the distribution of materials within the cell.

Intermediate Filaments

Intermediate filaments are named for their diameter, which is larger than the diameter of microfilaments but smaller than that of microtubules (see Table 4.1). Specialized for bearing tension (like microfilaments), intermediate filaments are a diverse class of cytoskeletal elements. Each type is constructed from a particular molecular subunit belonging to a family of proteins whose members include the keratins in hair and nails.

Intermediate filaments are more permanent fixtures of cells than are microfilaments and microtubules, which are often disassembled and reassembled in various parts of a cell. Even after cells die, intermediate filament networks often persist; for example, the outer layer of our skin consists of dead skin cells full of keratin filaments. Intermediate filaments are especially sturdy and play an important role in reinforcing the shape of a cell and fixing the position of certain organelles. For instance, the nucleus typically sits within a cage made of intermediate filaments. Other intermediate filaments make up the nuclear lamina, which lines the interior of the nuclear envelope (see Figure 4.8). In general, the various kinds of intermediate filaments seem to function together as the permanent framework of the entire cell.

CONCEPT CHECK 4.6

1. How do cilia and flagella bend?
2. **WHAT IF?** Males afflicted with Kartagener's syndrome are sterile because of immotile sperm, and they tend to suffer from lung infections. This disorder has a genetic basis. Suggest what the underlying defect might be.

For suggested answers, see Appendix A.

CONCEPT 4.7

Extracellular components and connections between cells help coordinate cellular activities

Having crisscrossed the cell to explore its interior components, we complete our tour of the cell by returning to the surface of this microscopic world, where there are additional structures with important functions. The plasma membrane is usually regarded as the boundary of the living cell, but most cells synthesize and secrete materials to their extracellular side, external to the plasma membrane. Although these materials and the structures they form are outside the cell, their study is important to cell biology because they are involved in a great many cellular functions.

Cell Walls of Plants

The **cell wall** is an extracellular structure of plant cells that distinguishes them from animal cells (see Figure 4.7). The wall protects the plant cell, maintains its shape, and prevents excessive uptake of water. On the level of the whole plant, the strong walls of specialized cells hold the plant up against the force of gravity. Prokaryotes, fungi, and some protists also have cell walls, as you saw in Figures 4.4 and 4.7, but we will postpone discussion of them until Chapters 24–26.

Plant cell walls are much thicker than the plasma membrane, ranging from 0.1 µm to several micrometers. The exact chemical composition of the wall varies from species to species and even from one cell type to another in the same plant, but the basic design of the wall is consistent. Microfibrils made of the polysaccharide cellulose (see Figure 3.10) are synthesized by an enzyme called cellulose synthase and secreted to the extracellular space, where they become embedded in a matrix of other polysaccharides and proteins. This combination of materials, strong fibers in a "ground substance" (matrix), is the same basic architectural design found in steel-reinforced concrete and in fiberglass.

A young plant cell first secretes a relatively thin and flexible wall called the **primary cell wall (Figure 4.25)**. Between primary walls of adjacent cells is the **middle lamella**, a thin layer rich in sticky polysaccharides called pectins. The middle lamella glues adjacent cells together. (Pectin is used as a thickening agent in fruit jellies.) When the cell matures and stops growing, it strengthens its wall. Some plant cells do this simply by secreting hardening substances into the primary wall. Other cells add a **secondary cell wall** between the plasma membrane and the primary wall. The secondary wall, often deposited in several laminated layers, has a strong and durable matrix that affords the cell protection and support. Wood, for example, consists mainly of secondary walls. Plant cell walls are usually perforated by channels between adjacent cells called plasmodesmata, which will be discussed shortly.

The Extracellular Matrix (ECM) of Animal Cells

Although animal cells lack walls akin to those of plant cells, they do have an elaborate **extracellular matrix (ECM)**. The main ingredients of the ECM are glycoproteins and other carbohydrate-containing molecules secreted by the cells. (Recall that glycoproteins are proteins with covalently bonded carbohydrates.) The most abundant glycoprotein in the ECM of most animal cells is **collagen**, which forms strong fibers outside the cells (see Figure 3.21, carbohydrate not shown). In fact, collagen accounts for about 40% of the total protein in the human

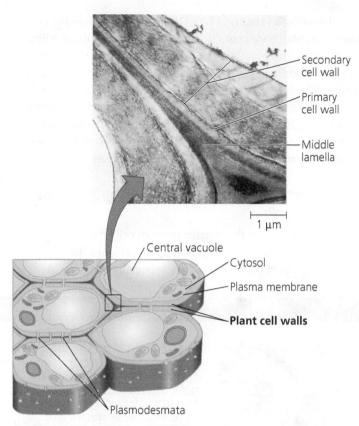

Secondary
cell wall

Primary
cell wall

Middle
lamella

1 μm

Central vacuole

Cytosol

Plasma membrane

Plant cell walls

Plasmodesmata

▲ **Figure 4.25 Plant cell walls.** The drawing shows several cells, each with a large vacuole, a nucleus, and several chloroplasts and mitochondria. The transmission electron micrograph shows the cell walls where two cells come together. The multilayered partition between plant cells consists of adjoining walls individually secreted by the cells. Plasmodesmata are channels through cell walls that connect the cytoplasm of adjacent plant cells.

body. The collagen fibers are embedded in a network woven of secreted **proteoglycans (Figure 4.26)**. A proteoglycan molecule consists of a small protein with many carbohydrate chains covalently attached; it may be up to 95% carbohydrate. Large proteoglycan complexes can form when hundreds of proteoglycan molecules become noncovalently attached to a single long polysaccharide molecule, as shown in Figure 4.26. Some cells are attached to the ECM by ECM glycoproteins such as **fibronectin**. Fibronectin and other ECM proteins bind to cell-surface receptor proteins called **integrins** that are built into the plasma membrane. Integrins span the membrane and bind on their cytoplasmic side to associated proteins attached to microfilaments of the cytoskeleton. The name *integrin* is based on the word *integrate*: Integrins are in a position to transmit signals between the ECM and the cytoskeleton and thus to integrate changes occurring outside and inside the cell.

Current research is revealing the influential role of the ECM in the lives of cells. By communicating with a cell through integrins, the ECM can regulate a cell's behavior. For example, some cells in a developing embryo migrate along specific pathways by matching the orientation of their microfilaments to the "grain" of fibers in the extracellular matrix. Researchers have also learned that the extracellular matrix around a cell can influence the activity of genes in the nucleus. Information about the ECM probably reaches the nucleus by a combination of mechanical and chemical signaling pathways. Mechanical signaling involves fibronectin, integrins, and microfilaments of the cytoskeleton. Changes in the cytoskeleton may in turn trigger chemical signaling pathways inside the cell, leading to changes in the set of proteins being made by

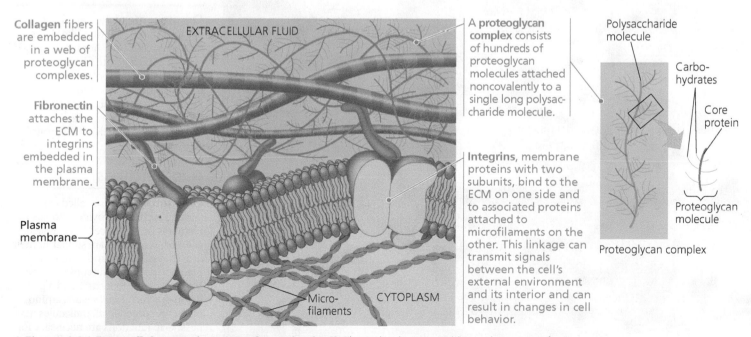

Collagen fibers are embedded in a web of proteoglycan complexes.

EXTRACELLULAR FLUID

A **proteoglycan complex** consists of hundreds of proteoglycan molecules attached noncovalently to a single long polysaccharide molecule.

Polysaccharide molecule

Carbohydrates

Core protein

Proteoglycan molecule

Fibronectin attaches the ECM to integrins embedded in the plasma membrane.

Plasma membrane

Integrins, membrane proteins with two subunits, bind to the ECM on one side and to associated proteins attached to microfilaments on the other. This linkage can transmit signals between the cell's external environment and its interior and can result in changes in cell behavior.

Microfilaments

CYTOPLASM

Proteoglycan complex

▲ **Figure 4.26 Extracellular matrix (ECM) of an animal cell.** The molecular composition and structure of the ECM vary from one cell type to another. In this example, three different types of ECM molecules are present: proteoglycans, collagen, and fibronectin.

the cell and therefore changes in the cell's function. In this way, the extracellular matrix of a particular tissue may help coordinate the behavior of all the cells of that tissue. Direct connections between cells also function in this coordination, as we discuss next.

Cell Junctions

Neighboring cells in an animal or plant often adhere, interact, and communicate via sites of direct physical contact.

Plasmodesmata in Plant Cells

It might seem that the nonliving cell walls of plants would isolate plant cells from one another. But in fact, as shown in Figure 4.25, cell walls are perforated with **plasmodesmata** (singular, *plasmodesma*; from the Greek *desma*, bond),

membrane-lined channels filled with cytosol. By joining adjacent cells, plasmodesmata unify most of a plant into one living continuum. The plasma membranes of adjacent cells line the channel of each plasmodesma and thus are continuous. Water and small solutes can pass freely from cell to cell, and recent experiments have shown that in some circumstances, certain proteins and RNA molecules can as well. The macromolecules transported to neighboring cells appear to reach the plasmodesmata by moving along fibers of the cytoskeleton.

Tight Junctions, Desmosomes, and Gap Junctions in Animal Cells

In animals, there are three main types of cell junctions: *tight junctions*, *desmosomes*, and *gap junctions* (**Figure 4.27**). All three types of cell junctions are especially common in

---

▼ **Figure 4.27**   **Exploring Cell Junctions in Animal Tissues**

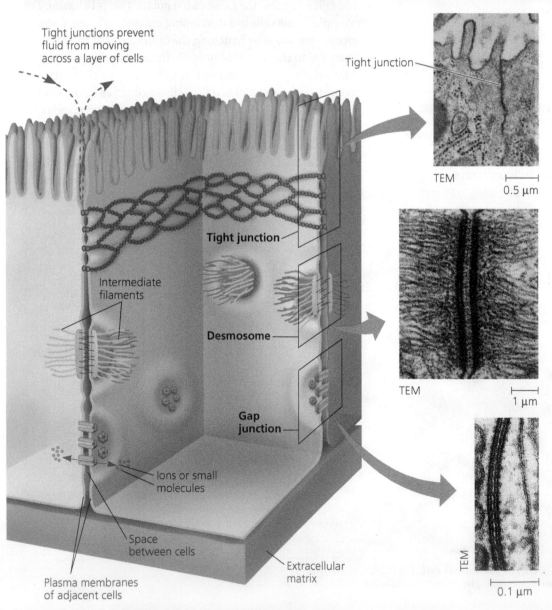

Tight junctions prevent fluid from moving across a layer of cells

Tight junction

Intermediate filaments

Tight junction

Desmosome

Gap junction

Ions or small molecules

Space between cells

Plasma membranes of adjacent cells

Extracellular matrix

Tight Junctions

Tight junction

TEM   0.5 μm

At **tight junctions**, the plasma membranes of neighboring cells are very tightly pressed against each other, bound together by specific proteins (purple). Forming continuous seals around the cells, tight junctions prevent leakage of extracellular fluid across a layer of epithelial cells. For example, tight junctions between skin cells make us watertight by preventing leakage between cells in our sweat glands.

Desmosomes

TEM   1 μm

Desmosomes (also called anchoring junctions) function like rivets, fastening cells together into strong sheets. Intermediate filaments made of sturdy keratin proteins anchor desmosomes in the cytoplasm. Desmosomes attach muscle cells to each other in a muscle. Some "muscle tears" involve the rupture of desmosomes.

Gap Junctions

TEM   0.1 μm

Gap junctions (also called communicating junctions) provide cytoplasmic channels from one cell to an adjacent cell and in this way are similar in their function to the plasmodesmata in plants. Gap junctions consist of membrane proteins that surround a pore through which ions, sugars, amino acids, and other small molecules may pass. Gap junctions are necessary for communication between cells in many types of tissues, such as heart muscle, and in animal embryos.

epithelial tissue, which lines the external and internal surfaces of the body. Figure 4.27 uses epithelial cells of the intestinal lining to illustrate these junctions. (Gap junctions are most like the plasmodesmata of plants, although gap junction pores are not lined with membrane.)

CONCEPT CHECK 4.7

1. In what way are the cells of plants and animals structurally different from single-celled eukaryotes?
2. **WHAT IF?** If the plant cell wall or the animal extracellular matrix were impermeable, what effect would this have on cell function?
3. **MAKE CONNECTIONS** The polypeptide chain that makes up a tight junction weaves back and forth through the membrane four times, with two extracellular loops, and one loop plus short C-terminal and N-terminal tails in the cytoplasm. Looking at Figure 3.17, what would you predict about the amino acids making up the tight-junction protein?

For suggested answers, see Appendix A.

The Cell: A Living Unit Greater Than the Sum of Its Parts

From our panoramic view of the cell's compartmental organization to our close-up inspection of each organelle's architecture, this tour of the cell has provided many opportunities to correlate structure with function. But even as we dissect the cell, remember that none of its components works alone. As an example of cellular integration, consider the microscopic scene in **Figure 4.28**. The large cell is a macrophage (see Figure 4.12). It helps defend the mammalian body against infections by ingesting bacteria (the smaller cells) into

phagocytic vesicles. The macrophage crawls along a surface and reaches out to the bacteria with thin cell extensions called pseudopodia (specifically, filopodia). Actin filaments interact with other elements of the cytoskeleton in these movements. After the macrophage engulfs the bacteria, they are destroyed by lysosomes. The elaborate endomembrane system produces the lysosomes. The digestive enzymes of the lysosomes and the proteins of the cytoskeleton are all made on ribosomes. And the synthesis of these proteins is programmed by genetic messages dispatched from the DNA in the nucleus. All these processes require energy, which mitochondria supply in the form of ATP. Cellular functions arise from cellular order: The cell is a living unit greater than the sum of its parts.

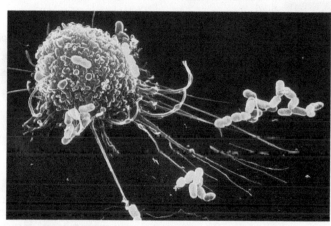

▲ **Figure 4.28 The emergence of cellular functions.** The ability of this macrophage (brown) to recognize, apprehend, and destroy bacteria (yellow) is a coordinated activity of the whole cell. Its cytoskeleton, lysosomes, and plasma membrane are among the components that function in phagocytosis (colorized SEM).

4 Chapter Review

SUMMARY OF KEY CONCEPTS

CONCEPT 4.1

Biologists use microscopes and the tools of biochemistry to study cells (pp. 67–69)

- Improvements in microscopy that affect the parameters of magnification, resolution, and contrast have catalyzed progress in the study of cell structure. The **light microscope** (LM) and **electron microscope** (EM), as well as other types, remain important tools.
- Cell biologists can obtain pellets enriched in particular cellular components by centrifuging disrupted cells at sequential speeds, a process known as **cell fractionation**. Larger cellular components are in the pellet after lower-speed centrifugation, and smaller components are in the pellet after higher-speed centrifugation.

? *How do microscopy and biochemistry complement each other to reveal cell structure and function?*

CONCEPT 4.2

Eukaryotic cells have internal membranes that compartmentalize their functions (pp. 69–74)

- All cells are bounded by a **plasma membrane**.
- **Prokaryotic cells** lack nuclei and other membrane-enclosed **organelles**, while **eukaryotic cells** have internal membranes that compartmentalize cellular functions.
- The surface-to-volume ratio is an important parameter affecting cell size and shape.
- Plant and animal cells have most of the same organelles: a nucleus, endoplasmic reticulum, Golgi apparatus, and mitochondria. Some organelles are found only in plant or in animal cells. Chloroplasts are present only in cells of photosynthetic eukaryotes.

? *Explain how the compartmental organization of a eukaryotic cell contributes to its biochemical functioning.*

	Cell Component	Structure	Function
CONCEPT 4.3 **The eukaryotic cell's genetic instructions are housed in the nucleus and carried out by the ribosomes (pp. 74–76)** **?** *Describe the relationship between the nucleus and ribosomes.*	Nucleus 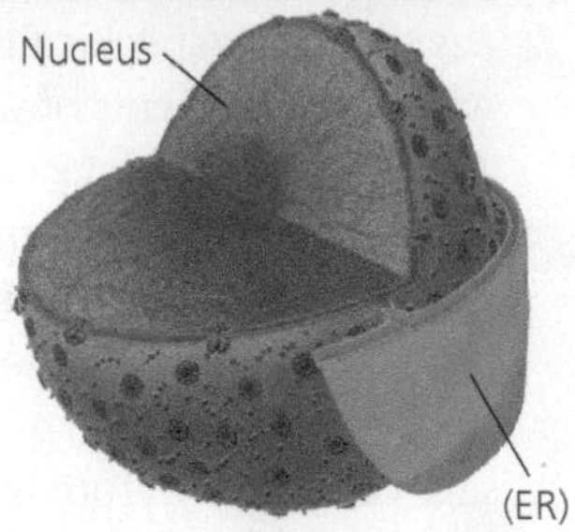 (ER)	Surrounded by nuclear envelope (double membrane) perforated by nuclear pores; nuclear envelope continuous with endoplasmic reticulum (ER)	Houses chromosomes, which are made of chromatin (DNA and proteins); contains nucleoli, where ribosomal subunits are made; pores regulate entry and exit of materials
	Ribosome 	Two subunits made of ribosomal RNA and proteins; can be free in cytosol or bound to ER	Protein synthesis
CONCEPT 4.4 **The endomembrane system regulates protein traffic and performs metabolic functions in the cell (pp. 76–81)** **?** *Describe the key role played by transport vesicles in the endomembrane system.*	Endoplasmic reticulum 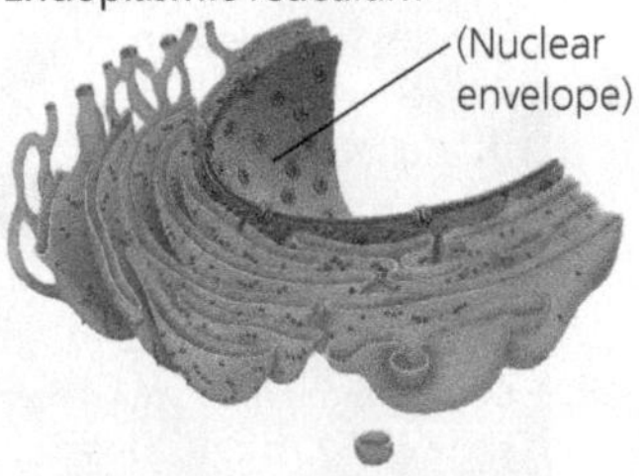 (Nuclear envelope)	Extensive network of membrane-bounded tubules and sacs; membrane separates lumen from cytosol; continuous with nuclear envelope	Smooth ER: synthesis of lipids, metabolism of carbohydrates, Ca^{2+} storage, detoxification of drugs and poisons Rough ER: aids in synthesis of secretory and other proteins from bound ribosomes; adds carbohydrates to proteins to make glycoproteins; produces new membrane
	Golgi apparatus 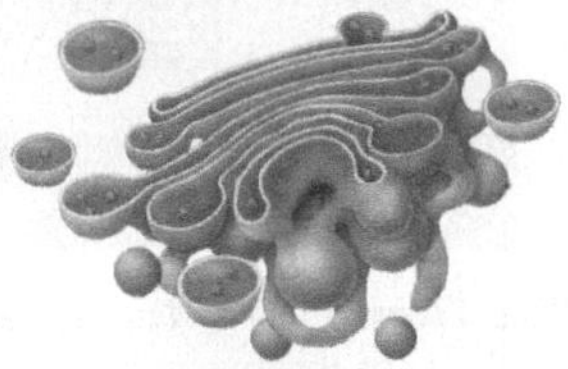	Stacks of flattened membranous sacs; has polarity (*cis* and *trans* faces)	Modification of proteins, carbohydrates on proteins, and phospholipids; synthesis of many polysaccharides; sorting of Golgi products, which are then released in vesicles
	Lysosome 	Membranous sac of hydrolytic enzymes (in animal cells)	Breakdown of ingested substances, cell macromolecules, and damaged organelles for recycling
	Vacuole 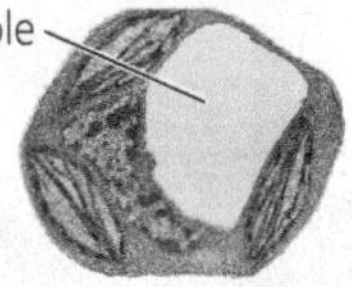	Large membrane-bounded vesicle	Digestion, storage, waste disposal, water balance, plant cell growth and protection
CONCEPT 4.5 **Mitochondria and chloroplasts change energy from one form to another (pp. 81–84)** **?** *What is the endosymbiont theory?*	Mitochondrion 	Bounded by double membrane; inner membrane has infoldings (cristae)	Cellular respiration
	Chloroplast 	Typically two membranes around fluid stroma, which contains thylakoids stacked into grana (in cells of photosynthetic eukaryotes, including plants)	Photosynthesis
	Peroxisome 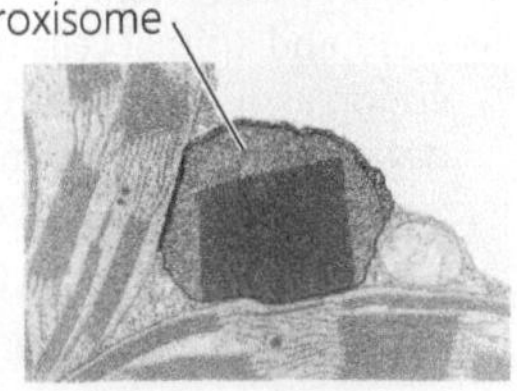	Specialized metabolic compartment bounded by a single membrane	Contains enzymes that transfer hydrogen atoms from certain molecules to oxygen, producing hydrogen peroxide (H_2O_2) as a by-product; H_2O_2 is converted to water by another enzyme

The cytoskeleton is a network of fibers that organizes structures and activities in the cell (pp. 84–88)

- The **cytoskeleton** functions in structural support for the cell and in motility and signal transmission.
- **Microtubules** shape the cell, guide organelle movement, and separate chromosomes in dividing cells. **Cilia** and **flagella** are motile appendages containing microtubules. *Primary cilia* play sensory and signaling roles. **Microfilaments** are thin rods functioning in muscle contraction, amoeboid movement, cytoplasmic streaming, and support of microvilli. **Intermediate filaments** support cell shape and fix organelles in place.

> **?** *Describe the role of motor proteins inside the eukaryotic cell and in whole-cell movement.*

Extracellular components and connections between cells help coordinate cellular activities (pp. 88–91)

- Plant **cell walls** are made of cellulose fibers embedded in other polysaccharides and proteins.
- Animal cells secrete glycoproteins and proteoglycans that form the **extracellular matrix (ECM)**, which functions in support, adhesion, movement, and regulation.
- Cell junctions connect neighboring cells in plants and animals. Plants have **plasmodesmata** that pass through adjoining cell walls. Animal cells have **tight junctions**, **desmosomes**, and **gap junctions**.

> **?** *Compare the composition and functions of a plant cell wall and the extracellular matrix of an animal cell.*

TEST YOUR UNDERSTANDING

Level 1: Knowledge/Comprehension

1. Which structure is *not* part of the endomembrane system?
   a. nuclear envelope
   b. chloroplast
   c. Golgi apparatus
   d. plasma membrane
   e. ER

2. Which structure is common to plant *and* animal cells?
   a. chloroplast
   b. wall made of cellulose
   c. central vacuole
   d. mitochondrion
   e. centriole

3. Which of the following is present in a prokaryotic cell?
   a. mitochondrion
   b. ribosome
   c. nuclear envelope
   d. chloroplast
   e. ER

4. Which structure-function pair is *mismatched*?
   a. nucleolus; production of ribosomal subunits
   b. lysosome; intracellular digestion
   c. ribosome; protein synthesis
   d. Golgi; protein trafficking
   e. microtubule; muscle contraction

Level 2: Application/Analysis

5. Cyanide binds to at least one molecule involved in producing ATP. If a cell is exposed to cyanide, most of the cyanide will be found within the
   a. mitochondria.
   b. ribosomes.
   c. peroxisomes.
   d. lysosomes.
   e. endoplasmic reticulum.

6. What is the most likely pathway taken by a newly synthesized protein that will be secreted by a cell?
   a. ER → Golgi → nucleus
   b. Golgi → ER → lysosome
   c. nucleus → ER → Golgi
   d. ER → Golgi → vesicles that fuse with plasma membrane
   e. ER → lysosomes → vesicles that fuse with plasma membrane

7. Which cell would be best for studying lysosomes?
   a. muscle cell
   b. nerve cell
   c. phagocytic white blood cell
   d. leaf cell of a plant
   e. bacterial cell

8. **DRAW IT** From memory, draw two eukaryotic cells, labeling the structures listed here and showing any physical connections between the internal structures of each cell: nucleus, rough ER, smooth ER, mitochondrion, centrosome, chloroplast, vacuole, lysosome, microtubule, cell wall, ECM, microfilament, Golgi apparatus, intermediate filament, plasma membrane, peroxisome, ribosome, nucleolus, nuclear pore, vesicle, flagellum, microvilli, plasmodesma.

Level 3: Synthesis/Evaluation

9. **SCIENTIFIC INQUIRY**
   In studying micrographs of an unusual protist (single-celled eukaryote) that you found in a sample of pond water, you spot an organelle that you can't recognize. You successfully develop a method for growing this organism in liquid in the laboratory. Describe how you would go about finding out what this organelle is and what it does in the cell. Assume that you would make use of additional microscopy, cell fractionation, and biochemical tests.

10. **FOCUS ON EVOLUTION**
    Which aspects of cell structure best reveal evolutionary unity? What are some examples of specialized modifications?

11. **FOCUS ON ORGANIZATION**
    Considering some of the characteristics that define life and drawing on your new knowledge of cellular structures and functions, write a short essay (100–150 words) that discusses this statement: Life is an emergent property that appears at the level of the cell. (Review the section on emergent properties in Concept 1.1.)

For selected answers, see Appendix A.

MasteringBiology®

Students Go to **MasteringBiology** for assignments, the eText, and the Study Area with practice tests, animations, and activities.

Instructors Go to **MasteringBiology** for automatically graded tutorials and questions that you can assign to your students, plus Instructor Resources.

5

Membrane Transport and Cell Signaling

KEY CONCEPTS

5.1 Cellular membranes are fluid mosaics of lipids and proteins

5.2 Membrane structure results in selective permeability

5.3 Passive transport is diffusion of a substance across a membrane with no energy investment

5.4 Active transport uses energy to move solutes against their gradients

5.5 Bulk transport across the plasma membrane occurs by exocytosis and endocytosis

5.6 The plasma membrane plays a key role in most cell signaling

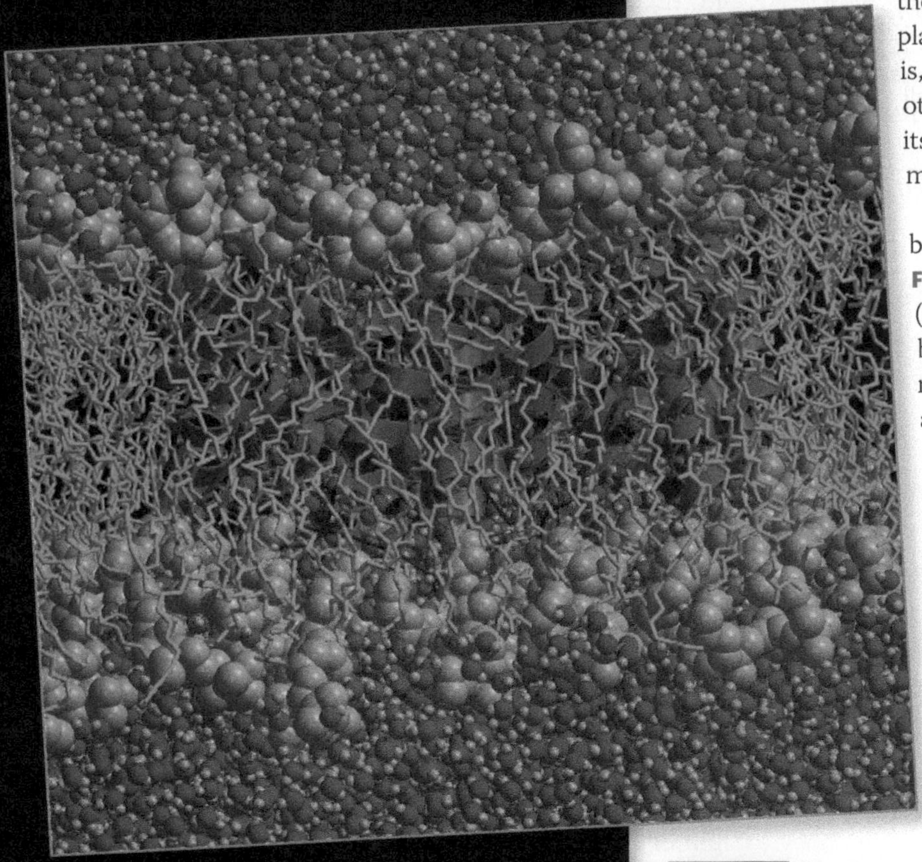

OVERVIEW

Life at the Edge

The plasma membrane is the edge of life, the boundary that separates the living cell from its surroundings. A remarkable film only about 8 nm thick—it would take over 8,000 plasma membranes to equal the thickness of this page—the plasma membrane controls traffic into and out of the cell it surrounds. Like all biological membranes, the plasma membrane exhibits **selective permeability**; that is, it allows some substances to cross it more easily than others. The resulting ability of the cell to discriminate in its chemical exchanges with its environment is fundamental to life.

Most of this chapter is devoted to how cellular membranes control the passage of substances through them. **Figure 5.1** shows a computer model of water molecules (red and gray) passing through a short section of membrane. The blue ribbons within the lipid bilayer (green) represent helical regions of a membrane protein called an aquaporin. One molecule of this protein enables billions of water molecules to pass through the membrane every second, many more than could cross on their own. Found in many kinds of cells, aquaporins are but one example of how the plasma membrane and its proteins enable cells to survive and function.

To understand how membranes work, we'll begin by examining their molecular structure. Then we'll describe in some detail how plasma membranes control transport into and out of cells. Finally, we'll discuss cell signaling, emphasizing the role of the plasma membrane in cell communication.

◀ **Figure 5.1** How do cell membrane proteins help regulate chemical traffic?

CONCEPT 5.1

Cellular membranes are fluid mosaics of lipids and proteins

Figure 5.2 shows the currently accepted model of the arrangement of molecules in the plasma membrane. Lipids and proteins are the staple ingredients of membranes, although carbohydrates are also important. The most abundant lipids in most membranes are phospholipids. The ability of phospholipids to

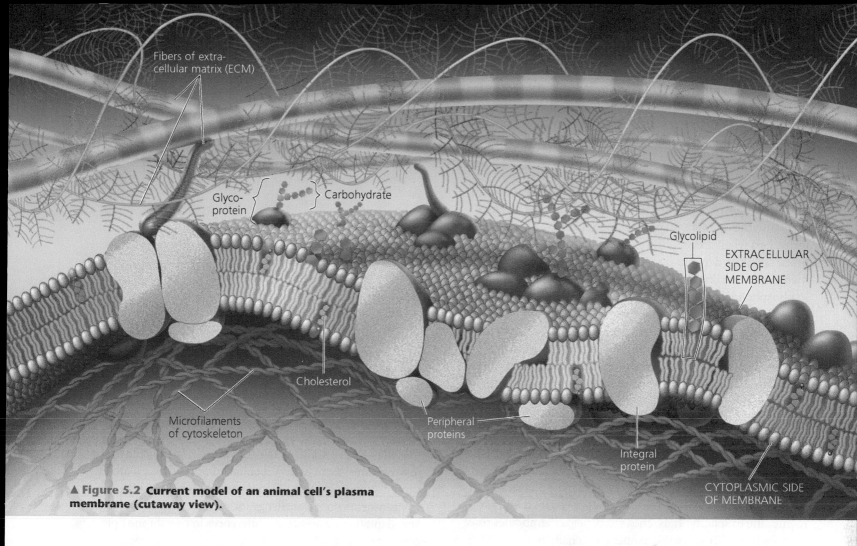

▲ Figure 5.2 Current model of an animal cell's plasma membrane (cutaway view).

Labels in figure:
- Fibers of extra-cellular matrix (ECM)
- Glyco-protein
- Carbohydrate
- Glycolipid
- EXTRACELLULAR SIDE OF MEMBRANE
- Cholesterol
- Microfilaments of cytoskeleton
- Peripheral proteins
- Integral protein
- CYTOPLASMIC SIDE OF MEMBRANE

form membranes is inherent in their molecular structure. A phospholipid is an **amphipathic** molecule, meaning it has both a hydrophilic region and a hydrophobic region (see Figure 3.14). A phospholipid bilayer can exist as a stable boundary between two aqueous compartments because the molecular arrangement shelters the hydrophobic tails of the phospholipids from water while exposing the hydrophilic heads to water **(Figure 5.3)**.

▼ Figure 5.3 Phospholipid bilayer (cross section).

- Hydrophilic head
- WATER
- Hydrophobic tail
- WATER

MAKE CONNECTIONS *Refer to Figure 3.14b, and then circle the hydrophilic and hydrophobic portions of one of the enlarged phospholipid molecules on the right in Figure 5.3. Explain what each portion contacts when the phospholipid is in the plasma membrane.*

Like phospholipids, most membrane proteins are amphipathic. Such proteins can reside in the phospholipid bilayer with their hydrophilic regions protruding. This molecular orientation maximizes contact of the hydrophilic regions of a protein with water in the cytosol and extracellular fluid, while providing its hydrophobic parts with a nonaqueous environment.

In the **fluid mosaic model** in Figure 5.2, the membrane is a mosaic of protein molecules bobbing in a fluid bilayer of phospholipids. The proteins are not randomly distributed in the membrane, however. Groups of proteins are often associated in long-lasting, specialized patches, as are certain lipids. In some regions, the membrane may be much more packed with proteins than shown in Figure 5.2. Like all models, the fluid mosaic model is continually being refined as new research reveals more about membrane structure.

The Fluidity of Membranes

Membranes are not static sheets of molecules locked rigidly in place. A membrane is held together primarily by hydrophobic interactions, which are much weaker than covalent bonds (see Figure 3.21). Most of the lipids and some of the proteins can shift about laterally—that is, in the plane of the membrane—like partygoers elbowing their way through a crowded room.

The lateral movement of phospholipids within the membrane is rapid. Proteins are much larger than lipids and move more slowly, but some membrane proteins do drift, as shown in a classic experiment described in **Figure 5.4**. And some membrane proteins seem to move in a highly directed manner, perhaps driven along cytoskeletal fibers by motor proteins. However, many other membrane proteins seem to be held immobile by their attachment to the cytoskeleton or to the extracellular matrix (see Figure 5.2).

A membrane remains fluid as temperature decreases until finally the phospholipids settle into a closely packed arrangement and the membrane solidifies, much as bacon grease forms lard when it cools. The temperature at which a membrane solidifies depends on the types of lipids it is made of. The membrane remains fluid to a lower temperature if it is rich in phospholipids with unsaturated hydrocarbon tails (see Figures 3.13 and 3.14). Because of kinks in the tails where double bonds are located, unsaturated hydrocarbon tails cannot pack together as closely as saturated hydrocarbon tails, and this looseness makes the membrane more fluid (**Figure 5.5a**).

The steroid cholesterol, which is wedged between phospholipid molecules in the plasma membranes of animal cells, has different effects on membrane fluidity at different temperatures (**Figure 5.5b**). At relatively high temperatures—at 37°C, the body temperature of humans, for example—cholesterol makes the membrane less fluid by restraining phospholipid movement. However, because cholesterol also hinders the close packing of phospholipids, it lowers the temperature required for the membrane to solidify. Thus, cholesterol helps membranes resist changes in fluidity when the temperature changes.

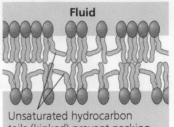
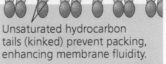
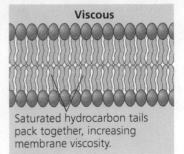

Fluid — Unsaturated hydrocarbon tails (kinked) prevent packing, enhancing membrane fluidity.

Viscous — Saturated hydrocarbon tails pack together, increasing membrane viscosity.

(a) Unsaturated versus saturated hydrocarbon tails.

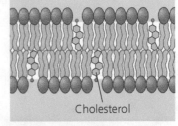

(b) Cholesterol within the animal cell membrane. Cholesterol reduces membrane fluidity at moderate temperatures by reducing phospholipid movement, but at low temperatures it hinders solidification by disrupting the regular packing of phospholipids.

Cholesterol

▲ **Figure 5.5 Factors that affect membrane fluidity.**

Membranes must be fluid to work properly; they are usually about as fluid as salad oil. When a membrane solidifies, its permeability changes, and enzymatic proteins in the membrane may become inactive. However, membranes that are too fluid cannot support protein function either. Therefore, extreme environments pose a challenge for life, resulting in evolutionary adaptations that include differences in membrane lipid composition.

Evolution of Differences in Membrane Lipid Composition

EVOLUTION Variations in the cell membrane lipid compositions of many species appear to be evolutionary adaptations that maintain the appropriate membrane fluidity under specific environmental conditions. For instance, fishes that live in extreme cold have membranes with a high proportion of unsaturated hydrocarbon tails, enabling their membranes to remain fluid (see Figure 5.5a). At the other extreme, some bacteria and archaea thrive at temperatures greater than 90°C (194°F) in thermal hot springs and geysers. Their membranes include unusual lipids that help prevent excessive fluidity at such high temperatures.

The ability to change the lipid composition of cell membranes in response to changing temperatures has evolved in organisms that live where temperatures vary. In many plants that tolerate extreme cold, such as winter wheat, the percentage of unsaturated phospholipids increases in autumn, keeping the membranes from solidifying during winter. Some bacteria and archaea can also change the proportion of unsaturated phospholipids in their cell membranes, depending on the temperature at which they are growing. Overall, natural selection has apparently favored organisms whose mix of membrane lipids ensures an appropriate level of membrane fluidity for their environment.

▼ **Figure 5.4** **Inquiry**

Do membrane proteins move?

Experiment Larry Frye and Michael Edidin, at Johns Hopkins University, labeled the plasma membrane proteins of a mouse cell and a human cell with two different markers and fused the cells. Using a microscope, they observed the markers on the hybrid cell.

Results

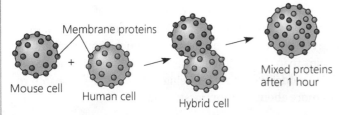

Membrane proteins

Mouse cell + Human cell → Hybrid cell → Mixed proteins after 1 hour

Conclusion The mixing of the mouse and human membrane proteins indicates that at least some membrane proteins move sideways within the plane of the plasma membrane.

Source L. D. Frye and M. Edidin, The rapid intermixing of cell surface antigens after formation of mouse-human heterokaryons, *Journal of Cell Science* 7:319 (1970).

WHAT IF? Suppose the proteins did not mix in the hybrid cell, even many hours after fusion. Would you be able to conclude that proteins don't move within the membrane? What other explanation could there be?

Membrane Proteins and Their Functions

Now we return to the *mosaic* aspect of the fluid mosaic model. Somewhat like a tile mosaic, a membrane is a collage of different proteins embedded in the fluid matrix of the lipid bilayer (see Figure 5.2). More than 50 kinds of proteins have been found so far in the plasma membrane of red blood cells, for example. Phospholipids form the main fabric of the membrane, but proteins determine most of the membrane's functions. Different types of cells contain different sets of membrane proteins, and the various membranes within a cell each have a unique collection of proteins.

Notice in Figure 5.2 that there are two major populations of membrane proteins: integral proteins and peripheral proteins. **Integral proteins** penetrate the hydrophobic interior of the lipid bilayer. The majority are *transmembrane proteins*, which span the membrane; other integral proteins extend only partway into the hydrophobic interior. The hydrophobic regions of an integral protein consist of one or more stretches of nonpolar amino acids (see Figure 3.17), usually coiled into α helices **(Figure 5.6)**. The hydrophilic parts of the molecule are exposed to the aqueous solutions on either side of the membrane. Some proteins also have one or more hydrophilic channels that allow passage of hydrophilic substances (even water itself, see Figure 5.1). **Peripheral proteins** are not embedded in the lipid bilayer at all; they are appendages loosely bound to the surface of the membrane, often to exposed parts of integral proteins (see Figure 5.2).

On the cytoplasmic side of the plasma membrane, some membrane proteins are held in place by attachment to the cytoskeleton. And on the extracellular side, certain membrane proteins are attached to fibers of the extracellular matrix (see Figure 4.26). These attachments combine to give animal cells a stronger framework than the plasma membrane alone could provide.

Figure 5.7 gives an overview of six major functions performed by proteins of the plasma membrane. A single cell may have membrane proteins carrying out several of these

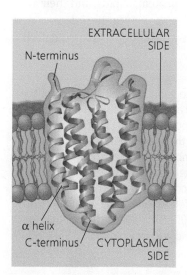

◀ **Figure 5.6 The structure of a transmembrane protein.** Bacteriorhodopsin (a bacterial transport protein) has a distinct orientation in the membrane, with its N-terminus outside the cell and its C-terminus inside. This ribbon model highlights the α-helical secondary structure of the hydrophobic parts, which lie mostly within the hydrophobic interior of the membrane. The protein includes seven transmembrane helices. The nonhelical hydrophilic segments are in contact with the aqueous solutions on the extracellular and cytoplasmic sides of the membrane.

(a) Transport. *Left:* A protein that spans the membrane may provide a hydrophilic channel across the membrane that is selective for a particular solute. *Right:* Other transport proteins shuttle a substance from one side to the other by changing shape. Some of these proteins hydrolyze ATP as an energy source to actively pump substances across the membrane.

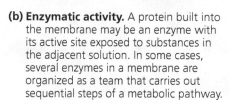

(b) Enzymatic activity. A protein built into the membrane may be an enzyme with its active site exposed to substances in the adjacent solution. In some cases, several enzymes in a membrane are organized as a team that carries out sequential steps of a metabolic pathway.

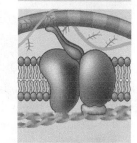

(c) Attachment to the cytoskeleton and extracellular matrix (ECM). Microfilaments or other elements of the cytoskeleton may be noncovalently bound to membrane proteins, a function that helps maintain cell shape and stabilizes the location of certain membrane proteins. Proteins that can bind to ECM molecules can coordinate extracellular and intracellular changes.

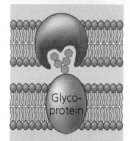

(d) Cell-cell recognition. Some glycoproteins serve as identification tags that are specifically recognized by membrane proteins of other cells. This type of cell-cell binding is usually short-lived compared to that shown in (e).

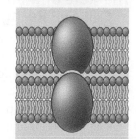

(e) Intercellular joining. Membrane proteins of adjacent cells may hook together in various kinds of junctions, such as gap junctions or tight junctions. This type of binding is more long-lasting than that shown in (d).

(f) Signal transduction. A membrane protein (receptor) may have a binding site with a specific shape that fits the shape of a chemical messenger, such as a hormone. The external messenger (signaling molecule) may cause the protein to change shape, allowing it to relay the message to the inside of the cell, usually by binding to a cytoplasmic protein.

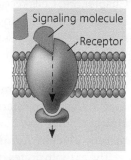

▲ **Figure 5.7 Some functions of membrane proteins.** In many cases, a single protein performs multiple tasks.

? *Some transmembrane proteins can bind to a particular ECM molecule and, when bound, transmit a signal into the cell. Use the proteins shown here to explain how this might occur.*

▼ **Figure 5.8 Synthesis of membrane components and their orientation in the membrane.** The cytoplasmic (orange) face of the plasma membrane differs from the extracellular (aqua) face. The latter arises from the inside face of ER, Golgi, and vesicle membranes.

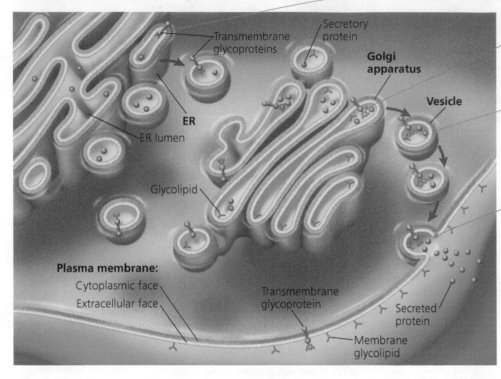

1 Membrane proteins and lipids are synthesized in the endoplasmic reticulum (ER). Carbohydrates (green) are added to the transmembrane proteins (purple dumbbells), making them glycoproteins. The carbohydrate portions may then be modified.

2 Inside the Golgi apparatus, the glycoproteins undergo further carbohydrate modification, and some lipids acquire carbohydrates, becoming glycolipids.

3 The glycoproteins, glycolipids, and secretory proteins (purple spheres) are transported in vesicles to the plasma membrane.

4 As vesicles fuse with the plasma membrane, the outside face of the vesicle becomes continuous with the inside (cytoplasmic) face of the plasma membrane. This releases the secretory proteins from the cell, a process called *exocytosis*, and positions the carbohydrates of membrane glycoproteins and glycolipids on the outside (extracellular) face of the plasma membrane.

DRAW IT *Draw an integral membrane protein extending from partway through the ER membrane into the ER lumen. Next, draw the protein where it would be located in a series of numbered steps ending at the plasma membrane. Would the protein contact the cytoplasm or the extracellular fluid?*

functions, and a single membrane protein may have multiple functions. In this way, the membrane is a functional mosaic as well as a structural one.

The Role of Membrane Carbohydrates in Cell-Cell Recognition

Cell-cell recognition, a cell's ability to distinguish one type of neighboring cell from another, is crucial to the functioning of an organism. It is important, for example, in the sorting of cells into tissues and organs in an animal embryo. It is also the basis for the rejection of foreign cells by the immune system, an important line of defense in vertebrate animals (see Chapter 35). Cells recognize other cells by binding to molecules, often containing carbohydrates, on the extracellular surface of the plasma membrane (see Figure 5.7d).

Membrane carbohydrates are usually short, branched chains of fewer than 15 sugar units. Some are covalently bonded to lipids, forming molecules called **glycolipids**. (Recall that *glyco* refers to the presence of carbohydrate.) However, most are covalently bonded to proteins, which are thereby **glycoproteins**.

The carbohydrates on the extracellular side of the plasma membrane vary from species to species, among individuals of the same species, and even from one cell type to another in a single individual. The diversity of the molecules and their location on the cell's surface enable membrane carbohydrates to

function as markers that distinguish one cell from another. For example, the four human blood types designated A, B, AB, and O reflect variation in the carbohydrate part of glycoproteins on the surface of red blood cells.

Synthesis and Sidedness of Membranes

A membrane has two distinct faces. The two lipid layers may differ in specific lipid composition, and each protein has directional orientation in the membrane (see Figure 5.6, for example). **Figure 5.8** shows how membrane sidedness arises: The asymmetric arrangement of proteins, lipids, and their associated carbohydrates in the plasma membrane is determined as the membrane is being built by the endoplasmic reticulum (ER) and Golgi apparatus.

CONCEPT CHECK 5.1

1. The carbohydrates attached to some proteins and lipids of the plasma membrane are added as the membrane is made and refined in the ER and Golgi apparatus. The new membrane then forms transport vesicles that travel to the cell surface. On which side of the vesicle membrane are the carbohydrates?

2. **WHAT IF?** The soil immediately around hot springs is much warmer than that in neighboring regions. Two closely related species of native grasses are found, one in the warmer region and one in the cooler region. If you analyzed their membrane lipid compositions, what would you expect to find? Explain.

For suggested answers, see Appendix A.

Membrane structure results in selective permeability

The biological membrane is an exquisite example of a supramolecular structure—many molecules ordered into a higher level of organization—with emergent properties beyond those of the individual molecules. We now focus on one of the most important of those properties: the ability to regulate transport across cellular boundaries, a function essential to the cell's existence. We will see once again that form fits function: The fluid mosaic model helps explain how membranes regulate the cell's molecular traffic.

A steady traffic of small molecules and ions moves across the plasma membrane in both directions. Consider the chemical exchanges between a muscle cell and the extracellular fluid that bathes it. Sugars, amino acids, and other nutrients enter the cell, and metabolic waste products leave it. The cell takes in O_2 for use in cellular respiration and expels CO_2. Also, the cell regulates its concentrations of inorganic ions, such as Na^+, K^+, Ca^{2+}, and Cl^-, by shuttling them one way or the other across the plasma membrane. In spite of heavy traffic through them, cell membranes are selectively permeable, and substances do not cross the barrier indiscriminately. The cell is able to take up some small molecules and ions and exclude others. Also, substances that move through the membrane do so at different rates.

The Permeability of the Lipid Bilayer

Nonpolar molecules, such as hydrocarbons, carbon dioxide, and oxygen, are hydrophobic and can therefore dissolve in the lipid bilayer of the membrane and cross it easily, without the aid of membrane proteins. However, the hydrophobic interior of the membrane impedes the direct passage of ions and polar molecules, which are hydrophilic, through the membrane. Polar molecules such as glucose and other sugars pass only slowly through a lipid bilayer, and even water, an extremely small polar molecule, does not cross very rapidly. A charged atom or molecule and its surrounding shell of water (see Figure 2.21) find the hydrophobic interior of the membrane even more difficult to penetrate. Furthermore, the lipid bilayer is only one aspect of the gatekeeper system responsible for the selective permeability of a cell. Proteins built into the membrane play key roles in regulating transport.

Transport Proteins

Cell membranes *are* permeable to specific ions and a variety of polar molecules. These hydrophilic substances can avoid contact with the lipid bilayer by passing through **transport proteins** that span the membrane.

Some transport proteins, called *channel proteins*, function by having a hydrophilic channel that certain molecules or atomic ions use as a tunnel through the membrane (see Figure 5.7a, left). For example, as you read earlier, the passage of water molecules through the plasma membrane of certain cells is greatly facilitated by channel proteins called **aquaporins** (see

Figure 5.1). Most aquaporin proteins consist of four identical subunits (see Figure 3.21). The polypeptide making up each subunit forms a channel that allows single-file passage of up to 3 billion (3×10^9) water molecules per second, many more than would cross the membrane without aquaporin. Other transport proteins, called *carrier proteins*, hold onto their passengers and change shape in a way that shuttles them across the membrane (see Figure 5.7a, right).

A transport protein is specific for the substance it translocates (moves), allowing only a certain substance (or a small group of related substances) to cross the membrane. For example, a specific carrier protein in the plasma membrane of red blood cells transports glucose across the membrane 50,000 times faster than glucose can pass through on its own. This "glucose transporter" is so selective that it even rejects fructose, which has the same molecular formula as glucose.

Thus, the selective permeability of a membrane depends on both the discriminating barrier of the lipid bilayer and the specific transport proteins built into the membrane. But what establishes the *direction* of traffic across a membrane? At a given time, what determines whether a particular substance will enter the cell or leave the cell? And what mechanisms actually drive molecules across membranes? We will address these questions next as we explore two modes of membrane traffic: passive transport and active transport.

CONCEPT CHECK 5.2

1. Two molecules that can cross a lipid bilayer without help from membrane proteins are O_2 and CO_2. What property allows this to occur?
2. Why is a transport protein needed to move water molecules rapidly and in large quantities across a membrane?
3. **MAKE CONNECTIONS** Aquaporins exclude passage of hydronium ions (H_3O^+; see Concept 2.5). Recent research on fat metabolism has shown that some aquaporins allow passage of glycerol, a three-carbon alcohol (see Figure 3.12), as well as H_2O. Since H_3O^+ is much closer in size to water than is glycerol, what do you suppose is the basis of this selectivity?

For suggested answers, see Appendix A.

Passive transport is diffusion of a substance across a membrane with no energy investment

Molecules have a type of energy called thermal energy, which is associated with their constant motion (see Concept 2.5). One result of this motion is **diffusion**, the movement of particles of any substance so that they tend to spread out into the available space. Each molecule moves randomly, yet diffusion of a *population* of molecules may be directional. To understand this process, let's imagine a synthetic membrane separating pure water from a solution of a dye in water. Study

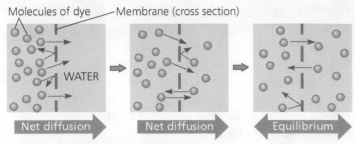

(a) Diffusion of one solute. The membrane has pores large enough for molecules of dye to pass through. Random movement of dye molecules will cause some to pass through the pores; this will happen more often on the side with more dye molecules. The dye diffuses from where it is more concentrated to where it is less concentrated (called diffusing down a concentration gradient). This leads to a dynamic equilibrium: The solute molecules continue to cross the membrane, but at equal rates in both directions.

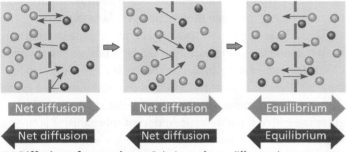

(b) Diffusion of two solutes. Solutions of two different dyes are separated by a membrane that is permeable to both. Each dye diffuses down its own concentration gradient. There will be a net diffusion of the purple dye toward the left, even though the *total* solute concentration was initially greater on the left side.

▲ **Figure 5.9 The diffusion of solutes across a synthetic membrane.** Each of the large arrows under the diagrams shows the net diffusion of the dye molecules of that color.

Figure 5.9a to appreciate how diffusion would result in both solutions having equal concentrations of the dye molecules. Once that point is reached, there will be a dynamic equilibrium, with as many dye molecules crossing the membrane each second in one direction as in the other.

We can now state a simple rule of diffusion: In the absence of other forces, a substance will diffuse from where it is more concentrated to where it is less concentrated. Put another way, any substance will diffuse down its **concentration gradient**, the region along which the density of a substance increases or decreases (in this case, decreases). No work must be done to make this happen; diffusion is a spontaneous process, needing no input of energy. Note that each substance diffuses down its *own* concentration gradient, unaffected by the concentration gradients of other substances **(Figure 5.9b)**.

Much of the traffic across cell membranes occurs by diffusion. When a substance is more concentrated on one side of a membrane than on the other, there is a tendency for the substance to diffuse across the membrane down its concentration gradient (assuming that the membrane is permeable to that substance). One important example is the uptake of oxygen by a cell performing cellular respiration. Dissolved oxygen diffuses

into the cell across the plasma membrane. As long as cellular respiration consumes the O_2 as it enters, diffusion into the cell will continue because the concentration gradient favors movement in that direction.

The diffusion of a substance across a biological membrane is called **passive transport** because the cell does not have to expend energy to make it happen. The concentration gradient itself represents potential energy and drives diffusion. Remember, however, that membranes are selectively permeable and therefore have different effects on the rates of diffusion of various molecules. In the case of water, aquaporins allow water to diffuse very rapidly across the membranes of certain cells. As we'll see next, the movement of water across the plasma membrane has important consequences for cells.

Effects of Osmosis on Water Balance

To see how two solutions with different solute concentrations interact, picture a U-shaped glass tube with a selectively permeable artificial membrane separating two sugar solutions **(Figure 5.10)**. Pores in this synthetic membrane are too small for sugar molecules to pass through but large enough

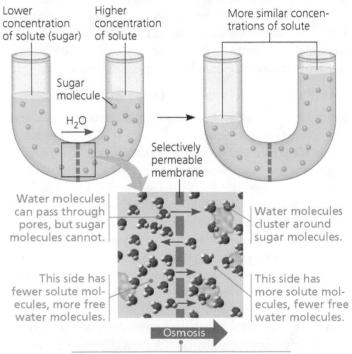

Water moves from an area of higher to lower free water concentration (lower to higher solute concentration).

▲ **Figure 5.10 Osmosis.** Two sugar solutions of different concentrations are separated by a membrane that the solvent (water) can pass through but the solute (sugar) cannot. Water molecules move randomly and may cross in either direction, but overall, water diffuses from the solution with less concentrated solute to that with more concentrated solute. This passive transport of water, called osmosis, reduces the difference in sugar concentrations.

WHAT IF? *If an orange dye capable of passing through the membrane was added to the left side of the tube above, how would it be distributed at the end of the experiment? (See Figure 5.9.) Would the final solution levels in the tube be affected?*

for water molecules. How does this affect the *water* concentration? It seems logical that the solution with the higher concentration of solute would have the lower concentration of water and that water would diffuse into it from the other side for that reason. However, for a dilute solution like most biological fluids, solutes do not affect the water concentration significantly. Instead, tight clustering of water molecules around the hydrophilic solute molecules makes some of the water unavailable to cross the membrane. It is the difference in *free* water concentration that is important. In the end, the effect is the same: Water diffuses across the membrane from the region of lower solute concentration (higher free water concentration) to that of higher solute concentration (lower free water concentration) until the solute concentrations on both sides of the membrane are more nearly equal. The diffusion of free water across a selectively permeable membrane, whether artificial or cellular, is called **osmosis**. The movement of water across cell membranes and the balance of water between the cell and its environment are crucial to organisms. Let's now apply to living cells what you have learned about osmosis in an artificial system.

Water Balance of Cells Without Walls

To explain the behavior of a cell in a solution, we must consider both solute concentration and membrane permeability. Both factors are taken into account in the concept of **tonicity**, the ability of a surrounding solution to cause a cell to gain or lose water. The tonicity of a solution depends in part on its concentration of solutes that cannot cross the membrane (nonpenetrating solutes) relative to that inside the cell. If there is a higher concentration of nonpenetrating solutes in the surrounding solution, water will tend to leave the cell, and vice versa.

If a cell without a wall, such as an animal cell, is immersed in an environment that is **isotonic** to the cell (*iso* means "same"), there will be no *net* movement of water across the plasma membrane. Water diffuses across the membrane, but at the same rate in both directions. In an isotonic environment, the volume of an animal cell is stable **(Figure 5.11a)**.

Now let's transfer the cell to a solution that is **hypertonic** to the cell (*hyper* means "more," in this case referring to nonpenetrating solutes). The cell will lose water, shrivel, and probably die. This is one way an increase in the salinity (saltiness) of a lake can kill the animals there; if the lake water becomes hypertonic to the animals' cells, the cells might shrivel and die. However, taking up too much water can be just as hazardous to an animal cell as losing water. If we place the cell in a solution that is **hypotonic** to the cell (*hypo* means "less"), water will enter the

(a) Animal cell. An animal cell fares best in an isotonic environment unless it has special adaptations that offset the osmotic uptake or loss of water.

(b) Plant cell. Plant cells are turgid (firm) and generally healthiest in a hypotonic environment, where the uptake of water is eventually balanced by the wall pushing back on the cell.

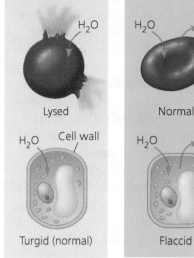
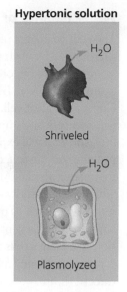

▲ **Figure 5.11 The water balance of living cells.** How living cells react to changes in the solute concentration of their environment depends on whether or not they have cell walls. **(a)** Animal cells, such as this red blood cell, do not have cell walls. **(b)** Plant cells do. (Arrows indicate net water movement after the cells were first placed in these solutions.)

cell faster than it leaves, and the cell will swell and lyse (burst) like an overfilled water balloon.

A cell without rigid walls can tolerate neither excessive uptake nor excessive loss of water. This problem of water balance is automatically solved if such a cell lives in isotonic surroundings. Seawater is isotonic to many marine invertebrates. The cells of most terrestrial (land-dwelling) animals are bathed in an extracellular fluid that is isotonic to the cells. In hypertonic or hypotonic environments, however, organisms that lack rigid cell walls must have other adaptations for **osmoregulation**, the control of solute concentrations and water balance. For example, the unicellular protist *Paramecium caudatum* lives in pond water, which is hypotonic to the cell. Water continually enters the cell. The *P. caudatum* cell doesn't burst because it is equipped with a contractile vacuole, an organelle that functions as a bilge pump to force water out of the cell as fast as it enters by osmosis **(Figure 5.12)**. We will examine other evolutionary adaptations for osmoregulation in Chapter 32.

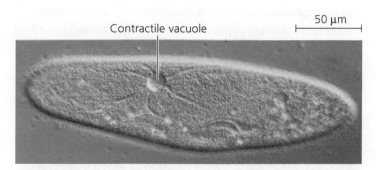

Contractile vacuole

50 μm

▲ **Figure 5.12 The contractile vacuole of *Paramecium caudatum*.** The vacuole collects fluid from a system of canals in the cytoplasm. When full, the vacuole and canals contract, expelling fluid from the cell (LM).

Water Balance of Cells with Walls

The cells of plants, prokaryotes, fungi, and some protists are surrounded by walls (see Figure 4.25). When such a cell is immersed in a hypotonic solution—bathed in rainwater, for example—the wall helps maintain the cell's water balance. Consider a plant cell. Like an animal cell, the plant cell swells as water enters by osmosis (**Figure 5.11b**). However, the relatively inelastic wall will expand only so much before it exerts a back pressure on the cell, called *turgor pressure*, that opposes further water uptake. At this point, the cell is **turgid** (very firm), which is the healthy state for most plant cells. Plants that are not woody, such as most houseplants, depend for mechanical support on cells kept turgid by a surrounding hypotonic solution. If a plant's cells and their surroundings are isotonic, there is no net tendency for water to enter, and the cells become **flaccid** (limp).

However, a wall is of no advantage if the cell is immersed in a hypertonic environment. In this case, a plant cell, like an animal cell, will lose water to its surroundings and shrink. As the plant cell shrivels, its plasma membrane pulls away from the wall. This phenomenon, called **plasmolysis**, causes the plant to wilt and can lead to plant death. The walled cells of bacteria and fungi also plasmolyze in hypertonic environments.

Facilitated Diffusion:
Passive Transport Aided by Proteins

Let's look more closely at how water and certain hydrophilic solutes cross a membrane. As mentioned earlier, many polar molecules and ions impeded by the lipid bilayer of the membrane diffuse passively with the help of transport proteins that span the membrane. This phenomenon is called **facilitated diffusion**. Cell biologists are still trying to learn exactly how various transport proteins facilitate diffusion. Most transport proteins are very specific: They transport some substances but not others.

As mentioned earlier, the two types of transport proteins are channel proteins and carrier proteins. Channel proteins simply provide corridors that allow specific molecules or ions to cross the membrane (**Figure 5.13a**). The hydrophilic passageways provided by these proteins can allow water molecules or small ions to diffuse very quickly from one side of the membrane to the other. Aquaporins, the water channel proteins, facilitate the massive amounts of diffusion that occur in plant cells and in animal cells such as red blood cells. Certain kidney cells also have many aquaporin molecules, allowing them to reclaim water from urine before it is excreted. If the kidneys did not perform this function, you would excrete about 180 L of urine per day—and have to drink an equal volume of water!

Channel proteins that transport ions are called **ion channels**. Many ion channels function as **gated channels**, which open or close in response to a stimulus. For some gated channels, the stimulus is electrical. Certain ion channels in

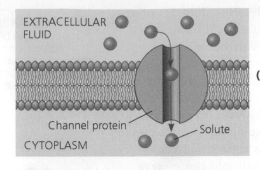

EXTRACELLULAR FLUID

Channel protein

Solute

CYTOPLASM

(a) A channel protein (purple) has a channel through which water molecules or a specific solute can pass.

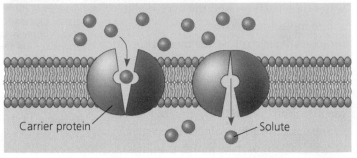

Carrier protein

Solute

(b) A carrier protein alternates between two shapes, moving a solute across the membrane during the shape change.

▲ **Figure 5.13 Two types of transport proteins that carry out facilitated diffusion.** In both cases, the protein can transport the solute in either direction, but the net movement is down the concentration gradient of the solute.

nerve cells, for example, open in response to an electrical stimulus, allowing potassium ions to leave the cell. Other gated channels open or close when a specific substance other than the one to be transported binds to the channel. Both types of gated channels are important in the functioning of the nervous system (as you'll learn in Chapter 37).

Carrier proteins, such as the glucose transporter mentioned earlier, seem to undergo a subtle change in shape that somehow translocates the solute-binding site across the membrane (**Figure 5.13b**). Such a change in shape may be triggered by the binding and release of the transported molecule. Like ion channels, carrier proteins involved in facilitated diffusion result in the net movement of a substance down its concentration gradient. No energy input is required: This is passive transport. The **Scientific Skills Exercise** gives you an opportunity to work with data from an experiment related to glucose transport.

CONCEPT CHECK 5.3

1. How do you think a cell performing cellular respiration rids itself of the resulting CO_2?
2. In the supermarket, produce is often sprayed with water. Explain why this makes vegetables look crisp.
3. **WHAT IF?** If a *Paramecium caudatum* cell swims from a hypotonic to an isotonic environment, will its contractile vacuole become more active or less? Why?

For suggested answers, see Appendix A.

Interpreting a Graph with Two Sets of Data

Is Glucose Uptake into Cells Affected by Age? Glucose, an important energy source for animals, is transported into cells by facilitated diffusion using protein carriers. In this exercise, you will interpret a graph with two sets of data from an experiment that examined glucose uptake over time in red blood cells from guinea pigs of different ages. You will determine if the age of the guinea pigs affected their cells' rate of glucose uptake.

How the Experiment Was Done Researchers incubated guinea pig red blood cells in a 300 mM (millimolar) radioactive glucose solution at pH 7.4 at 25°C. Every 10 or 15 minutes, they removed a sample of cells from the solution and measured the concentration of radioactive glucose inside those cells. The cells came from either a 15-day-old guinea pig or a 1-month-old guinea pig.

Data from the Experiment When you have multiple sets of data, it can be useful to plot them on the same graph for comparison. In the graph here, each set of dots (dots of the same color) forms a *scatter plot,* in which every data point represents two numerical values, one for each variable. For each data set, a curve that best fits the points has been drawn to make it easier to see the trends. (For additional information about graphs, see the Scientific Skills Review in Appendix F and in the Study Area in MasteringBiology.)

Interpret the Data
1. First make sure you understand the parts of the graph. (a) Which variable is the independent variable—the variable that was controlled by the researchers? (b) Which variable is the dependent variable—the variable that depended on the treatment and was measured by the researchers? (c) What do the red dots represent? (d) What do the blue dots represent?
2. From the data points on the graph, construct a table of the data. Put "Incubation Time (min)" in the left column of the table.
3. What does the graph show? Compare and contrast glucose uptake in red blood cells from 15-day-old guinea pigs and from 1-month-old guinea pigs.
4. Develop a hypothesis to explain the difference between glucose uptake in red blood cells from 15-day-old guinea pigs

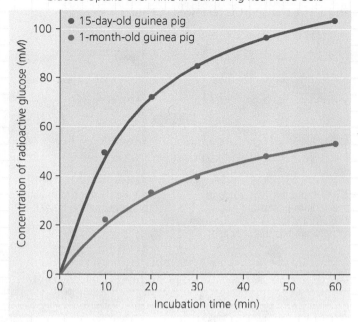

Glucose Uptake Over Time in Guinea Pig Red Blood Cells

and from 1-month-old guinea pigs. Think about how glucose gets into cells.

5. Design an experiment to test your hypothesis.

Data from T. Kondo and E. Beutler, Developmental changes in glucose transport of guinea pig erythrocytes, *Journal of Clinical Investigation* 65:1-4 (1980).

(MB) A version of this Scientific Skills Exercise can be assigned in MasteringBiology.

CONCEPT 5.4

Active transport uses energy to move solutes against their gradients

Despite the help of transport proteins, facilitated diffusion is considered passive transport because the solute is moving down its concentration gradient, a process that requires no energy. Facilitated diffusion speeds transport of a solute by providing efficient passage through the membrane, but it does not alter the direction of transport. Some transport proteins, however, can move solutes against their concentration gradients, across the plasma membrane from the side where they are less concentrated (whether inside or outside) to the side where they are more concentrated.

The Need for Energy in Active Transport

To pump a solute across a membrane against its gradient requires work; the cell must expend energy. Therefore, this type of membrane traffic is called **active transport**. Active transport enables a cell to maintain internal concentrations of small solutes that differ from concentrations in its environment. For example, compared with its surroundings, an animal cell has a much higher concentration of potassium ions (K^+) and a much lower concentration of sodium ions (Na^+). The plasma membrane helps maintain these steep gradients by pumping Na^+ out of the cell and K^+ into the cell.

As in other types of cellular work, ATP supplies the energy for most active transport. One way ATP can power active transport is by transferring its terminal phosphate group directly to the transport protein. This can induce the protein to change its shape in a manner that translocates a solute bound to the protein across the membrane. One transport system that works this way is the **sodium-potassium pump**, which exchanges Na$^+$ for K$^+$ across the plasma membrane of animal cells **(Figure 5.14)**. The distinction between passive transport and active transport is reviewed in **Figure 5.15**.

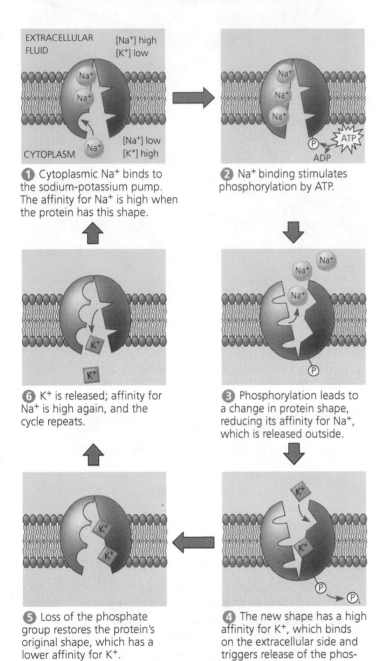

① Cytoplasmic Na$^+$ binds to the sodium-potassium pump. The affinity for Na$^+$ is high when the protein has this shape.

[Na$^+$] high
[K$^+$] low

EXTRACELLULAR FLUID

[Na$^+$] low
[K$^+$] high

CYTOPLASM

② Na$^+$ binding stimulates phosphorylation by ATP.

③ Phosphorylation leads to a change in protein shape, reducing its affinity for Na$^+$, which is released outside.

④ The new shape has a high affinity for K$^+$, which binds on the extracellular side and triggers release of the phosphate group.

⑤ Loss of the phosphate group restores the protein's original shape, which has a lower affinity for K$^+$.

⑥ K$^+$ is released; affinity for Na$^+$ is high again, and the cycle repeats.

▲ **Figure 5.14 The sodium-potassium pump: a specific case of active transport.** This transport system pumps ions against steep concentration gradients: Sodium ion concentration ([Na$^+$]) is high outside the cell and low inside, while potassium ion concentration ([K$^+$]) is low outside the cell and high inside. The pump oscillates between two shapes in a cycle that moves 3 Na$^+$ out of the cell for every 2 K$^+$ pumped into the cell. The two shapes have different affinities for Na$^+$ and K$^+$. ATP powers the shape change by transferring a phosphate group to the transport protein (phosphorylating the protein).

▼ **Figure 5.15   Review: passive and active transport.**

Passive transport. Substances diffuse spontaneously down their concentration gradients, crossing a membrane with no expenditure of energy by the cell. The rate of diffusion can be greatly increased by transport proteins in the membrane.

Active transport. Some transport proteins act as pumps, moving substances across a membrane against their concentration (or electrochemical) gradients. Energy for this work is usually supplied by ATP.

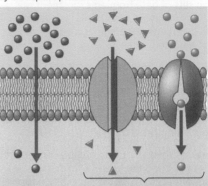

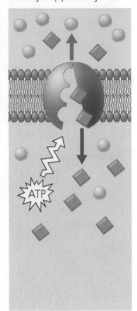

Diffusion. Hydrophobic molecules and (at a slow rate) very small uncharged polar molecules can diffuse through the lipid bilayer.

Facilitated diffusion. Many hydrophilic substances diffuse through membranes with the assistance of transport proteins, either channel proteins (left) or carrier proteins (right).

? *For each of the two solutes in the right panel, describe its direction of movement, and state whether it is going with or against its concentration gradient.*

How Ion Pumps Maintain Membrane Potential

All cells have voltages across their plasma membranes. Voltage is electrical potential energy—a separation of opposite charges. The cytoplasmic side of the membrane is negative in charge relative to the extracellular side because of an unequal distribution of anions and cations on the two sides. The voltage across a membrane, called a **membrane potential**, ranges from about −50 to −200 millivolts (mV). (The minus sign indicates that the inside of the cell is negative relative to the outside.)

The membrane potential acts like a battery, an energy source that affects the traffic of all charged substances across the membrane. Because the inside of the cell is negative compared with the outside, the membrane potential favors the passive transport of cations into the cell and anions out of the cell. Thus, *two* forces drive the diffusion of ions across a membrane: a chemical force (the ion's concentration gradient) and an electrical force (the effect of the membrane potential on the ion's movement). This combination of forces acting on an ion is called the **electrochemical gradient**.

In the case of ions, then, we must refine our concept of passive transport: An ion diffuses not simply down its concentration gradient but, more exactly, down its *electrochemical* gradient. For example, consider the cation Na^+. The concentration of Na^+ inside a resting nerve cell is much lower than outside it. When the cell is stimulated, gated channels open that facilitate Na^+ diffusion. Sodium ions then "fall" down their electrochemical gradient, driven by the concentration gradient of Na^+ and by the attraction of these cations to the negative side (inside) of the membrane. In this example, both electrical and chemical contributions to the electrochemical gradient act in the same direction across the membrane, but this is not always so. In cases where electrical forces due to the membrane potential oppose the simple diffusion of an ion down its concentration gradient, active transport may be necessary. Electrochemical gradients and membrane potentials are important in the transmission of nerve impulses (as you'll learn in Chapter 37).

Some membrane proteins that actively transport ions contribute to the membrane potential. An example is the sodium-potassium pump. Notice in Figure 5.14 that the pump does not translocate Na^+ and K^+ one for one, but pumps three sodium ions out of the cell for every two potassium ions it pumps into the cell. With each "crank" of the pump, there is a net transfer of one positive charge from the cytoplasm to the extracellular fluid, a process that stores energy as voltage. A transport protein that generates voltage across a membrane is called an **electrogenic pump**. The sodium-potassium pump appears to be the major electrogenic pump of animal cells. The main electrogenic pump of plants, fungi, and bacteria is a **proton pump**, which actively transports protons (hydrogen ions, H^+) out of the cell. The pumping of H^+ transfers positive charge from the cytoplasm to the extracellular solution **(Figure 5.16)**. By generating voltage across membranes, electrogenic pumps help store energy that can be tapped for cellular work. One important use of proton gradients in the cell is for ATP synthesis during cellular respiration (as you will see in Chapter 7). Another is a type of membrane traffic called cotransport.

Cotransport: Coupled Transport by a Membrane Protein

A single ATP-powered pump that transports a specific solute can indirectly drive the active transport of several other solutes in a mechanism called **cotransport**. A substance that has been pumped across a membrane can do work as it moves back across the membrane by diffusion, analogous to water that has been pumped uphill and performs work as it flows back down. Another transport protein, a cotransporter separate from the pump, can couple the "downhill" diffusion of this substance to the "uphill" transport of a second substance against its own concentration (or electrochemical) gradient. For example, a plant cell uses the gradient of H^+ generated by its proton pumps to drive the active transport of sugars, amino acids, and several other nutrients into the cell. One transport protein couples the return of H^+ to the transport of sucrose into the cell **(Figure 5.17)**. This protein can translocate sucrose into the cell against a concentration gradient, but only if the sucrose molecule travels in the company of a hydrogen ion. The hydrogen ion uses the transport protein as an avenue to diffuse down the electrochemical gradient maintained by the proton pump. Plants use sucrose-H^+ cotransport to load sucrose produced by photosynthesis into cells in the veins of leaves. The vascular tissue of the plant can then distribute the sugar to nonphotosynthetic organs, such as roots.

What we know about cotransport proteins in animal cells has helped us find more effective treatments for diarrhea, a serious problem in developing countries. Normally, sodium in waste is reabsorbed in the colon, maintaining constant levels

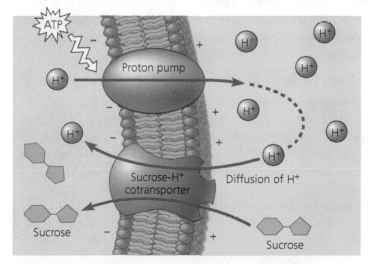

▲ **Figure 5.17 Cotransport: active transport driven by a concentration gradient.** A carrier protein, such as this sucrose-H^+ cotransporter in a plant cell, is able to use the diffusion of H^+ down its electrochemical gradient into the cell to drive the uptake of sucrose. The H^+ gradient is maintained by an ATP-driven proton pump that concentrates H^+ outside the cell, thus storing potential energy that can be used for active transport, in this case of sucrose. Thus, ATP indirectly provides the energy necessary for cotransport. (The cell wall is not shown.)

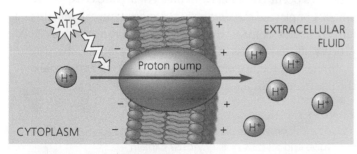

▲ **Figure 5.16 A proton pump.** Proton pumps are electrogenic pumps that store energy by generating voltage (charge separation) across membranes. A proton pump translocates positive charge in the form of hydrogen ions (that is, protons). The voltage and H^+ concentration gradient represent a dual energy source that can drive other processes, such as the uptake of nutrients. Most proton pumps are powered by ATP.

in the body, but diarrhea expels waste so rapidly that reabsorption is not possible, and sodium levels fall precipitously. To treat this life-threatening condition, patients are given a solution to drink containing high concentrations of salt (NaCl) and glucose. The solutes are taken up by sodium-glucose cotransporters on the surface of intestinal cells and passed through the cells into the blood. This simple treatment has lowered infant mortality worldwide.

CONCEPT CHECK 5.4

1. Sodium-potassium pumps help nerve cells establish a voltage across their plasma membranes. Do these pumps use ATP or produce ATP? Explain.
2. Explain why the sodium-potassium pump in Figure 5.14 would not be considered a cotransporter.
3. **MAKE CONNECTIONS** Review the characteristics of the lysosome discussed in Concept 4.4. Given the internal environment of a lysosome, what transport protein might you expect to see in its membrane?

For suggested answers, see Appendix A.

CONCEPT 5.5

Bulk transport across the plasma membrane occurs by exocytosis and endocytosis

Water and small solutes enter and leave the cell by diffusing through the lipid bilayer of the plasma membrane or by being moved across the membrane by transport proteins. However, large molecules, such as proteins and polysaccharides, as well as larger particles, generally cross the membrane in bulk by mechanisms that involve packaging in vesicles. Like active transport, these processes require energy.

Exocytosis

The cell secretes certain biological molecules by the fusion of vesicles with the plasma membrane; this process is called **exocytosis**. A transport vesicle that has budded from the Golgi apparatus moves along microtubules of the cytoskeleton to the plasma membrane. When the vesicle membrane and plasma membrane come into contact, specific proteins rearrange the lipid molecules of the two bilayers so that the two membranes fuse. The contents of the vesicle then spill to the outside of the cell, and the vesicle membrane becomes part of the plasma membrane (see Figure 5.8, step 4).

Many secretory cells use exocytosis to export products. For example, the cells in the pancreas that make insulin secrete it into the extracellular fluid by exocytosis. In another example, nerve cells use exocytosis to release neurotransmitters that signal other neurons or muscle cells. When plant cells are making walls, exocytosis delivers proteins and carbohydrates from Golgi vesicles to the outside of the cell.

Endocytosis

In **endocytosis**, the cell takes in molecules and particulate matter by forming new vesicles from the plasma membrane. Although the proteins involved in the two processes are different, the events of endocytosis look like the reverse of exocytosis. A small area of the plasma membrane sinks inward to form a pocket. As the pocket deepens, it pinches in, forming a vesicle containing material that had been outside the cell. Study **Figure 5.18** carefully to understand three types of endocytosis: phagocytosis ("cellular eating"), pinocytosis ("cellular drinking"), and receptor-mediated endocytosis.

Human cells use receptor-mediated endocytosis to take in cholesterol for membrane synthesis and the synthesis of other steroids. Cholesterol travels in the blood in particles called low-density lipoproteins (LDLs), each a complex of lipids and a protein. LDLs bind to LDL receptors on plasma membranes and then enter the cells by endocytosis. In the inherited disease familial hypercholesterolemia, LDLs cannot enter cells because the LDL receptor proteins are defective or missing:

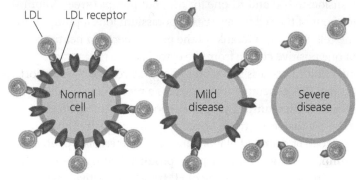

Consequently, in people with the disease, a large amount of cholesterol accumulates in the blood, where it contributes to early atherosclerosis, the buildup of lipid deposits within the walls of blood vessels. This buildup narrows the space in the vessels and impedes blood flow.

Endocytosis and exocytosis also provide mechanisms for rejuvenating or remodeling the plasma membrane. These processes occur continually in most eukaryotic cells, yet the amount of plasma membrane in a nongrowing cell remains fairly constant. Apparently, the addition of membrane by one process offsets the loss of membrane by the other.

In the final section of this chapter, we'll look at the role of the plasma membrane and its proteins in cell signaling.

CONCEPT CHECK 5.5

1. As a cell grows, its plasma membrane expands. Does this involve endocytosis or exocytosis? Explain.
2. **DRAW IT** Return to Figure 5.8, and circle a patch of plasma membrane that is coming from a vesicle involved in exocytosis.
3. **MAKE CONNECTIONS** In Concept 4.7, you learned that animal cells make an extracellular matrix (ECM). Describe the cellular pathway of synthesis and deposition of an ECM glycoprotein.

For suggested answers, see Appendix A.

Phagocytosis

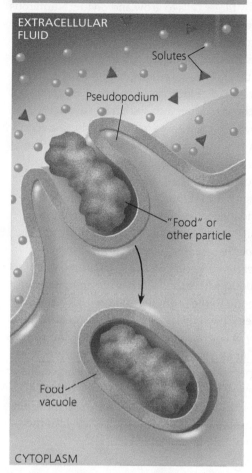

EXTRACELLULAR FLUID

Solutes

Pseudopodium

"Food" or other particle

Food vacuole

CYTOPLASM

In **phagocytosis**, a cell engulfs a particle by wrapping pseudopodia (singular, *pseudopodium*) around it and packaging it within a membranous sac called a food vacuole. The particle will be digested after the food vacuole fuses with a lysosome containing hydrolytic enzymes (see Figure 4.12).

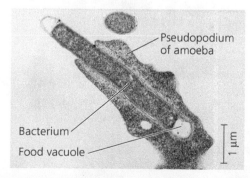

Pseudopodium of amoeba

Bacterium

Food vacuole

1 μm

An amoeba engulfing a bacterium via phago-cytosis (TEM).

Pinocytosis

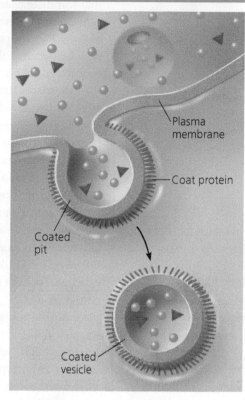

Plasma membrane

Coat protein

Coated pit

Coated vesicle

In **pinocytosis**, a cell continually "gulps" droplets of extracellular fluid into tiny vesicles. In this way, the cell obtains molecules dissolved in the droplets. Because any and all solutes are taken into the cell, pinocytosis as shown here is nonspecific for the substances it transports. In many cases, as above, the parts of the plasma membrane that form vesicles are lined on their cytoplasmic side by a fuzzy layer of coat protein; the "pits" and resulting vesicles are said to be "coated."

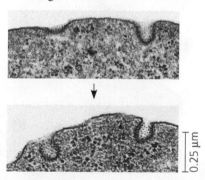

0.25 μm

Pinocytotic vesicles forming (TEMs).

 ANIMATION *BioFlix* Visit the Study Area in **MasteringBiology** for the BioFlix® 3-D Animation on Membrane Transport.

Receptor-Mediated Endocytosis

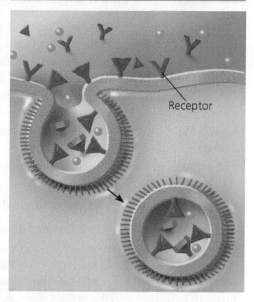

Receptor

Receptor-mediated endocytosis is a specialized type of pinocytosis that enables the cell to acquire bulk quantities of specific substances, even though those substances may not be very concentrated in the extracellular fluid. Embedded in the plasma membrane are proteins with receptor sites exposed to the extracellular fluid. Specific solutes bind to the sites. The receptor proteins then cluster in coated pits, and each coated pit forms a vesicle containing the bound molecules. Notice that there are relatively more bound molecules (purple triangles) inside the vesicle, but other molecules (green balls) are also present. After the ingested material is liberated from the vesicle, the emptied receptors are recycled to the plasma membrane by the same vesicle (not shown).

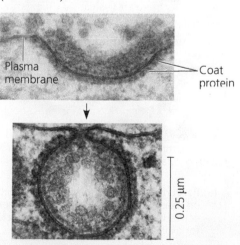

Plasma membrane

Coat protein

0.25 μm

Top: A coated pit. *Bottom*: A coated vesicle forming during receptor-mediated endocytosis (TEMs).

The plasma membrane plays a key role in most cell signaling

In a multicellular organism, whether a human being or an oak tree, it is cell-to-cell communication that allows the trillions of cells of the body to coordinate their activities, and the communication process usually involves the cells' plasma membranes. In fact, communication between cells is also essential for many unicellular organisms, including prokaryotes. However, here we will focus on cell signaling in animals and plants. We'll describe the main mechanisms by which cells receive, process, and respond to chemical signals sent from other cells.

Local and Long-Distance Signaling

The chemical messages sent out from cells are targeted for other cells that may or may not be immediately adjacent. As discussed earlier in this chapter and in Chapter 4, eukaryotic cells may communicate by direct contact, a type of local signaling. Both animals and plants have cell junctions that, where present, directly connect the cytoplasms of adjacent cells; in animals, these are gap junctions (see Figure 4.27), and in plants, plasmodesmata (see Figure 4.25). In these cases, signaling substances dissolved in the cytosol can pass freely between adjacent cells. Also, animal cells may communicate via direct contact between membrane-bound cell-surface molecules in cell-cell recognition (see Figure 5.7d). This sort of local signaling is important in embryonic development and in the immune response.

In many other cases of local signaling, the signaling cell secretes messenger molecules. Some of these, which are called **local regulators**, travel only short distances. One class of local regulators in animals, *growth factors*, consists of compounds that stimulate nearby target cells to grow and divide. Numerous cells can simultaneously receive and respond to the molecules of growth factor produced by a nearby cell. This type of local signaling in animals is called *paracrine signaling* (**Figure 5.19a**). (Local signaling in plants is discussed in Chapter 31.)

A more specialized type of local signaling called *synaptic signaling* occurs in the animal nervous system (**Figure 5.19b**). An electrical signal moving along a nerve cell triggers the secretion of neurotransmitter molecules carrying a chemical signal. These molecules diffuse across the synapse, the narrow space between the nerve cell and its target cell (often another nerve cell), triggering a response in the target cell.

Both animals and plants use chemicals called **hormones** for long-distance signaling. In hormonal signaling in animals, also known as *endocrine signaling*, specialized cells release hormone molecules, which travel via the circulatory system to other parts of the body, where they reach target cells that can recognize and respond to the hormones (**Figure 5.19c**). Most plant hormones (see Chapter 31) reach distant targets via plant vascular tissues (xylem or phloem; see Chapter 28), but some travel through the air as a gas. Hormones vary widely in molecular size and type, as do local regulators. For instance, the plant hormone ethylene, a gas that promotes fruit ripening, is a hydrocarbon of only six atoms (C_2H_4). In contrast, the mammalian hormone insulin, which regulates sugar levels in the blood, is a protein with thousands of atoms.

▼ **Figure 5.19 Local and long-distance cell signaling by secreted molecules in animals.** In both local and long-distance signaling, only specific target cells that can recognize a given signaling molecule will respond to it.

Local signaling

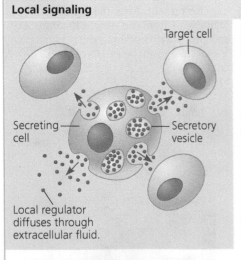

(a) Paracrine signaling. A secreting cell acts on nearby target cells by discharging molecules of a local regulator (a growth factor, for example) into the extracellular fluid.

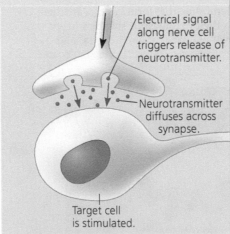

(b) Synaptic signaling. A nerve cell releases neurotransmitter molecules into a synapse, stimulating the target cell.

Long-distance signaling

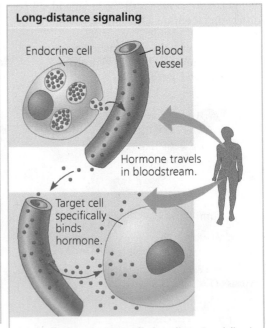

(c) Endocrine (hormonal) signaling. Specialized endocrine cells secrete hormones into body fluids, often blood. Hormones reach virtually all body cells, but are bound only by some cells.

The transmission of an electrical signal along the length of a single nerve cell can also be long-distance signaling, because nerve cells can be quite long. Jumping from cell to cell via synapses, a nerve signal can quickly travel great distances—from your brain to your big toe, for example. (This type of long-distance signaling is covered in detail in Chapter 37.)

What happens when a cell encounters a secreted signaling molecule? We will now consider this question, beginning with a bit of historical background.

The Three Stages of Cell Signaling: *A Preview*

Our current understanding of how chemical messengers act on cells had its origins in the pioneering work of the American Earl W. Sutherland about a half-century ago. He was investigating how the animal hormone epinephrine (also called adrenaline) stimulates the breakdown of the storage polysaccharide glycogen within liver cells and skeletal muscle cells. (This breakdown yields glucose molecules for use by the body.)

Sutherland's research team discovered that epinephrine never actually enters the glycogen-containing cells, and this discovery provided two insights. First, epinephrine does not interact directly with the enzyme responsible for glycogen breakdown; an intermediate step or series of steps must be occurring inside the cell. Second, the plasma membrane must somehow be involved in transmitting the signal. Sutherland's research suggested that the process going on at the receiving end of a cell-to-cell message can be divided into three stages: reception, transduction, and response **(Figure 5.20)**: **❶ Reception** is the target cell's detection of a signaling molecule coming from outside the cell. A chemical signal is "detected" when the signaling molecule binds to a receptor protein located at the cell's surface or, in some cases, inside the cell. **❷ Transduction** is a step or series of steps that converts the signal to a form that can bring about a specific cellular response. Transduction usually requires a sequence of changes in a series of different molecules—a **signal transduction pathway**. The molecules in the pathway are often called relay molecules. **❸** In the third stage of cell signaling, the transduced signal finally triggers a specific cellular **response**. The response may be almost any imaginable cellular activity—such as catalysis by an enzyme (for example, the enzyme that breaks down glycogen), rearrangement of the cytoskeleton, or activation of specific genes in the nucleus. The cell-signaling process helps ensure that crucial activities like these occur in the right cells, at the right time, and in proper coordination with the activities of other cells of the organism. We'll now explore the mechanisms of cell signaling in more detail.

Reception, the Binding of a Signaling Molecule to a Receptor Protein

A radio station broadcasts its signal indiscriminately, but it can be picked up only by radios tuned to the right wavelength; reception of the signal depends on the receiver. Similarly, in the case of epinephrine, the hormone encounters many types of cells as it circulates in the blood, but only certain target cells detect and react to the hormone molecule. A receptor protein on or in the target cell allows the cell to detect the signal and respond to it. The signaling molecule is complementary in shape to a specific site on the receptor and attaches there, like a key in a lock. The signaling molecule behaves as a **ligand**, a molecule that specifically binds to another molecule, often a larger one. (LDLs, mentioned in Concept 5.5, act as ligands when they bind to their receptors, as do the molecules that bind to enzymes; see Figure 3.16.) Ligand binding generally causes a receptor protein to undergo a change in shape. For many receptors, this shape change directly activates the receptor, enabling it to interact with other cellular molecules.

Most signal receptors are plasma membrane proteins. Their ligands are water-soluble and generally too large to pass freely through the plasma membrane. Other signal receptors, however, are located inside the cell. We discuss both of these types next.

Receptors in the Plasma Membrane

Most water-soluble signaling molecules bind to specific sites on receptor proteins that span the cell's plasma membrane. Such a transmembrane receptor transmits information from the extracellular environment to the inside of the cell by changing shape when a specific ligand binds to it. We can see how transmembrane receptors work by looking at two major types: G protein-coupled receptors and ligand-gated ion channels.

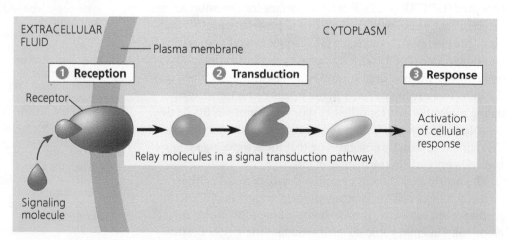

▲ **Figure 5.20 Overview of cell signaling.** From the perspective of the cell receiving the message, cell signaling can be divided into three stages: signal reception, signal transduction, and cellular response. When reception occurs at the plasma membrane, as shown here, the transduction stage is usually a pathway of several steps, with each relay molecule in the pathway bringing about a change in the next molecule. The final molecule in the pathway triggers the cell's response. The three stages are explained in more detail in the text.

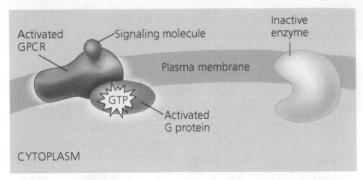

① When the appropriate signaling molecule binds to the extracellular side of the receptor, the receptor is activated and changes shape. Its cytoplasmic side then binds and activates a G protein. The activated G protein carries a GTP molecule.

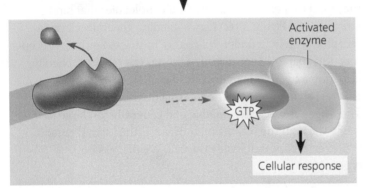

② The activated G protein leaves the receptor, diffuses along the membrane, and then binds to an enzyme, altering the enzyme's shape and activity. Once activated, the enzyme can trigger the next step leading to a cellular response. Binding of signaling molecules is reversible. The activating change in the GPCR, as well as the changes in the G protein and enzyme, are only temporary; these molecules soon become available for reuse.

© 1996 Pearson Education, Inc.

▲ **Figure 5.21 A G protein-coupled receptor (GPCR) in action.**

Figure 5.21 shows the functioning of a **G protein-coupled receptor (GPCR)**. A GPCR is a cell-surface transmembrane receptor that works with the help of a **G protein**, a protein that binds the energy-rich molecule GTP, which is similar to ATP (see end of Concept 3.1). Many signaling molecules, including epinephrine, many other hormones, and neurotransmitters, use GPCRs. These receptors vary in the binding sites for their signaling molecules and for different types of G proteins inside the cell. Nevertheless, all GPCRs and many G proteins are remarkably similar in structure, suggesting that these signaling systems evolved very early in the history of life.

The nearly 1,000 GPCRs examined to date make up the largest family of cell-surface receptors in mammals. GPCR pathways are extremely diverse in their functions, which include roles in embryonic development and the senses of smell and taste. They are also involved in many human diseases. For example, cholera, pertussis (whooping cough), and botulism are caused by bacterial toxins that interfere with G protein function. Up to 60% of all medicines used today exert their effects by influencing G protein pathways.

A **ligand-gated ion channel** is a membrane receptor that has a region that can act as a "gate" for ions when the receptor assumes a certain shape **(Figure 5.22)**. When a signaling molecule binds as a ligand to the receptor protein, the gate opens or closes, allowing or blocking the diffusion of specific ions, such as Na^+ or Ca^{2+}, through a channel in the protein. Like other membrane receptors, these proteins bind the ligand at a specific site on their extracellular side.

Ligand-gated ion channels are very important in the nervous system. For example, the neurotransmitter molecules released at a synapse between two nerve cells (see Figure 5.19b) bind as ligands to ion channels on the receiving cell, causing the channels to open. The diffusion of ions through the open channels may trigger an electrical signal that propagates down the length of the receiving cell. (You'll learn more about ion channels in Chapter 37.)

Intracellular Receptors

Intracellular receptor proteins are found in either the cytoplasm or nucleus of target cells. To reach such a receptor, a chemical messenger passes through the target cell's plasma membrane. A

① Here we show a ligand-gated ion channel receptor in which the gate remains closed until a ligand binds to the receptor.

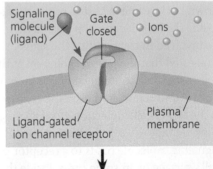

② When the ligand binds to the receptor and the gate opens, specific ions can flow through the channel and rapidly change the concentration of that particular ion inside the cell. This change may directly affect the activity of the cell in some way.

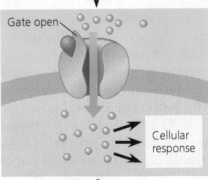

③ When the ligand dissociates from this receptor, the gate closes and ions no longer enter the cell.

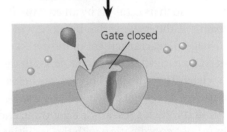

▲ **Figure 5.22 Ion channel receptor.** This is a ligand-gated ion channel, a type of receptor protein that regulates the passage of specific ions across the membrane. Whether the channel is open or closed depends on whether a specific ligand is bound to the protein.

number of important signaling molecules can do this because they are hydrophobic enough to cross the hydrophobic interior of the membrane. Such hydrophobic chemical messengers include the steroid hormones and thyroid hormones of animals. In both animals and plants, another chemical signaling molecule with an intracellular receptor is nitric oxide (NO), a gas; its very small, hydrophobic molecules can easily pass between the membrane phospholipids.

The behavior of testosterone is representative of steroid hormones. In males, the hormone is secreted by cells of the testes. It then travels through the blood and enters cells all over the body. However, only cells that contain receptors for testosterone respond. In these cells, the hormone binds to the receptor protein, activating it **(Figure 5.23)**. With the hormone attached, the active form of the receptor protein then enters the nucleus and turns on specific genes that control male sex characteristics.

How does the activated hormone-receptor complex turn on genes? Recall that the genes in a cell's DNA function by being transcribed and processed into messenger RNA (mRNA), which leaves the nucleus and is translated into a specific protein by ribosomes in the cytoplasm. Special proteins called *transcription factors* control which genes are turned on—that is,

which genes are transcribed into mRNA—in a particular cell at a particular time. The testosterone receptor, when activated, acts as a transcription factor that turns on specific genes.

By acting as a transcription factor, the testosterone receptor itself carries out the complete transduction of the signal. Most other intracellular receptors function in the same way, although many of them, such as the thyroid hormone receptor, are already in the nucleus before the signaling molecule reaches them. Interestingly, many of these intracellular receptor proteins are structurally similar, suggesting an evolutionary kinship.

Transduction by Cascades of Molecular Interactions

When receptors for signaling molecules are plasma membrane proteins, like most of those we have discussed, the transduction stage of cell signaling is usually a multistep pathway. Steps often include activation of proteins by addition or removal of phosphate groups or release of other small molecules or ions that act as messengers. One benefit of multiple steps is the possibility of greatly amplifying a signal. If some of the molecules in a pathway transmit the signal to numerous molecules at the next step in the series, the result can be a large number of activated molecules at the end of the pathway. Moreover, multistep pathways provide more opportunities for coordination and regulation than simpler systems do.

The binding of a specific signaling molecule to a receptor in the plasma membrane triggers the first step in the chain of molecular interactions—the signal transduction pathway—that leads to a particular response within the cell. Like falling dominoes, the signal-activated receptor activates another molecule, which activates yet another molecule, and so on, until the protein that produces the final cellular response is activated. The molecules that relay a signal from receptor to response, which we call relay molecules in this book, are often proteins. The interaction of proteins is a major theme of cell signaling.

Keep in mind that the original signaling molecule is not physically passed along a signaling pathway; in most cases, it never even enters the cell. When we say that the signal is relayed along a pathway, we mean that certain information is passed on. At each step, the signal is transduced into a different form, commonly via a shape change in a protein. Very often, the shape change is brought about by phosphorylation, the addition of phosphate groups to a protein (see Figure 3.5).

Protein Phosphorylation and Dephosphorylation

The phosphorylation of proteins and its reverse, dephosphorylation, are a widespread cellular mechanism for regulating protein activity. An enzyme that transfers phosphate groups from ATP to a protein is known as a **protein kinase**. Such enzymes are widely involved in signaling pathways in animals, plants, and fungi.

Many of the relay molecules in signal transduction pathways are protein kinases, and they often act on other protein kinases in the pathway. A hypothetical pathway containing

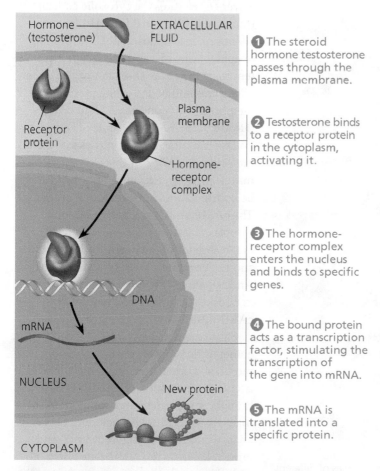

Hormone (testosterone) — **EXTRACELLULAR FLUID**

Receptor protein

Plasma membrane

Hormone-receptor complex

DNA

mRNA

NUCLEUS

New protein

CYTOPLASM

❶ The steroid hormone testosterone passes through the plasma membrane.

❷ Testosterone binds to a receptor protein in the cytoplasm, activating it.

❸ The hormone-receptor complex enters the nucleus and binds to specific genes.

❹ The bound protein acts as a transcription factor, stimulating the transcription of the gene into mRNA.

❺ The mRNA is translated into a specific protein.

▲ **Figure 5.23 Steroid hormone interacting with an intracellular receptor.**

? *Why is a cell-surface receptor protein not required for this steroid hormone to enter the cell?*

two different protein kinases that form a short "phosphorylation cascade" is depicted in **Figure 5.24**. The sequence shown is similar to many known pathways, although typically three protein kinases are involved. The signal is transmitted by a cascade of protein phosphorylations, each bringing with it a shape change. Each such shape change results from the interaction of the newly added phosphate groups with charged or polar amino acids (see Figure 3.17). The addition of phosphate groups often changes the form of a protein from inactive to active.

The importance of protein kinases can hardly be overstated. About 2% of our own genes are thought to code for protein kinases. A single cell may have hundreds of different kinds, each specific for a different protein. Together, they probably regulate a large proportion of the thousands of proteins in a cell. Among these are most of the proteins that, in turn, regulate cell division. Abnormal activity of such a kinase can cause abnormal cell growth and contribute to the development of cancer.

Equally important in the phosphorylation cascade are the **protein phosphatases**, enzymes that can rapidly remove phosphate groups from proteins, a process called dephosphorylation. By dephosphorylating and thus inactivating protein kinases, phosphatases provide the mechanism for turning off the signal transduction pathway when the initial signal is no longer present. Phosphatases also make the protein kinases available for reuse, enabling the cell to respond again to an extracellular signal. A phosphorylation-dephosphorylation system acts as a molecular switch in the cell, turning an activity on or off, or up or down, as required. At any given moment, the activity of a protein regulated by phosphorylation depends on the balance in the cell between active kinase molecules and active phosphatase molecules.

Small Molecules and Ions as Second Messengers

Not all components of signal transduction pathways are proteins. Many signaling pathways also involve small, nonprotein, water-soluble molecules or ions called **second messengers**. (The pathway's "first messenger" is considered to be the extracellular signaling molecule that binds to the membrane receptor.) Because they are small, second messengers can readily spread throughout the cell by diffusion. The two most common second messengers are cyclic AMP and calcium ions, Ca^{2+}. Here we'll limit our discussion to cyclic AMP.

In his research on epinephrine, Earl Sutherland discovered that the binding of epinephrine to the plasma membrane of a liver cell elevates the cytosolic concentration of **cyclic AMP (cAMP)** (cyclic adenosine monophosphate). The binding of epinephrine to a specific receptor protein leads to activation of adenylyl cyclase, an enzyme embedded in the plasma membrane that converts ATP to cAMP **(Figure 5.25)**. Each molecule of adenylyl cyclase can catalyze the synthesis of many molecules of cAMP. In this way, the normal cellular concentration of cAMP can be boosted 20-fold in a matter of seconds. The cAMP broadcasts the signal to the cytoplasm. It does not persist for long in the absence of the hormone because another enzyme converts cAMP to AMP. Another surge of epinephrine is needed to boost the cytosolic concentration of cAMP again.

Subsequent research has revealed that epinephrine is only one of many hormones and other signaling molecules that trigger the formation of cAMP. It has also brought to light the other components of many cAMP pathways, including G proteins, G protein-coupled receptors, and protein kinases. The immediate effect of cAMP is usually the activation of a protein kinase called *protein kinase A*. The activated protein kinase A then phosphorylates various other proteins.

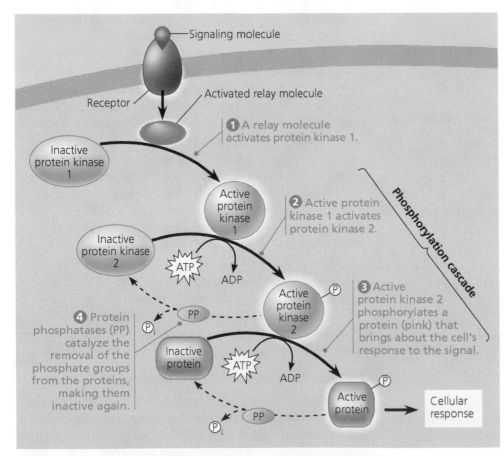

▲ **Figure 5.24 A phosphorylation cascade.** In a phosphorylation cascade, a series of different molecules in a pathway are phosphorylated in turn, each molecule adding a phosphate group to the next one in line. Dephosphorylation returns the molecule to its inactive form.

? *Which protein is responsible for activation of protein kinase 2?*

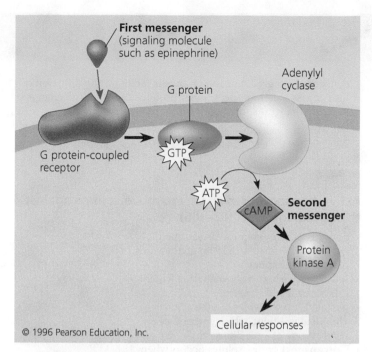

© 1996 Pearson Education, Inc.

▲ **Figure 5.25 cAMP as a second messenger in a G protein signaling pathway.** The first messenger activates a G protein-coupled receptor, which activates a specific G protein. In turn, the G protein activates adenylyl cyclase, which catalyzes the conversion of ATP to cAMP. The cAMP then acts as a second messenger and activates another protein, usually protein kinase A, leading to cellular responses.

Response: Regulation of Transcription or Cytoplasmic Activities

What is the nature of the final step in a signaling pathway—the *response* to an external signal? Ultimately, a signal transduction pathway leads to the regulation of one or more cellular activities. The response may occur in the nucleus of the cell or in the cytoplasm.

Many signaling pathways ultimately regulate protein synthesis, usually by turning specific genes on or off in the nucleus. Like an activated steroid receptor (see Figure 5.23), the final activated molecule in a signaling pathway may function as a transcription factor. **Figure 5.26** shows an example in which a signaling pathway activates a transcription factor that turns a gene on: The response to the growth factor signal is transcription, the synthesis of mRNA, which will be translated in the cytoplasm into a specific protein. In other cases, the transcription factor might regulate a gene by turning it off. Often a transcription factor regulates several different genes.

Sometimes a signaling pathway may regulate the *activity* of a protein rather than its synthesis, directly affecting a cellular activity outside the nucleus. For example, a signal may cause the opening or closing of an ion channel in the plasma membrane or a change in cell metabolism. As we have discussed, the response of cells to the hormone epinephrine helps regulate cellular energy metabolism by affecting the activity of an enzyme: The final step in the signaling pathway that begins with epinephrine binding activates the enzyme that catalyzes the breakdown of glycogen.

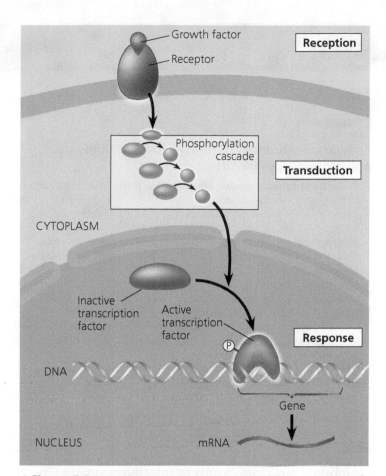

▲ **Figure 5.26 Nuclear response to a signal: the activation of a specific gene by a growth factor.** This diagram shows a typical signaling pathway that leads to the regulation of gene activity in the cell nucleus. The initial signaling molecule, a local regulator called a growth factor, triggers a phosphorylation cascade. (The ATP molecules and phosphate groups are not shown.) Once phosphorylated, the last kinase in the sequence enters the nucleus and activates a transcription factor, which stimulates transcription of a specific gene. The resulting mRNA then directs the synthesis of a particular protein in the cytoplasm.

The Evolution of Cell Signaling

EVOLUTION In studying how cells signal to each other and how they interpret the signals they receive, biologists have discovered some universal mechanisms of cellular regulation, additional evidence for the evolutionary relatedness of all life. The same small set of cell-signaling mechanisms shows up again and again in diverse species, in biological processes ranging from hormone action to embryonic development to cancer. Scientists think that early versions of today's cell-signaling mechanisms evolved well before the first multicellular creatures appeared on Earth.

CONCEPT CHECK 5.6

1. Explain how nerve cells provide examples of both local and long-distance signaling.
2. When a signal transduction pathway involves a phosphorylation cascade, what turns off the cell's response?
3. **WHAT IF?** How can a target cell's response to a single hormone molecule result in a response that affects a million other molecules?

   For suggested answers, see Appendix A.

5 Chapter Review

SUMMARY OF KEY CONCEPTS

CONCEPT 5.1

Cellular membranes are fluid mosaics of lipids and proteins (pp. 94–98)

- In the **fluid mosaic model**, **amphipathic** proteins are embedded in the phospholipid bilayer.
- Phospholipids and some proteins move laterally within the membrane. The unsaturated hydrocarbon tails of some phospholipids keep membranes fluid at lower temperatures, while cholesterol helps membranes resist changes in fluidity caused by temperature changes.
- Membraine proteins function in transport, enzymatic activity, attachment to the cytoskeleton and extracellular matrix, cell-cell recognition, intercellular joining, and signal transduction. Short chains of sugars linked to proteins (in **glycoproteins**) and lipids (in **glycolipids**) on the exterior side of the plasma membrane interact with surface molecules of other cells.
- Membrane proteins and lipids are synthesized in the ER and modified in the ER and Golgi apparatus. The inside and outside faces of membranes differ in molecular composition.

? *In what ways are membranes crucial to life?*

CONCEPT 5.2

Membrane structure results in selective permeability (p. 99)

- A cell must exchange substances with its surroundings, a process controlled by the **selective permeability** of the plasma membrane. Hydrophobic molecules pass through membranes rapidly, whereas polar molecules and ions usually need specific **transport proteins**.

? *How do aquaporins affect the permeability of a membrane?*

CONCEPT 5.3

Passive transport is diffusion of a substance across a membrane with no energy investment (pp. 99–103)

- Diffusion is the spontaneous movement of a substance down its **concentration gradient**. Water diffuses out through the permeable membrane of a cell (**osmosis**) if the solution outside has a higher solute concentration than the cytosol (is **hypertonic**); water enters the cell if the solution has a lower solute concentration (is **hypotonic**). If the concentrations are equal (**isotonic**), no net osmosis occurs. Cell survival depends on balancing water uptake and loss.
- In **facilitated diffusion**, a transport protein speeds the movement of water or a solute across a membrane down its

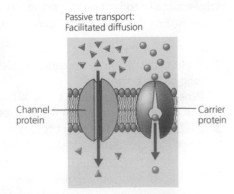

Passive transport:
Facilitated diffusion

Channel protein — Carrier protein

concentration gradient. **Ion channels** facilitate the diffusion of ions across a membrane. Carrier proteins can undergo changes in shape that transport bound solutes.

? *What happens to a cell placed in a hypertonic solution? Describe the free water concentration inside and out.*

CONCEPT 5.4

Active transport uses energy to move solutes against their gradients (pp. 103–106)

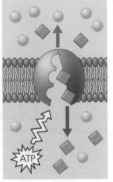

Active transport

- Specific membrane proteins use energy, usually in the form of ATP, to do the work of **active transport**.
- Ions can have both a concentration (chemical) gradient and an electrical gradient (voltage). These combine in the **electrochemical gradient**, which determines the net direction of ionic diffusion.
- **Cotransport** of two solutes occurs when a membrane protein enables the "downhill" diffusion of one solute to drive the "uphill" transport of the other.

? *ATP is not directly involved in the functioning of a cotransporter. Why, then, is cotransport considered active transport?*

CONCEPT 5.5

Bulk transport across the plasma membrane occurs by exocytosis and endocytosis (pp. 106–107)

- Three main types of **endocytosis** are **phagocytosis**, **pinocytosis**, and **receptor-mediated endocytosis**.

? *Which type of endocytosis involves the binding of specific substances in the extracellular fluid to membrane proteins? What does this type of transport enable a cell to do?*

CONCEPT 5.6

The plasma membrane plays a key role in most cell signaling (pp. 108–113)

- In local signaling, animal cells may communicate by direct contact or by secreting **local regulators**. For long-distance signaling, both animals and plants use **hormones**; animals also signal electrically.
- Signaling molecules that bind to membrane receptors trigger a three-stage cell-signaling pathway:

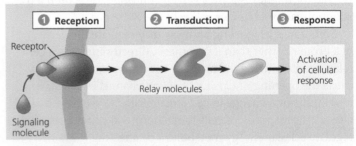

① Reception **② Transduction** **③ Response**

Receptor

Signaling molecule

Relay molecules

Activation of cellular response

- In **reception**, a signaling molecule binds to a receptor protein, causing the protein to change shape. Two major types of membrane receptors are **G protein-coupled receptors (GPCRs)**, which work with the help of cytoplasmic **G proteins**, and **ligand-gated ion channels**, which open or close in response to binding by signaling molecules. Signaling molecules that are hydrophobic cross the plasma membrane and bind to receptors inside the cell.

- At each step in a **signal transduction pathway**, the signal is *transduced* into a different form, which commonly involves a change in a protein's shape. Many pathways include phosphorylation cascades, in which a series of **protein kinases** each add a phosphate group to the next one in line, activating it. The balance between phosphorylation and dephosphorylation, by **protein phosphatases**, regulates the activity of proteins in the pathway.

- **Second messengers**, such as the small molecule **cyclic AMP (cAMP)**, diffuse readily through the cytosol and thus help broadcast signals quickly. Many G proteins activate the enzyme that makes cAMP from ATP.

- The cell's **response** to a signal may be the regulation of transcription in the nucleus or of an activity in the cytoplasm.

? *What determines* whether *a cell responds to a hormone such as epinephrine? What determines* how *the cell responds?*

TEST YOUR UNDERSTANDING

Level 1: Knowledge/Comprehension

1. In what way do the membranes of a eukaryotic cell vary?
   a. Phospholipids are found only in certain membranes.
   b. Certain proteins are unique to each kind of membrane.
   c. Only certain membranes of the cell are selectively permeable.
   d. Only certain membranes are constructed from amphipathic molecules.
   e. Some membranes have hydrophobic surfaces exposed to the cytoplasm, while others have hydrophilic surfaces facing the cytoplasm.

2. Which of the following factors would tend to increase membrane fluidity?
   a. a greater proportion of unsaturated phospholipids
   b. a greater proportion of saturated phospholipids
   c. a lower temperature
   d. a relatively high protein content in the membrane
   e. a greater proportion of relatively large glycolipids compared with lipids having smaller molecular masses

3. Phosphorylation cascades involving a series of protein kinases are useful for cellular signal transduction because
   a. they are species specific.
   b. they always lead to the same cellular response.
   c. they amplify the original signal manyfold.
   d. they counter the harmful effects of phosphatases.
   e. the number of molecules used is small and fixed.

4. Lipid-soluble signaling molecules, such as testosterone, cross the membranes of all cells but affect only target cells because
   a. only target cells retain the appropriate DNA segments.
   b. intracellular receptors are present only in target cells.
   c. most cells lack the Y chromosome required.
   d. only target cells possess the cytosolic enzymes that transduce the testosterone.
   e. only in target cells is testosterone able to initiate a phosphorylation cascade.

Level 2: Application/Analysis

5. Which of the following processes includes all the others?
   a. osmosis
   b. diffusion of a solute across a membrane
   c. facilitated diffusion
   d. passive transport
   e. transport of an ion down its electrochemical gradient

6. Based on Figure 5.17, which of these experimental treatments would increase the rate of sucrose transport into the cell?
   a. decreasing extracellular sucrose concentration
   b. decreasing extracellular pH
   c. decreasing cytoplasmic pH
   d. adding an inhibitor that blocks regeneration of ATP
   e. adding a substance that makes the membrane more permeable to hydrogen ions

Level 3: Synthesis/Evaluation

7. **SCIENTIFIC INQUIRY**
   An experiment is designed to study the mechanism of sucrose uptake by plant cells. Cells are immersed in a sucrose solution, and the pH of the solution is monitored. Samples of the cells are taken at intervals and their sucrose concentration measured. After a decrease in the pH of the solution to a steady, slightly acidic level, sucrose uptake begins. Propose a hypothesis for these results. What do you think would happen if an inhibitor of ATP regeneration by the cell were added to the beaker once the pH was at a steady level? Explain.

8. **SCIENCE, TECHNOLOGY, AND SOCIETY**
   Extensive irrigation in arid regions causes salts to accumulate in the soil. (When water evaporates, salts that were dissolved in the water are left behind in the soil.) Based on what you have learned about water balance in plant cells, explain why increased soil salinity (saltiness) might be harmful to crops. Suggest ways to minimize damage. What costs are attached to your solutions?

9. **FOCUS ON EVOLUTION**
   Paramecium and other protists that live in hypotonic environments have cell membranes that limit water uptake, while those living in isotonic environments have membranes that are more permeable to water. What water regulation adaptations might have evolved in protists in hypertonic habitats such as Great Salt Lake? In habitats with changing salt concentration?

10. **FOCUS ON INTERACTIONS**
    A human pancreatic cell obtains O_2, fuel molecules such as glucose, and building materials such as amino acids and cholesterol from its environment, and it releases CO_2 as a waste product of cellular respiration. In response to hormonal signals, the cell secretes digestive enzymes. It also regulates its ion concentrations by exchange with its environment. Based on what you have just learned about the structure and function of cellular membranes, write a short essay (100–150 words) that describes how such a cell accomplishes these interactions with its environment.

For selected answers, see Appendix A.

6
An Introduction to Metabolism

KEY CONCEPTS

6.1 An organism's metabolism transforms matter and energy

6.2 The free-energy change of a reaction tells us whether or not the reaction occurs spontaneously

6.3 ATP powers cellular work by coupling exergonic reactions to endergonic reactions

6.4 Enzymes speed up metabolic reactions by lowering energy barriers

6.5 Regulation of enzyme activity helps control metabolism

OVERVIEW

The Energy of Life

The living cell is a chemical factory in miniature, where thousands of reactions occur within a microscopic space. Sugars can be converted to amino acids that are linked together into proteins when needed, and when food is digested, proteins are dismantled into amino acids that can be converted to sugars. The process called cellular respiration drives the cellular economy by extracting the energy stored in sugars and other fuels. Cells apply this energy to perform various types of work. Cells of the two firefly squid (*Watasenia scintillans*) shown mating in **Figure 6.1** convert the energy stored in organic molecules to light, a process called bioluminescence that aids in mate recognition. Such metabolic activities are precisely coordinated and controlled in the cell. In its complexity, its efficiency, and its responsiveness to subtle changes, the cell is peerless as a chemical factory. The concepts of metabolism that you learn in this chapter will help you understand how matter and energy flow during life's processes and how that flow is regulated.

▼ **Figure 6.1** What causes these two squid to glow?

CONCEPT 6.1

An organism's metabolism transforms matter and energy

The totality of an organism's chemical reactions is called **metabolism** (from the Greek *metabole*, change). Metabolism is an emergent property of life that arises from orderly interactions between molecules.

Metabolic Pathways

We can picture a cell's metabolism as an elaborate road map of many chemical reactions, arranged as intersecting metabolic pathways. In a **metabolic pathway**, a specific molecule is altered in a series of defined steps, resulting in a product. Each step of the pathway is catalyzed by a specific enzyme:

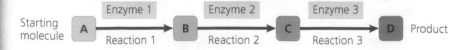

Starting molecule	Enzyme 1		Enzyme 2		Enzyme 3		
	A	→	B	→	C	→	D	Product
		Reaction 1		Reaction 2		Reaction 3	

Analogous to the red, yellow, and green stoplights that control the flow of automobile traffic, mechanisms that regulate enzymes balance metabolic supply and demand.

Metabolism as a whole manages the material and energy resources of the cell. Some metabolic pathways release energy by breaking down complex molecules to simpler compounds. These degradative processes are called **catabolic pathways**, or breakdown pathways. A major pathway of catabolism is cellular respiration, in which the sugar glucose and other organic fuels are broken down in the presence of oxygen to carbon dioxide and water. Energy stored in the organic molecules becomes available to do the work of the cell, such as ciliary beating or membrane transport. **Anabolic pathways**, in contrast, consume energy to build complicated molecules from simpler ones; they are sometimes called biosynthetic pathways. Examples of anabolism are the synthesis of an amino acid from simpler molecules and the synthesis of a protein from amino acids. Catabolic and anabolic pathways are the "downhill" and "uphill" avenues of the metabolic landscape. Energy released from the downhill reactions of catabolic pathways can be stored and then used to drive the uphill reactions of anabolic pathways.

In this chapter, we will focus on mechanisms common to metabolic pathways. Because energy is fundamental to all metabolic processes, a basic knowledge of energy is necessary to understand how the living cell works. Although we will use some nonliving examples to study energy, the concepts demonstrated by these examples also apply to **bioenergetics**, the study of how energy flows through living organisms.

Forms of Energy

Energy is the capacity to cause change. In everyday life, energy is important because some forms of energy can be used to do work—that is, to move matter against opposing forces, such as gravity and friction. Put another way, energy is the ability to rearrange a collection of matter. For example, you expend energy to turn the pages of this book, and your cells expend energy in transporting certain substances across membranes. Energy exists in various forms, and the work of life depends on the ability of cells to transform energy from one form to another.

Energy can be associated with the relative motion of objects; this energy is called **kinetic energy**. Moving objects can perform work by imparting motion to other matter: Water gushing through a dam turns turbines, and the contraction of leg muscles pushes bicycle pedals. **Thermal energy** is kinetic energy associated with the random movement of atoms or molecules; thermal energy in transfer from one object to another is called **heat**. Light is also a type of energy that can be harnessed to perform work, such as powering photosynthesis in green plants.

An object not presently moving may still possess energy. Energy that is not kinetic is called **potential energy**; it is energy that matter possesses because of its location or structure. Water behind a dam, for instance, possesses energy because of

its altitude above sea level. Molecules possess energy because of the arrangement of electrons in the bonds between their atoms. **Chemical energy** is a term used by biologists to refer to the potential energy available for release in a chemical reaction. Recall that catabolic pathways release energy by breaking down complex molecules. Biologists say that these complex molecules, such as glucose, are high in chemical energy. During a catabolic reaction, some bonds are broken and others formed, releasing energy and resulting in lower-energy breakdown products. This transformation also occurs, for example, in the engine of a car when the hydrocarbons of gasoline react explosively with oxygen, releasing the energy that pushes the pistons and producing exhaust. Although less explosive, a similar reaction of food molecules with oxygen provides chemical energy in biological systems, producing carbon dioxide and water as waste products. Biochemical pathways, carried out in the context of cellular structures, enable cells to release chemical energy from food molecules and use the energy to power life processes.

How is energy converted from one form to another? Consider the divers in **Figure 6.2**. The young woman climbing the ladder to the diving platform is releasing chemical energy from the food she ate for lunch and using some of that energy to perform the work of climbing. The kinetic energy of muscle movement is thus being transformed into potential energy due to her increasing height above the water. The young man diving is converting his potential energy to kinetic energy, which is then transferred to the water as he enters it. A small amount of energy is lost as heat due to friction.

Now let's go back one step and consider the original source of the organic food molecules that provided the necessary

A diver has more potential energy on the platform than in the water.

Diving converts potential energy to kinetic energy.

Climbing up converts the kinetic energy of muscle movement to potential energy.

A diver has less potential energy in the water than on the platform.

▲ **Figure 6.2 Transformations between potential and kinetic energy.**

chemical energy for the diver to climb the steps. This chemical energy was itself derived from light energy by plants during photosynthesis. Organisms are energy transformers.

The Laws of Energy Transformation

The study of the energy transformations that occur in a collection of matter is called **thermodynamics**. Scientists use the word *system* to denote the matter under study; they refer to the rest of the universe—everything outside the system—as the *surroundings*. An *isolated system*, such as that approximated by liquid in a thermos bottle, is unable to exchange either energy or matter with its surroundings. In an *open system*, energy and matter can be transferred between the system and its surroundings. Organisms are open systems. They absorb energy—for instance, light energy or chemical energy in the form of organic molecules—and release heat and metabolic waste products, such as carbon dioxide, to the surroundings. Two laws of thermodynamics govern energy transformations in organisms and all other collections of matter.

The First Law of Thermodynamics

According to the **first law of thermodynamics**, the energy of the universe is constant: *Energy can be transferred and transformed, but it cannot be created or destroyed.* The first law is also known as the *principle of conservation of energy*. The electric company does not make energy, but merely converts it to a form that is convenient for us to use. By converting sunlight to chemical energy, a plant acts as an energy transformer, not an energy producer.

The brown bear in **Figure 6.3a** will convert the chemical energy of the organic molecules in its food to kinetic and other forms of energy as it carries out biological processes. What happens to this energy after it has performed work? The second law of thermodynamics helps to answer this question.

The Second Law of Thermodynamics

If energy cannot be destroyed, why can't organisms simply recycle their energy over and over again? It turns out that during every energy transfer or transformation, some energy is converted to thermal energy and released as heat, becoming unavailable to do work. Only a small fraction of the chemical energy from the food in Figure 6.3a is transformed into the motion of the brown bear shown in **Figure 6.3b**; most is lost as heat, which dissipates rapidly through the surroundings.

A system can put thermal energy to work only when there is a temperature difference that results in the thermal energy flowing from a warmer location to a cooler one. If temperature is uniform, as it is in a living cell, then the heat generated during a chemical reaction will simply warm a body of matter, such as the organism. (This can make a room crowded with people uncomfortably warm, as each person is carrying out a multitude of chemical reactions!)

A logical consequence of the loss of usable energy during energy transfer or transformation is that each such event makes the universe more disordered. Scientists use a quantity called **entropy** as a measure of disorder, or randomness. The more randomly arranged a collection of matter is, the greater its entropy. We can now state the **second law of thermodynamics**: *Every energy transfer or transformation increases the entropy of the universe.* Although order can increase locally, there is an unstoppable trend toward randomization of the universe as a whole.

In many cases, increased entropy is evident in the physical disintegration of a system's organized structure. For example, you can observe increasing entropy in the gradual decay of an unmaintained building. Much of the increasing entropy of the universe is less apparent, however, because it appears as increasing amounts of heat and less ordered forms of matter. As the bear in Figure 6.3b converts chemical energy to kinetic energy, it is also increasing the disorder of its surroundings by

(a) First law of thermodynamics: Energy can be transferred or transformed but neither created nor destroyed. For example, chemical reactions in this brown bear will convert the chemical (potential) energy in the fish to the kinetic energy of running.

(b) Second law of thermodynamics: Every energy transfer or transformation increases the disorder (entropy) of the universe. For example, as the bear runs, disorder is increased around the bear by the release of heat and small molecules that are the by-products of metabolism. A brown bear can run at speeds up to 35 miles per hour (56 km/hr)—as fast as a racehorse.

▲ **Figure 6.3 The two laws of thermodynamics.**

producing heat and small molecules, such as the CO_2 it exhales, that are the breakdown products of food.

The concept of entropy helps us understand why certain processes occur without any input of energy. It turns out that for a process to occur on its own, without outside help, it must increase the entropy of the universe. A process that can occur without an input of energy is called a **spontaneous process**. Note that as we're using it here, the word *spontaneous* does not imply that such a process would occur quickly; rather, the word signifies that the process is energetically favorable. (In fact, it may be helpful for you to think of the phrase "energetically favorable" when you read the formal term "spontaneous.") Some spontaneous processes, such as an explosion, may be virtually instantaneous, while others, such as the rusting of an old car over time, are much slower. A process that cannot occur on its own is said to be nonspontaneous; it will happen only if energy is added to the system. We know from experience that certain events occur spontaneously and others do not. For instance, we know that water flows downhill spontaneously but moves uphill only with an input of energy, such as when a machine pumps the water against gravity. This understanding gives us another way to state the second law: *For a process to occur spontaneously, it must increase the entropy of the universe.*

Biological Order and Disorder

Living systems increase the entropy of their surroundings, as predicted by thermodynamic law. It is true that cells create ordered structures from less organized starting materials. For example, simpler molecules are ordered into the more complex structure of an amino acid, and amino acids are ordered into polypeptide chains. At the organismal level as well, complex and beautifully ordered structures result from biological processes that use simpler starting materials **(Figure 6.4)**. However, an organism also takes in organized forms of matter and energy from the surroundings and replaces them with less ordered forms. For example, an animal obtains starch, proteins, and other complex molecules from the food it eats. As catabolic pathways break these molecules down, the animal releases carbon dioxide and water—small molecules that possess less chemical energy than the food did. The depletion of chemical energy is accounted for by heat generated during metabolism. On a larger scale, energy flows into most ecosystems in the form of light and exits in the form of heat.

During the early history of life, complex organisms evolved from simpler ancestors. For example, we can trace the ancestry of the plant kingdom from much simpler organisms called green algae to more complex flowering plants. However, this increase in organization over time in no way violates the second law. The entropy of a particular system, such as an organism, may actually decrease as long as the total entropy of the *universe*—the system plus its surroundings—increases. Thus, organisms are islands of low entropy in an increasingly random universe. The evolution of biological order is perfectly consistent with the laws of thermodynamics.

▲ **Figure 6.4 Order as a characteristic of life.** Order is evident in the detailed structures of the sea urchin skeleton and the succulent plant shown here. As open systems, organisms can increase their order as long as the order of their surroundings decreases.

CONCEPT CHECK 6.1

1. **MAKE CONNECTIONS** How does the second law of thermodynamics help explain the diffusion of a substance across a membrane? (See Figure 5.9.)
2. Describe the forms of energy found in an apple as it grows on a tree, then falls, then is digested by someone who eats it.

For suggested answers, see Appendix A.

CONCEPT 6.2

The free-energy change of a reaction tells us whether or not the reaction occurs spontaneously

The laws of thermodynamics that we've just discussed apply to the universe as a whole. As biologists, we want to understand the chemical reactions of life—for example, which reactions occur spontaneously and which ones require some input of energy from outside. But how can we know this without assessing the energy and entropy changes in the entire universe for each separate reaction?

Free-Energy Change (ΔG), Stability, and Equilibrium

Recall that the universe is really equivalent to "the system" plus "the surroundings." In 1878, J. Willard Gibbs, a professor at Yale, defined a very useful function called the Gibbs free energy of a system (without considering its surroundings), symbolized by the letter G. We'll refer to the Gibbs free energy simply as free energy. **Free energy** is the portion of a system's energy that can perform work when temperature and pressure are uniform throughout the system, as in a living cell. Biologists find it most informative to focus on the *change* in free energy (ΔG) during the chemical reactions of life. ΔG represents the difference between the free energy of the final state and the free energy of the initial state:

$$\Delta G = G_{\text{final state}} - G_{\text{initial state}}$$

Using chemical methods, we can measure ΔG for any reaction. More than a century of experiments has shown that only reactions with a negative ΔG can occur with no input of energy, so the value of ΔG tells us whether a particular reaction is a spontaneous one. This principle is very important in the study of metabolism, where a major goal is to determine which reactions occur spontaneously and can be harnessed to supply energy for cellular work.

For a reaction to have a negative ΔG, the system must lose free energy during the change from initial state to final state. Because it has less free energy, the system in its final state is less likely to change and is therefore more stable than it was previously. We can think of free energy as a measure of a system's instability—its tendency to change to a more stable state. Unstable systems (higher G) tend to change in such a way that they become more stable (lower G), as shown in **Figure 6.5**.

Another term that describes a state of maximum stability is chemical *equilibrium*. At equilibrium, the forward and reverse reactions occur at the same rate, and there is no further net change in the relative concentration of products and reactants. For a system at equilibrium, G is at its lowest possible value in that system. We can think of the equilibrium state as a free-energy valley. Any change from the equilibrium position will have a positive ΔG and will not be spontaneous. For this reason, systems never spontaneously move away from equilibrium. Because a system at equilibrium cannot spontaneously change, it can do no work. *A process is spontaneous and can perform work only when it is moving toward equilibrium.*

Free Energy and Metabolism

We can now apply the free-energy concept more specifically to the chemistry of life's processes.

Exergonic and Endergonic Reactions in Metabolism

Based on their free-energy changes, chemical reactions can be classified as either exergonic ("energy outward") or endergonic ("energy inward"). An **exergonic reaction** proceeds with a net release of free energy **(Figure 6.6a)**. Because the chemical mixture loses free energy (G decreases), ΔG is negative for an exergonic reaction. Using ΔG as a standard for spontaneity, exergonic reactions are those that occur spontaneously. (Remember, the word *spontaneous* implies that it is energetically favorable, not that it will occur rapidly.) The magnitude of ΔG for an exergonic reaction represents the maximum amount of work the reaction can perform (some of the free energy is released as heat and cannot do work). The greater the decrease in free energy, the greater the amount of work that can be done.

Consider the overall reaction for cellular respiration:

$$C_6H_{12}O_6 + 6\,O_2 \rightarrow 6\,CO_2 + 6\,H_2O$$
$$\Delta G = -686 \text{ kcal/mol } (-2{,}870 \text{ kJ/mol})$$

686 kcal (2,870 kJ) of energy are made available for work for each mole (180 g) of glucose broken down by respiration under "standard conditions" (1 M of each reactant and product, 25°C, pH 7). Because energy must be conserved, the products of respiration store 686 kcal less free energy per mole than the reactants. The products are the "exhaust" of a process that tapped the free energy stored in the bonds of the sugar molecules.

- More free energy (higher G)
- Less stable
- Greater work capacity

In a **spontaneous change**
- The free energy of the system decreases ($\Delta G < 0$)
- The system becomes more stable
- The released free energy can be harnessed to do work

- Less free energy (lower G)
- More stable
- Less work capacity

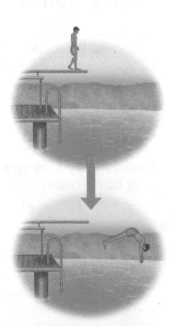

(a) Gravitational motion. Objects move spontaneously from a higher altitude to a lower one.

(b) Diffusion. Molecules in a drop of dye diffuse until they are randomly dispersed.

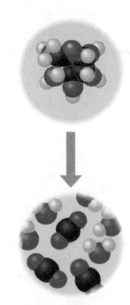

(c) Chemical reaction. In a cell, a glucose molecule is broken down into simpler molecules.

▲ **Figure 6.5 The relationship of free energy to stability, work capacity, and spontaneous change.** Unstable systems (top) are rich in free energy, G. They have a tendency to change spontaneously to a more stable state (bottom), and it is possible to harness this "downhill" change to perform work.

Figure 6.6 Free energy changes (ΔG) in exergonic and endergonic reactions.

(a) Exergonic reaction: energy released, spontaneous

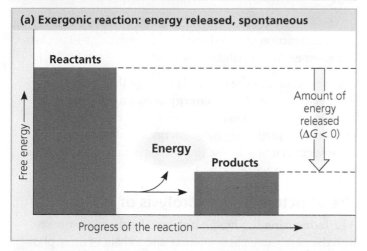

Amount of energy released (ΔG < 0)

Free energy — Reactants, Energy, Products

Progress of the reaction ——→

(b) Endergonic reaction: energy required, nonspontaneous

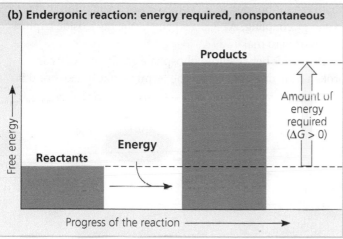

Amount of energy required (ΔG > 0)

Free energy — Products, Energy, Reactants

Progress of the reaction ——→

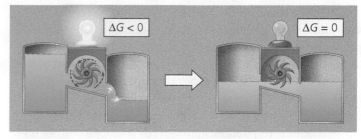

ΔG < 0   ΔG = 0

(a) An isolated hydroelectric system. Water flowing downhill turns a turbine that drives a generator providing electricity to a lightbulb, but only until the system reaches equilibrium.

(b) An open hydroelectric system. Flowing water keeps driving the generator because intake and outflow of water keep the system from reaching equilibrium.

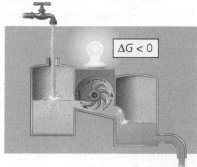

ΔG < 0

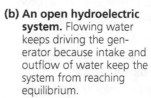

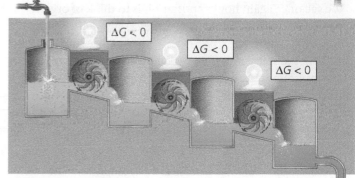

ΔG < 0   ΔG < 0   ΔG < 0

(c) A multistep open hydroelectric system. Cellular respiration is analogous to this system: Glucose is broken down in a series of exergonic reactions that power the work of the cell. The product of each reaction is used as the reactant for the next, so no reaction reaches equilibrium.

▲ **Figure 6.7 Equilibrium and work in isolated and open systems.**

It is important to realize that the breaking of bonds does not release energy; on the contrary, as you will soon see, it requires energy. The phrase "energy stored in bonds" is shorthand for the potential energy that can be released when new bonds are formed after the original bonds break, as long as the products are of lower free energy than the reactants.

An **endergonic reaction** is one that absorbs free energy from its surroundings (**Figure 6.6b**). Because this kind of reaction essentially *stores* free energy in molecules (*G* increases), ΔG is positive. Such reactions are nonspontaneous, and the magnitude of ΔG is the quantity of energy required to drive the reaction. If a chemical process is exergonic (downhill), releasing energy in one direction, then the reverse process must be endergonic (uphill), using energy. A reversible process cannot be downhill in both directions. If ΔG = −686 kcal/mol for respiration, which converts glucose and oxygen to carbon dioxide and water, then the reverse process—the conversion of carbon dioxide and water to glucose and oxygen—must be strongly endergonic, with ΔG = +686 kcal/mol. Such a reaction would never happen by itself.

How, then, do plants make the sugar that organisms use for energy? Plants get the required energy—686 kcal to make a mole of glucose—from the environment by capturing light and converting its energy to chemical energy. Next, in a long series of exergonic steps, they gradually spend that chemical energy to assemble glucose molecules.

Equilibrium and Metabolism

Reactions in an isolated system eventually reach equilibrium and can then do no work, as illustrated by the isolated hydroelectric system in **Figure 6.7a**. The chemical reactions of metabolism are reversible, and they, too, would reach equilibrium if they occurred in the isolation of a test tube. Because systems at equilibrium are at a minimum of *G* and can do no work, a cell that has reached metabolic equilibrium is dead! The fact that metabolism as a whole is never at equilibrium is one of the defining features of life.

Like most systems, a living cell is not in equilibrium. The constant flow of materials in and out of the cell keeps the metabolic pathways from ever reaching equilibrium, and the cell continues to do work throughout its life. This principle is illustrated by the open (and more realistic) hydroelectric system in **Figure 6.7b**. However, unlike this simple single-step system, a catabolic pathway in a cell releases free energy in a series of reactions. An example is cellular respiration, illustrated by analogy in **Figure 6.7c**. Some of the reversible reactions of respiration are constantly "pulled" in one direction—that is, they are kept out of equilibrium. The key to maintaining this lack of equilibrium is that the product of a reaction does not accumulate but instead becomes a reactant in the next step; finally, waste products are expelled from the cell. The overall sequence of reactions is kept going by the huge free-energy difference between glucose and oxygen at the top of the energy "hill" and carbon dioxide and water at the "downhill" end. As long as our cells have a steady supply of glucose or other fuels and oxygen and are able to expel waste products to the surroundings, their metabolic pathways never reach equilibrium and can continue to do the work of life.

We see once again how important it is to think of organisms as open systems. Sunlight provides a daily source of free energy for an ecosystem's plants and other photosynthetic organisms. Animals and other nonphotosynthetic organisms in an ecosystem must have a source of free energy in the form of the organic products of photosynthesis. Now that we have applied the free-energy concept to metabolism, we are ready to see how a cell actually performs the work of life.

CONCEPT CHECK 6.2

1. Cellular respiration uses glucose and oxygen, which have high levels of free energy, and releases CO_2 and water, which have low levels of free energy. Is cellular respiration spontaneous or not? Is it exergonic or endergonic? What happens to the energy released from glucose?
2. **WHAT IF?** Some nighttime partygoers wear glow-in-the-dark necklaces. The necklaces start glowing once they are "activated" by snapping the necklace in a way that allows two chemicals to react and emit light in the form of chemiluminescence. Is this chemical reaction exergonic or endergonic? Explain your answer.

For suggested answers, see Appendix A.

CONCEPT 6.3

ATP powers cellular work by coupling exergonic reactions to endergonic reactions

A cell does three main kinds of work:

- *Chemical work*, the pushing of endergonic reactions that would not occur spontaneously, such as the synthesis of polymers from monomers (chemical work will be discussed further in this chapter and in Chapters 7 and 8)

- *Transport work*, the pumping of substances across membranes against the direction of spontaneous movement (see Chapter 5)
- *Mechanical work*, such as the beating of cilia (see Chapter 4), the contraction of muscle cells, and the movement of chromosomes during cellular reproduction

A key feature in the way cells manage their energy resources to do this work is **energy coupling**, the use of an exergonic process to drive an endergonic one. ATP is responsible for mediating most energy coupling in cells, and in most cases it acts as the immediate source of energy that powers cellular work.

The Structure and Hydrolysis of ATP

ATP (adenosine triphosphate) contains the sugar ribose, with the nitrogenous base adenine and a chain of three phosphate groups bonded to it **(Figure 6.8a)**. In addition to its role in energy coupling, ATP is also one of the nucleoside triphosphates used to make RNA.

The bonds between the phosphate groups of ATP can be broken by hydrolysis. When the terminal phosphate bond is

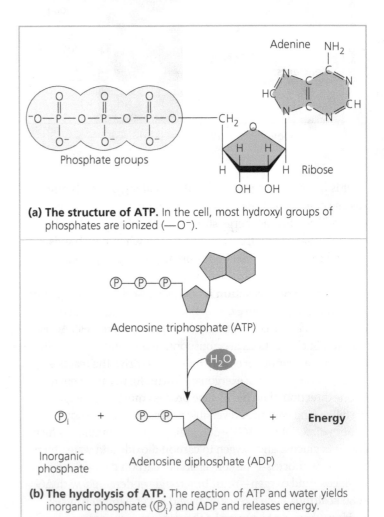

(a) The structure of ATP. In the cell, most hydroxyl groups of phosphates are ionized (—O⁻).

(b) The hydrolysis of ATP. The reaction of ATP and water yields inorganic phosphate ((P)$_i$) and ADP and releases energy.

▲ **Figure 6.8 The structure and hydrolysis of adenosine triphosphate (ATP).**

broken by the addition of a water molecule, a molecule of inorganic phosphate ($HOPO_3^{2-}$, which is abbreviated P_i throughout this book) leaves the ATP. In this way, adenosine *tri*phosphate becomes adenosine *di*phosphate, or ADP **(Figure 6.8b)**. The reaction is exergonic and releases 7.3 kcal of energy per mole of ATP hydrolyzed:

$$ATP + H_2O \rightarrow ADP + P_i$$
$$\Delta G = -7.3 \text{ kcal/mol } (-30.5 \text{ kJ/mol})$$

This is the free-energy change measured under standard conditions. In the cell, conditions do not conform to standard conditions, primarily because reactant and product concentrations differ from 1 M. For example, when ATP hydrolysis occurs under cellular conditions, the actual ΔG is about -13 kcal/mol, 78% greater than the energy released by ATP hydrolysis under standard conditions.

Because their hydrolysis releases energy, the phosphate bonds of ATP are sometimes referred to as high-energy phosphate bonds, but the term is misleading. The phosphate bonds of ATP are not unusually strong bonds, as "high-energy" may imply; rather, the reactants (ATP and water) themselves have high energy relative to the energy of the products (ADP and P_i). The release of energy during the hydrolysis of ATP comes from the chemical change to a state of lower free energy, not from the phosphate bonds themselves.

ATP is useful to the cell because the energy it releases on losing a phosphate group is somewhat greater than the energy most other molecules could deliver. But why does this hydrolysis release so much energy? If we reexamine the ATP molecule in Figure 6.8a, we can see that all three phosphate groups are negatively charged. These like charges are crowded together, and their mutual repulsion contributes to the instability of this region of the ATP molecule. The triphosphate tail of ATP is the chemical equivalent of a compressed spring.

How the Hydrolysis of ATP Performs Work

When ATP is hydrolyzed in a test tube, the release of free energy merely heats the surrounding water. In an organism, this same generation of heat can sometimes be beneficial. For instance, the process of shivering uses ATP hydrolysis during muscle contraction to warm the body. In most cases in the cell, however, the generation of heat alone would be an inefficient (and potentially dangerous) use of a valuable energy resource. Instead, the cell's proteins harness the energy released during ATP hydrolysis in several ways to perform the three types of cellular work—chemical, transport, and mechanical.

For example, with the help of specific enzymes, the cell is able to use the energy released by ATP hydrolysis directly to drive chemical reactions that, by themselves, are endergonic **(Figure 6.9)**. If the ΔG of an endergonic reaction is less than

(a) Glutamic acid conversion to glutamine. Glutamine synthesis from glutamic acid (Glu) by itself is endergonic (ΔG is positive), so it is not spontaneous.

$\Delta G_{Glu} = +3.4$ kcal/mol

Glutamic acid   Ammonia   Glutamine

(b) Conversion reaction coupled with ATP hydrolysis. In the cell, glutamine synthesis occurs in two steps, coupled by a phosphorylated intermediate. ❶ ATP phosphorylates glutamic acid, making it less stable. ❷ Ammonia displaces the phosphate group, forming glutamine.

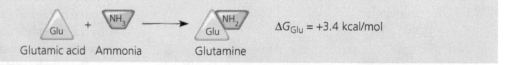

Glutamic acid   Phosphorylated intermediate   Glutamine

(c) Free-energy change for coupled reaction. ΔG for the glutamic acid conversion to glutamine (+3.4 kcal/mol) plus ΔG for ATP hydrolysis (−7.3 kcal/mol) gives the free-energy change for the overall reaction (−3.9 kcal/mol). Because the overall process is exergonic (net ΔG is negative), it occurs spontaneously.

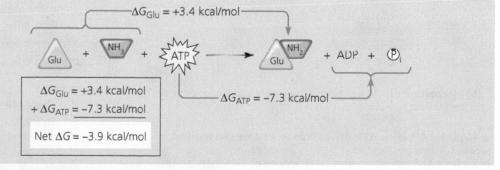

$\Delta G_{Glu} = +3.4$ kcal/mol

$\Delta G_{ATP} = -7.3$ kcal/mol

$\Delta G_{Glu} = +3.4$ kcal/mol
$+ \Delta G_{ATP} = -7.3$ kcal/mol

Net $\Delta G = -3.9$ kcal/mol

▲ **Figure 6.9 How ATP drives chemical work: energy coupling using ATP hydrolysis.** In this example, the exergonic process of ATP hydrolysis is used to drive an endergonic process—the cellular synthesis of the amino acid glutamine from glutamic acid and ammonia.

MAKE CONNECTIONS *Explain why glutamine is drawn as it is in this figure. (See Figure 3.17.)*

the amount of energy released by ATP hydrolysis, then the two reactions can be coupled so that, overall, the coupled reactions are exergonic. This usually involves phosphorylation, the transfer of a phosphate group from ATP to some other molecule, such as the reactant. The recipient with the phosphate group covalently bonded to it is then called a **phosphorylated intermediate**. The key to coupling exergonic and endergonic reactions is the formation of this phosphorylated intermediate, which is more reactive (less stable) than the original unphosphorylated molecule.

Transport and mechanical work in the cell are also nearly always powered by the hydrolysis of ATP. In these cases, ATP hydrolysis leads to a change in a protein's shape and often its ability to bind another molecule. Sometimes this occurs via a phosphorylated intermediate, as seen for the transport protein in **Figure 6.10a**. In most instances of mechanical work involving motor proteins "walking" along cytoskeletal elements **(Figure 6.10b)**, a cycle occurs in which ATP is first bound noncovalently to the motor protein. Next, ATP is hydrolyzed, releasing ADP and P_i. Another ATP molecule can then bind. At each stage, the motor protein changes its shape and ability to bind the cytoskeleton, resulting in movement of the protein along the cytoskeletal track. Phosphorylation and dephosphorylation also promote crucial protein shape changes during cell signaling (see Figure 5.24).

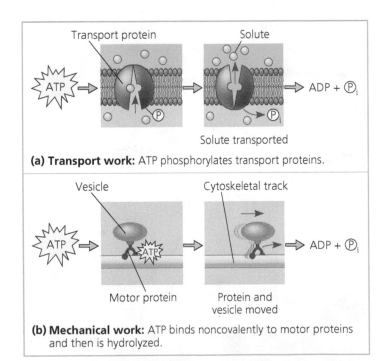

Transport protein   Solute

ATP   ADP + P_i

P

Solute transported

(a) Transport work: ATP phosphorylates transport proteins.

Vesicle   Cytoskeletal track

ATP   ATP   ADP + P_i

Motor protein   Protein and vesicle moved

(b) Mechanical work: ATP binds noncovalently to motor proteins and then is hydrolyzed.

▲ **Figure 6.10 How ATP drives transport and mechanical work.** ATP hydrolysis causes changes in the shapes and binding affinities of proteins. This can occur either **(a)** directly, by phosphorylation, as shown for a membrane protein carrying out active transport of a solute (see also Figure 5.14), or **(b)** indirectly, via noncovalent binding of ATP and its hydrolytic products, as is the case for motor proteins that move vesicles (and other organelles) along cytoskeletal "tracks" in the cell (see also Figure 4.20).

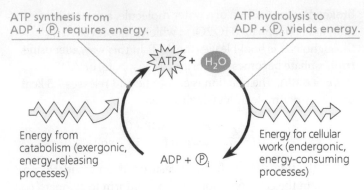

ATP synthesis from ADP + P_i requires energy.

ATP hydrolysis to ADP + P_i yields energy.

ATP + H_2O

Energy from catabolism (exergonic, energy-releasing processes)

ADP + P_i

Energy for cellular work (endergonic, energy-consuming processes)

▲ **Figure 6.11 The ATP cycle.** Energy released by breakdown reactions (catabolism) in the cell is used to phosphorylate ADP, regenerating ATP. Chemical potential energy stored in ATP drives most cellular work.

The Regeneration of ATP

An organism at work uses ATP continuously, but ATP is a renewable resource that can be regenerated by the addition of phosphate to ADP **(Figure 6.11)**. The free energy required to phosphorylate ADP comes from exergonic breakdown reactions (catabolism) in the cell. This shuttling of inorganic phosphate and energy is called the ATP cycle, and it couples the cell's energy-yielding (exergonic) processes to the energy-consuming (endergonic) ones. The ATP cycle proceeds at an astonishing pace. For example, a working muscle cell recycles its entire pool of ATP in less than a minute. That turnover represents 10 million molecules of ATP consumed and regenerated per second per cell. If ATP could not be regenerated by the phosphorylation of ADP, humans would use up nearly their body weight in ATP each day.

Because both directions of a reversible process cannot be downhill, the regeneration of ATP from ADP and P_i is necessarily endergonic:

$$ADP + P_i \rightarrow ATP + H_2O$$
$$\Delta G = +7.3 \text{ kcal/mol } (+30.5 \text{ kJ/mol}) \text{ (standard conditions)}$$

Since ATP formation from ADP and P_i is not spontaneous, free energy must be spent to make it occur. Catabolic (exergonic) pathways, especially cellular respiration, provide the energy for the endergonic process of making ATP. Plants also use light energy to produce ATP. Thus, the ATP cycle is a revolving door through which energy passes during its transfer from catabolic to anabolic pathways.

CONCEPT CHECK 6.3

1. How does ATP typically transfer energy from exergonic to endergonic reactions in the cell?
2. Which of the following combinations has more free energy: glutamic acid + ammonia + ATP or glutamine + ADP + P_i? Explain your answer.
3. **MAKE CONNECTIONS** Does Figure 6.10a show passive or active transport? Explain. (See Concepts 5.3 and 5.4.)

For suggested answers, see Appendix A.

Enzymes speed up metabolic reactions by lowering energy barriers

The laws of thermodynamics tell us what will and will not happen under given conditions but say nothing about the rate of these processes. A spontaneous chemical reaction occurs without any requirement for outside energy, but it may occur so slowly that it is imperceptible. For example, even though the hydrolysis of sucrose (table sugar) to glucose and fructose is exergonic, occurring spontaneously with a release of free energy ($\Delta G = -7$ kcal/mol), a solution of sucrose dissolved in sterile water will sit for years at room temperature with no appreciable hydrolysis. However, if we add a small amount of the enzyme sucrase to the solution, then all the sucrose may be hydrolyzed within seconds, as shown below:

Sucrase

Sucrose ($C_{12}H_{22}O_{11}$) + H_2O → Glucose ($C_6H_{12}O_6$) + Fructose ($C_6H_{12}O_6$)

How does the enzyme do this?

An **enzyme** is a macromolecule that acts as a **catalyst**, a chemical agent that speeds up a reaction without being consumed by the reaction. (In this chapter, we are focusing on enzymes that are proteins. Some RNA molecules, called ribozymes, can function as enzymes; these will be discussed in Chapters 14 and 24.) Without regulation by enzymes, chemical traffic through the pathways of metabolism would become terribly congested because many chemical reactions would take such a long time. In the next two sections, we will see what prevents a spontaneous reaction from occurring faster and how an enzyme changes the situation.

The Activation Energy Barrier

Every chemical reaction between molecules involves both bond breaking and bond forming. For example, the hydrolysis of sucrose involves breaking the bond between glucose and fructose and one of the bonds of a water molecule and then forming two new bonds, as shown above. Changing one molecule into another generally involves contorting the starting molecule into a highly unstable state before the reaction can proceed. This contortion can be compared to the bending of a metal key ring when you pry it open to add a new key. The key ring is highly unstable in its opened form but returns to a stable state once the key is threaded all the way onto the ring. To reach the contorted state where bonds can change, reactant molecules must absorb energy from their surroundings. When the new bonds of the product molecules form, energy

is released as heat, and the molecules return to stable shapes with lower energy than the contorted state.

The initial investment of energy for starting a reaction—the energy required to contort the reactant molecules so the bonds can break—is known as the *free energy of activation*, or **activation energy**, abbreviated E_A in this book. We can think of activation energy as the amount of energy needed to push the reactants to the top of an energy barrier, or uphill, so that the "downhill" part of the reaction can begin. Activation energy is often supplied by heat in the form of thermal energy that the reactant molecules absorb from the surroundings. The absorption of thermal energy accelerates the reactant molecules, so they collide more often and more forcefully. It also agitates the atoms within the molecules, making the breakage of bonds more likely. When the molecules have absorbed enough energy for the bonds to break, the reactants are in an unstable condition known as the *transition state*.

Figure 6.12 graphs the energy changes for a hypothetical exergonic reaction that swaps portions of two reactant molecules:

$$AB + CD \rightarrow AC + BD$$
Reactants   Products

The reactants AB and CD must absorb enough energy from the surroundings to reach the unstable transition state, where bonds can break.

After bonds have broken, new bonds form, releasing energy to the surroundings.

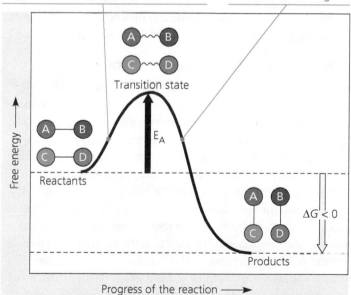

▲ **Figure 6.12 Energy profile of an exergonic reaction.** The "molecules" are hypothetical, with A, B, C, and D representing portions of the molecules. Thermodynamically, this is an exergonic reaction, with a negative ΔG, and the reaction occurs spontaneously. However, the activation energy (E_A) provides a barrier that determines the rate of the reaction.

DRAW IT *Graph the progress of an endergonic reaction in which EF and GH form products EG and FH, assuming that the reactants must pass through a transition state.*

The activation of the reactants is represented by the uphill portion of the graph, in which the free-energy content of the reactant molecules is increasing. At the summit, when energy equivalent to E_A has been absorbed, the reactants are in the transition state: They are activated, and their bonds can be broken. As the atoms then settle into their new, more stable bonding arrangements, energy is released to the surroundings. This corresponds to the downhill part of the curve, which shows the loss of free energy by the molecules. The overall decrease in free energy means that E_A is repaid with interest, as the formation of new bonds releases more energy than was invested in the breaking of old bonds.

The reaction shown in Figure 6.12 is exergonic and occurs spontaneously ($\Delta G < 0$). However, the activation energy provides a barrier that determines the rate of the reaction. The reactants must absorb enough energy to reach the top of the activation energy barrier before the reaction can occur. For some reactions, E_A is modest enough that even at room temperature there is sufficient thermal energy for many of the reactant molecules to reach the transition state in a short time. In most cases, however, E_A is so high and the transition state is reached so rarely that the reaction will hardly proceed at all. In these cases, the reaction will occur at a noticeable rate only if the reactants are heated. For example, the reaction of gasoline and oxygen is exergonic and will occur spontaneously, but energy is required for the molecules to reach the transition state and react. Only when the spark plugs fire in an automobile engine can there be the explosive release of energy that pushes the pistons. Without a spark, a mixture of gasoline hydrocarbons and oxygen will not react because the E_A barrier is too high.

How Enzymes Speed Up Reactions

Proteins, DNA, and other complex molecules of the cell are rich in free energy and have the potential to decompose spontaneously; that is, the laws of thermodynamics favor their breakdown. These molecules persist only because at temperatures typical for cells, few molecules can make it over the hump of activation energy. However, the barriers for selected reactions must occasionally be surmounted for cells to carry out the processes needed for life. Heat speeds a reaction by allowing reactants to attain the transition state more often, but this solution would be inappropriate for biological systems. First, high temperature denatures proteins and kills cells. Second, heat would speed up *all* reactions, not just those that are needed. Instead of heat, organisms use catalysis to speed up reactions.

An enzyme catalyzes a reaction by lowering the E_A barrier **(Figure 6.13)**, enabling the reactant molecules to absorb enough energy to reach the transition state even at moderate temperatures. An enzyme cannot change the ΔG for a reaction; it cannot make an endergonic reaction exergonic. Enzymes can only hasten reactions that would eventually occur anyway, but this function makes it possible for the cell to have a dynamic metabolism, routing chemicals smoothly through the cell's

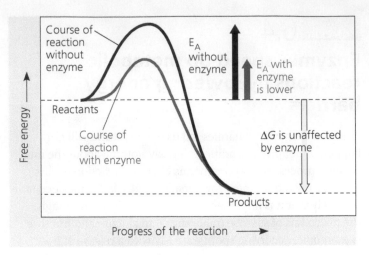

▲ **Figure 6.13 The effect of an enzyme on activation energy.** Without affecting the free-energy change (ΔG) for a reaction, an enzyme speeds the reaction by reducing its activation energy (E_A).

metabolic pathways. And because enzymes are very specific for the reactions they catalyze, they determine which chemical processes will be going on in the cell at any particular time.

Substrate Specificity of Enzymes

The reactant an enzyme acts on is referred to as the enzyme's **substrate**. The enzyme binds to its substrate (or substrates, when there are two or more reactants), forming an **enzyme-substrate complex**. While enzyme and substrate are joined, the catalytic action of the enzyme converts the substrate to the product (or products) of the reaction. The overall process can be summarized as follows:

$$\text{Enzyme} + \text{Substrate(s)} \rightleftharpoons \text{Enzyme-substrate complex} \rightleftharpoons \text{Enzyme} + \text{Product(s)}$$

For example, the enzyme sucrase (most enzyme names end in *-ase*) catalyzes the hydrolysis of the disaccharide sucrose into its two monosaccharides, glucose and fructose (see the illustrated equation on the previous page):

$$\text{Sucrase} + \text{Sucrose} + H_2O \rightleftharpoons \text{Sucrase-sucrose-}H_2O \text{ complex} \rightleftharpoons \text{Sucrase} + \text{Glucose} + \text{Fructose}$$

The reaction catalyzed by each enzyme is very specific; an enzyme can recognize its specific substrate even among closely related compounds. For instance, sucrase will act only on sucrose and will not bind to other disaccharides, such as maltose. What accounts for this molecular recognition? Recall that most enzymes are proteins, and proteins are macromolecules with unique three-dimensional configurations. The specificity of an enzyme results from its shape, which is a consequence of its amino acid sequence.

Only a restricted region of the enzyme molecule actually binds to the substrate. This region, called the **active site**, is typically a pocket or groove on the surface of the enzyme where catalysis occurs (**Figure 6.14a**). Usually, the active site is formed by only a few of the enzyme's amino acids, with the rest of the protein molecule providing a framework that determines the configuration of the active site. The specificity of an enzyme is attributed to a complementary fit between the shape of its active site and the shape of the substrate, like that seen in the binding of a signaling molecule to a receptor protein (see Concept 5.6).

An enzyme is not a stiff structure locked into a given shape. In fact, recent work by biochemists has shown clearly that enzymes (and other proteins as well) seem to "dance" between subtly different shapes in a dynamic equilibrium, with slight differences in free energy for each "pose." The shape that best fits the substrate isn't necessarily the one with the lowest energy, but during the very short time the enzyme takes on this shape, its active site can bind to the substrate. It has been known for more than 50 years that the active site itself is also not a rigid receptacle for the substrate. As the substrate enters the active site, the enzyme changes shape slightly due to interactions between the substrate's chemical groups and chemical groups on the side chains of the amino acids that form the active site. This shape change makes the active site fit even more snugly around the substrate (**Figure 6.14b**). This **induced fit** is like a clasping handshake. Induced fit brings chemical groups of the active site into positions that enhance their ability to catalyze the chemical reaction.

Catalysis in the Enzyme's Active Site

In most enzymatic reactions, the substrate is held in the active site by so-called weak interactions, such as hydrogen bonds and ionic bonds. R groups of a few of the amino acids that make up the active site catalyze the conversion of substrate to product, and the product departs from the active site. The enzyme is then free to take another substrate molecule into its active site. The entire cycle happens so fast that a single enzyme molecule typically acts on about a thousand substrate molecules per second, and some enzymes are even faster. Enzymes, like other catalysts, emerge from the reaction in their original form. Therefore, very small amounts of enzyme can have a huge metabolic impact by functioning over and over again in catalytic cycles. **Figure 6.15** shows a catalytic cycle involving two substrates and two products.

Most metabolic reactions are reversible, and an enzyme can catalyze either the forward or the reverse reaction, depending on which direction has a negative ΔG. This in turn depends mainly on the relative concentrations of reactants and products. The net effect is always in the direction of equilibrium.

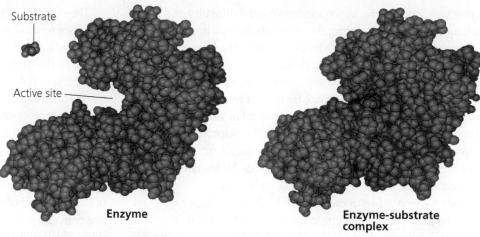

Enzyme

Enzyme-substrate complex

(a) In this computer graphic model, the active site of this enzyme (hexokinase, shown in blue) forms a groove on its surface. Its substrate is glucose (red).

(b) When the substrate enters the active site, it forms weak bonds with the enzyme, inducing a change in the shape of the enzyme. This change allows additional weak bonds to form, causing the active site to enfold the substrate and hold it in place.

▲ **Figure 6.14 Induced fit between an enzyme and its substrate.**

Enzymes use a variety of mechanisms that lower activation energy and speed up a reaction. First, in reactions involving two or more reactants, the active site provides a template on which the substrates can come together in the proper orientation for a reaction to occur between them (see Figure 6.15, step ❷). Second, as the active site of an enzyme clutches the bound substrates, the enzyme may stretch the substrate molecules toward their transition-state form, stressing and bending critical chemical bonds that must be broken during

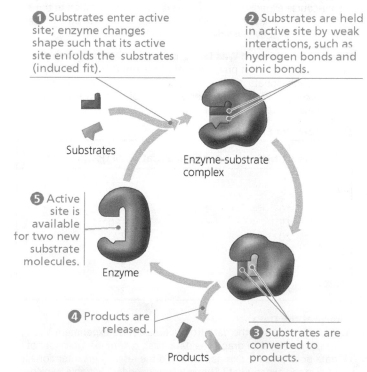

❶ Substrates enter active site; enzyme changes shape such that its active site enfolds the substrates (induced fit).

❷ Substrates are held in active site by weak interactions, such as hydrogen bonds and ionic bonds.

Substrates

Enzyme-substrate complex

❺ Active site is available for two new substrate molecules.

Enzyme

❹ Products are released.

❸ Substrates are converted to products.

Products

▲ **Figure 6.15 The active site and catalytic cycle of an enzyme.** An enzyme can convert one or more reactant molecules to one or more product molecules. The enzyme shown here converts two substrate molecules to two product molecules.

the reaction. Because E_A is proportional to the difficulty of breaking the bonds, distorting the substrate helps it approach the transition state and thus reduces the amount of free energy that must be absorbed to achieve that state.

Third, the active site may also provide a microenvironment that is more conducive to a particular type of reaction than the solution itself would be without the enzyme. For example, if the active site has amino acids with acidic R groups, the active site may be a pocket of low pH in an otherwise neutral cell. In such cases, an acidic amino acid may facilitate H+ transfer to the substrate as a key step in catalyzing the reaction.

A fourth mechanism of catalysis is the direct participation of the active site in the chemical reaction. Sometimes this process even involves brief covalent bonding between the substrate and the side chain of an amino acid of the enzyme. Subsequent steps of the reaction restore the side chains to their original states, so that the active site is the same after the reaction as it was before.

The rate at which a particular amount of enzyme converts substrate to product is partly a function of the initial concentration of the substrate: The more substrate molecules that are available, the more frequently they access the active sites of the enzyme molecules. However, there is a limit to how fast the reaction can be pushed by adding more substrate to a fixed concentration of enzyme. At some point, the concentration of substrate will be high enough that all enzyme molecules have their active sites engaged. As soon as the product exits an active site, another substrate molecule enters. At this substrate concentration, the enzyme is said to be *saturated*, and the rate of the reaction is determined by the speed at which the active site converts substrate to product. When an enzyme population is saturated, the only way to increase the rate of product formation is to add more enzyme. Cells often increase the rate of a reaction by producing more enzyme molecules. You can graph the progress of an enzymatic reaction in the **Scientific Skills Exercise**.

Scientific Skills Exercise

Making a Line Graph and Calculating a Slope

Does the Rate of Glucose 6-Phosphatase Activity Change over Time in Isolated Liver Cells? Glucose 6-phosphatase, which is found in mammalian liver cells, is a key enzyme in control of blood glucose levels. The enzyme catalyzes the breakdown of glucose 6-phosphate into glucose and inorganic phosphate (Ⓟi). These products are transported out of liver cells into the blood, increasing blood glucose levels. In this exercise, you will graph data from a time-course experiment that measured Ⓟi concentration in the buffer outside isolated liver cells, thus indirectly measuring glucose 6-phosphatase activity inside the cells.

How the Experiment Was Done Isolated rat liver cells were placed in a dish with buffer at physiological conditions (pH 7.4, 37°C). Glucose 6-phosphate (the substrate) was added to the dish, where it was taken up by the cells. Then a sample of buffer was removed every 5 minutes and the concentration of Ⓟi determined.

Data from the Experiment

Time (min)	Concentration of Ⓟi (µmol/mL)
0	0
5	10
10	90
15	180
20	270
25	330
30	355
35	355
40	355

Interpret the Data

1. To see patterns in the data from a time-course experiment like this, it is helpful to graph the data. First, determine which set of data goes on each axis. (a) What did the researchers intentionally vary in the experiment? This is the independent variable, which goes on the x-axis. (b) What are the units (abbreviated) for the independent variable? Explain in words what the abbreviation stands for. (c) What was measured by the researchers? This is the dependent variable, which goes on the y-axis. (d) What does the units abbreviation stand for? Label each axis, including the units.

2. Next, you'll want to mark off the axes with just enough evenly spaced tick marks to accommodate the full set of data. Determine the range of data values for each axis. (a) What is the largest value to go on the x-axis? What is a reasonable spacing for the tick marks, and what should be the highest one? (b) What is the largest value to go on the y-axis? What is a reasonable spacing for the tick marks, and what should be the highest one?

3. Plot the data points on your graph. Match each x-value with its partner y-value and place a point on the graph at that coordinate. Draw a line that connects the points. (For additional information about graphs, see the Scientific Skills Review in Appendix F and in the Study Area in MasteringBiology.)

4. Examine your graph and look for patterns in the data. (a) Does the concentration of Ⓟi increase evenly through the course of the experiment? To answer this question, describe the pattern you see in the graph. (b) What part of the graph shows the highest rate of enzyme activity? Consider that the rate of enzyme activity is related to the slope of the line, $\Delta y / \Delta x$ (the "rise" over the "run"), in µmol/mL·min, with the steepest slope indicating the highest rate of enzyme activity. Calculate the rate of enzyme activity (slope) where the graph is steepest. (c) Can you think of a biological explanation for the pattern you see?

5. If your blood sugar level is low from skipping lunch, what reaction (discussed in this exercise) will occur in your liver cells? Write out the reaction and put the name of the enzyme over the reaction arrow. How will this reaction affect your blood sugar level?

Data from S. R. Commerford et al., Diets enriched in sucrose or fat increase gluconeogenesis and G-6-Pase but not basal glucose production in rats, *American Journal of Physiology— Endocrinology and Metabolism* 283:E545–E555 (2002).

(MB) A version of this Scientific Skills Exercise can be assigned in MasteringBiology.

Effects of Local Conditions on Enzyme Activity

The activity of an enzyme—how efficiently the enzyme functions—is affected by general environmental factors, such as temperature and pH. It can also be affected by chemicals that specifically influence that enzyme. In fact, researchers have learned much about enzyme function by employing such chemicals.

Effects of Temperature and pH

The three-dimensional structures of proteins are sensitive to their environment (see Chapter 3). As a consequence, each enzyme works better under some conditions than under other conditions, because these *optimal conditions* favor the most active shape for the enzyme molecule.

Temperature and pH are environmental factors important in the activity of an enzyme. Up to a point, the rate of an enzymatic reaction increases with increasing temperature, partly because substrates collide with active sites more frequently when the molecules move rapidly. Above that temperature, however, the speed of the enzymatic reaction drops sharply. The thermal agitation of the enzyme molecule disrupts the hydrogen bonds, ionic bonds, and other weak interactions that stabilize the active shape of the enzyme, and the protein molecule eventually denatures. Each enzyme has an optimal temperature at which its reaction rate is greatest. Without denaturing the enzyme, this temperature allows the greatest number of molecular collisions and the fastest conversion of the reactants to product molecules. Most human enzymes have optimal temperatures of about 35–40°C (close to human body temperature). The thermophilic bacteria that live in hot springs contain enzymes with optimal temperatures of 70°C or higher **(Figure 6.16a)**.

Just as each enzyme has an optimal temperature, it also has a pH at which it is most active. The optimal pH values for most enzymes fall in the range of pH 6–8, but there are exceptions. For example, pepsin, a digestive enzyme in the human stomach, works best at pH 2. Such an acidic environment denatures most enzymes, but pepsin is adapted to maintain its functional three-dimensional structure in the acidic environment of the stomach. In contrast, trypsin, a digestive enzyme residing in the alkaline environment of the human intestine, has an optimal pH of 8 and would be denatured in the stomach **(Figure 6.16b)**.

Cofactors

Many enzymes require nonprotein helpers for catalytic activity. These adjuncts, called **cofactors**, may be bound tightly to the enzyme as permanent residents, or they may bind loosely and reversibly along with the substrate. The cofactors of some enzymes are inorganic, such as the metal atoms zinc, iron, and copper in ionic form. If the cofactor is an organic molecule, it is more specifically called a **coenzyme**. Most vitamins are important in nutrition because they act as coenzymes or raw materials from which coenzymes are made. Cofactors function

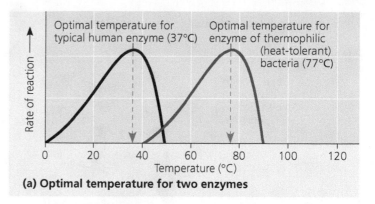

(a) **Optimal temperature for two enzymes**

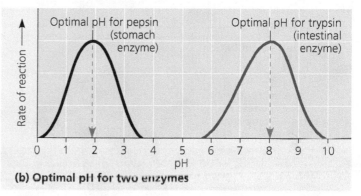

(b) **Optimal pH for two enzymes**

▲ **Figure 6.16 Environmental factors affecting enzyme activity.** Each enzyme has an optimal **(a)** temperature and **(b)** pH that favor the most active shape of the protein molecule.

DRAW IT *Given that a mature lysosome has an internal pH of around 4.5, draw a curve in (b) showing what you would predict for a lysosomal enzyme, labeling its optimal pH.*

in various ways, but in all cases where they are used, they perform a crucial chemical function in catalysis. You'll encounter examples of cofactors later in the book.

Enzyme Inhibitors

Certain chemicals selectively inhibit the action of specific enzymes, and we have learned a lot about enzyme function by studying the effects of these molecules. If the inhibitor attaches to the enzyme by covalent bonds, inhibition is usually irreversible.

Many enzyme inhibitors, however, bind to the enzyme by weak interactions, in which case inhibition is reversible. Some reversible inhibitors resemble the normal substrate molecule and compete for admission into the active site **(Figure 6.17a and b)**. These mimics, called **competitive inhibitors**, reduce the productivity of enzymes by blocking substrates from entering active sites. This kind of inhibition can be overcome by increasing the concentration of substrate so that as active sites become available, more substrate molecules than inhibitor molecules are around to gain entry to the sites.

In contrast, **noncompetitive inhibitors** do not directly compete with the substrate to bind to the enzyme at the active site. Instead, they impede enzymatic reactions by binding to another part of the enzyme. This interaction causes the

(a) Normal binding

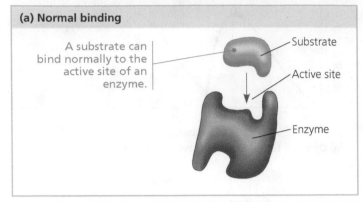

A substrate can bind normally to the active site of an enzyme.

Substrate

Active site

Enzyme

(b) Competitive inhibition

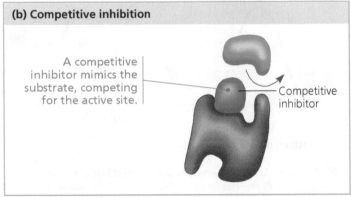

A competitive inhibitor mimics the substrate, competing for the active site.

Competitive inhibitor

(c) Noncompetitive inhibition

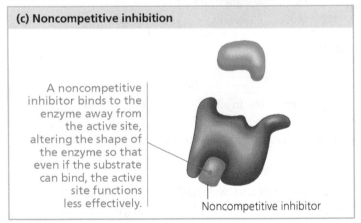

A noncompetitive inhibitor binds to the enzyme away from the active site, altering the shape of the enzyme so that even if the substrate can bind, the active site functions less effectively.

Noncompetitive inhibitor

enzyme molecule to change its shape in such a way that the active site becomes less effective at catalyzing the conversion of substrate to product **(Figure 6.17c)**.

Toxins and poisons are often irreversible enzyme inhibitors. An example is sarin, a nerve gas that caused the death of several people and injury to many others when it was released by terrorists in the Tokyo subway in 1995. This small molecule binds covalently to the R group on the amino acid serine, which is found in the active site of acetylcholinesterase, an enzyme important in the nervous system. Other examples include the pesticides DDT and parathion, inhibitors of key enzymes in the nervous system. Finally, many antibiotics are inhibitors of specific enzymes in bacteria. For instance, penicillin blocks the active site of an enzyme that many bacteria use to make their cell walls.

Citing enzyme inhibitors that are metabolic poisons may give the impression that enzyme inhibition is generally abnormal and harmful. In fact, molecules naturally present in the cell often regulate enzyme activity by acting as inhibitors. Such regulation—selective inhibition—is essential to the control of cellular metabolism, as we will discuss in Concept 6.5.

The Evolution of Enzymes

EVOLUTION Thus far, biochemists have discovered and named more than 4,000 different enzymes in various species, and this list probably represents the tip of the proverbial iceberg. How did this grand profusion of enzymes arise? Recall that most enzymes are proteins, and proteins are encoded by genes. A permanent change in a gene, known as a *mutation*, can result in a protein with one or more changed amino acids. In the case of an enzyme, if the changed amino acids are in the active site or some other crucial region, the altered enzyme might have a novel activity or might bind to a different substrate. Under environmental conditions where the new function benefits the organism, natural selection would tend to favor the mutated form of the gene, causing it to persist in the population. This simplified model is generally accepted as the main way in which the multitude of different enzymes arose over the past few billion years of life's history.

CONCEPT CHECK 6.4

1. Many spontaneous reactions occur very slowly. Why don't all spontaneous reactions occur instantly?
2. Why do enzymes act only on very specific substrates?
3. **WHAT IF?** Malonate is an inhibitor of the enzyme succinate dehydrogenase. How would you determine whether malonate is a competitive or noncompetitive inhibitor?

For suggested answers, see Appendix A.

CONCEPT 6.5

Regulation of enzyme activity helps control metabolism

Chemical chaos would result if all of a cell's metabolic pathways were operating simultaneously. Intrinsic to life's processes is a cell's ability to tightly regulate its metabolic pathways by controlling when and where its various enzymes are active. It does this either by switching on and off the genes that encode specific enzymes (as we will discuss in Unit Two) or, as we discuss next, by regulating the activity of enzymes once they are made.

Allosteric Regulation of Enzymes

In many cases, the molecules that naturally regulate enzyme activity in a cell behave something like reversible noncompetitive inhibitors (see Figure 6.17c): These regulatory molecules change an enzyme's shape and the functioning of its active site by binding to a site elsewhere on the molecule, via noncovalent interactions. **Allosteric regulation** is the term used to describe any case in which a protein's function at one site is affected by the binding of a regulatory molecule to a separate site. It may result in either inhibition or stimulation of an enzyme's activity.

Allosteric Activation and Inhibition

Most enzymes known to be allosterically regulated are constructed from two or more subunits, each composed of a polypeptide chain with its own active site. The entire complex oscillates between two different shapes, one catalytically active and the other inactive **(Figure 6.18a)**. In the simplest kind of allosteric regulation, an activating or inhibiting regulatory molecule binds to a regulatory site (sometimes called an allosteric site), often located where subunits join. The binding of an *activator* to a regulatory site stabilizes the shape that has functional active sites, whereas the binding of an *inhibitor* stabilizes the inactive form of the enzyme. The subunits of an allosteric enzyme fit together in such a way that a shape change in one subunit is transmitted to all others. Through this interaction of subunits, a single activator or inhibitor molecule that binds to one regulatory site will affect the active sites of all subunits.

Fluctuating concentrations of regulators can cause a sophisticated pattern of response in the activity of cellular enzymes. The products of ATP hydrolysis (ADP and P_i), for example, play a complex role in balancing the flow of traffic between anabolic and catabolic pathways by their effects on key enzymes. ATP binds to several catabolic enzymes allosterically, lowering their affinity for substrate and thus inhibiting their activity. ADP, however, functions as an activator of the same enzymes. This is logical because catabolism functions in regenerating ATP. If ATP production lags behind its use, ADP accumulates and activates the enzymes that speed up catabolism, producing more ATP. If the supply of ATP exceeds demand, then catabolism slows down as ATP molecules accumulate and bind to the same enzymes, inhibiting them. (You'll see specific examples of this type of regulation when you learn about cellular respiration in the next chapter.) ATP, ADP, and other related molecules also affect key enzymes in anabolic pathways. In this way, allosteric enzymes control the rates of important reactions in both sorts of metabolic pathways.

In another kind of allosteric activation, a *substrate* molecule binding to one active site in a multisubunit enzyme triggers a shape change in all the subunits, thereby increasing catalytic activity at the other active sites **(Figure 6.18b)**. Called **cooperativity**, this mechanism amplifies the response of enzymes to substrates: One substrate molecule primes an enzyme to act on additional substrate molecules more readily. Cooperativity is considered "allosteric" regulation because binding of the substrate to one active site affects catalysis in another active site.

▼ **Figure 6.18 Allosteric regulation of enzyme activity.**

(a) Allosteric activators and inhibitors

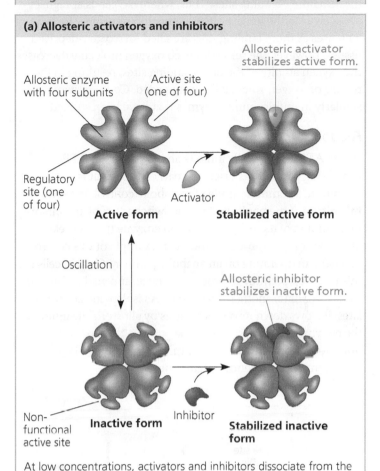

At low concentrations, activators and inhibitors dissociate from the enzyme. The enzyme can then oscillate again.

(b) Cooperativity: another type of allosteric activation

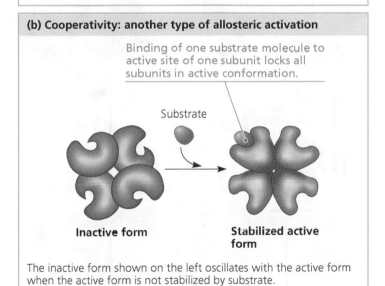

The inactive form shown on the left oscillates with the active form when the active form is not stabilized by substrate.

Although the vertebrate oxygen transport protein hemoglobin is not an enzyme, classic studies of cooperative binding in this protein have elucidated the principle of cooperativity. Hemoglobin is made up of four subunits, each of which has an oxygen-binding site (see Figure 3.21). The binding of an oxygen molecule to one binding site increases the affinity for oxygen of the remaining binding sites. Thus, where oxygen is at high levels, such as in the lungs or gills, hemoglobin's affinity for oxygen increases as more binding sites are filled. In oxygen-deprived tissues, however, the release of each oxygen molecule decreases the oxygen affinity of the other binding sites, resulting in the release of oxygen where it is most needed. Cooperativity works similarly in multisubunit enzymes that have been studied.

Feedback Inhibition

When ATP allosterically inhibits an enzyme in an ATP-generating pathway, as discussed earlier, the result is feedback inhibition, a common mode of metabolic control. In **feedback inhibition**, a metabolic pathway is switched off by the inhibitory binding of its end product to an enzyme that acts early in the pathway. **Figure 6.19** shows an example of this control mechanism operating on an anabolic pathway. Certain cells use this five-step pathway to synthesize the amino acid isoleucine from threonine, another amino acid. As isoleucine accumulates, it slows down its own synthesis by allosterically inhibiting the enzyme for the first step of the pathway. Feedback inhibition thereby prevents the cell from wasting chemical resources by making more isoleucine than is necessary.

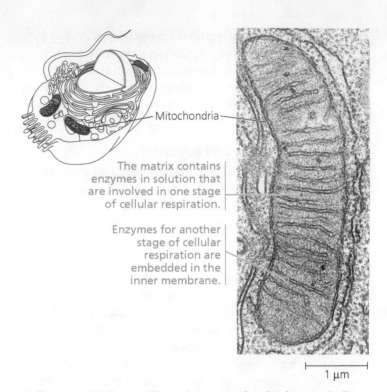

▲ **Figure 6.20 Organelles and structural order in metabolism.** Organelles such as the mitochondrion (TEM) contain enzymes that carry out specific functions, in this case cellular respiration.

Specific Localization of Enzymes Within the Cell

The cell is not just a bag of chemicals with thousands of different kinds of enzymes and substrates in a random mix. The cell is compartmentalized, and cellular structures help bring order to metabolic pathways. In some cases, a team of enzymes for several steps of a metabolic pathway is assembled into a multienzyme complex. The arrangement facilitates the sequence of reactions, with the product from the first enzyme becoming the substrate for an adjacent enzyme in the complex, and so on, until the end product is released. Some enzymes and enzyme complexes have fixed locations within the cell and act as structural components of particular membranes. Others are in solution within particular membrane-enclosed eukaryotic organelles, each with its own internal chemical environment. For example, in eukaryotic cells, the enzymes for cellular respiration reside in specific locations within mitochondria **(Figure 6.20)**.

In this chapter, you learned that metabolism, the intersecting set of chemical pathways characteristic of life, is a choreographed interplay of thousands of different kinds of cellular molecules. In the next chapter, we'll explore cellular respiration, the major catabolic pathway that breaks down organic molecules, releasing energy for the crucial processes of life.

CONCEPT CHECK 6.5

1. How do an activator and an inhibitor have different effects on an allosterically regulated enzyme?

For suggested answers, see Appendix A.

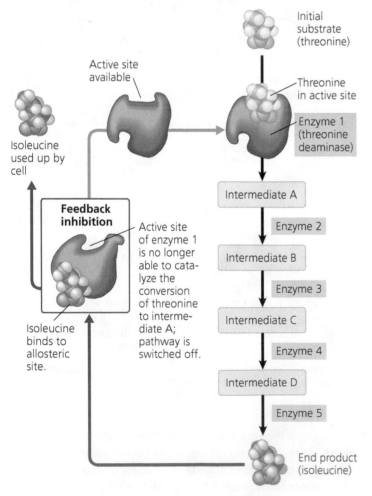

▲ **Figure 6.19 Feedback inhibition in isoleucine synthesis.**

6 Chapter Review

SUMMARY OF KEY CONCEPTS

CONCEPT 6.1

An organism's metabolism transforms matter and energy (pp. 116–119)

- **Metabolism** is the collection of chemical reactions that occur in an organism. Enzymes catalyze reactions in intersecting **metabolic pathways**, which may be **catabolic** (breaking down molecules, releasing energy) or **anabolic** (building molecules, consuming energy).
- **Energy** is the capacity to cause change; some forms of energy do work by moving matter. **Kinetic energy** is associated with motion and includes **thermal energy**, associated with the random motion of atoms or molecules. **Heat** is thermal energy in transfer from one object to another. **Potential energy** is related to the location or structure of matter and includes **chemical energy** possessed by a molecule due to its structure.
- The **first law of thermodynamics**, conservation of energy, states that energy cannot be created or destroyed, only transferred or transformed. The **second law of thermodynamics** states that **spontaneous processes**, those requiring no outside input of energy, increase the **entropy** (disorder) of the universe.

? *Explain how the highly ordered structure of a cell does not conflict with the second law of thermodynamics.*

CONCEPT 6.2

The free-energy change of a reaction tells us whether or not the reaction occurs spontaneously (pp. 119–122)

- A living system's **free energy** is energy that can do work under cellular conditions. Organisms live at the expense of free energy. The change in free energy (ΔG) during a biological process tells us if the process is spontaneous. During a spontaneous process, free energy decreases and the stability of a system increases. At maximum stability, the system is at equilibrium and can do no work.
- In an **exergonic** (spontaneous) chemical reaction, the products have less free energy than the reactants ($-\Delta G$). **Endergonic** (nonspontaneous) reactions require an input of energy ($+\Delta G$). The addition of starting materials and the removal of end products prevent metabolism from reaching equilibrium.

? *Why are spontaneous reactions important in the metabolism of a cell?*

CONCEPT 6.3

ATP powers cellular work by coupling exergonic reactions to endergonic reactions (pp. 122–124)

- **ATP** is the cell's energy shuttle. Hydrolysis of its terminal phosphate yields ADP and Ⓟ$_i$ and releases free energy.
- Through **energy coupling**, the exergonic process of ATP hydrolysis drives endergonic reactions by transfer of a phosphate group to specific reactants, forming a **phosphorylated intermediate** that is more reactive. ATP hydrolysis (sometimes with protein phosphorylation) also causes changes in the shape and binding affinities of transport and motor proteins.
- Catabolic pathways drive regeneration of ATP from ADP + ⓅATP.

? *Describe the ATP cycle: How is ATP used and regenerated in a cell?*

CONCEPT 6.4

Enzymes speed up metabolic reactions by lowering energy barriers (pp. 125–130)

- In a chemical reaction, the energy necessary to break the bonds of the reactants is the **activation energy**, E_A.
- **Enzymes** lower the E_A barrier:

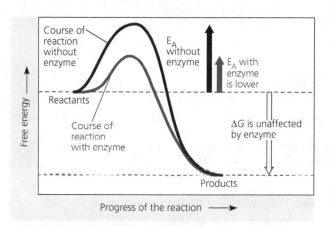

- Each type of enzyme has a unique **active site** that combines specifically with its **substrate(s)**, the reactant molecule(s) on which it acts. The enzyme changes shape slightly when it binds the substrate(s) (**induced fit**).
- The active site can lower an E_A barrier by orienting substrates correctly, straining their bonds, providing a favorable microenvironment, or even covalently bonding with the substrate.
- Each enzyme has an optimal temperature and pH. Inhibitors reduce enzyme function. A **competitive inhibitor** binds to the active site, whereas a **noncompetitive inhibitor** binds to a different site on the enzyme.
- Natural selection, acting on organisms with mutant genes encoding altered enzymes, is a major evolutionary force responsible for the diverse array of enzymes found in organisms.

? *How do both activation energy barriers and enzymes help maintain the structural and metabolic order of life?*

CONCEPT 6.5

Regulation of enzyme activity helps control metabolism (pp. 130–132)

- Many enzymes are subject to **allosteric regulation**: Regulatory molecules, either activators or inhibitors, bind to specific regulatory sites, affecting the shape and function of the enzyme. In **cooperativity**, binding of one substrate molecule can stimulate binding or activity at other active sites. In **feedback inhibition**, the end product of a metabolic pathway allosterically inhibits the enzyme for a previous step in the pathway.
- Some enzymes are grouped into complexes, some are incorporated into membranes, and some are contained inside organelles, increasing the efficiency of metabolic processes.

? *What roles do allosteric regulation and feedback inhibition play in the metabolism of a cell?*

TEST YOUR UNDERSTANDING

Level 1: Knowledge/Comprehension

1. Choose the pair of terms that correctly completes this sentence: Catabolism is to anabolism as _____ is to _____.
   a. exergonic; spontaneous
   b. exergonic; endergonic
   c. free energy; entropy
   d. work; energy
   e. entropy; heat

2. Most cells cannot harness heat to perform work because
   a. heat does not involve a transfer of energy.
   b. cells do not have much heat; they are relatively cool.
   c. temperature is usually uniform throughout a cell.
   d. heat can never be used to do work.
   e. heat must remain constant during work.

3. Which of the following metabolic processes can occur without a net influx of energy from some other process?
   a. $ADP + Ⓟi → ATP + H_2O$
   b. $C_6H_{12}O_6 + 6 O_2 → 6 CO_2 + 6 H_2O$
   c. $6 CO_2 + 6 H_2O → C_6H_{12}O_6 + 6 O_2$
   d. amino acids → protein
   e. glucose + fructose → sucrose

4. If an enzyme in solution is saturated with substrate, the most effective way to obtain a faster yield of products is to
   a. add more of the enzyme.
   b. heat the solution to 90°C.
   c. add more substrate.
   d. add an allosteric inhibitor.
   e. add a noncompetitive inhibitor.

5. Some bacteria are metabolically active in hot springs because
   a. they are able to maintain a lower internal temperature.
   b. high temperatures make catalysis unnecessary.
   c. their enzymes have high optimal temperatures.
   d. their enzymes are completely insensitive to temperature.
   e. they use molecules other than proteins or RNAs as their main catalysts.

Level 2: Application/Analysis

6. If an enzyme is added to a solution where its substrate and product are in equilibrium, what will occur?
   a. Additional product will be formed.
   b. Additional substrate will be formed.
   c. The reaction will change from endergonic to exergonic.
   d. The free energy of the system will change.
   e. Nothing; the reaction will stay at equilibrium.

Level 3: Synthesis/Evaluation

7. **DRAW IT** Using a series of arrows, draw the branched metabolic reaction pathway described by the following statements. Then answer the question at the end. Use red arrows and minus signs to indicate inhibition.

   L can form either M or N.
   M can form O.
   O can form either P or R.
   P can form Q.
   R can form S.
   O inhibits the reaction of L to form M.
   Q inhibits the reaction of O to form P.
   S inhibits the reaction of O to form R.

   Which reaction would prevail if both Q and S were present in the cell at high concentrations?
   a. $L → M$ c. $L → N$ e. $R → S$
   b. $M → O$ d. $O → P$

8. **SCIENTIFIC INQUIRY**
   DRAW IT A researcher has developed an assay to measure the activity of an important enzyme present in liver cells growing in culture. She adds the enzyme's substrate to a dish of cells and then measures the appearance of reaction products. The results are graphed as the amount of product on the y-axis versus time on the x-axis. The researcher notes four sections of the graph. For a short period of time, no products appear (section A). Then (section B) the reaction rate is quite high (the slope of the line is steep). Next, the reaction gradually slows down (section C). Finally, the graph line becomes flat (section D). Draw and label the graph, and propose a model to explain the molecular events occurring at each stage of this reaction profile.

9. **SCIENCE, TECHNOLOGY, AND SOCIETY**
   Organophosphates (organic compounds containing phosophate groups) are commonly used as insecticides to improve crop yield. Organophosphates typically interfere with nerve signal transmission by inhibiting the enzymes that degrade transmitter molecules. They affect humans and other vertebrates as well as insects. Thus, the use of organophosphate pesticides poses some health risks. On the other hand, these molecules break down rapidly upon exposure to air and sunlight. As a consumer, what level of risk are you willing to accept in exchange for an abundant and affordable food supply?

10. **FOCUS ON EVOLUTION**
    A recent revival of the antievolutionary "intelligent design" argument holds that biochemical pathways are too complex to have evolved, because all intermediate steps in a given pathway must be present to produce the final product. Critique this argument. How could you use the diversity of metabolic pathways that produce the same or similar products to support your case?

11. **FOCUS ON ENERGY AND MATTER**
    Life requires energy. In a short essay (100–150 words), describe the basic principles of bioenergetics in an animal cell. How is the flow and transformation of energy different in a photosynthesizing cell? Include the role of ATP and enzymes in your discussion.

For selected answers, see Appendix A.

MasteringBiology®

Students Go to **MasteringBiology** for assignments, the eText, and the Study Area with practice tests, animations, and activities.

Instructors Go to **MasteringBiology** for automatically graded tutorials and questions that you can assign to your students, plus Instructor Resources.

7

Cellular Respiration and Fermentation

KEY CONCEPTS

7.1 Catabolic pathways yield energy by oxidizing organic fuels

7.2 Glycolysis harvests chemical energy by oxidizing glucose to pyruvate

7.3 After pyruvate is oxidized, the citric acid cycle completes the energy-yielding oxidation of organic molecules

7.4 During oxidative phosphorylation, chemiosmosis couples electron transport to ATP synthesis

7.5 Fermentation and anaerobic respiration enable cells to produce ATP without the use of oxygen

7.6 Glycolysis and the citric acid cycle connect to many other metabolic pathways

OVERVIEW

Life Is Work

Living cells require transfusions of energy from outside sources to perform their many tasks—for example, assembling polymers, pumping substances across membranes, moving, and reproducing. The giraffe in **Figure 7.1** obtains energy for its cells by eating plants; some animals feed on other organisms that eat plants. The energy stored in the organic molecules of food ultimately comes from the sun. Energy flows into an ecosystem as sunlight and exits as heat; in contrast, the chemical elements essential to life are recycled **(Figure 7.2)**. Photosynthesis generates oxygen and organic molecules used by the mitochondria of eukaryotes (including plants and algae) as fuel for cellular respiration. Respiration breaks this fuel down, generating ATP. The waste products of this type of respiration, carbon dioxide and water, are the raw materials for photosynthesis.

In this chapter, we'll consider how cells harvest the chemical energy stored in organic molecules and use it to generate ATP, the molecule that drives most cellular work. After presenting some basics about respiration,

▼ **Figure 7.1** How do these leaves power the work of life for this giraffe?

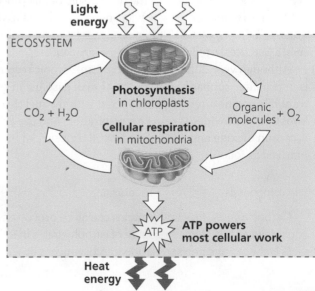

▶ **Figure 7.2**
Energy flow and chemical recycling in ecosystems.
Energy flows into an ecosystem as sunlight and ultimately leaves as heat, while the chemical elements essential to life are recycled.

we'll focus on three key pathways of respiration: glycolysis, the citric acid cycle, and oxidative phosphorylation. We'll also consider fermentation, a somewhat simpler pathway coupled to glycolysis that has deep evolutionary roots.

Catabolic pathways yield energy by oxidizing organic fuels

Metabolic pathways that release stored energy by breaking down complex molecules are called catabolic pathways (see Chapter 6). Electron transfer plays a major role in these pathways. In this section, we'll consider these processes, which are central to cellular respiration.

Catabolic Pathways and Production of ATP

Organic compounds possess potential energy as a result of the arrangement of electrons in the bonds between their atoms. Compounds that can participate in exergonic reactions can act as fuels. With the help of enzymes, a cell systematically degrades complex organic molecules that are rich in potential energy to simpler waste products that have less energy. Some of the energy taken out of chemical storage can be used to do work; the rest is dissipated as heat.

One catabolic process, **fermentation**, is a partial degradation of sugars or other organic fuel that occurs without the use of oxygen. However, the most efficient catabolic pathway is **aerobic respiration**, in which oxygen is consumed as a reactant along with the organic fuel (*aerobic* is from the Greek *aer*, air, and *bios*, life). The cells of most eukaryotic and many prokaryotic organisms can carry out aerobic respiration. Some prokaryotes use substances other than oxygen as reactants in a similar process that harvests chemical energy without oxygen; this process is called *anaerobic respiration* (the prefix *an-* means "without"). Technically, the term **cellular respiration** includes both aerobic and anaerobic processes. However, it originated as a synonym for aerobic respiration because of the relationship of that process to organismal respiration, in which an animal breathes in oxygen. Thus, *cellular respiration* is often used to refer to the aerobic process, a practice we follow in most of this chapter.

Although very different in mechanism, aerobic respiration is in principle similar to the combustion of gasoline in an automobile engine after oxygen is mixed with the fuel (hydrocarbons). Food provides the fuel for respiration, and the exhaust is carbon dioxide and water. The overall process can be summarized as follows:

$$\text{Organic compounds} + \text{Oxygen} \rightarrow \text{Carbon dioxide} + \text{Water} + \text{Energy}$$

Carbohydrates, fats, and proteins can all be processed and consumed as fuel. A major source of carbohydrates in animal diets is the storage polysaccharide starch, which is broken down into glucose ($C_6H_{12}O_6$) subunits. We will learn the steps of cellular respiration by tracking the degradation of the sugar glucose:

$$C_6H_{12}O_6 + 6\,O_2 \rightarrow 6\,CO_2 + 6\,H_2O + \text{Energy (ATP + heat)}$$

This breakdown of glucose is exergonic, having a free-energy change of –686 kcal (2,870 kJ) per mole of glucose decomposed ($\Delta G = -686$ kcal/mol). Recall that a negative ΔG indicates that the products of the chemical process store less energy than the reactants and that the reaction can happen spontaneously—in other words, without an input of energy.

Catabolic pathways do not directly move flagella, pump solutes across membranes, polymerize monomers, or perform other cellular work. Catabolism is linked to work by a chemical drive shaft—ATP (which you learned about in Chapters 3 and 6). To keep working, the cell must regenerate its supply of ATP from ADP and $℗_i$ (see Figure 6.11). To understand how cellular respiration accomplishes this, let's examine the fundamental chemical processes known as oxidation and reduction.

Redox Reactions: Oxidation and Reduction

How do the catabolic pathways that decompose glucose and other organic fuels yield energy? The answer is based on the transfer of electrons during the chemical reactions. The relocation of electrons releases energy stored in organic molecules, and this energy ultimately is used to synthesize ATP.

The Principle of Redox

In many chemical reactions, there is a transfer of one or more electrons (e^-) from one reactant to another. These electron transfers are called oxidation-reduction reactions, or **redox reactions** for short. In a redox reaction, the loss of electrons from one substance is called **oxidation**, and the addition of electrons to another substance is known as **reduction**. (Note that *adding* electrons is called *reduction*; adding negatively charged electrons to an atom *reduces* the amount of positive charge of that atom.) To take a simple, nonbiological example, consider the reaction between the elements sodium (Na) and chlorine (Cl) that forms table salt:

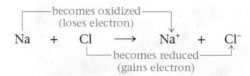

We could generalize a redox reaction this way:

$$\text{Xe}^- + \text{Y} \longrightarrow \text{X} + \text{Ye}^-$$
becomes oxidized / becomes reduced

In the generalized reaction, substance Xe^-, the electron donor, is called the **reducing agent**; it reduces Y, which accepts the donated electron. Substance Y, the electron acceptor, is the **oxidizing agent**; it oxidizes Xe^- by removing its electron. Because an electron transfer requires both a donor and an acceptor, oxidation and reduction always go together.

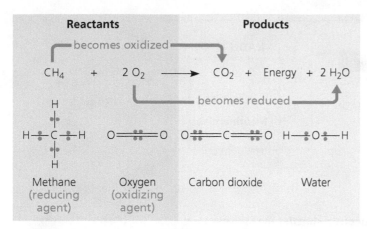

Reactants		Products

becomes oxidized

$$CH_4 \ + \ 2 \, O_2 \ \longrightarrow \ CO_2 \ + \ Energy \ + \ 2 \, H_2O$$

becomes reduced

Methane (reducing agent)   Oxygen (oxidizing agent)   Carbon dioxide   Water

▲ **Figure 7.3 Methane combustion as an energy-yielding redox reaction.** The reaction releases energy to the surroundings because the electrons lose potential energy when they end up being shared unequally, spending more time near electronegative atoms such as oxygen.

Not all redox reactions involve the complete transfer of electrons from one substance to another; some change the degree of electron sharing in covalent bonds. The reaction between methane and oxygen, shown in **Figure 7.3**, is an example. The covalent electrons in methane are shared nearly equally between the bonded atoms because carbon and hydrogen have about the same affinity for valence electrons; they are about equally electronegative. But when methane reacts with oxygen, forming carbon dioxide, electrons end up shared less equally between the carbon atom and its new covalent partners, the oxygen atoms, which are very electronegative. In effect, the carbon atom has partially "lost" its shared electrons; thus, methane has been oxidized.

Now let's examine the fate of the reactant O_2. The two atoms of the oxygen molecule (O_2) share their electrons equally. But when oxygen reacts with the hydrogen from methane, forming water, the electrons of the covalent bonds spend more time near the oxygen (see Figure 7.3). In effect, each oxygen atom has partially "gained" electrons, so the oxygen molecule has been reduced. Because oxygen is so electronegative, it is one of the most potent of all oxidizing agents.

Energy must be added to pull an electron away from an atom, just as energy is required to push a ball uphill. The more electronegative the atom (the stronger its pull on electrons), the more energy is required to take an electron away from it. An electron loses potential energy when it shifts from a less electronegative atom toward a more electronegative one, just as a ball loses potential energy when it rolls downhill. A redox reaction that moves electrons closer to oxygen, such as the burning (oxidation) of methane, therefore releases chemical energy that can be put to work.

Oxidation of Organic Fuel Molecules During Cellular Respiration

The oxidation of methane by oxygen is the main combustion reaction that occurs at the burner of a gas stove. The combustion of gasoline in an automobile engine is also a redox reaction; the energy released pushes the pistons. But the energy-yielding redox process of greatest interest to biologists is respiration: the oxidation of glucose and other molecules in food. Examine again the summary equation for cellular respiration, but this time think of it as a redox process:

becomes oxidized

$$C_6H_{12}O_6 \ + \ 6 \, O_2 \ \longrightarrow \ 6 \, CO_2 \ + \ 6 \, H_2O \ + \ Energy$$

becomes reduced

As in the combustion of methane or gasoline, the fuel (glucose) is oxidized and oxygen is reduced. The electrons lose potential energy along the way, and energy is released.

In general, organic molecules that have an abundance of hydrogen are excellent fuels because their bonds are a source of "hilltop" electrons, whose energy may be released as these electrons "fall" down an energy gradient when they are transferred to oxygen. The summary equation for respiration indicates that hydrogen is transferred from glucose to oxygen. But the important point, not visible in the summary equation, is that the energy state of the electron changes as hydrogen (with its electron) is transferred to oxygen. In respiration, the oxidation of glucose transfers electrons to a lower energy state, liberating energy that becomes available for ATP synthesis.

The main energy-yielding foods, carbohydrates and fats, are reservoirs of electrons associated with hydrogen. Only the barrier of activation energy holds back the flood of electrons to a lower energy state (see Figure 6.12). Without this barrier, a food substance like glucose would combine almost instantaneously with O_2. If we supply the activation energy by igniting glucose, it burns in air, releasing 686 kcal (2,870 kJ) of heat per mole of glucose (about 180 g). Body temperature is not high enough to initiate burning, of course. Instead, if you swallow some glucose, enzymes in your cells will lower the barrier of activation energy, allowing the sugar to be oxidized in a series of steps.

Stepwise Energy Harvest via NAD⁺ and the Electron Transport Chain

If energy is released from a fuel all at once, it cannot be harnessed efficiently for constructive work. For example, if a gasoline tank explodes, it cannot drive a car very far. Cellular respiration does not oxidize glucose in a single explosive step either. Rather, glucose and other organic fuels are broken down in a series of steps, each one catalyzed by an enzyme. At key steps, electrons are stripped from the glucose. As is often the case in oxidation reactions, each electron travels with a proton—thus, as a hydrogen atom. The hydrogen atoms are not transferred directly to oxygen, but instead are usually passed first to an electron carrier, a coenzyme called **NAD⁺** (nicotinamide adenine dinucleotide, a derivative of the vitamin niacin). NAD⁺ is well suited as an electron carrier because it can cycle easily between oxidized (NAD⁺) and reduced (NADH) states. As an electron acceptor, NAD⁺ functions as an oxidizing agent during respiration.

How does NAD⁺ trap electrons from glucose and other organic molecules? Enzymes called dehydrogenases remove a

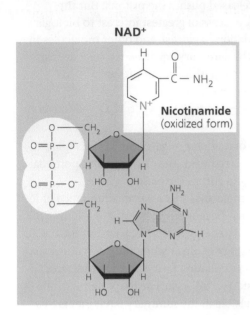

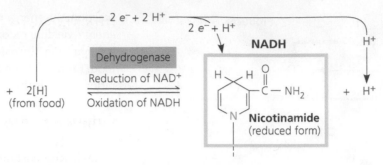

NAD⁺

Nicotinamide
(oxidized form)

Dehydrogenase

Reduction of NAD⁺

Oxidation of NADH

$+ \quad 2[H]$
(from food)

$2\,e^- + 2\,H^+$

$2\,e^- + H^+$

NADH

Nicotinamide
(reduced form)

$+ \quad H^+$

H^+

◄ **Figure 7.4 NAD⁺ as an electron shuttle.** The full name for NAD⁺, nicotinamide adenine dinucleotide, describes its structure: The molecule consists of two nucleotides joined together at their phosphate groups (shown in yellow). (Nicotinamide is a nitrogenous base, although not one that is present in DNA or RNA.) The enzymatic transfer of 2 electrons and 1 proton (H⁺) from an organic molecule in food to NAD⁺ reduces the NAD⁺ to NADH; the second proton (H⁺) is released. Most of the electrons removed from food are transferred initially to NAD⁺.

pair of hydrogen atoms (2 electrons and 2 protons) from the substrate (glucose, in this example), thereby oxidizing it. The enzyme delivers the 2 electrons along with 1 proton to its coenzyme, NAD⁺ **(Figure 7.4)**. The other proton is released as a hydrogen ion (H⁺) into the surrounding solution:

$$\mathrm{H-\overset{|}{\underset{|}{C}}-OH + NAD^+} \xrightarrow{\text{Dehydrogenase}} \mathrm{\overset{|}{C}=O + NADH + H^+}$$

By receiving 2 negatively charged electrons but only 1 positively charged proton, the nicotinamide portion of NAD⁺ has its charge neutralized when NAD⁺ is reduced to NADH. The name NADH shows the hydrogen that has been received in the reaction. NAD⁺ is the most versatile electron acceptor in cellular respiration and functions in several of the redox steps during the breakdown of glucose.

Electrons lose very little of their potential energy when they are transferred from glucose to NAD⁺. Each NADH molecule formed during respiration represents stored energy that can be tapped to make ATP when the electrons complete their "fall" down an energy gradient from NADH to oxygen.

How do electrons that are extracted from glucose and stored as potential energy in NADH finally reach oxygen? It will help to compare the redox chemistry of cellular respiration to a much simpler reaction: the reaction between hydrogen and oxygen to form water **(Figure 7.5a)**. Mix H₂ and O₂, provide a spark for activation

energy, and the gases combine explosively. In fact, combustion of liquid H₂ and O₂ was harnessed to help power the main engines of the Space Shuttle, boosting it into orbit. The explosion represents a release of energy as the electrons of hydrogen "fall" closer to the electronegative oxygen atoms. Cellular respiration also brings hydrogen and oxygen together to form water, but there are two important differences. First, in cellular respiration, the hydrogen that reacts with oxygen is derived from organic molecules rather than H₂. Second, instead of occurring

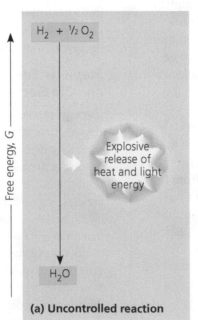

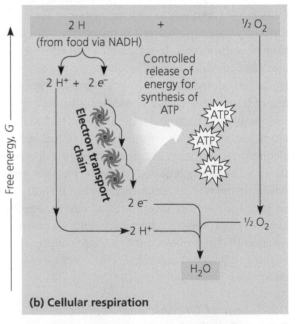

▲ **Figure 7.5 An introduction to electron transport chains. (a)** The one-step exergonic reaction of hydrogen with oxygen to form water releases a large amount of energy in the form of heat and light: an explosion. **(b)** In cellular respiration, the same reaction occurs in stages: An electron transport chain breaks the "fall" of electrons in this reaction into a series of smaller steps and stores some of the released energy in a form that can be used to make ATP. (The rest of the energy is released as heat.)

in one explosive reaction, respiration uses an electron transport chain to break the fall of electrons to oxygen into several energy-releasing steps (**Figure 7.5b**). An **electron transport chain** consists of a number of molecules, mostly proteins, built into the inner membrane of the mitochondria of eukaryotic cells and the plasma membrane of aerobically respiring prokaryotes. Electrons removed from glucose are shuttled by NADH to the "top," higher-energy end of the chain. At the "bottom," lower-energy end, O_2 captures these electrons along with hydrogen nuclei (H^+), forming water.

Electron transfer from NADH to oxygen is an exergonic reaction with a free-energy change of −53 kcal/mol (−222 kJ/mol). Instead of this energy being released and wasted in a single explosive step, electrons cascade down the chain from one carrier molecule to the next in a series of redox reactions, losing a small amount of energy with each step until they finally reach oxygen, the terminal electron acceptor, which has a very great affinity for electrons. Each "downhill" carrier is more electronegative than, and thus capable of oxidizing, its "uphill" neighbor, with oxygen at the bottom of the chain. Therefore, the electrons transferred from glucose to NAD^+ fall down an energy gradient in the electron transport chain to a far more stable location in the electronegative oxygen atom. Put another way, oxygen pulls electrons down the chain in an energy-yielding tumble analogous to gravity pulling objects downhill.

In summary, during cellular respiration, most electrons travel the following "downhill" route: glucose → NADH → electron transport chain → oxygen. Later in this chapter, you will learn more about how the cell uses the energy released from this exergonic electron fall to regenerate its supply of ATP. For now, having covered the basic redox mechanisms of cellular respiration, let's look at the entire process by which energy is harvested from organic fuels.

The Stages of Cellular Respiration: *A Preview*

The harvesting of energy from glucose by cellular respiration is a cumulative function of three metabolic stages. We list them here along with a color-coding scheme that we will use throughout the chapter to help you keep track of the big picture.

1. Glycolysis (color-coded teal throughout the chapter)
2. Pyruvate oxidation and the citric acid cycle (color-coded salmon)
3. Oxidative phosphorylation: electron transport and chemiosmosis (color-coded violet)

Biochemists usually reserve the term *cellular respiration* for stages 2 and 3 together. In this text, we include glycolysis, however, because most respiring cells deriving energy from glucose use glycolysis to produce the starting material for the citric acid cycle.

As diagrammed in **Figure 7.6**, glycolysis and pyruvate oxidation followed by the citric acid cycle are the catabolic pathways that break down glucose and other organic fuels. **Glycolysis**, which occurs in the cytosol, begins the degradation process by breaking glucose into two molecules of a compound called pyruvate. In eukaryotes, pyruvate enters the mitochondrion and is oxidized to a compound called acetyl CoA, which enters the **citric acid cycle** (also called the Krebs cycle). There, the breakdown of glucose to carbon dioxide is completed. (In prokaryotes, these processes take place in the cytosol.) Thus, the carbon dioxide produced by respiration represents fragments of oxidized organic molecules.

Some of the steps of glycolysis and the citric acid cycle are redox reactions in which dehydrogenases transfer electrons from substrates to NAD^+, forming NADH. In the third stage of respiration, the electron transport chain accepts electrons

▶ **Figure 7.6 An overview of cellular respiration.** During glycolysis, each glucose molecule is broken down into two molecules of the compound pyruvate. In eukaryotic cells, as shown here, the pyruvate enters the mitochondrion. There it is oxidized to acetyl CoA, which is further oxidized to CO_2 in the citric acid cycle. NADH and a similar electron carrier, a coenzyme called $FADH_2$, transfer electrons derived from glucose to electron transport chains, which are built into the inner mitochondrial membrane. (In prokaryotes, the electron transport chains are located in the plasma membrane.) During oxidative phosphorylation, electron transport chains convert the chemical energy to a form used for ATP synthesis in the process called chemiosmosis.

 ANIMATION **BioFlix** Visit the Study Area in **MasteringBiology** for the BioFlix® 3-D Animation on Cellular Respiration.

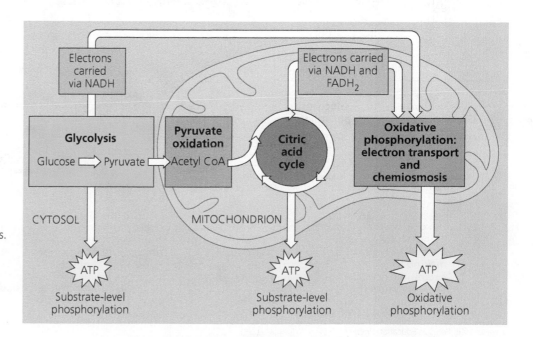

(most often via NADH) from the breakdown products of the first two stages and passes these electrons from one molecule to another. At the end of the chain, the electrons are combined with molecular oxygen and hydrogen ions (H^+), forming water (see Figure 7.5b). The energy released at each step of the chain is stored in a form the mitochondrion (or prokaryotic cell) can use to make ATP from ADP. This mode of ATP synthesis is called **oxidative phosphorylation** because it is powered by the redox reactions of the electron transport chain.

In eukaryotic cells, the inner membrane of the mitochondrion is the site of electron transport and chemiosmosis, the processes that together constitute oxidative phosphorylation. (In prokaryotes, these processes take place in the plasma membrane.) Oxidative phosphorylation accounts for almost 90% of the ATP generated by respiration. A smaller amount of ATP is formed directly in a few reactions of glycolysis and the citric acid cycle by a mechanism called **substrate-level phosphorylation (Figure 7.7)**. This mode of ATP synthesis occurs when an enzyme transfers a phosphate group from a substrate molecule to ADP, rather than adding an inorganic phosphate to ADP as in oxidative phosphorylation. "Substrate molecule" here refers to an organic molecule generated as an intermediate during the catabolism of glucose.

For each molecule of glucose degraded to carbon dioxide and water by respiration, the cell makes up to about 32 molecules of ATP, each with 7.3 kcal/mol of free energy. Respiration cashes in the large denomination of energy banked in a single molecule of glucose (686 kcal/mol) for the small change

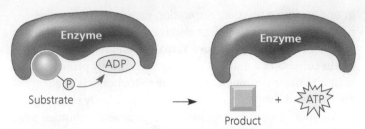

▲ **Figure 7.7 Substrate-level phosphorylation.** Some ATP is made by direct transfer of a phosphate group from an organic substrate to ADP by an enzyme. (For examples in glycolysis, see Figure 7.9, steps 7 and 10.)

MAKE CONNECTIONS *Review Figure 6.8. Do you think the potential energy is higher for the reactants or the products in the reaction shown above? Explain.*

of many molecules of ATP, which is more practical for the cell to spend on its work.

This preview has introduced you to how glycolysis, the citric acid cycle, and oxidative phosphorylation fit into the process of cellular respiration. We are now ready to take a closer look at each of these three stages of respiration.

CONCEPT CHECK 7.1

1. Compare and contrast aerobic and anaerobic respiration.
2. Name and describe the two ways in which ATP is made during cellular respiration. During what stage(s) in the process does each type occur?
3. **WHAT IF?** If the following redox reaction occurred, which compound would be oxidized? Which reduced?

$$C_4H_6O_5 + NAD^+ \rightarrow C_4H_4O_5 + NADH + H^+$$

For suggested answers, see Appendix A.

▼ **Figure 7.9 A closer look at glycolysis.** Note that glycolysis is a source of ATP and NADH.

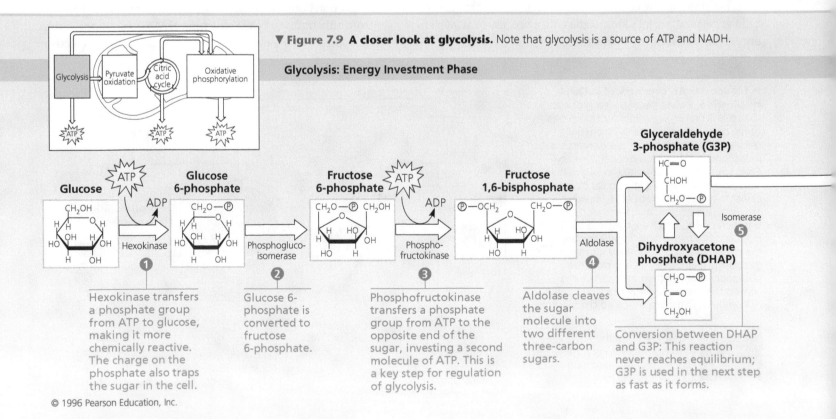

Glycolysis: Energy Investment Phase

Glucose

Glucose 6-phosphate

Fructose 6-phosphate

Fructose 1,6-bisphosphate

Glyceraldehyde 3-phosphate (G3P)

Dihydroxyacetone phosphate (DHAP)

① Hexokinase transfers a phosphate group from ATP to glucose, making it more chemically reactive. The charge on the phosphate also traps the sugar in the cell.

② Phosphogluco-isomerase Glucose 6-phosphate is converted to fructose 6-phosphate.

③ Phosphofructokinase transfers a phosphate group from ATP to the opposite end of the sugar, investing a second molecule of ATP. This is a key step for regulation of glycolysis.

④ Aldolase cleaves the sugar molecule into two different three-carbon sugars.

⑤ Isomerase Conversion between DHAP and G3P: This reaction never reaches equilibrium; G3P is used in the next step as fast as it forms.

© 1996 Pearson Education, Inc.

CONCEPT 7.2

Glycolysis harvests chemical energy by oxidizing glucose to pyruvate

The word *glycolysis* means "sugar splitting," and that is exactly what happens during this pathway. Glucose, a six-carbon sugar, is split into two three-carbon sugars. These smaller sugars are then oxidized and their remaining atoms rearranged to form two molecules of pyruvate. (Pyruvate is the ionized form of pyruvic acid.)

As summarized in **Figure 7.8**, glycolysis can be divided into two phases: energy investment and energy payoff. During the energy investment phase, the cell actually spends ATP. This investment is repaid with interest during the energy payoff phase, when ATP is produced by substrate-level phosphorylation and NAD⁺ is reduced to NADH by electrons released from the oxidation of glucose. The net energy yield from glycolysis, per glucose molecule, is 2 ATP plus 2 NADH.

Because glycolysis is a fundamental core process shared by bacteria, archaea, and eukaryotes alike, we will use it as an example of a biochemical pathway, detailing each of the enzyme-catalyzed reactions. The ten steps of the glycolytic pathway are shown in **Figure 7.9**, which begins on the previous page.

All of the carbon originally present in glucose is accounted for in the two molecules of pyruvate; no carbon is released as CO_2 during glycolysis. Glycolysis occurs whether or not O_2 is present. However, if O_2 *is* present, the chemical energy stored in pyruvate and NADH can be extracted by pyruvate oxidation, the citric acid cycle, and oxidative phosphorylation.

▼ **Figure 7.8 The energy input and output of glycolysis.**

Energy Investment Phase

Glucose

$2\ ADP + 2\ \text{P} \Longleftarrow\qquad 2\ ATP$ used

Energy Payoff Phase

$4\ ADP + 4\ \text{P} \Longrightarrow 4\ ATP$ formed

$2\ NAD^+ +\ 4\ e^- + 4\ H^+ \Longrightarrow 2\ NADH + 2\ H^+$

$\longrightarrow 2$ Pyruvate $+ 2\ H_2O$

Net

Glucose $\longrightarrow 2$ Pyruvate $+ 2\ H_2O$

4 ATP formed – 2 ATP used $\longrightarrow$ 2 ATP

$2\ NAD^+ + 4\ e^- + 4\ H^+ \longrightarrow 2\ NADH + 2\ H^+$

CONCEPT CHECK 7.2

1. During step 6 in Figure 7.9, which molecule acts as the oxidizing agent? The reducing agent?

   For suggested answers, see Appendix A.

The energy payoff phase occurs after glucose is split into two three-carbon sugars. Thus, the coefficient 2 precedes all molecules in this phase.

Glycolysis: Energy Payoff Phase

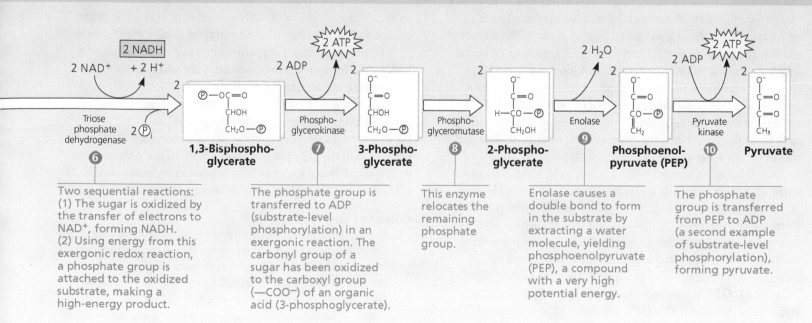

6 Triose phosphate dehydrogenase

1,3-Bisphospho-glycerate

7 Phospho-glycerokinase

3-Phospho-glycerate

8 Phospho-glyceromutase

2-Phospho-glycerate

9 Enolase

Phosphoenol-pyruvate (PEP)

10 Pyruvate kinase

Pyruvate

Two sequential reactions: (1) The sugar is oxidized by the transfer of electrons to NAD⁺, forming NADH. (2) Using energy from this exergonic redox reaction, a phosphate group is attached to the oxidized substrate, making a high-energy product.

The phosphate group is transferred to ADP (substrate-level phosphorylation) in an exergonic reaction. The carbonyl group of a sugar has been oxidized to the carboxyl group (—COO⁻) of an organic acid (3-phosphoglycerate).

This enzyme relocates the remaining phosphate group.

Enolase causes a double bond to form in the substrate by extracting a water molecule, yielding phosphoenolpyruvate (PEP), a compound with a very high potential energy.

The phosphate group is transferred from PEP to ADP (a second example of substrate-level phosphorylation), forming pyruvate.

CONCEPT 7.3

After pyruvate is oxidized, the citric acid cycle completes the energy-yielding oxidation of organic molecules

Glycolysis releases less than a quarter of the chemical energy in glucose that can be harvested by cells; most of the energy remains stockpiled in the two molecules of pyruvate. If molecular oxygen is present, the pyruvate enters a mitochondrion (in eukaryotic cells), where the oxidation of glucose is completed. (In prokaryotic cells, this process occurs in the cytosol.)

Once inside the mitochondrion, pyruvate undergoes a series of enzymatic reactions that remove CO_2 and oxidizes the remaining fragment, forming NADH from NAD^+. The product is a highly reactive compound called acetyl coenzyme A, or **acetyl CoA**, which will feed its acetyl group into the citric acid cycle for further oxidation **(Figure 7.10)**.

The citric acid cycle (also known as the Krebs cycle) functions as a metabolic furnace that oxidizes organic fuel derived from pyruvate. Figure 7.10 summarizes the inputs and outputs as pyruvate is broken down to three CO_2 molecules, including the molecule of CO_2 released during the conversion of pyruvate to acetyl CoA. The cycle generates 1 ATP per turn by substrate-level phosphorylation, but most of the chemical energy is transferred to NAD^+ and a related electron carrier, the coenzyme FAD (flavin adenine dinucleotide, derived from riboflavin, a B vitamin), during the redox reactions. The reduced coenzymes, NADH and $FADH_2$, shuttle their cargo of high-energy electrons into the electron transport chain.

Now let's look at the citric acid cycle in more detail. The cycle has eight steps, each catalyzed by a specific enzyme. You can see in **Figure 7.11** that for each turn of the citric acid cycle, two carbons (red type) enter in the relatively reduced form of an acetyl group (step 1), and two different carbons (blue type) leave in the completely oxidized form of CO_2 molecules (steps 3 and 4). The acetyl group of acetyl CoA joins the cycle by combining with the compound oxaloacetate, forming citrate (step 1). (Citrate is the ionized form of citric acid, for which the cycle is named.) The next seven steps decompose the citrate back to oxaloacetate. It is this regeneration of oxaloacetate that makes this process a *cycle*.

Now let's tally the energy-rich molecules produced by the citric acid cycle. For each acetyl group entering the cycle, 3 NAD^+ are reduced to NADH (steps 3, 4, and 8). In step 6, electrons are transferred not to NAD^+, but to FAD, which accepts 2 electrons and 2 protons to become $FADH_2$. In many animal tissue cells, step 5 produces a guanosine triphosphate (GTP) molecule by substrate-level phosphorylation, as shown in Figure 7.11. GTP is a molecule similar to ATP in its structure and cellular function. This GTP may be used to make an ATP

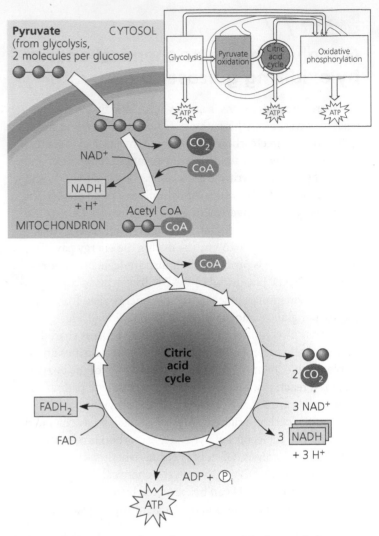

▲ **Figure 7.10 An overview of pyruvate oxidation and the citric acid cycle.** The inputs and outputs per pyruvate molecule are shown. To calculate on a per-glucose basis, multiply by 2, because each glucose molecule is split during glycolysis into two pyruvate molecules.

molecule (as shown) or directly power work in the cell. In the cells of plants, bacteria, and some animal tissues, step 5 forms an ATP molecule directly by substrate-level phosphorylation. The output from step 5 represents the only ATP generated during the citric acid cycle.

Most of the ATP produced by respiration results from oxidative phosphorylation, when the NADH and $FADH_2$ produced by the citric acid cycle relay the electrons extracted from food to the electron transport chain. In the process, they supply the necessary energy for the phosphorylation of ADP to ATP. We will explore this process in the next section.

CONCEPT CHECK 7.3

1. Name the molecules that conserve most of the energy from the citric acid cycle's redox reactions. How is this energy converted to a form that can be used to make ATP?
2. What processes in your cells produce the CO_2 that you exhale?

For suggested answers, see Appendix A.

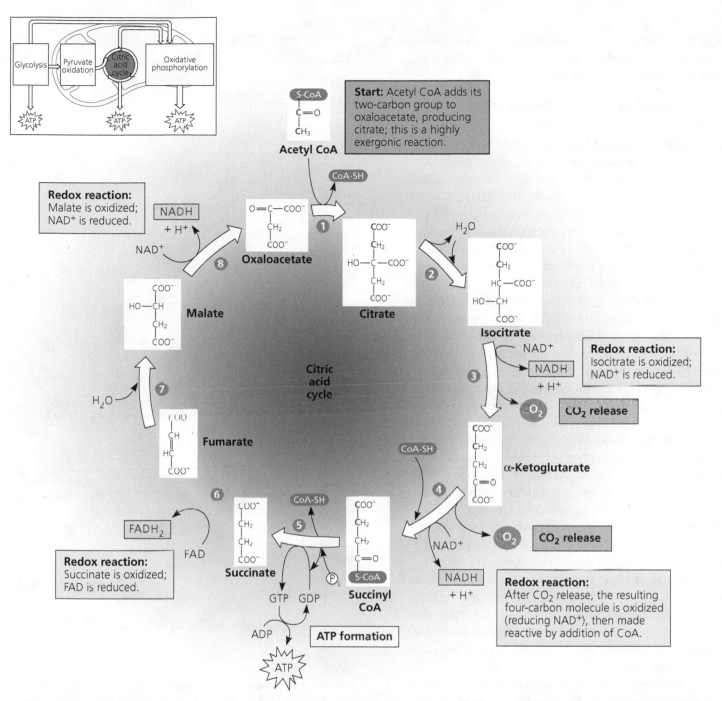

Start: Acetyl CoA adds its two-carbon group to oxaloacetate, producing citrate; this is a highly exergonic reaction.

Acetyl CoA

CoA-SH

Redox reaction: Malate is oxidized; NAD⁺ is reduced.

NADH + H⁺

NAD⁺

Oxaloacetate

8

Malate

H₂O

7

Fumarate

FADH₂

FAD

Redox reaction: Succinate is oxidized; FAD is reduced.

6

Succinate

CoA-SH

5

GTP GDP

ADP

ATP

ATP formation

Pᵢ

Succinyl CoA

S-CoA

CoA-SH

Citric acid cycle

1

Citrate

2

H₂O

Isocitrate

NAD⁺

NADH + H⁺

3

O₂

CO₂ release

Redox reaction: Isocitrate is oxidized; NAD⁺ is reduced.

α-Ketoglutarate

4

NAD⁺

O₂

CO₂ release

NADH + H⁺

Redox reaction: After CO₂ release, the resulting four-carbon molecule is oxidized (reducing NAD⁺), then made reactive by addition of CoA.

▲ **Figure 7.11 A closer look at the citric acid cycle.** Key steps (redox reactions, CO₂ release, and ATP formation) are labeled. In the chemical structures, red type traces the fate of the two carbon atoms that enter the cycle via acetyl CoA (step 1), and blue type indicates the two carbons that exit the cycle as CO₂ in steps 3 and 4. (The red labeling goes only through step 5 because the succinate molecule is symmetrical; the two ends cannot be distinguished from each other.) Notice that the carbon atoms that enter the cycle from acetyl CoA do not leave the cycle in the same turn. They remain in the cycle, occupying a different location in the molecules on their next turn, after another acetyl group is added. As a consequence, the oxaloacetate that is regenerated at step 8 is composed of different carbon atoms each time around.

CONCEPT 7.4

During oxidative phosphorylation, chemiosmosis couples electron transport to ATP synthesis

Our main objective in this chapter is to learn how cells harvest the energy of glucose and other nutrients in food to make ATP.

But the metabolic components of respiration we have dissected so far, glycolysis and the citric acid cycle, produce only 4 ATP molecules per glucose molecule, all by substrate-level phosphorylation: 2 net ATP from glycolysis and 2 ATP from the citric acid cycle. At this point, molecules of NADH (and FADH₂) account for most of the energy extracted from the glucose. These electron escorts link glycolysis and the citric acid cycle to the machinery of oxidative phosphorylation, which uses energy released by the electron transport chain to power

ATP synthesis. In this section, you will learn first how the electron transport chain works and then how electron flow down the chain is coupled to ATP synthesis.

The Pathway of Electron Transport

The electron transport chain is a collection of molecules embedded in the inner membrane of the mitochondrion in eukaryotic cells. (In prokaryotes, these molecules reside in the plasma membrane.) The folding of the inner membrane to form cristae increases its surface area, providing space for thousands of copies of the chain in each mitochondrion. (Once again, we see that structure fits function—the infolded membrane with its placement of electron carrier molecules in a chain, one after the other, is well-suited for the series of sequential redox reactions that take place along the chain.) Most components of the chain are proteins, which exist in multiprotein complexes numbered I through IV. Tightly bound to these proteins are *prosthetic groups*, nonprotein components essential for the catalytic functions of certain enzymes.

Figure 7.12 shows the sequence of electron carriers in the electron transport chain and the drop in free energy as electrons travel down the chain. During electron transport along the chain, electron carriers alternate between reduced and oxidized states as they accept and donate electrons. Each component of the chain becomes reduced when it accepts electrons from its "uphill" neighbor, which has a lower affinity for electrons (is less electronegative). It then returns to its oxidized form as it passes electrons to its "downhill," more electronegative neighbor.

Now let's take a closer look at the electron transport chain in Figure 7.12. We'll first describe the passage of electrons through complex I in some detail as an illustration of the general principles involved in electron transport. Electrons removed from glucose by NAD$^+$ during glycolysis and the citric acid cycle are transferred from NADH to the first molecule of the electron transport chain in complex I. This molecule is a flavoprotein, so named because it has a prosthetic group called flavin mononucleotide (FMN). In the next redox reaction, the flavoprotein returns to its oxidized form as it passes electrons to an iron-sulfur protein (Fe • S in complex I), one of a family of proteins with both iron and sulfur tightly bound. The iron-sulfur protein then passes the electrons to a compound called ubiquinone (Q in Figure 7.12). This electron carrier is a small hydrophobic molecule, the only member of the electron transport chain that is not a protein. Ubiquinone is individually mobile within the membrane rather than residing in a particular complex. (Another name for ubiquinone is coenzyme Q, or CoQ; you may have seen it sold as a nutritional supplement.)

Most of the remaining electron carriers between ubiquinone and oxygen are proteins called **cytochromes**. Their prosthetic group, called a heme group, has an iron atom that accepts and donates electrons. (It is similar to the heme group in hemoglobin, the protein of red blood cells, except that the

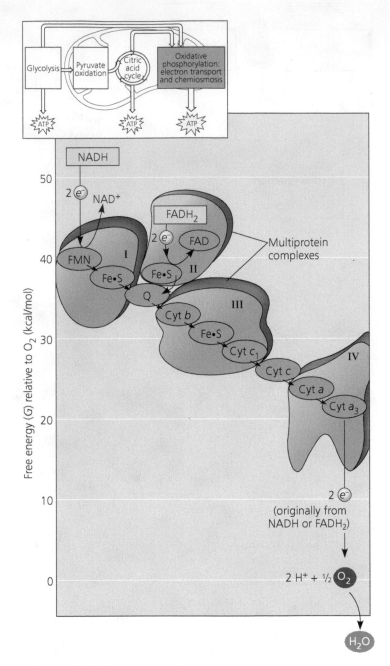

▲ **Figure 7.12 Free-energy change during electron transport.** The overall energy drop (ΔG) for electrons traveling from NADH to oxygen is 53 kcal/mol, but this "fall" is broken up into a series of smaller steps by the electron transport chain. (An oxygen atom is represented here as $^1/_2$ O_2 to emphasize that the electron transport chain reduces molecular oxygen, O_2, not individual oxygen atoms.)

iron in hemoglobin carries oxygen, not electrons.) The electron transport chain has several types of cytochromes, each a different protein with a slightly different electron-carrying heme group. The last cytochrome of the chain, cyt a_3, passes its electrons to oxygen, which is *very* electronegative. Each oxygen atom also picks up a pair of hydrogen ions from the aqueous solution, forming water.

Another source of electrons for the transport chain is FADH$_2$, the other reduced product of the citric acid cycle.

Notice in Figure 7.12 that $FADH_2$ adds its electrons to the electron transport chain from within complex II, at a lower energy level than NADH does. Consequently, although NADH and $FADH_2$ each donate an equivalent number of electrons (2) for oxygen reduction, the electron transport chain provides about one-third less energy for ATP synthesis when the electron donor is $FADH_2$ rather than NADH. We'll see why in the next section.

The electron transport chain makes no ATP directly. Instead, it eases the fall of electrons from food to oxygen, breaking a large free-energy drop into a series of smaller steps that release energy in manageable amounts. How does the mitochondrion (or the prokaryotic plasma membrane) couple this electron transport and energy release to ATP synthesis? The answer is a mechanism called chemiosmosis.

Chemiosmosis: The Energy-Coupling Mechanism

Populating the inner membrane of the mitochondrion or the prokaryotic plasma membrane are many copies of a protein complex called **ATP synthase**, the enzyme that actually makes ATP from ADP and inorganic phosphate. ATP synthase works like an ion pump running in reverse. Ion pumps usually use ATP as an energy source to transport ions against their gradients. (In fact, the proton pump shown in Figure 5.16 is an ATP synthase.) Enzymes can catalyze a reaction in either direction, depending on the ΔG for the reaction, which is affected by the local concentrations of reactants and products (see Chapter 6). Rather than hydrolyzing ATP to pump protons against their concentration gradient, under the conditions of cellular respiration ATP synthase uses the energy of an existing ion gradient to power ATP synthesis. The power source for the ATP synthase is a difference in the concentration of H^+ on opposite sides of the inner mitochondrial membrane. (We can also think of this gradient as a difference in pH, since pH is a measure of H^+ concentration.) This process, in which energy stored in the form of a hydrogen ion gradient across a membrane is used to drive cellular work such as the synthesis of ATP, is called **chemiosmosis** (from the Greek *osmos*, push). We have previously used the word *osmosis* in discussing water transport, but here it refers to the flow of H^+ across a membrane.

From studying the structure of ATP synthase, scientists have learned how the flow of H^+ through this large enzyme powers ATP generation. ATP synthase is a multisubunit complex with four main parts, each made up of multiple polypeptides. Protons move one by one into binding sites on one of the parts (the rotor), causing it to spin in a way that catalyzes ATP production from ADP and inorganic phosphate **(Figure 7.13)**. The flow of protons thus behaves somewhat like a rushing stream that turns a waterwheel. ATP synthase is the smallest molecular rotary motor known in nature.

How does the inner mitochondrial membrane or the prokaryotic plasma membrane generate and maintain the H^+

INTERMEMBRANE SPACE

1 H^+ ions flowing down their gradient enter a half channel in a **stator**, which is anchored in the membrane.

2 H^+ ions enter binding sites within a **rotor**, changing the shape of each subunit so that the rotor spins within the membrane.

3 Each H^+ ion makes one complete turn before leaving the rotor and passing through a second half channel in the stator into the mitochondrial matrix.

4 Spinning of the rotor causes an internal rod to spin as well. This rod extends like a stalk into the **knob** below it, which is held stationary by part of the stator.

5 Turning of the rod activates catalytic sites in the knob that produce ATP from ADP and P_i.

H^+ · Stator · Rotor · Internal rod · Catalytic knob · ADP + P_i · ATP

MITOCHONDRIAL MATRIX

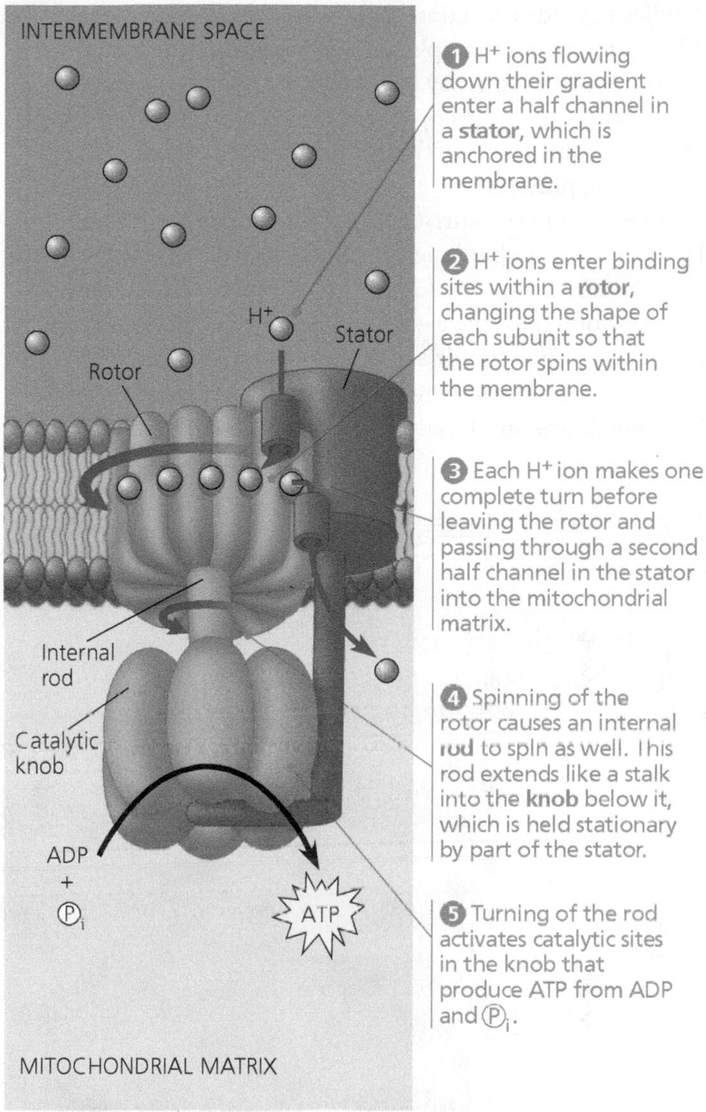

▲ **Figure 7.13 ATP synthase, a molecular mill.** The ATP synthase protein complex functions as a mill, powered by the flow of hydrogen ions. Multiple copies of this complex reside in mitochondrial and chloroplast membranes of eukaryotes and in the plasma membranes of prokaryotes. Each of the four parts of ATP synthase consists of a number of polypeptide subunits.

gradient that drives ATP synthesis by the ATP synthase protein complex? Establishing the H^+ gradient across the inner mitochondrial membrane is a major function of the electron transport chain **(Figure 7.14)**. The chain is an energy converter that uses the exergonic flow of electrons from NADH and $FADH_2$ to pump H^+ across the membrane, from the mitochondrial matrix into the intermembrane space. The H^+ has a tendency to move back across the membrane, diffusing down its gradient. And the ATP synthases are the only sites that provide a route through the membrane for H^+. As we described previously, the passage of H^+ through ATP synthase uses the exergonic flow of H^+ to drive the phosphorylation of ADP. Thus, the energy stored in an H^+ gradient across a membrane couples the redox reactions of the electron transport chain to ATP synthesis, an example of chemiosmosis (see Figure 7.14).

At this point, you may be wondering how the electron transport chain pumps hydrogen ions. Researchers have found that certain members of the electron transport chain accept and release protons (H^+) along with electrons. (The aqueous solutions inside and surrounding the cell are a ready source of H^+.) At certain steps along the chain, electron transfers cause H^+ to be taken up and released into the surrounding solution. In eukaryotic cells, the electron carriers are spatially arranged in the inner mitochondrial membrane in such a way that H^+ is accepted from the mitochondrial matrix and deposited in the intermembrane space (see Figure 7.14). The H^+ gradient that

results is referred to as a **proton-motive force**, emphasizing the capacity of the gradient to perform work. The force drives H^+ back across the membrane through the H^+ channels provided by ATP synthases.

In general terms, *chemiosmosis is an energy-coupling mechanism that uses energy stored in the form of an H^+ gradient across a membrane to drive cellular work*. In mitochondria, the energy for gradient formation comes from exergonic redox reactions, and ATP synthesis is the work performed. But chemiosmosis also occurs elsewhere and in other variations. Chloroplasts use chemiosmosis to generate ATP during photosynthesis; in these organelles, light (rather than chemical energy) drives both electron flow down an electron

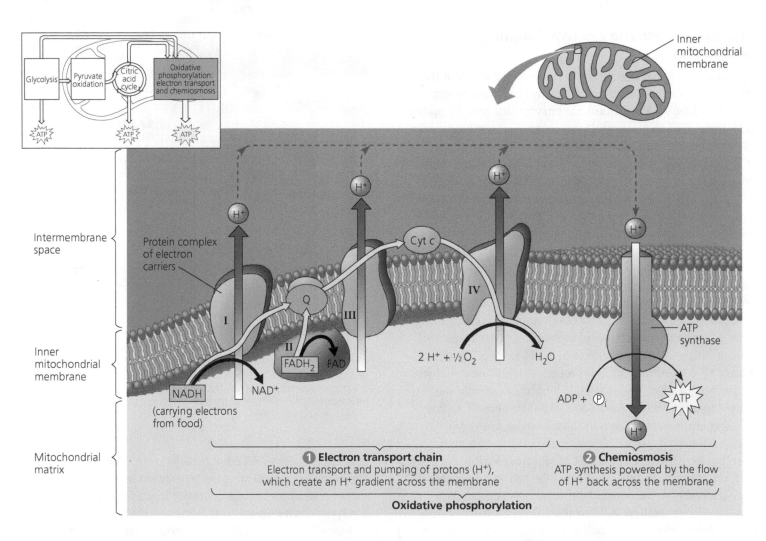

▲ **Figure 7.14 Chemiosmosis couples the electron transport chain to ATP synthesis.** ❶ NADH and FADH₂ shuttle high-energy electrons extracted from food during glycolysis and the citric acid cycle into an electron transport chain built into the inner mitochondrial membrane. The gold arrows trace the transport of electrons, which are finally passed to oxygen at the "downhill" end of the chain, forming water. Most of the electron carriers of the chain are grouped into four complexes. Two mobile

carriers, ubiquinone (Q) and cytochrome *c* (Cyt *c*), move rapidly, ferrying electrons between the large complexes. As complexes shuttle electrons, they pump protons from the mitochondrial matrix into the intermembrane space. FADH₂ deposits its electrons via complex II and so results in fewer protons being pumped into the intermembrane space than occurs with NADH. Chemical energy originally harvested from food is transformed into a proton-motive force, a gradient of H^+ across the membrane.

❷ During chemiosmosis, the protons flow back down their gradient via ATP synthase, which is built into the membrane nearby. The ATP synthase harnesses the proton-motive force to phosphorylate ADP, forming ATP. Together, electron transport and chemiosmosis make up oxidative phosphorylation.

WHAT IF? *If complex IV were nonfunctional, could chemiosmosis produce any ATP, and if so, how would the rate of synthesis differ?*

transport chain and the resulting H⁺ gradient formation. Prokaryotes, as already mentioned, generate H⁺ gradients across their plasma membranes. They then tap the proton-motive force not only to make ATP inside the cell but also to rotate their flagella and to pump nutrients and waste products across the membrane. Because of its central importance to energy conversions in prokaryotes and eukaryotes, chemiosmosis has helped unify the study of bioenergetics. Peter Mitchell was awarded the Nobel Prize in 1978 for originally proposing the chemiosmotic model.

An Accounting of ATP Production by Cellular Respiration

In the last few sections, we have looked rather closely at the key processes of cellular respiration. Now let's take a step back and remind ourselves of its overall function: harvesting the energy of glucose for ATP synthesis.

During respiration, most energy flows in this sequence: glucose → NADH → electron transport chain → proton-motive force → ATP. We can do some bookkeeping to calculate the ATP profit when cellular respiration oxidizes a molecule of glucose to six molecules of carbon dioxide. The three main departments of this metabolic enterprise are glycolysis, the citric acid cycle, and the electron transport chain, which drives oxidative phosphorylation. **Figure 7.15** gives a detailed accounting of the ATP yield per glucose molecule oxidized. The tally adds

the 4 ATP produced directly by substrate-level phosphorylation during glycolysis and the citric acid cycle to the many more molecules of ATP generated by oxidative phosphorylation. Each NADH that transfers a pair of electrons from glucose to the electron transport chain contributes enough to the proton-motive force to generate a maximum of about 3 ATP.

Why are the numbers in Figure 7.15 inexact? There are three reasons we cannot state an exact number of ATP molecules generated by the breakdown of one molecule of glucose. First, phosphorylation and the redox reactions are not directly coupled to each other, so the ratio of the number of NADH molecules to the number of ATP molecules is not a whole number. We know that 1 NADH results in 10 H⁺ being transported out across the inner mitochondrial membrane, but the exact number of H⁺ that must reenter the mitochondrial matrix via ATP synthase to generate 1 ATP has long been debated. Based on experimental data, however, most biochemists now agree that the most accurate number is 4 H⁺. Therefore, a single molecule of NADH generates enough proton-motive force for the synthesis of 2.5 ATP. The citric acid cycle also supplies electrons to the electron transport chain via FADH₂, but since its electrons enter later in the chain, each molecule of this electron carrier is responsible for transport of only enough H⁺ for the synthesis of 1.5 ATP. These numbers also take into account the slight energetic cost of moving the ATP formed in the mitochondrion out into the cytosol, where it will be used.

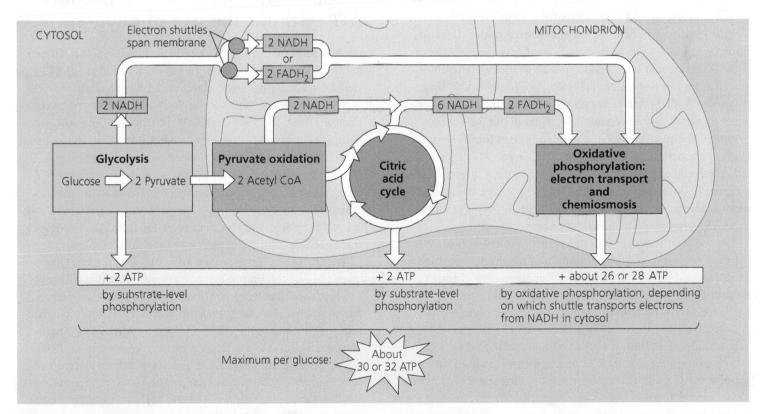

▲ **Figure 7.15 ATP yield per molecule of glucose at each stage of cellular respiration.**

[?] *Explain exactly how the numbers "26 or 28" in the yellow bar were calculated.*

Second, the ATP yield varies slightly depending on the type of shuttle used to transport electrons from the cytosol into the mitochondrion. The mitochondrial inner membrane is impermeable to NADH, so NADH in the cytosol is segregated from the machinery of oxidative phosphorylation. The 2 electrons of NADH captured in glycolysis must be conveyed into the mitochondrion by one of several electron shuttle systems. Depending on the kind of shuttle in a particular cell type, the electrons are passed either to NAD⁺ or to FAD in the mitochondrial matrix (see Figure 7.15). If the electrons are passed to FAD, as in brain cells, only about 1.5 ATP can result from each NADH that was originally generated in the cytosol. If the electrons are passed to mitochondrial NAD⁺, as in liver cells and heart cells, the yield is about 2.5 ATP per NADH.

A third variable that reduces the yield of ATP is the use of the proton-motive force generated by the redox reactions of respiration to drive other kinds of work. For example, the proton-motive force powers the mitochondrion's uptake of pyruvate from the cytosol. However, if *all* the proton-motive force generated by the electron transport chain were used to drive ATP synthesis, one glucose molecule could generate a maximum of 28 ATP produced by oxidative phosphorylation plus 4 ATP (net) from substrate-level phosphorylation to give a total yield of about 32 ATP (or only about 30 ATP if the less efficient shuttle were functioning).

We can now roughly estimate the efficiency of respiration—that is, the percentage of chemical energy in glucose that has been transferred to ATP. Recall that the complete oxidation of a mole of glucose releases 686 kcal of energy under standard conditions ($\Delta G = -686$ kcal/mol). Phosphorylation of ADP to form ATP stores at least 7.3 kcal per mole of ATP. Therefore, the efficiency of respiration is 7.3 kcal per mole of ATP times 32 moles of ATP per mole of glucose divided by 686 kcal per mole of glucose, which equals 0.34. Thus, about 34% of the potential chemical energy in glucose has been transferred to ATP; the actual percentage is bound to vary as ΔG varies under different cellular conditions. Cellular respiration is remarkably efficient in its energy conversion. By comparison, the most efficient automobile converts only about 25% of the energy stored in gasoline to energy that moves the car.

The rest of the energy stored in glucose is lost as heat. We humans use some of this heat to maintain our relatively high body temperature (37°C), and we dissipate the rest through sweating and other cooling mechanisms.

Surprisingly, perhaps, it is beneficial under certain conditions to reduce the efficiency of cellular respiration. A remarkable adaptation is shown by hibernating mammals, which overwinter in a state of inactivity and lowered metabolism. Although their internal body temperature is lower than normal, it still must be kept significantly higher than the external air temperature. One type of tissue, called brown fat, is made up of cells packed full of mitochondria. The inner mitochondrial membrane contains a channel protein called the uncoupling protein, which allows protons to flow back down their concentration gradient without generating ATP. Activation of these proteins in hibernating mammals results in ongoing oxidation of stored fuel stores (fats), generating heat without any ATP production. In the absence of such an adaptation, the ATP level would build up to a point that cellular respiration would be shut down due to regulatory mechanisms in the cell. In the **Scientific Skills Exercise**, you can work with data in a different case where a decrease in metabolic efficiency in cells is used to generate heat.

CONCEPT CHECK 7.4

1. What effect would an absence of O_2 have on the process shown in Figure 7.14?
2. **WHAT IF?** In the absence of O_2, as in question 1, what do you think would happen if you decreased the pH of the intermembrane space of the mitochondrion? Explain your answer.
3. **MAKE CONNECTIONS** Membranes must be fluid to function properly (as you learned in Concept 5.1). How does the operation of the electron transport chain support that assertion?

For suggested answers, see Appendix A.

CONCEPT 7.5

Fermentation and anaerobic respiration enable cells to produce ATP without the use of oxygen

Because most of the ATP generated by cellular respiration is due to the work of oxidative phosphorylation, our estimate of ATP yield from aerobic respiration is contingent on an adequate supply of oxygen to the cell. Without the electronegative oxygen to pull electrons down the transport chain, oxidative phosphorylation eventually ceases. However, there are two general mechanisms by which certain cells can oxidize organic fuel and generate ATP *without* the use of oxygen: anaerobic respiration and fermentation. The distinction between these two is that an electron transport chain is used in anaerobic respiration but not in fermentation. (The electron transport chain is also called the respiratory chain because of its role in both types of cellular respiration.)

We have already mentioned anaerobic respiration, which takes place in certain prokaryotic organisms that live in environments without oxygen. These organisms have an electron transport chain but do not use oxygen as a final electron acceptor at the end of the chain. Oxygen performs this function very well because it is extremely electronegative, but other, less electronegative substances can also serve as final electron acceptors. Some "sulfate-reducing" marine bacteria, for instance, use the sulfate ion (SO_4^{2-}) at the end of their respiratory chain. Operation of the chain builds up a proton-motive force used to produce ATP, but H_2S (hydrogen sulfide) is produced as a

Making a Bar Graph and Evaluating a Hypothesis

Does Thyroid Hormone Level Affect Oxygen Consumption in Cells? Some animals, such as mammals and birds, maintain a relatively constant body temperature, above that of their environment, using heat produced as a by-product of metabolism. When the core temperature of these animals drops below an internal set point, their cells are triggered to reduce the efficiency of ATP produced by the electron transport chains in mitochondria. At lower efficiency, extra fuel must be consumed to produce the same number of ATPs, generating additional heat. Because this response is moderated by the endocrine system, researchers hypothesized that thyroid hormone might trigger this cellular response. In this exercise, you will use a bar graph to visualize data from an experiment that compared the metabolic rate (by measuring oxygen consumption) in mitochondria of cells from animals with different levels of thyroid hormone.

How the Experiment Was Done Liver cells were isolated from sibling rats that had low, normal, or elevated thyroid hormone levels. The oxygen consumption rate due to activity of the mitochondrial electron transport chains of each type of cell was measured under controlled conditions.

Data from the Experiment

Thyroid Hormone Level	Oxygen Consumption Rate ($nmol\ O_2/min \cdot mg$ cells)
Low	4.3
Normal	4.8
Elevated	8.7

Interpret the Data

1. To visualize any differences in oxygen consumption between cell types, it will be useful to graph the data in a bar graph. First, you'll set up the axes. (a) What is the independent variable (intentionally varied by the researchers), which goes on the x-axis? List the categories along the x-axis; because they are discrete rather than continuous, you can list them in any order. (b) What is the dependent variable (measured by the researchers), which goes on the y-axis? (c) What units (abbreviated) should go on the y-axis? Label the y-axis, including the units specified in the data table.

Determine the range of values of the data that will need to go on the y-axis. What is the largest value? Draw evenly spaced tick marks and label them, starting with 0 at the bottom.

2. Graph the data for each sample. Match each x-value with its y-value and place a mark on the graph at that coordinate, then draw a bar from the x-axis up to the correct height for each sample. Why is a bar graph more appropriate than a scatter plot or line graph? (For additional information about graphs, see the Scientific Skills Review in Appendix F and in the Study Area in MasteringBiology.)

3. Examine your graph and look for a pattern in the data. (a) Which cell type had the highest rate of oxygen consumption, and which had the lowest? (b) Does this support the researchers' hypothesis? Explain. (c) Based on what you know about mitochondrial electron transport and heat production, predict which rats had the highest, and which had the lowest, body temperature.

Data from M. E. Harper and M. D. Brand, The quantitative contributions of mitochondrial proton leak and ATP turnover reactions to the changed respiration rates of hepatocytes from rats of different thyroid status, *Journal of Biological Chemistry* 268:14850–14860 (1993).

(MB) A version of this Scientific Skills Exercise can be assigned in MasteringBiology.

---

by-product rather than water. The rotten-egg odor you may have smelled while walking through a salt marsh or a mudflat signals the presence of sulfate-reducing bacteria.

Fermentation is a way of harvesting chemical energy without using either oxygen or any electron transport chain—in other words, without cellular respiration. How can food be oxidized without cellular respiration? Remember, oxidation simply refers to the loss of electrons to an electron acceptor, so it does not need to involve oxygen. Glycolysis oxidizes glucose to two molecules of pyruvate. The oxidizing agent of glycolysis is NAD^+, and neither oxygen nor any electron transfer chain is involved. Overall, glycolysis is exergonic, and some of the energy made available is used to produce 2 ATP (net) by substrate-level phosphorylation. If oxygen *is* present, then additional ATP is made by oxidative phosphorylation when NADH passes

electrons removed from glucose to the electron transport chain. But glycolysis generates 2 ATP whether oxygen is present or not—that is, whether conditions are aerobic or anaerobic.

As an alternative to respiratory oxidation of organic nutrients, fermentation is an extension of glycolysis that allows continuous generation of ATP by the substrate-level phosphorylation of glycolysis. For this to occur, there must be a sufficient supply of NAD^+ to accept electrons during the oxidation step of glycolysis. Without some mechanism to recycle NAD^+ from NADH, glycolysis would soon deplete the cell's pool of NAD^+ by reducing it all to NADH and would shut itself down for lack of an oxidizing agent. Under aerobic conditions, NAD^+ is recycled from NADH by the transfer of electrons to the electron transport chain. An anaerobic alternative is to transfer electrons from NADH to pyruvate, the end product of glycolysis.

Types of Fermentation

Fermentation consists of glycolysis plus reactions that regenerate NAD^+ by transferring electrons from NADH to pyruvate or derivatives of pyruvate. The NAD^+ can then be reused to oxidize sugar by glycolysis, which nets two molecules of ATP by substrate-level phosphorylation. There are many types of fermentation, differing in the end products formed from pyruvate. Two common types are alcohol fermentation and lactic acid fermentation.

In **alcohol fermentation (Figure 7.16a)**, pyruvate is converted to ethanol (ethyl alcohol) in two steps. The first step releases carbon dioxide from the pyruvate, which is converted to the two-carbon compound acetaldehyde. In the second step, acetaldehyde is reduced by NADH to ethanol. This regenerates the supply of NAD^+ needed for the continuation of glycolysis. Many bacteria carry out alcohol fermentation under anaerobic conditions. Yeast (a fungus) also carries out alcohol fermentation. For thousands of years, humans have used yeast in brewing, winemaking, and baking. The CO_2 bubbles generated by baker's yeast during alcohol fermentation allow bread to rise.

During **lactic acid fermentation (Figure 7.16b)**, pyruvate is reduced directly by NADH to form lactate as an end product, with no release of CO_2. (Lactate is the ionized form of lactic acid.) Lactic acid fermentation by certain fungi and bacteria is used in the dairy industry to make cheese and yogurt.

Human muscle cells make ATP by lactic acid fermentation when oxygen is scarce. This occurs during strenuous exercise, when sugar catabolism for ATP production outpaces the muscle's supply of oxygen from the blood. Under these conditions, the cells switch from aerobic respiration to fermentation. The lactate that accumulates was previously thought to cause muscle fatigue and pain, but recent research suggests instead that increased levels of potassium ions (K^+) may be to blame, while lactate appears to enhance muscle performance. In any case, the excess lactate is gradually carried away by the blood to the liver, where it is converted back to pyruvate by liver cells. Because oxygen is available, this pyruvate can then enter the mitochondria in liver cells and complete cellular respiration.

Comparing Fermentation with Anaerobic and Aerobic Respiration

Fermentation, anaerobic respiration, and aerobic respiration are three alternative cellular pathways for producing ATP by harvesting the chemical energy of food. All three use glycolysis to oxidize glucose and other organic fuels to pyruvate, with a net production of 2 ATP by substrate-level phosphorylation. And in all three pathways, NAD^+ is the oxidizing agent that accepts electrons from food during glycolysis.

A key difference is the contrasting mechanisms for oxidizing NADH back to NAD^+, which is required to sustain glycolysis. In fermentation, the final electron acceptor is an organic molecule such as pyruvate (lactic acid fermentation) or acetaldehyde (alcohol fermentation). In cellular respiration,

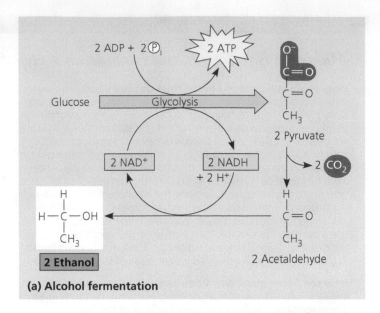

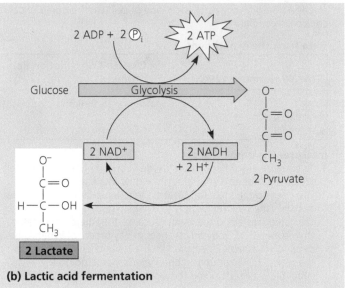

(a) Alcohol fermentation

(b) Lactic acid fermentation

▲ **Figure 7.16 Fermentation.** In the absence of oxygen, many cells use fermentation to produce ATP by substrate-level phosphorylation. Pyruvate, the end product of glycolysis, serves as an electron acceptor for oxidizing NADH back to NAD^+, which can then be reused in glycolysis. Two of the common end products formed from fermentation are **(a)** ethanol and **(b)** lactate, the ionized form of lactic acid.

by contrast, electrons carried by NADH are transferred to an electron transport chain, which generates the NAD^+ required for glycolysis.

Another major difference is the amount of ATP produced. Fermentation yields two molecules of ATP, produced by substrate-level phosphorylation. In the absence of an electron transport chain, the energy stored in pyruvate is unavailable. In cellular respiration, however, pyruvate is completely oxidized in the mitochondrion. Most of the chemical energy from this process is shuttled by NADH and $FADH_2$ in the form of the electrons to the electron transport chain. There, the electrons move stepwise down a series of redox reactions to a final electron acceptor. (In aerobic respiration, the final electron is oxygen;

in anaerobic respiration, the final acceptor is another molecule that is electronegative, although less so than oxygen.) Stepwise electron transport drives oxidative phosphorylation, yielding ATPs. Thus, cellular respiration harvests much more energy from each sugar molecule than fermentation can. In fact, aerobic respiration yields up to 32 molecules of ATP per glucose molecule—up to 16 times as much as does fermentation.

Some organisms, called **obligate anaerobes**, carry out only fermentation or anaerobic respiration. In fact, these organisms cannot survive in the presence of oxygen. A few cell types can carry out only aerobic oxidation of pyruvate, not fermentation. Other organisms, including yeasts and many bacteria, can make enough ATP to survive using either fermentation or respiration. Such species are called **facultative anaerobes**. On the cellular level, our muscle cells behave as facultative anaerobes. In such cells, pyruvate is a fork in the metabolic road that leads to two alternative catabolic routes **(Figure 7.17)**. Under aerobic conditions, pyruvate can be converted to acetyl CoA, which enters the citric acid cycle. Under anaerobic conditions, lactic acid fermentation occurs: Pyruvate is diverted from the citric acid cycle, serving instead as an electron acceptor to recycle NAD+. To make the same amount of ATP, a facultative anaerobe has to consume sugar at a much faster rate when fermenting than when respiring.

The Evolutionary Significance of Glycolysis

EVOLUTION The role of glycolysis in both fermentation and respiration has an evolutionary basis. Ancient prokaryotes are thought to have used glycolysis to make ATP long before oxygen was present in Earth's atmosphere. The oldest known fossils of bacteria date back 3.5 billion years, but appreciable quantities of oxygen probably did not begin to accumulate in the atmosphere until about 2.7 billion years ago, produced by photosynthesizing cyanobacteria. Therefore, early prokaryotes may have generated ATP exclusively from glycolysis. The fact that glycolysis is today the most widespread metabolic pathway among Earth's organisms suggests that it evolved very early in the history of life. The cytosolic location of glycolysis also implies great antiquity; the pathway does not require any of the membrane-enclosed organelles of the eukaryotic cell, which evolved approximately 1 billion years after the prokaryotic cell. Glycolysis is a metabolic heirloom from early cells that continues to function in fermentation and as the first stage in the breakdown of organic molecules by respiration.

CONCEPT CHECK 7.5

1. Consider the NADH formed during glycolysis. What is the final acceptor for its electrons during fermentation? What is the final acceptor for its electrons during aerobic respiration?
2. **WHAT IF?** A glucose-fed yeast cell is moved from an aerobic environment to an anaerobic one. How would its rate of glucose consumption change if ATP were to be generated at the same rate?

For suggested answers, see Appendix A.

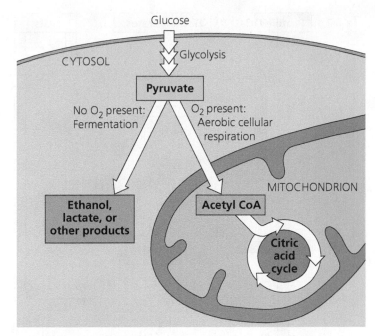

▲ **Figure 7.17 Pyruvate as a key juncture in catabolism.** Glycolysis is common to fermentation and cellular respiration. The end product of glycolysis, pyruvate, represents a fork in the catabolic pathways of glucose oxidation. In a facultative anaerobe or a muscle cell, which are capable of both aerobic cellular respiration and fermentation, pyruvate is committed to one of those two pathways, usually depending on whether or not oxygen is present.

CONCEPT 7.6

Glycolysis and the citric acid cycle connect to many other metabolic pathways

So far, we have treated the oxidative breakdown of glucose in isolation from the cell's overall metabolic economy. In this section, you will learn that glycolysis and the citric acid cycle are major intersections of the cell's catabolic and anabolic (biosynthetic) pathways.

The Versatility of Catabolism

Throughout this chapter, we have used glucose as an example of a fuel for cellular respiration. But free glucose molecules are not common in the diets of humans and other animals. We obtain most of our calories in the form of fats, proteins, sucrose and other disaccharides, and starch, a polysaccharide. All these organic molecules in food can be used by cellular respiration to make ATP **(Figure 7.18)**.

Glycolysis can accept a wide range of carbohydrates for catabolism. In the digestive tract, starch is hydrolyzed to glucose, which can then be broken down in the cells by glycolysis and the citric acid cycle. Similarly, glycogen, the polysaccharide that humans and many other animals store in their liver and muscle cells, can be hydrolyzed to glucose between meals as fuel for respiration. The digestion of

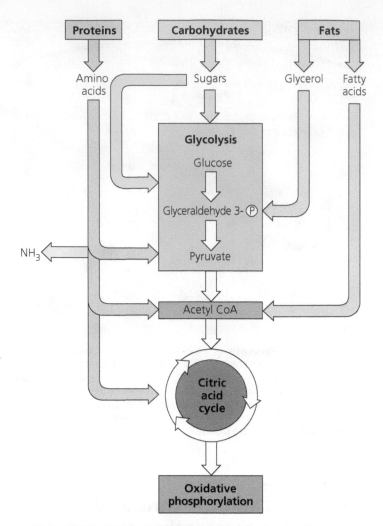

▲ Figure 7.18 The catabolism of various molecules from food. Carbohydrates, fats, and proteins can all be used as fuel for cellular respiration. Monomers of these molecules enter glycolysis or the citric acid cycle at various points. Glycolysis and the citric acid cycle are catabolic funnels through which electrons from all kinds of organic molecules flow on their exergonic fall to oxygen.

disaccharides, including sucrose, provides glucose and other monosaccharides as fuel for respiration.

Proteins can also be used for fuel, but first they must be digested to their constituent amino acids. Many of the amino acids are used by the organism to build new proteins. Amino acids present in excess are converted by enzymes to intermediates of glycolysis and the citric acid cycle. Before amino acids can feed into glycolysis or the citric acid cycle, their amino groups must be removed, a process called *deamination*. The nitrogenous refuse is excreted from the animal in the form of ammonia (NH_3), urea, or other waste products.

Catabolism can also harvest energy stored in fats obtained either from food or from storage cells in the body. After fats are digested to glycerol and fatty acids, the glycerol is converted to glyceraldehyde 3-phosphate, an intermediate of glycolysis. Most of the energy of a fat is stored in the fatty acids. A metabolic sequence called **beta oxidation** breaks the fatty acids down to two-carbon fragments, which enter the citric acid cycle as

acetyl CoA. NADH and $FADH_2$ are also generated during beta oxidation, resulting in further ATP production. Fats make excellent fuel, in large part due to their chemical structure and the high energy level of their electrons compared to those of carbohydrates. A gram of fat oxidized by respiration produces more than twice as much ATP as a gram of carbohydrate.

Biosynthesis (Anabolic Pathways)

Cells need substance as well as energy. Not all the organic molecules of food are destined to be oxidized as fuel to make ATP. In addition to calories, food must also provide the carbon skeletons that cells require to make their own molecules. Some organic monomers obtained from digestion can be used directly. For example, as previously mentioned, amino acids from the hydrolysis of proteins in food can be incorporated into the organism's own proteins. Often, however, the body needs specific molecules that are not present as such in food. Compounds formed as intermediates of glycolysis and the citric acid cycle can be diverted into anabolic pathways as precursors from which the cell can synthesize the molecules it requires. For example, humans can make about half of the 20 amino acids in proteins by modifying compounds siphoned away from the citric acid cycle; the rest are "essential amino acids" that must be obtained in the diet. Also, glucose can be made from pyruvate, and fatty acids can be synthesized from acetyl CoA. Of course, these anabolic, or biosynthetic, pathways do not generate ATP, but instead consume it.

In addition, glycolysis and the citric acid cycle function as metabolic interchanges that enable our cells to convert some kinds of molecules to others as we need them. For example, an intermediate compound generated during glycolysis, dihydroxyacetone phosphate (see Figure 7.9, step 5), can be converted to one of the major precursors of fats. If we eat more food than we need, we store fat even if our diet is fat-free. Metabolism is remarkably versatile and adaptable.

Cellular respiration and metabolic pathways play a role of central importance in organisms. Examine Figure 7.2 again to put cellular respiration into the broader context of energy flow and chemical cycling in ecosystems. The energy that keeps us alive is *released*, not *produced*, by cellular respiration. We are tapping energy that was stored in food by photosynthesis, which captures light and converts it to chemical energy, a process you will learn about in Chapter 8.

CONCEPT CHECK 7.6

1. **MAKE CONNECTIONS** Compare the structure of a fat (see Figure 3.12) with that of a carbohydrate (see Figure 3.7). What features of their structures make fat a much better fuel?

2. Under what circumstances might your body synthesize fat molecules?

3. **WHAT IF?** During intense exercise, can a muscle cell use fat as a concentrated source of chemical energy? Explain. (Review Figures 7.17 and 7.18.)

For suggested answers, see Appendix A.

SUMMARY OF KEY CONCEPTS

CONCEPT 7.1

Catabolic pathways yield energy by oxidizing organic fuels (pp. 136–140)

- Cells break down glucose and other organic fuels to yield chemical energy in the form of ATP. **Fermentation** is a partial degradation of glucose without the use of oxygen. **Cellular respiration** is a more complete breakdown of glucose; in **aerobic respiration**, oxygen is used as a reactant. The cell taps the energy stored in food molecules through **redox reactions**, in which one substance partially or totally shifts electrons to another. **Oxidation** is the loss of electrons from one substance, while **reduction** is the addition of electrons to the other.
- During aerobic respiration, glucose ($C_6H_{12}O_6$) is oxidized to CO_2, and O_2 is reduced to H_2O. Electrons lose potential energy during their transfer from glucose or other organic compounds to oxygen. Electrons are usually passed first to **NAD⁺**, reducing it to NADH, and then from NADH to an **electron transport chain**, which conducts them to O_2 in energy-releasing steps. The energy is used to make ATP.
- Aerobic respiration occurs in three stages: (1) **glycolysis**, (2) pyruvate oxidation and the **citric acid cycle**, and (3) **oxidative phosphorylation** (electron transport and chemiosmosis).

> **?** *Describe the difference between the two processes in cellular respiration that produce ATP: oxidative phosphorylation and substrate-level phosphorylation.*

CONCEPT 7.2

Glycolysis harvests chemical energy by oxidizing glucose to pyruvate (pp. 140–141)

Inputs	Outputs
Glucose →(Glycolysis)→	2 **Pyruvate** + 2 **ATP** + 2 **NADH**

> **?** *What is the source of energy for the formation of ATP and NADH in glycolysis?*

CONCEPT 7.3

After pyruvate is oxidized, the citric acid cycle completes the energy-yielding oxidation of organic molecules (pp. 142–143)

- In eukaryotic cells, pyruvate enters the mitochondrion and is oxidized to **acetyl CoA**, which is further oxidized in the citric acid cycle.

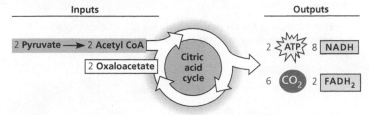

> **?** *What molecular products indicate the complete oxidation of glucose during cellular respiration?*

CONCEPT 7.4

During oxidative phosphorylation, chemiosmosis couples electron transport to ATP synthesis (pp. 143–148)

- NADH and $FADH_2$ transfer electrons to the electron transport chain. Electrons move down the chain, losing energy in several energy-releasing steps. Finally, electrons are passed to O_2, reducing it to H_2O.

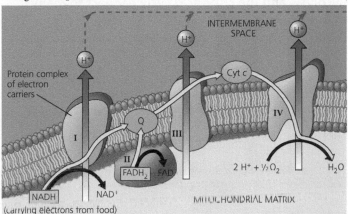

- At certain steps along the electron transport chain, electron transfer causes protein complexes to move H⁺ from the mitochondrial matrix (in eukaryotes) to the intermembrane space, storing energy as a **proton-motive force** (H⁺ gradient). As H⁺ diffuses back into the matrix through **ATP synthase**, its passage drives the phosphorylation of ADP, a process called **chemiosmosis**.
- About 34% of the energy stored in a glucose molecule is transferred to ATP during cellular respiration, producing a maximum of about 32 ATP.

> **?** *Briefly explain the mechanism by which ATP synthase produces ATP. List three locations in which ATP synthases are found.*

CONCEPT 7.5

Fermentation and anaerobic respiration enable cells to produce ATP without the use of oxygen (pp. 148–151)

- Glycolysis nets 2 ATP by substrate-level phosphorylation, whether oxygen is present or not. Under anaerobic conditions, either anaerobic respiration or fermentation can take place. In anaerobic respiration, an electron transport chain is present with a final electron acceptor other than oxygen. In fermentation, the electrons from NADH are passed to pyruvate or a derivative of pyruvate, regenerating the NAD⁺ required to oxidize more glucose. Two common types of fermentation are **alcohol fermentation** and **lactic acid fermentation**.
- Fermentation, anaerobic respiration, and aerobic respiration all use glycolysis to oxidize glucose, but they differ in their final

electron acceptor and whether an electron transport chain is used (respiration) or not (fermentation). Respiration yields more ATP; aerobic respiration, with O_2 as the final electron acceptor, yields about 16 times as much ATP as does fermentation.

- Glycolysis occurs in nearly all organisms and is thought to have evolved in ancient prokaryotes before there was O_2 in the atmosphere.

? *Which process yields more ATP, fermentation or anaerobic respiration? Explain.*

CONCEPT 7.6

Glycolysis and the citric acid cycle connect to many other metabolic pathways (pp. 151–152)

- Catabolic pathways funnel electrons from many kinds of organic molecules into cellular respiration. Many carbohydrates can enter glycolysis, most often after conversion to glucose. Amino acids of proteins must be deaminated before being oxidized. The fatty acids of fats undergo **beta oxidation** to two-carbon fragments and then enter the citric acid cycle as acetyl CoA. Anabolic pathways can use small molecules from food directly or build other substances using intermediates of glycolysis or the citric acid cycle.

? *Describe how the catabolic pathways of glycolysis and the citric acid cycle intersect with anabolic pathways in the metabolism of a cell.*

TEST YOUR UNDERSTANDING

Level 1: Knowledge/Comprehension

1. The *immediate* energy source that drives ATP synthesis by ATP synthase during oxidative phosphorylation is the
   a. oxidation of glucose and other organic compounds.
   b. flow of electrons down the electron transport chain.
   c. affinity of oxygen for electrons.
   d. H^+ movement down its concentration gradient.
   e. transfer of phosphate to ADP.

2. Which metabolic pathway is common to both fermentation and cellular respiration of a glucose molecule?
   a. the citric acid cycle
   b. the electron transport chain
   c. glycolysis
   d. synthesis of acetyl CoA from pyruvate
   e. reduction of pyruvate to lactate

3. In mitochondria, exergonic redox reactions
   a. are the source of energy driving prokaryotic ATP synthesis.
   b. are directly coupled to substrate-level phosphorylation.
   c. provide the energy that establishes the proton gradient.
   d. reduce carbon atoms to carbon dioxide.
   e. use ATP to pump H^+ out of the mitochondrion.

4. The final electron acceptor of the electron transport chain that functions in aerobic oxidative phosphorylation is
   a. oxygen.   d. pyruvate.
   b. water.   e. ADP.
   c. NAD^+.

Level 2: Application/Analysis

5. What is the oxidizing agent in the following reaction?

   Pyruvate + NADH + H^+ → Lactate + NAD^+

   a. oxygen   d. lactate
   b. NADH   e. pyruvate
   c. NAD^+

6. When electrons flow along the electron transport chains of mitochondria, which of the following changes occurs?
   a. The pH of the matrix increases.
   b. ATP synthase pumps protons by active transport.
   c. The electrons gain free energy.
   d. The cytochromes phosphorylate ADP to form ATP.
   e. NAD^+ is oxidized.

7. Most CO_2 from catabolism is released during
   a. glycolysis.
   b. the citric acid cycle.
   c. lactate fermentation.
   d. electron transport.
   e. oxidative phosphorylation.

Level 3: Synthesis/Evaluation

8. **DRAW IT** The graph here shows the pH difference across the inner mitochondrial membrane over time in an actively respiring cell. At the time indicated by the vertical arrow, a metabolic poison is added that specifically and completely inhibits all function of mitochondrial ATP synthase. Draw what you would expect to see for the rest of the graphed line.

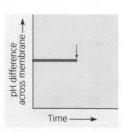

9. **SCIENTIFIC INQUIRY**
   In the 1930s, some physicians prescribed low doses of a compound called dinitrophenol (DNP) to help patients lose weight. This unsafe method was abandoned after some patients died. DNP uncouples the chemiosmotic machinery by making the lipid bilayer of the inner mitochondrial membrane leaky to H^+. Explain how this could cause weight loss and death.

10. **FOCUS ON EVOLUTION**
    ATP synthases are found in the prokaryotic plasma membrane and in mitochondria and chloroplasts. What does this suggest about the evolutionary relationship of these eukaryotic organelles to prokaryotes? How might the amino acid sequences of the ATP synthases from the different sources support or refute your hypothesis?

11. **FOCUS ON ENERGY AND MATTER**
    In a short essay (100–150 words), explain how oxidative phosphorylation—the production of ATP using energy derived from the redox reactions of a spatially organized electron transport chain followed by chemiosmosis—is an example of how new properties emerge at each level of the biological hierarchy.

For selected answers, see Appendix A.

MasteringBiology®

Students Go to **MasteringBiology** for assignments, the eText, and the Study Area with practice tests, animations, and activities.

Instructors Go to **MasteringBiology** for automatically graded tutorials and questions that you can assign to your students, plus Instructor Resources.

Photosynthesis

8

KEY CONCEPTS

8.1 Photosynthesis converts light energy to the chemical energy of food

8.2 The light reactions convert solar energy to the chemical energy of ATP and NADPH

8.3 The Calvin cycle uses the chemical energy of ATP and NADPH to reduce CO_2 to sugar

OVERVIEW

The Process That Feeds the Biosphere

Life on Earth is solar powered. The chloroplasts of plants capture light energy that has traveled 150 million kilometers from the sun and convert it to chemical energy that is stored in sugar and other organic molecules. This conversion process is called **photosynthesis**. Let's begin by placing photosynthesis in its ecological context.

Photosynthesis nourishes almost the entire living world directly or indirectly. An organism acquires the organic compounds it uses for energy and carbon skeletons by one of two major modes: autotrophic nutrition or heterotrophic nutrition. **Autotrophs** are "self-feeders" (*auto* means "self," and *trophos* means "feeder"); they sustain themselves without eating anything derived from other living beings. Autotrophs produce their organic molecules from CO_2 and other inorganic raw materials obtained from the environment. They are the ultimate sources of organic compounds for all nonautotrophic organisms, and for this reason, biologists refer to autotrophs as the *producers* of the biosphere.

Almost all plants are autotrophs; the only nutrients they require are water and minerals from the soil and carbon dioxide from the air. Specifically, plants are *photo*autotrophs, organisms that use light as a source of energy to synthesize organic substances **(Figure 8.1)**. Photosynthesis also occurs in algae, certain other unicellular eukaryotes, and some prokaryotes.

Heterotrophs are unable to make their own food; they live on compounds produced by other organisms (*hetero-* means "other"). Heterotrophs are the biosphere's *consumers*. This "other-feeding" is most obvious when an animal eats plants or other animals, but heterotrophic nutrition may be more subtle. Some heterotrophs decompose and feed on the remains of dead organisms and organic litter such as feces and fallen leaves; these types of heterotrophs are known as decomposers. Most fungi and many types of prokaryotes get their nourishment this way. Almost all heterotrophs, including humans, are completely dependent, either directly or indirectly, on photoautotrophs for food—and also for oxygen, a by-product of photosynthesis.

▼ **Figure 8.1** How can sunlight, seen here as a spectrum of colors in a rainbow, power the synthesis of organic substances?

In this chapter, you'll learn how photosynthesis works. A variety of photosynthetic organisms are shown in **Figure 8.2**, including both eukaryotes and prokaryotes. Our discussion here will focus mainly on plants. (Variations in autotrophic nutrition that occur in prokaryotes and algae will be described in Chapters 24 and 25.) After discussing the general principles of photosynthesis, we'll consider the two stages of photosynthesis: the light reactions, which capture solar energy and transform it into chemical energy; and the Calvin cycle, which uses the chemical energy to make organic molecules of food. Finally, we'll consider a few aspects of photosynthesis from an evolutionary perspective.

Photosynthesis converts light energy to the chemical energy of food

The remarkable ability of an organism to harness light energy and use it to drive the synthesis of organic compounds emerges from structural organization in the cell: Photosynthetic enzymes and other molecules are grouped together in a biological membrane, enabling the necessary series of chemical reactions to be carried out efficiently. The process of photosynthesis most likely originated in a group of bacteria that had infolded regions of the plasma membrane containing clusters of such molecules. In photosynthetic bacteria that exist today, infolded photosynthetic membranes function similarly to the internal membranes of the chloroplast, a eukaryotic organelle. According to the endosymbiont theory, the original chloroplast was a photosynthetic prokaryote that lived inside an ancestor of eukaryotic cells. (You learned about this theory in Chapter 4, and it will be described more fully in Chapter 25.) Chloroplasts are present in a variety of photosynthesizing organisms, but here we focus on chloroplasts in plants.

Chloroplasts: The Sites of Photosynthesis in Plants

All green parts of a plant, including green stems and unripened fruit, have chloroplasts, but the leaves are the major sites of photosynthesis in most plants **(Figure 8.3)**. There are about half a million chloroplasts in a chunk of leaf with a top surface area of 1 mm^2. Chloroplasts are found mainly in the cells of the **mesophyll**, the tissue in the interior of the leaf. Carbon dioxide enters the leaf, and oxygen exits, by way of microscopic pores called **stomata** (singular, *stoma*; from the Greek, meaning "mouth"). Water absorbed by the roots is delivered to the leaves in veins. Leaves also use veins to export sugar to roots and other nonphotosynthetic parts of the plant.

A typical mesophyll cell has about 30–40 chloroplasts, each organelle measuring about 2–4 μm by 4–7 μm. A chloroplast has an envelope of two membranes surrounding a dense fluid called the **stroma**. Suspended within the stroma is a third membrane system, made up of sacs called **thylakoids**, which segregates the stroma from the *thylakoid space* inside

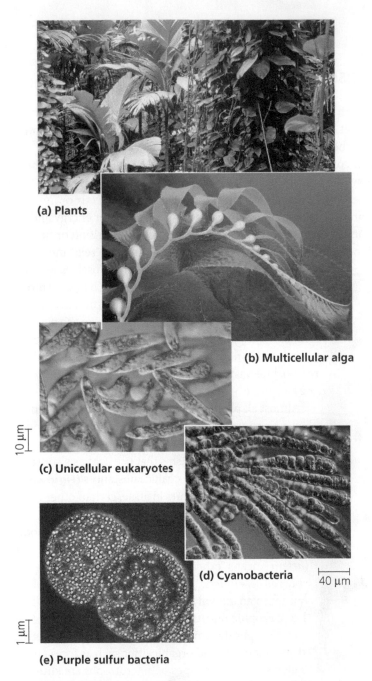

(a) Plants

(b) Multicellular alga

10 μm

(c) Unicellular eukaryotes

(d) Cyanobacteria   40 μm

1 μm

(e) Purple sulfur bacteria

▲ **Figure 8.2 Photoautotrophs.** These organisms use light energy to drive the synthesis of organic molecules from carbon dioxide and (in most cases) water. They feed themselves and the entire living world. **(a)** On land, plants are the predominant producers of food. In aquatic environments, photoautotrophs include unicellular and **(b)** multicellular algae, such as this kelp; **(c)** some non-algal unicellular eukaryotes, such as *Euglena*; **(d)** the prokaryotes called cyanobacteria; and **(e)** other photosynthetic prokaryotes, such as these purple sulfur bacteria, which produce sulfur (the yellow globules within the cells) (c–e, LMs).

these sacs. In some places, thylakoid sacs are stacked in columns called *grana* (singular, *granum*). **Chlorophyll**, the green pigment that gives leaves their color, resides in the thylakoid membranes of the chloroplast. (The internal photosynthetic membranes of some prokaryotes are also called thylakoid membranes; see Figure 24.11b.) It is the light energy absorbed by chlorophyll that drives the synthesis of organic molecules in the chloroplast. Now that we have looked at the sites of photosynthesis in plants, we are ready to look more closely at the process of photosynthesis.

Tracking Atoms Through Photosynthesis: *Scientific Inquiry*

Scientists have tried for centuries to piece together the process by which plants make food. Although some of the steps are still not completely understood, the overall photosynthetic equation has been known since the 1800s: In the presence of light, the green parts of plants produce organic compounds and oxygen from carbon dioxide and water. Using molecular formulas, we can summarize the complex series of chemical reactions in photosynthesis with this chemical equation:

$$6\ CO_2 + 12\ H_2O + \text{Light energy} \rightarrow C_6H_{12}O_6 + 6\ O_2 + 6\ H_2O$$

We use glucose ($C_6H_{12}O_6$) here to simplify the relationship between photosynthesis and respiration, but the direct product of photosynthesis is actually a three-carbon sugar that can be used to make glucose. Water appears on both sides of the equation because 12 molecules are consumed and 6 molecules are newly formed during photosynthesis. We can simplify the equation by indicating only the net consumption of water:

$$6\ CO_2 + 6\ H_2O + \text{Light energy} \rightarrow C_6H_{12}O_6 + 6\ O_2$$

Writing the equation in this form, we can see that the overall chemical change during photosynthesis is the reverse of the one that occurs during cellular respiration. Both of these metabolic processes occur in plant cells. However, as you will soon learn, chloroplasts do not synthesize sugars by simply reversing the steps of respiration.

Now let's divide the photosynthetic equation by 6 to put it in its simplest possible form:

$$CO_2 + H_2O \rightarrow [CH_2O] + O_2$$

Here, the brackets indicate that CH_2O is not an actual sugar but represents the general formula for a carbohydrate. In other words, we are imagining the synthesis of a sugar molecule one carbon at a time. Six repetitions would theoretically produce a glucose molecule. Let's now use this simplified formula to see how researchers tracked the elements C, H, and O from the reactants of photosynthesis to the products.

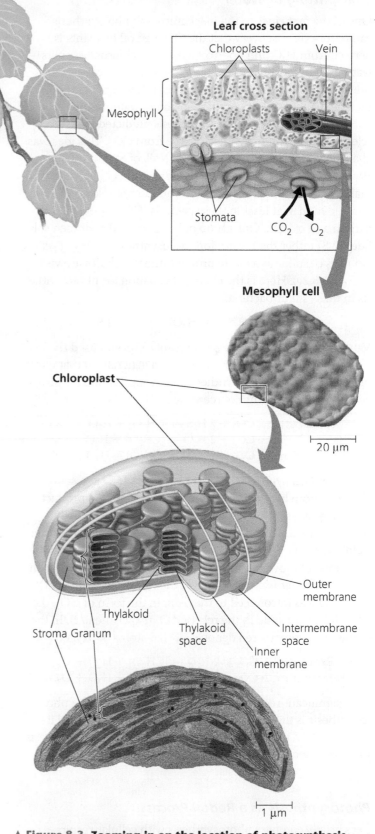

▲ **Figure 8.3 Zooming in on the location of photosynthesis in a plant.** Leaves are the major organs of photosynthesis in plants. These pictures take you into a leaf, then into a cell, and finally into a chloroplast, the organelle where photosynthesis occurs (middle, LM; bottom, TEM).

The Splitting of Water

One of the first clues to the mechanism of photosynthesis came from the discovery that the O_2 given off by plants is derived from H_2O and not from CO_2. The chloroplast splits water into hydrogen and oxygen. Before this discovery, the prevailing hypothesis was that photosynthesis split carbon dioxide ($CO_2 \rightarrow C + O_2$) and then added water to the carbon ($C + H_2O \rightarrow [CH_2O]$). This hypothesis predicted that the O_2 released during photosynthesis came from CO_2. This idea was challenged in the 1930s by C. B. van Niel, of Stanford University. Van Niel was investigating photosynthesis in bacteria that make their carbohydrate from CO_2 but do not release O_2. He concluded that, at least in these bacteria, CO_2 is not split into carbon and oxygen. One group of bacteria used hydrogen sulfide (H_2S) rather than water for photosynthesis, forming yellow globules of sulfur as a waste product (these globules are visible in Figure 8.2e). Here is the chemical equation for photosynthesis in these sulfur bacteria:

$$CO_2 + 2 H_2S \rightarrow [CH_2O] + H_2O + 2 S$$

Van Niel reasoned that the bacteria split H_2S and used the hydrogen atoms to make sugar. He then generalized that idea, proposing that all photosynthetic organisms require a hydrogen source but that the source varies:

Sulfur bacteria: $CO_2 + 2 H_2S \rightarrow [CH_2O] + H_2O + 2 S$
Plants: $CO_2 + 2 H_2O \rightarrow [CH_2O] + H_2O + O_2$
General: $CO_2 + 2 H_2X \rightarrow [CH_2O] + H_2O + 2 X$

Thus, van Niel hypothesized that plants split H_2O as a source of electrons from hydrogen atoms, releasing O_2 as a by-product.

Nearly 20 years later, scientists confirmed van Niel's hypothesis by using oxygen-18 (^{18}O), a heavy isotope, as a tracer to follow the fate of oxygen atoms during photosynthesis. The experiments showed that the O_2 from plants was labeled with ^{18}O *only* if water was the source of the tracer (experiment 1). If the ^{18}O was introduced to the plant in the form of CO_2, the label did not turn up in the released O_2 (experiment 2). In the following summary, red denotes labeled atoms of oxygen (^{18}O):

Experiment 1: $CO_2 + 2 H_2O \rightarrow [CH_2O] + H_2O + O_2$
Experiment 2: $CO_2 + 2 H_2O \rightarrow [CH_2O] + H_2O + O_2$

A significant result of the shuffling of atoms during photosynthesis is the extraction of hydrogen from water and its incorporation into sugar. The waste product of photosynthesis, O_2, is released to the atmosphere. **Figure 8.4** shows the fates of all atoms in photosynthesis.

Photosynthesis as a Redox Process

Let's briefly compare photosynthesis with cellular respiration. Both processes involve redox reactions. During cellular respiration, energy is released from sugar when electrons associated with hydrogen are transported by carriers to oxygen, forming water as a by-product (see Figure 7.3). The electrons

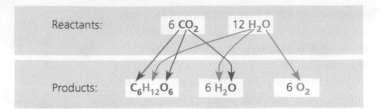

▲ **Figure 8.4 Tracking atoms through photosynthesis.** The atoms from CO_2 are shown in magenta, and the atoms from H_2O are shown in blue.

lose potential energy as they "fall" down the electron transport chain toward electronegative oxygen, and the mitochondrion harnesses that energy to synthesize ATP (see Figure 7.14). Photosynthesis reverses the direction of electron flow. Water is split, and electrons are transferred along with hydrogen ions from the water to carbon dioxide, reducing it to sugar.

$$\text{Energy} + 6 CO_2 + 6 H_2O \longrightarrow C_6H_{12}O_6 + 6 O_2$$

becomes reduced
becomes oxidized

Because the electrons increase in potential energy as they move from water to sugar, this process requires energy—in other words, is endergonic. This energy boost is provided by light.

The Two Stages of Photosynthesis: *A Preview*

The equation for photosynthesis is a deceptively simple summary of a very complex process. Actually, photosynthesis is not a single process, but two processes, each with multiple steps. These two stages of photosynthesis are known as the **light reactions** (the *photo* part of photosynthesis) and the **Calvin cycle** (the *synthesis* part) **(Figure 8.5)**.

The light reactions are the steps of photosynthesis that convert solar energy to chemical energy. Water is split, providing a source of electrons and protons (hydrogen ions, H^+) and giving off O_2 as a by-product. Light absorbed by chlorophyll drives a transfer of the electrons and hydrogen ions from water to an acceptor called **$NADP^+$** (nicotinamide adenine dinucleotide phosphate), where they are temporarily stored. The electron acceptor $NADP^+$ is first cousin to NAD^+, which functions as an electron carrier in cellular respiration; the two molecules differ only by the presence of an extra phosphate group in the $NADP^+$ molecule. The light reactions use solar power to reduce $NADP^+$ to NADPH by adding a pair of electrons along with an H^+. The light reactions also generate ATP, using chemiosmosis to power the addition of a phosphate group to ADP, a process called **photophosphorylation**. Thus, light energy is initially converted to chemical energy in the form of two compounds: NADPH and ATP. NADPH, a source of electrons, acts as "reducing power" that can be passed along to an electron acceptor, reducing it; ATP is the versatile energy currency of cells. Notice that the light reactions produce no sugar; that happens in the second stage of photosynthesis, the Calvin cycle.

▶ **Figure 8.5 An overview of photosynthesis: cooperation of the light reactions and the Calvin cycle.** In the chloroplast, the thylakoid membranes (green) are the sites of the light reactions, whereas the Calvin cycle occurs in the stroma (gray). The light reactions use solar energy to make ATP and NADPH, which supply chemical energy and reducing power, respectively, to the Calvin cycle. The Calvin cycle incorporates CO_2 into organic molecules, which are converted to sugar. (Recall that most simple sugars have formulas that are some multiple of CH_2O.)

 ANIMATION **BioFlix** Visit the Study Area in **MasteringBiology** for the BioFlix® 3-D Animation on Photosynthesis.

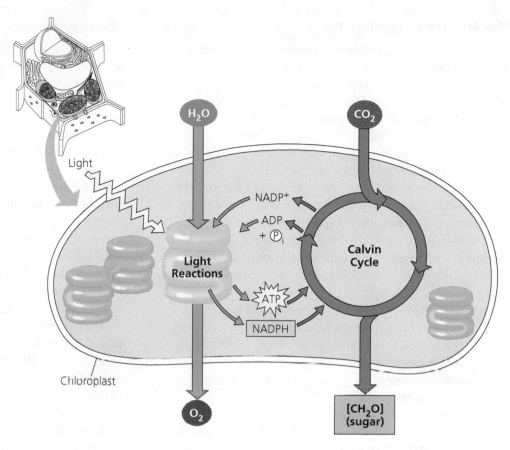

The Calvin cycle is named for Melvin Calvin, who, along with his colleagues, began to elucidate its steps in the late 1940s. The cycle begins by incorporating CO_2 from the air into organic molecules already present in the chloroplast. This initial incorporation of carbon into organic compounds is known as **carbon fixation**. The Calvin cycle then reduces the fixed carbon to carbohydrate by the addition of electrons. The reducing power is provided by NADPH, which acquired its cargo of electrons in the light reactions. To convert CO_2 to carbohydrate, the Calvin cycle also requires chemical energy in the form of ATP, which is also generated by the light reactions. Thus, it is the Calvin cycle that makes sugar, but it can do so only with the help of the NADPH and ATP produced by the light reactions. The metabolic steps of the Calvin cycle are sometimes referred to as the dark reactions, or light-independent reactions, because none of the steps requires light *directly*. Nevertheless, the Calvin cycle in most plants occurs during daylight, for only then can the light reactions provide the NADPH and ATP that the Calvin cycle requires. In essence, the chloroplast uses light energy to make sugar by coordinating the two stages of photosynthesis.

As Figure 8.5 indicates, the thylakoids of the chloroplast are the sites of the light reactions, while the Calvin cycle occurs in the stroma. On the outside of the thylakoids, molecules of $NADP^+$ and ADP pick up electrons and phosphate, respectively, and NADPH and ATP are then released to the stroma, where they play crucial roles in the Calvin cycle. The two

stages of photosynthesis are treated in this figure as metabolic modules that take in ingredients and crank out products. In the next two sections, we'll look more closely at how the two stages work, beginning with the light reactions.

CONCEPT CHECK 8.1

1. How do the reactant molecules of photosynthesis reach the chloroplasts in leaves?
2. How did the use of an oxygen isotope help elucidate the chemistry of photosynthesis?
3. **WHAT IF?** The Calvin cycle requires ATP and NADPH, products of the light reactions. If a classmate asserted that the light reactions don't depend on the Calvin cycle and, with continual light, could just keep on producing ATP and NADPH, how would you respond?

For suggested answers, see Appendix A.

CONCEPT 8.2

The light reactions convert solar energy to the chemical energy of ATP and NADPH

Chloroplasts are chemical factories powered by the sun. Their thylakoids transform light energy into the chemical energy of ATP and NADPH. To understand this conversion better, we need to know about some important properties of light.

The Nature of Sunlight

Light is a form of energy known as electromagnetic energy, also called electromagnetic radiation. Electromagnetic energy travels in rhythmic waves analogous to those created by dropping a pebble into a pond. Electromagnetic waves, however, are disturbances of electric and magnetic fields rather than disturbances of a material medium such as water.

The distance between the crests of electromagnetic waves is called the **wavelength**. Wavelengths range from less than a nanometer (for gamma rays) to more than a kilometer (for radio waves). This entire range of radiation is known as the **electromagnetic spectrum (Figure 8.6)**. The segment most important to life is the narrow band from about 380 nm to 750 nm in wavelength. This radiation is known as **visible light** because it can be detected as various colors by the human eye.

The model of light as waves explains many of light's properties, but in certain respects light behaves as though it consists of discrete particles, called **photons**. Photons are not tangible objects, but they act like objects in that each of them has a fixed quantity of energy. The amount of energy is inversely related to the wavelength of the light: The shorter the wavelength, the greater the energy of each photon of that light. Thus, a photon of violet light packs nearly twice as much energy as a photon of red light.

Although the sun radiates the full spectrum of electromagnetic energy, the atmosphere acts like a selective window, allowing visible light to pass through while screening out a substantial fraction of other radiation. The part of the spectrum we can see—visible light—is also the radiation that drives photosynthesis.

Photosynthetic Pigments: The Light Receptors

When light meets matter, it may be reflected, transmitted, or absorbed. Substances that absorb visible light are known as *pigments*. Different pigments absorb light of different wavelengths, and the wavelengths that are absorbed disappear. If a pigment is illuminated with white light, the color we see is the color most reflected or transmitted by the pigment. (If a pigment absorbs all wavelengths, it appears black.) We see green when we look at a leaf because chlorophyll absorbs violet-blue and red light while transmitting and reflecting green light **(Figure 8.7)**. The ability of a pigment to absorb various wavelengths of light can be measured with an instrument called a **spectrophotometer**. This machine directs beams of light of different wavelengths through a solution of the pigment and measures the fraction of the light transmitted at each wavelength. A graph plotting a pigment's light absorption versus wavelength is called an **absorption spectrum (Figure 8.8)**.

The absorption spectra of chloroplast pigments provide clues to the relative effectiveness of different wavelengths for driving photosynthesis, since light can perform work in chloroplasts only if it is absorbed. **Figure 8.9a** shows the absorption spectra of three types of pigments in chloroplasts: **chlorophyll a**, which participates directly in the light reactions; the accessory pigment **chlorophyll b**; and a group of accessory pigments called carotenoids. The spectrum of chlorophyll *a* suggests that violet-blue and red light work best for photosynthesis, since they are absorbed, while green is the least effective color. This is confirmed by an **action spectrum** for photosynthesis **(Figure 8.9b)**, which profiles the relative effectiveness of different wavelengths of

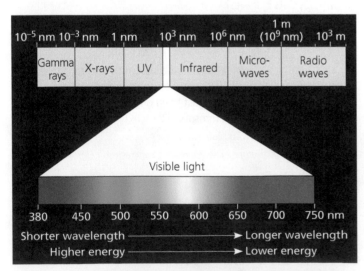

▲ **Figure 8.6 The electromagnetic spectrum.** White light is a mixture of all wavelengths of visible light. A prism can sort white light into its component colors by bending light of different wavelengths at different angles. (Droplets of water in the atmosphere can act as prisms, forming a rainbow; see Figure 8.1.) Visible light drives photosynthesis.

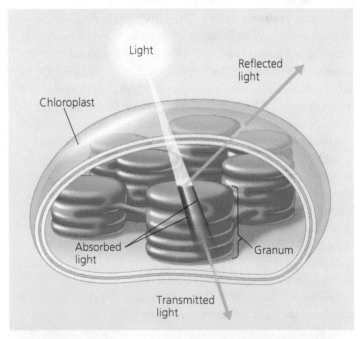

▲ **Figure 8.7 Why leaves are green: interaction of light with chloroplasts.** The chlorophyll molecules of chloroplasts absorb violet-blue and red light (the colors most effective in driving photosynthesis) and reflect or transmit green light. This is why leaves appear green.

Determining an Absorption Spectrum

Application An absorption spectrum is a visual representation of how well a particular pigment absorbs different wavelengths of visible light. Absorption spectra of various chloroplast pigments help scientists decipher each pigment's role in a plant.

Technique A spectrophotometer measures the relative amounts of light of different wavelengths absorbed and transmitted by a pigment solution.

1 White light is separated into colors (wavelengths) by a prism.

2 One by one, the different colors of light are passed through the sample (chlorophyll in this example). Green light and blue light are shown here.

3 The transmitted light strikes a photoelectric tube, which converts the light energy to electricity.

4 The electric current is measured by a galvanometer. The meter indicates the fraction of light transmitted through the sample, from which we can determine the amount of light absorbed.

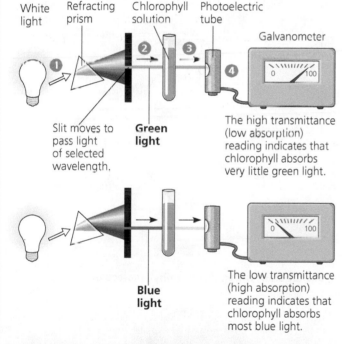

White light · Refracting prism · Chlorophyll solution · Photoelectric tube · Galvanometer

Slit moves to pass light of selected wavelength.

Green light

The high transmittance (low absorption) reading indicates that chlorophyll absorbs very little green light.

Blue light

The low transmittance (high absorption) reading indicates that chlorophyll absorbs most blue light.

Results See Figure 8.9a for absorption spectra of three types of chloroplast pigments.

radiation in driving the process. An action spectrum is prepared by illuminating chloroplasts with light of different colors and then plotting wavelength against some measure of photosynthetic rate, such as CO_2 consumption or O_2 release. The action spectrum for photosynthesis was first demonstrated by Theodor W. Engelmann, a German botanist, in 1883. Before equipment for measuring O_2 levels had even been invented, Engelmann performed a clever experiment in which he used bacteria to measure rates of photosynthesis in filamentous algae **(Figure 8.9c)**. His results are a striking match to the modern action spectrum shown in Figure 8.9b.

Which wavelengths of light are most effective in driving photosynthesis?

Experiment Absorption and action spectra, along with a classic experiment by Theodor W. Engelmann, reveal which wavelengths of light are photosynthetically important.

Results

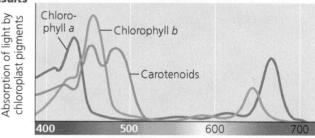

(a) Absorption spectra. The three curves show the wavelengths of light best absorbed by three types of chloroplast pigments.

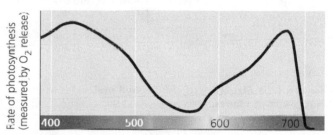

(b) Action spectrum. This graph plots the rate of photosynthesis versus wavelength. The resulting action spectrum resembles the absorption spectrum for chlorophyll *a* but does not match exactly (see part a). This is partly due to the absorption of light by accessory pigments such as chlorophyll *b* and carotenoids.

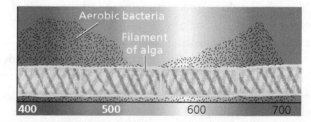

(c) Engelmann's experiment. In 1883, Theodor W. Engelmann illuminated a filamentous alga with light that had been passed through a prism, exposing different segments of the alga to different wavelengths. He used aerobic bacteria, which concentrate near an oxygen source, to determine which segments of the alga were releasing the most O_2 and thus photosynthesizing most. Bacteria congregated in greatest numbers around the parts of the alga illuminated with violet-blue or red light.

Conclusion Light in the violet-blue and red portions of the spectrum is most effective in driving photosynthesis.

Source T. W. Engelmann, *Bacterium photometricum*. Ein Beitrag zur vergleichenden Physiologie des Licht-und Farbensinnes, *Archiv. für Physiologie* 30:95–124 (1883).

(MB) A related Experimental Inquiry Tutorial can be assigned in MasteringBiology.

WHAT IF? If Engelmann had used a filter that allowed only red light to pass through, how would the results have differed?

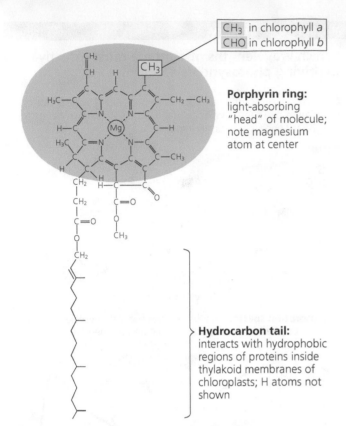

CH₃ in chlorophyll *a*
CHO in chlorophyll *b*

Porphyrin ring: light-absorbing "head" of molecule; note magnesium atom at center

Hydrocarbon tail: interacts with hydrophobic regions of proteins inside thylakoid membranes of chloroplasts; H atoms not shown

▲ **Figure 8.10 Structure of chlorophyll molecules in chloroplasts of plants.** Chlorophyll *a* and chlorophyll *b* differ only in one of the functional groups bonded to the porphyrin ring. (Also see the space-filling model of chlorophyll in Figure 1.3.)

Notice by comparing Figures 8.9a and 8.9b that the action spectrum for photosynthesis is much broader than the absorption spectrum of chlorophyll *a*. The absorption spectrum of chlorophyll *a* alone underestimates the effectiveness of certain wavelengths in driving photosynthesis. This is partly because accessory pigments with different absorption spectra are also photosynthetically important in chloroplasts and broaden the spectrum of colors that can be used for photosynthesis. **Figure 8.10** shows the structure of chlorophyll *a* compared with that of chlorophyll *b*. A slight structural difference between them is enough to cause the two pigments to absorb at slightly different wavelengths in the red and blue parts of the spectrum (see Figure 8.9a). As a result, chlorophyll *a* appears blue green and chlorophyll *b* is olive green in visible light.

Other accessory pigments include **carotenoids**, hydrocarbons that are various shades of yellow and orange because they absorb violet and blue-green light (see Figure 8.9a). Carotenoids may broaden the spectrum of colors that can drive photosynthesis. However, a more important function of at least some carotenoids seems to be *photoprotection*: These compounds absorb and dissipate excessive light energy that would otherwise damage chlorophyll or interact with oxygen, forming reactive oxidative molecules that are dangerous to the cell. Interestingly, carotenoids similar to the photoprotective ones in chloroplasts have a photoprotective role in the human eye.

Excitation of Chlorophyll by Light

What exactly happens when chlorophyll and other pigments absorb light? The colors corresponding to the absorbed wavelengths disappear from the spectrum of the transmitted and reflected light, but energy cannot disappear. When a molecule absorbs a photon of light, one of the molecule's electrons is elevated to an electron shell where it has more potential energy. When the electron is in its normal shell, the pigment molecule is said to be in its ground state. Absorption of a photon boosts an electron to a higher-energy electron shell, and the pigment molecule is then said to be in an excited state **(Figure 8.11a)**. The only photons absorbed are those whose energy is exactly equal to the energy difference between the ground state and an excited state, and this energy difference varies from one kind of molecule to another. Thus, a particular compound absorbs only photons corresponding to specific wavelengths, which is why each pigment has a unique absorption spectrum.

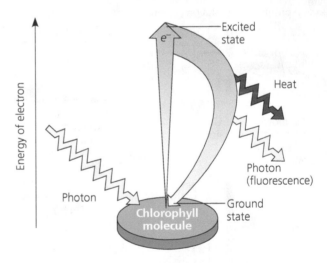

(a) Excitation of isolated chlorophyll molecule

(b) Fluorescence

▲ **Figure 8.11 Excitation of isolated chlorophyll by light.**
(a) Absorption of a photon causes a transition of the chlorophyll molecule from its ground state to its excited state. The photon boosts an electron to an orbital where it has more potential energy. If the illuminated molecule exists in isolation, its excited electron immediately drops back down to the ground-state orbital, and its excess energy is given off as heat and fluorescence (light). **(b)** A chlorophyll solution excited with ultraviolet light fluoresces with a red-orange glow.

Once absorption of a photon raises an electron from the ground state to an excited state, the electron cannot remain there long. The excited state, like all high-energy states, is unstable. Generally, when isolated pigment molecules absorb light, their excited electrons drop back down to the ground-state electron shell in a billionth of a second, releasing their excess energy as heat. This conversion of light energy to heat is what makes the top of an automobile so hot on a sunny day. (White cars are coolest because their paint reflects all wavelengths of visible light, although it may absorb ultraviolet and other invisible radiation.) In isolation, some pigments, including chlorophyll, emit light as well as heat after absorbing photons. As excited electrons fall back to the ground state, photons are given off. This afterglow is called fluorescence. If a solution of chlorophyll isolated from chloroplasts is illuminated, it will fluoresce in the red-orange part of the spectrum and also give off heat (**Figure 8.11b**). If the same flask were viewed under visible light, it would appear green.

A Photosystem: A Reaction-Center Complex Associated with Light-Harvesting Complexes

Chlorophyll molecules excited by the absorption of light energy produce very different results in an intact chloroplast than they do in isolation. In their native environment of the thylakoid membrane, chlorophyll molecules are organized along with other small organic molecules and proteins into complexes called photosystems.

A **photosystem** is composed of a **reaction-center complex** surrounded by several light-harvesting complexes (**Figure 8.12**). The reaction-center complex is an organized association of proteins holding a special pair of chlorophyll *a* molecules. Each **light-harvesting complex** consists of various pigment molecules (which may include chlorophyll *a*, chlorophyll *b*, and carotenoids) bound to proteins. The number and variety of pigment molecules enable a photosystem to harvest light over a larger surface area and a larger portion of the spectrum than could any single pigment molecule alone. Together, these light-harvesting complexes act as an antenna for the reaction-center complex. When a pigment molecule absorbs a photon, the energy is transferred from pigment molecule to pigment molecule within a light-harvesting complex, somewhat like a human "wave" at a sports arena, until it is passed into the reaction-center complex. The reaction-center complex also contains a molecule capable of accepting electrons and becoming reduced; this is called the **primary electron acceptor**. The pair of chlorophyll *a* molecules in the reaction-center complex are special because their molecular environment— their location and the other molecules with which they are associated—enables them to use the energy from light not only to boost one of their electrons to a higher energy level, but also to transfer it to a different molecule—the primary electron acceptor.

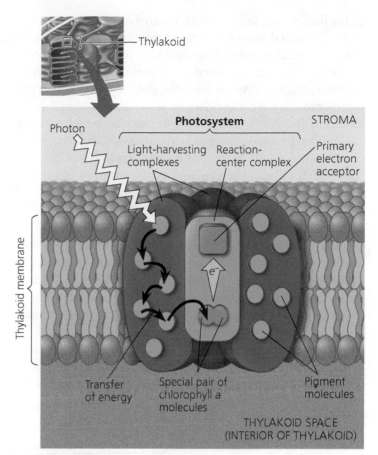

(a) **How a photosystem harvests light.** When a photon strikes a pigment molecule in a light-harvesting complex, the energy is passed from molecule to molecule until it reaches the reaction-center complex. Here, an excited electron from the special pair of chlorophyll *a* molecules is transferred to the primary electron acceptor.

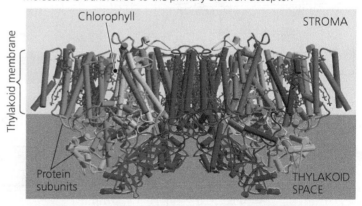

(b) **Structure of a photosystem.** This computer model, based on X-ray crystallography, shows two photosystem complexes side by side. Chlorophyll molecules (small green ball-and-stick models) are interspersed with protein subunits (cylinders and ribbons). For simplicity, a photosystem will be shown as a single complex in the rest of the chapter.

© 2004 AAAS

▲ **Figure 8.12 The structure and function of a photosystem.**

The solar-powered transfer of an electron from the reaction-center chlorophyll *a* pair to the primary electron acceptor is one of the first steps of the light reactions. As soon as the chlorophyll electron is excited to a higher energy level, the primary electron acceptor captures it; this is a redox reaction.

In the flask shown in Figure 8.11, isolated chlorophyll fluoresces because there is no electron acceptor, so electrons of photoexcited chlorophyll drop right back to the ground state. In the structured environment of a chloroplast, however, an electron acceptor is readily available, and the potential energy represented by the excited electron is not dissipated as light and heat. Thus, each photosystem—a reaction-center complex surrounded by light-harvesting complexes—functions in the chloroplast as a unit. It converts light energy to chemical energy, which will ultimately be used for the synthesis of sugar.

The thylakoid membrane is populated by two types of photosystems that cooperate in the light reactions of photosynthesis. They are called **photosystem II (PS II)** and **photosystem I (PS I)**. (They were named in order of their discovery, but photosystem II functions first in the light reactions.) Each has a characteristic reaction-center complex—a particular kind of primary electron acceptor next to a special pair of chlorophyll *a* molecules associated with specific proteins. The reaction-center chlorophyll *a* of photosystem II is known as P680 because this pigment is best at absorbing light having a wavelength of 680 nm (in the red part of the spectrum). The chlorophyll *a* at the reaction-center complex of photosystem I

is called P700 because it most effectively absorbs light of wavelength 700 nm (in the far-red part of the spectrum). These two pigments, P680 and P700, are nearly identical chlorophyll *a* molecules. However, their association with different proteins in the thylakoid membrane affects the electron distribution in the two pigments and accounts for the slight differences in their light-absorbing properties. Now let's see how the two photosystems work together in using light energy to generate ATP and NADPH, the two main products of the light reactions.

Linear Electron Flow

Light drives the synthesis of ATP and NADPH by energizing the two photosystems embedded in the thylakoid membranes of chloroplasts. The key to this energy transformation is a flow of electrons through the photosystems and other molecular components built into the thylakoid membrane. This is called **linear electron flow**, and it occurs during the light reactions of photosynthesis, as shown in **Figure 8.13**. The following steps correspond to the numbered steps in the figure.

1 A photon of light strikes a pigment molecule in a light-harvesting complex of PS II, boosting one of its electrons to

▼ **Figure 8.13 How linear electron flow during the light reactions generates ATP and NADPH.** The gold arrows trace the current of light-driven electrons from water to NADPH.

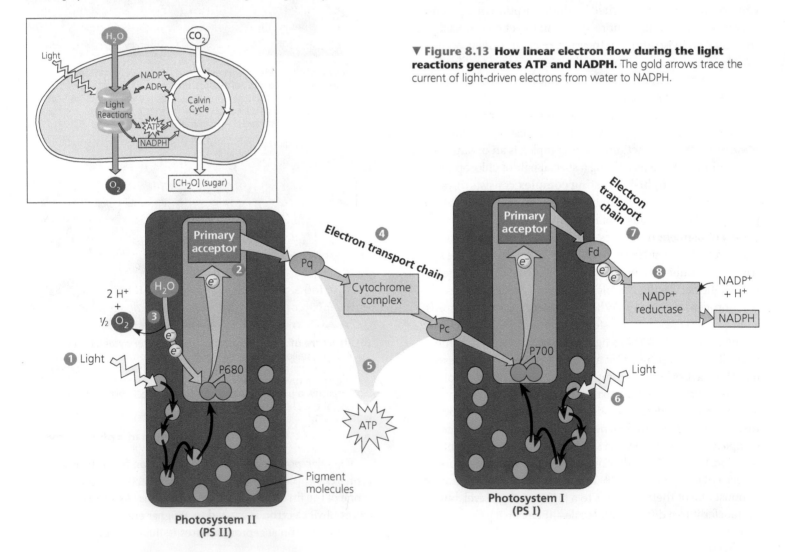

a higher energy level. As this electron falls back to its ground state, an electron in a nearby pigment molecule is simultaneously raised to an excited state. The process continues, with the energy being relayed to other pigment molecules until it reaches the P680 pair of chlorophyll *a* molecules in the PS II reaction-center complex. It excites an electron in this pair of chlorophylls to a higher energy state.

2 This electron is transferred from the excited P680 to the primary electron acceptor. We can refer to the resulting form of P680, missing an electron, as $P680^+$.

3 An enzyme catalyzes the splitting of a water molecule into two electrons, two hydrogen ions (H^+), and an oxygen atom. The electrons are supplied one by one to the $P680^+$ pair, each electron replacing one transferred to the primary electron acceptor. ($P680^+$ is the strongest biological oxidizing agent known; its electron "hole" must be filled. This greatly facilitates the transfer of electrons from the split water molecule.) The H^+ are released into the thylakoid space. The oxygen atom immediately combines with an oxygen atom generated by the splitting of another water molecule, forming O_2.

4 Each photoexcited electron passes from the primary electron acceptor of PS II to PS I via an electron transport chain, the components of which are similar to those of the electron transport chain that functions in cellular respiration. The electron transport chain between PS II and PS I is made up of the electron carrier plastoquinone (Pq), a cytochrome complex, and a protein called plastocyanin (Pc).

5 The exergonic "fall" of electrons to a lower energy level provides energy for the synthesis of ATP. As electrons pass through the cytochrome complex, H^+ are pumped into the thylakoid space, contributing to the proton gradient that is subsequently used in chemiosmosis.

6 Meanwhile, light energy has been transferred via light-harvesting complex pigments to the PS I reaction-center complex, exciting an electron of the P700 pair of chlorophyll *a* molecules located there. The photoexcited electron is then transferred to PS I's primary electron acceptor, creating an electron "hole" in the P700—which we now can call $P700^+$. In other words, $P700^+$ can now act as an electron acceptor, accepting an electron that reaches the bottom of the electron transport chain from PS II.

7 Photoexcited electrons are passed in a series of redox reactions from the primary electron acceptor of PS I down a second electron transport chain through the protein ferredoxin (Fd). (This chain does not create a proton gradient and thus does not produce ATP.)

8 The enzyme $NADP^+$ reductase catalyzes the transfer of electrons from Fd to $NADP^+$. Two electrons are required for its reduction to NADPH. This molecule is at a higher energy level than water, and its electrons are more readily available for the reactions of the Calvin cycle than were those of water. This process also removes an H^+ from the stroma.

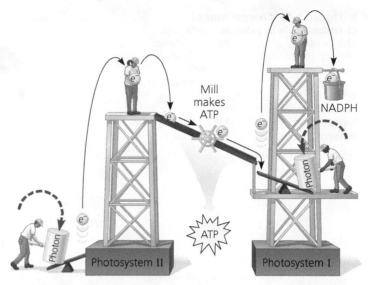

▲ **Figure 8.14 A mechanical analogy for linear electron flow during the light reactions.**

The energy changes of electrons during their linear flow through the light reactions are shown in a mechanical analogy in **Figure 8.14**. Although the scheme shown in Figures 8.13 and 8.14 may seem complicated, do not lose track of the big picture. The light reactions use solar power to generate ATP and NADPH, which provide chemical energy and reducing power, respectively, to the carbohydrate-synthesizing reactions of the Calvin cycle. Before we move on to consider the Calvin cycle, let's review chemiosmosis, the process that uses membranes to couple redox reactions to ATP production.

A Comparison of Chemiosmosis in Chloroplasts and Mitochondria

Chloroplasts and mitochondria generate ATP by the same basic mechanism: chemiosmosis. An electron transport chain assembled in a membrane pumps protons across the membrane as electrons are passed through a series of carriers that are progressively more electronegative. In this way, electron transport chains transform redox energy to a proton-motive force, potential energy stored in the form of an H^+ gradient across a membrane. Built into the same membrane is an ATP synthase complex that couples the diffusion of hydrogen ions down their gradient to the phosphorylation of ADP. Some of the electron carriers, including the iron-containing proteins called cytochromes, are very similar in chloroplasts and mitochondria. The ATP synthase complexes of the two organelles are also very much alike. But there are noteworthy differences between oxidative phosphorylation in mitochondria and photophosphorylation in chloroplasts. In mitochondria, the high-energy electrons dropped down the transport chain are extracted from organic molecules (which are thus oxidized), while in chloroplasts, the source of electrons is water. Chloroplasts do not need molecules from food to make ATP; their photosystems capture light energy

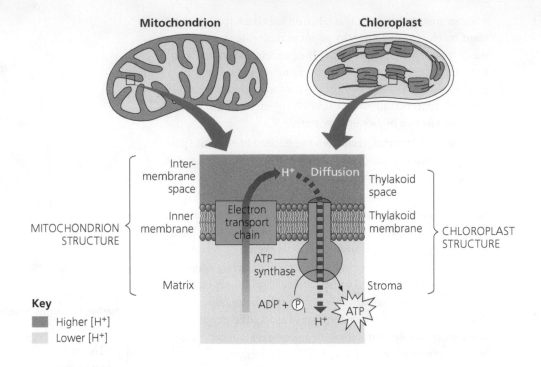

► **Figure 8.15 Comparison of chemiosmosis in mitochondria and chloroplasts.** In both kinds of organelles, electron transport chains pump protons (H⁺) across a membrane from a region of low H⁺ concentration (light gray in this diagram) to one of high H⁺ concentration (dark gray). The protons then diffuse back across the membrane through ATP synthase, driving the synthesis of ATP.

Mitochondrion

Chloroplast

Inter-membrane space

Inner membrane

Matrix

MITOCHONDRION STRUCTURE

H⁺   Diffusion

Electron transport chain

ATP synthase

ADP + P_i

H⁺

ATP

Thylakoid space

Thylakoid membrane

Stroma

CHLOROPLAST STRUCTURE

Key

Higher [H⁺]

Lower [H⁺]

and use it to drive the electrons from water to the top of the transport chain. In other words, mitochondria use chemiosmosis to transfer chemical energy from food molecules to ATP, whereas chloroplasts transform light energy into chemical energy in ATP.

Although the spatial organization of chemiosmosis differs slightly between chloroplasts and mitochondria, it is easy to see similarities in the two **(Figure 8.15)**. The inner membrane of the mitochondrion pumps protons from the mitochondrial matrix out to the intermembrane space, which then serves as a reservoir of hydrogen ions. The thylakoid membrane of the chloroplast pumps protons from the stroma into the thylakoid space (interior of the thylakoid), which functions as the H⁺ reservoir. If you imagine the cristae of mitochondria pinching off from the inner membrane, this may help you see how the thylakoid space and the intermembrane space are comparable spaces in the two organelles, while the mitochondrial matrix is analogous to the stroma of the chloroplast. In the mitochondrion, protons diffuse down their concentration gradient from the intermembrane space through ATP synthase to the matrix, driving ATP synthesis. In the chloroplast, ATP is synthesized as the hydrogen ions diffuse from the thylakoid space back to the stroma through ATP synthase complexes, whose catalytic knobs are on the stroma side of the membrane. Thus, ATP forms in the stroma, where it is used to help drive sugar synthesis during the Calvin cycle.

The proton (H⁺) gradient, or pH gradient, across the thylakoid membrane is substantial. When chloroplasts in an experimental setting are illuminated, the pH in the thylakoid space drops to about 5 (the H⁺ concentration increases), and the pH in the stroma increases to about 8 (the H⁺ concentration decreases). This gradient of three pH units corresponds to a

thousandfold difference in H⁺ concentration. If in the laboratory the lights are turned off, the pH gradient is abolished, but it can quickly be restored by turning the lights back on. Experiments such as this provided strong evidence in support of the chemiosmotic model.

Based on studies in several laboratories, **Figure 8.16** shows a current model for the organization of the light-reaction "machinery" within the thylakoid membrane. Each of the molecules and molecular complexes in the figure is present in numerous copies in each thylakoid. Notice that NADPH, like ATP, is produced on the side of the membrane facing the stroma, where the Calvin cycle reactions take place.

Let's summarize the light reactions. Electron flow pushes electrons from water, where they are at a low state of potential energy, ultimately to NADPH, where they are stored at a high state of potential energy. The light-driven electron current also generates ATP. Thus, the equipment of the thylakoid membrane converts light energy to chemical energy stored in ATP and NADPH. (Oxygen is a by-product.) Let's now see how the Calvin cycle uses the products of the light reactions to synthesize sugar from CO_2.

CONCEPT CHECK 8.2

1. What color of light is *least* effective in driving photosynthesis? Explain.

2. In the light reactions, what is the initial electron donor? At the end of the light reactions, where are the electrons?

3. **WHAT IF?** In an experiment, isolated chloroplasts placed in an illuminated solution with the appropriate chemicals can carry out ATP synthesis. Predict what will happen to the rate of synthesis if a compound is added to the solution that makes membranes freely permeable to hydrogen ions.

For suggested answers, see Appendix A.

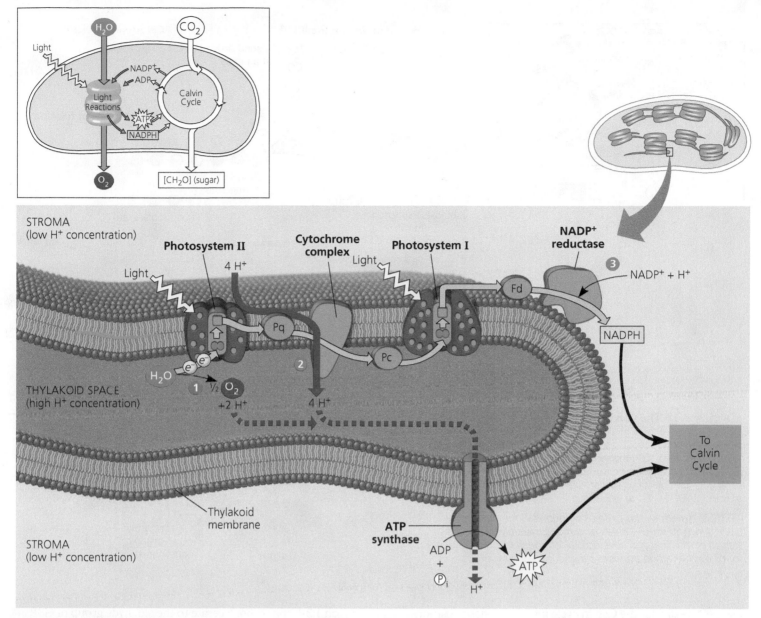

▲ **Figure 8.16 The light reactions and chemiosmosis: the organization of the thylakoid membrane.** This diagram shows the current model for the organization of the thylakoid membrane. The gold arrows track the linear electron flow outlined in Figure 8.13. At least three steps contribute to the H⁺ gradient by increasing H⁺ concentration in the thylakoid space: ❶ Water is split by photosystem II on the side of the membrane facing the thylakoid space; ❷ as plastoquinone (Pq) transfers electrons to the cytochrome complex, four protons are translocated across the membrane into the thylakoid space; and ❸ a hydrogen ion is removed from the stroma when it is taken up by NADP⁺. Notice that in step 2, hydrogen ions are being pumped from the stroma into the thylakoid space, as in Figure 8.15. The diffusion of H⁺ from the thylakoid space back to the stroma (along the H⁺ concentration gradient) powers the ATP synthase.

CONCEPT 8.3

The Calvin cycle uses the chemical energy of ATP and NADPH to reduce CO_2 to sugar

The Calvin cycle is similar to the citric acid cycle in that a starting material is regenerated after molecules enter and leave the cycle. However, while the citric acid cycle is catabolic, oxidizing acetyl CoA and using the energy to synthesize ATP, the Calvin cycle is anabolic, building carbohydrates from smaller molecules and consuming energy. Carbon enters the Calvin cycle in the form of CO_2 and leaves in the form of sugar. The cycle spends ATP as an energy source and consumes NADPH as reducing power for adding high-energy electrons to make the sugar.

As we mentioned previously, the carbohydrate produced directly from the Calvin cycle is actually not glucose, but a three-carbon sugar named **glyceraldehyde 3-phosphate (G3P)**. For net synthesis of one molecule of G3P, the cycle must take place three times, fixing three molecules of CO_2. (Recall that carbon fixation refers to the initial incorporation of CO_2 into organic material.) As we trace the steps of the cycle, keep in mind that we are following three molecules of CO_2 through the reactions.

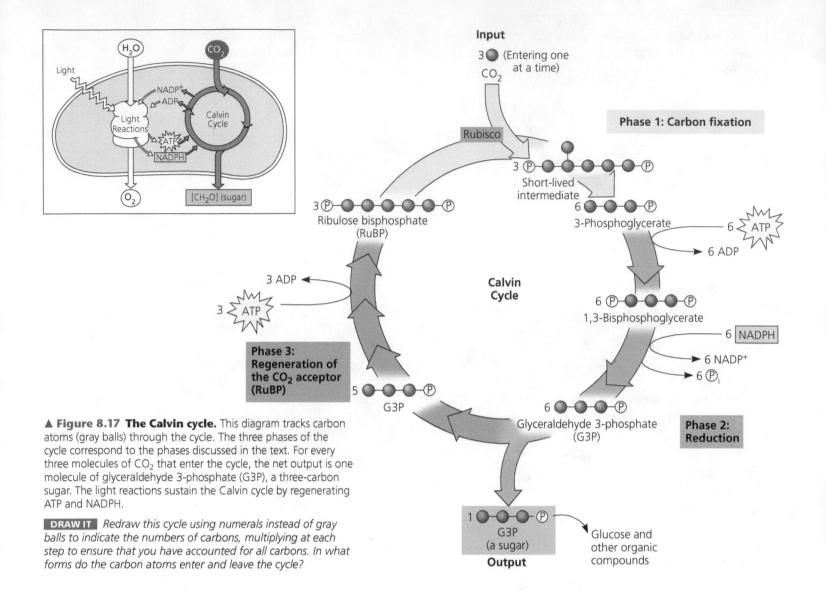

▲ Figure 8.17 The Calvin cycle. This diagram tracks carbon atoms (gray balls) through the cycle. The three phases of the cycle correspond to the phases discussed in the text. For every three molecules of CO_2 that enter the cycle, the net output is one molecule of glyceraldehyde 3-phosphate (G3P), a three-carbon sugar. The light reactions sustain the Calvin cycle by regenerating ATP and NADPH.

DRAW IT *Redraw this cycle using numerals instead of gray balls to indicate the numbers of carbons, multiplying at each step to ensure that you have accounted for all carbons. In what forms do the carbon atoms enter and leave the cycle?*

Figure 8.17 divides the Calvin cycle into three phases: carbon fixation, reduction, and regeneration of the CO_2 acceptor.

Phase 1: Carbon fixation. The Calvin cycle incorporates each CO_2 molecule, one at a time, by attaching it to a five-carbon sugar named ribulose bisphosphate (abbreviated RuBP). The enzyme that catalyzes this first step is RuBP carboxylase, or **rubisco**. (This is the most abundant protein in chloroplasts and is also thought to be the most abundant protein on Earth.) The product of the reaction is a six-carbon intermediate so unstable that it immediately splits in half, forming two molecules of 3-phosphoglycerate (for each CO_2 fixed).

Phase 2: Reduction. Each molecule of 3-phosphoglycerate receives an additional phosphate group from ATP, becoming 1,3-bisphosphoglycerate. Next, a pair of electrons donated from NADPH reduces 1,3-bisphosphoglycerate, which also loses a phosphate group, becoming G3P. Specifically, the electrons from NADPH reduce a carboyxl group

on 1,3-bisphosphoglycerate to the aldehyde group of G3P, which stores more potential energy. G3P is a sugar—the same three-carbon sugar formed in glycolysis by the splitting of glucose (see Figure 7.9). Notice in Figure 8.17 that for every *three* molecules of CO_2 that enter the cycle, there are *six* molecules of G3P formed. But only one molecule of this three-carbon sugar can be counted as a net gain of carbohydrate. The cycle began with 15 carbons' worth of carbohydrate in the form of three molecules of the five-carbon sugar RuBP. Now there are 18 carbons' worth of carbohydrate in the form of six molecules of G3P. One molecule exits the cycle to be used by the plant cell, but the other five molecules must be recycled to regenerate the three molecules of RuBP.

Phase 3: Regeneration of the CO_2 acceptor (RuBP). In a complex series of reactions, the carbon skeletons of five molecules of G3P are rearranged by the last steps of the Calvin cycle into three molecules of RuBP. To accomplish this, the cycle spends three more molecules of ATP. The RuBP is now prepared to receive CO_2 again, and the cycle continues.

For the net synthesis of one G3P molecule, the Calvin cycle consumes a total of nine molecules of ATP and six molecules of NADPH. The light reactions regenerate the ATP and NADPH. The G3P spun off from the Calvin cycle becomes the starting material for metabolic pathways that synthesize other organic compounds, including glucose and other carbohydrates. Neither the light reactions nor the Calvin cycle alone can make sugar from CO_2. Photosynthesis is an emergent property of the intact chloroplast, which integrates the two stages of photosynthesis.

Evolution of Alternative Mechanisms of Carbon Fixation in Hot, Arid Climates

EVOLUTION Ever since plants first moved onto land about 475 million years ago, they have been adapting to the problem of dehydration. The solutions often involve trade-offs. An example is the compromise between photosynthesis and the prevention of excessive water loss from the plant. The CO_2 required for photosynthesis enters a leaf (and the resulting O_2 exits) via stomata, the pores on the leaf surface (see Figure 8.3). However, stomata are also the main avenues of the evaporative loss of water from leaves and may be partially or fully closed on hot, dry days. This prevents water loss, but it also reduces CO_2 levels.

In most plants, initial fixation of carbon occurs via rubisco, the Calvin cycle enzyme that adds CO_2 to ribulose bisphosphate. Such plants are called **C_3 plants** because the first organic product of carbon fixation is a three-carbon compound, 3-phosphoglycerate (see Figure 8.17). C_3 plants include important agricultural plants such as rice, wheat, and soybeans. When their stomata close on hot, dry days, C_3 plants produce less sugar because the declining level of CO_2 in the leaf starves the Calvin cycle. In addition, rubisco is capable of binding O_2 in place of CO_2. As CO_2 becomes scarce and O_2 builds up, rubisco adds O_2 to the Calvin cycle instead of CO_2. The product splits, forming a two-carbon compound that leaves the chloroplast and is broken down in the cell, releasing CO_2. The process is called **photorespiration** because it occurs in the light (*photo*) and consumes O_2 while producing CO_2 (*respiration*). However, unlike normal cellular respiration, photorespiration uses ATP rather than generating it. And unlike photosynthesis, photorespiration produces no sugar. In fact, photorespiration *decreases* photosynthetic output by siphoning organic material from the Calvin cycle and releasing CO_2 that would otherwise be fixed.

According to one hypothesis, photorespiration is evolutionary baggage—a metabolic relic from a much earlier time when the atmosphere had less O_2 and more CO_2 than it does today. In the ancient atmosphere that prevailed when rubisco first evolved, the inability of the enzyme's active site to exclude O_2 would have made little difference. The hypothesis suggests that modern rubisco retains some of its chance affinity for O_2, which is now so concentrated in the atmosphere that a certain amount of photorespiration is inevitable. There is also some evidence that photorespiration may provide protection against damaging products of the light reactions that build up when the Calvin cycle slows due to low CO_2.

In some plant species, alternate modes of carbon fixation have evolved that minimize photorespiration and optimize the Calvin cycle—even in hot, arid climates. The two most important of these photosynthetic adaptations are C_4 photosynthesis and crassulacean acid metabolism (CAM).

C_4 Plants

The **C_4 plants** are so named because they carry out a modified pathway for sugar synthesis that first fixes CO_2 into a four-carbon compound. When the weather is hot and dry, a C_4 plant partially closes its stomata, thus conserving water. Sugar continues to be made, though, through the function of two different types of photosynthetic cells: mesophyll cells and bundle-sheath cells **(Figure 8.18a)**. An enzyme in the mesophyll cells has a high affinity for CO_2 and can fix carbon even when the CO_2 concentration in the leaf is low. The resulting four-carbon compound then acts as a carbon shuttle; it moves into bundle-sheath cells, which are packed around the veins of the leaf, and releases CO_2. Thus, the CO_2 concentration in these cells remains high enough for the Calvin cycle to make

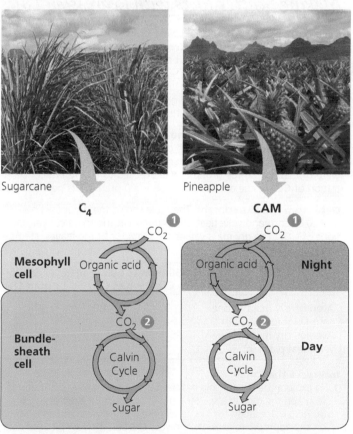

(a) Spatial separation of steps. In C_4 plants, carbon fixation and the Calvin cycle occur in different types of cells.

(b) Temporal separation of steps. In CAM plants, carbon fixation and the Calvin cycle occur in the same cell at different times.

▲ **Figure 8.18 C_4 and CAM photosynthesis compared.** Both adaptations are characterized by ❶ preliminary incorporation of CO_2 into organic acids, followed by ❷ transfer of CO_2 to the Calvin cycle. The C_4 and CAM pathways are two evolutionary solutions to the problem of maintaining photosynthesis with stomata partially or completely closed on hot, dry days.

sugars and avoid photorespiration. The C_4 pathway is believed to have evolved independently at least 45 times and is used by several thousand species in at least 19 plant families. Among the C_4 plants important to agriculture are sugarcane and corn (maize), members of the grass family. In the **Scientific Skills Exercise**, you will work with data to see how different concentrations of CO_2 affect growth in plants that use the C_4 pathway versus those that use the C_3 pathway.

CAM Plants

A second photosynthetic adaptation to arid conditions has evolved in pineapples, many cacti, and other succulent (water-storing) plants, such as aloe and jade plants **(Figure 8.18b)**. These plants open their stomata during the night and close them during the day, the reverse of how other plants behave. Closing stomata during the day helps desert plants conserve water, but it also prevents CO_2 from entering the leaves.

During the night, when their stomata are open, these plants take up CO_2 and incorporate it into a variety of organic acids. This mode of carbon fixation is called **crassulacean acid metabolism (CAM)** after the plant family Crassulaceae, the succulents in which the process was first discovered. The mesophyll cells of **CAM plants** store the organic acids they make during the night in their vacuoles until morning, when the stomata close. During the day, when the light reactions can supply ATP and NADPH for the Calvin cycle, CO_2 is released from the organic acids made the night before to become incorporated into sugar in the chloroplasts.

Notice in Figure 8.18 that the CAM pathway is similar to the C_4 pathway in that carbon dioxide is first incorporated into organic intermediates before it enters the Calvin cycle. The difference is that in C_4 plants, the initial steps of carbon fixation are separated structurally from the Calvin cycle, whereas in CAM plants, the two steps occur at separate times but within the same

---

Scientific Skills Exercise

Making Scatter Plots with Regression Lines

Does Atmospheric Carbon Dioxide Concentration Affect the Productivity of Agricultural Crops? Atmospheric concentration of carbon dioxide (CO_2) has been rising globally, and scientists wondered whether this would affect C_3 and C_4 plants differently. In this exercise, you will make a scatter plot to examine the relationship between CO_2 concentration and growth of corn (maize), a C_4 crop plant, and velvetleaf, a C_3 weed found in cornfields.

How the Experiment Was Done Researchers grew corn and velvetleaf plants under controlled conditions for 45 days, where all plants received the same amount of water and light. The plants were divided into three groups, each exposed to a different concentration of CO_2 in the air: 350, 600, or 1,000 ppm (parts per million).

Data from the Experiment The table shows the dry mass (in grams) of corn and velvetleaf plants grown at the three concentrations of CO_2. The dry mass values are averages of the leaves, stems, and roots of eight plants.

	350 ppm CO_2	600 ppm CO_2	1,000 ppm CO_2
Average dry mass of one corn plant (g)	91	89	80
Average dry mass of one velvetleaf plant (g)	35	48	54

Interpret the Data

1. To explore the relationship between the two variables, it is useful to graph the data in a scatter plot, and then draw a regression line. (a) First, place labels for the dependent and independent variables on the appropriate axes. Explain your choices. (b) Now plot the data points for corn and velvetleaf using different symbols for each set of data, and add a key for the two symbols. (For additional information about graphs, see the Scientific Skills Review in Appendix F and in the Study Area in MasteringBiology.)

2. Draw a "best-fit" line for each set of points. A best-fit line does not necessarily pass through all or even most points. Instead, it is a straight line that passes as close as possible to all data points from that set. Use your eye to draw a best-fit line for each set of

data. Because this is a matter of judgment, two individuals may draw two slightly different lines for a given set of points. The line that actually fits best, a regression line, can be identified by adding up the distances of all points to any candidate line, then selecting the line that minimizes the summed distances. (See the graph in the Scientific Skills Exercise in Chapter 2 for an example of a linear regression line.) Excel or other software programs, including those on a graphing calculator, can plot a regression line once data points are entered. Using either Excel or a graphing calculator, enter the data points for each data set and have the program draw the two regression lines. Compare them to the lines you drew by eye.

3. Describe the trends shown by the regression lines in your scatter plot. (a) Compare the relationship between increasing concentration of CO_2 and the dry mass of corn to that of velvetleaf. (b) Considering that velvetleaf is a weed invasive to cornfields, predict how increased CO_2 concentration may affect interactions between the two species.

4. Based on the data in the scatter plot, estimate the percentage change in dry mass of corn and velvetleaf plants if atmospheric CO_2 concentration increases from 390 ppm (current levels) to 800 ppm. (a) First draw vertical lines on your graph at 390 ppm and 800 ppm. Next, where each vertical line intersects a regression line, draw a horizontal line to the y-axis. What is the estimated dry mass of corn and velvetleaf plants at 390 ppm? 800 ppm? (b) To calculate the percentage change in mass for each plant, subtract the mass at 390 ppm from the mass at 800 ppm, divide by the mass at 390 ppm, and multiply by 100. What is the estimated percentage change in dry mass for corn? For velvetleaf? (c) Do these results support the conclusion from other experiments that C_3 plants grow better than C_4 plants under increased CO_2 concentration? Why or why not?

Data from D. T. Patterson and E. P. Flint, Potential effects of global atmospheric CO_2 enrichment on the growth and competitiveness of C_3 and C_4 weed and crop plants, *Weed Science* 28(1): 71–75 (1980).

(MB) A version of this Scientific Skills Exercise can be assigned in MasteringBiology.

cell. (Keep in mind that CAM, C$_4$, and C$_3$ plants all eventually use the Calvin cycle to make sugar from carbon dioxide.)

CONCEPT CHECK 8.3

1. **MAKE CONNECTIONS** How are the large numbers of ATP and NADPH molecules used during the Calvin cycle consistent with the high value of glucose as an energy source? (Compare Figures 7.15 and 8.17.)

2. **WHAT IF?** Explain why a poison that inhibits an enzyme of the Calvin cycle will also inhibit the light reactions.

3. Describe how photorespiration lowers photosynthetic output for plants.

For suggested answers, see Appendix A.

The Importance of Photosynthesis: *A Review*

In this chapter, we have followed photosynthesis from photons to food. The light reactions capture solar energy and use it to make ATP and transfer electrons from water to NADP$^+$, forming NADPH. The Calvin cycle uses the ATP and NADPH to produce sugar from carbon dioxide. The energy that enters the chloroplasts as sunlight becomes stored as chemical energy in organic compounds. See **Figure 8.19** for a review of the entire process.

What are the fates of photosynthetic products? The sugar made in the chloroplasts supplies the entire plant with chemical energy and carbon skeletons for the synthesis of all the major organic molecules of plant cells. About 50% of the organic material made by photosynthesis is consumed as fuel for cellular respiration in the mitochondria of the plant cells. Sometimes there is a loss of photosynthetic products to photorespiration.

Technically, green cells are the only autotrophic parts of the plant. The rest of the plant depends on organic molecules exported from leaves via veins. In most plants, carbohydrate is transported out of the leaves in the form of sucrose, a disaccharide. After arriving at nonphotosynthetic cells, the sucrose provides raw material for cellular respiration and a multitude of anabolic pathways that synthesize proteins, lipids, and other products. A considerable amount of sugar in the form of glucose is

linked together to make the polysaccharide cellulose, especially in plant cells that are still growing and maturing. Cellulose, the main ingredient of cell walls, is the most abundant organic molecule in the plant—and probably on the surface of the planet.

Most plants manage to make more organic material each day than they need to use as respiratory fuel and precursors for biosynthesis. They stockpile the extra sugar by synthesizing starch, storing some in the chloroplasts themselves and some in storage cells of roots, tubers, seeds, and fruits. In accounting for the consumption of the food molecules produced by photosynthesis, let's not forget that most plants lose leaves, roots, stems, fruits, and sometimes their entire bodies to heterotrophs, including humans.

On a global scale, photosynthesis is the process responsible for the presence of oxygen in our atmosphere. Furthermore, while each chloroplast is minuscule, their collective productivity in terms of food production is prodigious: Photosynthesis makes an estimated 160 billion metric tons of carbohydrate per year (a metric ton is 1,000 kg, about 1.1 tons). That's organic matter equivalent in mass to a stack of about 60 trillion copies of this textbook—17 stacks of books reaching from Earth to the sun! No other chemical process on the planet can match the output of photosynthesis. In fact, researchers are seeking ways to capitalize on photosynthetic production to produce alternative fuels. No process is more important than photosynthesis to the welfare of life on Earth.

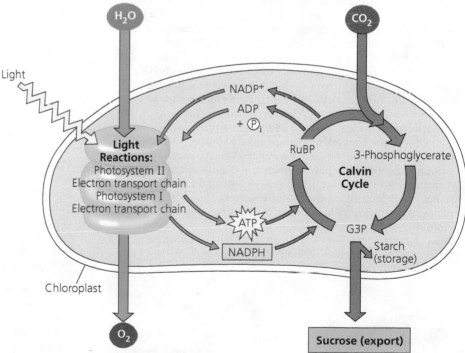

▶ **Figure 8.19 A review of photosynthesis.** This diagram outlines the main reactants and products of the light reactions and the Calvin cycle as they occur in the chloroplasts of plant cells. The entire ordered operation depends on the structural integrity of the chloroplast and its membranes. Enzymes in the chloroplast and cytosol convert glyceraldehyde 3-phosphate (G3P), the direct product of the Calvin cycle, to many other organic compounds.

Light Reactions:
- Are carried out by molecules in the thylakoid membranes
- Convert light energy to the chemical energy of ATP and NADPH
- Split H$_2$O and release O$_2$ to the atmosphere

Calvin Cycle Reactions:
- Take place in the stroma
- Use ATP and NADPH to convert CO$_2$ to the sugar G3P
- Return ADP, inorganic phosphate, and NADP$^+$ to the light reactions

8 Chapter Review

SUMMARY OF KEY CONCEPTS

CONCEPT 8.1

Photosynthesis converts light energy to the chemical energy of food (pp. 156–159)

- In **autotrophic** eukaryotes, photosynthesis occurs in **chloroplasts**, organelles containing **thylakoids**. Stacks of thylakoids form grana. **Photosynthesis** is summarized as

  $$6 \, CO_2 + 12 \, H_2O + \text{Light energy} \rightarrow C_6H_{12}O_6 + 6 \, O_2 + 6 \, H_2O.$$

  Chloroplasts split water into hydrogen and oxygen, incorporating the electrons of hydrogen into sugar molecules. Photosynthesis is a redox process: H_2O is oxidized, and CO_2 is reduced. The **light reactions** in the thylakoid membranes split water, releasing O_2, producing ATP, and forming **NADPH**. The **Calvin cycle** in the **stroma** forms sugars from CO_2, using ATP for energy and NADPH for reducing power.

  ? *Compare and describe the roles of CO_2 and H_2O in respiration and photosynthesis.*

CONCEPT 8.2

The light reactions convert solar energy to the chemical energy of ATP and NADPH (pp. 159–167)

- Light is a form of electromagnetic energy. The colors we see as **visible light** include those **wavelengths** that drive photosynthesis. A pigment absorbs light of specific wavelengths; **chlorophyll *a*** is the main photosynthetic pigment in plants. Other accessory pigments absorb different wavelengths of light and pass the energy on to chlorophyll *a*.
- A pigment goes from a ground state to an excited state when a **photon** of light boosts one of the pigment's electrons to a higher-energy electron shell. Electrons from isolated pigments tend to fall back to the ground state, giving off heat and/or light.
- A **photosystem** is composed of a **reaction-center complex** surrounded by **light-harvesting complexes** that funnel the energy of photons to the reaction-center complex. When a special pair of reaction-center chlorophyll *a* molecules absorbs energy, one of its electrons is boosted to a higher energy level and transferred to the **primary electron acceptor**. **Photosystem II** contains P680 chlorophyll *a* molecules in the reaction-center complex; **photosystem I** contains P700 molecules.
- **Linear electron flow** during the light reactions uses both photosystems and produces NADPH, ATP, and oxygen:

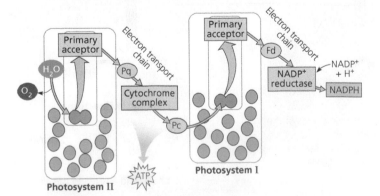

- During chemiosmosis in both mitochondria and chloroplasts, electron transport chains generate an H^+ (proton) gradient across a membrane. ATP synthase uses this proton-motive force to synthesize ATP.

  ? *The absorption spectrum of chlorophyll a differs from the action spectrum of photosynthesis. Explain this observation.*

CONCEPT 8.3

The Calvin cycle uses the chemical energy of ATP and NADPH to reduce CO_2 to sugar (pp. 167–171)

- The Calvin cycle occurs in the stroma, using electrons from NADPH and energy from ATP. One molecule of **G3P** exits the cycle per three CO_2 molecules fixed and is converted to glucose and other organic molecules.

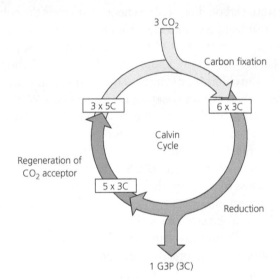

- On hot, dry days, **C_3 plants** close their stomata, conserving water but keeping CO_2 out and O_2 in. Under these conditions, **photorespiration** can occur: Rubisco binds O_2 instead of CO_2, leading to consumption of ATP and release of CO_2 without the production of sugar. Photorespiration may be an evolutionary relic and it may also play a protective role.
- **C_4 plants** are adapted to hot, dry climates. Even with their stomata partially or completely closed, they minimize the cost of photorespiration by incorporating CO_2 into four-carbon compounds in mesophyll cells. These compounds are exported to bundle-sheath cells, where they release carbon dioxide for use in the Calvin cycle.
- **CAM plants** are also adapted to hot, dry climates. They open their stomata at night, incorporating CO_2 into organic acids, which are stored in mesophyll cells. During the day, the stomata close, and the CO_2 is released from the organic acids for use in the Calvin cycle.
- Organic compounds produced by photosynthesis provide the energy and building material for ecosystems.

  DRAW IT *On the diagram above, draw where ATP and NADPH are used and where rubisco functions. Describe these steps.*

TEST YOUR UNDERSTANDING

Level 1: Knowledge/Comprehension

1. The light reactions of photosynthesis supply the Calvin cycle with
   a. light energy.
   b. CO_2 and ATP.
   c. H_2O and NADPH.
   d. ATP and NADPH.
   e. sugar and O_2.

2. Which of the following sequences correctly represents the flow of electrons during photosynthesis?
   a. NADPH $\rightarrow$ O_2 $\rightarrow$ CO_2
   b. H_2O $\rightarrow$ NADPH $\rightarrow$ Calvin cycle
   c. NADPH $\rightarrow$ chlorophyll $\rightarrow$ Calvin cycle
   d. H_2O $\rightarrow$ photosystem I $\rightarrow$ photosystem II
   e. NADPH $\rightarrow$ electron transport chain $\rightarrow$ O_2

3. How is photosynthesis similar in C_4 plants and CAM plants?
   a. In both cases, electron transport is not used.
   b. Both types of plants make sugar without the Calvin cycle.
   c. In both cases, rubisco is not used to fix carbon initially.
   d. Both types of plants make most of their sugar in the dark.
   e. In both cases, thylakoids are not involved in photosynthesis.

4. Which of the following statements is a correct distinction between autotrophs and heterotrophs?
   a. Only heterotrophs require chemical compounds from the environment.
   b. Cellular respiration is unique to heterotrophs.
   c. Only heterotrophs have mitochondria.
   d. Autotrophs, but not heterotrophs, can nourish themselves beginning with CO_2 and other nutrients that are inorganic.
   e. Only heterotrophs require oxygen.

5. Which of the following does *not* occur during the Calvin cycle?
   a. carbon fixation
   b. oxidation of NADPH
   c. release of oxygen
   d. regeneration of the CO_2 acceptor
   e. consumption of ATP

Level 2: Application/Analysis

6. In mechanism, photophosphorylation is most similar to
   a. substrate-level phosphorylation in glycolysis.
   b. oxidative phosphorylation in cellular respiration.
   c. the Calvin cycle.
   d. carbon fixation.
   e. reduction of $NADP^+$.

7. Which process is most directly driven by light energy?
   a. creation of a pH gradient by pumping protons across the thylakoid membrane
   b. carbon fixation in the stroma
   c. reduction of $NADP^+$ molecules
   d. removal of electrons from chlorophyll molecules
   e. ATP synthesis

8. To synthesize one glucose molecule, the Calvin cycle uses _____ molecules of CO_2, _____ molecules of ATP, and _____ molecules of NADPH.

Level 3: Synthesis/Evaluation

9. **SCIENTIFIC INQUIRY**
   MAKE CONNECTIONS The following diagram represents an experiment with isolated thylakoids. The thylakoids were first made acidic by soaking them in a solution at pH 4. After the thylakoid space reached pH 4, the thylakoids were transferred to a basic solution at pH 8. The thylakoids then made ATP in the dark. (See Concept 2.5 to review pH.)

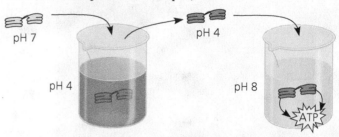

Draw an enlargement of part of the thylakoid membrane in the beaker with the solution at pH 8. Draw ATP synthase. Label the areas of high H^+ concentration and low H^+ concentration. Show the direction protons flow through the enzyme, and show the reaction where ATP is synthesized. Would ATP end up in the thylakoid or outside of it? Explain why the thylakoids in the experiment were able to make ATP in the dark.

10. **SCIENCE, TECHNOLOGY, AND SOCIETY**
   Scientific evidence indicates that the CO_2 added to the air by the burning of wood and fossil fuels is contributing to global warming, a rise in global temperature. Tropical rain forests are estimated to be responsible for approximately 20% of global photosynthesis, yet the consumption of large amounts of CO_2 by living trees is thought to make little or no *net* contribution to reduction of global warming. Why might this be? (*Hint:* What processes in both living and dead trees produce CO_2?)

11. **FOCUS ON EVOLUTION**
   Consider the endosymbiont theory (see Concept 4.5) and the fact that chloroplasts contain DNA molecules. Given that chloroplast DNA has genes, what would you expect if you compared the sequence of a chloroplast gene in a plant cell to the same gene in other plant species or bacteria? Would it be more similar to a plant or a bacterial gene sequence?

12. **FOCUS ON ENERGY AND MATTER**
   Life is solar powered. Almost all the producers of the biosphere depend on energy from the sun to produce the organic molecules that supply the energy and carbon skeletons needed for life. In a short essay (100–150 words), describe how the process of photosynthesis in the chloroplasts of plants transforms the energy of sunlight into the chemical energy of sugar molecules.

For selected answers, see Appendix A.

MasteringBiology®

Students Go to **MasteringBiology** for assignments, the eText, and the Study Area with practice tests, animations, and activities.

Instructors Go to **MasteringBiology** for automatically graded tutorials and questions that you can assign to your students, plus Instructor Resources.

9 The Cell Cycle

KEY CONCEPTS

9.1 Most cell division results in genetically identical daughter cells
9.2 The mitotic phase alternates with interphase in the cell cycle
9.3 The eukaryotic cell cycle is regulated by a molecular control system

▼ **Figure 9.1** How do a cell's chromosomes change during cell division?

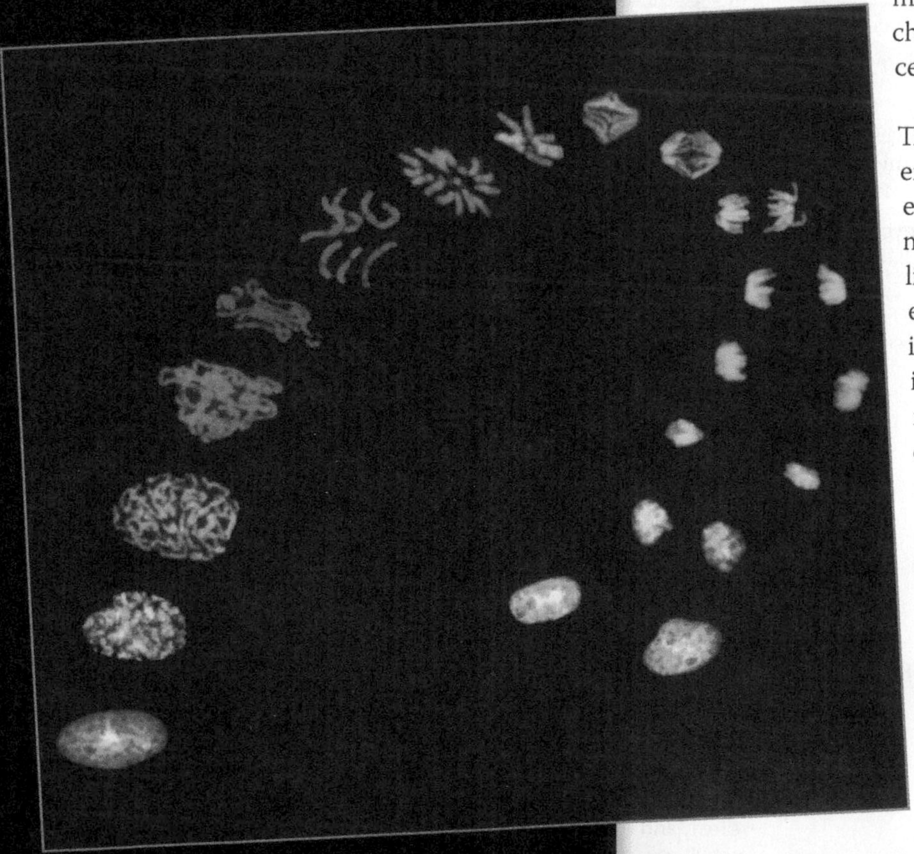

OVERVIEW

The Key Roles of Cell Division

The ability of organisms to produce more of their own kind is the one characteristic that best distinguishes living things from non-living matter. This unique capacity to procreate, like all biological functions, has a cellular basis. Rudolf Virchow, a German physician, put it this way in 1855: "Where a cell exists, there must have been a preexisting cell, just as the animal arises only from an animal and the plant only from a plant." He summarized this concept with the Latin axiom "*Omnis cellula e cellula*," meaning "Every cell from a cell." The continuity of life is based on the reproduction of cells, or **cell division**. The series of fluorescence micrographs in **Figure 9.1** follows an animal cell's chromosomes, from lower left to lower right, as one cell divides into two.

Cell division plays several important roles in life. The division of one prokaryotic cell reproduces an entire organism. The same is true of a unicellular eukaryote **(Figure 9.2a)**. Cell division also enables multicellular eukaryotes to develop from a single cell, like the fertilized egg that gave rise to the two-celled embryo in **Figure 9.2b**. And after such an organism is fully grown, cell division continues to function in renewal and repair, replacing cells that die from normal wear and tear or accidents. For example, dividing cells in your bone marrow continuously make new blood cells **(Figure 9.2c)**.

The cell division process is an integral part of the **cell cycle**, the life of a cell from the time it is first formed from a dividing parent cell until its own division into two daughter cells. (Our use of the words *daughter* or *sister* in relation to cells is not meant to imply gender.) Passing identical genetic material to cellular offspring is a crucial function of cell division. In this chapter, you'll learn how this process occurs. After studying the mechanics of cell division in eukaryotes and bacteria, you'll learn about the molecular control system that regulates progress through the eukaryotic cell cycle and what happens when the control system malfunctions. Because a breakdown in cell cycle control plays a major role in cancer development, this aspect of cell biology is an active area of research.

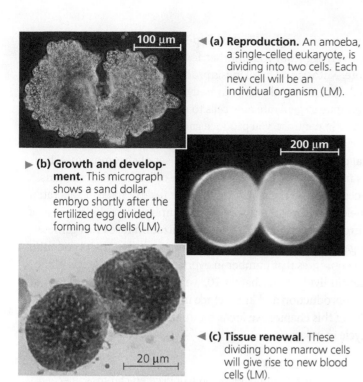

◀ **(a) Reproduction.** An amoeba, a single-celled eukaryote, is dividing into two cells. Each new cell will be an individual organism (LM).

▶ **(b) Growth and development.** This micrograph shows a sand dollar embryo shortly after the fertilized egg divided, forming two cells (LM).

◀ **(c) Tissue renewal.** These dividing bone marrow cells will give rise to new blood cells (LM).

▲ **Figure 9.2 The functions of cell division.**

CONCEPT 9.1

Most cell division results in genetically identical daughter cells

The reproduction of a cell, with all its complexity, cannot occur by a mere pinching in half; a cell is not like a soap bubble that simply enlarges and splits in two. In both prokaryotes and eukaryotes, most cell division involves the distribution of identical genetic material—DNA—to two daughter cells. (The exception is meiosis, the special type of eukaryotic cell division that can produce sperm and eggs.) What is most remarkable about cell division is the fidelity with which the DNA is passed along from one generation of cells to the next. A dividing cell duplicates its DNA, allocates the two copies to opposite ends of the cell, and only then splits into daughter cells. After we describe the distribution of DNA during cell division in animal and plant cells, we'll consider the process in other eukaryotes as well as in bacteria.

Cellular Organization of the Genetic Material

A cell's endowment of DNA, its genetic information, is called its **genome**. Although a prokaryotic genome is often a single DNA molecule, eukaryotic genomes usually consist of a number of DNA molecules. The overall length of DNA in a eukaryotic cell is enormous. A typical human cell, for example, has about 2 m of DNA—a length about 250,000 times greater than the cell's diameter. Before the cell can divide to form genetically identical daughter cells, all of this DNA must be copied, or replicated, and then the two copies must be separated so that each daughter cell ends up with a complete genome.

The replication and distribution of so much DNA are manageable because the DNA molecules are packaged into structures called **chromosomes** (from the Greek *chroma*, color, and *soma*, body), so named because they take up certain dyes used in microscopy **(Figure 9.3)**. Each eukaryotic chromosome consists of one very long, linear DNA molecule associated with many proteins (see Figure 4.8). The DNA molecule carries several hundred to a few thousand genes, the units of information that specify an organism's inherited traits. The associated proteins maintain the structure of the chromosome and help control the activity of the genes. Together, the entire complex of DNA and proteins that is the building material of chromosomes is referred to as **chromatin**. As you will soon see, the chromatin of a chromosome varies in its degree of condensation during the process of cell division.

Every eukaryotic species has a characteristic number of chromosomes in each cell nucleus. For example, the nuclei of human **somatic cells** (all body cells except the reproductive cells) each contain 46 chromosomes, made up of two sets of 23, one set inherited from each parent. Reproductive cells, or **gametes**—sperm and eggs—have half as many chromosomes as somatic cells, or one set of 23 chromosomes in humans. The number of chromosomes in somatic cells varies widely among species: 18 in cabbage plants, 48 in chimpanzees, 56 in elephants, 90 in hedgehogs, and 148 in one species of alga. We'll now consider how these chromosomes behave during cell division.

Distribution of Chromosomes During Eukaryotic Cell Division

When a cell is not dividing, and even as it replicates its DNA in preparation for cell division, each chromosome is in the form of a long, thin chromatin fiber. After DNA replication, however, the chromosomes condense as a part of cell division: Each chromatin fiber becomes densely coiled and folded,

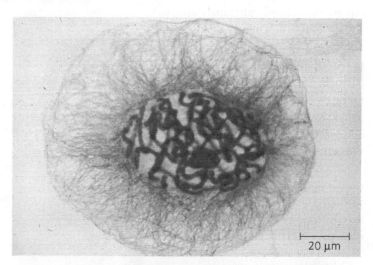

▲ **Figure 9.3 Eukaryotic chromosomes.** Chromosomes (stained purple) are visible within the nucleus of this cell from an African blood lily. The thinner red threads in the surrounding cytoplasm are the cytoskeleton. The cell is preparing to divide (LM).

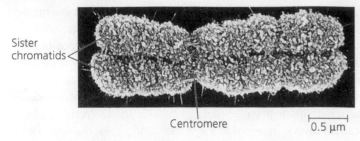

Sister chromatids

Centromere

0.5 μm

▲ **Figure 9.4 A highly condensed, duplicated human chromosome (SEM).**

DRAW IT *Circle one sister chromatid of the chromosome in this micrograph.*

making the chromosomes much shorter and so thick that we can see them with a light microscope.

Each duplicated chromosome has two **sister chromatids**, which are joined copies of the original chromosome **(Figure 9.4)**. The two chromatids, each containing an identical DNA molecule, are initially attached all along their lengths by protein complexes called *cohesins*; this attachment is known as *sister chromatid cohesion*. Each sister chromatid has a **centromere**, a region containing specific DNA sequences where the chromatid is attached most closely to its sister chromatid. This attachment is mediated by proteins bound to the centromeric DNA sequences and gives the condensed, duplicated chromosome a narrow "waist." The part of a chromatid on either side of the centromere is referred to as an *arm* of the chromatid. (An uncondensed, unduplicated chromosome has a single centromere and two arms.)

Later in the cell division process, the two sister chromatids of each duplicated chromosome separate and move into two new nuclei, one forming at each end of the cell. Once the sister chromatids separate, they are no longer called sister chromatids but are considered individual chromosomes. Thus, each new nucleus receives a collection of chromosomes identical to that of the parent cell **(Figure 9.5)**. **Mitosis**, the division of the genetic material in the nucleus, is usually followed immediately by **cytokinesis**, the division of the cytoplasm. One cell has become two, each the genetic equivalent of the parent cell.

What happens to the chromosome number as we follow the human life cycle through the generations? You inherited 46 chromosomes, one set of 23 from each parent. They were combined in the

nucleus of a single cell when a sperm from your father united with an egg from your mother, forming a fertilized egg, or zygote. Mitosis and cytokinesis produced the 200 trillion somatic cells that now make up your body, and the same processes continue to generate new cells to replace dead and damaged ones. In contrast, you produce gametes—eggs or sperm—by a variation of cell division called *meiosis*, which yields nonidentical daughter cells that have only one set of chromosomes, half as many chromosomes as the parent cell. Meiosis in humans occurs only in the gonads (ovaries or testes). In each generation, meiosis reduces the chromosome number from 46 (two sets of chromosomes) to 23 (one set). Fertilization fuses two gametes together and returns the chromosome number to 46, and mitosis conserves that number in every somatic cell nucleus of the new individual. In Chapter 10, we'll examine the role of meiosis in reproduction and inheritance in more detail. In the remainder of this chapter, we focus on mitosis and the rest of the cell cycle in eukaryotes.

CONCEPT CHECK 9.1

1. How many chromatids are in a duplicated chromosome?
2. **WHAT IF?** A chicken has 78 chromosomes in its somatic cells. How many chromosomes did the chicken inherit from each parent? How many chromosomes are in each of the chicken's gametes? How many chromosomes will be in each somatic cell of the chicken's offspring?

For suggested answers, see Appendix A.

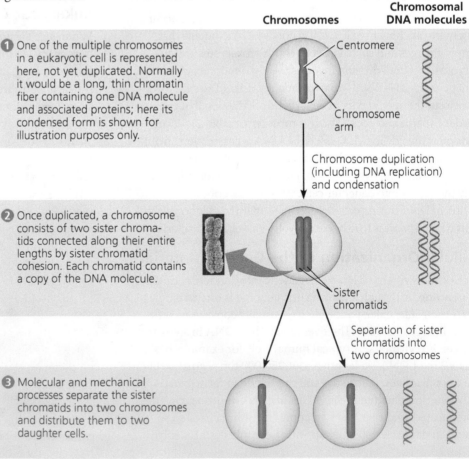

Chromosomes **Chromosomal DNA molecules**

1 One of the multiple chromosomes in a eukaryotic cell is represented here, not yet duplicated. Normally it would be a long, thin chromatin fiber containing one DNA molecule and associated proteins; here its condensed form is shown for illustration purposes only.

Centromere

Chromosome arm

Chromosome duplication (including DNA replication) and condensation

2 Once duplicated, a chromosome consists of two sister chromatids connected along their entire lengths by sister chromatid cohesion. Each chromatid contains a copy of the DNA molecule.

Sister chromatids

Separation of sister chromatids into two chromosomes

3 Molecular and mechanical processes separate the sister chromatids into two chromosomes and distribute them to two daughter cells.

▶ **Figure 9.5 Chromosome duplication and distribution during cell division.**

? *How many chromatid arms does the chromosome in step 2 have?*

The mitotic phase alternates with interphase in the cell cycle

In 1882, a German anatomist named Walther Flemming developed dyes that allowed him to observe, for the first time, the behavior of chromosomes during mitosis and cytokinesis. (In fact, Flemming coined the terms *mitosis* and *chromatin*.) It appeared to Flemming that during the period between one cell division and the next, the cell was simply growing larger. But we now know that many critical events occur during this stage in the life of a cell.

Phases of the Cell Cycle

Mitosis is just one part of the cell cycle **(Figure 9.6)**. In fact, the **mitotic (M) phase**, which includes both mitosis and cytokinesis, is usually the shortest part of the cell cycle. Mitotic cell division alternates with a much longer stage called **interphase**, which often accounts for about 90% of the cycle. Interphase can be divided into subphases: the **G_1 phase** ("first gap"), the **S phase** ("synthesis"), and the **G_2 phase** ("second gap"). During all three subphases, a cell that will eventually divide grows by producing proteins and cytoplasmic organelles such as mitochondria and endoplasmic reticulum. However, chromosomes are duplicated only during the S phase. (We will discuss synthesis of DNA in Chapter 13.) Thus, a cell grows (G_1), continues to grow as it copies its chromosomes (S), grows more as it completes preparations for cell division (G_2), and divides (M). The daughter cells may then repeat the cycle.

A particular human cell might undergo one division in 24 hours. Of this time, the M phase would occupy less than 1 hour, while the S phase might occupy about 10–12 hours, or about half the cycle. The rest of the time would be apportioned

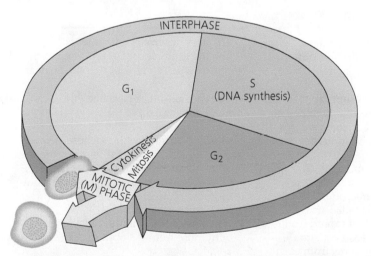

▲ **Figure 9.6 The cell cycle.** In a dividing cell, the mitotic (M) phase alternates with interphase, a growth period. The first part of interphase (G_1) is followed by the S phase, when the chromosomes duplicate; G_2 is the last part of interphase. In the M phase, mitosis distributes the daughter chromosomes to daughter nuclei, and cytokinesis divides the cytoplasm, producing two daughter cells. The relative durations of G_1, S, and G_2 may vary.

between the G_1 and G_2 phases. The G_2 phase usually takes 4–6 hours; in our example, G_1 would occupy about 5–6 hours. G_1 is the most variable in length in different types of cells. Some cells in a multicellular organism divide very infrequently or not at all. These cells spend their time in G_1 (or a related phase called G_0) doing their job in the organism—a nerve cell carries impulses, for example.

Mitosis is conventionally broken down into five stages: **prophase**, **prometaphase**, **metaphase**, **anaphase**, and **telophase**. Overlapping with the latter stages of mitosis, cytokinesis completes the mitotic phase. **Figure 9.7** describes these stages in an animal cell. Study this figure thoroughly before progressing to the next two sections, which examine mitosis and cytokinesis more closely.

The Mitotic Spindle: *A Closer Look*

Many of the events of mitosis depend on the **mitotic spindle**, which begins to form in the cytoplasm during prophase. This structure consists of fibers made of microtubules and associated proteins. While the mitotic spindle assembles, the other microtubules of the cytoskeleton partially disassemble, providing the material used to construct the spindle. The spindle microtubules elongate (polymerize) by incorporating more subunits of the protein tubulin (see Table 4.1) and shorten (depolymerize) by losing subunits.

In animal cells, the assembly of spindle microtubules starts at the **centrosome**, a subcellular region containing material that functions throughout the cell cycle to organize the cell's microtubules. (It is also a type of *microtubule-organizing center*.) A pair of centrioles is located at the center of the centrosome, but they are not essential for cell division: If the centrioles are destroyed with a laser microbeam, a spindle nevertheless forms during mitosis. In fact, centrioles are not even present in plant cells, which do form mitotic spindles.

During interphase in animal cells, the single centrosome duplicates, forming two centrosomes, which remain together near the nucleus (see Figure 9.7). The two centrosomes move apart during prophase and prometaphase of mitosis as spindle microtubules grow out from them. By the end of prometaphase, the two centrosomes, one at each pole of the spindle, are at opposite ends of the cell. An **aster**, a radial array of short microtubules, extends from each centrosome. The spindle includes the centrosomes, the spindle microtubules, and the asters.

Each of the two sister chromatids of a duplicated chromosome has a **kinetochore**, a structure made up of proteins that have assembled on specific sections of chromosomal DNA at each centromere. The chromosome's two kinetochores face in opposite directions. During prometaphase, some of the spindle microtubules attach to the kinetochores; these are called kinetochore microtubules. (The number of microtubules attached to a kinetochore varies among species, from one microtubule in yeast cells to 40 or so in some mammalian cells.) When one of a chromosome's kinetochores is "captured" by microtubules, the chromosome begins to move toward the pole from which those

Exploring Mitosis in an Animal Cell

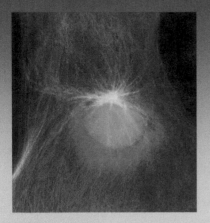

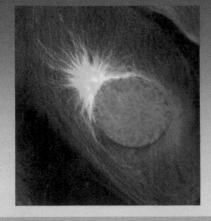

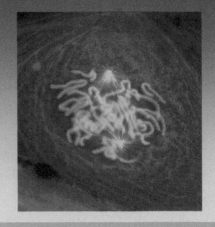

G₂ of Interphase

Centrosomes (with centriole pairs)

Chromosomes (duplicated, uncondensed)

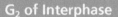

Nucleolus   Nuclear envelope   Plasma membrane

Prophase

Early mitotic spindle   Aster   Centromere

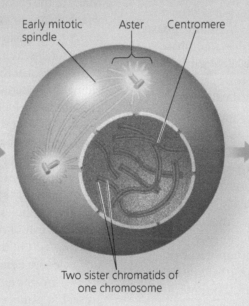

Two sister chromatids of one chromosome

Prometaphase

Fragments of nuclear envelope   Nonkinetochore microtubules

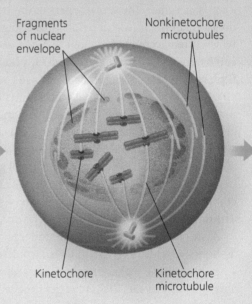

Kinetochore   Kinetochore microtubule

G₂ of Interphase

- A nuclear envelope encloses the nucleus.
- The nucleus contains one or more nucleoli (singular, *nucleolus*).
- Two centrosomes have formed by duplication of a single centrosome. Centrosomes are regions in animal cells that organize the microtubules of the spindle. Each centrosome contains two centrioles.
- Chromosomes, duplicated during S phase, cannot be seen individually because they have not yet condensed.

The fluorescence micrographs show dividing lung cells from a newt; this species has 22 chromosomes. Chromosomes appear blue, microtubules green, and intermediate filaments red. For simplicity, the drawings show only 6 chromosomes.

Prophase

- The chromatin fibers become more tightly coiled, condensing into discrete chromosomes observable with a light microscope.
- The nucleoli disappear.
- Each duplicated chromosome appears as two identical sister chromatids joined at their centromeres and, in some species, all along their arms by cohesins (sister chromatid cohesion).
- The mitotic spindle (named for its shape) begins to form. It is composed of the centrosomes and the microtubules that extend from them. The radial arrays of shorter microtubules that extend from the centrosomes are called asters ("stars").
- The centrosomes move away from each other, propelled partly by the lengthening microtubules between them.

Prometaphase

- The nuclear envelope fragments.
- The microtubules extending from each centrosome can now invade the nuclear area.
- The chromosomes have become even more condensed.
- Each of the two chromatids of each chromosome now has a kinetochore, a specialized protein structure at the centromere.
- Some of the microtubules attach to the kinetochores, becoming kinetochore microtubules, which jerk the chromosomes back and forth.
- Nonkinetochore microtubules interact with those from the opposite pole of the spindle.

? *How many molecules of DNA are in the prometaphase drawing? How many molecules per chromosome? How many double helices are there per chromosome? Per chromatid?*

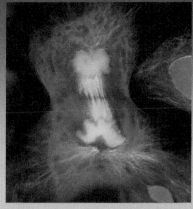
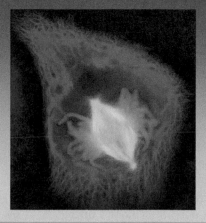
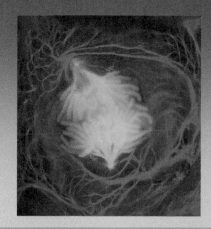

10 µm

| **Metaphase** | **Anaphase** | **Telophase and Cytokinesis** |

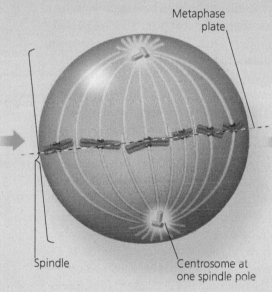
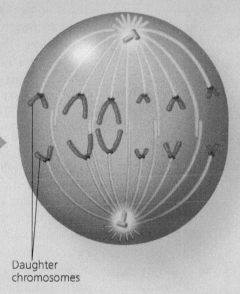
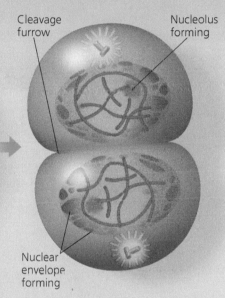

Metaphase plate

Spindle

Centrosome at one spindle pole

Daughter chromosomes

Cleavage furrow

Nucleolus forming

Nuclear envelope forming

Metaphase

- The centrosomes are now at opposite poles of the cell.
- The chromosomes convene at the metaphase plate, a plane that is equidistant between the spindle's two poles. The chromosomes' centromeres lie at the metaphase plate.
- For each chromosome, the kinetochores of the sister chromatids are attached to kinetochore microtubules coming from opposite poles.

Anaphase

- Anaphase is the shortest stage of mitosis, often lasting only a few minutes.
- Anaphase begins when the cohesin proteins are cleaved. This allows the two sister chromatids of each pair to part suddenly. Each chromatid thus becomes a full-fledged chromosome.
- The two liberated daughter chromosomes begin moving toward opposite ends of the cell as their kinetochore microtubules shorten. Because these microtubules are attached at the centromere region, the chromosomes move centromere first (at about 1 µm/min).
- The cell elongates as the nonkinetochore microtubules lengthen.
- By the end of anaphase, the two ends of the cell have equivalent—and complete— collections of chromosomes.

Telophase

- Two daughter nuclei form in the cell. Nuclear envelopes arise from the fragments of the parent cell's nuclear envelope and other portions of the endomembrane system.
- Nucleoli reappear.
- The chromosomes become less condensed.
- Any remaining spindle microtubules are depolymerized.
- Mitosis, the division of one nucleus into two genetically identical nuclei, is now complete.

Cytokinesis

- The division of the cytoplasm is usually well under way by late telophase, so the two daughter cells appear shortly after the end of mitosis.
- In animal cells, cytokinesis involves the formation of a cleavage furrow, which pinches the cell in two.

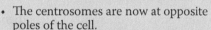

ANIMATION

BioFlix Visit the Study Area in **MasteringBiology** for the BioFlix® 3-D Animation on Mitosis.

microtubules extend. However, this movement is checked as soon as microtubules from the opposite pole attach to the other kinetochore. What happens next is like a tug-of-war that ends in a draw. The chromosome moves first in one direction, then the other, back and forth, finally settling midway between the two ends of the cell. At metaphase, the centromeres of all the duplicated chromosomes are on a plane midway between the spindle's two poles. This plane is called the **metaphase plate**, which is an imaginary rather than an actual cellular structure **(Figure 9.8)**. Meanwhile, microtubules that do not attach to kinetochores have been elongating, and by metaphase they

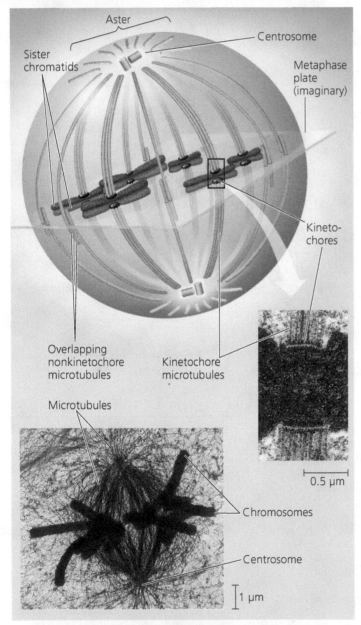

▲ Figure 9.8 The mitotic spindle at metaphase. The kinetochores of each chromosome's two sister chromatids face in opposite directions. Here, each kinetochore is attached to a cluster of kinetochore microtubules extending from the nearest centrosome. Nonkinetochore microtubules overlap at the metaphase plate (TEMs).

DRAW IT *On the lower micrograph, draw a line indicating the position of the metaphase plate. Circle the asters. Draw arrows indicating the directions of chromosome movement once anaphase begins.*

overlap and interact with other nonkinetochore microtubules from the opposite pole of the spindle. (These are sometimes called "polar" microtubules.) By metaphase, the microtubules of the asters have also grown and are in contact with the plasma membrane. The spindle is now complete.

The structure of the completed spindle correlates well with its function during anaphase. Anaphase commences suddenly when the cohesins holding together the sister chromatids of each chromosome are cleaved by an enzyme called *separase*. Once separated, the chromatids become full-fledged chromosomes that move toward opposite ends of the cell.

How do the kinetochore microtubules function in this pole-ward movement of chromosomes? Apparently, two mechanisms are in play, both involving motor proteins. (To review how motor proteins move an object along a microtubule, see Figure 4.21.) A clever experiment carried out in 1987 suggested that motor proteins on the kinetochores "walk" the chromosomes along the microtubules, which depolymerize at their kinetochore ends after the motor proteins have passed **(Figure 9.9)**. (This is referred to as the "Pacman" mechanism because of its resemblance to the arcade game character that moves by eating all the dots in its path.) However, other researchers, working with different cell types or cells from other species, have shown that chromosomes are "reeled in" by motor proteins at the spindle poles and that the microtubules depolymerize after they pass by these motor proteins. The general consensus now is that both mechanisms are used and that their relative contributions vary among cell types.

In a dividing animal cell, the nonkinetochore microtubules are responsible for elongating the whole cell during anaphase. Nonkinetochore microtubules from opposite poles overlap each other extensively during metaphase (see Figure 9.8). During anaphase, the region of overlap is reduced as motor proteins attached to the microtubules walk them away from one another, using energy from ATP. As the microtubules push apart from each other, their spindle poles are pushed apart, elongating the cell. At the same time, the microtubules lengthen somewhat by the addition of tubulin subunits to their overlapping ends. As a result, the microtubules continue to overlap.

At the end of anaphase, duplicate groups of chromosomes have arrived at opposite ends of the elongated parent cell. Nuclei re-form during telophase. Cytokinesis generally begins during anaphase or telophase, and the spindle eventually disassembles by depolymerization of microtubules.

Cytokinesis: *A Closer Look*

In animal cells, cytokinesis occurs by a process known as **cleavage**. The first sign of cleavage is the appearance of a **cleavage furrow**, a shallow groove in the cell surface near the old metaphase plate **(Figure 9.10a)**. On the cytoplasmic side of the furrow is a contractile ring of actin microfilaments associated with molecules of the protein myosin. The actin microfilaments interact with the myosin molecules, causing the ring to contract. The contraction of the dividing cell's ring of

At which end do kinetochore microtubules shorten during anaphase?

Experiment Gary Borisy and colleagues at the University of Wisconsin wanted to determine whether kinetochore microtubules depolymerize at the kinetochore end or the pole end as chromosomes move toward the poles during mitosis. First they labeled the microtubules of a pig kidney cell in early anaphase with a yellow fluorescent dye.

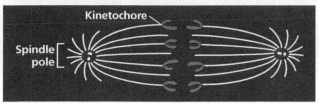

Then they marked a region of the kinetochore microtubules between one spindle pole and the chromosomes by using a laser to eliminate the fluorescence from that region, leaving the microtubules intact (see below). As anaphase proceeded, they monitored the changes in microtubule length on either side of the mark.

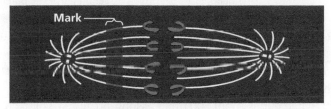

Results As the chromosomes moved poleward, the microtubule segments on the kinetochore side of the mark shortened, while those on the spindle pole side stayed the same length.

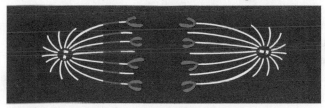

Conclusion During anaphase in this cell type, chromosome movement is correlated with kinetochore microtubules shortening at their kinetochore ends and not at their spindle pole ends. This experiment supports the hypothesis that during anaphase, a chromosome is walked along a microtubule as the microtubule depolymerizes at its kinetochore end, releasing tubulin subunits.

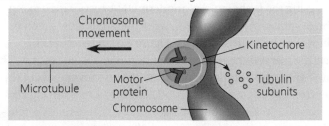

Source G. J. Gorbsky, P. J. Sammak, and G. G. Borisy, Chromosomes move poleward in anaphase along stationary microtubules that coordinately disassemble from their kinetochore ends, *Journal of Cell Biology* 104:9–18 (1987).

WHAT IF? If this experiment had been done on a cell type in which "reeling in" at the poles was the main cause of chromosome movement, how would the mark have moved relative to the poles? How would the microtubule lengths have changed?

▼ **Figure 9.10 Cytokinesis in animal and plant cells.**

(a) Cleavage of an animal cell (SEM)

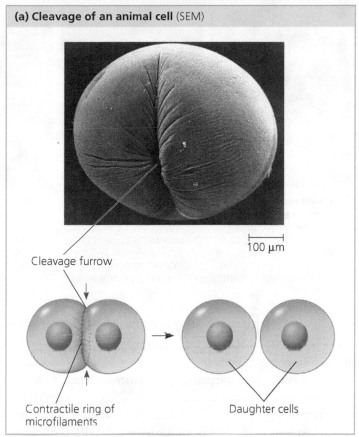

(b) Cell plate formation in a plant cell (TEM)

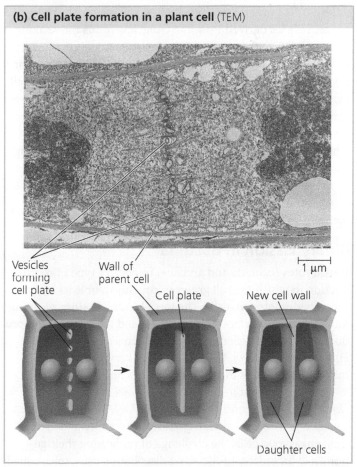

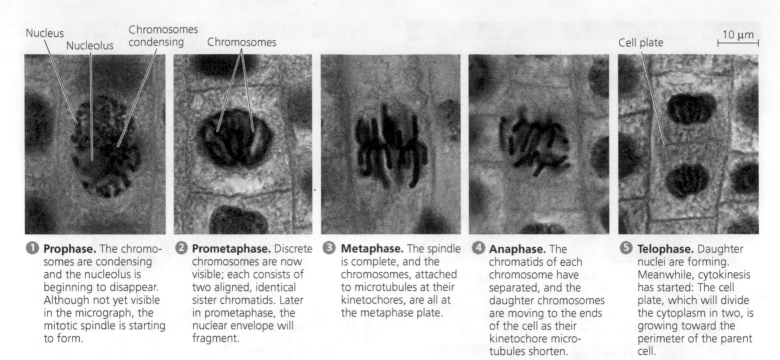

Nucleus · Nucleolus · Chromosomes condensing · Chromosomes · Cell plate · 10 μm

① Prophase. The chromosomes are condensing and the nucleolus is beginning to disappear. Although not yet visible in the micrograph, the mitotic spindle is starting to form.

② Prometaphase. Discrete chromosomes are now visible; each consists of two aligned, identical sister chromatids. Later in prometaphase, the nuclear envelope will fragment.

③ Metaphase. The spindle is complete, and the chromosomes, attached to microtubules at their kinetochores, are all at the metaphase plate.

④ Anaphase. The chromatids of each chromosome have separated, and the daughter chromosomes are moving to the ends of the cell as their kinetochore microtubules shorten.

⑤ Telophase. Daughter nuclei are forming. Meanwhile, cytokinesis has started: The cell plate, which will divide the cytoplasm in two, is growing toward the perimeter of the parent cell.

▲ **Figure 9.11 Mitosis in a plant cell.** These light micrographs show mitosis in cells of an onion root.

microfilaments is like the pulling of a drawstring. The cleavage furrow deepens until the parent cell is pinched in two, producing two completely separated cells, each with its own nucleus and share of cytosol, organelles, and other subcellular structures.

Cytokinesis in plant cells, which have cell walls, is markedly different. There is no cleavage furrow. Instead, during telophase, vesicles derived from the Golgi apparatus move along microtubules to the middle of the cell, where they coalesce, producing a **cell plate (Figure 9.10b)**. Cell wall materials carried in the vesicles collect in the cell plate as it grows. The cell plate enlarges until its surrounding membrane fuses with the plasma membrane along the perimeter of the cell. Two daughter cells result, each with its own plasma membrane. Meanwhile, a new cell wall arising from the contents of the cell plate has formed between the daughter cells.

Figure 9.11 is a series of micrographs of a dividing plant cell. Examining this figure will help you review mitosis and cytokinesis.

Binary Fission in Bacteria

Prokaryotes (bacteria and archaea) undergo a type of reproduction in which the cell grows to roughly double its size and then divides into two cells. The term **binary fission**, meaning "division in half," refers to this process and to the asexual reproduction of single-celled eukaryotes, such as the amoeba in Figure 9.2a. However, the process in eukaryotes involves mitosis; the process in prokaryotes does not.

In bacteria, most genes are carried on a single *bacterial chromosome* that consists of a circular DNA molecule and associated proteins. Although bacteria are smaller and simpler than eukaryotic cells, the challenge of replicating their genomes in an orderly fashion and distributing the copies equally

to two daughter cells is still formidable. The chromosome of the bacterium *Escherichia coli*, for example, when it is fully stretched out, is about 500 times as long as the cell. For such a long chromosome to fit within the cell requires that it be highly coiled and folded.

In *E. coli*, the process of cell division is initiated when the DNA of the bacterial chromosome begins to replicate at a specific place on the chromosome called the **origin of replication**, producing two origins. As the chromosome continues to replicate, one origin moves rapidly toward the opposite end of the cell **(Figure 9.12)**. While the chromosome is replicating, the cell elongates. When replication is complete and the bacterium has reached about twice its initial size, its plasma membrane pinches inward, dividing the parent *E. coli* cell into two daughter cells. In this way, each cell inherits a complete genome.

Using the techniques of modern DNA technology to tag the origins of replication with molecules that glow green in fluorescence microscopy (see Figure 4.3), researchers have directly observed the movement of bacterial chromosomes. This movement is reminiscent of the poleward movements of the centromere regions of eukaryotic chromosomes during anaphase of mitosis, but bacteria don't have visible mitotic spindles or even microtubules. In most bacterial species studied, the two origins of replication end up at opposite ends of the cell or in some other very specific location, possibly anchored there by one or more proteins. How bacterial chromosomes move and how their specific location is established and maintained are still not fully understood. However, several proteins have been identified that play important roles: One resembling eukaryotic actin apparently functions in bacterial chromosome movement during cell division, and another that is related to

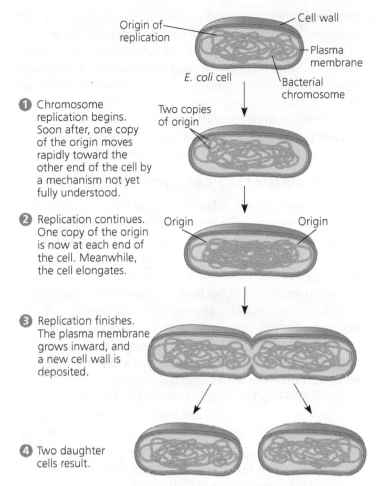

1 **Chromosome replication begins.** Soon after, one copy of the origin moves rapidly toward the other end of the cell by a mechanism not yet fully understood.

2 **Replication continues.** One copy of the origin is now at each end of the cell. Meanwhile, the cell elongates.

3 **Replication finishes.** The plasma membrane grows inward, and a new cell wall is deposited.

4 **Two daughter cells result.**

Origin of replication — Cell wall — Plasma membrane — *E. coli* cell — Bacterial chromosome

Two copies of origin

Origin — Origin

▲ **Figure 9.12 Bacterial cell division by binary fission.** The bacterium *E. coli*, shown here, has a single, circular chromosome.

tubulin seems to help pinch the plasma membrane inward, separating the two bacterial daughter cells.

The Evolution of Mitosis

EVOLUTION Given that prokaryotes preceded eukaryotes on Earth by more than a billion years, we might hypothesize that mitosis evolved from simpler prokaryotic mechanisms of cell reproduction. The fact that some of the proteins involved in bacterial binary fission are related to eukaryotic proteins that function in mitosis supports that hypothesis.

As eukaryotes with nuclear envelopes and larger genomes evolved, the ancestral process of binary fission, seen today in bacteria, somehow gave rise to mitosis. Possible intermediate stages are suggested by two unusual types of nuclear division found today in certain unicellular eukaryotes—dinoflagellates, diatoms, and some yeasts. **(Figure 9.13)**. These processes may be similar to mechanisms used by ancestral species and thus may resemble steps in the evolution of mitosis from a binary fission-like process presumably carried out by very early bacteria. The two modes of nuclear division shown in Figure 9.13 are thought to be cases where ancestral mechanisms have remained relatively unchanged over evolutionary time. In both types, the nuclear envelope remains intact, in contrast to what happens in most eukaryotic cells.

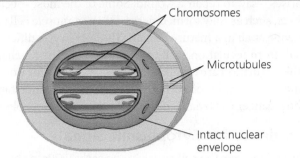

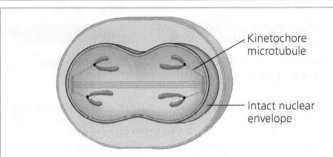

(a) Dinoflagellates. In unicellular eukaryotes called dinoflagellates, the chromosomes attach to the nuclear envelope, which remains intact during cell division. Microtubules pass through the nucleus inside cytoplasmic tunnels, reinforcing the spatial orientation of the nucleus, which then divides in a process reminiscent of bacterial binary fission.

(b) Diatoms and some yeasts. In two other groups of unicellular eukaryotes, diatoms and some yeasts, the nuclear envelope also remains intact during cell division. In these organisms, the microtubules form a spindle *within* the nucleus. Microtubules separate the chromosomes, and the nucleus splits into two daughter nuclei.

▲ **Figure 9.13 Mechanisms of cell division.** Some unicellular eukaryotes existing today have mechanisms of cell division that may resemble intermediate steps in the evolution of mitosis.

CONCEPT CHECK 9.2

1. How many chromosomes are shown in the diagram in Figure 9.8? Are they duplicated? How many chromatids are shown?
2. Compare cytokinesis in animal cells and plant cells.
3. What is the function of nonkinetochore microtubules?
4. During which stages of the cell cycle does a chromosome consist of two identical chromatids?

For suggested answers, see Appendix A.

CONCEPT 9.3

The eukaryotic cell cycle is regulated by a molecular control system

The timing and rate of cell division in different parts of a plant or animal are crucial to normal growth, development, and maintenance. The frequency of cell division varies with the type of cell. For example, human skin cells divide frequently throughout life, whereas liver cells maintain the ability to divide but keep it in reserve until an appropriate need arises—say, to repair a wound. Some of the most specialized cells, such as fully formed nerve cells and muscle cells, do not divide at all in a mature human. These cell cycle differences result from regulation at the molecular level. The mechanisms of this regulation are of intense interest, not only for understanding the life cycles of normal cells but also for understanding how cancer cells manage to escape the usual controls.

Evidence for Cytoplasmic Signals

What controls the cell cycle? In the early 1970s, a variety of experiments led to the hypothesis that the cell cycle is driven by specific signaling molecules present in the cytoplasm. Some of the first strong evidence for this hypothesis came from experiments with mammalian cells grown in culture (Figure 9.14). In

these experiments, two cells in different phases of the cell cycle were fused to form a single cell with two nuclei. If one of the original cells was in the S phase and the other was in G_1, the G_1 nucleus immediately entered the S phase, as though stimulated by signaling molecules present in the cytoplasm of the first cell. Similarly, if a cell undergoing mitosis (M phase) was fused with another cell in any stage of its cell cycle, even G_1, the second nucleus immediately entered mitosis, with condensation of the chromatin and formation of a mitotic spindle.

Checkpoints of the Cell Cycle Control System

The experiment shown in Figure 9.14 and other experiments on animal cells and yeasts demonstrated that the sequential events of the cell cycle are directed by a distinct **cell cycle control system**, a cyclically operating set of molecules in the cell that both triggers and coordinates key events in the cell cycle. The cell cycle control system has been compared to the control device of a washing machine (Figure 9.15). Like the washer's timing device, the cell cycle control system proceeds on its own, according to a built-in clock. However, just as a washer's cycle is subject to both internal control (such as the sensor that detects when the tub is filled with water) and external adjustment (such as starting the machine), the cell cycle is regulated at certain checkpoints by both internal and external signals.

A **checkpoint** in the cell cycle is a control point where stop and go-ahead signals can regulate the cycle. (The signals are transmitted within the cell by the kinds of signal transduction pathways discussed in Concept 5.6.) Animal cells generally have built-in stop signals that halt the cell cycle at checkpoints until overridden by go-ahead signals. Many signals registered at checkpoints come from cellular surveillance mechanisms inside the cell. These signals report whether crucial cellular

---

▼ Figure 9.14 Inquiry

Do molecular signals in the cytoplasm regulate the cell cycle?

Experiment Researchers at the University of Colorado wondered whether a cell's progression through the cell cycle is controlled by cytoplasmic molecules. To investigate this, they selected cultured mammalian cells that were at different phases of the cell cycle and induced them to fuse. Two such experiments are shown here.

Experiment 1   **Experiment 2**

| S | G_1 |

| M | G_1 |

| S | S |

| M | M |

When a cell in the S phase was fused with a cell in G_1, the G_1 nucleus immediately entered the S phase—DNA was synthesized.

When a cell in the M phase was fused with a cell in G_1, the G_1 nucleus immediately began mitosis—a spindle formed and the chromosomes condensed, even though the chromosomes had not been duplicated.

Conclusion The results of fusing a G_1 cell with a cell in the S or M phase of the cell cycle suggest that molecules present in the cytoplasm during the S or M phase control the progression to those phases.

Source R. T. Johnson and P. N. Rao, Mammalian cell fusion: Induction of premature chromosome condensation in interphase nuclei, *Nature* 226:717–722 (1970).

WHAT IF? If the progression of phases did not depend on cytoplasmic molecules and each phase began when the previous one was complete, how would the results have differed?

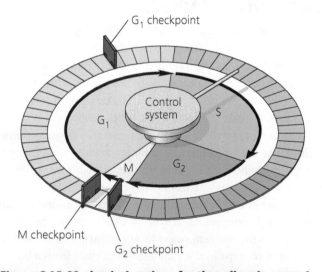

▲ **Figure 9.15 Mechanical analogy for the cell cycle control system.** In this diagram of the cell cycle, the flat "stepping stones" around the perimeter represent sequential events. Like the control device of an automatic washer, the cell cycle control system proceeds on its own, driven by a built-in clock. However, the system is subject to internal and external regulation at various checkpoints, of which three are shown (red).

processes that should have occurred by that point have in fact been completed correctly and thus whether or not the cell cycle should proceed. Checkpoints also register signals from outside the cell, as we'll discuss later. Three major checkpoints are found in the G_1, G_2, and M phases (see Figure 9.15).

For many cells, the G_1 checkpoint—dubbed the "restriction point" in mammalian cells—seems to be the most important. If a cell receives a go-ahead signal at the G_1 checkpoint, it will usually complete the G_1, S, G_2, and M phases and divide **(Figure 9.16a)**. If it does not receive a go-ahead signal at that point, it will exit the cycle, switching into a nondividing state called the **G_0 phase**. Most cells of the human body are actually in the G_0 phase. As mentioned earlier, mature nerve cells and muscle cells never divide. Other cells, such as liver cells, can be "called back" from the G_0 phase to the cell cycle by external cues, such as growth factors released during injury.

The cell cycle is regulated at the molecular level by a set of regulatory proteins and protein complexes, including kinases (enzymes that activate or inactivate other proteins by phosphorylating them; see Figure 5.24) and proteins called *cyclins*. To understand how a cell progresses through the cycle, let's consider the checkpoint signals that can make the cell cycle clock pause or continue.

Biologists are currently working out the pathways that link signals originating inside and outside the cell with the responses by kinases, cyclins, and other proteins. An example of an internal signal occurs at the third important checkpoint, the M phase checkpoint **(Figure 9.16b)**. Anaphase, the separation of sister chromatids, does not begin until all the chromosomes are properly attached to the spindle at the metaphase plate. Researchers have learned that as long as some kinetochores are unattached to spindle microtubules, the sister chromatids remain together, delaying anaphase. Only when the kinetochores of all the chromosomes are properly attached to the spindle does the appropriate regulatory protein complex become activated. Once activated, the complex sets off a chain of molecular events that activates the enzyme separase, which cleaves the cohesins, allowing the sister chromatids to separate. This mechanism ensures that daughter cells do not end up with missing or extra chromosomes.

Studies using animal cells in culture have led to the identification of many external factors, both chemical and physical, that can influence cell division. For example, cells fail to divide if an essential nutrient is lacking in the culture medium. (This is analogous to trying to run a washing machine without the water supply hooked up; an internal sensor won't allow the machine to continue past the point where water is needed.) And even if all other conditions are favorable, most types of mammalian cells divide in culture only if the growth medium includes specific growth factors. A **growth factor** is a protein released by certain cells that stimulates other cells to divide. Different cell types respond specifically to different growth factors or combinations of growth factors.

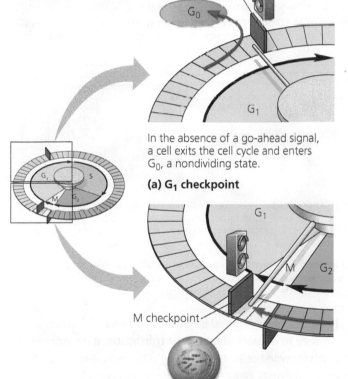

In the absence of a go-ahead signal, a cell exits the cell cycle and enters G_0, a nondividing state.

(a) G_1 checkpoint

M checkpoint

Prometaphase

A cell in mitosis receives a stop signal when any of its chromosomes are not attached to spindle fibers.

(b) M checkpoint

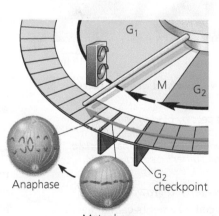

If a cell receives a go-ahead signal, the cell continues on in the cell cycle.

Anaphase

Metaphase

When all chromosomes are attached to spindle fibers from both poles, a go-ahead signal allows the cell to proceed into anaphase.

▲ **Figure 9.16 Two important checkpoints.** At certain points in the cell cycle, cells can do different things depending on the signals they receive. Events of the G_1 and M checkpoints are shown. In part (b), the G_2 checkpoint has already been passed by the cell.

WHAT IF? *In (a), what might be the result if the cell ignored the checkpoint and progressed through the cell cycle?*

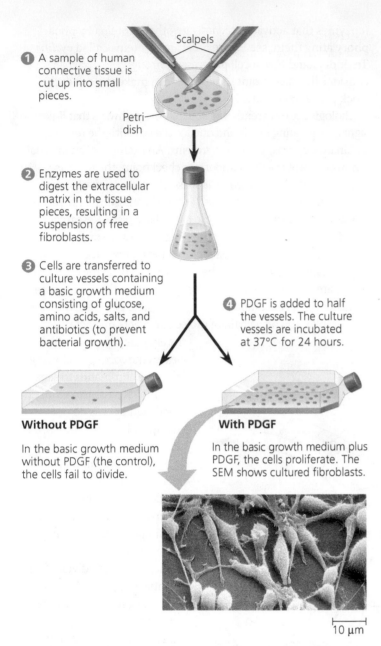

① A sample of human connective tissue is cut up into small pieces.

Scalpels

Petri dish

② Enzymes are used to digest the extracellular matrix in the tissue pieces, resulting in a suspension of free fibroblasts.

③ Cells are transferred to culture vessels containing a basic growth medium consisting of glucose, amino acids, salts, and antibiotics (to prevent bacterial growth).

④ PDGF is added to half the vessels. The culture vessels are incubated at 37°C for 24 hours.

Without PDGF

In the basic growth medium without PDGF (the control), the cells fail to divide.

With PDGF

In the basic growth medium plus PDGF, the cells proliferate. The SEM shows cultured fibroblasts.

10 μm

▲ **Figure 9.17 The effect of platelet-derived growth factor (PDGF) on cell division.**

MAKE CONNECTIONS *PDGF signals cells by binding to a cell-surface receptor that then becomes phosphorylated, activating it so that it transduces a signal. If you added a chemical that blocked phosphorylation, how would the results differ? (See Figure 5.24.)*

Consider, for example, *platelet-derived growth factor (PDGF)*, which is made by blood cell fragments called platelets. The experiment illustrated in **Figure 9.17** demonstrates that PDGF is required for the division of cultured fibroblasts, a type of connective tissue cell. Fibroblasts have PDGF receptors on their plasma membranes. The binding of PDGF molecules to these receptors triggers a signal transduction pathway that allows the cells to pass the G_1 checkpoint and divide. PDGF stimulates fibroblast division not only in the artificial conditions of cell culture, but also in an animal's body. When an injury occurs, platelets release PDGF in the vicinity. The resulting proliferation of fibroblasts helps heal the wound.

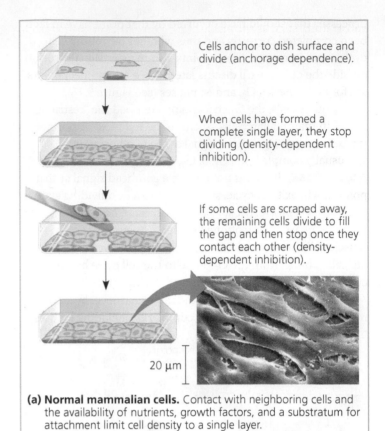

Cells anchor to dish surface and divide (anchorage dependence).

When cells have formed a complete single layer, they stop dividing (density-dependent inhibition).

If some cells are scraped away, the remaining cells divide to fill the gap and then stop once they contact each other (density-dependent inhibition).

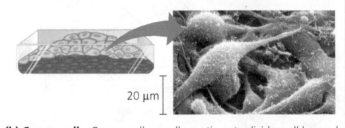

20 μm

(a) Normal mammalian cells. Contact with neighboring cells and the availability of nutrients, growth factors, and a substratum for attachment limit cell density to a single layer.

20 μm

(b) Cancer cells. Cancer cells usually continue to divide well beyond a single layer, forming a clump of overlapping cells. They do not exhibit anchorage dependence or density-dependent inhibition.

▲ **Figure 9.18 Density-dependent inhibition and anchorage dependence of cell division.** Individual cells are shown disproportionately large in the drawings.

The effect of an external physical factor on cell division is clearly seen in **density-dependent inhibition**, a phenomenon in which crowded cells stop dividing **(Figure 9.18a)**. As first observed many years ago, cultured cells normally divide until they form a single layer of cells on the inner surface of the culture container, at which point the cells stop dividing. If some cells are removed, those bordering the open space begin dividing again and continue until the vacancy is filled. Follow-up studies revealed that the binding of a cell-surface protein to its counterpart on an adjoining cell sends a cell division-inhibiting signal to both cells, preventing them from moving forward in the cell cycle. Growth factors also have a role in determining the density that cells attain before ceasing division.

Most animal cells also exhibit **anchorage dependence** (see Figure 9.18a). To divide, they must be attached to a substratum, such as the inside of a culture flask or the extracellular matrix of a tissue. Experiments suggest that like cell density, anchorage is signaled to the cell cycle control system via pathways involving plasma membrane proteins and elements of the cytoskeleton linked to them.

Density-dependent inhibition and anchorage dependence appear to function not only in cell culture but also in the body's tissues, checking the growth of cells at some optimal density and location during embryonic development and throughout an organism's life. Cancer cells, which we discuss next, exhibit neither density-dependent inhibition nor anchorage dependence **(Figure 9.18b)**.

Loss of Cell Cycle Controls in Cancer Cells

Cancer cells do not heed the normal signals that regulate the cell cycle. They divide excessively and invade other tissues. If unchecked, they can kill the organism.

Cancer cells in culture do not stop dividing when growth factors are depleted. A logical hypothesis is that cancer cells do not need growth factors in their culture medium to grow and divide. They may make a required growth factor themselves, or they may have an abnormality in the signaling pathway that conveys the growth factor's signal to the cell cycle control system even in the absence of that factor. Another possibility is an abnormal cell cycle control system. In these scenarios, the underlying basis of the abnormality is almost always a change in one or more genes that alters the function of their protein products, resulting in faulty cell cycle control. (You will learn more in Chapter 16 about the genetic bases of these changes and how these conditions may lead to cancer.)

There are other important differences between normal cells and cancer cells that reflect derangements of the cell cycle. If and when they stop dividing, cancer cells do so at random points in the cycle, rather than at the normal checkpoints. Moreover, cancer cells can go on dividing indefinitely in culture if they are given a continual supply of nutrients; in essence, they are "immortal." A striking example is a cell line that has been reproducing in culture since 1951. Cells of this line are called HeLa cells because their original source was a tumor removed from a woman named *He*nrietta *La*cks. By contrast, nearly all normal mammalian cells growing in culture divide only about 20 to 50 times before they stop dividing, age, and die. (We'll see a possible reason for this phenomenon when we discuss DNA replication in Chapter 13.) Finally, cancer cells evade the normal controls that trigger a cell to undergo a type of programmed cell death called *apoptosis* when something is wrong—for example, when an irreparable mistake has occurred during DNA replication preceding mitosis.

The abnormal behavior of cancer cells can be catastrophic when it occurs in the body. The problem begins when a single cell in a tissue undergoes **transformation**, the process that converts a normal cell to a cancer cell. The body's immune system normally recognizes a transformed cell as an insurgent and destroys it. However, if the cell evades destruction, it may proliferate and form a tumor, a mass of abnormal cells within otherwise normal tissue. The abnormal cells may remain at the original site if they have too few genetic and cellular changes to survive at another site. In that case, the tumor is called a **benign tumor**. Most benign tumors do not cause serious problems and can be completely removed by surgery. In contrast, a **malignant tumor** has cells whose genetic and cellular changes enable them to spread to new tissues and impair the functions of one or more organs. An individual with a malignant tumor is said to have cancer; **Figure 9.19** shows the development of breast cancer.

The changes that have occurred in cells of malignant tumors show up in many ways besides excessive proliferation. These cells may have unusual numbers of chromosomes, though whether this is a cause or an effect of transformation is an ongoing topic of debate. Their metabolism may be disabled, and they may cease to function in any constructive way. Abnormal changes on the cell surface cause cancer cells to lose attachments to neighboring cells and the extracellular matrix, allowing

▼ **Figure 9.19 The growth and metastasis of a malignant breast tumor.** The cells of malignant (cancerous) tumors grow in an uncontrolled way and can spread to neighboring tissues and, via lymph and blood vessels, to other parts of the body (metastasis).

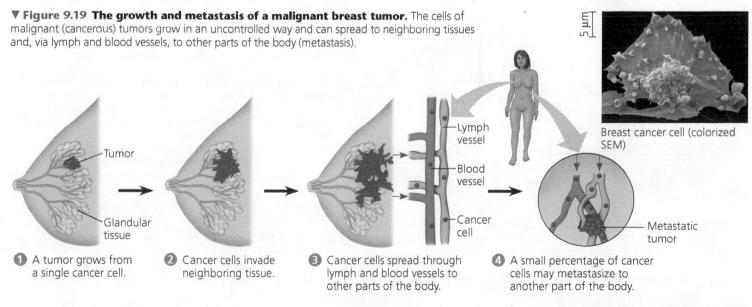

5 μm

Breast cancer cell (colorized SEM)

❶ A tumor grows from a single cancer cell.

❷ Cancer cells invade neighboring tissue.

❸ Cancer cells spread through lymph and blood vessels to other parts of the body.

❹ A small percentage of cancer cells may metastasize to another part of the body.

Lymph vessel

Blood vessel

Cancer cell

Tumor

Glandular tissue

Metastatic tumor

them to spread into nearby tissues. Cancer cells may also secrete signaling molecules that cause blood vessels to grow toward the tumor. A few tumor cells may separate from the original tumor, enter blood vessels and lymph vessels, and travel to other parts of the body. There, they may proliferate and form a new tumor. This spread of cancer cells to locations distant from their original site is called **metastasis** (see Figure 9.19).

A tumor that appears to be localized may be treated with high-energy radiation, which damages DNA in cancer cells much more than it does in normal cells, apparently because the majority of cancer cells have lost the ability to repair such damage. To treat known or suspected metastatic tumors, chemotherapy is used, in which drugs that are toxic to actively dividing cells are administered through the circulatory system. As you might expect, chemotherapeutic drugs interfere with specific steps in the cell cycle. For example, the drug Taxol freezes the

mitotic spindle by preventing microtubule depolymerization; this stops actively dividing cells from proceeding past metaphase and leads to their destruction. In the **Scientific Skills Exercise**, you'll work with data from an experiment involving a potential chemotherapeutic agent. The side effects of chemotherapy are due to the drugs' effects on normal cells that divide often. For example, nausea results from chemotherapy's effects on intestinal cells, hair loss from effects on hair follicle cells, and susceptibility to infection from effects on immune system cells.

Over the past several decades, researchers have produced a flood of valuable information about cell-signaling pathways and how their malfunction contributes to the development of cancer through effects on the cell cycle. Coupled with new molecular techniques, such as the ability to rapidly sequence the DNA of cells in a particular tumor, medical treatments for cancer are beginning to become more "personalized" to a

Scientific Skills Exercise

Interpreting Histograms

At What Phase Is the Cell Cycle Arrested by an Inhibitor? Many medical treatments are aimed at stopping cancer cell proliferation by blocking the cell cycle of cancerous tumor cells. One potential treatment is a cell cycle inhibitor derived from human umbilical cord stem cells. In this exercise, you will compare two histograms to determine where in the cell cycle the inhibitor blocks the division of cancer cells.

How the Experiment Was Done In the treated sample, human glioblastoma (brain cancer) cells were grown in tissue culture in the presence of the inhibitor, while control sample cells were grown in its absence. After 72 hours of growth, the two cell samples were harvested. To get a "snapshot" of the phase of the cell cycle each cell was in at that time, the samples were treated with a fluorescent chemical that binds to DNA and then run through a flow cytometer, an instrument that records the fluorescence level of each cell. Computer software then graphed the number of cells in each sample with a particular fluorescence level, as shown below.

Data from the Experiment

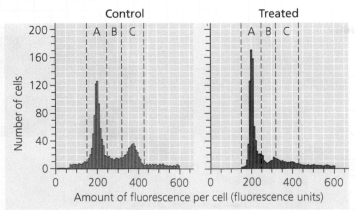

The data are plotted in a type of graph called a histogram (above), which groups values for a numeric variable on the x-axis into intervals. A histogram allows you to see how an entire group of experimental

subjects (cells, in this case) are distributed along a continuous variable (amount of fluorescence). In these histograms, the bars are so narrow that the data appear to follow a curve for which you can detect peaks and dips. Each narrow bar represents the number of cells observed to have a level of fluorescence in the range of that interval. This in turn indicates the relative amount of DNA in those cells. Overall, comparing histograms allows you to see how the DNA content of this cell population is altered by the treatment.

Interpret the Data

1. Familiarize yourself with the data shown in the histograms. (a) Which axis indirectly shows the relative amount of DNA per cell? Explain your answer. (b) In the control sample, compare the first peak in the histogram (in region A) to the second peak (in region C). Which peak shows the population of cells with the higher amount of DNA per cell? Explain. (For additional information about graphs, see the Scientific Skills Review in Appendix F and in the Study Area in MasteringBiology.)

2. (a) In the control sample histogram, identify the phase of the cell cycle (G_1, S, or G_2) of the population of cells in each region delineated by vertical lines. Label the histogram with these phases and explain your answer. (b) Does the S phase population of cells show a distinct peak in the histogram? Why or why not?

3. The histogram representing the treated sample shows the effect of growing the cancer cells alongside human umbilical cord stem cells. (a) Label the histogram with the cell cycle phases. Which phase of the cell cycle has the greatest number of cells in the treated sample? Explain. (b) Compare the distribution of cells among G_1, S, and G_2 phases in the control and treated samples. What does this tell you about the cells in the treated sample? (c) Based on what you learned in Concept 9.3, propose a mechanism by which the stem cell–derived inhibitor might arrest the cancer cell cycle at this stage. (More than one answer is possible.)

Data from K. K. Velpula et al., Regulation of glioblastoma progression by cord blood stem cells is mediated by downregulation of cyclin D1, *PLoS ONE* 6(3): e18017 (2011). doi:10.1371/journal.pone.0018017

(MB) A version of this Scientific Skills Exercise can be assigned in MasteringBiology.

particular patient's tumor. Breast cancer provides a good example. Basic research on cell signaling and the cell cycle has augmented our understanding of the molecular events underlying the development of breast cancer. Proteins functioning in cell-signaling pathways that affect the cell cycle are often found to be altered in breast cancer cells. Analyzing the level and sequences of such proteins has allowed physicians to better tailor the treatment to the cancers of some individuals.

One of the big lessons we've learned about the development of cancer, though, is how very complex the process is. There are many areas that remain to be explored. Perhaps the reason we have so many unanswered questions about cancer cells is that there is still so much to learn about how normal cells function. The cell, life's basic unit of structure and function, holds enough secrets to engage researchers well into the future.

CONCEPT CHECK 9.3

1. In Figure 9.14, why do the nuclei resulting from experiment 2 contain different amounts of DNA?
2. What phase are most of your body cells in?
3. Compare and contrast a benign tumor and a malignant tumor.
4. **WHAT IF?** What would happen if you performed the experiment in Figure 9.17 with cancer cells?

For suggested answers, see Appendix A.

9 Chapter Review

SUMMARY OF KEY CONCEPTS

- Unicellular organisms reproduce by **cell division**; multicellular organisms depend on cell division for their development from a fertilized egg and for growth and repair. Cell division is part of the **cell cycle**, an ordered sequence of events in the life of a cell from its origin until it divides into daughter cells.

CONCEPT 9.1

Most cell division results in genetically identical daughter cells (pp. 175–176)

- The genetic material (DNA) of a cell—its **genome**—is partitioned among **chromosomes**. Each eukaryotic chromosome consists of one DNA molecule associated with many proteins that maintain chromosome structure and help control the activity of genes. Together, the complex of DNA and associated proteins is called **chromatin**. The chromatin of a chromosome exists in different states of condensation at different times. In animals, gametes have one set of chromosomes and **somatic cells** have two sets.
- Cells replicate their genetic material before they divide, ensuring that each daughter cell can receive a copy of the DNA. In preparation for cell division, chromosomes are duplicated, each one then consisting of two identical **sister chromatids** joined along their lengths by sister chromatid cohesion and held most tightly together at a constricted region at the **centromeres** of the chromatids. When this cohesion is broken, the chromatids separate during cell division, becoming the chromosomes of the new daughter cells. Eukaryotic cell division consists of **mitosis** (division of the nucleus) and **cytokinesis** (division of the cytoplasm).

? *Differentiate between these terms: chromosome, chromatin, and chromatid.*

CONCEPT 9.2

The mitotic phase alternates with interphase in the cell cycle (pp. 177–183)

- Between divisions, a cell is in **interphase**: the G_1, **S**, and G_2 phases. The cell grows throughout interphase, but DNA is replicated only during the synthesis (S) phase. Mitosis and cytokinesis make up the **mitotic (M) phase** of the cell cycle.

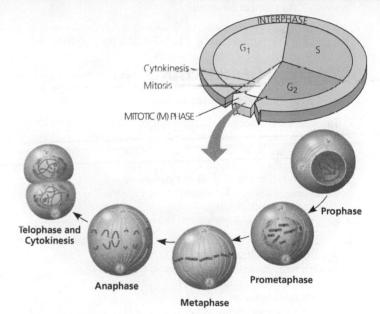

- The **mitotic spindle** is an apparatus of microtubules that controls chromosome movement during mitosis. In animal cells, the spindle arises from the **centrosomes** and includes spindle microtubules and **asters**. Some spindle microtubules attach to the **kinetochores** of chromosomes and move the chromosomes to the **metaphase plate**. In anaphase, sister chromatids separate, and motor proteins move them along the kinetochore microtubules toward opposite ends of the cell. Meanwhile, motor proteins push nonkinetochore microtubules from opposite poles away from each other, elongating the cell. In telophase, genetically identical daughter nuclei form at opposite ends of the cell.
- Mitosis is usually followed by cytokinesis. Animal cells carry out cytokinesis by **cleavage**, and plant cells form a **cell plate**.
- During **binary fission** in bacteria, the chromosome replicates and the two daughter chromosomes actively move apart. Some of the proteins involved in bacterial binary fission are related to eukaryotic actin and tubulin. Since prokaryotes preceded eukaryotes by more than a billion years, it is likely that mitosis evolved from prokaryotic cell division.

? *In which of the three subphases of interphase and the stages of mitosis do chromosomes exist as single DNA molecules?*

The eukaryotic cell cycle is regulated by a molecular control system (pp. 183–189)

- Signaling molecules present in the cytoplasm regulate progress through the cell cycle.
- The **cell cycle control system** is molecularly based; key regulatory proteins are kinases and cyclins. The cell cycle clock has specific **checkpoints** where the cell cycle stops until a go-ahead signal is received. Cell culture has enabled researchers to study the molecular details of cell division. Both internal signals and external signals control the cell cycle checkpoints via signal transduction pathways. Most cells exhibit **density-dependent inhibition** of cell division as well as **anchorage dependence**.
- Cancer cells elude normal cell cycle regulation and divide out of control, forming tumors. **Malignant tumors** invade surrounding tissues and can undergo **metastasis**, exporting cancer cells to other parts of the body, where they may form secondary tumors. Recent advances in understanding the cell cycle and cell signaling, as well as techniques for sequencing DNA, have allowed improvements in cancer treatment.

? *Explain the significance of the G_1 and M checkpoints and the go-ahead signals involved in the cell cycle control system.*

TEST YOUR UNDERSTANDING

Level 1: Knowledge/Comprehension

1. Through a microscope, you can see a cell plate beginning to develop across the middle of a cell and nuclei forming on either side of the cell plate. This cell is most likely
   a. an animal cell in the process of cytokinesis.
   b. a plant cell in the process of cytokinesis.
   c. an animal cell in the S phase of the cell cycle.
   d. a bacterial cell dividing.
   e. a plant cell in metaphase.

2. In the cells of some organisms, mitosis occurs without cytokinesis. This will result in
   a. cells with more than one nucleus.
   b. cells that are unusually small.
   c. cells lacking nuclei.
   d. destruction of chromosomes.
   e. cell cycles lacking an S phase.

3. Which of the following does *not* occur during mitosis?
   a. condensation of the chromosomes
   b. replication of the DNA
   c. separation of sister chromatids
   d. spindle formation
   e. separation of the spindle poles

Level 2: Application/Analysis

4. A particular cell has half as much DNA as some other cells in a mitotically active tissue. The cell in question is most likely in
   a. G_1.   **d.** metaphase.
   b. G_2.   **e.** anaphase.
   c. prophase.

5. The drug cytochalasin B blocks the function of actin. Which of the following aspects of the animal cell cycle would be most disrupted by cytochalasin B?
   a. spindle formation
   b. spindle attachment to kinetochores
   c. DNA synthesis
   d. cell elongation during anaphase
   e. cleavage furrow formation and cytokinesis

6. In the light micrograph below of dividing cells near the tip of an onion root, identify a cell in each of the following stages: prophase, prometaphase, metaphase, anaphase, and telophase. Describe the major events occurring at each stage.

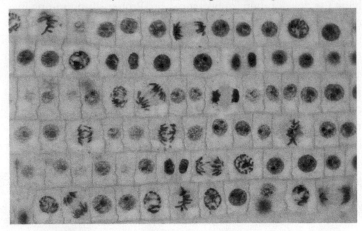

7. **DRAW IT** Draw one eukaryotic chromosome as it would appear during interphase, during each of the stages of mitosis, and during cytokinesis. Also draw and label the nuclear envelope and any microtubules attached to the chromosome(s).

Level 3: Synthesis/Evaluation

8. **SCIENTIFIC INQUIRY**
   Although both ends of a microtubule can gain or lose subunits, one end (called the plus end) polymerizes and depolymerizes at a higher rate than the other end (the minus end). For spindle microtubules, the plus ends are in the center of the spindle, and the minus ends are at the poles. Motor proteins that move along microtubules specialize in walking either toward the plus end or toward the minus end; the two types are called plus end–directed and minus end–directed motor proteins, respectively. Given what you know about chromosome movement and spindle changes during anaphase, predict which type of motor proteins would be present on (a) kinetochore microtubules and (b) nonkinetochore microtubules.

9. **FOCUS ON EVOLUTION**
   The result of mitosis is that the daughter cells end up with the same number of chromosomes that the parent cell had. Another way to maintain the number of chromosomes would be to carry out cell division first and then duplicate the chromosomes in each daughter cell. Do you think this would be an equally good way of organizing the cell cycle? Why do you suppose that evolution has not led to this alternative?

10. **FOCUS ON INFORMATION**
    The continuity of life is based on heritable information in the form of DNA. In a short essay (100–150 words), explain how the process of mitosis faithfully parcels out exact copies of this heritable information in the production of genetically identical daughter cells.

For selected answers, see Appendix A.

MasteringBiology®

Students Go to **MasteringBiology** for assignments, the eText, and the Study Area with practice tests, animations, and activities.

Instructors Go to **MasteringBiology** for automatically graded tutorials and questions that you can assign to your students, plus Instructor Resources.

Unit 2 Genetics

10 Meiosis and Sexual Life Cycles

Sexually reproducing species alternate fertilization with **meiosis**, accurately passing on genetic information while generating genetic diversity.

11 Mendel and the Gene Idea

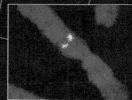

Although unaware of meiosis, Mendel did experiments that enabled him to describe the behavior of **genes**.

12 The Chromosomal Basis of Inheritance

Genes are located on **chromosomes**, and chromosomal behavior underlies genetic inheritance.

13 The Molecular Basis of Inheritance

The nucleotide sequence of the DNA in chromosomes provides the **molecular basis for inheritance**.

14 Gene Expression: From Gene to Protein

An organism's characteristics emerge from **gene expression**, the process in which information in **genes** is transcribed into **RNAs** that can be translated into **proteins**.

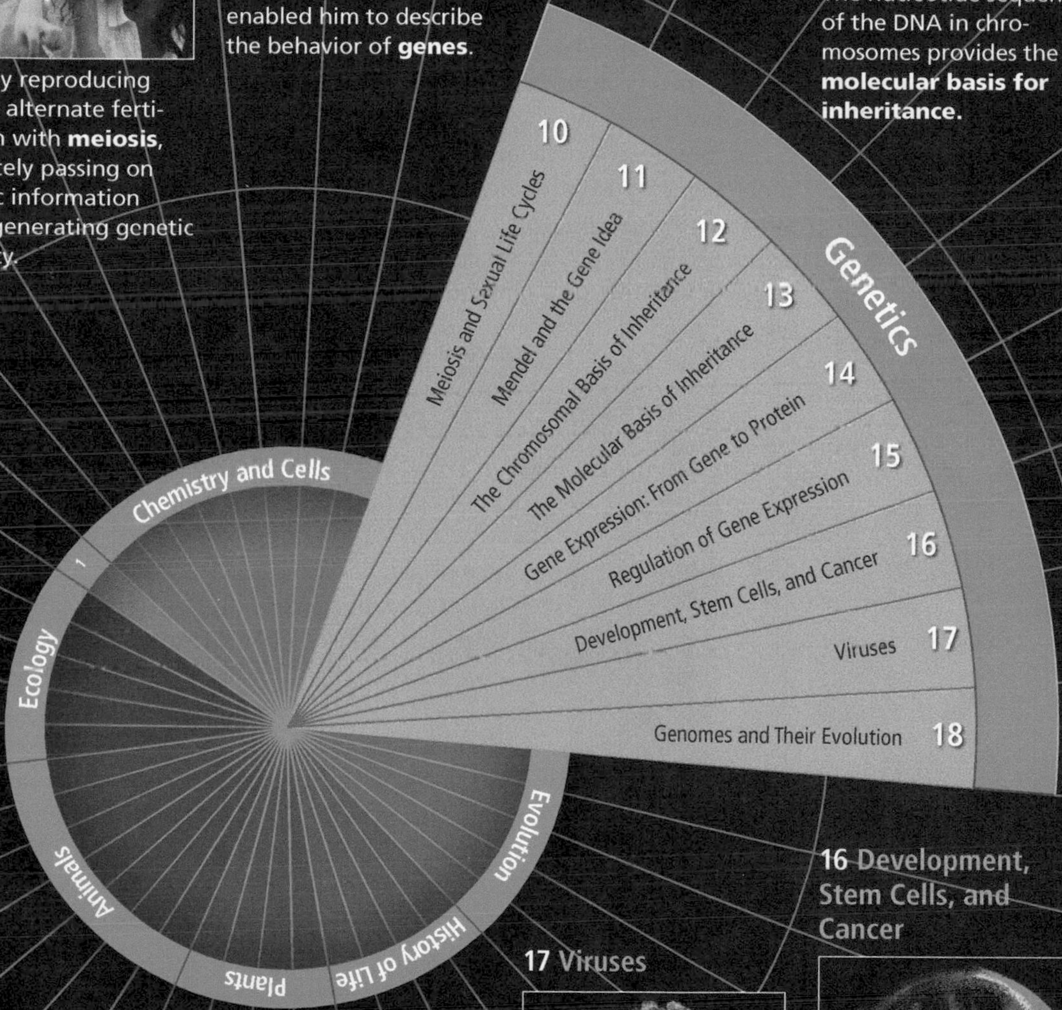

15 Regulation of Gene Expression

An organism's different cell types and responses to its environment depend on **regulation of gene expression**.

16 Development, Stem Cells, and Cancer

Coordinated gene regulation underlies **embryonic development**, while misregulation can contribute to **cancer**.

17 Viruses

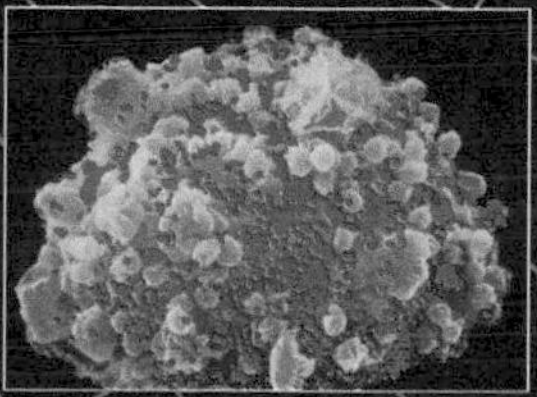

Our understanding of gene expression is informed by studying **viruses**, protein-coated packets of genetic information that hijack cellular resources and replicate themselves.

18 Genomes and Their Evolution

The **evolution of genomes** is the basis of life's diversity.

Chemistry and Cells

Ecology

Animals

Plants

History of Life

Evolution

Genetics

10 Meiosis and Sexual Life Cycles
11 Mendel and the Gene Idea
12 The Chromosomal Basis of Inheritance
13 The Molecular Basis of Inheritance
14 Gene Expression: From Gene to Protein
15 Regulation of Gene Expression
16 Development, Stem Cells, and Cancer
17 Viruses
18 Genomes and Their Evolution

10

Meiosis and Sexual Life Cycles

KEY CONCEPTS

10.1 Offspring acquire genes from parents by inheriting chromosomes

10.2 Fertilization and meiosis alternate in sexual life cycles

10.3 Meiosis reduces the number of chromosome sets from diploid to haploid

10.4 Genetic variation produced in sexual life cycles contributes to evolution

OVERVIEW

Variations on a Theme

Most people who send out birth announcements mention the sex of the baby, but they don't feel the need to specify that their offspring is a human being! One of the characteristics of life is the ability of organisms to reproduce their own kind—elephants produce little elephants, and oak trees generate oak saplings. Exceptions to this rule show up only as sensational but highly suspect stories in tabloid newspapers.

Another rule often taken for granted is that offspring resemble their parents more than they do unrelated individuals. If you examine the family members shown in **Figure 10.1**, you can pick out some similar features among them. The transmission of traits from one generation to the next is called inheritance, or **heredity** (from the Latin *heres*, heir). However, sons and daughters are not identical copies of either parent or of their siblings. Along with inherited similarity, there is also **variation**. Farmers have exploited the principles of heredity and variation for thousands of years, breeding plants and animals for desired traits. But what are the biological mechanisms leading to the hereditary similarity and variation that we call a "family resemblance"? The answer to this question eluded biologists until the advance of genetics in the 20th century.

Genetics is the scientific study of heredity and hereditary variation. In this unit, you'll learn about genetics at multiple levels, from organisms to cells to molecules. On the practical side, you'll see how genetics continues to revolutionize medicine, and you'll be asked to consider some social and ethical questions raised by our ability to manipulate DNA, the genetic material. At the end of the unit, you'll be able to stand back and consider the whole genome, an organism's entire complement of DNA. Rapid acquisition and analysis of the genome sequences of many species, including our own, have taught us a great deal about evolution on the molecular level—in other words, evolution of the genome itself. In fact, genetic methods and discoveries are catalyzing progress in all areas of biology, from cell biology to physiology, developmental biology, behavior, and even ecology.

We begin our study of genetics in this chapter by examining how chromosomes pass from parents to offspring in sexually reproducing organisms. The processes of meiosis (a special type of cell division) and fertilization (the fusion of sperm and egg) maintain a species' chromosome count during the sexual life cycle. We'll describe the cellular mechanics of meiosis and explain how this process differs from mitosis. Finally, we'll consider how both meiosis and fertilization contribute to genetic variation, such as the variation obvious in the family shown in Figure 10.1.

▼ **Figure 10.1** What accounts for family resemblance?

Offspring acquire genes from parents by inheriting chromosomes

Family friends may tell you that you have your mother's freckles or your father's eyes. Of course, parents do not, in any literal sense, give their children freckles, eyes, hair, or any other traits. What, then, *is* actually inherited?

Inheritance of Genes

Parents endow their offspring with coded information in the form of hereditary units called **genes**. The genes we inherit from our mothers and fathers are our genetic link to our parents, and they account for family resemblances such as shared eye color or freckles. Our genes program the specific traits that emerge as we develop from fertilized eggs into adults.

The genetic program is written in the language of DNA, the polymer of four different nucleotides (see Chapter 3). Inherited information is passed on in the form of each gene's specific sequence of DNA nucleotides, much as printed information is communicated in the form of meaningful sequences of letters. In both cases, the language is symbolic. Just as your brain translates the word *apple* into a mental image of the fruit, cells translate genes into freckles and other features. Most genes program cells to synthesize specific enzymes and other proteins, whose cumulative action produces an organism's inherited traits. The programming of these traits in the form of DNA is one of the unifying themes of biology.

The transmission of hereditary traits has its molecular basis in the precise replication of DNA, which produces copies of genes that can be passed from parents to offspring. In animals and plants, reproductive cells called **gametes** are the vehicles that transmit genes from one generation to the next. During fertilization, male and female gametes (sperm and eggs) unite, thereby passing on genes of both parents to their offspring.

Except for small amounts of DNA in mitochondria and chloroplasts, the DNA of a eukaryotic cell is packaged into chromosomes within the nucleus. Every species has a characteristic number of chromosomes. For example, humans have 46 chromosomes in their **somatic cells**—all the cells of the body except the gametes and their precursors. Each chromosome consists of a single long DNA molecule elaborately coiled in association with various proteins. One chromosome includes several hundred to a few thousand genes, each of which is a specific sequence of nucleotides within the DNA molecule. A gene's specific location along the length of a chromosome is called the gene's **locus** (plural, *loci*; from the Latin, meaning "place"). Our genetic endowment consists of the genes that are part of the chromosomes we inherited from our parents.

Comparison of Asexual and Sexual Reproduction

Only organisms that reproduce asexually have offspring that are exact genetic copies of themselves. In **asexual reproduction**, a single individual is the sole parent and passes copies of all its genes to its offspring without the fusion of gametes. For example, single-celled eukaryotic organisms can reproduce asexually by mitotic cell division, in which DNA is copied and allocated equally to two daughter cells. The genomes of the offspring are virtually exact copies of the parent's genome. Some multicellular organisms are also capable of reproducing asexually **(Figure 10.2)**. Because the cells of the offspring are derived by mitosis in the parent, the "chip off the old block" is usually genetically identical to its parent. An individual that reproduces asexually gives rise to a **clone**, a group of genetically identical individuals. Genetic differences occasionally arise in asexually reproducing organisms as a result of changes in the DNA called mutations, which we will discuss in Chapter 14.

In **sexual reproduction**, two parents give rise to offspring that have unique combinations of genes inherited from the two parents. In contrast to a clone, offspring of sexual reproduction vary genetically from their siblings and both parents: They are variations on a common theme of family resemblance, not exact replicas. Genetic variation like that shown in Figure 10.1 is an important consequence of sexual reproduction. What mechanisms generate this genetic variation? The key is the behavior of chromosomes during the sexual life cycle.

0.5 mm

Parent

Bud

(a) Hydra

(b) Redwoods

▲ **Figure 10.2 Asexual reproduction in two multicellular organisms. (a)** This relatively simple animal, a hydra, reproduces by budding. The bud, a localized mass of mitotically dividing cells, develops into a small hydra, which detaches from the parent (LM). **(b)** All the trees in this circle of redwoods arose asexually from a single parent tree, whose stump is in the center of the circle.

1. Explain what causes the traits of parents (such as hair color) to show up in their offspring.
2. How do asexually reproducing organisms produce offspring that are genetically identical to each other and to their parent?
3. **WHAT IF?** A horticulturalist breeds orchids, trying to obtain a plant with a unique combination of desirable traits. After many years, she finally succeeds. To produce more plants like this one, should she cross-breed it with another plant or clone it? Why?

For suggested answers, see Appendix A.

CONCEPT 10.2

Fertilization and meiosis alternate in sexual life cycles

A **life cycle** is the generation-to-generation sequence of stages in the reproductive history of an organism, from conception to production of its own offspring. In this section, we use humans as an example to track the behavior of chromosomes through the sexual life cycle. We begin by considering the chromosome count in human somatic cells and gametes. We will then explore how the behavior of chromosomes relates to the human life cycle and other types of sexual life cycles.

Sets of Chromosomes in Human Cells

In humans, each somatic cell has 46 chromosomes, usually found in a diffused state throughout the nucleus. During mitosis, however, the chromosomes become condensed enough to be distinguished microscopically from each other. They differ in size, centromere position, and the pattern of bands produced by certain chromatin-binding stains.

Careful examination of a micrograph of the 46 human chromosomes from a single cell in mitosis reveals that there are two chromosomes of each of 23 types. This becomes clear when images of the chromosomes are arranged in pairs, starting with the longest chromosomes. The resulting ordered display is called a **karyotype (Figure 10.3)**. The two chromosomes of a pair have the same length, centromere position, and staining pattern: These are called **homologous chromosomes**, or homologs. Both chromosomes of each pair carry genes controlling the same inherited characters. For example, if a gene for eye color is situated at a particular locus on a certain chromosome, then its homolog will also have a version of the same gene specifying eye color at the equivalent locus.

The two chromosomes referred to as X and Y are an important exception to the general pattern of homologous chromosomes in human somatic cells. Human females have a homologous pair of X chromosomes (XX), but males have one X and one Y chromosome (XY). Only small parts of the X and Y are homologous. Most of the genes carried on the X chromosome do not have counterparts on the tiny Y, and the Y chromosome has genes not present on the X. Because they determine an

▼ **Figure 10.3** **Research Method**

Preparing a Karyotype

Application A karyotype is a display of condensed chromosomes arranged in pairs. Karyotyping can be used to screen for defective chromosomes or abnormal numbers of chromosomes associated with certain congenital disorders, such as Down syndrome.

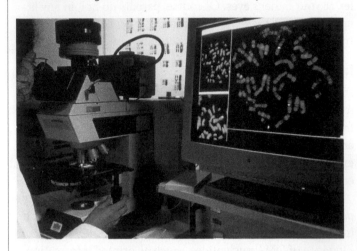

Technique Karyotypes are prepared from isolated somatic cells, which are treated with a drug to stimulate mitosis and then grown in culture for several days. Cells arrested in metaphase, when chromosomes are most highly condensed, are stained and then viewed with a microscope equipped with a digital camera. A photograph of the chromosomes is displayed on a computer monitor, and the images of the chromosomes are arranged into pairs according to their appearance.

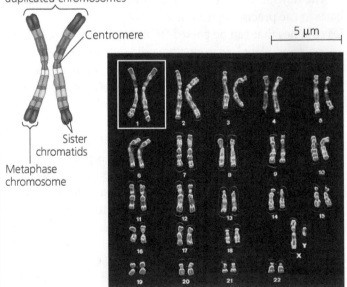

Results This karyotype shows the chromosomes from a normal human male. The size of the chromosome, position of the centromere, and pattern of stained bands help identify specific chromosomes. Although difficult to discern in the karyotype, each metaphase chromosome consists of two closely attached sister chromatids (see the diagram of a pair of homologous duplicated chromosomes).

individual's sex, the X and Y chromosomes are called **sex chromosomes**. The other chromosomes are called **autosomes**.

The occurrence of pairs of homologous chromosomes in each human somatic cell is a consequence of our sexual origins. We inherit one chromosome of each pair from each parent. Thus, the 46 chromosomes in our somatic cells are actually two sets of 23 chromosomes—a maternal set (from our mother) and a paternal set (from our father). The number of chromosomes in a single set is represented by n. Any cell with two chromosome sets is called a **diploid cell** and has a diploid number of chromosomes, abbreviated $2n$. For humans, the diploid number is 46 ($2n = 46$), the number of chromosomes in our somatic cells. In a cell in which DNA synthesis has occurred, all the chromosomes are duplicated, and therefore each consists of two identical sister chromatids, associated closely at the centromere and along the arms. **Figure 10.4** helps clarify the various terms that we use to describe duplicated chromosomes in a diploid cell. Study this figure so that you understand the differences between homologous chromosomes, sister chromatids, nonsister chromatids, and chromosome sets.

Unlike somatic cells, gametes contain a single set of chromosomes. Such cells are called **haploid cells**, and each has a haploid number of chromosomes (n). For humans, the haploid number is 23 ($n = 23$). The set of 23 consists of the 22 autosomes plus a single sex chromosome. An unfertilized egg contains an X chromosome, but a sperm may contain an X or a Y chromosome.

Note that each sexually reproducing species has a characteristic diploid number and haploid number. For example, the fruit fly, *Drosophila melanogaster*, has a diploid number ($2n$)

of 8 and a haploid number (n) of 4, while dogs have a diploid number of 78 and a haploid number of 39.

Now that you have learned the concepts of diploid and haploid numbers of chromosomes, let's consider chromosome behavior during sexual life cycles. We'll use the human life cycle as an example.

Behavior of Chromosome Sets in the Human Life Cycle

The human life cycle begins when a haploid sperm from the father fuses with a haploid egg from the mother. This union of gametes, culminating in fusion of their nuclei, is called **fertilization**. The resulting fertilized egg, or **zygote**, is diploid because it contains two haploid sets of chromosomes bearing genes representing the maternal and paternal family lines. As a human develops into a sexually mature adult, mitosis of the zygote and its descendant cells generates all the somatic cells of the body. Both chromosome sets in the zygote and all the genes they carry are passed with precision to the somatic cells.

The only cells of the human body not produced by mitosis are the gametes, which develop from specialized cells called *germ cells* in the gonads—ovaries in females and testes in males **(Figure 10.5)**. Imagine what would happen if human gametes

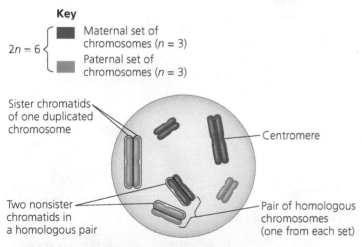

Key

$2n = 6$ {
■ Maternal set of chromosomes ($n = 3$)
■ Paternal set of chromosomes ($n = 3$)
}

Sister chromatids of one duplicated chromosome

Centromere

Two nonsister chromatids in a homologous pair

Pair of homologous chromosomes (one from each set)

▲ **Figure 10.4 Describing chromosomes.** A cell from an organism with a diploid number of 6 ($2n = 6$) is depicted here following chromosome duplication and condensation. Each of the six duplicated chromosomes consists of two sister chromatids associated closely along their lengths. Each homologous pair is composed of one chromosome from the maternal set (red) and one from the paternal set (blue). Each set is made up of three chromosomes in this example. Nonsister chromatids are any two chromatids in a pair of homologous chromosomes that are not sister chromatids—in other words, one maternal and one paternal chromatid.

? *What is the haploid number of this cell? Is a "set" of chromosomes haploid or diploid? How many sets are present in this cell? In the karyotype in Figure 10.3?*

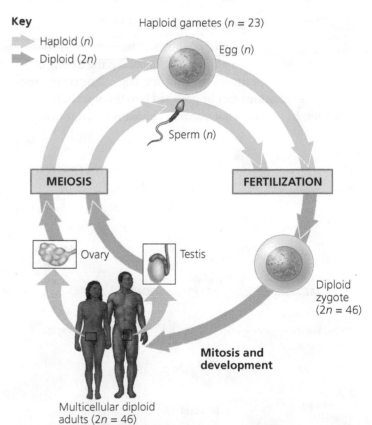

Key
→ Haploid (n)
→ Diploid ($2n$)

Haploid gametes ($n = 23$)

Egg (n)

Sperm (n)

MEIOSIS

FERTILIZATION

Ovary

Testis

Diploid zygote ($2n = 46$)

Mitosis and development

Multicellular diploid adults ($2n = 46$)

▲ **Figure 10.5 The human life cycle.** In each generation, the number of chromosome sets doubles at fertilization but is halved during meiosis. For humans, the number of chromosomes in a haploid cell is 23, consisting of one set ($n = 23$); the number of chromosomes in the diploid zygote and all somatic cells arising from it is 46, consisting of two sets ($2n = 46$).

This figure introduces a color code that will be used for other life cycles later in this book. The aqua arrows identify haploid stages of a life cycle, and the tan arrows identify diploid stages.

were made by mitosis: They would be diploid like the somatic cells. At the next round of fertilization, when two gametes fused, the normal chromosome number of 46 would double to 92, and each subsequent generation would double the number of chromosomes yet again. This does not happen, however, because in sexually reproducing organisms, gamete formation involves a sort of cell division called **meiosis**. This type of cell division reduces the number of sets of chromosomes from two to one in the gametes, counterbalancing the doubling that occurs at fertilization. In animals, meiosis occurs only in germ cells, which are in the ovaries or testes. As a result of meiosis, each human sperm and egg is haploid ($n = 23$). Fertilization restores the diploid condition by combining two haploid sets of chromosomes, and the human life cycle is repeated, generation after generation (see Figure 10.5). You will learn more about the production of sperm and eggs in Chapter 36.

In general, the steps of the human life cycle are typical of many sexually reproducing animals. Indeed, the processes of fertilization and meiosis are the hallmarks of sexual reproduction in plants, fungi, and protists as well as in animals. Fertilization and meiosis alternate in sexual life cycles, maintaining a constant number of chromosomes in each species from one generation to the next.

The Variety of Sexual Life Cycles

Although the alternation of meiosis and fertilization is common to all organisms that reproduce sexually, the timing of these two events in the life cycle varies, depending on the species. These variations can be grouped into three main types of life cycles. In the type that occurs in humans and most other animals, gametes are the only haploid cells. Meiosis occurs in germ cells during the production of gametes, which undergo no further cell division prior to fertilization. After fertilization, the diploid zygote divides by mitosis, producing a multicellular organism that is diploid **(Figure 10.6a)**.

Plants and some species of algae exhibit a second type of life cycle called **alternation of generations**. This type includes both diploid and haploid stages that are multicellular. The multicellular diploid stage is called the *sporophyte*. Meiosis in the sporophyte produces haploid cells called *spores*. Unlike a gamete, a haploid spore doesn't fuse with another cell but divides mitotically, generating a multicellular haploid stage called the *gametophyte*. Cells of the gametophyte give rise to gametes by mitosis. Fusion of two haploid gametes at fertilization results in a diploid zygote, which develops into the next sporophyte generation. Therefore, in this type of life cycle, the sporophyte generation produces a gametophyte as its offspring, and the gametophyte generation produces the next sporophyte generation **(Figure 10.6b)**. The term *alternation of generations* fits well as a name for this type of life cycle.

A third type of life cycle occurs in most fungi and some protists, including some algae. After gametes fuse and form a diploid zygote, meiosis occurs without a multicellular diploid offspring developing. Meiosis produces not gametes but haploid cells that then divide by mitosis and give rise to either unicellular descendants or a haploid multicellular adult organism. Subsequently, the haploid organism carries out further mitoses, producing the cells that develop into gametes. The only diploid stage found in these species is the single-celled zygote **(Figure 10.6c)**.

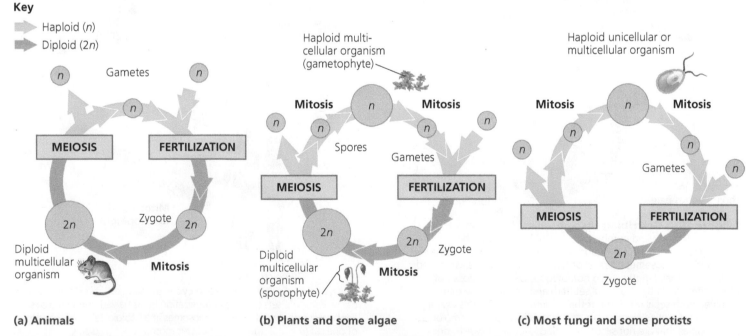

Key

➤ Haploid (n)
➤ Diploid ($2n$)

(a) Animals

(b) Plants and some algae

(c) Most fungi and some protists

▲ **Figure 10.6 Three types of sexual life cycles.** The common feature of all three cycles is the alternation of meiosis and fertilization, key events that contribute to genetic variation among offspring. The cycles differ in the timing of these two key events.

Note that *either* haploid or diploid cells can divide by mitosis, depending on the type of life cycle. Only diploid cells, however, can undergo meiosis because haploid cells have a single set of chromosomes that cannot be further reduced. Though the three types of sexual life cycles differ in the timing of meiosis and fertilization, they share a fundamental result: genetic variation among offspring. A closer look at meiosis will reveal the sources of this variation.

CONCEPT CHECK 10.2

1.  In Figure 10.4, how many DNA molecules (double helices) are present (see Figure 9.5)?
2. How does the alternation of meiosis and fertilization in the life cycles of sexually reproducing organisms maintain the normal chromosome count for each species?
3. Each sperm of a pea plant contains seven chromosomes. What are the haploid and diploid numbers for this species?
4. **WHAT IF?** A certain eukaryote lives as a unicellular organism, but during environmental stress, it produces gametes. The gametes fuse, and the resulting zygote undergoes meiosis, generating new single cells. What type of organism could this be?

For suggested answers, see Appendix A.

CONCEPT 10.3

Meiosis reduces the number of chromosome sets from diploid to haploid

Many of the steps of meiosis closely resemble corresponding steps in mitosis. Meiosis, like mitosis, is preceded by the duplication of chromosomes. However, this single duplication is followed not by one but by two consecutive cell divisions, called **meiosis I** and **meiosis II**. These two divisions result in four daughter cells (rather than the two daughter cells of mitosis), each with only half as many chromosomes as the parent cell.

The Stages of Meiosis

The overview of meiosis in **Figure 10.7** shows, for a single pair of homologous chromosomes in a diploid cell, that both members of the pair are duplicated and the copies sorted into four haploid daughter cells. Recall that sister chromatids are two copies of *one* chromosome, closely associated all along their lengths; this association is called *sister chromatid cohesion*. Together, the sister chromatids make up one duplicated chromosome (see Figure 10.4). In contrast, the two chromosomes of a homologous pair are individual chromosomes that were inherited from different parents. Homologs appear alike in the microscope, but they may have different versions of genes, each called an *allele*, at corresponding loci (for example, an allele for freckles on one chromosome and an allele for the absence of freckles at the same locus on the homolog).

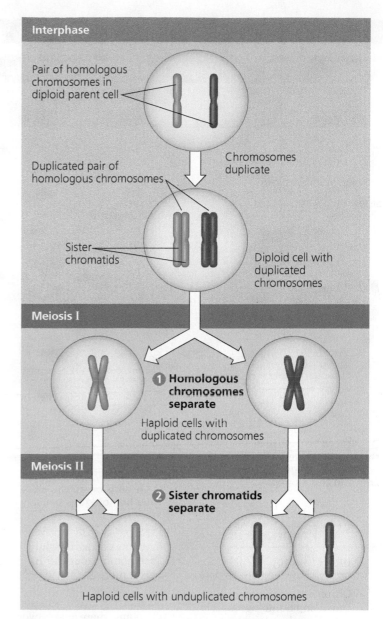

▲ **Figure 10.7 Overview of meiosis: how meiosis reduces chromosome number.** After the chromosomes duplicate in interphase, the diploid cell divides *twice*, yielding four haploid daughter cells. This overview tracks just one pair of homologous chromosomes, which for the sake of simplicity are drawn in the condensed state throughout. (They would not normally be condensed during interphase.) The red chromosome was inherited from the female parent, the blue chromosome from the male parent.

DRAW IT *Redraw the cells in this figure using a simple double helix to represent each DNA molecule.*

Homologs are not associated with each other in any obvious way except during meiosis, as you will soon see.

Figure 10.8, on the next two pages, describes in detail the stages of the two divisions of meiosis for an animal cell whose diploid number is 6. Meiosis halves the total number of chromosomes in a very specific way, reducing the number of sets from two to one, with each daughter cell receiving one set of chromosomes. Study Figure 10.8 thoroughly before going on.

MEIOSIS I: Separates homologous chromosomes

| Prophase I | Metaphase I | Anaphase I | Telophase I and Cytokinesis |

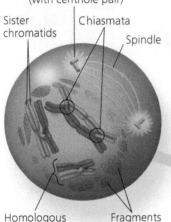

Centrosome (with centriole pair)

Sister chromatids

Chiasmata

Spindle

Homologous chromosomes

Fragments of nuclear envelope

Duplicated homologous chromosomes (red and blue) pair and exchange segments; 2*n* = 6 in this example.

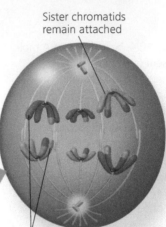

Centromere (with kinetochore)

Metaphase plate

Microtubule attached to kinetochore

Chromosomes line up by homologous pairs.

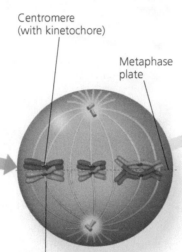

Sister chromatids remain attached

Homologous chromosomes separate

Each pair of homologous chromosomes separates.

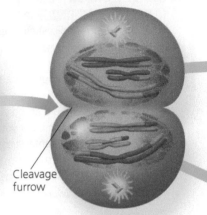

Cleavage furrow

Two haploid cells form; each chromosome still consists of two sister chromatids.

Prophase I

During early prophase I, before the stage shown above:

- Chromosomes begin to condense, and homologs loosely pair along their lengths, aligned gene by gene.
- Paired homologs become physically connected to each other along their lengths by a zipper-like protein structure, the *synaptonemal complex*; this state is called **synapsis**.
- **Crossing over**, a genetic rearrangement between nonsister chromatids involving the exchange of corresponding segments of DNA molecules, begins during pairing and synaptonemal complex formation and is completed while homologs are in synapsis.

At the stage shown above:

- Synapsis has ended with the disassembly of the synaptonemal complex in mid-prophase, and the

chromosomes in each pair have moved apart slightly.

- Each homologous pair has one or more X-shaped regions called **chiasmata** (singular, *chiasma*). A chiasma exists at the point where a crossover has occurred. It appears as a cross because sister chromatid cohesion still holds the two original sister chromatids together, even in regions beyond the crossover point, where one chromatid is now part of the other homolog.
- Centrosome movement, spindle formation, and nuclear envelope breakdown occur as in mitosis.

Later in prophase I, after the stage shown above:

- Microtubules from one pole or the other attach to the two kinetochores, protein structures at the centromeres of the two homologs. The homologous pairs then move toward the metaphase plate.

Metaphase I

- Pairs of homologous chromosomes are now arranged at the metaphase plate, with one chromosome in each pair facing each pole.
- Both chromatids of one homolog are attached to kinetochore microtubules from one pole; those of the other homolog are attached to microtubules from the opposite pole.

Anaphase I

- Breakdown of proteins responsible for sister chromatid cohesion along chromatid arms allows homologs to separate.
- The homologs move toward opposite poles, guided by the spindle apparatus.
- Sister chromatid cohesion persists at the centromere, causing chromatids to move as a unit toward the same pole.

Telophase I and Cytokinesis

- At the beginning of telophase I, each half of the cell has a complete haploid set of duplicated chromosomes. Each chromosome is composed of two sister chromatids; one or both chromatids include regions of nonsister chromatid DNA.
- Cytokinesis (division of the cytoplasm) usually occurs simultaneously with telophase I, forming two haploid daughter cells.
- In animal cells like these, a cleavage furrow forms. (In plant cells, a cell plate forms.)
- In some species, chromosomes decondense and nuclear envelopes form.
- No chromosome duplication occurs between meiosis I and meiosis II.

MEIOSIS II: Separates sister chromatids

Prophase II	Metaphase II	Anaphase II	Telophase II and Cytokinesis

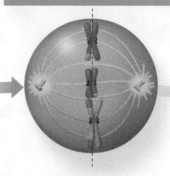 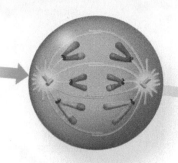 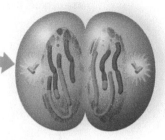

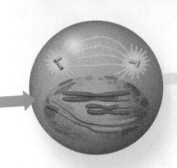

During another round of cell division, the sister chromatids finally separate; four haploid daughter cells result, containing unduplicated chromosomes.

 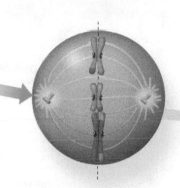 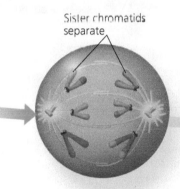

Sister chromatids separate

Haploid daughter cells forming

Prophase II

- A spindle apparatus forms.
- In late prophase II (not shown here), chromosomes, each still composed of two chromatids associated at the centromere, move toward the metaphase II plate.

Metaphase II

- The chromosomes are positioned at the metaphase plate as in mitosis.
- Because of crossing over in meiosis I, the two sister chromatids of each chromosome are not genetically identical.
- The kinetochores of sister chromatids are attached to microtubules extending from opposite poles.

Anaphase II

- Breakdown of proteins holding the sister chromatids together at the centromere allows the chromatids to separate. The chromatids move toward opposite poles as individual chromosomes.

Telophase II and Cytokinesis

- Nuclei form, the chromosomes begin decondensing, and cytokinesis occurs.
- The meiotic division of one parent cell produces four daughter cells, each with a haploid set of (unduplicated) chromosomes.
- The four daughter cells are genetically distinct from one another and from the parent cell.

MAKE CONNECTIONS *Imagine the two daughter cells in Figure 9.7 undergoing another round of mitosis, yielding four cells. Compare the number of chromosomes in each of those four cells, after mitosis, with the number in each cell in Figure 10.8, after meiosis. What is it about the process of meiosis that accounts for this difference, even though meiosis also includes two cell divisions?*

ANIMATION

BioFlix Visit the Study Area in **MasteringBiology** for the BioFlix® 3-D Animation on Meiosis.

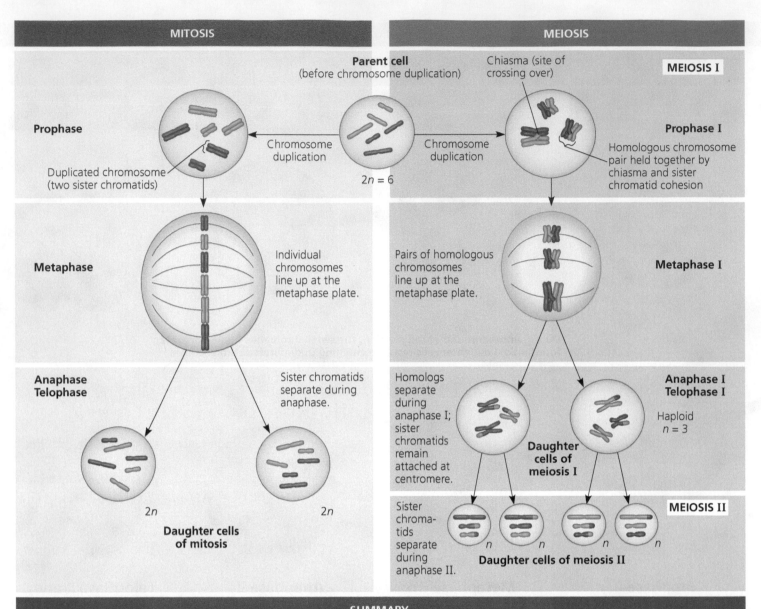

MITOSIS	MEIOSIS

Parent cell
(before chromosome duplication)

Chiasma (site of crossing over)

MEIOSIS I

Prophase

Chromosome duplication

Chromosome duplication

Prophase I

Duplicated chromosome
(two sister chromatids)

$2n = 6$

Homologous chromosome pair held together by chiasma and sister chromatid cohesion

Metaphase

Individual chromosomes line up at the metaphase plate.

Pairs of homologous chromosomes line up at the metaphase plate.

Metaphase I

**Anaphase
Telophase**

Sister chromatids separate during anaphase.

Homologs separate during anaphase I; sister chromatids remain attached at centromere.

**Anaphase I
Telophase I**

Haploid
$n = 3$

Daughter cells of meiosis I

$2n$

$2n$

Daughter cells of mitosis

Sister chromatids separate during anaphase II.

MEIOSIS II

n n n n

Daughter cells of meiosis II

SUMMARY		
Property	**Mitosis**	**Meiosis**
DNA replication	Occurs during interphase before mitosis begins	Occurs during interphase before meiosis I begins
Number of divisions	One, including prophase, prometaphase, metaphase, anaphase, and telophase	Two, each including prophase, metaphase, anaphase, and telophase
Synapsis of homologous chromosomes	Does not occur	Occurs during prophase I along with crossing over between nonsister chromatids; resulting chiasmata hold pairs together due to sister chromatid cohesion
Number of daughter cells and genetic composition	Two, each diploid ($2n$) and genetically identical to the parent cell	Four, each haploid (n), containing half as many chromosomes as the parent cell; genetically different from the parent cell and from each other
Role in the animal body	Enables multicellular adult to arise from zygote; produces cells for growth, repair, and, in some species, asexual reproduction	Produces gametes; reduces number of chromosome sets by half and introduces genetic variability among the gametes

▲ **Figure 10.9 A comparison of mitosis and meiosis in diploid cells.**

DRAW IT *Could any other combinations of chromosomes be generated during meiosis II from the specific cells shown in telophase I? Explain. (Hint: Draw all possible arrangements of chromosomes at metaphase II.)*

A Comparison of Mitosis and Meiosis

Figure 10.9 summarizes the key differences between meiosis and mitosis in diploid cells. Basically, meiosis reduces the number of chromosome sets from two to one, whereas mitosis conserves the number. Meiosis produces cells that differ genetically from their parent cell and from each other, whereas mitosis produces daughter cells that are genetically identical to their parent cell and to each other.

Three events unique to meiosis occur during meiosis I:

1. **Synapsis and crossing over.** During prophase I, duplicated homologs pair up, and the formation of the synaptonemal complex between them holds them in synapsis. Crossing over also occurs during prophase I. Synapsis and crossing over normally do not occur during prophase of mitosis.
2. **Homologous pairs at the metaphase plate.** At metaphase I of meiosis, chromosomes are positioned at the metaphase plate as pairs of homologs, rather than individual chromosomes, as in metaphase of mitosis.
3. **Separation of homologs.** At anaphase I of meiosis, the duplicated chromosomes of each homologous pair move toward opposite poles, but the sister chromatids of each duplicated chromosome remain attached. In anaphase of mitosis, by contrast, sister chromatids separate.

How do sister chromatids stay together through meiosis I but separate from each other in meiosis II and mitosis? Sister chromatids are attached along their lengths by protein complexes called *cohesins* and are said to exhibit *sister chromatid cohesion*. In mitosis, this attachment lasts until the end of metaphase, when enzymes cleave the cohesins, freeing the sister chromatids to move to opposite poles of the cell. In meiosis, sister chromatid cohesion is released in two steps, one at the start of anaphase I and one at anaphase II. In metaphase I, homologs are held together by cohesion between sister chromatid arms in regions beyond points of crossing over, where stretches of sister chromatids now belong to different chromosomes. As shown in Figure 10.8, the combination of crossing over and sister chromatid cohesion along the arms results in the formation of a chiasma. Chiasmata hold homologs together as the spindle forms for the first meiotic division. At the onset of anaphase I, the release of cohesion along sister chromatid arms allows homologs to separate. At anaphase II, the release of sister chromatid cohesion at the centromeres allows the sister chromatids to separate. Thus, sister chromatid cohesion and crossing over, acting together, play an essential role in the lining up of chromosomes by homologous pairs at metaphase I.

Meiosis I is called the *reductional division* because it halves the number of chromosome sets per cell—a reduction from two sets (the diploid state) to one set (the haploid state). During the second meiotic division, meiosis II (sometimes called the *equational division*), the sister chromatids separate,

producing haploid daughter cells. The mechanism for separating sister chromatids is virtually identical in meiosis II and mitosis. The molecular basis of chromosome behavior during meiosis continues to be a focus of intense research. In the **Scientific Skills Exercise**, you can work with data from an experiment that tracked the amount of DNA in cells as they proceeded through the stages of meiosis.

CONCEPT CHECK 10.3

1. **MAKE CONNECTIONS** How are the chromosomes in a cell at metaphase of mitosis similar to and different from the chromosomes in a cell at metaphase of meiosis II? (Compare Figures 9.7 and 10.8.)
2. **WHAT IF?** Given that the synaptonemal complex has disappeared by the end of prophase, how would the two homologs be associated if crossing over did not occur? What effect might this ultimately have on gamete formation?

For suggested answers, see Appendix A.

CONCEPT 10.4

Genetic variation produced in sexual life cycles contributes to evolution

How do we account for the genetic variation among the family members illustrated in Figure 10.1? As you'll learn later in more detail, mutations are the original source of all genetic diversity. These changes in an organism's DNA create the different versions of genes known as *alleles*. Once these differences arise, reshuffling of the alleles during sexual reproduction produces the variation that results in each member of a sexually reproducing population having a unique combination of traits.

Origins of Genetic Variation Among Offspring

In species that reproduce sexually, the behavior of chromosomes during meiosis and fertilization is responsible for most of the variation that arises in each generation. Let's examine three mechanisms that contribute to the genetic variation arising from sexual reproduction: independent assortment of chromosomes, crossing over, and random fertilization.

Independent Assortment of Chromosomes

One aspect of sexual reproduction that generates genetic variation is the random orientation of pairs of homologous chromosomes at metaphase of meiosis I. At metaphase I, the homologous pairs, each consisting of one maternal and one paternal chromosome, are situated at the metaphase plate. (Note that the terms *maternal* and *paternal* refer, respectively, to the mother and father of the individual whose cells are undergoing meiosis.) Each pair may orient with either its maternal or paternal homolog closer to a given pole—its orientation is as random as the flip of a coin. Thus, there is a 50% chance that a given daughter cell of meiosis I will get the maternal

Making a Line Graph and Converting Between Units of Data

How Does DNA Content Change as Budding Yeast Cells Proceed Through Meiosis? When nutrients are low, cells of the budding yeast (*Saccharomyces cerevisiae*) exit the mitotic cell cycle and enter meiosis. In this exercise you will track the DNA content of a population of yeast cells as they progress through meiosis.

How the Experiment Was Done Researchers grew a culture of yeast cells in a nutrient-rich medium and then transferred them to a nutrient-poor medium to induce meiosis. At different times after induction, the DNA content per cell was measured in a sample of the cells, and the average DNA content per cell was recorded in femtograms (fg; 1 femtogram = 1×10^{-15} gram).

Data from the Experiment

Time after Induction (hours)	Average Amount of DNA per Cell (fg)
0.0	24.0
1.0	24.0
2.0	40.0
3.0	47.0
4.0	47.5
5.0	48.0
6.0	48.0
7.0	47.5
7.5	25.0
8.0	24.0
9.0	23.5
9.5	14.0
10.0	13.0
11.0	12.5
12.0	12.0
13.0	12.5
14.0	12.0

Interpret the Data

1. First, set up your graph. (a) Place the labels for the independent variable and the dependent variable on the appropriate axes, followed by units of measurement in parentheses. Explain your choices. (b) Add tick marks and values for each axis in your graph. Note that while the timed samples were not all taken at equal intervals, the tick marks signifying the elapsed times along the x-axis should be regularly spaced and labeled. Explain your choices. (For additional information about graphs, see the Scientific Skills Review in Appendix F and in the Study Area in MasteringBiology.)

2. Because the variable on the x-axis varies continuously, it makes sense to plot the data on a line graph. (a) Plot each data point from the table onto the graph by placing a dot at the appropriate (x, y) coordinate. (b) Connect the data points with line segments.

3. Most of the yeast cells in the culture were in G_1 of the cell cycle before being moved to the nutrient-poor medium. (a) How many femtograms of DNA are there in each yeast cell in G_1? Estimate this value from the data in your graph. (b) How many femtograms of DNA should be present in each cell in G_2? (See Concept 9.2 and Figure 9.6.) At the end of meiosis I (MI)? At the end of meiosis II (MII)? (See Figure 10.7.) (c) Using these values as a guideline, distinguish the different phases by inserting vertical dashed lines in the graph between phases and label each phase (G_1, S, G_2, MI, MII). You can figure out where to put the dividing lines based on what you know about the DNA content of each phase (see Figure 10.7). (d) Think carefully about the point where the line at the highest value begins to slope downward. What specific point of meiosis does this "corner" represent? What stage(s) correspond to the downward sloping line?

4. Given the fact that 1 fg of DNA = 9.78×10^5 base pairs (on average), you can convert the amount of DNA per cell to the length of DNA in numbers of base pairs. (a) Calculate the number of base pairs of DNA in the haploid yeast genome. Express your answer in millions of base pairs (Mb), a standard unit for expressing genome size. Show your work. (b) How many base pairs per minute were synthesized during the S phase of these yeast cells?

Further Reading G. Simchen, Commitment to meiosis: what determines the mode of division in budding yeast? *BioEssays* 31:169–177 (2009). doi 10.1002/bies.200800124

(MB) A version of this Scientific Skills Exercise can be assigned in MasteringBiology.

---

chromosome of a certain homologous pair and a 50% chance that it will get the paternal chromosome.

Because each pair of homologous chromosomes is positioned independently of the other pairs at metaphase I, the first meiotic division results in each pair sorting its maternal and paternal homologs into daughter cells independently of every other pair. This is called *independent assortment*. Each daughter cell represents one outcome of all possible combinations of maternal and paternal chromosomes. As shown in **Figure 10.10**, the number of combinations possible for daughter cells formed by meiosis of a diploid cell with *n* = 2 (two pairs of homologous chromosomes) is four: two possible arrangements for the first pair times two possible arrangements for the second pair. Note that only two of the four combinations of daughter cells shown in the figure

would result from meiosis of a *single* diploid cell, because a single parent cell would have one or the other possible chromosomal arrangement at metaphase I, but not both. However, the population of daughter cells resulting from meiosis of a large number of diploid cells contains all four types in approximately equal numbers. In the case of *n* = 3, eight combinations of chromosomes are possible for daughter cells. More generally, the number of possible combinations when chromosomes sort independently during meiosis is 2^n, where *n* is the haploid number of the organism.

In the case of humans (*n* = 23), the number of possible combinations of maternal and paternal chromosomes in the resulting gametes is 2^{23}, or about 8.4 million. Each gamete that you produce in your lifetime contains one of roughly 8.4 million possible combinations of chromosomes.

Crossing Over

As a consequence of the independent assortment of chromosomes during meiosis, each of us produces a collection of gametes differing greatly in their combinations of the chromosomes we inherited from our two parents. Figure 10.10 suggests that each chromosome in a gamete is exclusively maternal or paternal in origin. In fact, this is *not* the case, because crossing over produces **recombinant chromosomes**, individual chromosomes that carry genes (DNA) derived from two different parents **(Figure 10.11)**. In meiosis in humans, an average of one to three crossover events occur per chromosome pair, depending on the size of the chromosomes and the position of their centromeres.

Crossing over begins very early in prophase I as homologous chromosomes pair loosely along their lengths. Each gene on one homolog is aligned precisely with the corresponding gene on the other homolog. In a single crossover event, the DNA of two *nonsister* chromatids—one maternal and one paternal chromatid of a homologous pair—is broken by specific proteins at precisely corresponding points, and the two segments beyond the crossover point are each joined to the other chromatid. Thus, a paternal (blue) chromatid is joined to a piece of maternal (red) chromatid beyond the crossover point, and vice versa. In this way, crossing over produces chromosomes with new combinations of maternal and paternal alleles (see Figure 10.11).

At metaphase II, chromosomes that contain one or more recombinant chromatids can be oriented in two alternative, nonequivalent ways with respect to other chromosomes, because their sister chromatids are no longer identical. The different possible arrangements of nonidentical sister chromatids during meiosis II further increase the number of genetic types of daughter cells that can result from meiosis.

You'll learn more about crossing over in Chapter 12. The important point for now is that crossing over, by combining DNA inherited from two parents into a single chromosome, is an important source of genetic variation in sexual life cycles.

Random Fertilization

The random nature of fertilization adds to the genetic variation arising from meiosis. In humans, each male and female gamete represents one of about 8.4 million (2^{23}) possible chromosome combinations due to independent assortment. The fusion of a male gamete with a female gamete during fertilization will produce a zygote with any of about 70 trillion ($2^{23} \times 2^{23}$) diploid combinations. If we factor in the variation brought about by crossing over, the number of possibilities is truly astronomical. It may sound trite, but you really *are* unique.

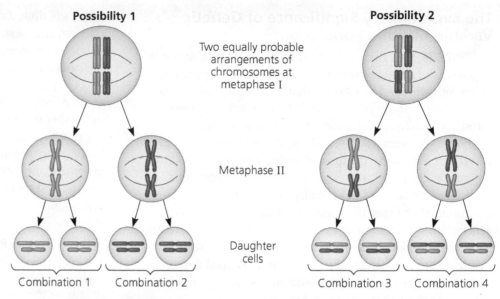

Two equally probable arrangements of chromosomes at metaphase I

Metaphase II

Daughter cells

Possibility 1 — Combination 1 — Combination 2

Possibility 2 — Combination 3 — Combination 4

▲ Figure 10.10 **The independent assortment of homologous chromosomes in meiosis.**

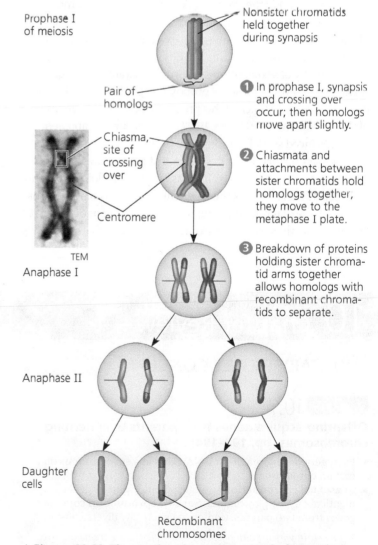

Prophase I of meiosis

Pair of homologs

Chiasma, site of crossing over

Centromere

TEM

Anaphase I

Anaphase II

Daughter cells

Nonsister chromatids held together during synapsis

1 In prophase I, synapsis and crossing over occur; then homologs move apart slightly.

2 Chiasmata and attachments between sister chromatids hold homologs together; they move to the metaphase I plate.

3 Breakdown of proteins holding sister chromatid arms together allows homologs with recombinant chromatids to separate.

Recombinant chromosomes

▲ Figure 10.11 **The results of crossing over during meiosis.**

The Evolutionary Significance of Genetic Variation Within Populations

EVOLUTION Now that you've learned how new combinations of genes arise among offspring in a sexually reproducing population, let's see how the genetic variation in a population relates to evolution. Darwin recognized that a population evolves through the differential reproductive success of its variant members. On average, those individuals best suited to the local environment leave the most offspring, thereby transmitting their genes. Thus, natural selection results in the accumulation of genetic variations favored by the environment. As the environment changes, the population may survive if, in each generation, at least some of its members can cope effectively with the new conditions. Mutations are the original source of different alleles, which are then mixed and matched during meiosis. New and different combinations of alleles may work better than those that previously prevailed.

In a stable environment, though, sexual reproduction seems as if it would be less advantageous than asexual reproduction, which ensures perpetuation of successful combinations of alleles. Furthermore, sexual reproduction is more expensive, energetically, than asexual reproduction. In spite of these apparent disadvantages, sexual reproduction is almost universal among animals. Why is this?

The ability of sexual reproduction to generate genetic diversity is the most commonly proposed explanation for the evolutionary persistence of this process. Consider the rare case of the bdelloid rotifer **(Figure 10.12)**. This group has apparently not reproduced sexually throughout the 40 million years of its evolutionary history. Does this mean that genetic diversity is not advantageous in this species? It turns out that bdelloid rotifers are an exception that proves the rule: This group has mechanisms other than sexual reproduction for generating genetic diversity. For example, they live in environments that can dry up for long periods of time, during which they can enter a state of suspended animation. In this state, their cell membranes may crack in places, allowing entry of DNA from other rotifers and even other species. Evidence suggests that this DNA can become incorporated into the genome of the rotifer, leading to increased genetic diversity. This supports the idea that genetic diversity is advantageous, and that sexual reproduction has persisted because it generates such diversity.

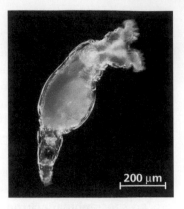

200 μm

▲ **Figure 10.12 A bdelloid rotifer, an animal that reproduces only asexually.**

In this chapter, we have seen how sexual reproduction greatly increases the genetic variation present in a population. Although Darwin realized that heritable variation is what makes evolution possible, he could not explain why offspring resemble—but are not identical to—their parents. Ironically, Gregor Mendel, a contemporary of Darwin, published a theory of inheritance that helps explain genetic variation, but his discoveries had no impact on biologists until 1900, more than 15 years after Darwin (1809–1882) and Mendel (1822–1884) had died. In the next chapter, you'll learn how Mendel discovered the basic rules governing the inheritance of specific traits.

CONCEPT CHECK 10.4

1. What is the original source of variation among the different alleles of a gene?
2. **WHAT IF?** Under what circumstances would crossing over during meiosis *not* contribute to genetic variation among daughter cells?

   For suggested answers, see Appendix A.

---

10 Chapter Review

SUMMARY OF KEY CONCEPTS

CONCEPT 10.1

Offspring acquire genes from parents by inheriting chromosomes (pp. 193–194)

- Each **gene** in an organism's DNA exists at a specific **locus** on a certain chromosome.
- In **asexual reproduction**, a single parent produces genetically identical offspring by mitosis. **Sexual reproduction** combines genes from two parents, leading to genetically diverse offspring.

> **?** *Explain why human offspring resemble their parents but are not identical to them.*

CONCEPT 10.2

Fertilization and meiosis alternate in sexual life cycles (pp. 194–197)

- Normal human **somatic cells** are **diploid**. They have 46 chromosomes made up of two sets of 23 chromosomes, one set from each parent. Human diploid cells have 22 **homologous** pairs of **autosomes** and one pair of **sex chromosomes**; the latter determines whether the person is female (XX) or male (XY).
- In humans, ovaries and testes produce **haploid gametes** by **meiosis**, each gamete containing a single set of 23 chromosomes ($n = 23$). During **fertilization**, an egg and sperm unite, forming a diploid ($2n = 46$) single-celled **zygote**, which develops into a multicellular organism by mitosis.

- Sexual **life cycles** differ in the timing of meiosis relative to fertilization and in the point(s) of the cycle at which a multicellular organism is produced by mitosis.

? *Compare the life cycles of animals and plants, mentioning their similarities and differences.*

CONCEPT 10.3

Meiosis reduces the number of chromosome sets from diploid to haploid (pp. 197–201)

- **Meiosis I** and **meiosis II** produce four haploid daughter cells. The number of chromosome sets is reduced from two (diploid) to one (haploid) during meiosis I, the reductional division.
- Meiosis is distinguished from mitosis by three events of meiosis I:

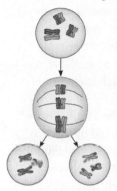

Prophase I: Each homologous pair undergoes **synapsis** and **crossing over** between nonsister chromatids with the subsequent appearance of **chiasmata**.

Metaphase I: Chromosomes line up as homologous pairs on the metaphase plate.

Anaphase I: Homologs separate from each other, sister chromatids remain joined at the centromere.

Meiosis II then separates the sister chromatids.

- Sister chromatid cohesion and crossing over allow chiasmata to hold homologs together until anaphase I. Cohesins are cleaved along the arms at anaphase I, allowing homologs to separate, and at the centromeres in anaphase II, releasing sister chromatids.

? *In prophase I, homologous chromosomes pair up and undergo crossing over. Can this also occur during prophase II? Explain.*

CONCEPT 10.4

Genetic variation produced in sexual life cycles contributes to evolution (pp. 201–204)

- Three events in sexual reproduction contribute to genetic variation in a population: independent assortment of chromosomes during meiosis, crossing over during meiosis I, and random fertilization of egg cells by sperm. During crossing over, DNA of nonsister chromatids in a homologous pair is broken and rejoined.
- Genetic variation is the raw material for evolution by natural selection. Mutations are the original source of this variation; recombination of variant genes generates additional diversity.

? *Explain how three processes unique to meiosis generate a great deal of genetic variation.*

TEST YOUR UNDERSTANDING

Level 1: Knowledge/Comprehension

1. A human cell containing 22 autosomes and a Y chromosome is
   a. a sperm.
   b. an egg.
   c. a zygote.
   d. a somatic cell of a male.
   e. a somatic cell of a female.

2. Homologous chromosomes move toward opposite poles of a dividing cell during
   a. mitosis.
   b. meiosis I.
   c. meiosis II.
   d. fertilization.
   e. binary fission.

Level 2: Application/Analysis

3. If the DNA content of a diploid cell in the G_1 phase of the cell cycle is x, then the DNA content of the same cell at metaphase of meiosis I would be
   a. $0.25x$.　**b.** $0.5x$.　**c.** x.　**d.** $2x$.　**e.** $4x$.

4. If we continued to follow the cell lineage from question 3, then the DNA content of a single cell at metaphase of meiosis II would be
   a. $0.25x$.　**b.** $0.5x$.　**c.** x.　**d.** $2x$.　**e.** $4x$.

5. How many different combinations of maternal and paternal chromosomes can be packaged in gametes made by an organism with a diploid number of 8 ($2n = 8$)?
   a. 2　**b.** 4　**c.** 8　**d.** 16　**e.** 32

6. **DRAW IT** The diagram at right shows a cell in meiosis.

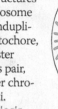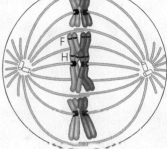

   (a) Label the appropriate structures with these terms: chromosome (label as duplicated or unduplicated), centromere, kinetochore, sister chromatids, nonsister chromatids, homologous pair, homologs, chiasma, sister chromatid cohesion, gene loci.
   (b) Identify the stage of meiosis shown.
   (c) Describe the makeup of a haploid set and a diploid set.

Level 3: Synthesis/Evaluation

7. How can you tell that the cell in question 6 is undergoing meiosis, not mitosis?

8. **SCIENTIFIC INQUIRY**
   The diagram above represents a meiotic cell. A previous study has shown that the freckles gene is located at the locus marked F, and the hair-color gene is located at the locus marked H, both on the long chromosome. The individual from whom this cell was taken has inherited different alleles for each gene ("freckles" and "black hair" from one parent and "no freckles" and "blond hair" from the other). Predict allele combinations in the gametes resulting from this meiotic event. List other possible combinations of these alleles in this individual's gametes.

9. **FOCUS ON EVOLUTION**
   Many species can reproduce either asexually or sexually. What might be the evolutionary significance of the switch from asexual to sexual reproduction that occurs in some organisms when the environment becomes unfavorable?

10. **FOCUS ON INFORMATION**
    The continuity of life is based on heritable information in the form of DNA. In a short essay (100–150 words), explain how chromosome behavior during sexual reproduction in animals ensures perpetuation of parental traits in offspring and, at the same time, genetic variation among offspring.

For selected answers, see Appendix A.

MasteringBiology®

Students Go to **MasteringBiology** for assignments, the eText, and the Study Area with practice tests, animations, and activities.

Instructors Go to **MasteringBiology** for automatically graded tutorials and questions that you can assign to your students, plus Instructor Resources.

11

Mendel and the Gene Idea

▼ **Figure 11.1** What principles of inheritance did Gregor Mendel discover by breeding garden pea plants?

KEY CONCEPTS

11.1 Mendel used the scientific approach to identify two laws of inheritance

11.2 The laws of probability govern Mendelian inheritance

11.3 Inheritance patterns are often more complex than predicted by simple Mendelian genetics

11.4 Many human traits follow Mendelian patterns of inheritance

OVERVIEW

Drawing from the Deck of Genes

Scanning the crowd at a soccer match attests to the marvelous variety and diversity of humankind. Brown, blue, green, or gray eyes; black, brown, blond, or red hair—these are just a few examples of heritable variations that we may observe among individuals in a population. What are the genetic principles that account for the transmission of such traits from parents to offspring in humans and other organisms?

The explanation of heredity most widely in favor during the 1800s was the "blending" hypothesis, the idea that genetic material contributed by the two parents mixes in a manner analogous to the way blue and yellow paints blend to make green. This hypothesis predicts that over many generations, a freely mating population will give rise to a uniform population of individuals. However, our everyday observations and the results of breeding experiments with animals and plants contradict that prediction. The blending hypothesis also fails to explain other phenomena of inheritance, such as traits reappearing after skipping a generation.

An alternative to the blending model is a "particulate" hypothesis of inheritance: the gene idea. According to this model, parents pass on discrete heritable units—genes—that retain their separate identities in offspring. An organism's collection of genes is more like a deck of cards than a bucket of paint. Like playing cards, genes can be shuffled and passed along, generation after generation, in undiluted form.

Modern genetics had its genesis in an abbey garden, where a monk named Gregor Mendel documented a particulate mechanism for inheritance. **Figure 11.1** shows Mendel (back row, holding a sprig of fuchsia) with his fellow monks. Mendel developed his theory of inheritance several decades before chromosomes were observed under the microscope and well before the significance of their behavior was understood. In this chapter, we will step into Mendel's garden to re-create his experiments and explain how he arrived at his theory of inheritance. We'll also explore inheritance patterns more complex than those observed by Mendel in garden peas. Finally, we'll see how the Mendelian model applies to the inheritance of human variations, including hereditary disorders such as sickle-cell disease.

CONCEPT 11.1

Mendel used the scientific approach to identify two laws of inheritance

Mendel discovered the basic principles of heredity by breeding garden peas in carefully planned experiments. As we retrace his work, you'll recognize the key elements of the scientific process that were introduced in Chapter 1.

Mendel's Experimental, Quantitative Approach

One reason Mendel probably chose to work with peas is that they are available in many varieties. For example, one variety has purple flowers, while another variety has white flowers. A heritable feature that varies among individuals, such as flower color, is called a **character**. Each variant for a character, such as purple or white color for flowers, is called a **trait**.

Mendel could strictly control mating between plants. Each pea flower has both pollen-producing organs (stamens) and an egg-bearing organ (carpel). In nature, pea plants usually self-fertilize: Pollen grains from the stamens land on the carpel of the same flower, and sperm released from the pollen grains fertilize eggs present in the carpel. To achieve cross-pollination (fertilization between different plants), Mendel removed the immature stamens of a plant before they produced pollen and then dusted pollen from another plant onto the altered flowers **(Figure 11.2)**. Each resulting zygote then developed into a plant embryo encased in a seed (a pea). Mendel could thus always be sure of the parentage of new seeds.

Mendel chose to track only those characters that occurred in two distinct, alternative forms: purple or white flower color. He also made sure that he started his experiments with varieties that, over many generations of self-pollination, had produced only the same variety as the parent plant. Such plants are said to be **true-breeding**. For example, a plant with purple flowers is true-breeding if the seeds produced by self-pollination in successive generations all give rise to plants that also have purple flowers.

In a typical breeding experiment, Mendel cross-pollinated two contrasting, true-breeding pea varieties—for example, purple-flowered plants and white-flowered plants (see Figure 11.2). This mating, or *crossing*, of two true-breeding varieties is called **hybridization**. The true-breeding parents are referred to as the **P generation** (parental generation), and their hybrid offspring are the **F₁ generation** (first filial generation, the word *filial* from the Latin word for "son"). Allowing these F₁ hybrids to self-pollinate (or to cross-pollinate with other F₁ hybrids) produces an **F₂ generation** (second filial generation). Mendel usually followed traits for at least the P, F₁, and F₂ generations. Had Mendel stopped his experiments with the F₁ generation, the basic patterns of inheritance would have escaped him.

Crossing Pea Plants

Application By crossing (mating) two true-breeding varieties of an organism, scientists can study patterns of inheritance. In this example, Mendel crossed pea plants that varied in flower color.

Technique

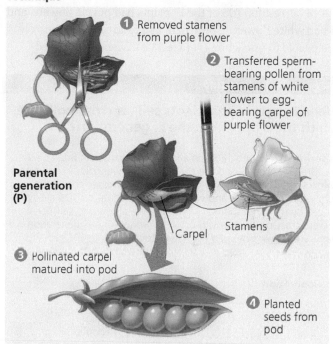

1 Removed stamens from purple flower

2 Transferred sperm-bearing pollen from stamens of white flower to egg-bearing carpel of purple flower

Parental generation (P)

Carpel   Stamens

3 Pollinated carpel matured into pod

4 Planted seeds from pod

Results When pollen from a white flower was transferred to a purple flower, the first-generation hybrids all had purple flowers. The result was the same for the reciprocal cross, which involved the transfer of pollen from purple flowers to white flowers.

First filial generation offspring (F₁)

5 Examined offspring: all purple flowers

Mendel's quantitative analysis of the F₂ plants from thousands of genetic crosses like these allowed him to deduce two fundamental principles of heredity, which have come to be called the law of segregation and the law of independent assortment.

The Law of Segregation

If the blending model of inheritance were correct, the F₁ hybrids from a cross between purple-flowered and white-flowered pea plants would have pale purple flowers, a trait intermediate between those of the P generation. Notice in Figure 11.2 that the experiment produced a very different result: All the F₁ offspring had flowers just as purple as the purple-flowered parents. What happened to the white-flowered plants' genetic contribution to the hybrids? If it were lost, then the F₁ plants could produce only purple-flowered offspring in the F₂ generation. But when Mendel allowed the F₁ plants to self-pollinate and planted their seeds, the white-flower trait reappeared in the F₂ generation.

Mendel used very large sample sizes and kept accurate records of his results: 705 of the F₂ plants had purple flowers, and 224 had white flowers. These data fit a ratio of approximately three purple to one white **(Figure 11.3)**. Mendel reasoned

that the heritable factor for white flowers did not disappear in the F₁ plants, but was somehow hidden, or masked, when the purple-flower factor was present. In Mendel's terminology, purple flower color is a *dominant* trait, and white flower color is a *recessive* trait. The reappearance of white-flowered plants in the F₂ generation was evidence that the heritable factor causing white flowers had not been diluted or destroyed by coexisting with the purple-flower factor in the F₁ hybrids.

Mendel observed the same pattern of inheritance in six other characters, each represented by two distinctly different traits **(Table 11.1)**. For example, when Mendel crossed a true-breeding variety that produced smooth, round pea seeds with one that produced wrinkled seeds, all the F₁ hybrids produced round seeds; this is the dominant trait for seed shape. In the F₂ generation, approximately 75% of the seeds were round and 25% were wrinkled—a 3:1 ratio, as in Figure 11.3. Now let's see how Mendel deduced the law of segregation from his experimental results. In the discussion that follows, we will use modern terms instead of some of the terms used by Mendel. (For example, we'll use "gene" instead of Mendel's "heritable factor.")

▼ **Figure 11.3** **Inquiry**

When F₁ hybrid pea plants self- or cross-pollinate, which traits appear in the F₂ generation?

Experiment Around 1860, in a monastery garden in Brünn, Austria, Gregor Mendel used the character of flower color in pea plants to follow traits through two generations. He crossed true-breeding purple-flowered plants and white-flowered plants (crosses are symbolized by ×). The resulting F₁ hybrids were allowed to self-pollinate or were cross-pollinated with other F₁ hybrids. The F₂ generation plants were then observed for flower color.

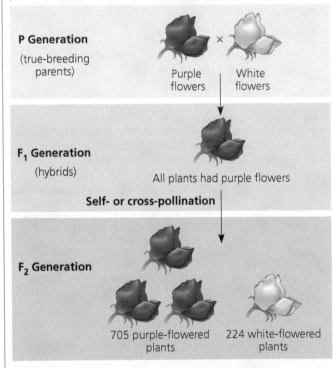

P Generation
(true-breeding parents)

Purple flowers   White flowers

F₁ Generation
(hybrids)

All plants had purple flowers

Self- or cross-pollination

F₂ Generation

705 purple-flowered plants   224 white-flowered plants

Results Both purple-flowered and white-flowered plants appeared in the F₂ generation, in a ratio of approximately 3:1.

Conclusion The "heritable factor" for the recessive trait (white flowers) had not been destroyed, deleted, or "blended" in the F₁ generation but was merely masked by the presence of the factor for purple flowers, which is the dominant trait.

Source G. Mendel, Experiments in plant hybridization, *Proceedings of the Natural History Society of Brünn* 4:3–47 (1866).

WHAT IF? If you mated two purple-flowered plants from the P generation, what ratio of traits would you expect to observe in the offspring? Explain.

Table 11.1 The Results of Mendel's F₁ Crosses for Seven Characters in Pea Plants

Character	Dominant Trait	×	Recessive Trait	F₂ Generation Dominant: Recessive	Ratio
Flower color	Purple	×	White	705:224	3.15:1
Seed color	Yellow	×	Green	6,022:2,001	3.01:1
Seed shape	Round	×	Wrinkled	5,474:1,850	2.96:1
Pod shape	Inflated	×	Constricted	882:299	2.95:1
Pod color	Green	×	Yellow	428:152	2.82:1
Flower position	Axial	×	Terminal	651:207	3.14:1
Stem length	Tall	×	Dwarf	787:277	2.84:1

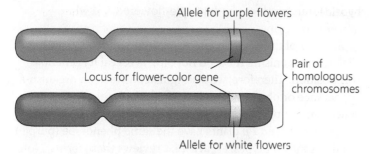

Allele for purple flowers

Locus for flower-color gene

Pair of homologous chromosomes

Allele for white flowers

▲ **Figure 11.4 Alleles, alternative versions of a gene.** A somatic cell has two copies of each chromosome (forming a homologous pair) and thus two versions of each gene; the alleles may be identical or different. This figure depicts a pair of homologous chromosomes in an F_1 hybrid pea plant. The paternally inherited chromosome (blue), which was present in the sperm within a pollen grain, has an allele for purple flowers, and the maternally inherited chromosome (red), which was present in an egg within a carpel, has an allele for white flowers.

Mendel's Model

Mendel developed a model to explain the 3:1 inheritance pattern that he consistently observed among the F_2 offspring in his pea experiments. We describe four related concepts making up this model, the fourth of which is the law of segregation.

First, *alternative versions of genes account for variations in inherited characters.* The gene for flower color in pea plants, for example, exists in two versions, one for purple flowers and the other for white flowers. These alternative versions of a gene are called **alleles (Figure 11.4)**. Today, we can relate this concept to chromosomes and DNA: Each gene is a sequence of nucleotides at a specific place, or locus, along a particular chromosome. The DNA at that locus, however, can vary slightly in its nucleotide sequence and hence in its information content. The purple-flower allele and the white-flower allele are two DNA sequence variations possible at the flower-color locus on one of a pea plant's chromosomes.

▶ **Figure 11.5 Mendel's law of segregation.** This diagram shows the genetic makeup of the generations in Figure 11.3. It illustrates Mendel's model for inheritance of the alleles of a single gene. Each plant has two alleles for the gene controlling flower color, one allele inherited from each of the plant's parents. To construct a Punnett square that predicts the F_2 generation offspring, we list all the possible gametes from one parent (here, the F_1 female) along the left side of the square and all the possible gametes from the other parent (here, the F_1 male) along the top. The boxes represent the offspring resulting from all the possible unions of male and female gametes.

Second, *for each character, an organism inherits two copies (that is, two alleles) of a gene, one from each parent.* Remarkably, Mendel made this deduction without knowing about the role, or even the existence, of chromosomes. Each somatic cell in a diploid organism has two sets of chromosomes, one set inherited from each parent (see Chapter 10). Thus, a genetic locus is actually represented twice in a diploid cell, once on each homolog of a specific pair of chromosomes. The two alleles at a particular locus may be identical, as in the true-breeding plants of Mendel's P generation. Or the alleles may differ, as in the F_1 hybrids (see Figure 11.4).

Third, *if the two alleles at a locus differ, then one, the **dominant allele**, determines the organism's appearance; the other, the **recessive allele**, has no noticeable effect on the organism's appearance.* Accordingly, Mendel's F_1 plants had purple flowers because the allele for that trait is dominant and the allele for white flowers is recessive.

The fourth and final part of Mendel's model, the **law of segregation**, states that *the two alleles for a heritable character segregate (separate from each other) during gamete formation and end up in different gametes* **(Figure 11.5)**. Thus, an egg or a sperm gets only one of the two alleles that are present in the somatic cells of the organism making the gamete. In terms of

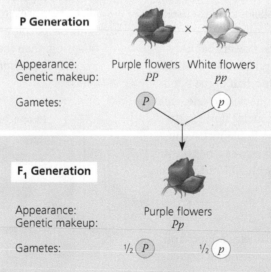

P Generation

Appearance: Purple flowers   White flowers
Genetic makeup: *PP*   *pp*

Gametes: *P*   *p*

Each true-breeding plant of the parental generation has two identical alleles, denoted as either *PP* or *pp*.

Gametes (circles) each contain only one allele for the flower-color gene. In this case, every gamete produced by a given parent has the same allele.

F₁ Generation

Appearance: Purple flowers
Genetic makeup: *Pp*

Gametes: ½ *P*   ½ *p*

Union of parental gametes produces F_1 hybrids having a *Pp* combination. Because the purple-flower allele is dominant, all these hybrids have purple flowers.

When the hybrid plants produce gametes, the two alleles segregate. Half of the gametes receive the *P* allele and the other half the *p* allele.

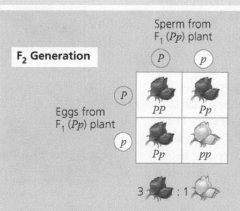

F₂ Generation

Sperm from F_1 (*Pp*) plant

   P   *p*

Eggs from F_1 (*Pp*) plant

P — *PP*   *Pp*
p — *Pp*   *pp*

3 : 1

This box, a Punnett square, shows all possible combinations of alleles in offspring that result from an $F_1 \times F_1$ (*Pp* × *Pp*) cross. Each square represents an equally probable product of fertilization. For example, the bottom left box shows the genetic combination resulting from a *p* egg fertilized by a *P* sperm.

Random combination of the gametes results in the 3:1 ratio that Mendel observed in the F_2 generation.

chromosomes, this segregation corresponds to the distribution of the two members of a pair of homologous chromosomes to different gametes in meiosis (see Figure 10.7). Note that if an organism has identical alleles for a particular character—that is, the organism is true-breeding for that character—then that allele is present in all gametes. But if different alleles are present, as in the F_1 hybrids, then 50% of the gametes receive the dominant allele and 50% receive the recessive allele.

Does Mendel's segregation model account for the 3:1 ratio he observed in the F_2 generation of his numerous crosses? For the flower-color character, the model predicts that the two different alleles present in an F_1 individual will segregate into gametes such that half the gametes will have the purple-flower allele and half will have the white-flower allele. During self-pollination, gametes of each class unite randomly. An egg with a purple-flower allele has an equal chance of being fertilized by a sperm with a purple-flower allele or one with a white-flower allele. Since the same is true for an egg with a white-flower allele, there are four equally likely combinations of sperm and egg. Figure 11.5 illustrates these combinations using a **Punnett square**, a handy diagrammatic device for predicting the allele composition of all offspring resulting from a cross between individuals of known genetic makeup. Notice that we use a capital letter to symbolize a dominant allele and a lower-case letter for a recessive allele. In our example, P is the purple-flower allele, and p is the white-flower allele; the gene itself is sometimes referred to as the P/p gene.

In the F_2 offspring, what color will the flowers be? One-fourth of the plants have inherited two purple-flower alleles; these plants will have purple flowers. One-half of the F_2 offspring have inherited one purple-flower allele and one white-flower allele; these plants will also have purple flowers, the dominant trait. Finally, one-fourth of the F_2 plants have inherited two white-flower alleles and will express the recessive trait. Thus, Mendel's model accounts for the 3:1 ratio of traits that he observed in the F_2 generation.

Useful Genetic Vocabulary

An organism that has a pair of identical alleles for a character is said to be **homozygous** for the gene controlling that character. In the parental generation in Figure 11.5, the purple pea plant is homozygous for the dominant allele (PP), while the white plant is homozygous for the recessive allele (pp). Homozygous plants "breed true" because all of their gametes contain the same allele—either P or p in this example. If we cross dominant homozygotes with recessive homozygotes, every offspring will have two different alleles—Pp in the case of the F_1 hybrids of our flower-color experiment (see Figure 11.5). An organism that has two different alleles for a gene is said to be **heterozygous** for that gene. Unlike homozygotes, heterozygotes produce gametes with different alleles, so they are not true-breeding. For example, P- and p-containing gametes are both produced by our F_1 hybrids. Self-pollination of the F_1

hybrids thus produces both purple-flowered and white-flowered offspring.

Because of the different effects of dominant and recessive alleles, an organism's traits do not always reveal its genetic composition. Therefore, we distinguish between an organism's appearance or observable traits, called its **phenotype**, and its genetic makeup, its **genotype**. In the case of flower color in pea plants, PP and Pp plants have the same phenotype (purple) but different genotypes. **Figure 11.6** reviews these terms. Note that "phenotype" refers to physiological traits as well as traits that relate directly to appearance. For example, there is a pea variety that lacks the normal ability to self-pollinate. This physiological variation (non-self-pollination) is a phenotypic trait.

The Testcross

Suppose we have a "mystery" pea plant that has purple flowers. We cannot tell from its flower color if this plant is homozygous (PP) or heterozygous (Pp) because both genotypes result in the same purple phenotype. To determine the genotype, we can cross this plant with a white-flowered plant (pp), which will make only gametes with the recessive allele (p). The allele in the gamete contributed by the mystery plant will therefore determine the appearance of the offspring (**Figure 11.7**). If all the offspring of the cross have purple flowers, then the purple-flowered mystery plant must be homozygous for the dominant allele, because a $PP \times pp$ cross produces all Pp offspring. But if both the purple and the white phenotypes appear among the offspring, then the purple-flowered parent must be heterozygous. The offspring of a $Pp \times pp$ cross will be expected to

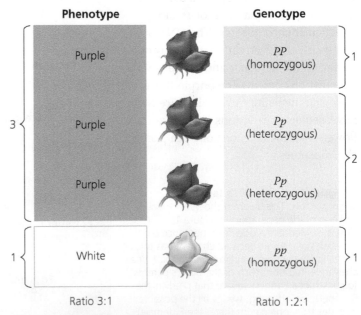

▲ **Figure 11.6 Phenotype versus genotype.** Grouping F_2 offspring from a cross for flower color according to phenotype results in the typical 3:1 phenotypic ratio. In terms of genotype, however, there are actually two categories of purple-flowered plants, PP (homozygous) and Pp (heterozygous), giving a 1:2:1 genotypic ratio.

The Testcross

Application An organism that exhibits a dominant trait, such as purple flowers in pea plants, can be either homozygous for the dominant allele or heterozygous. To determine the organism's genotype, geneticists can perform a testcross.

Technique In a testcross, the individual with the unknown genotype is crossed with a homozygous individual expressing the recessive trait (white flowers in this example), and Punnett squares are used to predict the possible outcomes.

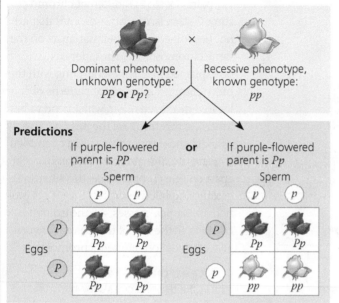

Dominant phenotype, unknown genotype: *PP* or *Pp*?

×

Recessive phenotype, known genotype: *pp*

Predictions

If purple-flowered parent is *PP*   **or**   If purple-flowered parent is *Pp*

Results Matching the results to either prediction identifies the unknown parental genotype (either *PP* or *Pp* in this example). In this testcross, we transferred pollen from a white-flowered plant to the carpels of a purple-flowered plant; the opposite (reciprocal) cross would have led to the same results.

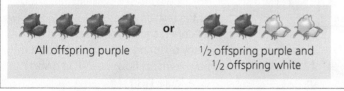

All offspring purple   **or**   1/2 offspring purple and 1/2 offspring white

---

have a 1:1 phenotypic ratio. Breeding an organism of unknown genotype with a recessive homozygote is called a **testcross** because it can reveal the genotype of that organism. The testcross was devised by Mendel and continues to be an important tool of geneticists.

The Law of Independent Assortment

Mendel derived the law of segregation from experiments in which he followed only a *single* character, such as flower color. All the F₁ progeny produced in his crosses of true-breeding parents were **monohybrids**, meaning that they were heterozygous for the one particular character being followed in the cross. We refer to a cross between such heterozygotes as a **monohybrid cross**.

Mendel identified his second law of inheritance by following *two* characters at the same time, such as seed color and seed shape. Seeds (peas) may be either yellow or green. They also may be either round (smooth) or wrinkled. From single-character crosses, Mendel knew that the allele for yellow seeds (*Y*) is dominant and the allele for green seeds (*y*) is recessive. For the seed-shape character, the allele for round (*R*) is dominant, and the allele for wrinkled (*r*) is recessive.

Imagine crossing two true-breeding pea varieties that differ in *both* of these characters—a cross between a plant with yellow-round seeds (*YYRR*) and a plant with green-wrinkled seeds (*yyrr*). The F₁ plants will be **dihybrids**, individuals heterozygous for the two characters being followed in the cross (*YyRr*). But are these two characters transmitted from parents to offspring as a package? That is, will the *Y* and *R* alleles always stay together, generation after generation? Or are seed color and seed shape inherited independently? **Figure 11.8** shows how a **dihybrid cross**, a cross between F₁ dihybrids, can determine which of these two hypotheses is correct.

The F₁ plants, of genotype *YyRr*, exhibit both dominant phenotypes, yellow seeds with round shapes, no matter which hypothesis is correct. The key step in the experiment is to see what happens when F₁ plants self-pollinate and produce F₂ offspring. If the hybrids must transmit their alleles in the same combinations in which the alleles were inherited from the P generation, then the F₁ hybrids will produce only two classes of gametes: *YR* and *yr*. This "dependent assortment" hypothesis predicts that the phenotypic ratio of the F₂ generation will be 3:1, just as in a monohybrid cross (see Figure 11.8, left side).

The alternative hypothesis is that the two pairs of alleles segregate independently of each other. In other words, genes are packaged into gametes in all possible allelic combinations, as long as each gamete has one allele for each gene. In our example, an F₁ plant will produce four classes of gametes in equal quantities: *YR*, *Yr*, *yR*, and *yr*. If sperm of the four classes fertilize eggs of the four classes, there will be 16 (4 × 4) equally probable ways in which the alleles can combine in the F₂ generation, as shown in Figure 11.8, right side. These combinations result in four phenotypic categories with a ratio of 9:3:3:1 (nine yellow-round to three green-round to three yellow-wrinkled to one green-wrinkled). When Mendel did the experiment and classified the F₂ offspring, his results were close to the predicted 9:3:3:1 phenotypic ratio, supporting the hypothesis that the alleles for one gene—controlling seed color or seed shape, in this example—segregate into gametes independently of the alleles of other genes.

Mendel tested his seven pea characters in various dihybrid combinations and always observed a 9:3:3:1 phenotypic ratio in the F₂ generation. Does this override the 3:1 phenotypic ratio seen for the monohybrid cross shown in Figure 11.5? To investigate this question, let's consider one of the two dihybrid characters by itself: Looking only at pea color, we see that there are 416 yellow and 140 green peas—a 2.97:1 ratio, or roughly

▼ Figure 11.8 Inquiry

Do the alleles for one character segregate into gametes dependently or independently of the alleles for a different character?

Experiment Gregor Mendel followed the characters of seed color and seed shape through the F_2 generation. He crossed two true-breeding plants, one with yellow-round seeds and one with green-wrinkled seeds, producing dihybrid F_1 plants. Self-pollination of the F_1 dihybrids produced the F_2 generation. The two hypotheses (dependent and independent "assortment" of the two genes) predict different phenotypic ratios.

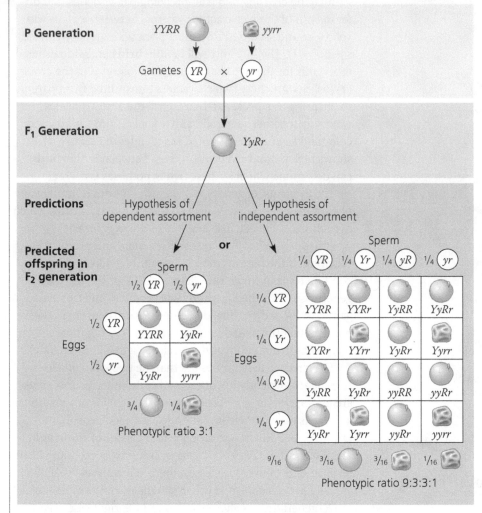

Results

315 ◯ 108 ◯ 101 🟫 32 🟫   Phenotypic ratio approximately 9:3:3:1

Conclusion Only the hypothesis of independent assortment predicts two of the observed phenotypes: green-round seeds and yellow-wrinkled seeds (see the right-hand Punnett square). The alleles for each gene segregate independently, and the two genes are said to assort independently.

Source G. Mendel, Experiments in plant hybridization, *Proceedings of the Natural History Society of Brünn* 4:3–47 (1866).

WHAT IF? Suppose Mendel had transferred pollen from an F_1 plant to the carpel of a plant that was homozygous recessive for both genes. Set up the cross and draw Punnett squares that predict the offspring for both hypotheses. Would this cross have supported the hypothesis of independent assortment equally well?

3:1. In the dihybrid cross, the pea color alleles segregate as if this were a monohybrid cross. The results of Mendel's dihybrid experiments are the basis for what we now call the **law of independent assortment**, which states that *two or more genes assort independently—that is, each pair of alleles segregates independently of each other pair during gamete formation.*

This law applies only to genes (allele pairs) located on different chromosomes—that is, on chromosomes that are not homologous—or very far apart on the same chromosome. (The latter case will be explained in Chapter 12, along with the more complex inheritance patterns of genes located near each other, which tend to be inherited together.) All the pea characters Mendel chose for analysis were controlled by genes on different chromosomes or far apart on one chromosome; this situation greatly simplified interpretation of his multicharacter pea crosses. All the examples we consider in the rest of this chapter involve genes located on different chromosomes.

CONCEPT CHECK 11.1

1. In the dihybrid cross shown in Figure 11.8, calculate the phenotypic ratio in the F_2 generation, considering only the character of pea shape.

2. **DRAW IT** Pea plants heterozygous for flower position and stem length (*AaTt*) are allowed to self-pollinate, and 400 of the resulting seeds are planted. Draw a Punnett square for this cross. How many offspring would be predicted to have terminal flowers and be dwarf? (See Table 11.1.)

3. List all gametes that could be made by a pea plant heterozygous for seed color, seed shape, and pod shape (*YyRrIi;* see Table 11.1). How large a Punnett square would you need to draw to predict the offspring of a self-pollination of this "trihybrid"?

4. **MAKE CONNECTIONS** In some pea plant crosses, the plants are self-pollinated. Explain whether self-pollination is considered asexual or sexual reproduction (refer back to Concept 10.1).

For suggested answers, see Appendix A.

The laws of probability govern Mendelian inheritance

Mendel's laws of segregation and independent assortment reflect the same rules of probability that apply to tossing coins, rolling dice, and drawing cards from a deck. The probability scale ranges from 0 to 1. An event that is certain to occur has a probability of 1, while an event that is certain *not* to occur has a probability of 0. With a coin that has heads on both sides, the probability of tossing heads is 1, and the probability of tossing tails is 0. With a normal coin, the chance of tossing heads is 1/2, and the chance of tossing tails is 1/2. The probability of drawing the ace of spades from a 52-card deck is 1/52. The probabilities of all possible outcomes for an event must add up to 1. With a deck of cards, the chance of picking a card other than the ace of spades is 51/52.

Tossing a coin illustrates an important lesson about probability. For every toss, the probability of heads is 1/2. The outcome of any particular toss is unaffected by what has happened on previous trials. We refer to phenomena such as coin tosses as independent events. Each toss of a coin, whether done sequentially with one coin or simultaneously with many, is independent of every other toss. And like two separate coin tosses, the alleles of one gene segregate into gametes independently of another gene's alleles (the law of independent assortment). Two basic rules of probability, described below, can help us predict the outcome of the fusion of such gametes in simple monohybrid crosses and more complicated crosses.

The Multiplication and Addition Rules Applied to Monohybrid Crosses

How do we determine the probability that two or more independent events will occur together in some specific combination? For example, what is the chance that two coins tossed simultaneously will both land heads up? The **multiplication rule** states that to determine this probability, we multiply the probability of one event (one coin coming up heads) by the probability of the other event (the other coin coming up heads). By the multiplication rule, then, the probability that both coins will land heads up is $1/2 \times 1/2 = 1/4$.

We can apply the same reasoning to an F_1 monohybrid cross. With seed shape in pea plants as the heritable character, the genotype of F_1 plants is Rr. Segregation in a heterozygous plant is like flipping a coin in terms of calculating the probability of each outcome: Each egg produced has a 1/2 chance of carrying the dominant allele (R) and a 1/2 chance of carrying the recessive allele (r). The same odds apply to each sperm cell produced. For a particular F_2 plant to have wrinkled seeds, the recessive trait, both the egg and the sperm that come together must carry the r allele. The probability that an r allele will be

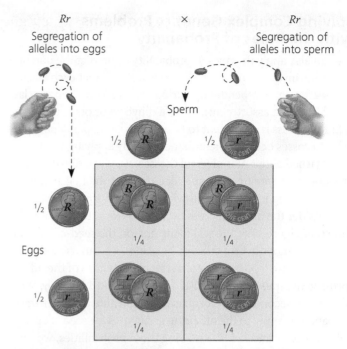

▲ **Figure 11.9 Segregation of alleles and fertilization as chance events.** When a heterozygote (Rr) forms gametes, whether a particular gamete ends up with an R or an r is like the toss of a coin. We can determine the probability for any genotype among the offspring of two heterozygotes by multiplying together the individual probabilities of an egg and sperm having a particular allele (R or r in this example).

present in both gametes at fertilization is found by multiplying 1/2 (the probability that the egg will have an r) $\times$ 1/2 (the probability that the sperm will have an r). Thus, the multiplication rule tells us that the probability of an F_2 plant having wrinkled seeds (rr) is 1/4 **(Figure 11.9)**. Likewise, the probability of an F_2 plant carrying both dominant alleles for seed shape (RR) is 1/4.

To figure out the probability that an F_2 plant from a monohybrid cross will be heterozygous rather than homozygous, we need to invoke a second rule. Notice in Figure 11.9 that the dominant allele can come from the egg and the recessive allele from the sperm, or vice versa. That is, F_1 gametes can combine to produce Rr offspring in two *mutually exclusive* ways: For any particular heterozygous F_2 plant, the dominant allele can come from the egg *or* the sperm, but not from both. According to the **addition rule**, the probability that any one of two or more mutually exclusive events will occur is calculated by adding their individual probabilities. As we have just seen, the multiplication rule gives us the individual probabilities that we will now add together. The probability for one possible way of obtaining an F_2 heterozygote—the dominant allele from the egg and the recessive allele from the sperm—is 1/4. The probability for the other possible way—the recessive allele from the egg and the dominant allele from the sperm—is also 1/4 (see Figure 11.9). Using the rule of addition, then, we can calculate the probability of an F_2 heterozygote as $1/4 + 1/4 = 1/2$.

Solving Complex Genetics Problems with the Rules of Probability

We can also apply the rules of probability to predict the outcome of crosses involving multiple characters. Recall that each allelic pair segregates independently during gamete formation (the law of independent assortment). Thus, a dihybrid or other multi-character cross is equivalent to two or more independent monohybrid crosses occurring simultaneously. By applying what we have learned about monohybrid crosses, we can determine the probability of specific genotypes occurring in the F_2 generation without having to construct unwieldy Punnett squares.

Consider the dihybrid cross between *YyRr* heterozygotes shown in Figure 11.8. We will focus first on the seed-color character. For a monohybrid cross of *Yy* plants, we can use a simple Punnett square to determine that the probabilities of the offspring genotypes are 1/4 for *YY*, 1/2 for *Yy*, and 1/4 for *yy*. We can draw a second Punnett square to determine that the same probabilities apply to the offspring genotypes for seed shape: 1/4 *RR*, 1/2 *Rr*, and 1/4 *rr*. Knowing these probabilities, we can simply use the multiplication rule to determine the probability of each of the genotypes in the F_2 generation. To give two examples, the calculations for finding the probabilities of two of the possible F_2 genotypes (*YYRR* and *YyRR*) are shown below:

Probability of *YYRR* = ¼ (probability of *YY*) × ¼ (*RR*) = ¹⁄₁₆

Probability of *YyRR* =   ½ (*Yy*) × ¼ (*RR*) = ⅛

The *YYRR* genotype corresponds to the upper left box in the larger Punnett square in Figure 11.8 (one box = 1/16). Looking closely at the larger Punnett square in Figure 11.8, you will see that 2 of the 16 boxes (1/8) correspond to the *YyRR* genotype.

Now let's see how we can combine the multiplication and addition rules to solve even more complex problems in Mendelian genetics. Imagine a cross of two pea varieties in which we track the inheritance of three characters. Let's cross a trihybrid with purple flowers and yellow, round seeds (heterozygous for all three genes) with a plant with purple flowers and green, wrinkled seeds (heterozygous for flower color but homozygous recessive for the other two characters). Using Mendelian symbols, our cross is *PpYyRr* × *Ppyyrr*. What fraction of offspring from this cross is predicted to exhibit the recessive phenotypes for *at least two* of the three characters?

To answer this question, we can start by listing all genotypes we could get that fulfill this condition: *ppyyRr*, *ppYyrr*, *Ppyyrr*, *PPyyrr*, and *ppyyrr*. (Because the condition is *at least two* recessive traits, it includes the last genotype, which shows all three recessive traits.) Next, we calculate the probability for each of these genotypes resulting from our *PpYyRr* × *Ppyyrr* cross by multiplying together the individual probabilities for the allele pairs, just as we did in our dihybrid example. Note that in a cross involving heterozygous and homozygous allele pairs (for example, *Yy* × *yy*), the probability of heterozygous offspring is 1/2 and the probability of homozygous offspring is 1/2. Finally, we use the addition rule to add the probabilities for all the different genotypes that fulfill the condition of at least two recessive traits, as shown below:

ppyyRr	¼ (probability of *pp*) × ½ (*yy*) × ½ (*Rr*)	= ¹⁄₁₆
ppYyrr	¼ × ½ × ½	= ¹⁄₁₆
Ppyyrr	½ × ½ × ½	= ²⁄₁₆
PPyyrr	¼ × ½ × ½	= ¹⁄₁₆
ppyyrr	¼ × ½ × ½	= ¹⁄₁₆
Chance of *at least two* recessive traits		= ⁶⁄₁₆ or ⅜

In time, you'll be able to solve genetics problems faster by using the rules of probability than by filling in Punnett squares.

We cannot predict with certainty the exact numbers of progeny of different genotypes resulting from a genetic cross. But the rules of probability give us the *chance* of various outcomes. Usually, the larger the sample size, the closer the results will conform to our predictions. The reason Mendel counted so many offspring from his crosses is that he understood this statistical feature of inheritance and had a keen sense of the rules of chance.

CONCEPT CHECK 11.2

1. For any gene with a dominant allele *A* and recessive allele *a*, what proportions of the offspring from an *AA* × *Aa* cross are expected to be homozygous dominant, homozygous recessive, and heterozygous?

2. Two organisms, with genotypes *BbDD* and *BBDd*, are mated. Assuming independent assortment of the *B/b* and *D/d* genes, write the genotypes of all possible offspring from this cross and use the rules of probability to calculate the chance of each genotype occurring.

3. **WHAT IF?** Three characters (flower color, seed color, and pod shape) are considered in a cross between two pea plants (*PpYyIi* × *ppYyii*). What fraction of offspring are predicted to be homozygous recessive for at least two of the three characters?

For suggested answers, see Appendix A.

CONCEPT 11.3

Inheritance patterns are often more complex than predicted by simple Mendelian genetics

In the 20th century, geneticists extended Mendelian principles not only to diverse organisms, but also to patterns of inheritance more complex than those described by Mendel. For the work that led to his two laws of inheritance, Mendel chose pea plant characters that turn out to have a relatively simple genetic basis: Each character is determined by one gene, for which there are only two alleles, one completely dominant and the other completely recessive. (There is one exception: Mendel's pod-shape character is actually determined by two genes.) Not all heritable characters are determined so simply, and the relationship between genotype and phenotype is rarely so straightforward. Mendel himself realized that he could not explain the more complicated patterns he observed in crosses involving other pea characters or other plant species. This does not diminish the utility of Mendelian genetics, however, because the basic principles of segregation and independent assortment apply even to more complex patterns of inheritance. In this section, we'll extend Mendelian genetics to hereditary patterns that were not reported by Mendel.

Extending Mendelian Genetics for a Single Gene

The inheritance of characters determined by a single gene deviates from simple Mendelian patterns when alleles are not completely dominant or recessive, when a particular gene has more than two alleles, or when a single gene produces multiple phenotypes. We'll describe examples of each of these situations in this section.

Degrees of Dominance

Alleles can show different degrees of dominance and recessiveness in relation to each other. In Mendel's classic pea crosses, the F_1 offspring always looked like one of the two parental varieties because one allele in a pair showed **complete dominance** over the other. In such situations, the phenotypes of the heterozygote and the dominant homozygote are indistinguishable.

For some genes, however, neither allele is completely dominant, and the F_1 hybrids have a phenotype somewhere between those of the two parental varieties. This phenomenon, called **incomplete dominance**, is seen when red snapdragons are crossed with white snapdragons: All the F_1 hybrids have pink flowers **(Figure 11.10)**. This third, intermediate phenotype results from flowers of the heterozygotes having less red pigment than the red homozygotes. (This is unlike the case of Mendel's pea plants, where the Pp heterozygotes make enough pigment for the flowers to be purple, indistinguishable from those of PP plants.)

At first glance, incomplete dominance of either allele seems to provide evidence for the blending hypothesis of inheritance, which would predict that the red or white trait could never be retrieved from the pink hybrids. In fact, interbreeding F_1 hybrids produces F_2 offspring with a phenotypic ratio of one red to two pink to one white. (Because heterozygotes have a separate phenotype, the genotypic and phenotypic ratios for the F_2 generation are the same, 1:2:1.) The segregation of the red-flower and white-flower alleles in the gametes produced by the pink-flowered plants confirms that the alleles for flower color are heritable factors that maintain their identity in the hybrids; that is, inheritance is particulate.

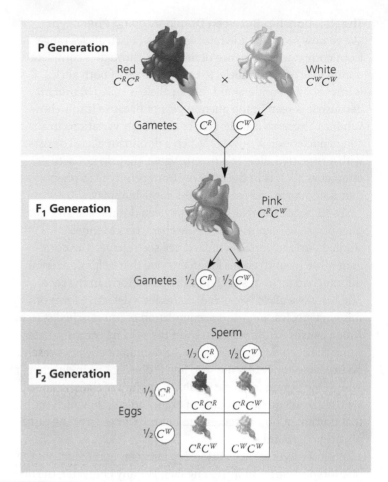

▲ **Figure 11.10 Incomplete dominance in snapdragon color.** When red snapdragons are crossed with white ones, the F_1 hybrids have pink flowers. Segregation of alleles into gametes of the F_1 plants results in an F_2 generation with a 1:2:1 ratio for both genotype and phenotype. Neither allele is dominant, so rather than using upper- and lowercase letters, we use the letter C with a superscript to indicate an allele for flower color: C^R for red and C^W for white.

? *Suppose a classmate argues that this figure supports the blending hypothesis for inheritance. What might your classmate say, and how would you respond?*

Another variation on dominance relationships between alleles is called **codominance**; in this variation, the two alleles each affect the phenotype in separate, distinguishable ways. For example, the human MN blood group is determined by codominant alleles for two specific molecules located on the surface of red blood cells, the M and N molecules. A single gene locus, at which two allelic variations are possible, determines the phenotype of this blood group. Individuals homozygous for the M allele (MM) have red blood cells with only M molecules; individuals homozygous for the N allele (NN) have red blood cells with only N molecules. But *both* M and N molecules are present on the red blood cells of individuals heterozygous for the M and N alleles (MN). Note that the MN phenotype is *not* intermediate between the M and N phenotypes, which distinguishes codominance from incomplete dominance. Rather, *both* M and N phenotypes are exhibited by heterozygotes, since both molecules are present.

The Relationship Between Dominance and Phenotype

We've now seen that the relative effects of two alleles range from complete dominance of one allele, through incomplete dominance of either allele, to codominance of both alleles. It is important to understand that an allele is called *dominant* because it is seen in the phenotype, not because it somehow subdues a recessive allele. Alleles are simply variations in a gene's nucleotide sequence. When a dominant allele coexists with a recessive allele in a heterozygote, they do not actually interact at all. It is in the pathway from genotype to phenotype that dominance and recessiveness come into play.

To illustrate the relationship between dominance and phenotype, we can use one of the characters Mendel studied—round versus wrinkled pea seed shape. The dominant allele (round) codes for an enzyme that helps convert an unbranched form of starch to a branched form in the seed. The recessive allele (wrinkled) codes for a defective form of this enzyme, leading to an accumulation of unbranched starch, which causes excess water to enter the seed by osmosis. Later, when the seed dries, it wrinkles. If a dominant allele is present, no excess water enters the seed and it does not wrinkle when it dries. One dominant allele results in enough of the enzyme to synthesize adequate amounts of branched starch, which means that dominant homozygotes and heterozygotes have the same phenotype: round seeds.

A closer look at the relationship between dominance and phenotype reveals an intriguing fact: For any character, the observed dominant/recessive relationship of alleles depends on the level at which we examine the phenotype. **Tay-Sachs disease**, an inherited disorder in humans, provides an example. The brain cells of a child with Tay-Sachs disease cannot metabolize certain lipids because a crucial enzyme does not work properly. As these lipids accumulate in brain cells, the child begins to suffer seizures, blindness, and degeneration of motor and mental performance and dies within a few years.

Only children who inherit two copies of the Tay-Sachs allele (homozygotes) have the disease. Thus, at the *organismal* level, the Tay-Sachs allele qualifies as recessive. However, the activity level of the lipid-metabolizing enzyme in heterozygotes is intermediate between that in individuals homozygous for the normal allele and that in individuals with Tay-Sachs disease. The intermediate phenotype observed at the *biochemical* level is characteristic of incomplete dominance of either allele. Fortunately, the heterozygote condition does not lead to disease symptoms, apparently because half the normal enzyme activity is sufficient to prevent lipid accumulation in the brain. Extending our analysis to yet another level, we find that heterozygous individuals produce equal numbers of normal and dysfunctional enzyme molecules. Thus, at the *molecular* level, the normal allele and the Tay-Sachs allele are codominant. As you can see, whether alleles appear to be completely dominant, incompletely dominant, or codominant depends on the level at which the phenotype is analyzed.

Frequency of Dominant Alleles While you might assume that the dominant allele for a particular character would be more common in a population than the recessive one, this is not always so. For example, about one baby out of 400 in the United States is born with extra digits (fingers or toes), a condition known as polydactyly. Some cases are caused by the presence of a dominant allele. The low frequency of polydactyly indicates that the recessive allele, which results in five digits per appendage when homozygous, is far more prevalent than the dominant allele. In Chapter 21, you'll learn how relative frequencies of alleles in a population are affected by natural selection.

Multiple Alleles

Only two alleles exist for each of the seven pea characters that Mendel studied, but most genes exist in more than two allelic forms. The ABO blood groups in humans, for instance, are determined by three alleles of a single gene: I^A, I^B, and i. Each person has two alleles of the three for the blood group gene, which determines his or her blood group (phenotype): A, B, AB, or O. These letters refer to two carbohydrates—A and B—that may be found on the surface of red blood cells. A person's blood cells may have carbohydrate A (type A blood), carbohydrate B (type B), both (type AB), or neither (type O), as shown schematically in **Figure 11.11**. Matching compatible blood groups is critical for safe blood transfusions (see Chapter 35).

(a) The three alleles for the ABO blood groups and their carbohydrates. Each allele codes for an enzyme that may add a specific carbohydrate (designated by the superscript on the allele and shown as a triangle or circle) to red blood cells.

Allele	I^A	I^B	i
Carbohydrate	A △	B ○	none

(b) Blood group genotypes and phenotypes. There are six possible genotypes, resulting in four different phenotypes.

Genotype	$I^A I^A$ or $I^A i$	$I^B I^B$ or $I^B i$	$I^A I^B$	ii
Red blood cell appearance				
Phenotype (blood group)	A	B	AB	O

▲ **Figure 11.11 Multiple alleles for the ABO blood groups.** The four blood groups result from different combinations of three alleles.

? *Based on the surface carbohydrate phenotypes in (b), what are the dominance relationships among the alleles?*

Pleiotropy

So far, we have treated Mendelian inheritance as though each gene affects only one phenotypic character. Most genes, however, have multiple phenotypic effects, a property called **pleiotropy** (from the Greek *pleion*, more). In humans, for example, pleiotropic alleles are responsible for the multiple symptoms associated with certain hereditary diseases, such as cystic fibrosis and sickle-cell disease, discussed later in this chapter. In the garden pea, the gene that determines flower color also affects the color of the coating on the outer surface of the seed, which can be gray or white. Given the intricate molecular and cellular interactions responsible for an organism's development and physiology, it isn't surprising that a single gene can affect a number of characteristics in an organism.

Extending Mendelian Genetics for Two or More Genes

Dominance relationships, multiple alleles, and pleiotropy all have to do with the effects of the alleles of a single gene. We now consider two situations in which two or more genes are involved in determining a particular phenotype. In the first case, one gene affects the phenotype of another because the two gene products interact, whereas in the second, multiple genes independently affect a single trait.

Epistasis

In **epistasis** (from the Greek for "standing upon"), the phenotypic expression of a gene at one locus alters that of a gene at a second locus. An example will help clarify this concept. In Labrador retrievers (commonly called Labs), black coat color is dominant to brown. Let's designate *B* and *b* as the two alleles for this character. For a Lab to have brown fur, its genotype must be *bb*; these dogs are called chocolate Labs. But there is more to the story. A second gene determines whether or not pigment will be deposited in the hair. The dominant allele, symbolized by *E*, results in the deposition of either black or brown pigment, depending on the genotype at the first locus. But if the Lab is homozygous recessive for the second locus (*ee*), then the coat is yellow, regardless of the genotype at the black/brown locus. In this case, the gene for pigment deposition (*E/e*) is said to be epistatic to the gene that codes for black or brown pigment (*B/b*).

What happens if we mate black Labs that are heterozygous for both genes (*BbEe*)? Although the two genes affect the same phenotypic character (coat color), they follow the law of independent assortment. Thus, our breeding experiment represents an F_1 dihybrid cross, like those that produced a 9:3:3:1 ratio in Mendel's experiments. We can use a Punnett square to represent the genotypes of the F_2 offspring **(Figure 11.12)**. As a result of epistasis, the phenotypic ratio among the F_2 offspring is nine black to three chocolate (brown) to four yellow. Other types of epistatic interactions produce different ratios, but all are modified versions of 9:3:3:1.

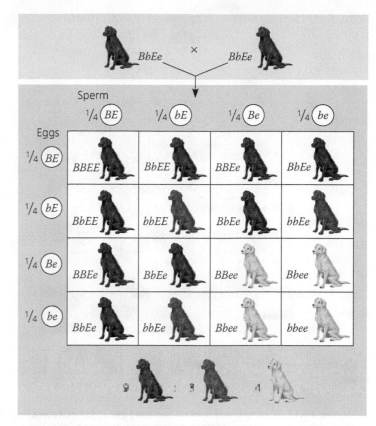

▲ **Figure 11.12 An example of epistasis.** This Punnett square illustrates the genotypes and phenotypes predicted for offspring of matings between two black Labrador retrievers of genotype *BbEe*. The *E/e* gene, which is epistatic to the *B/b* gene coding for hair pigment, controls whether or not pigment of any color will be deposited in the hair.

Polygenic Inheritance

Mendel studied characters that could be classified on an either-or basis, such as purple versus white flower color. But for many characters, such as human skin color and height, an either-or classification is impossible because the characters vary in the population in gradations along a continuum. These are called **quantitative characters**. Quantitative variation usually indicates **polygenic inheritance**, the additive effect of two or more genes on a single phenotypic character. (In a way, this is the converse of pleiotropy, where a single gene affects several phenotypic characters.)

There is evidence, for instance, that skin pigmentation in humans is controlled by at least three separately inherited genes (probably more, but we will simplify). Let's consider three genes, with a dark-skin allele for each gene (*A*, *B*, or *C*) contributing one "unit" of darkness (also a simplification) to the phenotype and being incompletely dominant to the other, light-skin allele (*a*, *b*, or *c*). An *AABBCC* person would be very dark, while an *aabbcc* individual would be very light. An *AaBbCc* person would have skin of an intermediate shade. Because the alleles have a cumulative effect, the genotypes *AaBbCc* and *AABbcc* would make the same genetic contribution (three units) to skin darkness. There are seven skin-color phenotypes that could result from a mating between *AaBbCc* heterozygotes. In a large

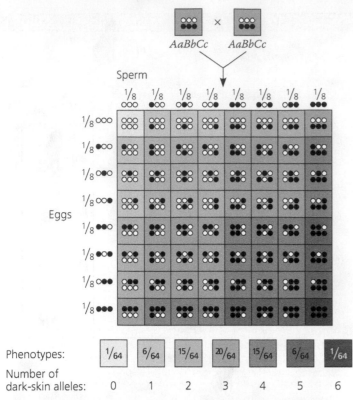

Phenotypes:

| $\frac{1}{64}$ | $\frac{6}{64}$ | $\frac{15}{64}$ | $\frac{20}{64}$ | $\frac{15}{64}$ | $\frac{6}{64}$ | $\frac{1}{64}$ |

Number of
dark-skin alleles: 0   1   2   3   4   5   6

▲ **Figure 11.13 A simplified model for polygenic inheritance of skin color.** In this model, three separately inherited genes affect skin color. The heterozygous individuals (*AaBbCc*) represented by the two rectangles at the top of this figure each carry three dark-skin alleles (black circles, representing *A*, *B*, or *C*) and three light-skin alleles (white circles, representing *a*, *b*, or *c*). The Punnett square shows all the possible genetic combinations in gametes and offspring of many hypothetical matings between these heterozygotes. The results are summarized by the phenotypic frequencies (fractions) under the Punnett square. (The phenotypic ratio is 1:6:15:20:15:6:1.)

number of such matings, the majority of offspring would be expected to have intermediate phenotypes (skin color in the middle range), as shown in **Figure 11.13**. You can graph the predictions from the Punnett square in the **Scientific Skills Exercise**. Environmental factors, such as exposure to the sun, also affect the skin-color phenotype.

Nature and Nurture: The Environmental Impact on Phenotype

Another departure from simple Mendelian genetics arises when the phenotype for a character depends on environment as well as genotype. A single tree, locked into its inherited genotype, has leaves that vary in size, shape, and greenness, depending on their exposure to wind and sun. In humans, nutrition influences height, exercise alters build, sun-tanning darkens the skin, and experience improves performance on intelligence tests. Even identical twins, who are genetic equals, accumulate phenotypic differences as a result of their unique experiences.

Whether human characteristics are more influenced by genes or the environment— in everyday terms, nature versus nurture—is a very old and hotly contested debate that we will

not attempt to settle here. We can say, however, that a genotype generally is not associated with a rigidly defined phenotype, but rather with a range of phenotypic possibilities due to environmental influences. For some characters, such as the ABO blood group system, the range is extremely narrow; that is, a given genotype mandates a very specific phenotype. Other characteristics, such as a person's blood count of red and white cells, vary quite a bit, depending on such factors as the altitude, the customary level of physical activity, and the presence of infectious agents.

Generally, the phenotypic range is broadest for polygenic characters. Environment contributes to the quantitative nature of these characters, as we have seen in the continuous variation of skin color. Geneticists refer to such characters as **multifactorial**, meaning that many factors, both genetic and environmental, collectively influence phenotype.

Integrating a Mendelian View of Heredity and Variation

We have now broadened our view of Mendelian inheritance by exploring the degrees of dominance as well as multiple alleles, pleiotropy, epistasis, polygenic inheritance, and the phenotypic impact of the environment. Stepping back to see the big picture, how can we integrate these refinements into a comprehensive theory of Mendelian genetics? The key is to make the transition from the reductionist emphasis on single genes and phenotypic characters to the emergent properties of the organism as a whole, one of the themes of this book.

The term *phenotype* can refer not only to specific characters, such as flower color and blood group, but also to an organism in its entirety—*all* aspects of its physical appearance, internal anatomy, physiology, and behavior. Similarly, the term *genotype* can refer to an organism's entire genetic makeup, not just its alleles for a single genetic locus. In most cases, a gene's impact on phenotype is affected by other genes and by the environment. In this integrated view of heredity and variation, an organism's phenotype reflects its overall genotype and unique environmental history.

Considering all that can occur in the pathway from genotype to phenotype, it is indeed impressive that Mendel could uncover the fundamental principles governing the transmission of individual genes from parents to offspring. Mendel's two laws, those of segregation and independent assortment, explain heritable variations in terms of alternative forms of genes (hereditary "particles," now known as the alleles of genes) that are passed along, generation after generation, according to simple rules of probability. This theory of inheritance is equally valid for peas, flies, fishes, birds, and human beings—indeed, for any organism with a sexual life cycle. Furthermore, by extending the principles of segregation and independent assortment to help explain such hereditary patterns as epistasis and quantitative characters, we begin to see how broadly Mendelian genetics applies. From Mendel's abbey

Making a Histogram and Analyzing a Distribution Pattern

What Is the Distribution of Phenotypes Among Offspring of Two Parents Who Are Both Heterozygous for Three Additive Genes? Human skin color is a polygenic trait that is determined by the additive effects of several different genes. In this exercise, you will work with a simplified model of skin-color genetics where three genes are assumed to affect the darkness of skin color and where each gene has two alleles—dark or light. In this model, each dark allele contributes equally to the darkness of skin color, and each pair of alleles segregates independently of each other pair. Using a type of graph called a histogram, you will determine the distribution of phenotypes of offspring with different numbers of dark-skin alleles. (For additional information about graphs, see the Scientific Skills Review in Appendix F and in the Study Area in MasteringBiology.)

How This Model Is Analyzed To predict the phenotypes of the offspring of heterozygous parents, the ratios of the genes for this trait must be calculated. Figure 11.13 shows a simplified model for polygenic inheritance of skin color that includes three of the known genes. According to this model, three separately inherited genes affect the darkness of skin. The heterozygous individuals (*AaBbCc*) represented by the two rectangles at the top of this figure each carry three dark-skin alleles (black circles, which represent *A*, *B*, or *C*) and three light-skin alleles (white circles, which represent *a*, *b*, or *c*). The Punnett square shows all the possible genetic combinations in gametes and in offspring of a large number of hypothetical matings between these heterozygotes. The possible phenotypes are shown under the Punnett square.

Predictions from the Punnett Square If we assume that each square in the Punnett square represents one offspring of the heterozygous *AaBbCc* parents, then the squares below show the phenotypic frequencies of individuals with the same number of dark-skin alleles.

Phenotypes:

Number of dark-skin alleles: 0 1 2 3 4 5 6

Interpret the Data

1. A histogram is a bar graph that shows the distribution of numeric data (here, the number of dark skin alleles). To make a histogram of the allele distribution, put skin color (as the number of dark-skin alleles) along the *x*-axis and number of offspring (out of 64) with each phenotype on the *y*-axis. There are no gaps in our allele data, so draw the bars side-to-side with no space in between.

2. You can see that the skin-color phenotypes are not distributed uniformly. (a) Which phenotype has the highest frequency? Draw a vertical dotted line through that bar. (b) Distributions of values like this one tend to show one of several common patterns. Sketch a rough curve that approximates the values and look at its shape. Is it symmetrically distributed around a central peak value (a "normal distribution," sometimes called a bell curve); is it skewed to one end of the *x*-axis or the other (a "skewed distribution"); or does it show two apparent groups of frequencies (a "bimodal distribution")? Explain the reason for the curve's shape. (It will help to read the text description that supports Figure 11.13.)

3. If one of the three genes were lethal when homozygous recessive, what would happen to the distribution of phenotype frequencies? To determine this, use *bb* as an example of a lethal genotype. Using Figure 11.13, identify offspring where the center circle (the *B/b* gene) in both the top and bottom rows of the square is white, representing the homozygous state *bb*. Because *bb* individuals would not survive, cross out those squares, then count the phenotype frequencies of the surviving offspring according to the number of dark-skin alleles (0–6) and graph the new data. What happens to the shape of the curve compared with the curve in question 2? What does this indicate about the distribution of phenotype frequencies?

Further Reading R.A. Sturm, A golden age of human pigmentation genetics, *Trends in Genetics* 22: 464–468 (2006). doi:10.1016/j.tig.2006.06.010

(MB) A version of this Scientific Skills Exercise can be assigned in MasteringBiology.

---

garden came a theory of particulate inheritance that anchors modern genetics. In the last section of this chapter, we'll apply Mendelian genetics to human inheritance, with emphasis on the transmission of hereditary diseases.

CONCEPT CHECK 11.3

1. *Incomplete dominance* and *epistasis* are both terms that define genetic relationships. What is the most basic distinction between these terms?

2. If a man with type AB blood marries a woman with type O, what blood types would you expect in their children? What fraction would you expect of each type?

3. **WHAT IF?** A rooster with gray feathers and a hen of the same phenotype produce 15 gray, 6 black, and 8 white chicks. What is the simplest explanation for the inheritance of these colors in chickens? What phenotypes would you expect in the offspring of a cross between a gray rooster and a black hen?

For suggested answers, see Appendix A.

CONCEPT 11.4

Many human traits follow Mendelian patterns of inheritance

Peas are convenient subjects for genetic research, but humans are not. The human generation span is long—about 20 years—and human parents produce many fewer offspring than peas and most other species. Even more important, it wouldn't be ethical to ask pairs of humans to breed so that the phenotypes of their offspring could be analyzed! In spite of these constraints, the study of human genetics continues, spurred on by our desire to understand our own inheritance. New molecular biological techniques have led to many breakthrough discoveries, but basic Mendelian genetics endures as the foundation of human genetics.

Unable to manipulate the matings of people, geneticists instead analyze results that have already occurred by collecting information about a family's history for a particular trait.

Pedigree Analysis

Geneticists assemble information about members of a family into a tree diagram that describes the traits of parents and children across the generations, called a **pedigree**.

Figure 11.14a shows a three-generation pedigree that traces the occurrence of a pointed contour of the hairline on the forehead. This trait, called a widow's peak, is due to a dominant allele, W. Because the widow's-peak allele is dominant, all individuals who lack a widow's peak must be homozygous recessive (ww). The two grandparents with widow's peaks must have the Ww genotype, since some of their offspring are homozygous recessive. The offspring in the second generation who *do* have widow's peaks must also be heterozygous, because they are the products of $Ww \times ww$ matings. The third generation in this pedigree consists of two sisters. The one who has a widow's peak could be either homozygous (WW) or heterozygous (Ww), given what we know about the genotypes of her parents (both Ww).

Figure 11.14b is a pedigree of the same family, but this time we focus on a recessive trait, attached earlobes. We'll use f for the recessive allele and F for the dominant allele, which results in free earlobes. As you work your way through the pedigree, notice once again that you can apply what you have learned about Mendelian inheritance to understand the genotypes shown for the family members.

An important application of a pedigree is to help us calculate the probability that a future child will have a particular genotype and phenotype. Suppose that the couple represented in the second generation of Figure 11.14 decides to have one more child. What is the probability that the child will have a widow's peak? This is equivalent to a Mendelian F_1 monohybrid cross ($Ww \times Ww$), and thus the probability that a child will inherit a dominant allele and have a widow's peak is 3/4 (1/4 WW + 1/2 Ww). What is the probability that the child will have attached earlobes? Again, we can treat this as a monohybrid cross ($Ff \times Ff$), but this time we want to know the chance that the offspring will be homozygous recessive (ff). That probability is 1/4. Finally, what is the chance that the child will have a widow's peak *and* attached earlobes? Assuming that the genes for these two characters are on different chromosomes, the two pairs of alleles will assort independently in this dihybrid cross ($WwFf \times WwFf$). Thus, we can use the multiplication rule: 3/4 (chance of widow's peak) $\times$ 1/4 (chance of attached earlobes) = 3/16 (chance of widow's peak and attached earlobes).

Pedigrees are a more serious matter when the alleles in question cause disabling or deadly diseases instead of innocuous human variations such as hairline or earlobe configuration. However, for disorders inherited as simple Mendelian traits, the same techniques of pedigree analysis apply.

Recessively Inherited Disorders

Thousands of genetic disorders are known to be inherited as simple recessive traits. These disorders range in severity from

Key

☐ Male	■ Affected male	Mating
○ Female	● Affected female	Offspring, in birth order (first-born on left)

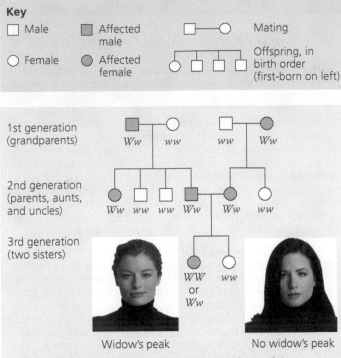

Widow's peak   No widow's peak

(a) Is a widow's peak a dominant or recessive trait?

Tips for pedigree analysis: Notice in the third generation that the second-born daughter lacks a widow's peak, although both of her parents had the trait. Such a pattern of inheritance supports the hypothesis that the trait is due to a dominant allele. If the trait were due to a *recessive* allele, and both parents had the recessive phenotype, then *all* of their offspring would also have the recessive phenotype.

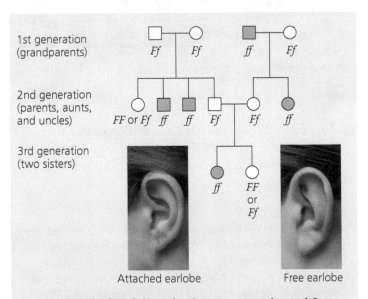

Attached earlobe   Free earlobe

(b) Is an attached earlobe a dominant or recessive trait?

Tips for pedigree analysis: Notice that the first-born daughter in the third generation has attached earlobes, although both of her parents lack that trait (they have free earlobes). Such a pattern is easily explained if the attached-lobe phenotype is due to a recessive allele. If it were due to a *dominant* allele, then at least one parent would also have had the trait.

▲ **Figure 11.14 Pedigree analysis.** Each of these pedigrees traces a trait through three generations of the same family. The two traits have different inheritance patterns, as seen by analysis of the pedigrees.

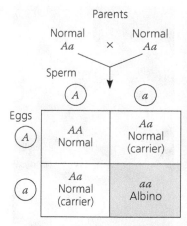

Parents

Normal
Aa × Normal
Aa

Sperm

	A	a
Eggs A	*AA* Normal	*Aa* Normal (carrier)
a	*Aa* Normal (carrier)	*aa* Albino

▲ **Figure 11.15 Albinism: a recessive trait.** One of the two sisters shown here has normal coloration; the other is albino. Most recessive homozygotes are born to parents who are carriers of the disorder but themselves have a normal phenotype, the case shown in the Punnett square.

? *What is the probability that the sister with normal coloration is a carrier of the albinism allele?*

relatively mild, such as albinism (lack of pigmentation, which results in susceptibility to skin cancers and vision problems), to life-threatening, such as cystic fibrosis.

The Behavior of Recessive Alleles

How can we account for the behavior of alleles that cause recessively inherited disorders? Recall that genes code for proteins of specific function. An allele that causes a genetic disorder (let's call it allele *a*) codes for either a malfunctioning protein or no protein at all. In the case of disorders classified as recessive, heterozygotes (*Aa*) are typically normal in phenotype because one copy of the normal allele (*A*) produces a sufficient amount of the specific protein. Thus, a recessively inherited disorder shows up only in the homozygous individuals (*aa*) who inherit one recessive allele from each parent. Although phenotypically normal with regard to the disorder, heterozygotes may transmit the recessive allele to their offspring and thus are called **carriers**. **Figure 11.15** illustrates these ideas using albinism as an example.

Most people who have recessive disorders are born to parents who are carriers of the disorder but have a normal phenotype, as is the case shown in the Punnett square in Figure 11.15. A mating between two carriers corresponds to a Mendelian F₁ monohybrid cross, so the predicted genotypic ratio for the offspring is 1 *AA* : 2 *Aa* : 1 *aa*. Thus, each child has a 1/4 chance of inheriting a double dose of the recessive allele; in the case of albinism, such a child will be albino. From the genotypic ratio, we also can see that out of three offspring with the *normal* phenotype (one *AA* plus two *Aa*), two are predicted to be heterozygous carriers, a 2/3 chance. Recessive homozygotes could also result from *Aa* × *aa* and *aa* × *aa* matings, but if the disorder is lethal before reproductive age or results in sterility (neither of which is true for albinism), no *aa* individuals will reproduce. Even if recessive homozygotes

are able to reproduce, such individuals will still account for a much smaller percentage of the population than heterozygous carriers (for reasons we will examine in Chapter 21).

In general, genetic disorders are not evenly distributed among all groups of people. For example, the incidence of Tay-Sachs disease, which we described earlier in this chapter, is disproportionately high among Ashkenazic Jews, Jewish people whose ancestors lived in central Europe. In that population, Tay-Sachs disease occurs in one out of 3,600 births, an incidence about 100 times greater than that among non-Jews or Mediterranean (Sephardic) Jews. This uneven distribution results from the different genetic histories of the world's peoples during less technological times, when populations were more geographically (and hence genetically) isolated.

When a disease-causing recessive allele is rare, it is relatively unlikely that two carriers of the same harmful allele will meet and mate. However, if the man and woman are close relatives (for example, siblings or first cousins), the probability of passing on recessive traits increases greatly. These are called consanguineous ("same blood") matings, and they are indicated in pedigrees by double lines. Because people with recent common ancestors are more likely to carry the same recessive alleles than are unrelated people, it is more likely that a mating of close relatives will produce offspring homozygous for recessive traits—including harmful ones. Such effects can be observed in many types of domesticated and zoo animals that have become inbred.

There is debate among geneticists about the extent to which human consanguinity increases the risk of inherited diseases. Many deleterious alleles have such severe effects that a homozygous embryo spontaneously aborts long before birth. Still, most societies and cultures have laws or taboos forbidding marriages between close relatives. These rules may have evolved out of empirical observation that in most populations, stillbirths and birth defects are more common when parents are closely related. Social and economic factors have also influenced the development of customs and laws against consanguineous marriages.

Cystic Fibrosis

The most common lethal genetic disease in the United States is **cystic fibrosis**, which strikes one out of every 2,500 people of European descent but is much rarer in other groups. Among people of European descent, one out of 25 (4%) are carriers of the cystic fibrosis allele. The normal allele for this gene codes for a membrane protein that functions in the transport of chloride ions between certain cells and the extracellular fluid. These chloride transport channels are defective or absent in the plasma membranes of children who inherit two recessive alleles for cystic fibrosis. The result is an abnormally high concentration of extracellular chloride, which causes the mucus that coats certain cells to become thicker and stickier than normal. The mucus builds up in the pancreas, lungs, digestive tract, and other organs, leading to multiple (pleiotropic) effects, including poor absorption of nutrients from the intestines, chronic bronchitis, and recurrent bacterial infections.

Untreated, cystic fibrosis can cause death by the age of 5. Daily doses of antibiotics to stop infection, gentle pounding on the chest to clear mucus from clogged airways, and other therapies can prolong life. In the U.S., more than half of those with cystic fibrosis now survive into their 30s and beyond. Recent research on gene-based treatments also shows much promise.

Sickle-Cell Disease: A Genetic Disorder with Evolutionary Implications

EVOLUTION The most common inherited disorder among people of African descent is **sickle-cell disease**, which affects one out of 400 African-Americans. Sickle-cell disease is caused by the substitution of a single amino acid in the hemoglobin protein of red blood cells; in homozygous individuals, all hemoglobin is of the sickle-cell (abnormal) variety. When the oxygen content of an affected individual's blood is low (at high altitudes or under physical stress, for instance), the sickle-cell hemoglobin molecules aggregate into long rods that deform the red cells into a sickle shape (see Figure 3.22). Sickled cells may clump and clog small blood vessels, often leading to other symptoms throughout the body, including physical weakness, pain, organ damage, and even paralysis. Regular blood transfusions can ward off brain damage in children with sickle-cell disease, and new drugs can help prevent or treat other problems, but there is no cure.

Although two sickle-cell alleles are necessary for an individual to manifest full-blown sickle-cell disease, the presence of one sickle-cell allele can affect the phenotype. Thus, at the organismal level, the normal allele is incompletely dominant to the sickle-cell allele. Heterozygotes (carriers), said to have *sickle-cell trait*, are usually healthy, but they may suffer some sickle-cell symptoms during prolonged periods of reduced blood oxygen. At the molecular level, the two alleles are codominant; both normal and abnormal (sickle-cell) hemoglobins are made in heterozygotes.

About one out of ten African-Americans have sickle-cell trait, an unusually high frequency of heterozygotes for an allele with severe detrimental effects in homozygotes. Why haven't evolutionary processes resulted in the disappearance of this allele from this population? One explanation is that having a single copy of the sickle-cell allele reduces the frequency and severity of malaria attacks, especially among young children. The malaria parasite spends part of its life cycle in red blood cells (see Figure 25.26), and the presence of even heterozygous amounts of sickle-cell hemoglobin results in lower parasite densities and hence reduced malaria symptoms. Thus, in tropical Africa, where infection with the malaria parasite is common, the sickle-cell allele confers an advantage to heterozygotes even though it is harmful in the homozygous state. (The balance between these two effects will be discussed in Chapter 21.) The relatively high frequency of African-Americans with sickle-cell trait is a vestige of their African roots.

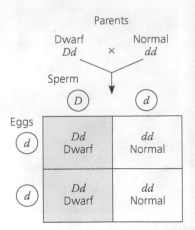

▲ **Figure 11.16 Achondroplasia: a dominant trait.** Dr. Michael C. Ain has achondroplasia, a form of dwarfism caused by a dominant allele. This has inspired his work: He is a specialist in the repair of bone defects caused by achondroplasia and other disorders. The dominant allele (*D*) might have arisen as a mutation in the egg or sperm of a parent or could have been inherited from an affected parent, as shown for an affected father in the Punnett square.

Dominantly Inherited Disorders

Although many harmful alleles are recessive, a number of human disorders are due to dominant alleles. One example is *achondroplasia*, a form of dwarfism that occurs in one of every 25,000 people. Heterozygous individuals have the dwarf phenotype **(Figure 11.16)**. Therefore, all people who are not achondroplastic dwarfs—99.99% of the population—are homozygous for the recessive allele. Like the presence of extra fingers or toes mentioned earlier, achondroplasia is a trait for which the recessive allele is much more prevalent than the corresponding dominant allele.

Dominant alleles that cause a lethal disease are much less common than recessive alleles that have lethal effects. All lethal alleles arise by mutations (changes to the DNA) in cells that produce sperm or eggs; presumably, such mutations are equally likely to be recessive or dominant. A lethal recessive allele can be passed from one generation to the next by heterozygous carriers because the carriers themselves have normal phenotypes. A lethal dominant allele, however, often causes the death of afflicted individuals before they can mature and reproduce, so the allele is not passed on to future generations.

In cases of late-onset diseases, however, a lethal dominant allele may be passed on. If symptoms first appear after reproductive age, the individual may already have transmitted the allele to his or her children. For example, **Huntington's disease**, a degenerative disease of the nervous system, is caused by a lethal dominant allele that has no obvious phenotypic effect until the individual is about 35 to 45 years old. Once the deterioration of the nervous system begins, it is irreversible and inevitably fatal. As with other dominant traits, a child born to a parent with the Huntington's disease allele has a 50% chance of inheriting the allele and the disorder (see the Punnett square in Figure 11.16). In the United States, this devastating disease afflicts about one in 10,000 people.

At one time, the onset of symptoms was the only way to know if a person had inherited the Huntington's allele, but this is no longer the case. By analyzing DNA samples from a large family with a high incidence of the disorder, geneticists tracked the allele for Huntington's disease to a locus near the tip of chromosome 4, and the gene was sequenced in 1993. This information led to the development of a genetic test that could detect the presence of the Huntington's allele in an individual's genome. The availability of this test poses an agonizing dilemma for those with a family history of Huntington's disease. Some individuals may want to be tested for this disease, whereas others may decide it would be too stressful to find out.

Multifactorial Disorders

The hereditary diseases we have discussed so far are sometimes described as simple Mendelian disorders because they result from an abnormality of one or both alleles at a single genetic locus. Many more people are susceptible to diseases that have a multifactorial basis—a genetic component plus a significant environmental influence. Heart disease, diabetes, cancer, alcoholism, certain mental illnesses such as schizophrenia and bipolar disorder, and many other diseases are multifactorial. In many cases, the hereditary component is polygenic. For example, many genes affect cardiovascular health, making some of us more prone than others to heart attacks and strokes. No matter what our genotype, however, our lifestyle has a tremendous effect on phenotype for cardiovascular health and other multifactorial characters. Exercise, a healthful diet, abstinence from smoking, and an ability to handle stressful situations all reduce our risk of heart disease and some types of cancer.

Genetic Counseling Based on Mendelian Genetics

Avoiding simple Mendelian disorders is possible when the risk of a particular genetic disorder can be assessed before a child is conceived or during the early stages of the pregnancy. Many hospitals have genetic counselors who can provide information to prospective parents concerned about a family history for a specific disease.

Consider the case of a hypothetical couple, John and Carol. Each had a brother who died from the same recessively inherited lethal disease. Before conceiving their first child, John and Carol seek genetic counseling to determine the risk of having a child with the disease. From the information about their brothers, we know that both parents of John and both parents of Carol must have been carriers of the recessive allele. Thus, John and Carol are both products of $Aa \times Aa$ crosses, where a symbolizes the allele that causes this particular disease. We also know that John and Carol are not homozygous recessive (aa), because they do not have the disease. Therefore, their genotypes are either AA or Aa.

Given a genotypic ratio of 1 AA : 2 Aa : 1 aa for offspring of an $Aa \times Aa$ cross, John and Carol each have a 2/3 chance of being carriers (Aa). According to the rule of multiplication, the overall probability of their firstborn having the disorder is 2/3 (the chance that John is a carrier) times 2/3 (the chance that Carol is a carrier) times 1/4 (the chance of two carriers having a child with the disease), which equals 1/9. Suppose that Carol and John decide to have a child—after all, there is an 8/9 chance that their baby will not have the disorder. If, despite these odds, their child is born with the disease, then we would know that *both* John and Carol are, in fact, carriers (Aa genotype). If both John and Carol are carriers, there is a 1/4 chance that any subsequent child this couple has will have the disease. The probability is higher for subsequent children because the diagnosis of the disease in the first child established that both parents are carriers, not because the genotype of the first child affects in any way that of future children.

When we use Mendel's laws to predict possible outcomes of matings, it is important to remember that each child represents an independent event in the sense that its genotype is unaffected by the genotypes of older siblings. Suppose that John and Carol have three more children, and *all three* have the hypothetical hereditary disease. There is only one chance in 64 ($1/4 \times 1/4 \times 1/4$) that such an outcome will occur. Despite this run of misfortune, the chance that still another child of this couple will have the disease remains 1/4.

Genetic counseling like this relies on the Mendelian model of inheritance. We owe the "gene idea"—the concept of heritable factors transmitted according to simple rules of chance—to the elegant quantitative experiments of Gregor Mendel. The importance of his discoveries was overlooked by most biologists until early in the 20th century, decades after he reported his findings. In the next chapter, you'll learn how Mendel's laws have their physical basis in the behavior of chromosomes during sexual life cycles and how the synthesis of Mendelian genetics and a chromosome theory of inheritance catalyzed progress in genetics.

CONCEPT CHECK 11.4

1. Beth and Tom each have a sibling with cystic fibrosis, but neither Beth nor Tom nor any of their parents have the disease. Calculate the probability that if this couple has a child, the child will have cystic fibrosis. What would be the probability if a test revealed that Tom is a carrier but Beth is not? Explain your answers.

2. **MAKE CONNECTIONS** In Table 11.1, note the phenotypic ratio of the dominant to recessive trait in the F_2 generation for the monohybrid cross involving flower color. Then determine the phenotypic ratio for the offspring of the second-generation couple in Figure 11.14b. What accounts for the difference in the two ratios?

For suggested answers, see Appendix A.

SUMMARY OF KEY CONCEPTS

CONCEPT 11.1

Mendel used the scientific approach to identify two laws of inheritance (pp. 207–212)

- Gregor Mendel formulated a theory of inheritance based on experiments with garden peas, proposing that parents pass on to their offspring discrete genes that retain their identity through generations. This theory includes two "laws."

- The **law of segregation** states that genes have alternative forms, or **alleles**. In a diploid organism, the two alleles of a gene segregate (separate) during meiosis and gamete formation; each sperm or egg carries only one allele of each pair. This law explains the 3:1 ratio of F$_2$ phenotypes observed when **monohybrids** self-pollinate. Each organism inherits one allele for each gene from each parent. In **heterozygotes**, the two alleles are different, and expression of one (the **dominant allele**) masks the phenotypic effect of the other (the **recessive allele**). **Homozygotes** have identical alleles of a given gene and are **true-breeding**.

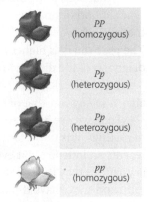

PP (homozygous)

Pp (heterozygous)

Pp (heterozygous)

pp (homozygous)

- The **law of independent assortment** states that the pair of alleles for a given gene segregates into gametes independently of the pair of alleles for any other gene. In a cross between **dihybrids** (individuals heterozygous for two genes), the offspring have four phenotypes in a 9:3:3:1 ratio.

? *When Mendel did crosses of true-breeding purple- and white-flowered pea plants, the white-flowered trait disappeared from the F$_1$ generation but reappeared in the F$_2$ generation. Use genetic terms to explain why that happened.*

CONCEPT 11.2

The laws of probability govern Mendelian inheritance (pp. 213–214)

- The **multiplication rule** states that the probability of two or more events occurring together is equal to the product of the individual probabilities of the independent single events. The **addition rule** states that the probability of an event that can occur in two or more independent, mutually exclusive ways is the sum of the individual probabilities.

- The rules of probability can be used to solve complex genetics problems. A dihybrid or other multicharacter cross is equivalent to two or more independent monohybrid crosses occurring simultaneously. In calculating the chances of the various offspring genotypes from such crosses, each character is first considered separately and then the individual probabilities are multiplied.

DRAW IT *Redraw the Punnett square on the right side of Figure 11.8 as two smaller monohybrid Punnett squares, one for each gene. Below each square, list the fraction of each phenotype produced. Use the rule of multiplication to compute the overall fraction of each possible dihybrid phenotype. Write the phenotypic ratio.*

CONCEPT 11.3

Inheritance patterns are often more complex than predicted by simple Mendelian genetics (pp. 214–219)

- Extensions of Mendelian genetics for a single gene:

Relationship among alleles of a single gene	Description	Example
Complete dominance of one allele	Heterozygous phenotype same as that of homozygous dominant	*PP* *Pp*
Incomplete dominance of either allele	Heterozygous phenotype intermediate between the two homozygous phenotypes	$C^R C^R$ $C^R C^W$ $C^W C^W$
Codominance	Both phenotypes expressed in heterozygotes	$I^A I^B$
Multiple alleles	In the whole population, some genes have more than two alleles	ABO blood group alleles I^A, I^B, i
Pleiotropy	One gene is able to affect multiple phenotypic characters	Sickle-cell disease

- Extensions of Mendelian genetics for two or more genes:

Relationship among two or more genes	Description	Example
Epistasis	The phenotypic expression of one gene affects the expression of another gene	*BbEe* × *BbEe* ... 9 : 3 : 4
Polygenic inheritance	A single phenotypic character is affected by two or more genes	*AaBbCc* × *AaBbCc*

- The expression of a genotype can be affected by environmental influences. Polygenic characters that are also influenced by the environment are called **multifactorial** characters.

- An organism's overall phenotype reflects its complete genotype and unique environmental history. Even in more complex inheritance patterns, Mendel's fundamental laws still apply.

? *Which relationships (in the first column of the two tables above) are demonstrated by the inheritance patterns of the ABO blood group alleles? Explain why or why not, for each genetic relationship.*

CONCEPT 11.4

Many human traits follow Mendelian patterns of inheritance (pp. 219–223)

- Analysis of family **pedigrees** can be used to deduce the possible genotypes of individuals and make predictions about future offspring. Predictions are statistical probabilities rather than certainties.

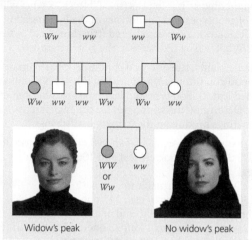

Widow's peak   No widow's peak

- Many genetic disorders are inherited as simple recessive traits, ranging from relatively mild disorders (albinism, for example) to life-threatening ones such as sickle-cell disease and cystic fibrosis. Most affected (homozygous recessive) individuals are children of phenotypically normal, heterozygous **carriers**.
- The sickle-cell allele has probably persisted for evolutionary reasons: Heterozygotes have an advantage because one copy of the sickle-cell allele reduces both the frequency and severity of malaria attacks.
- Lethal dominant alleles are eliminated from the population if affected people die before reproducing. Nonlethal dominant alleles and lethal alleles that are expressed relatively late in life are inherited in a Mendelian way.
- Many human diseases are multifactorial—that is, they have both genetic and environmental components and do not follow simple Mendelian inheritance patterns.
- Using family histories, genetic counselors help couples determine the probability of their children having genetic disorders.

> **?** *Both members of a couple know that they are carriers of the cystic fibrosis allele. None of their three children have cystic fibrosis, but any one of them might be a carrier. The couple would like to have a fourth child but are worried that he or she would very likely have the disease, since the first three do not. What would you tell the couple? Would it remove some more uncertainty in their prediction if they could find out from genetic tests whether the three children are carriers?*

TIPS FOR GENETICS PROBLEMS

1. Write down symbols for the alleles. (These may be given in the problem.) When represented by single letters, the dominant allele is uppercase and the recessive allele is lowercase.

2. Write down the possible genotypes, as determined by the phenotype.
   a. If the phenotype is that of the dominant trait (for example, purple flowers), then the genotype is either homozygous dominant or heterozygous (*PP* or *Pp*, in this example).
   b. If the phenotype is that of the recessive trait, the genotype must be homozygous recessive (for example, *pp*).
   c. If the problem says "true-breeding," the genotype is homozygous.

3. Determine what the problem is asking. If asked to do a cross, write it out in the form [Genotype] × [Genotype], using the alleles you've decided on.

4. To figure out the outcome of a cross, set up a Punnett square.
   a. Put the gametes of one parent at the top and those of the other on the left. To determine the allele(s) in each gamete for a given genotype, set up a systematic way to list all the possibilities. (Remember, each gamete has one allele of each gene.) Note that there are 2^n possible types of gametes, where n is the number of gene loci that are heterozygous. For example, an individual with genotype *AaBbCc* would produce $2^3 = 8$ types of gametes. Write the genotypes of the gametes in circles above the columns and to the left of the rows.
   b. Fill in the Punnett square as if each possible sperm were fertilizing each possible egg, making all of the possible offspring. In a cross of *AaBbCc* × *AaBbCc*, for example, the Punnett square would have 8 columns and 8 rows, so there are 64 different offspring; you would know the genotype of each and thus the phenotype. Count genotypes and phenotypes to obtain the genotypic and phenotypic ratios. Because the Punnett square is so large, this method is not the most efficient. Instead, see tip 5.

5. You can use the rules of probability if the Punnett square would be too big. (For example, see the question at the end of the summary for Concept 11.2 and question 7.) You can consider each gene separately (see the section Solving Complex Genetics Problems with the Rules of Probability in Concept 11.2).

6. If the problem gives you the phenotypic ratios of offspring, but not the genotypes of the parents in a given cross, the phenotypes can help you deduce the parents' unknown genotypes.
   a. For example, if 1/2 of the offspring have the recessive phenotype and 1/2 the dominant, you know that the cross was between a heterozygote and a homozygous recessive.
   b. If the ratio is 3:1, the cross was between two heterozygotes.
   c. If two genes are involved and you see a 9:3:3:1 ratio in the offspring, you know that each parent is heterozygous for both genes. *Caution*: Don't assume that the reported numbers will exactly equal the predicted ratios. For example, if there are 13 offspring with the dominant trait and 11 with the recessive, assume that the ratio is one dominant to one recessive.

7. For pedigree problems, use the tips in Figure 11.14 and below to determine what kind of trait is involved.
   a. If parents without the trait have offspring with the trait, the trait must be recessive and both of the parents must be carriers.
   b. If the trait is seen in every generation, it is most likely dominant (see the next possibility, though).
   c. If both parents have the trait, then in order for it to be recessive, all offspring must show the trait.
   d. To determine the likely genotype of a certain individual in a pedigree, first label the genotypes of all individuals in the pedigree as well as you can. If an individual has the dominant phenotype, the genotype must be *AA* or *Aa*; you can write this as *A-*; the recessive phenotype means the genotype must be *aa*. Try different possibilities to see how well each fits the results. Use Mendel's laws and the rules of probability to calculate the probability of each possible genotype being the correct one.

TEST YOUR UNDERSTANDING

Level 1: Knowledge/Comprehension

1. Match each term on the left with a statement on the right.

Term	Statement
— Gene	a. Has no effect on phenotype in a heterozygote
— Allele	b. A variant for a character
— Character	c. Having two identical alleles for a gene
— Trait	d. A cross between individuals heterozygous for a single character
— Dominant allele	e. An alternative version of a gene
— Recessive allele	f. Having two different alleles for a gene
— Genotype	g. A heritable feature that varies among individuals
— Phenotype	h. An organism's appearance or observable traits
— Homozygous	i. A cross between an individual with an unknown genotype and a homozygous recessive individual
— Heterozygous	j. Determines phenotype in a heterozygote
— Testcross	k. The genetic makeup of an individual
— Monohybrid cross	l. A heritable unit that determines a character; can exist in different forms

2. **DRAW IT** Two pea plants heterozygous for the characters of pod color and pod shape are crossed. Draw a Punnett square to determine the phenotypic ratios of the offspring.

3. A man with type A blood marries a woman with type B blood. Their child has type O blood. What are the genotypes of these three individuals? What genotypes, and in what frequencies, would you expect in future offspring from this marriage?

4. A man has six fingers on each hand and six toes on each foot. His wife and their daughter have the normal number of digits. Remember that extra digits is a dominant trait. What fraction of this couple's children would be expected to have extra digits?

5. **DRAW IT** A pea plant heterozygous for inflated pods (Ii) is crossed with a plant homozygous for constricted pods (ii). Draw a Punnett square for this cross. Assume that pollen comes from the ii plant.

Level 2: Application/Analysis

6. Flower position, stem length, and seed shape are three characters that Mendel studied. Each is controlled by an independently assorting gene and has dominant and recessive expression as follows:

Character	Dominant	Recessive
Flower position	Axial (A)	Terminal (a)
Stem length	Tall (T)	Dwarf (t)
Seed shape	Round (R)	Wrinkled (r)

If a plant that is heterozygous for all three characters is allowed to self-fertilize, what proportion of the offspring would you expect to be as follows? (*Note*: Use the rules of probability instead of a huge Punnett square.)

(a) homozygous for the three dominant traits
(b) homozygous for the three recessive traits
(c) heterozygous for all three characters
(d) homozygous for axial and tall, while heterozygous for seed shape

7. A black guinea pig crossed with an albino guinea pig produces 12 black offspring. When the albino is crossed with a second black one, 7 blacks and 5 albinos are obtained. What is the best explanation for this genetic outcome? Write genotypes for the parents, gametes, and offspring.

8. In some plants, a true-breeding, red-flowered strain gives all pink flowers when crossed with a white-flowered strain: $C^R C^R$ (red) × $C^W C^W$ (white) → $C^R C^W$ (pink). If flower position (axial or terminal) is inherited as it is in peas (see Table 11.1), what will be the ratios of genotypes and phenotypes of the F_1 generation resulting from the following cross: axial-red (true-breeding) × terminal-white? What will be the ratios in the F_2 generation?

9. In sesame plants, the one-pod condition (P) is dominant to the three-pod condition (p), and normal leaf (L) is dominant to wrinkled leaf (l). Pod type and leaf type are inherited independently. Determine the genotypes for the two parents for all possible matings producing the following offspring:

(a) 318 one-pod, normal leaf and 98 one-pod, wrinkled leaf
(b) 323 three-pod, normal leaf and 106 three-pod, wrinkled leaf
(c) 401 one-pod, normal leaf
(d) 150 one-pod, normal leaf, 147 one-pod, wrinkled leaf, 51 three-pod, normal leaf, and 48 three-pod, wrinkled leaf
(e) 223 one-pod, normal leaf, 72 one-pod, wrinkled leaf, 76 three-pod, normal leaf, and 27 three-pod, wrinkled leaf

10. Phenylketonuria (PKU) is an inherited disease caused by a recessive allele. If a woman and her husband, who are both carriers, have three children, what is the probability of each of the following?

(a) All three children are of normal phenotype.
(b) One or more of the three children have the disease.
(c) All three children have the disease.
(d) At least one child is phenotypically normal.

(*Note*: It will help to remember that the probabilities of all possible outcomes always add up to 1.)

11. The genotype of F_1 individuals in a tetrahybrid cross is *AaBbCcDd*. Assuming independent assortment of these four genes, what are the probabilities that F_2 offspring will have the following genotypes?

(a) *aabbccdd*
(b) *AaBbCcDd*
(c) *AABBCCDD*
(d) *AaBBccDd*
(e) *AaBBCCdd*

12. What is the probability that each of the following pairs of parents will produce the indicated offspring? (Assume independent assortment of all gene pairs.)

(a) *AABBCC* × *aabbcc* → *AaBbCc*
(b) *AABbCc* × *AaBbCc* → *AAbbCC*
(c) *AaBbCc* × *AaBbCc* → *AaBbCc*
(d) *aaBbCC* × *AABbcc* → *AaBbCc*

13. Karen and Steve each have a sibling with sickle-cell disease. Neither Karen nor Steve nor any of their parents have the disease, and none of them have been tested to see if they have the sickle-cell trait. Based on this incomplete information, calculate the probability that if this couple has a child, the child will have sickle-cell disease.

14. In tigers, a recessive allele that is pleiotropic causes an absence of fur pigmentation (a white tiger) and a cross-eyed condition. If two phenotypically normal tigers that are heterozygous at this locus are mated, what percentage of their offspring will be cross-eyed? What percentage of cross-eyed tigers will be white?

15. In 1981, a stray black cat with unusual rounded, curled-back ears was adopted by a family in California. Hundreds of descendants of the cat have since been born, and cat fanciers hope to develop the curl cat into a show breed. Suppose you owned the first curl cat and wanted to develop  a true-breeding variety. How would you determine whether the curl allele is dominant or recessive? How would you obtain true-breeding curl cats? How could you be sure they are true-breeding?

16. Imagine that a newly discovered, recessively inherited disease is expressed only in individuals with type O blood, although the disease and blood group are independently inherited. A normal man with type A blood and a normal woman with type B blood have already had one child with the disease. The woman is now pregnant for a second time. What is the probability that the second child will also have the disease? Assume that both parents are heterozygous for the gene that causes the disease.

17. In maize (corn) plants, a dominant allele *I* inhibits kernel color, while the recessive allele *i* permits color when homozygous. At a different locus, the dominant allele *P* causes purple kernel color, while the homozygous recessive genotype *pp* causes red kernels. If plants heterozygous at both loci are crossed, what will be the genotypic and phenotypic ratios of the offspring?

18. The pedigree below traces the inheritance of alkaptonuria, a biochemical disorder. Affected individuals, indicated here by the colored circles and squares, are unable to metabolize a substance called alkapton, which colors the urine and stains body tissues. Does alkaptonuria appear to be caused by a dominant allele or by a recessive allele? Fill in the genotypes of the individuals whose genotypes can be deduced. What genotypes are possible for each of the other individuals?

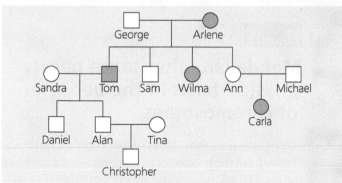

19. Imagine that you are a genetic counselor, and a couple planning to start a family comes to you for information. Charles was married once before, and he and his first wife had a child with cystic fibrosis. The brother of his current wife, Elaine, died of cystic fibrosis. What is the probability that Charles and Elaine will have a baby with cystic fibrosis? (Neither Charles, Elaine, nor their parents have cystic fibrosis.)

20. In mice, black fur (*B*) is dominant to white (*b*). At a different locus, a dominant allele (*A*) produces a band of yellow just below the tip of each hair in mice with black fur. This gives a frosted appearance known as agouti. Expression of the recessive allele (*a*) results in a solid coat color. If mice that are heterozygous at both loci are crossed, what are the expected genotypic and phenotypic ratios of their offspring?

Level 3: Synthesis/Evaluation

21. SCIENTIFIC INQUIRY
You are handed a mystery pea plant with tall stems and axial flowers and asked to determine its genotype as quickly as possible. You know that the allele for tall stems (*T*) is dominant to that for dwarf stems (*t*) and that the allele for axial flowers (*A*) is dominant to that for terminal flowers (*a*).

(a) What are *all* the possible genotypes for your mystery plant?
(b) Describe the *one* cross you would do, out in your garden, to determine the exact genotype of your mystery plant.
(c) While waiting for the results of your cross, you predict the results for each possible genotype listed in part a. How do you do this? Why is this not called "performing a cross"?
(d) Explain how the results of your cross and your predictions will help you learn the genotype of your mystery plant.

22. SCIENCE, TECHNOLOGY, AND SOCIETY
Imagine that one of your parents has Huntington's disease. What is the probability that you, too, will someday manifest the disease? There is no cure for Huntington's. Would you want to be tested for the Huntington's allele? Why or why not?

23. FOCUS ON EVOLUTION
Over the past half century, there has been a trend in the United States and other developed countries for people to marry and start families later in life than did their parents and grandparents. What effects might this trend have on the incidence (frequency) of late-acting dominant lethal alleles in the population?

24. FOCUS ON INFORMATION
The continuity of life is based on heritable information in the form of DNA. In a short essay (100–150 words), explain how the passage of genes from parents to offspring, in the form of particular alleles, ensures perpetuation of parental traits in offspring and, at the same time, genetic variation among offspring. Use genetic terms in your explanation.

For selected answers, see Appendix A.

MasteringBiology®

Students Go to **MasteringBiology** for assignments, the eText, and the Study Area with practice tests, animations, and activities.

Instructors Go to **MasteringBiology** for automatically graded tutorials and questions that you can assign to your students, plus Instructor Resources.

12

The Chromosomal Basis of Inheritance

▼ **Figure 12.1** Where are Mendel's hereditary factors located in the cell?

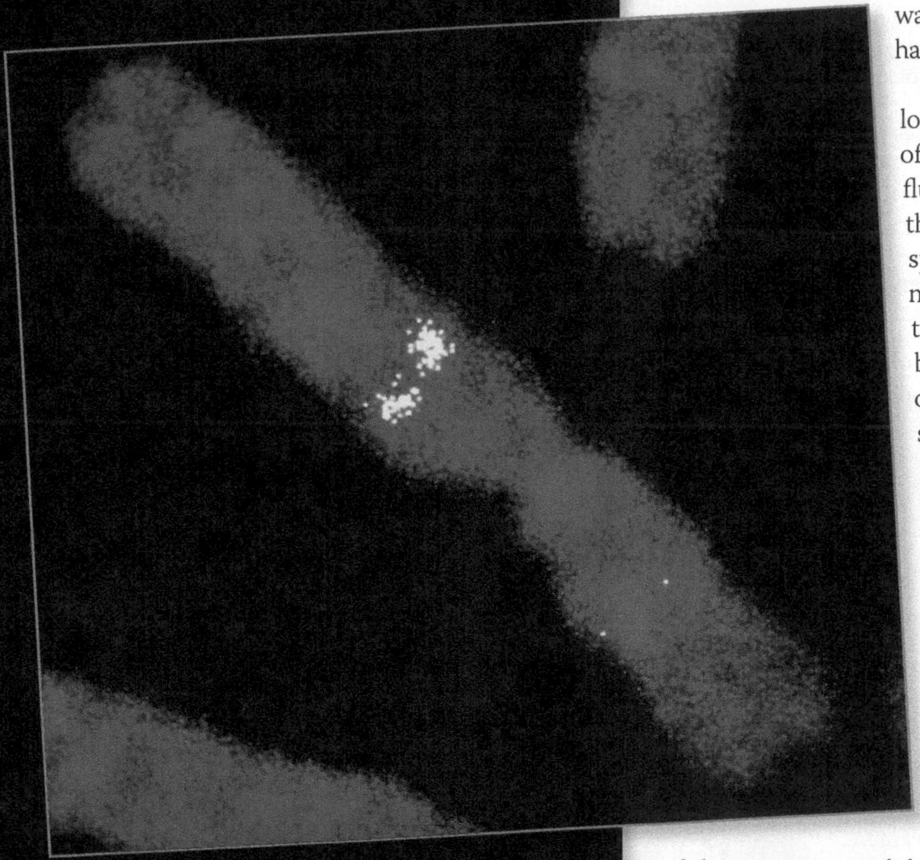

KEY CONCEPTS

12.1 Mendelian inheritance has its physical basis in the behavior of chromosomes

12.2 Sex-linked genes exhibit unique patterns of inheritance

12.3 Linked genes tend to be inherited together because they are located near each other on the same chromosome

12.4 Alterations of chromosome number or structure cause some genetic disorders

OVERVIEW

Locating Genes Along Chromosomes

Gregor Mendel's "hereditary factors" were purely an abstract concept when he proposed their existence in 1860. At that time, no cellular structures were known that could house these imaginary units. Even after chromosomes were first observed, many biologists remained skeptical about Mendel's laws of segregation and independent assortment until there was sufficient evidence that these principles of heredity had a physical basis in chromosomal behavior.

Today, we know that genes—Mendel's "factors"—are located along chromosomes. We can see the location of a particular gene by tagging chromosomes with a fluorescent dye that highlights that gene. For example, the two yellow spots in **Figure 12.1** mark the locus of a specific gene on the sister chromatids of human chromosome 6. This chapter will extend what you learned in the past two chapters. We'll describe the chromosomal basis for the transmission of genes from parents to offspring, along with some important exceptions to the standard mode of inheritance.

CONCEPT 12.1

Mendelian inheritance has its physical basis in the behavior of chromosomes

Using improved techniques of microscopy, cytologists worked out the process of mitosis in 1875 and meiosis in the 1890s. Cytology and genetics converged when biologists began to see parallels between the behavior of chromosomes and the behavior of Mendel's proposed hereditary factors during sexual life cycles: Chromosomes and genes are both present in pairs in diploid cells; homologous chromosomes separate and alleles segregate during the process of meiosis; and fertilization restores the paired condition for both chromosomes and genes. Around 1902, Walter S. Sutton, Theodor Boveri, and others independently noted these parallels, and the **chromosome theory of inheritance** began to take form. According to this theory, Mendelian genes have specific loci (positions) along chromosomes, and it is the chromosomes that undergo segregation and independent assortment.

Figure 12.2 shows that the behavior of homologous chromosomes during meiosis can account for the segregation of the alleles at each genetic locus to

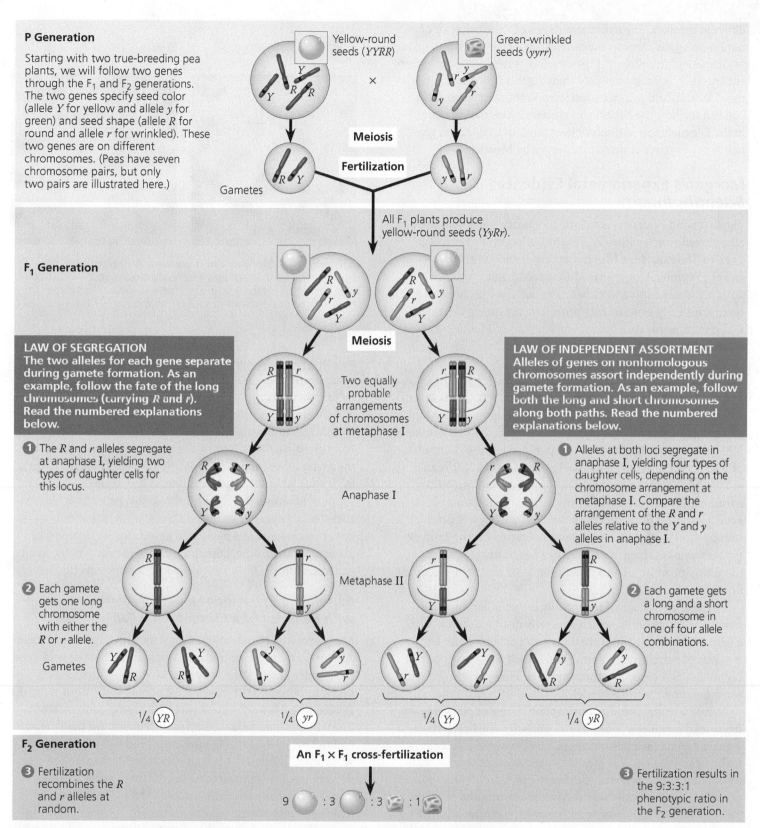

P Generation

Starting with two true-breeding pea plants, we will follow two genes through the F_1 and F_2 generations. The two genes specify seed color (allele Y for yellow and allele y for green) and seed shape (allele R for round and allele r for wrinkled). These two genes are on different chromosomes. (Peas have seven chromosome pairs, but only two pairs are illustrated here.)

Yellow-round seeds (*YYRR*)

Green-wrinkled seeds (*yyrr*)

×

Meiosis

Fertilization

Gametes

All F_1 plants produce yellow-round seeds (*YyRr*).

F_1 Generation

LAW OF SEGREGATION
The two alleles for each gene separate during gamete formation. As an example, follow the fate of the long chromosomes (carrying R and r). Read the numbered explanations below.

Meiosis

Two equally probable arrangements of chromosomes at metaphase I

LAW OF INDEPENDENT ASSORTMENT
Alleles of genes on nonhomologous chromosomes assort independently during gamete formation. As an example, follow both the long and short chromosomes along both paths. Read the numbered explanations below.

1 The R and r alleles segregate at anaphase I, yielding two types of daughter cells for this locus.

Anaphase I

1 Alleles at both loci segregate in anaphase I, yielding four types of daughter cells, depending on the chromosome arrangement at metaphase I. Compare the arrangement of the R and r alleles relative to the Y and y alleles in anaphase I.

Metaphase II

2 Each gamete gets one long chromosome with either the R or r allele.

Gametes

2 Each gamete gets a long and a short chromosome in one of four allele combinations.

¼ *YR*   ¼ *yr*   ¼ *Yr*   ¼ *yR*

F_2 Generation

3 Fertilization recombines the R and r alleles at random.

An $F_1 \times F_1$ cross-fertilization

9 : 3 : 3 : 1

3 Fertilization results in the 9:3:3:1 phenotypic ratio in the F_2 generation.

▲ **Figure 12.2 The chromosomal basis of Mendel's laws.** Here we correlate a dihybrid cross that Mendel performed (see Figure 11.8) with the behavior of chromosomes during meiosis (see Figure 10.8). The arrangement of chromosomes at metaphase I of meiosis and their movement during anaphase I account, respectively, for the independent assortment and segregation of the alleles for seed color and shape. Each cell that undergoes meiosis in an F_1 plant produces two kinds of gametes. If we count the results for all cells, however, each F_1 plant produces equal numbers of all four kinds of gametes because the alternative chromosome arrangements at metaphase I are equally likely.

? *If you crossed an F_1 plant with a plant that was homozygous recessive for both genes (yyrr), how would the phenotypic ratio of the offspring compare with the 9:3:3:1 ratio seen here?*

different gametes. The figure also shows that the behavior of nonhomologous chromosomes can account for the independent assortment of the alleles for two or more genes located on different chromosomes. By carefully studying this figure, which traces the same dihybrid pea cross you learned about in Figure 11.8, you can see how the behavior of chromosomes during meiosis in the F_1 generation and subsequent random fertilization give rise to the F_2 phenotypic ratio observed by Mendel.

Morgan's Experimental Evidence: *Scientific Inquiry*

The first solid evidence associating a specific gene with a specific chromosome came early in the 20th century from the work of Thomas Hunt Morgan, an experimental embryologist at Columbia University. Although Morgan was initially skeptical about both Mendelian genetics and the chromosome theory, his early experiments provided convincing evidence that chromosomes are indeed the location of Mendel's heritable factors.

Morgan's Choice of Experimental Organism

Many times in the history of biology, important discoveries have come to those insightful or lucky enough to choose an experimental organism suitable for the research problem being tackled. Mendel chose the garden pea because a number of distinct varieties were available. For his work, Morgan selected a species of fruit fly, *Drosophila melanogaster*, a common insect that feeds on the fungi growing on fruit. Fruit flies are prolific breeders; a single mating will produce hundreds of offspring, and a new generation can be bred every two weeks. Morgan's laboratory began using this convenient organism for genetic studies in 1907 and soon became known as "the fly room."

Another advantage of the fruit fly is that it has only four pairs of chromosomes, which are easily distinguishable with a light microscope. There are three pairs of autosomes and one pair of sex chromosomes. Female fruit flies have a pair of homologous X chromosomes, and males have one X chromosome and one Y chromosome.

While Mendel could readily obtain different pea varieties from seed suppliers, Morgan was probably the first person to want different varieties of the fruit fly. He faced the tedious task of carrying out many matings and then microscopically inspecting large numbers of offspring in search of naturally occurring variant individuals. After many months of this, he lamented, "Two years' work wasted. I have been breeding those flies for all that time and I've got nothing out of it." Morgan persisted, however, and was finally rewarded with the discovery of a single male fly with white eyes instead of the usual red. The phenotype for a character most commonly observed in natural populations, such as red eyes in *Drosophila*, is called the **wild type (Figure 12.3)**. Traits that are alternatives to the wild type, such as white eyes in *Drosophila*,

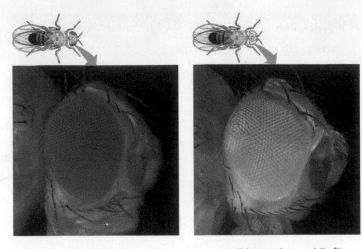

▲ **Figure 12.3 Morgan's first mutant.** Wild-type *Drosophila* flies have red eyes (left). Among his flies, Morgan discovered a mutant male with white eyes (right). This variation made it possible for Morgan to trace a gene for eye color to a specific chromosome (LMs).

are called *mutant phenotypes* because they are due to alleles assumed to have originated as changes, or mutations, in the wild-type allele.

Morgan and his students invented a notation for symbolizing alleles in *Drosophila* that is still widely used for fruit flies. For a given character in flies, the gene takes its symbol from the first mutant (non-wild type) discovered. Thus, the allele for white eyes in *Drosophila* is symbolized by w. The superscript + identifies the allele for the wild-type trait: w^+ for the allele for red eyes, for example. Over the years, a variety of gene notation systems have been developed for different organisms. For example, human genes are usually written in all capitals, such as *HD* for the allele for Huntington's disease.

Correlating Behavior of a Gene's Alleles with Behavior of a Chromosome Pair

Morgan mated his white-eyed male fly with a red-eyed female. All the F_1 offspring had red eyes, suggesting that the wild-type allele is dominant. When Morgan bred the F_1 flies to each other, he observed the classical 3:1 phenotypic ratio among the F_2 offspring. However, there was a surprising additional result: The white-eye trait showed up only in males. All the F_2 females had red eyes, while half the males had red eyes and half had white eyes. Therefore, Morgan concluded that somehow a fly's eye color was linked to its sex. (If the eye-color gene were unrelated to sex, half of the white-eyed flies would have been male and half female.)

Recall that a female fly has two X chromosomes (XX), while a male fly has an X and a Y (XY). The correlation between the trait of white eye color and the male sex of the affected F_2 flies suggested to Morgan that the gene involved in his white-eyed mutant was located exclusively on the X chromosome, with no corresponding allele present on the Y chromosome. His reasoning can be followed in **Figure 12.4**. For a male, a single

In a cross between a wild-type female fruit fly and a mutant white-eyed male, what color eyes will the F₁ and F₂ offspring have?

Experiment Thomas Hunt Morgan wanted to analyze the behavior of two alleles of a fruit fly eye-color gene. In crosses similar to those done by Mendel with pea plants, Morgan and his colleagues mated a wild-type (red-eyed) female with a mutant white-eyed male.

P Generation ♀ × ♂

F₁ Generation — All offspring had red eyes.

Morgan then bred an F₁ red-eyed female to an F₁ red-eyed male to produce the F₂ generation.

Results The F₂ generation showed a typical Mendelian ratio of three red-eyed flies to one white-eyed fly. However, no females displayed the white-eye trait; all white-eyed flies were males.

F₂ Generation ♀ ♀ ♂ ♂

Conclusion All F₁ offspring had red eyes, so the mutant white-eye trait (w) must be recessive to the wild-type red-eye trait (w^+). Since the recessive trait—white eyes—was expressed only in males in the F₂ generation, Morgan deduced that this eye-color gene is located on the X chromosome and that there is no corresponding locus on the Y chromosome.

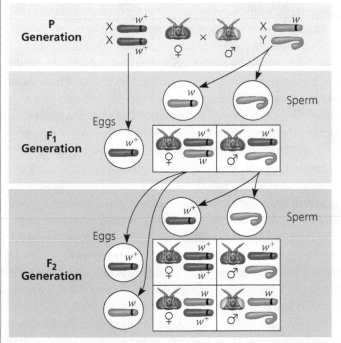

Source T. H. Morgan, Sex-limited inheritance in *Drosophila, Science* 32:120–122 (1910).

A related Experimental Inquiry Tutorial can be assigned in MasteringBiology.

WHAT IF? Suppose this eye-color gene were located on an autosome. Predict the phenotypes (including gender) of the F₂ flies in this hypothetical cross. (*Hint*: Draw a Punnett square.)

---

copy of the mutant allele would confer white eyes; since a male has only one X chromosome, there can be no wild-type allele (w^+) present to mask the recessive allele. On the other hand, a female could have white eyes only if both her X chromosomes carried the recessive mutant allele (w). This was impossible for the F₂ females in Morgan's experiment because all the F₁ fathers had red eyes.

Morgan's finding of the correlation between a particular trait and an individual's sex provided support for the chromosome theory of inheritance: namely, that a specific gene is carried on a specific chromosome (in this case, an eye-color gene on the X chromosome). In addition, Morgan's work indicated that genes located on a sex chromosome exhibit unique inheritance patterns, which we'll discuss in the next section. Recognizing the importance of Morgan's early work, many bright students were attracted to his fly room.

CONCEPT CHECK 12.1

1. Which one of Mendel's laws relates to the inheritance of alleles for a single character? Which law relates to the inheritance of alleles for two characters in a dihybrid cross?
2. **MAKE CONNECTIONS** Review the description of meiosis (see Figure 10.8) and Mendel's laws of segregation and independent assortment (see Concept 11.1). What is the physical basis for each of Mendel's laws?
3. **WHAT IF?** Propose a possible reason that the first naturally occurring mutant fruit fly Morgan saw involved a gene on a sex chromosome.

For suggested answers, see Appendix A.

CONCEPT 12.2

Sex-linked genes exhibit unique patterns of inheritance

As you just learned, Morgan's discovery of a trait (white eyes) that correlated with a fly's sex was a key episode in the development of the chromosome theory of inheritance. Because the identity of a fly's sex chromosomes could be inferred by observing the sex of the fly, the behavior of the two members of the pair of sex chromosomes could be correlated with the behavior of the two alleles of the eye-color gene. In this section, we'll take a closer look at the role of sex chromosomes in inheritance. We'll begin by reviewing the chromosomal basis of sex determination in humans and some other animals.

The Chromosomal Basis of Sex

Whether we are male or female is one of our more obvious phenotypic characters. Although the anatomical and physiological differences between women and men are numerous, the chromosomal basis for determining sex is rather simple. In humans and other mammals, there are two varieties of sex chromosomes, designated X and Y. The Y chromosome is much smaller than the X chromosome **(Figure 12.5)**. A person who inherits two X chromosomes, one from each parent, usually develops as a female. A male develops from a zygote containing one X chromosome and one Y chromosome **(Figure 12.6)**. Short segments at either end of the

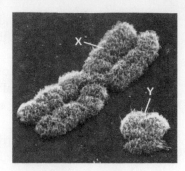

▲ **Figure 12.5 Human sex chromosomes.**

Y chromosome are the only regions that are homologous with corresponding regions of the X. These homologous regions allow the X and Y chromosomes in males to pair and behave like homologous chromosomes during meiosis in the testes.

In mammalian testes and ovaries, the two sex chromosomes segregate during meiosis. Each egg receives one X chromosome. In contrast, sperm fall into two categories: Half the sperm cells a male produces receive an X chromosome, and half receive a Y chromosome. We can trace the sex of each offspring to the events of conception: If a sperm cell bearing an X chromosome happens to fertilize an egg, the zygote is XX, a female; if a sperm cell containing a Y chromosome fertilizes an egg, the zygote is XY, a male (see Figure 12.6). Thus, mammalian sex determination is a matter of chance—a fifty-fifty chance. In *Drosophila*, males are XY, but sex depends on the ratio between the number of X chromosomes and the number of autosome sets, not simply on the presence of the Y. There are other chromosomal systems as well, besides the X-Y system, for determining sex.

In humans, the anatomical signs of sex begin to emerge when the embryo is about 2 months old. Before then, the rudiments of the gonads are generic—they can develop into either testes or ovaries, depending on whether or not a Y chromosome is present. In 1990, a British research team identified a

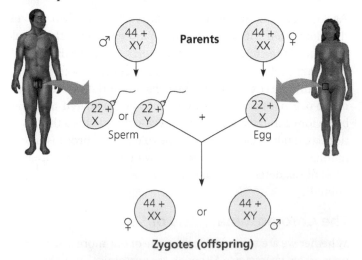

Parents

Sperm

Egg

Zygotes (offspring)

▲ **Figure 12.6 The mammalian X-Y chromosomal system of sex determination.** In mammals, the sex of an offspring depends on whether the sperm cell contains an X chromosome or a Y. Numerals indicate the number of autosomes.

gene on the Y chromosome required for the development of testes. They named the gene *SRY*, for sex-determining region of Y. In the absence of *SRY*, the gonads develop into ovaries. The biochemical, physiological, and anatomical features that distinguish males and females are complex, and many genes are involved in their development. In fact, *SRY* codes for a protein that regulates other genes.

Researchers have sequenced the human Y chromosome and have identified 78 genes that code for about 25 proteins (some genes are duplicates). About half of these genes are expressed only in the testis, and some are required for normal testicular functioning and the production of normal sperm. A gene located on either sex chromosome is called a **sex-linked gene**; those located on the Y chromosome are called *Y-linked genes*. The Y chromosome is passed along virtually intact from a father to all his sons. Because there are so few Y-linked genes, very few disorders are transferred from father to son on the Y chromosome. A rare example is that in the absence of certain Y-linked genes, an XY individual is male but does not produce normal sperm.

The human X chromosome contains approximately 1,100 genes, which are called **X-linked genes**. The fact that males and females inherit a different number of X chromosomes leads to a pattern of inheritance different from that produced by genes located on autosomes.

Inheritance of X-Linked Genes

While most Y-linked genes help determine sex, the X chromosomes have genes for many characters unrelated to sex. X-linked genes in humans follow the same pattern of inheritance that Morgan observed for the eye-color locus he studied in *Drosophila* (see Figure 12.4). Fathers pass X-linked alleles to all of their daughters but to none of their sons. In contrast, mothers can pass X-linked alleles to both sons and daughters, as shown in **Figure 12.7** for the inheritance of a mild X-linked disorder, color blindness.

If an X-linked trait is due to a recessive allele, a female will express the phenotype only if she is homozygous for that allele. Because males have only one locus, the terms *homozygous* and *heterozygous* are meaningless when describing their X-linked genes; the term *hemizygous* is used in such cases. Any male receiving the recessive allele from his mother will express the trait. For this reason, far more males than females have X-linked recessive disorders. However, even though the chance of a female inheriting a double dose of the mutant allele is much less than the probability of a male inheriting a single dose, there *are* females with X-linked disorders. For instance, color blindness is almost always inherited as an X-linked trait. A color-blind daughter may be born to a color-blind father whose mate is a carrier (see Figure 12.7c). Because the X-linked allele for color blindness is relatively rare, however, the probability that such a man and woman will mate is low.

A number of human X-linked disorders are much more serious than color blindness. An example is **Duchenne muscular dystrophy**, which affects about one out of every 3,500 males

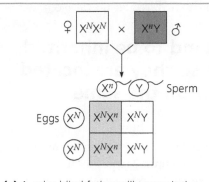
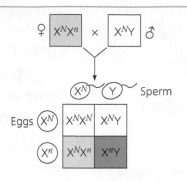
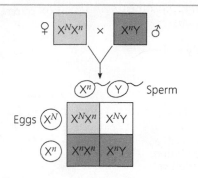

(a) A color-blind father will transmit the mutant allele to all daughters but to no sons. When the mother is a dominant homozygote, the daughters will have the normal phenotype but will be carriers of the mutation.

(b) If a carrier mates with a male who has normal color vision, there is a 50% chance that each daughter will be a carrier like her mother and a 50% chance that each son will have the disorder.

(c) If a carrier mates with a color-blind male, there is a 50% chance that each child born to them will have the disorder, regardless of sex. Daughters who have normal color vision will be carriers, whereas males who have normal color vision will be free of the recessive allele.

▲ **Figure 12.7 The transmission of X-linked recessive traits.** In this diagram, color blindness is used as an example. The superscript N represents the dominant allele for normal color vision carried on the X chromosome, and the superscript n represents the recessive allele, which has a mutation causing color blindness. White boxes indicate unaffected individuals, light orange boxes indicate carriers, and dark orange boxes indicate color-blind individuals.

? *If a color-blind woman married a man who had normal color vision, what would be the probable phenotypes of their children?*

born in the United States. The disease is characterized by a progressive weakening of the muscles and loss of coordination. Affected individuals rarely live past their early 20s. Researchers have traced the disorder to the absence of a key muscle protein called dystrophin and have mapped the gene for this protein to a specific locus on the X chromosome.

Hemophilia is an X-linked recessive disorder defined by the absence of one or more of the proteins required for blood clotting. When a person with hemophilia is injured, bleeding is prolonged because a firm clot is slow to form. Small cuts in the skin are usually not a problem, but bleeding in the muscles or joints can be painful and can lead to serious damage. In the 1800s, hemophilia was widespread among the royal families of Europe. Queen Victoria of England is known to have passed the allele to several of her descendants. Subsequent intermarriage with royal family members of other nations, such as Spain and Russia, further spread this X-linked trait, and its incidence is well documented in royal pedigrees. Today, people with hemophilia are treated as needed with intravenous injections of the protein that is missing.

X Inactivation in Female Mammals

Female mammals, including humans, inherit two X chromosomes—twice the number inherited by males—so you may wonder if females make twice as much of the proteins encoded by X-linked genes. In fact, most of one X chromosome in each cell in female mammals becomes inactivated during early embryonic development. As a result, the cells of females and males have the same effective dose (one copy) of most X-linked genes. The inactive X in each cell of a female condenses into a compact object called a

Barr body (discovered by Canadian anatomist Murray Barr), which lies along the inside of the nuclear envelope. Most of the genes of the X chromosome that forms the Barr body are not expressed. In the ovaries, Barr-body chromosomes are reactivated in the cells that give rise to eggs, so every female gamete has an active X.

British geneticist Mary Lyon demonstrated that the selection of which X chromosome will form the Barr body occurs randomly and independently in each embryonic cell present at the time of X inactivation. As a consequence, females consist of a *mosaic* of two types of cells: those with the active X derived from the father and those with the active X derived from the mother. After an X chromosome is inactivated in a particular cell, all mitotic descendants of that cell have the same inactive X. Thus, if a female is heterozygous for a sex-linked trait, about half her cells will express one allele, while the others will express the alternate allele. **Figure 12.8** shows how this mosaicism results in the mottled coloration of a tortoiseshell cat. In humans, mosaicism can be observed in a recessive X-linked mutation that prevents the development of sweat glands. A woman who is heterozygous for this trait has patches of normal skin and patches of skin lacking sweat glands.

Inactivation of an X chromosome involves modification of the DNA and the histone proteins bound to it, including attachment of methyl groups ($-CH_3$) to one of the nitrogenous bases of DNA nucleotides. (The regulatory role of DNA methylation is discussed further in Chapter 15.) A particular region of each X chromosome contains several genes involved in the inactivation process. The two regions, one on each X chromosome, associate briefly with each other in each cell at an early stage of embryonic development. Then one of the

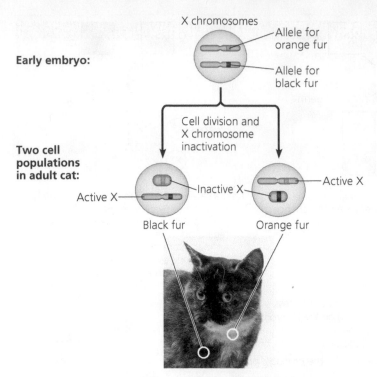

▲ **Figure 12.8 X inactivation and the tortoiseshell cat.** The tortoiseshell gene is on the X chromosome, and the tortoiseshell phenotype requires the presence of two different alleles, one for orange fur and one for black fur. Normally, only females can have both alleles, because only they have two X chromosomes. If a female cat is heterozygous for the tortoiseshell gene, she is tortoiseshell. Orange patches are formed by populations of cells in which the X chromosome with the orange allele is active; black patches have cells in which the X chromosome with the black allele is active. ("Calico" cats also have white areas, which are determined by yet another gene.)

genes, called *XIST* (for X-inactive specific transcript), becomes active *only* on the chromosome that will become the Barr body. Multiple copies of the RNA product of this gene apparently attach to the X chromosome on which they are made, eventually almost covering it. Interaction of this RNA with the chromosome seems to initiate X inactivation, and the RNA products of other genes nearby on the X chromosome help to regulate the process.

CONCEPT CHECK 12.2

1. A white-eyed *Drosophila* female is mated with a red-eyed (wild-type) male, the reciprocal of the cross shown in Figure 12.4. What phenotypes and genotypes do you predict for the offspring?

2. Neither Tim nor Rhoda has Duchenne muscular dystrophy, but their firstborn son does have it. What is the probability that a second child of this couple will have the disease? What is the probability if the second child is a boy? A girl?

3. **MAKE CONNECTIONS** Consider what you learned about dominant and recessive alleles in Concept 11.1. If a disorder were caused by a dominant X-linked allele, how would the inheritance pattern differ from what we see for recessive X-linked disorders?

For suggested answers, see Appendix A.

Linked genes tend to be inherited together because they are located near each other on the same chromosome

The number of genes in a cell is far greater than the number of chromosomes; in fact, each chromosome has hundreds or thousands of genes. (The small Y chromosome is an exception.) Genes located near each other on the same chromosome tend to be inherited together in genetic crosses; such genes are said to be genetically linked and are called **linked genes**. (Note the distinction between the terms *sex-linked gene*, referring to a single gene on a sex chromosome, and *linked genes*, referring to two or more genes on the same chromosome that tend to be inherited together.) When geneticists follow linked genes in breeding experiments, the results deviate from those expected from Mendel's law of independent assortment.

How Linkage Affects Inheritance

To see how linkage between genes affects the inheritance of two different characters, let's examine another of Morgan's *Drosophila* experiments. In this case, the characters are body color and wing size, each with two different phenotypes. Wild-type flies have gray bodies and normal-sized wings. In addition to these flies, Morgan had managed to obtain, through breeding, doubly mutant flies with black bodies and wings much smaller than normal, called vestigial wings. The mutant alleles are recessive to the wild-type alleles, and neither gene is on a sex chromosome. In his investigation of these two genes, Morgan carried out the crosses shown in **Figure 12.9**. The first was a P generation cross to generate F₁ dihybrid flies, and the second was a testcross.

The resulting flies had a much higher proportion of the combinations of traits seen in the P generation flies (called parental phenotypes) than would be expected if the two genes assorted independently. Morgan thus concluded that body color and wing size are usually inherited together in specific (parental) combinations because the genes for these characters are near each other on the same chromosome:

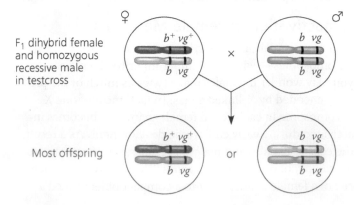

F₁ dihybrid female and homozygous recessive male in testcross

Most offspring

How does linkage between two genes affect inheritance of characters?

Experiment Morgan wanted to know whether the genes for body color and wing size are genetically linked, and if so, how this affects their inheritance. The alleles for body color are b^+ (gray) and b (black), and those for wing size are vg^+ (normal) and vg (vestigial).

Morgan mated true-breeding P (parental) generation flies—wild-type flies with black, vestigial-winged flies—to produce heterozygous F_1 dihybrids ($b^+ b\ vg^+ vg$), all of which are wild-type in appearance.

He then mated wild-type F_1 dihybrid females with homozygous recessive males. This testcross will reveal the genotype of the eggs made by the dihybrid female.

The male's sperm contributes only recessive alleles, so the phenotype of the offspring reflects the genotype of the female's eggs.

Note: Although only females (with pointed abdomens) are shown, half the offspring in each class would be males (with rounded abdomens).

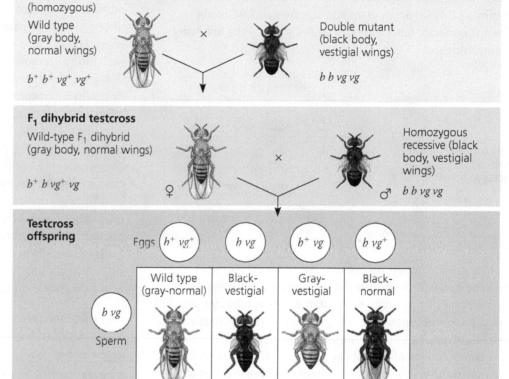

P Generation
(homozygous)

Wild type
(gray body,
normal wings)

$b^+ b^+ vg^+ vg^+$

Double mutant
(black body,
vestigial wings)

$b\ b\ vg\ vg$

F_1 dihybrid testcross

Wild-type F_1 dihybrid
(gray body, normal wings)

$b^+ b\ vg^+ vg$ ♀

Homozygous recessive (black body, vestigial wings)

♂ $b\ b\ vg\ vg$

Testcross offspring

Eggs $b^+ vg^+$ $b\ vg$ $b^+ vg$ $b\ vg^+$

Sperm $b\ vg$

Wild type (gray-normal)	Black-vestigial	Gray-vestigial	Black-normal
$b^+ b\ vg^+ vg$	$b\ b\ vg\ vg$	$b^+ b\ vg\ vg$	$b\ b\ vg^+ vg$

PREDICTED RATIOS

If genes are located on different chromosomes:	1 :	1 :	1 :	1
If genes are located on the same chromosome *and* parental alleles are always inherited together:	1 :	1 :	0 :	0

Results   965 : 944 : 206 : 185

Conclusion Since most offspring had a parental (P generation) phenotype, Morgan concluded that the genes for body color and wing size are genetically linked on the same chromosome. However, the production of a relatively small number of offspring with nonparental phenotypes indicated that some mechanism occasionally breaks the linkage between specific alleles of genes on the same chromosome.

Source T. H. Morgan and C. J. Lynch, The linkage of two factors in *Drosophila* that are not sex-linked, *Biological Bulletin* 23:174–182 (1912).

WHAT IF? *If the parental (P generation) flies had been true-breeding for gray body with vestigial wings and true-breeding for black body with normal wings, which phenotypic class(es) would be largest among the testcross offspring?*

However, as Figure 12.9 shows, both of the combinations of traits not seen in the P generation (called nonparental phenotypes) were also produced in Morgan's experiments, suggesting that the body-color and wing-size alleles are not always linked genetically. To understand this conclusion, we need to further explore **genetic recombination**, the production of offspring with combinations of traits that differ from those found in either P generation parent.

Genetic Recombination and Linkage

Meiosis and random fertilization generate genetic variation among offspring of sexually reproducing organisms due to independent assortment of chromosomes and crossing over in meiosis I, and the possibility of any sperm fertilizing any egg (see Chapter 10). Here we'll examine the chromosomal basis of recombination in relation to the genetic findings of Mendel and Morgan.

Recombination of Unlinked Genes: Independent Assortment of Chromosomes

Mendel learned from crosses in which he followed two characters that some offspring have combinations of traits that do not match those of either parent. For example, we can represent the cross between a pea plant with yellow-round seeds that is heterozygous for both seed color and seed shape (a dihybrid, $YyRr$) and a plant with green-wrinkled seeds (homozygous for both recessive alleles, $yyrr$) by the following Punnett square:

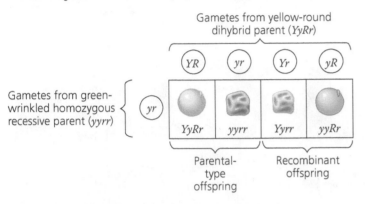

Gametes from yellow-round dihybrid parent ($YyRr$)

Gametes from green-wrinkled homozygous recessive parent ($yyrr$)

Parental-type offspring

Recombinant offspring

Notice in this Punnett square that one-half of the offspring are expected to inherit a phenotype that matches either of the parental (P generation) phenotypes. These offspring are called **parental types**. But two nonparental phenotypes are also found among the offspring. Because these offspring have new combinations of seed shape and color, they are called **recombinant types**, or **recombinants** for short. When 50% of all offspring are recombinants, as in this example, geneticists say that there is a 50% frequency of recombination. The predicted phenotypic ratios among the offspring are similar to what Mendel actually found in $YyRr \times yyrr$ crosses (a type of testcross because it reveals the genotype of the gametes made by the dihybrid $YyRr$ plant).

A 50% frequency of recombination in such testcrosses is observed for any two genes that are located on different chromosomes and thus unlinked. The physical basis of recombination between unlinked genes is the random orientation of homologous chromosomes at metaphase I of meiosis, which leads to the independent assortment of the two unlinked genes (see Figure 10.10 and the question in the Figure 12.2 legend).

Recombination of Linked Genes: Crossing Over

Now let's return to Morgan's fly room to see how we can explain the results of the *Drosophila* testcross illustrated in Figure 12.9. Recall that most of the offspring from the testcross for body color and wing size had parental phenotypes. That suggested that the two genes were on the same chromosome, since the occurrence of parental types with a frequency greater than 50% indicates that the genes are linked. About 17% of offspring, however, were recombinants.

Faced with these results, Morgan proposed that some process must occasionally break the physical connection between specific genes on the same chromosome. Subsequent experiments demonstrated that this process, now called **crossing over**, accounts for the recombination of linked genes. In crossing over, which occurs while replicated homologous chromosomes are paired during prophase of meiosis I, a set of proteins orchestrates an exchange of corresponding segments of one maternal and one paternal chromatid (see Figure 10.11). In effect, end portions of two nonsister chromatids trade places each time a crossover occurs.

Figure 12.10 shows how crossing over in a dihybrid female fly resulted in recombinant eggs and ultimately recombinant offspring in Morgan's testcross. Most of the eggs had a chromosome with either the $b^+ vg^+$ or $b vg$ parental genotype for body color and wing size, but some eggs had a recombinant chromosome ($b^+ vg$ or $b vg^+$). Fertilization of these various classes of eggs by homozygous recessive sperm ($b vg$) produced an offspring population in which 17% exhibited a nonparental, recombinant phenotype, reflecting combinations of alleles not seen before in either P generation parent. In the **Scientific Skills Exercise**, you can use a statistical test to analyze the results from another F_1 dihybrid testcross to see whether the two genes are assorting independently or are linked.

New Combinations of Alleles: Variation for Natural Selection

EVOLUTION The physical behavior of chromosomes during meiosis contributes to the generation of variation in offspring (see Chapter 10). Each pair of homologous chromosomes lines up independently of other pairs during metaphase I, and crossing over prior to that, during prophase I, can mix and match parts of maternal and paternal homologs. Mendel's elegant experiments show that the behavior of the abstract entities known as genes—or, more concretely, alleles of genes—also leads to variation in offspring (see Chapter 11). Now, putting these different ideas together, you can see that the recombinant chromosomes resulting from crossing over may bring alleles together in new combinations, and the subsequent events of meiosis distribute to gametes the recombinant chromosomes in a multitude of combinations, such as the new variants seen in Figures 12.9 and 12.10. Random fertilization then increases even further the number of variant allele combinations that can be created.

This abundance of genetic variation provides the raw material on which natural selection works. If the traits conferred by particular combinations of alleles are better suited for a given environment, organisms possessing those genotypes will be expected to thrive and leave more offspring, ensuring the continuation of their genetic complement. In the next generation, of course, the alleles will be shuffled anew. Ultimately, the interplay between environment and genotype will determine which genetic combinations persist over time.

▶ **Figure 12.10 Chromosomal basis for recombination of linked genes.** In these diagrams re-creating the testcross in Figure 12.9, we track chromosomes as well as genes. The maternal chromosomes are color-coded red and pink to distinguish one homolog from the other before any meiotic crossing over has taken place. Because crossing over between the b^+/b and vg^+/vg loci occurs in some, but not all, egg-producing cells, more eggs with parental-type chromosomes than with recombinant ones are produced in the mating females. Fertilization of the eggs by sperm of genotype $b\ vg$ gives rise to some recombinant offspring. The recombination frequency is the percentage of recombinant flies in the total pool of offspring.

DRAW IT *Suppose, as in the question at the bottom of Figure 12.9, the parental (P generation) flies were true-breeding for gray body with vestigial wings and black body with normal wings. Draw the chromosomes in each of the four possible kinds of eggs from an F₁ female, and label each chromosome as "parental" or "recombinant."*

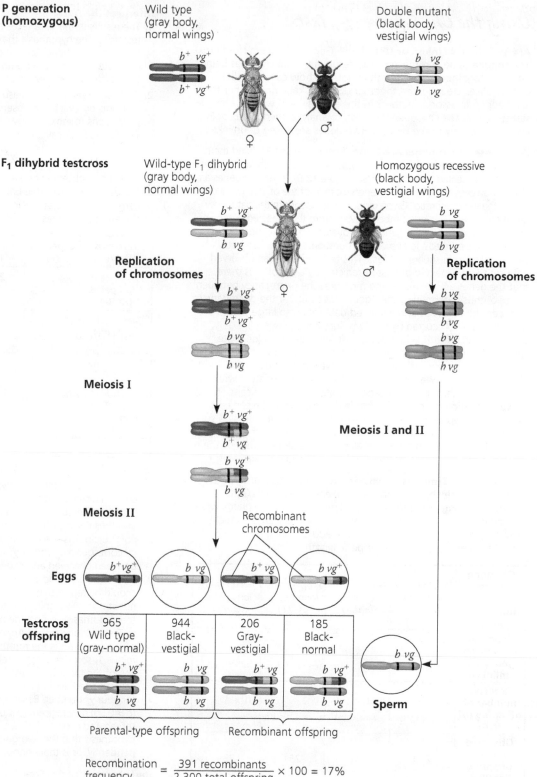

$$\text{Recombination frequency} = \frac{391 \text{ recombinants}}{2{,}300 \text{ total offspring}} \times 100 = 17\%$$

Mapping the Distance Between Genes Using Recombination Data: *Scientific Inquiry*

The discovery of linked genes and recombination due to crossing over led one of Morgan's students, Alfred H. Sturtevant, to a method for constructing a **genetic map**, an ordered list of the genetic loci along a particular chromosome.

Sturtevant hypothesized that the percentage of recombinant offspring, the *recombination frequency*, calculated from experiments like that in Figures 12.9 and 12.10, depends on the distance between genes on a chromosome. He assumed that crossing over is a random event, with the chance of crossing over approximately equal at all points along a chromosome.

Using the Chi-Square (χ^2) Test

Are Two Genes Linked or Unlinked? Genes that are in close proximity on the same chromosome will result in the linked alleles being inherited together more often than not. But how can you tell if certain alleles are inherited together due to linkage or whether they just happen to assort together? In this exercise, you will use a simple statistical test, the chi-square (χ^2) test, to analyze phenotypes of F_1 testcross progeny to see whether two genes are linked or unlinked.

How These Experiments Are Done If genes are unlinked and therefore assort independently, the phenotypic ratio of offspring from an F_1 testcross is expected to be 1:1:1:1 (see Figure 12.9). If the two genes are linked, however, the observed phenotypic ratio of the offspring will not match the expected ratio. Given random fluctuations in the data, how much must the observed numbers deviate from the expected numbers for us to conclude that the genes are not assorting independently but may instead be linked? To answer this question, scientists use a statistical test called a chi-square (χ^2) test. This test compares an observed data set to an expected data set predicted by a hypothesis (here, that the genes are unlinked) and measures the discrepancy between the two, thus determining the "goodness of fit." If the discrepancy between the observed and expected data sets is so large that it is unlikely to have occurred by random fluctuation, we say there is statistically significant evidence against the hypothesis (or, more specifically, evidence for the genes being linked). If the discrepancy is small, then our observations are well explained by random variation alone. In this case, we say the observed data are consistent with our hypothesis, or that the discrepancy is statistically insignificant. Note, however, that consistency with our hypothesis is not the same as proof of our hypothesis. Also, the size of the experimental data set is important: With small data sets like this one, even if the genes are linked, discrepancies might be small by chance alone if the linkage is weak. (For simplicity, we overlook the effect of sample size here.)

Data from the Simulated Experiment In cosmos plants, purple stem (A) is dominant to green stem (a), and short petals (B) is dominant to long petals (b). In a simulated cross, $AABB$ plants were crossed with $aabb$ plants to generate F_1 dihybrids ($AaBb$), which were then test crossed ($AaBb \times aabb$). 900 offspring plants were scored for stem color and flower petal length.

Offspring from testcross of $AaBb$ (F_1) × $aabb$	Purple stem/ short petals (A–B–)	Green stem/ short petals (aaB–)	Purple stem/ long petals (A–bb)	Green stem/ long petals ($aabb$)
Expected ratio if the genes are unlinked	1	1	1	1
Expected number of offspring (of 900)				
Observed number of offspring (of 900)	220	210	231	239

Interpret the Data

1. The results in the data table are from a simulated F_1 dihybrid testcross. The hypothesis that the two genes are unlinked predicts the offspring phenotypic ratio will be 1:1:1:1. Using this ratio, calculate the expected number of each phenotype out of the 900 total offspring, and enter the values in the data table.

2. The goodness of fit is measured by χ^2. This statistic measures the amounts by which the observed values differ from their respective predictions to indicate how closely the two sets of values match. The formula for calculating this value is

$$\chi^2 = \Sigma \frac{(o - e)^2}{e}$$

where o = observed and e = expected. Calculate the χ^2 value for the data using the table below. Enter the data into the table, and carry out the operations indicated in the top row. Then add up the entries in the last column to find the χ^2 value.

Testcross offspring	Expected (e)	Observed (o)	Deviation ($o - e$)	$(o - e)^2$	$(o - e)^2/e$
Purple stem/ short petals (A–B–)		220			
Green stem/ short petals (aaB–)		210			
Purple stem/ long petals (A–bb)		231			
Green stem/ long petals ($aabb$)		239			
				χ^2 = Sum	

3. The χ^2 value means nothing on its own—it is used to find the probability that, assuming the hypothesis is true, the observed data set could have resulted from random fluctuations. A low probability suggests the observed data is not consistent with the hypothesis, and thus the hypothesis should be rejected. A standard cut-off point biologists use is a probability of 0.05 (5%). If the probability corresponding to the χ^2 value is 0.05 or less, the differences between observed and expected values are considered statistically significant and the hypothesis (that the genes are unlinked) should be rejected. If the probability is above 0.05, the results are not statistically significant; the observed data is consistent with the hypothesis. To find the probability, locate your χ^2 value in the χ^2 Distribution Table in Appendix F. The "degrees of freedom" (df) of your data set is the number of categories (here, 4 phenotypes) minus 1, so df = 3. (a) Determine which values on the df = 3 line of the table your calculated χ^2 value lies between. (b) The column headings for these values show the probability range for your χ^2 number. Based on whether there are nonsignificant ($p > 0.05$) or significant ($p \leq 0.05$) differences between the observed and expected values, are the data consistent with the hypothesis that the two genes are unlinked and assorting independently, or is there enough evidence to reject this hypothesis?

(MB) A version of this Scientific Skills Exercise can be assigned in MasteringBiology.

Based on these assumptions, Sturtevant predicted that *the farther apart two genes are, the higher the probability that a crossover will occur between them and therefore the higher the recombination frequency.* His reasoning was simple: The greater the distance between two genes, the more points there are between them where crossing over can occur. Using recombination data from various fruit fly crosses, Sturtevant proceeded to assign relative positions to genes on the same chromosomes—that is, to *map* genes to their locations on the chromosomes.

▼ Figure 12.11 **Research Method**

Constructing a Linkage Map

Application A linkage map shows the relative locations of genes along a chromosome.

Technique A linkage map is based on the assumption that the probability of a crossover between two genetic loci is proportional to the distance separating the loci. The recombination frequencies used to construct a linkage map for a particular chromosome are obtained from experimental crosses, such as the cross depicted in Figures 12.9 and 12.10. The distances between genes are expressed as map units, with one map unit equivalent to a 1% recombination frequency. Genes are arranged on the chromosome in the order that best fits the data.

Results In this example, the observed recombination frequencies between three *Drosophila* gene pairs (*b–cn* 9%, *cn–vg* 9.5%, and *b–vg* 17%) best fit a linear order in which *cn* is positioned about halfway between the other two genes:

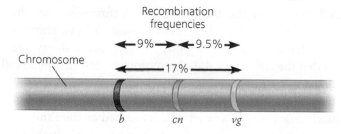

The *b–vg* recombination frequency (17%) is slightly less than the sum of the *b–cn* and *cn–vg* frequencies (9 + 9.5 = 18.5%) because of the few times that one crossover occurs between *b* and *cn* and another crossover occurs between *cn* and *vg*. The second crossover would "cancel out" the first, reducing the observed *b–vg* recombination frequency while contributing to the frequency between each of the closer pairs of genes. The value of 18.5% (18.5 map units) is closer to the actual distance between the genes, so a geneticist would add the smaller distances in constructing a map.

A genetic map based on recombination frequencies is called a **linkage map**. **Figure 12.11** shows Sturtevant's linkage map of three genes: the body-color (*b*) and wing-size (*vg*) genes depicted in Figure 12.10 and a third gene, called cinnabar (*cn*). Cinnabar is one of many *Drosophila* genes affecting eye color. Cinnabar eyes, a mutant phenotype, are a brighter red than the wild-type color. The recombination frequency between *cn* and *b* is 9%; that between *cn* and *vg*, 9.5%; and that between *b* and *vg*, 17%. In other words, crossovers between *cn* and *b* and between *cn* and *vg* are about half as frequent as crossovers between *b* and *vg*. Only a map that locates *cn* about midway between *b* and *vg* is consistent with these data, as you can prove to yourself by drawing alternative maps. Sturtevant expressed the distances between genes in **map units**, defining one map unit as equivalent to a 1% recombination frequency.

In practice, the interpretation of recombination data is more complicated than this example suggests. Some genes on a chromosome are so far from each other that a crossover between them is virtually certain. The observed frequency of recombination in crosses involving two such genes can have

a maximum value of 50%, a result indistinguishable from that for genes on different chromosomes. In this case, the physical connection between genes on the same chromosome is not reflected in the results of genetic crosses. Despite being on the same chromosome and thus being *physically connected*, the genes are *genetically unlinked*; alleles of such genes assort independently, as if they were on different chromosomes. In fact, at least two of the genes for pea characters that Mendel studied are now known to be on the same chromosome, but the distance between them is so great that linkage is not observed in genetic crosses. Consequently, the two genes behaved as if they were on different chromosomes in Mendel's experiments. Genes located far apart on a chromosome are mapped by adding the recombination frequencies from crosses involving closer pairs of genes lying between the two distant genes.

Using recombination data, Sturtevant and his colleagues were able to map numerous *Drosophila* genes in linear arrays. They found that the genes clustered into four groups of linked genes (*linkage groups*). Light microscopy had revealed four pairs of chromosomes in *Drosophila*, so the linkage map provided additional evidence that genes are located on chromosomes. Each chromosome has a linear array of specific genes, each gene with its own locus **(Figure 12.12)**.

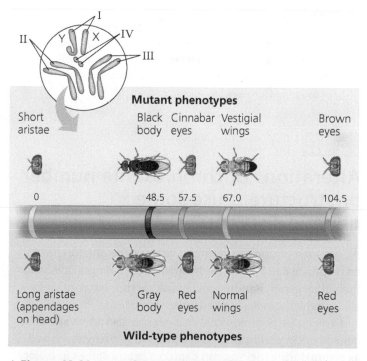

▲ **Figure 12.12 A partial genetic (linkage) map of a *Drosophila* chromosome.** This simplified map shows just a few of the many genes that have been mapped on *Drosophila* chromosome II. The number at each gene locus indicates the number of map units between that locus and the locus for arista length (left). Notice that more than one gene can affect a given phenotypic characteristic, such as eye color. Also, note that in contrast to the homologous autosomes (II–IV), the X and Y sex chromosomes (I) have distinct shapes.

Because a linkage map is based strictly on recombination frequencies, it gives only an approximate picture of a chromosome. The frequency of crossing over is not actually uniform over the length of a chromosome, as Sturtevant assumed, and therefore map units do not correspond to actual physical distances (in nanometers, for instance). A linkage map does portray the order of genes along a chromosome, but it does not accurately portray the precise locations of those genes. Other methods enable geneticists to construct **cytogenetic maps** of chromosomes, which locate genes with respect to chromosomal features, such as stained bands, that can be seen in the microscope. The ultimate maps display the physical distances between gene loci in DNA nucleotides (see Chapter 18). Comparing a linkage map with such a physical map or with a cytogenetic map of the same chromosome, we find that the linear order of genes is identical in all the maps, but the spacing between genes is not.

CONCEPT CHECK 12.3

1. When two genes are located on the same chromosome, what is the physical basis for the production of recombinant offspring in a testcross between a dihybrid parent and a double-mutant (recessive) parent?

2. For each type of offspring of the testcross in Figure 12.9, explain the relationship between its phenotype and the alleles contributed by the female parent. (It will be useful to draw out the chromosomes of each fly and follow the alleles throughout the cross.)

3. **WHAT IF?** Genes *A*, *B*, and *C* are located on the same chromosome. Testcrosses show that the recombination frequency between *A* and *B* is 28% and between *A* and *C* is 12%. Can you determine the linear order of these genes? Explain.

For suggested answers, see Appendix A.

CONCEPT 12.4

Alterations of chromosome number or structure cause some genetic disorders

As you have learned so far in this chapter, the phenotype of an organism can be affected by small-scale changes involving individual genes. Random mutations are the source of all new alleles, which can lead to new phenotypic traits.

Large-scale chromosomal changes can also affect an organism's phenotype. Physical and chemical disturbances, as well as errors during meiosis, can damage chromosomes in major ways or alter their number in a cell. Large-scale chromosomal alterations in humans and other mammals often lead to spontaneous abortion (miscarriage) of a fetus, and individuals born with these types of genetic defects commonly exhibit various developmental disorders. Plants may tolerate such genetic defects better than animals do.

Abnormal Chromosome Number

Ideally, the meiotic spindle distributes chromosomes to daughter cells without error. But there is an occasional mishap, called a **nondisjunction**, in which the members of a pair of homologous chromosomes do not move apart properly during meiosis I or sister chromatids fail to separate during meiosis II **(Figure 12.13)**. In these cases, one gamete receives two of the same type of chromosome and another gamete receives no copy. The other chromosomes are usually distributed normally.

If either of the aberrant gametes unites with a normal one at fertilization, the zygote will also have an abnormal number of a particular chromosome, a condition known as **aneuploidy**. (Aneuploidy may involve more than one chromosome.) Fertilization involving a gamete that has no copy of a particular chromosome will lead to a missing chromosome in the zygote (so that the cell has $2n - 1$ chromosomes); the aneuploid zygote is said to be **monosomic** for that chromosome. If a chromosome is present in triplicate in the zygote (so that the cell has $2n + 1$ chromosomes), the aneuploid cell is **trisomic** for that chromosome. Mitosis will subsequently transmit the anomaly to all embryonic cells. If the organism survives, it usually has a set of traits caused by the abnormal dose of the genes associated with the extra or missing chromosome. Down syndrome is an example of trisomy in humans that will be discussed shortly. Nondisjunction can also

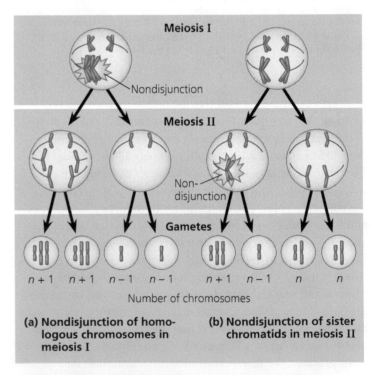

Figure 12.13 Meiotic nondisjunction. Gametes with an abnormal chromosome number can arise by nondisjunction in either meiosis I or meiosis II. For simplicity, the figure does not show the spores formed by meiosis in plants. Ultimately, spores form gametes that have the defects shown. (See Figure 10.6.)

Within the figure:
Meiosis I
Nondisjunction
Meiosis II
Non-disjunction
Gametes
$n + 1$ $n + 1$ $n - 1$ $n - 1$ $n + 1$ $n - 1$ n n
Number of chromosomes

(a) Nondisjunction of homologous chromosomes in meiosis I

(b) Nondisjunction of sister chromatids in meiosis II

occur during mitosis. If such an error takes place early in embryonic development, then the aneuploid condition is passed along by mitosis to a large number of cells and is likely to have a substantial effect on the organism.

Some organisms have more than two complete chromosome sets in all somatic cells. The general term for this chromosomal alteration is **polyploidy**; the specific terms *triploidy* (3*n*) and *tetraploidy* (4*n*) indicate three or four chromosomal sets, respectively. One way a triploid cell may arise is by the fertilization of an abnormal diploid egg produced by nondisjunction of all its chromosomes. Tetraploidy could result from the failure of a 2*n* zygote to divide after replicating its chromosomes. Subsequent normal mitotic divisions would then produce a 4*n* embryo.

Polyploidy is fairly common in plants; the spontaneous origin of polyploid individuals plays an important role in the evolution of plants (see Chapter 22). Many of the plant species we eat are polyploid; for example, bananas are triploid, wheat is hexaploid (6*n*), and strawberries are octoploid (8*n*).

Alterations of Chromosome Structure

Errors in meiosis or damaging agents such as radiation can cause breakage of a chromosome, which can lead to four types of changes in chromosome structure **(Figure 12.14)**. A **deletion** occurs when a chromosomal fragment is lost. The affected chromosome is then missing certain genes. The "deleted" fragment may become attached as an extra segment to a sister chromatid, producing a **duplication**. Alternatively, a detached fragment could attach to a nonsister chromatid of a homologous chromosome. In that case, though, the "duplicated" segments might not be identical because the homologs could carry different alleles of certain genes. A chromosomal fragment may also reattach to the original chromosome but in the reverse orientation, producing an **inversion**. A fourth possible result of chromosomal breakage is for the fragment to join a nonhomologous chromosome, a rearrangement called a **translocation**.

Deletions and duplications are especially likely to occur during meiosis. In crossing over, nonsister chromatids sometimes exchange unequal-sized segments of DNA, so that one partner gives up more genes than it receives. The products of such an unequal crossover are one chromosome with a deletion and one chromosome with a duplication.

A diploid embryo that is homozygous for a large deletion (or has a single X chromosome with a large deletion, in a male) is usually missing a number of essential genes, a condition that is ordinarily lethal. Duplications and translocations also tend to be harmful. In reciprocal translocations, in which segments are exchanged between nonhomologous chromosomes, and in inversions, the balance of genes is not abnormal—all genes are present in their normal doses. Nevertheless, translocations and inversions can alter phenotype because a gene's expression can be influenced by its location among neighboring genes; such events sometimes have devastating effects.

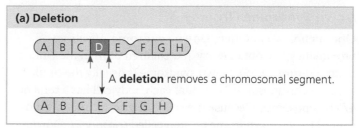

▼ **Figure 12.14 Alterations of chromosome structure.** Red arrows indicate breakage points. Dark purple highlights the chromosomal parts affected by the rearrangements.

(a) Deletion

A **deletion** removes a chromosomal segment.

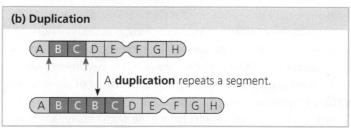

(b) Duplication

A **duplication** repeats a segment.

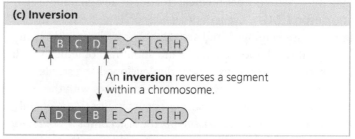

(c) Inversion

An **inversion** reverses a segment within a chromosome.

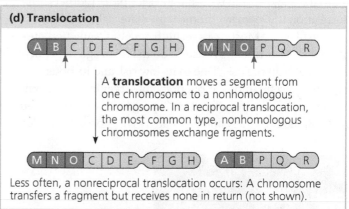

(d) Translocation

A **translocation** moves a segment from one chromosome to a nonhomologous chromosome. In a reciprocal translocation, the most common type, nonhomologous chromosomes exchange fragments.

Less often, a nonreciprocal translocation occurs: A chromosome transfers a fragment but receives none in return (not shown).

Human Disorders Due to Chromosomal Alterations

Alterations of chromosome number and structure are associated with a number of serious human disorders. As described earlier, nondisjunction in meiosis results in aneuploidy in gametes and subsequently in any resulting zygotes. Although the frequency of aneuploid zygotes may be quite high in humans, most of these chromosomal alterations are so disastrous to development that the affected embryos are spontaneously aborted long before birth. However, some types of aneuploidy appear to upset the genetic balance less than others, with the result that individuals with certain aneuploid conditions can survive to birth and beyond. These individuals have a set of traits—a *syndrome*—characteristic of the type of aneuploidy. Genetic disorders caused by aneuploidy can be diagnosed before birth by genetic testing of the fetus.

Down Syndrome (Trisomy 21)

One aneuploid condition, **Down syndrome**, affects approximately one out of every 700 children born in the United States **(Figure 12.15)**. Down syndrome is usually the result of an extra chromosome 21, so that each body cell has a total of 47 chromosomes. Because the cells are trisomic for chromosome 21, Down syndrome is often called *trisomy 21*. Down syndrome includes characteristic facial features, short stature, correctable heart defects, and developmental delays. Individuals with Down syndrome have an increased chance of developing leukemia and Alzheimer's disease but have a lower rate of high blood pressure, atherosclerosis (hardening of the arteries), stroke, and many types of solid tumors. Although people with Down syndrome, on average, have a life span shorter than normal, most, with proper medical treatment, live to middle age and beyond. Many live independently or at home with their families, are employed, and are valuable contributors to their communities. Almost all males and about half of females with Down syndrome are sexually underdeveloped and sterile.

The frequency of Down syndrome increases with the age of the mother. While the disorder occurs in just 0.04% of children born to women under age 30, the risk climbs to 0.92% for mothers at age 40 and is even higher for older mothers. The correlation of Down syndrome with maternal age has not yet been explained. Most cases result from nondisjunction during meiosis I, and some research points to an age-dependent abnormality in meiosis. Due to its low risk and its potential for providing useful information, prenatal screening for trisomies in the embryo is now offered to all pregnant women. Passed in 2008, the Prenatally and Postnatally Diagnosed Conditions Awareness Act stipulates that medical practitioners give accurate, up-to-date information about any prenatal or postnatal diagnosis received by parents and that they connect parents with appropriate support services.

Aneuploidy of Sex Chromosomes

Aneuploid conditions involving sex chromosomes appear to upset the genetic balance less than those involving autosomes. This may be because the Y chromosome carries relatively few genes. Also, extra copies of the X chromosome simply become inactivated as Barr bodies in somatic cells.

An extra X chromosome in a male, producing XXY, occurs approximately once in every 500 to 1,000 live male births. People with this disorder, called *Klinefelter syndrome*, have male sex organs, but the testes are abnormally small and the man is sterile. Even though the extra X is inactivated, some breast enlargement and other female body characteristics are common. Affected individuals may have subnormal intelligence. About one of every 1,000 males is born with an extra Y chromosome (XYY). These males undergo normal sexual development and do not exhibit any well-defined syndrome.

Females with trisomy X (XXX), which occurs once in approximately 1,000 live female births, are healthy and have no unusual physical features other than being slightly taller than average. Triple-X females are at risk for learning disabilities but are fertile. Monosomy X, called *Turner syndrome*, occurs about once in every 2,500 female births and is the only known viable monosomy in humans. Although these X0 individuals are phenotypically female, they are sterile because their sex organs do not mature. When provided with estrogen replacement therapy, girls with Turner syndrome do develop secondary sex characteristics. Most have normal intelligence.

Disorders Caused by Structurally Altered Chromosomes

Many deletions in human chromosomes, even in a heterozygous state, cause severe problems. One such syndrome, known as *cri du chat* ("cry of the cat"), results from a specific deletion in chromosome 5. A child born with this deletion is severely intellectually disabled, has a small head with unusual facial features, and has a cry that sounds like the mewing of a distressed cat. Such individuals usually die in infancy or early childhood.

Chromosomal translocations have been implicated in certain cancers, including *chronic myelogenous leukemia* (*CML*). This disease occurs when a reciprocal translocation happens during mitosis of cells that will become white blood cells. In these cells, the exchange of a large portion of chromosome 22 with a small fragment from a tip of chromosome 9 produces a much shortened, easily recognized chromosome 22, called the *Philadelphia chromosome* **(Figure 12.16)**. Such an exchange causes cancer by activating a gene that leads to uncontrolled cell cycle progression. (The mechanism of gene activation will be discussed in Chapter 16.)

▲ **Figure 12.15 Down syndrome.** The karyotype shows trisomy 21, the most common cause of Down syndrome. The child exhibits the facial features characteristic of this disorder.

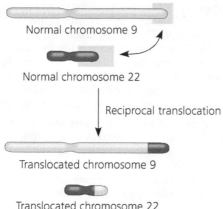

▲ Figure 12.16 Translocation associated with chronic myelogenous leukemia (CML). The cancerous cells in nearly all CML patients contain an abnormally short chromosome 22, the so-called Philadelphia chromosome, and an abnormally long chromosome 9. These altered chromosomes result from the reciprocal translocation shown here, which presumably occurred in a single white blood cell precursor undergoing mitosis and was then passed along to all descendant cells.

CONCEPT CHECK 12.4

1. About 5% of individuals with Down syndrome have a chromosomal translocation in which a third copy of chromosome 21 is attached to chromosome 14. If this translocation occurred in a parent's gonad, how could it lead to Down syndrome in a child?
2. **WHAT IF?** The ABO blood type locus has been mapped on chromosome 9. A father who has type AB blood and a mother who has type O blood have a child with trisomy 9 and type A blood. Using this information, can you tell in which parent the nondisjunction occurred? Explain your answer.
3. **MAKE CONNECTIONS** The gene that is activated on the Philadelphia chromosome codes for an intracellular kinase. Review the discussion of cell cycle control and cancer in Concept 9.3, and explain how the activation of this gene could contribute to the development of cancer.
4. Women born with an extra X chromosome (XXX) are generally healthy and indistinguishable in appearance from normal XX women. What is a likely explanation for this finding? How could you test this explanation?

For suggested answers, see Appendix A.

12 Chapter Review

SUMMARY OF KEY CONCEPTS

CONCEPT 12.1

Mendelian inheritance has its physical basis in the behavior of chromosomes (pp. 228–231)

- The **chromosome theory of inheritance** states that genes are located on chromosomes and that the behavior of chromosomes during meiosis accounts for Mendel's laws of segregation and independent assortment.
- Morgan's discovery that transmission of the X chromosome in *Drosophila* correlates with inheritance of an eye-color trait was the first solid evidence indicating that a specific gene is associated with a specific chromosome.

> **?** *What characteristic of the sex chromosomes allowed Morgan to correlate their behavior with that of the alleles of the eye-color gene?*

CONCEPT 12.2

Sex-linked genes exhibit unique patterns of inheritance (pp. 231–234)

- Sex is often chromosomally based. Humans and other mammals have an X-Y system in which sex is determined by whether a Y chromosome is present.
- The sex chromosomes carry **sex-linked genes**, virtually all of which are on the X chromosome (X-linked). Any male who inherits a recessive X-linked allele (from his mother) will express the trait, such as color blindness.
- In mammalian females, one of the two X chromosomes in each cell is randomly inactivated during early embryonic development, becoming highly condensed into a **Barr body**.

> **?** *Why are males affected much more often than females by X-linked disorders?*

CONCEPT 12.3

Linked genes tend to be inherited together because they are located near each other on the same chromosome (pp. 234–240)

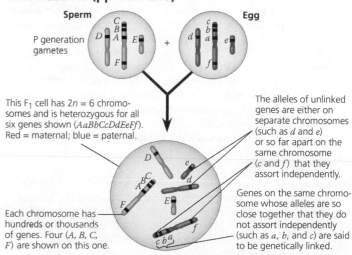

This F$_1$ cell has 2n = 6 chromosomes and is heterozygous for all six genes shown (*AaBbCcDdEeFf*). Red = maternal; blue = paternal.

Each chromosome has hundreds or thousands of genes. Four (*A, B, C, F*) are shown on this one.

The alleles of unlinked genes are either on separate chromosomes (such as *d* and *e*) or so far apart on the same chromosome (*c* and *f*) that they assort independently.

Genes on the same chromosome whose alleles are so close together that they do not assort independently (such as *a, b,* and *c*) are said to be genetically linked.

- An F$_1$ testcross yields **parental types** with the same combination of traits as those in the P generation parents and **recombinant types** with new combinations of traits. Unlinked genes exhibit a 50% frequency of recombination in the gametes. For genetically **linked genes**, crossing over accounts for the observed recombinants, always less than 50%.
- Recombination frequencies observed in genetic crosses allow construction of a **linkage map** (a type of **genetic map**).

> **?** *Why are specific alleles of two genes that are farther apart more likely to show recombination than those of two closer genes?*

Alterations of chromosome number or structure cause some genetic disorders (pp. 240–243)

- **Aneuploidy**, an abnormal chromosome number, results from **nondisjunction** during meiosis. When a normal gamete unites with one containing two copies or no copies of a particular chromosome, the resulting zygote and its descendant cells either have one extra copy of that chromosome (**trisomy**, $2n + 1$) or are missing a copy (**monosomy**, $2n - 1$). **Polyploidy** (extra sets of chromosomes) can result from complete nondisjunction.
- Chromosome breakage can result in alterations of chromosome structure: **deletions**, **duplications**, **inversions**, and **translocations**.

? *Why are inversions and reciprocal translocations less likely to be lethal than are aneuploidy, duplications, and deletions?*

TEST YOUR UNDERSTANDING

Level 1: Knowledge/Comprehension

1. A man with hemophilia (a recessive, sex-linked condition) has a normal daughter, who marries a normal man. What is the probability that a daughter will be a hemophiliac? A son? If the couple has four sons, that all will be affected?

2. Pseudohypertrophic muscular dystrophy is an inherited disorder that causes gradual deterioration of the muscles. It is seen almost exclusively in boys born to apparently normal parents and usually results in death in the early teens. Is this disorder caused by a dominant or a recessive allele? Is its inheritance sex-linked or autosomal? How do you know? Explain why this disorder is almost never seen in girls.

3. A space probe discovers a planet inhabited by creatures that reproduce with the same hereditary patterns seen in humans. Three phenotypic characters are height (T = tall, t = dwarf), head appendages (A = antennae, a = no antennae), and nose morphology (S = upturned snout, s = downturned snout). Since the creatures are not "intelligent," Earth scientists are able to do some controlled breeding experiments using various heterozygotes in testcrosses. For tall heterozygotes with antennae, the offspring are tall-antennae, 46; dwarf-antennae, 7; dwarf–no antennae, 42; tall–no antennae, 5. For heterozygotes with antennae and an upturned snout, the offspring are antennae–upturned snout, 47; antennae–downturned snout, 2; no antennae–downturned snout, 48; no antennae–upturned snout, 3. Calculate the recombination frequencies for both experiments.

Level 2: Application/Analysis

4. Using the information from problem 3, scientists do a further testcross using a heterozygote for height and nose morphology. The offspring are tall–upturned snout, 40; dwarf–upturned snout, 9; dwarf–downturned snout, 42; tall–downturned snout, 9. Calculate the recombination frequency from these data; then use your answer from problem 3 to determine the correct sequence of the three linked genes.

5. A man with red-green color blindness (a recessive, sex-linked condition) marries a woman with normal vision whose father was color-blind. What is the probability that they will have a color-blind daughter? That their first son will be color-blind? (Note the different wording in the two questions.)

6. You design *Drosophila* crosses to provide recombination data for gene *a*, which is located on the chromosome shown in Figure 12.12. Gene *a* has recombination frequencies of 14% with the vestigial-wing locus and 26% with the brown-eye locus. Approximately where is gene *a* located along the chromosome?

7. A wild-type fruit fly (heterozygous for gray body color and red eyes) is mated with a black fruit fly with purple eyes. The offspring are wild-type, 721; black-purple, 751; gray-purple, 49; black-red, 45. What is the recombination frequency between these genes for body color and eye color? Using information from Figure 12.9, what fruit flies (genotypes and phenotypes) would you mate to determine the sequence of the body-color, wing-size, and eye-color genes on the chromosome?

8. Assume that genes *A* and *B* are 50 map units apart on the same chromosome. An animal heterozygous at both loci is crossed with one that is homozygous recessive at both loci. What percentage of the offspring will show recombinant phenotypes? Without knowing that these genes are on the same chromosome, how would you interpret the results of this cross?

9. Two genes of a flower, one controlling blue (*B*) versus white (*b*) petals and the other controlling round (*R*) versus oval (*r*) stamens, are linked and are 10 map units apart. You cross a homozygous blue-oval plant with a homozygous white-round plant. The resulting F_1 progeny are crossed with homozygous white-oval plants, and 1,000 F_2 progeny are obtained. How many F_2 plants of each of the four phenotypes do you expect?

Level 3: Synthesis/Evaluation

10. **SCIENTIFIC INQUIRY**
Butterflies have an X-Y sex determination system that is different from that of flies or humans. Female butterflies may be either XY or XO, while butterflies with two or more X chromosomes are males. This photograph shows a tiger swallowtail *gynandromorph*, an individual that is half male (left side) and half female (right side). Given that the first division of the zygote divides the embryo into the future right and left halves of the butterfly, propose a hypothesis that explains how nondisjunction during the first mitosis might have produced this unusual-looking butterfly.

11. **FOCUS ON EVOLUTION**
Crossing over, or recombination, is thought to be evolutionarily advantageous because it continually shuffles genetic alleles into novel combinations. Until recently, it was thought that Y-linked genes might degenerate because they have no homologous genes on the X chromosome with which to recombine. However, when the Y chromosome was sequenced, eight large regions were found to be internally homologous to each other, and quite a few of the 78 genes represent duplicates. How might this be beneficial?

12. **FOCUS ON INFORMATION**
The continuity of life is based on heritable information in the form of DNA. In a short essay (100–150 words), relate the structure and behavior of chromosomes to inheritance in both asexually and sexually reproducing species.

For selected answers, see Appendix A.

The Molecular Basis of Inheritance

13

KEY CONCEPTS

13.1 DNA is the genetic material

13.2 Many proteins work together in DNA replication and repair

13.3 A chromosome consists of a DNA molecule packed together with proteins

13.4 Understanding DNA structure and replication makes genetic engineering possible

Life's Operating Instructions

▼ **Figure 13.1** How was the structure of DNA determined?

In April 1953, James Watson and Francis Crick shook the scientific world with an elegant double-helical model for the three-dimensional structure of deoxyribonucleic acid, or DNA. **Figure 13.1** shows Watson (left) and Crick admiring their DNA model, which they built from tin and wire. Over the past 60 years or so, their model has evolved from a novel proposition to an icon of modern biology. Mendel's heritable factors and Morgan's genes on chromosomes are, in fact, composed of DNA. Chemically speaking, your genetic endowment is the DNA you inherited from your parents. DNA, the substance of inheritance, is the most celebrated molecule of our time.

Of all nature's molecules, nucleic acids are unique in their ability to direct their own replication from monomers. Indeed, the resemblance of offspring to their parents has its basis in the precise replication of DNA and its transmission from one generation to the next. Hereditary information is encoded in the chemical language of DNA and reproduced in all the cells of your body. It is this DNA program that directs the development of your biochemical, anatomical, physiological, and, to some extent, behavioral traits. In this chapter, you'll discover how biologists deduced that DNA is the genetic material and how Watson and Crick worked out its structure. You'll also learn how a molecule of DNA is copied during **DNA replication** and how cells repair their DNA. Next, you'll see how DNA is packaged with proteins in a chromosome. Finally, you'll explore how an understanding of DNA-related processes has allowed scientists to directly manipulate genes for practical purposes.

CONCEPT 13.1

DNA is the genetic material

Today, even schoolchildren have heard of DNA, and scientists routinely manipulate DNA in the laboratory, often to change the heritable traits of cells in their experiments. Early in the 20th century, however, identifying the molecules of inheritance loomed as a major challenge to biologists.

The Search for the Genetic Material: *Scientific Inquiry*

Once T. H. Morgan's group showed that genes exist as parts of chromosomes (described in Chapter 12), the two chemical components of chromosomes—DNA and protein—emerged as the leading candidates for the genetic material. Until the 1940s, the case for proteins seemed stronger, especially since biochemists had identified them as a class of macromolecules with great heterogeneity and specificity of function, essential requirements for the hereditary material. Moreover, little was known about nucleic acids, whose physical and chemical properties seemed far too uniform to account for the multitude of specific inherited traits exhibited by every organism. This view gradually changed as experiments with microorganisms yielded unexpected results. As with the work of Mendel and Morgan, a key factor in determining the identity of the genetic material was the choice of appropriate experimental organisms. The role of DNA in heredity was first worked out while studying bacteria and the viruses that infect them, which are far simpler than pea plants, fruit flies, or humans. In this section, we'll trace the search for the genetic material in some detail as a case study in scientific inquiry.

Evidence That DNA Can Transform Bacteria

In 1928, a British medical officer named Frederick Griffith was trying to develop a vaccine against pneumonia. He was studying *Streptococcus pneumoniae*, a bacterium that causes pneumonia in mammals. Griffith had two strains (varieties) of the bacterium, one pathogenic (disease-causing) and one nonpathogenic (harmless). He was surprised to find that when he killed the pathogenic bacteria with heat and then mixed the cell remains with living bacteria of the nonpathogenic strain, some of the living cells became pathogenic **(Figure 13.2)**. Furthermore, this newly acquired trait of pathogenicity was inherited by all the descendants of the transformed bacteria. Clearly, some chemical component of the dead pathogenic cells caused this heritable change, although the identity of the substance was not known. Griffith called the phenomenon **transformation**, now defined as a change in genotype and phenotype due to the assimilation of external DNA by a cell. Later work by Oswald Avery and others identified the transforming substance as DNA.

Scientists remained skeptical, however, many viewing proteins as better candidates for the genetic material. Moreover, many biologists were not convinced that the genes of bacteria would be similar in composition and function to those of more complex organisms. But the major reason for the continued doubt was that so little was known about DNA.

Evidence That Viral DNA Can Program Cells

Additional evidence for DNA as the genetic material came from studies of viruses that infect bacteria. These viruses are called **bacteriophages** (meaning "bacteria-eaters"), or **phages**

▼ **Figure 13.2** **Inquiry**

Can a genetic trait be transferred between different bacterial strains?

Experiment Frederick Griffith studied two strains of the bacterium *Streptococcus pneumoniae*. Bacteria of the S (smooth) strain can cause pneumonia in mice; they are pathogenic because they have an outer capsule that protects them from an animal's immune system. Bacteria of the R (rough) strain lack a capsule and are nonpathogenic. To test for the trait of pathogenicity, Griffith injected mice with the two strains:

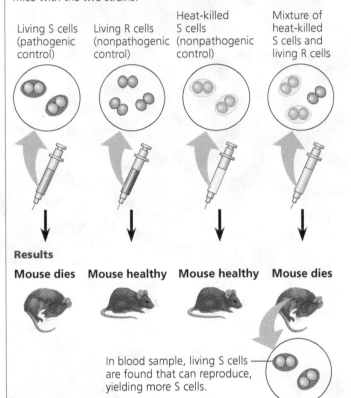

Living S cells (pathogenic control)

Living R cells (nonpathogenic control)

Heat-killed S cells (nonpathogenic control)

Mixture of heat-killed S cells and living R cells

Results

Mouse dies   **Mouse healthy**   **Mouse healthy**   **Mouse dies**

In blood sample, living S cells are found that can reproduce, yielding more S cells.

Conclusion Griffith concluded that the living R bacteria had been transformed into pathogenic S bacteria by an unknown, heritable substance from the dead S cells that allowed the R cells to make capsules.

Source F. Griffith, The significance of pneumococcal types, *Journal of Hygiene* 27:113–159 (1928).

WHAT IF? How did this experiment rule out the possibility that the R cells could have simply used the capsules of the dead S cells to become pathogenic?

---

for short. Viruses are much simpler than cells. A **virus** is little more than DNA (or sometimes RNA) enclosed by a protective coat, which is often simply protein **(Figure 13.3)**. To produce more viruses, a virus must infect a cell and take over the cell's metabolic machinery.

Phages have been widely used as tools by researchers in molecular genetics. In 1952, Alfred Hershey and Martha Chase performed experiments showing that DNA is the genetic material of a phage known as T2. This is one of many phages that

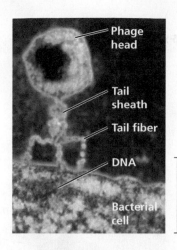

► **Figure 13.3 Viruses infecting a bacterial cell.** Phages called T2 attach to the host cell and inject their genetic material through the plasma membrane while the head and tail parts remain on the outer bacterial surface (colorized TEM).

Phage head

Tail sheath

Tail fiber

DNA

Bacterial cell

100 nm

infect *Escherichia coli* (*E. coli*), a bacterium that normally lives in the intestines of mammals and is a model organism for molecular biologists. At that time, biologists already knew that T2, like many other phages, was composed almost entirely of DNA and protein. They also knew that the T2 phage could quickly turn an *E. coli* cell into a T2-producing factory that released many copies when the cell ruptured. Somehow, T2 could reprogram its host cell to produce viruses. But which viral component—protein or DNA—was responsible?

Hershey and Chase answered this question by devising an experiment showing that only one of the two components of T2 actually enters the *E. coli* cell during infection (**Figure 13.4**).

▼ **Figure 13.4** Inquiry

Is protein or DNA the genetic material of phage T2?

Experiment Alfred Hershey and Martha Chase used radioactive sulfur and phosphorus to trace the fates of protein and DNA, respectively, of T2 phages that infected bacterial cells. They wanted to see which of these molecules entered the cells and could reprogram them to make more phages.

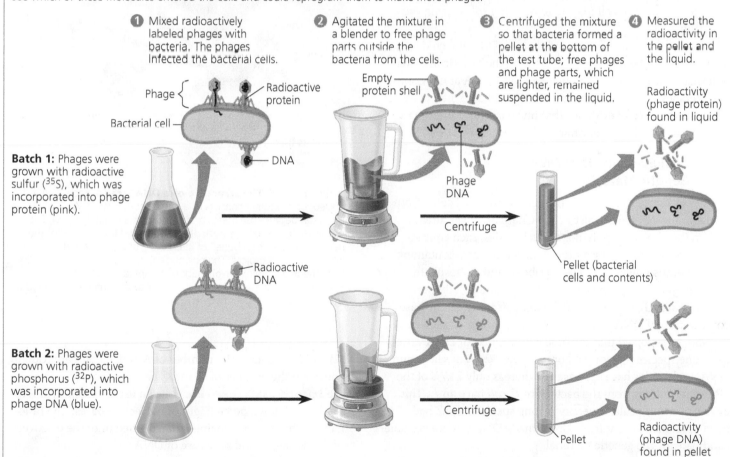

❶ Mixed radioactively labeled phages with bacteria. The phages infected the bacterial cells.

❷ Agitated the mixture in a blender to free phage parts outside the bacteria from the cells.

❸ Centrifuged the mixture so that bacteria formed a pellet at the bottom of the test tube; free phages and phage parts, which are lighter, remained suspended in the liquid.

❹ Measured the radioactivity in the pellet and the liquid.

Phage

Radioactive protein

Bacterial cell

Empty protein shell

Radioactivity (phage protein) found in liquid

Batch 1: Phages were grown with radioactive sulfur (³⁵S), which was incorporated into phage protein (pink).

DNA

Phage DNA

Centrifuge

Pellet (bacterial cells and contents)

Radioactive DNA

Batch 2: Phages were grown with radioactive phosphorus (³²P), which was incorporated into phage DNA (blue).

Centrifuge

Pellet

Radioactivity (phage DNA) found in pellet

Results When proteins were labeled (batch 1), radioactivity remained outside the cells; but when DNA was labeled (batch 2), radioactivity was found inside the cells. Bacterial cells with radioactive phage DNA released new phages with some radioactive phosphorus.

Conclusion Phage DNA entered bacterial cells, but phage proteins did not. Hershey and Chase concluded that DNA, not protein, functions as the genetic material of phage T2.

Source A. D. Hershey and M. Chase, Independent functions of viral protein and nucleic acid in growth of bacteriophage, *Journal of General Physiology* 36:39–56 (1952).

WHAT IF? *How would the results have differed if proteins carried the genetic information?*

In their experiment, they used a radioactive isotope of sulfur to tag protein in one batch of T2 and a radioactive isotope of phosphorus to tag DNA in a second batch. Because protein, but not DNA, contains sulfur, radioactive sulfur atoms were incorporated only into the protein of the phage. In a similar way, the atoms of radioactive phosphorus labeled only the DNA, not the protein, because nearly all the phage's phosphorus is in its DNA. In the experiment, separate samples of nonradioactive *E. coli* cells were allowed to be infected by the protein-labeled and DNA-labeled batches of T2. The researchers then tested the two samples shortly after the onset of infection to see which type of molecule—protein or DNA—had entered the bacterial cells and would therefore have been capable of reprogramming them.

Hershey and Chase found that the phage DNA entered the host cells but the phage protein did not. Moreover, when these bacteria were returned to a culture medium, the infection ran its course, and the *E. coli* released phages that contained some radioactive phosphorus, further showing that the DNA inside the cell played an ongoing role during the infection process.

Hershey and Chase concluded that the DNA injected by the phage must be the molecule carrying the genetic information that makes the cells produce new viral DNA and proteins. The Hershey-Chase experiment was a landmark study because it provided powerful evidence that nucleic acids, rather than proteins, are the hereditary material, at least for viruses.

Additional Evidence That DNA Is the Genetic Material

Further evidence that DNA is the genetic material came from the laboratory of biochemist Erwin Chargaff. It was already known that DNA is a polymer of nucleotides, each consisting of three components: a nitrogenous (nitrogen-containing) base, a pentose sugar called deoxyribose, and a phosphate group (Figure 13.5). The base can be adenine (A), thymine (T), guanine (G), or cytosine (C). Chargaff analyzed the base composition of DNA from a number of different organisms. In 1950, he reported that the base composition of DNA varies from one species to another. For example, 32.8% of sea urchin DNA nucleotides have the base A, whereas only 24.7% of the DNA nucleotides from the bacterium *E. coli* have an A. This evidence of molecular diversity among species, which had been presumed absent from DNA, made DNA a more credible candidate for the genetic material.

Chargaff also noticed a peculiar regularity in the ratios of nucleotide bases. In the DNA of each species he studied, the number of adenines approximately equaled the number of thymines, and the number of guanines approximately equaled the number of cytosines. In sea urchin DNA, for example, the four bases are present in these percentages: A = 32.8% and T = 32.1%; G = 17.7% and C = 17.3%.

These two findings became known as *Chargaff's rules*: (1) the base composition varies between species, and

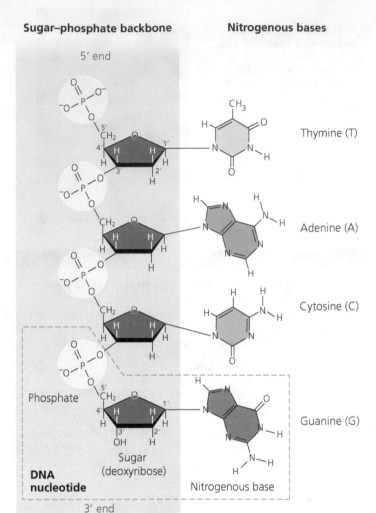

Sugar–phosphate backbone   **Nitrogenous bases**

5' end

Phosphate

DNA nucleotide

Sugar (deoxyribose)

3' end

Thymine (T)

Adenine (A)

Cytosine (C)

Guanine (G)

Nitrogenous base

▲ **Figure 13.5 The structure of a DNA strand.** Each DNA nucleotide monomer consists of a nitrogenous base (T, A, C, or G), the sugar deoxyribose (blue), and a phosphate group (yellow). The phosphate group of one nucleotide is attached to the sugar of the next, forming a "backbone" of alternating phosphates and sugars from which the bases project. The polynucleotide strand has directionality, from the 5' end (with the phosphate group) to the 3' end (with the —OH group of the sugar). 5' and 3' refer to the numbers assigned to the carbons in the sugar ring.

(2) within a species, the number of A and T bases are roughly equal and the number of G and C bases are roughly equal. In the **Scientific Skills Exercise,** you can use Chargaff's rules to predict unknown percentages of nucleotide bases. The rationale for these rules remained unexplained until the discovery of the double helical structure of DNA.

Building a Structural Model of DNA: *Scientific Inquiry*

Once most biologists were convinced that DNA was the genetic material, the challenge was to determine how the structure of DNA could account for its role in inheritance. By the early 1950s, the arrangement of covalent bonds in a single nucleic acid polymer was well established (see Figure 13.5), and researchers focused on discovering the three-dimensional

Working with Data in a Table

Given the Percentage Composition of One Nucleotide in a Genome, Can We Predict the Percentages of the Other Three Nucleotides? Even before the structure of DNA was elucidated, Erwin Chargaff and his coworkers noticed a pattern in the base composition of nucleotides from different organisms: the number of adenine (A) bases roughly equaled the number of thymine (T) bases, and the number of cytosine (C) bases roughly equaled the number of guanine (G) bases. Further, each species they studied had a different distribution of A/T and C/G bases. We now know that these consistent ratios are due to complementary base pairing between A and T and between C and G in the DNA double helix, and interspecies differences are due to the unique sequences of bases along a DNA strand. In this exercise, you will apply Chargaff's rules to predict the composition of nucleotide bases in a genome.

How the Experiments Were Done In Chargaff's experiments, DNA was extracted from the given organism, denatured, and hydrolyzed to break apart the individual nucleotides before analyzing them chemically. These experiments provided approximate values for each type of nucleotide. Today, the availability of whole-genome sequencing has allowed base composition analysis to be done more precisely directly from the sequence data.

Data from the Experiments Tables are useful for organizing sets of data representing a common set of values (here percentages of A, G, C, and T) for a number of different samples (in this case, species). You can apply the patterns that you see in the known data to predict unknown values. In the table in the upper right, complete base distribution data are given for sea urchin DNA and salmon DNA; you will use Chargaff's rules to fill in the rest of the table with predicted values.

Source of DNA	Adenine	Guanine	Cytosine	Thymine
Sea urchin	32.8%	17.7%	17.3%	32.1%
Salmon	29.7	20.8	20.4	29.1
Wheat	28.1	21.8	22.7	
E. coli	24.7	26.0		
Human	30.4			30.1
Ox	29.0			

Interpret the Data

1. Explain how the sea urchin and salmon data demonstrate both of Chargaff's rules.
2. Based on Chargaff's rules, fill in the table with your predictions of the missing percentages of bases, starting with the wheat genome and proceeding through *E. coli*, human, and ox. Show how you arrived at your answers.
3. If Chargaff's rule is valid, that the amount of A equals the amount of T and the amount of C equals the amount of G, then hypothetically we could extrapolate this to the combined DNA of all species on Earth (like one huge Earth genome). To see whether the data in the table support this hypothesis, calculate the average percentage for each base in your completed table by averaging the values in each column. Does Chargaff's equivalence rule still hold true?

Data from several papers by Chargaff: for example, E. Chargaff et al., Composition of the desoxypentose nucleic acids of four genera of sea-urchin, *Journal of Biological Chemistry* 195:155–160 (1952).

(MB) A version of this Scientific Skills Exercise can be assigned in MasteringBiology.

---

structure of DNA. Among the scientists working on the problem were Linus Pauling, at the California Institute of Technology, and Maurice Wilkins and Rosalind Franklin, at King's College in London. First to come up with the correct answer, however, were two scientists who were relatively unknown at the time—the American James Watson and the Englishman Francis Crick.

The brief but celebrated partnership that solved the puzzle of DNA structure began soon after Watson journeyed to Cambridge University, where Crick was studying protein structure with a technique called X-ray crystallography (see Figure 3.24). While visiting the laboratory of Maurice Wilkins, Watson saw an X-ray diffraction image of DNA produced by Wilkins's accomplished colleague Rosalind Franklin (**Figure 13.6a**). Images produced by X-ray crystallography are not actually pictures of molecules. The spots and smudges in **Figure 13.6b** were produced by X-rays that were diffracted (deflected) as they passed through aligned fibers of purified DNA. Watson was familiar with the type of X-ray diffraction pattern that helical molecules produce, and an examination of the photo that Wilkins showed him confirmed that DNA was helical in shape. It also augmented earlier data obtained by Franklin and others suggesting the

width of the helix and the spacing of the nitrogenous bases along it. The pattern in this photo implied that the helix was made up of two strands, contrary to a three-stranded model that Linus Pauling had proposed a short time earlier. The

(a) Rosalind Franklin

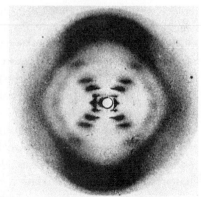

(b) Franklin's X-ray diffraction photograph of DNA

▲ **Figure 13.6 Rosalind Franklin and her X-ray diffraction photo of DNA.** Franklin, a very accomplished X-ray crystallographer, conducted critical experiments resulting in the photograph that allowed Watson and Crick to deduce the double-helical structure of DNA.

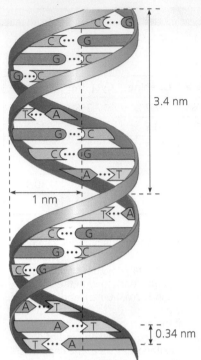

(a) Key features of DNA structure. The "ribbons" in this diagram represent the sugar-phosphate backbones of the two DNA strands. The helix is "right-handed," curving up to the right. The two strands are held together by hydrogen bonds (dotted lines) between the nitrogenous bases, which are paired in the interior of the double helix.

(b) Partial chemical structure. For clarity, the two DNA strands are shown untwisted in this partial chemical structure. Strong covalent bonds link the units of each strand, while weaker hydrogen bonds between the bases hold one strand to the other. Notice that the strands are antiparallel, meaning that they are oriented in opposite directions.

(c) Space-filling model. The tight stacking of the base pairs is clear in this computer model. Van der Waals interactions between the stacked pairs play a major role in holding the molecule together.

▲ **Figure 13.7 The double helix.**

presence of two strands accounts for the now-familiar term **double helix (Figure 13.7)**.

Watson and Crick began building models of a double helix that would conform to the X-ray measurements and what was then known about the chemistry of DNA, including Chargaff's rules. They knew that Franklin had concluded that the sugar-phosphate backbones were on the outside of the DNA molecule. This arrangement was appealing because it put the negatively charged phosphate groups facing the aqueous surroundings, while the relatively hydrophobic nitrogenous bases were hidden in the interior. Watson constructed such a model (see Figure 13.1). In this model, the two sugar-phosphate backbones are **antiparallel**—that is, their subunits run in opposite directions (see Figure 13.7b). You can imagine the overall arrangement as a rope ladder with rigid rungs. The side ropes represent the sugar-phosphate backbones, and the rungs represent pairs of nitrogenous bases. Now imagine twisting the ladder to form a helix. Franklin's X-ray data indicated that the helix makes one full turn every 3.4 nm along its length. With the bases stacked just 0.34 nm apart, there are ten "rungs" of base pairs in each full turn of the helix.

The nitrogenous bases of the double helix are paired in specific combinations: adenine (A) with thymine (T), and guanine (G) with cytosine (C). It was mainly by trial and error that Watson and Crick arrived at this key feature of DNA. At first, Watson imagined that the bases paired like with like—for example, A with A and C with C. But this model did not fit the X-ray data, which suggested that the double helix had a uniform diameter. Why is this requirement inconsistent with like-with-like pairing of bases? Adenine and guanine are purines, nitrogenous bases with two organic rings, while cytosine and thymine are nitrogenous bases called pyrimidines, which have a single ring. Thus, purines (A and G) are about twice as wide as pyrimidines (C and T). A purine-purine pair is too wide and a pyrimidine-pyrimidine pair too narrow to account for the 2-nm diameter of the double helix. Always pairing a purine with a pyrimidine, however, results in a uniform diameter:

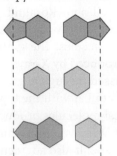

Purine + purine: too wide

Pyrimidine + pyrimidine: too narrow

Purine + pyrimidine: width consistent with X-ray data

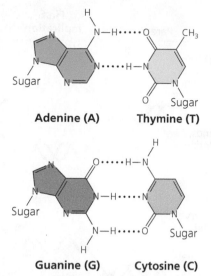

▲ Figure 13.8 Base pairing in DNA. The pairs of nitrogenous bases in a DNA double helix are held together by hydrogen bonds, shown here as black dotted lines.

Watson and Crick reasoned that there must be additional specificity of pairing dictated by the structure of the bases. Each base has chemical side groups that can form hydrogen bonds with its appropriate partner: Adenine forms two hydrogen bonds with thymine and only thymine; guanine forms three hydrogen bonds with cytosine and only cytosine. In shorthand, A pairs with T, and G pairs with C **(Figure 13.8)**.

The Watson-Crick model took into account Chargaff's ratios and ultimately explained them. Wherever one strand of a DNA molecule has an A, the partner strand has a T. Similarly, a G in one strand is always paired with a C in the complementary strand. Therefore, in the DNA of any organism, the amount of adenine equals the amount of thymine, and the amount of guanine equals the amount of cytosine. Although the base-pairing rules dictate the combinations of nitrogenous bases that form the "rungs" of the double helix, they do not restrict the sequence of nucleotides *along* each DNA strand. The

linear sequence of the four bases can be varied in countless ways, and each gene has a unique order, or base sequence.

In April 1953, Watson and Crick surprised the scientific world with a succinct, one-page paper that reported their molecular model for DNA: the double helix, which has since become the symbol of molecular biology. Watson and Crick, along with Maurice Wilkins, were awarded the Nobel Prize in 1962 for this work. (Sadly, Rosalind Franklin had died at the age of 38 in 1958 and was thus ineligible for the prize.) The beauty of the double helix model was that the structure of DNA suggested the basic mechanism of its replication.

CONCEPT CHECK 13.1

1. Given a polynucleotide sequence such as GAATTC, can you tell which is the 5' end? If not, what further information do you need to identify the ends? (See Figure 13.5.)
2. **WHAT IF?** Griffith did not expect transformation to occur in his experiment. What results was he expecting? Explain.

For suggested answers, see Appendix A.

CONCEPT 13.2

Many proteins work together in DNA replication and repair

The relationship between structure and function is manifest in the double helix. The idea that there is specific pairing of nitrogenous bases in DNA was the flash of inspiration that led Watson and Crick to the double helix. At the same time, they saw the functional significance of the base-pairing rules. They ended their classic paper with this wry statement: "It has not escaped our notice that the specific pairing we have postulated immediately suggests a possible copying mechanism for the genetic material." In this section, you'll learn about the basic principle of DNA replication **(Figure 13.9)**, as well as some important details of the process.

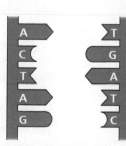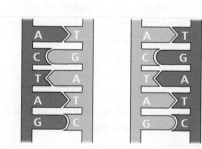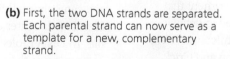

(a) The parental molecule (dark blue) has two complementary strands of DNA. Each base is paired by hydrogen bonding with its specific partner, A with T and G with C.

(b) First, the two DNA strands are separated. Each parental strand can now serve as a template for a new, complementary strand.

(c) Nucleotides complementary to the parental strands are connected to form the sugar-phosphate backbones of the new (light blue) strands.

▲ Figure 13.9 A model for DNA replication: the basic concept. In this simplified illustration, a short segment of DNA has been untwisted. Simple shapes symbolize the four kinds of bases, here represented as ladder rungs. Dark blue represents DNA strands present in the parental molecule; light blue represents newly synthesized DNA.

The Basic Principle: Base Pairing to a Template Strand

In a second paper, Watson and Crick stated their hypothesis for how DNA replicates:

> Now our model for deoxyribonucleic acid is, in effect, a pair of templates, each of which is complementary to the other. We imagine that prior to duplication the hydrogen bonds are broken, and the two chains unwind and separate. Each chain then acts as a template for the formation onto itself of a new companion chain, so that eventually we shall have two pairs of chains, where we only had one before. Moreover, the sequence of the pairs of bases will have been duplicated exactly.*

Figure 13.9 illustrates Watson and Crick's basic idea. To make it easier to follow, only a short section of double helix is shown, in untwisted form. Notice that if you cover one of the two DNA strands of Figure 13.9a, you can still determine its linear sequence of nucleotides by referring to the uncovered strand and applying the base-pairing rules. The two strands are complementary; each stores the information necessary to reconstruct the other. When a cell copies a DNA molecule, each strand serves as a template for ordering nucleotides into a new, complementary strand. Nucleotides line up along the template strand according to the base-pairing rules and are linked to form the new strands. Where there was one double-stranded DNA molecule at the beginning of the process, there are soon two, each an exact replica of the "parental" molecule. The copying mechanism is analogous to using a photographic negative to make a positive image, which can in turn be used to make another negative, and so on.

This model of DNA replication remained untested for several years following publication of the DNA structure. The requisite experiments were simple in concept but difficult to perform. Watson and Crick's model predicts that when a double helix replicates, each of the two daughter molecules will have one old strand, from the parental molecule, and one newly made strand. This **semiconservative model** can be distinguished from a conservative model of replication, in which the two parental strands somehow come back together after the process (that is, the parental molecule is conserved). In yet a third model, called the dispersive model, all four strands of DNA following replication have a mixture of old and new DNA. These three models are shown in **Figure 13.10**. Although mechanisms for conservative or dispersive DNA replication are not easy to come up with, these models remained possibilities until they could be ruled out. After two years of preliminary work in the late 1950s, Matthew Meselson and Franklin Stahl devised a clever experiment that distinguished between the three models, described in detail in **Figure 13.11**. Their experiment supported the semiconservative model of DNA replication, as predicted by Watson and Crick,

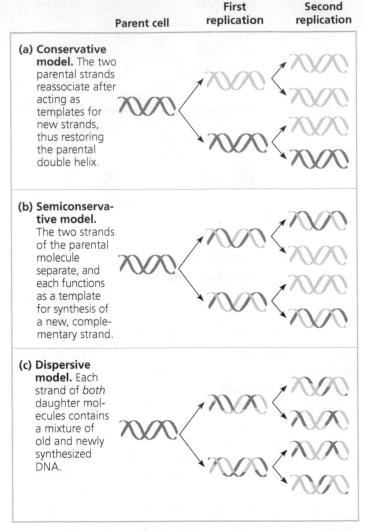

Parent cell — **First replication** — **Second replication**

(a) Conservative model. The two parental strands reassociate after acting as templates for new strands, thus restoring the parental double helix.

(b) Semiconservative model. The two strands of the parental molecule separate, and each functions as a template for synthesis of a new, complementary strand.

(c) Dispersive model. Each strand of *both* daughter molecules contains a mixture of old and newly synthesized DNA.

▲ **Figure 13.10 Three alternative models of DNA replication.** Each short segment of double helix symbolizes the DNA within a cell. Beginning with a parent cell, we follow the DNA for two more generations of cells—two rounds of DNA replication. Newly made DNA is light blue.

and is widely acknowledged among biologists to be a classic example of elegant experimental design.

The basic principle of DNA replication is conceptually simple. However, the actual process involves some complicated biochemical gymnastics, as we will now see.

DNA Replication: *A Closer Look*

The bacterium *E. coli* has a single chromosome of about 4.6 million nucleotide pairs. In a favorable environment, an *E. coli* cell can copy all this DNA and divide to form two genetically identical daughter cells in less than an hour. Each of *your* cells has 46 DNA molecules in its nucleus, one long double-helical molecule per chromosome. In all, that represents about 6 billion nucleotide pairs, or over a thousand times more DNA than is found in a bacterial cell. If we were to print the one-letter symbols for these bases (A, G, C, and T) the size of the

---

*F. H. C. Crick and J. D. Watson, The complementary structure of deoxyribonucleic acid, *Proceedings of the Royal Society of London A* 223:80 (1954).

Does DNA replication follow the conservative, semiconservative, or dispersive model?

Experiment At the California Institute of Technology, Matthew Meselson and Franklin Stahl cultured *E. coli* for several generations in a medium containing nucleotide precursors labeled with a heavy isotope of nitrogen, ^{15}N. They then transferred the bacteria to a medium with only ^{14}N, a lighter isotope. A sample was taken after DNA replicated once; another sample was taken after DNA replicated again. They extracted DNA from the bacteria in the samples and then centrifuged each DNA sample to separate DNA of different densities.

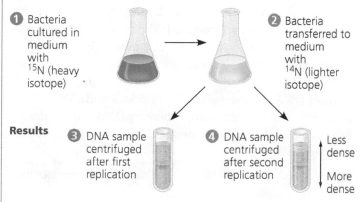

1 Bacteria cultured in medium with ^{15}N (heavy isotope)

2 Bacteria transferred to medium with ^{14}N (lighter isotope)

Results

3 DNA sample centrifuged after first replication

4 DNA sample centrifuged after second replication

Less dense

More dense

Conclusion Meselson and Stahl compared their results with those predicted by each of the three models in Figure 13.10, as shown below. The first replication in the ^{14}N medium produced a band of hybrid (^{15}N–^{14}N) DNA. This result eliminated the conservative model. The second replication produced both light and hybrid DNA, a result that refuted the dispersive model and supported the semiconservative model. They therefore concluded that DNA replication is semiconservative.

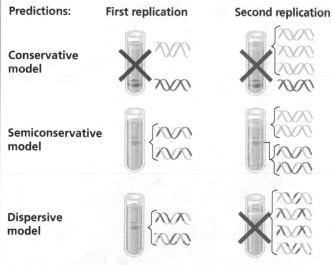

Predictions:	First replication	Second replication

Conservative model

Semiconservative model

Dispersive model

Source M. Meselson and F. W. Stahl, The replication of DNA in *Escherichia coli*, *Proceedings of the National Academy of Sciences USA* 44:671–682 (1958).

Inquiry in Action Read and analyze the original paper in *Inquiry in Action: Interpreting Scientific Papers*.

(MB) A related Experimental Inquiry Tutorial can be assigned in MasteringBiology.

WHAT IF? If Meselson and Stahl had first grown the cells in ^{14}N-containing medium and then moved them into ^{15}N-containing medium before taking samples, what would have been the result?

---

type you are now reading, the 6 billion nucleotide pairs of information in a diploid human cell would fill about 1,200 books as thick as this text. Yet it takes one of your cells just a few hours to copy all of this DNA. This replication of an enormous amount of genetic information is achieved with very few errors—only about one per 10 billion nucleotides. The copying of DNA is remarkable in its speed and accuracy.

More than a dozen enzymes and other proteins participate in DNA replication. Much more is known about how this "replication machine" works in bacteria (such as *E. coli*) than in eukaryotes, and we will describe the basic steps of the process for *E. coli*, except where otherwise noted. What scientists have learned about eukaryotic DNA replication suggests, however, that most of the process is fundamentally similar for prokaryotes and eukaryotes.

Getting Started

The replication of a DNA molecule begins at particular sites called **origins of replication**, short stretches of DNA having a specific sequence of nucleotides. Proteins that initiate DNA replication recognize this sequence and attach to the DNA, separating the two strands and opening up a replication "bubble." At each end of a bubble is a **replication fork**, a Y-shaped region where the parental strands of DNA are being unwound. Several kinds of proteins participate in the unwinding **(Figure 13.12)**. **Helicases** are enzymes that untwist the double helix at the replication forks, separating the two parental strands and making them available as template strands. After the parental strands separate, **single-strand binding proteins** bind to the unpaired DNA strands, keeping them from re-pairing. The untwisting of the double helix causes tighter twisting and strain ahead of the replication fork. **Topoisomerase** helps relieve this strain by breaking, swiveling, and rejoining DNA strands.

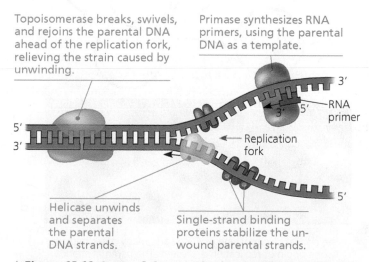

Topoisomerase breaks, swivels, and rejoins the parental DNA ahead of the replication fork, relieving the strain caused by unwinding.

Primase synthesizes RNA primers, using the parental DNA as a template.

3'

RNA primer

3' 5'

5'

3'

Replication fork

5'

Helicase unwinds and separates the parental DNA strands.

Single-strand binding proteins stabilize the unwound parental strands.

▲ **Figure 13.12 Some of the proteins involved in the initiation of DNA replication.** The same proteins function at both replication forks in a replication bubble. For simplicity, only the left-hand fork is shown, and the DNA bases are drawn much larger in relation to the proteins than they are in reality.

The *E. coli* chromosome, like many other bacterial chromosomes, is circular and has a single origin of replication, forming one replication bubble (**Figure 13.13a**). Replication of DNA then proceeds in both directions until the entire molecule is copied. In contrast to a bacterial chromosome, a eukaryotic chromosome may have hundreds or even a few thousand replication origins. Multiple replication bubbles form and eventually fuse, thus speeding up the copying of the very long DNA molecules (**Figure 13.13b**). As in bacteria, eukaryotic DNA replication proceeds in both directions from each origin.

Synthesizing a New DNA Strand

Within a bubble, the unwound sections of parental DNA strands are available to serve as templates for the synthesis of new complementary DNA strands. However, the enzymes that synthesize DNA cannot *initiate* the synthesis of a polynucleotide; they can only add nucleotides to the end of an already existing chain that is base-paired with the template strand. The initial nucleotide chain that is produced during DNA synthesis is actually a short stretch of RNA, not DNA. This RNA chain is called a **primer** and is synthesized by the enzyme **primase** (see Figure 13.12). Primase starts a complementary RNA chain from a single RNA nucleotide, adding RNA nucleotides one at a time, using the parental DNA strand as a template. The completed primer, generally 5–10 nucleotides long, is thus base-paired to the template strand. The new DNA strand will start from the 3′ end of the RNA primer.

Enzymes called **DNA polymerases** catalyze the synthesis of new DNA by adding nucleotides to a preexisting chain. In *E. coli*, there are several different DNA polymerases, but two appear to play the major roles in DNA replication: DNA polymerase III and DNA polymerase I. The situation in eukaryotes is more complicated, with at least 11 different DNA polymerases discovered so far; however, the general principles are the same.

Most DNA polymerases require a primer and a DNA template strand along which complementary DNA nucleotides line up. In *E. coli*, DNA polymerase III (abbreviated DNA pol

▼ **Figure 13.13 Origins of replication in *E. coli* and eukaryotes.** The red arrows indicate the movement of the replication forks and thus the overall directions of DNA replication within each bubble.

(a) Origin of replication in an *E. coli* cell

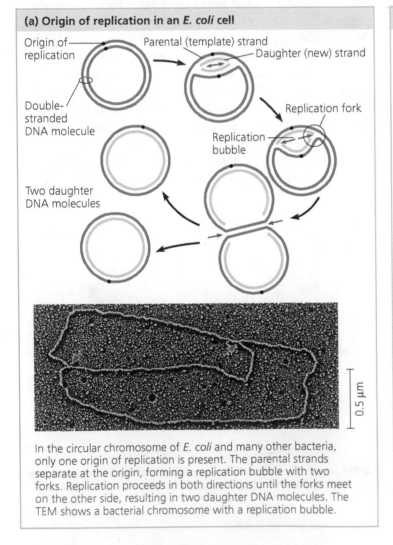

In the circular chromosome of *E. coli* and many other bacteria, only one origin of replication is present. The parental strands separate at the origin, forming a replication bubble with two forks. Replication proceeds in both directions until the forks meet on the other side, resulting in two daughter DNA molecules. The TEM shows a bacterial chromosome with a replication bubble.

(b) Origins of replication in a eukaryotic cell

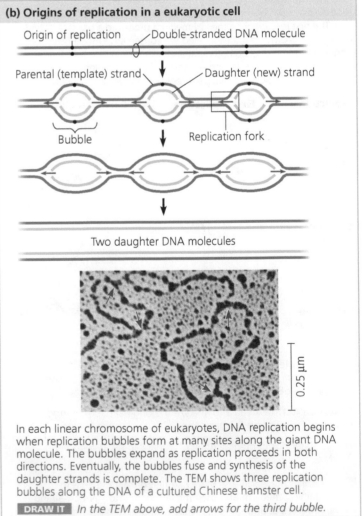

In each linear chromosome of eukaryotes, DNA replication begins when replication bubbles form at many sites along the giant DNA molecule. The bubbles expand as replication proceeds in both directions. Eventually, the bubbles fuse and synthesis of the daughter strands is complete. The TEM shows three replication bubbles along the DNA of a cultured Chinese hamster cell.

DRAW IT *In the TEM above, add arrows for the third bubble.*

III) adds a DNA nucleotide to the RNA primer and then continues adding DNA nucleotides, complementary to the parental DNA template strand, to the growing end of the new DNA strand. The rate of elongation is about 500 nucleotides per second in bacteria and 50 per second in human cells.

Each nucleotide to be added to a growing DNA strand consists of a sugar attached to a base and three phosphate groups. You have already encountered such a molecule—ATP (adenosine triphosphate; see Figure 6.8). The only difference between the ATP of energy metabolism and dATP, the adenine nucleotide used to make DNA, is the sugar component, which is deoxyribose in the building block of DNA but ribose in ATP. Like ATP, the nucleotides used for DNA synthesis are chemically reactive, partly because their triphosphate tails have an unstable cluster of negative charge. As each monomer joins the growing end of a DNA strand, two phosphate groups are lost as a molecule of pyrophosphate (P—P_i). Subsequent hydrolysis of the pyrophosphate to two molecules of inorganic phosphate P_i is a coupled exergonic reaction that helps drive the polymerization reaction **(Figure 13.14)**.

Antiparallel Elongation

As we have noted previously, the two ends of a DNA strand are different, giving each strand directionality, like a one-way street (see Figure 13.5). In addition, the two strands of DNA in a double helix are antiparallel, meaning that they are oriented in opposite directions to each other, like a divided highway (see Figure 13.14). Therefore, the two new strands formed during DNA replication must also end up antiparallel to their template strands.

How does the antiparallel arrangement of the double helix affect replication? Because of their structure, DNA polymerases can add nucleotides only to the free 3' end of a primer or growing DNA strand, never to the 5' end (see Figure 13.14). Thus, a new DNA strand can elongate only in the 5' → 3' direction. With this in mind, let's examine one of the two replication forks in a bubble **(Figure 13.15)**. Along one template strand, DNA polymerase III can synthesize a complementary strand continuously by elongating the new DNA in the mandatory 5' → 3' direction. DNA pol III remains in the replication fork on that template strand and continuously adds nucleotides to the new complementary strand as the fork progresses. The DNA strand made by this mechanism is called the **leading strand**. Only one primer is required for DNA pol III to synthesize the leading strand.

To elongate the other new strand of DNA in the mandatory 5' → 3' direction, DNA pol III must work along the other

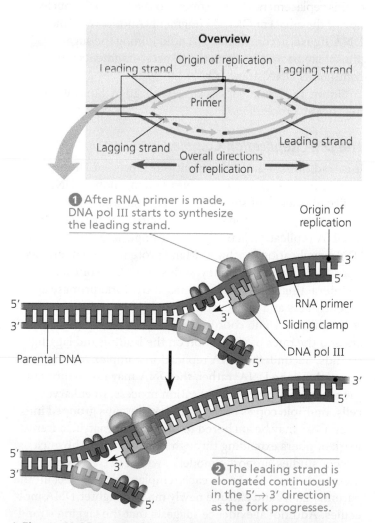

① After RNA primer is made, DNA pol III starts to synthesize the leading strand.

② The leading strand is elongated continuously in the 5' → 3' direction as the fork progresses.

▲ **Figure 13.15 Synthesis of the leading strand during DNA replication.** This diagram focuses on the left replication fork shown in the overview box. DNA polymerase III (DNA pol III), shaped like a cupped hand, is shown closely associated with a protein called the "sliding clamp" that encircles the newly synthesized double helix like a doughnut. The sliding clamp moves DNA pol III along the DNA template strand.

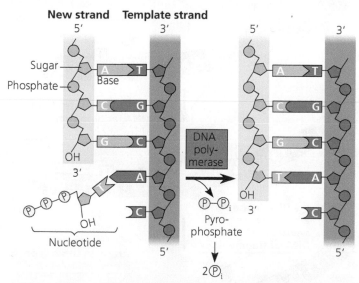

▲ **Figure 13.14 Addition of a nucleotide to a DNA strand.** DNA polymerase catalyzes the addition of a nucleotide to the 3' end of a growing DNA strand, with the release of two phosphates.

❓ *Use this diagram to explain what we mean when we say that each DNA strand has directionality.*

template strand in the direction *away from* the replication fork. The DNA strand elongating in this direction is called the **lagging strand**.* In contrast to the leading strand, which elongates continuously, the lagging strand is synthesized discontinuously, as a series of segments. These segments of the lagging strand are called **Okazaki fragments**, after the Japanese scientist who discovered them. The fragments are about 1,000–2,000 nucleotides long in *E. coli* and 100–200 nucleotides long in eukaryotes.

Figure 13.16 illustrates the steps in the synthesis of the lagging strand at one fork. Whereas only one primer is required on the leading strand, each Okazaki fragment on the lagging strand must be primed separately (**1** and **4**). After DNA pol III forms an Okazaki fragment (**2–4**), another DNA polymerase, DNA polymerase I (DNA pol I), replaces the RNA nucleotides of the adjacent primer with DNA nucleotides (**5**). But DNA pol I cannot join the final nucleotide of this replacement DNA segment to the first DNA nucleotide of the adjacent Okazaki fragment. Another enzyme, **DNA ligase**, accomplishes this task, joining the sugar-phosphate backbones of all the Okazaki fragments into a continuous DNA strand (**6**).

Figure 13.17 summarizes DNA replication. Study it carefully before proceeding.

The DNA Replication Complex

It is traditional—and convenient—to represent DNA polymerase molecules as locomotives moving along a DNA "railroad track," but such a model is inaccurate in two important ways. First, the various proteins that participate in DNA replication actually form a single large complex, a "DNA replication machine." Many protein-protein interactions facilitate the efficiency of this complex. For example, by interacting with other proteins at the fork, primase apparently acts as a molecular brake, slowing progress of the replication fork and coordinating the placement of primers and the rates of replication on the leading and lagging strands. Second, the DNA replication complex may not move along the DNA; rather, the DNA may move through the complex during the replication process. In eukaryotic cells, multiple copies of the complex, perhaps grouped into "factories," may be anchored to the nuclear matrix, a framework of fibers extending through the interior of the nucleus. Recent studies support a model in which two DNA polymerase molecules, one on each template strand, "reel in" the parental DNA and extrude newly made daughter DNA molecules. Additional evidence suggests that the lagging strand is looped back through the complex (**Figure 13.18**).

---

*Synthesis of the leading strand and synthesis of the lagging strand occur concurrently and at the same rate. The lagging strand is so named because its synthesis is delayed slightly relative to synthesis of the leading strand; each new fragment of the lagging strand cannot be started until enough template has been exposed at the replication fork.

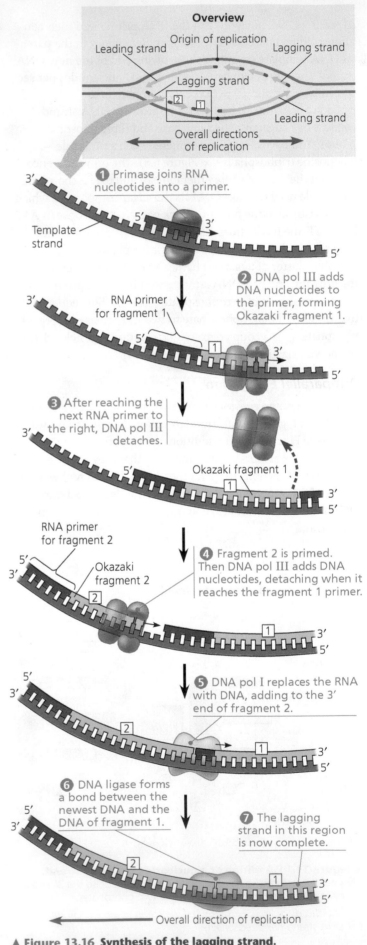

▲ **Figure 13.16 Synthesis of the lagging strand.**

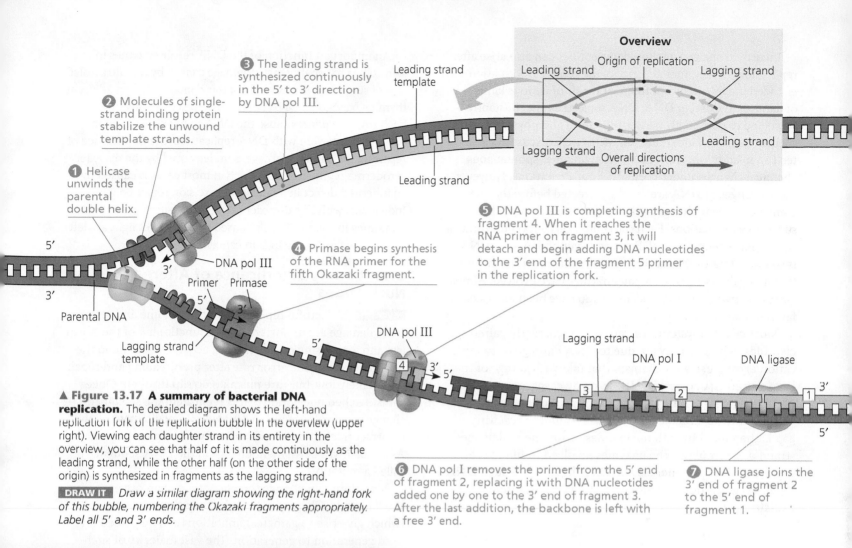

Overview

Leading strand — Origin of replication — Lagging strand

Lagging strand — Leading strand

Overall directions of replication

① Helicase unwinds the parental double helix.

② Molecules of single-strand binding protein stabilize the unwound template strands.

③ The leading strand is synthesized continuously in the 5′ to 3′ direction by DNA pol III.

Leading strand template

Leading strand

DNA pol III

Primer Primase

④ Primase begins synthesis of the RNA primer for the fifth Okazaki fragment.

⑤ DNA pol III is completing synthesis of fragment 4. When it reaches the RNA primer on fragment 3, it will detach and begin adding DNA nucleotides to the 3′ end of the fragment 5 primer in the replication fork.

Parental DNA

Lagging strand template

DNA pol III

Lagging strand

DNA pol I

DNA ligase

5′

3′

▲ **Figure 13.17 A summary of bacterial DNA replication.** The detailed diagram shows the left-hand replication fork of the replication bubble in the overview (upper right). Viewing each daughter strand in its entirety in the overview, you can see that half of it is made continuously as the leading strand, while the other half (on the other side of the origin) is synthesized in fragments as the lagging strand.

DRAW IT *Draw a similar diagram showing the right-hand fork of this bubble, numbering the Okazaki fragments appropriately. Label all 5′ and 3′ ends.*

⑥ DNA pol I removes the primer from the 5′ end of fragment 2, replacing it with DNA nucleotides added one by one to the 3′ end of fragment 3. After the last addition, the backbone is left with a free 3′ end.

⑦ DNA ligase joins the 3′ end of fragment 2 to the 5′ end of fragment 1.

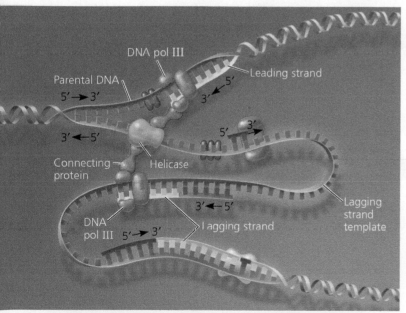

▲ **Figure 13.18 A current model of the DNA replication complex.** Two DNA polymerase III molecules work together in a complex, one on each template strand. The lagging strand template DNA loops through the complex.

 ANIMATION ***BioFlix*** Visit the Study Area in **MasteringBiology** for the BioFlix ® 3-D Animation on DNA Replication.

Proofreading and Repairing DNA

We cannot attribute the accuracy of DNA replication solely to the specificity of base pairing. Initial pairing errors between incoming nucleotides and those in the template strand occur at a rate of one in 10^5 nucleotides. However, errors in the completed DNA molecule amount to only one in 10^{10} (10 billion) nucleotides, an error rate that is 100,000 times lower. This is because during DNA replication, DNA polymerases proofread each nucleotide against its template as soon as it is added to the growing strand. Upon finding an incorrectly paired nucleotide, the polymerase removes the nucleotide and then resumes synthesis. (This action is similar to fixing a typing error by deleting the wrong letter and then entering the correct letter.)

Mismatched nucleotides sometimes do evade proofreading by a DNA polymerase. In **mismatch repair**, other enzymes remove and replace incorrectly paired nucleotides resulting from replication errors. Researchers spotlighted the importance of such repair enzymes when they found that a hereditary defect in one of them is associated with a form of colon cancer. Apparently, this defect allows cancer-causing errors to accumulate in the DNA faster than normal.

Incorrectly paired or altered nucleotides can also arise after replication. In fact, maintenance of the genetic information encoded in DNA requires frequent repair of various kinds of damage to existing DNA. DNA molecules are constantly subjected to potentially harmful chemical and physical agents, such as cigarette smoke and X-rays (as we'll discuss in Chapter 14). In addition, DNA bases often undergo spontaneous chemical changes under normal cellular conditions. However, these changes in DNA are usually corrected before they become permanent changes—*mutations*—perpetuated through successive replications. Each cell continuously monitors and repairs its genetic material. Because repair of damaged DNA is so important to the survival of an organism, it is no surprise that many different DNA repair enzymes have evolved. Almost 100 are known in *E. coli*, and about 130 have been identified so far in humans.

Most cellular systems for repairing incorrectly paired nucleotides, whether they are due to DNA damage or to replication errors, use a mechanism that takes advantage of the base-paired structure of DNA. In many cases, a segment of the strand containing the damage is cut out (excised) by a DNA-cutting enzyme—a **nuclease**—and the resulting gap is then filled in with nucleotides, using the undamaged strand as a template. The enzymes involved in filling the gap are a DNA polymerase and DNA ligase. One such DNA repair system, shown in **Figure 13.19**, is called **nucleotide excision repair**.

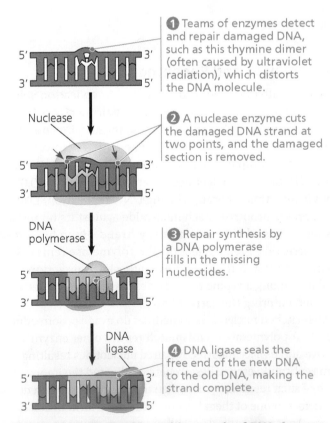

❶ Teams of enzymes detect and repair damaged DNA, such as this thymine dimer (often caused by ultraviolet radiation), which distorts the DNA molecule.

Nuclease

❷ A nuclease enzyme cuts the damaged DNA strand at two points, and the damaged section is removed.

DNA polymerase

❸ Repair synthesis by a DNA polymerase fills in the missing nucleotides.

DNA ligase

❹ DNA ligase seals the free end of the new DNA to the old DNA, making the strand complete.

▲ **Figure 13.19 Nucleotide excision repair of DNA damage.**

An important function of the DNA repair enzymes in our skin cells is to repair genetic damage caused by the ultraviolet rays of sunlight. One example of this damage is when adjacent thymine bases on a DNA strand become covalently linked. Such *thymine dimers* cause the DNA to buckle (see Figure 13.19) and interfere with DNA replication. The importance of repairing this kind of damage is underscored by the disorder xeroderma pigmentosum, which in most cases is caused by an inherited defect in a nucleotide excision repair enzyme. Individuals with this disorder are hypersensitive to sunlight; mutations in their skin cells caused by ultraviolet light are left uncorrected, resulting in skin cancer.

Evolutionary Significance of Altered DNA Nucleotides

EVOLUTION Faithful replication of the genome and repair of DNA damage are important for the functioning of the organism and for passing on a complete, accurate genome to the next generation. The error rate after proofreading and repair is extremely low, but rare mistakes do slip through. Once a mismatched nucleotide pair is replicated, the sequence change is permanent in the daughter molecule that has the incorrect nucleotide as well as in any subsequent copies. As you know, a permanent change in the DNA sequence is called a mutation.

Mutations can change the phenotype of an organism (as you'll learn in Chapter 14). And if they occur in germ cells (which give rise to gametes), mutations can be passed on from generation to generation. The vast majority of such changes are harmful, but a very small percentage can be beneficial. In either case, mutations are the source of the variation on which natural selection operates during evolution and are ultimately responsible for the appearance of new species. (You'll learn more about this process in Unit Three.) The balance between complete fidelity of DNA replication or repair and a low mutation rate has, over long periods of time, allowed the evolution of the rich diversity of species we see on Earth today.

Replicating the Ends of DNA Molecules

For linear DNA, such as the DNA of eukaryotic chromosomes, the usual replication machinery cannot complete the 5′ ends of daughter DNA strands. (This is a consequence of the fact that a DNA polymerase can add nucleotides only to the 3′ ends.) As a result, repeated rounds of replication produce shorter and shorter DNA molecules with uneven ends.

What protects the genes near the ends of eukaryotic chromosomes from being eroded away during successive replications? Eukaryotic chromosomal DNA molecules have special nucleotide sequences called telomeres at their ends **(Figure 13.20)**. Telomeres do not contain genes; instead, the DNA typically consists of multiple repetitions of one short nucleotide sequence. In each human telomere, for example, the sequence

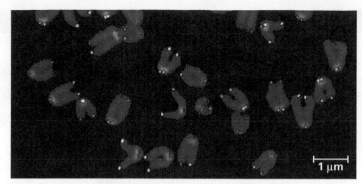

▲ **Figure 13.20 Telomeres.** Eukaryotes have repetitive, noncoding sequences called telomeres at the ends of their DNA. Telomeres are stained orange in these mouse chromosomes (LM).

TTAGGG is repeated 100 to 1,000 times. Telomeric DNA acts as a buffer zone that protects the organism's genes.

Telomeres do not prevent the erosion of genes near ends of chromosomes; they merely postpone it. As you would expect, telomeres tend to be shorter in cultured cells that have divided many times and in dividing somatic cells of older individuals. Shortening of telomeres is proposed to play a role in the aging process of some tissues and even of the organism as a whole.

If the chromosomes of germ cells became shorter in every cell cycle, essential genes would eventually be missing from the gametes they produce. However, this does not occur: An enzyme called **telomerase** catalyzes the lengthening of telomeres in eukaryotic germ cells, thus restoring their original length and compensating for the shortening that occurs during DNA replication. Telomerase is not active in most human somatic cells, but shows inappropriate activity in some cancer cells that may remove limits to a cell's normal life span. Thus, telomerase is under study as a target for cancer therapies.

CONCEPT CHECK 13.2

1. What role does base pairing play in the replication of DNA?
2. Make a table listing the functions of seven proteins involved in DNA replication in *E. coli*.
3. **MAKE CONNECTIONS** What is the relationship between DNA replication and the S phase of the cell cycle? See Figure 9.6.

For suggested answers, see Appendix A.

CONCEPT 13.3

A chromosome consists of a DNA molecule packed together with proteins

Now that you have learned about the structure and replication of DNA, let's take a step back and examine how DNA is packaged into chromosomes, the structures that carry genetic information. The main component of the genome in most bacteria is one double-stranded, circular DNA molecule that is associated with a small amount of protein. Although we refer to this structure as a bacterial chromosome, it is very different from a eukaryotic chromosome, which consists of one linear DNA molecule associated with a large amount of protein. In *E. coli*, the chromosomal DNA consists of about 4.6 million nucleotide pairs, representing about 4,400 genes. This is 100 times more DNA than is found in a typical virus, but only about one-thousandth as much DNA as in a human somatic cell. Still, that is a lot of DNA to be packaged in such a small container.

Stretched out, the DNA of an *E. coli* cell would measure about a millimeter in length, 500 times longer than the cell. Within a bacterium, however, certain proteins cause the chromosome to coil and "supercoil," densely packing it so that it fills only part of the cell. Unlike the nucleus of a eukaryotic cell, this dense region of DNA in a bacterium, called the **nucleoid**, is not surrounded by membrane (see Figure 4.5).

Each eukaryotic chromosome contains a single linear DNA double helix that, in humans, averages about 1.5×10^8 nucleotide pairs. This is an enormous amount of DNA relative to a chromosome's condensed length. If completely stretched out, such a DNA molecule would be about 4 cm long, thousands of times the diameter of a cell nucleus—and that's not even considering the DNA of the other 45 human chromosomes!

In the cell, eukaryotic DNA is precisely combined with a large amount of protein. Together, this complex of DNA and protein, called **chromatin**, fits into the nucleus through an elaborate, multilevel system of packing.

Chromatin undergoes striking changes in its degree of packing during the course of the cell cycle (see Figure 9.7). In interphase cells stained for light microscopy, the chromatin usually appears as a diffuse mass within the nucleus, suggesting that the chromatin is highly extended. As a cell prepares for mitosis, its chromatin coils and folds up (condenses), eventually forming a characteristic number of short, thick metaphase chromosomes that are distinguishable from each other with the light microscope. Our current view of the successive levels of DNA packing in a chromosome is outlined in **Figure 13.21**. Study this figure carefully before reading further.

Though interphase chromatin is generally much less condensed than the chromatin of mitotic chromosomes, it shows several of the same levels of higher-order packing. Some of the chromatin comprising a chromosome seems to be present as a 10-nm fiber, but much is compacted into a 30-nm fiber, which in some regions is further folded into looped domains. Even during interphase, the centromeres of chromosomes, as well as other chromosomal regions in some cells, exist in a highly condensed state similar to that seen in a metaphase chromosome. This type of interphase chromatin, visible as irregular clumps with a light microscope, is called **heterochromatin**, to distinguish it from the less compacted, more dispersed **euchromatin** ("true chromatin"). Because of its compaction, heterochromatic DNA is largely inaccessible to the machinery

This series of diagrams and transmission electron micrographs depicts a current model for the progressive levels of DNA coiling and folding. The illustration zooms out from a single molecule of DNA to a metaphase chromosome, which is large enough to be seen with a light microscope.

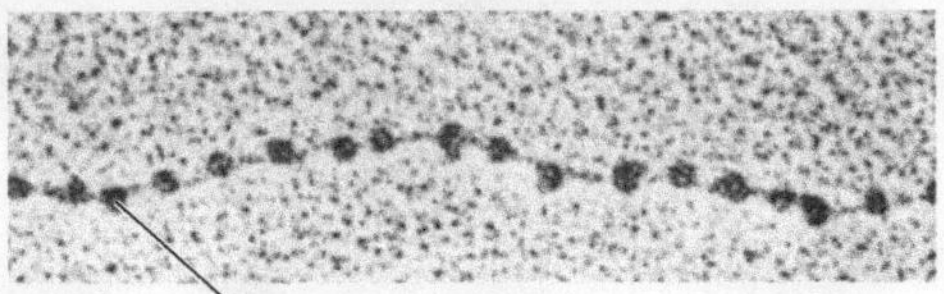

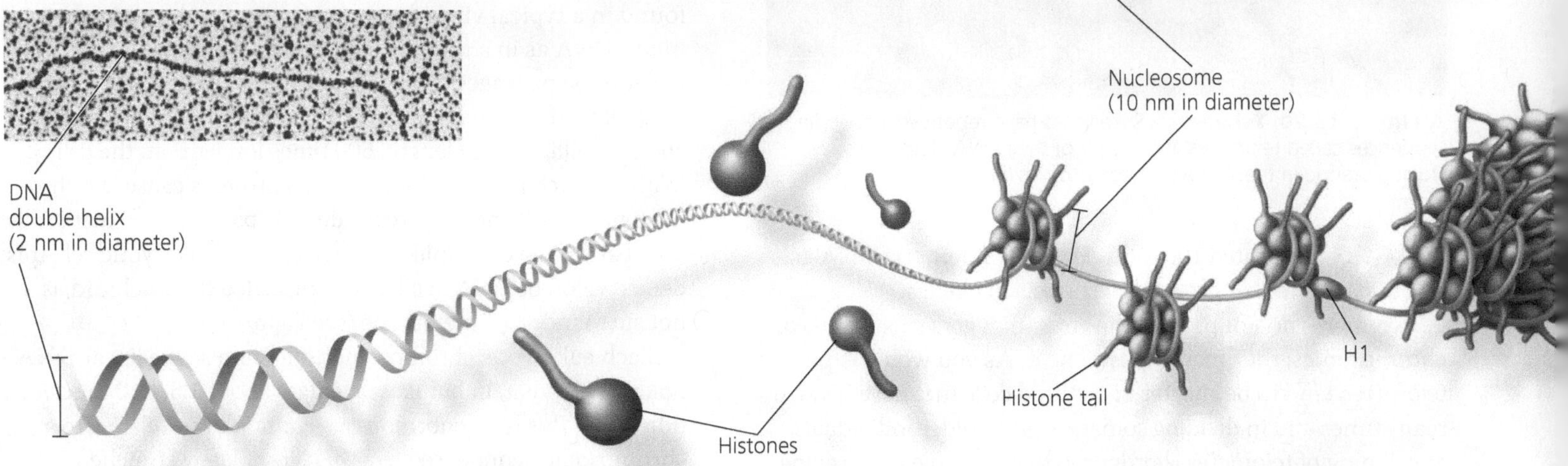

DNA double helix (2 nm in diameter)

Nucleosome (10 nm in diameter)

H1

Histone tail

Histones

DNA, the double helix

Shown here is a ribbon model of DNA, with each ribbon representing one of the sugar-phosphate backbones. As you will recall from Figure 13.7, the phosphate groups along the backbone contribute a negative charge along the outside of each strand. The TEM shows a molecule of naked DNA; the double helix alone is 2 nm across.

Histones

Proteins called **histones** are responsible for the first level of DNA packing in chromatin. Although each histone is small—containing only about 100 amino acids—the total mass of histone in chromatin approximately equals the mass of DNA. More than a fifth of a histone's amino acids are positively charged (lysine or arginine) and therefore bind tightly to the negatively charged DNA.

Four types of histones are most common in chromatin: H2A, H2B, H3, and H4. The histones are very similar among eukaryotes; for example, all but two of the amino acids in cow H4 are identical to those in pea H4. The apparent conservation of histone genes during evolution probably reflects the important role of histones in organizing DNA within cells.

The four main types of histones are critical to the next level of DNA packing. (A fifth type of histone, called H1, is involved in a further stage of packing.)

Nucleosomes, or "beads on a string" (10-nm fiber)

In electron micrographs, unfolded chromatin is 10 nm in diameter (the *10-nm fiber*). Such chromatin resembles beads on a string (see the TEM). Each "bead" is a **nucleosome**, the basic unit of DNA packing; the "string" between beads is called *linker DNA*.

A nucleosome consists of DNA wound twice around a protein core composed of two molecules each of the four main histone types. The amino end (N-terminus) of each histone (the *histone tail*) extends outward from the nucleosome.

In the cell cycle, the histones leave the DNA only briefly during DNA replication. Generally, they do the same during transcription, another process that requires access to the DNA by the cell's molecular machinery. Chapter 18 will discuss some recent findings about the role of histone tails and nucleosomes in the regulation of gene expression.

---

in the cell responsible for transcribing the genetic information coded in the DNA, a crucial early step in gene expression. In contrast, the looser packing of euchromatin makes its DNA accessible to this machinery, so the genes present in euchromatin can be transcribed.

The chromosome is a dynamic structure that is condensed, loosened, modified, and remodeled as necessary for various cell processes, including mitosis, meiosis, and gene activity. Certain chemical modifications of histones affect the state of

chromatin condensation and also have multiple effects on gene activity (as you'll see in Chapter 15).

CONCEPT CHECK 13.3

1. Describe the structure of a nucleosome, the basic unit of DNA packing in eukaryotic cells.
2. What two properties, one structural and one functional, distinguish heterochromatin from euchromatin?

For suggested answers, see Appendix A.

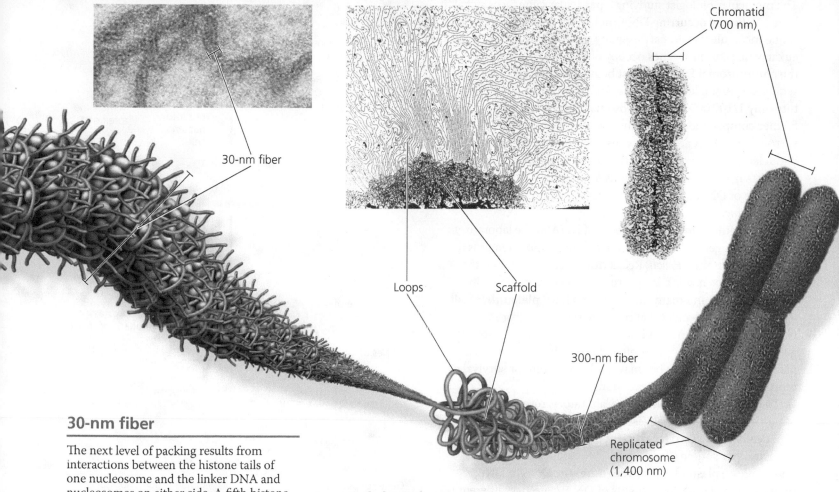

Chromatid
(700 nm)

30-nm fiber

Loops   Scaffold

300-nm fiber

Replicated
chromosome
(1,400 nm)

30-nm fiber

The next level of packing results from interactions between the histone tails of one nucleosome and the linker DNA and nucleosomes on either side. A fifth histone, H1, is involved at this level. These interactions cause the extended 10-nm fiber to coil or fold, forming a chromatin fiber roughly 30 nm in thickness, the *30-nm fiber*. Although the 30-nm fiber is quite prevalent in the interphase nucleus, the packing arrangement of nucleosomes in this form of chromatin is still a matter of some debate.

Looped domains
(300-nm fiber)

The 30-nm fiber, in turn, forms loops called *looped domains* attached to a chromosome scaffold composed of proteins, thus making up a *300-nm fiber*. The scaffold is rich in one type of topoisomerase, and H1 molecules also appear to be present.

Metaphase chromosome

In a mitotic chromosome, the looped domains themselves coil and fold in a manner not yet fully understood, further compacting all the chromatin to produce the characteristic metaphase chromosome shown in the micrograph above. The width of one chromatid is 700 nm. Particular genes always end up located at the same places in metaphase chromosomes, indicating that the packing steps are highly specific and precise.

CONCEPT 13.4

Understanding DNA structure and replication makes genetic engineering possible

The discovery of the structure of DNA marked a milestone in biology and changed the course of biological research. Most notable was the realization that the two strands of a DNA molecule are complementary to each other. This fundamental structural property of DNA is the basis for **nucleic acid hybridization**, the base pairing of one strand of a nucleic acid to a complementary sequence on another strand. Nucleic acid hybridization forms the foundation of virtually every technique used in **genetic engineering**, the direct manipulation of genes for practical purposes. Genetic engineering has launched a revolution in fields ranging from agriculture to criminal law to medical and basic biological research. In this section, we'll describe several of the most important techniques and their uses.

DNA Cloning: Making Multiple Copies of a Gene or Other DNA Segment

The molecular biologist studying a particular gene faces a challenge. Naturally occurring DNA molecules are very long, and a single molecule usually carries many genes. Moreover, in many eukaryotic genomes, genes occupy only a small proportion of the chromosomal DNA, the rest being noncoding nucleotide sequences. A single human gene, for example, might constitute only 1/100,000 of a chromosomal DNA molecule. As a further complication, the distinctions between a gene and the surrounding DNA are subtle, consisting only of differences in nucleotide sequence. To work directly with specific genes, scientists have developed methods for preparing well-defined segments of DNA in multiple identical copies, a process called *DNA cloning*.

Most methods for cloning pieces of DNA in the laboratory share certain general features. One common approach uses bacteria, most often *E. coli*. Recall from Figure 13.13 that the *E. coli* chromosome is a large circular molecule of DNA. In addition, *E. coli* and many other bacteria have **plasmids**, small circular DNA molecules that replicate separately from the bacterial chromosome. A plasmid has only a small number of genes; these genes may be useful when the bacterium is in a particular environment but may not be required for survival or reproduction under most conditions.

To clone pieces of DNA in the laboratory, researchers first obtain a plasmid (originally isolated from a bacterial cell and genetically engineered for efficient cloning) and insert DNA from another source ("foreign" DNA) into it **(Figure 13.22)**. The resulting plasmid is now **recombinant DNA**, a DNA molecule formed when segments of DNA from two different sources—often different species—are combined *in vitro* (in a test tube). The plasmid is then returned to a bacterial cell, producing a *recombinant bacterium*. This single cell reproduces through repeated cell divisions to form a clone of cells, a population of genetically identical cells. Because the dividing bacteria replicate the recombinant plasmid and pass it on to their descendants, the foreign DNA and any genes it carries are cloned at the same time. The production of multiple copies of a single gene is called **gene cloning**.

Gene cloning is useful for two basic purposes: to make many copies of, or *amplify*, a particular gene and to produce a protein product. Researchers can isolate copies of a cloned gene from bacteria for use in basic research or to endow an organism with a new metabolic trait, such as pest resistance. For example, a resistance gene present in one crop species might be cloned and transferred into plants of another species. Alternatively, a protein with medical uses, such as human growth hormone, can be harvested in large quantities from cultures of bacteria carrying the cloned gene for the protein. Since a single gene is usually a very small part of the total DNA in a cell, the ability to amplify such rare DNA fragments is therefore crucial for any application involving a single gene.

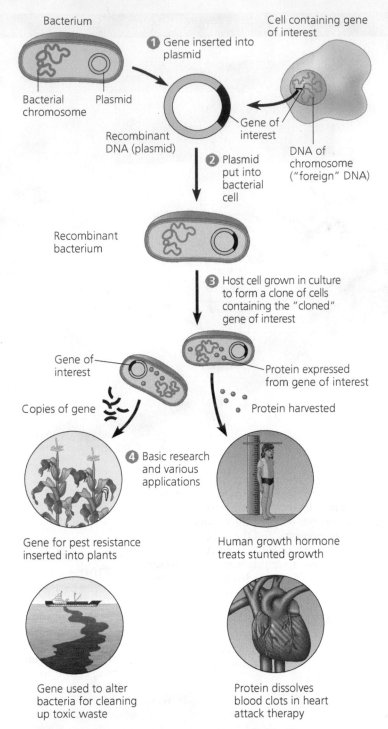

Bacterium

Cell containing gene of interest

1 Gene inserted into plasmid

Bacterial chromosome   Plasmid

Recombinant DNA (plasmid)

Gene of interest

DNA of chromosome ("foreign" DNA)

2 Plasmid put into bacterial cell

Recombinant bacterium

3 Host cell grown in culture to form a clone of cells containing the "cloned" gene of interest

Gene of interest

Protein expressed from gene of interest

Copies of gene

Protein harvested

4 Basic research and various applications

Gene for pest resistance inserted into plants

Human growth hormone treats stunted growth

Gene used to alter bacteria for cleaning up toxic waste

Protein dissolves blood clots in heart attack therapy

▲ **Figure 13.22 An overview of gene cloning and some uses of cloned genes.** In this simplified diagram of gene cloning, we start with a plasmid (originally isolated from a bacterial cell) and a gene of interest from another organism. Only one plasmid and one copy of the gene of interest are shown at the top of the figure, but the starting materials would include many of each.

Using Restriction Enzymes to Make Recombinant DNA

Gene cloning and genetic engineering rely on the use of enzymes that cut DNA molecules at a limited number of specific locations. These enzymes, called restriction endonucleases, or **restriction enzymes**, were discovered in the late 1960s by biologists doing basic research on bacteria. Restriction enzymes protect the bacterial cell by cutting up foreign DNA from other organisms or phages.

Hundreds of different restriction enzymes have been identified and isolated. Each restriction enzyme is very specific, recognizing a particular short DNA sequence, or **restriction site**, and cutting both DNA strands at precise points within this restriction site. The DNA of a bacterial cell is protected from the cell's own restriction enzymes by the addition of methyl groups ($-CH_3$) to adenines or cytosines within the sequences recognized by the enzymes.

The top of **Figure 13.23** illustrates a restriction site recognized by a particular restriction enzyme from *E. coli*. As shown

in this example, most restriction sites are symmetric. That is, the sequence of nucleotides is the same on both strands when read in the 5′ → 3′ direction. The most commonly used restriction enzymes recognize sequences containing 4–8 nucleotides. Because any sequence this short usually occurs (by chance) many times in a long DNA molecule, a restriction enzyme will make many cuts in a DNA molecule, yielding a set of **restriction fragments**. All copies of a particular DNA molecule always yield the same set of restriction fragments when exposed to the same restriction enzyme. To see the fragments, researchers carry out a technique called **gel electrophoresis**, which can separate a mixture of nucleic acid fragments by length **(Figure 13.24)**.

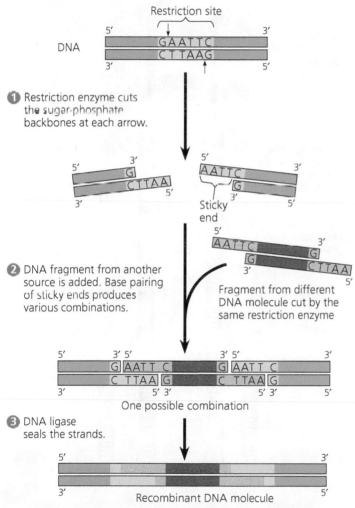

▲ **Figure 13.23 Using a restriction enzyme and DNA ligase to make recombinant DNA.** The restriction enzyme in this example (called *Eco*RI) recognizes a specific six-base-pair sequence, the restriction site, and makes staggered cuts in the sugar-phosphate backbones within this sequence, producing fragments with sticky ends. Any fragments with complementary sticky ends can base-pair, including the two original fragments. If the fragments come from different DNA molecules, the ligated product is recombinant DNA.

DRAW IT *The restriction enzyme HindIII recognizes the sequence 5′-AAGCTT-3′, cutting between the two A's. Draw the double-stranded sequence before and after the enzyme cuts.*

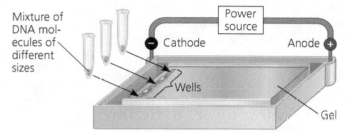

(a) Each sample, a mixture of DNA molecules, is placed in a separate well near one end of a thin slab of agarose gel. The gel is set into a small plastic support and immersed in an aqueous, buffered solution in a tray with electrodes at each end. The current is then turned on, causing the negatively charged DNA molecules to move toward the positive electrode.

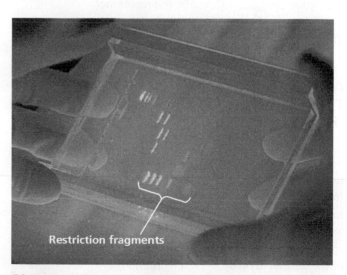

(b) Shorter molecules are impeded less than longer ones, so they move faster through the gel. After the current is turned off, a DNA-binding dye is added that fluoresces pink in ultraviolet light. Each pink band corresponds to many thousands of DNA molecules of the same length. The horizontal ladder of bands at the bottom of the gel is a set of restriction fragments used as size standards.

▲ **Figure 13.24 Gel electrophoresis.** A gel made of a polymer acts as a molecular sieve to separate nucleic acids or proteins differing in size, electrical charge, or other physical properties as they move in an electric field. In the example shown here, DNA molecules are separated by length in a gel made of the polysaccharide agarose.

The most useful restriction enzymes cleave the sugar-phosphate backbones in the two DNA strands in a staggered manner, as indicated in Figure 13.23. The resulting double-stranded restriction fragments have at least one single-stranded end, called a **sticky end**. These short extensions can form hydrogen-bonded base pairs (hybridize) with complementary sticky ends on any other DNA molecules cut with the same enzyme. The associations formed in this way are only temporary but can be made permanent by DNA ligase. As you saw in Figure 13.16, this enzyme catalyzes the formation of covalent bonds that close up the sugar-phosphate backbones of DNA strands; for example, it joins Okazaki fragments during replication.

You can see at the bottom of Figure 13.23 that the ligase-catalyzed joining of DNA from two different sources produces a stable recombinant DNA molecule. In gene cloning, the two DNA molecules to be joined are a **cloning vector**—a DNA molecule that can carry foreign DNA into a host cell and replicate there—and the gene to be cloned (see Figure 13.22). The cloning vector is often a bacterial plasmid that has one copy of a restriction site recognized by a particular restriction enzyme, selected by the researcher and purchased from a commercial source. The most common way to obtain many copies of the gene to be cloned is described next.

Amplifying DNA *in Vitro*: The Polymerase Chain Reaction (PCR) and Its Use in Cloning

Today, most researchers have some information about the sequence of the gene or DNA fragment they want to clone. Using this information, they can start with the entire collection of genomic DNA from the particular species of interest and obtain enough copies of the desired gene by using a technique called the **polymerase chain reaction**, or **PCR**. **Figure 13.25** illustrates the steps in PCR. Within a few hours, this technique can make billions of copies of a specific target DNA segment in a sample, even if that segment makes up less than 0.001% of the total DNA in the sample.

In the PCR procedure, a three-step cycle brings about a chain reaction that produces

▼ **Figure 13.25**　**Research Method**

The Polymerase Chain Reaction (PCR)

Application　With PCR, any specific segment—the target sequence—within a DNA sample can be copied many times (amplified), completely *in vitro*.

Technique　PCR requires double-stranded DNA containing the target sequence, a heat-resistant DNA polymerase, all four nucleotides, and two 15- to 20-nucleotide DNA strands that serve as primers. One primer is complementary to one end of the target sequence on one strand; the second primer is complementary to the other end of the sequence on the other strand.

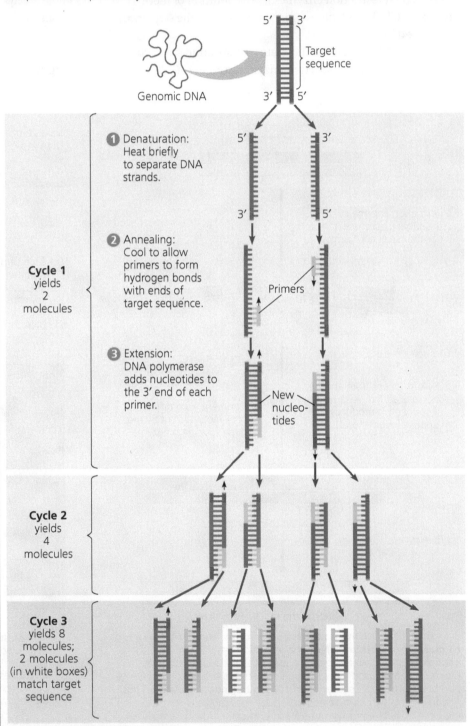

© 1996 Pearson Education, Inc.

Results　After 3 cycles, two molecules match the target sequence exactly. After 30 more cycles, over 1 billion (10^9) molecules match the target sequence.

an exponentially growing population of identical DNA molecules. During each cycle, the reaction mixture is heated to denature (separate) the DNA strands and then cooled to allow annealing (hybridization) of short, single-stranded DNA primers complementary to sequences on opposite strands at each end of the target segment; finally, a DNA polymerase extends the primers in the $5' \rightarrow 3'$ direction. If a standard DNA polymerase were used, the protein would be denatured along with the DNA during the first heating step and would have to be replaced after each cycle. The key to automating PCR was the discovery of an unusually heat-stable DNA polymerase called Taq polymerase, named after the bacterial species from which it was first isolated. This bacterial species, *Thermus aquaticus*, lives in hot springs, and the stability of its DNA polymerase at high temperatures is an evolutionary adaptation that enables the bacterium to survive at temperatures up to 95°C.

PCR is speedy and very specific. Only minuscule amounts of DNA need be present in the starting material, and this DNA can be partially degraded, as long as a few molecules contain the complete target segment. The key to this high specificity is the primers, the sequences of which are chosen so they hybridize *only* with complementary sequences at opposite ends of the target segment. (For high specificity, the primers must be at least 15 or so nucleotides long.) By the end of the third cycle, one-fourth of the molecules are identical to the target segment, with both strands the appropriate length. With each successive cycle, the number of target segment molecules of the correct length doubles, so the number of molecules equals 2^n, where n is the number of cycles. After 30 more cycles, about a billion copies of the target sequence are present!

Despite its speed and specificity, PCR amplification alone cannot substitute for gene cloning in cells to make large amounts of a gene. This is because occasional errors during PCR replication limit the number of good copies and the length of DNA fragments that can be copied. Instead, PCR is used to provide the specific DNA fragment for cloning. PCR primers are synthesized to include a restriction site at each end of the DNA fragment that matches the site in the cloning vector, and the fragment and vector are cut and ligated together **(Figure 13.26)**. The resulting clones are sequenced so that clones with error-free inserts can be selected.

Devised in 1985, PCR has had a major impact on biological research and genetic engineering. PCR has been used to amplify DNA from a wide variety of sources: a 40,000-year-old frozen woolly mammoth; fingerprints or tiny amounts of blood, tissue, or semen found at crime scenes; single embryonic cells for rapid prenatal diagnosis of genetic disorders; and cells infected with viruses that are difficult to detect, such as HIV (in the latter case, viral genes are amplified).

DNA Sequencing

Once a gene is cloned, researchers can exploit the principle of complementary base pairing to determine the gene's complete

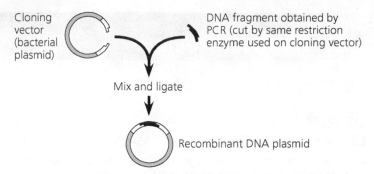

▲ **Figure 13.26 Use of restriction enzymes and PCR in gene cloning.** In a closer look at the process shown at the top of Figure 13.22, PCR is used to produce the DNA fragment or gene of interest that will be ligated into a cloning vector, in this case a bacterial plasmid. Both the plasmid and the DNA fragments are cut with the same restriction enzyme, combined so the sticky ends can hybridize, ligated together, and introduced into bacterial cells.

nucleotide sequence, a process called **DNA sequencing**. In the last ten years, "next-generation" sequencing techniques have been developed that are rapid and inexpensive. A single template strand is immobilized, and DNA polymerase and other reagents are added that allow so-called *sequencing by synthesis* of the complementary strand, one nucleotide at a time. A chemical trick enables electronic monitors to identify which of the four nucleotides is being added, allowing determination of the sequence. Technical advances continue to produce "third-generation" sequencing techniques, with each new technique being faster and less expensive than the previous. In Chapter 18, you'll learn more about how this rapid acceleration of sequencing technology has enhanced our study of genes and whole genomes.

In this chapter, you've learned how DNA molecules are arranged in chromosomes and how DNA replication provides the copies of genes that parents pass to offspring. However, it is not enough that genes be copied and transmitted; the information they carry must be used by the cell. In other words, genes must also be "expressed." In the next few chapters, we'll examine how the cell expresses the genetic information encoded in DNA. We'll also return to the subject of genetic engineering by exploring a few techniques for analyzing gene expression.

CONCEPT CHECK 13.4

1. The restriction site for an enzyme called *PvuI* is the following sequence:

   5'-C G A T C G-3'
   3'-G C T A G C-5'

   Staggered cuts are made between the T and C on each strand. What type of bonds are being cleaved?

2. **DRAW IT** One strand of a DNA molecule has the following sequence: 5'-CCTTGACGATCGTTACCG-3'. Draw the other strand. Will *PvuI* cut this molecule? If so, draw the products.

3. Describe the role of complementary base pairing during cloning, DNA sequencing, and PCR.

For suggested answers, see Appendix A.

SUMMARY OF KEY CONCEPTS

CONCEPT 13.1

DNA is the genetic material (pp. 245–251)

- Experiments with bacteria and **phages** provided the first strong evidence that the genetic material is DNA.
- Watson and Crick deduced that DNA is a **double helix** and built a structural model. Two **antiparallel** sugar-phosphate chains wind around the outside of the molecule; the nitrogenous bases project into the interior, where they hydrogen-bond in specific pairs: A with T, G with C.

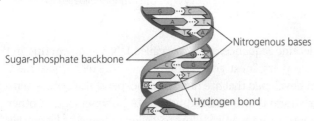

Sugar-phosphate backbone — Nitrogenous bases

Hydrogen bond

> **?** *What does it mean when we say that the two DNA strands in the double helix are antiparallel? What would an end of the double helix look like if the strands were parallel?*

CONCEPT 13.2

Many proteins work together in DNA replication and repair (pp. 251–259)

- The Meselson-Stahl experiment showed that **DNA replication** is **semiconservative**: The parental molecule unwinds, and each strand then serves as a template for the synthesis of a new strand according to base-pairing rules.
- DNA replication at one **replication fork** is summarized here:

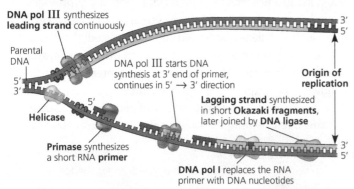

DNA pol III synthesizes **leading strand** continuously

Parental DNA

DNA pol III starts DNA synthesis at 3' end of primer, continues in 5' → 3' direction

Origin of replication

Helicase

Lagging strand synthesized in short **Okazaki fragments**, later joined by **DNA ligase**

Primase synthesizes a short RNA **primer**

DNA pol I replaces the RNA primer with DNA nucleotides

- DNA polymerases proofread new DNA, replacing incorrect nucleotides. In **mismatch repair**, enzymes correct errors that persist. **Nucleotide excision repair** is a general process by which **nucleases** cut out and replace damaged stretches of DNA.

> **?** *Compare DNA replication on the leading and lagging strands, including both similarities and differences.*

CONCEPT 13.3

A chromosome consists of a DNA molecule packed together with proteins (pp. 259–261)

- The chromosome of most bacterial species is a circular DNA molecule with some associated proteins, making up the **nucleoid**

of the cell. The **chromatin** making up a eukaryotic chromosome is composed of DNA, **histones**, and other proteins. The histones bind to each other and to the DNA to form **nucleosomes**, the most basic units of DNA packing. Additional coiling and folding lead ultimately to the highly condensed chromatin of the metaphase chromosome. In interphase cells, most chromatin is less compacted (**euchromatin**), but some remains highly condensed (**heterochromatin**). Euchromatin, but not heterochromatin, is generally accessible for transcription of genes.

> **?** *Describe the levels of chromatin packing you would expect to see in an interphase nucleus.*

CONCEPT 13.4

Understanding DNA structure and replication makes genetic engineering possible (pp. 261–265)

- **Gene cloning** (or DNA cloning) produces multiple copies of a gene (or DNA fragment) that can be used to manipulate and analyze DNA and to produce useful new products or organisms with beneficial traits.
- In **genetic engineering**, bacterial **restriction enzymes** are used to cut DNA molecules within short, specific nucleotide sequences (**restriction sites**), yielding a set of double-stranded **restriction fragments** with single-stranded **sticky ends**.

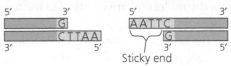

Sticky end

- DNA fragments of different lengths can be separated and their lengths assessed by **gel electrophoresis**.
- The sticky ends on restriction fragments from one DNA source—such as a bacterial **plasmid** or other **cloning vector**—can base-pair with complementary sticky ends on fragments from other DNA molecules; sealing the base-paired fragments with DNA ligase produces **recombinant DNA** molecules.
- The **polymerase chain reaction (PCR)** can produce many copies of (amplify) a specific target segment of DNA *in vitro* for use as a DNA fragment for cloning. PCR uses primers that bracket the desired segment and requires a heat-resistant DNA polymerase.
- The rapid development of fast, inexpensive techniques for **DNA sequencing** is based on *sequencing by synthesis*: DNA polymerase is used to replicate a stretch of DNA from a single-stranded template, and the order in which nucleotides are added reveals the sequence.

> **?** *Describe how the process of gene cloning results in a cell clone containing a recombinant plasmid.*

TEST YOUR UNDERSTANDING

Level 1: Knowledge/Comprehension

1. In his work with pneumonia-causing bacteria and mice, Griffith found that
   a. the protein coat from pathogenic cells was able to transform nonpathogenic cells.
   b. heat-killed pathogenic cells caused pneumonia.
   c. some substance from pathogenic cells was transferred to nonpathogenic cells, making them pathogenic.
   d. the polysaccharide coat of bacteria caused pneumonia.
   e. bacteriophages injected DNA into bacteria.

2. What is the basis for the difference in how the leading and lagging strands of DNA molecules are synthesized?
   a. The origins of replication occur only at the 5′ end.
   b. Helicases and single-strand binding proteins work at the 5′ end.
   c. DNA polymerase can join new nucleotides only to the 3′ end of a growing strand.
   d. DNA ligase works only in the 3′ → 5′ direction.
   e. Polymerase can work on only one strand at a time.

3. In analyzing the number of different bases in a DNA sample, which result would be consistent with the base-pairing rules?
   a. A = G
   b. A + G = C + T
   c. A + T = G + T
   d. A = C
   e. G = T

4. The elongation of the leading strand during DNA synthesis
   a. progresses away from the replication fork.
   b. occurs in the 3′ → 5′ direction.
   c. produces Okazaki fragments.
   d. depends on the action of DNA polymerase.
   e. does not require a template strand.

5. In a nucleosome, the DNA is wrapped around
   a. polymerase molecules.
   b. ribosomes.
   c. histones.
   d. a thymine dimer.
   e. satellite DNA.

6. Which of the following sequences in double-stranded DNA is most likely to be recognized as a cutting site for a restriction enzyme?
   a. AAGG   b. AGTC   c. GGCC   d. ACCA   e. AAAA
      TTCC   TCAG   CCGG   TGGT   TTTT

Level 2: Application/Analysis

7. *E. coli* cells grown on ^{15}N medium are transferred to ^{14}N medium and allowed to grow for two more generations (two rounds of DNA replication). DNA extracted from these cells is centrifuged. What density distribution of DNA would you expect in this experiment?
   a. one high-density and one low-density band
   b. one intermediate-density band
   c. one high-density and one intermediate-density band
   d. one low-density and one intermediate-density band
   e. one low-density band

8. A biochemist isolates, purifies, and combines in a test tube a variety of molecules needed for DNA replication. When she adds some DNA to the mixture, replication occurs, but each DNA molecule consists of a normal strand paired with numerous segments of DNA a few hundred nucleotides long. What has she probably left out of the mixture?
   a. DNA polymerase
   b. DNA ligase
   c. nucleotides
   d. Okazaki fragments
   e. primase

9. The spontaneous loss of amino groups from adenine in DNA results in hypoxanthine, an uncommon base, opposite thymine. What combination of proteins could repair such damage?
   a. nuclease, DNA polymerase, DNA ligase
   b. topoisomerase, primase, DNA polymerase
   c. topoisomerase, helicase, single-strand binding protein
   d. DNA ligase, replication fork proteins, adenylyl cyclase
   e. nuclease, topoisomerase, primase

10. **MAKE CONNECTIONS** Although the proteins that cause the *E. coli* chromosome to coil are not histones, what property would you expect them to share with histones, given their ability to bind to DNA (see Figure 3.17)?

Level 3: Synthesis/Evaluation

11. **SCIENTIFIC INQUIRY**

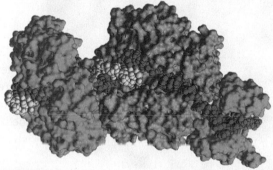

DRAW IT Model building can be an important part of the scientific process. The illustration shown above is a computer-generated model of a DNA replication complex. The parental and newly synthesized DNA strands are color-coded differently, as are each of the following three proteins: DNA pol III, the sliding clamp, and single-strand binding protein. Use what you've learned in this chapter to clarify this model by labeling each DNA strand and each protein and indicating the overall direction of DNA replication.

12. **FOCUS ON EVOLUTION**
Some bacteria may be able to respond to environmental stress by increasing the rate at which mutations occur during cell division. How might this be accomplished? Might there be an evolutionary advantage of this ability? Explain.

13. **FOCUS ON ORGANIZATION**
The continuity of life is based on heritable information in the form of DNA, and structure and function are correlated at all levels of biological organization. In a short essay (100–150 words), describe how the structure of DNA is correlated with its role as the molecular basis of inheritance.

For selected answers, see Appendix A.

MasteringBiology®

Students Go to **MasteringBiology** for assignments, the eText, and the Study Area with practice tests, animations, and activities.

Instructors Go to **MasteringBiology** for automatically graded tutorials and questions that you can assign to your students, plus Instructor Resources.

Gene Expression: From Gene to Protein

KEY CONCEPTS

14.1 Genes specify proteins via transcription and translation

14.2 Transcription is the DNA-directed synthesis of RNA: *a closer look*

14.3 Eukaryotic cells modify RNA after transcription

14.4 Translation is the RNA-directed synthesis of a polypeptide: *a closer look*

14.5 Mutations of one or a few nucleotides can affect protein structure and function

OVERVIEW

The Flow of Genetic Information

In 2006, a young albino deer seen frolicking with several brown deer in the mountains of eastern Germany elicited a public outcry **(Figure 14.1)**. A local hunting organization announced that the albino deer suffered from a "genetic disorder" and should be shot. Some argued that the deer should merely be prevented from mating with other deer to safeguard the population's gene pool. Others favored relocating the albino deer to a nature reserve because they worried that it might be more noticeable to predators if left in the wild. A German rock star even held a benefit concert to raise funds for the relocation. What led to the striking phenotype of this deer, the cause of this lively debate?

Inherited traits are determined by genes, and the trait of albinism is caused by a recessive allele of a pigmentation gene (see Chapter 11). The information content of genes is in the form of specific sequences of nucleotides along strands of DNA, the genetic material. But how does this information determine an organism's traits? Put another way, what does a gene actually say? And how is its message translated by cells into a specific trait, such as brown hair, type A blood, or, in the case of an albino deer, a total lack of pigment? The albino deer has a faulty version of a key protein, an enzyme required for pigment synthesis, and this protein is faulty because the gene that codes for it contains incorrect information.

This example illustrates the main point of this chapter: The DNA inherited by an organism leads to specific traits by dictating the synthesis of proteins and of RNA molecules involved in protein synthesis. In other words, proteins are the link between genotype and phenotype. **Gene expression** is the process by which DNA directs the synthesis of proteins (or, in some cases, just RNAs). The expression of genes that code for proteins includes two stages: transcription and translation. This chapter describes the flow of information from gene to protein in detail and explains how genetic mutations affect organisms through their proteins. Understanding the processes of gene expression, which are similar in all three domains of life, will allow us to revisit the concept of the gene in more detail at the end of the chapter.

▼ **Figure 14.1** How does a single faulty gene result in the dramatic appearance of an albino deer?

CONCEPT 14.1

Genes specify proteins via transcription and translation

Before going into the details of how genes direct protein synthesis, let's step back and examine how the fundamental relationship between genes and proteins was discovered.

Evidence from the Study of Metabolic Defects

In 1902, British physician Archibald Garrod was the first to suggest that genes dictate phenotypes through enzymes that catalyze specific chemical reactions in the cell. Garrod postulated that the symptoms of an inherited disease reflect a person's inability to make a particular enzyme. He later referred to such diseases as "inborn errors of metabolism." Garrod gave as one example the hereditary condition called alkaptonuria. In this disorder, the urine is black because it contains the chemical alkapton, which darkens upon exposure to air. Garrod reasoned that most people have an enzyme that metabolizes alkapton, whereas people with alkaptonuria have inherited an inability to make that enzyme.

Garrod may have been the first to recognize that Mendel's principles of heredity apply to humans as well as peas. Garrod's realization was ahead of its time, but research several decades later supported his hypothesis that a gene dictates the production of a specific enzyme. Biochemists accumulated much evidence that cells synthesize and degrade most organic molecules via metabolic pathways, in which each chemical reaction in a sequence is catalyzed by a specific enzyme (see Concept 6.1). Such metabolic pathways lead, for instance, to the synthesis of the pigments that give the brown deer in Figure 14.1 their fur

color or fruit flies (*Drosophila*) their eye color (see Figure 12.3). In the 1930s, the American geneticist George Beadle and his French colleague Boris Ephrussi speculated that in *Drosophila*, each of the mutations affecting eye color blocks pigment synthesis at a specific step by preventing production of the enzyme that catalyzes that step. But neither the chemical reactions nor the enzymes that catalyze them were known at the time.

Nutritional Mutants in Neurospora: Scientific Inquiry

A breakthrough in demonstrating the relationship between genes and enzymes came a few years later at Stanford University, where Beadle and Edward Tatum began working with the bread mold *Neurospora crassa* to investigate the role of genes in this organism's metabolic pathways. Their experimental approach still plays a central role in genetic research today: They disabled genes one by one and looked for changes in each mutant's phenotype, thereby revealing the normal function of the gene.

Like Mendel and T. H. Morgan before them, Beadle and Tatum chose their experimental organism carefully. They elected to work with *Neurospora*, a haploid species. They realized that it would be easier to detect a disabled gene in a haploid species than in a diploid species like *Drosophila*. In a diploid species, two copies of each gene are present, and both would need to be disabled for an effect to be seen on the organism's phenotype. In *Neurospora*, though, disabling a single gene would allow them to see the consequences and thus to deduce what the function of the wild-type gene might be. (In other words, haploidy makes it easier to detect recessive mutations.)

What was known about metabolism in *Neurospora* also made it a good choice. Wild-type *Neurospora* has modest food requirements. It can grow in the laboratory on a simple solution of inorganic salts, glucose, and the vitamin biotin (**Figure 14.2**). From this *minimal medium*, the mold cells use their metabolic

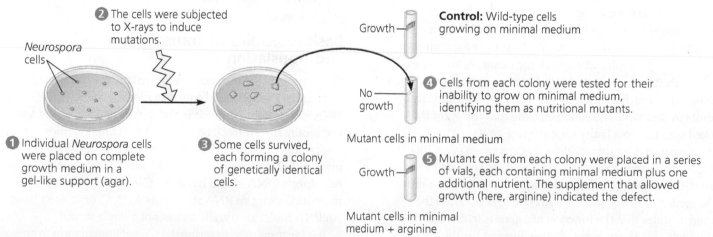

1 Individual *Neurospora* cells were placed on complete growth medium in a gel-like support (agar).

2 The cells were subjected to X-rays to induce mutations.

3 Some cells survived, each forming a colony of genetically identical cells.

Control: Wild-type cells growing on minimal medium — Growth

4 Cells from each colony were tested for their inability to grow on minimal medium, identifying them as nutritional mutants. — No growth

Mutant cells in minimal medium

5 Mutant cells from each colony were placed in a series of vials, each containing minimal medium plus one additional nutrient. The supplement that allowed growth (here, arginine) indicated the defect. — Growth

Mutant cells in minimal medium + arginine

▲ **Figure 14.2 The experimental approach of Beadle and Tatum.** To obtain nutritional mutants, Beadle and Tatum exposed *Neurospora* cells to X-rays to induce mutations. They then screened mutants with new nutritional requirements, such as arginine, as shown here.

WHAT IF? *What do you predict would happen if the same mutant were tested in a tube of minimal medium plus the amino acid glycine?*

pathways to produce all the other molecules they need. Wild-type cells grow and divide repeatedly on this medium.

Neurospora cells can be plated individually on a petri dish containing minimal medium embedded in a gel-like substance called agar. Although single cells are microscopic, after many divisions the resulting daughter cells can be seen by eye on the surface of the agar. Thus their ability to grow and divide on a particular medium can easily be monitored. Any change in their ability to grow on minimal medium would be easy to recognize in the lab simply as a new food requirement for growth (cell division).

Because any mutant that could not synthesize an essential nutrient would be unable to grow on minimal medium, Beadle and Tatum placed single *Neurospora* cells on a *complete growth medium*, which consisted of minimal medium supplemented with all 20 amino acids and a few other nutrients. As diagrammed in Figure 14.2 they bombarded the cells with X-rays, shown in the 1920s to cause mutations. Each of the surviving cells formed a visible colony of genetically identical cells. Next, the researchers screened the surviving colonies for "nutritional mutants" that grew well on complete medium but not at all on minimal medium. Apparently, each nutritional mutant was unable to synthesize a certain essential molecule from the minimal ingredients. In the final step of this experimental approach, Beadle and Tatum took cells from each mutant colony growing on complete medium and distributed them to a number of different vials. Each vial contained minimal medium plus a single additional nutrient. The particular supplement that allowed growth indicated the nutrient that the mutant could not synthesize.

Thus, the researchers amassed a valuable collection of mutant strains of *Neurospora*, catalogued by their defects. The collection would prove useful for focusing in on particular metabolic pathways in which the individual steps were either known or strongly suspected. For example, a series of experiments on mutants requiring the amino acid arginine revealed that they could be grouped into classes, each corresponding to a particular step in the biochemical pathway for arginine synthesis. These results, along with the results of similar experiments with other nutritional mutants, suggested that each class was blocked at a different step in the pathway because mutants in that class lacked the enzyme that catalyzes the blocked step due to a faulty gene **(Figure 14.3)**.

Because each mutant was defective in a single gene, Beadle and Tatum saw that, taken together, the collected results provided strong support for a working hypothesis they had proposed earlier. The *one gene–one enzyme hypothesis*, as they dubbed it, states that the function of a gene is to dictate the production of a specific enzyme. Further support for this hypothesis came from experiments that identified the specific enzymes lacking in the mutants. Beadle and Tatum shared a Nobel Prize in 1958 for "their discovery that genes act by regulating definite chemical events," in the words of the Nobel committee.

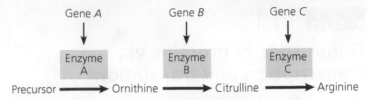

▲ **Figure 14.3 The one gene–one protein hypothesis.** Based on results from work in their lab on nutritional mutants, Beadle and Tatum proposed that the function of a specific gene is to dictate production of a specific enzyme that catalyzes a particular reaction. The model shown here for the arginine-synthesizing pathway illustrates their hypothesis.

The Products of Gene Expression: A Developing Story

As researchers learned more about proteins, they made revisions to the one gene–one enzyme hypothesis. First of all, not all proteins are enzymes. Keratin, the structural protein of animal hair, and the hormone insulin are two examples of nonenzyme proteins. Because proteins that are not enzymes are nevertheless gene products, molecular biologists began to think in terms of one gene–one protein. However, many proteins are constructed from two or more different polypeptide chains, and each polypeptide is specified by its own gene. For example, hemoglobin, the oxygen-transporting protein of vertebrate red blood cells, contains two kinds of polypeptides, and thus two genes code for this protein (see Figure 3.21). Beadle and Tatum's idea was therefore restated as the *one gene–one polypeptide hypothesis*. Even this description is not entirely accurate, though. First, many eukaryotic genes can each code for a set of closely related polypeptides via a process called alternative splicing, which you will learn about later in this chapter. Second, quite a few genes code for RNA molecules that have important functions in cells even though they are never translated into protein. For now, we will focus on genes that do code for polypeptides. (Note that it is common to refer to these gene products as proteins—a practice you'll encounter in this text—rather than more precisely as polypeptides.)

Basic Principles of Transcription and Translation

Genes provide the instructions for making specific proteins. But a gene does not build a protein directly. The bridge between DNA and protein synthesis is the nucleic acid RNA. RNA is chemically similar to DNA except that it contains ribose instead of deoxyribose as its sugar and has the nitrogenous base uracil rather than thymine (see Chapter 3). Thus, each nucleotide along a DNA strand has A, G, C, or T as its base, and each nucleotide along an RNA strand has A, G, C, or U as its base. An RNA molecule usually consists of a single strand.

It is customary to describe the flow of information from gene to protein in linguistic terms because both nucleic acids and proteins are polymers with specific sequences of monomers that convey information, much as specific sequences of letters communicate information in a language like English. In

DNA or RNA, the monomers are the four types of nucleotides, which differ in their nitrogenous bases. Genes are typically hundreds or thousands of nucleotides long, each gene having a specific sequence of nucleotides. Each polypeptide of a protein also has monomers arranged in a particular linear order (the protein's primary structure), but its monomers are amino acids. Thus, nucleic acids and proteins contain information written in two different chemical languages. Getting from DNA to protein requires two major stages: transcription and translation.

Transcription is the synthesis of RNA using information in the DNA. The two nucleic acids are written in different forms of the same language, and the information is simply transcribed, or "rewritten," from DNA to RNA. Just as a DNA strand provides a template for making a new complementary strand during DNA replication, it also can serve as a template for assembling a complementary sequence of RNA nucleotides. For a protein-coding gene, the resulting RNA molecule is a faithful transcript of the gene's protein-building instructions. This type of RNA molecule is called **messenger RNA (mRNA)** because it carries a genetic message from the DNA to the protein-synthesizing machinery of the cell. (Transcription is the general term for the synthesis of *any* kind of RNA on a DNA template. Later, you'll learn about some other types of RNA produced by transcription.)

Translation is the synthesis of a polypeptide using the information in the mRNA. During this stage, there is a change in language: The cell must translate the nucleotide sequence of an mRNA molecule into the amino acid sequence of a polypeptide. The sites of translation are **ribosomes**, complex particles that facilitate the orderly linking of amino acids into polypeptide chains.

Transcription and translation occur in all organisms—those that lack a membrane-enclosed nucleus (bacteria and archaea) and those that have one (eukaryotes). Because most studies of transcription and translation have used bacteria and eukaryotic cells, they are our main focus in this chapter. While our understanding of transcription and translation in archaea lags behind, we do know that archaeal cells share some features of gene expression with bacteria, and others with eukaryotes.

The basic mechanics of transcription and translation are similar for bacteria and eukaryotes, but there is an important difference in the flow of genetic information within the cells. Because bacteria do not have nuclei, their DNA is not separated by nuclear membranes from ribosomes and the other protein-synthesizing equipment **(Figure 14.4a)**. As you will see later, this lack of compartmentalization allows translation of an mRNA to begin while its transcription is still in progress. In a eukaryotic cell, by contrast, the nuclear envelope separates transcription from translation in space and time **(Figure 14.4b)**. Transcription occurs in the nucleus, and mRNA is then transported to the cytoplasm, where translation occurs. But before eukaryotic RNA transcripts from protein-coding genes can leave the nucleus, they are modified in various ways to produce the final, functional mRNA.

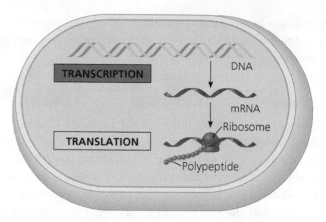

(a) Bacterial cell. In a bacterial cell, which lacks a nucleus, mRNA produced by transcription is immediately translated without additional processing.

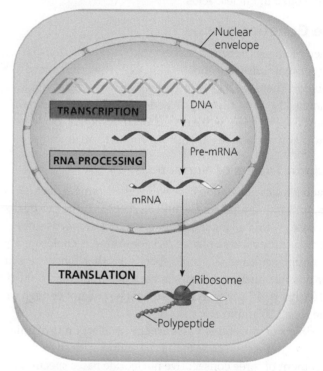

(b) Eukaryotic cell. The nucleus provides a separate compartment for transcription. The original RNA transcript, called pre-mRNA, is processed in various ways before leaving the nucleus as mRNA.

▲ **Figure 14.4 Overview: the roles of transcription and translation in the flow of genetic information.** In a cell, inherited information flows from DNA to RNA to protein. The two main stages of information flow are transcription and translation. A miniature version of part (a) or (b) accompanies several figures later in the chapter as an orientation diagram to help you see where a particular figure fits into the overall scheme.

The transcription of a protein-coding eukaryotic gene results in *pre-mRNA*, and further processing yields the finished mRNA. The initial RNA transcript from any gene, including those specifying RNA that is not translated into protein, is more generally called a **primary transcript**.

To summarize: Genes program protein synthesis via genetic messages in the form of messenger RNA. Put another way, cells are governed by a molecular chain of command with a directional flow of genetic information, shown here by arrows:

This concept was dubbed the *central dogma* by Francis Crick in 1956. How has the concept held up over time? In the 1970s, scientists were surprised to discover that some RNA molecules can act as templates for DNA synthesis (a process you'll read about in Chapter 17). However, these exceptions do not invalidate the idea that, in general, genetic information flows from DNA to RNA to protein. Now let's discuss how the instructions for assembling amino acids into a specific order are encoded in nucleic acids.

The Genetic Code

When biologists began to suspect that the instructions for protein synthesis were encoded in DNA, they recognized a problem: There are only four nucleotide bases to specify 20 amino acids. Thus, the genetic code cannot be a language like Chinese, where each written symbol corresponds to a word. How many nucleotides, then, correspond to an amino acid?

Codons: Triplets of Nucleotides

If each kind of nucleotide base were translated into an amino acid, only four amino acids could be specified, one per nucleotide base. Would a language of two-letter code words suffice? The two-nucleotide sequence AG, for example, could specify one amino acid, and GT could specify another. Since there are four possible nucleotide bases in each position, this would give us 16 (that is, 4^2) possible arrangements—still not enough to code for all 20 amino acids.

Triplets of nucleotide bases are the smallest units of uniform length that can code for all the amino acids. If each arrangement of three consecutive nucleotide bases specifies an amino acid, there can be 64 (that is, 4^3) possible code words—more than enough to specify all the amino acids. Experiments have verified that the flow of information from gene to protein is based on a **triplet code**: The genetic instructions for a polypeptide chain are written in the DNA as a series of nonoverlapping, three-nucleotide words. The series of words in a gene is transcribed into a complementary series of nonoverlapping, three-nucleotide words in mRNA, which is then translated into a chain of amino acids **(Figure 14.5)**.

During transcription, the gene determines the sequence of nucleotide bases along the length of the RNA molecule that is being synthesized. For each gene, only one of the two DNA strands is transcribed. This strand is called the **template strand** because it provides the pattern, or template, for the sequence of nucleotides in an RNA transcript. For any given gene, the same strand is used as the template every time the gene is transcribed. For other genes on the same DNA molecule, however, the opposite strand may be the one that always functions as the template.

▶ **Figure 14.5 The triplet code.** For each gene, one DNA strand functions as a template for transcription of RNAs, such as mRNA. The base-pairing rules for DNA synthesis also guide transcription, except that uracil (U) takes the place of thymine (T) in RNA. During translation, the mRNA is read as a sequence of nucleotide triplets, called codons. Each codon specifies an amino acid to be added to the growing polypeptide chain. The mRNA is read in the 5′ → 3′ direction.

? *Compare the sequence of the mRNA to that of the nontemplate DNA strand, in both cases reading from 5′ → 3′.*

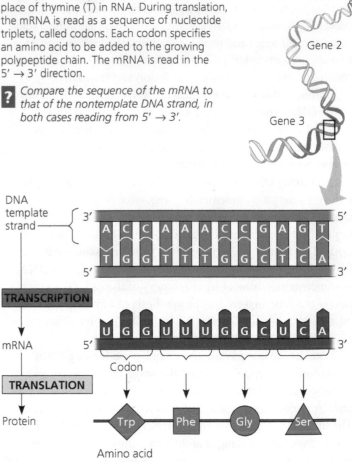

An mRNA molecule is complementary rather than identical to its DNA template because RNA nucleotides are assembled on the template according to base-pairing rules (see Figure 14.5). The pairs are similar to those that form during DNA replication, except that U, the RNA substitute for T, pairs with A and the mRNA nucleotides contain ribose instead of deoxyribose. Like a new strand of DNA, the RNA molecule is synthesized in an antiparallel direction to the template strand of DNA. (To review what is meant by "antiparallel" and the 5′ and 3′ ends of a nucleic acid chain, see Figure 13.7.) In the example in Figure 14.5, the nucleotide triplet ACC along the DNA (written as 3′-ACC-5′) provides a template for 5′-UGG-3′ in the mRNA molecule. The mRNA nucleotide triplets are called **codons**, and they are customarily written in the 5′ → 3′ direction. In our example, UGG is the codon for the amino acid tryptophan (abbreviated Trp). The term *codon* is also used for the DNA nucleotide triplets along the *nontemplate* strand. These codons are complementary to the template strand and thus identical in sequence to the mRNA, except that they have T instead of U. (For this reason, the nontemplate DNA strand is sometimes called the "coding strand.")

During translation, the sequence of codons along an mRNA molecule is decoded, or translated, into a sequence of amino acids making up a polypeptide chain. The codons are read by the translation machinery in the 5′ → 3′ direction along the mRNA. Each codon specifies which one of the 20 amino acids will be incorporated at the corresponding position along a polypeptide. Because codons are nucleotide triplets, the number of nucleotides making up a genetic message must be three times the number of amino acids in the protein product. For example, it takes 300 nucleotides along an mRNA strand to code for the amino acids in a polypeptide that is 100 amino acids long.

Cracking the Code

Molecular biologists cracked the genetic code of life in the early 1960s when a series of elegant experiments disclosed the amino acid translations of each of the RNA codons. The first codon was deciphered in 1961 by Marshall Nirenberg, of the National Institutes of Health, and his colleagues. Nirenberg synthesized an artificial mRNA by linking identical RNA nucleotides containing uracil as their base. No matter where this message started or stopped, it could contain only one codon in repetition: UUU. Nirenberg added this "poly-U" to a test-tube mixture containing amino acids, ribosomes, and the other components required for protein synthesis. His artificial system translated the poly-U into a polypeptide containing many units of the amino acid phenylalanine (Phe), strung together as a long polyphenylalanine chain. Thus, Nirenberg determined that the mRNA codon UUU specifies the amino acid phenylalanine. Soon, the amino acids specified by the codons AAA, GGG, and CCC were determined in the same way.

Although more elaborate techniques were required to decode mixed triplets such as AUA and CGA, all 64 codons were deciphered by the mid-1960s. As **Figure 14.6** shows, 61 of the 64 triplets code for amino acids. The three codons that do not designate amino acids are "stop" signals, or termination codons, marking the end of translation. Notice that the codon AUG has a dual function: It codes for the amino acid methionine (Met) and also functions as a "start" signal, or initiation codon. Genetic messages usually begin with the mRNA codon AUG, which signals the protein-synthesizing machinery to begin translating the mRNA at that location. (Because AUG also stands for methionine, polypeptide chains begin with methionine when they are synthesized. However, an enzyme may subsequently remove this starter amino acid from the chain.)

Notice in Figure 14.6 that there is redundancy in the genetic code, but no ambiguity. For example, although codons GAA and GAG both specify glutamic acid (redundancy), neither of them ever specifies any other amino acid (no ambiguity). The redundancy in the code is not altogether random. In many cases, codons that are synonyms for a particular amino acid differ only in the third nucleotide base of the triplet. We'll consider a possible benefit of this redundancy later in the chapter.

Our ability to extract the intended message from a written language depends on reading the symbols in the correct

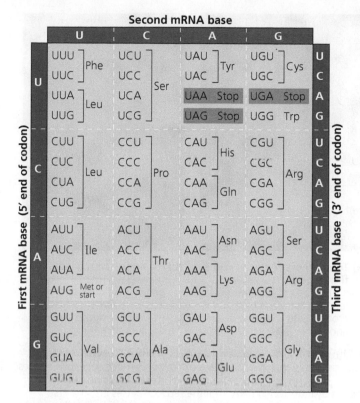

▲ **Figure 14.6 The codon table for mRNA.** The three nucleotide bases of an mRNA codon are designated here as the first, second, and third bases, reading in the 5′ → 3′ direction along the mRNA. (Practice using this table by finding the codons in Figure 14.5.) The codon AUG not only stands for the amino acid methionine (Met) but also functions as a "start" signal for ribosomes to begin translating the mRNA at that point. Three of the 64 codons function as "stop" signals, marking where ribosomes end translation. See Figure 3.17 for a list of the full names of all the amino acids.

groupings—that is, in the correct **reading frame**. Consider this statement: "The red dog ate the bug." Group the letters incorrectly by starting at the wrong point, and the result will probably be gibberish: for example, "her edd oga tet heb ug." The reading frame is also important in the molecular language of cells. The short stretch of polypeptide shown in Figure 14.5, for instance, will be made correctly only if the mRNA nucleotides are read from left to right (5′ → 3′) in the groups of three shown in the figure: UGG UUU GGC UCA. Although a genetic message is written with no spaces between the codons, the cell's protein-synthesizing machinery reads the message as a series of nonoverlapping three-letter words. The message is *not* read as a series of overlapping words—UGGUUU, and so on—which would convey a very different message.

Evolution of the Genetic Code

EVOLUTION The genetic code is nearly universal, shared by organisms from the simplest bacteria to the most complex plants and animals. The RNA codon CCG, for instance, is translated as the amino acid proline in all organisms whose genetic code has been examined. In laboratory experiments, genes can be transcribed and translated after being transplanted from one species to another, sometimes with quite

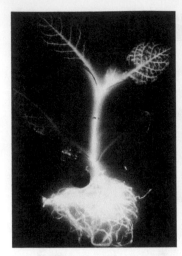

(a) Tobacco plant expressing a firefly gene. The yellow glow is produced by a chemical reaction catalyzed by the protein product of the firefly gene.

(b) Pig expressing a jellyfish gene. Researchers injected the gene for a fluorescent protein into fertilized pig eggs. One of the eggs developed into this fluorescent pig.

▲ **Figure 14.7 Expression of genes from different species.** Because diverse forms of life share a common genetic code, one species can be programmed to produce proteins characteristic of a second species by introducing DNA from the second species into the first.

striking results, as shown in **Figure 14.7**. Bacteria can be programmed by the insertion of human genes to synthesize certain human proteins for medical use, such as insulin. Such applications have produced many exciting developments in the area of genetic engineering (see Chapter 13).

Despite a small number of exceptions in which a few codons differ from the standard ones, the evolutionary significance of the code's *near* universality is clear. A language shared by all living things must have been operating very early in the history of life—early enough to be present in the common ancestor of all present-day organisms. A shared genetic vocabulary is a reminder of the kinship that bonds all life on Earth.

CONCEPT CHECK 14.1

1. **MAKE CONNECTIONS** In a research article about alkaptonuria published in 1902, Garrod suggested that humans inherit two "characters" (alleles) for a particular enzyme and that both parents must contribute a faulty version for the offspring to have the disorder. Today, would this disorder be called dominant or recessive? (See Concept 11.4.)

2. What polypeptide product would you expect from a poly-G mRNA that is 30 nucleotides long?

3. **DRAW IT** The template strand of a gene contains the sequence 3'-TTCAGTCGT-5'. Imagine that the nontemplate sequence was transcribed instead of the template sequence. Draw the mRNA sequence and translate it using Figure 14.6. (Be sure to pay attention to the 5' and 3' ends.) Predict how well the protein synthesized from the nontemplate strand would function, if at all.

For suggested answers, see Appendix A.

Transcription is the DNA-directed synthesis of RNA: *a closer look*

Now that we have considered the linguistic logic and evolutionary significance of the genetic code, we are ready to reexamine transcription, the first stage of gene expression, in more detail.

Molecular Components of Transcription

Messenger RNA, the carrier of information from DNA to the cell's protein-synthesizing machinery, is transcribed from the template strand of a gene. An enzyme called an **RNA polymerase** pries the two strands of DNA apart and joins together RNA nucleotides complementary to the DNA template strand, thus elongating the RNA polynucleotide (**Figure 14.8**). Like the DNA polymerases that function in DNA replication, RNA polymerases can assemble a polynucleotide only in its 5' → 3' direction. Unlike DNA polymerases, however, RNA polymerases are able to start a chain from scratch; they don't need a primer.

Specific sequences of nucleotides along the DNA mark where transcription of a gene begins and ends. The DNA sequence where RNA polymerase attaches and initiates transcription is known as the **promoter**; in bacteria, the sequence that signals the end of transcription is called the **terminator**. (The termination mechanism is different in eukaryotes; we'll describe it later.) Molecular biologists refer to the direction of transcription as "downstream" and the other direction as "upstream." These terms are also used to describe the positions of nucleotide sequences within the DNA or RNA. Thus, the promoter sequence in DNA is said to be upstream from the terminator. The stretch of DNA that is transcribed into an RNA molecule is called a **transcription unit**.

Bacteria have a single type of RNA polymerase that synthesizes not only mRNA but also other types of RNA that function in protein synthesis, such as ribosomal RNA. In contrast, eukaryotes have at least three types of RNA polymerase in their nuclei; the one used for mRNA synthesis is called RNA polymerase II. In the discussion of transcription that follows, we start with the features of mRNA synthesis common to both bacteria and eukaryotes and then describe some key differences.

Synthesis of an RNA Transcript

The three stages of transcription, as shown in Figure 14.8 and described next, are initiation, elongation, and termination of the RNA chain. Study Figure 14.8 to familiarize yourself with the stages and the terms used to describe them.

RNA Polymerase Binding and Initiation of Transcription

The promoter of a gene includes within it the transcription **start point** (the nucleotide where RNA synthesis actually begins) and typically extends several dozen or more nucleotide pairs upstream from the start point. RNA polymerase binds in a precise location and orientation on the promoter, thereby determining where transcription starts and which of the two strands of the DNA helix is used as the template.

Certain sections of a promoter are especially important for binding RNA polymerase. In bacteria, the RNA polymerase itself specifically recognizes and binds to the promoter. In eukaryotes, a collection of proteins called **transcription factors** mediate the binding of RNA polymerase and the initiation of transcription. Only after transcription factors are attached to the promoter does RNA polymerase II bind to it. The whole complex of transcription factors and RNA polymerase II bound to the promoter is called a **transcription initiation complex**. **Figure 14.9** shows the role of transcription factors and a crucial

Promoter

Transcription unit

1 Initiation. After RNA polymerase binds to the promoter, the DNA strands unwind, and the polymerase initiates RNA synthesis at the start point on the template strand.

Start point

RNA polymerase

DNA

Nontemplate strand of DNA

Unwound DNA

RNA transcript

Template strand of DNA

2 Elongation. The polymerase moves downstream, unwinding the DNA and elongating the RNA transcript 5' → 3'. In the wake of transcription, the DNA strands re-form a double helix.

Rewound DNA

RNA transcript

3 Termination. Eventually, the RNA transcript is released, and the polymerase detaches from the DNA.

Completed RNA transcript

Direction of transcription ("downstream")

▲ **Figure 14.8 The stages of transcription: initiation, elongation, and termination.** This general depiction of transcription applies to both bacteria and eukaryotes, but the details of termination differ, as described in the text. Also, in a bacterium, the RNA transcript is immediately usable as mRNA; in a eukaryote, the RNA transcript must first undergo processing.

MAKE CONNECTIONS *Compare the use of a template strand during transcription and replication. See Figure 13.17.*

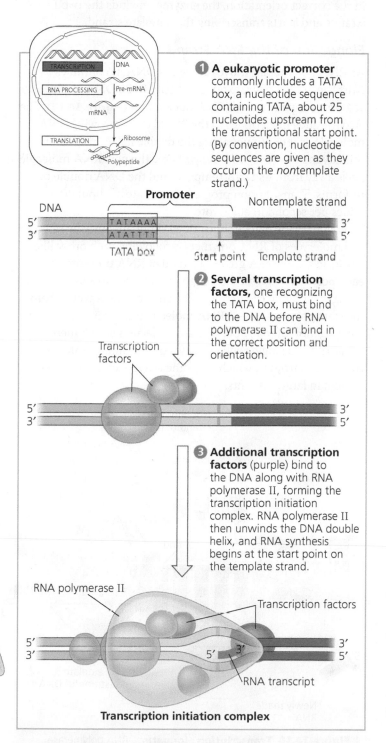

1 A eukaryotic promoter commonly includes a TATA box, a nucleotide sequence containing TATA, about 25 nucleotides upstream from the transcriptional start point. (By convention, nucleotide sequences are given as they occur on the *non*template strand.)

Promoter

Nontemplate strand

DNA

TATAAAA

ATATTTT

TATA box

Start point

Template strand

2 Several transcription factors, one recognizing the TATA box, must bind to the DNA before RNA polymerase II can bind in the correct position and orientation.

Transcription factors

3 Additional transcription factors (purple) bind to the DNA along with RNA polymerase II, forming the transcription initiation complex. RNA polymerase II then unwinds the DNA double helix, and RNA synthesis begins at the start point on the template strand.

RNA polymerase II

Transcription factors

RNA transcript

Transcription initiation complex

▲ **Figure 14.9 The initiation of transcription at a eukaryotic promoter.** In eukaryotic cells, proteins called transcription factors mediate the initiation of transcription by RNA polymerase II.

? *Explain how the interaction of RNA polymerase with the promoter would differ if the figure showed transcription initiation for bacteria.*

DNA sequence in the promoter—the so-called **TATA box**—in forming the initiation complex at a eukaryotic promoter.

The interaction between eukaryotic RNA polymerase II and transcription factors is an example of the importance of protein-protein interactions in controlling eukaryotic transcription. Once the appropriate transcription factors are firmly attached to the promoter DNA and the polymerase is bound in the correct orientation, the enzyme unwinds the two DNA strands and starts transcribing the template strand.

Elongation of the RNA Strand

As RNA polymerase moves along the DNA, it continues to untwist the double helix, exposing about 10–20 DNA nucleotides at a time for pairing with RNA nucleotides **(Figure 14.10)**. The enzyme adds nucleotides to the 3′ end of the growing RNA molecule as it continues along the double helix. In the wake of this advancing wave of RNA synthesis, the new RNA molecule peels away from its DNA template, and the DNA double helix re-forms. Transcription progresses at a rate of about 40 nucleotides per second in eukaryotes.

A single gene can be transcribed simultaneously by several molecules of RNA polymerase following each other like trucks in a convoy. A growing strand of RNA trails off from each polymerase, with the length of each new strand reflecting how far along the template the enzyme has traveled from the start point (see the mRNA molecules in Figure 14.23). The congregation of many polymerase molecules simultaneously transcribing a single gene increases the amount of mRNA transcribed from it, which helps the cell make the encoded protein in large amounts.

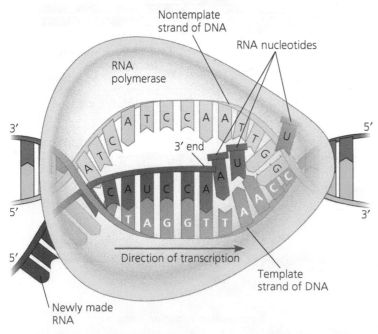

▲ **Figure 14.10 Transcription elongation.** RNA polymerase moves along the DNA template strand, joining complementary RNA nucleotides to the 3′ end of the growing RNA transcript. Behind the polymerase, the new RNA peels away from the template strand, which re-forms a double helix with the nontemplate strand.

Termination of Transcription

The mechanism of termination differs between bacteria and eukaryotes. In bacteria, transcription proceeds through a terminator sequence in the DNA. The transcribed terminator (an RNA sequence) functions as the termination signal, causing the polymerase to detach from the DNA and release the transcript, which requires no further modification before translation. In eukaryotes, RNA polymerase II transcribes a sequence on the DNA called the polyadenylation signal sequence, which codes for a polyadenylation signal (AAUAAA) in the pre-mRNA. Then, at a point about 10–35 nucleotides downstream from the AAUAAA signal, proteins associated with the growing RNA transcript cut it free from the polymerase, releasing the pre-mRNA. The pre-mRNA then undergoes processing, the topic of the next section.

CONCEPT CHECK 14.2

1. What is a promoter, and is it located at the upstream or downstream end of a transcription unit?
2. What enables RNA polymerase to start transcribing a gene at the right place on the DNA in a bacterial cell? In a eukaryotic cell?
3. **WHAT IF?** Suppose X-rays caused a sequence change in the TATA box of a particular gene's promoter. How would that affect transcription of the gene? (See Figure 14.9.)

For suggested answers, see Appendix A.

CONCEPT 14.3

Eukaryotic cells modify RNA after transcription

Enzymes in the eukaryotic nucleus modify pre-mRNA in specific ways before the genetic messages are dispatched to the cytoplasm. During this **RNA processing**, both ends of the primary transcript are altered. Also, in most cases, certain interior sections of the RNA molecule are cut out and the remaining parts spliced together. These modifications produce an mRNA molecule ready for translation.

Alteration of mRNA Ends

Each end of a pre-mRNA molecule is modified in a particular way **(Figure 14.11)**. The 5′ end is synthesized first; it receives a **5′ cap**, a modified form of a guanine (G) nucleotide added onto the 5′ end after transcription of the first 20–40 nucleotides. The 3′ end of the pre-mRNA molecule is also modified before the mRNA exits the nucleus. Recall that the pre-mRNA is released soon after the polyadenylation signal, AAUAAA, is transcribed. At the 3′ end, an enzyme adds 50–250 more adenine (A) nucleotides, forming a **poly-A tail**. The 5′ cap and poly-A tail share several important functions. First, they seem to facilitate the export of the mature mRNA from the nucleus. Second, they help protect the mRNA from degradation by hydrolytic enzymes. And third, they help ribosomes attach to the

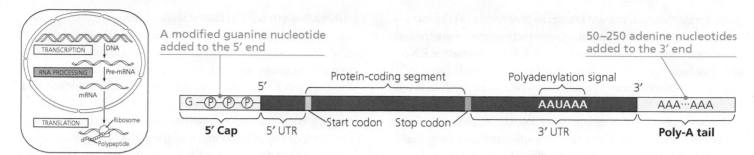

A modified guanine nucleotide added to the 5' end

50–250 adenine nucleotides added to the 3' end

Protein-coding segment

Polyadenylation signal

5'   Start codon   Stop codon   AAUAAA   3'   AAA···AAA

5' Cap   5' UTR   3' UTR   **Poly-A tail**

▲ **Figure 14.11   RNA processing: Addition of the 5' cap and poly-A tail.** Enzymes modify the two ends of a eukaryotic pre-mRNA molecule. The modified ends may promote the export of mRNA from the nucleus, and they help protect the mRNA from degradation. When the mRNA reaches the cytoplasm, the modified ends, in conjunction with certain cytoplasmic proteins, facilitate ribosome attachment. The 5' cap and poly-A tail are not translated into protein, nor are the regions called the 5' untranslated region (5' UTR) and 3' untranslated region (3' UTR).

5' end of the mRNA once the mRNA reaches the cytoplasm. Figure 14.11 shows a diagram of a eukaryotic mRNA molecule with cap and tail. The figure also shows the untranslated regions (UTRs) at the 5' and 3' ends of the mRNA (referred to as the 5' UTR and 3' UTR). The UTRs are parts of the mRNA that will not be translated into protein, but they have other functions, such as ribosome binding.

Split Genes and RNA Splicing

A remarkable stage of RNA processing in the eukaryotic nucleus is the removal of large portions of the RNA molecule that is initially synthesized—a cut-and-paste job called **RNA splicing**, similar to editing a video **(Figure 14.12)**. The average length of a transcription unit along a human DNA molecule is about 27,000 nucleotide pairs, so the primary RNA transcript is also that long. However, the average-sized protein of 400 amino acids requires only 1,200 nucleotides in RNA to code for it. (Remember, each amino acid is encoded by a *triplet* of nucleotides.) This means that most eukaryotic genes and their RNA transcripts have long noncoding stretches of nucleotides, regions that are not translated. Even more surprising is that most of these noncoding sequences are interspersed between

coding segments of the gene and thus between coding segments of the pre-mRNA. In other words, the sequence of DNA nucleotides that codes for a eukaryotic polypeptide is usually not continuous; it is split into segments. The noncoding segments of nucleic acid that lie between coding regions are called *in*tervening sequences, or **introns**. The other regions are called **exons**, because they are eventually *ex*pressed, usually by being translated into amino acid sequences. (Exceptions include the UTRs of the exons at the ends of the RNA, which make up part of the mRNA but are not translated into protein. Because of these exceptions, you may find it helpful to think of exons as sequences of RNA that *exit* the nucleus.) The terms *intron* and *exon* are used for both RNA sequences and the DNA sequences that encode them.

In making a primary transcript from a gene, RNA polymerase II transcribes both introns and exons from the DNA, but the mRNA molecule that enters the cytoplasm is an abridged version. The introns are cut out from the molecule and the exons joined together, forming an mRNA molecule with a continuous coding sequence. This is the process of RNA splicing.

One important consequence of the presence of introns in genes is that a single gene can encode more than one kind of

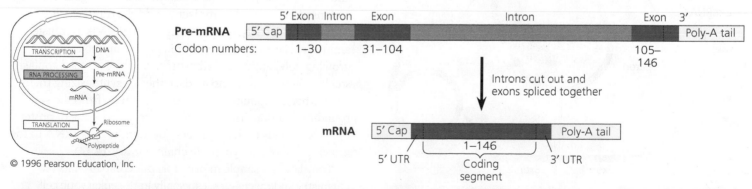

© 1996 Pearson Education, Inc.

Pre-mRNA: 5' Exon   Intron   Exon   Intron   Exon   3'
5' Cap ... Poly-A tail
Codon numbers: 1–30   31–104   105–146

Introns cut out and exons spliced together

mRNA: 5' Cap   1–146   Poly-A tail
5' UTR   Coding segment   3' UTR

▲ **Figure 14.12 RNA processing: RNA splicing.** The RNA molecule shown here codes for β-globin, one of the polypeptides of hemoglobin. The numbers under the RNA refer to codons; β-globin is 146 amino acids long. The β-globin gene and its pre-mRNA transcript have three exons, corresponding to sequences that will leave the nucleus as mRNA. (The 5' UTR and 3' UTR are parts of exons because they are included in the mRNA; however, they do not code for protein.) During RNA processing, the introns are cut out and the exons spliced together. In many genes, the introns are much larger than the exons.

polypeptide. Many genes are known to give rise to two or more different polypeptides, depending on which segments are treated as exons during RNA processing; this is called **alternative RNA splicing** (see Figure 15.12). Because of alternative splicing, the number of different protein products an organism produces can be much greater than its number of genes.

How is pre-mRNA splicing carried out? The removal of introns is accomplished by a large complex made of proteins and small RNAs called a **spliceosome**. This complex binds to several short nucleotide sequences along the intron, including key sequences at each end **(Figure 14.13)**. The intron is then released (and rapidly degraded), and the spliceosome joins together the two exons that flanked the intron. It turns out that the small RNAs in the spliceosome catalyze these processes, as well as participating in spliceosome assembly and splice site recognition.

Ribozymes

The idea of a catalytic role for the RNAs in the spliceosome arose from the discovery of **ribozymes**, RNA molecules that function as enzymes. In some organisms, RNA splicing can occur without proteins or even additional RNA molecules: The intron RNA functions as a ribozyme and catalyzes its own excision! For example, in the ciliate protist *Tetrahymena*, self-splicing occurs in the production of ribosomal RNA (rRNA), a component of the organism's ribosomes. The pre-rRNA actually removes its own introns. The discovery of ribozymes rendered obsolete the idea that all biological catalysts are proteins.

Three properties of RNA enable some RNA molecules to function as enzymes. First, because RNA is single-stranded, a region of an RNA molecule may base-pair with a complementary region elsewhere in the same molecule, which gives the molecule a particular three-dimensional structure. A specific

structure is essential to the catalytic function of ribozymes, just as it is for enzymatic proteins. Second, like certain amino acids in an enzymatic protein, some of the bases in RNA contain functional groups that may participate in catalysis. Third, the ability of RNA to hydrogen-bond with other nucleic acid molecules (either RNA or DNA) adds specificity to its catalytic activity. For example, complementary base pairing between the RNA of the spliceosome and the RNA of a primary RNA transcript precisely locates the region where the ribozyme catalyzes splicing. Later in this chapter, you'll see how these properties of RNA also allow it to perform important noncatalytic roles in the cell, such as recognition of the three-nucleotide codons on mRNA.

CONCEPT CHECK 14.3

1. How can human cells make 75,000–100,000 different proteins, given that there are about 20,000 human genes?
2. How is RNA splicing similar to editing a video? What would introns correspond to in this analogy?
3. **WHAT IF?** What would be the effect of treating cells with an agent that removed the cap from mRNAs?

For suggested answers, see Appendix A.

CONCEPT 14.4

Translation is the RNA-directed synthesis of a polypeptide: *a closer look*

We will now examine in greater detail how genetic information flows from mRNA to protein—the process of translation. As we did for transcription, we'll concentrate on the basic steps of translation that occur in both bacteria and eukaryotes, while pointing out key differences.

Molecular Components of Translation

In the process of translation, a cell "reads" a genetic message and builds a polypeptide accordingly. The message is a series of codons along an mRNA molecule, and the translator is called **transfer RNA (tRNA)**. The function of tRNA is to transfer amino acids from the cytoplasmic pool of amino acids to a growing polypeptide in a ribosome. A cell keeps its cytoplasm stocked with all 20 amino acids, either by synthesizing them from other compounds or by taking them up from the surrounding solution. The ribosome, a structure made of proteins and RNAs, adds each amino acid brought to it by tRNA to the growing end of a polypeptide chain **(Figure 14.14)**.

Translation is simple in principle but complex in its biochemistry and mechanics, especially in the eukaryotic cell. In dissecting translation, we'll concentrate on the slightly less complicated version of the process that occurs in bacteria. We'll begin by looking at the major players in this cellular process and then see how they act together in making a polypeptide.

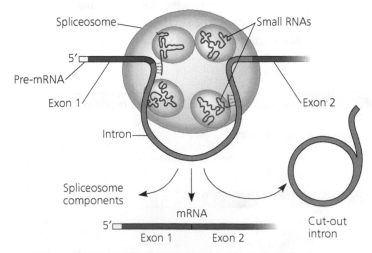

▲ **Figure 14.13 A spliceosome splicing a pre-mRNA.** The diagram shows a portion of a pre-mRNA transcript, with an intron (pink) flanked by two exons (red). Small RNAs within the spliceosome base-pair with nucleotides at specific sites along the intron. Next, the spliceosome catalyzes cutting of the pre-mRNA and the splicing together of the exons, releasing the intron for rapid degradation.

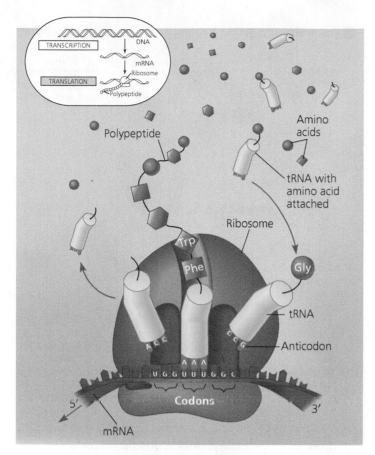

▲ **Figure 14.14 Translation: the basic concept.** As a molecule of mRNA is moved through a ribosome, codons are translated into amino acids, one by one. The interpreters are tRNA molecules, each type with a specific nucleotide triplet called an anticodon at one end and a corresponding amino acid at the other end. A tRNA adds its amino acid cargo to a growing polypeptide chain after the anticodon hydrogen-bonds to a complementary codon on the mRNA. The figures that follow show some of the details of translation in a bacterial cell.

 ANIMATION **BioFlix** Visit the Study Area in **MasteringBiology** for the BioFlix® 3-D Animation on Protein Synthesis.

The Structure and Function of Transfer RNA

The key to translating a genetic message into a specific amino acid sequence is the fact that each tRNA can translate a particular mRNA codon into a given amino acid. This is possible because a tRNA bears a specific amino acid at one end, while at the other end is a nucleotide triplet that can base-pair with the complementary codon on mRNA.

A tRNA molecule consists of a single RNA strand that is only about 80 nucleotides long (compared to hundreds of nucleotides for most mRNA molecules). Because of the presence of complementary stretches of nucleotide bases that can hydrogen-bond to each other, this single strand can fold back on itself and form a molecule with a three-dimensional structure. Flattened into one plane to clarify this base pairing, a tRNA molecule looks like a cloverleaf **(Figure 14.15a).** The tRNA actually twists and folds into a

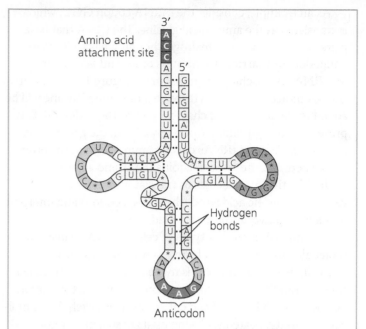

(a) Two-dimensional structure. The four base-paired regions and three loops are characteristic of all tRNAs, as is the base sequence of the amino acid attachment site at the 3' end. The anticodon triplet is unique to each tRNA type, as are some sequences in the other two loops. (The asterisks mark bases that have been chemically modified, a characteristic of tRNA. The modified bases contribute to tRNA function in a way that is not yet understood.)

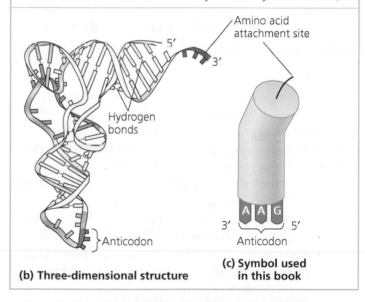

(b) Three-dimensional structure

(c) Symbol used in this book

▲ **Figure 14.15 The structure of transfer RNA (tRNA).** Anticodons are conventionally written 3' → 5' to align properly with codons written 5' → 3' (see Figure 14.14). For base pairing, RNA strands must be antiparallel, like DNA. For example, anticodon 3'-AAG-5' pairs with mRNA codon 5'-UUC-3'.

compact three-dimensional structure that is roughly L-shaped **(Figure 14.15b).** The loop extending from one end of the L includes the **anticodon**, the particular nucleotide triplet that base-pairs to a specific mRNA codon. From the other end of the L-shaped tRNA molecule protrudes its 3' end, which is the attachment site for an amino acid. Thus, the structure of a tRNA molecule fits its function.

As an example, consider the mRNA codon GGC, which is translated as the amino acid glycine. The tRNA that base-pairs with this codon by hydrogen bonding has CCG as its anticodon and carries glycine at its other end (see the incoming tRNA approaching the ribosome in Figure 14.14). As an mRNA molecule is moved through a ribosome, glycine will be added to the polypeptide chain whenever the codon GGC is presented for translation. Codon by codon, the genetic message is translated as tRNAs deposit amino acids in the order prescribed, and the ribosome joins the amino acids into a chain. The tRNA molecule is a translator in the sense that it can read a nucleic acid word (the mRNA codon) and interpret it as a protein word (the amino acid).

Like mRNA and other types of cellular RNA, transfer RNA molecules are transcribed from DNA templates. In a eukaryotic cell, tRNA, like mRNA, is made in the nucleus and then travels from the nucleus to the cytoplasm, where translation occurs. In both bacterial and eukaryotic cells, each tRNA molecule is used repeatedly, picking up its designated amino acid in the cytosol, depositing this cargo onto a polypeptide chain at the ribosome, and then leaving the ribosome, ready to pick up another of the same amino acid.

The accurate translation of a genetic message requires two instances of molecular recognition. First, a tRNA that binds to an mRNA codon specifying a particular amino acid must carry that amino acid, and no other, to the ribosome. The correct matching up of tRNA and amino acid is carried out by a family of related enzymes called **aminoacyl-tRNA synthetases (Figure 14.16)**. The active site of each type of aminoacyl-tRNA synthetase fits only a specific combination of amino acid and tRNA. (Regions of both the amino acid attachment end and the anticodon end of the tRNA are instrumental in ensuring the specific fit.) There are 20 different synthetases, one for each amino acid; each synthetase is able to bind to all the different tRNAs that match the codons for its particular amino acid. The synthetase catalyzes the covalent attachment of the amino acid to its tRNA in a process driven by the hydrolysis of ATP. The resulting aminoacyl tRNA, also called a charged tRNA, is released from the enzyme and is then available to deliver its amino acid to a growing polypeptide chain on a ribosome.

The second instance of molecular recognition is the pairing of the tRNA anticodon with the appropriate mRNA codon. If one tRNA variety existed for each mRNA codon specifying an amino acid, there would be 61 tRNAs (see Figure 14.6). In fact, there are only about 45, signifying that some tRNAs must be able to bind to more than one codon. Such versatility is possible because the rules for base pairing between the third nucleotide base of a codon and the corresponding base of a tRNA anticodon are relaxed compared to those at other codon positions. For example, the nucleotide base U at the 5′ end of a tRNA anticodon can pair with either A or G in the third position (at the 3′ end) of an mRNA codon. The flexible base pairing at

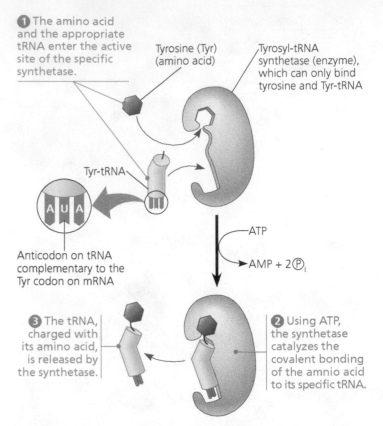

▲ **Figure 14.16 Aminoacyl-tRNA synthetases provide specificity in joining amino acids to their tRNAs.** Linkage of a tRNA to its amino acid is an endergonic process that occurs at the expense of ATP (which loses two phosphate groups, becoming AMP). Shown as an example is the joining of the amino acid tyrosine (Tyr) to the appropriate tRNA by the aminoacyl-tRNA synthetase specific for tyrosine.

this codon position is called **wobble**. Wobble explains why the synonymous codons for a given amino acid most often differ in their third nucleotide base, but not in the other bases. For example, a tRNA with the anticodon 3′-UCU-5′ can base-pair with either the mRNA codon 5′-AGA-3′ or 5′-AGG-3′, both of which code for arginine (see Figure 14.6).

Ribosomes

Ribosomes facilitate the specific coupling of tRNA anticodons with mRNA codons during protein synthesis. A ribosome consists of a large subunit and a small subunit, each made up of proteins and one or more **ribosomal RNAs (rRNAs) (Figure 14.17)**. In eukaryotes, the subunits are made in the nucleolus. Ribosomal RNA genes are transcribed, and the RNA is processed and assembled with proteins imported from the cytoplasm. The resulting ribosomal subunits are then exported via nuclear pores to the cytoplasm. In both bacteria and eukaryotes, large and small subunits join to form a functional ribosome only when they attach to an mRNA molecule. About one-third of the mass of a ribosome is made up of proteins; the rest consists of rRNAs, either three molecules (in bacteria) or four

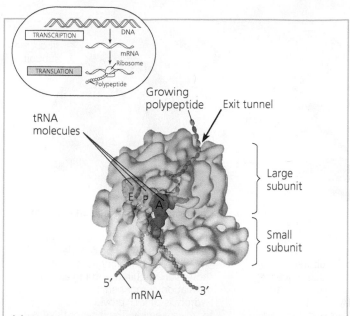

Growing
polypeptide

Exit tunnel

tRNA
molecules

E P A

Large
subunit

Small
subunit

5'   3'
mRNA

(a) Computer model of functioning ribosome. This is a model of a bacterial ribosome, showing its overall shape. The eukaryotic ribosome is roughly similar. A ribosomal subunit is a complex of ribosomal RNA molecules and proteins.

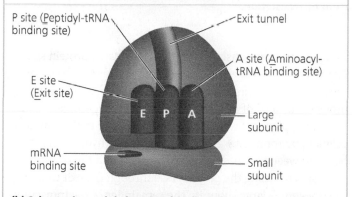

P site (Peptidyl-tRNA
binding site)

Exit tunnel

A site (Aminoacyl-
tRNA binding site)

E site
(Exit site)

E P A

Large
subunit

mRNA
binding site

Small
subunit

(b) Schematic model showing binding sites. A ribosome has an mRNA binding site and three tRNA binding sites, known as the A, P, and E sites. This schematic ribosome will appear in later diagrams.

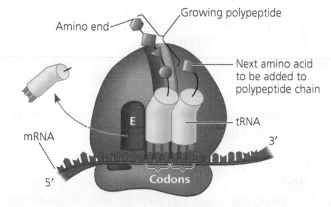

Growing polypeptide

Amino end

Next amino acid
to be added to
polypeptide chain

E

tRNA

mRNA

3'

5'

Codons

(c) Schematic model with mRNA and tRNA. A tRNA fits into a binding site when its anticodon base-pairs with an mRNA codon. The P site holds the tRNA attached to the growing polypeptide. The A site holds the tRNA carrying the next amino acid to be added to the polypeptide chain. Discharged tRNAs leave from the E site. The polypeptide grows at its carboxyl end.

© 1996 Pearson Education, Inc.

▲ **Figure 14.17   The anatomy of a functioning ribosome.**

(in eukaryotes). Because most cells contain thousands of ribosomes, rRNA is the most abundant type of cellular RNA.

Although the ribosomes of bacteria and eukaryotes are very similar in structure and function, eukaryotic ribosomes are slightly larger, and they differ somewhat from bacterial ribosomes in their molecular composition. The differences are medically significant. Certain antibiotic drugs can inactivate bacterial ribosomes without inhibiting the ability of eukaryotic ribosomes to make proteins. These drugs, including tetracycline and streptomycin, are used to combat bacterial infections.

The structure of a ribosome reflects its function of bringing mRNA together with tRNAs carrying amino acids. In addition to a binding site for mRNA, each ribosome has three binding sites for tRNA, as described in Figure 14.17. The **P site** (peptidyl-tRNA binding site) holds the tRNA carrying the growing polypeptide chain, while the **A site** (aminoacyl-tRNA binding site) holds the tRNA carrying the next amino acid to be added to the chain. Discharged tRNAs leave the ribosome from the **E site** (exit site). The ribosome holds the tRNA and mRNA in close proximity and positions the new amino acid so it can be added to the carboxyl end of the growing polypeptide. It then catalyzes the formation of the peptide bond. As the polypeptide becomes longer, it passes through an *exit tunnel* in the ribosome's large subunit. When the polypeptide is complete, it is released through the exit tunnel.

There is strong evidence supporting the hypothesis that rRNA, not protein, is primarily responsible for both the structure and the function of the ribosome. The proteins, which are largely on the exterior, support the shape changes of the rRNA molecules as they carry out catalysis during translation. Ribosomal RNA is the main constituent of the A and P sites and of the interface between the two ribosomal subunits; it also acts as the catalyst of peptide bond formation. Thus, a ribosome can be regarded as one colossal ribozyme!

Building a Polypeptide

We can divide translation, the synthesis of a polypeptide chain, into three stages (analogous to those of transcription): initiation, elongation, and termination. All three stages require protein "factors" that aid in the translation process. For certain aspects of chain initiation and elongation, energy is also required. It is provided by the hydrolysis of guanosine triphosphate (GTP), a molecule closely related to ATP.

Ribosome Association and Initiation of Translation

The initiation stage of translation brings together mRNA, a tRNA bearing the first amino acid of the polypeptide, and the two subunits of a ribosome. First, a small ribosomal subunit binds to both mRNA and a specific initiator tRNA, which carries the amino acid methionine. In bacteria, the small subunit can bind these two in either order; it binds the mRNA at a specific RNA sequence, just upstream of the start codon, AUG. In eukaryotes, the small subunit, with the initiator tRNA already bound, binds to the 5′ cap of the mRNA and then moves, or *scans*, downstream along the mRNA until it reaches the start codon; the initiator tRNA then hydrogen-bonds to the AUG start codon **(Figure 14.18)**. In either case, the start codon signals the start of translation; this is important because it establishes the codon reading frame for the mRNA. In the **Scientific Skills Exercise**, you can work with DNA sequences encoding the ribosomal binding sites on the mRNAs of a group of *E. coli* genes.

The union of mRNA, initiator tRNA, and a small ribosomal subunit is followed by the attachment of a large ribosomal subunit, completing the *translation initiation complex*. Proteins called *initiation factors* are required to bring all these components together. The cell also expends energy obtained by hydrolysis of a GTP molecule to form the initiation complex. At the completion of the initiation process, the initiator tRNA sits in the P site of the ribosome, and the vacant A site is ready for the next aminoacyl tRNA. Note that a polypeptide is always synthesized in one direction, from the initial methionine at the amino end, also called the N-terminus, toward the final amino acid at the carboxyl end, also called the C-terminus (see Figure 3.18).

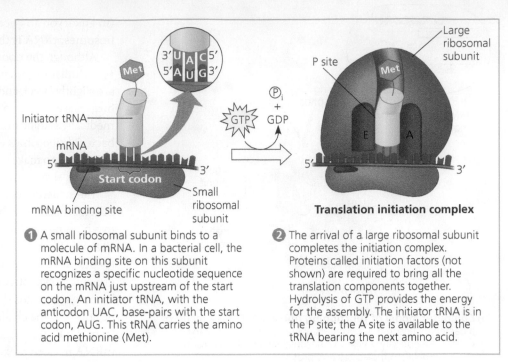

① A small ribosomal subunit binds to a molecule of mRNA. In a bacterial cell, the mRNA binding site on this subunit recognizes a specific nucleotide sequence on the mRNA just upstream of the start codon. An initiator tRNA, with the anticodon UAC, base-pairs with the start codon, AUG. This tRNA carries the amino acid methionine (Met).

② The arrival of a large ribosomal subunit completes the initiation complex. Proteins called initiation factors (not shown) are required to bring all the translation components together. Hydrolysis of GTP provides the energy for the assembly. The initiator tRNA is in the P site; the A site is available to the tRNA bearing the next amino acid.

▲ **Figure 14.18 The initiation of translation.**

Elongation of the Polypeptide Chain

In the elongation stage of translation, amino acids are added one by one to the previous amino acid at the C-terminus of the growing chain. Each addition involves the participation of several proteins called *elongation factors* and occurs in a three-step cycle described in **Figure 14.19**. Energy expenditure occurs in the first and third steps. Codon recognition requires hydrolysis of one molecule of GTP, which increases the accuracy and efficiency of this step. One more GTP is hydrolyzed to provide energy for the translocation step.

The mRNA is moved through the ribosome in one direction only, 5′ end first; this is equivalent to the ribosome moving 5′ → 3′ on the mRNA. The important point is that the ribosome and the mRNA move relative to each other, unidirectionally, codon by codon. The elongation cycle takes less than a tenth of a second in bacteria and is repeated as each amino acid is added to the chain until the polypeptide is completed.

Termination of Translation

The final stage of translation is termination **(Figure 14.20)**. Elongation continues until a stop codon in the mRNA reaches the A site of the ribosome. The nucleotide base triplets UAG, UAA, and UGA do not code for amino acids but instead act as signals to stop translation. A *release factor*, a protein shaped like an aminoacyl tRNA, binds directly to the stop codon in the A site. The release factor causes the addition of a water molecule instead of an amino acid to the polypeptide chain. (There are plenty of water molecules available in the aqueous cellular environment.) This reaction breaks (hydrolyzes) the bond between the completed polypeptide and the tRNA in the P site, releasing the polypeptide through the exit tunnel of the ribosome's large subunit. The remainder of the translation assembly then comes apart in a multistep process, aided by other protein factors. Breakdown of the translation assembly requires the hydrolysis of two more GTP molecules.

Completing and Targeting the Functional Protein

The process of translation is often not sufficient to make a functional protein. In this section, you'll learn about modifications that polypeptide chains undergo after the translation process as well as some of the mechanisms used to target completed proteins to specific sites in the cell.

Protein Folding and Post-Translational Modifications

During its synthesis, a polypeptide chain begins to coil and fold spontaneously as a consequence of its amino acid sequence (primary structure), forming a protein with a specific shape: a three-dimensional molecule with secondary and tertiary structure (see Figure 3.21). Thus, a gene determines primary structure, and primary structure in turn determines shape. In many cases, a chaperone protein helps the polypeptide fold correctly.

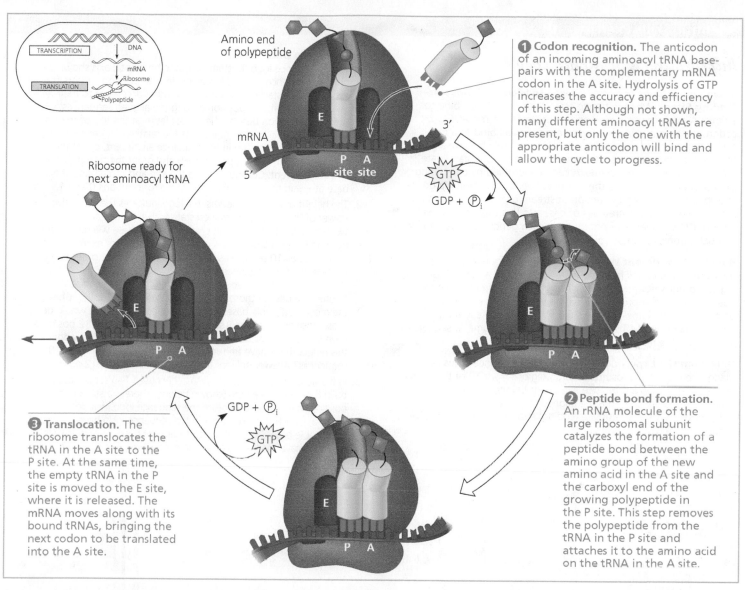

Amino end of polypeptide

mRNA

Ribosome ready for next aminoacyl tRNA

① Codon recognition. The anticodon of an incoming aminoacyl tRNA base-pairs with the complementary mRNA codon in the A site. Hydrolysis of GTP increases the accuracy and efficiency of this step. Although not shown, many different aminoacyl tRNAs are present, but only the one with the appropriate anticodon will bind and allow the cycle to progress.

GTP
GDP + Ⓟ$_i$

② Peptide bond formation. An rRNA molecule of the large ribosomal subunit catalyzes the formation of a peptide bond between the amino group of the new amino acid in the A site and the carboxyl end of the growing polypeptide in the P site. This step removes the polypeptide from the tRNA in the P site and attaches it to the amino acid on the tRNA in the A site.

GDP + Ⓟ$_i$
GTP

③ Translocation. The ribosome translocates the tRNA in the A site to the P site. At the same time, the empty tRNA in the P site is moved to the E site, where it is released. The mRNA moves along with its bound tRNAs, bringing the next codon to be translated into the A site.

▲ **Figure 14.19 The elongation cycle of translation.** The hydrolysis of GTP plays an important role in the elongation process. Not shown are the proteins called elongation factors.

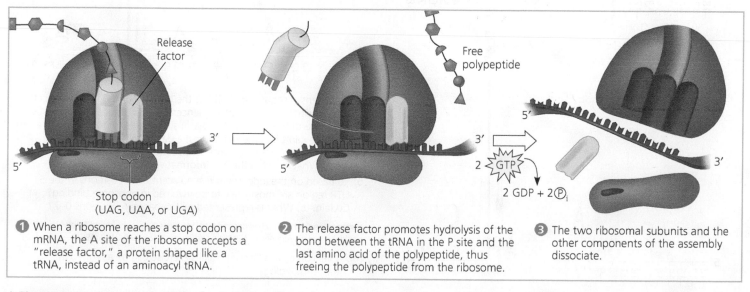

Release factor

Free polypeptide

Stop codon (UAG, UAA, or UGA)

2 GTP
2 GDP + 2 Ⓟ$_i$

① When a ribosome reaches a stop codon on mRNA, the A site of the ribosome accepts a "release factor," a protein shaped like a tRNA, instead of an aminoacyl tRNA.

② The release factor promotes hydrolysis of the bond between the tRNA in the P site and the last amino acid of the polypeptide, thus freeing the polypeptide from the ribosome.

③ The two ribosomal subunits and the other components of the assembly dissociate.

▲ **Figure 14.20 The termination of translation.** Like elongation, termination requires GTP hydrolysis as well as additional protein factors, which are not shown here.

Interpreting a Sequence Logo

How Can a Sequence Logo Be Used to Identify Ribosome-Binding Sites?
When initiating translation, ribosomes bind to an mRNA at a ribosome-binding site upstream of the AUG start codon. Because mRNAs from different genes all bind to a ribosome, the genes encoding these mRNAs are likely to have a similar base sequence where the ribosomes bind. Therefore, candidate ribosome-binding sites on mRNA can be identified by comparing DNA sequences (and thus the mRNA sequences) of multiple genes in a species, searching the region upstream of the start codon for shared ("conserved") stretches of bases. In this exercise you will analyze DNA sequences from multiple such genes, represented by a visual graphic called a sequence logo.

How the Experiment Was Done
The DNA sequences of 149 genes from the *E. coli* genome were aligned and analyzed using computer software. The aim was to identify similar base sequences—at the appropriate location in each gene—that might be potential ribosome-binding sites. Rather than presenting the data as a series of 149 sequences aligned in a column (a sequence alignment), the researchers used a sequence logo.

Data from the Experiment
To show how sequence logos are made, the potential ribosome-binding regions from 10 of the *E. coli* genes are shown in a sequence alignment below, followed by the sequence logo derived from the aligned sequences. Note that the DNA shown is the nontemplate (coding) strand, which is how DNA sequences are typically given.

thrA	G G T A A C G A G G T A A C A A C C A T G C G A G T G
lacA	C A T A A C G G A G T G A T C G C A T T G A A C A T G
lacY	C G C G T A A G G A A A T C C A T T A T G T A C T A T
lacZ	T T C A C A C A G G A A A C A G C T A T G A C C A T G
lacI	C A A T T C A G G G T G G T G A A T G T G A A A C C A
recA	G G C A T G A C A G G A G T A A A A A T G G C T A T C
galR	A C C C A C T A A G G T A T T T T C A T G G C G A C C
metJ	A A G A G G A T T A A G T A T C T C A T G G C T G A A
lexA	A T A C A C C C A G G G G G C G G A A T G A A A G C G
trpR	T A A C A A T G G C G A C A T A T T A T G G C C C A A

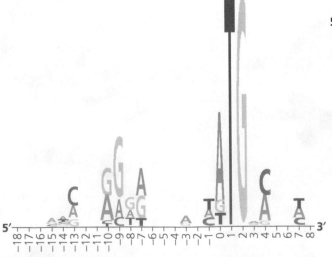

Interpret the Data

1. In the sequence logo (bottom, left), the horizontal axis shows the primary sequence of the DNA by nucleotide position. Letters for each base are stacked on top of each other according to their relative frequency at that position among the aligned sequences, with the most common base as the largest letter at the top of the stack. The height of each letter represents the relative frequency of that base *at that position*. (a) In the sequence alignment, count the number of each base at position −9 and order them from most to least frequent. Compare this to the size and placement of each base at −9 in the logo. (b) Do the same for positions 0 and 1.

2. The height of a stack of letters in a logo indicates the predictive power of that stack (determined statistically). If the stack is tall, we can be more confident in predicting what base will be in that position if a new sequence is added to the logo. For example, at position 2, all 10 sequences have a G; the probability of finding a G there in a new sequence is very high, as is the stack. For short stacks, the bases all have about the same frequency, and so it's hard to predict a base at those positions. (a) Which two positions have the most predictable bases? What bases do you predict would be at those positions in a newly sequenced gene? (b) Which 12 positions have the least predictable bases? How do you know? How does this reflect the relative frequencies of the bases shown in the 10 sequences? Answer only for the two left-most of the 12 positions.

3. In the actual experiment, the researchers used 149 sequences to build their sequence logo (shown below). There is a stack at each position, even if short, because the sequence logo includes more data. (a) Which three positions in the sequence logo have the most predictable bases? Name the most frequent base at each. (b) Which positions have the least predictable bases? How can you tell?

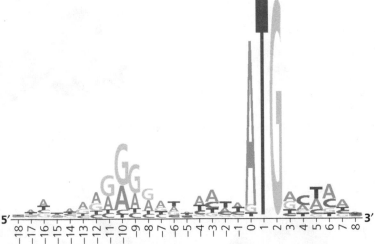

4. A consensus sequence identifies the base occurring most often at each position in the set of sequences. (a) Write out the consensus sequence of this (the nontemplate) strand. In any position where the base can't be determined, put a dash. (b) Which provides more information—the consensus sequence or the sequence logo? What is lost in the less informative method?

5. (a) Based on the logo, what five adjacent base positions in the 5′ UTR region are most likely to be involved in ribosome binding? Explain. (b) What is represented by the bases in positions 0–2?

Further Reading T.D. Schneider and R.M. Stephens, Sequence logos: A new way to display consensus sequences, *Nucleic Acids Research* 18:6097–6100 (1990).

(MB) A version of this Scientific Skills Exercise can be assigned in MasteringBiology.

Additional steps—*post-translational modifications*—may be required before the protein can begin doing its particular job in the cell. Certain amino acids may be chemically modified by the attachment of sugars, lipids, phosphate groups, or other additions. Enzymes may remove one or more amino acids from the leading (amino) end of the polypeptide chain. In some cases, a polypeptide chain may be enzymatically cleaved into two or more pieces. For example, the protein insulin is first synthesized as a single polypeptide chain but becomes active only after an enzyme cuts out a central part of the chain, leaving a protein made up of two shorter polypeptide chains connected by disulfide bridges. In other cases, two or more polypeptides that are synthesized separately may come together, becoming the subunits of a protein that has quaternary structure. A familiar example is hemoglobin (see Figure 3.21).

Targeting Polypeptides to Specific Locations

In electron micrographs of eukaryotic cells active in protein synthesis, two populations of ribosomes (and polyribosomes) are evident: free and bound (see Figure 4.9). Free ribosomes are suspended in the cytosol and mostly synthesize proteins that stay in the cytosol and function there. In contrast, bound ribosomes are attached to the cytosolic side of the endoplasmic reticulum (ER) or to the nuclear envelope. Bound ribosomes make proteins of the endomembrane system (the nuclear envelope, ER, Golgi apparatus, lysosomes, vacuoles, and plasma membrane) as well as proteins secreted from the cell, such as insulin. It is important to note that the ribosomes themselves are identical and can alternate between being free and bound.

What determines whether a ribosome is free in the cytosol or bound to rough ER? Polypeptide synthesis always begins in the cytosol as a free ribosome starts to translate an mRNA molecule. There the process continues to completion—*unless* the growing polypeptide itself cues the ribosome to attach to the ER. The polypeptides of proteins destined for the endomembrane system or for secretion are marked by a **signal peptide**, which targets the protein to the ER **(Figure 14.21)**. The signal peptide, a sequence of about 20 amino acids at or near the leading end (N-terminus) of the polypeptide, is recognized as it emerges from the ribosome by a protein-RNA complex called a **signal-recognition particle (SRP)**. This particle functions as an escort that brings the ribosome to a receptor protein built into the ER membrane. The receptor is part of a multiprotein translocation complex. Polypeptide synthesis continues there, and the growing polypeptide snakes across the membrane into the ER lumen via a protein pore. The signal peptide is usually removed by an enzyme. The rest of the completed polypeptide, if it is to be secreted from the cell, is released into solution within the ER lumen (as in Figure 14.21). Alternatively, if the polypeptide is to be a membrane protein,

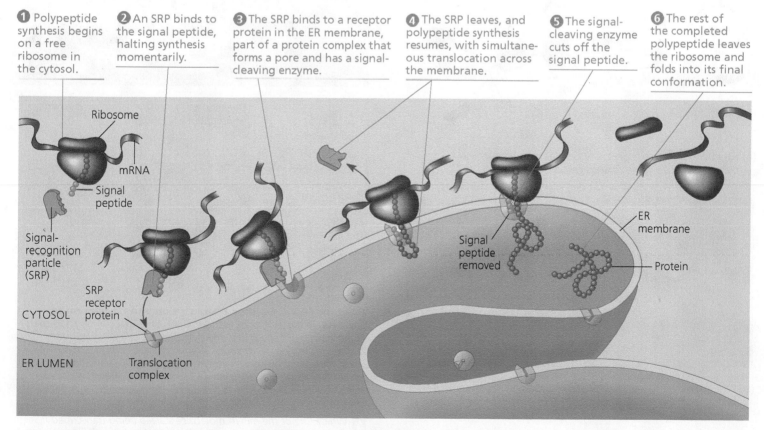

❶ Polypeptide synthesis begins on a free ribosome in the cytosol.

❷ An SRP binds to the signal peptide, halting synthesis momentarily.

❸ The SRP binds to a receptor protein in the ER membrane, part of a protein complex that forms a pore and has a signal-cleaving enzyme.

❹ The SRP leaves, and polypeptide synthesis resumes, with simultaneous translocation across the membrane.

❺ The signal-cleaving enzyme cuts off the signal peptide.

❻ The rest of the completed polypeptide leaves the ribosome and folds into its final conformation.

Ribosome
mRNA
Signal peptide
Signal-recognition particle (SRP)
SRP receptor protein
CYTOSOL
ER LUMEN
Translocation complex
Signal peptide removed
ER membrane
Protein

▲ **Figure 14.21** **The signal mechanism for targeting proteins to the ER.**

it remains partially embedded in the ER membrane. In either case, it travels in a transport vesicle to the plasma membrane (see Figure 5.8).

Other kinds of signal peptides are used to target polypeptides to mitochondria, chloroplasts, the interior of the nucleus, and other organelles that are not part of the endomembrane system. The critical difference in these cases is that translation is completed in the cytosol before the polypeptide is imported into the organelle. Translocation mechanisms also vary, but in all cases studied to date, the "postal zip codes" that address proteins for secretion or to cellular locations are signal peptides of some sort. Bacteria also employ signal peptides to target proteins to the plasma membrane for secretion.

Making Multiple Polypeptides in Bacteria and Eukaryotes

In previous sections, you have learned how a single polypeptide is synthesized using the information encoded in an mRNA molecule. When a polypeptide is required in a cell, though, the need is for many copies, not just one.

In both bacteria and eukaryotes, multiple ribosomes translate an mRNA at the same time (**Figure 14.22**); that is, a single mRNA is used to make many copies of a polypeptide simultaneously. Once a ribosome is far enough past the start codon, a second ribosome can attach to the mRNA, eventually resulting in a number of ribosomes trailing along the mRNA. Such strings of ribosomes, called polyribosomes (or polysomes), can be seen with an electron microscope (see Figure 14.22). They enable a cell to make many copies of a polypeptide very quickly.

Another way both bacteria and eukaryotes augment the number of copies of a polypeptide is by transcribing multiple mRNAs from the same gene, as we mentioned earlier. However, the coordination of the two processes—transcription and translation—differ in the two groups. The most important differences between bacteria and eukaryotes arise from the bacterial cell's lack of compartmental organization. Like a one-room workshop, a bacterial cell ensures a streamlined operation by coupling the two processes. In the absence of a nuclear envelope, it can simultaneously transcribe and translate the same gene (**Figure 14.23**), and the newly made protein can then diffuse to its site of function.

In contrast, the eukaryotic cell's nuclear envelope segregates transcription from translation and provides a compartment for extensive RNA processing. This processing stage includes additional steps whose regulation can help coordinate the eukaryotic cell's elaborate activities (see Chapter 15). **Figure 14.24** summarizes the path from gene to polypeptide in a eukaryotic cell.

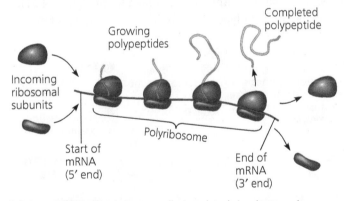

(a) An mRNA molecule is generally translated simultaneously by several ribosomes in clusters called polyribosomes.

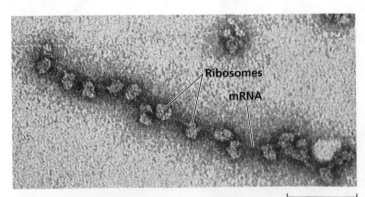

(b) This micrograph shows a large polyribosome in a bacterial cell. Growing polypeptides are not visible here (TEM).

0.1 μm

▲ **Figure 14.22 Polyribosomes.**

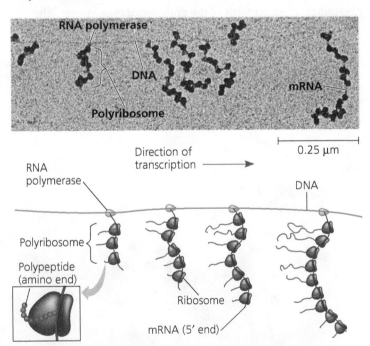

▲ **Figure 14.23 Coupled transcription and translation in bacteria.** In bacterial cells, the translation of mRNA can begin as soon as the leading (5') end of the mRNA molecule peels away from the DNA template. The micrograph (TEM) shows a stretch of *E. coli* DNA being transcribed by RNA polymerase molecules. Attached to each RNA polymerase molecule is a growing strand of mRNA, which is already being translated by ribosomes. The newly synthesized polypeptides are not visible in the micrograph but are shown in the diagram.

? *Which one of the mRNA molecules started being transcribed first? On that mRNA, which ribosome started translating first?*

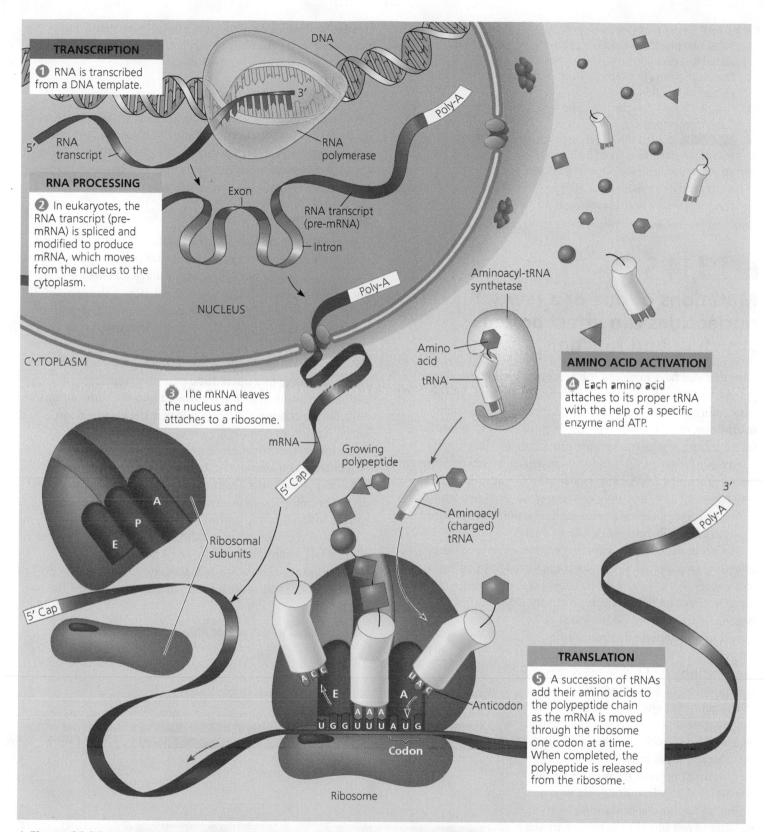

TRANSCRIPTION

1 RNA is transcribed from a DNA template.

DNA

3'

RNA transcript

5'

RNA polymerase

Poly-A

RNA PROCESSING

2 In eukaryotes, the RNA transcript (pre-mRNA) is spliced and modified to produce mRNA, which moves from the nucleus to the cytoplasm.

Exon

RNA transcript (pre-mRNA)

Intron

Poly-A

NUCLEUS

CYTOPLASM

3 The mRNA leaves the nucleus and attaches to a ribosome.

mRNA

5' Cap

Ribosomal subunits

5' Cap

A

P

E

Aminoacyl-tRNA synthetase

Amino acid

tRNA

AMINO ACID ACTIVATION

4 Each amino acid attaches to its proper tRNA with the help of a specific enzyme and ATP.

Growing polypeptide

Aminoacyl (charged) tRNA

Poly-A

3'

Anticodon

TRANSLATION

5 A succession of tRNAs add their amino acids to the polypeptide chain as the mRNA is moved through the ribosome one codon at a time. When completed, the polypeptide is released from the ribosome.

A C C

E

A

U A C

A A A

U G G U U U A U G

Codon

Ribosome

▲ **Figure 14.24 A summary of transcription and translation in a eukaryotic cell.** This diagram shows the path from one gene to one polypeptide. Keep in mind that each gene in the DNA can be transcribed repeatedly into many identical RNA molecules and that each mRNA can be translated repeatedly to yield many identical polypeptide molecules. (Also, remember that the final products of some genes are not polypeptides but RNA molecules, including tRNA and rRNA.) In general, the steps of transcription and translation are similar in bacterial, archaeal, and eukaryotic cells. The major difference is the occurrence of RNA processing in the eukaryotic nucleus. Other significant differences are found in the initiation stages of both transcription and translation and in the termination of transcription.

CONCEPT CHECK 14.4

1. What two processes ensure that the correct amino acid is added to a growing polypeptide chain?
2. Discuss the ways in which rRNA structure likely contributes to ribosomal function.
3. Describe how a polypeptide to be secreted is transported to the endomembrane system.
4. **DRAW IT** Draw a tRNA with the anticodon 3'-CGU-5'. What two different codons could it bind to? Draw each codon on an mRNA, labeling all 5' and 3' ends, the tRNA, and the amino acid it carries.

For suggested answers, see Appendix A.

CONCEPT 14.5

Mutations of one or a few nucleotides can affect protein structure and function

Now that you have explored the process of gene expression, you are ready to understand the effects of changes to the genetic information of a cell (or virus). These changes, called **mutations**, are responsible for the huge diversity of genes found among organisms because mutations are the ultimate source of new genes. Chromosomal rearrangements that affect long segments of DNA are considered large-scale mutations (see Figure 12.14). Here we'll examine small-scale mutations of one or a few nucleotide pairs, including **point mutations**, changes in a single nucleotide pair of a gene.

If a point mutation occurs in a gamete or in a cell that gives rise to gametes, it may be transmitted to offspring and to a succession of future generations. If the mutation has an adverse effect on the phenotype of an organism, the mutant condition is referred to as a genetic disorder or hereditary disease. For example, we can trace the genetic basis of sickle-cell disease to the mutation of a single nucleotide pair in the gene that encodes the β-globin polypeptide of hemoglobin. The change of a single nucleotide in the DNA's template strand leads to the production of an abnormal protein **(Figure 14.25;** also see Figure 3.22). In individuals who are homozygous for the mutant allele, the sickling of red blood cells caused by the altered hemoglobin produces the multiple symptoms associated with sickle-cell disease (see Chapter 11). Another disorder caused by a point mutation is a heart condition, familial cardio-myopathy, that is responsible for some

incidents of sudden death in young athletes. Point mutations in several genes have been identified, any of which can lead to this disorder.

Types of Small-Scale Mutations

Let's now consider how small-scale mutations affect proteins. Small-scale mutations within a gene can be divided into two general categories: (1) single nucleotide-pair substitutions and (2) nucleotide-pair insertions or deletions. Insertions and deletions can involve one or more nucleotide pairs.

Substitutions

A **nucleotide-pair substitution** is the replacement of one nucleotide and its partner with another pair of nucleotides **(Figure 14.26a)**. Some substitutions have no effect on the encoded protein, owing to the redundancy of the genetic code. For example, if 3'-CCG-5' on the template strand mutated to 3'-CCA-5', the mRNA codon that used to be GGC would become GGU, but a glycine would still be inserted at the proper location in the protein (see Figure 14.6). In other words, a change in a nucleotide pair may transform one codon into another that is translated into the same amino acid. Such a change is an example of a **silent mutation**, which has no observable effect on the phenotype. (Silent mutations can occur outside genes as well.) Substitutions that change one amino acid to another one are called **missense mutations**. Such a mutation may have little effect on the protein: The new amino acid may have properties similar to those of the amino acid it replaces, or it may be in a region of the protein where the exact sequence of amino acids is not essential to the protein's function.

However, the nucleotide-pair substitutions of greatest interest are those that cause a major change in a protein. The

▼ **Figure 14.25 The molecular basis of sickle-cell disease: a point mutation.** The allele that causes sickle-cell disease differs from the wild-type (normal) allele by a single DNA nucleotide pair.

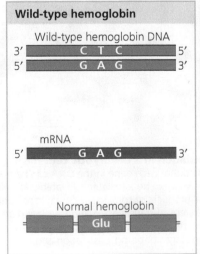

Wild-type hemoglobin		

Wild-type hemoglobin DNA
3' ▆▆▆ C T C ▆▆▆ 5'
5' ▆▆▆ G A G ▆▆▆ 3'

mRNA
5' ▆▆▆ G A G ▆▆▆ 3'

Normal hemoglobin
Glu

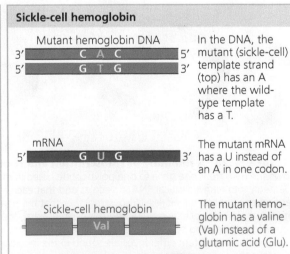

Sickle-cell hemoglobin		

Mutant hemoglobin DNA
3' ▆▆▆ C A C ▆▆▆ 5'
5' ▆▆▆ G T G ▆▆▆ 3'

In the DNA, the mutant (sickle-cell) template strand (top) has an A where the wild-type template has a T.

mRNA
5' ▆▆▆ G U G ▆▆▆ 3'

The mutant mRNA has a U instead of an A in one codon.

Sickle-cell hemoglobin
Val

The mutant hemoglobin has a valine (Val) instead of a glutamic acid (Glu).

alteration of a single amino acid in a crucial area of a protein—such as in the part of hemoglobin shown in Figure 14.25 or in the active site of an enzyme—will significantly alter protein activity. Occasionally, such a mutation leads to an improved protein or one with novel capabilities, but much more often such mutations are detrimental, leading to a useless or less active protein that impairs cellular function.

Substitution mutations are usually missense mutations; that is, the altered codon still codes for an amino acid and thus makes sense, although not necessarily the *right* sense. But a point mutation can also change a codon for an amino acid

into a stop codon. This is called a **nonsense mutation**, and it causes translation to be terminated prematurely; the resulting polypeptide will be shorter than the polypeptide encoded by the normal gene. Nearly all nonsense mutations lead to non-functional proteins.

Insertions and Deletions

Insertions and **deletions** are additions or losses of nucleotide pairs in a gene **(Figure 14.26b)**. These mutations have a disastrous effect on the resulting protein more often than substitutions do. Insertion or deletion of nucleotides

▼ **Figure 14.26 Types of small-scale mutations that affect mRNA sequence.** All but one of the types shown here also affect the amino acid sequence of the encoded polypeptide.

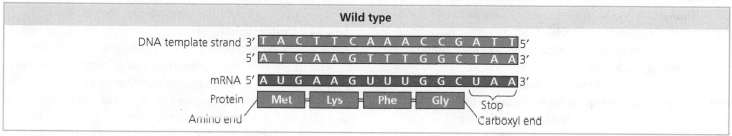

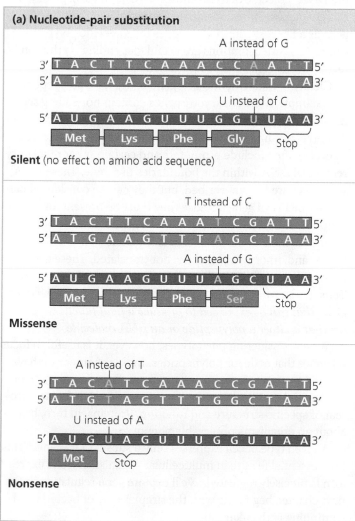

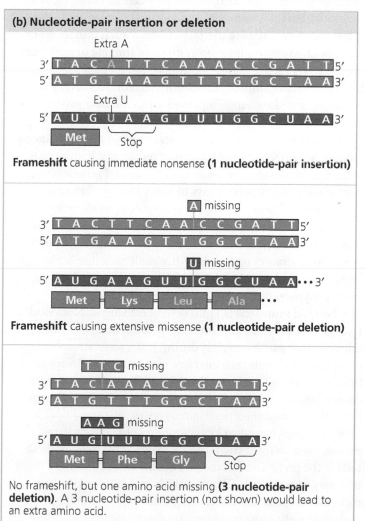

may alter the reading frame of the genetic message, the triplet grouping of nucleotides on the mRNA that is read during translation. Such a mutation, called a **frameshift mutation**, will occur whenever the number of nucleotides inserted or deleted is not a multiple of three. All the nucleotides that are downstream of the deletion or insertion will be improperly grouped into codons, and the result will be extensive missense, usually ending sooner or later in nonsense and premature termination. Unless the frameshift is very near the end of the gene, the protein is almost certain to be nonfunctional.

Mutagens

Mutations can arise in a number of ways. Errors during DNA replication or recombination can lead to nucleotide-pair substitutions, insertions, or deletions, as well as to mutations affecting longer stretches of DNA. If an incorrect nucleotide is added to a growing chain during replication, for example, the base on that nucleotide will then be mismatched with the nucleotide base on the other strand. In many cases, the error will be corrected by proofreading and repair systems (see Chapter 13). Otherwise, the incorrect base will be used as a template in the next round of replication, resulting in a mutation. Such mutations are called *spontaneous mutations*. It is difficult to calculate the rate at which such mutations occur. Rough estimates have been made of the rate of mutation during DNA replication for both *E. coli* and eukaryotes, and the numbers are similar: About one nucleotide in every 10^{10} is altered, and the change is passed on to the next generation of cells.

A number of physical and chemical agents, called **mutagens**, interact with DNA in ways that cause mutations. In the 1920s, Hermann Muller discovered that X-rays caused genetic changes in fruit flies, and he used X-rays to make *Drosophila* mutants for his genetic studies. But he also recognized an alarming implication of his discovery: X-rays and other forms of high-energy radiation pose hazards to the genetic material of people as well as laboratory organisms. Mutagenic radiation, a physical mutagen, includes ultraviolet (UV) light, which can cause disruptive thymine dimers in DNA (see Figure 13.19).

Chemical mutagens fall into several categories. Nucleotide analogs are chemicals that are similar to normal DNA nucleotides but that pair incorrectly during DNA replication. Some other chemical mutagens interfere with correct DNA replication by inserting themselves into the DNA and distorting the double helix. Still other mutagens cause chemical changes in bases that change their pairing properties.

Researchers have developed a variety of methods to test the mutagenic activity of chemicals. A major application of these tests is the preliminary screening of chemicals to identify those that may cause cancer. This approach makes sense because most carcinogens (cancer-causing chemicals) are mutagenic, and conversely, most mutagens are carcinogenic.

What Is a Gene? *Revisiting the Question*

Our definition of a gene has evolved over the past few chapters, as it has through the history of genetics. We began with the Mendelian concept of a gene as a discrete unit of inheritance that affects a phenotypic character (Chapter 11). We saw that Morgan and his colleagues assigned such genes to specific loci on chromosomes (Chapter 12). We went on to view a gene as a region of specific nucleotide sequence along the length of the DNA molecule of a chromosome (Chapter 13). Finally, in this chapter, we have considered a functional definition of a gene as a DNA sequence that codes for a specific polypeptide chain. All these definitions are useful, depending on the context in which genes are being studied.

Clearly, the statement that a gene codes for a polypeptide is too simple. Most eukaryotic genes contain noncoding segments (such as introns), so large portions of these genes have no corresponding segments in polypeptides. Molecular biologists also often include promoters and certain other regulatory regions of DNA within the boundaries of a gene. These DNA sequences are not transcribed, but they can be considered part of the functional gene because they must be present for transcription to occur. Our definition of a gene must also be broad enough to include the DNA that is transcribed into rRNA, tRNA, and other RNAs that are not translated. These genes have no polypeptide products but play crucial roles in the cell. Thus, we arrive at the following definition: *A gene is a region of DNA that can be expressed to produce a final functional product that is either a polypeptide or an RNA molecule.*

When considering phenotypes, however, it is useful to focus on genes that code for polypeptides. In this chapter, you have learned how a typical gene is expressed—by transcription into RNA and then translation into a polypeptide that forms a protein of specific structure and function. Proteins, in turn, bring about an organism's observable phenotype.

A given type of cell expresses only a subset of its genes. This is an essential feature in multicellular organisms: Gene expression is precisely regulated. We'll explore gene regulation in the next chapter, beginning with the simpler case of bacteria and continuing with eukaryotes.

SUMMARY OF KEY CONCEPTS

CONCEPT 14.1

Genes specify proteins via transcription and translation (pp. 269–274)

- DNA controls metabolism by directing cells to make specific enzymes and other proteins, via the process of **gene expression**. Beadle and Tatum's studies of mutant strains of *Neurospora* led to the one gene–one polypeptide hypothesis. Genes code for polypeptide chains or specify RNA molecules.
- **Transcription** is the synthesis of RNA complementary to a **template strand** of DNA, providing a nucleotide-to-nucleotide transfer of information. **Translation** is the synthesis of a polypeptide whose amino acid sequence is specified by the nucleotide sequence in **mRNA**; this informational transfer involves a change of language, from that of nucleotides to that of amino acids.
- Genetic information is encoded as a sequence of nonoverlapping nucleotide triplets, or **codons**. A codon in messenger RNA (mRNA) either is translated into an amino acid (61 of the 64 codons) or serves as a stop signal (3 codons). Codons must be read in the correct **reading frame**.

> **?** *Describe the process of gene expression, by which a gene affects the phenotype of an organism.*

CONCEPT 14.2

Transcription is the DNA-directed synthesis of RNA: *a closer look* (pp. 274–276)

- RNA synthesis is catalyzed by **RNA polymerase**, which links together RNA nucleotides complementary to a DNA template strand. This process follows the same base-pairing rules as DNA replication, except that in RNA, uracil substitutes for thymine.

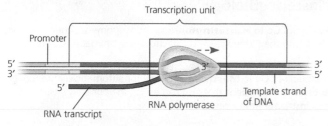

- The three stages of transcription are initiation, elongation, and termination. A **promoter**, often including a **TATA box** in eukaryotes, establishes where RNA synthesis is initiated. **Transcription factors** help eukaryotic RNA polymerase recognize promoter sequences, forming a **transcription initiation complex**. The mechanisms of termination are different in bacteria and eukaryotes.

> **?** *What are the similarities and differences in the initiation of gene transcription in bacteria and eukaryotes?*

CONCEPT 14.3

Eukaryotic cells modify RNA after transcription (pp. 276–278)

- Eukaryotic pre-mRNAs undergo **RNA processing**, which includes RNA splicing, the addition of a

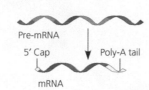

modified nucleotide **5′ cap** to the 5′ end, and the addition of a **poly-A tail** to the 3′ end.

- Most eukaryotic genes are split into segments: They have **introns** interspersed among the **exons** (regions included in the mRNA). In **RNA splicing**, introns are removed and exons joined. RNA splicing is typically carried out by **spliceosomes**, but in some cases, RNA alone catalyzes its own splicing. The catalytic ability of some RNA molecules, called **ribozymes**, derives from the properties of RNA. The presence of introns allows for **alternative RNA splicing**.

> **?** *What function do the 5′ cap and the poly-A tail serve on a eukaryotic mRNA?*

CONCEPT 14.4

Translation is the RNA-directed synthesis of a polypeptide: *a closer look* (pp. 278–288)

- A cell translates an mRNA message into protein using **transfer RNAs (tRNAs)**. After being bound to a specific amino acid by an **aminoacyl-tRNA synthetase**, a tRNA lines up via its **anticodon** at the complementary codon on mRNA. A **ribosome**, made up of **ribosomal RNAs (rRNAs)** and proteins, facilitates this coupling with binding sites for mRNA and tRNA.
- Ribosomes coordinate the three stages of translation: initiation, elongation, and termination. The formation of peptide bonds between amino acids is catalyzed by rRNA as tRNAs move through the **A** and **P sites** and exit through the **E site**.

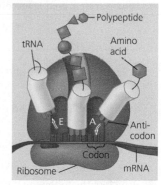

- After translation, modifications to proteins can affect their shape. Free ribosomes in the cytosol initiate synthesis of all proteins, but proteins with a **signal peptide** are synthesized on the ER.
- A gene can be transcribed by multiple RNA polymerases simultaneously. A single mRNA molecule can be translated simultaneously by a number of ribosomes, forming a **polyribosome**. In bacteria, these processes are coupled, but in eukaryotes they are separated in time and space by the nuclear membrane.

> **?** *What function do tRNAs serve in the process of translation?*

CONCEPT 14.5

Mutations of one or a few nucleotides can affect protein structure and function (pp. 288–290)

- Small-scale **mutations** include **point mutations**, changes in one DNA nucleotide pair, which may lead to production of nonfunctional proteins. **Nucleotide-pair substitutions** can cause **missense** or **nonsense mutations**. Nucleotide-pair **insertions** or **deletions** may produce **frameshift mutations**.
- Spontaneous mutations can occur during DNA replication, recombination, or repair. Chemical and physical **mutagens** cause DNA damage that can alter genes.

> **?** *What will be the results of chemically modifying one nucleotide base of a gene? What role is played by DNA repair systems in the cell?*

TEST YOUR UNDERSTANDING

Level 1: Knowledge/Comprehension

1. In eukaryotic cells, transcription cannot begin until
   a. the two DNA strands have completely separated and exposed the promoter.
   b. several transcription factors have bound to the promoter.
   c. the 5′ caps are removed from the mRNA.
   d. the DNA introns are removed from the template.
   e. DNA nucleases have isolated the transcription unit.

2. Which of the following is *not* true of a codon?
   a. It consists of three nucleotides.
   b. It may code for the same amino acid as another codon.
   c. It never codes for more than one amino acid.
   d. It extends from one end of a tRNA molecule.
   e. It is the basic unit of the genetic code.

3. The anticodon of a particular tRNA molecule is
   a. complementary to the corresponding mRNA codon.
   b. complementary to the corresponding triplet in rRNA.
   c. the part of tRNA that bonds to a specific amino acid.
   d. changeable, depending on the amino acid that attaches to the tRNA.
   e. catalytic, making the tRNA a ribozyme.

4. Which of the following is *not* true of RNA processing?
   a. Exons are cut out before mRNA leaves the nucleus.
   b. Nucleotides may be added at both ends of the RNA.
   c. Ribozymes may function in RNA splicing.
   d. RNA splicing can be catalyzed by spliceosomes.
   e. A primary transcript is often much longer than the final RNA molecule that leaves the nucleus.

5. Which component is *not* directly involved in translation?
   a. mRNA
   b. DNA
   c. tRNA
   d. ribosomes
   e. GTP

Level 2: Application/Analysis

6. Using Figure 14.6, identify a 5′ → 3′ sequence of nucleotides in the DNA template strand for an mRNA coding for the polypeptide sequence Phe-Pro-Lys.
   a. 5′-UUUGGGAAA-3′
   b. 5′-GAACCCCTT-3′
   c. 5′-AAAACCTTT-3′
   d. 5′-CTTCGGGAA-3′
   e. 5′-AAACCCUUU-3′

7. Which of the following mutations would be *most* likely to have a harmful effect on an organism?
   a. a nucleotide-pair substitution
   b. a deletion of three nucleotides near the middle of a gene
   c. a single nucleotide deletion in the middle of an intron
   d. a single nucleotide deletion near the end of the coding sequence
   e. a single nucleotide insertion downstream of, and close to, the start of the coding sequence

8. Fill in the following table:

Type of RNA	Functions
Messenger RNA (mRNA)	
Transfer RNA (tRNA)	
	Plays catalytic (ribozyme) roles and structural roles in ribosomes
Primary transcript	
Small RNAs in spliceosome	

Level 3: Synthesis/Evaluation

9. **SCIENTIFIC INQUIRY**
   Knowing that the genetic code is almost universal, a scientist uses molecular biological methods to insert the human β-globin gene (shown in Figure 14.12) into bacterial cells, hoping the cells will express it and synthesize functional β-globin protein. Instead, the protein produced is nonfunctional and is found to contain many fewer amino acids than does β-globin made by a eukaryotic cell. Explain why.

10. **FOCUS ON EVOLUTION**
    Most amino acids are coded for by a set of similar codons (see Figure 14.6). What evolutionary explanations can you give for this pattern? (*Hint*: There is one explanation relating to ancestry, and some less obvious ones of a "form-fits-function" type.)

11. **FOCUS ON INFORMATION**
    Evolution accounts for the unity and diversity of life, and the continuity of life is based on heritable information in the form of DNA. In a short essay (100–150 words), discuss how the fidelity with which DNA is inherited is related to the processes of evolution. (Review the discussion of proofreading and DNA repair in Concept 13.2.)

For selected answers, see Appendix A.

MasteringBiology®

Students Go to **MasteringBiology** for assignments, the eText, and the Study Area with practice tests, animations, and activities.

Instructors Go to **MasteringBiology** for automatically graded tutorials and questions that you can assign to your students, plus Instructor Resources.

15
Regulation of Gene Expression

KEY CONCEPTS

15.1 Bacteria often respond to environmental change by regulating transcription

15.2 Eukaryotic gene expression is regulated at many stages

15.3 Noncoding RNAs play multiple roles in controlling gene expression

15.4 Researchers can monitor expression of specific genes

OVERVIEW

Differential Expression of Genes

The fish shown in **Figure 15.1** is keeping an eye out for predators—or, more precisely, half of each eye! *Anableps anableps* is commonly known as "cuatro ojos" ("four eyes") where it lives in regions of southern Mexico and Central and South America. The fish glides through freshwater lakes and ponds with the upper half of each eye protruding from the water. The eye's upper half is particularly well suited for aerial vision and the lower half for aquatic vision. The molecular basis of this specialization has recently been revealed: The cells of the two parts of the eye express a slightly different set of genes involved in vision, even though these two groups of cells are quite similar and contain identical genomes.

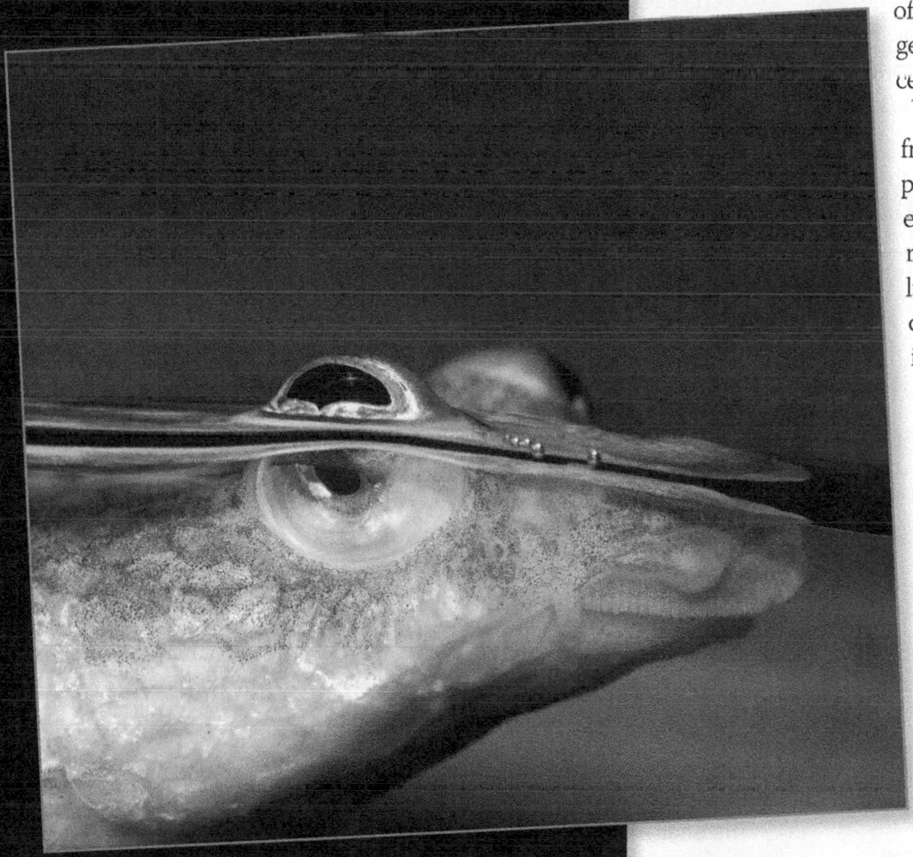

A hallmark of prokaryotic and eukaryotic cells alike—from bacteria to the cells of a fish—is their intricate and precise regulation of gene expression. Both prokaryotes and eukaryotes must alter their patterns of gene expression in response to changes in environmental conditions. Multicellular eukaryotes must also develop and maintain multiple cell types, each expressing a different subset of genes. This is a significant challenge in gene regulation.

In this chapter, we'll first explore how bacteria regulate expression of their genes in response to different environmental conditions. We'll then examine how eukaryotes regulate gene expression to maintain different cell types. Gene expression in eukaryotes, as in bacteria, is often regulated at the stage of transcription, but control at other stages is also important. In recent years, researchers have been surprised to discover the many roles played by RNA molecules in regulating eukaryotic gene expression, a topic we'll touch on next. Finally, we'll describe a few techniques related to those in Chapter 13 that have been developed to investigate gene expression. Elucidating how gene expression is regulated in different cells is crucial to our understanding of living systems.

▼ **Figure 15.1** How can this fish's eyes see equally well in both air and water?

CONCEPT 15.1

Bacteria often respond to environmental change by regulating transcription

Bacterial cells that can conserve resources and energy have a selective advantage over cells that are unable to do so. Thus, natural selection has favored bacteria that express only the genes whose products are needed by the cell.

Consider, for instance, an individual *E. coli* cell living in the erratic environment of a human colon, dependent for its nutrients on the whimsical eating habits of its host. If the environment is lacking in the amino acid tryptophan, which the bacterium needs to survive, the cell responds by activating a metabolic pathway that makes tryptophan from another compound. Later, if the human host eats a tryptophan-rich meal, the bacterial cell stops producing tryptophan, thus avoiding wasting its resources to produce a substance that is available from the surrounding solution in prefabricated form. This is just one example of how bacteria tune their metabolism to changing environments.

Metabolic control occurs on two levels, as shown for the synthesis of tryptophan in **Figure 15.2**. First, cells can adjust the activity of enzymes already present. This is a fairly fast response, which relies on the sensitivity of many enzymes to chemical cues that increase or decrease their catalytic activity (see Chapter 6). The activity of the first enzyme in the tryptophan synthesis pathway is inhibited by the pathway's end product **(Figure 15.2a)**. Thus, if tryptophan accumulates in a cell, it shuts down the synthesis of more tryptophan by inhibiting enzyme activity. Such *feedback inhibition*, typical of anabolic (biosynthetic) pathways, allows a cell to adapt to short-term fluctuations in the supply of a substance it needs.

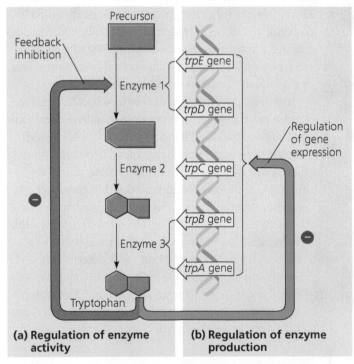

▲ **Figure 15.2 Regulation of a metabolic pathway.** In the pathway for tryptophan synthesis, an abundance of tryptophan can both **(a)** inhibit the activity of the first enzyme in the pathway (feedback inhibition), a rapid response, and **(b)** repress expression of the genes encoding all subunits of the enzymes in the pathway, a longer-term response. Genes *trpE* and *trpD* encode the two subunits of enzyme 1, and genes *trpB* and *trpA* encode the two subunits of enzyme 3. (The genes were named before the order in which they functioned in the pathway was determined.) The ⊖ symbol stands for inhibition.

Second, cells can adjust the production level of certain enzymes; that is, they can regulate the expression of the genes encoding the enzymes. If, in our example, the environment provides all the tryptophan the cell needs, the cell stops making the enzymes that catalyze the synthesis of tryptophan **(Figure 15.2b)**. In this case, the control of enzyme production occurs at the level of transcription, the synthesis of messenger RNA coding for these enzymes. More generally, many genes of the bacterial genome are switched on or off by changes in the metabolic status of the cell. One basic mechanism for this control of gene expression in bacteria, described as the *operon model*, was discovered in 1961 by François Jacob and Jacques Monod at the Pasteur Institute in Paris. Let's see what an operon is and how it works, using the control of tryptophan synthesis as our first example.

Operons: The Basic Concept

E. coli synthesizes the amino acid tryptophan from a precursor molecule in the multistep pathway shown in Figure 15.2. Each reaction in the pathway is catalyzed by a specific enzyme, and the five genes that code for the subunits of these enzymes are clustered together on the bacterial chromosome. A single promoter serves all five genes, which together constitute a transcription unit. (Recall from Chapter 14 that a promoter is a site where RNA polymerase can bind to DNA and begin transcription.) Thus, transcription gives rise to one long mRNA molecule that codes for the five polypeptides making up the enzymes in the tryptophan pathway. The cell can translate this one mRNA into five separate polypeptides because the mRNA is punctuated with start and stop codons that signal where the coding sequence for each polypeptide begins and ends.

A key advantage of grouping genes of related function into one transcription unit is that a single "on-off switch" can control the whole cluster of functionally related genes; in other words, these genes are *coordinately controlled*. When an *E. coli* cell must make tryptophan for itself because the nutrient medium lacks this amino acid, all the enzymes for the metabolic pathway are synthesized at one time. The switch is a segment of DNA called an **operator**. Both its location and name suit its function: Positioned within the promoter or, in some cases, between the promoter and the enzyme-coding genes, the operator controls the access of RNA polymerase to the genes. All together, the operator, the promoter, and the genes they control—the entire stretch of DNA required for enzyme production for the tryptophan pathway—constitute an **operon**. The *trp* operon (*trp* for tryptophan) is one of many operons in the *E. coli* genome **(Figure 15.3)**.

If the operator is the operon's switch for controlling transcription, how does this switch work? By itself, the *trp* operon is turned on; that is, RNA polymerase can bind to the promoter and transcribe the genes of the operon. The operon

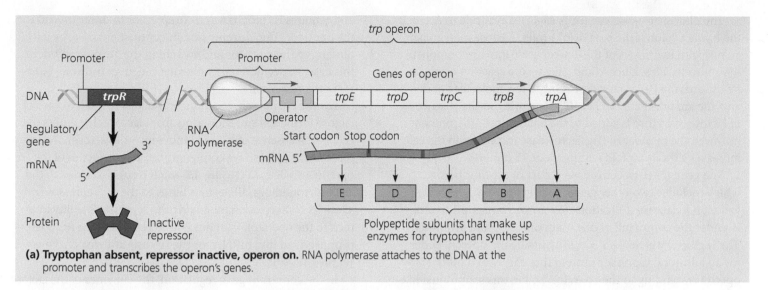

(a) **Tryptophan absent, repressor inactive, operon on.** RNA polymerase attaches to the DNA at the promoter and transcribes the operon's genes.

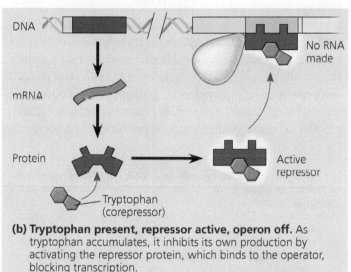

(b) **Tryptophan present, repressor active, operon off.** As tryptophan accumulates, it inhibits its own production by activating the repressor protein, which binds to the operator, blocking transcription.

▲ **Figure 15.3 The *trp* operon in *E. coli*: regulated synthesis of repressible enzymes.** Tryptophan is an amino acid produced by an anabolic pathway catalyzed by repressible enzymes. **(a)** The five genes encoding the polypeptide subunits of the enzymes in this pathway (see Figure 15.2) are grouped, along with a promoter, into the *trp* operon. The *trp* operator (the repressor binding site) is located within the *trp* promoter (the RNA polymerase binding site). **(b)** Accumulation of tryptophan, the end product of the pathway, represses transcription of the *trp* operon, thus blocking synthesis of all the enzymes in the pathway and shutting down tryptophan production.

? *Describe what happens to the* trp *operon as the cell uses up its store of tryptophan.*

can be switched off by a protein called the *trp* **repressor**. The repressor binds to the operator and blocks attachment of RNA polymerase to the promoter, preventing transcription of the genes. A repressor protein is specific for the operator of a particular operon. For example, the repressor that switches off the *trp* operon by binding to the *trp* operator has no effect on other operons in the *E. coli* genome.

The *trp* repressor is the protein product of a **regulatory gene** called *trpR*, which is located some distance from the *trp* operon and has its own promoter. Regulatory genes are expressed continuously, although at a low rate, and a few *trp* repressor molecules are always present in *E. coli* cells. Why, then, is the *trp* operon not switched off permanently? First, the binding of repressors to operators is reversible. An operator alternates between two states: one with the repressor bound and one without. The relative duration of the repressor-bound state is higher when more active repressor molecules are present. Second, the *trp* repressor, like most regulatory proteins, is an allosteric protein, with two alternative shapes, active and

inactive (see Figure 6.18). The *trp* repressor is synthesized in an inactive form with little affinity for the *trp* operator. Only if tryptophan binds to the *trp* repressor at an allosteric site does the repressor protein change to the active form that can attach to the operator, turning the operon off.

Tryptophan functions in this system as a **corepressor**, a small molecule that cooperates with a repressor protein to switch an operon off. As tryptophan accumulates, more tryptophan molecules associate with *trp* repressor molecules, which can then bind to the *trp* operator and shut down production of the tryptophan pathway enzymes. If the cell's tryptophan level drops, transcription of the operon's genes resumes. The *trp* operon is one example of how gene expression can respond to changes in the cell's internal and external environment.

Repressible and Inducible Operons: Two Types of Negative Gene Regulation

The *trp* operon is said to be a *repressible operon* because its transcription is usually on but can be inhibited (repressed) when a specific small molecule (in this case, tryptophan) binds allosterically to a regulatory protein. In contrast, an *inducible operon* is usually off but can be stimulated (induced) when a specific small molecule interacts with a regulatory protein. The classic example of an inducible operon is the *lac* operon (*lac* for lactose), which was the subject of Jacob and Monod's pioneering research.

The disaccharide lactose (milk sugar) is available to *E. coli* in the human colon if the host drinks milk. Lactose metabolism begins with hydrolysis of the disaccharide into its component monosaccharides, glucose and galactose, a reaction catalyzed by the enzyme β-galactosidase. Only a few molecules of this enzyme are present in an *E. coli* cell growing in the absence of lactose. If lactose is added to the bacterium's environment, however, the number of β-galactosidase molecules in the cell increases a thousandfold within about 15 minutes.

The gene for β-galactosidase is part of the *lac* operon, which includes two other genes coding for enzymes that function in lactose utilization. The entire transcription unit is under the command of one main operator and promoter. The regulatory gene, *lacI*, located outside the operon, codes for an allosteric repressor protein that can switch off the *lac* operon by binding to the operator. So far, this sounds just like regulation of the *trp* operon, but there is one important difference. Recall that the *trp* repressor protein is inactive by itself and requires tryptophan as a corepressor in order to bind to the operator. The *lac* repressor, in contrast, is active by itself, binding to the operator and switching the *lac* operon off. In this case, a specific small molecule, called an **inducer**, *inactivates* the repressor.

For the *lac* operon, the inducer is allolactose, an isomer of lactose formed in small amounts from lactose that enters the cell. In the absence of lactose (and hence allolactose), the *lac* repressor is in its active configuration, and the genes of the *lac* operon are silenced **(Figure 15.4a)**. If lactose is added to the cell's surroundings, allolactose binds to the *lac* repressor and alters its conformation, nullifying the repressor's ability to attach to the operator. Without bound repressor, the *lac* operon is transcribed into mRNA for the lactose-utilizing enzymes **(Figure 15.4b)**.

In the context of gene regulation, the enzymes of the lactose pathway are referred to as *inducible enzymes* because their synthesis is induced by a chemical signal (allolactose, in this case). Analogously, the enzymes for tryptophan synthesis are said to be repressible. *Repressible enzymes* generally function in anabolic pathways, which synthesize essential end products from raw materials (precursors). By suspending production of an end product when it is already present in sufficient quantity, the cell can allocate its organic precursors and energy for

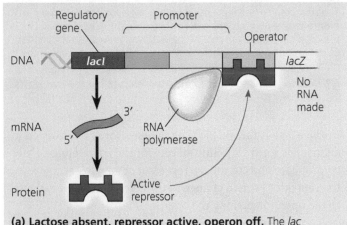

(a) Lactose absent, repressor active, operon off. The *lac* repressor is innately active, and in the absence of lactose it switches off the operon by binding to the operator.

▼ **Figure 15.4 The *lac* operon in *E. coli*: regulated synthesis of inducible enzymes.** *E. coli* uses three enzymes to take up and metabolize lactose. The genes for these three enzymes are clustered in the *lac* operon. One gene, *lacZ*, codes for β-galactosidase, which hydrolyzes lactose to glucose and galactose. The second gene, *lacY*, codes for a permease, the membrane protein that transports lactose into the cell. The third gene, *lacA*, codes for an enzyme called transacetylase, whose function in lactose metabolism is still unclear. The gene for the *lac* repressor, *lacI*, happens to be adjacent to the *lac* operon, an unusual situation. The function of the teal region at the upstream end of the promoter (the left end in these diagrams) will be revealed in Figure 15.5.

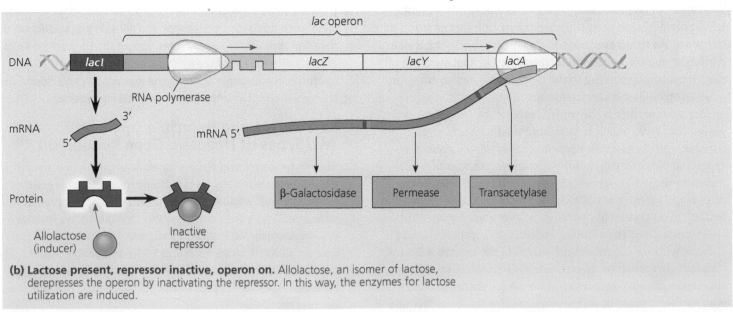

(b) Lactose present, repressor inactive, operon on. Allolactose, an isomer of lactose, derepresses the operon by inactivating the repressor. In this way, the enzymes for lactose utilization are induced.

other uses. In contrast, inducible enzymes usually function in catabolic pathways, which break down a nutrient to simpler molecules. By producing the appropriate enzymes only when the nutrient is available, the cell avoids wasting energy and precursors making proteins that are not needed.

Regulation of both the *trp* and *lac* operons involves the *negative* control of genes, because the operons are switched off by the active form of the repressor protein. It may be easier to see this for the *trp* operon, but it is also true for the *lac* operon. Allolactose induces enzyme synthesis not by acting directly on the genome, but by freeing the *lac* operon from the negative effect of the repressor. Gene regulation is said to be *positive* only when a regulatory protein interacts directly with the genome to switch transcription on. Let's look at an example of the positive control of genes, again involving the *lac* operon.

Positive Gene Regulation

When glucose and lactose are both present in its environment, *E. coli* preferentially uses glucose. The enzymes for glucose breakdown in glycolysis (see Figure 7.9) are continually present. Only when lactose is present *and* glucose is in short supply does *E. coli* use lactose as an energy source, and only then does it synthesize appreciable quantities of the enzymes for lactose breakdown.

How does the *E. coli* cell sense the glucose concentration and relay this information to the genome? Again, the mechanism depends on the interaction of an allosteric regulatory protein with a small organic molecule, in this case **cyclic AMP (cAMP)**, which accumulates when glucose is scarce. The regulatory protein, called *catabolite activator protein (CAP)*, is an **activator**, a protein that binds to DNA and stimulates transcription of a gene. When cAMP binds to this regulatory protein, CAP assumes its active shape and can attach to a specific site at the upstream end of the *lac* promoter **(Figure 15.5a)**. This attachment increases the affinity of RNA polymerase for the promoter, which is actually rather low even when no repressor is bound to the operator. By facilitating the binding of RNA polymerase to the promoter and thereby increasing the rate of transcription, the attachment of CAP to the promoter directly stimulates gene expression. Therefore, this mechanism qualifies as positive regulation.

If the amount of glucose in the cell increases, the cAMP concentration falls, and without cAMP, CAP detaches from the operon. Because CAP is inactive, RNA polymerase binds less efficiently to the promoter, and transcription of the *lac* operon proceeds at only a low level, even in the presence of lactose **(Figure 15.5b)**. Thus, the *lac* operon is under dual control: negative control by the *lac* repressor and positive control by CAP. The state of the *lac* repressor (with or without bound allolactose) determines whether or not transcription of the *lac* operon's genes occurs at all; the state of CAP (with or without bound cAMP) controls the *rate* of transcription if the operon

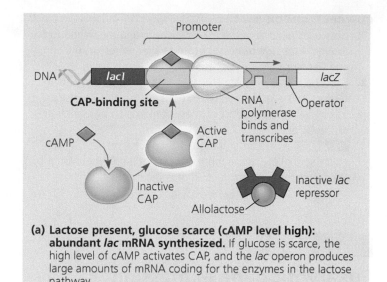

(a) Lactose present, glucose scarce (cAMP level high): abundant *lac* mRNA synthesized. If glucose is scarce, the high level of cAMP activates CAP, and the *lac* operon produces large amounts of mRNA coding for the enzymes in the lactose pathway.

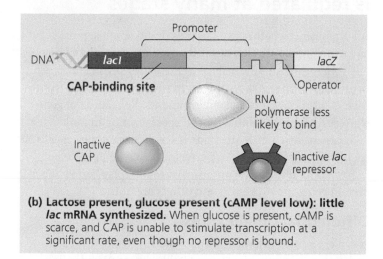

(b) Lactose present, glucose present (cAMP level low): little *lac* mRNA synthesized. When glucose is present, cAMP is scarce, and CAP is unable to stimulate transcription at a significant rate, even though no repressor is bound.

▲ **Figure 15.5 Positive control of the lac operon by catabolite activator protein (CAP).** RNA polymerase has high affinity for the *lac* promoter only when catabolite activator protein (CAP) is bound to a DNA site at the upstream end of the promoter. CAP attaches to its DNA site only when associated with cyclic AMP (cAMP), whose concentration in the cell rises when the glucose concentration falls. Thus, when glucose is present, even if lactose also is available, the cell preferentially catabolizes glucose and makes very little of the lactose-utilizing enzymes.

is repressor-free. It is as though the operon has both an on-off switch and a volume control.

In addition to regulating the *lac* operon, CAP helps regulate other operons that encode enzymes used in catabolic pathways. All told, it may affect the expression of more than 100 genes in *E. coli*. When glucose is plentiful and CAP is inactive, the synthesis of enzymes that catabolize compounds other than glucose generally slows down. The ability to catabolize other compounds, such as lactose, enables a cell deprived of glucose to survive. The compounds present in the cell at the moment determine which operons are switched on—the result of simple interactions of activator and repressor proteins with the promoters of the genes in question.

1. How does binding of the *trp* corepressor and the *lac* inducer to their respective repressor proteins alter repressor function and transcription in each case?

2. Describe the binding of RNA polymerase, repressors, and activators to the *lac* operon when both lactose and glucose are scarce. What is the effect of these scarcities on transcription of the *lac* operon?

3. **WHAT IF?** A certain mutation in *E. coli* changes the *lac* operator so that the active repressor cannot bind. How would this affect the cell's production of β-galactosidase?

For suggested answers, see Appendix A.

CONCEPT 15.2

Eukaryotic gene expression is regulated at many stages

All organisms, whether prokaryotes or eukaryotes, must regulate which genes are expressed at any given time. Both unicellular organisms and the cells of multicellular organisms must continually turn genes on and off in response to signals from their external and internal environments. Regulation of gene expression is also essential for cell specialization in multicellular organisms, which are made up of different types of cells, each with a distinct role. To perform its role, each cell type must maintain a specific program of gene expression in which certain genes are expressed and others are not.

Differential Gene Expression

A typical human cell might express about 20% of its protein-coding genes at any given time. Highly differentiated cells, such as muscle or nerve cells, express an even smaller fraction of their genes. Almost all the cells in an organism contain an identical genome. (Cells of the immune system are one exception, as you will see in Chapter 35.) However, the subset of genes expressed in the cells of each type is unique, allowing these cells to carry out their specific function. The differences between cell types, therefore, are due not to different genes being present, but to **differential gene expression**, the expression of different genes by cells with the same genome.

The function of any cell, whether a single-celled eukaryote or a particular cell type in a multicellular organism, depends on the appropriate set of genes being expressed. The transcription factors of a cell must locate the right genes at the right time, a task on a par with finding a needle in a haystack. When gene expression proceeds abnormally, serious imbalances and diseases, including cancer, can arise.

Figure 15.6 summarizes the process of gene expression in a eukaryotic cell, highlighting key stages in the expression of a protein-coding gene. Each stage depicted in Figure 15.6 is a potential control point at which gene expression can be turned on or off, accelerated, or slowed down.

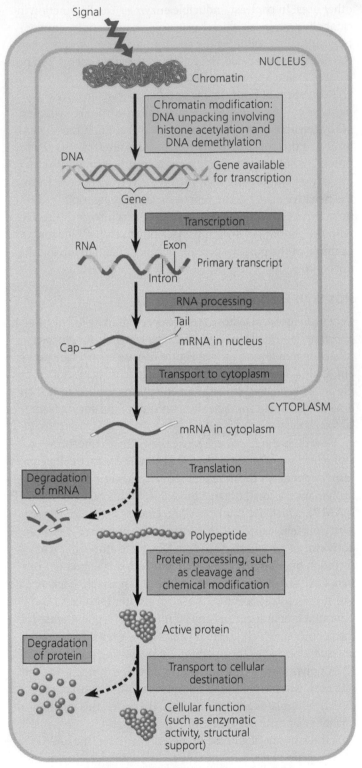

▲ **Figure 15.6 Stages in gene expression that can be regulated in eukaryotic cells.** In this diagram, the colored boxes indicate the processes most often regulated; each color indicates the type of molecule that is affected (blue = DNA, orange = RNA, purple = protein). The nuclear envelope separating transcription from translation in eukaryotic cells offers an opportunity for post-transcriptional control in the form of RNA processing that is absent in prokaryotes. In addition, eukaryotes have a greater variety of control mechanisms operating before transcription and after translation. The expression of any given gene, however, does not necessarily involve every stage shown; for example, not every polypeptide is cleaved.

Fifty years ago, an understanding of the mechanisms that control gene expression in eukaryotes seemed almost hopelessly out of reach. Since then, new research methods, notably advances in DNA technology (see Concept 13.4), have enabled molecular biologists to uncover many of the details of eukaryotic gene regulation. In all organisms, a common control point for gene expression is at transcription; regulation at this stage often occurs in response to signals coming from outside the cell, such as hormones or other signaling molecules. For this reason, the term *gene expression* is often equated with transcription for both bacteria and eukaryotes. While this is most often the case for bacteria, the greater complexity of eukaryotic cell structure and function provides opportunities for regulating gene expression at many additional stages (see Figure 15.6). In the remainder of this section, we'll examine some of the important control points of eukaryotic gene expression more closely.

Regulation of Chromatin Structure

Recall that the DNA of eukaryotic cells is packaged with proteins in an elaborate complex known as chromatin, the basic unit of which is the nucleosome (see Figure 13.21). The structural organization of chromatin not only packs a cell's DNA into a compact form that fits inside the nucleus, but also helps regulate gene expression in several ways. The location of a gene's promoter relative to nucleosomes and to the sites where the DNA attaches to the chromosome scaffold or nuclear lamina can affect whether the gene is transcribed. In addition, genes within heterochromatin, which is highly condensed, are usually not expressed. Lastly, certain chemical modifications to the histone proteins and to the DNA of chromatin can influence both chromatin structure and gene expression. Here we examine the effects of these modifications, which are catalyzed by specific enzymes.

Histone Modifications and DNA Methylation

There is abundant evidence that chemical modifications to histones, the proteins around which the DNA is wrapped in nucleosomes, play a direct role in the regulation of gene transcription. The N-terminus of each histone molecule in a nucleosome protrudes outward from the nucleosome. These histone tails are accessible to various modifying enzymes that catalyze the addition or removal of specific chemical groups, such as acetyl ($-COCH_3$), methyl, and phosphate groups. Generally, **histone acetylation** appears to promote transcription by opening up the chromatin structure **(Figure 15.7)**, while addition of methyl groups can lead to condensation of chromatin and reduced transcription.

While some enzymes methylate the tails of histone proteins, a different set of enzymes can methylate certain bases in the DNA itself, usually cytosine. Such **DNA methylation** occurs in most plants, animals, and fungi. Long stretches of inactive DNA, such as that of inactivated mammalian X chromosomes (see Figure 12.8), are generally more methylated than regions of actively transcribed DNA, although there are exceptions.

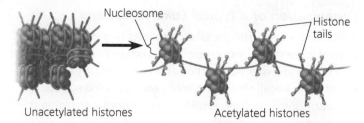

▲ **Figure 15.7 A simple model of the effect of histone acetylation.** The amino acids in the N-terminal tails of histones are accessible for chemical modification such as addition of acetyl groups (green balls). A region of chromatin in which nucleosomes are unacetylated forms a compact structure (left) in which the DNA is not transcribed. Highly acetylated nucleosomes (right) cause the chromatin to be less compact and the DNA accessible for transcription.

On a smaller scale, individual genes are usually more heavily methylated in cells in which they are not expressed. Removal of the extra methyl groups can turn on some of these genes. Once methylated, genes usually stay that way through successive cell divisions in a given individual. At DNA sites where one strand is already methylated, enzymes methylate the correct daughter strand after each round of DNA replication. In this way, methylation patterns can be inherited.

Epigenetic Inheritance

The chromatin modifications discussed above do not entail a change in the DNA sequence, yet they may be passed along to future generations of cells. Inheritance of traits transmitted by mechanisms not directly involving the nucleotide sequence is called **epigenetic inheritance**. Whereas mutations in DNA are permanent, modifications to the chromatin can be reversed.

Researchers are amassing more and more evidence for the importance of epigenetic information in the regulation of gene expression. Epigenetic variations might help explain why one identical twin acquires a genetically based disease, such as schizophrenia, but the other does not, despite their identical genomes. Alterations in normal patterns of DNA methylation are also seen in some cancers, where the alterations are associated with inappropriate gene expression. Evidently, enzymes that modify chromatin structure are integral parts of the eukaryotic cell's machinery for regulating transcription.

Regulation of Transcription Initiation

Chromatin-modifying enzymes provide initial control of gene expression by making a region of DNA either more or less able to bind the transcription machinery. Once the chromatin of a gene is optimally modified for expression, the initiation of transcription is the next major step at which gene expression is regulated. As in bacteria, the regulation of transcription initiation in eukaryotes involves proteins that bind to DNA and either facilitate or inhibit binding of RNA polymerase. The process is more complicated in eukaryotes, however. Before looking at how eukaryotic cells control their transcription, let's review the structure of a typical eukaryotic gene and its transcript.

Organization of a Typical Eukaryotic Gene

A eukaryotic gene and the DNA elements (segments) that control it are typically organized as shown in **Figure 15.8**, which extends what you learned about eukaryotic genes in Chapter 14. Recall that a cluster of proteins called a *transcription initiation complex* assembles on the promoter sequence at the "upstream" end of the gene. One of these proteins, RNA polymerase II, then proceeds to transcribe the gene, synthesizing a primary RNA transcript (more specifically, pre-mRNA). RNA processing includes enzymatic addition of a 5′ cap and a poly-A tail, as well as splicing out of introns, to yield a mature mRNA. Associated with most eukaryotic genes are multiple **control elements**, segments of noncoding DNA having particular nucleotide sequences that serve as binding sites for the proteins called transcription factors, which in turn regulate transcription. Control elements on the DNA and the transcription factors they bind are critical to the precise regulation of gene expression seen in different cell types.

The Roles of Transcription Factors

To initiate transcription, eukaryotic RNA polymerase requires the assistance of transcription factors. Some transcription factors, such as those illustrated in Figure 14.9, are essential for the transcription of *all* protein-coding genes; therefore, they are often called *general transcription factors*. Only a few general transcription factors independently bind a DNA sequence, such as

the TATA box within the promoter; the others primarily bind proteins, including each other and RNA polymerase II. Protein-protein interactions are crucial to the initiation of eukaryotic transcription. Only when the complete initiation complex has assembled can the polymerase begin to move along the DNA template strand, producing a complementary strand of RNA.

The interaction of general transcription factors and RNA polymerase II with a promoter usually leads to only a low rate of initiation and production of few RNA transcripts. In eukaryotes, high levels of transcription of particular genes at the appropriate time and place depend on the interaction of control elements with another set of proteins, which can be thought of as *specific transcription factors*.

Enhancers and Specific Transcription Factors As you can see in Figure 15.8, some control elements, named *proximal control elements*, are located close to the promoter. (Although some biologists consider proximal control elements part of the promoter, in this book we do not.) The more distant *distal control elements*, groupings of which are called **enhancers**, may be thousands of nucleotides upstream or downstream of a gene or even within an intron. A given gene may have multiple enhancers, each active at a different time or in a different cell type or location in the organism. Each enhancer, however, is generally associated with only that gene and no other.

In eukaryotes, the rate of gene expression can be strongly increased or decreased by the binding of specific transcription

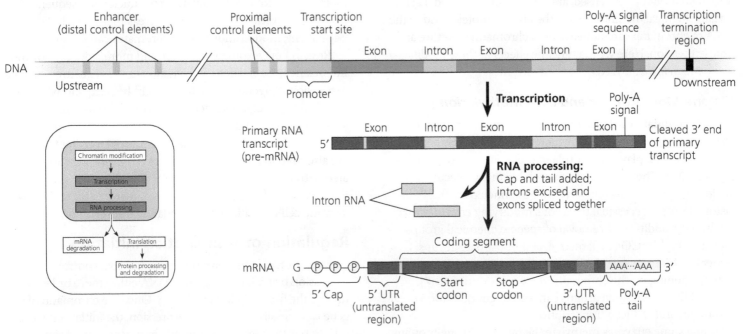

▲ **Figure 15.8 A eukaryotic gene and its transcript.** Each eukaryotic gene has a promoter, a DNA sequence where RNA polymerase binds and starts transcription, proceeding "downstream." A number of control elements (gold) are involved in regulating the initiation of transcription; these are DNA sequences located near (proximal to) or far from (distal to) the promoter. Distal control elements can be grouped together as enhancers, one of which is shown for this gene. A polyadenylation (poly-A) signal sequence in the last exon of the gene is transcribed into an RNA sequence that signals where the transcript is cleaved and the poly-A tail added. Transcription may continue for hundreds of nucleotides beyond the poly-A signal before terminating. RNA processing of the primary transcript into a functional mRNA involves three steps: addition of the 5′ cap, addition of the poly-A tail, and splicing. In the cell, the 5′ cap is added soon after transcription is initiated; splicing and poly-A tail addition may also occur while transcription is under way (see Figure 14.11).

factors, either activators or repressors, to the control elements of enhancers. Hundreds of transcription activators have been discovered in eukaryotes; the structure of one example is shown in **Figure 15.9**. In a large number of activator proteins, researchers have identified two common structural elements: a DNA-binding domain—a part of the protein's three-dimensional structure that binds to DNA—and one or more activation domains. Activation domains bind other regulatory proteins or components of the transcription machinery, facilitating a series of protein-protein interactions that result in transcription of a given gene.

Figure 15.10 shows a current model for how binding of activators to an enhancer located far from the promoter can influence transcription. Protein-mediated bending of the DNA is thought to bring the bound activators into contact with a group of *mediator proteins*, which in turn interact with proteins at the promoter. These protein-protein interactions help assemble and position the initiation complex on the promoter.

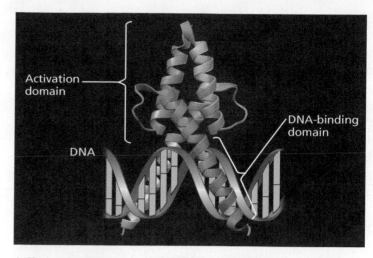

▲ **Figure 15.9 MyoD, a transcription activator.** The MyoD protein is made up of two subunits (purple and salmon) with extensive regions of α helix. Each subunit has one DNA-binding domain and one activation domain. The latter includes binding sites for the other subunit and other proteins. MyoD is involved in muscle development in vertebrate embryos.

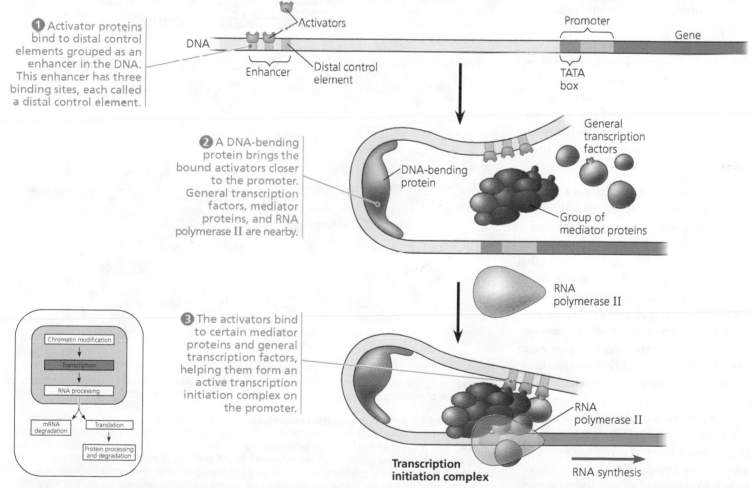

1 Activator proteins bind to distal control elements grouped as an enhancer in the DNA. This enhancer has three binding sites, each called a distal control element.

2 A DNA-bending protein brings the bound activators closer to the promoter. General transcription factors, mediator proteins, and RNA polymerase II are nearby.

3 The activators bind to certain mediator proteins and general transcription factors, helping them form an active transcription initiation complex on the promoter.

▲ **Figure 15.10   A model for the action of enhancers and transcription activators.** Bending of the DNA by a protein enables enhancers to influence a promoter hundreds or even thousands of nucleotides away. Specific transcription factors called activators bind to the enhancer DNA sequences and then to a group of mediator proteins, which in turn bind to general transcription factors, assembling the transcription initiation complex. These protein-protein interactions facilitate the correct positioning of the complex on the promoter and the initiation of RNA synthesis. Only one enhancer (with three gold control elements) is shown here, but a gene may have several enhancers that act at different times or in different cell types.

Support for this model includes a study showing that the proteins regulating a mouse globin gene contact both the gene's promoter and an enhancer located about 50,000 nucleotides upstream. Evidently, these two regions in the DNA must come together in a very specific fashion for this interaction to occur.

Specific transcription factors that function as repressors can inhibit gene expression in several different ways. Some repressors bind directly to control element DNA (in enhancers or elsewhere), blocking activator binding or, in some cases, turning off transcription even when activators are bound. Other repressors block the binding of activators to proteins that allow the activators to bind to DNA. In the **Scientific Skills Exercise**, you can work with data from an experiment that identified the control elements in the enhancer of a particular human gene.

In addition to influencing transcription directly, some activators and repressors act indirectly by affecting chromatin structure. Studies using yeast and mammalian cells show that some activators recruit proteins that acetylate histones near the promoters of specific genes, thus promoting transcription (see Figure 15.7). Similarly, some repressors recruit proteins that remove acetyl groups from histones, leading to reduced transcription, a phenomenon called *silencing*. Indeed, the recruitment of proteins that modify chromatin seems to be the most common mechanism of repression in eukaryotes.

Combinatorial Control of Gene Activation In eukaryotes, the precise control of transcription depends largely on the binding of activators to DNA control elements. Considering the great number of genes that must be regulated in a typical animal or plant cell, the number of completely different nucleotide sequences found in control elements is surprisingly small. A dozen or so short nucleotide sequences appear again and again in the control elements for different genes. On average, each enhancer is composed of about ten control elements, each of which can bind only one or two specific transcription factors. It is the particular *combination* of control elements in an enhancer associated with a gene, rather than the presence of a single unique control element, that is important in regulating transcription of the gene.

Even with only a dozen control element sequences available, a very large number of combinations are possible. A particular

combination of control elements will be able to activate transcription only when the appropriate activator proteins are present, which may occur at a precise time during development or in a particular cell type. **Figure 15.11** illustrates how the use of different combinations of just a few control elements can allow differential regulation of transcription in two cell types. This can occur because each cell type contains a different group of activator proteins. How these groups came to differ during embryonic development will be explored in Chapter 16.

Coordinately Controlled Genes in Eukaryotes

How does the eukaryotic cell deal with a group of genes of related function that need to be turned on or off at the same time? Earlier in this chapter, you learned that in bacteria, such

DNA in both cells:

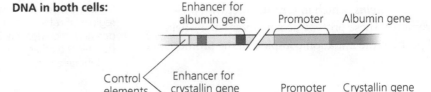

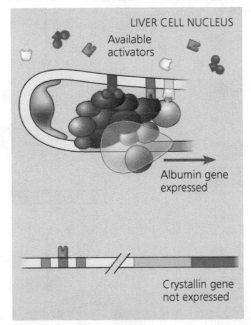

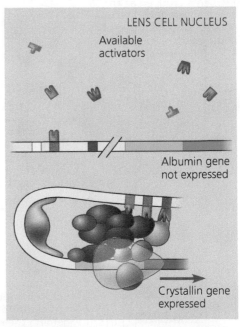

(a) Liver cell. The albumin gene is expressed, and the crystallin gene is not.

(b) Lens cell. The crystallin gene is expressed, and the albumin gene is not.

▲ **Figure 15.11 Cell type–specific transcription.** Both liver cells and lens cells have the genes for making the proteins albumin and crystallin, but only liver cells make albumin (a blood protein) and only lens cells make crystallin (the main protein of the lens of the eye). The specific transcription factors made in a cell determine which genes are expressed. In this example, the genes for albumin and crystallin are shown at the top, each with an enhancer made up of three different control elements. Although the enhancers for the two genes share one control element (gray), each enhancer has a unique combination of elements. All the activators required for high-level expression of the albumin gene are present only in liver cells **(a)**, whereas the activators needed for expression of the crystallin gene are present only in lens cells **(b)**. For simplicity, we consider only the role of activators here, although the presence or absence of repressors may also influence transcription in certain cell types.

? *Describe the enhancer for the albumin gene in each cell. How would the nucleotide sequence of this enhancer in the liver cell compare with that in the lens cell?*

Analyzing DNA Deletion Experiments

What Control Elements Regulate Expression of the *mPGES-1* Gene?

The promoter of a gene includes the DNA immediately upstream of the transcription start site, but expression of the gene can also be affected by control elements. These can be thousands of base pairs upstream of the promoter, grouped in an enhancer. Since the distance and spacing of these control elements make them difficult to identify, scientists begin by deleting possible control elements and measuring the effect on gene expression. In this exercise, you will analyze data obtained from DNA deletion experiments that tested possible control elements for the human gene *mPGES-1*. This gene codes for an enzyme that synthesizes a type of prostaglandin, a chemical made during inflammation.

How the Experiment Was Done

The researchers hypothesized that there were three possible control elements in an enhancer region 8–9 kilobases upstream of the *mPGES-1* gene. Control elements regulate whatever gene is in the appropriate downstream location. Thus, to test the activity of the possible elements, researchers first synthesized molecules of DNA ("constructs") with the intact enhancer region upstream of a "reporter gene," a gene whose mRNA product could be easily measured experimentally. Next, they synthesized three more DNA constructs but deleted one of the three proposed control elements in each (see left side of figure). The researchers then introduced each DNA construct into a separate human cell culture, where the cells took up the artificial DNA molecules. After 48 hours the amount of reporter gene mRNA made by the cells was measured. Comparing these amounts allowed researchers to determine if any of the deletions had an effect on expression of the reporter gene, mimicking the effect that deletions would have had on *mPGES-1* gene expression. (The *mPGES-1* gene itself couldn't be used to measure expression levels because the cells express their own *mPGES-1* gene, so expression of the reporter gene is used to mimic expression of the *mPGES-1* gene.)

Data from the Experiment

The diagrams on the left side of the figure show the intact DNA sequence (top) and the three experimental DNA sequences. A red X indicates the possible control element (1, 2, or 3) that was deleted in each experimental DNA sequence. The area between the slashes represents the approximately 8 kilobases of DNA located between the promoter and the enhancer region. The horizontal bar graph on the right shows the amount of reporter gene mRNA that was present in each cell culture after 48 hours relative to the amount that was in the culture containing the intact enhancer region (top bar = 100%).

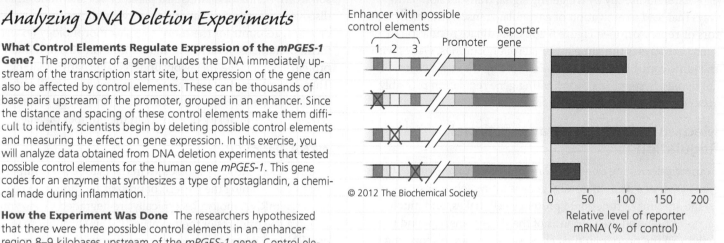

Enhancer with possible control elements — Promoter — Reporter gene
1 2 3

© 2012 The Biochemical Society

Relative level of reporter mRNA (% of control)

Interpret the Data

1. (a) What is the independent variable in the graph (that is, what variable was manipulated by the scientists)? (b) What is the dependent variable (that is, what variable responded to the changes in the independent variable)? (c) What was the control treatment in this experiment? Label it on the diagram.

2. Do the data suggest that any of these possible control elements are actual control elements? Explain.

3. (a) Did deletion of any of the possible control elements cause a *reduction* in reporter gene expression? If so, which one(s), and how can you tell? (b) If loss of a control element causes a reduction in gene expression, what must be the normal role of that control element? Provide a biological explanation for how the loss of such a control element could lead to a reduction in gene expression.

4. (a) Did deletion of any of the possible control elements cause an *increase* in reporter gene expression relative to the control? If so, which one(s), and how can you tell? (b) If loss of a control element causes an increase in gene expression, what must be the normal role of that control element? Propose a biological explanation for how the loss of such a control element could lead to an increase in gene expression.

Data from J. N. Walters et al., Regulation of human microsomal prostaglandin E synthase-1 by IL-1β requires a distal enhancer element with a unique role for C/EBPβ, *Biochemical Journal* (2012). doi:10.1042/BJ20111801

MB A version of this Scientific Skills Exercise can be assigned in MasteringBiology.

---

coordinately controlled genes are often clustered into an operon, which is regulated by a single promoter and transcribed into a single mRNA molecule. Thus, the genes are expressed together, and the encoded proteins are produced concurrently. With a few minor exceptions, operons that work in this way have *not* been found in eukaryotic cells.

Co-expressed eukaryotic genes, such as genes coding for the enzymes of a metabolic pathway, are typically scattered over different chromosomes. In these cases, coordinate gene expression depends on the association of a specific combination of control elements with every gene of a dispersed group. The presence of these elements can be compared to the raised flags on a few mailboxes out of many, signaling to the mail carrier to check those boxes. Copies of the activators that recognize the control

elements bind to them, promoting simultaneous transcription of the genes, no matter where they are in the genome.

Coordinate control of dispersed genes in a eukaryotic cell often occurs in response to chemical signals from outside the cell. A steroid hormone, for example, enters a cell and binds to a specific intracellular receptor protein, forming a hormone-receptor complex that serves as a transcription activator (see Figure 5.23). Every gene whose transcription is stimulated by a particular steroid hormone, regardless of its chromosomal location, has a control element recognized by that hormone-receptor complex. This is how estrogen activates a group of genes that stimulate cell division in uterine cells, preparing the uterus for pregnancy.

Many signaling molecules, such as nonsteroid hormones and growth factors, bind to receptors on a cell's surface and never actually enter the cell. Such molecules can control gene expression indirectly by triggering signal transduction pathways that lead to activation of particular transcription activators or repressors (see Figure 5.26). Coordinate regulation in such pathways is the same as for steroid hormones: Genes with the same control elements are activated by the same chemical signals. Systems for coordinating gene regulation probably arose early in evolutionary history.

Mechanisms of Post-Transcriptional Regulation

Transcription alone does not constitute gene expression. The expression of a protein-coding gene is ultimately measured by the amount of functional protein a cell makes, and much happens between the synthesis of the RNA transcript and the activity of the protein in the cell. Researchers are discovering more and more regulatory mechanisms that operate at various stages after transcription (see Figure 15.6). These mechanisms allow a cell to fine-tune gene expression rapidly in response to environmental changes without altering its transcription patterns. Here we discuss how cells can regulate gene expression once a gene has been transcribed.

RNA Processing

RNA processing in the nucleus and the export of mature RNA to the cytoplasm provide opportunities for regulating gene expression that are not available in prokaryotes. One example of regulation at the RNA-processing level is **alternative RNA splicing**, in which different mRNA molecules are produced from the same primary transcript, depending on which RNA segments are treated as exons and which as introns. Regulatory proteins specific to a cell type control intron-exon choices by binding to RNA sequences within the primary transcript.

A simple example of alternative RNA splicing is shown in **Figure 15.12** for the troponin T gene, which encodes two different (though related) proteins. Other genes offer possibilities for far greater numbers of products. For instance, researchers have found a gene in *Drosophila* with enough alternatively spliced exons to generate about 19,000 membrane proteins with different extracellular domains. At least 17,500 (94%) of the alternative mRNAs are actually synthesized. Each developing nerve cell in the fly appears to synthesize a unique form of the protein, which acts as an identification badge on the cell surface.

It is clear that alternative RNA splicing can significantly expand the repertoire of a eukaryotic genome. In fact, alternative splicing was proposed as one explanation for the surprisingly low number of human genes counted when the human genome was sequenced about ten years ago. The

number of human genes was found to be similar to that of a soil worm (nematode), mustard plant, or sea anemone. This discovery prompted questions about what, if not the number of genes, accounts for the more complex morphology (external form) of humans. It turns out that more than 90% of human protein-coding genes probably undergo alternative splicing. Thus, the extent of alternative splicing greatly multiplies the number of possible human proteins, which may be better correlated with complexity of form than the number of genes.

mRNA Degradation

The life span of mRNA molecules in the cytoplasm is important in determining the pattern of protein synthesis in a cell. Bacterial mRNA molecules typically are degraded by enzymes within a few minutes of their synthesis. This short life span of mRNAs is one reason bacteria can change their patterns of protein synthesis so quickly in response to environmental changes. In contrast, mRNAs in multicellular eukaryotes typically survive for hours, days, or even weeks. For instance, the mRNAs for the hemoglobin polypeptides (α-globin and β-globin) in developing red blood cells are unusually stable, and these long-lived mRNAs are translated repeatedly in these cells. Nucleotide sequences that affect how long an mRNA remains intact are often found in the untranslated region (UTR) at the 3' end of the molecule (see Figure 15.8).

During the past few years, other mechanisms that degrade or block expression of mRNA molecules have come to light. These mechanisms involve an important group of newly discovered RNA molecules that regulate gene expression at several levels, and we'll discuss them later in this chapter.

Initiation of Translation

Translation presents another opportunity for regulating gene expression; such regulation occurs most commonly at the initiation stage (see Figure 14.18). For some mRNAs, the initiation of translation can be blocked by regulatory proteins that bind

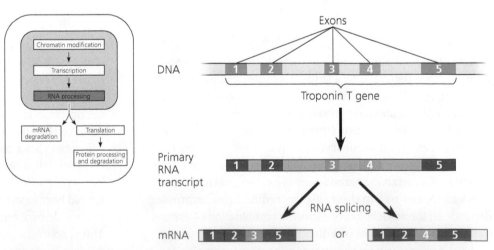

▲ **Figure 15.12 Alternative RNA splicing of the troponin T gene.** The primary transcript of this gene can be spliced in more than one way, generating different mRNA molecules. Notice that one mRNA molecule has ended up with exon 3 (green) and the other with exon 4 (purple). These two mRNAs are translated into different but related muscle proteins.

to specific sequences or structures within the 5′ or 3′ UTR, preventing the attachment of ribosomes. (Recall from Chapter 14 that both the 5′ cap and the poly-A tail of an mRNA molecule are important for ribosome binding.) A different mechanism for blocking translation is seen in a variety of mRNAs present in the eggs of many organisms: Initially, these stored mRNAs lack poly-A tails of sufficient length to allow translation initiation. At the appropriate time during embryonic development, however, a cytoplasmic enzyme adds more adenine (A) nucleotides, prompting translation to begin.

Alternatively, translation of *all* the mRNAs in a cell may be regulated simultaneously. In a eukaryotic cell, such "global" control usually involves the activation or inactivation of one or more of the protein factors required to initiate translation. This mechanism plays a role in starting translation of mRNAs that are stored in eggs. Just after fertilization, translation is triggered by the sudden activation of translation initiation factors. The response is a burst of synthesis of the proteins encoded by the stored mRNAs. Some plants and algae store mRNAs during periods of darkness; light then triggers the reactivation of the translational apparatus.

Protein Processing and Degradation

The final opportunities for controlling gene expression occur after translation. Often, eukaryotic polypeptides must be processed to yield functional protein molecules. For instance, cleavage of the initial insulin polypeptide (pro-insulin) forms the active hormone. In addition, many proteins undergo chemical modifications that make them functional. Regulatory proteins are commonly activated or inactivated by the reversible addition of phosphate groups, and proteins destined for the surface of animal cells acquire sugars. Cell-surface proteins and many others must also be transported to target destinations in the cell in order to function. Regulation might occur at any of the steps involved in modifying or transporting a protein.

Finally, the length of time each protein functions in the cell is strictly regulated by means of selective degradation. Many proteins, such as the cyclins involved in regulating the cell cycle, must be relatively short-lived if the cell is to function appropriately. To mark a particular protein for destruction, the cell commonly attaches molecules of a small protein called ubiquitin to the protein, which triggers its destruction by protein complexes in the cell.

CONCEPT CHECK 15.2

1. In general, what is the effect of histone acetylation and DNA methylation on gene expression?
2. Compare the roles of general and specific transcription factors in regulating gene expression.
3. Suppose you compared the nucleotide sequences of the distal control elements in the enhancers of three genes that are expressed only in muscle cells. What would you expect to find? Why?

For suggested answers, see Appendix A.

Noncoding RNAs play multiple roles in controlling gene expression

Genome sequencing has revealed that protein-coding DNA accounts for only 1.5% of the human genome and a similarly small percentage of the genomes of many other multicellular eukaryotes. A very small fraction of the non-protein-coding DNA consists of genes for RNAs such as ribosomal RNA and transfer RNA. Until recently, most of the remaining DNA was assumed to be untranscribed. The idea was that since it didn't specify proteins or the few known types of RNA, such DNA didn't contain meaningful genetic information. However, a flood of recent data has contradicted this idea. For example, an in-depth study of a region comprising 1% of the human genome showed that more than 90% of that region was transcribed. Introns accounted for only a fraction of this transcribed, nontranslated RNA. These and other results suggest that a significant amount of the genome may be transcribed into non-protein-coding RNAs (also called *noncoding RNAs*, or *ncRNAs*), including a variety of small RNAs and longer RNA transcripts. While many questions about the functions of these RNAs remain unanswered, researchers are uncovering more evidence of their biological roles every day.

Biologists are excited about these recent discoveries, which hint at a large, diverse population of RNA molecules in the cell that play crucial roles in regulating gene expression—and have gone largely unnoticed until now. Clearly, we must revise our long-standing view that because mRNAs code for proteins, they are the most important RNAs functioning in the cell. This represents a major shift in the thinking of biologists, one that you are witnessing as students entering this field of study. It's as if our exclusive focus on a famous rock star has blinded us to the many backup musicians and songwriters working behind the scenes.

Regulation by both small and large ncRNAs is known to occur at several points in the pathway of gene expression, including mRNA translation and chromatin modification. We'll focus mainly on two types of small ncRNAs that have been extensively studied in the past few years; the importance of these RNAs was acknowledged when they were the focus of the 2006 Nobel Prize in Physiology or Medicine.

Effects on mRNAs by MicroRNAs and Small Interfering RNAs

Since 1993, a number of research studies have uncovered small single-stranded RNA molecules, called **microRNAs (miRNAs)**, that are capable of binding to complementary sequences in mRNA molecules. A longer RNA precursor is processed by cellular enzymes into an miRNA, a single-stranded RNA of about 22 nucleotides that forms a complex with one or more proteins. The miRNA allows the complex to bind to any mRNA molecule with 7–8 nucleotides of complementary sequence. The miRNA-protein complex then either degrades the target mRNA or blocks its translation (**Figure 15.13**). It has been estimated that expression of at least one-half of all human genes may be regulated by miRNAs, a remarkable figure given that the existence of miRNAs was unknown a mere two decades ago.

Another class of small RNAs are called **small interfering RNAs (siRNAs)**. These are similar in size and function to miRNAs—both can associate with the same proteins, producing similar results. The distinction between miRNAs and siRNAs is based on subtle differences in the structure of their double-stranded RNA precursor molecules. If researchers inject siRNA precursor molecules into a cell, the cell's machinery can process them into siRNAs that turn off expression of genes with related sequences. The blocking of gene expression by siRNAs is called **RNA interference (RNAi)**; it is used in the laboratory as a means of disabling specific genes to investigate their function.

EVOLUTION How did the RNAi pathway evolve? As you will learn in Chapter 17, some viruses have double-stranded RNA genomes. Because the cellular RNAi pathway can process double-stranded RNAs into homing devices that lead to the destruction of RNAs with complementary sequences, this pathway may have evolved as a natural defense against infection by such viruses. However, the fact that RNAi can also affect the expression of nonviral cellular genes may reflect a different evolutionary origin for the RNAi pathway. Moreover, many species, including mammals, apparently produce their own long, double-stranded RNA precursors to small RNAs such as siRNAs. Once produced, these RNAs can interfere with gene expression at stages other than translation, as we'll discuss next.

Chromatin Remodeling and Effects on Transcription by ncRNAs

In addition to affecting mRNAs, small RNAs can cause remodeling of chromatin structure. In some yeasts, siRNAs produced by the yeast cells themselves are required for the formation of heterochromatin at the centromeres of chromosomes. According to one model, an RNA transcript produced from DNA in the centromeric region of the chromosome is copied into double-stranded RNA by a yeast enzyme and then processed into siRNAs. These siRNAs associate with a complex of proteins (different from the one shown in Figure 15.13) and act as a homing device, targeting the complex back to RNA transcripts being made from the centromeric sequences of DNA. Once there, proteins in the complex recruit enzymes that modify the chromatin, turning it into the highly condensed heterochromatin found at the centromere.

A newly discovered class of small ncRNAs called *piwi-associated RNAs* (*piRNAs*) also induce formation of heterochromatin, blocking expression of some parasitic DNA elements in the genome known as transposons. (Transposons are discussed in Chapter 18.) Usually 24–31 nucleotides in length, piRNAs are probably processed from single-stranded RNA precursors. They play an indispensable role in the germ cells of many animal species, where they appear to help reestablish appropriate methylation patterns in the genome during gamete formation.

The role of ncRNAs in regulation of gene expression adds yet another layer to the complex and intricate process described in the previous section. As more is learned about the multiple, interacting ways a cell can fine-tune expression of its genes, the goal is to understand how a specific set of genes is expressed in a particular cell. In the next section, we'll describe a few methods that researchers use to monitor expression of specific genes.

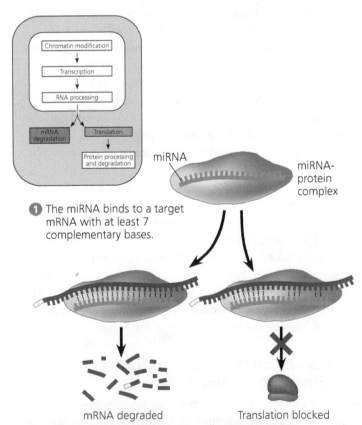

1 The miRNA binds to a target mRNA with at least 7 complementary bases.

miRNA

miRNA-protein complex

mRNA degraded   Translation blocked

2 If miRNA and mRNA bases are complementary all along their length, the mRNA is degraded (left); if the match is less complete, translation is blocked (right).

▲ **Figure 15.13 Regulation of gene expression by miRNAs.** A 22-nucleotide miRNA, formed by enzymatic processing of an RNA precursor, associates with one or more proteins in a complex that can affect target mRNAs.

CONCEPT CHECK 15.3

1. **WHAT IF?** If the mRNA being degraded in Figure 15.13 coded for a protein that promotes cell division in a multicellular organism, what would happen if a mutation disabled the gene encoding the miRNA that triggers this degradation?

2. **MAKE CONNECTIONS** Inactivation of one of the X chromosomes in female mammals results in a Barr body (see Concept 12.2). Suggest a model for how the *XIST* noncoding RNA functions to cause Barr body formation.

For suggested answers, see Appendix A.

CONCEPT 15.4

Researchers can monitor expression of specific genes

The diverse mechanisms of regulating gene expression discussed in this chapter underlie one basic generality: Cells of a given multicellular organism differ from each other because they express different genes from an identical genome. Biologists driven to understand the assorted cell types of a multicellular organism, cancer cells, or the developing tissues of an embryo first try to discover which genes are expressed by the cells of interest. The most straightforward way to do this is usually to identify the mRNAs being made. Techniques related to those developed for genetic engineering (see Concept 13.4) are widely used to track expression of mRNAs. In this section we'll first examine techniques that look for patterns of expression of specific individual genes. Next, we'll explore techniques that characterize groups of genes being expressed by cells or tissues of interest. As you will see, all of these techniques depend in some way on base pairing between complementary nucleotide sequences.

Studying the Expression of Single Genes

Suppose we have cloned a gene that may play an important role in the embryonic development of *Drosophila* (the fruit fly). The first thing we might want to know is which embryonic cells express the gene—in other words, where in the embryo is the corresponding mRNA found? We can detect the mRNA using the technique of **nucleic acid hybridization**, the base pairing of one strand of a nucleic acid to the complementary sequence on another strand. The complementary molecule, a short single-stranded nucleic acid that can be either RNA or DNA, is called a **nucleic acid probe**. Using our cloned gene as a template, we can synthesize a probe complementary to the mRNA. For example, if part of the sequence on the mRNA were

5′ ···CUCAUCACCGGC··· 3′

then we would synthesize this single-stranded DNA probe:

3′ GAGTAGTGGCCG 5′

Each probe molecule is labeled during synthesis with a fluorescent tag so we can follow it. A solution with the probe is applied to *Drosophila* embryos, allowing probe molecules to hybridize specifically to any complementary sequences on the many mRNAs in embryonic cells that are transcribing the gene. Because this technique allows us to see the mRNA in place (or *in situ*) in the intact organism, this technique is called *in situ* **hybridization**. Different probes can be labeled with different fluorescent dyes, sometimes with strikingly beautiful results **(Figure 15.14)**.

Other mRNA detection techniques may be preferable for comparing the amounts of a specific mRNA in several samples at the same time—for example, in different cell types or in embryos of different stages. One method that is widely

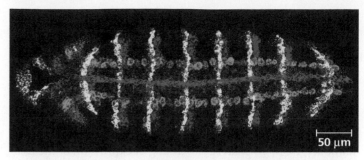

▲ **Figure 15.14 Determining where genes are expressed by *in situ* hybridization analysis.** This *Drosophila* embryo was incubated in a solution containing probes for five different mRNAs, each probe labeled with a different fluorescently colored tag. The embryo was then viewed using fluorescence microscopy. Each color marks cells in which a specific gene is expressed as mRNA.

used is called the **reverse transcriptase–polymerase chain reaction (RT-PCR)**. RT-PCR begins by turning sample sets of mRNAs into double-stranded DNAs with the corresponding sequences. This feat is accomplished by an enzyme called *reverse transcriptase*, isolated in the late 1980s from a type of virus called a retrovirus. (You'll learn more about retroviruses, including HIV, in Chapter 17.) Reverse transcriptase is able to synthesize a complementary DNA copy of an mRNA, thus making a *reverse transcript* **(Figure 15.15)**. Recall that the 3′

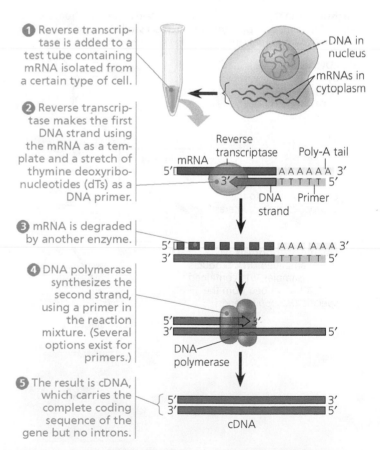

1. Reverse transcriptase is added to a test tube containing mRNA isolated from a certain type of cell.

DNA in nucleus
mRNAs in cytoplasm

2. Reverse transcriptase makes the first DNA strand using the mRNA as a template and a stretch of thymine deoxyribonucleotides (dTs) as a DNA primer.

Reverse transcriptase
Poly-A tail
mRNA
5′
3′
A A A A A A 3′
T T T T T 5′
DNA strand
Primer

3. mRNA is degraded by another enzyme.

5′
3′
A A A A A A 3′
T T T T T 5′

4. DNA polymerase synthesizes the second strand, using a primer in the reaction mixture. (Several options exist for primers.)

5′
3′
3′
5′
DNA polymerase

5. The result is cDNA, which carries the complete coding sequence of the gene but no introns.

5′
3′
3′
5′
cDNA

▲ **Figure 15.15 Making complementary DNA (cDNA) from eukaryotic genes.** Complementary DNA is DNA made *in vitro* using mRNA as a template for the first strand. Although only one mRNA is shown here, the final collection of cDNAs would reflect all the mRNAs that were present in the cell.

end of the mRNA has a stretch of adenine (A) ribonucleotides called a poly-A tail. This feature allows use of a short strand of thymine deoxyribonucleotides (dT's) as a primer for the reverse transcriptase. Following enzymatic degradation of the mRNA, a second DNA strand, complementary to the first, is synthesized by DNA polymerase. The resulting double-stranded DNA is called **complementary DNA (cDNA)**, and the reverse transcription step accounts for the "RT" in the name RT-PCR. To analyze the timing of expression of the *Drosophila* gene of interest, for example, we would first isolate all the mRNAs from different stages of *Drosophila* embryos and then make cDNA from each stage **(Figure 15.16)**.

Next in RT-PCR is the PCR step (see Figure 13.25). As you may recall, PCR is a way of rapidly making many copies of one specific stretch of double-stranded DNA, using primers that hybridize to the opposite ends of the region of interest. In our case, we would add primers corresponding to a region of our *Drosophila* gene, using the cDNA from each sample as a template for PCR amplification. When the products are run on a gel, copies of the amplified region will be observed as bands only in samples that originally contained mRNA from the gene we're focusing on. RT-PCR can also be carried out with mRNAs collected from different tissues at one time to discover which tissue is producing a specific mRNA.

Studying the Expression of Groups of Genes

A major goal of biologists is to learn how genes act together to produce and maintain a functioning organism. Now that the entire genomes of a number of organisms have been sequenced, it is possible to study the expression of large groups of genes—the so-called systems approach. Researchers use what is known about the whole genome to investigate which groups of genes are transcribed in different tissues or at different stages of development. One of their aims is to identify networks of gene expression across an entire genome.

Genome-wide expression studies can be carried out using **DNA microarray assays**. A DNA microarray consists of tiny amounts of a large number of single-stranded DNA fragments representing different genes fixed to a glass slide in a tightly spaced array, or grid. (The microarray is also called a *DNA chip* by analogy to a computer chip.) Ideally, these fragments represent all the genes in the genome of an organism.

The basic strategy in such studies is to isolate the mRNAs made in a cell of interest and use these mRNAs as templates for making the corresponding cDNAs by reverse transcription. In microarray assays, these cDNAs are labeled with fluorescent molecules and then allowed to hybridize to a microarray slide. Most often, the cDNAs from two samples are labeled with molecules that emit different colors and tested on the same microarray. **Figure 15.17** shows the result of such an experiment, identifying the subsets of genes in the genome that are being expressed in one tissue compared with another. DNA technology makes such studies possible; with automation, they are easily performed on a large scale. Scientists can now measure the expression of thousands of genes at one time.

Alternatively, with the advent of rapid, inexpensive DNA sequencing methods (see Chapter 13), researchers can now afford to simply sequence the cDNA samples from different tissues or different embryonic stages in order to discover which genes are expressed. This straightforward method is called *RNA sequencing* or *RNA-seq*, even though it is the cDNA that is actually sequenced. As the price of sequencing plummets, this method is growing more widespread.

By characterizing sets of genes that are expressed together in some tissues but not others, genome-wide gene expression studies may contribute to a better understanding of diseases and suggest new diagnostic techniques or therapies. For instance, comparing patterns of gene expression in breast cancer tumors and noncancerous breast tissue has already resulted in

---

▼ **Figure 15.16** **Research Method**

RT-PCR Analysis of the Expression of Single Genes

Application RT-PCR uses the enzyme reverse transcriptase (RT) in combination with PCR and gel electrophoresis. RT-PCR can be used to compare gene expression in different embryonic stages, in different tissues, or in the same type of cell under different conditions.

Technique In this example, samples containing mRNAs from six embryonic stages of *Drosophila* were processed as shown below. (The mRNAs from only one stage are shown here.)

1 **cDNA synthesis** is carried out by incubating the mRNAs with reverse transcriptase and other necessary components.

mRNAs

cDNAs

2 **PCR amplification** of the sample is performed using primers specific to the *Drosophila* gene of interest.

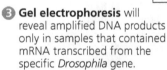

Primers

β-globin gene

3 **Gel electrophoresis** will reveal amplified DNA products only in samples that contained mRNA transcribed from the specific *Drosophila* gene.

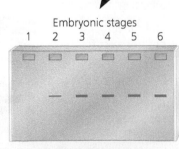

Embryonic stages
1 2 3 4 5 6

Results The mRNA for this gene is expressed from stage 2 through stage 6. The size of the amplified fragment (shown by its position on the gel) depends on the distance between the primers that were used.

Each dot is a well containing identical copies of DNA fragments that carry a specific gene.

The genes in the red wells are expressed in one tissue and bind the red cDNAs.

The genes in the green wells are expressed in the other tissue and bind the green cDNAs.

The genes in the yellow wells are expressed in both tissues and bind both red and green cDNAs, appearing yellow.

The genes in the black wells are not expressed in either tissue and do not bind either cDNA.

DNA microarray (actual size)

▲ **Figure 15.17 DNA microarray assay of gene expression levels.** Researchers synthesized two sets of cDNAs, fluorescently labeled red or green, from mRNAs from two different human tissues. These cDNAs were hybridized with a microarray containing 5,760 human genes (about 25% of human genes), resulting in the pattern shown here. The intensity of fluorescence at each spot measures the relative expression in the two samples of the gene represented by that spot: Red indicates expression in one sample, green in the other, yellow in both, and black in neither.

more informed and effective treatment protocols. Ultimately, information from genome-wide studies should provide a grander view of how ensembles of genes interact to form an organism and maintain its vital systems. The genetic basis of embryonic development and disease will be considered in the next chapter.

CONCEPT CHECK 15.4

1. Describe the role of complementary base pairing during RT-PCR and microarray analysis.

2. **WHAT IF?** Consider the microarray in Figure 15.17. If a sample from normal tissue is labeled with a green fluorescent dye, and a sample from cancerous tissue is labeled red, what color spots would represent genes you would be interested in if you were studying cancer? Explain.

For suggested answers, see Appendix A.

15 Chapter Review

SUMMARY OF KEY CONCEPTS

CONCEPT 15.1

Bacteria often respond to environmental change by regulating transcription (pp. 293–298)

- In bacteria, certain groups of genes are clustered into an operon with a single promoter. An operator site on the DNA switches the operon on or off, resulting in coordinate regulation of the genes.
- Both repressible and inducible operons are examples of negative gene regulation. Binding of a specific **repressor** protein to the operator shuts off transcription. (The repressor is encoded by a separate **regulatory gene**.) In a repressible operon, the repressor is active when bound to a **corepressor**.

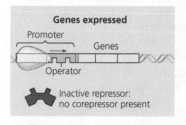
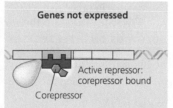

Genes expressed — Promoter, Genes, Operator. Inactive repressor: no corepressor present

Genes not expressed — Active repressor: corepressor bound. Corepressor

In an inducible operon, binding of an **inducer** to an innately active repressor inactivates the repressor and turns on transcription. Inducible enzymes usually function in catabolic pathways.

- Some operons have positive gene regulation. A stimulatory **activator** protein (such as CAP, when activated by **cyclic AMP**), binds to a site within the promoter and stimulates transcription.

? Compare and contrast the roles of the corepressor and the inducer in negative regulation of an operon.

CONCEPT 15.2

Eukaryotic gene expression is regulated at many stages (pp. 298–305)

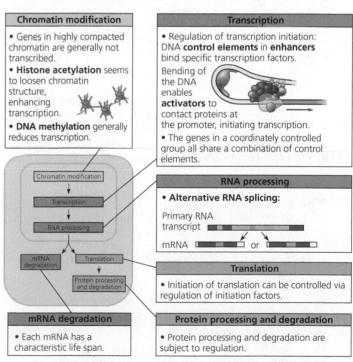

Chromatin modification
- Genes in highly compacted chromatin are generally not transcribed.
- **Histone acetylation** seems to loosen chromatin structure, enhancing transcription.
- **DNA methylation** generally reduces transcription.

Transcription
- Regulation of transcription initiation: DNA **control elements** in **enhancers** bind specific transcription factors.
- Bending of the DNA enables **activators** to contact proteins at the promoter, initiating transcription.
- The genes in a coordinately controlled group all share a combination of control elements.

RNA processing
- **Alternative RNA splicing:**
- Primary RNA transcript
- mRNA ⟶ or ⟶

Translation
- Initiation of translation can be controlled via regulation of initiation factors.

mRNA degradation
- Each mRNA has a characteristic life span.

Protein processing and degradation
- Protein processing and degradation are subject to regulation.

(Flowchart: Chromatin modification → Transcription → RNA processing → mRNA degradation / Translation → Protein processing and degradation)

? Describe what must happen for a cell type–specific gene to be transcribed in a cell of that type.

CONCEPT 15.3

Noncoding RNAs play multiple roles in controlling gene expression (pp. 305–306)

- Noncoding RNAs (e.g., **miRNAs** and **siRNAs**) can block translation or cause degradation of mRNAs.

> **?** *Why are miRNAs called noncoding RNAs? Explain how they participate in gene regulation.*

CONCEPT 15.4

Researchers can monitor expression of specific genes (pp. 307–309)

- In **nucleic acid hybridization**, a **nucleic acid probe** is used to detect the presence of a specific mRNA.
- *In situ* hybridization and **RT-PCR** can detect the presence of a given mRNA in a tissue or an RNA sample, respectively.
- **DNA microarrays** are used to identify sets of genes co-expressed by a group of cells. Their cDNAs can also be sequenced.

> **?** *What useful information is obtained by detecting expression of specific genes?*

TEST YOUR UNDERSTANDING

Level 1: Knowledge/Comprehension

1. If a particular operon encodes enzymes for making an essential amino acid and is regulated like the *trp* operon, then
   a. the amino acid inactivates the repressor.
   b. the enzymes produced are called inducible enzymes.
   c. the repressor is active in the absence of the amino acid.
   d. the amino acid acts as a corepressor.
   e. the amino acid turns on transcription of the operon.

2. The functioning of enhancers is an example of
   a. transcriptional control of gene expression.
   b. a post-transcriptional mechanism to regulate mRNA.
   c. the stimulation of translation by initiation factors.
   d. post-translational control that activates certain proteins.
   e. a eukaryotic equivalent of prokaryotic promoter functioning.

3. Which of the following is an example of post-transcriptional control of gene expression?
   a. the addition of methyl groups to cytosine bases of DNA
   b. the binding of transcription factors to a promoter
   c. the removal of introns and alternative splicing of exons
   d. the binding of RNA polymerase to transcription factors
   e. the folding of DNA to form heterochromatin

Level 2: Application/Analysis

4. What would occur if the repressor of an inducible operon were mutated so it could not bind the operator?
   a. irreversible binding of the repressor to the promoter
   b. reduced transcription of the operon's genes
   c. buildup of a substrate for the pathway controlled by the operon
   d. continuous transcription of the operon's genes
   e. overproduction of catabolite activator protein (CAP)

5. Which of the following statements about the DNA in one of your brain cells is true?
   a. Most of the DNA codes for protein.
   b. The majority of genes are likely to be transcribed.
   c. Each gene lies immediately adjacent to an enhancer.
   d. Many genes are grouped into operon-like clusters.
   e. It is the same as the DNA in one of your kidney cells.

6. Which of the following would *not* be true of cDNA produced using human brain tissue as the starting material?
   a. It could be amplified by the polymerase chain reaction.
   b. It would contain sequences representing all the genes in the genome.
   c. It was produced from mRNA using reverse transcriptase.
   d. It could be used as a probe to detect genes expressed in the brain.
   e. It lacks the introns of the human genes.

Level 3: Synthesis/Evaluation

7. **DRAW IT** The diagram to the right shows five genes, including their enhancers, from the genome of a certain species. Imagine that orange, blue, green, black, red, and purple activator proteins exist that can bind to the appropriately color-coded control elements in the enhancers of these genes.

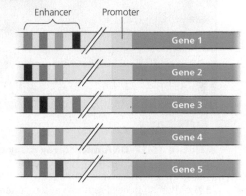

   (a) Draw an X above enhancer elements (of all the genes) that would have activators bound in a cell in which only gene 5 is transcribed. Which colored activators would be present?
   (b) Draw a dot above all enhancer elements that would have activators bound in a cell in which the green, blue, and orange activators are present. Which gene(s) would be transcribed?
   (c) Imagine that genes 1, 2, and 4 code for nerve-specific proteins, and genes 3 and 5 are skin specific. Which activators would have to be present in each cell type to ensure transcription of the appropriate genes?

8. **SCIENTIFIC INQUIRY**
   Imagine you want to study one of the mouse crystallins, proteins present in the lens of the eye. Assuming the gene has been cloned, describe two ways you could investigate expression of this gene in the developing mouse embryo.

9. **FOCUS ON EVOLUTION**
   DNA sequences can act as "tape measures of evolution" (see Concept 3.6). Scientists analyzing the human genome sequence were surprised to find that some of the regions of the human genome that are most highly conserved (similar to comparable regions in other species) don't code for proteins. Propose a possible explanation for this observation.

10. **FOCUS ON INTERACTIONS**
    In a short essay (100–150 words), discuss how the processes shown in Figure 15.2 are examples of feedback mechanisms regulating biological systems in bacterial cells.

For selected answers, see Appendix A.

MasteringBiology®

Students Go to **MasteringBiology** for assignments, the eText, and the Study Area with practice tests, animations, and activities.

Instructors Go to **MasteringBiology** for automatically graded tutorials and questions that you can assign to your students, plus Instructor Resources.

16

Development, Stem Cells, and Cancer

KEY CONCEPTS

16.1 A program of differential gene expression leads to the different cell types in a multicellular organism

16.2 Cloning of organisms showed that differentiated cells could be "reprogrammed" and ultimately led to the production of stem cells

16.3 Abnormal regulation of genes that affect the cell cycle can lead to cancer

OVERVIEW

Orchestrating Life's Processes

The development of the fertilized egg, a single cell, into an embryo and later an adult is an astounding transformation that requires a precisely regulated program of gene expression. All of the levels of eukaryotic gene regulation you learned about in the previous chapter come into play during embryonic development. The elaborate sequence of genes being turned on and off in different cells is the ultimate example of regulation of gene expression.

Understanding the genetic underpinnings of development has progressed mainly by studying the process in **model organisms**, species that are easy to raise in the lab and use in experiments. A prime example is the fruit fly *Drosophila melanogaster*. An adult fruit fly develops from a fertilized egg, passing through a wormlike stage called a larva. At every stage, gene expression is carefully regulated, ensuring that the right genes are expressed only at the correct time and place. In the larva, the adult wing forms in a disk-shaped pocket of several thousand cells, shown in **Figure 16.1**. The tissue in this image has been analyzed by *in situ* hybridization (see Figure 15.14) to reveal the mRNA for three genes—labeled red, blue, and green. (Red and green together appear yellow.) The intricate pattern of expression for each gene is the same from larva to larva at this stage, and it provides a graphic display of the precision of gene regulation. But what is the molecular basis for this pattern? Why is one particular gene expressed only in the few hundred cells that appear blue in this image and not in the other cells?

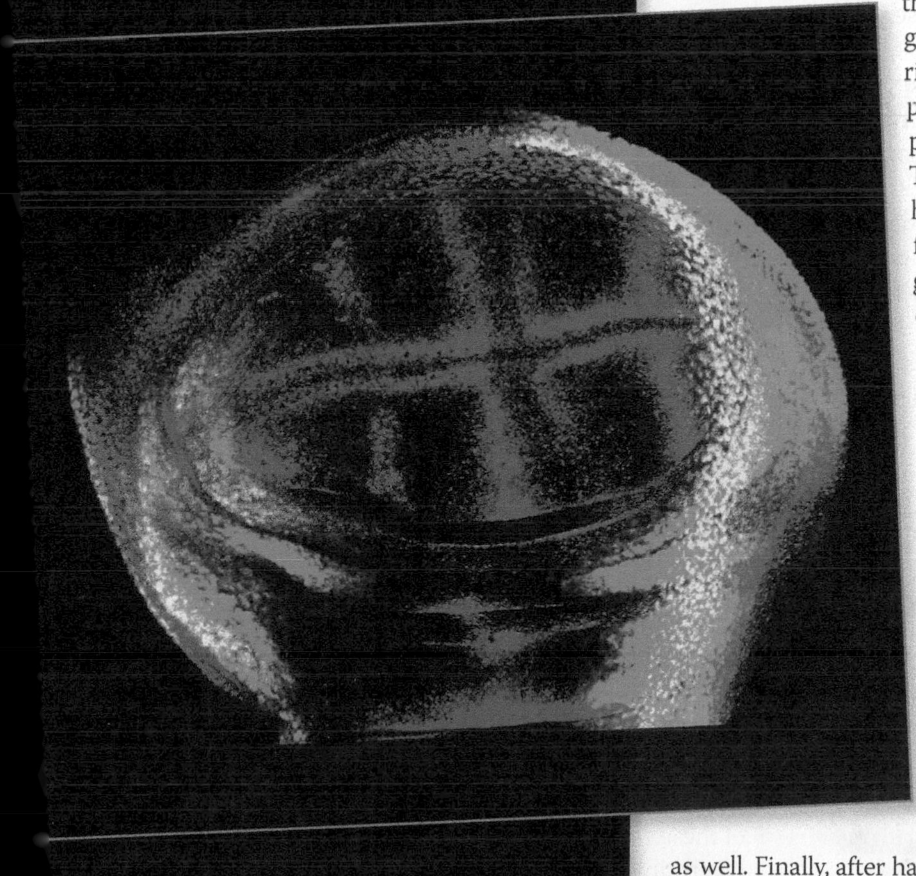

Part of the answer involves the transcription factors and other regulatory molecules you learned about in the previous chapter. But how do they come to be different in distinct cell types? In this chapter, we'll first explain the mechanisms that send cells down diverging genetic pathways to adopt different fates. Then we'll take a closer look at *Drosophila* development. Next, we'll describe the discovery of stem cells, a powerful cell type that is key to the developmental process. These cells offer hope for medical treatments as well. Finally, after having explored embryonic development and stem cells, we will underscore the crucial role played by regulation of gene expression by investigating how cancer can result when this regulation goes awry. Orchestrating proper gene expression by all cells is crucial to the functions of life.

◀ **Figure 16.1** What regulates the precise pattern of gene expression in the developing wing of a fly embryo?

A program of differential gene expression leads to the different cell types in a multicellular organism

In the embryonic development of multicellular organisms, a fertilized egg (a zygote) gives rise to cells of many different types, each with a different structure and corresponding function. Typically, cells are organized into tissues, tissues into organs, organs into organ systems, and organ systems into the whole organism. Thus, any developmental program must produce cells of different types that form higher-level structures arranged in a particular way in three dimensions. Here, we'll focus on the program of regulation of gene expression that orchestrates development using a few animal species as examples.

A Genetic Program for Embryonic Development

The photos in **Figure 16.2** illustrate the dramatic difference between a zygote and the organism it becomes. This remarkable transformation results from three interrelated processes: cell division, cell differentiation, and morphogenesis. Through a succession of mitotic cell divisions, the zygote gives rise to a large number of cells. Cell division alone, however, would merely produce a great ball of identical cells, nothing like a tadpole. During embryonic development, cells not only increase in number, but also undergo cell **differentiation**, the process by which cells become specialized in structure and function. Moreover, the different kinds of cells are not randomly distributed but are organized into tissues and organs in a particular three-dimensional arrangement. The physical processes that give an organism its shape constitute **morphogenesis**, the development of the form of an organism and its structures.

All three processes have their basis in cellular behavior. Even morphogenesis, the shaping of the organism, can be traced back to changes in the motility, shape, and other characteristics of the cells that make up various regions of the embryo. As you have seen, the activities of a cell depend on the genes it expresses and the proteins it produces. Almost all cells in an organism have the same genome; therefore, differential gene expression results from the genes being regulated differently in each cell type.

In Figure 15.11, you saw a simplified view of how differential gene expression occurs in two cell types, a liver cell and a lens cell. Each of these fully differentiated cells has a particular mix of specific activators that turn on the collection of genes whose products are required in the cell. The fact that both cells arose through a series of mitoses from a common fertilized egg inevitably leads to a question: How do different sets of activators come to be present in the two cells?

It turns out that materials placed into the egg by the mother set up a sequential program of gene regulation that is carried out as cells divide, and this program makes the cells become different from each other in a coordinated fashion. To understand how this works, we'll consider two basic developmental processes: First, we'll explore how cells that arise from early embryonic mitoses develop the differences that start each cell along its own differentiation pathway. Second, we'll see how cellular differentiation leads to one particular cell type, using muscle development as an example.

Cytoplasmic Determinants and Inductive Signals

What generates the first differences among cells in an early embryo? And what controls the differentiation of all the various cell types as development proceeds? You can probably deduce the answer: The specific genes expressed in any particular cell of a developing organism determine its path. Two sources of information, used to varying extents in different species, "tell" a cell which genes to express at any given time during embryonic development.

One important source of information early in development is the egg's cytoplasm, which contains both RNA and proteins encoded by the mother's DNA. The cytoplasm of an unfertilized egg is not homogeneous. Messenger RNA, proteins, other substances, and organelles are distributed unevenly in the unfertilized egg, and this unevenness has a profound impact on the development of the future embryo in many species. Maternal substances in the egg that influence the course of early development are called **cytoplasmic determinants** (**Figure 16.3a**). After fertilization, early mitotic divisions distribute the zygote's cytoplasm into separate cells. The nuclei of these cells may thus be exposed to different cytoplasmic determinants, depending on which portions of the zygotic cytoplasm a cell received. The combination of cytoplasmic determinants in a cell helps determine its developmental fate by regulating expression of the cell's genes during the course of cell differentiation.

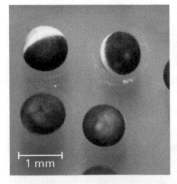 

1 mm

2 mm

(a) Fertilized eggs of a frog　　　**(b) Newly hatched tadpole**

▲ **Figure 16.2 From fertilized egg to animal: What a difference four days makes.** It takes just four days for cell division, differentiation, and morphogenesis to transform each of the fertilized frog eggs shown in (a) into a tadpole like the one in (b).

▼ Figure 16.3 Sources of developmental information for the early embryo.

(a) Cytoplasmic determinants in the egg

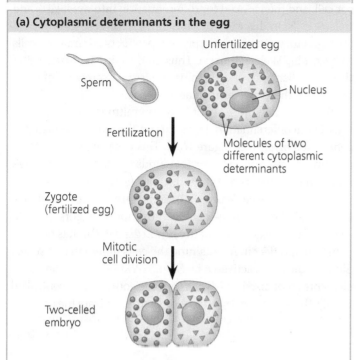

Sperm

Unfertilized egg

Nucleus

Fertilization

Molecules of two different cytoplasmic determinants

Zygote (fertilized egg)

Mitotic cell division

Two-celled embryo

The unfertilized egg has molecules in its cytoplasm, encoded by the mother's genes, that influence development. Many of these cytoplasmic determinants, like the two shown here, are unevenly distributed in the egg. After fertilization and mitotic division, the cell nuclei of the embryo are exposed to different sets of cytoplasmic determinants and, as a result, express different genes.

(b) Induction by nearby cells

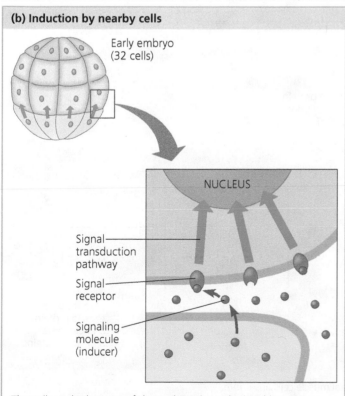

Early embryo (32 cells)

NUCLEUS

Signal transduction pathway

Signal receptor

Signaling molecule (inducer)

The cells at the bottom of the early embryo depicted here are releasing chemicals that signal nearby cells to change their gene expression.

The other major source of developmental information, which becomes increasingly important as the number of embryonic cells increases, is the environment around a particular cell. Most influential are the signals communicated to an embryonic cell from other embryonic cells in the vicinity, including contact with cell-surface molecules on neighboring cells and the binding of growth factors secreted by neighboring cells (see Concept 5.6). Such signals cause changes in the target cells, a process called **induction (Figure 16.3b)**. The molecules conveying these signals within the target cell are cell-surface receptors and other proteins expressed by the embryo's own genes. In general, the signaling molecules send a cell down a specific developmental path by causing changes in its gene expression that eventually result in observable cellular changes. Thus, interactions between embryonic cells help induce differentiation of the many specialized cell types making up a new organism.

Sequential Regulation of Gene Expression during Cellular Differentiation

As the tissues and organs of an embryo develop and their cells differentiate, the cells become noticeably different in structure and function. These observable changes are actually the outcome of a cell's developmental history, which begins at the first mitotic division of the zygote, as we have just seen. The earliest changes that set a cell on a path to specialization are subtle ones, showing up only at the molecular level. Before biologists knew much about the molecular changes occurring in embryos, they coined the term **determination** to refer to the unseen events that lead to the observable differentiation of a cell. Once it has undergone determination, an embryonic cell is irreversibly committed to its final fate. If a committed cell is experimentally placed in another location in the embryo, it will still differentiate into the cell type that is its normal fate.

Differentiation of Cell Types

Today we understand determination in terms of molecular changes. The outcome of determination, observable cell differentiation, is marked by the expression of genes for *tissue-specific proteins*. These proteins are found only in a specific cell type and give the cell its characteristic structure and function. The first evidence of differentiation is the appearance of mRNAs for these proteins. Eventually, differentiation is observable with a microscope as changes in cellular structure. On the molecular level, different sets of genes are sequentially expressed in a regulated manner as new cells arise from division of their precursors. A number of the steps in gene expression may be regulated during differentiation, with transcription among the most important. At the end of the process, in the fully differentiated cell, transcription remains the principal regulatory point for maintaining appropriate gene expression.

Differentiated cells are specialists at making tissue-specific proteins. For example, as a result of transcriptional regulation, liver cells specialize in making albumin, and lens cells specialize in making crystallin (see Figure 15.11). Skeletal muscle cells in vertebrates are another instructive example. Each of these cells is a long fiber containing many nuclei within a single plasma membrane. Skeletal muscle cells have high concentrations of muscle-specific versions of the contractile proteins myosin and actin, as well as membrane receptor proteins that detect signals from nerve cells.

Muscle cells develop from embryonic precursor cells that have the potential to develop into a number of cell types, including cartilage cells and fat cells, but particular conditions commit them to becoming muscle cells. Although the committed cells appear unchanged under the microscope, determination has occurred, and they are now *myoblasts*. Eventually, myoblasts start to churn out large amounts of muscle-specific proteins and fuse to form mature, elongated, multinucleate skeletal muscle cells.

Researchers have worked out what happens at the molecular level during muscle cell determination (**Figure 16.4**). To do so, they grew embryonic precursor cells in culture and analyzed them using molecular biological techniques like those described in Chapters 13 and 15. They isolated different genes one by one, caused each to be expressed in a separate precursor cell, and then looked for differentiation into myoblasts and muscle cells. In this way, they identified several so-called "master regulatory genes" whose protein products commit the cells to becoming skeletal muscle. Thus, in the case of muscle cells, the molecular basis of determination is the expression of one or more of these master regulatory genes.

To understand more about how commitment occurs in muscle cell differentiation, let's focus on the master regulatory gene called *myoD* (see Figure 16.4). This gene encodes MyoD protein, a transcription factor that binds to specific control elements in the enhancers of various target genes and stimulates their expression (see Figure 15.9). Some target genes for MyoD encode still other muscle-specific transcription factors. MyoD also stimulates expression of the *myoD* gene itself, thus perpetuating its effect in maintaining the cell's differentiated state. Since all the genes activated by MyoD have enhancer control elements recognized by MyoD, they are coordinately controlled. Finally, the secondary transcription factors activate the genes

① **Determination.** Signals from other cells lead to activation of a master regulatory gene called *myoD*, and the cell makes MyoD protein, a specific transcription factor that acts as an activator. The cell, now called a myoblast, is irreversibly committed to becoming a skeletal muscle cell.

② **Differentiation.** MyoD protein stimulates the *myoD* gene further and activates genes encoding other muscle-specific transcription factors, which in turn activate genes for muscle proteins. MyoD also turns on genes that block the cell cycle, thus stopping cell division. The nondividing myoblasts fuse to become mature multinucleate muscle cells, also called muscle fibers.

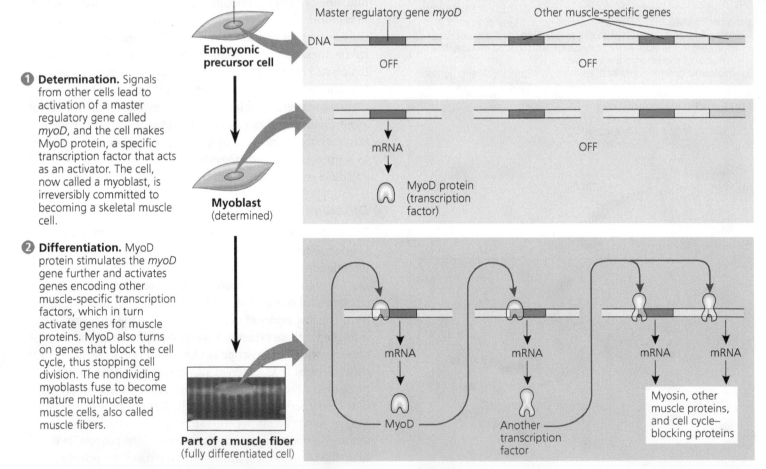

▲ **Figure 16.4 Determination and differentiation of muscle cells.** Skeletal muscle cells arise from embryonic cells as a result of changes in gene expression. (In this depiction, the process of gene activation is greatly simplified.)

WHAT IF? *What would happen if a mutation in the* myoD *gene resulted in a MyoD protein that could not activate the* myoD *gene?*

for proteins such as myosin and actin that confer the unique properties of skeletal muscle cells. The MyoD protein deserves its designation as a master regulatory gene.

The regulation of genes that play important roles in development of embryonic tissues and structures is often complex. In the **Scientific Skills Exercise**, you'll work with data from an experiment that tested how different regulatory regions in the DNA affect expression of a gene that helps establish the pattern of the different digits in a mouse's paw.

Apoptosis: A Type of Programmed Cell Death

During the time when most cells are differentiating, some cells in the developing organism are genetically programmed to die. The best-understood type of "programmed cell death" is **apoptosis** (from the Greek, meaning "falling off," and used in a classic Greek poem to refer to leaves falling from a tree). Apoptosis also occurs in cells of the mature organism that are infected, damaged, or have reached the end of their functional life span. During this process, cellular agents chop up the DNA and fragment the organelles and other cytoplasmic components. The cell becomes multilobed, a change called "blebbing" **(Figure 16.5)**, and the cell's parts are packaged up in vesicles. These "blebs" are then engulfed by scavenger cells, leaving no trace. Apoptosis protects neighboring cells from damage that they would otherwise suffer if a dying cell merely leaked out all its contents, including its many digestive enzymes.

Apoptosis plays a crucial role in the developing embryo. The molecular mechanisms underlying apoptosis were worked out in detail by researchers studying embryonic development of a small soil worm, a nematode called *Caenorhabditis elegans* that has now become a popular model organism for genetic studies. Because the adult worm has only about a thousand cells, the researchers were able to work out the complete ancestry of each cell. The timely suicide of cells occurs exactly 131 times during normal development of *C. elegans*, at precisely the same points in the cell lineage of each worm. In worms and other species, apoptosis is triggered by signal transduction pathways (see Figure 5.20). These activate a cascade of apoptotic "suicide"

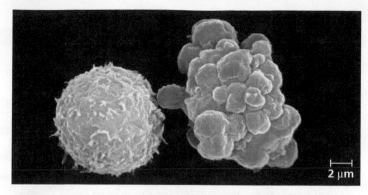

▲ **Figure 16.5 Apoptosis of a human white blood cell.** We can compare a normal white blood cell (left) with a white blood cell undergoing apoptosis (right). The apoptotic cell is shrinking and forming lobes ("blebs"), which eventually are shed as membrane-enclosed cell fragments (colorized SEMs).

proteins in the cells destined to die, including the enzymes that break down and package cellular molecules in the "blebs."

Apoptosis is essential to development and maintenance in all animals. There are similarities in genes encoding apoptotic proteins in nematodes and mammals, and apoptosis is known to occur as well in multicellular fungi and single-celled yeasts, evidence that the basic mechanism evolved early among eukaryotes. In vertebrates, apoptosis is essential for normal development of the nervous system and for normal morphogenesis of hands and feet in humans and paws in other mammals **(Figure 16.6)**. The level of apoptosis between the developing digits is lower in the webbed feet of ducks and other water birds than in the nonwebbed feet of land birds, such as chickens. In the case of humans, the failure of appropriate apoptosis can result in webbed fingers and toes.

We have seen how different programs of gene expression that are activated in the fertilized egg can result in differentiated cells and tissues as well as the death of some cells. But for tissues to function properly in the organism as a whole, the organism's *body plan*—its overall three-dimensional arrangement—must be established and superimposed on the differentiation process. Next we'll look at the molecular basis for establishing the body plan, using the well-studied *Drosophila* as an example.

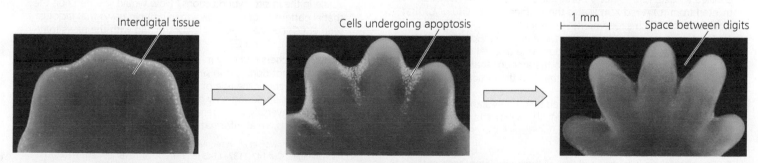

▲ **Figure 16.6 Effect of apoptosis during paw development in the mouse.** In mice, humans, other mammals, and land birds, the embryonic region that develops into feet or hands initially has a solid, platelike structure. Apoptosis eliminates the cells in the interdigital regions, thus forming the digits. The embryonic mouse paws shown in these fluorescence light micrographs are stained so that cells undergoing apoptosis appear a bright yellowish green. Apoptosis of cells begins at the margin of each interdigital region (left), peaks as the tissue in these regions is reduced (middle), and is no longer visible when the interdigital tissue has been eliminated (right). (Note that the Scientific Skills Exercise shows a different genetic process involved in mouse paw development.)

Analyzing Quantitative and Spatial Gene Expression Data

How Is a Particular *Hox* Gene Regulated During Paw Development? *Hox* genes code for transcription factor proteins, which in turn control sets of genes important for animal development (see Concept 18.6 for more information on *Hox* genes). One group of *Hox* genes, the *Hoxd* genes, plays a role in establishing the pattern of the different digits (fingers and toes) at the end of a limb. Unlike the *mPGES-1* gene mentioned in the last chapter, *Hox* genes have very large, complicated regulatory regions, including control elements that may be hundreds of kilobases (kb; thousands of nucleotides) away from the gene.

In cases like this, how do biologists narrow down the segments that contain important elements? They begin by removing (deleting) large segments of DNA and studying the effect on gene expression. In this exercise, you'll compare data from two different but complementary approaches that look at the expression of a specific *Hoxd* gene (*Hoxd13*). One approach quantifies overall expression; the other approach is less quantitative but gives important spatial localization information.

How the Experiment Was Done Researchers interested in the regulation of *Hoxd13* gene expression genetically engineered a set of mice (*transgenic* mice) that had different segments of DNA deleted upstream of the gene. They then compared levels and patterns of *Hoxd13* gene expression in the developing paws of 12.5-day-old transgenic mouse embryos with those seen in wild-type mouse embryos of the same age.

They used two different approaches: In some mice, they extracted the mRNA from the embryonic paws and quantified the overall level of *Hoxd13* mRNA in the whole paw. In another set of the same transgenic mice, they used *in situ* hybridization (see Concept 15.4) to pinpoint exactly where in the paws the *Hoxd13* gene was expressed as mRNA. The particular technique that was used causes the *Hoxd13* mRNA to appear blue.

Data from the Experiment The top diagram (upper right) depicts the very large regulatory region upstream of the *Hoxd13* gene. The area between the slashes represents the DNA located between the promoter and the regulatory region.

The diagrams to the left of the bar graph show, first, the intact DNA (830 kb) and, next, the three altered DNA sequences. (Each is called a "deletion," since a particular section of DNA has been deleted from it.) A red X indicates the segment (A, B, and/or C) that was deleted in each experimental treatment.

The horizontal bar graph shows the amount of *Hoxd13* mRNA that was present in the digit-formation zone of each mutant 12.5-day-old embryo paw relative to the amount that was in the digit-formation zone of the mouse that had the intact regulatory region (top bar = 100%).

The images on the right are the embryo paws showing the location of the *Hoxd13* mRNA (blue stain). The white triangles show the location where the thumb will form.

Interpret the Data

1. The researchers hypothesized that all three regulatory segments (A, B, and C) were required for full expression of the *Hoxd13* gene. By measuring the amount of *Hoxd13* mRNA in the embryo paw

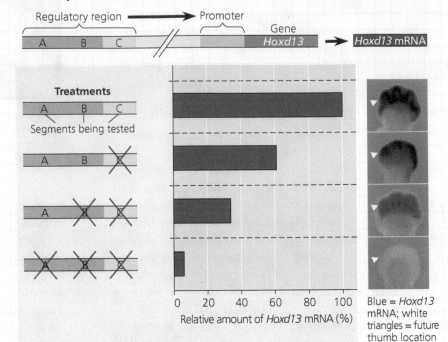

Blue = *Hoxd13* mRNA; white triangles = future thumb location

zones where digits will develop, they could measure the effect of the regulatory segments singly and in combination. Refer to the graph to answer these questions, noting that the segments being tested are shown on the vertical axis and the relative amount of *Hoxd13* mRNA is shown on the horizontal axis. (a) Which of the four treatments was used as a control for the experiment? (b) Their hypothesis is that all three segments are required for highest expression of the *Hoxd13* gene. Is this supported by their results? Explain your answer.

2. (a) What is the effect on the amount of *Hoxd13* mRNA when segments B and C are both deleted, compared with the control? (b) Is this effect visible in the blue-stained regions of the *in situ* hybridizations? How would you describe the spatial pattern of gene expression in the embryo paws that lack segments B and C?

3. (a) What is the effect on the amount of *Hoxd13* mRNA when just segment C is deleted, compared with the control? (b) Is this effect visible in the *in situ* hybridizations? How would you describe the spatial pattern of gene expression in embryo paws that lack just segment C, compared with the control and with the paws that lack segments B and C?

4. If the researchers had only measured the amount of *Hoxd13* mRNA and not done the *in situ* hybridizations, what important information about the role of the regulatory segments in *Hoxd13* gene expression during paw development would have been missed? Conversely, if the researchers had only done the *in situ* hybridizations, what information would have been inaccessible?

Data from T. Montavon et al., A regulatory archipelago controls *Hox* genes transcription in digits, *Cell* 147:1132–1145 (2011). doi 10.1016/j.cell.2011.10.023

(MB) A version of this Scientific Skills Exercise can be assigned in MasteringBiology.

Pattern Formation: Setting Up the Body Plan

Cytoplasmic determinants and inductive signals both contribute to the development of a spatial organization in which the tissues and organs of an organism are all in their characteristic places. This process is called **pattern formation**.

Just as the locations of the front, back, and sides of a new building are determined before construction begins, pattern formation in animals begins in the early embryo, when the major axes of an animal are established. In a bilaterally symmetric animal, the relative positions of head and tail, right and left sides, and back and front—the three major body axes—are set up before the tissues and organs appear. The molecular cues that control pattern formation, collectively called **positional information**, are provided by cytoplasmic determinants and inductive signals (see Figure 16.3). These cues tell a cell its location relative to the body axes and to neighboring cells and determine how the cell and its progeny will respond to future molecular signals.

During the first half of the 20th century, classical embryologists made detailed anatomical observations of embryonic development in a number of species and performed experiments in which they manipulated embryonic tissues. This research laid the groundwork for understanding the mechanisms of development, but it did not reveal the specific molecules that guide development or determine how patterns are established.

In the 1940s, scientists began using the genetic approach—the study of mutants—to investigate *Drosophila* development. That approach has had spectacular success and continues today. Genetic studies have established that genes control development and have led to an understanding of the key roles that specific molecules play in defining position and directing differentiation. By combining anatomical, genetic, and biochemical approaches to the study of *Drosophila* development, researchers have discovered developmental principles common to many other species, including humans.

The Life Cycle of Drosophila

Fruit flies and other arthropods have a modular construction, an ordered series of segments. These segments make up the body's three major parts: the head, the thorax (the midbody, from which the wings and legs extend), and the abdomen **(Figure 16.7a)**. Like other bilaterally symmetric animals, *Drosophila* has an anterior-posterior (head-to-tail) axis, a dorsal-ventral (back-to-belly) axis, and a right-left axis. In *Drosophila*, cytoplasmic determinants that are localized in the unfertilized egg provide positional information for the placement of anterior-posterior and dorsal-ventral axes even before fertilization. We'll focus here on the molecules involved in establishing the anterior-posterior axis as a case in point.

The *Drosophila* egg develops in the female's ovary, surrounded by ovarian cells called nurse cells and follicle cells **(Figure 16.7b**, top). These support cells supply the egg with

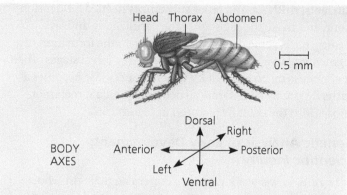

(a) Adult. The adult fly is segmented, and multiple segments make up each of the three main body parts—head, thorax, and abdomen. The body axes are shown by arrows.

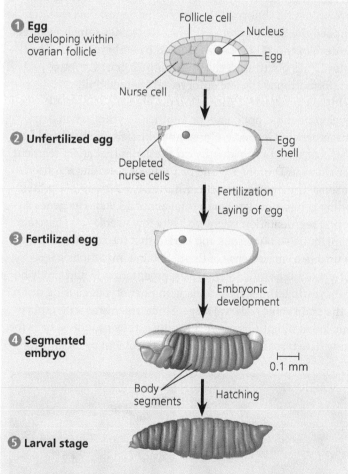

(b) Development from egg to larva. ❶ The egg (yellow) is surrounded by other cells that form a structure called the follicle within one of the mother's ovaries. ❷ The nurse cells shrink as they supply nutrients and mRNAs to the developing egg, which grows larger. Eventually, the mature egg fills the egg shell that is secreted by the follicle cells. ❸ The egg is fertilized within the mother and then laid. ❹ Embryonic development forms ❺ a larva, which goes through three stages. The third stage forms a cocoon (not shown), within which the larva metamorphoses into the adult shown in (a).

▲ **Figure 16.7 Key developmental events in the life cycle of *Drosophila*.**

nutrients, mRNAs, and other substances needed for development, and make the egg shell. After fertilization and laying of the egg, embryonic development results in the formation of a segmented larva, which goes through three larval stages. Then, in a process much like that by which a caterpillar becomes a butterfly, the fly larva forms a cocoon in which it metamorphoses into the adult fly pictured in Figure 16.7a.

Genetic Analysis of Early Development: *Scientific Inquiry*

Edward B. Lewis was a visionary American biologist who, in the 1940s, first showed the value of the genetic approach to studying embryonic development in *Drosophila*. Lewis studied bizarre mutant flies with developmental defects that led to extra wings or legs in the wrong place **(Figure 16.8)**. He located the mutations on the fly's genetic map, thus connecting the developmental abnormalities to specific genes. This research supplied the first concrete evidence that genes somehow direct the developmental processes studied by embryologists. The genes Lewis discovered, called **homeotic genes**, control pattern formation in the late embryo, larva, and adult.

Insight into pattern formation during early embryonic development did not come for another 30 years, when two researchers in Germany, Christiane Nüsslein-Volhard and Eric Wieschaus, set out to identify *all* the genes that affect segment formation in *Drosophila*. The project was daunting for three reasons. The first was the sheer number of *Drosophila* protein-coding genes, now known to total about 13,900. The genes affecting segmentation might be just a few needles in a haystack or might be so numerous and varied that the scientists would be unable to make sense of them. Second, mutations affecting a process as fundamental as segmentation would surely be **embryonic lethals**, mutations with phenotypes causing death at the embryonic or larval stage. Since organisms with embryonic lethal mutations never reproduce, they cannot be bred for study. The researchers dealt with this problem by looking for

recessive mutations, which can be propagated in heterozygous flies that act as genetic carriers. Third, cytoplasmic determinants in the egg were known to play a role in axis formation, so the researchers knew they would have to study the mother's genes as well as those of the embryo. It is the mother's genes that we will discuss further as we focus on how the anterior-posterior body axis is set up in the developing egg.

Nüsslein-Volhard and Wieschaus began their search for segmentation genes by exposing flies to a mutagenic chemical that affected the flies' gametes. They mated the mutagenized flies and then scanned their descendants for dead embryos or larvae with abnormal segmentation or other defects. For example, to find genes that might set up the anterior-posterior axis, they looked for embryos or larvae with abnormal ends, such as two heads or two tails, predicting that such abnormalities would arise from mutations in maternal genes required for correctly setting up the offspring's head or tail end.

Using this approach, Nüsslein-Volhard and Wieschaus eventually identified about 1,200 genes essential for pattern formation during embryonic development. Of these, about 120 were essential for normal segmentation. Over several years, the researchers were able to group these segmentation genes by general function, to map them, and to clone many of them for further study in the lab. The result was a detailed molecular understanding of the early steps in pattern formation in *Drosophila*.

When the results of Nüsslein-Volhard and Wieschaus were combined with Lewis's earlier work, a coherent picture of *Drosophila* development emerged. In recognition of their discoveries, the three researchers were awarded a Nobel Prize in 1995.

Let's consider further the genes that Nüsslein-Volhard, Wieschaus, and co-workers found for cytoplasmic determinants deposited in the egg by the mother. These genes set up the initial pattern of the embryo by regulating gene expression in broad regions of the early embryo.

Axis Establishment

As we mentioned earlier, cytoplasmic determinants in the egg are the substances that initially establish the axes of the *Drosophila* body. These substances are encoded by genes of the mother, fittingly called maternal effect genes. A **maternal effect gene** is a gene that, when mutant in the mother, results in a mutant phenotype in the offspring, regardless of the offspring's own genotype. In fruit fly development, the mRNA or protein products of maternal effect genes are placed in the egg while it is still in the mother's ovary. When the mother has a mutation in such a gene, she makes a defective gene product (or none at all), and her eggs are defective; when these eggs are fertilized, they fail to develop properly.

Because they control the orientation (polarity) of the egg and consequently of the fly, maternal effect genes are also called **egg-polarity genes**. One group of these genes sets up the anterior-posterior axis of the embryo, while a second group establishes the dorsal-ventral axis. Like mutations in

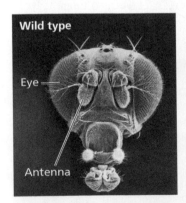 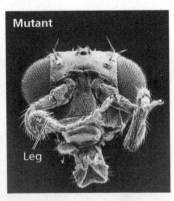

▲ **Figure 16.8 Abnormal pattern formation in *Drosophila*.** Mutations in certain regulatory genes, called homeotic genes, cause a misplacement of structures in an animal. These scanning electron micrographs contrast the head of a wild-type fly, bearing a pair of small antennae, with that of a homeotic mutant (a fly with a mutation in a single gene), bearing a pair of legs in place of antennae.

segmentation genes, mutations in maternal effect genes are generally embryonic lethals.

Bicoid: A Morphogen Determining Head Structures

To see how maternal effect genes determine the body axes of the offspring, we'll focus on one such gene, called *bicoid*, a term meaning "two-tailed." An embryo whose mother has two mutant alleles of the *bicoid* gene lacks the front half of its body and has posterior structures at both ends **(Figure 16.9)**. This phenotype suggested to Nüsslein-Volhard and her colleagues that the product of the mother's *bicoid* gene is essential for setting up the anterior end of the fly and might be concentrated at the future anterior end of the embryo. This hypothesis is an example of the *morphogen gradient hypothesis* first proposed by embryologists a century ago; in this hypothesis, gradients of substances called **morphogens** establish an embryo's axes and other features of its form.

DNA technology and other modern biochemical methods enabled the researchers to test whether the *bicoid* product, a protein called Bicoid, is in fact a morphogen that determines the anterior end of the fly. The first question they asked was whether the mRNA and protein products of these genes are located in the egg in a position consistent with the hypothesis. They found that *bicoid* mRNA is highly concentrated at the extreme anterior end of the mature egg, as predicted by the hypothesis **(Figure 16.10)**. After the egg is fertilized, the mRNA is translated into protein. The Bicoid protein then diffuses from the anterior end toward the posterior, resulting in a gradient of protein within the early embryo, with the highest concentration at the anterior end. These results are consistent with the hypothesis that Bicoid protein specifies the fly's anterior end. To test the hypothesis more specifically, scientists injected

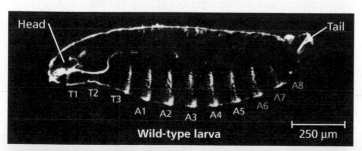

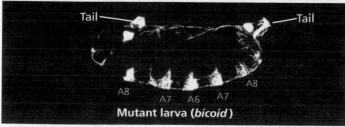

▲ **Figure 16.9 Effect of the *bicoid* gene on *Drosophila* development.** A wild-type fruit fly larva has a head, three thoracic (T) segments, eight abdominal (A) segments, and a tail. A larva whose mother has two mutant alleles of the *bicoid* gene has two tails and lacks all anterior structures (LMs).

▼ **Figure 16.10** **Inquiry**

Could Bicoid be a morphogen that determines the anterior end of a fruit fly?

Experiment Using a genetic approach to study *Drosophila* development, Christiane Nüsslein-Volhard and colleagues at the European Molecular Biology Laboratory in Heidelberg, Germany, analyzed expression of the *bicoid* gene. The researchers hypothesized that *bicoid* normally codes for a morphogen that specifies the head (anterior) end of the embryo. To begin to test this hypothesis, they used molecular techniques to determine whether the mRNA and protein encoded by this gene were found in the anterior end of the fertilized egg and early embryo of wild-type flies.

Results Bicoid mRNA (dark blue) was confined to the anterior end of the unfertilized egg. Later in development, Bicoid protein (dark orange) was seen to be concentrated in cells at the anterior end of the embryo.

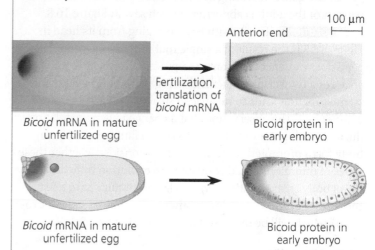

100 µm

Anterior end

Fertilization, translation of *bicoid* mRNA

Bicoid mRNA in mature unfertilized egg

Bicoid protein in early embryo

Bicoid mRNA in mature unfertilized egg

Bicoid protein in early embryo

Conclusion The location of *bicoid* mRNA and the diffuse gradient of Bicoid protein seen later are consistent with the hypothesis that Bicoid protein is a morphogen specifying formation of head-specific structures.

Source C. Nüsslein-Volhard et al., Determination of anteroposterior polarity in *Drosophila*, *Science* 238:1675–1681 (1987); W. Driever and C. Nüsslein-Volhard, A gradient of *bicoid* protein in *Drosophila* embryos, *Cell* 54:83–93 (1988); T. Berleth et al., The role of localization of *bicoid* RNA in organizing the anterior pattern of the *Drosophila* embryo, *EMBO Journal* 7:1749–1756 (1988).

WHAT IF? *The researchers needed further evidence, so they injected* bicoid *mRNA into the anterior end of an egg from a female with a mutation disabling the* bicoid *gene. Given that the hypothesis was supported, predict what happened.*

pure *bicoid* mRNA into various regions of early embryos. The protein that resulted from its translation caused anterior structures to form at the injection sites.

The *bicoid* research was groundbreaking for several reasons. First, it led to the identification of a specific protein required for some of the earliest steps in pattern formation. It thus helped us understand how different regions of the egg can give rise to cells that go down different developmental pathways. Second, it increased our understanding of the mother's critical role in the initial phases of embryonic development. Finally, the principle that a gradient of morphogens can determine polarity and position has proved to be a key developmental concept for a number of species, just as early embryologists had hypothesized.

Maternal mRNAs are crucial during development of many species. In *Drosophila*, gradients of specific proteins encoded by maternal mRNAs determine the posterior and anterior ends and establish the dorsal-ventral axis. As the fly embryo grows, it reaches a point when the embryonic program of gene expression takes over, and the maternal mRNAs must be destroyed. (This process involves miRNAs in *Drosophila* and other species.) Later, positional information encoded by the embryo's genes, operating on an ever finer scale, establishes a specific number of correctly oriented segments and triggers the formation of each segment's characteristic structures. When the genes operating in this final step are abnormal, the pattern of the adult is abnormal, as you saw in Figure 16.8.

EVOLUTION The fly with legs emerging from its head in Figure 16.8 is the result of a single mutation in one gene. The gene does not encode an antenna protein, however. Instead, it encodes a transcription factor that regulates other genes, and its malfunction leads to misplaced structures like legs instead of antennae. The observation that a change in gene regulation during development could lead to such a fantastic change in body form prompted some scientists to consider whether these types of mutations could contribute to evolution by generating novel body shapes. Ultimately this line of inquiry gave rise to the field of evolutionary developmental biology, so-called "evo-devo," which will be discussed further in Chapter 18.

CONCEPT CHECK 16.1

1. **MAKE CONNECTIONS** As you learned in Chapter 9, mitosis gives rise to two daughter cells that are genetically identical to the parent cell. Yet you, the product of many mitotic divisions, are not composed of identical cells. Why?

2. **MAKE CONNECTIONS** Explain how the signaling molecules released by an embryonic cell can induce changes in a neighboring cell without entering the cell. (See Figure 5.26.)

3. Why are fruit fly maternal effect genes also called egg-polarity genes?

For suggested answers, see Appendix A.

CONCEPT 16.2

Cloning of organisms showed that differentiated cells could be "reprogrammed" and ultimately led to the production of stem cells

When the field of developmental biology (then called embryology) was first taking shape at the beginning of the 20th century, a major question was whether all the cells of an organism have the same genes (a concept called *genomic equivalence*) or whether cells lose genes during the process of differentiation.

Today, we know that genes are not lost—but the question that remains is whether each cell is able to express all of its genes.

One way to answer this question is to see whether a differentiated cell has the potential to generate a whole organism. Because the organism develops from a single cell without either meiosis or fertilization, this is called "cloning." In this context, cloning produces one or more organisms genetically identical to the "parent" that donated the single cell. This is often called *organismal cloning* to differentiate it from gene cloning and, more significantly, from cell cloning—the division of an asexually reproducing cell into a collection of genetically identical cells. (The common theme for all types of cloning is that the product is genetically identical to the parent. In fact, the word *clone* comes from the Greek *klon*, meaning "twig.")

The current interest in organismal cloning arises primarily from its potential to generate stem cells, which can in turn generate many different tissues. Conceptually, though, the series of experiments discussed here provides a context for thinking about how regulation of gene expression genetically programs the overall potential of a cell—what genes it can express. Let's discuss early organismal cloning experiments before we consider more recent progress in cloning and procedures for producing stem cells.

Cloning Plants and Animals

The successful cloning of whole plants from single differentiated cells was accomplished during the 1950s by F. C. Steward and his students at Cornell University, who worked with carrot plants. They found that single differentiated cells taken from the root (the carrot) and incubated in culture medium could grow into normal adult plants, each genetically identical to the parent plant. These results showed that differentiation does not necessarily involve irreversible changes in the DNA. In plants, at least, mature cells can "dedifferentiate" and then give rise to all the specialized cell types of the organism. Any cell with this potential is said to be **totipotent**.

Differentiated cells from animals generally do not divide in culture, much less develop into the multiple cell types of a new organism. Therefore, early researchers had to use a different approach to the question of whether differentiated animal cells can be totipotent. Their approach was to remove the nucleus of an unfertilized or fertilized egg and replace it with the nucleus of a differentiated cell, a procedure called *nuclear transplantation*. If the nucleus from the differentiated donor cell retains its full genetic capability, then it should be able to direct development of the recipient cell into all the tissues and organs of an organism.

Such experiments were conducted on one species of frog (*Rana pipiens*) by Robert Briggs and Thomas King in the 1950s and on another (*Xenopus laevis*) by John Gurdon in the 1970s. These researchers transplanted a nucleus from an embryonic or tadpole cell into an enucleated (nucleus-lacking) egg of the same species. In Gurdon's experiments, the transplanted nucleus was often able to support normal development of the egg into a tadpole (**Figure 16.11**). However, he found that the potential of a transplanted nucleus to direct normal development was inversely related to the age of the donor: the older the donor nucleus, the lower the percentage of normally developing tadpoles.

From these results, Gurdon concluded that something in the nucleus *does* change as animal cells differentiate. In frogs and most other animals, nuclear potential tends to be restricted more and more as embryonic development and cell differentiation progress.

Reproductive Cloning of Mammals

In addition to cloning frogs, researchers have long been able to clone mammals by transplanting nuclei or cells from a variety of early embryos. But until about 15 years ago, it was not known whether a nucleus from a fully differentiated cell could be reprogrammed to successfully act as a donor nucleus. In 1997, however, researchers at the Roslin Institute in Scotland captured newspaper headlines when they announced the birth of Dolly, a lamb cloned from an adult sheep by nuclear transplantation from a differentiated cell (**Figure 16.12**). These researchers achieved

---

▼ Figure 16.11 Inquiry

Can the nucleus from a differentiated animal cell direct development of an organism?

Experiment John Gurdon and colleagues at Oxford University, in England, destroyed the nuclei of frog (*Xenopus laevis*) eggs by exposing the eggs to ultraviolet light. They then transplanted nuclei from cells of frog embryos and tadpoles into the enucleated eggs.

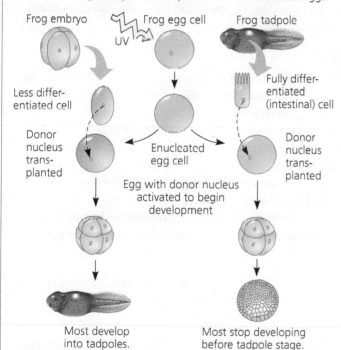

Most develop into tadpoles.

Most stop developing before tadpole stage.

Results When the transplanted nuclei came from an early embryo, whose cells are relatively undifferentiated, most of the recipient eggs developed into tadpoles. But when the nuclei came from the fully differentiated intestinal cells of a tadpole, fewer than 2% of the eggs developed into normal tadpoles, and most of the embryos stopped developing at a much earlier stage.

Conclusion The nucleus from a differentiated frog cell can direct development of a tadpole. However, its ability to do so decreases as the donor cell becomes more differentiated, presumably because of changes in the nucleus.

Source J. B. Gurdon et al., The developmental capacity of nuclei transplanted from keratinized cells of adult frogs, *Journal of Embryology and Experimental Morphology* 34:93–112 (1975).

WHAT IF? *If each cell in a four-cell embryo was already so specialized that it was not totipotent, what results would you predict for the experiment on the left side of the figure?*

▼ Figure 16.12 Research Method

Reproductive Cloning of a Mammal by Nuclear Transplantation

Application This method produces cloned animals with nuclear genes identical to those of the animal supplying the nucleus.

Technique The procedure below produced Dolly, the first case of a mammal cloned using the nucleus of a differentiated cell.

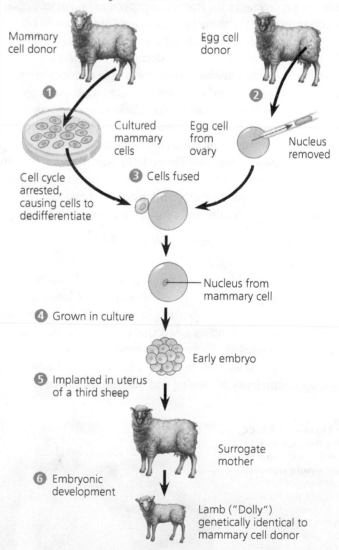

Results The cloned animal is genetically identical to the animal supplying the nucleus but differs from the egg donor and surrogate mother. (The latter two are "Scottish blackface" sheep.)

the necessary dedifferentiation of donor nuclei by culturing mammary cells in nutrient-poor medium. They then fused these cells with enucleated sheep eggs. The resulting diploid cells divided to form early embryos, which were implanted into surrogate mothers. Out of several hundred embryos, one successfully completed normal development, and Dolly was born.

Later analyses showed that Dolly's chromosomal DNA was indeed identical to that of the nucleus donor. (Her mitochondrial DNA came from the egg donor, as expected.) At the age of 6, Dolly suffered complications from a lung disease usually seen only in much older sheep and was euthanized. Dolly's premature death, as well as an arthritic condition, led to speculation that her cells were in some way not quite as healthy as those of a normal sheep, possibly reflecting incomplete reprogramming of the original transplanted nucleus.

Since that time, researchers have cloned numerous other mammals, including mice, cats, cows, horses, pigs, dogs, and monkeys. In most cases, their goal has been the production of new individuals; this is known as *reproductive cloning*. We have already learned a lot from such experiments. For example, cloned animals of the same species do *not* always look or behave identically. In a herd of cows cloned from the same line of cultured cells, certain cows are dominant in behavior and others are more submissive. Another example of nonidentity in clones is the first cloned cat, named CC for Carbon Copy **(Figure 16.13)**. She has a calico coat, like her single female parent, but the color and pattern are different because of random X chromosome inactivation, which is a normal occurrence during embryonic development (see Figure 12.8). And identical human twins, which are naturally occurring "clones," are always slightly different. Clearly, environmental influences and random phenomena can play a significant role during development.

Faulty Gene Regulation in Cloned Animals

In most nuclear transplantation studies thus far, only a small percentage of cloned embryos develop normally to birth. And like Dolly, many cloned animals exhibit defects. Cloned mice, for instance, are prone to obesity, pneumonia, liver failure, and premature death. Scientists assert that even cloned animals that appear normal are likely to have subtle defects.

In recent years, we have begun to uncover some reasons for the low efficiency of cloning and the high incidence of abnormalities. In the nuclei of fully differentiated cells, a small subset of genes is turned on and expression of the rest is repressed. This regulation often is the result of epigenetic changes in chromatin, such as acetylation of histones or methylation of DNA (see Figure 15.7). During the nuclear transfer procedure, many of these changes must be reversed in the later-stage nucleus from a donor animal for genes to be expressed or repressed appropriately in early stages of development. Researchers have found that the DNA in cells from cloned embryos, like that of differentiated cells, often has more methyl groups than does the DNA in equivalent cells from normal embryos of the same species. This finding suggests that the reprogramming of donor nuclei requires more accurate and complete chromatin restructuring than occurs during cloning procedures. Because DNA methylation helps regulate gene expression, misplaced or extra methyl groups in the DNA of donor nuclei may interfere with the pattern of gene expression necessary for normal embryonic development. In fact, the success of a cloning attempt may depend in large part on whether or not the chromatin in the donor nucleus can be artificially "rejuvenated" to resemble that of a newly fertilized egg.

Stem Cells of Animals

The successful cloning of many mammals, including primates, has heightened speculation about the cloning of humans, which has not yet been achieved. The main reason researchers are trying to clone human embryos is not for reproduction, but for the production of stem cells to treat human diseases. A **stem cell** is a relatively unspecialized cell that can both reproduce itself indefinitely and, under appropriate conditions, differentiate into specialized cells of one or more types **(Figure 16.14)**. Thus, stem cells can both replenish their own undifferentiated population and generate cells that travel down specific differentiation pathways.

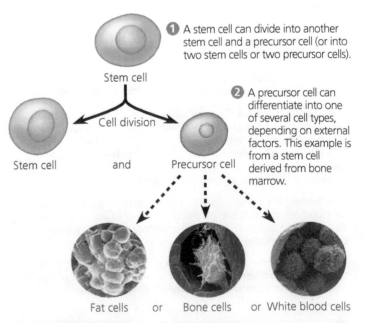

❶ A stem cell can divide into another stem cell and a precursor cell (or into two stem cells or two precursor cells).

Stem cell

Cell division

Stem cell   and   Precursor cell

❷ A precursor cell can differentiate into one of several cell types, depending on external factors. This example is from a stem cell derived from bone marrow.

Fat cells   or   Bone cells   or   White blood cells

▲ **Figure 16.14 How stem cells maintain their own population and generate differentiated cells.**

▶ **Figure 16.13 CC, the first cloned cat (right), and her single parent.** Rainbow (left) donated the nucleus in a cloning procedure that resulted in CC. However, the two cats are not identical: Rainbow has orange patches on her fur, but CC does not.

Many early animal embryos contain stem cells capable of giving rise to differentiated embryonic cells of any type. Stem cells can be isolated from early embryos at a stage called the blastula stage or its human equivalent, the blastocyst stage (**Figure 16.15**). In culture, these *embryonic stem (ES) cells* reproduce indefinitely; and depending on culture conditions, they can be made to differentiate into a wide variety of specialized cells, including even eggs and sperm.

The adult body also has stem cells, which serve to replace nonreproducing specialized cells as needed. In contrast to ES cells, *adult stem cells* are not able to give rise to all cell types in the organism, though in many cases they can generate multiple types. For example, one of the several types of stem cells in bone marrow can generate all the different kinds of blood cells (see Figure 16.15), and another can differentiate into bone, cartilage, fat, muscle, and the linings of blood vessels. To the surprise of many, the adult brain has been found to contain

stem cells that continue to produce certain kinds of nerve cells there. Researchers have also reported finding stem cells in skin, hair, eyes, and dental pulp. Although adult animals have only tiny numbers of stem cells, scientists are learning to identify and isolate these cells from various tissues and, in some cases, to grow them in culture. With the right culture conditions (often including the addition of specific growth factors), cultured stem cells from adult animals have been made to differentiate into multiple types of specialized cells, although none are as versatile as ES cells.

Research with embryonic or adult stem cells is a source of valuable data about differentiation and has enormous potential for medical applications. The ultimate aim is to supply cells for the repair of damaged or diseased organs: for example, insulin-producing pancreatic cells for people with type 1 diabetes or certain kinds of brain cells for people with Parkinson's disease or Huntington's disease. Adult stem cells from bone marrow have long been used as a source of immune system cells in patients whose own immune systems are nonfunctional because of genetic disorders or radiation treatments for cancer.

The developmental potential of adult stem cells is limited to certain tissues. ES cells hold more promise than adult stem cells for most medical applications because ES cells are **pluripotent**, capable of differentiating into many different cell types. The only way to obtain ES cells thus far, however, has been to harvest them from human embryos, which raises ethical and political issues.

ES cells are currently obtained from embryos donated (with informed consent) by patients undergoing infertility treatment or from long-term cell cultures originally established with cells isolated from donated embryos. If scientists were able to clone human embryos to the blastocyst stage, they might be able to use such clones as the source of ES cells in the future. Furthermore, with a donor nucleus from a person with a particular disease, they might be able to produce ES cells for treatment that match the patient and are thus not rejected by his or her immune system. When the main aim of cloning is to produce ES cells to treat disease, the process is called *therapeutic cloning*. Although most people believe that reproductive cloning of humans is unethical, opinions vary about the morality of therapeutic cloning.

Resolving the debate now seems less imperative because researchers have been able to turn back the clock in fully differentiated cells, reprogramming them to act like ES cells. The accomplishment of this feat, which posed formidable obstacles, was announced in 2007, first by labs using mouse skin cells and then by additional groups using cells from human skin and other organs or tissues. In all these cases, researchers transformed the differentiated cells into ES cells by using types of viruses called retroviruses to introduce extra cloned copies of four "stem cell" master regulatory genes. All the tests that were carried out at the time indicated that the transformed cells, known as *induced pluripotent stem (iPS) cells*, could do everything ES cells can do. More recently, however, several research groups have uncovered differences between iPS and ES cells in

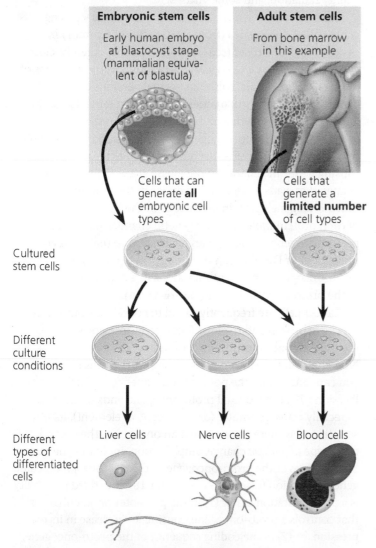

Embryonic stem cells

Early human embryo at blastocyst stage (mammalian equivalent of blastula)

Cells that can generate **all** embryonic cell types

Adult stem cells

From bone marrow in this example

Cells that generate a **limited number** of cell types

Cultured stem cells

Different culture conditions

Different types of differentiated cells

Liver cells   Nerve cells   Blood cells

▲ **Figure 16.15 Working with stem cells.** Animal stem cells, which can be isolated from early embryos or adult tissues and grown in culture, are self-perpetuating, relatively undifferentiated cells. Embryonic stem cells are easier to grow than adult stem cells and can theoretically give rise to *all* types of cells in an organism. The range of cell types that can arise from adult stem cells is not yet fully understood.

gene expression and other cellular functions, such as cell division. At least until these differences are fully understood, the study of ES cells will continue to make important contributions to the development of stem cell therapies. (In fact, ES cells will likely always be a focus of basic research as well.) In the meantime, work is proceeding using the iPS cells in hand.

There are two major potential uses for human iPS cells. First, cells from patients suffering from diseases can be reprogrammed to become iPS cells, which can act as model cells for studying the disease and potential treatments. Human iPS cell lines have already been developed from individuals with type 1 diabetes, Parkinson's disease, and at least a dozen other diseases. Second, in the field of regenerative medicine, a patient's own cells could be reprogrammed into iPS cells and then used to replace nonfunctional tissues. Developing techniques that direct iPS cells to become specific cell types for this purpose is an area of intense research, one that has already seen some success. The iPS cells created in this way could eventually provide tailor-made replacement cells for patients without using any human eggs or embryos, thus circumventing most ethical objections.

The research described in this and the preceding section on stem cells and cell differentiation has underscored the key role of gene regulation in embryonic development. The genetic program is carefully balanced between turning on the genes for differentiation in the right place and turning off other genes. Even when an organism is fully developed, gene expression is regulated in a similarly fine-tuned manner. In the final section of the chapter, we'll consider how fine this tuning is by looking at how specific changes in expression of one or a few genes can lead to the development of cancer.

CONCEPT CHECK 16.2

1. Based on current knowledge, how would you explain the difference in the percentage of tadpoles that developed from the two kinds of donor nuclei in Figure 16.11?

2. If you were to clone a sheep using the technique shown in Figure 16.12, would all the progeny sheep ("clones") look identical? Why or why not?

3. **WHAT IF?** If you were a doctor who wanted to use iPS cells to treat a patient with severe type 1 diabetes, what new technique would have to be developed?

For suggested answers, see Appendix A.

CONCEPT 16.3

Abnormal regulation of genes that affect the cell cycle can lead to cancer

In Chapter 9, we considered cancer as a set of diseases in which cells escape from the control mechanisms that normally limit their growth. Now that we have discussed the molecular basis of gene expression and its regulation, we are ready to look at

cancer more closely. The gene regulation systems that go wrong during cancer turn out to be the very same systems that play important roles in embryonic development, the maintenance of stem cell populations, and many other biological processes. Thus, research into the molecular basis of cancer has both benefited from and informed many other fields of biology.

Types of Genes Associated with Cancer

The genes that normally regulate cell growth and division during the cell cycle include genes for growth factors, their receptors, and the intracellular molecules of signaling pathways. (To review the cell cycle, see Chapter 9; for cell signaling, see Concept 5.6.) Mutations that alter any of these genes in somatic cells can lead to cancer. The agent of such change can be random spontaneous mutation. However, it is likely that many cancer-causing mutations result from environmental influences, such as chemical carcinogens, X-rays and other high-energy radiation, and some viruses.

Cancer research led to the discovery of cancer-causing genes called **oncogenes** (from the Greek *onco*, tumor) in certain types of viruses (see Chapter 17). Subsequently, close counterparts of viral oncogenes were found in the genomes of humans and other animals. The normal versions of the cellular genes, called **proto-oncogenes**, code for proteins that stimulate normal cell growth and division.

How might a proto-oncogene—a gene that has an essential function in normal cells—become an oncogene, a cancer-causing gene? In general, an oncogene arises from a genetic change that leads to an increase either in the amount of the proto-oncogene's protein product or in the intrinsic activity of each protein molecule. The genetic changes that convert proto-oncogenes to oncogenes fall into three main categories: movement of DNA within the genome, amplification of a proto-oncogene, and point mutations in a control element or in the proto-oncogene itself **(Figure 16.16)**.

Cancer cells are frequently found to contain chromosomes that have broken and rejoined incorrectly, translocating fragments from one chromosome to another (see Figure 12.14). Now that you have learned how gene expression is regulated, you can understand the possible consequences of such translocations. If a translocated proto-oncogene ends up near an especially active promoter (or other control element), its transcription may increase, making it an oncogene. The second main type of genetic change, amplification, increases the number of copies of the proto-oncogene in the cell through repeated gene duplication (discussed in Chapter 18). The third possibility is a point mutation either (1) in the promoter or an enhancer that controls a proto-oncogene, causing an increase in its expression, or (2) in the coding sequence of the proto-oncogene, changing the gene's product to a protein that is more active or more resistant to degradation than the normal protein. All these mechanisms can lead to abnormal stimulation of the cell cycle and put the cell on the path to becoming malignant.

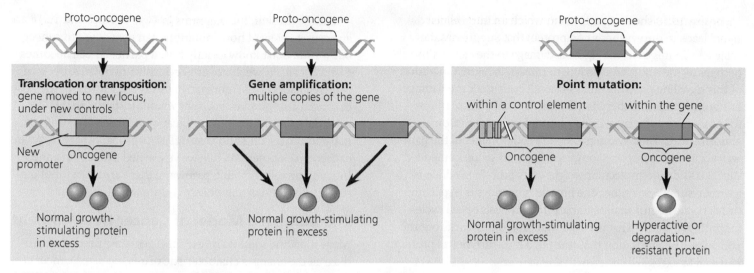

Translocation or transposition: gene moved to new locus, under new controls

New promoter
Oncogene

Normal growth-stimulating protein in excess

Gene amplification: multiple copies of the gene

Normal growth-stimulating protein in excess

Point mutation:

within a control element
Oncogene

within the gene
Oncogene

Normal growth-stimulating protein in excess

Hyperactive or degradation-resistant protein

▲ **Figure 16.16 Genetic changes that can turn proto-oncogenes into oncogenes.**

In addition to genes whose products normally promote cell division, cells contain genes whose normal products *inhibit* cell division. Such **tumor-suppressor genes** encode proteins that help prevent uncontrolled cell growth. Any mutation that decreases the normal activity of a tumor-suppressor protein may contribute to the onset of cancer, in effect stimulating growth through the absence of suppression.

Tumor-suppressor gene products have various functions. Some tumor-suppressor proteins repair damaged DNA, a function that prevents the cell from accumulating cancer-causing mutations. Other tumor-suppressor proteins control the adhesion of cells to each other or to the extracellular matrix; proper cell anchorage is crucial in normal tissues—and is often absent in cancers. Still other tumor-suppressor proteins are components of cell-signaling pathways that inhibit the cell cycle.

Interference with Cell-Signaling Pathways

The proteins encoded by many proto-oncogenes and tumor-suppressor genes are components of cell-signaling pathways.

Let's take a closer look at how such proteins function in normal cells and what goes wrong with their function in cancer cells. We'll focus on the products of two key genes, the *ras* proto-oncogene and the *p53* tumor-suppressor gene. Mutations in *ras* occur in about 30% of human cancers, and mutations in *p53* in more than 50%.

The Ras protein, encoded by the ***ras* gene** (named for r̲a̲t̲ s̲arcoma, a connective tissue cancer), is a G protein that relays a signal from a growth factor receptor on the plasma membrane to a cascade of protein kinases (see Figure 5.21). The cellular response at the end of the pathway is the synthesis of a protein that stimulates the cell cycle **(Figure 16.17)**. Normally, such a pathway will not operate unless triggered by the appropriate growth factor. But certain mutations in the *ras* gene can lead to production of a hyperactive Ras protein that triggers the kinase cascade even in the absence of growth factor, resulting in increased cell division. In fact, hyperactive versions or excess amounts of any of the pathway's components can have the same outcome: excessive cell division.

▶ **Figure 16.17 Normal and mutant cell cycle–stimulating pathway. (a)** The normal pathway is triggered by ❶ a growth factor that binds to ❷ its receptor in the plasma membrane. The signal is relayed to ❸ a G protein called Ras. Like all G proteins, Ras is active when GTP is bound to it. Ras passes the signal to ❹ a series of protein kinases. The last kinase activates ❺ a transcription factor (activator) that turns on one or more genes for ❻ a protein that stimulates the cell cycle. **(b)** If a mutation makes Ras or any other pathway component abnormally active, excessive cell division and cancer may result.

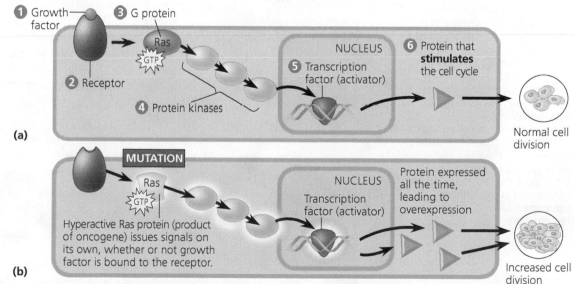

❶ Growth factor
❸ G protein
Ras
GTP
❷ Receptor
❹ Protein kinases
NUCLEUS
❺ Transcription factor (activator)
❻ Protein that **stimulates** the cell cycle
Normal cell division

(a)

MUTATION

Ras
GTP
Hyperactive Ras protein (product of oncogene) issues signals on its own, whether or not growth factor is bound to the receptor.
NUCLEUS
Transcription factor (activator)
Protein expressed all the time, leading to overexpression
Increased cell division

(b)

Figure 16.18 shows a pathway in which an intracellular signal leads to the synthesis of a protein that suppresses the cell cycle. In this case, the signal is damage to the cell's DNA, perhaps as the result of exposure to ultraviolet light. Operation of this signaling pathway blocks the cell cycle until the damage has been repaired. Otherwise, the damage might contribute to tumor formation by causing mutations or chromosomal abnormalities. Thus, the genes for the components of the pathway act as tumor-suppressor genes. The *p53 gene*, named for the 53,000-dalton molecular weight of its protein product, is a tumor-suppressor gene. The protein it encodes is a specific transcription factor that promotes the synthesis of cell cycle–inhibiting proteins. That is why a mutation that knocks out the *p53* gene, like a mutation that leads to a hyperactive Ras protein, can lead to excessive cell growth and cancer.

The *p53* gene has been called the "guardian angel of the genome." Once the gene is activated—for example, by DNA damage—the p53 protein functions as an activator for several other genes. Often it activates a gene called *p21*, whose product halts the cell cycle by binding to cyclin-dependent kinases, allowing time for the cell to repair the DNA. Researchers recently showed that p53 also activates expression of a group of miRNAs, which in turn inhibit the cell cycle. In addition, the p53 protein can turn on genes directly involved in DNA repair. Finally, when DNA damage is irreparable, p53 activates "suicide" genes, whose protein products bring about apoptosis, as described in the first section of this chapter. Thus, p53 acts in several ways to prevent a cell from passing on mutations due to DNA damage. If mutations do accumulate and the cell survives through many divisions—as is more likely if the *p53* tumor-suppressor gene is defective or missing—cancer may ensue. The many functions of p53 suggest a complex picture of regulation in normal cells, one that we do not yet fully understand.

For the present, the diagrams in Figures 16.17 and 16.18 are an accurate view of how mutations can contribute to cancer, but we still don't know exactly how a particular cell becomes a cancer cell. As we discover previously unknown aspects of gene regulation, it is informative to study their role in the onset of cancer. Such studies have shown, for instance, that DNA methylation and histone modification patterns differ in normal and cancer cells and that miRNAs probably participate in cancer development. While we've learned a lot about cancer by studying cell-signaling pathways, there are still a lot of outstanding questions that need to be answered.

The Multistep Model of Cancer Development

More than one somatic mutation is generally needed to produce all the changes characteristic of a full-fledged cancer cell. This may help explain why the incidence of cancer increases greatly with age. If cancer results from an accumulation of mutations and if mutations occur throughout life, then the longer we live, the more likely we are to develop cancer.

The model of a multistep path to cancer is well supported by studies of one of the best-understood types of human cancer, colorectal cancer. About 140,000 new cases of colorectal cancer are diagnosed each year in the United States, and the disease causes 50,000 deaths each year. Like most cancers, colorectal cancer develops gradually **(Figure 16.19)**. The first sign is often a polyp, a small, benign growth in the colon lining. The cells of the polyp look normal, although they divide unusually frequently. The tumor grows and may eventually become malignant, invading other tissues. The development of a malignant tumor is paralleled by a gradual accumulation of mutations that convert proto-oncogenes to oncogenes and knock out tumor-suppressor genes. A *ras* oncogene and a mutated *p53* tumor-suppressor gene are often involved.

▶ **Figure 16.18 Normal and mutant cell cycle–inhibiting pathway. (a)** In the normal pathway, ❶ DNA damage is an intracellular signal that is passed via ❷ protein kinases and leads to activation of ❸ p53. Activated p53 promotes transcription of the gene for a protein that inhibits the cell cycle. The resulting suppression of cell division ensures that the damaged DNA is not replicated. If the DNA damage is irreparable, the p53 signal leads to programmed cell death (apoptosis). **(b)** Mutations causing deficiencies in any pathway component can contribute to the development of cancer.

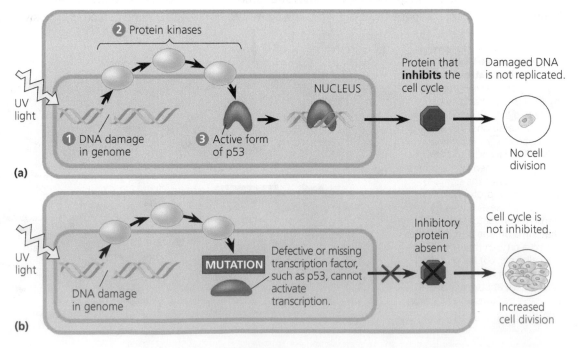

About half a dozen changes must occur at the DNA level for a cell to become fully cancerous. These changes usually include the appearance of at least one active oncogene and the mutation or loss of several tumor-suppressor genes. Furthermore, since mutant tumor-suppressor alleles are usually recessive, in most cases mutations must knock out *both* alleles in a cell's genome to block tumor suppression. (Most oncogenes, on the other hand, behave as dominant alleles.) The order in which these changes must occur is still under investigation, as is the relative importance of different mutations.

Since we understand the progression of this type of cancer, routine screenings are recommended to identify and remove any suspicious polyps. The colorectal cancer rate has been declining for the past 20 years, due in part to increased screening and in part to improved treatments. Treatments for other cancers have improved as well. Technical advances in the sequencing of DNA and mRNA have allowed medical researchers to compare the genes expressed by different types of tumors and by the same type in different individuals. These comparisons have led to personalized cancer treatments based on the molecular characteristics of an individual's tumor.

Inherited Predisposition and Other Factors Contributing to Cancer

The fact that multiple genetic changes are required to produce a cancer cell helps explain the observation that cancers can run in families. An individual inheriting an oncogene or a mutant allele of a tumor-suppressor gene is one step closer to accumulating the necessary mutations for cancer to develop than is an individual without any such mutations.

Geneticists are devoting much effort to identifying inherited cancer alleles so that predisposition to certain cancers can be detected early in life. About 15% of colorectal cancers, for example, involve inherited mutations. Many of these affect the tumor-suppressor gene called *adenomatous polyposis coli*, or *APC* (see Figure 16.19). This gene has multiple functions in the cell, including regulation of cell migration and adhesion. Even in patients with no family history of the disease, the *APC* gene is mutated in 60% of colorectal cancers. In these individuals, new mutations must occur in both *APC* alleles before the gene's function is lost. Since only 15% of colorectal cancers are associated with known inherited mutations, researchers continue in their efforts to identify "markers" that could predict the risk of developing this type of cancer.

There is evidence of a strong inherited predisposition in 5–10% of patients with breast cancer. This is the second most common type of cancer in the United States, striking over 230,000 women (and some men) annually and killing 40,000 each year. In 1990, after 16 years of research, geneticist Mary-Claire King convincingly demonstrated that mutations in one gene—*BRCA1*—were associated with increased susceptibility to breast cancer, a finding that flew in the face of medical opinion at the time. (*BRCA* stands for breast cancer.) Mutations in that gene or the related *BRCA2* gene are found in at least half of inherited breast cancers, and tests using DNA sequencing can detect these mutations. A woman who inherits one mutant *BRCA1* allele has a 60% probability of developing breast cancer before the age of 50, compared with only a 2% probability for an individual homozygous for the normal allele. Both *BRCA1* and *BRCA2* are considered tumor-suppressor genes because their wild-type alleles protect against breast cancer and their mutant alleles are recessive. Apparently, the BRCA1 and BRCA2 proteins both function in the cell's DNA damage repair pathway. More is known about BRCA2, which, in association with another protein, helps repair breaks that occur in both strands of DNA; it is crucial for maintaining undamaged DNA in a cell's nucleus.

Because DNA breakage can contribute to cancer, it makes sense that the risk of cancer can be lowered by minimizing exposure to DNA-damaging agents, such as the ultraviolet radiation in sunlight and chemicals found in cigarette smoke. Novel methods for early diagnosis and treatment of specific cancers

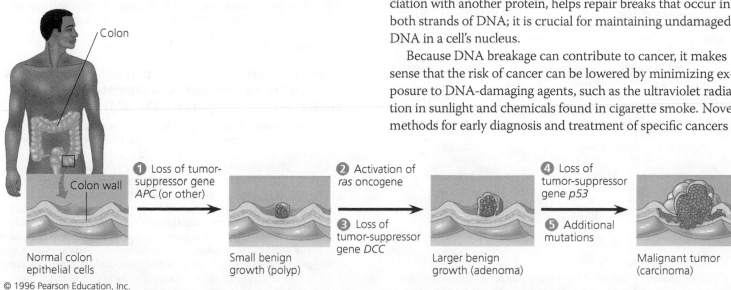

© 1996 Pearson Education, Inc.

▲ **Figure 16.19 A multistep model for the development of colorectal cancer.** Affecting the colon and/or rectum, this type of cancer is one of the best understood. Changes in a tumor parallel a series of genetic changes, including mutations affecting several tumor-suppressor genes (such as *p53*) and the *ras* proto-oncogene. Mutations of tumor-suppressor genes often entail loss (deletion) of the gene. *APC* stands for "adenomatous polyposis coli," and *DCC* stands for "deleted in colorectal cancer." Other mutation sequences can also lead to colorectal cancer.

are being developed that rely on new techniques for analyzing, and perhaps interfering with, gene expression in tumors. Ultimately, such approaches may lower the death rate from cancer.

The study of genes associated with cancer, inherited or not, increases our basic understanding of how disruption of normal gene regulation can result in this disease. In addition to the mutations and other genetic alterations described in this section, a number of *tumor viruses* can cause cancer in various animals, including humans. In fact, one of the earliest breakthroughs in understanding cancer came in 1911, when Peyton Rous, an American pathologist, discovered a virus that causes cancer in chickens. The Epstein-Barr virus, which causes infectious mononucleosis, has been linked to several types of cancer in humans, notably Burkitt's lymphoma. Papillomaviruses are associated with cancer of the cervix, and a virus called HTLV-1 causes a type of adult leukemia. Worldwide, viruses seem to play a role in about 15% of the cases of human cancer.

Viruses may at first seem very different from mutations as a cause of cancer. However, we now know that viruses can interfere with gene regulation in several ways if they integrate their genetic material into the DNA of a cell. Viral integration may donate an oncogene to the cell, disrupt a tumor-suppressor gene, or convert a proto-oncogene to an oncogene. In addition, some viruses produce proteins that inactivate p53 and other tumor-suppressor proteins, making the cell more prone to becoming cancerous. Viruses are powerful biological agents, and you'll learn more about their function in Chapter 17.

CONCEPT CHECK 16.3

1. The p53 protein can activate genes involved in apoptosis, or programmed cell death. Review Concept 16.1 and discuss how mutations in genes coding for proteins that function in apoptosis could contribute to cancer.
2. Under what circumstances is cancer considered to have a hereditary component?
3. **WHAT IF?** Explain how the types of mutations that lead to cancer are different for a proto-oncogene and a tumor-suppressor gene in terms of the effect of the mutation on the activity of the gene product.

For suggested answers, see Appendix A.

16 Chapter Review

SUMMARY OF KEY CONCEPTS

CONCEPT 16.1

A program of differential gene expression leads to the different cell types in a multicellular organism (pp. 312–320)

- Embryonic cells undergo **differentiation**, becoming specialized in structure and function. **Morphogenesis** encompasses the processes that give shape to the organism and its various structures. Cells differ in structure and function not because they contain different genes but because they express different portions of a common genome.
- Localized **cytoplasmic determinants** in the unfertilized egg are distributed differentially to daughter cells, where they regulate the expression of genes that affect those cells' developmental fates. In the process called **induction**, signaling molecules from embryonic cells cause transcriptional changes in nearby target cells.

Cytoplasmic determinants   Induction

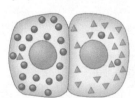 

- Differentiation is heralded by the appearance of tissue-specific proteins, which enable differentiated cells to carry out their specialized roles.
- **Apoptosis** is a type of programmed cell death in which cell components are disposed of in an orderly fashion, without damage to neighboring cells. Studies of the soil worm *Caenorhabditis elegans* showed that apoptosis occurs at defined times during embryonic development. Related apoptotic signaling pathways exist in the cells of humans and other mammals, as well as yeasts.
- In animals, **pattern formation**, the development of a spatial organization of tissues and organs, begins in the early embryo. **Positional information**, the molecular cues that control pattern formation, tells a cell its location relative to the body's axes and to other cells. In *Drosophila*, gradients of **morphogens** encoded by **maternal effect genes** determine the body axes. For example, the gradient of **Bicoid** protein determines the anterior-posterior axis.

? *Describe the two main processes that cause embryonic cells to head down different pathways to their final fates.*

CONCEPT 16.2

Cloning of organisms showed that differentiated cells could be "reprogrammed" and ultimately led to the production of stem cells (pp. 320–324)

- Studies showing genomic equivalence (that an organism's cells all have the same genome) provided the first examples of organismal cloning.
- Single differentiated cells from plants are often **totipotent**: capable of generating all the tissues of a complete new plant.
- Transplantation of the nucleus from a differentiated animal cell into an enucleated egg can sometimes give rise to a new animal.
- Certain embryonic **stem cells** (ES cells) from animal embryos or adult stem cells from adult tissues can reproduce and differentiate *in vitro* as well as *in vivo*, offering the potential for medical use. ES cells are **pluripotent** but difficult to acquire. Induced pluripotent stem (iPS) cells resemble ES cells in their capacity to differentiate; they can be generated by reprogramming differentiated cells. iPS cells hold promise for medical research and regenerative medicine.

? *Describe how a researcher could carry out organismal cloning, production of ES cells, and generation of iPS cells, focusing on how the cells are reprogrammed and using mice as an example. (The procedures are basically the same in humans and mice.)*

CONCEPT 16.3

Abnormal regulation of genes that affect the cell cycle can lead to cancer (pp. 324–328)

- The products of **proto-oncogenes** and **tumor-suppressor genes** control cell division. A DNA change that makes a proto-oncogene excessively active converts it to an **oncogene**, which may promote excessive cell division and cancer. A tumor-suppressor gene encodes a protein that inhibits abnormal cell division. A mutation in such a gene that reduces the activity of its protein product may also lead to excessive cell division and possibly to cancer.

- Many proto-oncogenes and tumor-suppressor genes encode components of growth-stimulating and growth-inhibiting signaling pathways, respectively, and mutations in these genes can interfere with normal cell-signaling pathways. A hyperactive version of a protein in a stimulatory pathway, such as **Ras** (a G protein), functions as an oncogene protein. A defective version of a protein in an inhibitory pathway, such as **p53** (a transcription activator), fails to function as a tumor suppressor.

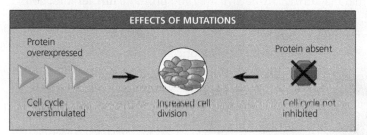

EFFECTS OF MUTATIONS

Protein overexpressed → Cell cycle overstimulated → Increased cell division ← Cell cycle not inhibited ← Protein absent

- In the multistep model of cancer development, normal cells are converted to cancer cells by the accumulation of mutations affecting proto-oncogenes and tumor-suppressor genes. Technical advances in DNA and mRNA sequencing are enabling cancer treatments that are more individually based.

- Individuals who inherit a mutant oncogene or tumor-suppressor allele have a predisposition to develop a particular cancer. Certain viruses promote cancer by integration of viral DNA into a cell's genome.

> **?** *Compare the usual functions of proteins encoded by proto-oncogenes with the functions of proteins encoded by tumor-suppressor genes.*

TEST YOUR UNDERSTANDING

Level 1: Knowledge/Comprehension

1. Muscle cells differ from nerve cells mainly because they
   a. express different genes.
   b. contain different genes.
   c. use different genetic codes.
   d. have unique ribosomes.
   e. have different chromosomes.

2. Cell differentiation always involves
   a. the production of tissue-specific proteins, such as muscle actin.
   b. the movement of cells.
   c. the transcription of the *myoD* gene.
   d. the selective loss of certain genes from the genome.
   e. the cell's sensitivity to environmental cues, such as light or heat.

Level 2: Application/Analysis

3. Apoptosis involves all but which of the following?
   a. fragmentation of the DNA
   b. cell-signaling pathways
   c. activation of cellular enzymes
   d. lysis of the cell
   e. digestion of cellular contents by scavenger cells

4. Absence of *bicoid* mRNA from a *Drosophila* egg leads to the absence of anterior larval body parts and mirror-image duplication of posterior parts. This is evidence that the product of the *bicoid* gene
   a. is transcribed in the early embryo.
   b. normally leads to formation of tail structures.
   c. normally leads to formation of head structures.
   d. is a protein present in all head structures.
   e. leads to programmed cell death.

5. Proto-oncogenes can change into oncogenes that cause cancer. Which of the following best explains the presence of these potential time bombs in eukaryotic cells?
   a. Proto-oncogenes first arose from viral infections.
   b. Proto-oncogenes normally help regulate cell division.
   c. Proto-oncogenes are genetic "junk."
   d. Proto-oncogenes are mutant versions of normal genes.
   e. Cells produce proto-oncogenes as they age.

Level 3: Synthesis/Evaluation

6. **SCIENTIFIC INQUIRY**
   Prostate cells usually require testosterone and other androgens to survive. But some prostate cancer cells thrive despite treatments that eliminate androgens. One hypothesis is that estrogen, often considered a female hormone, may be activating genes normally controlled by an androgen in these cancer cells. Describe one or more experiments to test this hypothesis. (See Figure 5.23 to review the action of these steroid hormones.)

7. **FOCUS ON EVOLUTION**
   Cancer cells can be considered a population that undergoes evolutionary processes such as random mutation and natural selection. Apply what you learned about evolution in Chapter 1 and about cancer in this chapter to discuss this concept.

8. **FOCUS ON ORGANIZATION**
   The property of life emerges at the biological level of the cell. The highly regulated process of apoptosis is not simply the destruction of a cell; it is also an emergent property. In a short essay (about 100–150 words), briefly explain the role of apoptosis in the development and proper functioning of an animal and describe how this form of programmed cell death is a process that emerges from the orderly integration of signaling pathways.

For selected answers, see Appendix A.

MasteringBiology®

Students Go to **MasteringBiology** for assignments, the eText, and the Study Area with practice tests, animations, and activities.

Instructors Go to **MasteringBiology** for automatically graded tutorials and questions that you can assign to your students, plus Instructor Resources.

Viruses

KEY CONCEPTS

17.1 A virus consists of a nucleic acid surrounded by a protein coat

17.2 Viruses replicate only in host cells

17.3 Viruses are formidable pathogens in animals and plants

OVERVIEW

A Borrowed Life

The photo in **Figure 17.1** shows a remarkable event: a cell under siege, releasing thousands more of its attackers, each capable of infecting another cell. The attackers (red) are human immunodeficiency viruses (HIV) emerging from a human immune cell. By injecting its genetic information into the infected cell, a single virus hijacks the cell, recruiting cellular machinery to manufacture many new viruses and promote further infection. Left untreated, HIV causes acquired immunodeficiency syndrome (AIDS) by destroying the immune system.

Compared to eukaryotic and even prokaryotic cells, viruses are much smaller and simpler in structure. Lacking the metabolic machinery found in a cell, a **virus** is an infectious particle consisting of little more than genes packaged in a protein coat.

Are viruses living or nonliving? Because viruses are capable of causing many diseases, researchers in the late 1800s saw a parallel with bacteria and proposed that viruses were the simplest of living forms. However, viruses cannot reproduce or carry out metabolism outside of a host cell. Most biologists studying viruses today would likely agree that they are not alive but exist in a shady area between life-forms and chemicals. The simple phrase used recently by two researchers describes them aptly enough: Viruses lead "a kind of borrowed life."

In this chapter, we'll explore the biology of viruses, beginning with their structure and then describing how they replicate. We'll end the chapter with a look at the role of viruses as disease-causing agents, or pathogens, of plants and animals.

▼ **Figure 17.1** Are the tiny viruses (red) budding from this cell alive?

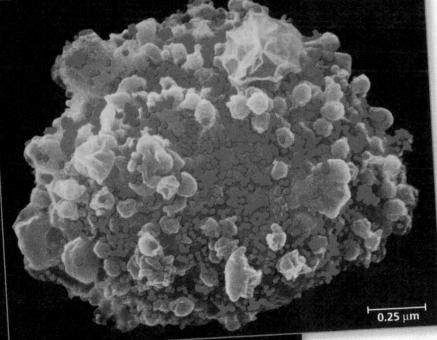

0.25 μm

CONCEPT 17.1

A virus consists of a nucleic acid surrounded by a protein coat

The tiniest viruses are only 20 nm in diameter—smaller than a ribosome. Millions could easily fit on a pinhead. Even the largest known virus, which has a diameter of several hundred nanometers, is barely visible under the light microscope. An early discovery that some viruses could be crystallized was exciting and puzzling news. Not even the simplest of cells can aggregate into regular crystals. But if viruses are not cells, then what are they? Examining the structure of a virus more closely reveals that it is an infectious particle consisting of nucleic acid enclosed in a protein coat and, for some viruses, surrounded by a membranous envelope.

Viral Genomes

We usually think of genes as being made of double-stranded DNA—the conventional double helix—but many viruses defy this convention. Their genomes may consist of double-stranded DNA, single-stranded DNA, double-stranded RNA, or single-stranded RNA, depending on the type of virus. A virus is called a DNA virus or an RNA virus, based on the kind of nucleic acid that makes up its genome. In either case, the genome is usually organized as a single linear or circular molecule of nucleic acid, although the genomes of some viruses consist of multiple molecules of nucleic acid. The smallest viruses known have only four genes in their genome, while the largest have several hundred to a thousand. For comparison, bacterial genomes contain about 200 to a few thousand genes.

Capsids and Envelopes

The protein shell enclosing the viral genome is called a **capsid**. Depending on the type of virus, the capsid may be rod-shaped, polyhedral, or more complex in shape. Capsids are built from a large number of protein subunits called *capsomeres*, but the number of different *kinds* of proteins in a capsid is usually small. Tobacco mosaic virus (TMV), for example, has a rigid, rod-shaped capsid made from over a thousand molecules of a single type of protein arranged in a helix; rod-shaped viruses are commonly called *helical viruses* for this reason **(Figure 17.2a)**. Adenoviruses, which infect the respiratory tracts of animals, have 252 identical protein molecules arranged in a polyhedral capsid with 20 triangular facets—an icosahedron; thus, these and other similarly shaped viruses are referred to as *icosahedral viruses* **(Figure 17.2b)**.

Some viruses have accessory structures that help them infect their hosts. For instance, a membranous envelope surrounds

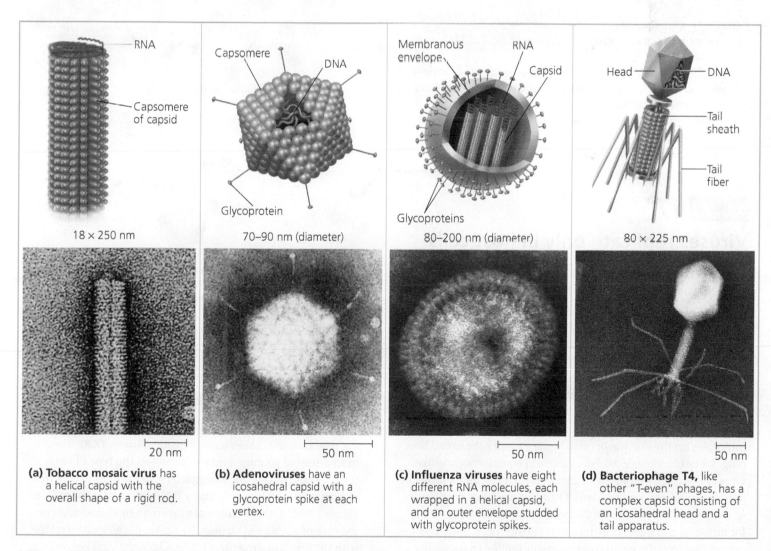

(a) **Tobacco mosaic virus** has a helical capsid with the overall shape of a rigid rod.

18 × 250 nm · 20 nm

(b) **Adenoviruses** have an icosahedral capsid with a glycoprotein spike at each vertex.

70–90 nm (diameter) · 50 nm

(c) **Influenza viruses** have eight different RNA molecules, each wrapped in a helical capsid, and an outer envelope studded with glycoprotein spikes.

80–200 nm (diameter) · 50 nm

(d) **Bacteriophage T4**, like other "T-even" phages, has a complex capsid consisting of an icosahedral head and a tail apparatus.

80 × 225 nm · 50 nm

▲ **Figure 17.2 Viral structure.** Viruses are made up of nucleic acid (DNA or RNA) enclosed in a protein coat (the capsid) and sometimes further wrapped in a membranous envelope. The individual protein subunits making up the capsid are called capsomeres. Although diverse in size and shape, viruses have many common structural features. (All micrographs are colorized TEMs.)

the capsids of influenza viruses and many other viruses found in animals **(Figure 17.2c)**. These **viral envelopes**, which are derived from the membranes of the host cell, contain host cell phospholipids and membrane proteins. They also contain proteins and glycoproteins of viral origin. (Glycoproteins are proteins with carbohydrates covalently attached.) Some viruses carry a few viral enzyme molecules within their capsids.

Many of the most complex capsids are found among the viruses that infect bacteria, called **bacteriophages**, or simply **phages**. The first phages studied included seven that infect *E. coli.* These seven phages were named type 1 (T1), type 2 (T2), and so forth, in the order of their discovery. The three T-even phages (T2, T4, and T6) turned out to be very similar in structure. Their capsids have elongated icosahedral heads enclosing their DNA. Attached to the head is a protein tail piece with fibers by which the phages attach to a bacterium **(Figure 17.2d)**. In the next section, we'll examine how these few viral parts function together with cellular components to produce large numbers of viral progeny.

CONCEPT CHECK 17.1

1. Compare the structures of tobacco mosaic virus and influenza virus (see Figure 17.2).
2. **MAKE CONNECTIONS** Bacteriophages were used to provide evidence that DNA carries genetic information (see Figure 13.4). Briefly describe the experiment carried out by Hershey and Chase, including in your description why the researchers chose to use phages.

For suggested answers, see Appendix A.

CONCEPT 17.2

Viruses replicate only in host cells

Viruses lack metabolic enzymes and equipment for making proteins, such as ribosomes. They are obligate intracellular parasites; in other words, they can replicate only within a host cell. It is fair to say that viruses in isolation are merely packaged sets of genes in transit from one host cell to another.

Each particular virus can infect cells of only a limited number of host species, called the **host range** of the virus. This host specificity results from the evolution of recognition systems by the virus. Viruses usually identify host cells by a "lock-and-key" fit between viral surface proteins and specific receptor molecules on the outside of cells. Some viruses have broad host ranges. For example, West Nile virus and equine encephalitis virus are distinctly different viruses that can each infect mosquitoes, birds, horses, and humans. Other viruses have host ranges so narrow that they infect only a single species. Measles virus, for instance, can infect only humans. Furthermore, viral infection of multicellular eukaryotes is usually limited to particular tissues. Human cold viruses infect only the cells lining the upper respiratory tract, and the HIV virus binds to receptors present only on certain types of white blood cells (see Figure 17.1).

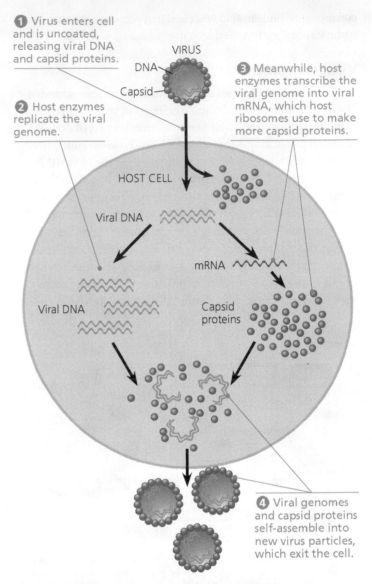

1 Virus enters cell and is uncoated, releasing viral DNA and capsid proteins.

2 Host enzymes replicate the viral genome.

3 Meanwhile, host enzymes transcribe the viral genome into viral mRNA, which host ribosomes use to make more capsid proteins.

VIRUS
DNA
Capsid

HOST CELL

Viral DNA

Viral DNA

mRNA

Capsid proteins

4 Viral genomes and capsid proteins self-assemble into new virus particles, which exit the cell.

▲ **Figure 17.3 A simplified viral replicative cycle.** A virus is an obligate intracellular parasite that uses the equipment and small molecules of its host cell to replicate. In this simplest of viral cycles, the parasite is a DNA virus with a capsid consisting of a single type of protein.

MAKE CONNECTIONS *Label each of the straight black arrows with one word representing the name of the process that is occurring. (Review Figure 14.24.)*

General Features of Viral Replicative Cycles

A viral infection begins when a virus binds to a host cell and the viral genome makes its way inside **(Figure 17.3)**. The mechanism of genome entry depends on the type of virus and the type of host cell. For example, T-even phages use their elaborate tail apparatus to inject DNA into a bacterium (see Figure 17.2d). Other viruses are taken up by endocytosis or, in the case of enveloped viruses, by fusion of the viral envelope with the host's plasma membrane. Once the viral genome is inside, the proteins it encodes can commandeer the host, reprogramming the cell to copy the viral nucleic acid and manufacture viral proteins. The host provides the nucleotides for making viral nucleic acids, as well as enzymes, ribosomes, tRNAs, amino acids, ATP, and other components needed for making the viral proteins. Many DNA viruses use the DNA polymerases of the host cell to synthesize new genomes along the templates provided by the viral DNA. In contrast, to replicate their genomes, RNA viruses use virally encoded RNA polymerases that can use RNA as a template. (Uninfected cells generally make no enzymes for carrying out this process.)

After the viral nucleic acid molecules and capsomeres are produced, they spontaneously self-assemble into new viruses. In fact, researchers can separate the RNA and capsomeres of TMV and then reassemble complete viruses simply by mixing the components together under the right conditions. The simplest type of viral replicative cycle ends with the exit of hundreds or thousands of viruses from the infected host cell, a process that often damages or destroys the cell. Such cellular damage and death, as well as the body's responses to this destruction, cause many of the symptoms associated with viral infections. The viral progeny that exit a cell have the potential to infect additional cells, spreading the viral infection.

There are many variations on the simplified viral replicative cycle we have just described. We'll now take a look at some of these variations in bacterial viruses (phages) and animal viruses; later in the chapter, we'll consider plant viruses.

Replicative Cycles of Phages

Phages are the best understood of all viruses, although some of them are also among the most complex. Research on phages led to the discovery that some double-stranded DNA viruses can replicate by two alternative mechanisms: the lytic cycle and the lysogenic cycle.

The Lytic Cycle

A phage replicative cycle that culminates in death of the host cell is known as a **lytic cycle**. The term refers to the last stage of infection, during which the bacterium lyses (breaks open) and releases the phages that were produced within the cell. Each of these phages can then infect a healthy cell, and a few successive lytic cycles can destroy an entire bacterial population in just a few hours. A phage that replicates only by a lytic cycle is called a **virulent phage**. **Figure 17.4** illustrates the major steps in the lytic cycle of T4, a typical virulent phage. Study this figure before proceeding.

After reading about the lytic cycle, you may wonder why phages haven't exterminated all bacteria. The reason is that bacteria have their own defenses. First, natural selection favors bacterial mutants with receptors that are no longer recognized by a particular type of phage. Second, when phage DNA does

▶ **Figure 17.4 The lytic cycle of phage T4, a virulent phage.** Phage T4 has almost 300 genes, which are transcribed and translated using the host cell's machinery. One of the first phage genes translated after the viral DNA enters the host cell codes for an enzyme that degrades the host cell's DNA (step 2); the phage DNA is protected from breakdown because it contains a modified form of cytosine that is not recognized by the enzyme. The entire lytic cycle, from the phage's first contact with the cell surface to cell lysis, takes only 20–30 minutes at 37°C.

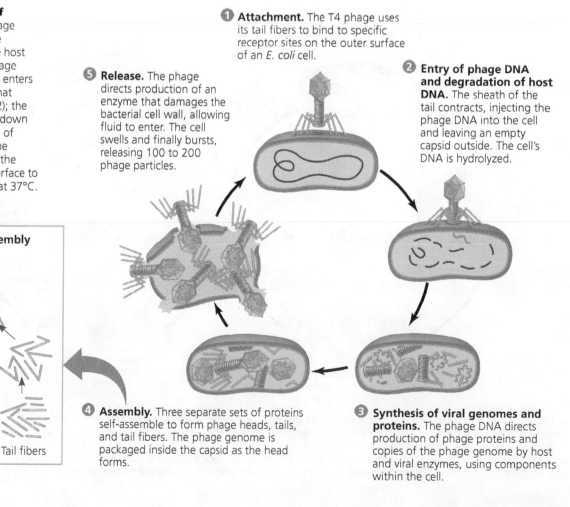

1 Attachment. The T4 phage uses its tail fibers to bind to specific receptor sites on the outer surface of an *E. coli* cell.

2 Entry of phage DNA and degradation of host DNA. The sheath of the tail contracts, injecting the phage DNA into the cell and leaving an empty capsid outside. The cell's DNA is hydrolyzed.

3 Synthesis of viral genomes and proteins. The phage DNA directs production of phage proteins and copies of the phage genome by host and viral enzymes, using components within the cell.

4 Assembly. Three separate sets of proteins self-assemble to form phage heads, tails, and tail fibers. The phage genome is packaged inside the capsid as the head forms.

5 Release. The phage directs production of an enzyme that damages the bacterial cell wall, allowing fluid to enter. The cell swells and finally bursts, releasing 100 to 200 phage particles.

Phage assembly

Head Tail Tail fibers

enter a bacterium, the DNA often is identified as foreign and cut up by cellular enzymes called **restriction enzymes**, which are so named because their activity *restricts* the ability of the phage to infect the bacterium. (These enzymes are used in molecular biology and DNA cloning techniques; see Concept 13.4.) The bacterial cell's own DNA is methylated in a way that prevents attack by its own restriction enzymes. But just as natural selection favors bacteria with mutant receptors or effective restriction enzymes, it also favors phage mutants that can bind the altered receptors or are resistant to particular restriction enzymes. Thus, the parasite-host relationship is in constant evolutionary flux.

There is yet a third important reason bacteria have been spared from extinction as a result of phage activity. Instead of lysing their host cells, many phages coexist with them in a state called lysogeny, which we'll now discuss.

The Lysogenic Cycle

In contrast to the lytic cycle, which kills the host cell, the **lysogenic cycle** allows replication of the phage genome without destroying the host. Phages capable of using both modes of replicating within a bacterium are called **temperate phages**. A temperate phage called lambda, written with the Greek letter λ, is widely used in biological research. Phage λ resembles T4, but its tail has only one short tail fiber.

Infection of an *E. coli* cell by phage λ begins when the phage binds to the surface of the cell and injects its linear DNA genome **(Figure 17.5)**. Within the host, the λ DNA molecule forms a circle. What happens next depends on the replicative mode: lytic cycle or lysogenic cycle. During a lytic cycle, the viral genes immediately turn the host cell into a λ-producing factory, and the cell soon lyses and releases its viral products. During a lysogenic cycle, however, the λ DNA molecule is incorporated into a specific site on the *E. coli* chromosome by viral proteins that break both circular DNA molecules and join them to each other. When integrated into the bacterial chromosome in this way, the viral DNA is known as a **prophage**. One prophage gene codes for a protein that prevents transcription of most of the other prophage genes. Thus, the phage genome is mostly silent within the bacterium. Every time the *E. coli* cell prepares to divide, it replicates the phage DNA along with its own and passes the copies on to daughter cells. A single infected cell can quickly give rise to a large population of bacteria carrying the virus in prophage form. This mechanism enables viruses to propagate without killing the host cells on which they depend.

The term *lysogenic* implies that prophages are capable of generating active phages that lyse their host cells. This occurs when the λ genome is induced to exit the bacterial chromosome and initiate a lytic cycle. An environmental signal, such

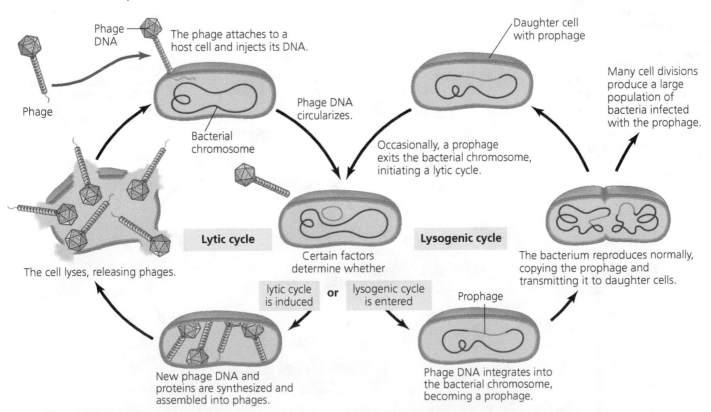

▲ **Figure 17.5 The lytic and lysogenic cycles of phage λ, a temperate phage.** After entering the bacterial cell and circularizing, the λ DNA can immediately initiate the production of a large number of progeny phages (lytic cycle) or integrate into the bacterial chromosome (lysogenic cycle). In most cases, phage λ follows the lytic pathway, which is similar to that detailed in Figure 17.4. However, once a lysogenic cycle begins, the prophage may be carried in the host cell's chromosome for many generations. Phage λ has one main tail fiber, which is short.

as a certain chemical or high-energy radiation, usually triggers the switchover from the lysogenic to the lytic mode.

In addition to the gene for the transcription-preventing protein, a few other prophage genes may be expressed during lysogeny. Expression of these genes may alter the host's phenotype, a phenomenon that can have important medical significance. For example, the three species of bacteria that cause the human diseases diphtheria, botulism, and scarlet fever would not be so harmful to humans without certain prophage genes that cause the host bacteria to make toxins. And the difference between the *E. coli* strain that resides in our intestines and the O157:H7 strain that has caused several deaths by food poisoning appears to be the presence of prophages in the O157:H7 strain.

Replicative Cycles of Animal Viruses

Everyone has suffered from viral infections, whether cold sores, influenza, or the common cold. Like all viruses, those that cause illness in humans and other animals can replicate only inside host cells. Many variations on the basic scheme of viral infection and replication are represented among the animal viruses. Key variables are the nature of the viral genome

(double- or single-stranded DNA or RNA) and the presence or absence of an envelope.

Whereas few bacteriophages have an RNA genome or envelope, many animal viruses have both. In fact, nearly all animal viruses with RNA genomes have an envelope, as do some with DNA genomes. Rather than consider all the mechanisms of viral infection and replication, we'll focus on the roles of viral envelopes and on the functioning of RNA as the genetic material of many animal viruses.

Viral Envelopes

An animal virus equipped with an envelope—that is, an outer membrane—uses it to enter the host cell. Protruding from the outer surface of this envelope are viral glycoproteins that bind to specific receptor molecules on the surface of a host cell. **Figure 17.6** outlines the events in the replicative cycle of an enveloped virus with an RNA genome. Ribosomes bound to the endoplasmic reticulum (ER) of the host cell make the protein parts of the envelope glycoproteins; cellular enzymes in the ER and Golgi apparatus then add the sugars. The resulting viral glycoproteins, embedded in membrane derived from the

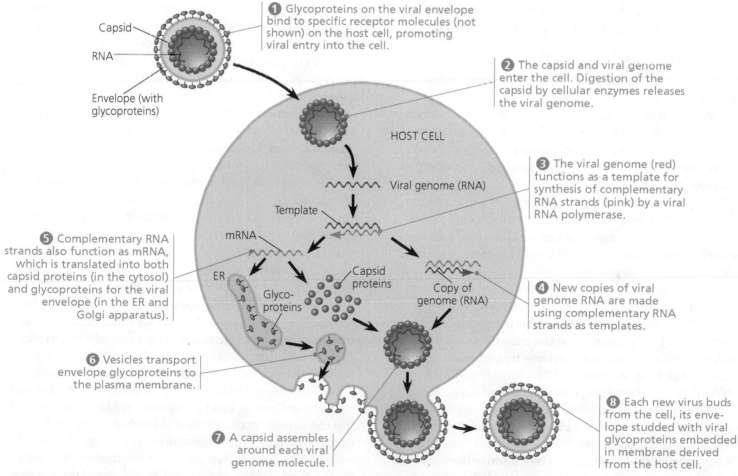

1 Glycoproteins on the viral envelope bind to specific receptor molecules (not shown) on the host cell, promoting viral entry into the cell.

Capsid

RNA

Envelope (with glycoproteins)

2 The capsid and viral genome enter the cell. Digestion of the capsid by cellular enzymes releases the viral genome.

HOST CELL

Viral genome (RNA)

3 The viral genome (red) functions as a template for synthesis of complementary RNA strands (pink) by a viral RNA polymerase.

Template

mRNA

5 Complementary RNA strands also function as mRNA, which is translated into both capsid proteins (in the cytosol) and glycoproteins for the viral envelope (in the ER and Golgi apparatus).

ER

Glyco-proteins

Capsid proteins

Copy of genome (RNA)

4 New copies of viral genome RNA are made using complementary RNA strands as templates.

6 Vesicles transport envelope glycoproteins to the plasma membrane.

7 A capsid assembles around each viral genome molecule.

8 Each new virus buds from the cell, its envelope studded with viral glycoproteins embedded in membrane derived from the host cell.

▲ **Figure 17.6 The replicative cycle of an enveloped RNA virus.** Shown here is a virus with a single-stranded RNA genome that functions as a template for synthesis of mRNA. Some enveloped viruses enter the host cell by fusion of the envelope with the cell's plasma membrane; others enter by endocytosis. For all enveloped RNA viruses, the formation of new envelopes for progeny viruses occurs by the mechanism depicted in this figure.

host cell, are transported to the cell surface. In a process much like exocytosis, new viral capsids are wrapped in membrane as they bud from the cell. In other words, the viral envelope is derived from the host cell's plasma membrane, although some of the molecules of this membrane are specified by viral genes. The enveloped viruses are now free to infect other cells. This replicative cycle does not necessarily kill the host cell, in contrast to the lytic cycles of phages.

Some viruses have envelopes that are not derived from plasma membrane. Herpesviruses, for example, are temporarily cloaked in membrane derived from the nuclear envelope of the host; they then shed this membrane in the cytoplasm and acquire a new envelope made from membrane of the Golgi apparatus. These viruses have a double-stranded DNA genome and replicate within the host cell nucleus, using a combination of viral and cellular enzymes to replicate and transcribe their DNA. In the case of herpesviruses, copies of the viral DNA can remain behind as mini-chromosomes in the nuclei of certain nerve cells. There they remain latent until some sort of physical or emotional stress triggers a new round of active virus production. The infection of other cells by these new viruses causes the blisters characteristic of herpes, such as cold sores or genital sores. Once someone acquires a herpesvirus infection, flare-ups may recur throughout the person's life.

RNA as Viral Genetic Material

Although some phages and most plant viruses are RNA viruses, the broadest variety of RNA genomes is found among the viruses that infect animals. There are three types of single-stranded RNA genomes found in animal viruses. In the first type, the viral genome can directly serve as mRNA and thus can be translated into viral protein immediately after infection. In a second type, the RNA genome serves as a *template* for mRNA synthesis. The RNA genome is transcribed into complementary RNA strands, which function both as mRNA and as templates for the synthesis of additional copies of genomic RNA. All viruses that require RNA → RNA synthesis to make mRNA use a viral enzyme capable of carrying out this process; there are no such enzymes in most cells. The viral enzyme is packaged with the genome inside the viral capsid.

The RNA animal viruses with the most complicated replicative cycles are the third type, the **retroviruses**. These viruses are equipped with an enzyme called **reverse transcriptase**, which transcribes an RNA template into DNA, providing an RNA → DNA information flow, the opposite of the usual direction. (Reverse transcriptase is the enzyme used in the technique called RT-PCR, described in Concept 15.4.) This unusual phenomenon is the source of the name retroviruses (*retro* means "backward"). Of particular medical importance is **HIV (human immunodeficiency virus)**, the retrovirus that causes **AIDS (acquired immunodeficiency syndrome)**. HIV and other retroviruses are enveloped viruses that contain two identical molecules of single-stranded RNA and two molecules of reverse transcriptase.

The HIV replicative cycle (traced in **Figure 17.7**) is typical of a retrovirus. After HIV enters a host cell, its reverse transcriptase molecules are released into the cytoplasm, where they catalyze synthesis of viral DNA. The newly made viral DNA then enters the cell's nucleus and integrates into the DNA of a chromosome. The integrated viral DNA, called a **provirus**, never leaves the host's genome, remaining a permanent resident of the cell. (Recall that a prophage, in contrast, leaves the host's genome at the start of a lytic cycle.) The host's RNA polymerase transcribes the proviral DNA into RNA molecules, which can function both as mRNA for the synthesis of viral proteins and as genomes for the new viruses that will be assembled and released from the cell. In Chapter 35, we'll describe how HIV causes the deterioration of the immune system that occurs in AIDS.

Evolution of Viruses

EVOLUTION We began this chapter by asking whether or not viruses are alive. Viruses do not really fit our definition of living organisms. An isolated virus is biologically inert, unable to replicate its genes or regenerate its own supply of ATP. Yet it has a genetic program written in the universal language of life. Do we think of viruses as nature's most complex associations of molecules or as the simplest forms of life? Either way, we must bend our usual definitions. Although viruses cannot replicate or carry out metabolic activities independently, their use of the genetic code makes it hard to deny their evolutionary connection to the living world.

How did viruses originate? Viruses have been found that infect every form of life—not just bacteria, animals, and plants, but also archaea, fungi, and algae and other protists. Because they depend on cells for their own propagation, it seems likely that viruses are not the descendants of precellular forms of life but evolved—possibly multiple times—*after* the first cells appeared. Most molecular biologists favor the hypothesis that viruses originated from naked bits of cellular nucleic acids that moved from one cell to another, perhaps via injured cell surfaces. The evolution of genes coding for capsid proteins may have facilitated the infection of uninjured cells.

Candidates for the original sources of viral genomes include plasmids and transposons. *Plasmids* are small, circular DNA molecules found in bacteria and in the unicellular eukaryotes called yeasts. Plasmids exist apart from the cell's genome, can replicate independently of the genome, and are occasionally transferred between cells. (Use of plasmids in gene cloning was discussed in Concept 13.4.) *Transposons* are DNA segments that can move from one location to another within a cell's genome. Thus, plasmids, transposons, and viruses all share an important feature: They are *mobile genetic elements*. (We'll discuss plasmids in more detail in Chapter 24 and transposons in Chapter 18.)

Consistent with this vision of pieces of DNA shuttling from cell to cell is the observation that a viral genome can have more

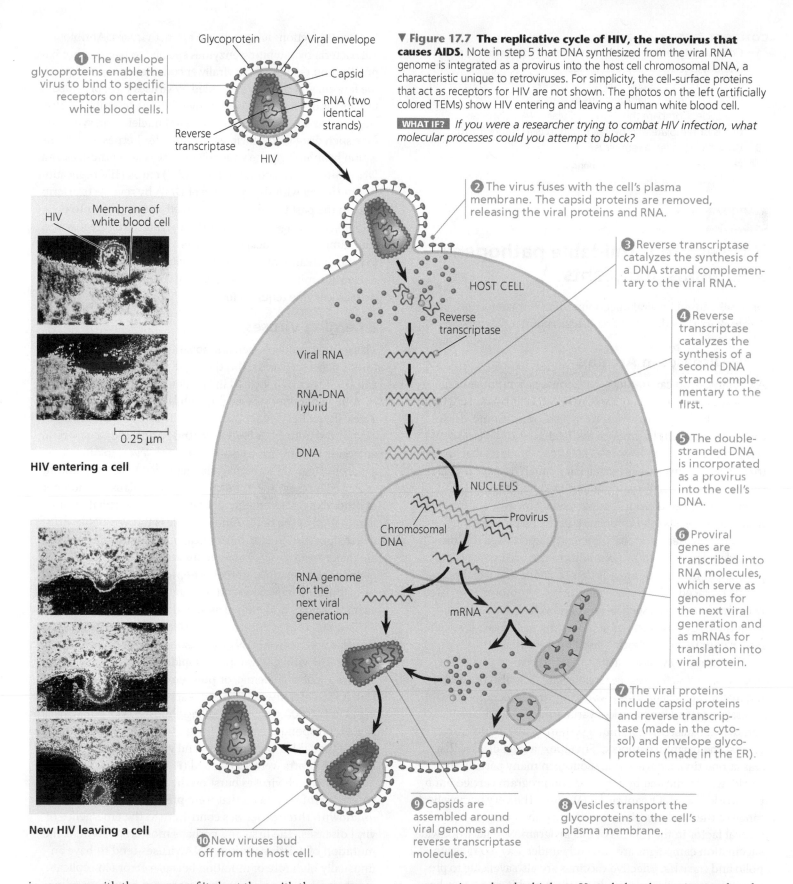

① The envelope glycoproteins enable the virus to bind to specific receptors on certain white blood cells.

Glycoprotein

Viral envelope

Capsid

RNA (two identical strands)

Reverse transcriptase

HIV

▼ **Figure 17.7 The replicative cycle of HIV, the retrovirus that causes AIDS.** Note in step 5 that DNA synthesized from the viral RNA genome is integrated as a provirus into the host cell chromosomal DNA, a characteristic unique to retroviruses. For simplicity, the cell-surface proteins that act as receptors for HIV are not shown. The photos on the left (artificially colored TEMs) show HIV entering and leaving a human white blood cell.

WHAT IF? *If you were a researcher trying to combat HIV infection, what molecular processes could you attempt to block?*

② The virus fuses with the cell's plasma membrane. The capsid proteins are removed, releasing the viral proteins and RNA.

③ Reverse transcriptase catalyzes the synthesis of a DNA strand complementary to the viral RNA.

④ Reverse transcriptase catalyzes the synthesis of a second DNA strand complementary to the first.

⑤ The double-stranded DNA is incorporated as a provirus into the cell's DNA.

⑥ Proviral genes are transcribed into RNA molecules, which serve as genomes for the next viral generation and as mRNAs for translation into viral protein.

⑦ The viral proteins include capsid proteins and reverse transcriptase (made in the cytosol) and envelope glycoproteins (made in the ER).

⑧ Vesicles transport the glycoproteins to the cell's plasma membrane.

⑨ Capsids are assembled around viral genomes and reverse transcriptase molecules.

⑩ New viruses bud off from the host cell.

HOST CELL

Reverse transcriptase

Viral RNA

RNA-DNA hybrid

DNA

NUCLEUS

Chromosomal DNA

Provirus

RNA genome for the next viral generation

mRNA

HIV

Membrane of white blood cell

0.25 μm

HIV entering a cell

New HIV leaving a cell

in common with the genome of its host than with the genomes of viruses that infect other hosts. The ongoing evolutionary relationship between viruses and the genomes of their host cells is an association that makes viruses very useful experimental systems in molecular biology. Knowledge about viruses also allows many practical applications, since viruses have a tremendous impact on all organisms through their ability to cause disease.

1. Compare the effect on the host cell of a lytic (virulent) phage and a lysogenic (temperate) phage.
2. **MAKE CONNECTIONS** The RNA virus in Figure 17.6 has a viral RNA polymerase that functions in step 3 of the virus's replicative cycle. Compare this with a cellular RNA polymerase in terms of template and overall function (see Figure 14.10).
3. Why is HIV called a retrovirus?

For suggested answers, see Appendix A.

CONCEPT 17.3

Viruses are formidable pathogens in animals and plants

Diseases caused by viral infections afflict humans, crops, and livestock worldwide. We'll first discuss animal viruses.

Viral Diseases in Animals

A viral infection can produce symptoms by a number of different routes. Viruses may damage or kill cells by causing the release of hydrolytic enzymes from lysosomes. Some viruses cause infected cells to produce toxins that lead to disease symptoms, and some have molecular components that are toxic, such as envelope proteins. How much damage a virus causes depends partly on the ability of the infected tissue to regenerate by cell division. People usually recover completely from colds because the epithelium of the respiratory tract, which the viruses infect, can efficiently repair itself. In contrast, damage inflicted by poliovirus to mature nerve cells is permanent because these cells do not divide and usually cannot be replaced. Many of the temporary symptoms associated with viral infections, such as fever and aches, actually result from the body's own efforts at defending itself against infection rather than from cell death caused by the virus.

The immune system is a complex and critical part of the body's natural defenses (see Chapter 35). It is also the basis for the major medical tool for preventing viral infections—vaccines. A **vaccine** is a harmless variant or derivative of a pathogen that stimulates the immune system to mount defenses against the harmful pathogen. Smallpox, a viral disease that was at one time a devastating scourge in many parts of the world, was eradicated by a vaccination program carried out by the World Health Organization (WHO). The very narrow host range of the smallpox virus—it infects only humans—was a critical factor in the success of this program. Similar worldwide vaccination campaigns are currently under way to eradicate polio and measles. Effective vaccines are also available to protect against rubella, mumps, hepatitis A and B, and a number of other viral diseases.

Although vaccines can prevent certain viral illnesses, medical technology can do little, at present, to cure most viral infections once they occur. The antibiotics that help us recover from bacterial infections are powerless against viruses. Antibiotics kill bacteria by inhibiting enzymes specific to bacteria but have no effect on eukaryotic or virally encoded enzymes. However, the few enzymes that are encoded by viruses have provided targets for other drugs. Most antiviral drugs resemble nucleosides and as a result interfere with viral nucleic acid synthesis. One such drug is acyclovir, which impedes herpesvirus replication by inhibiting the viral polymerase that synthesizes viral DNA. Similarly, azidothymidine (AZT) curbs HIV replication by interfering with the synthesis of DNA by reverse transcriptase. In the past two decades, much effort has gone into developing drugs against HIV. Currently, multidrug treatments, sometimes called "cocktails," have been found to be most effective. Such treatments commonly include a combination of two nucleoside mimics and a protease inhibitor, which interferes with an enzyme required for assembly of the viruses.

Emerging Viruses

Viruses that suddenly become apparent are often referred to as *emerging viruses*. HIV, the AIDS virus, is a classic example: This virus appeared in San Francisco in the early 1980s, seemingly out of nowhere, although later studies uncovered a case in the Belgian Congo in 1959. The deadly Ebola virus, recognized initially in 1976 in central Africa, is one of several emerging viruses that cause *hemorrhagic fever*, an often fatal syndrome (set of symptoms) characterized by fever, vomiting, massive bleeding, and circulatory system collapse. A number of other dangerous emerging viruses cause encephalitis, inflammation of the brain. One example is the West Nile virus, which appeared in North America for the first time in 1999 and has spread to all 48 contiguous states in the United States.

In 2009, a general outbreak, or **epidemic**, of a flu-like illness appeared in Mexico and the United States. The infectious agent was quickly identified as an influenza virus related to viruses that cause the seasonal flu **(Figure 17.8a)**. This particular virus was named H1N1 for reasons that will be explained shortly. The viral disease spread rapidly, prompting WHO to declare a global epidemic, or **pandemic**, shortly thereafter. Half a year later, the disease had reached 207 countries, infecting over 600,000 people and killing almost 8,000. Public health agencies responded rapidly with guidelines for shutting down schools and other public places, and vaccine development and screening efforts were accelerated **(Figure 17.8b)**.

How do such viruses burst on the human scene, giving rise to harmful diseases that were previously rare or even unknown? Three processes contribute to the emergence of viral diseases. The first, and perhaps most important, is the mutation of existing viruses. RNA viruses tend to have an unusually high rate of mutation because errors in replicating their RNA genomes are not corrected by proofreading. Some mutations change existing viruses into new genetic varieties (strains) that can cause disease, even in individuals who are immune to the ancestral virus. For instance, seasonal flu epidemics are caused by new strains of influenza virus

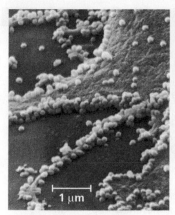

(a) 2009 pandemic H1N1 influenza A virus. Viruses (blue) are seen on an infected cell (green) in this colorized SEM.

(b) 2009 pandemic screening. At a South Korean airport, thermal scans were used to detect passengers with a fever who might have the H1N1 flu.

▲ **Figure 17.8 Influenza in humans.**

genetically different enough from earlier strains that people have little immunity to them. You'll see an example of this process in the **Scientific Skills Exercise**, where you'll analyze genetic changes in variants of the 2009 flu virus and correlate them with spread of the disease.

A second process that can lead to the emergence of viral diseases is the dissemination of a viral disease from a small, isolated human population. For instance, AIDS went unnamed and virtually unnoticed for decades before it began to spread around the world. In this case, technological and social factors, including affordable international travel, blood transfusions, sexual promiscuity, and the abuse of intravenous drugs, allowed a previously rare human disease to become a global scourge.

A third source of new viral diseases in humans is the spread of existing viruses from other animals. Scientists estimate that about three-quarters of new human diseases originate in this way. Animals that harbor and can transmit a particular virus but are generally unaffected by it are said to act as a natural reservoir for that virus. For example, the 2009 flu pandemic mentioned earlier was likely passed to humans from pigs; for this reason, it was originally called "swine flu."

In general, flu epidemics provide an instructive example of the effects of viruses moving between species. There are three types of influenza virus: types B and C, which infect only humans and have never caused an epidemic, and type A, which infects a wide range of animals, including birds, pigs, horses, and humans. Influenza A strains have caused four major flu epidemics among humans in the last 100 years. The worst was the first one, the "Spanish flu" pandemic of 1918–1919, which killed between 10 and 20% of those infected—about 40 million people, including many World War I soldiers.

Different strains of influenza A are given standardized names; for example, both the strain that caused the 1918 flu and the one that caused the 2009 pandemic flu are called H1N1. The name identifies which forms of two viral surface proteins are present: hemagglutinin (H) and neuraminidase (N). There are 16 different types of hemagglutinin, a protein that helps the flu virus attach to host cells, and 9 types of neuraminidase, an enzyme that helps release new virus particles from infected cells. Waterbirds have been found that carry viruses with all possible combinations of H and N.

A likely scenario for the 1918 pandemic is that the virus mutated as it passed from one host species to another. When an animal like a pig or a bird is infected with more than one strain of flu virus, the different strains can undergo genetic recombination if the RNA molecules of their genomes mix and match during viral assembly. Pigs were probably the breeding ground for the 2009 flu virus, which contains sequences from bird, pig, and human flu viruses. Coupled with mutation, these reassortments can lead to the emergence of a viral strain capable of infecting human cells. People who have never been exposed to that particular strain before will lack immunity, and the recombinant virus has the potential to be highly pathogenic. If such a flu virus recombines with viruses that circulate widely among humans, it may acquire the ability to spread easily from person to person, dramatically increasing the potential for a major human outbreak.

One potential long-term threat is the avian flu caused by an H5N1 virus carried by wild and domestic birds. The first documented transmission from birds to humans occurred in Hong Kong in 1997. Since then, the overall mortality rate of the H5N1 virus has been greater than 50% of those infected, an alarming number. Also, the host range of H5N1 is expanding, which provides increasing chances for reassortment between different strains. If the H5N1 avian flu virus evolves so that it can spread easily from person to person, it could represent a major global health threat akin to that of the 1918 pandemic.

How easily could this happen? Recently, scientists working with ferrets, small mammals that are animal models for human flu, found that only a few mutations of the avian flu virus would allow infection of cells in the human nasal cavity and windpipe. Furthermore, when the scientists transferred nasal swabs serially from ferret to ferret, the virus became transmissible through the air. Reports of this startling discovery at a scientific conference in 2011 ignited a firestorm of debate about whether to publish the results. Ultimately, the scientific community decided the benefits of potentially understanding how to prevent pandemics would outweigh the risks of the information being used for harmful purposes, and the work was published in 2012.

As we have seen, emerging viruses are generally not new; rather, they are existing viruses that mutate, disseminate more widely in the current host species, or spread to new host species. Changes in host behavior or environmental changes can increase the viral traffic responsible for emerging diseases. For instance, new roads built through remote areas can allow viruses to spread between previously isolated human populations. Also, the destruction of forests to expand cropland can bring humans into contact with other animals that may host viruses capable of infecting humans.

Analyzing a DNA Sequence-Based Phylogenetic Tree to Understand Viral Evolution

How Can DNA Sequence Data Be Used to Track Flu Virus Evolution During Pandemic Waves? In 2009, an influenza A H1N1 virus caused a pandemic, and the virus has continued to resurface in outbreaks across the world. Researchers in Taiwan were curious about why the virus kept appearing despite widespread flu vaccine initiatives. They hypothesized that newly evolved variants of the H1N1 virus were able to evade human immune system defenses. To test this hypothesis, they needed to determine if each wave of the flu outbreak was caused by a different H1N1 variant.

How the Experiment Was Done Scientists obtained the genome sequences for 4,703 virus isolates collected from patients with H1N1 flu in Taiwan. They compared the sequences in different strains for the viral hemagglutinin (HA) gene, and based on mutations that had occurred, arranged the isolates into a phylogenetic tree (see Figure 20.5 for information on how to read phylogenetic trees).

Data from the Experiment The figure below, left, shows a phylogenetic tree; each branch tip is one variant of the H1N1 virus with a unique HA gene sequence. The tree is a way to visualize a working hypothesis about the evolutionary relationships between H1N1 variants.

Interpret the Data

1. The phylogenetic tree shows the hypothesized evolutionary relationship between the variant strains of H1N1 virus. The more closely connected two variants are, the more alike they are in terms of HA gene sequence. Each fork in a branch, called a node, shows where two lineages separate due to different accumulated mutations. The length of the branches is a measure of how many DNA sequence differences there are between the variants, thus how distantly related they are. Referring to the phylogenetic tree, which variants are more closely related to each other: A/Taiwan1018/2011 and A/Taiwan/552/2011 or A/Taiwan1018/2011 and A/Taiwan/8542/2009? Explain your answer.

2. The scientists arranged the branches into groups made up of one ancestral variant and all of its descendant, mutated variants. They are color-coded in the figure. Using Group 11 as an example, trace the lineage of its variants. (a) Do all of the nodes have the same number of branches or branch tips? (b) Are all of the branches in the group the same length? (c) What do these results indicate?

3. The graph shows the number of isolates collected from ill patients on the *y*-axis and the month and year that the isolates were collected on the *x*-axis. Each group of variants is plotted separately with a line color that matches the tree diagram. (a) Which group of variants was the earliest to cause H1N1 flu in over 100 patients in Taiwan?
(b) Once a group of variants had a peak number of infections, did members of that same group cause another wave of infection?
(c) One variant in Group 1 was used to make a vaccine very early in the pandemic. Based on the graphed data, does it look like the vaccine was effective?

4. Groups 9, 10, and 11 all had H1N1 variants that caused a large number of infections at the same time in Taiwan. Does this mean that the scientists' hypothesis, that new variants cause new waves of infection, was incorrect? Explain your answer.

Data from J-R. Yang et al., New variants and age shift to high fatality groups contribute to severe successive waves in the 2009 influenza pandemic in Taiwan, *PLoS ONE* 6(11): e28288 (2011). doi:10.1371/journal.pone.0028288.

(MB) A version of this Scientific Skills Exercise can be assigned in MasteringBiology.

Scientists also graphed the isolates by the month and year of isolate collection, which reflects the time period in which each viral variant was actively causing illness in people.

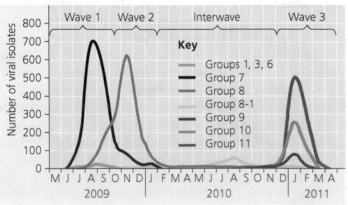

Viral Diseases in Plants

More than 2,000 types of viral diseases of plants are known, and together they account for an estimated annual loss of $15 billion worldwide due to their destruction of agricultural and horticultural crops. Common signs of viral infection include bleached or brown spots on leaves and fruits (as on the squash to the right), stunted growth, and damaged flowers or roots, all tending to diminish the yield and quality of crops.

Plant viruses have the same basic structure and mode of replication as animal viruses. Most plant viruses discovered thus far, including tobacco mosaic virus (TMV), have an RNA genome. Many have a helical capsid, like TMV, while others have an icosahedral capsid (see Figure 17.2).

Viral diseases of plants spread by two major routes. In the first route, called *horizontal transmission*, a plant is infected from an external source of the virus. Because the invading virus must get past the plant's outer protective layer of cells (the epidermis), a plant becomes more susceptible to viral infections if it has been damaged by wind, injury, or herbivores. Herbivores, especially insects, pose a double threat because they can also act as carriers of viruses, transmitting disease from plant to plant. Moreover, farmers and gardeners may transmit plant viruses inadvertently on pruning shears and other tools. The other route of viral infection is *vertical transmission*, in which a plant inherits a viral infection from a parent. Vertical transmission can occur in asexual propagation (for example, through cuttings) or in sexual reproduction via infected seeds.

Once a virus enters a plant cell and begins replicating, viral genomes and associated proteins can spread throughout the plant by means of plasmodesmata, the cytoplasmic connections that penetrate the walls between adjacent plant cells (see Figure 4.25). The passage of viral macromolecules from cell to cell is facilitated by virally encoded proteins that cause enlargement of plasmodesmata. Scientists have not yet devised cures for most viral plant diseases. Consequently, research efforts are focused largely on reducing the transmission of such diseases and on breeding resistant varieties of crop plants.

Earlier in this chapter, we mentioned the ongoing evolutionary relationship between viruses and the genomes of their host cells. In fact, the original source of viral genetic material may have been transposons, mobile genetic elements that are present in multiple copies in many genomes. In the next chapter, we'll discuss the structure of genomes and how they evolve.

CONCEPT CHECK 17.3

1. Describe two ways a preexisting virus can become an emerging virus.
2. Contrast horizontal and vertical transmission of viruses in plants.
3. **WHAT IF?** TMV has been isolated from virtually all commercial tobacco products. Why, then, is TMV infection not an additional hazard for smokers?

For suggested answers, see Appendix A.

---

17 Chapter Review

SUMMARY OF KEY CONCEPTS

CONCEPT 17.1

A virus consists of a nucleic acid surrounded by a protein coat (pp. 330–332)

- A **virus** is a small nucleic acid genome enclosed in a protein **capsid** and sometimes a membranous **viral envelope** containing viral proteins that help the virus enter a cell. The genome may be single- or double-stranded DNA or RNA.

> **?** *Are viruses generally considered living or nonliving? Explain.*

CONCEPT 17.2

Viruses replicate only in host cells (pp. 332–338)

- Viruses use enzymes, ribosomes, and small molecules of host cells to synthesize progeny viruses during replication. Each type of virus has a characteristic **host range**.
- **Phages** (viruses that infect bacteria) can replicate by two alternative mechanisms: the **lytic cycle** and the **lysogenic cycle**.

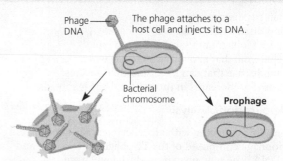

Lytic cycle
- **Virulent** or **temperate phage**
- Destruction of host DNA
- Production of new phages
- Lysis of host cell causes release of progeny phages

Lysogenic cycle
- **Temperate phage** only
- Genome integrates into bacterial chromosome as **prophage**, which (1) is replicated and passed on to daughter cells and (2) can be induced to leave the chromosome and initiate a lytic cycle

- Many animal viruses have an envelope. **Retroviruses** (such as **HIV**) use the enzyme **reverse transcriptase** to copy their RNA genome into DNA, which can be integrated into the host genome as a **provirus**.

- Since viruses can replicate only within cells, they probably evolved after the first cells appeared, perhaps as packaged fragments of cellular nucleic acid. The origin of viruses is still being debated.

? *Describe enzymes that are not found in most cells but are necessary for the replication of certain viruses.*

CONCEPT 17.3

Viruses are formidable pathogens in animals and plants (pp. 338–341)

- Symptoms of viral diseases in animals may be caused by direct viral harm to cells or by the body's immune response. **Vaccines** stimulate the immune system to defend the host against specific viruses.
- Outbreaks of "new" viral diseases in humans are usually caused by existing viruses that expand their host territory. The H1N1 2009 flu virus was a new combination of pig, human, and avian viral genes that caused a pandemic.
- Viruses enter plant cells through damaged cell walls (horizontal transmission) or are inherited from a parent (vertical transmission).

? *What aspect of an RNA virus makes it more likely than a DNA virus to become an emerging virus?*

TEST YOUR UNDERSTANDING

Level 1: Knowledge/Comprehension

1. Which of the following characteristics, structures, or processes is common to both bacteria and viruses?
   a. metabolism
   b. ribosomes
   c. genetic material composed of nucleic acid
   d. cell division
   e. independent existence

2. Emerging viruses arise by
   a. mutation of existing viruses.
   b. the spread of existing viruses to new host species.
   c. the spread of existing viruses more widely within their host species.
   d. all of the above
   e. none of the above

3. A human pandemic is
   a. a viral disease that infects all humans.
   b. a flu that kills more than 1 million people.
   c. an epidemic that extends around the world.
   d. a viral disease that can infect multiple species.
   e. a virus that increases in mortality rate as it spreads.

Level 2: Application/Analysis

4. A bacterium is infected with an experimentally constructed bacteriophage composed of the T2 phage protein coat and T4 phage DNA. The new phages produced will have
   a. T2 protein and T4 DNA.
   b. T2 protein and T2 DNA.
   c. a mixture of the DNA and proteins of both phages.
   d. T4 protein and T4 DNA.
   e. T4 protein and T2 DNA.

5. RNA viruses require their own supply of certain enzymes because
   a. host cells rapidly destroy the viruses.
   b. host cells lack enzymes that can replicate the viral genome.
   c. these enzymes translate viral mRNA into proteins.
   d. these enzymes penetrate host cell membranes.
   e. these enzymes cannot be made in host cells.

6. **DRAW IT** Redraw Figure 17.6 to show the replicative cycle of a virus with a single-stranded genome that can function as mRNA.

Level 3: Synthesis/Evaluation

7. **SCIENTIFIC INQUIRY**
   When bacteria infect an animal, the number of bacteria in the body increases in an exponential fashion (graph A). After infection by a virulent animal virus with a lytic replicative cycle, there is no evidence of infection for a while. Then the number of viruses rises suddenly and subsequently increases in a series of steps (graph B). Explain the difference in the curves.

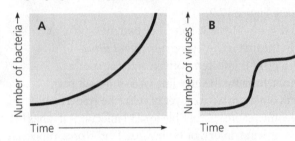

8. **FOCUS ON EVOLUTION**
   The success of some viruses lies in their ability to evolve rapidly within the host. Such a virus evades the host's defenses by mutating and producing many altered progeny viruses before the body can mount an attack. Thus, the viruses present late in infection differ from those that initially infected the body. Discuss this as an example of evolution in microcosm. Which viral lineages tend to predominate?

9. **FOCUS ON ORGANIZATION**
   While viruses are considered by most scientists to be nonliving, they do show some characteristics of life, including the correlation of structure and function. In a short essay (100–150 words), discuss how the structure of a virus correlates with its function.

For selected answers, see Appendix A.

MasteringBiology®

Students Go to **MasteringBiology** for assignments, the eText, and the Study Area with practice tests, animations, and activities.

Instructors Go to **MasteringBiology** for automatically graded tutorials and questions that you can assign to your students, plus Instructor Resources.

Genomes and Their Evolution

KEY CONCEPTS

18.1 The Human Genome Project fostered development of faster, less expensive sequencing techniques

18.2 Scientists use bioinformatics to analyze genomes and their functions

18.3 Genomes vary in size, number of genes, and gene density

18.4 Multicellular eukaryotes have much noncoding DNA and many multigene families

18.5 Duplication, rearrangement, and mutation of DNA contribute to genome evolution

18.6 Comparing genome sequences provides clues to evolution and development

OVERVIEW

Reading the Leaves from the Tree of Life

The chimpanzee (*Pan troglodytes*) is our closest living relative on the evolutionary tree of life. The boy in **Figure 18.1** and his chimpanzee companion are intently studying the same leaf, but only one of them is able to talk about what he sees. What accounts for this difference between two primates that share so much of their evolutionary history? With the advent of recent techniques for rapidly sequencing complete genomes, we can now start to address the genetic basis of intriguing questions like this.

The chimpanzee genome was sequenced two years after sequencing of the human genome was largely completed. Now that we can compare our genome, base by base, with that of the chimpanzee, we can tackle the more general issue of what differences in genetic information account for the distinct characteristics of these two species of primates.

In addition to determining the sequences of the human and chimpanzee genomes, researchers have obtained complete genome sequences for *E. coli* and numerous other prokaryotes, as well as many eukaryotes, including *Zea mays* (corn), *Drosophila melanogaster* (fruit fly), *Mus musculus* (house mouse), and *Pongo pigmaeus* (orangutan). In 2010, a draft sequence was announced for the genome of *Homo neanderthalensis*, an extinct species closely related to present-day humans. These whole and partial genomes are of great interest in their own right and are also providing important insights into evolution and other biological processes. Broadening the human-chimpanzee comparison to the genomes of other primates and more distantly related animals should reveal the sets of genes that control group-defining characteristics. Beyond that, comparisons with the genomes of bacteria, archaea, fungi, protists, and plants will enlighten us about the long evolutionary history of shared ancient genes and their products.

With the genomes of many species fully sequenced, scientists can study whole sets of genes and their interactions, an approach called **genomics**. The sequencing efforts that feed this approach have generated, and continue to generate, enormous volumes of data. The need to deal with this

▼ **Figure 18.1** What genomic information distinguishes a human from a chimpanzee?

ever-increasing flood of information has spawned the field of **bioinformatics**, the application of computational methods to the storage and analysis of biological data.

We'll begin this chapter by discussing genome sequencing and some of the advances in bioinformatics and its applications. We'll then summarize what has been learned from the genomes that have been sequenced thus far. Next, we'll describe the composition of the human genome as a representative genome of a complex multicellular eukaryote. Finally, we'll explore current ideas about how genomes evolve and about how the evolution of developmental mechanisms could have generated the great diversity of life on Earth today.

CONCEPT 18.1

The Human Genome Project fostered development of faster, less expensive sequencing techniques

Sequencing of the human genome, an ambitious undertaking, officially began as the **Human Genome Project** in 1990. Organized by an international, publicly funded consortium of scientists at universities and research institutes, the project involved 20 large sequencing centers in six countries plus a host of other labs working on small projects.

After sequencing of the human genome was reported in 2003, the sequence of each chromosome was analyzed and described in a series of papers, the last of which covered chromosome 1 and was published in 2006. With this refinement, researchers termed the sequencing "virtually complete."

The ultimate goal in mapping any genome is to determine the complete nucleotide sequence of each chromosome. For the human genome, this was accomplished by sequencing machines. Even with automation, the sequencing of all 3 billion base pairs in a haploid set of human chromosomes presented a formidable challenge. In fact, a major thrust of the Human Genome Project was the development of technology for faster sequencing. Improvements over the years chipped away at each time-consuming step, enabling the rate of sequencing to accelerate impressively: Whereas a productive lab could typically sequence 1,000 base pairs a day in the 1980s, by the year 2000 each research center working on the Human Genome Project was sequencing 1,000 base pairs *per second*, 24 hours a day, seven days a week. Methods that can analyze biological materials very rapidly and produce enormous volumes of data are said to be "high-throughput." Sequencing machines are an example of high-throughput devices.

Two approaches complemented each other in obtaining the complete sequence. The initial approach was a methodical one that built on an earlier storehouse of human genetic information. In 1998, however, molecular biologist J. Craig Venter set up a company (Celera Genomics) and declared his intention to

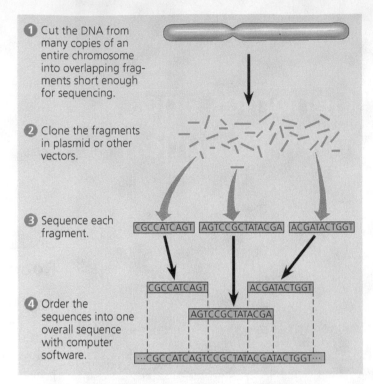

1. Cut the DNA from many copies of an entire chromosome into overlapping fragments short enough for sequencing.

2. Clone the fragments in plasmid or other vectors.

3. Sequence each fragment.

CGCCATCAGT   AGTCCGCTATACGA   ACGATACTGGT

4. Order the sequences into one overall sequence with computer software.

CGCCATCAGT   ACGATACTGGT
         AGTCCGCTATACGA
···CGCCATCAGTCCGCTATACGATACTGGT···

▲ **Figure 18.2 Whole-genome shotgun approach to sequencing.** In this approach, developed by J. Craig Venter and colleagues at the company he founded, Celera Genomics, random DNA fragments are sequenced and then ordered relative to each other.

? *The fragments in stage 2 of this figure are depicted as scattered, rather than being in an ordered array. How does this depiction accurately reflect the approach?*

sequence the entire human genome using an alternative strategy. The **whole-genome shotgun approach** starts with the cloning and sequencing of DNA fragments from randomly cut DNA. Powerful computer programs then assemble the resulting very large number of overlapping short sequences into a single continuous sequence **(Figure 18.2)**.

Today, the whole-genome shotgun approach is widely used. Also, the development of newer sequencing techniques, generally called *sequencing by synthesis* (see Chapter 13), has resulted in massive increases in speed and decreases in the cost of sequencing entire genomes. In these new techniques, many very small fragments (fewer than 100 base pairs) are sequenced at the same time, and computer software rapidly assembles the complete sequence. Because of the sensitivity of these techniques, the fragments can be sequenced directly; the cloning step (stage ② in Figure 18.2) is unnecessary. By 2010, the worldwide output was astronomical: close to 100 *billion* bases per day, with the rate estimated to double every 9 months. Whereas sequencing the first human genome took 13 years and cost $100 million, biologist James Watson's genome was sequenced during 4 months in 2007 for about $1 million, and we are rapidly approaching the day when an individual's genome can be sequenced in a matter of hours for less than $1,000!

These technological advances have also facilitated an approach called **metagenomics** (from the Greek *meta*, beyond), in which DNA from a group of species (a *metagenome*) is collected from an environmental sample and sequenced. Again, computer software accomplishes the task of sorting out the partial sequences and assembling them into specific genomes. So far, this approach has been applied to microbial communities found in environments as diverse as the Sargasso Sea and the human intestine. The ability to sequence the DNA of mixed populations eliminates the need to culture each species separately in the lab, a difficulty that has limited the study of many microbial species.

At first glance, genome sequences of humans and other organisms are simply dry lists of nucleotide bases—millions of A's, T's, C's, and G's in mind-numbing succession. Crucial to making sense of this massive amount of data have been new analytical approaches, which we discuss next.

CONCEPT CHECK 18.1

1. Describe the whole-genome shotgun approach to genome sequencing.

   For suggested answers, see Appendix A.

CONCEPT 18.2

Scientists use bioinformatics to analyze genomes and their functions

Each of the 20 or so sequencing centers around the world working on the Human Genome Project in the 1990s churned out voluminous amounts of DNA sequence day after day. As the data began to accumulate, the need to coordinate efforts to keep track of all the sequences became clear. Thanks to the foresight of research scientists and government officials involved in the Human Genome Project, its goals included the establishment of banks of data, or databases, and the refining of analytical software. These databases and software programs would then be centralized and made readily accessible on the Internet. Accomplishing this aim has accelerated progress in DNA sequence analysis by making bioinformatics resources available to researchers worldwide and by speeding up the dissemination of information.

Centralized Resources for Analyzing Genome Sequences

Government-funded agencies carried out their mandate to establish databases and provide software with which scientists could analyze the sequence data. For example, in the United States, a joint endeavor between the National Library of Medicine and the National Institutes of Health (NIH) created the National Center for Biotechnology Information (NCBI), which maintains a website (www.ncbi.nlm.nih.gov) with extensive bioinformatics resources. On this site are links to databases,

software, and a wealth of information about genomics and related topics. Similar websites have also been established by the European Molecular Biology Laboratory, the DNA Data Bank of Japan, and BGI (formerly known as the Beijing Genome Institute) in Shenzhen, China, three genome centers with which NCBI collaborates. These large, comprehensive websites are complemented by others maintained by individual or small groups of laboratories. Smaller websites often provide databases and software designed for a narrower purpose, such as studying genetic and genomic changes in one particular type of cancer.

The NCBI database of sequences is called GenBank. As of August 2012, it included the sequences of 156 million fragments of genomic DNA, totaling 143 billion base pairs! GenBank is constantly updated, and the amount of data it contains is estimated to double approximately every 18 months. Any sequence in the database can be retrieved and analyzed using software from the NCBI website or elsewhere.

One software program available on the NCBI website, called BLAST, allows the visitor to compare a DNA sequence with every sequence in GenBank, base by base, to look for similar regions. Another program allows comparison of predicted protein sequences. Yet a third can search any protein sequence for common stretches of amino acids (domains) for which a function is known or suspected, and it can show a three-dimensional model of the domain alongside other relevant information (**Figure 18.3**). There is even a software program that can compare a collection of sequences, either nucleic acids or polypeptides, and diagram them in the form of an evolutionary tree based on the sequence relationships. (One such diagram is shown in Figure 18.15.)

Two research institutions, Rutgers University and the University of California, San Diego, also maintain a worldwide Protein Data Bank, a database of all three-dimensional protein structures that have been determined. (The database is accessible at www.wwpdb.org.) These structures can be rotated by the viewer to show all sides of the protein.

There is a vast array of resources available for researchers anywhere in the world to use. Now let's consider the types of questions scientists can address using these resources.

Understanding the Functions of Protein-Coding Genes

The identities of about half of the human genes were known before the Human Genome Project began. But what about the others, the previously unknown genes revealed by analysis of DNA sequences? Clues about their identities and functions come from comparing sequences that might be genes with known genes from other organisms, using the software described previously. Due to redundancy in the genetic code, the DNA sequence itself may vary more than the protein sequence does. Thus, scientists interested in proteins often compare the predicted amino acid sequence of a protein with that of other proteins.

Sometimes a newly identified sequence will match, at least partially, the sequence of a gene or protein whose function

In this window, a partial amino acid sequence from an unknown muskmelon protein ("Query") is aligned with sequences from other proteins that the computer program found to be similar. Each sequence represents a domain called WD40.

Four hallmarks of the WD40 domain are highlighted in yellow. (Sequence similarity is based on chemical aspects of the amino acids, so the amino acids in each hallmark region are not always identical.)

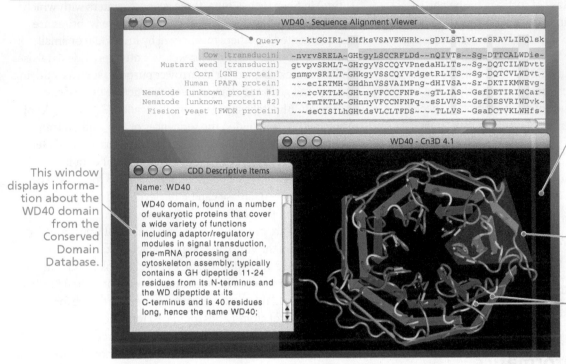

WD40 - Sequence Alignment Viewer

Query	~~~ktGGIRL~RHfksVSAVEWHRk~~~gDYLSTlvLreSRAVLIHQlsk
Cow [transducin]	~nvrvSRELA~GHtgyLSCCRFLDd~~nQIVTs~~Sg~DTTCALWDie~
Mustard weed [transducin]	gtvpvSRMLT~GHrgyVSCCQYVPnedaHLITs~~Sg~DQTCILWDvtt
Corn [GNB protein]	gnmpvSRILT~GHkgyVSSCQYVPdgetRLITS~~Sg~DQTCVLWDvt~
Human [PAFA protein]	~~~ecIRTMH~GHdhnVSSVAIMPng~dHIVSA~~Sr~DKTIKMWEvg~
Nematode [unknown protein #1]	~~~rcVKTLK~GHtnyVFCCFNPs~~gTLIAS~~GsfDETIRIWCar~
Nematode [unknown protein #2]	~~~rmTKTLK~GHnnyVFCCNFNPq~~sSLVVS~~GsfDESVRIWDvk~
Fission yeast [FWDR protein]	~~~seCISILhGHtdsVLCLTFDS~~~~TLLVS~~GsaDCTVKLWHfs~

The Cn3D program displays a three-dimensional ribbon model of cow transducin (the protein highlighted in purple in the Sequence Alignment Viewer). This protein is the only one of those shown for which a structure has been determined. The sequence similarity of the other proteins to cow transducin suggests that their structures are likely to be similar.

This window displays information about the WD40 domain from the Conserved Domain Database.

WD40 - Cn3D 4.1

CDD Descriptive Items

Name: WD40

WD40 domain, found in a number of eukaryotic proteins that cover a wide variety of functions including adaptor/regulatory modules in signal transduction, pre-mRNA processing and cytoskeleton assembly; typically contains a GH dipeptide 11-24 residues from its N-terminus and the WD dipeptide at its C-terminus and is 40 residues long, hence the name WD40;

Cow transducin contains seven WD40 domains, one of which is highlighted here in gray.

The yellow segments correspond to the WD40 hallmarks highlighted in yellow in the window above.

▲ **Figure 18.3 Bioinformatics tools available on the Internet.** A website maintained by the National Center for Biotechnology Information allows scientists and the public to access DNA and protein sequences and other stored data. The site includes a link to a protein structure database (Conserved Domain Database, CDD) that can find and describe similar domains in related proteins, as well as software (Cn3D, "See in 3D") that displays three-dimensional models of domains for which the structure has been determined. Some results are shown from a search for regions of proteins similar to an amino acid sequence in a muskmelon protein.

is well known. For example, part of a new gene may match a known gene that encodes an important signaling pathway protein such as a protein kinase (see Chapter 5), suggesting that the new gene does, too. Alternatively, the new gene sequence may be similar to a previously encountered sequence whose function is still unknown. Another possibility is that the sequence is entirely unlike anything ever seen before. This was true for about a third of the genes of *E. coli* when its genome was sequenced. In the last case, protein function is usually deduced through a combination of biochemical and functional studies. The biochemical approach aims to determine the three-dimensional structure of the protein as well as other attributes, such as potential binding sites for other molecules. Functional studies usually involve blocking or disabling the gene to see how the phenotype is affected.

Understanding Genes and Gene Expression at the Systems Level

The impressive computational power provided by the tools of bioinformatics allows the study of whole sets of genes and their interactions, as well as the comparison of genomes from different species. Genomics is a rich source of new insights into fundamental questions about genome organization, regulation of gene expression, growth and development, and evolution.

One informative approach has been taken by a research project called ENCODE (Encyclopedia of DNA Elements). First, researchers focused intensively on 1% of the human genome and attempted to learn all they could about the functionally important elements in that sequence. They looked for protein-coding genes and genes for noncoding RNAs as well as sequences that regulate DNA replication, gene expression (such as enhancers and promoters), and chromatin modifications. This pilot project, completed in 2007, yielded a wealth of information. One big surprise, discussed in Concept 18.3, was that over 90% of the region was transcribed into RNA, even though less than 2% codes for proteins. The success of this approach led to two follow-up studies, one extending the analysis to the entire human genome and the other analyzing in a similar fashion the genomes of two model organisms, the soil nematode *Caenorhabditis elegans* and the fruit fly *Drosophila melanogaster*. Because genetic and molecular biological experiments can be performed on these species, testing the activities of potentially functional DNA elements in their genomes is expected to reveal much about how the human genome works.

Systems Biology

The success in sequencing genomes and studying entire sets of genes has encouraged scientists to attempt similar systematic study of the full protein sets (*proteomes*) encoded by genomes, an approach called **proteomics**. Proteins, not the genes that encode them, actually carry out most of the activities of the cell. Therefore, we must study when and where proteins are produced in an organism if we are to understand the functioning of cells and organisms.

Genomics and proteomics are enabling molecular biologists to approach the study of life from an increasingly global perspective. Using the tools we have described, biologists have begun to compile catalogs of genes and proteins—listings of all the "parts" that contribute to the operation of cells, tissues, and organisms. With such catalogs in hand, researchers have shifted their attention from the individual parts to their functional integration in biological systems. This is called the **systems biology** approach, which aims to model the dynamic behavior of whole biological systems based on the study of the interactions among the system's parts. Because of the vast amounts of data generated in these types of studies, the systems biology approach has really been made possible by advances in computer technology and bioinformatics.

Application of Systems Biology to Medicine

The Cancer Genome Atlas is an example of systems biology in which a large group of interacting genes and gene products are analyzed together. This project, under the joint leadership of the National Cancer Institute and NIH, aims to determine how changes in biological systems lead to cancer. A three-year pilot project that ended in 2010 set out to find all the common mutations in three types of cancer—lung cancer, ovarian cancer, and glioblastoma of the brain—by comparing gene sequences and patterns of gene expression in cancer cells with those in normal cells. Work on glioblastoma has confirmed the role of several suspected genes and identified a few unknown ones, suggesting possible new targets for therapies. The approach has proved so fruitful for these three types of cancer that it has been extended to ten other types, chosen because they are common and often lethal in humans.

Systems biology has tremendous potential in human medicine that is just starting to be explored. Silicon and glass "chips" have been developed that hold a microarray of most of the known human genes **(Figure 18.4)**. Such chips are being used to analyze gene expression patterns in patients suffering from various cancers and other diseases, with the eventual aim of tailoring their treatment to their unique genetic makeup and the specifics of their cancers. This approach has had modest success in characterizing subsets of several cancers.

Ultimately, people may carry with their medical records a catalog of their DNA sequence, a sort of genetic bar code, with regions highlighted that predispose them to specific diseases. The use of such sequences for personalized medicine—disease prevention and treatment—has great potential.

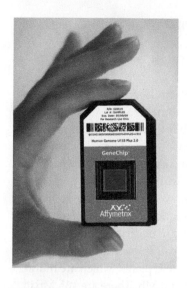

◀ **Figure 18.4 A human gene microarray chip.** Tiny spots of DNA arranged in a grid on this silicon wafer represent almost all of the genes in the human genome. Using this chip, researchers can analyze expression patterns for all these genes at the same time.

Systems biology is a very efficient way to study emergent properties at the molecular level. Novel properties emerge at each successive level of biological complexity as a result of the arrangement of building blocks at the underlying level (see Chapter 1). The more we can learn about the arrangement and interactions of the components of genetic systems, the deeper will be our understanding of whole organisms. The rest of this chapter will survey what we've learned from genomic studies thus far.

CONCEPT CHECK 18.2

1. What role does the Internet play in current genomics and proteomics research?
2. Explain the advantage of the systems biology approach to studying cancer versus the approach of studying a single gene at a time.
3. **MAKE CONNECTIONS** The ENCODE pilot project found that more than 90% of the genomic region being studied was transcribed into RNAs, far more than could be accounted for by protein-coding genes. Suggest some roles that these RNAs might play. (Review Concept 15.3.)

For suggested answers, see Appendix A.

CONCEPT 18.3

Genomes vary in size, number of genes, and gene density

By August 2012, the sequencing of about 3,700 genomes had been completed and that of over 7,500 genomes and about 340 metagenomes was in progress. In the completely sequenced group, about 3,300 are genomes of bacteria, and 160 are archaeal genomes. Among the 183 eukaryotic species in the group are vertebrates, invertebrates, protists, fungi, and plants. The accumulated genome sequences contain a wealth of information that we are now beginning to mine. What have we learned so far by comparing the genomes that have been sequenced? In this section, we'll examine the characteristics of genome size, number of genes, and gene density. Because these characteristics are so broad, we'll focus on general trends, for which there are often exceptions.

Genome Size

Comparing the three domains (Bacteria, Archaea, and Eukarya), we find a general difference in genome size between prokaryotes and eukaryotes (Table 18.1). While there are some exceptions, most bacterial genomes have between 1 and 6 million base pairs (Mb); the genome of *E. coli*, for instance, has 4.6 Mb. Genomes of archaea are, for the most part, within the size range of bacterial genomes. (Keep in mind, however, that many fewer archaeal genomes have been completely sequenced, so this picture may change.) Eukaryotic genomes tend to be larger: The genome of the single-celled yeast *Saccharomyces cerevisiae* (a fungus) has about 12 Mb, while most animals and plants, which are multicellular, have genomes of at least 100 Mb. There are 165 Mb in the fruit fly genome, while humans have 3,000 Mb, about 500 to 3,000 times as many as a typical bacterium.

Aside from this general difference between prokaryotes and eukaryotes, a comparison of genome sizes among eukaryotes fails to reveal any systematic relationship between genome size and the organism's phenotype. For instance, the genome of *Fritillaria assyriaca*, a flowering plant in the lily family, contains 124 billion base pairs (124,000 Mb), about 40 times the size of the human genome. On a finer scale, comparing two insect species, the cricket (*Anabrus simplex*) genome turns out to have 11 times as many base pairs as the *Drosophila melanogaster* genome. There is a wide range of genome sizes within the groups of protists, insects, amphibians, and plants and less of a range within mammals and reptiles.

Number of Genes

The number of genes also varies between prokaryotes and eukaryotes: Bacteria and archaea, in general, have fewer genes than eukaryotes. Free-living bacteria and archaea have from 1,500 to 7,500 genes, while the number of genes in eukaryotes ranges from about 5,000 for unicellular fungi to at least 40,000 for some multicellular eukaryotes (see Table 18.1).

Within the eukaryotes, the number of genes in a species is often lower than expected from simply considering the size of its genome. Looking at Table 18.1, you can see that the genome of the nematode *C. elegans* is 100 Mb in size and contains 20,100 genes. The *Drosophila* genome, in comparison, is much bigger (165 Mb) but has about two-thirds the number of genes—only 13,900 genes.

Considering an example closer to home, we noted that the human genome contains 3,000 Mb, well over ten times the size of either the *Drosophila* or *C. elegans* genome. At the outset of the Human Genome Project, biologists expected somewhere between 50,000 and 100,000 genes to be identified in the completed sequence, based on the number of known human proteins. As the project progressed, the estimate was revised downward several times, and currently, the most reliable count

Table 18.1 Genome Sizes and Estimated Numbers of Genes*

Organism	Haploid Genome Size (Mb)	Number of Genes	Genes per Mb
Bacteria			
Haemophilus influenzae	1.8	1,700	940
Escherichia coli	4.6	4,400	950
Archaea			
Archaeoglobus fulgidus	2.2	2,500	1,130
Methanosarcina barkeri	4.8	3,600	750
Eukaryotes			
Saccharomyces cerevisiae (yeast, a fungus)	12	6,300	525
Caenorhabditis elegans (nematode)	100	22,000	200
Arabidopsis thaliana (mustard family plant)	120	27,400	228
Drosophila melanogaster (fruit fly)	165	17,000	84
Oryza sativa (rice)	430	40,600	95
Zea mays (corn)	2,300	32,000	14
Mus musculus (house mouse)	2,600	22,000	11
Ailuropoda melanoleuca (giant panda)	2,400	21,000	9
Homo sapiens (human)	3,000	<21,000	7
Fritillaria assyriaca (lily family plant)	124,000	ND	ND

*Some values given here are likely to be revised as genome analysis continues. Mb = million base pairs. ND = not determined.

has placed the number at fewer than 21,000. This relatively low number, similar to the number of genes in the nematode *C. elegans*, has surprised biologists, who had clearly expected many more human genes.

What genetic attributes allow humans (and other vertebrates) to get by with no more genes than nematodes? An important factor is that vertebrate genomes "get more bang for the buck" from their coding sequences because of extensive alternative splicing of RNA transcripts. Recall that this process generates more than one functional protein from a single gene (see Figure 15.12). A typical human gene contains about ten exons, and an estimated 90% or more of these multi-exon genes are spliced in at least two different ways. Some genes are expressed in hundreds of alternatively spliced forms, others in just two. It is not yet possible to catalog all of the different forms, but it is clear that the number of different proteins encoded in the human genome far exceeds the proposed number of genes.

Additional polypeptide diversity could result from posttranslational modifications such as cleavage or the addition of carbohydrate groups in different cell types or at different developmental stages. Finally, the discovery of miRNAs and other small RNAs that play regulatory roles have added a new variable to the mix (see Concept 15.3). Some scientists think that this added level of regulation, when present, may contribute to greater organismal complexity for a given number of genes.

Gene Density and Noncoding DNA

In addition to genome size and number of genes, we can compare gene density in different species—in other words, how many genes there are in a given length of DNA. When we compare the genomes of bacteria, archaea, and eukaryotes, we see that eukaryotes generally have larger genomes but fewer genes in a given number of base pairs. Humans have hundreds or thousands of times as many base pairs in their genome as most bacteria, as we already noted, but only 5 to 15 times as many genes; thus, gene density is lower in humans (see Table 18.1). Even unicellular eukaryotes, such as yeasts, have fewer genes per million base pairs than bacteria and archaea. Among the genomes that have been sequenced completely thus far, humans and other mammals have the lowest gene density.

In all bacterial genomes studied so far, most of the DNA consists of genes for protein, tRNA, or rRNA, the small amount remaining consists mainly of nontranscribed regulatory sequences, such as promoters. The sequence of nucleotides along a bacterial protein-coding gene proceeds from start to finish without interruption by noncoding sequences (introns). In eukaryotic genomes, by contrast, most of the DNA neither encodes protein nor is transcribed into RNA molecules of known function, and the DNA includes more complex regulatory sequences. In fact, humans have 10,000 times as much noncoding DNA as bacteria. Some of this DNA in multicellular eukaryotes is present as introns within genes. Indeed, introns account for most of the difference in average length between human genes (27,000 base pairs) and bacterial genes (1,000 base pairs).

In addition to introns, multicellular eukaryotes have a vast amount of non-protein-coding DNA between genes. In the next section, we'll explore the composition and arrangement of these great stretches of DNA in the human genome.

CONCEPT 18.4

Multicellular eukaryotes have much noncoding DNA and many multigene families

We have spent most of this chapter, and indeed this unit, focusing on genes that code for proteins. Yet the coding regions of these genes and the genes for RNA products such as rRNA, tRNA, and miRNA make up only a small portion of the genomes of most multicellular eukaryotes. For example, once the sequencing of the human genome was completed, it became clear that only a tiny part—1.5%—codes for proteins or is transcribed into rRNAs or tRNAs. **Figure 18.5** shows what is known about the makeup of the remaining 98.5% of the genome.

Gene-related regulatory sequences and introns account, respectively, for 5% and about 20% of the human genome. The rest, located between functional genes, includes some unique noncoding DNA, such as gene fragments and **pseudogenes**, former genes that have accumulated mutations over a long time

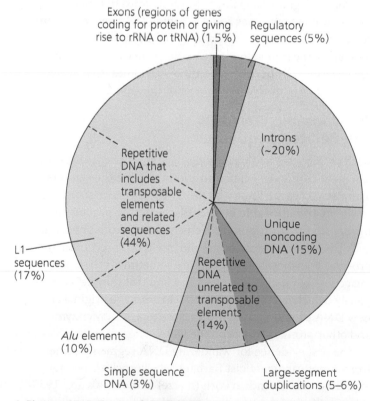

▲ **Figure 18.5 Types of DNA sequences in the human genome.** The gene sequences that code for proteins or are transcribed into rRNA or tRNA molecules make up only about 1.5% of the human genome (dark purple in the pie chart), while introns and regulatory sequences associated with genes (lighter purple) make up about a quarter. The vast majority of the human genome does not code for proteins or give rise to known RNAs, and much of it is repetitive DNA (dark and light green and teal). Because repetitive DNA is the most difficult to sequence and analyze, classification of some portions is tentative, and the percentages given here may shift slightly as genome analysis proceeds.

and no longer produce functional proteins. (The genes that produce small noncoding RNAs are a tiny percentage of the genome, distributed between the 20% introns and the 15% unique noncoding DNA.) Most intergenic DNA, however, is **repetitive DNA**, which consists of sequences that are present in multiple copies in the genome. Somewhat surprisingly, about 75% of this repetitive DNA (44% of the entire human genome) is made up of units called transposable elements and sequences related to them.

The bulk of many eukaryotic genomes likewise consists of DNA sequences that neither code for proteins nor are transcribed to produce RNAs with known functions; this noncoding DNA was often described in the past as "junk DNA." However, much evidence is accumulating that this DNA plays important roles in the cell. One measure of its importance is the high degree of sequence conservation between species that diverged many hundreds of generations ago. For example, comparison of the genomes of humans, rats, and mice has revealed the presence of almost 500 regions of noncoding DNA that are *identical* in sequence in all three species. This is a higher level of sequence conservation than is seen for protein-coding regions in these species, strongly suggesting that the noncoding regions have important functions. In this section, we'll examine how genes and noncoding DNA sequences are currently organized within genomes of multicellular eukaryotes, using the human genome as our main example. Genome organization tells us much about how genomes have evolved and continue to evolve, the subject of Concept 18.5.

Transposable Elements and Related Sequences

Both prokaryotes and eukaryotes have stretches of DNA that can move from one location to another within the genome. These stretches are known as *transposable genetic elements*, or simply **transposable elements**. During the process called *transposition*, a transposable element moves from one site in a cell's DNA to a different target site by a type of recombination process. Transposable elements are sometimes called "jumping genes," but it should be kept in mind that they never completely detach from the cell's DNA. Instead, the original and new DNA sites are brought very close together by enzymes and other proteins that bend the DNA.

The first evidence for wandering DNA segments came from American geneticist Barbara McClintock's breeding experiments with Indian corn (maize) in the 1940s and 1950s **(Figure 18.6)**. As she tracked corn plants through multiple generations, McClintock identified changes in the color of corn kernels that made sense only if she postulated the existence of genetic elements capable of moving from other locations in the genome into the genes for kernel color, disrupting the genes so that the kernel color was changed. McClintock's discovery was met with great skepticism and virtually discounted at the time. Her careful work and insightful ideas

▲ **Figure 18.6 The effect of transposable elements on corn kernel color.** Barbara McClintock first proposed the idea of mobile genetic elements after observing variegations in corn kernel color (right). She received the Nobel Prize in 1983.

were finally validated many years later when transposable elements were found in bacteria. In 1983, at the age of 81, McClintock received the Nobel Prize for her pioneering research.

Movement of Transposons and Retrotransposons

Eukaryotic transposable elements are of two types. The first type, **transposons**, can move within a genome by means of a DNA intermediate. Transposons can move by a "cut-and-paste" mechanism, which removes the element from the original site, or by a "copy-and-paste" mechanism, which leaves a copy behind **(Figure 18.7)**. Both mechanisms require an enzyme called *transposase*, which is generally encoded by the transposon.

Most transposable elements in eukaryotic genomes are of the second type, **retrotransposons**, which move by means of an RNA intermediate that is a transcript of the retrotransposon DNA. Retrotransposons always leave a copy at the original site during transposition, since they are initially transcribed

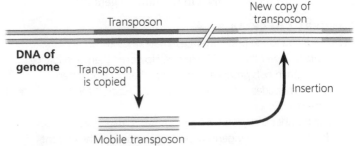

© 1996 Pearson Education, Inc.

▲ **Figure 18.7 Transposon movement.** Movement of transposons by either the copy-and-paste mechanism (shown here) or the cut-and-paste mechanism involves a double-stranded DNA intermediate that is inserted into the genome.

? *How would this figure differ if it showed the cut-and-paste mechanism?*

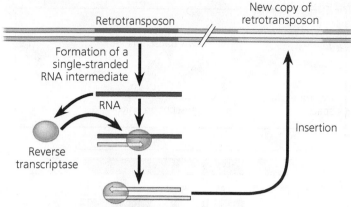

Formation of a
single-stranded
RNA intermediate

Retrotransposon

New copy of
retrotransposon

RNA

Reverse
transcriptase

Insertion

© 1996 Pearson Education, Inc.

▲ **Figure 18.8 Retrotransposon movement.** Movement begins with formation of a single-stranded RNA intermediate. The remaining steps are essentially identical to part of the retrovirus replicative cycle (see Figure 17.7).

into an RNA intermediate (**Figure 18.8**). To insert at another site, the RNA intermediate is first converted back to DNA by reverse transcriptase, an enzyme encoded by the retrotransposon. (Reverse transcriptase is also encoded by retroviruses, as you learned in Chapter 17. In fact, retroviruses may have evolved from retrotransposons.) Another cellular enzyme catalyzes insertion of the reverse-transcribed DNA at a new site.

Sequences Related to Transposable Elements

Multiple copies of transposable elements and sequences related to them are scattered throughout eukaryotic genomes. A single unit is usually hundreds to thousands of base pairs long, and the dispersed "copies" are similar but usually not identical to each other. Some of these are transposable elements that can move; the enzymes required for this movement may be encoded by any transposable element, including the one that is moving. Others are related sequences that have lost the ability to move altogether. Transposable elements and related sequences make up 25–50% of most mammalian genomes (see Figure 18.5) and even higher percentages in amphibians and many plants. In fact, the very large size of some plant genomes is accounted for not by extra genes, but by extra transposable elements. For example, such sequences make up 85% of the corn genome!

In humans and other primates, a large portion of transposable element–related DNA consists of a family of similar sequences called *Alu elements*. These sequences alone account for approximately 10% of the human genome. *Alu* elements are about 300 nucleotides long, much shorter than most functional transposable elements, and they do not code for any protein. However, many *Alu* elements are transcribed into RNA; its cellular function, if any, is currently unknown.

An even larger percentage (17%) of the human genome is made up of a type of retrotransposon called *LINE-1*, or *L1*. These sequences are much longer than *Alu* elements—about 6,500 base pairs—and have a low rate of transposition. An accompanying genomic analysis found L1 sequences within the introns of nearly 80% of the human genes that were analyzed, suggesting that L1 may help regulate gene expression. Other researchers have proposed that L1 retrotransposons may have differential effects on gene expression in developing neurons, contributing to the great diversity of neuronal cell types (see Chapter 37).

Although many transposable elements encode proteins, these proteins do not carry out normal cellular functions. Therefore, transposable elements are usually included in the "noncoding" DNA category, along with other repetitive sequences.

Other Repetitive DNA, Including Simple Sequence DNA

Repetitive DNA that is not related to transposable elements probably arises due to mistakes during DNA replication or recombination. Such DNA accounts for about 14% of the human genome (see Figure 18.5). About a third of this (5–6% of the human genome) consists of duplications of long stretches of DNA, with each unit ranging from 10,000 to 300,000 base pairs. The large segments seem to have been copied from one chromosomal location to another site on the same or a different chromosome and probably include some functional genes.

In contrast to scattered copies of long sequences, **simple sequence DNA** contains many copies of tandemly repeated short sequences, as in the following example (showing one DNA strand only):

… GTTACGTTACGTTACGTTACGTTACGTTAC …

In this case, the repeated unit (GTTAC) consists of 5 nucleotides. Repeated units may contain as many as 500 nucleotides, but often contain fewer than 15 nucleotides, as in this example. When the unit contains 2–5 nucleotides, the series of repeats is called a **short tandem repeat (STR)**. The number of copies of the repeated unit can vary from site to site within a given genome. There could be as many as several hundred thousand repetitions of the GTTAC unit at one site, but only half that number at another.

The repeat number also varies from person to person, and since humans are diploid, each person has two alleles per site, which can differ. This diversity produces variation that can be used to identify a unique set of genetic markers for each individual, his or her **genetic profile**. Forensic scientists can use STR analysis on DNA extracted from samples of tissues or body fluids to identify victims of a crime scene or natural disaster. In such an application, STR analysis is performed on STR sites selected because they have relatively few repeats and are easily sequenced. This technique has also been used by The Innocence Project, a nonprofit organization, to free more than 250 wrongly convicted people from prison.

Altogether, simple sequence DNA makes up 3% of the human genome. Much of a genome's simple sequence DNA is located at chromosomal telomeres and centromeres, suggesting that this DNA plays a structural role for chromosomes. The DNA at centromeres is essential for the separation of chromatids in cell division (see Chapter 9). Centromeric DNA, along with simple sequence DNA located elsewhere, may also help organize the chromatin within the interphase nucleus. The simple sequence DNA located at telomeres, at the tips of chromosomes, binds proteins that protect the ends of a chromosome from degradation and from joining to other chromosomes.

Genes and Multigene Families

We finish our discussion of the various types of DNA sequences in eukaryotic genomes with a closer look at genes. Recall that DNA sequences that code for proteins or give rise to tRNA or rRNA compose a mere 1.5% of the human genome (see Figure 18.5). If we include introns and regulatory sequences associated with genes, the total amount of DNA that is gene related—coding and noncoding—constitutes about 25% of the human genome. Put another way, only about 6% (1.5% out of 25%) of the length of the average gene is represented in the final gene product.

Like the genes of bacteria, many eukaryotic genes are present as unique sequences, with only one copy per haploid set of chromosomes. But in the human genome and the genomes of many other animals and plants, solitary genes make up less than half of the total gene-related DNA. The rest occur in **multigene families**, collections of two or more identical or very similar genes.

In multigene families consisting of *identical* DNA sequences, those sequences are usually clustered tandemly and, with the notable exception of the genes for histone proteins, have RNAs as their final products. An example is the family of identical DNA sequences that are the genes for the three largest rRNA molecules **(Figure 18.9a)**. These rRNA molecules are transcribed from a single transcription unit that is repeated tandemly hundreds to thousands of times in one or several clusters in the genome of a multicellular eukaryote. The many copies of this rRNA transcription unit help cells to quickly make the millions of ribosomes needed for active protein synthesis. The primary transcript is cleaved to yield the three rRNA molecules, which combine with proteins and one other kind of rRNA (5S rRNA) to form ribosomal subunits.

The classic examples of multigene families of *nonidentical* genes are two related families of genes that encode globins, a group of proteins that include the α and β polypeptide subunits of hemoglobin. One family, located on chromosome 16 in humans, encodes various forms of α-globin; the other, on chromosome 11, encodes forms of β-globin **(Figure 18.9b)**. The different forms of each globin subunit are expressed at different times in development, allowing hemoglobin to function effectively in the changing environment of the developing animal. In humans, for example, the embryonic and fetal forms

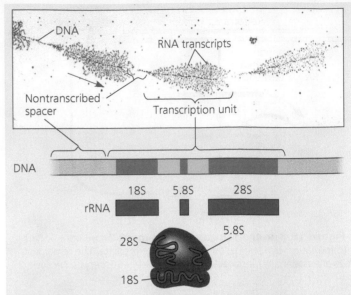

(a) Part of the ribosomal RNA gene family. The TEM at the top shows three of the hundreds of copies of rRNA transcription units in a salamander genome. Each "feather" corresponds to a single unit being transcribed by about 100 molecules of RNA polymerase (dark dots along the DNA), moving left to right (red arrow). The growing RNA transcripts extend from the DNA. In the diagram of a transcription unit below the TEM, the genes for three types of rRNA (blue) are adjacent to regions that are transcribed but later removed (yellow). A single transcript is processed to yield one of each of the three rRNAs (red), key components of the ribosome.

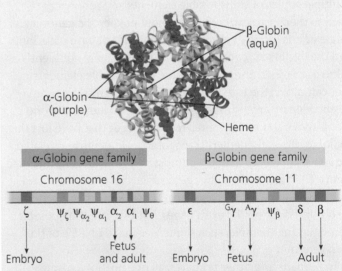

(b) The human α-globin and β-globin gene families. Adult hemoglobin is composed of two α-globin and two β-globin polypeptide subunits, as shown in the molecular model. The genes (dark blue) encoding α- and β-globins are found in two families, organized as shown here. The noncoding DNA separating the functional genes within each family includes pseudogenes (ψ; green), versions of the original genes that no longer produce functional proteins. Genes and pseudogenes are named with Greek letters. Some genes are expressed only in the embryo or fetus.

▲ **Figure 18.9 Gene families.**

? *In (a), how could you determine the direction of transcription if it weren't indicated by the red arrow?*

of hemoglobin have a higher affinity for oxygen than the adult forms, ensuring the efficient transfer of oxygen from mother to fetus. Also found in the globin gene family clusters are several pseudogenes (green in Figure 18.9b).

We'll return to the globin gene family to consider the evolutionary history of these gene clusters in the next section. We'll also consider some of the processes that have shaped the genomes of different species over evolutionary time.

CONCEPT CHECK 18.4

1. Discuss the characteristics of mammalian genomes that make them larger than prokaryotic genomes.
2. Which of the three mechanisms described in Figures 18.7 and 18.8 result(s) in a copy remaining at the original site as well as appearing in a new location?
3. Contrast the organizations of the rRNA gene family and the globin gene families. For each, explain how the existence of a family of genes benefits the organism.

For suggested answers, see Appendix A.

CONCEPT 18.5

Duplication, rearrangement, and mutation of DNA contribute to genome evolution

EVOLUTION The basis of change at the genomic level is mutation, which underlies much of genome evolution. It seems likely that the earliest forms of life had a minimal number of genes—those necessary for survival and reproduction. If this were indeed the case, one aspect of evolution must have been an increase in the size of the genome, with the extra genetic material providing the raw material for gene diversification. In this section, we'll first describe how extra copies of all or part of a genome can arise and then consider subsequent processes that can lead to the evolution of proteins (or RNA products) with slightly different or entirely new functions.

Duplication of Entire Chromosome Sets

An accident in meiosis can result in one or more extra sets of chromosomes, a condition known as polyploidy. Although such accidents would most often be lethal, in rare cases they could facilitate the evolution of genes. In a polyploid organism, one set of genes can provide essential functions for the organism. The genes in the one or more extra sets can diverge by accumulating mutations; these variations may persist if the organism carrying them survives and reproduces. In this way, genes with novel functions can evolve. As long as one copy of an essential gene is expressed, the divergence of another copy can lead to its encoded protein acting in a novel way, thereby changing the organism's phenotype. The outcome of this accumulation of mutations may be the branching off of a new species, as happens often in flowering plants (see Chapter 22).

Polyploid animals also exist, but they are much rarer; the tetraploid model organism *Xenopus laevis*, the African clawed frog, is an example.

Alterations of Chromosome Structure

Scientists have long known that sometime in the last 6 million years, when the ancestors of humans and chimpanzees diverged as species, the fusion of two ancestral chromosomes in the human line led to different haploid numbers for humans ($n = 23$) and chimpanzees ($n = 24$). The banding patterns in stained chromosomes suggested that the ancestral versions of current chimp chromosomes 12 and 13 fused end to end, forming chromosome 2 in an ancestor of the human lineage. With the recent explosion in genomic sequence information, we can now compare the chromosomal organizations of many different species on a much finer scale. This information allows us to make inferences about the evolutionary processes that shape chromosomes and may drive speciation. Sequencing and analysis of human chromosome 2 provided very strong supporting evidence for the model we have just described **(Figure 18.10)**.

In another study of broader scope, researchers compared the DNA sequence of each human chromosome with the whole-genome sequence of the mouse. For human chromosome 16, the comparison revealed that large blocks of genes on this chromosome are found on four mouse chromosomes. This

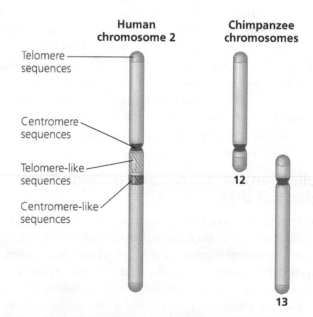

▲ **Figure 18.10 Related human and chimpanzee chromosomes.** The positions of telomere-like and centromere-like sequences on human chromosome 2 (left) match those of telomeres on chimp chromosome 13 (right). This suggests that chromosomes 12 and 13 in a human ancestor fused end-to-end to form human chromosome 2. The centromere from ancestral chromosome 12 remained functional on human chromosome 2, while the one from ancestral chromosome 13 did not. (Chimp chromosomes 12 and 13 have since been renamed 2a and 2b, respectively.)

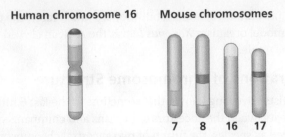

Human chromosome 16 **Mouse chromosomes**

7 8 16 17

▲ **Figure 18.11 A comparison of human and mouse chromosomes.** DNA sequences very similar to large blocks of human chromosome 16 (colored areas in this diagram) are found on mouse chromosomes 7, 8, 16, and 17. This suggests that the DNA sequence in each block has stayed together in the mouse and human lineages since the time they diverged from a common ancestor.

observation suggests that the genes in each block stayed together during the evolution of the mouse and human lineages **(Figure 18.11).**

Performing the same comparative analysis between chromosomes of humans and six other mammalian species allowed the researchers to reconstruct the evolutionary history of chromosomal rearrangements in these eight species. They found many duplications and inversions of large portions of chromosomes, the result of mistakes during meiotic recombination in which the DNA broke and was rejoined incorrectly. The rate of these events seems to have accelerated about 100 million years ago, around the time large dinosaurs became extinct and the number of mammalian species increased rapidly. The apparent coincidence is interesting because chromosomal rearrangements are thought to contribute to the generation of new species. Although two individuals with different arrangements could still mate and produce offspring, the offspring would have two nonequivalent sets of chromosomes, making meiosis inefficient or even impossible. Thus, chromosomal rearrangements would reduce the success of matings between members of the two populations, a step on the way to the populations becoming two separate species. (You'll learn more about this in Chapter 22.)

Duplication and Divergence of Gene-Sized Regions of DNA

Errors during meiosis can also lead to the duplication of chromosomal regions that are smaller than the ones we've just discussed, including segments the length of individual genes. Unequal crossing over during prophase I of meiosis, for instance, can result in one chromosome with a deletion and another with a duplication of a particular gene. As illustrated in **Figure 18.12,** transposable elements can provide homologous sites where nonsister chromatids can cross over, even when other chromatid regions are not correctly aligned.

Also, slippage can occur during DNA replication, such that the template shifts with respect to the new complementary strand, and a part of the template strand is either skipped by the replication machinery or used twice as a template. As a result,

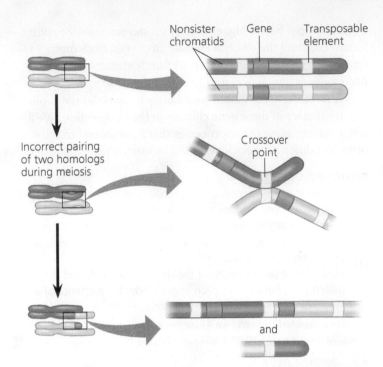

▲ **Figure 18.12 Gene duplication due to unequal crossing over.** One mechanism by which a gene (or other DNA segment) can be duplicated is recombination during meiosis between copies of a transposable element flanking the gene. Such recombination between misaligned nonsister chromatids of homologous chromosomes produces one chromatid with two copies of the gene and one chromatid with no copy.

MAKE CONNECTIONS *Recall how crossing over occurs (see Figure 10.11). In the middle panel above, draw a line along the portions that result in the upper chromatid in the bottom panel. Use a different color to do the same for the other chromatid.*

a segment of DNA is deleted or duplicated. It is easy to imagine how such errors could occur in regions of repeats. The variable number of repeated units of simple sequence DNA at a given site, used for STR analysis, is probably due to errors like these. Evidence that unequal crossing over and template slippage during DNA replication lead to duplication of genes is found in the existence of multigene families, such as the globin family.

Evolution of Genes with Related Functions: The Human Globin Genes

Duplication events can lead to the evolution of genes with related functions, such as those of the α-globin and β-globin gene families (see Figure 18.9b). A comparison of gene sequences within a multigene family can suggest the order in which the genes arose. This approach to re-creating the evolutionary history of the globin genes indicates that they all evolved from one common ancestral globin gene that underwent duplication and divergence into the α-globin and β-globin ancestral genes about 450–500 million years ago **(Figure 18.13).** Each of these genes was later duplicated several times, and the copies then diverged from each other in sequence, yielding the current family members. In fact, the common ancestral globin gene also gave rise to the oxygen-binding muscle protein myoglobin and to the plant protein

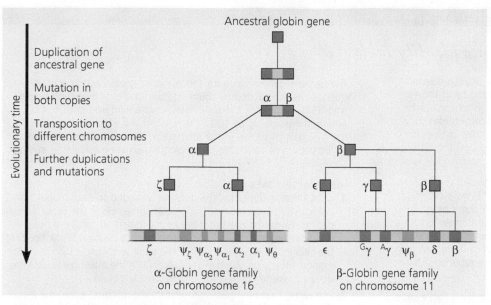

▲ **Figure 18.13 A model for the evolution of the human α-globin and β-globin gene families from a single ancestral globin gene.**

? *The green elements are pseudogenes. Explain how they could have arisen after gene duplication.*

leghemoglobin. The latter two proteins function as monomers, and their genes are included in a "globin superfamily."

After the duplication events, the differences between the genes in the globin families undoubtedly arose from mutations that accumulated in the gene copies over many generations. The current model is that the necessary function provided by an α-globin protein, for example, was fulfilled by one gene, while other copies of the α-globin gene accumulated random mutations. Many mutations may have had an adverse effect on the organism and others may have had no effect, but a few mutations must have altered the function of the protein product in a way that was advantageous to the organism at a particular life stage without substantially changing the protein's oxygen-carrying function. Presumably, natural selection acted on these altered genes, maintaining them in the population.

In the **Scientific Skills Exercise**, you can compare amino acid sequences of the globin family members and see how such comparisons were used to generate the model for globin gene evolution shown in Figure 18.13. The existence of several pseudogenes among the functional globin genes provides additional evidence for this model (see Figure 18.9b): Random mutations in these "genes" over evolutionary time have destroyed their function.

Evolution of Genes with Novel Functions

In the evolution of the globin gene families, gene duplication and subsequent divergence produced family members whose protein products performed similar functions (oxygen transport). Alternatively, one copy of a duplicated gene can undergo alterations that lead to a completely new function for the protein product. The genes for lysozyme and α-lactalbumin are good examples.

Lysozyme is an enzyme that helps protect animals against bacterial infection by hydrolyzing bacterial cell walls; α-lactalbumin is a nonenzymatic protein that plays a role in milk production in mammals. The two proteins are quite similar in their amino acid sequences and three-dimensional structures. Both genes are found in mammals, whereas only the lysozyme gene is present in birds. These findings suggest that at some time after the lineages leading to mammals and birds had separated, the lysozyme gene was duplicated in the mammalian lineage but not in the avian lineage. Subsequently, one copy of the duplicated lysozyme gene evolved into a gene encoding α-lactalbumin, a protein with a completely different function.

Besides the duplication and divergence of whole genes, rearrangement of existing DNA sequences within genes has also contributed to genome evolution. The presence of introns may have promoted the evolution of new proteins by facilitating the duplication or shuffling of exons, as we'll discuss next.

Rearrangements of Parts of Genes: Exon Duplication and Exon Shuffling

Proteins often have a modular architecture consisting of discrete structural and functional regions called **domains**. One domain of an enzyme, for example, might include the active site, while another might allow the enzyme to bind to a cellular membrane. In quite a few cases, different exons code for the different domains of a protein.

We've already seen that unequal crossing over during meiosis can lead to duplication of a gene on one chromosome and its loss from the homologous chromosome (see Figure 18.12). By a similar process, a particular exon within a gene could be duplicated on one chromosome and deleted from the other. The gene with the duplicated exon would code for a protein containing a second copy of the encoded domain. This change in the protein's structure could augment its function by increasing its stability, enhancing its ability to bind a particular ligand, or altering some other property. Quite a few protein-coding genes have multiple copies of related exons, which presumably arose by duplication and then diverged. The gene encoding the extracellular matrix protein collagen is a good example. Collagen is a structural protein with a highly repetitive amino acid sequence, which is reflected in the repetitive pattern of exons in the collagen gene.

Alternatively, we can imagine the occasional mixing and matching of different exons either within a gene or between two different (nonallelic) genes owing to errors in meiotic recombination. This process, termed *exon shuffling*, could lead

Reading an Amino Acid Sequence Identity Table

How Have Amino Acid Sequences of Human Globin Genes Diverged During Their Evolution? To build a model of the evolutionary history of the globin genes (see Figure 18.13), researchers compared the amino acid sequences of the polypeptides they encode. In this exercise, you will analyze comparisons of the amino acid sequences of globin polypeptides to shed light on their evolutionary relationships.

How the Experiment Was Done Scientists obtained the DNA sequences for each of the eight globin genes and "translated" them into amino acid sequences. They then used a computer program to align the sequences and calculate a percent identity value for each pair. The percent identity reflects the number of positions with identical amino acids relative to the total number of amino acids in a globin polypeptide. The data were arranged in a table to show the pairwise comparisons.

Data from the Experiment The following table shows an example of a pairwise alignment—that of the α_1-globin (alpha-1 globin) and ζ-globin (zeta globin) amino acid sequences—using the standard single-letter codes for amino acids. To the left of each line of amino acid sequence is the number of the first amino acid in that line.

Globin	Alignment of Globin Amino Acid Sequences
α_1	1 MVLSPADKTNVKAAWGKVGAHAGEYGAEAL
ζ	1 MSLTKTERTIIVSMWAKISTQADTIGTETL
α_1	31 ERMFLSFPTTKTYFPHFDLSH–GSAQVKGH
ζ	31 ERLFLSHPQTKTYFPHFDL–HPGSAQLRAH
α_1	61 GKKVADALTNAVAHVDDMPNALSALSDLHA
ζ	61 GSKVVAAVGDAVKSIDDIGGALSKLSELHA
α_1	91 HKLRVDPVNFKLLSHCLLVTLAAHLPAEFT
ζ	91 YILRVDPVNFKLLSHCLLVTLAARFPADFT
α_1	121 PAVHASLDKFLASVSTVLTSKYR
ζ	121 AEAHAAWDKFLSVVSSVLTEKYR

The percent identity value for the α_1- and ζ-globin amino acid sequences was calculated by counting the number of matching amino acids (87), dividing by the total number of amino acid positions (143), and then multiplying by 100. This resulted in a 61% identity value for the α_1-ζ pair, as shown in the amino acid identity table at the bottom of the page. The values for other globin pairs were calculated in the same way.

Interpret the Data

1. Notice that in the table, the data are arranged so each globin pair can be compared. (a) Notice that some cells in the table have dashes. Given the pairs that are being compared for these cells, what percent identity value is implied by the dashes? (b) Notice that the cells in the lower left half of the table are blank. Using the information already provided in the table, fill in the missing values. Why does it make sense that these cells were left blank?

2. The earlier that two genes arose from a duplicated gene, the more divergent their bases can become, which may result in amino acid differences in the protein products. (a) Based on that premise, identify which two genes are most divergent from each other. What is the percent identity between them? (b) Using the same approach, identify which two globin genes are the most recently duplicated. What is the percent identity between them?

3. The model of globin gene evolution shown in Figure 18.13 suggests that an ancestral gene duplicated into α- and β-globin genes and then each one was further duplicated and modified by mutation. What features of the data set support the model?

4. Make a list of all the percent identity values from the table, starting with 100% at the top. Next to each number write the globin pair(s) with that percent identity value. Use one color for the globins from the α family and a different color for the globins from the β family. (a) Compare the order of pairs on your list with their position in the model shown in Figure 18.13. Does the order of pairs describe the same relative "closeness" of globin family members seen in the model? (b) Compare the percent identity values for pairs within the α or β group to the values for between-group pairs.

Further Reading R. C. Hardison, Globin genes on the move, *Journal of Biology* 7:35.1–35.5 (2008). doi:10.1186/jbiol92

(MB) A version of this Scientific Skills Exercise can be assigned in MasteringBiology.

Amino Acid Identity Table								
	α Family			**β Family**				
	α_1 (alpha 1)	α_2 (alpha 2)	ζ (zeta)	β (beta)	δ (delta)	ϵ (epsilon)	$^A\gamma$ (gamma A)	$^G\gamma$ (gamma G)
α_1	------	100	61	45	44	39	42	42
α_2		------	61	45	44	39	42	42
ζ			------	38	40	41	41	41
β				------	93	76	73	73
δ					------	73	71	72
ϵ						------	80	80
$^A\gamma$							------	99
$^G\gamma$								------

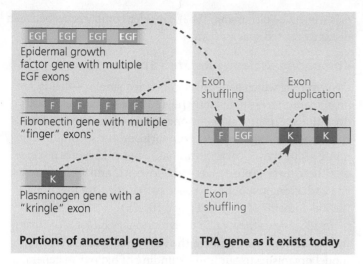

| Portions of ancestral genes | TPA gene as it exists today |

▲ Figure 18.14 Evolution of a new gene by exon shuffling. Exon shuffling could have moved exons, each encoding a particular domain, from ancestral forms of the genes for epidermal growth factor, fibronectin, and plasminogen (left) into the evolving gene for tissue plasminogen activator, TPA (right). Duplication of the "kringle" exon from the plasminogen gene after its movement could account for the two copies of this exon in the TPA gene.

? *How could the presence of transposable elements in introns have facilitated the exon shuffling shown here?*

to new proteins with novel combinations of functions. As an example, let's consider the gene for tissue plasminogen activator (TPA). The TPA protein is an extracellular protein that helps control blood clotting. It has four domains of three types, each encoded by an exon; one exon is present in two copies. Because each type of exon is also found in other proteins, the gene for TPA is thought to have arisen by several instances of exon shuffling and duplication **(Figure 18.14)**.

How Transposable Elements Contribute to Genome Evolution

The persistence of transposable elements as a large fraction of some eukaryotic genomes is consistent with the idea that they play an important role in shaping a genome over evolutionary time. These elements can contribute to the evolution of the genome in several ways. They can promote recombination, disrupt cellular genes or control elements, and carry entire genes or individual exons to new locations.

Transposable elements of similar sequence scattered throughout the genome facilitate recombination between different chromosomes by providing homologous regions for crossing over. Most such recombination events are probably detrimental, causing chromosomal translocations and other changes in the genome that may be lethal to the organism. But over the course of evolutionary time, an occasional recombination event of this sort may be advantageous to the organism. (For the change to be heritable, of course, it must happen in a cell that will give rise to a gamete.)

The movement of a transposable element can have a variety of consequences. For instance, if a transposable element "jumps" into the middle of a protein-coding sequence, it will

prevent the production of a normal transcript of the gene. If a transposable element inserts within a regulatory sequence, the transposition may lead to increased or decreased production of one or more proteins. Transposition caused both types of effects on the genes coding for pigment-synthesizing enzymes in McClintock's corn kernels. Again, while such changes are usually harmful, in the long run some may prove beneficial by providing a survival advantage.

During transposition, a transposable element may carry along a gene or group of genes to a new position in the genome. This mechanism probably accounts for the location of the α-globin and β-globin gene families on different human chromosomes, as well as the dispersion of the genes of certain other gene families. By a similar tag-along process, an exon from one gene may be inserted into another gene in a mechanism similar to that of exon shuffling during recombination. For example, an exon may be inserted by transposition into the intron of a protein-coding gene. If the inserted exon is retained in the RNA transcript during RNA splicing, the protein that is synthesized will have an additional domain, which may confer a new function on the protein.

All the processes discussed in this section most often produce either harmful effects, which may be lethal, or no effect at all. In a few cases, however, small beneficial heritable changes may occur. Over many generations, the resulting genetic diversity provides valuable raw material for natural selection. Diversification of genes and their products is an important factor in the evolution of new species. Thus, the accumulation of changes in the genome of each species provides a record of its evolutionary history. To read this record, we must be able to identify genomic changes. Comparing the genomes of different species allows us to do that and has increased our understanding of how genomes evolve. You'll learn more about these topics in the final section.

CONCEPT CHECK 18.5

1. Describe three examples of errors in cellular processes that lead to DNA duplications.
2. Explain how multiple exons might have arisen in the ancestral EGF and fibronectin genes shown in Figure 18.14 (left).
3. What are three ways that transposable elements are thought to contribute to genome evolution?

For suggested answers, see Appendix A.

CONCEPT 18.6

Comparing genome sequences provides clues to evolution and development

EVOLUTION One researcher has likened the current state of biology to the Age of Exploration in the 15th century after major improvements in navigation and the building of faster ships. In the last 25 years, we have seen rapid advances in genome sequencing and data collection, new techniques for assessing gene activity across the whole genome, and refined approaches for understanding how genes and their products work together in complex systems. We are truly poised on the brink of a new world.

Comparisons of genome sequences from different species reveal much about the evolutionary history of life, from very ancient to more recent. Similarly, comparative studies of the genetic programs that direct embryonic development in different species are beginning to clarify the mechanisms that generated the great diversity of life-forms present today. In this final section of the chapter, we'll discuss what has been learned from these two approaches.

Comparing Genomes

The more similar in sequence the genes and genomes of two species are, the more closely related those species are in their evolutionary history. Comparing genomes of closely related species sheds light on more recent evolutionary events, whereas comparing genomes of very distantly related species helps us understand ancient evolutionary history. In either case, learning about characteristics that are shared or divergent between groups enhances our picture of the evolution of life-forms and biological processes. Evolutionary relationships between species can be represented by a diagram in the form of a tree (often turned sideways), where each branch point marks the divergence of two lineages (see Chapter 1). **Figure 18.15** shows the evolutionary relationships of some groups and spe-

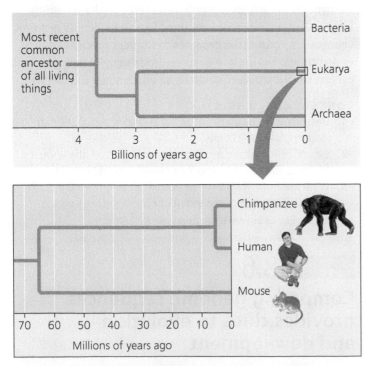

▲ **Figure 18.15 Evolutionary relationships of the three domains of life.** This tree diagram shows the ancient divergence of bacteria, archaea, and eukaryotes. A portion of the eukaryote lineage is expanded in the inset to show the more recent divergence of three mammalian species discussed in this chapter.

cies we will be discussing. We'll consider comparisons between distantly related species first.

Comparing Distantly Related Species

Determining which genes have remained similar—that is, are *highly conserved*—in distantly related species can help clarify evolutionary relationships among species that diverged from each other long ago. Indeed, comparisons of the complete genome sequences of bacteria, archaea, and eukaryotes indicate that these three groups diverged between 2 and 4 billion years ago and strongly support the theory that they are the fundamental domains of life (see Figure 18.15).

In addition to their value in evolutionary biology, comparative genomic studies confirm the relevance of research on model organisms to our understanding of biology in general and human biology in particular. Genes that evolved a very long time ago can still be surprisingly similar in disparate species. As a case in point, several genes in yeast are so similar to certain human disease genes that researchers have deduced the functions of the disease genes by studying their yeast counterparts. This striking similarity underscores the common origin of these two distantly related species.

Comparing Closely Related Species

The genomes of two closely related species are likely to be organized similarly because of their relatively recent divergence. This allows the fully sequenced genome of one species to be used as a scaffold for assembling the genomic sequences of a closely related species, accelerating mapping of the second genome. For instance, using the human genome sequence as a guide, researchers were able to quickly sequence the entire chimpanzee genome.

The recent divergence of two closely related species also underlies the small number of gene differences that are found when their genomes are compared. The particular genetic differences can therefore be more easily correlated with phenotypic differences between the two species. An exciting application of this type of analysis is seen as researchers compare the human genome with the genomes of the chimpanzee, mouse, rat, and other mammals. Identifying the genes shared by all of these species but not by nonmammals should give clues about what it takes to make a mammal, while finding the genes shared by chimpanzees and humans but not by rodents should tell us something about primates. And, of course, comparing the human genome with that of the chimpanzee should help us answer the tantalizing question we asked at the beginning of the chapter: What genomic information makes a human or a chimpanzee?

An analysis of the overall composition of the human and chimpanzee genomes, which are thought to have diverged only about 6 million years ago (see Figure 18.15), reveals some general differences. Considering single nucleotide substitutions, the two genomes differ by only 1.2%. When researchers looked at longer stretches of DNA, however, they were surprised to find a further 2.7% difference due to insertions or deletions of larger regions in the genome of one or the other species; many of the insertions were duplications or other repetitive DNA. In fact, a third of the human duplications are not present in the chimpanzee genome, and some of these duplications contain regions associated with human diseases. There are more *Alu* elements in the human genome than in the chimpanzee genome, and the latter contains many copies of a retroviral provirus not present in humans. All of these observations provide clues to the forces that might have swept the two genomes along different paths, but we don't have a complete picture yet. We also don't know how these differences might account for the distinct characteristics of each species.

To discover the basis for the phenotypic differences between the two species, biologists are studying specific genes and types of genes that differ between humans and chimpanzees and comparing them with their counterparts in other mammals. This approach has revealed a number of genes that are apparently changing (evolving) faster in the human than in either the chimpanzee or the mouse. Among them are genes involved in defense against malaria and tuberculosis and at least one gene that regulates brain size. When genes are classified by function, the genes that seem to be evolving the fastest are those that code for transcription factors. This discovery makes sense because transcription factors regulate gene expression and thus play a key role in orchestrating the overall genetic program.

One transcription factor whose gene shows evidence of rapid change in the human lineage is called *FOXP2*. Several lines of evidence suggest that the *FOXP2* gene functions in vocalization in vertebrates. For one thing, mutations in this gene can produce severe speech and language impairment in humans. Moreover, the *FOXP2* gene is expressed in the brains of zebra finches and canaries at the time when these songbirds are learning their songs. But perhaps the strongest evidence comes from a "knock-out" experiment in which researchers disrupted the *FOXP2* gene in mice and analyzed the resulting phenotype. Normal mice produce ultrasonic squeaks (whistles) to communicate stress, but mice that were homozygous for a mutated form of *FOXP2* had malformed brains and failed to vocalize normally **(Figure 18.16)**. Heterozygous mice, with one faulty copy of the gene, also showed vocalization defects. These results augmented the evidence from birds and humans, supporting the idea that the *FOXP2* gene product turns on genes involved in vocalization.

The *FOXP2* story is an excellent example of how different approaches can complement each other in uncovering biological phenomena of widespread importance. The *FOXP2* experiments used mice as a model for humans because it would be unethical (as well as impractical) to carry out such experiments in humans. Mice and humans diverged about 65.5 million years ago (see Figure 18.15) and share about 85% of their genes. This genetic similarity can be exploited in studying

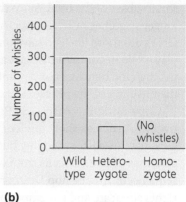

(a)   (b)

▲ **Figure 18.16 The function of *FOXP2*, a gene that is rapidly evolving in the human lineage. (a)** Wild-type mice emit ultrasonic squeaks (whistles) to communicate stress. **(b)** Researchers used genetic engineering to produce mice in which one or both copies of *FOXP2* were disrupted, separated each newborn pup from its mother, and recorded the number of ultrasonic whistles produced by the pup. No vocalization was observed in homozygous mutants, and the effect on heterozygotes was also extreme.

human genetic disorders. If researchers know the organ or tissue that is affected by a particular genetic disorder, they can look for genes that are expressed in these locations in mice.

Further research efforts are under way to extend genomic studies to many more microbial species, additional primates, and neglected species from diverse branches of the tree of life. These studies will advance our understanding of all aspects of biology, including health and ecology as well as evolution.

Comparing Genomes Within a Species

Another exciting consequence of our ability to analyze genomes is our growing understanding of the spectrum of genetic variation in humans. Because the history of the human species is so short—probably about 200,000 years—the amount of DNA variation among humans is small compared with that of many other species. Much of our diversity seems to be in the form of **single nucleotide polymorphisms** (**SNPs**, pronounced "snips"), defined as single base-pair sites where variation is found in at least 1% of the population. Usually detected by DNA sequencing, SNPs occur on average about once in 100–300 base pairs in the human genome. Scientists have already identified the location of several million human SNP sites and continue to find more.

In the course of this search, they have also found other variations—including inversions, deletions, and duplications. The most surprising discovery has been the widespread occurrence of *copy-number variants* (*CNVs*), loci where some individuals have one or multiple copies of a particular gene or genetic region, rather than the standard two copies (one on each homolog). CNVs result from regions of the genome being duplicated or deleted inconsistently within the population. One study of 40 people found more than 8,000 CNVs involving 13% of the genes in the genome, and these CNVs probably represent just a small subset of the total. Since these variants encompass much longer stretches of DNA than the single nucleotides of SNPs, CNVs are more likely to have phenotypic consequences and to play a role in complex diseases and disorders. At the very least, the high incidence of copy-number variation casts doubt on the meaning of the phrase "a normal human genome."

Copy-number variants, SNPs, and variations in repetitive DNA such as short tandem repeats (STRs) will be useful genetic markers for studying human evolution. In 2010, the genomes of two Africans from different communities were sequenced: Archbishop Desmond Tutu, the South African civil rights advocate and a member of the Bantu tribe, the majority population in southern Africa; and !Gubi, a hunter-gatherer from the Khoisan community in Namibia, a minority African population that is probably the human group with the oldest known lineage. The comparison revealed many differences, as you might expect. The analysis was then broadened to compare the protein-coding regions of !Gubi's genome with those of three other Khoisan community members (self-identified Bushmen) living nearby. Remarkably, these four Khoisan genomes differed more from each other than a European would from an Asian. These data highlight the extensive diversity among African genomes. Extending this approach will help us answer important questions about the differences between human populations and the migratory routes of human populations throughout history.

Comparing Developmental Processes

Biologists in the field of evolutionary developmental biology, or **evo-devo** as it is often called, compare developmental processes of different multicellular organisms. Their aim is to understand how these processes have evolved and how changes in them can modify existing organismal features or lead to new ones. With the advent of molecular techniques and the recent flood of genomic information, we are beginning to realize that the genomes of related species with strikingly different forms may have only minor differences in gene sequence or regulation. Discovering the molecular basis of these differences in turn helps us understand the origins of the myriad diverse forms that cohabit this planet, thus informing our study of evolution.

Widespread Conservation of Developmental Genes Among Animals

You may recall that the homeotic genes in *Drosophila* specify the identity of body segments in the fruit fly (see Figure 16.8). Molecular analysis of the homeotic genes in *Drosophila* has shown that they all include a 180-nucleotide sequence called a **homeobox**, which specifies a 60-amino-acid *homeodomain* in the encoded proteins. An identical or very similar nucleotide sequence has been discovered in the homeotic genes of many invertebrates and vertebrates. The sequences are so similar between humans and fruit flies, in fact, that one researcher has whimsically referred to flies as "little people with wings." The

resemblance even extends to the organization of these genes: The vertebrate genes homologous to the homeotic genes of fruit flies have kept the same chromosomal arrangement **(Figure 18.17)**. Homeobox-containing sequences have also been found in regulatory genes of much more distantly related eukaryotes, including plants and yeasts. From these similarities, we can deduce that the homeobox DNA sequence evolved very early in the history of life and was sufficiently beneficial to organisms to have been conserved in animals and plants virtually unchanged for hundreds of millions of years.

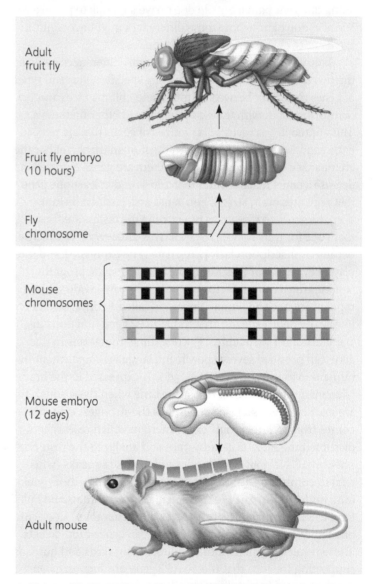

▲ **Figure 18.17 Conservation of homeotic genes in a fruit fly and a mouse.** Homeotic genes that control the form of anterior and posterior structures of the body occur in the same linear sequence on chromosomes in *Drosophila* and mice. Each colored band on the chromosomes shown here represents a homeotic gene. In fruit flies, all homeotic genes are found on one chromosome. The mouse and other mammals have the same or similar sets of genes on four chromosomes. The color code indicates the parts of the embryos in which these genes are expressed and the adult body regions that result. All of these genes are essentially identical in flies and mice, except for those represented by black bands, which are less similar in the two animals.

Homeotic genes in animals were named *Hox* genes, short for *homeobox*-containing genes, because homeotic genes were the first genes found to have this sequence. Other homeobox-containing genes were later found that do not act as homeotic genes; that is, they do not directly control the identity of body parts. However, most of these genes, in animals at least, are associated with development, suggesting their ancient and fundamental importance in that process. In *Drosophila*, for example, homeoboxes are present not only in the homeotic genes but also in the egg-polarity gene *bicoid* (see Figures 16.9 and 16.10), in several of the segmentation genes, and in a master regulatory gene for eye development.

Researchers have discovered that the homeobox-encoded homeodomain is the part of a protein that binds to DNA when the protein functions as a transcriptional regulator. However, the shape of the homeodomain allows it to bind to any DNA segment; its own structure is not specific for a particular sequence. Instead, other, more variable domains in a homeodomain-containing protein determine which genes the protein regulates. Interaction of these variable domains with still other transcription factors helps a homeodomain-containing protein recognize specific enhancers in the DNA. Proteins with homeodomains probably regulate development by coordinating the transcription of batteries of developmental genes, switching them on or off. In embryos of *Drosophila* and other animal species, different combinations of homeobox genes are active in different parts of the embryo. This selective expression of regulatory genes, varying over time and space, is central to pattern formation.

Developmental biologists have found that in addition to homeotic genes, many other genes involved in development are highly conserved from species to species. These include numerous genes encoding components of signaling pathways. The extraordinary similarity among particular developmental genes in different animal species raises a question: How can the same genes be involved in the development of animals whose forms are so very different from each other?

Ongoing studies are suggesting answers to this question. In some cases, small changes in regulatory sequences of particular genes cause changes in gene expression patterns that can lead to major changes in body form. For example, the differing patterns of expression of the *Hox* genes along the body axis in insects and crustaceans can explain the variation in the number of leg-bearing segments among these segmented animals (**Figure 18.18**). Also, recent research suggests that the same *Hox* gene product may have subtly dissimilar effects in different species, turning on new genes or turning on the same genes at higher or lower levels. In other cases, similar genes direct different developmental processes in different organisms, resulting in diverse body shapes. Several *Hox* genes, for instance, are expressed in the embryonic and larval stages of the sea urchin, a nonsegmented animal that has a body plan quite different from those of insects and mice. Sea urchin adults make the pincushion-shaped shells you may have seen on the

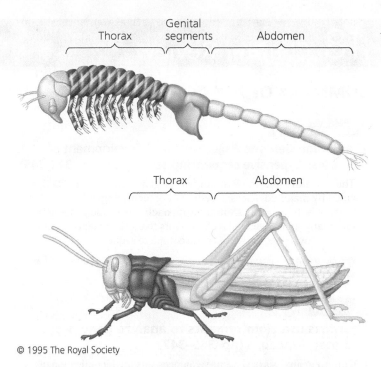

© 1995 The Royal Society

▲ **Figure 18.18 Effect of differences in *Hox* gene expression in crustaceans and insects.** Changes in the expression patterns of *Hox* genes have occurred over evolutionary time. These changes account in part for the different body plans of the brine shrimp *Artemia*, a crustacean (top), and the grasshopper, an insect. Shown here are regions of the adult body color-coded for expression of four *Hox* genes that determine formation of particular body parts during embryonic development. Each color represents a specific *Hox* gene. Colored stripes on the thorax of *Artemia* indicate co-expression of three *Hox* genes.

beach (see Figure 6.4). They are among the organisms long used in classical embryological studies (see Chapter 36).

In this final chapter of the genetics unit, you have learned how studying genomic composition and comparing the genomes of different species can disclose much about how genomes evolve. Further, comparing developmental programs, we can see that the unity of life is reflected in the similarity of molecular and cellular mechanisms used to establish body pattern, although the genes directing development may differ among organisms. The similarities between genomes reflect the common ancestry of life on Earth. But the differences are also crucial, for they have created the huge diversity of organisms that have evolved. In the remainder of the book, we expand our perspective beyond the level of molecules, cells, and genes to explore this diversity on the organismal level.

CONCEPT CHECK 18.6

1. Would you expect the genome of the macaque (a monkey) to be more similar to the mouse genome or the human genome? Why?

2. The DNA sequences called homeoboxes, which help homeotic genes in animals direct development, are common to flies and mice. Given this similarity, explain why these animals are so different.

For suggested answers, see Appendix A.

18 Chapter Review

SUMMARY OF KEY CONCEPTS

CONCEPT 18.1

The Human Genome Project fostered development of faster, less expensive sequencing techniques (pp. 344–345)

- The **Human Genome Project** was largely completed in 2003, aided by major advances in sequencing technology.
- In the **whole-genome shotgun approach**, the whole genome is cut into many small, overlapping fragments that are sequenced; computer software then assembles the complete sequence.

? *How did the Human Genome Project result in more rapid, less expensive DNA sequencing?*

CONCEPT 18.2

Scientists use bioinformatics to analyze genomes and their functions (pp. 345–347)

- Computer analysis of genome sequences aids the identification of protein-coding sequences. Methods for determining gene function include comparing the sequences of newly discovered genes with those of known genes in other species, and also observing the phenotypic effects of experimentally inactivating genes whose functions are unknown.
- In **systems biology**, researchers aim to model the dynamic behavior of whole biological systems based on the study of the interactions among the system's parts. For example, scientists use the computer-based tools of **bioinformatics** to compare genomes and to study sets of genes and proteins as whole systems (**genomics** and **proteomics**). These studies include large-scale analyses of functional DNA elements.

? *What was the most significant finding of the ENCODE pilot project? Why has the project been expanded to include other species?*

CONCEPT 18.3

Genomes vary in size, number of genes, and gene density (pp. 347–349)

	Bacteria	Archaea	Eukarya
Genome size	Most are 1–6 Mb		Most are 10–4,000 Mb, but a few are much larger
Number of genes	1,500–7,500		5,000–40,000
Gene density	Higher than in eukaryotes		Lower than in prokaryotes (Within eukaryotes, lower density is correlated with larger genomes.)
Introns	None in protein-coding genes	Present in some genes	Present in most genes of multicellular eukaryotes, but only in some genes of unicellular eukaryotes
Other noncoding DNA	Very little		Can be large amounts; generally more repetitive noncoding DNA in multicellular eukaryotes

? *Compare genome size, gene number, and gene density (a) in the three domains and (b) among eukaryotes.*

CONCEPT 18.4

Multicellular eukaryotes have much noncoding DNA and many multigene families (pp. 349–353)

- Only 1.5% of the human genome codes for proteins or gives rise to rRNAs or tRNAs; the rest is noncoding DNA, including **pseudogenes** and **repetitive DNA** of unknown function.
- The most abundant type of repetitive DNA in multicellular eukaryotes consists of **transposable elements** and related sequences. In eukaryotes, there are two types of transposable elements: **transposons**, which move via a DNA intermediate, and **retrotransposons**, which are more prevalent and move via an RNA intermediate.
- Other repetitive DNA includes short noncoding sequences that are tandemly repeated thousands of times (**simple sequence DNA**, which includes **STRs**); these sequences are especially prominent in centromeres and telomeres, where they probably play structural roles in the chromosome.
- Though many eukaryotic genes are present in one copy per haploid chromosome set, others are members of a family of related genes, such as the human globin gene families:

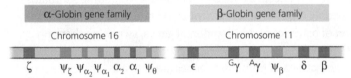

? *Explain how the function of transposable elements might account for their prevalence in human noncoding DNA.*

CONCEPT 18.5

Duplication, rearrangement, and mutation of DNA contribute to genome evolution (pp. 353–357)

- Accidents in cell division can lead to extra copies of all or part of entire chromosome sets, which may then diverge if one set accumulates sequence changes.
- The chromosomal organization of genomes can be compared among species, providing information about evolutionary relationships. Within a given species, rearrangements of chromosomes are thought to contribute to the emergence of new species.
- The genes encoding the various globin proteins evolved from one common ancestral globin gene, which duplicated and diverged into α-globin and β-globin ancestral genes. Subsequent duplication and random mutation gave rise to the present globin genes, all of which code for oxygen-binding proteins. The copies of some duplicated genes have diverged so much that the functions of their encoded proteins (such as lysozyme and α-lactalbumin) are now substantially different.
- Each exon may code for a **domain**, a discrete structural and functional region of a protein. Rearrangement of exons within and between genes during evolution has led to genes containing multiple copies of similar exons and/or several different exons derived from other genes.
- Movement of transposable elements or recombination between copies of the same element can generate new sequence combinations that are beneficial to the organism, which can alter the functions of genes or their patterns of expression and regulation.

? *How could chromosomal rearrangements lead to the emergence of new species?*

CONCEPT 18.6

Comparing genome sequences provides clues to evolution and development (pp. 357–361)

- Comparative studies of genomes from widely divergent and closely related species provide valuable information about ancient and more recent evolutionary history, respectively. Human and chimpanzee sequences are about 4% different. Along with nucleotide variations in specific genes, these differences may account for the distinct characteristics of the two species. Analysis of **single nucleotide polymorphisms (SNPs)** and copy-number variants (CNVs) within a species can also shed light on the evolution of that species.
- Evolutionary developmental (**evo-devo**) biologists have shown that homeotic genes and some other genes associated with animal development contain a **homeobox** region whose sequence is highly conserved among diverse species. Related sequences are present in the genes of plants and yeasts.

? *What type of information can be obtained by comparing the genomes of closely related species? Of very distantly related species?*

TEST YOUR UNDERSTANDING

Level 1: Knowledge/Comprehension

1. Bioinformatics includes all of the following except
   a. using computer programs to align DNA sequences.
   b. analyzing protein interactions in a species.
   c. using molecular biology to combine DNA from two different sources in a test tube.
   d. developing computer-based tools for genome analysis.
   e. using mathematical tools to make sense of biological systems.

2. One of the characteristics of retrotransposons is that
   a. they code for an enzyme that synthesizes DNA using an RNA template.
   b. they are found only in animal cells.
   c. they generally move by a cut-and-paste mechanism.
   d. they contribute a significant portion of the genetic variability seen within a population of gametes.
   e. their amplification is dependent on a retrovirus.

3. Homeotic genes
   a. encode transcription factors that control the expression of genes responsible for specific anatomical structures.
   b. are found only in *Drosophila* and other arthropods.
   c. are the only genes that contain the homeobox domain.
   d. encode proteins that form anatomical structures in the fly.
   e. are responsible for differentiation in muscle cells.

Level 2: Application/Analysis

4. Two eukaryotic proteins have one domain in common but are otherwise very different. Which of the following processes is most likely to have contributed to this similarity?
   a. gene duplication
   b. alternative splicing
   c. exon shuffling
   d. histone modification
   e. random point mutations

5. Two eukaryotic proteins are identical except for one domain in each protein, and these two domains are completely different from each other. Which of the following processes is most likely to have contributed to this difference?
   a. gene duplication
   b. alternative splicing
   c. exon shuffling
   d. histone modification
   e. random point mutations

6. **DRAW IT** Below are the amino acid sequences (using the single-letter code; see Figure 3.17) of four short segments of the *FOXP2* protein from six species: chimpanzee (C), orangutan (O), gorilla (G), rhesus macaque (R), mouse (M), and human (H). These segments contain all of the amino acid differences between the *FOXP2* proteins of these species.

   1. ATETI...PKSSD...TSSTT...NARRD
   2. ATETI...PKSSE...TSSTT...NARRD
   3. ATETI...PKSSD...TSSTT...NARRD
   4. ATETI...PKSSD...TSSNT...SARRD
   5. ATETI...PKSSD...TSSTT...NARRD
   6. VTETI...PKSSD...TSSTT...NARRD

   Use a highlighter to color any amino acid that varies among the species. (Color that amino acid in all sequences.)

   (a) The C, G, R sequences are identical. Which lines correspond to those sequences?

   (b) The H sequence differs from that of the C, G, R species at two amino acids. Underline the two differences in the H sequence.

   (c) The O sequence differs from the C, G, R sequences at one amino acid (having V instead of A) and from the H sequence at three amino acids. Which line is the O sequence?

   (d) In the M sequence, circle the amino acid(s) that differ from the C, G, R sequences, and draw a square around those that differ from the H sequence. Describe these differences.

   (e) Primates and rodents diverged between 60 and 100 million years ago, and chimpanzees and humans, about 6 million years ago. What can you conclude by comparing the amino acid differences between the mouse and the C, G, R species with those between the human and the C, G, R species?

Level 3: Synthesis/Evaluation

7. **SCIENTIFIC INQUIRY**
   The scientists mapping human SNPs noticed that groups of SNPs tended to be inherited together, in blocks known as haplotypes, ranging from 5,000 to 200,000 base pairs long. There are only four or five commonly occurring combinations of SNPs per haplotype. Propose an explanation, integrating what you've learned throughout this chapter and this unit.

8. **FOCUS ON EVOLUTION**
   Genes important in the embryonic development of animals, such as homeobox-containing genes, have been relatively well conserved during evolution; that is, they are more similar among different species than are many other genes. Why is this?

9. **FOCUS ON INFORMATION**
   The continuity of life is based on heritable information in the form of DNA. In a short essay (100–150 words), explain how mutations in protein-coding genes and regulatory DNA contribute to evolution.

For selected answers, see Appendix A.

Unit 3 Evolution

20 Phylogeny

As organisms adapt to their environments over time, they become increasingly different from their ancestors. To reconstruct an organism's evolutionary history, or **phylogeny**, biologists use data ranging from fossils to molecules.

19 Descent with Modification

Darwin proposed that the diversity of life and the match between organisms and their environments arose through **natural selection** over time, as species adapted to their environments.

21 The Evolution of Populations

The evolutionary impact of natural selection appears in the genetic changes of a **population** of organisms over time.

22 The Origin of Species

Evolutionary changes in a population ultimately can result in **speciation**, a process in which one species gives rise to two or more species.

23 Broad Patterns of Evolution

As speciation occurs again and again, new groups of organisms arise while others disappear. These changes make up the **broad patterns of evolutionary change** documented in the fossil record.

Evolution

19 Descent with Modification
20 Phylogeny
21 The Evolution of Populations
22 The Origin of Species
23 Broad Patterns of Evolution

Genetics

History of Life

Plants

Animals

Ecology

Chemistry and Cells

19

Descent with Modification

KEY CONCEPTS

19.1 The Darwinian revolution challenged traditional views of a young Earth inhabited by unchanging species

19.2 Descent with modification by natural selection explains the adaptations of organisms and the unity and diversity of life

19.3 Evolution is supported by an overwhelming amount of scientific evidence

OVERVIEW

Endless Forms Most Beautiful

A hungry bird would have to look very closely to spot this caterpillar of the moth *Synchlora aerata*, which blends in well with the flowers on which it feeds **(Figure 19.1)**. The disguise is enhanced by the caterpillar's flair for "decorating"—it glues pieces of flower petals to its body, transforming itself into its own background.

This striking caterpillar is a member of a diverse group, the more than 120,000 species of lepidopteran insects (moths and butterflies). All lepidopteran species go through a juvenile stage characterized by a well-developed head with chewing mouthparts: the ravenous, efficient feeding machines we call caterpillars. As adults, all lepidopterans share other features, such as three pairs of legs and two pairs of wings covered with small scales. But the many lepidopteran species also differ from one another, in both their caterpillar and adult forms. How did there come to be so many different moths and butterflies, and what causes their similarities and differences?

The self-decorating caterpillar and its many close relatives illustrate three key observations about life:

- the striking ways in which organisms are suited for life in their environments*
- the many shared characteristics (unity) of life
- the rich diversity of life

A century and a half ago, Charles Darwin was inspired to develop a scientific explanation for these three broad observations. When he published his hypothesis in *The Origin of Species*, Darwin ushered in a scientific revolution—the era of evolutionary biology.

For now, we will define **evolution** as *descent with modification*, a phrase Darwin used in proposing that Earth's many species are descendants of ancestral species that were different from the present-day species. Evolution can also be defined more narrowly as a change in the genetic composition of a population from generation to generation (as discussed further in Chapter 21).

Whether it is defined broadly or narrowly, we can view evolution in two related but different ways: as a pattern and as a process. The *pattern* of evolutionary change is revealed by data from a range of scientific disciplines, including biology, geology, physics, and chemistry. These data are

▼ **Figure 19.1** How is this caterpillar protecting itself from predators?

*Here and throughout this book, the term *environment* refers to other organisms as well as to the physical aspects of an organism's surroundings.

365

facts—they are observations about the natural world. The *process* of evolution consists of the mechanisms that produce the observed pattern of change. These mechanisms represent natural causes of the natural phenomena we observe. Indeed, the power of evolution as a unifying theory is its ability to explain and connect a vast array of observations about the living world.

As with all general theories in science, we continue to test our understanding of evolution by examining whether it can account for new observations and experimental results. In this and the following chapters, we'll examine how ongoing discoveries shape what we know about the pattern and process of evolution. To set the stage, we'll first retrace Darwin's quest to explain the adaptations, unity, and diversity of what he called life's "endless forms most beautiful."

CONCEPT 19.1

The Darwinian revolution challenged traditional views of a young Earth inhabited by unchanging species

What impelled Darwin to challenge the prevailing views about Earth and its life? Darwin's revolutionary proposal developed over time, influenced by the work of others and by his travels **(Figure 19.2)**. As we'll see, his ideas had deep historical roots.

Scala Naturae and Classification of Species

Long before Darwin was born, several Greek philosophers suggested that life might have changed gradually over time. But one philosopher who greatly influenced early Western science, Aristotle (384–322 BCE), viewed species as fixed (unchanging). Through his observations of nature, Aristotle recognized certain "affinities" among organisms. He concluded that life-forms could be arranged on a ladder, or scale, of increasing complexity, later called the *scala naturae* ("scale of nature"). Each form of life, perfect and permanent, had its allotted rung on this ladder.

These ideas were generally consistent with the Old Testament account of creation, which holds that species were individually designed by God and therefore perfect. In the 1700s, many scientists interpreted the often remarkable match of organisms to their environment as evidence that the Creator had designed each species for a particular purpose.

One such scientist was Carolus Linnaeus (1707–1778), a Swedish physician and botanist who sought to classify life's diversity, in his words, "for the greater glory of God." Linnaeus developed the two-part, or *binomial*, format for naming species (such as *Homo sapiens* for humans) that is still used today. In contrast to the linear hierarchy of the *scala naturae*, Linnaeus adopted a nested classification system, grouping similar species into increasingly general categories. For example, similar species are grouped in the same genus, similar genera (plural of genus) are grouped in the same family, and so on.

Linnaeus did not ascribe the resemblances among species to evolutionary kinship, but rather to the pattern of their creation. A century later, however, Darwin argued that classification should be based on evolutionary relationships. He also noted that scientists using the Linnaean system often grouped organisms in ways that reflected those relationships.

Ideas About Change over Time

Among other sources of information, Darwin drew from the work of scientists studying **fossils**, the remains or traces of organisms from the past. As depicted in **Figure 19.3**, many fossils are found in sedimentary rocks formed from the

▲ **Figure 19.2 Unusual species inspired novel ideas.** Darwin observed this species of marine iguana and many other unique animals when he visited the Galápagos Islands in 1835.

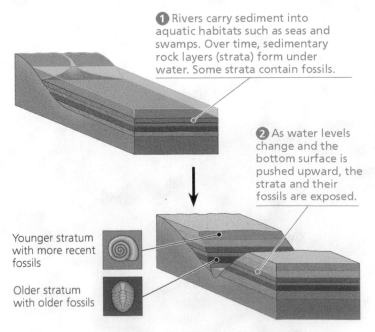

❶ Rivers carry sediment into aquatic habitats such as seas and swamps. Over time, sedimentary rock layers (strata) form under water. Some strata contain fossils.

❷ As water levels change and the bottom surface is pushed upward, the strata and their fossils are exposed.

Younger stratum with more recent fossils

Older stratum with older fossils

▲ **Figure 19.3 Formation of sedimentary strata with fossils.**

sand and mud that settle to the bottom of seas, lakes, and swamps. New layers of sediment cover older ones and compress them into layers of rock called **strata** (singular, *stratum*). The fossils in a particular stratum provide a glimpse of some of the organisms that populated Earth at the time that layer formed. Later, erosion may carve through upper (younger) strata, revealing deeper (older) strata that had been buried.

Paleontology, the study of fossils, was developed in large part by French scientist Georges Cuvier (1769–1832). In examining strata near Paris, Cuvier noted that the older the stratum, the more dissimilar its fossils were to current life-forms. He also observed that from one layer to the next, some new species appeared while others disappeared. He inferred that extinctions must have been a common occurrence, but he staunchly opposed the idea of evolution. Cuvier speculated that each boundary between strata represented a sudden catastrophic event, such as a flood, that had destroyed many of the species living in that area. Such regions, he reasoned, were later repopulated by different species immigrating from other areas.

In contrast, other scientists suggested that profound change could take place through the cumulative effect of slow but continuous processes. In 1795, Scottish geologist James Hutton (1726–1797) proposed that Earth's geologic features could be explained by gradual mechanisms, such as valleys being formed by rivers wearing through rocks. The leading geologist of Darwin's time, Charles Lyell (1797–1875), incorporated Hutton's thinking into his proposal that the same geologic processes are operating today as in the past, and at the same rate.

Hutton and Lyell's ideas strongly influenced Darwin's thinking. Darwin agreed that if geologic change results from slow, continuous actions rather than from sudden events, then Earth must be much older than the widely accepted age of a few thousand years. It would, for example, take a very long time for a river to carve a canyon by erosion. He later reasoned that perhaps similarly slow and subtle processes could produce substantial biological change. Darwin was not the first to apply the idea of gradual change to biological evolution, however.

Lamarck's Hypothesis of Evolution

Although some 18th-century naturalists suggested that life evolves as environments change, only one of Charles Darwin's predecessors proposed a mechanism for *how* life changes over time: French biologist Jean-Baptiste de Lamarck (1744–1829). Alas, Lamarck is primarily remembered today *not* for his visionary recognition that evolutionary change explains patterns in fossils and the match of organisms to their environments, but for the incorrect mechanism he proposed.

Lamarck published his hypothesis in 1809, the year Darwin was born. By comparing living species with fossil forms, Lamarck had found what appeared to be several lines

◀ **Figure 19.4 Acquired traits cannot be inherited.** This bonsai tree was "trained" to grow as a dwarf by pruning and shaping. However, seeds from this tree would produce offspring of normal size.

of descent, each a chronological series of older to younger fossils leading to a living species. He explained his findings using two principles that were widely accepted at the time. The first was *use and disuse*, the idea that parts of the body that are used extensively become larger and stronger, while those that are not used deteriorate. Among many examples, he cited a giraffe stretching its neck to reach leaves on high branches. The second principle, *inheritance of acquired characteristics*, stated that an organism could pass these modifications to its offspring. Lamarck reasoned that the long, muscular neck of the living giraffe had evolved over many generations as giraffes stretched their necks ever higher.

Lamarck also thought that evolution happens because organisms have an innate drive to become more complex. Darwin rejected this idea, but he, too, thought that variation was introduced into the evolutionary process in part through inheritance of acquired characteristics. Today, however, our understanding of genetics refutes this mechanism: Experiments show that traits acquired by use during an individual's life are not inherited in the way proposed by Lamarck **(Figure 19.4)**.

Lamarck was vilified in his own time, especially by Cuvier, who denied that species ever evolve. In retrospect, however, Lamarck did recognize that the match of organisms to their environments can be explained by gradual evolutionary change, and he did propose a testable explanation for how this change occurs.

CONCEPT CHECK 19.1

1. How did Hutton's and Lyell's ideas influence Darwin's thinking about evolution?

2. **MAKE CONNECTIONS** Scientific hypotheses must be testable and falsifiable (see Concept 1.3). Applying these criteria, are Cuvier's explanation of the fossil record and Lamarck's hypothesis of evolution scientific? Explain your answer in each case.

For suggested answers, see Appendix A.

Descent with modification by natural selection explains the adaptations of organisms and the unity and diversity of life

As the 19th century dawned, it was generally thought that species had remained unchanged since their creation. A few clouds of doubt about the permanence of species were beginning to gather, but no one could have forecast the thundering storm just beyond the horizon. How did Charles Darwin become the lightning rod for a revolutionary view of life?

Darwin's Research

Charles Darwin (1809–1882) was born in Shrewsbury, England. He had a consuming interest in nature—reading nature books, fishing, hunting, and collecting insects. Darwin's father, a physician, could see no future for his son as a naturalist and sent him to medical school in Edinburgh. But Charles found medicine boring and surgery before the days of anesthesia horrifying. He enrolled at Cambridge University, intending to become a clergyman. (At that time many scholars of science belonged to the clergy.)

At Cambridge, Darwin became the protégé of John Henslow, a botany professor. Henslow recommended him to Captain Robert FitzRoy, who was preparing the survey ship HMS *Beagle* for a voyage around the world. FitzRoy, who was himself an accomplished scientist, accepted Darwin because he was a skilled naturalist and because they were of similar age and social class.

The Voyage of the Beagle

Darwin embarked on the *Beagle* in December 1831. The primary mission of the voyage was to chart poorly known stretches of the South American coastline. Darwin spent most of his time on shore, observing and collecting thousands of plants and animals. He noted the characteristics that made organisms well suited to such diverse environments as Brazil's humid jungles, Argentina's broad grasslands, and the Andes' towering peaks.

Darwin observed that the plants and animals in temperate regions of South America more closely resembled species living in the South American tropics than species living in temperate regions of Europe. Furthermore, the fossils he found, though clearly different from living species, distinctly resembled the living organisms of South America.

Darwin also read Lyell's *Principles of Geology* during the voyage. He experienced geologic change firsthand when a violent earthquake shook the coast of Chile, and he observed afterward that rocks along the coast had been thrust upward by several feet. Finding fossils of ocean organisms high in the Andes, Darwin inferred that the rocks containing the fossils must have been raised there by many similar earthquakes. These observations reinforced what he had learned from Lyell: Physical evidence did not support the traditional view that Earth was only a few thousand years old.

Darwin's interest in the geographic distribution of species was further stimulated by the *Beagle*'s stop at the Galápagos, a group of volcanic islands located near the equator about 900 km west of South America (**Figure 19.5**). Darwin was fascinated by the unusual organisms there. The birds he collected included several kinds of mockingbirds. These mockingbirds, though similar to each other, seemed to be different species.

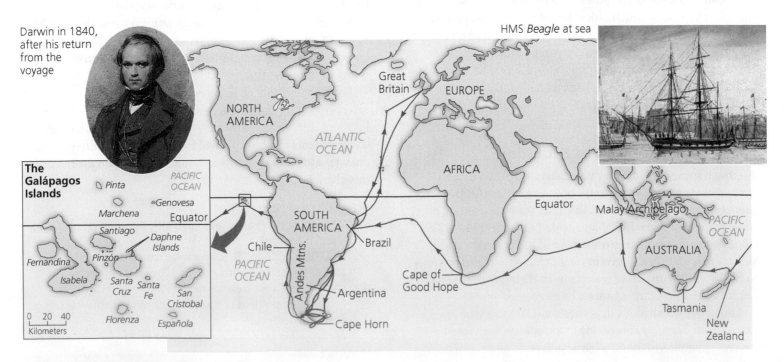

▲ **Figure 19.5 The voyage of HMS *Beagle*.**

Some were unique to individual islands, while others lived on two or more adjacent islands. Furthermore, although the animals on the Galápagos resembled species living on the South American mainland, most of the Galápagos species were not known from anywhere else in the world. Darwin hypothesized that the Galápagos had been colonized by organisms that had strayed from South America and then diversified, giving rise to new species on the various islands.

Darwin's Focus on Adaptation

During the voyage of the *Beagle*, Darwin observed many examples of **adaptations**, inherited characteristics of organisms that enhance their survival and reproduction in specific environments. Later, as he reassessed his observations, he began to perceive adaptation to the environment and the origin of new species as closely related processes. Could a new species arise from an ancestral form by the gradual accumulation of adaptations to a different environment? From studies made years after Darwin's voyage, biologists have concluded that this is indeed what happened to the diverse group of Galápagos finches (see Figure 1.16). The finches' various beaks and behaviors are adapted to the specific foods available on their home islands **(Figure 19.6)**. Darwin realized that explaining such adaptations was essential to understanding evolution. As we'll explore further, his explanation of how adaptations arise centered on **natural selection**, a process in which individuals that have certain inherited traits tend to survive and reproduce at higher rates than other individuals *because of* those traits.

By the early 1840s, Darwin had worked out the major features of his hypothesis. He set these ideas on paper in 1844, when he wrote a long essay on descent with modification and its underlying mechanism, natural selection. Yet he was still reluctant to publish his ideas, apparently because he anticipated the uproar they would cause. During this time, Darwin continued to compile evidence in support of his hypothesis. By the mid-1850s, he had described his ideas to Lyell and a few others. Lyell, who was not yet convinced of evolution, nevertheless urged Darwin to publish on the subject before someone else came to the same conclusions and published first.

In June 1858, Lyell's prediction came true. Darwin received a manuscript from Alfred Russel Wallace (1823–1913), a British naturalist working in the South Pacific islands of the Malay Archipelago **(Figure 19.7)**. Wallace had developed a hypothesis of natural selection nearly identical to Darwin's. He asked Darwin to evaluate his paper and forward it to Lyell if it merited publication. Darwin complied, writing to Lyell: "Your words have come

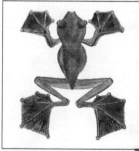

(a) Cactus-eater. The long, sharp beak of the cactus ground finch (*Geospiza scandens*) helps it tear and eat cactus flowers and pulp.

(b) Seed-eater. The large ground finch (*Geospiza magnirostris*) has a large beak adapted for cracking seeds on the ground.

(c) Insect-eater. The green warbler finch (*Certhidea olivacea*) uses its narrow, pointed beak to grasp insects.

▲ **Figure 19.6 Three examples of beak variation in Galápagos finches.** The Galápagos Islands are home to more than a dozen species of closely related finches, some found only on a single island. A striking difference among them is their beaks, which are adapted to specific diets.

MAKE CONNECTIONS *Review Figure 1.16. To which of the other two species shown above is the cactus-eater more closely related (that is, with which does it share a more recent common ancestor)?*

▶ **Figure 19.7 Alfred Russel Wallace.** The inset is a painting Wallace made of a flying tree frog from the Malay Archipelago.

true with a vengeance. . . . I never saw a more striking coincidence . . . so all my originality, whatever it may amount to, will be smashed." On July 1, 1858, Lyell and a colleague presented Wallace's paper, along with extracts from Darwin's unpublished 1844 essay, to the Linnean Society of London. Darwin quickly finished his book, titled *On the Origin of Species by Means of Natural Selection* (commonly referred to as *The Origin of Species*), and published it the next year. Although Wallace had submitted his ideas for publication first, he admired Darwin and thought that Darwin had developed the idea of natural selection so extensively that he should be known as its main architect.

Within a decade, Darwin's book and its proponents had convinced most scientists that life's diversity is the product of evolution. Darwin succeeded where previous evolutionists had failed, mainly by presenting a plausible scientific mechanism with immaculate logic and an avalanche of evidence.

Ideas from *The Origin of Species*

In his book, Darwin amassed evidence that descent with modification by natural selection explains the three broad observations about nature listed in the Overview: the unity of life, the diversity of life, and the match between organisms and their environments.

Descent with Modification

In the first edition of *The Origin of Species*, Darwin never used the word *evolution* (although the final word of the book is "evolved"). Rather, he discussed *descent with modification*, a phrase that summarized his view of life. Organisms share many characteristics, leading Darwin to perceive unity in life. He attributed the unity of life to the descent of all organisms from an ancestor that lived in the remote past. He also thought that as the descendants of that ancestral organism lived in various habitats over millions of years, they accumulated diverse modifications, or adaptations, that fit them to specific ways of life. Darwin reasoned that over a long time, descent with modification eventually led to the rich diversity of life today.

Darwin viewed the history of life as a tree, with multiple branchings from a common trunk out to the tips of the youngest twigs **(Figure 19.8)**. In his diagram, the tips of the twigs that are labeled A, B, C, and D represent several groups of organisms living in the present day, while the unlabeled branches represent groups that are extinct. Each fork of the tree represents the most recent common ancestor of all the lines of evolution that subsequently branch from that point. Darwin reasoned that such a branching process, along with past extinction events, could explain the large morphological gaps (differences in form) that sometimes exist in between related groups of organisms.

As an example, consider the three living species of elephants: the Asian elephant (*Elephas maximus*) and two species of African elephants (*Loxodonta africana* and *L. cyclotis*). As shown in the tree diagram in **Figure 19.9**, these closely related species are very similar because they shared the same line of

▶ **Figure 19.8 "I think . . ."** In this 1837 sketch, Darwin envisioned the branching pattern of evolution. Branches that end in twigs labeled A–D represent particular groups of living organisms; all other branches represent extinct groups.

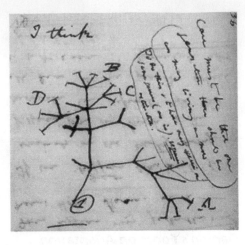

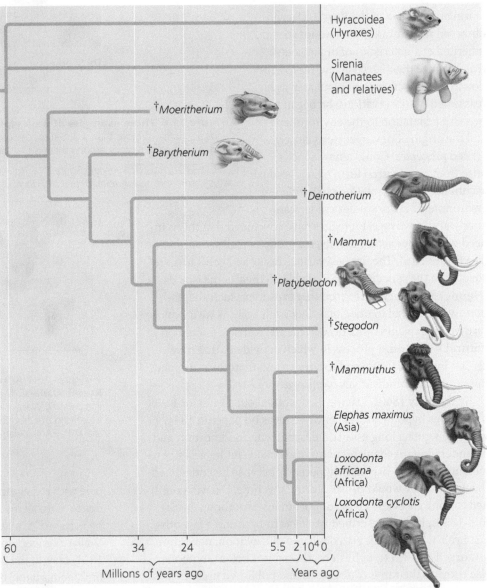

▲ **Figure 19.9 Descent with modification.** This evolutionary tree of elephants and their relatives is based mainly on fossils—their anatomy, order of appearance in strata, and geographic distribution. Note that most branches of descent ended in extinction (denoted by the dagger symbol †). (Time line not to scale.)

? *Based on the tree shown here, approximately when did the most recent ancestor shared by* Mammuthus *(woolly mammoths), Asian elephants, and African elephants live?*

descent until a relatively recent split from their common ancestor.

Note that seven lineages related to elephants have become extinct over the past 32 million years. As a result, there are no living species that fill the morphological gap between elephants and their nearest relatives today, the hyraxes and the manatees and their relatives. Such extinctions are not uncommon. In fact, many evolutionary branches, even some major ones, are dead ends: Scientists estimate that over 99% of all species that have ever lived are now extinct. As in Figure 19.9, fossils of extinct species can document the divergence of present-day groups by "filling in" gaps between them.

Artificial Selection, Natural Selection, and Adaptation

Darwin proposed the mechanism of natural selection to explain the observable patterns of evolution. He crafted his argument carefully, hoping to persuade even the most skeptical readers. First he discussed familiar examples of selective breeding of domesticated plants and animals. Humans have modified other species over many generations by selecting and breeding individuals that possess desired traits, a process called **artificial selection (Figure 19.10)**. As a result of artificial selection, crops, livestock animals, and pets often bear little resemblance to their wild ancestors.

Darwin then argued that a similar process occurs in nature. He based his argument on two observations, from which he drew two inferences.

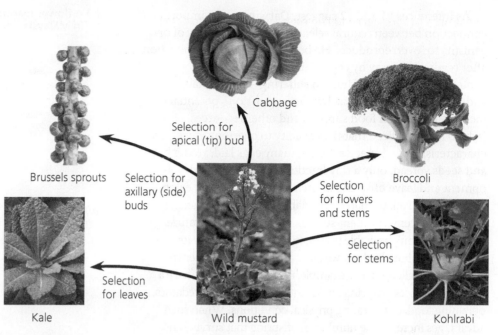

▲ **Figure 19.10 Artificial selection.** These different vegetables have all been selected from one species of wild mustard. By selecting variations in different parts of the plant, breeders have obtained these divergent results.

Observation #1: Members of a population often vary in their inherited traits **(Figure 19.11)**.

Observation #2: All species can produce more offspring than their environment can support **(Figure 19.12)**, and many of these offspring fail to survive and reproduce.

Inference #1: Individuals whose inherited traits give them a higher probability of surviving and reproducing in a given environment tend to leave more offspring than other individuals.

Inference #2: This unequal ability of individuals to survive and reproduce will lead to the accumulation of favorable traits in the population over generations.

▲ **Figure 19.11 Variation in a population.** Individuals in this population of Asian ladybird beetles vary in color and spot pattern. Natural selection may act on these variations only if (1) they are heritable and (2) they affect the beetles' ability to survive and reproduce.

◀ **Figure 19.12 Overproduction of offspring.** A single puffball fungus can produce billions of offspring. If all of these offspring and their descendants survived to maturity, they would carpet the surrounding land surface.

As inferences #1 and #2 suggest, Darwin saw an important connection between natural selection and the capacity of organisms to "overreproduce." He began to make this connection after reading an essay by economist Thomas Malthus, who contended that much of human suffering—disease, famine, and war—resulted from the human population's potential to increase faster than food supplies and other resources. Similarly, Darwin realized that the capacity to overreproduce was characteristic of all species. Of the many eggs laid, young born, and seeds spread, only a tiny fraction complete their development and leave offspring of their own. The rest are eaten, starved, diseased, unmated, or unable to tolerate physical conditions of the environment such as salinity or temperature.

An organism's heritable traits can influence not only its own performance, but also how well its offspring cope with environmental challenges. For example, an organism might have a trait that gives its offspring an advantage in escaping predators, obtaining food, or tolerating physical conditions. When such advantages increase the number of offspring that survive and reproduce, the traits that are favored will likely appear at a greater frequency in the next generation. Thus, over time, natural selection resulting from factors such as predators, lack of food, or adverse physical conditions can lead to an increase in the proportion of favorable traits in a population.

How rapidly do such changes occur? Darwin reasoned that if artificial selection can bring about dramatic change in a relatively short period of time, then natural selection should be capable of substantial modification of species over many hundreds of generations. Even if the advantages of some heritable traits over others are slight, the advantageous variations will gradually accumulate in the population, and less favorable ones will diminish. Over time, this process will increase the frequency of individuals with favorable adaptations and refine the match between organisms and their environment.

Natural Selection: A Summary

Let's now recap the main ideas of natural selection:

- Natural selection is a process in which individuals that have certain heritable traits survive and reproduce at a higher rate than other individuals because of those traits.
- Over time, natural selection can increase the match between organisms and their environment **(Figure 19.13)**.
- If an environment changes, or if individuals move to a new habitat, natural selection may result in adaptation to these new conditions, sometimes giving rise to new species.

One subtle but important point is that although natural selection occurs through interactions between individual organisms and their environment, *individuals do not evolve*. Rather, it is the population that evolves over time.

A second key point is that natural selection can amplify or diminish only those heritable traits that differ among the individuals in a population. Thus, even if a trait is heritable, if all

(a) A flower mantid in Malaysia

(b) A leaf mantid in Borneo

▲ **Figure 19.13 Camouflage as an example of evolutionary adaptation.** Related species of the insects called mantids have diverse shapes and colors that evolved in different environments.

? *Explain how these mantids demonstrate the three key observations about life introduced in the Overview: the match between organisms and their environments, unity, and diversity.*

the individuals in a population are genetically identical for that trait, evolution by natural selection cannot occur.

Third, remember that environmental factors vary from place to place and over time. A trait that is favorable in one place or time may be useless—or even detrimental—in other places or times. Natural selection is always operating, but which traits are favored depends on the context in which a species lives and mates.

Next, we'll survey the wide range of observations that support a Darwinian view of evolution by natural selection.

CONCEPT CHECK 19.2

1. How does the concept of descent with modification explain both the unity and diversity of life?
2. **WHAT IF?** Predict whether a fossil of an extinct mammal that lived high in the Andes would more closely resemble present-day mammals that live in South American jungles or present-day mammals that live high in African mountains? Explain.
3. **MAKE CONNECTIONS** Review the relationship between genotype and phenotype (see Figure 11.6). Suppose that in a particular pea population, flowers with the white phenotype are favored by natural selection. Predict what would happen over time to the frequency of the *p* allele in the population, and explain your reasoning.

For suggested answers, see Appendix A.

CONCEPT 19.3

Evolution is supported by an overwhelming amount of scientific evidence

In *The Origin of Species*, Darwin marshaled a broad range of evidence to support the concept of descent with modification. Still—as he readily acknowledged—there were instances in which key evidence was lacking. For example, Darwin referred to the origin of flowering plants as an "abominable mystery," and he lamented the lack of fossils showing how earlier groups of organisms gave rise to new groups.

In the last 150 years, new discoveries have filled many of the gaps that Darwin identified. The origin of flowering plants, for example, is much better understood (see Chapter 26), and many fossils have been discovered that signify the origin of new groups of organisms (see Chapter 23). In this section, we'll consider four types of data that document the pattern of evolution and illuminate the processes by which it occurs.

Direct Observations of Evolutionary Change

Biologists have documented evolutionary change in thousands of scientific studies. We'll examine many such studies throughout this unit, but let's look at two examples here.

Natural Selection in Response to Introduced Plant Species

Animals that eat plants, called herbivores, often have adaptations that help them feed efficiently on their primary food sources. What happens when herbivores begin to feed on a plant species with different characteristics than their usual food source?

An opportunity to study this question in nature is provided by soapberry bugs, which use their "beak," a hollow, needlelike mouthpart, to feed on seeds located within the fruits of various plants. In southern Florida, the soapberry bug (*Jadera haematoloma*) feeds on the seeds of a native plant, the balloon vine (*Cardiospermum corindum*). In central Florida, however, balloon vines have become rare. Instead, soapberry bugs in that region now feed on seeds of the goldenrain tree (*Koelreuteria elegans*), a species recently introduced from Asia.

Soapberry bugs feed most effectively when their beak length closely matches the depth at which the seeds are found within the fruit. Goldenrain tree fruit consists of three flat lobes, and its seeds are much closer to the fruit surface than are the seeds of the plump, round native balloon vine fruit. Researchers at the University of Utah predicted that in populations that feed on goldenrain tree, natural selection would result in beaks that are *shorter* than those in populations that feed on balloon vine **(Figure 19.14)**. Indeed, beak lengths are shorter in the populations that feed on goldenrain tree.

▼ **Figure 19.14** **Inquiry**

Can a change in a population's food source result in evolution by natural selection?

Field Study Soapberry bugs feed most effectively when the length of their "beak" closely matches the depth of the seeds within the fruit. Scott Carroll and his colleagues measured beak lengths in soapberry bug populations feeding on the native balloon vine. They also measured beak lengths in populations feeding on the introduced goldenrain tree. The researchers then compared the measurements with those of museum specimens collected in the two areas before the goldenrain tree was introduced.

Soapberry bug with beak inserted in balloon vine fruit

Results Beak lengths were shorter in populations feeding on the introduced species than in populations feeding on the native species, in which the seeds are buried more deeply. The average beak length in museum specimens from each population (indicated by red arrows) was similar to beak lengths in populations feeding on native species.

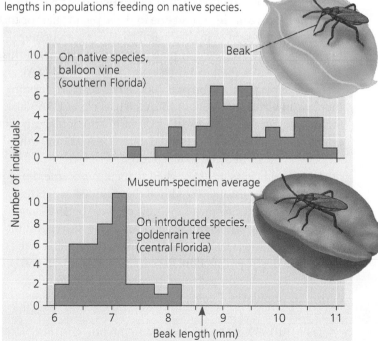

Conclusion Museum specimens and contemporary data suggest that a change in the size of the soapberry bug's food source can result in evolution by natural selection for matching beak size.

Source S. P. Carroll and C. Boyd, Host race radiation in the soapberry bug: natural history with the history, *Evolution* 46:1052–1069 (1992).

WHAT IF? Additional results showed that when soapberry bug eggs from a population fed on balloon vine fruits were reared on goldenrain tree fruits (or vice versa), the beak lengths of the adult insects matched those in the population from which the eggs were obtained. Interpret these results.

Researchers have also studied beak length evolution in soapberry bug populations that feed on plants introduced to Louisiana, Oklahoma, and Australia. In each of these locations, the fruit of the introduced plants is larger than the fruit of the native plant. Thus, in populations feeding on introduced species in these regions, the researchers predicted that natural selection would result in the evolution of *longer* beaks. Again, data from field studies upheld this prediction.

The adaptation observed in these soapberry bug populations had important consequences: In Australia, for example, the increase in beak length nearly doubled the success with which soapberry bugs could eat the seeds of the introduced species. Furthermore, since historical data show that the goldenrain tree reached central Florida just 35 years before the scientific studies were initiated, the results demonstrate that natural selection can cause rapid evolution in a wild population.

The Evolution of Drug-Resistant Bacteria

An example of ongoing natural selection that dramatically affects humans is the evolution of drug-resistant pathogens (disease-causing organisms and viruses). This is a particular problem with bacteria and viruses because resistant strains of these pathogens can proliferate very quickly.

Consider the evolution of drug resistance in the bacterium *Staphylococcus aureus*. About one in three people harbor this species on their skin or in their nasal passages with no negative effects. However, certain genetic varieties (strains) of this species, known as methicillin-resistant *S. aureus* (MRSA), are formidable pathogens. The past decade has seen an alarming increase in virulent forms of MRSA such as clone USA300, a strain that can cause "flesh-eating disease" and potentially fatal infections **(Figure 19.15)**. How did clone USA300 and other strains of MRSA become so dangerous?

The story begins in 1943, when penicillin became the first widely used antibiotic. Although penicillin and other antibiotics have since saved millions of lives, by 1945, over 20% of the *S. aureus* strains seen in hospitals were resistant to penicillin. These bacteria had an enzyme, penicillinase, that could destroy penicillin. Researchers developed antibiotics that were not destroyed by penicillinase, but some *S. aureus* populations developed resistance to each new drug within a few years.

Then, in 1959, doctors began using the powerful antibiotic methicillin. But within two years, methicillin-resistant strains of *S. aureus* appeared. How did these resistant strains emerge? Methicillin works by deactivating a protein that bacteria use to synthesize their cell walls. However, *S. aureus* populations exhibited variations in how strongly their members were affected by the drug. In particular, some individuals were able to synthesize their cell walls using a different protein that was not affected by methicillin. These individuals survived the methicillin treatments and reproduced at higher rates than did other individuals. Over time, these resistant individuals became increasingly common, leading to the spread of MRSA.

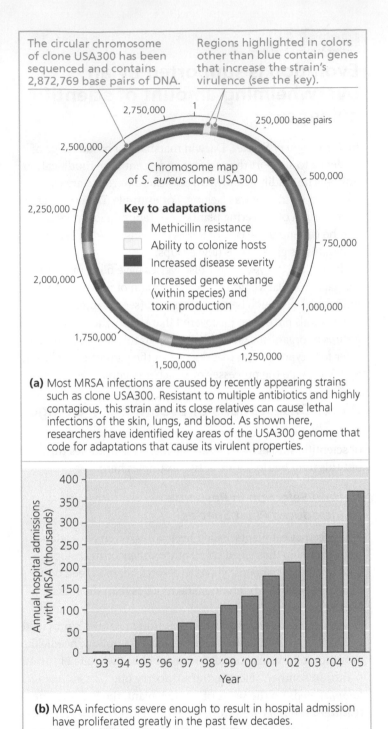

(a) Most MRSA infections are caused by recently appearing strains such as clone USA300. Resistant to multiple antibiotics and highly contagious, this strain and its close relatives can cause lethal infections of the skin, lungs, and blood. As shown here, researchers have identified key areas of the USA300 genome that code for adaptations that cause its virulent properties.

(b) MRSA infections severe enough to result in hospital admission have proliferated greatly in the past few decades.

▲ **Figure 19.15 The rise of methicillin-resistant *Staphylococcus aureus* (MRSA).**

Initially, MRSA could be controlled by antibiotics that work differently from the way methicillin works. But this has become increasingly difficult because some MRSA strains are resistant to multiple antibiotics—probably because bacteria can exchange genes with members of their own and other species (see Figure 24.17). Thus, the present-day multidrug-resistant strains may have emerged over time as MRSA strains that were resistant to different antibiotics exchanged genes.

The *S. aureus* and soapberry bug examples highlight two key points about natural selection. First, natural selection is a process of editing, not a creative mechanism. A drug does not *create* resistant pathogens; it *selects for* resistant individuals that are already present in the population. Second, natural selection depends on time and place. It favors those characteristics in a genetically variable population that provide advantage in the current, local environment. What is beneficial in one situation may be useless or even harmful in another. Beak lengths arise that match the size of the typical fruit eaten by a particular soapberry bug population. However, a beak length suitable for fruit of one size can be disadvantageous when the bug is feeding on fruit of another size.

Homology

A second type of evidence for evolution comes from analyzing similarities among different organisms. As we've discussed, evolution is a process of descent with modification: Characteristics present in an ancestral organism are altered (by natural selection) in its descendants over time as they face different environmental conditions. As a result, related species can have characteristics that have an underlying similarity yet function differently. Similarity resulting from common ancestry is known as **homology**. As we'll describe in this section, an understanding of homology can be used to make testable predictions and explain observations that are otherwise puzzling.

Anatomical and Molecular Homologies

The view of evolution as a remodeling process leads to the prediction that closely related species should share similar features—and they do. Of course, closely related species share the features used to determine their relationship, but they also share many other features. Some of these shared features make little sense except in the context of evolution. For example, the forelimbs of all mammals, including humans, cats, whales, and bats, show the same arrangement of bones from the shoulder to the tips of the digits, even though these appendages have very different functions: lifting, walking, swimming, and flying

(Figure 19.16). Such striking anatomical resemblances would be highly unlikely if these structures had arisen anew in each species. Rather, the underlying skeletons of the arms, forelegs, flippers, and wings of different mammals are **homologous structures** that represent variations on a structural theme that was present in their common ancestor.

Comparing early stages of development in different animal species reveals additional anatomical homologies not visible in adult organisms. For example, at some point in their development, all vertebrate embryos have a tail located posterior to (behind) the anus, as well as structures called pharyngeal (throat) arches (Figure 19.17). These homologous throat arches ultimately develop into structures with very different functions, such as gills in fishes and parts of the ears and throat in humans and other mammals.

Some of the most intriguing homologies concern "leftover" structures of marginal, if any, importance to the organism. These **vestigial structures** are remnants of features that served a function in the organism's ancestors. For instance, the

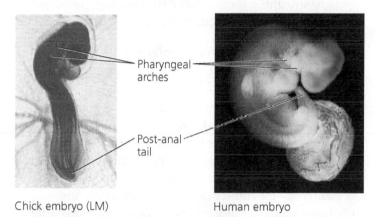

Chick embryo (LM)   Human embryo

▲ **Figure 19.17 Anatomical similarities in vertebrate embryos.** At some stage in their embryonic development, all vertebrates have a tail located posterior to the anus (referred to as a post-anal tail), as well as pharyngeal (throat) arches. Descent from a common ancestor can explain such similarities.

▶ **Figure 19.16 Mammalian forelimbs: homologous structures.** Even though they have become adapted for different functions, the forelimbs of all mammals are constructed from the same basic skeletal elements: one large bone (purple), attached to two smaller bones (orange and tan), attached to several small bones (gold), attached to several metacarpals (green), attached to approximately five digits, each of which is composed of phalanges (blue).

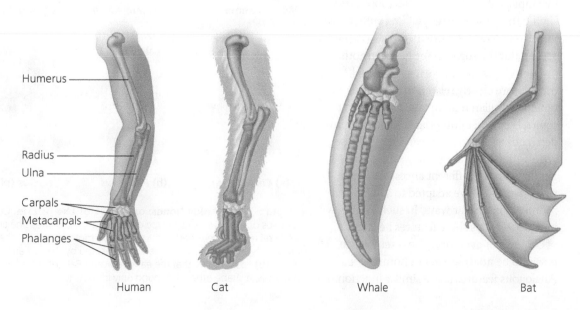

Human   Cat   Whale   Bat

skeletons of some snakes retain vestiges of the pelvis and leg bones of walking ancestors. Another example is provided by eye remnants that are buried under scales in blind species of cave fishes. We would not expect to see these vestigial structures if snakes and blind cave fishes had origins separate from other vertebrate animals.

Biologists also observe similarities among organisms at the molecular level. All forms of life use the same genetic language of DNA and RNA, and the genetic code is essentially universal. Thus, it is likely that all species descended from common ancestors that used this code. But molecular homologies go beyond a shared code. For example, organisms as dissimilar as humans and bacteria share genes inherited from a very distant common ancestor. Some of these homologous genes have acquired new functions, while others, such as those coding for the ribosomal subunits used in protein synthesis (see Figure 14.17), have retained their original functions. It is also common for organisms to have genes that have lost their function, even though the homologous genes in related species may be fully functional. Like vestigial structures, it appears that such inactive "pseudogenes" may be present simply because a common ancestor had them.

A Different Cause of Resemblance: Convergent Evolution

Although organisms that are closely related share characteristics because of common descent, distantly related organisms can resemble one another for a different reason: **convergent evolution**, the independent evolution of similar features in different lineages. Consider marsupial mammals, many of which live in Australia. Marsupials are distinct from another group of mammals—the eutherians—few of which live in Australia. (Eutherians complete their embryonic development in the uterus, whereas marsupials are born as embryos and complete their development in an external pouch.) Some Australian marsupials have eutherian look-alikes with superficially similar adaptations. For instance, a forest-dwelling Australian marsupial called the sugar glider looks very similar to flying squirrels, gliding eutherians that live in North American forests **(Figure 19.18)**. But the sugar glider has many other characteristics that make it a marsupial, much more closely related to kangaroos and other Australian marsupials than to flying squirrels or other eutherians. Again, our understanding of evolution can explain these observations: Although they evolved independently from different ancestors, these two mammals have adapted to similar environments in similar ways. In such examples in which species share features because of convergent evolution, the resemblance is said to be **analogous**, not homologous. Analogous features share similar function,

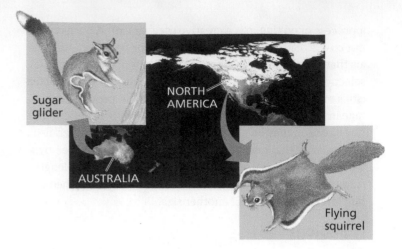

▲ **Figure 19.18 Convergent evolution.** The ability to glide through the air evolved independently in these two distantly related mammals.

but not common ancestry, while homologous features share common ancestry, but not necessarily similar function.

The Fossil Record

A third type of evidence for evolution comes from fossils. The fossil record documents the pattern of evolution, showing that past organisms differed from present-day organisms and that many species have become extinct. Fossils also show the evolutionary changes that have occurred in various groups of organisms. To give one of hundreds of examples, researchers found that the pelvic bone in fossil stickleback fish became greatly reduced in size over time in a number of different lakes. The consistent nature of this change suggests that the reduction in the size of the pelvic bone may have been driven by natural selection.

Fossils can also shed light on the origins of new groups of organisms. An example is the fossil record of cetaceans, the mammalian order that includes whales, dolphins, and porpoises. As shown in **Figure 19.19**, some of these fossils

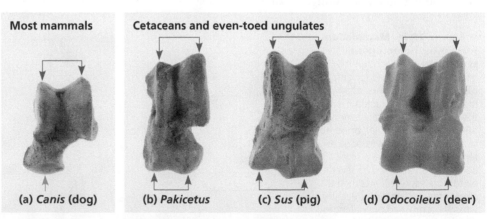

▲ **Figure 19.19 Ankle bones: one piece of the puzzle.** Comparing fossils and present-day examples of the astragalus (a type of ankle bone) provides one line of evidence that cetaceans are closely related to even-toed ungulates. **(a)** In most mammals, the astragalus is shaped like that of a dog, with a double hump on one end (indicated by the red arrows) but not at the opposite end (blue arrow). **(b)** Fossils show that the early cetacean *Pakicetus* had an astragalus with double humps at both ends, a shape otherwise found only in even-toed ungulates, such as **(c)** pigs and **(d)** deer.

provided an unexpected line of support for a hypothesis based on DNA sequence data: that cetaceans are closely related to even-toed ungulates, a group that includes deer, pigs, camels, and cows. What else can fossils tell us about cetacean origins? The earliest cetaceans lived 50–60 million years ago. The fossil record indicates that prior to that time, most mammals were terrestrial. Although scientists had long realized that whales and other cetaceans originated from land mammals, few fossils had been found that revealed how cetacean limb structure had changed over time, leading eventually to the loss of hind limbs and the development of flippers and tail flukes. In the past few decades, however, a series of remarkable fossils have been discovered in Pakistan, Egypt, and North America. These fossils document steps in the transition from life on land to life in the sea, filling in some of the gaps between ancestral and living cetaceans (Figure 19.20).

Collectively, the recent fossil discoveries document the formation of new species and the origin of a major new group of mammals, the cetaceans. These discoveries also show that cetaceans and their close living relatives (hippopotamuses, pigs, deer, and other even-toed ungulates) are much more different

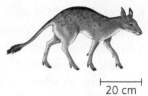

▲ *Diacodexis*, an early even-toed ungulate

from each other than were *Pakicetus* and early even-toed ungulates, such as *Diacodexis*. Similar patterns are seen in fossils documenting the origins of other major new groups of organisms, including mammals (see Chapter 23), flowering plants (see Chapter 26), and tetrapods (see Chapter 27). In each of these cases, the fossil record shows that over time, descent with modification produced increasingly large differences among related groups of organisms, ultimately resulting in the diversity of life we see today.

Biogeography

A fourth type of evidence for evolution has to do with **biogeography**, the scientific study of the geographic distributions of species. The geographic distributions of organisms are influenced by many factors, including *continental drift*, the slow movement of Earth's continents over time. About 250 million years ago, these movements united all of Earth's landmasses into a single large continent called **Pangaea** (see Figure 23.8).

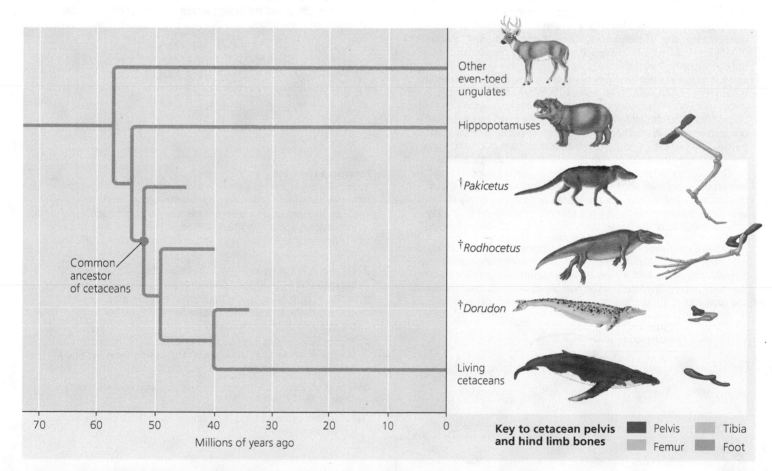

▲ **Figure 19.20 The transition to life in the sea.** Multiple lines of evidence support the hypothesis that cetaceans evolved from terrestrial mammals. Fossils document the reduction over time in the pelvis and hind limb bones of extinct cetacean ancestors, including *Pakicetus*, *Rodhocetus*, and *Dorudon*. DNA sequence data support the hypothesis that cetaceans are most closely related to hippopotamuses.

? *Which happened first during the evolution of cetaceans: changes in hind limb structure or the origin of tail flukes?*

Roughly 200 million years ago, Pangaea began to break apart; by 20 million years ago, the continents we know today were within a few hundred kilometers of their present locations.

We can use our understanding of evolution and continental drift to predict where fossils of different groups of organisms might be found. For example, scientists have constructed evolutionary trees for horses based on anatomical data. These trees and the ages of fossils of horse ancestors suggest that present-day horse species originated 5 million years ago in North America. At that time, North and South America were close to their present locations, but they were not yet connected, making it difficult for horses to travel between them. Thus, we would predict that the oldest horse fossils should be found only on the continent on which horses originated—North America. This prediction and others like it for differ-

ent groups of organisms have been upheld, providing more evidence for evolution.

We can also use our understanding of evolution to explain biogeographic data. For example, islands generally have many plant and animal species that are **endemic**—they are nowhere else in the world. Yet, as Darwin described in *The Origin of Species*, most island species are closely related to species from the nearest mainland or a neighboring island. He explained this observation by suggesting that islands are colonized by species from the nearest mainland. These colonists eventually give rise to new species as they adapt to their new environments. Such a process also explains why two islands with similar environments in distant parts of the world tend to be populated not by species that are closely related to each other, but rather by species related to those of the nearest mainland, where the environment is often quite different.

Scientific Skills Exercise

Making and Testing Predictions

Can Predation Result in Natural Selection for Color Patterns in Guppies? What we know about evolution changes constantly as new observations lead to new hypotheses—and hence to new ways to test our understanding of evolutionary theory. Consider the wild guppies (*Poecilia reticulata*) that live in pools connected by streams on the Caribbean island of Trinidad. Male guppies have highly varied color patterns, which are controlled by genes that are only expressed in adult males. Female guppies choose males with bright color patterns as mates more often than they choose males with drab coloring. But the bright colors that attract females also make the males more conspicuous to predators. Researchers observed that in pools with few predator species, the benefits of bright colors appear to "win out," and males are more brightly colored than in pools where predation is intense.

One guppy predator, the killifish, preys on juvenile guppies that have not yet displayed their adult coloration. Researchers predicted that if guppies with drab colors were transferred to a pool with only killifish, eventually the descendants of these guppies would be more brightly colored (because of the female preference for brightly colored males).

How the Experiment Was Done Researchers transplanted 200 guppies from pools containing pike-cichlid fish, intense guppy predators, to pools containing killifish, less active predators that prey mainly on juvenile guppies. They tracked the number of bright-colored spots and the total area of those spots on male guppies in each generation.

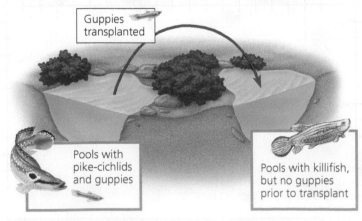

Guppies transplanted

Pools with pike-cichlids and guppies

Pools with killifish, but no guppies prior to transplant

Data from the Experiment After 22 months (15 generations), researchers compared the color pattern data for the source and transplanted populations.

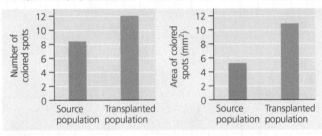

Interpret the Data

1. Identify the following elements of hypothesis-based science in this example: (a) question, (b) hypothesis, (c) prediction, (d) control group, and (e) experimental group. (For additional information about hypothesis-based science, see Chapter 1 and the Scientific Skills Review in Appendix F and in the Study Area in MasteringBiology.)

2. Explain how the types of data the researchers chose to collect enabled them to test their prediction.

3. (a) What conclusion would you draw from the data presented above? (b) What additional questions might you ask to determine the strength of this conclusion?

4. Predict what would happen if, after 22 months, guppies from the transplanted population were returned to the source pool. Describe an experiment to test your prediction.

Data from J.A. Endler, Natural selection on color patterns in *Poecilia reticulata*, *Evolution* 34:76–91 (1980).

(MB) A related version of this Scientific Skills Exercise can be assigned in MasteringBiology.

What Is Theoretical About Darwin's View of Life?

Some people dismiss Darwin's ideas as "just a theory." However, as we have seen, the *pattern* of evolution—the observation that life has evolved over time—has been documented directly and is supported by a great deal of evidence. In addition, Darwin's explanation of the *process* of evolution—that natural selection is the primary cause of the observed pattern of evolutionary change—makes sense of massive amounts of data. The effects of natural selection also can be observed and tested in nature.

What, then, is theoretical about evolution? Keep in mind that the scientific meaning of the term *theory* is very different from its meaning in everyday use. The colloquial use of the word *theory* comes close to what scientists mean by a hypothesis. In science, a theory is more comprehensive than a hypothesis. A theory, such as the theory of evolution by natural selection, accounts for many observations and explains and integrates a great variety of phenomena. Such a unifying theory does not become widely accepted unless its predictions stand up to thorough and continual testing by experiment and additional observation (see Chapter 1). As the rest of this unit demonstrates, this has certainly been the case with the theory of evolution by natural selection.

The skepticism of scientists as they continue to test theories prevents these ideas from becoming dogma. For example, although Darwin thought that evolution was a very slow process, we now know that this isn't always true. New species can form in relatively short periods of time—a few thousand years or less (see Chapter 22). Furthermore, evolutionary biologists now recognize that natural selection is not the only mechanism responsible for evolution. Indeed, the study of evolution today is livelier than ever as scientists use a wide range of experimental approaches and genetic analyses to test predictions based on natural selection and other evolutionary mechanisms. In the **Scientific Skills Exercise**, you'll work with data from an experiment on natural selection in wild guppies.

Although Darwin's theory attributes the diversity of life to natural processes, the diverse products of evolution nevertheless remain elegant and inspiring. As Darwin wrote in the final sentence of *The Origin of Species*, "There is grandeur in this view of life . . . [in which] endless forms most beautiful and most wonderful have been, and are being, evolved."

CONCEPT CHECK 19.3

1. Explain how the following statement is inaccurate: "Antibiotics have created drug resistance in MRSA."
2. How does evolution account for (a) the similar mammalian forelimbs with different functions shown in Figure 19.16 and (b) the similar forms of the two distantly related mammals shown in Figure 19.18?
3. **WHAT IF?** Fossils show that dinosaurs originated 250–200 million years ago. Would you expect the geographic distribution of early dinosaur fossils to be broad (on many continents) or narrow (on one or a few continents only)? Explain.

For suggested answers, see Appendix A.

---

19 Chapter Review

SUMMARY OF KEY CONCEPTS

CONCEPT 19.1

The Darwinian revolution challenged traditional views of a young Earth inhabited by unchanging species (pp. 366–367)

- Darwin proposed that life's diversity arose from ancestral species through natural selection, a departure from prevailing views.
- Cuvier studied fossils but denied that evolution occurs; he proposed that sudden catastrophic events in the past caused species to disappear from an area. Hutton and Lyell thought that geologic change could result from gradual, continuous mechanisms. Lamarck hypothesized that species evolve, but the underlying mechanisms he proposed are not supported by evidence.

? *Why was the age of Earth important for Darwin's ideas about evolution?*

CONCEPT 19.2

Descent with modification by natural selection explains the adaptations of organisms and the unity and diversity of life (pp. 368–372)

- Darwin's voyage on the *Beagle* gave rise to his idea that species originate from ancestral forms through the accumulation of

adaptations. He refined his theory for many years and finally published it in 1859 after learning that Wallace had come to the same idea. In *The Origin of Species*, Darwin proposed that evolution occurs by **natural selection**.

Observations

Individuals in a population vary in their heritable characteristics.	Organisms produce more offspring than the environment can support.

Inferences

Individuals that are well suited to their environment tend to leave more offspring than other individuals.

and

Over time, favorable traits accumulate in the population.

? *Describe how overreproduction and heritable variation relate to evolution by natural selection.*

Evolution is supported by an overwhelming amount of scientific evidence (pp. 373–379)

- Researchers have directly observed natural selection leading to adaptive evolution in many studies, including research on soapberry bug populations and on MRSA.
- Organisms share characteristics because of common descent (**homology**) or because natural selection affects independently evolving species in similar environments in similar ways (**convergent evolution**).
- Fossils show that past organisms differed from living organisms, that many species have become extinct, and that species have evolved over long periods of time; fossils also document the origin of major new groups of organisms.
- Evolutionary theory can explain biogeographic patterns.

? *Summarize the different lines of evidence supporting the hypothesis that cetaceans descended from land mammals and are closely related to even-toed ungulates.*

TEST YOUR UNDERSTANDING

Level 1: Knowledge/Comprehension

1. Which of the following is *not* an observation or inference on which natural selection is based?
   a. There is heritable variation among individuals.
   b. Poorly adapted individuals never produce offspring.
   c. Species produce more offspring than the environment can support.
   d. Individuals whose characteristics are best suited to the environment generally leave more offspring than those whose characteristics are less well suited.
   e. Only a fraction of an individual's offspring may survive.

2. Which of the following observations helped Darwin shape his concept of descent with modification?
   a. Species diversity declines farther from the equator.
   b. Fewer species live on islands than on the nearest continents.
   c. Birds live on islands located farther from the mainland than the birds' maximum nonstop flight distance.
   d. South American temperate plants are more similar to the tropical plants of South America than to the temperate plants of Europe.
   e. Earthquakes reshape life by causing mass extinctions.

Level 2: Application/Analysis

3. Within six months of effectively using methicillin to treat *S. aureus* infections in a community, all new infections were caused by MRSA. How can this result best be explained?
   a. *S. aureus* can resist vaccines.
   b. A patient must have become infected with MRSA from another community.
   c. In response to the drug, *S. aureus* began making drug-resistant versions of the protein targeted by the drug.
   d. Some drug-resistant bacteria were present at the start of treatment, and natural selection increased their frequency.
   e. The drug caused the *S. aureus* DNA to change.

4. The upper forelimbs of humans and bats have fairly similar skeletal structures, whereas the corresponding bones in whales have very different shapes and proportions. However, genetic data suggest that all three kinds of organisms diverged from a common ancestor at about the same time. Which of the following is the most likely explanation for these data?
   a. Humans and bats evolved by natural selection, and whales evolved by Lamarckian mechanisms.
   b. Forelimb evolution was adaptive in people and bats, but not in whales.
   c. Natural selection in an aquatic environment resulted in significant changes to whale forelimb anatomy.
   d. Genes mutate faster in whales than in humans or bats.
   e. Whales are not properly classified as mammals.

5. DNA sequences in many human genes are very similar to the sequences of corresponding genes in chimpanzees. The most likely explanation for this result is that
   a. humans and chimpanzees share a relatively recent common ancestor.
   b. humans evolved from chimpanzees.
   c. chimpanzees evolved from humans.
   d. convergent evolution led to the DNA similarities.
   e. humans and chimpanzees are not closely related.

Level 3: Synthesis/Evaluation

6. **SCIENTIFIC INQUIRY**
   DRAW IT Mosquitoes resistant to the pesticide DDT first appeared in India in 1959, but now are found throughout the world. (a) Graph the data in the table below. (b) Examining the graph, hypothesize why the percentage of mosquitoes resistant to DDT rose rapidly. (c) Suggest an explanation for the global spread of DDT resistance.

Month	0	8	12
Mosquitoes Resistant* to DDT	4%	45%	77%

Source C. F. Curtis et al., Selection for and against insecticide resistance and possible methods of inhibiting the evolution of resistance in mosquitoes, *Ecological Entomology* 3:273–287 (1978).

*Mosquitoes were considered resistant if they were not killed within 1 hour of receiving a dose of 4% DDT.

7. **FOCUS ON EVOLUTION**
   Explain why anatomical and molecular features often fit a similar nested pattern. In addition, describe a process that can cause this not to be the case.

8. **FOCUS ON INTERACTIONS**
   Write a short essay (about 100–150 words) evaluating whether changes to an organism's physical environment are likely to result in evolutionary change. Use an example to support your reasoning.

For selected answers, see Appendix A.

MasteringBiology®

Students Go to **MasteringBiology** for assignments, the eText, and the Study Area with practice tests, animations, and activities.

Instructors Go to **MasteringBiology** for automatically graded tutorials and questions that you can assign to your students, plus Instructor Resources.

21

The Evolution of Populations

KEY CONCEPTS

21.1 Genetic variation makes evolution possible

21.2 The Hardy-Weinberg equation can be used to test whether a population is evolving

21.3 Natural selection, genetic drift, and gene flow can alter allele frequencies in a population

21.4 Natural selection is the only mechanism that consistently causes adaptive evolution

OVERVIEW

The Smallest Unit of Evolution

One common misconception about evolution is that individual organisms evolve. It is true that natural selection acts on individuals: Each organism's traits affect its survival and reproductive success compared with that of other individuals. But the evolutionary impact of natural selection is only apparent in the changes in a *population* of organisms over time.

Consider the medium ground finch (*Geospiza fortis*), a seed-eating bird that inhabits the Galápagos Islands **(Figure 21.1)**. In 1977, the *G. fortis* population on the island of Daphne Major was decimated by a long period of drought: Of some 1,200 birds, only 180 survived. Researchers Peter and Rosemary Grant observed that during the drought, small, soft seeds were in short supply. The finches mostly fed on large, hard seeds that were more plentiful. Birds with larger, deeper beaks were better able to crack and eat these larger seeds, and they survived at a higher rate than finches with smaller beaks. Since beak depth is an inherited trait in these birds, the average beak depth in the next generation of *G. fortis* was greater than it had been in the pre-drought population **(Figure 21.2)**. The finch population had evolved by natural selection. However, the *individual* finches did not evolve. Each bird had a beak of a particular size, which did not grow larger during the drought. Rather, the proportion of large beaks in the population increased from generation to generation: The population evolved, not its individual members.

▼ **Figure 21.1** Is this finch evolving?

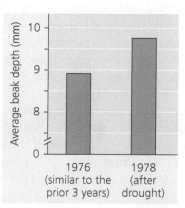

▶ **Figure 21.2 Evidence of selection by food source.** The data represent adult beak depth measurements of medium ground finches hatched in the generations before and after the 1977 drought. In a single generation, evolution by natural selection resulted in a larger average beak size in the population.

(MB) A related Experimental Inquiry Tutorial can be assigned in MasteringBiology.

Focusing on evolutionary change in populations, we can define evolution on its smallest scale, called **microevolution**, as a change in allele frequencies in a population over generations. As you will see in this chapter, natural selection is not the only cause of microevolution. In fact, there are three main mechanisms that can cause allele frequency change: natural selection, genetic drift (chance events that alter allele frequencies), and gene flow (the transfer of alleles between populations). Each of these mechanisms has distinctive effects on the genetic composition of populations. However, only natural selection consistently improves the match between organisms and their environment (adaptation). Before we examine natural selection and adaptation more closely, let's revisit a prerequisite for these processes in a population: genetic variation.

▲ **Figure 21.3 Phenotypic variation in horses.** In horses, coat color varies along a continuum and is influenced by multiple genes.

Genetic variation makes evolution possible

In *The Origin of Species*, Darwin provided abundant evidence that life on Earth has evolved over time, and he proposed natural selection as the primary mechanism for that change. He observed that individuals differ in their inherited traits and that selection acts on such differences, leading to evolutionary change. Although Darwin realized that variation in heritable traits is a prerequisite for evolution, he did not know precisely how organisms pass heritable traits to their offspring.

Just a few years after Darwin published *The Origin of Species*, Gregor Mendel wrote a groundbreaking paper on inheritance in pea plants (see Chapter 11). In that paper, Mendel proposed a model of inheritance in which organisms transmit discrete heritable units (now called genes) to their offspring. Although Darwin did not know about genes, Mendel's paper set the stage for understanding the genetic differences on which evolution is based. Here we'll examine such genetic differences and how they are produced.

Genetic Variation

Individuals within a species vary in their specific characteristics. Among humans, you can easily observe phenotypic variation in facial features, height, and voice. And though you cannot identify a person's blood group (A, B, AB, or O) from his or her appearance, this and many other molecular traits also vary extensively among individuals.

Such phenotypic variations often reflect **genetic variation**, differences among individuals in the composition of their genes or other DNA sequences. Some heritable phenotypic differences occur on an "either-or" basis, such as the flower colors of Mendel's pea plants: Each plant had flowers that were either purple or white (see Figure 11.3). Characters that vary in this

way are typically determined by a single gene locus, with different alleles producing distinct phenotypes. In contrast, other phenotypic differences vary in gradations along a continuum. Such variation usually results from the influence of two or more genes on a single phenotypic character. In fact, many phenotypic characters are influenced by multiple genes, including coat color in horses **(Figure 21.3)**, seed number in maize (corn), and height in humans.

How much do genes and other DNA sequences vary from one individual to another? Genetic variation at the whole-gene level (*gene variability*) can be quantified as the average percentage of loci that are heterozygous. (Recall that a heterozygous individual has two different alleles for a given locus, whereas a homozygous individual has two identical alleles for that locus.) As an example, on average the fruit fly *Drosophila melanogaster* is heterozygous for about 1,920 of its 13,700 loci (14%) and homozygous for all the rest.

Considerable genetic variation can also be measured at the molecular level of DNA (*nucleotide variability*). But little of this variation results in phenotypic variation because many of the differences occur within *introns*, noncoding segments of DNA lying between *exons*, the regions retained in mRNA after RNA processing (see Figure 14.12). And of the variations that occur within exons, most do not cause a change in the amino acid sequence of the protein encoded by the gene. In the sequence comparison summarized in **Figure 21.4**, there are 43 nucleotide sites with variable base pairs (where substitutions have occurred), as well as several sites where insertions or deletions have occurred. Although 18 variable sites occur within the four exons of the *Adh* gene, only one of these variations—at site 1,490—results in an amino acid change. Note, however, that this single variable site is enough to cause genetic variation at the level of the gene, and two different forms of the Adh enzyme are produced.

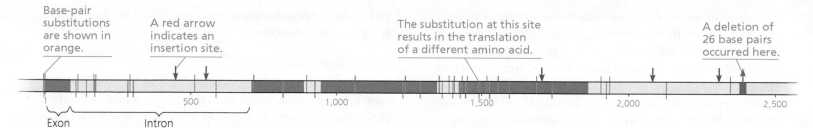

Base-pair substitutions are shown in orange.

A red arrow indicates an insertion site.

The substitution at this site results in the translation of a different amino acid.

A deletion of 26 base pairs occurred here.

Exon   Intron

▲ **Figure 21.4 Extensive genetic variation at the molecular level.** This diagram summarizes data from a study comparing the DNA sequence of the alcohol dehydrogenase (*Adh*) gene in several fruit flies (*Drosophila melanogaster*). The *Adh* gene has four exons (dark blue) separated by introns (light blue); the exons include the coding regions that are ultimately translated into the amino acids of the Adh enzyme. Only one substitution has a phenotypic effect, producing a different form of the Adh enzyme.

MAKE CONNECTIONS   *Review Figures 14.6 and 14.12. Explain how a base-pair substitution that alters a coding region of the* Adh *locus could have no effect on amino acid sequence. Then explain how an insertion in an exon could have no effect on the protein produced.*

It is important to bear in mind that some phenotypic variation is not heritable (**Figure 21.5** shows a striking example in a caterpillar of the southwestern United States). Phenotype is the product of an inherited genotype and many environmental influences (see Concept 11.3). In a human example, bodybuilders alter their phenotypes dramatically but do not pass their huge muscles on to the next generation. In general, only the genetically determined part of phenotypic variation can have evolutionary consequences. As such, genetic variation provides the raw material for evolutionary change: Without genetic variation, evolution cannot occur.

Sources of Genetic Variation

The genetic variation on which evolution depends originates when mutation, gene duplication, or other processes produce new alleles and new genes. Genetic variants can be produced in short periods of time in organisms that reproduce rapidly. Sexual reproduction can also result in genetic variation as existing genes are arranged in new ways.

Formation of New Alleles

New alleles can arise by *mutation*, a change in the nucleotide sequence of an organism's DNA. A mutation is like a shot in the dark—we cannot predict accurately which segments of DNA will be altered or in what way. In multicellular organisms, only mutations in cell lines that produce gametes can be passed to offspring. In plants and fungi, this is not as limiting as it may sound, since many different cell lines can produce gametes. But in most animals, the majority of mutations occur in somatic cells and are lost when the individual dies.

A change of as little as one base in a gene, called a "point mutation," can have a significant impact on phenotype, as in sickle-cell disease (see Figure 14.25). Organisms reflect many generations of past selection, and hence their phenotypes tend to be well matched to their environments. As a result, it's unlikely that a new mutation that alters a phenotype will improve it. In fact, most such mutations are at least slightly harmful. But since much of the DNA in eukaryotic genomes does not code for protein products, point mutations in these noncoding

▲ **Figure 21.5 Nonheritable variation.** These caterpillars of the moth *Nemoria arizonaria* owe their different appearances to chemicals in their diets, not to differences in their genotypes. **(a)** Caterpillars raised on a diet of oak flowers resemble the flowers, whereas **(b)** their siblings raised on oak leaves resemble oak twigs.

regions are generally harmless. Also, because of the redundancy in the genetic code, even a point mutation in a gene that encodes a protein will have no effect on the protein's function if the amino acid composition is not changed. And even where there is a change in the amino acid, it may not affect the protein's shape and function. However, as you will see later in this chapter, a mutant allele may on rare occasions actually make its bearer better suited to the environment, enhancing reproductive success.

Altering Gene Number or Position

Chromosomal changes that delete, disrupt, or rearrange many loci at once are usually harmful. However, when such large-scale changes leave genes intact, they may not affect the organisms' phenotypes. In rare cases, chromosomal rearrangements may even be beneficial. For example, the translocation of part of one chromosome to a different chromosome could link DNA segments in a way that produces a positive effect.

A key potential source of variation is the duplication of genes due to errors in meiosis (such as unequal crossing over), slippage during DNA replication, or the activities of transposable elements (see Concept 18.4). Duplications of large chromosome segments, like other chromosomal aberrations, are often harmful, but the duplication of smaller pieces of DNA may not be. Gene duplications that do not have severe effects can persist over generations, allowing mutations to accumulate. The result is an expanded genome with new genes that may take on new functions.

Such increases in gene number appear to have played a major role in evolution. For example, the remote ancestors of mammals had a single gene for detecting odors that has since been duplicated many times: Humans today have about 350 functional olfactory receptor genes, and mice have 1,000. This proliferation of olfactory genes probably helped mammals over the course of evolution, enabling them to detect faint odors and to distinguish among many different smells.

Rapid Reproduction

Mutation rates tend to be low in plants and animals, averaging about one mutation in every 100,000 genes per generation, and they are often even lower in prokaryotes. But prokaryotes have many more generations per unit of time, so mutations can quickly generate genetic variation in populations of these organisms. The same is true of viruses. For instance, HIV has a generation span of about two days. It also has an RNA genome, which has a much higher mutation rate than a typical DNA genome because of the lack of RNA repair mechanisms in host cells (see Chapter 17). For this reason, it is unlikely that a single-drug treatment would ever be effective against HIV; mutant forms of the virus that are resistant to a particular drug would no doubt proliferate in relatively short order. The most effective AIDS treatments to date have been drug "cocktails" that combine several medications. It is less likely that a set of mutations that together confer resistance to *all* the drugs will occur in a short time period.

Sexual Reproduction

In organisms that reproduce sexually, most of the genetic variation in a population results from the unique combination of alleles that each individual receives from its parents. Of course, at the nucleotide level, all the differences among these alleles have originated from past mutations. Sexual reproduction then shuffles existing alleles and deals them at random to produce individual genotypes.

Three mechanisms contribute to this shuffling: crossing over, independent assortment of chromosomes, and fertilization (see Chapter 10). During meiosis, homologous chromosomes, one inherited from each parent, trade some of their alleles by crossing over. These homologous chromosomes and the alleles they carry are then distributed at random into gametes. Then, because myriad possible mating combinations exist in a population, fertilization brings together gametes that are likely to have different genetic backgrounds. The combined effects of these three mechanisms ensure that sexual reproduction rearranges existing alleles into fresh combinations each generation, providing much of the genetic variation that makes evolution possible.

CONCEPT CHECK 21.1

1. Explain why genetic variation within a population is a prerequisite for evolution.
2. Of all the mutations that occur in a population, why do only a small fraction become widespread?
3. **MAKE CONNECTIONS** If a population stopped reproducing sexually (but still reproduced asexually), how would its genetic variation be affected over time? Explain. (See Concept 10.4.)

For suggested answers, see Appendix A.

CONCEPT 21.2

The Hardy-Weinberg equation can be used to test whether a population is evolving

Although the individuals in a population must differ genetically for evolution to occur, the presence of genetic variation does not guarantee that a population will evolve. For that to happen, one of the factors that cause evolution must be at work. In this section, we'll explore one way to test whether evolution is occurring in a population. First, let's clarify what we mean by a population.

Gene Pools and Allele Frequencies

A **population** is a group of individuals of the same species that live in the same area and interbreed, producing fertile offspring. Different populations of a single species may

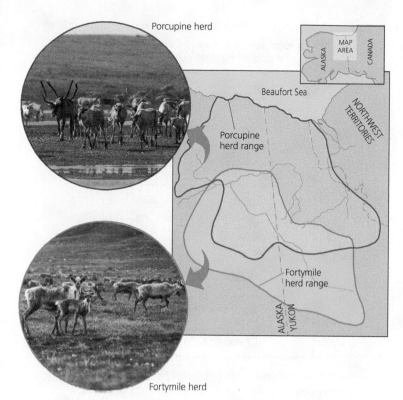

Porcupine herd

MAP AREA

ALASKA CANADA

Beaufort Sea

NORTHWEST TERRITORIES

Porcupine herd range

Fortymile herd range

ALASKA YUKON

Fortymile herd

▲ **Figure 21.6 One species, two populations.** These two caribou populations in the Yukon are not totally isolated; they sometimes share the same area. Still, members of either population are most likely to breed within their own population.

be isolated geographically from one another, exchanging genetic material only rarely. Such isolation is common for species that live on widely separated islands or in different lakes. But not all populations are isolated, nor must populations have sharp boundaries **(Figure 21.6)**. Still, members of a population typically breed with one another and thus on average are more closely related to each other than to members of other populations.

We can characterize a population's genetic makeup by describing its **gene pool**, which consists of all copies of every type of allele at every locus in all members of the population. If only one allele exists for a particular locus in a population, that allele is said to be *fixed* in the gene pool, and all individuals are homozygous for that allele. But if there are two or more alleles for a particular locus in a population, individuals may be either homozygous or heterozygous.

Each allele has a frequency (proportion) in the population. For example, imagine a population of 500 wild-flower plants with two alleles, C^R and C^W, for a locus that codes for flower pigment. These alleles show incomplete dominance (see Figure 11.10); thus, each genotype has a distinct phenotype. Plants homozygous for the C^R allele ($C^R C^R$) produce red pigment and have red flowers; plants homozygous for the C^W allele ($C^W C^W$) produce no red pigment and have

 $C^R C^R$

 $C^W C^W$

 $C^R C^W$

white flowers; and heterozygotes ($C^R C^W$) produce some red pigment and have pink flowers.

In our population, suppose there are 320 plants with red flowers, 160 with pink flowers, and 20 with white flowers. Because these are diploid organisms, these 500 individuals have a total of 1,000 copies of the gene for flower color. The C^R allele accounts for 800 of these copies ($320 \times 2 = 640$ for $C^R C^R$ plants, plus $160 \times 1 = 160$ for $C^R C^W$ plants). Thus, the frequency of the C^R allele is $800/1,000 = 0.8$ (80%).

When studying a locus with two alleles, the convention is to use p to represent the frequency of one allele and q to represent the frequency of the other allele. Thus, p, the frequency of the C^R allele in the gene pool of this population, is $p = 0.8$ (80%). And because there are only two alleles for this gene, the frequency of the C^W allele, represented by q, must be $q = 1 - p = 0.2$ (20%). For loci that have more than two alleles, the sum of all allele frequencies must still equal 1 (100%).

Next we'll see how allele and genotype frequencies can be used to test whether evolution is occurring in a population.

The Hardy-Weinberg Principle

One way to assess whether natural selection or other factors are causing evolution at a particular locus is to determine what the genetic makeup of a population would be if it were *not* evolving at that locus. We can then compare that scenario with the data that we actually observe for the population. If there are no differences, we can conclude that the population is not evolving. If there are differences, this suggests that the population may be evolving—and then we can try to figure out why.

Hardy-Weinberg Equilibrium

The gene pool of a population that is not evolving can be described by the **Hardy-Weinberg principle**, named for the British mathematician and German physician, respectively, who independently derived it in 1908. This principle states that the frequencies of alleles and genotypes in a population will remain constant from generation to generation, provided that only Mendelian segregation and recombination of alleles are at work. Such a gene pool is in *Hardy-Weinberg equilibrium*.

To use the Hardy-Weinberg principle, it is helpful to think about genetic crosses in a new way. Previously, we used Punnett squares to determine the genotypes of offspring in a genetic cross (see Figure 11.5). Here, instead of considering the possible allele combinations from one cross, we'll consider the combination of alleles in *all* of the crosses in a population.

Imagine that all the alleles for a given locus from all the individuals in a population are placed in a large bin. We can think of this bin as holding the population's gene pool for that locus. "Reproduction" occurs by selecting alleles at random from the bin; somewhat similar events occur in nature when fish release sperm and eggs into the water or when pollen (containing plant sperm) is blown about by the wind. By viewing reproduction as a process of randomly selecting and combining alleles from

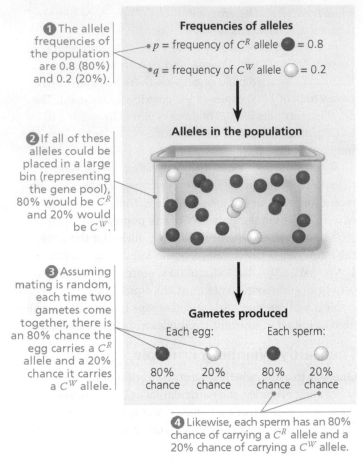

Frequencies of alleles

① The allele frequencies of the population are 0.8 (80%) and 0.2 (20%).

p = frequency of C^R allele ⚫ = 0.8

q = frequency of C^W allele ⚪ = 0.2

Alleles in the population

② If all of these alleles could be placed in a large bin (representing the gene pool), 80% would be C^R and 20% would be C^W.

Gametes produced

③ Assuming mating is random, each time two gametes come together, there is an 80% chance the egg carries a C^R allele and a 20% chance it carries a C^W allele.

Each egg:

80% chance	20% chance
Each sperm:	
80% chance	20% chance

④ Likewise, each sperm has an 80% chance of carrying a C^R allele and a 20% chance of carrying a C^W allele.

▲ **Figure 21.7 Selecting alleles at random from a gene pool.**

the bin (the gene pool), we are in effect assuming that mating occurs at random—that is, that all male-female matings are equally likely.

Let's apply the bin analogy to the hypothetical wildflower population discussed earlier **(Figure 21.7)**. In that population of 500 flowers, the frequency of the allele for red flowers (C^R) is $p = 0.8$, and the frequency of the allele for white flowers (C^W) is $q = 0.2$. In other words, a bin holding all 1,000 copies of the flower-color gene in the population would contain 800 C^R alleles and 200 C^W alleles. Assuming that gametes are formed by selecting alleles at random from the bin, the probability that an egg or sperm contains a C^R or C^W allele is equal to the frequency of these alleles in the bin. Thus, as shown in Figure 21.7, each egg has an 80% chance of containing a C^R allele and a 20% chance of containing a C^W allele; the same is true for each sperm.

Using the rule of multiplication (see Figure 11.9), we can now calculate the frequencies of the three possible genotypes, assuming random unions of sperm and eggs. The probability that two C^R alleles will come together is $p \times p = p^2 = 0.8 \times 0.8 = 0.64$. Thus, about 64% of the plants in the next generation will have the genotype $C^R C^R$. The frequency of $C^W C^W$ individuals is expected to be about $q \times q = q^2 = 0.2 \times 0.2 = 0.04$, or 4%. $C^R C^W$ heterozygotes can arise in two different

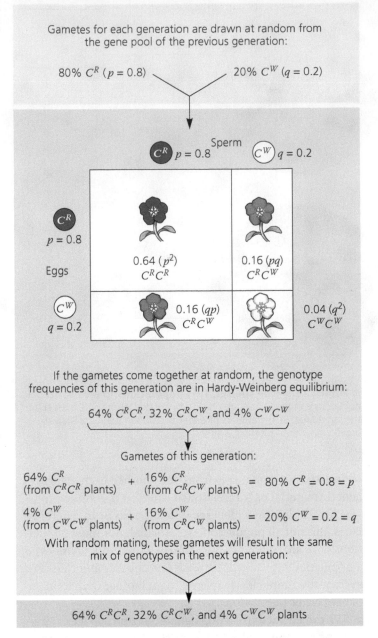

Gametes for each generation are drawn at random from the gene pool of the previous generation:

80% C^R ($p = 0.8$)   20% C^W ($q = 0.2$)

Sperm

C^R $p = 0.8$ C^W $q = 0.2$

C^R $p = 0.8$

Eggs

C^W $q = 0.2$

	Sperm C^R $p = 0.8$	Sperm C^W $q = 0.2$
C^R $p = 0.8$	0.64 (p^2) $C^R C^R$	0.16 (pq) $C^R C^W$
C^W $q = 0.2$	0.16 (qp) $C^R C^W$	0.04 (q^2) $C^W C^W$

If the gametes come together at random, the genotype frequencies of this generation are in Hardy-Weinberg equilibrium:

64% $C^R C^R$, 32% $C^R C^W$, and 4% $C^W C^W$

Gametes of this generation:

64% C^R (from $C^R C^R$ plants) + 16% C^R (from $C^R C^W$ plants) = 80% C^R = 0.8 = p

4% C^W (from $C^W C^W$ plants) + 16% C^W (from $C^R C^W$ plants) = 20% C^W = 0.2 = q

With random mating, these gametes will result in the same mix of genotypes in the next generation:

64% $C^R C^R$, 32% $C^R C^W$, and 4% $C^W C^W$ plants

▲ **Figure 21.8 The Hardy-Weinberg principle.** In our wildflower population, the gene pool remains constant from one generation to the next. Mendelian processes alone do not alter frequencies of alleles or genotypes.

? *If the frequency of the* C^R *allele is 0.6, predict the frequencies of the* $C^R C^R$, $C^R C^W$, *and* $C^W C^W$ *genotypes.*

ways. If the sperm provides the C^R allele and the egg provides the C^W allele, the resulting heterozygotes will be $p \times q = 0.8 \times 0.2 = 0.16$, or 16% of the total. If the sperm provides the C^W allele and the egg the C^R allele, the heterozygous offspring will make up $q \times p = 0.2 \times 0.8 = 0.16$, or 16%. The frequency of heterozygotes is thus the sum of these possibilities: $pq + qp = 2pq = 0.16 + 0.16 = 0.32$, or 32%.

As shown in **Figure 21.8**, the genotype frequencies in the next generation must add up to 1 (100%). Thus, the equation for Hardy-Weinberg equilibrium states that at a locus with two alleles, the three genotypes will appear in the following proportions:

$$\underbrace{p^2}_{\substack{\text{Expected}\\\text{frequency}\\\text{of genotype}\\C^R C^R}} + \underbrace{2pq}_{\substack{\text{Expected}\\\text{frequency}\\\text{of genotype}\\C^R C^W}} + \underbrace{q^2}_{\substack{\text{Expected}\\\text{frequency}\\\text{of genotype}\\C^W C^W}} = 1$$

Note that for a locus with two alleles, only three genotypes are possible (in this case, $C^R C^R$, $C^R C^W$, and $C^W C^W$). As a result, the sum of the frequencies of the three genotypes must equal 1 (100%) in *any* population—regardless of whether the population is in Hardy-Weinberg equilibrium. A population is in Hardy-Weinberg equilibrium only if the genotype frequencies are such that the actual frequency of one homozygote is p^2, the actual frequency of the other homozygote is q^2, and the actual frequency of heterozygotes is $2pq$. Finally, as suggested by Figure 21.8, if a population such as our wildflowers is in Hardy-Weinberg equilibrium and its members continue to mate randomly generation after generation, allele and genotype frequencies will remain constant. The system operates somewhat like a deck of cards: No matter how many times the deck is reshuffled to deal out new hands, the deck itself remains the same. Aces do not grow more numerous than jacks. And the repeated shuffling of a population's gene pool over the generations cannot, in itself, change the frequency of one allele relative to another.

Conditions for Hardy-Weinberg Equilibrium

The Hardy-Weinberg principle describes a hypothetical population that is not evolving. But in real populations, the allele and genotype frequencies often *do* change over time. Such changes can occur when at least one of the following five conditions of Hardy-Weinberg equilibrium is not met:

1. **No mutations.** The gene pool is modified if mutations alter alleles or if entire genes are deleted or duplicated.
2. **Random mating.** If individuals tend to mate within a subset of the population, such as their near neighbors or close relatives (inbreeding), random mixing of gametes does not occur, and genotype frequencies change.
3. **No natural selection.** Differences in the survival and reproductive success of individuals carrying different genotypes can alter allele frequencies.
4. **Extremely large population size.** The smaller the population, the more likely it is that allele frequencies will fluctuate by chance from one generation to the next (a process called genetic drift).
5. **No gene flow.** By moving alleles into or out of populations, gene flow can alter allele frequencies.

Departure from these conditions usually results in evolutionary change, which, as we've already described, is common in natural populations. But it is also common for natural populations to be in Hardy-Weinberg equilibrium for specific genes. This apparent contradiction occurs because a population can be evolving at some loci, yet simultaneously be in Hardy-Weinberg equilibrium at other loci. In addition, some populations evolve so slowly that the changes in their allele and genotype frequencies are difficult to distinguish from those predicted for a nonevolving population.

Applying the Hardy-Weinberg Principle

The Hardy-Weinberg equation is often used as an initial test of whether evolution is occurring in a population (you'll encounter an example in Concept Check 21.2, question 3). The equation also has medical applications, such as estimating the percentage of a population carrying the allele for an inherited disease. For example, consider phenylketonuria (PKU), a metabolic disorder that results from homozygosity for a recessive allele and occurs in about one out of every 10,000 babies born in the United States. Left untreated, PKU results in mental disability and other problems. (Newborns are now tested for PKU, and symptoms can be largely avoided with a diet very low in phenylalanine. For this reason, products that contain phenylalanine, such as diet colas, carry warning labels.)

To apply the Hardy-Weinberg equation, we must assume that no new PKU mutations are being introduced into the population (condition 1), and that people neither choose their mates on the basis of whether or not they carry this gene nor generally mate with close relatives (condition 2). We must also ignore any effects of differential survival and reproductive success among PKU genotypes (condition 3) and assume that there are no effects of genetic drift (condition 4) or of gene flow from other populations into the United States (condition 5). These assumptions are reasonable: The mutation rate for the PKU gene is low, inbreeding and other forms of nonrandom mating are not common in the United States, selection occurs only against the rare homozygotes (and then only if dietary restrictions are not followed), the U.S. population is very large, and populations outside the country have PKU allele frequencies similar to those seen in the United States. If all these assumptions hold, then the frequency of individuals in the population born with PKU will correspond to q^2 in the Hardy-Weinberg equation (q^2 = frequency of homozygotes). Because the allele is recessive, we must estimate the number of heterozygotes rather than counting them directly as we did with the pink flowers. Since we know there is one PKU occurrence per 10,000 births (q^2 = 0.0001), the frequency (q) of the recessive allele for PKU is

$$q = \sqrt{0.0001} = 0.01$$

and the frequency of the dominant allele is

$$p = 1 - q = 1 - 0.01 = 0.99$$

Using the Hardy-Weinberg Equation to Interpret Data and Make Predictions

Is Evolution Occurring in a Soybean Population? One way to test whether evolution is occurring in a population is to compare the observed genotype frequencies at a locus with those expected for a nonevolving population based on the Hardy-Weinberg equation. In this exercise, you'll test whether a soybean population is evolving at a locus with two alleles, C^G and C^Y, that affect chlorophyll production and hence leaf color.

How the Experiment Was Done Students planted soybean seeds and then counted the number of seedlings of each genotype at day 7 and again at day 21. Seedlings of each genotype could be distinguished visually because the C^G and C^Y alleles show incomplete dominance: $C^G C^G$ seedlings have green leaves, $C^G C^Y$ seedlings have green-yellow leaves, and $C^Y C^Y$ seedlings have yellow leaves.

Data from the Experiment

	Number of Seedlings			
Time (days)	Green ($C^G C^G$)	Green-yellow ($C^G C^Y$)	Yellow ($C^Y C^Y$)	Total
7	49	111	56	216
21	47	106	20	173

Interpret the Data

1. Use the observed genotype frequencies from the day 7 data to calculate the frequencies of the C^G allele (p) and the C^Y allele (q). (Remember that the frequency of an allele in a gene pool is the number of copies of that allele divided by the total number of copies of all alleles at that locus.)

2. Next, use the Hardy-Weinberg equation ($p^2 + 2pq + q^2 = 1$) to calculate the expected frequencies of genotypes $C^G C^G$, $C^G C^Y$, and $C^Y C^Y$ for a population in Hardy-Weinberg equilibrium.

3. Calculate the observed frequencies of genotypes $C^G C^G$, $C^G C^Y$, and $C^Y C^Y$ at day 7. (The observed frequency of a genotype in a gene pool is the number of individuals with that genotype divided by the total number of individuals.) Compare these frequencies to the expected frequencies calculated in step 2. Is the seedling population in Hardy-Weinberg equilibrium at day 7, or is evolution occurring? Explain your reasoning and identify which genotypes, if any, appear to be selected for or against.

4. Calculate the observed frequencies of genotypes $C^G C^G$, $C^G C^Y$, and $C^Y C^Y$ at day 21. Compare these frequencies to the expected frequencies calculated in step 2 and the observed frequencies at day 7. Is the seedling population in Hardy-Weinberg equilibrium at day 21, or is evolution occurring? Explain your reasoning and identify which genotypes, if any, appear to be selected for or against.

5. Homozygous $C^Y C^Y$ individuals cannot produce chlorophyll. The ability to photosynthesize becomes more critical as seedlings age and begin to exhaust the supply of food that was stored in the seed from which they emerged. Develop a hypothesis that explains the data for days 7 and 21. Based on this hypothesis, predict how the frequencies of the C^G and C^Y alleles will change beyond day 21.

(MB) A version of this Scientific Skills Exercise can be assigned in MasteringBiology.

The frequency of carriers, heterozygous people who do not have PKU but may pass the PKU allele to offspring, is

$$2pq = 2 \times 0.99 \times 0.01 = 0.0198$$
(approximately 2% of the U.S. population)

Remember, the assumption of Hardy-Weinberg equilibrium yields an approximation; the real number of carriers may differ. Still, our calculations suggest that harmful recessive alleles at this and other loci can be concealed in a population because they are carried by healthy heterozygotes. The **Scientific Skills Exercise** provides another opportunity for you to apply the Hardy-Weinberg principle to allele data for a population.

CONCEPT CHECK 21.2

1. A population has 700 individuals, 85 of genotype *AA*, 320 of genotype *Aa*, and 295 of genotype *aa*. What are the frequencies of alleles *A* and *a*?

2. The frequency of allele *a* is 0.45 for a population in Hardy-Weinberg equilibrium. What are the expected frequencies of genotypes *AA*, *Aa*, and *aa*?

3. **WHAT IF?** A locus that affects susceptibility to a degenerative brain disease has two alleles, *V* and *v*. In a population, 16 people have genotype *VV*, 92 have genotype *Vv*, and 12 have genotype *vv*. Is this population evolving? Explain.

For suggested answers, see Appendix A.

CONCEPT 21.3

Natural selection, genetic drift, and gene flow can alter allele frequencies in a population

Note again the five conditions required for a population to be in Hardy-Weinberg equilibrium. A deviation from any of these conditions is a potential cause of evolution. New mutations (violation of condition 1) can alter allele frequencies, but because mutations are rare, the change from one generation to the next is likely to be very small. Nonrandom mating (violation of condition 2) can affect the frequencies of homozygous and heterozygous genotypes but by itself has no effect on allele frequencies in the gene pool. (Allele frequencies can change if individuals with certain inherited traits are more likely than other individuals to obtain mates. However, such a situation not only causes a deviation from random mating; it also violates condition 3, no natural selection.) For the rest of this section, we will focus on the three mechanisms that alter allele frequencies directly and cause most evolutionary change: natural selection, genetic drift, and gene flow (violations of conditions 3–5).

Natural Selection

The concept of natural selection is based on differential success in survival and reproduction: Individuals in a population exhibit variations in their heritable traits, and those with traits that are better suited to their environment tend to produce more offspring than those with traits that are not as well suited (see Chapter 19).

In genetic terms, we now know that selection results in alleles being passed to the next generation in proportions that differ from those in the present generation. For example, the fruit fly *D. melanogaster* has an allele that confers resistance to several insecticides, including DDT. This allele has a frequency of 0% in laboratory strains of *D. melanogaster* established from flies collected in the wild in the early 1930s, prior to DDT use. However, in strains established from flies collected after 1960 (following 20 or more years of DDT use), the allele frequency is 37%. We can infer that this allele either arose by mutation between 1930 and 1960 or was present in 1930, but very rare. In any case, the rise in frequency of this allele most likely occurred because DDT is a powerful poison that is a strong selective force in exposed fly populations.

As the *D. melanogaster* example shows, an allele that confers resistance to an insecticide will increase in frequency in a population exposed to that insecticide. Such changes are not coincidental. By consistently favoring some alleles over others, natural selection can cause *adaptive evolution* (evolution that results in a better match between organisms and their environment). We'll explore this process in more detail a little later in this chapter.

Genetic Drift

If you flip a coin 1,000 times, a result of 700 heads and 300 tails might make you suspicious about that coin. But if you flip a coin only 10 times, an outcome of 7 heads and 3 tails would not be surprising. The smaller the number of coin flips, the more likely it is that chance alone will cause a deviation from the predicted result. (In this case, the prediction is an equal number of heads and tails.) Chance events can also cause allele frequencies to fluctuate unpredictably from one generation to the next, especially in small populations—a process called **genetic drift**.

Figure 21.9 models how genetic drift might affect a small population of our wildflowers. In this example, drift leads to the loss of an allele from the gene pool, but it is a matter of chance that the C^W allele is lost and not the C^R allele. Such unpredictable changes in allele frequencies can be caused by chance events associated with survival and reproduction. Perhaps a large animal such as a moose stepped on the three $C^W C^W$ individuals in generation 2, killing them and increasing the chance that only the C^R allele would be passed to the next generation. Allele frequencies can also be affected by chance events that occur during fertilization. For example, suppose two individuals

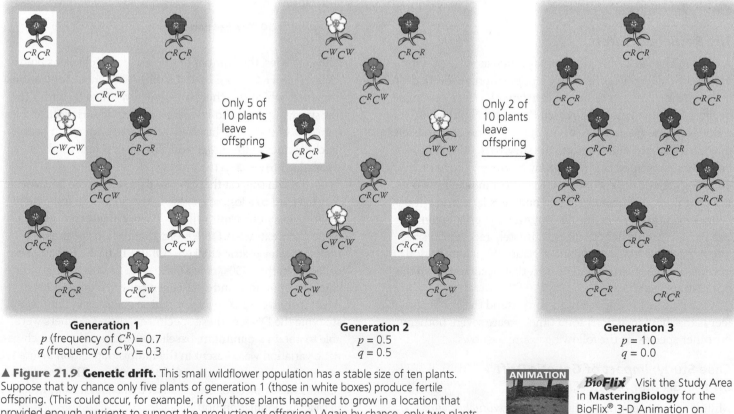

Generation 1
p (frequency of C^R) = 0.7
q (frequency of C^W) = 0.3

Generation 2
p = 0.5
q = 0.5

Generation 3
p = 1.0
q = 0.0

▲ **Figure 21.9 Genetic drift.** This small wildflower population has a stable size of ten plants. Suppose that by chance only five plants of generation 1 (those in white boxes) produce fertile offspring. (This could occur, for example, if only those plants happened to grow in a location that provided enough nutrients to support the production of offspring.) Again by chance, only two plants of generation 2 leave fertile offspring. As a result, by chance the frequency of the C^W allele first increases in generation 2, then falls to zero in generation 3.

ANIMATION *BioFlix* Visit the Study Area in **MasteringBiology** for the BioFlix® 3-D Animation on Mechanisms of Evolution.

of genotype C^RC^W had a small number of offspring. By chance alone, every egg and sperm pair that generated offspring could happen to have carried the C^R allele and not the C^W allele.

Certain circumstances can result in genetic drift having a significant impact on a population. Two examples are the founder effect and the bottleneck effect.

The Founder Effect

When a few individuals become isolated from a larger population, this smaller group may establish a new population whose gene pool differs from the source population; this is called the **founder effect**. The founder effect might occur, for example, when a few members of a population are blown by a storm to a new island. Genetic drift, in which chance events alter allele frequencies, will occur in such a case if the storm indiscriminately transports some individuals (and their alleles), but not others, from the source population.

The founder effect probably accounts for the relatively high frequency of certain inherited disorders among isolated human populations. For example, in 1814, 15 British colonists founded a settlement on Tristan da Cunha, a group of small islands in the Atlantic Ocean midway between Africa and South America. Apparently, one of the colonists carried a recessive allele for retinitis pigmentosa, a progressive form of blindness that afflicts homozygous individuals. Of the founding colonists' 240 descendants on the island in the late 1960s, 4 had retinitis pigmentosa. The frequency of the allele that causes this disease is ten times higher on Tristan da Cunha than in the populations from which the founders came.

The Bottleneck Effect

A sudden change in the environment, such as a fire or flood, may drastically reduce the size of a population. A severe drop in population size can cause the **bottleneck effect**, so named because the population has passed through a "bottleneck" that reduces its size **(Figure 21.10)**. By chance alone, certain alleles may be overrepresented among the survivors, others may be underrepresented, and some may be absent altogether. Ongoing genetic drift is likely to have substantial effects on the gene pool until the population becomes large enough that chance events have less impact. But even if a population that has passed through a bottleneck ultimately recovers in size, it may have low levels of genetic variation for a long period of time—a legacy of the genetic drift that occurred when the population was small.

One reason it is important to understand the bottleneck effect is that human actions sometimes create severe bottlenecks for other species, as the following example shows.

Case Study: *Impact of Genetic Drift on the Greater Prairie Chicken*

Millions of greater prairie chickens (*Tympanuchus cupido*) once lived on the prairies of Illinois. As these prairies were converted to farmland and other uses during the 19th and

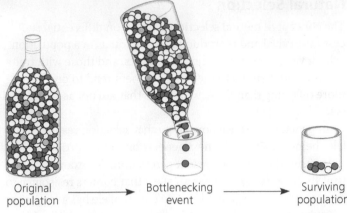

(a) Shaking just a few marbles through the narrow neck of a bottle is analogous to a drastic reduction in the size of a population. By chance, blue marbles are overrepresented in the surviving population, and gold marbles are absent.

(b) Similarly, bottlenecking a wild population tends to reduce genetic variation, as in the Florida panther (*Puma concolor coryi*), a subspecies in danger of extinction.

▲ **Figure 21.10 The bottleneck effect.**

20th centuries, the number of greater prairie chickens plummeted **(Figure 21.11a)**. By 1993, only two Illinois populations remained, which together harbored fewer than 50 birds. The few surviving birds had low levels of genetic variation, and less than 50% of their eggs hatched, compared with much higher hatching rates of the larger populations in Kansas and Nebraska **(Figure 21.11b)**.

These data suggest that genetic drift during the bottleneck may have led to a loss of genetic variation and an increase in the frequency of harmful alleles. To investigate this hypothesis, researchers extracted DNA from 15 museum specimens of Illinois greater prairie chickens. Of the 15 birds, 10 had been collected in the 1930s, when there were 25,000 greater prairie chickens in Illinois, and 5 had been collected in the 1960s, when there were 1,000 greater prairie chickens in Illinois. By studying the DNA of these specimens, the researchers were able to obtain a minimum, baseline estimate of how much genetic variation was present in the Illinois population *before* the population shrank to extremely low numbers. This baseline estimate is a key piece of information that is not usually available in cases of population bottlenecks.

The researchers surveyed six loci and found that the 1993 Illinois greater prairie chicken population had lost nine alleles

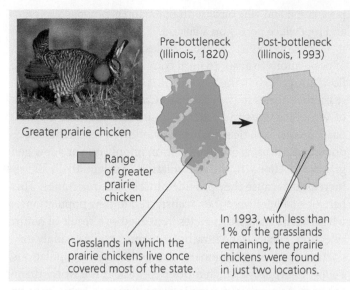

Pre-bottleneck
(Illinois, 1820)

Post-bottleneck
(Illinois, 1993)

Greater prairie chicken

■ Range
of greater
prairie
chicken

Grasslands in which the
prairie chickens live once
covered most of the state.

In 1993, with less than
1% of the grasslands
remaining, the prairie
chickens were found
in just two locations.

(a) The Illinois population of greater prairie chickens dropped from
millions of birds in the 1800s to fewer than 50 birds in 1993.

Location	Population size	Number of alleles per locus	Percentage of eggs hatched
Illinois			
1930–1960s	1,000–25,000	5.2	93
1993	<50	3.7	<50
Kansas, 1998 (no bottleneck)	750,000	5.8	99
Nebraska, 1998 (no bottleneck)	75,000–200,000	5.8	96

(b) As a consequence of the drastic reduction in the size of the Illinois
population, genetic drift resulted in a drop in the number of alleles
per locus (averaged across six loci studied) and a decrease in the
percentage of eggs that hatched.

▲ **Figure 21.11 Genetic drift and loss of genetic variation.**

that were present in the museum specimens. The 1993 population also had fewer alleles per locus than the pre-bottleneck
Illinois or the current Kansas and Nebraska populations (see
Figure 21.11b). Thus, as predicted, drift had reduced the genetic variation of the small 1993 population. Drift may also
have increased the frequency of harmful alleles, leading to
the low egg-hatching rate. To counteract these negative effects, 271 birds from neighboring states were added to the
Illinois population over four years. This strategy succeeded:
New alleles entered the population, and the egg-hatching rate
improved to over 90%. Overall, studies on the Illinois greater
prairie chicken illustrate the powerful effects of genetic drift
in small populations and provide hope that in at least some
populations, these effects can be reversed.

Effects of Genetic Drift: A Summary

The examples we've described highlight four key points:

1. **Genetic drift is significant in small populations.**
   Chance events can cause an allele to be disproportionately over- or underrepresented in the next generation.
   Although chance events occur in populations of all sizes,
   they tend to alter allele frequencies substantially only in
   small populations.
2. **Genetic drift can cause allele frequencies to change at
   random.** Because of genetic drift, an allele may increase
   in frequency one year, then decrease the next; the change
   from year to year is not predictable. Thus, unlike natural
   selection, which in a given environment consistently favors some alleles over others, genetic drift causes allele
   frequencies to change at random over time.
3. **Genetic drift can lead to a loss of genetic variation
   within populations.** By causing allele frequencies to
   fluctuate randomly over time, genetic drift can eliminate
   alleles from a population. Because evolution depends
   on genetic variation, such losses can influence how
   effectively a population can adapt to a change in the
   environment.
4. **Genetic drift can cause harmful alleles to become
   fixed.** Alleles that are neither harmful nor beneficial can
   be lost or become fixed entirely by chance through genetic
   drift. In very small populations, genetic drift can also
   cause alleles that are slightly harmful to become fixed.
   When this occurs, the population's survival can be threatened (as in the case of the greater prairie chicken).

Gene Flow

Natural selection and genetic drift are not the only phenomena
affecting allele frequencies. Allele frequencies can also change
by **gene flow**, the transfer of alleles into or out of a population
due to the movement of fertile individuals or their gametes. For
example, suppose that near our original hypothetical wildflower
population there is another population consisting primarily of
white-flowered individuals ($C^W C^W$). Insects carrying pollen
from these plants may fly to and pollinate plants in our original
population. The introduced C^W alleles would modify our original
population's allele frequencies in the next generation. Because
alleles are transferred between populations, gene flow tends to
reduce the genetic differences between populations. In fact, if
it is extensive enough, gene flow can result in two populations
combining into a single population with a common gene pool.

Alleles transferred by gene flow can also affect how well
populations are adapted to local environmental conditions.
Researchers studying the songbird *Parus major* (great tit) on
the small Dutch island of Vlieland noted survival differences
between two populations on the island. Females born in the
eastern population survive twice as well as females born in the
central population, regardless of where the females eventually

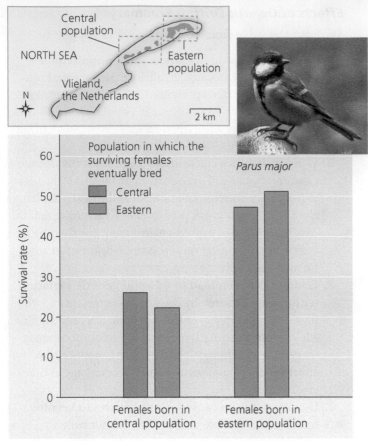

Population in which the surviving females eventually bred

■ Central
■ Eastern

Parus major

© 2005 Macmillan Publishers Ltd.

▲ **Figure 21.12 Gene flow and local adaptation.** In *Parus major* populations on Vlieland, the yearly survival rate of females born in the eastern population is higher than that of females born in the central population. Gene flow from the mainland to the central population is 3.3 times higher than gene flow to the eastern population, and birds from the mainland are selected against in both populations. These data suggest that gene flow from the mainland has prevented the central population from adapting fully to its local conditions.

settle and raise offspring **(Figure 21.12)**. This finding suggests that females born in the eastern population are better adapted to life on the island than females born in the central population. But extensive field studies also showed that the two populations are connected by high levels of gene flow (mating), which should reduce genetic differences between them.

So how can the eastern population be better adapted to life on Vlieland than the central population? The answer lies in the unequal amounts of gene flow from the mainland. In any given year, 43% of the first-time breeders in the central population are immigrants from the mainland, compared with only 13% in the eastern population. Birds with mainland genotypes survive and reproduce poorly on Vlieland, and in the eastern population, selection reduces the frequency of these genotypes. In the central population, however, gene flow from the mainland is so high that it overwhelms the effects of selection. As a result, females born in the central population have many immigrant genes, reducing the degree to which members of that population are adapted to life on the island. Researchers are currently investigating why gene flow is so much higher in the central

population and why birds with mainland genotypes survive and reproduce poorly on Vlieland.

Gene flow can also transfer alleles that improve the ability of populations to adapt to local conditions. For example, gene flow has resulted in the worldwide spread of some insecticide-resistance alleles in the mosquito *Culex pipiens*, a vector of West Nile virus and other diseases. Each of these alleles has a unique genetic signature that allowed researchers to document that it arose by mutation in only one or a few geographic locations. In their population of origin, these alleles increased because they provided insecticide resistance. These beneficial alleles were then transferred to new populations, where again, their frequencies increased as a result of natural selection. Finally, gene flow has become an increasingly important agent of evolutionary change in human populations. People today move much more freely about the world than in the past. As a result, mating is more common between members of populations that previously had very little contact, leading to an exchange of alleles and fewer genetic differences between those populations.

CONCEPT CHECK 21.3

1. In what sense is natural selection more "predictable" than genetic drift?
2. Distinguish genetic drift from gene flow in terms of (a) how they occur and (b) their implications for future genetic variation in a population.
3. **WHAT IF?** Suppose two plant populations exchange pollen and seeds. In one population, individuals of genotype *AA* are most common (9,000 *AA*, 900 *Aa*, 100 *aa*), while the opposite is true in the other population (100 *AA*, 900 *Aa*, 9,000 *aa*). If neither allele has a selective advantage, what will happen over time to the allele and genotype frequencies of these populations?

For suggested answers, see Appendix A.

CONCEPT 21.4

Natural selection is the only mechanism that consistently causes adaptive evolution

Evolution by natural selection is a blend of chance and "sorting": chance in the creation of new genetic variations (as in mutation) and sorting as natural selection favors some alleles over others. Because of this favoring process, the outcome of natural selection is *not* random. Instead, natural selection consistently increases the frequencies of alleles that provide reproductive advantage and thus leads to **adaptive evolution**.

Natural Selection: *A Closer Look*

In examining how natural selection brings about adaptive evolution, we'll begin with the concept of relative fitness and the different ways that an organism's phenotype is subject to natural selection.

Relative Fitness

The phrases "struggle for existence" and "survival of the fittest" are commonly used to describe natural selection, but these expressions are misleading if taken to mean direct competitive contests among individuals. There *are* animal species in which individuals, usually males, lock horns or otherwise spar to determine mating privilege. But reproductive success is generally more subtle and depends on many factors besides outright battle. For example, a barnacle that is more efficient at collecting food than its neighbors may have greater stores of energy and hence be able to produce more eggs. A moth may have more offspring than other moths in the same population because its body colors more effectively conceal it from predators, improving its chance of surviving long enough to produce more offspring. These examples illustrate how in a given environment, certain traits can lead to greater **relative fitness**: the contribution an individual makes to the gene pool of the next generation *relative to* the contributions of other individuals.

Although we often refer to the relative fitness of a genotype, remember that the entity that is subjected to natural selection is the whole organism, not the underlying genotype. Thus, selection acts more directly on the phenotype than on the genotype; it acts on the genotype indirectly, via how the genotype affects the phenotype.

Directional, Disruptive, and Stabilizing Selection

Natural selection can occur in three ways, depending on which phenotypes in a population are favored. These three modes of selection are called directional selection, disruptive selection, and stabilizing selection.

Directional selection occurs when conditions favor individuals at one extreme of a phenotypic range, thereby shifting a population's frequency curve for the phenotypic character in one direction or the other **(Figure 21.13a)**. Directional selection is common when a population's environment changes or when members of a population migrate to a different habitat. For instance, an increase in the relative abundance of large seeds over small seeds led to increased beak depth in a population of Galápagos finches (see Figure 21.2).

▼ **Figure 21.13 Modes of selection.** These cases describe three ways in which a hypothetical deer mouse population with heritable variation in fur coloration from light to dark might evolve. The graphs show how the frequencies of individuals with different fur colors change over time. The large white arrows symbolize selective pressures against certain phenotypes.

MAKE CONNECTIONS *Review Figure 19.14. Which mode of selection has occurred in soapberry bug populations that feed on the introduced goldenrain tree? Explain.*

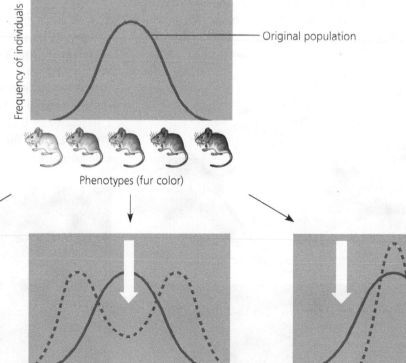

Original population

Phenotypes (fur color)

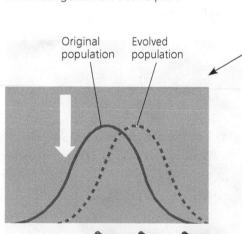

Original population   Evolved population

(a) Directional selection shifts the overall makeup of the population by favoring variants that are at one extreme of the distribution. In this case, lighter mice are selected against because they live among dark rocks, making it harder for them to hide from predators.

(b) Disruptive selection favors variants at both ends of the distribution. These mice have colonized a patchy habitat made up of light and dark rocks, with the result that mice of an intermediate color are selected against.

(c) Stabilizing selection removes extreme variants from the population and preserves intermediate types. If the environment consists of rocks of an intermediate color, both light and dark mice will be selected against.

Disruptive selection (Figure 21.13b) occurs when conditions favor individuals at both extremes of a phenotypic range over individuals with intermediate phenotypes. One example is a population of black-bellied seedcracker finches in Cameroon whose members display two distinctly different beak sizes. Small-billed birds feed mainly on soft seeds, whereas large-billed birds specialize in cracking hard seeds. It appears that birds with intermediate-sized bills are relatively inefficient at cracking both types of seeds and thus have lower relative fitness.

Stabilizing selection (Figure 21.13c) acts against both extreme phenotypes and favors intermediate variants. This mode of selection reduces variation and tends to maintain the status quo for a particular phenotypic character. For example, the birth weights of most human babies lie in the range of 3–4 kg (6.6–8.8 pounds); babies who are either much smaller or much larger suffer higher rates of mortality.

Regardless of the mode of selection, however, the basic mechanism remains the same. Selection favors individuals whose heritable phenotypic traits provide higher reproductive success than do the traits of other individuals.

The Key Role of Natural Selection in Adaptive Evolution

The adaptations of organisms include many striking examples. Certain octopuses, for instance, can change color rapidly, enabling them to blend into different backgrounds. Another example is the remarkable jaws of snakes **(Figure 21.14)**, which allow them to swallow prey much larger than their own head (a feat analogous to a person swallowing a whole watermelon). Other adaptations, such as a version of an enzyme that shows improved function in cold environments, may be less visually dramatic but just as important for survival and reproduction.

Such adaptations can arise gradually over time as natural selection increases the frequencies of alleles that enhance survival and reproduction. As the proportion of individuals that have favorable traits increases, the match between a species and its environment improves; that is, adaptive evolution occurs. Note, however, that the physical and biological components of an organism's environment may change over time. As a result, what constitutes a "good match" between an organism and its environment can be a moving target, making adaptive evolution a continuous, dynamic process.

And what about genetic drift and gene flow? Both can, in fact, increase the frequencies of alleles that improve the match between organisms and their environment, but neither does so consistently. Genetic drift can cause the frequency of a slightly beneficial allele to increase, but it also can cause the frequency of such an allele to decrease. Similarly, gene flow may introduce alleles that are advantageous or ones that are disadvantageous. Natural selection is the only evolutionary mechanism that consistently leads to adaptive evolution.

Sexual Selection

Charles Darwin was the first to explore the implications of **sexual selection**, a form of natural selection in which individuals with certain inherited characteristics are more likely than other individuals to obtain mates. Sexual selection can result in **sexual dimorphism**, a difference in secondary sexual characteristics between males and females of the same species **(Figure 21.15)**. These distinctions include differences in size, color, ornamentation, and behavior.

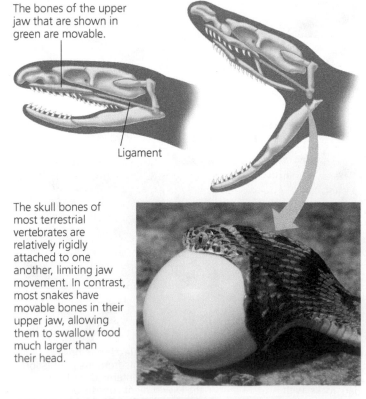

The bones of the upper jaw that are shown in green are movable.

Ligament

The skull bones of most terrestrial vertebrates are relatively rigidly attached to one another, limiting jaw movement. In contrast, most snakes have movable bones in their upper jaw, allowing them to swallow food much larger than their head.

▲ **Figure 21.14 Movable jaw bones in snakes.**

▲ **Figure 21.15 Sexual dimorphism and sexual selection.** Peacocks (above left) and peahens (above right) show extreme sexual dimorphism. There is intrasexual selection between competing males, followed by intersexual selection when the females choose among the showiest males.

How does sexual selection operate? There are several ways. In *intrasexual selection*, meaning selection within the same sex, individuals of one sex compete directly for mates of the opposite sex. In many species, intrasexual selection occurs among males. For example, a single male may patrol a group of females and prevent other males from mating with them. The patrolling male may defend his status by defeating smaller, weaker, or less fierce males in combat. More often, this male is the psychological victor in ritualized displays that discourage would-be competitors but do not risk injury that would reduce his own fitness. Intrasexual selection has also been observed among females in a variety of species, including ring-tailed lemurs and broad-nosed pipefish.

In *intersexual selection*, also called *mate choice*, individuals of one sex (usually the females) are choosy in selecting their mates from the other sex. In many cases, the female's choice depends on the showiness of the male's appearance or behavior (see Figure 21.15). What intrigued Darwin about mate choice is that male showiness may not seem adaptive in any other way and in fact pose some risk. For example, bright plumage may make male birds more visible to predators. But if such characteristics help a male gain a mate, and if this benefit outweighs the risk from predation, then both the bright plumage and the female preference for it will be reinforced because they enhance overall reproductive success.

How do female preferences for certain male characteristics evolve in the first place? One hypothesis is that females prefer male traits that are correlated with "good genes." If the trait preferred by females is indicative of a male's overall genetic quality, both the male trait and female preference for it should increase in frequency. **Figure 21.16** describes one experiment testing this hypothesis in gray tree frogs (*Hyla versicolor*).

Other researchers have shown that in several bird species, the traits preferred by females are related to overall male health. Here, too, female preference appears to be based on traits that reflect "good genes," in this case alleles indicative of a robust immune system.

The Preservation of Genetic Variation

Some of the genetic variation in populations represents **neutral variation**, differences in DNA sequence that do not confer a selective advantage or disadvantage. But variation is also found at loci affected by selection. What prevents natural selection from reducing genetic variation at those loci by culling all unfavorable alleles? The tendency for directional and stabilizing selection to reduce variation is countered by mechanisms that preserve or restore it, such as diploidy and balancing selection.

Diploidy

In diploid organisms, a considerable amount of genetic variation is hidden from selection in the form of recessive alleles. Recessive alleles that are less favorable than their dominant

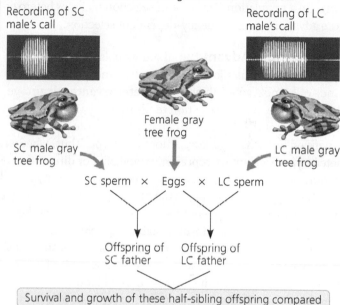

▼ **Figure 21.16** **Inquiry**

Do females select mates based on traits indicative of "good genes"?

Experiment Female gray tree frogs (*Hyla versicolor*) prefer to mate with males that give long mating calls. Allison Welch and colleagues, at the University of Missouri, tested whether the genetic makeup of long-calling (LC) males is superior to that of short-calling (SC) males. The researchers fertilized half the eggs of each female with sperm from an LC male and fertilized the remaining eggs with sperm from an SC male. In two separate experiments (one in 1995, the other in 1996), the resulting half-sibling offspring were raised in a common environment and their survival and growth were monitored.

Recording of SC male's call

Recording of LC male's call

SC male gray tree frog

Female gray tree frog

LC male gray tree frog

SC sperm × Eggs × LC sperm

Offspring of SC father

Offspring of LC father

Survival and growth of these half-sibling offspring compared

Results

Offspring Performance	1995	1996
Larval survival	LC better	NSD
Larval growth	NSD	LC better
Time to metamorphosis	LC better (shorter)	LC better (shorter)

NSD = no significant difference; LC better = offspring of LC males superior to offspring of SC males.

Conclusion Because offspring fathered by an LC male outperformed their half-siblings fathered by an SC male, the team concluded that the duration of a male's mating call is indicative of the male's overall genetic quality. This result supports the hypothesis that female mate choice can be based on a trait that indicates whether the male has "good genes."

Source A. M. Welch et al., Call duration as an indicator of genetic quality in male gray tree frogs, *Science* 280:1928–1930 (1998).

Inquiry in Action Read and analyze the original paper in *Inquiry in Action: Interpreting Scientific Papers*.

WHAT IF? Why did the researchers split each female frog's eggs into two batches for fertilization by different males? Why didn't they mate each female with a single male frog?

counterparts or even harmful in the current environment can persist by propagation in heterozygous individuals. This latent variation is exposed to natural selection only when both parents carry the same recessive allele and two copies end up in the same zygote. This happens only rarely if the frequency of the recessive allele is very low. Heterozygote protection maintains a huge pool of alleles that might not be favored under present conditions, but which could bring new benefits if the environment changes.

Balancing Selection

Selection itself may preserve variation at some loci. **Balancing selection** occurs when natural selection maintains two or more forms in a population. This type of selection includes heterozygote advantage and frequency-dependent selection.

Heterozygote Advantage If individuals who are heterozygous at a particular locus have greater fitness than do both kinds of homozygotes, they exhibit **heterozygote advantage**. In such a case, natural selection tends to maintain two or more alleles at that locus. Note that heterozygote advantage is defined in terms of *genotype*, not phenotype. Thus, whether heterozygote advantage represents stabilizing or directional selection depends on the relationship between the genotype and the phenotype. For example, if the phenotype of a heterozygote is intermediate to the phenotypes of both homozygotes, heterozygote advantage is a form of stabilizing selection.

An example of heterozygote advantage occurs at the locus in humans that codes for the β polypeptide subunit of hemoglobin, the oxygen-carrying protein of red blood cells. In homozygous individuals, a certain recessive allele at that locus causes sickle-cell disease. The red blood cells of people with sickle-cell disease become distorted in shape, or *sickled*, under low-oxygen conditions (see Figure 3.22), as occurs in the capillaries. These sickled cells can clump together and block the flow of blood in the capillaries, resulting in serious damage to organs such as the kidney, heart, and brain. Although some red blood cells become sickled in heterozygotes, not enough become sickled to cause sickle-cell disease.

Heterozygotes for the sickle-cell allele are protected against the most severe effects of malaria, a disease caused by a parasite that infects red blood cells (see Figure 25.26). One reason for this partial protection is that the body destroys sickled red blood cells rapidly, killing the parasites they harbor (but not affecting parasites inside normal red blood cells). Protection against malaria is important in tropical regions where the disease is a major killer. In such regions, selection favors heterozygotes over homozygous dominant individuals, who are more vulnerable to the effects of malaria, and also over homozygous recessive individuals, who develop sickle-cell disease. The frequency of the sickle-cell allele in Africa is generally highest in areas where the malaria parasite is most common **(Figure 21.17)**. In some populations, it accounts for 20% of the hemoglobin alleles in the gene pool, a very high frequency for such a harmful allele.

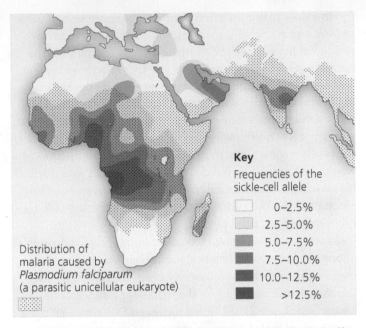

▲ **Figure 21.17 Mapping malaria and the sickle-cell allele.** The sickle-cell allele is most common in Africa, but it is not the only case of heterozygote advantage providing protection against malaria. Alleles at other loci (not shown on this map) are also favored by heterozygote advantage in populations near the Mediterranean Sea and in Southeast Asia where malaria is widespread.

Frequency-Dependent Selection In **frequency-dependent selection**, the fitness of a phenotype depends on how common it is in the population. Consider the scale-eating fish (*Perissodus microlepis*) of Lake Tanganyika, in Africa. These fish attack other fish from behind, darting in to remove a few scales from the flank of their prey. Of interest here is a peculiar feature of the scale-eating fish: Some are "left-mouthed" and some are "right-mouthed." Simple Mendelian inheritance determines these phenotypes, with the right-mouthed allele being dominant to the left-mouthed allele. Because their mouth twists to the left, left-mouthed fish always attack their prey's right flank **(Figure 21.18)**. (To see why, twist your lower jaw and lips to the left and imagine trying to take a bite from the left side of a fish, approaching it from behind.) Similarly, right-mouthed fish always attack from the left. Prey species guard against attack from whatever phenotype of scale-eating fish is most common in the lake. Thus, from year to year, selection favors whichever mouth phenotype is least common. As a result, the frequency of left- and right-mouthed fish oscillates over time, and balancing selection (due to frequency dependence) keeps the frequency of each phenotype close to 50%.

Why Natural Selection Cannot Fashion Perfect Organisms

Though natural selection leads to adaptation, nature abounds with examples of organisms that are less than ideally suited for their lifestyles. There are several reasons why.

1. **Selection can act only on existing variations.** Natural selection favors only the fittest phenotypes among those

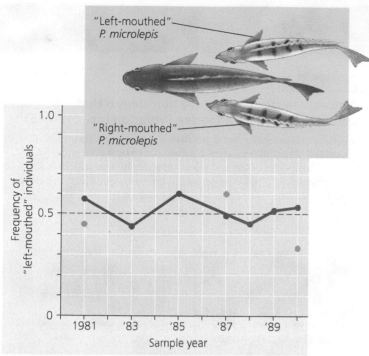

"Left-mouthed" *P. microlepis*

"Right-mouthed" *P. microlepis*

1.0

Frequency of
"left-mouthed" individuals

0.5

0

1981   '83   '85   '87   '89

Sample year

© 1993 AAAS

▲ **Figure 21.18 Frequency-dependent selection.** In a population of the scale-eating fish *Perissodus microlepis*, the frequency of left-mouthed individuals rises and falls in a regular manner (shown in red). At each of three time periods when the phenotypes of breeding adults were assessed, a majority of the adults that reproduced (represented by green dots) had the opposite phenotype of that which was most common in the population. Thus, it appears that right-mouthed individuals were favored by selection when left-mouthed individuals were more common, and vice versa.

? *What did the researchers measure to determine which phenotype was favored by selection? Are any assumptions implied by this choice? Explain.*

currently in the population, which may not be the ideal traits. New advantageous alleles do not arise on demand.

2. **Evolution is limited by historical constraints.** Each species has a legacy of descent with modification from ancestral forms. Evolution does not scrap the ancestral anatomy and build each new complex structure from scratch; rather, evolution co-opts existing structures and adapts them to new situations. We could imagine that if a terrestrial animal were to adapt to an environment in which flight would be advantageous, it might be best just to grow an extra pair of limbs that would serve as wings. However, evolution does not work this way; instead, it operates on the traits an organism already has. Thus, in birds and bats, an existing pair of limbs took on new functions for flight as these organisms evolved from nonflying ancestors.

3. **Adaptations are often compromises.** Each organism must do many different things. A seal spends part of its time on rocks; it could probably walk better if it had legs instead of flippers, but then it would not swim nearly as well. We humans owe much of our versatility and athleticism to our prehensile hands and flexible limbs, but these

▲ **Figure 21.19 Evolutionary compromise.** The loud call that enables a Túngara frog to attract mates also attracts more dangerous characters in the neighborhood—in this case, a bat about to seize a meal.

also make us prone to sprains, torn ligaments, and dislocations: Structural reinforcement has been compromised for agility. **Figure 21.19** depicts another example of evolutionary compromise.

4. **Chance, natural selection, and the environment interact.** Chance events can affect the subsequent evolutionary history of populations. For instance, when a storm blows insects or birds hundreds of kilometers over an ocean to an island, the wind does not necessarily transport those individuals that are best suited to the new environment. Thus, not all alleles present in the founding population's gene pool are better suited to the new environment than the alleles that are "left behind." In addition, the environment at a particular location may change unpredictably from year to year, again limiting the extent to which adaptive evolution results in a close match between the organism and current environmental conditions.

With these four constraints, evolution does not tend to craft perfect organisms. Natural selection operates on a "better than" basis. We can, in fact, see evidence for evolution in the many imperfections of the organisms it produces.

CONCEPT CHECK 21.4

1. What is the relative fitness of a sterile mule? Explain.
2. Explain why natural selection is the only evolutionary mechanism that consistently leads to adaptive evolution.
3. **WHAT IF?** Consider a population in which heterozygotes at a certain locus have an extreme phenotype (such as being larger than homozygotes) that confers a selective advantage. Does such a situation represent directional, disruptive, or stabilizing selection? Explain your answer.
4. **WHAT IF?** Would individuals who are heterozygous for the sickle-cell allele be selected for or against in a region free from malaria? Explain.

For suggested answers, see Appendix A.

SUMMARY OF KEY CONCEPTS

CONCEPT 21.1

Genetic variation makes evolution possible (pp. 400–402)

- **Genetic variation** refers to genetic differences among individuals within a population.
- The nucleotide differences that provide the basis of genetic variation originate when mutation and gene duplication produce new alleles and new genes.
- New genetic variants are produced rapidly in organisms with short generation times. In sexually reproducing organisms, most of the genetic differences among individuals result from crossing over, the independent assortment of chromosomes, and fertilization.

> **?** *Typically, most of the nucleotide variability that occurs within a genetic locus does not affect the phenotype. Explain why.*

CONCEPT 21.2

The Hardy-Weinberg equation can be used to test whether a population is evolving (pp. 402–406)

- A **population**, a localized group of organisms belonging to one species, is united by its **gene pool**, the aggregate of all the alleles in the population.
- The **Hardy-Weinberg principle** states that the allele and genotype frequencies of a population will remain constant if the population is large, mating is random, mutation is negligible, there is no gene flow, and there is no natural selection. For such a population, if p and q represent the frequencies of the only two possible alleles at a particular locus, then p^2 is the frequency of one kind of homozygote, q^2 is the frequency of the other kind of homozygote, and $2pq$ is the frequency of the heterozygous genotype.

> **?** *Is it circular reasoning to calculate p and q from observed genotype frequencies and then use those values of p and q to test if the population is in Hardy-Weinberg equilibrium? Explain your answer. (Hint: Consider a specific case, such as a population with 195 individuals of genotype AA, 10 of genotype Aa, and 195 of genotype aa.)*

CONCEPT 21.3

Natural selection, genetic drift, and gene flow can alter allele frequencies in a population (pp. 406–410)

- In natural selection, individuals that have certain inherited traits tend to survive and reproduce at higher rates than other individuals *because of* those traits.
- In **genetic drift**, chance fluctuations in allele frequencies over generations tend to reduce genetic variation.
- **Gene flow**, the transfer of alleles between populations, tends to reduce genetic differences between populations over time.

> **?** *Would two small, geographically isolated populations in very different environments be likely to evolve in similar ways? Explain.*

CONCEPT 21.4

Natural selection is the only mechanism that consistently causes adaptive evolution (pp. 410–415)

- One organism has greater **relative fitness** than a second organism if it leaves more fertile descendants than the second organism. The modes of natural selection differ in how selection acts on phenotype (the white arrows in the summary diagram below represent selective pressure on a population).

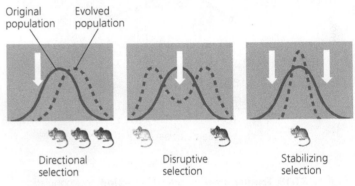

Original population   Evolved population

Directional selection   Disruptive selection   Stabilizing selection

- Unlike genetic drift and gene flow, natural selection consistently increases the frequencies of alleles that enhance survival and reproduction, thus improving the match between organisms and their environment.
- **Sexual selection** influences evolutionary change in secondary sex characteristics that can give individuals advantages in mating.
- Despite the winnowing effects of selection, populations have considerable genetic variation. Some of this variation represents **neutral variation**; additional variation can be maintained by diploidy and **balancing selection**.
- There are constraints to evolution: Natural selection can act only on available variation; structures result from modified ancestral anatomy; adaptations are often compromises; and chance, natural selection, and the environment interact.

> **?** *How might secondary sex characteristics differ between males and females in a species in which females compete for mates?*

TEST YOUR UNDERSTANDING

Level 1: Knowledge/Comprehension

1. Natural selection changes allele frequencies because some _____ survive and reproduce more successfully than others.
   - **a.** alleles
   - **b.** loci
   - **c.** gene pools
   - **d.** species
   - **e.** individuals

2. No two people are genetically identical, except for identical twins. The main source of genetic variation among human individuals is
   - **a.** new mutations that occurred in the preceding generation.
   - **b.** genetic drift due to the small size of the population.
   - **c.** the reshuffling of alleles in sexual reproduction.
   - **d.** natural selection.
   - **e.** environmental effects.

3. Sparrows with average-sized wings survive severe storms better than those with longer or shorter wings, illustrating
   - **a.** the bottleneck effect.
   - **b.** disruptive selection.
   - **c.** frequency-dependent selection.
   - **d.** neutral variation.
   - **e.** stabilizing selection.

Level 2: Application/Analysis

4. If the nucleotide variability of a locus equals 0%, what is the gene variability and number of alleles at that locus?
  a. gene variability = 0%; number of alleles = 0
  b. gene variability = 0%; number of alleles = 1
  c. gene variability = 0%; number of alleles = 2
  d. gene variability > 0%; number of alleles = 2
  e. Without more information, gene variability and number of alleles cannot be determined.

5. There are 25 individuals in population 1, all with genotype *AA*, and there are 40 individuals in population 2, all with genotype *aa*. Assume that these populations are located far from each other and that their environmental conditions are very similar. Based on the information given here, the observed genetic variation most likely resulted from
  a. genetic drift.   **d.** nonrandom mating.
  b. gene flow.   **e.** directional selection.
  c. disruptive selection.

6. A fruit fly population has a gene with two alleles, *A1* and *A2*. Tests show that 70% of the gametes produced in the population contain the *A1* allele. If the population is in Hardy-Weinberg equilibrium, what proportion of the flies carry both *A1* and *A2*?
  a. 0.7   **d.** 0.42
  b. 0.49   **e.** 0.09
  c. 0.21

Level 3: Synthesis/Evaulation

7. SCIENTIFIC INQUIRY
  DRAW IT Researchers studied genetic variation in the marine mussel *Mytilus edulis* around Long Island, New York. They measured the frequency of a particular allele (lap^{94}) for an enzyme involved in regulating the mussel's internal saltwater balance. The researchers presented their data as a series of pie charts linked to sampling sites within Long Island Sound, where the salinity is highly variable, and along the coast of the open ocean, where salinity is constant:

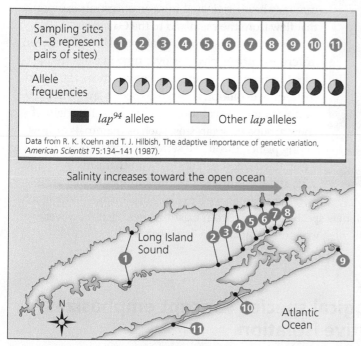

Data from R. K. Koehn and T. J. Hilbish, The adaptive importance of genetic variation, *American Scientist* 75:134–141 (1987).

Create a data table for the 11 sampling sites by estimating the frequency of lap^{94} from the pie charts. (*Hint*: Think of each pie chart as a clock face to help you estimate the proportion of the shaded area.) Then graph the frequencies for sites 1–8 to show how the frequency of this allele changes with increasing salinity in Long Island Sound (from southwest to northeast). How do the data from sites 9–11 compare with the data from the sites within the Sound?

Construct a hypothesis that explains the patterns you observe in the data and that accounts for the following observations: (1) The lap^{94} allele helps mussels maintain osmotic balance in water with a high salt concentration but is costly to use in less salty water; and (2) mussels produce larvae that can disperse long distances before they settle on rocks and grow into adults.

8. FOCUS ON EVOLUTION
  Using at least two examples, explain how the process of evolution is revealed by the imperfections of living organisms.

9. FOCUS ON ORGANIZATION
  Heterozygotes at the sickle-cell locus produce both normal and abnormal (sickle-cell) hemoglobin (see Concept 11.4). When hemoglobin molecules are packed into a heterozygote's red blood cells, some cells receive relatively large quantities of abnormal hemoglobin, making these cells prone to sickling. In a short essay (approximately 100–150 words), explain how these molecular and cellular events lead to emergent properties at the individual and population levels of biological organization.

For selected answers, see Appendix A.

MasteringBiology®

Students Go to **MasteringBiology** for assignments, the eText, and the Study Area with practice tests, animations, and activities.

Instructors Go to **MasteringBiology** for automatically graded tutorials and questions that you can assign to your students, plus Instructor Resources.

22

The Origin of Species

KEY CONCEPTS

22.1 The biological species concept emphasizes reproductive isolation

22.2 Speciation can take place with or without geographic separation

22.3 Hybrid zones reveal factors that cause reproductive isolation

22.4 Speciation can occur rapidly or slowly and can result from changes in few or many genes

OVERVIEW

That "Mystery of Mysteries"

When Darwin came to the Galápagos, he noted that these volcanic islands, despite their geologic youth, were teeming with plants and animals found nowhere else in the world **(Figure 22.1)**. Later he realized that these species had formed relatively recently. He wrote in his diary: "Both in space and time, we seem to be brought somewhat near to that great fact—that mystery of mysteries—the first appearance of new beings on this Earth."

The "mystery of mysteries" that captivated Darwin is **speciation**, the process by which one species splits into two or more species. Speciation fascinated Darwin (and many biologists since) because it leads to the tremendous diversity of life, repeatedly yielding new species that differ from existing ones. Speciation also explains the many features that organisms share (the unity of life). When a species splits, the species that result share many characteristics because they are descended from this common ancestor. At the DNA sequence level, such similarities indicate that the flightless cormorant (*Phalacrocorax harrisi*) in Figure 22.1 is closely related to flying cormorants found in the Americas. This suggests that the flightless cormorant may have originated from an ancestral cormorant that flew from the mainland to the Galápagos.

Speciation also forms a conceptual bridge between **microevolution**, changes over time in allele frequencies in a population, and **macroevolution**, the broad pattern of evolution above the species level. An example of macroevolutionary change is the origin of new groups of organisms, such as mammals or flowering plants, through a series of speciation events. We examined microevolutionary mechanisms in Chapter 21, and we'll turn to macroevolution in Chapter 23.

In this chapter, we'll explore the "bridge"—the mechanisms by which new species originate from existing ones. First, however, we need to establish what we actually mean by a "species."

▼ **Figure 22.1** How did this flightless bird come to live on the isolated Galápagos Islands?

CONCEPT 22.1

The biological species concept emphasizes reproductive isolation

The word *species* is Latin for "kind" or "appearance." In daily life, we commonly distinguish between various "kinds" of organisms—dogs and cats, for instance—from differences in their appearance. But are organisms truly divided into the discrete units we call species, or is this classification an arbitrary attempt to impose order on the natural world? To answer this question, biologists compare not only the morphology (body form) of different groups of organisms but also less obvious differences in physiology, biochemistry, and DNA sequences. The results generally confirm that morphologically distinct species are indeed discrete groups, differing in many ways besides their body forms.

The Biological Species Concept

The primary definition of species used in this textbook is the **biological species concept**. According to this concept, a **species** is a group of populations whose members have the potential to interbreed in nature and produce viable, fertile offspring—but do not produce viable, fertile offspring with members of other such groups **(Figure 22.2)**. Thus, the members of a biological species are united by being reproductively compatible, at least potentially. All human beings, for example, belong to the same species. A businesswoman in Manhattan may be unlikely to meet a dairy farmer in Mongolia, but if the two should happen to meet and mate, they could have viable babies that develop into fertile adults. In contrast, humans and chimpanzees remain distinct biological species even where they live in the same region, because many factors keep them from interbreeding and producing fertile offspring.

What holds the gene pool of a species together, causing its members to resemble each other more than they resemble members of other species? To answer this question, we need to return to the evolutionary mechanism called *gene flow*, the transfer of alleles into or out of a population (see Concept 21.3). Typically, gene flow occurs between the different populations of a species. This ongoing transfer of alleles tends to hold the populations together genetically. As we'll explore in the following sections, the absence of gene flow plays a key role in the formation of new species, as well as in keeping them apart once their potential to interbreed has been reduced.

Reproductive Isolation

Because biological species are defined in terms of reproductive compatibility, the formation of a new species hinges on **reproductive isolation**—the existence of biological barriers that impede members of two species from interbreeding and producing viable, fertile offspring. Such barriers block gene flow between the species and limit the formation of **hybrids**, offspring that result from an interspecific mating. Although a single barrier may not prevent all gene flow, a combination of several barriers can effectively isolate a species' gene pool.

Clearly, a fly cannot mate with a frog or a fern, but the reproductive barriers between more closely related species are not so obvious. These barriers can be classified according to whether they contribute to reproductive isolation before or after fertilization. **Prezygotic barriers** ("before the zygote") block fertilization from occurring. Such barriers typically act in one of three ways: by impeding members of different species

(a) Similarity between different species. The eastern meadowlark (*Sturnella magna*, left) and the western meadowlark (*Sturnella neglecta*, right) have similar body shapes and colorations. Nevertheless, they are distinct biological species because their songs and other behaviors are different enough to prevent interbreeding should they meet in the wild.

(b) Diversity within a species. As diverse as we may be in appearance, all humans belong to a single biological species (*Homo sapiens*), defined by our capacity to interbreed successfully.

▲ **Figure 22.2 The biological species concept is based on the potential to interbreed rather than on physical similarity.**

from attempting to mate, by preventing an attempted mating from being completed successfully, or by hindering fertilization if mating is completed successfully. If a sperm cell from one species overcomes prezygotic barriers and fertilizes an ovum from another species, a variety of **postzygotic barriers** ("after the zygote") may contribute to reproductive isolation after the hybrid zygote is formed. For example, developmental errors may reduce survival among hybrid embryos. Or problems after birth may cause hybrids to be infertile or may decrease their chance of surviving long enough to reproduce. **Figure 22.3** describes prezygotic and postzygotic barriers in more detail.

Prezygotic barriers impede mating or hinder fertilization if mating does occur

| Habitat Isolation | Temporal Isolation | Behavioral Isolation | Mechanical Isolation |

Individuals of different species

MATING ATTEMPT

Two species that occupy different habitats within the same area may encounter each other rarely, if at all, even though they are not isolated by obvious physical barriers, such as mountain ranges.

Species that breed during different times of the day, different seasons, or different years cannot mix their gametes.

Courtship rituals that attract mates and other behaviors unique to a species are effective reproductive barriers, even between closely related species. Such behavioral rituals enable *mate recognition*—a way to identify potential mates of the same species.

Mating is attempted, but morphological differences prevent its successful completion.

Example: Two species of garter snakes in the genus *Thamnophis* occur in the same geographic areas, but one lives mainly in water (a) while the other is primarily terrestrial (b).

Example: In North America, the geographic ranges of the eastern spotted skunk (*Spilogale putorius*) (c) and the western spotted skunk (*Spilogale gracilis*) (d) overlap, but *S. putorius* mates in late winter and *S. gracilis* mates in late summer.

Example: Blue-footed boobies, inhabitants of the Galápagos, mate only after a courtship display unique to their species. Part of the "script" calls for the male to high-step (e), a behavior that calls the female's attention to his bright blue feet.

Example: The shells of two species of snails in the genus *Bradybaena* spiral in different directions: Moving inward to the center, one spirals in a counterclockwise direction (f, left), the other in a clockwise direction (f, right). As a result, the snails' genital openings (indicated by arrows) are not aligned, and mating cannot be completed.

(a)

(b)

(c)

(d)

(e)

(f)

Gametic Isolation

Reduced Hybrid Viability

Reduced Hybrid Fertility

Hybrid Breakdown

FERTILIZATION

VIABLE, FERTILE OFFSPRING

Sperm of one species may not be able to fertilize the eggs of another species. For instance, sperm may not be able to survive in the reproductive tract of females of the other species, or biochemical mechanisms may prevent the sperm from penetrating the membrane surrounding the other species' eggs.

The genes of different parent species may interact in ways that impair the hybrid's development or survival in its environment.

Even if hybrids are vigorous, they may be sterile. If the chromosomes of the two parent species differ in number or structure, meiosis in the hybrids may fail to produce normal gametes. Since the infertile hybrids cannot produce offspring when they mate with either parent species, genes cannot flow freely between the species.

Some first-generation hybrids are viable and fertile, but when they mate with one another or with either parent species, offspring of the next generation are feeble or sterile.

Example: Gametic isolation separates certain closely related species of aquatic animals, such as sea urchins (g). Sea urchins release their sperm and eggs into the surrounding water, where they fuse and form zygotes. It is difficult for gametes of different species, such as the red and purple urchins shown here, to fuse because proteins on the surfaces of the eggs and sperm bind very poorly to each other.

Example: Some salamander subspecies of the genus *Ensatina* live in the same regions and habitats, where they may occasionally hybridize. But most of the hybrids do not complete development, and those that do are frail (h).

Example: The hybrid offspring of a male donkey (i) and a female horse (j) is a mule (k), which is robust but sterile. A "hinny" (not shown), the offspring of a female donkey and a male horse, is also sterile.

Example: Strains of cultivated rice have accumulated different mutant recessive alleles at two loci in the course of their divergence from a common ancestor. Hybrids between them are vigorous and fertile (l, left and right), but plants in the next generation that carry too many of these recessive alleles are small and sterile (l, center). Although these rice strains are not yet considered different species, they have begun to be separated by postzygotic barriers.

Limitations of the Biological Species Concept

One strength of the biological species concept is that it directs our attention to a way by which speciation can occur: by the evolution of reproductive isolation. However, the number of species to which this concept can be usefully applied is limited. There is, for example, no way to evaluate the reproductive isolation of fossils. The biological species concept also does not apply to organisms that reproduce asexually all or most of the time, such as prokaryotes. (Many prokaryotes do transfer genes among themselves, as we will discuss in Chapter 24, but this is not part of their reproductive process.) Furthermore, in the biological species concept, species are designated by the *absence* of gene flow. But there are many pairs of species that are morphologically and ecologically distinct, and yet gene flow occurs between them. An example is the grizzly bear (*Ursus arctos*) and polar bear (*Ursus maritimus*), whose hybrid offspring have been dubbed "grolar bears" **(Figure 22.4)**. As we'll discuss, natural selection can cause such species to remain distinct even though some gene flow occurs between them. This observation has led some researchers to argue that the biological species concept overemphasizes gene flow and downplays the role of natural selection. Because of the limitations to the biological species concept, alternative species concepts are useful in certain situations.

◀ Grizzly bear (*U. arctos*)

▼ Polar bear (*U. maritimus*)

▲ Hybrid "grolar bear"

▲ **Figure 22.4 Hybridization between two species of bears in the genus *Ursus*.**

Other Definitions of Species

While the biological species concept emphasizes the *separateness* of species from one another due to reproductive barriers, several other definitions emphasize the *unity within* a species. For example, the **morphological species concept** characterizes a species by body shape and other structural features. The morphological species concept can be applied to asexual and sexual organisms, and it can be useful even without information on the extent of gene flow. In practice, scientists often distinguish species using morphological criteria. A disadvantage of this approach, however, is that it relies on subjective criteria; researchers may disagree on which structural features distinguish a species.

The **ecological species concept** views a species in terms of its ecological niche, the sum of how members of the species interact with the nonliving and living parts of their environment (see Chapter 41). For example, two species of oak trees might differ in their size or in their ability to tolerate dry conditions, yet still occasionally interbreed. Because they occupy different ecological niches, these oaks would be considered two separate species even though some gene flow occurs between them. Unlike the biological species concept, the ecological species concept can accommodate asexual as well as sexual species. It also emphasizes the role of disruptive natural selection as organisms adapt to different environmental conditions.

The **phylogenetic species concept** defines a species as the smallest group of individuals that share a common ancestor, forming one branch on the tree of life. Biologists trace the phylogenetic history of a species by comparing its characteristics, such as morphology or molecular sequences, with those of other organisms. Such analyses can distinguish groups of individuals that are sufficiently different to be considered separate species. Of course, the difficulty with this species concept is determining the degree of difference required to indicate separate species.

In addition to those discussed here, more than 20 other species definitions have been proposed. The usefulness of each definition depends on the situation and the research questions being asked. For our purposes of studying how species originate, the biological species concept, with its focus on reproductive barriers, is particularly helpful.

CONCEPT CHECK 22.1

1. (a) Which species concept(s) could you apply to both asexual and sexual species? (b) Which would be most useful for identifying species in the field? Explain.
2. **WHAT IF?** Suppose you are studying two bird species that live in a forest and are not known to interbreed. One species feeds and mates in the treetops and the other on the ground. But in captivity, the birds can interbreed and produce viable, fertile offspring. What type of reproductive barrier most likely keeps these species separate in nature? Explain.

For suggested answers, see Appendix A.

Speciation can take place with or without geographic separation

Now that we have a clearer sense of what constitutes a unique species, let's return to our discussion of the process by which such species arise from existing species. Speciation can occur in two main ways, depending on how gene flow is interrupted between populations of the existing species **(Figure 22.5)**.

Allopatric ("Other Country") Speciation

In **allopatric speciation** (from the Greek *allos*, other, and *patra*, homeland), gene flow is interrupted when a population is divided into geographically isolated subpopulations. For example, the water level in a lake may subside, resulting in two or more smaller lakes that are now home to separated populations (see Figure 22.5a). Or a river may change course and divide a population of animals that cannot cross it. Allopatric speciation can also occur without geologic remodeling, such as when individuals colonize a remote area and their descendants become isolated from the parent population. The flightless cormorant in Figure 22.1 likely originated in this way from an ancestral flying species that reached the Galápagos Islands.

The Process of Allopatric Speciation

How formidable must a geographic barrier be to promote allopatric speciation? The answer depends on the ability of the organisms to move about. Birds, mountain lions, and coyotes can cross rivers and canyons—as can the windblown pollen of pine trees and the seeds of many flowering plants. In contrast, small rodents may find a wide river or deep canyon a formidable barrier.

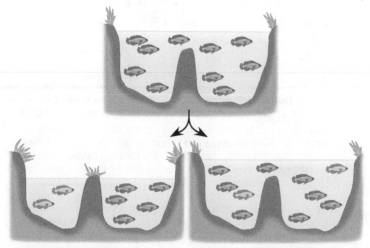

(a) Allopatric speciation. A population forms a new species while geographically isolated from its parent population.

(b) Sympatric speciation. A subset of a population forms a new species without geographic separation.

▲ **Figure 22.5 Two main modes of speciation.**

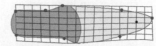

(a) Under high predation   **(b) Under low predation**

In ponds with predatory fishes, the head region of the mosquitofish is streamlined and the tail region is powerful, enabling rapid bursts of speed.

In ponds without predatory fishes, mosquitofish have a different body shape that favors long, steady swimming.

▲ **Figure 22.6 Reproductive isolation as a by-product of selection.** Bringing together mosquitofish from different ponds indicates that selection for traits that enable mosquitofish in high-predation ponds to avoid predators has isolated them reproductively from mosquitofish in low-predation ponds.

Once geographic separation has occurred, the separated gene pools may diverge. Different mutations arise, and natural selection and genetic drift may alter allele frequencies in different ways in the separated populations. Reproductive isolation may then arise as a by-product of the genetic divergence that results from selection or drift.

Let's consider an example. On Andros Island, in the Bahamas, populations of the mosquitofish *Gambusia hubbsi* colonized a series of ponds that later became isolated from one another. Genetic analyses indicate that little or no gene flow currently occurs between the ponds. The environments of these ponds are very similar except that some contain many predatory fishes, while others do not. In the "high-predation" ponds, selection has favored the evolution of a mosquitofish body shape that enables rapid bursts of speed **(Figure 22.6)**. In low-predation ponds, selection has favored a different body shape, one that improves the ability to swim for long periods of time. How have these different selective pressures affected the evolution of reproductive barriers? Researchers studied this question by bringing together mosquitofish from the two types of ponds. They found that female mosquitofish prefer to mate with males whose body shape is similar to their own. This preference establishes a behavioral barrier to reproduction between mosquitofish from high-predation and low-predation ponds. Thus, as a by-product of selection for avoiding predators, reproductive barriers have started to form in these allopatric populations.

Evidence of Allopatric Speciation

Many studies provide evidence that speciation can occur in allopatric populations. Consider the 30 species of snapping shrimp in the genus *Alpheus* that live off the Isthmus of Panama, the land bridge that connects South and North America. Fifteen of these species live on the Atlantic side of the isthmus, while the other 15 live on the Pacific side. Before the isthmus formed, gene flow could occur between the Atlantic and Pacific populations of snapping shrimp. Did the species

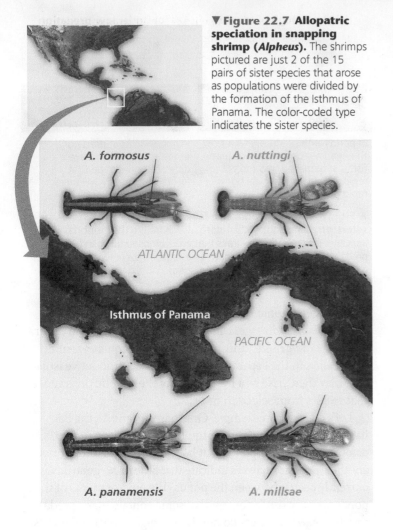

▼ **Figure 22.7 Allopatric speciation in snapping shrimp (Alpheus).** The shrimps pictured are just 2 of the 15 pairs of sister species that arose as populations were divided by the formation of the Isthmus of Panama. The color-coded type indicates the sister species.

A. formosus
A. nuttingi

ATLANTIC OCEAN

Isthmus of Panama

PACIFIC OCEAN

A. panamensis
A. millsae

on different sides of the isthmus originate by allopatric speciation? Morphological and genetic data group these shrimp into 15 pairs of *sister species*, pairs whose member species are each other's closest relative (see Figure 20.5). In each of these 15 pairs, one of the sister species lives on the Atlantic side of the isthmus, while the other lives on the Pacific side **(Figure 22.7)**, strongly suggesting that the two species arose as a consequence of geographic separation. Furthermore, genetic analyses indicate that the *Alpheus* species originated from 9 million to 3 million years ago, with the sister species that live in the deepest water diverging first. These divergence times are consistent with geologic evidence that the isthmus formed gradually, starting 10 million years ago and closing completely about 3 million years ago.

The importance of allopatric speciation is also suggested by the fact that regions that are isolated or highly subdivided by barriers typically have more species than do otherwise similar regions that lack such features. For example, many unique plants and animals are found on the geographically isolated Hawaiian Islands (we'll return to the origin of Hawaiian species in Chapter 23). Similarly, unusually high numbers of butterfly species are found in regions of South America that are subdivided by many rivers.

▼ **Figure 22.8 Inquiry**

Can divergence of allopatric populations lead to reproductive isolation?

Experiment A researcher divided a laboratory population of the fruit fly *Drosophila pseudoobscura*, raising some flies on a starch medium and others on a maltose medium. After one year (about 40 generations), natural selection resulted in divergent evolution: Populations raised on starch digested starch more efficiently, while those raised on maltose digested maltose more efficiently. The researcher then put flies from the same or different populations in mating cages and measured mating frequencies. All flies used in the mating preference tests were reared for one generation on a standard cornmeal medium.

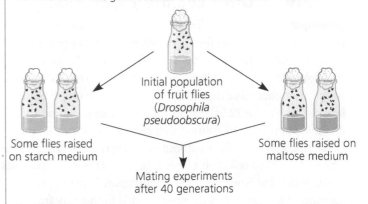

Initial population of fruit flies (*Drosophila pseudoobscura*)

Some flies raised on starch medium

Some flies raised on maltose medium

Mating experiments after 40 generations

Results Mating patterns among populations of flies raised on different media are shown below. When flies from "starch populations" were mixed with flies from "maltose populations," the flies tended to mate with like partners. But in the control group (shown on the right), flies from different populations adapted to starch were about as likely to mate with each other as with flies from their own population; similar results were obtained for control groups adapted to maltose.

		Female	
		Starch	Maltose
Male	Starch	22	9
	Maltose	8	20

Number of matings in experimental group

		Female	
		Starch population 1	Starch population 2
Male	Starch population 1	18	15
	Starch population 2	12	15

Number of matings in control group

Conclusion In the experimental group, the strong preference of "starch flies" and "maltose flies" to mate with like-adapted flies indicates that a reproductive barrier was forming between these fly populations. Although this reproductive barrier was not absolute (some mating between starch flies and maltose flies did occur), after 40 generations it appeared to be under way. This barrier may have been caused by differences in courtship behavior that arose as an incidental by-product of differing selective pressures as these allopatric populations adapted to different sources of food.

Source D. M. B. Dodd, Reproductive isolation as a consequence of adaptive divergence in *Drosophila pseudoobscura*, *Evolution* 43:1308–1311 (1989).

WHAT IF? Why were all flies used in the mating preference tests reared on a standard medium (rather than on a starch or maltose medium)?

Field observations show that reproductive isolation between two populations generally increases as the geographic distance between them increases. Researchers have also tested whether intrinsic reproductive barriers develop when populations are isolated experimentally and subjected to different environmental conditions. In such cases, too, the results provide strong support for allopatric speciation (**Figure 22.8**, on the preceding page).

We need to emphasize here that although geographic isolation prevents interbreeding between allopatric populations, physical separation is not a biological barrier to reproduction. Biological reproductive barriers such as those described in Figure 22.3 are intrinsic to the organisms themselves. Hence, it is biological barriers that can prevent interbreeding when members of different populations come into contact with one another.

Sympatric ("Same Country") Speciation

In **sympatric speciation** (from the Greek *syn*, together), speciation occurs in populations that live in the same geographic area. How can reproductive barriers form between sympatric populations while their members remain in contact with each other? Although such contact (and the ongoing gene flow that results) makes sympatric speciation less common than allopatric speciation, sympatric speciation can occur if gene flow is reduced by such factors as polyploidy, habitat differentiation, and sexual selection. (Note that these factors can also promote allopatric speciation.)

Polyploidy

A species may originate from an accident during cell division that results in extra sets of chromosomes, a condition called **polyploidy**. Polyploid speciation occasionally occurs in animals; for example, the gray tree frog *Hyla versicolor* (see Figure 21.16) is thought to have originated in this way. However, polyploidy is far more common in plants. Botanists estimate that more than 80% of the plant species alive today are descended from ancestors that formed by polyploid speciation.

Two distinct forms of polyploidy have been observed in plant (and a few animal) populations. An **autopolyploid** (from the Greek *autos*, self) is an individual that has more than two chromosome sets that are all derived from a single species. In plants, for example, a failure of cell division could double a cell's chromosome number from the diploid number (2n) to a tetraploid number (4n).

A tetraploid can produce fertile tetraploid offspring by

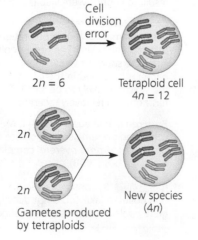

self-pollinating or by mating with other tetraploids. In addition, the tetraploids are reproductively isolated from diploid plants of the original population, because the triploid (3n) offspring of such unions have reduced fertility. Thus, in just one generation, autopolyploidy can generate reproductive isolation without any geographic separation.

A second form of polyploidy can occur when two different species interbreed and produce hybrid offspring. Most such hybrids are sterile because the set of chromosomes from one species cannot pair during meiosis with the set of chromosomes from the other species. However, an infertile hybrid may be able to propagate itself asexually (as many plants can do). In subsequent generations, various mechanisms can change a sterile hybrid into a fertile polyploid called an **allopolyploid** (**Figure 22.9**). The allopolyploids are fertile when mating with

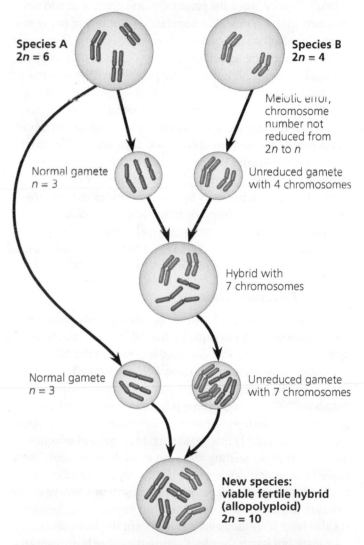

▲ **Figure 22.9 One mechanism for allopolyploid speciation in plants.** Most hybrids are sterile because their chromosomes are not homologous and cannot pair during meiosis. However, such a hybrid may be able to reproduce asexually. This diagram traces one mechanism that can produce fertile hybrids (allopolyploids) as new species. The new species has a diploid chromosome number equal to the sum of the diploid chromosome numbers of the two parent species.

each other but cannot interbreed with either parent species; thus, they represent a new biological species.

Although polyploid speciation is relatively rare, even in plants, scientists have documented that at least five new plant species have originated in this way since 1850. One of these examples involves the origin of a new species of goatsbeard plant (genus *Tragopogon*) in the Pacific Northwest. *Tragopogon* first arrived in the region when humans introduced three European species in the early 1900s. These three species are now common weeds in abandoned parking lots and other urban sites. In 1950, a new *Tragopogon* species was discovered near the Idaho-Washington border, a region where all three European species also were found. Genetic analyses revealed that this new species, *Tragopogon miscellus*, is a tetraploid hybrid of two of the European species. Although the *T. miscellus* population grows mainly by reproduction of its own members, additional episodes of hybridization between the parent species continue to add new members to the *T. miscellus* population—just one of many examples in which scientists have observed speciation in progress.

Many important agricultural crops—such as oats, cotton, potatoes, tobacco, and wheat—are polyploids. The wheat used for bread, *Triticum aestivum*, is an allohexaploid (six sets of chromosomes, two sets from each of three different species). The first of the polyploidy events that eventually led to modern wheat probably occurred about 8,000 years ago in the Middle East as a spontaneous hybrid of an early cultivated wheat species and a wild grass. Today, plant geneticists generate new polyploids in the laboratory by using chemicals that induce meiotic and mitotic errors. By harnessing the evolutionary process, researchers can produce new hybrid species with desired qualities, such as a hybrid that combines the high yield of wheat with the hardiness of rye.

Habitat Differentiation

Sympatric speciation can also occur when genetic factors enable a subpopulation to exploit a habitat or resource not used by the parent population. Such is the case with the North American apple maggot fly (*Rhagoletis pomonella*), a pest of apples. The fly's original habitat was the native hawthorn tree, but about 200 years ago, some populations colonized apple trees that had been introduced by European settlers. As apples mature more quickly than hawthorn fruit, natural selection has favored apple-feeding flies with rapid development. These apple-feeding populations now show temporal isolation from the hawthorn-feeding *R. pomonella*, providing a prezygotic restriction to gene flow between the two populations. Researchers also have identified alleles that benefit the flies that use one host plant but harm the flies that use the other host plant. As a result, natural selection operating on these alleles provides a postzygotic barrier to reproduction, further limiting gene flow. Altogether, although the two populations are still classified as subspecies rather than separate species, sympatric speciation appears to be well under way.

Sexual Selection

There is evidence that sympatric speciation can also be driven by sexual selection. Clues to how this can occur have been found in cichlid fishes from one of Earth's hot spots of animal speciation, East Africa's Lake Victoria. This lake was once home to as many as 600 species of cichlids. Genetic data indicate that these species originated within the last 100,000 years from a small number of colonizing species that arrived from rivers and lakes located elsewhere. How did so many species—more than double the number of freshwater fish species known in all of Europe—originate within a single lake?

One hypothesis is that subgroups of the original cichlid populations adapted to different food sources and that the resulting genetic divergence contributed to speciation in Lake Victoria.

▼ **Figure 22.10** **Inquiry**

Does sexual selection in cichlids result in reproductive isolation?

Experiment Researchers placed males and females of *Pundamilia pundamilia* and *P. nyererei* together in two aquarium tanks, one with natural light and one with a monochromatic orange lamp. Under normal light, the two species are noticeably different in male breeding coloration; under monochromatic orange light, the two species are very similar in color. The researchers then observed the mate choices of the females in each tank.

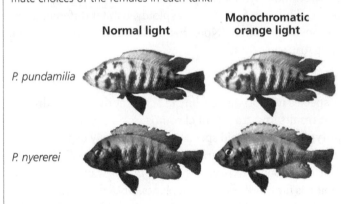

Results Under normal light, females of each species strongly preferred males of their own species. But under orange light, females of each species responded indiscriminately to males of both species. The resulting hybrids were viable and fertile.

Conclusion The researchers concluded that mate choice by females based on male breeding coloration is the main reproductive barrier that normally keeps the gene pools of these two species separate. Since the species can still interbreed when this prezygotic behavioral barrier is breached in the laboratory, the genetic divergence between the species is likely to be small. This suggests that speciation in nature has occurred relatively recently.

Source O. Seehausen and J. J. M. van Alphen, The effect of male coloration on female mate choice in closely related Lake Victoria cichlids (*Haplochromis nyererei* complex), *Behavioral Ecology and Sociobiology* 42:1–8 (1998).

WHAT IF? Suppose that female cichlids living in the murky waters of a polluted lake could not distinguish colors well. How might the gene pools of these species change over time?

But sexual selection, in which (typically) females select males based on their appearance (see Chapter 21), may also have been a factor. Researchers have studied two closely related sympatric species of cichlids that differ mainly in the coloration of breeding males: Breeding *Pundamilia pundamilia* males have a blue-tinged back, whereas breeding *Pundamilia nyererei* males have a red-tinged back (**Figure 22.10**, on the preceding page). Their results suggest that mate choice based on male breeding coloration is the main reproductive barrier that normally keeps the gene pools of these two species separate.

Allopatric and Sympatric Speciation: *A Review*

Now let's recap the two main modes by which new species form. In allopatric speciation, a new species forms in geographic isolation from its parent population. Geographic isolation severely restricts gene flow. As a result, other reproductive barriers from the ancestral species may arise as a by-product of genetic changes that occur within the isolated population. Many different processes can produce such genetic changes, including natural selection under different environmental conditions, genetic drift, and sexual selection. Once formed, intrinsic reproductive barriers that arise in allopatric populations can prevent interbreeding with the parent population even if the populations come back into contact. In the **Scientific Skills Exercise**, you will interpret data from a study of reproductive isolation in geographically separated salamander populations.

Sympatric speciation, in contrast, requires the emergence of a reproductive barrier that isolates a subset of a population from the remainder of the population in the same area. Though rarer than allopatric speciation, sympatric speciation can occur when gene flow to and from the isolated subpopulation is blocked. This can occur as a result of polyploidy, a condition in which an organism has extra sets of chromosomes.

Identifying Independent and Dependent Variables, Making a Scatter Plot, and Interpreting Data

Does Distance Between Salamander Populations Increase Their Reproductive Isolation? The process of allopatric speciation begins when populations become geographically isolated, preventing mating between individuals in different populations and thus stopping gene flow. It seems logical that as distance between populations increases, so will their degree of reproductive isolation. To test this hypothesis, researchers studied populations of the dusky salamander (*Desmognathus ochrophaeus*) living on different mountain ranges in the southern Appalachian Mountains.

How the Experiment Was Done The researchers tested the reproductive isolation of pairs of salamander populations by leaving one male and one female together and later checking the females for the presence of sperm. Four mating combinations were tested for each pair of populations (A and B)—two *within* the same population (female A with male A and female B with male B) and two *between* populations (female A with male B and female B with male A).

Data from the Experiment The researchers used an index of reproductive isolation that ranged from a value of 0 (no isolation) to a value of 2 (full isolation). The proportion of successful matings for each mating combination was measured, with 100% success = 1 and no success = 0. The reproductive isolation value for two populations is the sum of the proportion of successful matings of each type within populations (AA + BB) minus the sum of the proportion of successful matings of each type between populations (AB + BA). The following table provides data for 27 pairs of dusky salamander populations:

Interpret the Data

1. State the researchers' hypothesis, and identify the independent and dependent variables in this study. Explain why the researchers used four mating combinations for each pair of populations.

2. Calculate the value of the reproductive isolation index if (a) *all* of the matings within a population were successful, but *none* of the matings between populations were successful; (b) salamanders are equally successful in mating with members of their own population and members of another population.

3. Make a scatter plot of one variable against the other to help you visualize whether or not there is a relationship between the variables. (For additional information about graphs, see the Scientific Skills Review in Appendix F and in the Study Area in MasteringBiology.) Plot the dependent variable on the *y*-axis and the independent variable on the *x*-axis.

4. Interpret your graph by (a) explaining in words the relationship between the variables that can be visualized by graphing the data and (b) hypothesizing the possible cause of this relationship.

Data from S. G. Tilley, P. A. Verrell, and S. J. Arnold, Correspondence between sexual isolation and allozyme differentiation: A test in the salamander *Desmognathus ochrophaeus, Proceedings of the National Academy of Sciences USA.* 87:2715–2719 (1990).

MB A version of this Scientific Skills Exercise can be assigned in MasteringBiology.

Geographic Distance (km)	15	32	40	47	42	62	63	81	86	107	107	115	137	147
Reproductive Isolation Value	0.32	0.54	0.50	0.50	0.82	0.37	0.67	0.53	1.15	0.73	0.82	0.81	0.87	0.87
Distance (continued)	137	150	165	189	219	239	247	53	55	62	105	179	169	
Isolation (continued)	0.50	0.57	0.91	0.93	1.50	1.22	0.82	0.99	0.21	0.56	0.41	0.72	1.15	

Sympatric speciation also can occur when a subset of a population becomes reproductively isolated because of natural selection that results from a switch to a habitat or food source not used by the parent population. Finally, sympatric speciation can result from sexual selection.

Having reviewed the geographic context in which species originate, we'll next explore in more detail what can happen when new or partially formed species come into contact.

CONCEPT CHECK 22.2

1. Summarize key differences between allopatric and sympatric speciation. Which type of speciation is more common, and why?

2. Describe two mechanisms that can decrease gene flow in sympatric populations, thereby making sympatric speciation more likely to occur.

3. **WHAT IF?** Is allopatric speciation more likely to occur on an island close to a mainland or on a more isolated island of the same size? Explain your prediction.

4. **MAKE CONNECTIONS** Review meiosis in Figure 10.8. Describe how an error during meiosis could lead to polyploidy.

For suggested answers, see Appendix A.

Hybrid zones reveal factors that cause reproductive isolation

What happens if species with incomplete reproductive barriers come into contact with one another? One possible outcome is the formation of a **hybrid zone**, a region in which members of different species meet and mate, producing at least some offspring of mixed ancestry. In this section, we'll explore hybrid zones and what they reveal about factors that cause the evolution of reproductive isolation.

Patterns Within Hybrid Zones

Some hybrid zones form as narrow bands, such as the one depicted in **Figure 22.11** for two species of toads in the genus *Bombina*, the yellow-bellied toad (*B. variegata*) and the fire-bellied toad (*B. bombina*). This hybrid zone, represented by the red line on the map, extends for 4,000 km but is less than 10 km wide in most places. The hybrid zone occurs where the higher-altitude habitat of the yellow-bellied toad meets the

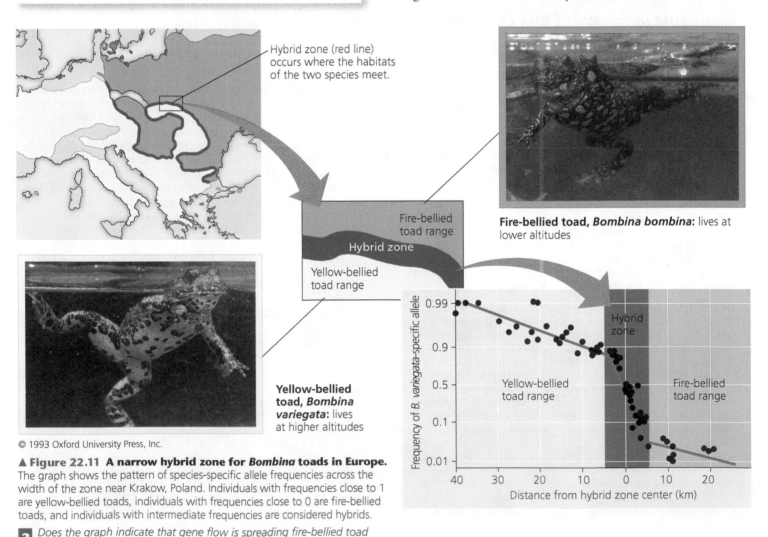

Hybrid zone (red line) occurs where the habitats of the two species meet.

Fire-bellied toad, *Bombina bombina*: lives at lower altitudes

Fire-bellied toad range

Hybrid zone

Yellow-bellied toad range

Yellow-bellied toad, *Bombina variegata*: lives at higher altitudes

© 1993 Oxford University Press, Inc.

▲ **Figure 22.11 A narrow hybrid zone for *Bombina* toads in Europe.** The graph shows the pattern of species-specific allele frequencies across the width of the zone near Krakow, Poland. Individuals with frequencies close to 1 are yellow-bellied toads, individuals with frequencies close to 0 are fire-bellied toads, and individuals with intermediate frequencies are considered hybrids.

? *Does the graph indicate that gene flow is spreading fire-bellied toad alleles into the range of the yellow-bellied toad? Explain.*

lowland habitat of the fire-bellied toad. Across a given "slice" of the zone, the frequency of alleles specific to yellow-bellied toads typically decreases from about 90% at the edge where only yellow-bellied toads are found, to 50% in the central portion of the zone, to less than 10% at the edge where only fire-bellied toads are found.

What causes such a pattern of allele frequencies across a hybrid zone? We can infer that there is an obstacle to gene flow—otherwise, alleles from one parent species would also be common in the gene pool of the other parent species. Are geographic barriers reducing gene flow? Not in this case, since the toads can move throughout the hybrid zone. A more important factor is that hybrid toads have increased rates of embryonic mortality and a variety of morphological abnormalities, including ribs that are fused to the spine and malformed tadpole mouthparts. Because the hybrids have poor survival and reproduction, they produce few viable offspring with members of the parent species. As a result, hybrid individuals rarely serve as a stepping-stone from which alleles are passed from one species to the other. Outside the hybrid zone, additional obstacles to gene flow may be provided by natural selection in the different environments in which the parent species live.

Hybrid zones typically are located wherever the habitats of the interbreeding species meet. Those regions often resemble a group of isolated patches scattered across the landscape—more like the complex pattern of spots on a Dalmatian than the continuous band shown in Figure 22.11. But regardless of whether they have complex or simple spatial patterns, hybrid zones form when two species lacking complete barriers to reproduction come into contact. Once formed, how does a hybrid zone change over time?

Hybrid Zones over Time

Studying a hybrid zone is like observing a naturally occurring experiment on speciation. Will the hybrids become reproductively isolated from their parents and form a new species, as occurred by polyploidy in the goatsbeard plant of the Pacific Northwest? If not, there are three possible outcomes for the hybrid zone over time: reinforcement of barriers, fusion of species, or stability (**Figure 22.12**). We'll discuss each of these outcomes in turn.

- **Reinforcement:** When hybrids are less fit than members of their parent species, natural selection tends to strengthen prezygotic barriers to reproduction, thus reducing the formation of unfit hybrids. Because this process involves *reinforcing* reproductive barriers, it is called reinforcement. If reinforcement is occurring, a logical prediction is that barriers to reproduction between species should be stronger for sympatric populations than for allopatric populations. Evidence in support of this prediction has been observed in birds, fishes, insects, plants, and other organisms.

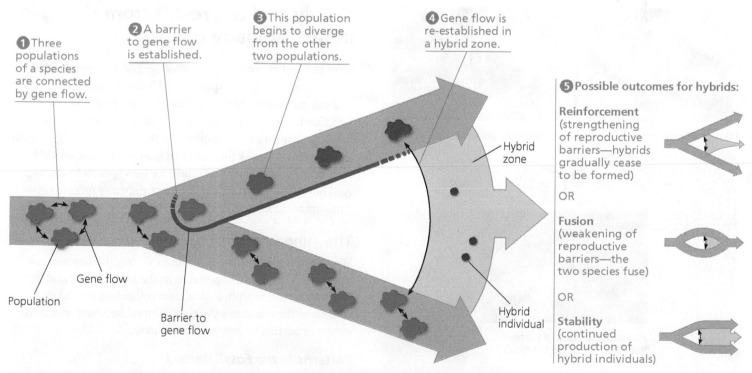

1 Three populations of a species are connected by gene flow.

2 A barrier to gene flow is established.

3 This population begins to diverge from the other two populations.

4 Gene flow is re-established in a hybrid zone.

Gene flow

Population

Barrier to gene flow

Hybrid zone

Hybrid individual

5 Possible outcomes for hybrids:

Reinforcement (strengthening of reproductive barriers—hybrids gradually cease to be formed)

OR

Fusion (weakening of reproductive barriers—the two species fuse)

OR

Stability (continued production of hybrid individuals)

▲ **Figure 22.12 Formation of a hybrid zone and possible outcomes for hybrids over time.** The thick colored arrows represent the passage of time.

WHAT IF? *Predict what might happen if gene flow were re-established at step 3 in this process.*

- **Fusion:** Barriers to reproduction may be weak when two species meet in a hybrid zone. Indeed, so much gene flow may occur that reproductive barriers weaken further and the gene pools of the two species become increasingly alike. In effect, the speciation process reverses, eventually causing the two hybridizing species to fuse into a single species. Such a situation may be occurring among Lake Victoria cichlids. Many pairs of ecologically similar cichlid species are reproductively isolated because the females of one species prefer to mate with males of one color, while females of the other species prefer to mate with males of a different color (see Figure 22.10). Murky waters caused by pollution may have reduced the ability of females to use color to distinguish males of their own species from males of closely related species. In some polluted waters, many hybrids have been produced, leading to fusion of the parent species' gene pools and a loss of species **(Figure 22.13)**.

- **Stability:** Many hybrid zones are stable in the sense that hybrids continue to be produced. In some cases, this occurs because the hybrids survive or reproduce better than members of either parent species, at least in certain habitats or years. But stable hybrid zones have also been observed in cases where the hybrids are selected *against*—an unexpected result. For example, hybrids continue to be formed in the *Bombina* hybrid zone even though they are strongly selected against. What could explain this finding? One possibility relates to the narrowness of the *Bombina* hybrid

Pundamilia nyererei

Pundamilia pundamilia

Pundamilia "turbid water," hybrid offspring from a location with turbid water

▲ **Figure 22.13 Fusion: The breakdown of reproductive barriers.** Increasingly cloudy water in Lake Victoria over the past 30 years may have weakened reproductive barriers between *P. nyererei* and *P. pundamilia*. In areas of cloudy water, the two species have hybridized extensively, causing their gene pools to fuse.

zone (see Figure 22.11). Evidence suggests that members of both parent species migrate into the zone from the parent populations located outside the zone, thus leading to the continued production of hybrids. If the hybrid zone were wider, this would be less likely to occur, since the center of the zone would receive little gene flow from distant parent populations located outside the hybrid zone.

As we've seen, events in hybrid zones can shed light on how barriers to reproduction between closely related species change over time. In the next section, we'll examine how interactions between hybridizing species can also provide a glimpse into the speed and genetic control of speciation.

CONCEPT CHECK 22.3

1. What are hybrid zones, and why can they be viewed as "natural laboratories" in which to study speciation?
2. **WHAT IF?** Consider two species that diverged while geographically separated but resumed contact before reproductive isolation was complete. Predict what would happen over time if the two species mated indiscriminately and (a) hybrid offspring survived and reproduced more poorly than offspring from intraspecific matings or (b) hybrid offspring survived and reproduced as well as offspring from intraspecific matings.

For suggested answers, see Appendix A.

CONCEPT 22.4

Speciation can occur rapidly or slowly and can result from changes in few or many genes

Darwin faced many questions when he began to ponder that "mystery of mysteries"—speciation. He found answers to some of those questions when he realized that evolution by natural selection helps explain both the diversity of life and the adaptations of organisms (see Chapter 19). But biologists since Darwin have continued to ask fundamental questions about speciation. For example, how long does it take for new species to form? And how many genes change when one species splits into two? Answers to these questions are also beginning to emerge.

The Time Course of Speciation

We can gather information about how long it takes new species to form from broad patterns in the fossil record and from studies that use morphological data (including fossils) or molecular data to assess the time interval between speciation events in particular groups of organisms.

Patterns in the Fossil Record

The fossil record includes many episodes in which new species appear suddenly in a geologic stratum, persist essentially unchanged through several strata, and then disappear. For example, there are dozens of species of marine invertebrates that make

(a) In a punctuated model, new species change most as they branch from a parent species and then change little for the rest of their existence.

Time

(b) In a gradual model, species diverge from one another more slowly and steadily over time.

▲ **Figure 22.14 Two models for the tempo of speciation, based on patterns observed in the fossil record.**

their debut in the fossil record with novel morphologies, but then change little for millions of years before becoming extinct. Paleontologists Niles Eldredge and Stephen Jay Gould coined the term **punctuated equilibria** to describe these patterns in the fossil record: periods of apparent stasis punctuated by sudden change **(Figure 22.14a)**. Other species do not show a punctuated pattern; instead, they appear to have changed more gradually over long periods of time **(Figure 22.14b)**. For example, the fossil record shows that many species of trilobites (early arthropods) changed gradually over the course of 10–20 million years.

What might punctuated and gradual patterns tell us about how long it takes new species to form? Suppose that a species survived for 5 million years, but most of the morphological changes that caused it to be designated a new species occurred during the first 50,000 years of its existence—just 1% of its total lifetime. Time periods this short (in geologic terms) often cannot be distinguished in fossil strata, in part because the rate of sediment accumulation may be too slow to separate layers formed so close together in time. Thus, based on its fossils, the species would seem to have appeared suddenly and then lingered with little or no change before becoming extinct. Even though such a species may have originated more slowly than its fossils suggest (in this case taking up to 50,000 years), a punctuated pattern indicates that speciation occurred relatively rapidly. For species whose fossils changed much more gradually, we also cannot tell exactly when a new biological species formed, since information about reproductive isolation does not fossilize. However, it is likely that speciation in such groups occurred relatively slowly, perhaps taking millions of years.

Speciation Rates

The existence of fossils that display a punctuated pattern suggests that once the process of speciation begins, it can be completed relatively rapidly—a suggestion supported by recent studies. For example, rapid speciation appears to have produced the wild sunflower *Helianthus anomalus*. Genetic evidence indicates that this species originated by the hybridization of two other sunflower species, *H. annuus* and *H. petiolaris*. The hybrid species *H. anomalus* is ecologically distinct and reproductively isolated from both parent species **(Figure 22.15)**. Unlike the outcome of allopolyploid speciation, in which there is a change in chromosome number after hybridization, in these sunflowers the two parent species and the hybrid all have the same number of chromosomes ($2n = 34$). How, then, did speciation occur? To study this question, researchers performed an experiment designed to mimic events in nature: They crossed the

▲ **Figure 22.15 A hybrid sunflower species and its dry sand dune habitat.** The wild sunflower *Helianthus anomalus* originated via the hybridization of two other sunflowers, *H. annuus* and *H. petiolaris*, which live in nearby but moister environments.

two parent species and followed the fate of the hybrid offspring over several generations (**Figure 22.16**). Their results indicated that natural selection could produce extensive genetic changes in hybrid populations over short periods of time. These changes appear to have caused the hybrids to diverge reproductively from their parents and form a new species, *H. anomalus*.

The sunflower example, along with the apple maggot fly, Lake Victoria cichlid, and fruit fly examples discussed earlier, suggests that new species can arise rapidly *once divergence begins*. But what is the total length of time between speciation events? This interval consists of the time that elapses before populations of a newly formed species start to diverge from one another plus the time it takes for speciation to be complete once divergence begins. It turns out that the total time between speciation events varies considerably. For example, in a survey of data from 84 groups of plants and animals, the interval between speciation events ranged from 4,000 years (in cichlids of Lake Nabugabo, Uganda) to 40 million years (in some beetles). Overall, the time between speciation events in the groups studied averaged 6.5 million years and was rarely less than 500,000 years.

These data suggest that on average, millions of years may pass before a newly formed plant or animal species will itself give rise to another new species. As we'll see in Chapter 23, this finding has implications for how long it takes life on Earth to recover from mass extinction events. Moreover, the extreme variability in the time it takes new species to form indicates that organisms do not have a "speciation clock" ticking inside them, causing them to produce new species at regular time intervals. Instead, speciation begins only after gene flow between populations is interrupted, perhaps by changing environmental conditions or by unpredictable events, such as a storm that transports a few individuals to an isolated area. Furthermore, once gene flow is interrupted, the populations must diverge genetically to such an extent that they become reproductively isolated—all before other events cause gene flow to resume, possibly reversing the speciation process (see Figure 22.13).

Studying the Genetics of Speciation

The central quest of studying the genetics of speciation is to identify genes that cause reproductive isolation. In general, genes that influence a particular trait can be identified by performing genetic crosses and analyzing gene linkages—but such studies are by definition hard to do when studying different species (since they do not interbreed). However, studies of ongoing speciation (as in hybrid zones) have uncovered specific traits that cause reproductive isolation. By identifying the genes that control those traits, scientists can explore a fundamental question of evolutionary biology: How many genes change when a new species forms?

In a few cases, the evolution of reproductive isolation is due to a change in a single gene. For example, in Japanese snails of the genus *Euhadra*, a change in a single gene results in a

How does hybridization lead to speciation in sunflowers?

Experiment Researchers crossed the two parent sunflower species, *H. annuus* and *H. petiolaris*, to produce experimental hybrids in the laboratory (for each gamete, only two of the *n* = 17 chromosomes are shown).

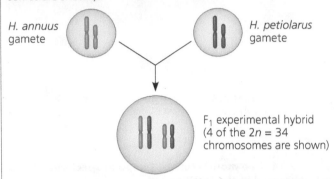

H. annuus gamete

H. petiolarus gamete

F₁ experimental hybrid (4 of the 2*n* = 34 chromosomes are shown)

Note that in the first (F₁) generation, each chromosome of the experimental hybrids consisted entirely of DNA from one or the other parent species. The researchers then tested whether the F₁ and subsequent generations of experimental hybrids were fertile. They also used species-specific genetic markers to compare the chromosomes in the experimental hybrids with the chromosomes in the naturally occurring hybrid *H. anomalus*.

Results Although only 5% of the F₁ experimental hybrids were fertile, after just four more generations the hybrid fertility rose to more than 90%. The chromosomes of individuals from this fifth hybrid generation differed from those in the F₁ generation (see above) but were similar to those in *H. anomalus* individuals from natural populations:

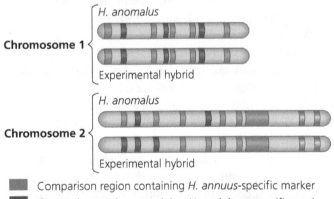

Chromosome 1

H. anomalus

Experimental hybrid

Chromosome 2

H. anomalus

Experimental hybrid

■ Comparison region containing *H. annuus*-specific marker
■ Comparison region containing *H. petiolarus*-specific marker
© 1996 AAAS

Conclusion Over time, the chromosomes in the population of experimental hybrids became similar to the chromosomes of *H. anomalus* individuals from natural populations. This suggests that the observed rise in the fertility of the experimental hybrids may have occurred as selection eliminated regions of DNA from the parent species that were not compatible with one another. Overall, it appeared that the initial steps of the speciation process occurred rapidly and could be mimicked in a laboratory experiment.

Source L. H. Rieseberg et al., Role of gene interactions in hybrid speciation: Evidence from ancient and experimental hybrids, *Science* 272:741–745 (1996).

WHAT IF? The increased fertility of the experimental hybrids could have resulted from natural selection for thriving under laboratory conditions. Evaluate this alternative explanation for the result.

mechanical barrier to reproduction. This gene controls the direction in which the shells spiral. When their shells spiral in different directions, the snails' genitalia are oriented in a manner that prevents mating (Figure 22.3f shows a similar example in a different genus of snail).

A major barrier to reproduction between two closely related species of monkey flower, *Mimulus cardinalis* and *M. lewisii*, also appears to be influenced by a relatively small number of genes. These two species are isolated by several prezygotic and postzygotic barriers. Of these, one prezygotic barrier, pollinator choice, accounts for most of the isolation: In a hybrid zone between *M. cardinalis* and *M. lewisii*, nearly 98% of pollinator visits were restricted to one species or the other.

The two monkey flower species are visited by different pollinators: Hummingbirds prefer the red-flowered *M. cardinalis*, and bumblebees prefer the pink-flowered *M. lewisii*. Pollinator choice is affected by at least two loci in the monkey flowers, one of which, the "yellow upper," or *yup*, locus, influences flower color **(Figure 22.17)**. By crossing the two parent species to produce F_1 hybrids and then performing repeated backcrosses of these F_1 hybrids to each parent species, researchers succeeded in transferring the *M. cardinalis* allele at this locus into *M. lewisii*, and vice versa. In a field experiment, *M. lewisii* plants with the *M. cardinalis yup* allele received 68-fold more visits from hummingbirds than did wild-type *M. lewisii*. Similarly, *M. cardinalis* plants with the *M. lewisii yup* allele received 74-fold more visits from bumblebees than did wild-type *M. cardinalis*. Thus, a mutation at a single locus can influence pollinator preference and hence contribute to reproductive isolation in monkey flowers.

In other organisms, the speciation process is influenced by larger numbers of genes and gene interactions. For example, hybrid sterility between two subspecies of the fruit fly *Drosophila pseudoobscura* results from gene interactions among at least four loci, and postzygotic isolation in the sunflower hybrid zone discussed earlier is influenced by at least 26 chromosome segments (and an unknown number of genes). Overall, studies suggest that few or many genes can influence the evolution of reproductive isolation and hence the emergence of a new species.

From Speciation to Macroevolution

As you've seen, speciation may begin with differences as seemingly small as the color on a cichlid's back. However, as speciation occurs again and again, such differences can accumulate and become more pronounced, eventually leading to the formation of new groups of organisms that differ greatly from their ancestors (as in the origin of whales from land-dwelling mammals; see Figure 19.20). Furthermore, as one group of organisms increases in size by producing many new species, another group of organisms may shrink, losing species to extinction. The cumulative effects of many such speciation and extinction events have helped shape the sweeping evolutionary

(a) Typical *Mimulus lewisii* | **(b) *M. lewisii* with an *M. cardinalis* flower-color allele**

(c) Typical *Mimulus cardinalis* | **(d) *M. cardinalis* with an *M. lewisii* flower-color allele**

▲ **Figure 22.17 A locus that influences pollinator choice.** Pollinator preferences provide a strong barrier to reproduction between *Mimulus lewisii* and *M. cardinalis*. After transferring the *M. lewisii* allele for a flower-color locus into *M. cardinalis* and vice versa, researchers observed a shift in some pollinators' preferences.

WHAT IF? *If* M. cardinalis *individuals that had the* M. lewisii yup *allele were planted in an area that housed both monkey flower species, how might the production of hybrid offspring be affected?*

changes that are documented in the fossil record. In the next chapter, we turn to such large-scale evolutionary changes as we begin our study of macroevolution.

CONCEPT CHECK 22.4

1. Speciation can occur rapidly between diverging populations, yet the length of time between speciation events is often more than a million years. Explain this apparent contradiction.
2. Summarize evidence that the *yup* locus acts as a prezygotic barrier to reproduction in two species of monkey flowers. Do these results demonstrate that the *yup* locus alone controls barriers to reproduction between these species? Explain.
3. **MAKE CONNECTIONS** Compare Figure 10.11 with Figure 22.16. What cellular process could cause the hybrid chromosomes in Figure 22.16 to contain DNA from both parent species? Explain.

For suggested answers, see Appendix A.

22 Chapter Review

SUMMARY OF KEY CONCEPTS

CONCEPT 22.1

The biological species concept emphasizes reproductive isolation (pp. 418–422)

- A biological **species** is a group of populations whose individuals have the potential to interbreed and produce viable, fertile offspring with each other but not with members of other species. The **biological species concept** emphasizes reproductive isolation through prezygotic and postzygotic barriers that separate gene pools.
- Although helpful in thinking about how speciation occurs, the biological species concept has limitations. For instance, it cannot be applied to organisms known only as fossils or to organisms that reproduce only asexually. Thus, scientists use other species concepts, such as the **morphological species concept**, in certain circumstances.

  ? *Explain the importance of gene flow to the biological species concept.*

CONCEPT 22.2

Speciation can take place with or without geographic separation (pp. 423–428)

- In **allopatric speciation**, gene flow is reduced when two populations of one species become geographically separated from each other. One or both populations may undergo evolutionary change during the period of separation, resulting in the establishment of prezygotic or postzygotic barriers to reproduction.
- In **sympatric speciation**, a new species originates while remaining in the same geographic area as the parent species. Plant species (and, more rarely, animal species) have evolved sympatrically through polyploidy. Sympatric speciation can also result from habitat shifts and sexual selection.

Original population

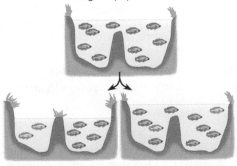

Allopatric speciation   Sympatric speciation

  ? *Can factors that cause sympatric speciation also cause allopatric speciation? Explain.*

CONCEPT 22.3

Hybrid zones reveal factors that cause reproductive isolation (pp. 428–430)

- Many groups of organisms form **hybrid zones** in which members of different species meet and mate, producing at least some offspring of mixed ancestry.

- Many hybrid zones exhibit **stability** in that hybrid offspring continue to be produced over time. In others, **reinforcement** strengthens prezygotic barriers to reproduction, thus decreasing the formation of unfit hybrids. In still other hybrid zones, barriers to reproduction may weaken over time, resulting in the **fusion** of the species' gene pools (reversing the speciation process).

  ? *What factors can support the long-term stability of a hybrid zone if the parent species live in different environments?*

CONCEPT 22.4

Speciation can occur rapidly or slowly and can result from changes in few or many genes (pp. 430–433)

- New species can form rapidly once divergence begins—but it can take millions of years for that to happen. The time interval between speciation events varies considerably, from a few thousand years to tens of millions of years.
- New developments in genetics have enabled researchers to identify specific genes involved in some cases of speciation. Results show that speciation can be driven by few or many genes.

  ? *Is speciation something that happened only in the distant past, or are new species continuing to arise today? Explain.*

TEST YOUR UNDERSTANDING

Level 1: Knowledge/Comprehension

1. The *largest* unit within which gene flow can readily occur is a
   a. population.
   b. species.
   c. genus.
   d. hybrid.
   e. phylum.

2. Males of different species of the fruit fly *Drosophila* that live in the same parts of the Hawaiian Islands have different elaborate courtship rituals. These rituals involve fighting other males and making stylized movements that attract females. What type of reproductive isolation does this represent?
   a. habitat isolation
   b. temporal isolation
   c. behavioral isolation
   d. gametic isolation
   e. postzygotic barriers

3. According to the punctuated equilibria model,
   a. natural selection is unimportant as a mechanism of evolution.
   b. given enough time, most existing species will branch gradually into new species.
   c. most new species accumulate their unique features relatively rapidly as they come into existence, then change little for the rest of their duration as a species.
   d. most evolution occurs in sympatric populations.
   e. speciation is usually due to a single mutation.

Level 2: Application/Analysis

4. Bird guides once listed the myrtle warbler and Audubon's warbler as distinct species. Recently, these birds have been reclassified as eastern and western forms of a single species, the

yellow-rumped warbler. Which of the following pieces of evidence, if true, would be cause for this reclassification?

a. The two forms interbreed often in nature, and their offspring survive and reproduce well.
b. The two forms live in similar habitats.
c. The two forms have many genes in common.
d. The two forms have similar food requirements.
e. The two forms are very similar in coloration.

5. Which of the following factors would *not* contribute to allopatric speciation?

a. A population becomes geographically isolated from the parent population.
b. The separated population is small, and genetic drift occurs.
c. The isolated population is exposed to different selection pressures than the ancestral population.
d. Different mutations begin to distinguish the gene pools of the separated populations.
e. Gene flow between the two populations is extensive.

6. Plant species A has a diploid number of 12. Plant species B has a diploid number of 16. A new species, C, arises as an allopolyploid from A and B. The diploid number for species C would probably be

a. 12.
b. 14.
c. 16.
d. 28.
e. 56.

Level 3: Synthesis/Evaluation

7. SCIENTIFIC INQUIRY

DRAW IT In this chapter, you read that bread wheat (*Triticum aestivum*) is an allohexaploid, containing two sets of chromosomes from each of three different parent species. Genetic analysis suggests that the three species pictured following this question each contributed chromosome sets to *T. aestivum*. (The capital letters here represent sets of chromosomes rather than individual genes.) Evidence also indicates that the first polyploidy event was a spontaneous hybridization of the early cultivated wheat species *T. monococcum* and a wild *Triticum* grass species. Based on this information, draw a diagram of one possible chain of events that could have produced the allohexaploid *T. aestivum*.

Ancestral species:

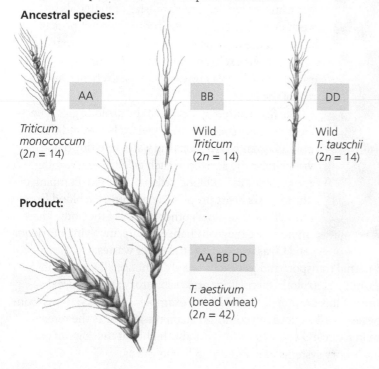

Triticum
monococcum
(2*n* = 14)

AA

Wild
Triticum
(2*n* = 14)

BB

Wild
T. tauschii
(2*n* = 14)

DD

Product:

AA BB DD

T. aestivum
(bread wheat)
(2*n* = 42)

8. SCIENCE, TECHNOLOGY, AND SOCIETY
In the United States, the rare red wolf (*Canis lupus*) has been known to hybridize with coyotes (*Canis latrans*), which are much more numerous. Although red wolves and coyotes differ in terms of morphology, DNA, and behavior, genetic evidence suggests that living red wolf individuals are actually hybrids. Red wolves are designated as an endangered species and hence receive legal protection under the Endangered Species Act. Some people think that their endangered status should be withdrawn because the remaining red wolves are hybrids, not members of a "pure" species. Do you agree? Why or why not?

9. FOCUS ON EVOLUTION
What is the biological basis for assigning all human populations to a single species? Can you think of a scenario by which a second human species could originate in the future?

10. FOCUS ON INFORMATION
In sexually reproducing species, each individual begins life with DNA inherited from both parent organisms. In a short essay (100–150 words), apply this idea to what occurs when organisms of two species that have homologous chromosomes mate and produce (F_1) hybrid offspring. What percentage of the DNA in the F_1 hybrids' chromosomes comes from each parent species? As the hybrids mate and produce F_2 and later-generation hybrid offspring, describe how recombination and natural selection may affect whether the DNA in hybrid chromosomes is derived from one parent species or the other.

For selected answers, see Appendix A.

MasteringBiology®

Students Go to **MasteringBiology** for assignments, the eText, and the Study Area with practice tests, animations, and activities.

Instructors Go to **MasteringBiology** for automatically graded tutorials and questions that you can assign to your students, plus Instructor Resources.

34

Circulation and Gas Exchange

KEY CONCEPTS

34.1 Circulatory systems link exchange surfaces with cells throughout the body

34.2 Coordinated cycles of heart contraction drive double circulation in mammals

34.3 Patterns of blood pressure and flow reflect the structure and arrangement of blood vessels

34.4 Blood components function in exchange, transport, and defense

34.5 Gas exchange occurs across specialized respiratory surfaces

34.6 Breathing ventilates the lungs

34.7 Adaptations for gas exchange include pigments that bind and transport gases

OVERVIEW

Trading Places

The animal in **Figure 34.1** may look like a creature from a science fiction film, but it's actually an axolotl, an amphibian native to shallow ponds in central Mexico. Unlike other amphibians, an axolotl doesn't develop into an air-breathing adult. Instead, it remains in a larval form indefinitely, using feathery gills behind its head to extract oxygen (O_2) from water.

Although external gills are uncommon among adult animals, they help the axolotl carry out a process common to all organisms—the exchange of substances between body cells and the environment. The resources that an animal cell requires, such as nutrients and O_2, enter the cytoplasm by crossing the plasma membrane. Metabolic by-products, such as carbon dioxide (CO_2), exit the cell by crossing the same membrane.

In unicellular organisms, exchange occurs directly with the external environment. For most multicellular organisms, however, direct transfer of materials between every cell and the environment is not possible. Instead, these organisms rely on specialized systems that carry out exchange with the environment and that transport materials between sites of exchange and the rest of the body.

The filamentous structure of the axolotl's gills reflects the intimate association between exchange and transport. Oxygen passes from the water into tiny blood vessels near the surface of each gill filament, turning the pigment in the blood cells a bright red. Pumping of the axolotl's heart propels the oxygen-rich blood from the gill filaments to all other tissues of the body. There, more short-range exchange occurs, involving nutrients and O_2 as well as CO_2 and other wastes.

Because internal transport and gas exchange are functionally related in most animals, not just axolotls, circulatory and respiratory systems are discussed together in this chapter. By considering examples of these systems from a range of species, we'll explore the common elements as well as the remarkable variation in form and organization. We'll also highlight the roles of circulatory and respiratory systems in maintaining homeostasis.

▼ **Figure 34.1** How does a feathery fringe help this animal survive?

Circulatory systems link exchange surfaces with cells throughout the body

The molecular trade that an animal carries out with its environment—gaining O_2 and nutrients while shedding CO_2 and other waste products—must ultimately involve every cell in the body. Small molecules, including O_2 and CO_2, can move between cells and their immediate surroundings by **diffusion** (see Chapter 5). When there is a difference in concentration, diffusion can result in net movement. But such movement is very slow for distances of more than a few millimeters. That's because the time it takes for a substance to diffuse from one place to another is proportional to the *square* of the distance. For example, a quantity of glucose that takes 1 second to diffuse 100 μm will take 100 seconds to diffuse 1 mm and almost 3 hours to diffuse 1 cm! This relationship between diffusion time and distance places a substantial constraint on the body plan of any animal.

Given that net movement by diffusion is rapid only over very small distances, how does each cell of an animal participate in exchange? Natural selection has resulted in two basic adaptations that allow for effective exchange for all of an animal's cells. One adaptation is a body size and shape that places many or all cells in direct contact with the environment. Each cell can thus exchange materials directly with the surrounding medium. This type of body plan is found only in certain invertebrates, including cnidarians and flatworms. The other adaptation, found in all other animals, is a circulatory system. Such systems move fluid between each cell's immediate surroundings and the body tissues where exchange with the environment occurs.

Gastrovascular Cavities

Let's begin by looking at some animals whose body shapes put many of their cells into contact with their environment; these animals lack a distinct circulatory system. In hydras and other cnidarians, a central **gastrovascular cavity** functions in the distribution of substances throughout the body and in digestion (see Figure 33.6). Planarians and most other flatworms also survive without a circulatory system. Their combination of a gastrovascular cavity and a flat body is well suited for exchange with the environment **(Figure 34.2)**.

▶ **Figure 34.2 Internal transport in the planarian *Dugesia* (LM).** The mouth, on the ventral (bottom) side, leads to a gastrovascular cavity (here stained red) that branches throughout the body.

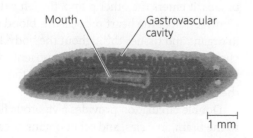

Mouth
Gastrovascular cavity

1 mm

Open and Closed Circulatory Systems

A circulatory system has three basic components: a circulatory fluid, a set of interconnecting vessels, and a muscular pump, the **heart**. The heart powers circulation by using metabolic energy to elevate the hydrostatic pressure of the circulatory fluid, which then flows through the vessels and back to the heart.

Circulatory systems are either open or closed. In an **open circulatory system**, the circulatory fluid, called **hemolymph**, is also the *interstitial fluid* that bathes body cells; arthropods and some molluscs, such as clams, have open systems. Heart contraction pumps the hemolymph through circulatory vessels into sinuses, spaces surrounding the organs **(Figure 34.3a)**. There, exchange with body cells occurs. Heart relaxation draws hemolymph back in through pores, which have valves that close when the heart contracts. Body movements periodically squeeze the sinuses, helping circulate the hemolymph.

In a **closed circulatory system**, a circulatory fluid called **blood** is confined to vessels and is distinct from the interstitial fluid **(Figure 34.3b)**. This type of circulatory system is

▼ **Figure 34.3 Open and closed circulatory systems.**

(a) An open circulatory system

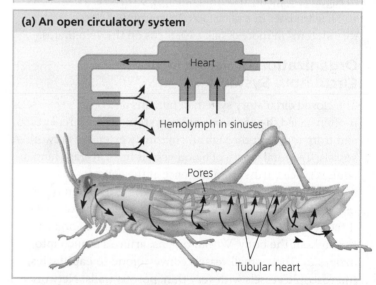

Heart
Hemolymph in sinuses
Pores
Tubular heart

(b) A closed circulatory system

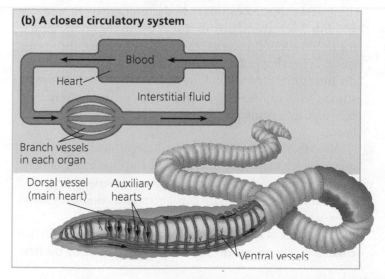

Blood
Heart
Interstitial fluid
Branch vessels in each organ
Dorsal vessel (main heart)
Auxiliary hearts
Ventral vessels

found in annelids (including earthworms), most cephalopod molluscs (including squids and octopuses), and all vertebrates. In closed circulatory systems, one or more hearts pump blood into large vessels that branch into smaller ones that infiltrate the organs. Chemical exchange occurs between the blood and the interstitial fluid, as well as between the interstitial fluid and body cells.

The fact that both open and closed circulatory systems are widespread suggests that each system offers evolutionary advantages. The lower hydrostatic pressures typically associated with open circulatory systems make them less costly than closed systems in terms of energy expenditure. In some invertebrates, open circulatory systems serve additional functions. For example, spiders use the hydrostatic pressure generated by their open circulatory system to extend their legs.

The benefits of closed circulatory systems include relatively high blood pressure, which enables the effective delivery of O_2 and nutrients to the cells of larger and more active animals. Among the molluscs, for instance, closed circulatory systems are found in the largest and most active species, the squids and octopuses. Closed systems are also particularly well suited to regulating the distribution of blood to different organs, as you'll learn later in this chapter. In examining closed circulatory systems in more detail, we'll focus on the vertebrates.

Organization of Vertebrate Circulatory Systems

The closed circulatory system of humans and other vertebrates is often called the **cardiovascular system**. Blood circulates to and from the heart through an amazingly extensive network of vessels: The total length of blood vessels in an average human adult is twice Earth's circumference at the equator!

Arteries, veins, and capillaries are the three main types of blood vessels. Within each type, blood flows in only one direction. **Arteries** carry blood from the heart to organs throughout the body. Within organs, arteries branch into *arterioles*. These small vessels convey blood to **capillaries**, microscopic vessels with very thin, porous walls. Networks of these vessels, called **capillary beds**, infiltrate tissues, passing within a few cell diameters of every cell in the body. Across the thin walls of capillaries, chemicals, including dissolved gases, are exchanged by net diffusion between the blood and the interstitial fluid around the tissue cells. At their "downstream" end, capillaries converge into *venules*, and venules converge into **veins**, the vessels that carry blood back to the heart.

Note that arteries and veins are distinguished by the *direction* in which they carry blood, not by the O_2 content or other characteristics of the blood they contain. Arteries carry blood *away* from the heart toward capillaries, and veins carry blood *toward* the heart from capillaries.

The hearts of all vertebrates contain two or more muscular chambers. The chambers that receive blood entering the heart are called **atria** (singular, *atrium*). The chambers responsible for pumping blood out of the heart are called **ventricles**. The number of chambers and the extent to which they are separated from one another differ substantially among vertebrate groups, as we'll discuss next. These differences reflect the close fit of form to function that arises from natural selection.

Single Circulation

In bony fishes, rays, and sharks, the heart consists of two chambers: an atrium and a ventricle **(Figure 34.4a)**. The blood passes through the heart once in each complete circuit, an arrangement called **single circulation**. Blood entering the heart collects in the atrium before transfer to the ventricle. Contraction of the ventricle pumps blood to the gills, where there is a net diffusion of O_2 into the blood and of CO_2 out of the blood. As blood leaves the gills, the capillaries converge into a vessel that carries oxygen-rich blood to capillary beds throughout the body. Blood then returns to the heart.

In single circulation, blood that leaves the heart passes through two sets of capillary beds before returning to the heart. When blood flows through a capillary bed, blood pressure drops substantially, for reasons we will explain shortly. The drop in blood pressure in the gills limits the velocity of blood flow in the rest of the animal's body. As the animal swims, however, the contraction and relaxation of its muscles help accelerate the relatively sluggish pace of circulation.

Double Circulation

The circulatory systems of amphibians, reptiles, and mammals have two circuits, an arrangement called **double circulation** **(Figure 34.4b** and **c)**. In animals with double circulation, the pumps for the two circuits are combined into a single organ, the heart. Having both pumps within a single heart simplifies coordination of the pumping cycles. One pump, the right side of the heart, delivers oxygen-poor blood to the capillary beds of the gas exchange tissues, where there is a net movement of O_2 into the blood and of CO_2 out of the blood. This **gas exchange circuit** is called a *pulmonary circuit* if the capillary beds involved are all in the lungs, as in reptiles and mammals. It is called a *pulmocutaneous circuit* if it includes capillaries in both the lungs and the skin, as in many amphibians.

After the oxygen-enriched blood leaves the gas exchange tissues, it enters the other pump, the left side of the heart. Contraction of the heart propels this blood to capillary beds in organs and tissues throughout the body. Following the exchange of O_2 and CO_2, as well as nutrients and waste products, the now oxygen-poor blood returns to the heart, completing the **systemic circuit**.

Double circulation provides a vigorous flow of blood to the brain, muscles, and other organs because the heart

▼ Figure 34.4 Generalized circulatory schemes of vertebrates. (a) Fishes provide an example of single circulation, in which blood passes through the heart once per complete circuit. In contrast, **(b)** amphibians and **(c)** mammals offer examples of double circulation, with separate gas exchange and systemic circuits. Comparing (b) and (c) indicates some variations of heart structure and gas exchange tissue types that have evolved in different groups of vertebrates with double circulation. (Note that circulatory systems are shown as if the body were facing you: The right side of the heart is shown on the left, and vice versa.)

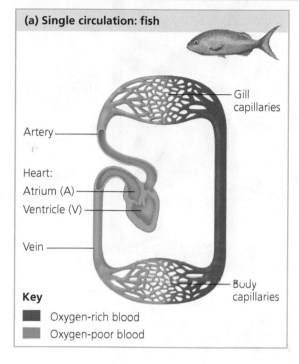

(a) Single circulation: fish

Gill capillaries

Artery

Heart:
Atrium (A)
Ventricle (V)

Vein

Body capillaries

Key
- Oxygen-rich blood
- Oxygen-poor blood

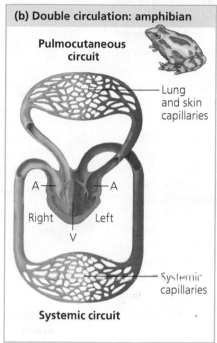

(b) Double circulation: amphibian

Pulmocutaneous circuit

Lung and skin capillaries

A —— A

Right | Left

V

Systemic capillaries

Systemic circuit

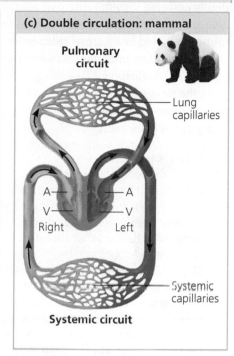

(c) Double circulation: mammal

Pulmonary circuit

Lung capillaries

A —— A
V —— V

Right | Left

Systemic capillaries

Systemic circuit

repressurizes the blood destined for these tissues after it passes through the capillary beds of the lungs or skin. Indeed, blood pressure is often much higher in the systemic circuit than in the gas exchange circuit. This contrasts sharply with single circulation, in which blood flows under reduced pressure directly from the gas exchange organs to other organs.

Evolutionary Variation in Double Circulation

EVOLUTION Some vertebrates with double circulation are intermittent breathers. For example, amphibians and many reptiles fill their lungs with air periodically, passing long periods of time without gas exchange or relying on another gas exchange tissue, typically the skin. These animals have adaptations that enable the circulatory system to temporarily bypass the lungs in part or in whole:

- Frogs and other amphibians have a heart with three chambers—two atria and one ventricle (see Figure 34.4b). A ridge within the ventricle diverts most (about 90%) of the oxygen-rich blood from the left atrium into the systemic circuit and most of the oxygen-poor blood from the right atrium into the gas exchange (pulmocutaneous) circuit. When a frog is underwater, the incomplete division of the ventricle allows the frog to adjust its circulation, shutting off most blood flow to its temporarily ineffective lungs. Blood

flow continues to the skin, which acts as the sole site of gas exchange while the frog is submerged.

- In the three-chambered heart of turtles, snakes, and lizards, an incomplete septum partially divides the single ventricle into separate right and left chambers. Two major arteries, called aortas, lead to the systemic circulation. As with amphibians, the circulatory system enables control of the relative amount of blood flowing to the lungs and the body.

- In alligators, caimans, and other crocodilians, the ventricles are divided by a complete septum, but the pulmonary and systemic circuits connect where the arteries exit the heart. This connection allows arterial valves to shunt blood flow away from the lungs temporarily, such as when the animal is underwater.

Double circulation in birds and mammals is quite different. As shown for a panda in Figure 34.4c, the heart has two atria and two completely divided ventricles. The left side of the heart receives and pumps only oxygen-rich blood, while the right side receives and pumps only oxygen-poor blood. Unlike amphibians and many reptiles, birds and mammals cannot vary blood flow to the lungs without altering blood flow throughout the body.

How has natural selection shaped the double circulation of birds and mammals? As endotherms, birds and mammals use about ten times as much energy as equal-sized ectotherms. Their circulatory systems therefore need to deliver about ten times as much fuel and O_2 to their tissues (and remove ten times as much CO_2 and other wastes). This large traffic of substances is made possible by the separate and independently powered systemic and pulmonary circuits and by large hearts that pump the necessary volume of blood. A powerful four-chambered heart arose independently in the different ancestors of birds and mammals and thus reflects convergent evolution (see Chapter 27).

In the next section, we'll restrict our focus to circulation in mammals and to the anatomy and physiology of the key circulatory organ—the heart.

CONCEPT CHECK 34.1

1. How is the flow of hemolymph through an open circulatory system similar to the flow of water through an outdoor fountain?
2. Three-chambered hearts with incomplete septa were once viewed as being less adapted to circulatory function than mammalian hearts. What advantage of such hearts did this viewpoint overlook?
3. **WHAT IF?** The heart of a normally developing human fetus has a hole between the left and right atria. In some cases, this hole does not close completely before birth. If the hole weren't surgically corrected, how would it affect the O_2 content of the blood entering the systemic circuit?

For suggested answers, see Appendix A.

CONCEPT 34.2

Coordinated cycles of heart contraction drive double circulation in mammals

Timely delivery of O_2 to body organs is critical: Some brain cells, for example, die if their O_2 supply is interrupted for even a few minutes. How does the mammalian cardiovascular system meet the body's continuous (although variable) demand for O_2? To answer this question, we need to consider how the parts of the system are arranged and how each part functions.

Mammalian Circulation

Let's first examine the overall organization of the mammalian cardiovascular system, beginning with the pulmonary circuit. (The circled numbers refer to corresponding locations in **Figure 34.5**.) Contraction of ❶ the right ventricle pumps blood to the lungs via ❷ the pulmonary arteries. As the blood flows through ❸ capillary beds in the left and right lungs, it loads O_2 and unloads CO_2. Oxygen-rich blood returns from the lungs via the pulmonary veins to ❹ the left atrium of the heart. Next, the oxygen-rich blood flows into ❺ the heart's left ventricle, which pumps the oxygen-rich blood out to body tissues through the systemic circuit. Blood leaves the left ventricle via ❻ the aorta,

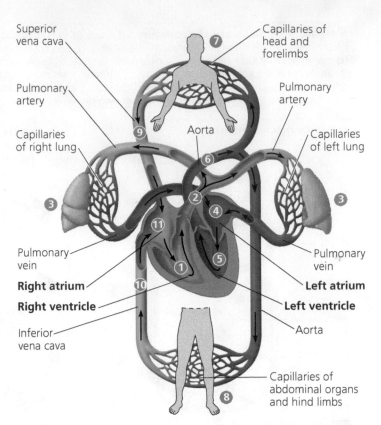

▲ **Figure 34.5 The mammalian cardiovascular system: an overview.** Note that the dual circuits operate simultaneously, not in the serial fashion that the numbering in the diagram suggests. The two ventricles pump almost in unison; while some blood is traveling in the pulmonary circuit, the rest of the blood is flowing in the systemic circuit.

which conveys blood to arteries leading throughout the body. The first branches leading from the aorta are the coronary arteries (not shown), which supply blood to the heart muscle itself. Then branches lead to ❼ capillary beds in the head and arms (forelimbs). The aorta then descends into the abdomen, supplying oxygen-rich blood to arteries leading to ❽ capillary beds in the abdominal organs and legs (hind limbs). Within the capillaries, there is a net diffusion of O_2 from the blood to the tissues and of CO_2 (produced by cellular respiration) into the blood. Capillaries rejoin, forming venules, which convey blood to veins. Oxygen-poor blood from the head, neck, and forelimbs is channeled into a large vein, ❾ the superior vena cava. Another large vein, ❿ the inferior vena cava, drains blood from the trunk and hind limbs. The two venae cavae empty their blood into ⓫ the right atrium, from which the oxygen-poor blood flows into the right ventricle.

The Mammalian Heart: *A Closer Look*

Located behind the sternum (breastbone), the human heart is about the size of a clenched fist and consists mostly of cardiac muscle. The two atria have relatively thin walls and serve as collection chambers for blood returning to the heart from the lungs or other body tissues **(Figure 34.6)**. Much of the blood that enters the atria flows into the ventricles while all

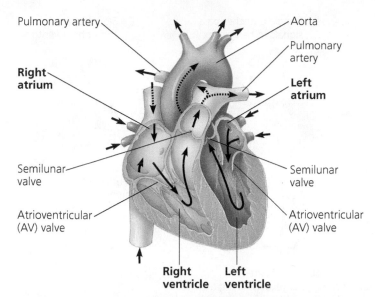

Right atrium

Pulmonary artery

Aorta

Pulmonary artery

Left atrium

Semilunar valve

Semilunar valve

Atrioventricular (AV) valve

Atrioventricular (AV) valve

Right ventricle

Left ventricle

▲ **Figure 34.6 The mammalian heart: a closer look.** Notice the locations of the valves, which prevent backflow of blood within the heart. Also notice how the atria and left and right ventricles differ in the thickness of their muscular walls.

heart chambers are relaxed. The remainder is transferred by contraction of the atria before the ventricles begin to contract. Compared to the atria, the ventricles have thicker walls and contract much more forcefully—especially the left ventricle, which pumps blood to all body organs through the systemic circuit. Although the left ventricle contracts with greater force than the right ventricle, it pumps the same volume of blood as the right ventricle during each contraction.

The heart contracts and relaxes in a rhythmic cycle. When it contracts, it pumps blood; when it relaxes, its chambers fill with blood. One complete sequence of pumping and filling is referred to as the **cardiac cycle (Figure 34.7)**. The contraction phase of the cycle is called **systole**, and the relaxation phase is called **diastole**.

The volume of blood each ventricle pumps per minute is the cardiac output. Two factors determine cardiac output: the rate of contraction, or heart rate (number of beats per minute), and the *stroke volume*, the amount of blood pumped by a ventricle in a single contraction. The average stroke volume in humans is about 70 mL. Multiplying this stroke volume by a resting heart rate of 72 beats per minute yields a cardiac output of 5 L/min—about equal to the total volume of blood in the human body. During heavy exercise, cardiac output increases as much as fivefold.

Four valves in the heart prevent backflow and keep blood moving in the correct direction (see Figures 34.6 and 34.7). Made of flaps of connective tissue, the valves open when pushed from one side and close when pushed from the other. An **atrioventricular (AV) valve** lies between each atrium and ventricle. The AV valves are anchored by strong fibers that prevent them from turning inside out. Pressure generated by the powerful contraction of the ventricles closes the AV valves, keeping blood from flowing back into the atria. **Semilunar valves** are located at

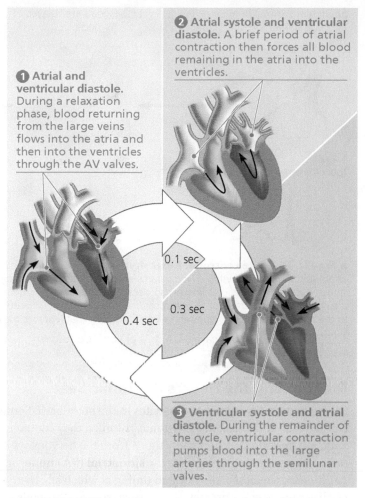

1 **Atrial and ventricular diastole.** During a relaxation phase, blood returning from the large veins flows into the atria and then into the ventricles through the AV valves.

2 **Atrial systole and ventricular diastole.** A brief period of atrial contraction then forces all blood remaining in the atria into the ventricles.

0.1 sec

0.4 sec

0.3 sec

3 **Ventricular systole and atrial diastole.** During the remainder of the cycle, ventricular contraction pumps blood into the large arteries through the semilunar valves.

▲ **Figure 34.7 The cardiac cycle.** Note that during all but 0.1 second of the cardiac cycle, the atria are relaxed and are filling with blood returning via the veins.

the two exits of the heart: where the aorta leaves the left ventricle and where the pulmonary artery leaves the right ventricle. These valves are pushed open by the pressure generated during contraction of the ventricles. When the ventricles relax, blood pressure built up in the aorta and pulmonary artery closes the semilunar valves and prevents significant backflow.

You can follow the closing of the two sets of heart valves either with a stethoscope or by pressing your ear tightly against the chest of a friend (or a friendly dog). The sound pattern is "lub-dup, lub-dup, lub-dup." The first heart sound ("lub") is created by the recoil of blood against the closed AV valves. The second sound ("dup") is due to the vibrations caused by closing of the semilunar valves.

If blood squirts backward through a defective valve, it may produce an abnormal sound called a heart murmur. Some people are born with heart murmurs; in others, the valves may be damaged by infection (from rheumatic fever, for instance). When a valve defect is severe enough to endanger health, surgeons may implant a mechanical replacement valve. However, not all heart murmurs are caused by a defect, and most valve defects do not reduce the efficiency of blood flow enough to warrant surgery.

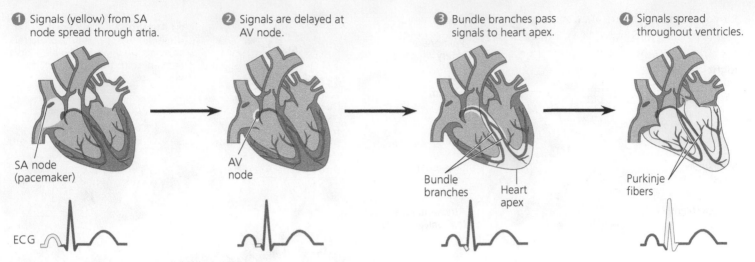

1 Signals (yellow) from SA node spread through atria.

2 Signals are delayed at AV node.

3 Bundle branches pass signals to heart apex.

4 Signals spread throughout ventricles.

SA node (pacemaker)

AV node

Bundle branches   Heart apex

Purkinje fibers

ECG

▲ **Figure 34.8 The control of heart rhythm.** Electrical signals follow a set path through the heart in establishing the heart rhythm. The diagrams at the top trace the movement of electrical signals (yellow) during the cardiac cycle; specialized cells involved in electrical control of the rhythm are indicated in orange. Under each step, the corresponding portion of an electrocardiogram (ECG) is highlighted in yellow. In step 4, the portion of the ECG to the right of the "spike" represents electrical activity that reprimes the ventricles for the next round of contraction.

WHAT IF? *If your doctor gave you a copy of your ECG recording, how could you determine what your heart rate had been during the test?*

Maintaining the Heart's Rhythmic Beat

In vertebrates, the heartbeat originates in the heart itself. Some cardiac muscle cells are autorhythmic, meaning they contract and relax repeatedly without any signal from the nervous system. A group of such cells forms the **sinoatrial (SA) node**, or *pacemaker*, which sets the rate and timing at which all other cardiac muscle cells contract. (In contrast, some arthropods have pacemakers located in the nervous system, outside the heart.)

The SA node produces electrical impulses that spread rapidly within heart tissue. Because body fluids can conduct electricity, currents generated by those impulses can be detected at the surface of the body in an **electrocardiogram** (**ECG** or, often, **EKG**, from the German spelling). The resulting graph of current against time has a characteristic shape that represents the stages in the cardiac cycle (**Figure 34.8**).

Impulses from the SA node first spread rapidly through the walls of the atria, causing both atria to contract in unison. During atrial contraction, the impulses originating at the SA node reach other autorhythmic cells located in the wall between the left and right atria. These cells form a relay point called the **atrioventricular (AV) node**. Here the impulses are delayed for about 0.1 second before spreading to the heart apex. This delay allows the atria to empty completely before the ventricles contract. Then the signals from the AV node are conducted to the heart apex and throughout the ventricular walls.

Physiological cues alter heart tempo by regulating the pacemaker function of the SA node. For example, when you stand up and start walking, the nervous system speeds up your pacemaker. The resulting increase in heart rate provides the additional O₂ needed by the muscles that are powering your activity. If you then sit down and relax, the nervous system slows down your pacemaker, decreasing your heart rate and thus conserving energy. Hormones and temperature also influence the pacemaker. For instance, epinephrine, the "fight-or-flight" hormone secreted by the adrenal glands, causes the heart rate to increase, as does an increase in body temperature.

Having examined the operation of the circulatory pump, we turn in the next section to the forces and structures that influence blood flow in the vessels of each circuit.

CONCEPT CHECK 34.2

1. Explain why blood in the pulmonary veins has a higher O₂ concentration than in the venae cavae, which are also veins.
2. Why is it important that the AV node delay the electrical impulse moving from the SA node and the atria to the ventricles?
3. **WHAT IF?** After you exercise regularly for several months, your resting heart rate decreases, but your cardiac output at rest is unchanged. What change in the function of your heart at rest could explain these findings?

For suggested answers, see Appendix A.

CONCEPT 34.3

Patterns of blood pressure and flow reflect the structure and arrangement of blood vessels

The vertebrate circulatory system enables blood to deliver oxygen and nutrients and remove wastes throughout the body. In doing so, the circulatory system relies on a branching network of vessels much like the plumbing system that delivers fresh water to a city and removes its wastes. In fact, the same physical principles that govern the operation of plumbing systems apply to the functioning of blood vessels.

Blood Vessel Structure and Function

Blood vessels contain a central lumen (cavity) lined with an **endothelium**, a single layer of flattened epithelial cells. The smooth surface of the endothelium minimizes resistance to the flow of blood. Surrounding the endothelium are layers of tissue that differ in capillaries, arteries, and veins, reflecting the specialized functions of these vessels.

Capillaries are the smallest blood vessels, having a diameter only slightly greater than that of a red blood cell **(Figure 34.9)**. Capillaries also have very thin walls, which consist of just the endothelium and its basal lamina. The exchange of substances between the blood and interstitial fluid occurs in capillaries because only there are blood vessel walls thin enough to permit this transfer.

The walls of other blood vessels have a more complex organization than those of capillaries. Both arteries and veins have two layers of tissue surrounding the endothelium. The outer layer is formed by connective tissue that contains elastic fibers, which allow the vessel to stretch and recoil, and collagen, which provides strength. The layer next to the endothelium contains smooth muscle and more elastic fibers.

While similar in organization, the walls of arteries and veins differ, reflecting adaptations to distinct functions. The walls of arteries are thick and strong, accommodating blood pumped at high pressure by the heart. Arterial walls also have an elastic recoil that helps maintain blood pressure and flow to capillaries when the heart relaxes between contractions. Nervous system signals and hormones in the blood act on the smooth muscle in arteries and arterioles, dilating or constricting these vessels and thus controlling blood flow to different body parts.

Because veins convey blood back to the heart at a lower pressure, they do not require thick walls. For a given blood vessel diameter, a vein has a wall only about a third as thick as that of an artery. Unlike arteries, veins contain valves, which maintain a unidirectional flow of blood despite the low blood pressure in these vessels.

We consider next how blood vessel diameter, vessel number, and pressure influence the velocity at which blood flows in different locations within the body.

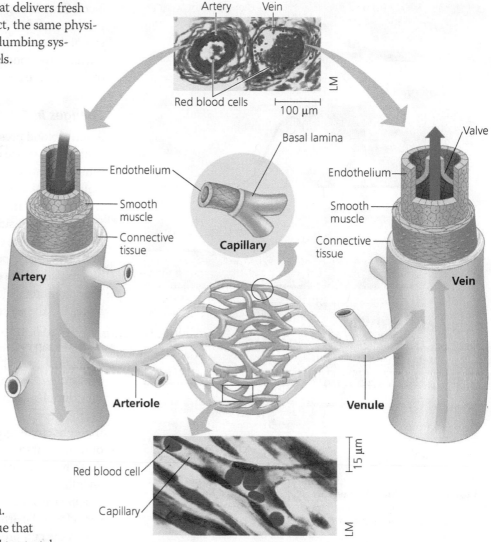

▲ Figure 34.9 **The structure of blood vessels.**

Blood Flow Velocity

To understand how blood vessel diameter influences blood flow, consider how water flows through a thick hose connected to a faucet. When the faucet is turned on, water flows at the same velocity at each point along the hose. However, if a narrow nozzle is attached to the end of the hose, the water will exit the nozzle at a much greater velocity. Because water doesn't compress under pressure, the volume of water moving through the nozzle in a given time must be the same as the volume moving through the rest of the hose. The cross-sectional area of the nozzle is smaller than that of the hose, so the water speeds up in the nozzle.

An analogous situation exists in the circulatory system, but blood *slows* as it moves from arteries to arterioles to the much narrower capillaries. Why? The reason is that the number of capillaries is enormous, roughly 7 billion in a human body. Each artery conveys blood to so many capillaries that the *total* cross-sectional area is much greater in

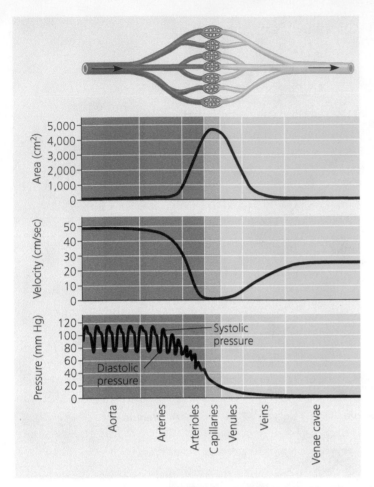

▲ Figure 34.10 The interrelationship of cross-sectional area of blood vessels, blood flow velocity, and blood pressure. Because total cross-sectional area increases in the arterioles and in the capillaries, blood flow velocity decreases markedly. Blood pressure, the main force driving blood from the heart to the capillaries, is highest in the aorta and other arteries.

blood flow, throughout the cardiac cycle. Once the blood enters the millions of tiny arterioles and capillaries, the narrow diameter of these vessels generates substantial resistance to flow. By the time the blood enters the veins, resistance dissipates much of the pressure generated by the pumping heart.

Changes in Blood Pressure During the Cardiac Cycle

Arterial blood pressure is highest during systole, the contraction phase of the cardiac cycle. The pressure at the time the ventricles contract is called *systolic pressure* (see Figure 34.10). Each spike in blood pressure caused by a contraction of a ventricle stretches the arteries. By placing your fingers on the inside of your wrist, you can feel a **pulse**—the rhythmic bulging of the artery walls with each heartbeat. The pressure surge is partly due to the narrow openings of arterioles impeding the exit of blood from the arteries. When the heart contracts, blood enters the arteries faster than it can leave, and the vessels stretch from the rise in pressure.

During diastole, the relaxation phase, the elastic walls of the arteries snap back. As a consequence, there is a lower but still substantial arterial blood pressure when the ventricles are relaxed (*diastolic pressure*). Before enough blood has flowed into the arterioles to completely relieve pressure in the arteries, the heart contracts again. Because the arteries remain pressurized throughout the cardiac cycle (see Figure 34.10), blood continuously flows into arterioles and capillaries.

To measure blood pressure, doctors or nurses often use an inflatable cuff attached to a pressure gauge. The cuff is wrapped around the upper arm and inflated until the pressure closes the artery; the cuff is then deflated gradually. When the cuff pressure drops just below that in the artery, blood begins to pulse past the cuff, making sounds that can be heard with a stethoscope. The pressure measured at this point equals the systolic pressure. As deflation continues, the cuff pressure at some point no longer constricts blood movement. The reading on the gauge when the blood begins to flow freely and silently equals the diastolic pressure. For a healthy 20-year-old human at rest, arterial blood pressure in the systemic circuit is typically about 120 millimeters of mercury (mm Hg) at systole and 70 mm Hg at diastole, expressed as 120/70. (Arterial blood pressure in the pulmonary circuit is six to ten times lower.)

capillary beds than in the arteries or any other part of the circulatory system **(Figure 34.10)**. The result is a dramatic decrease in velocity from the arteries to the capillaries: Blood travels 500 times more slowly in the capillaries (about 0.1 cm/sec) than in the aorta (about 48 cm/sec). After passing through the capillaries, the blood speeds up as it enters the venules and veins, which have smaller *total* cross-sectional areas than the capillaries.

Blood Pressure

Blood, like all fluids, flows from areas of higher pressure to areas of lower pressure. Contraction of a heart ventricle generates blood pressure, which exerts force in all directions. The force directed lengthwise in an artery causes the blood to flow away from the heart, the site of highest pressure. The force exerted sideways against the wall of an artery stretches the wall. Following ventricular contraction, the recoil of the elastic arterial walls plays a critical role in maintaining blood pressure, and hence

Maintenance of Blood Pressure

Homeostatic mechanisms regulate arterial blood pressure by altering the diameter of arterioles. As the smooth muscles in arteriole walls contract, the arterioles narrow, a process called **vasoconstriction**. Narrowing of the arterioles increases blood pressure upstream in the arteries. When the smooth muscles relax, the arterioles undergo **vasodilation**, an increase in diameter that causes blood pressure in the arteries to fall.

Researchers have identified nitric oxide (NO), a gas, as a major inducer of vasodilation and endothelin, a peptide, as the most potent inducer of vasoconstriction. Cues from the nervous and endocrine systems regulate production of NO and endothelin in blood vessels, where their activities regulate blood pressure.

Gravity also has a significant effect on blood pressure. When you are standing, for example, your head is roughly 0.35 m higher than your chest, and the arterial blood pressure in your brain is about 27 mm Hg less than that near your heart. If the blood pressure in your brain is too low to provide adequate blood flow, you will likely faint. By causing your body to collapse to the ground, fainting effectively places your head at the level of your heart, quickly increasing blood flow to your brain.

For animals with very long necks, the blood pressure required to overcome gravity is particularly high. A giraffe, for example, requires a systolic pressure of more than 250 mm Hg near the heart to get blood to its head. When a giraffe lowers its head to drink, one-way valves and sinuses, along with feedback mechanisms that reduce cardiac output, prevent this high pressure from damaging its brain.

Gravity is also a consideration for blood flow in veins, especially those in the legs. Although blood pressure in veins is relatively low, the valves inside the veins maintain a unidirectional flow of blood. The return of blood to the heart is further enhanced by rhythmic contractions of smooth muscles in the walls of venules and veins and by the contraction of skeletal muscles during exercise **(Figure 34.11)**.

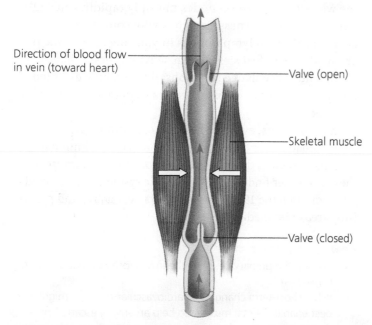

Direction of blood flow in vein (toward heart)

Valve (open)

Skeletal muscle

Valve (closed)

▲ **Figure 34.11 Blood flow in veins.** Skeletal muscle contraction squeezes and constricts veins. Flaps of tissue within the veins act as one-way valves that keep blood moving only toward the heart. If you sit or stand too long, the lack of muscular activity may cause your feet to swell as blood pools in your veins.

Capillary Function

At any given time, only about 5–10% of the body's capillaries have blood flowing through them. Capillaries in the brain, heart, kidneys, and liver are usually filled to capacity, but at many other sites the blood supply varies over time as blood is diverted from one destination to another. For example, blood flow to the skin is regulated to help control body temperature, and blood supply to the digestive tract increases after a meal. During strenuous exercise, blood is diverted from the digestive tract and supplied more generously to skeletal muscles and skin. This is one reason why exercising heavily immediately after eating a big meal may cause indigestion.

Given that capillaries lack smooth muscle, how is blood flow in capillary beds altered? One mechanism for altering blood flow is vasoconstriction or vasodilation of the arteriole that supplies a capillary bed. A second mechanism involves rings of smooth muscle located at the entrance to capillary beds. Opening and closing these muscular rings regulates and redirects the passage of blood into particular sets of capillaries.

As you have read, the critical exchange of substances between the blood and interstitial fluid takes place across the thin endothelial walls of the capillaries. Some substances are carried across the endothelium in vesicles that form on one side by endocytosis and release their contents on the opposite side by exocytosis. Small molecules, such as O_2 and CO_2, simply diffuse across the endothelial cells or, in some tissues, through microscopic pores in the capillary wall. These openings also provide the route for transport of small solutes such as sugars, salts, and urea, as well as for bulk flow of fluid into tissues driven by blood pressure within the capillary.

Two opposing forces control the movement of fluid between the capillaries and the surrounding tissues: Blood pressure tends to drive fluid out of the capillaries, and the presence of blood proteins tends to pull fluid back. Many blood proteins (and all blood cells) are too large to pass readily through the endothelium, and they remain in the capillaries. These dissolved proteins are responsible for much of the blood's *osmotic pressure* (the pressure produced by the difference in solute concentration across a membrane). The difference in osmotic pressure between the blood and the interstitial fluid opposes fluid movement out of the capillaries. On average, blood pressure is greater than the opposing forces, leading to a net loss of fluid from capillaries. The net loss is generally greatest at the arterial end of these vessels, where blood pressure is highest.

Fluid Return by the Lymphatic System

Each day, the adult human body loses approximately 4–8 L of fluid from capillaries to the surrounding tissues. There is also some leakage of blood proteins, even though the capillary wall is not very permeable to large molecules. The lost fluid and proteins return to the blood via the **lymphatic system**,

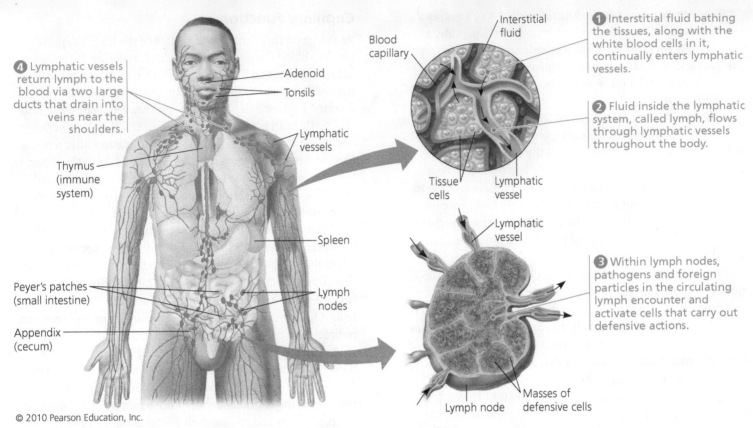

© 2010 Pearson Education, Inc.

▲ Figure 34.12 The human lymphatic system. Lymph flows through lymphatic vessels (shown in green). Foreign substances carried by the lymph are trapped in lymph nodes (orange) and lymphoid organs (yellow): the adenoids, tonsils, spleen, Peyer's patches, and appendix. Steps 1–4 trace the flow of lymph and illustrate the critical role of lymph nodes in activating immune responses.

which includes a network of tiny vessels intermingled among capillaries of the cardiovascular system **(Figure 34.12)**.

After entering the lymphatic system by diffusion, the fluid lost by capillaries is called **lymph**; its composition is about the same as that of interstitial fluid. The lymphatic system drains into large veins of the circulatory system at the base of the neck. This joining of the lymphatic and circulatory systems enables lipids to be transferred from the small intestine to the blood (see Chapter 33).

The movement of lymph from peripheral tissues to the heart relies on much the same mechanisms that assist blood flow in veins. Lymph vessels, like veins, have valves that prevent the backflow of fluid. Rhythmic contractions of the vessel walls help draw fluid into the small lymphatic vessels. In addition, skeletal muscle contractions play a role in moving lymph.

Disorders that interfere with the lymphatic system highlight its role in maintaining proper fluid distribution in the body. Disruptions in the movement of lymph often cause edema, swelling that results from the excessive accumulation of fluid in tissues. Severe blockage of lymph flow, as occurs when certain parasitic worms lodge in lymph vessels, results in extremely swollen limbs or other body parts, a condition known as elephantiasis.

Along a lymph vessel are organs called **lymph nodes**. By filtering the lymph and by housing cells that attack viruses and bacteria, lymph nodes play an important role in

the body's defense. When the body is fighting an infection, the white cells in lymph nodes multiply rapidly, causing swelling and tenderness (which is why your doctor may check for swollen lymph nodes in your neck, armpits, or groin when you feel sick). Because lymph nodes have filtering and surveillance functions, doctors may examine the lymph nodes of cancer patients to detect the spread of diseased cells.

In recent years, evidence has surfaced demonstrating that the lymphatic system also plays a role in harmful immune responses, such as those responsible for asthma. Because of these and other findings, the lymphatic system, given limited attention until the 1990s, has become a very active and promising area of biomedical research.

CONCEPT CHECK 34.3

1. What is the primary cause of the low velocity of blood flow in capillaries?
2. What short-term changes in cardiovascular function might best enable skeletal muscles to help an animal escape from a dangerous situation?
3. **WHAT IF?** If you had additional hearts distributed throughout your body, what would be one likely advantage and one likely disadvantage?

For suggested answers, see Appendix A.

CONCEPT 34.4

Blood components function in exchange, transport, and defense

As you read in Concept 34.1, the fluid transported by an open circulatory system is continuous with the fluid that surrounds all of the body cells and therefore has the same composition. In contrast, the fluid in a closed circulatory system can be more highly specialized, as is the case for the blood of vertebrates.

Blood Composition and Function

Vertebrate blood is a connective tissue consisting of cells suspended in a liquid matrix called **plasma**. Separating the components of blood using a centrifuge reveals that cellular elements (cells and cell fragments) occupy about 45% of the volume of blood **(Figure 34.13)**. The remainder is plasma. Dissolved in the plasma are ions and proteins that, together with the blood cells, function in osmotic regulation, transport, and defense.

Plasma

Among the many solutes in plasma are inorganic salts in the form of dissolved ions, sometimes referred to as blood electrolytes. Although plasma is about 90% water, the dissolved salts are an essential component of the blood. Some of these ions buffer the blood, which in humans normally has a pH of 7.4. Salts are also important in maintaining the osmotic balance of the blood. In addition, the concentration of ions in plasma directly affects the composition of the interstitial fluid, where many of these ions have a vital role in muscle and nerve activity. Serving all of these functions necessitates keeping plasma electrolytes within narrow concentration ranges via homeostatic mechanisms.

Plasma proteins act as buffers against pH changes and help maintain the osmotic balance between blood and interstitial fluid. Particular plasma proteins have additional functions. The immunoglobulins, or antibodies, help combat viruses and other foreign agents that invade the body (see Chapter 35). Other plasma proteins serve as escorts for lipids, which are insoluble in water and can travel in blood only when bound to proteins. Still other plasma proteins are clotting factors that help plug leaks when blood vessels are injured. (The term *serum* refers to blood plasma from which these clotting factors have been removed.)

Plasma also contains a wide variety of other substances in transit from one part of the body to another, including nutrients, metabolic wastes, respiratory gases, and hormones.

Plasma 55%	
Constituent	**Major functions**
Water	Solvent for carrying other substances
Ions (blood electrolytes) Sodium Potassium Calcium Magnesium Chloride Bicarbonate	Osmotic balance, pH buffering, and regulation of membrane permeability
Plasma proteins Albumin	Osmotic balance, pH buffering
Fibrinogen	Clotting
Immunoglobulins (antibodies)	Defense
Substances transported by blood Nutrients (such as glucose, fatty acids, vitamins) Waste products of metabolism Respiratory gases (O_2 and CO_2) Hormones	

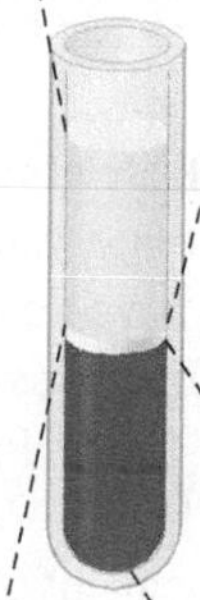

Separated blood elements

Cellular elements 45%		
Cell type	**Number** per µL (mm^3) of blood	**Functions**
Leukocytes (white blood cells) 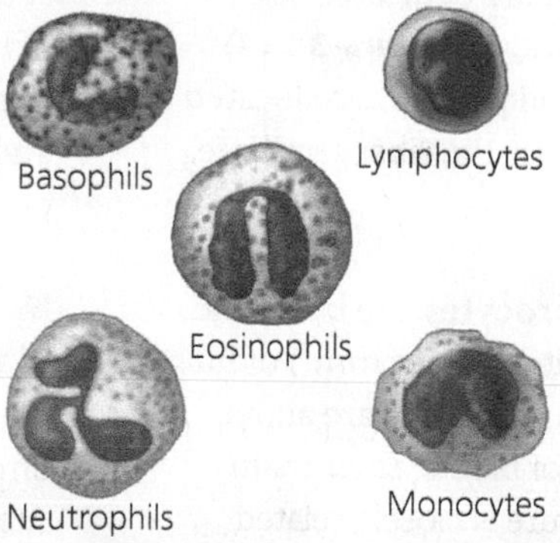 Basophils Lymphocytes Eosinophils Neutrophils Monocytes	5,000–10,000	Defense and immunity
Platelets 	250,000–400,000	Blood clotting
Erythrocytes (red blood cells) 	5,000,000–6,000,000	Transport of O_2 and some CO_2

▲ **Figure 34.13 The composition of mammalian blood.**

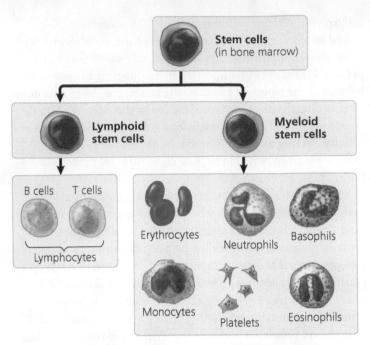

Stem cells
(in bone marrow)

Lymphoid
stem cells

Myeloid
stem cells

B cells   T cells

Lymphocytes

Erythrocytes

Neutrophils

Basophils

Monocytes

Platelets

Eosinophils

▲ **Figure 34.14 Differentiation of blood cells.** Some of the multipotent stem cells differentiate into lymphoid stem cells, which then develop into B cells and T cells, two types of lymphocytes that function in immunity (see Chapter 35). All other blood cells and platelets arise from myeloid stem cells.

Cellular Elements

Blood contains two classes of cells: red blood cells, which transport O_2, and white blood cells, which function in defense. Also suspended in blood plasma are **platelets**, cell fragments that are involved in the clotting process. All these cellular elements develop from multipotent **stem cells** that are dedicated to replenishing the body's blood cell populations **(Figure 34.14)**. The stem cells that produce blood cells and platelets are located in the red marrow inside bones, particularly the ribs, vertebrae, sternum, and pelvis.

Erythrocytes Red blood cells, or **erythrocytes**, are by far the most numerous blood cells. Each microliter (μL, or mm^3) of human blood contains 5–6 million red cells, and there are about 25 trillion of these cells in the body's 5 L of blood. Their main function is O_2 transport, and their structure is closely related to this function. Human erythrocytes are small disks (7–8 μm in diameter) that are biconcave—thinner in the center than at the edges. This shape increases surface area, enhancing the rate of diffusion of O_2 across the plasma membrane. Mature mammalian erythrocytes lack nuclei. This unusual characteristic leaves more space in these tiny cells for **hemoglobin**, the iron-containing protein that transports O_2 (see Figure 3.21).

Despite its small size, an erythrocyte contains about 250 million molecules of hemoglobin (Hb). Because each molecule of hemoglobin binds up to four molecules of O_2, one erythrocyte can transport about one billion O_2 molecules. As erythrocytes pass through the capillary beds of lungs, gills, or other respiratory organs, O_2 diffuses into the erythrocytes and binds to hemoglobin. In the systemic capillaries, O_2 dissociates from hemoglobin and diffuses into body cells.

In **sickle-cell disease**, an abnormal form of hemoglobin (Hb^S) polymerizes into aggregates. Because the concentration of hemoglobin in erythrocytes is so high, these aggregates are large enough to distort the erythrocyte into an elongated, curved shape that resembles a sickle. This abnormality results from an alteration in the amino acid sequence of hemoglobin at a single position (see Figure 3.22).

Throughout a person's life, stem cells replace the worn-out cellular elements of blood. Erythrocytes are the shortest lived, circulating for only 120 days on average before being replaced. A negative-feedback mechanism, sensitive to the amount of O_2 reaching the body's tissues via the blood, controls erythrocyte production. If the tissues do not receive enough O_2, the kidneys synthesize and secrete *erythropoietin* (*EPO*), a hormone that stimulates erythrocyte production. Today, EPO produced by recombinant DNA technology is used to treat health problems such as *anemia*, a condition of lower-than-normal erythrocyte or hemoglobin levels. Some athletes inject themselves with EPO to increase their erythrocyte levels, although this practice, a form of blood doping, has been banned by major sports organizations.

Leukocytes The blood contains five major types of white blood cells, or **leukocytes**. Their function is to fight infections. Some are phagocytic, engulfing and digesting microorganisms as well as debris from the body's own dead cells. Other leukocytes, called lymphocytes, develop into specialized B cells and T cells that mount immune responses against foreign substances (as will be discussed in Chapter 35). Normally, 1 μL of human blood contains about 5,000–10,000 leukocytes; their numbers increase temporarily whenever the body is fighting an infection. Unlike erythrocytes, leukocytes are also found outside the circulatory system, patrolling both interstitial fluid and the lymphatic system.

Platelets Platelets are pinched-off cytoplasmic fragments of specialized bone marrow cells. They are about 2–3 μm in diameter and have no nuclei. Platelets serve both structural and molecular functions in blood clotting.

Blood Clotting

The occasional cut or scrape is not life-threatening because blood components seal the broken blood vessels. A break in a blood vessel wall exposes proteins that attract platelets and initiate coagulation, the conversion of liquid components of blood to a solid clot. The coagulant, or sealant, circulates in an inactive form called fibrinogen. In response to a broken blood vessel, platelets release clotting factors that trigger reactions leading to the formation of thrombin, an enzyme that converts fibrinogen to fibrin. Newly formed fibrin aggregates into threads that form the framework of the clot. Thrombin also

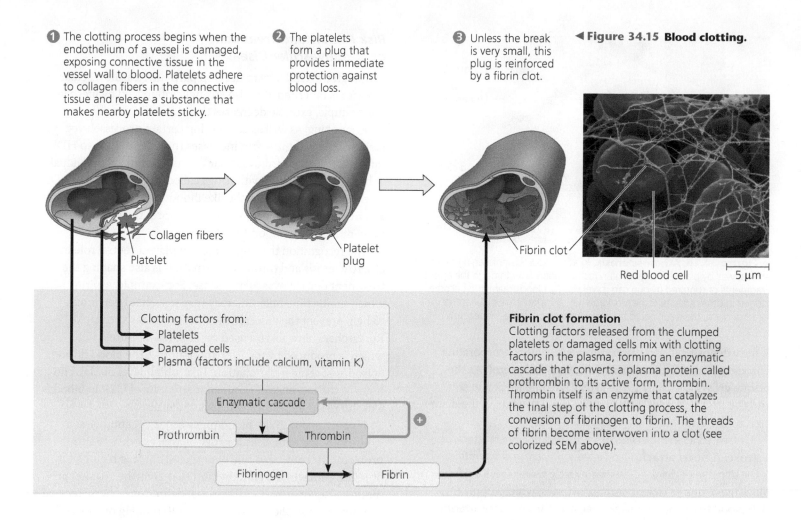

1 The clotting process begins when the endothelium of a vessel is damaged, exposing connective tissue in the vessel wall to blood. Platelets adhere to collagen fibers in the connective tissue and release a substance that makes nearby platelets sticky.

2 The platelets form a plug that provides immediate protection against blood loss.

3 Unless the break is very small, this plug is reinforced by a fibrin clot.

◀ **Figure 34.15 Blood clotting.**

Collagen fibers

Platelet

Platelet plug

Fibrin clot

Red blood cell

5 μm

Clotting factors from:
→ Platelets
→ Damaged cells
→ Plasma (factors include calcium, vitamin K)

Enzymatic cascade

Prothrombin → Thrombin

Fibrinogen → Fibrin

Fibrin clot formation
Clotting factors released from the clumped platelets or damaged cells mix with clotting factors in the plasma, forming an enzymatic cascade that converts a plasma protein called prothrombin to its active form, thrombin. Thrombin itself is an enzyme that catalyzes the final step of the clotting process, the conversion of fibrinogen to fibrin. The threads of fibrin become interwoven into a clot (see colorized SEM above).

activates a factor that catalyzes the formation of more thrombin, driving clotting to completion through positive feedback (see Chapter 32). The steps in the production of a blood clot are diagrammed in **Figure 34.15**. Any genetic mutation that blocks a step in the clotting process can cause hemophilia, a disease characterized by excessive bleeding and bruising from even minor cuts and bumps (see Chapter 12).

Anticlotting factors in the blood normally prevent spontaneous clotting in the absence of injury. Sometimes, however, clots form within a blood vessel, blocking the flow of blood. Such a clot is called a **thrombus**. We'll explore how a thrombus forms and the danger that it poses later in this chapter.

Cardiovascular Disease

Cardiovascular diseases—disorders of the heart and blood vessels—cause more than 750,000 human deaths each year in the United States. These diseases range from a minor disturbance of vein or heart valve function to a life-threatening disruption of blood flow to the heart or brain.

Cholesterol metabolism has a central role in cardiovascular disease. In animal cell membranes this steroid is important for maintaining normal membrane fluidity (see Figure 5.5). Cholesterol travels in plasma in particles that consist of thousands of cholesterol molecules and other lipids bound

to a protein. One type of particle—**low-density lipoprotein (LDL)**—delivers cholesterol to cells for membrane production. Another type—**high-density lipoprotein (HDL)**—scavenges excess cholesterol for return to the liver. A high ratio of LDL to HDL substantially increases the risk for atherosclerosis, a form of heart disease discussed below.

Another factor in cardiovascular disease is *inflammation*, the body's reaction to injury. Tissue damage leads to recruitment of two types of circulating immune cells, macrophages and leukocytes. Signals released by these cells trigger a flow of fluid out of blood vessels at the site of injury, resulting in the tissue swelling characteristic of inflammation (see Figure 35.5). Although inflammation is often a normal and healthy response to injury, it sometimes significantly disrupts circulatory function, as explained in the next section.

Atherosclerosis, Heart Attacks, and Stroke

Circulating cholesterol and inflammation can act together to produce a cardiovascular disease called **atherosclerosis**, the hardening of the arteries by fatty deposits. Healthy arteries have a smooth inner lining that reduces resistance to blood flow. Damage or infection can roughen the lining and lead to inflammation. Leukocytes are attracted to the damaged lining and begin to take up lipids, including cholesterol from LDL.

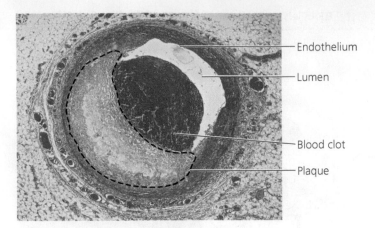

Endothelium

Lumen

Blood clot

Plaque

▲ **Figure 34.16 Atherosclerosis.** In atherosclerosis, thickening of an arterial wall by plaque formation can restrict blood flow through the artery. Fragments of ruptured plaque can travel via the bloodstream and become lodged in other arteries. If the blockage is in an artery that supplies the heart or brain, the result could be a heart attack or stroke, respectively.

A fatty deposit called a plaque, grows steadily, incorporating fibrous connective tissue and additional cholesterol. As the plaque grows, the walls of the artery become thick and stiff, and the obstruction of the artery increases, reducing the diameter available for blood flow **(Figure 34.16)**.

Untreated atherosclerosis often results in a heart attack or a stroke. A **heart attack**, also called a *myocardial infarction*, is the damage or death of cardiac muscle tissue resulting from blockage of one or more coronary arteries, which supply O_2-rich blood to the heart muscle. Because the coronary arteries are small in diameter, they are especially vulnerable to obstruction. Such blockage can destroy cardiac muscle quickly because the constantly beating heart muscle cannot survive long without O_2. If the heart stops beating, the victim may nevertheless survive if a heartbeat is restored within a few minutes by cardiopulmonary resuscitation (CPR) or some other emergency procedure.

A **stroke** is the death of nervous tissue in the brain due to a lack of O_2. Strokes usually result from rupture or blockage of arteries in the neck or head. The effects of a stroke and the individual's chance of survival depend on the extent and location of the damaged brain tissue. If a stroke results from a blocked artery, rapid administration of a clot-dissolving drug may help limit the damage.

Although atherosclerosis often isn't detected until critical blood flow is disrupted, there can be warning signs. Partial blockage of the coronary arteries may cause occasional chest pain, a condition known as angina pectoris, or more commonly angina. The pain is most likely to be felt when the heart is laboring hard during physical or emotional stress, and it signals that part of the heart is not receiving enough O_2. An obstructed coronary artery may be treated surgically, either by inserting a metal mesh tube called a stent to expand the artery or by transplanting a healthy blood vessel from the chest or a limb to bypass the blockage.

Risk Factors and Treatment of Cardiovascular Disease

Although the tendency to develop particular cardiovascular diseases is inherited, it is also strongly influenced by lifestyle. For example, exercise decreases the LDL/HDL ratio. In contrast, smoking, as well as consuming certain processed vegetable oils called *trans fats*, increases the ratio of LDL to HDL, raising the risk of cardiovascular disease. For many individuals at high risk, treatment with drugs called statins can lower LDL levels and thereby reduce the likelihood of heart attacks. In the **Scientific Skills Exercise**, you can interpret the effect of a genetic mutation on blood LDL levels.

The recognition that inflammation plays a central role in atherosclerosis and thrombus formation is also shaping the treatment of cardiovascular disease. For example, aspirin, which inhibits the inflammatory response, has been found to help prevent the recurrence of heart attacks and stroke. Researchers have also focused on C-reactive protein (CRP), which is produced by the liver and found in the blood during episodes of acute inflammation. Like a high level of LDL cholesterol, the presence of significant amounts of CRP in blood is a useful risk indicator for cardiovascular disease.

Hypertension (high blood pressure) is yet another contributor to heart attack and stroke as well as other health problems. According to one hypothesis, chronic high blood pressure damages the endothelium that lines the arteries, promoting plaque formation. The usual definition of hypertension in adults is a systolic pressure above 140 mm Hg or a diastolic pressure above 90 mm Hg. Fortunately, hypertension is simple to diagnose and can usually be controlled by dietary changes, exercise, medication, or a combination of these approaches.

CONCEPT CHECK 34.4

1. Explain why a physician might order a white cell count for a patient with symptoms of an infection.
2. Clots in arteries can cause heart attacks and strokes. Why, then, does it make sense to treat people with hemophilia by introducing clotting factors into their blood?
3. **WHAT IF?** Nitroglycerin (the key ingredient in dynamite) is sometimes prescribed for heart disease patients. Within the body, the nitroglycerin is converted to nitric oxide. Why would you expect nitroglycerin to relieve chest pain in these patients?

For suggested answers, see Appendix A.

CONCEPT 34.5

Gas exchange occurs across specialized respiratory surfaces

In the remainder of this chapter, we will focus on the process of **gas exchange**. Although this process is often called respiratory exchange or respiration, it should not be confused

Interpreting Data in Histograms

Does Inactivating the PCSK9 Enzyme Lower LDL Levels in Humans? Researchers interested in genetic factors affecting susceptibility to cardiovascular disease examined the DNA of 15,000 individuals. This screening revealed that 3% of the population sample had a mutation that inactivated one copy (allele) of the gene for PCSK9, a human liver enzyme. Because mutations that *increase* PCSK9 activity *increase* levels of LDL cholesterol in the blood, the researchers hypothesized that *inactivating* mutations in this gene would *lower* LDL levels. In this exercise, you will interpret the results of an experiment they carried out to test this hypothesis.

How the Experiment Was Done Researchers measured LDL cholesterol levels in blood plasma from 85 individuals with one copy of the *PCSK9* gene inactivated (the study group) and from 3,278 wild-type individuals (the control group).

Data from the Experiment

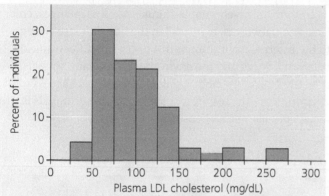

Individuals with an inactivating mutation in one copy of *PCSK9* gene (study group)

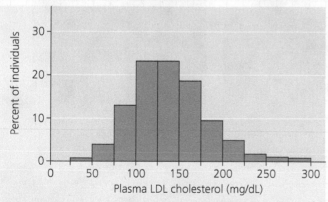

Individuals with two functional copies of *PCSK9* gene (control group)

Interpret the Data

1. The results are presented using a variant of bar graph called a *histogram*. In a histogram, the variable on the *x*-axis is grouped into ranges. The height of each bar in this histogram reflects the percentage of samples that fall into the range specified on the *x*-axis for that bar. For example, in the top histogram, about 4% of individuals studied had plasma LDL cholesterol levels in the 25–50 mg/dL (milligrams per deciliter) range. Add the percentages for the relevant bars to calculate the percentage of individuals in the study group and the control group with an LDL level below 100 mg/dL. (For additional information about histograms, see the Scientific Skills Review in Appendix F and in the Study Area in MasteringBiology.)

2. Given the differences between these two histograms, what conclusion can you draw?

3. Based on what you know about LDL cholesterol, would you predict that the individuals in the study group have an increased, unchanged, or reduced risk for cardiovascular disease relative to wild-type individuals? Explain.

4. Propose an explanation for the fact that the two histograms overlap as much as they do.

5. Comparing these two histograms allowed researchers to draw a conclusion regarding the effect of PCSK9 mutations on LDL cholesterol levels in blood. Suppose you now consider two individuals with a plasma LDL level of 160 mg/dL, one from the study group and one from the control group. What do you predict regarding their relative risk of cardiovascular disease? Explain how you arrived at your prediction. What role did the histograms play in making your prediction in this case?

Data from J. C. Cohen et al., Sequence variations in PCSK9, low LDL, and protection against coronary heart disease, *New England Journal of Medicine* 354:1264–1272 (2006).

(MB) A version of this Scientific Skills Exercise can be assigned in MasteringBiology.

---

with the energy transformations of cellular respiration. Gas exchange is the uptake of molecular O_2 from the environment and the discharge of CO_2 to the environment. The respiratory medium—the source of oxygen in the environment—is either water or air.

Partial Pressure Gradients in Gas Exchange

To understand the driving forces for gas exchange, we must calculate **partial pressure**, which is simply the pressure exerted by a particular gas in a mixture of gases. To do so, we need to know the pressure that the mixture exerts and the fraction of the mixture represented by a particular gas. Let's consider O_2 as an example. At sea level, the atmosphere exerts a downward force equal to that of a column of mercury (Hg) 760 mm high. Atmospheric pressure at sea level is thus 760 mm Hg. Since the atmosphere is 21% O_2 by volume, the partial pressure of O_2 is 0.21×760, or about 160 mm Hg. This value is called the *partial pressure* of O_2 (abbreviated P_{O_2}) because it is the part of atmospheric pressure contributed by O_2. The partial pressure of CO_2 (abbreviated P_{CO_2}) is much less, only 0.29 mm Hg at sea level.

Partial pressures also apply to gases dissolved in a liquid, such as water. When water is exposed to air, an equilibrium is reached in which the partial pressure of each gas in the water equals the partial pressure of that gas in the air. Thus, water exposed to air at sea level has a P_{O_2} of 160 mm Hg, the same as in the atmosphere. However, the *concentrations* of O_2 in the air and water differ substantially because O_2 is much less soluble in water than in air.

Once we have calculated partial pressures, we can readily predict the net result of diffusion at gas exchange surfaces: A gas always undergoes net diffusion from a region of higher partial pressure to a region of lower partial pressure.

Respiratory Media

The conditions for gas exchange vary considerably, depending on whether the respiratory medium—the source of O_2—is air or water. As already noted, O_2 is plentiful in air, making up about 21% of Earth's atmosphere by volume. Compared to water, air is much less dense and less viscous, so it is easier to move and to force through small passageways. As a result, breathing air is relatively easy and need not be particularly efficient. Humans, for example, extract only about 25% of the O_2 in inhaled air.

Gas exchange with water as the respiratory medium is much more demanding. The amount of O_2 dissolved in a given volume of water varies but is always less than in an equivalent volume of air: Water in many marine and freshwater habitats contains only 4–8 mL of dissolved O_2 per liter, a concentration roughly 40 times less than in air. The warmer and saltier the water is, the less dissolved O_2 it can hold. Water's lower O_2 content, greater density, and greater viscosity mean that aquatic animals such as fishes and lobsters must expend considerable energy to carry out gas exchange. In the context of these challenges, adaptations have evolved that enable most aquatic animals to be very efficient in gas exchange. Many of these adaptations involve the organization of the surfaces dedicated to exchange, as illustrated in a marine worm and a sea star **(Figure 34.17)**.

Respiratory Surfaces

Specialization for gas exchange is apparent in the structure of the respiratory surface, the part of an animal's body where gas exchange occurs. Like all living cells, the cells that carry out gas exchange have a plasma membrane that must be in contact with an aqueous solution. Respiratory surfaces are therefore always moist.

The movement of O_2 and CO_2 across respiratory surfaces takes place by diffusion. The rate of diffusion is proportional to the surface area across which it occurs and inversely proportional to the square of the distance through which molecules must move. In other words, gas exchange is fast when the area for diffusion is large and the path for diffusion is short. As a result, respiratory surfaces tend to be large and thin.

In some relatively simple animals, such as sponges, cnidarians, and flatworms, every cell in the body is close enough to the external environment that gases can diffuse quickly between any cell and the environment. In many animals, however, the bulk of the body's cells lack immediate access to the environment. The respiratory surface in these animals is a thin, moist epithelium that constitutes a respiratory organ.

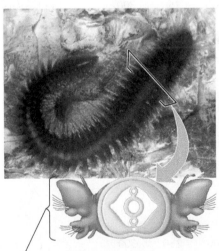

Parapodium (functions as gill)

(a) Marine worm. Many polychaetes (marine worms of the phylum Annelida) have a pair of flattened appendages called parapodia (singular, *parapodium*) on each body segment. The parapodia serve as gills and also function in crawling and swimming.

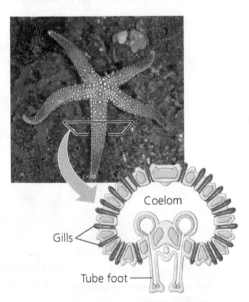

Coelom

Gills

Tube foot

(b) Sea star. The gills of a sea star are simple tubular projections of the skin. The hollow core of each gill is an extension of the coelom (body cavity). Gas exchange occurs by diffusion across the gill surfaces, and fluid in the coelom circulates in and out of the gills, aiding gas transport. The surfaces of a sea star's tube feet also function in gas exchange.

▲ **Figure 34.17 Diversity in the structure of gills, external body surfaces that function in gas exchange.**

In some animals, including earthworms and some amphibians, the skin serves as a respiratory organ. A dense network of capillaries just below the surface facilitates the exchange of gases between the circulatory system and the environment. For most animals, however, the general body surface lacks sufficient area to exchange gases for the whole organism. The evolutionary solution to this limitation is a respiratory organ that is extensively folded or branched, thereby enlarging the available surface area for gas exchange. Gills, tracheae, and lungs are three such organs.

Gills in Aquatic Animals

Gills are outfoldings of the body surface that are suspended in the water. They can be localized to specialized structures, as in an axolotl (see Figure 34.1), or distributed across the body, as in a sea star or marine worm (see Figure 34.17). Movement of the respiratory medium over the respiratory surface, a process called **ventilation**, maintains the partial pressure gradients of O_2 and CO_2 across the gill that are necessary for gas exchange.

To promote ventilation, most gill-bearing animals move their gills through the water or move water over their gills. For example, crayfish and lobsters have paddle-like appendages that drive a current of water over the gills, whereas mussels and clams move water with cilia. Octopuses and squids ventilate their gills by taking in and ejecting water, with the side benefit of locomotion by jet propulsion. Fishes use the motion of swimming or movements of the mouth and gill covers to ventilate their gills. In both cases, a water current enters the mouth of the fish, passes through slits in the pharynx, flows over the gills, and exits the body **(Figure 34.18)**.

In fishes, the efficiency of gas exchange is maximized by **countercurrent exchange**, the exchange of a substance or heat between two fluids flowing in opposite directions. (Recall from Chapter 32 that countercurrent exchange also contributes to temperature regulation and kidney function.) Because blood flows in the direction opposite to that of water passing over the gills, blood is less saturated with O_2 at each point in its travel than the water it meets (see Figure 34.18). As blood enters a gill capillary, it encounters water that is completing its passage through the gill. Depleted of much of its dissolved O_2, this water nevertheless has a higher P_{O_2} than the incoming blood, and O_2 transfer takes place. As the blood continues its passage, its P_{O_2} steadily increases, but so does that of the water it encounters, since each successive position in the blood's travel corresponds to an earlier position in the water's passage over the gills. Thus, a partial pressure gradient favoring the diffusion of O_2 from water to blood exists along the entire length of the capillary.

Countercurrent exchange mechanisms are remarkably efficient. More than 80% of the O_2 dissolved in the water entering the mouth and gills of the fish is removed as it passes over the respiratory surface.

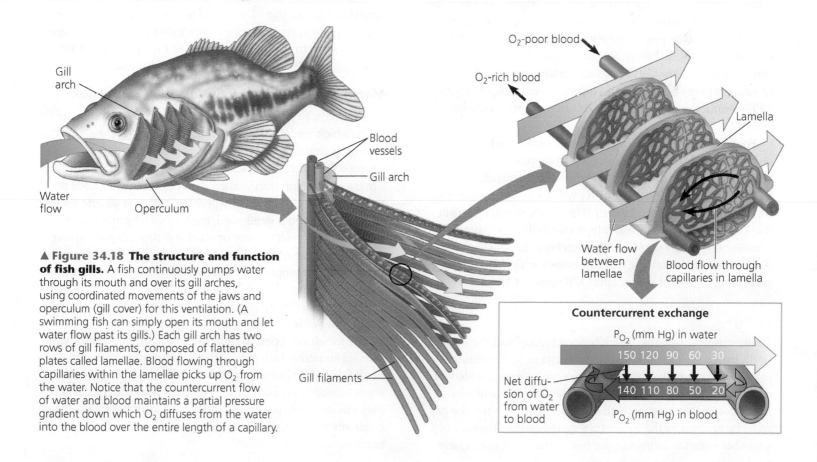

▲ **Figure 34.18 The structure and function of fish gills.** A fish continuously pumps water through its mouth and over its gill arches, using coordinated movements of the jaws and operculum (gill cover) for this ventilation. (A swimming fish can simply open its mouth and let water flow past its gills.) Each gill arch has two rows of gill filaments, composed of flattened plates called lamellae. Blood flowing through capillaries within the lamellae picks up O_2 from the water. Notice that the countercurrent flow of water and blood maintains a partial pressure gradient down which O_2 diffuses from the water into the blood over the entire length of a capillary.

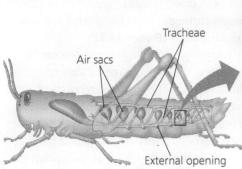
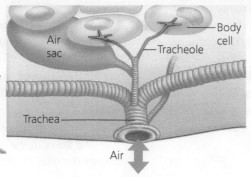
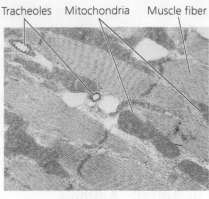

Tracheoles Mitochondria Muscle fiber

2.5 μm

(a) The respiratory system of an insect consists of branched internal tubes. The largest tubes, called tracheae, connect to external openings spaced along the insect's body surface. Air sacs formed from enlarged portions of the tracheae are found near organs that require a large supply of oxygen.

(b) Rings of chitin keep the tracheae open, allowing air to enter and pass into smaller tubes called tracheoles. The branched tracheoles deliver air directly to cells throughout the body. Tracheoles have closed ends filled with fluid (blue-gray). When the animal is active and using more O_2, most of the fluid is withdrawn into the body. This increases the surface area of air-filled tracheoles in contact with cells.

(c) The TEM above shows cross sections of tracheoles in a tiny piece of insect flight muscle. Each of the numerous mitochondria in the muscle cells lies within about 5 μm of a tracheole.

▲ **Figure 34.19 The insect tracheal system.**

Tracheal Systems in Insects

In most terrestrial animals, respiratory surfaces are enclosed within the body, exposed to the atmosphere only through narrow tubes. Although the most familiar such structure is the lung, the most common is the insect **tracheal system**, a network of air tubes that branch throughout the body. The largest tubes, called tracheae, open to the outside **(Figure 34.19a)**. The finest branches extend close to the surface of nearly every cell, where gas is exchanged by diffusion across the moist epithelium that lines the tips of the tracheal branches **(Figure 34.19b)**. Because the tracheal system brings air within a very short distance of virtually every body cell in an insect, it can transport O_2 and CO_2 without the participation of the animal's open circulatory system.

For small insects, diffusion through the tracheae brings in enough O_2 and removes enough CO_2 to support cellular respiration. Larger insects meet their higher energy demands by ventilating their tracheal systems with rhythmic body movements that compress and expand the air tubes like bellows. Insects in flight, which have a particularly high metabolic rate, have flight muscle cells that are packed with mitochondria. Tracheal tubes distributed throughout the flight muscles supply these ATP-generating organelles with ample O_2 **(Figure 34.19c)**.

Lungs

Unlike tracheal systems, which branch throughout the insect body, **lungs** are localized respiratory organs. Representing an infolding of the body surface, they are typically subdivided into numerous pockets. Because the respiratory surface of a lung is not in direct contact with all other parts of the body, the gap must be bridged by the circulatory system, which transports gases between the lungs and the rest of the body. Lungs have

evolved in organisms with open circulatory systems, such as spiders and land snails, as well as in vertebrates.

Among vertebrates that lack gills, the use of lungs for gas exchange varies. Amphibians rely heavily on diffusion across body surfaces, such as the skin, to carry out gas exchange; lungs, if present, are relatively small. In contrast, most reptiles (including all birds) and all mammals depend entirely on lungs for gas exchange. Lungs and air breathing have evolved in a few aquatic vertebrates as adaptations to living in oxygen-poor water or to spending part of their time exposed to air (for instance, when the water in a pond recedes).

Mammalian Respiratory Systems: A Closer Look

In mammals, a system of branching ducts conveys air to the lungs, which are located in the thoracic cavity **(Figure 34.20)**. Air enters through the nostrils and is then filtered by hairs, warmed, humidified, and sampled for odors as it flows through a maze of spaces in the nasal cavity. The nasal cavity leads to the pharynx, an intersection where the paths for air and food cross. When food is swallowed, the **larynx** (the upper part of the respiratory tract) moves upward and tips a flap of cartilage over the opening of the **trachea**, or windpipe. This allows food to go down the esophagus to the stomach. The rest of the time, the airway is open, enabling breathing.

From the larynx, air passes into the trachea. Cartilage reinforcing the walls of both the larynx and the trachea keeps this part of the airway open. Within the larynx of most mammals, exhaled air rushes by a pair of elastic bands of muscle called vocal folds, or, in humans, vocal cords. Sounds are produced when muscles in the larynx are tensed, stretching the cords so they vibrate. High-pitched sounds result from tightly stretched cords vibrating rapidly; low-pitched sounds come from less tense cords vibrating slowly.

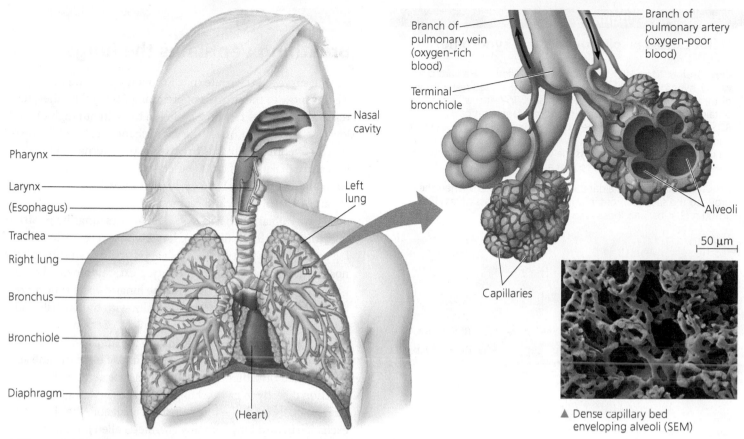

Branch of pulmonary vein (oxygen-rich blood)

Branch of pulmonary artery (oxygen-poor blood)

Terminal bronchiole

Alveoli

Capillaries

50 μm

Nasal cavity

Pharynx

Larynx

(Esophagus)

Trachea

Right lung

Bronchus

Bronchiole

Diaphragm

Left lung

(Heart)

▲ Dense capillary bed enveloping alveoli (SEM)

▲ **Figure 34.20 The mammalian respiratory system.** From the nasal cavity and pharynx, inhaled air passes through the larynx, trachea, and bronchi to the bronchioles, which end in microscopic alveoli lined by a thin, moist epithelium. Branches of the pulmonary arteries convey oxygen-poor blood to the alveoli; branches of the pulmonary veins transport oxygen-rich blood from the alveoli back to the heart.

The trachea branches into two **bronchi** (singular, *bronchus*), one leading to each lung. Within the lung, the bronchi branch repeatedly into finer and finer tubes called *bronchioles*. The entire system of air ducts has the appearance of an inverted tree, the trunk being the trachea. The epithelium lining the major branches of this respiratory tree is covered by cilia and a thin film of mucus. The mucus traps dust, pollen, and other particulate contaminants, and the beating cilia move the mucus upward to the pharynx, where it can be swallowed into the esophagus. This process, sometimes referred to as the "mucus escalator," plays a crucial role in cleansing the respiratory system.

Gas exchange in mammals occurs in **alveoli** (singular, *alveolus*; see Figure 34.20), air sacs clustered at the tips of the tiniest bronchioles. Human lungs contain millions of alveoli, which together have a surface area of about 100 m²—50 times that of the skin. Oxygen in the air entering the alveoli dissolves in the moist film lining their inner surfaces and rapidly undergoes net diffusion across the epithelium into a web of capillaries that surrounds each alveolus. Net diffusion of carbon dioxide occurs in the opposite direction, from the capillaries across the epithelium of the alveolus and into the air space.

Lacking cilia or significant air currents to remove particles from their surface, alveoli are highly susceptible to contamination. White blood cells patrol the alveoli, engulfing foreign particles. However, if too much particulate matter reaches the alveoli, the defenses can be overwhelmed, leading to inflammation and irreversible damage. For example, particulates from cigarette smoke that enter alveoli can cause a permanent reduction in lung capacity. For coal miners, inhalation of large amounts of coal dust can lead to silicosis, a disabling, irreversible, and sometimes fatal lung disease.

The film of liquid that lines alveoli is subject to surface tension, an attractive force that has the effect of minimizing a liquid's surface area (see Chapter 2). Given their tiny diameter (about 0.25 mm), why don't alveoli collapse under high surface tension? Researchers reasoned that alveoli must be coated with a material that reduces surface tension. In 1955, English biophysicist Richard Pattle obtained experimental evidence for such a material, now called a **surfactant**, for *surface-active* agent. In addition, he proposed that the absence of surfactant might cause respiratory distress syndrome (RDS), a disease common among preterm infants born 6 weeks or more before their due dates. In the 1950s, RDS killed 10,000 infants annually in the United States alone.

In the late 1950s, Mary Ellen Avery carried out the first experiment linking RDS to a surfactant deficiency

What causes respiratory distress syndrome?

Experiment Mary Ellen Avery, a research fellow at Harvard University, hypothesized that a lack of surfactant caused respiratory distress syndrome (RDS) in preterm infants. To test this idea, she obtained autopsy samples of lungs from infants that had died of RDS or from other causes. She extracted material from the samples and let it form a film on water. Avery then measured the tension (in dynes per centimeter) across the water surface and recorded the lowest surface tension observed for each sample.

Results In analyzing the data, Avery noted a pattern when she grouped the samples from infants with a body mass of less than 1,200 g (2.7 lbs) and those who had grown larger.

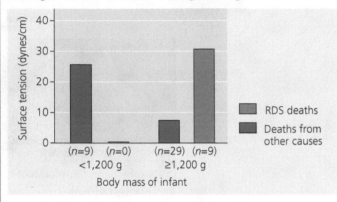

Conclusion For infants 1,200 g and above, the samples from those that had died of causes other than RDS exhibited much lower surface tension than samples from those that had died of RDS. Avery inferred that infants' lungs normally contain a surface-tension reducing material (now called surfactant), and that a lack of this material was a likely cause of RDS. The results from infants less than 1,200 g were similar to those of infants who had died from RDS, suggesting that surfactant is not normally produced until a fetus reaches this size.

Source M. E. Avery and J. Mead, Surface properties in relation to atelectasis and hyaline membrane disease, *American Journal of Diseases of Children* 97:517–523 (1959).

WHAT IF? Suppose researchers had measured the amount of surfactant in lung samples. Describe the graph you would expect if the amount of surfactant were plotted against infant weight.

(Figure 34.21). Subsequent studies revealed that surfactant contains a mixture of phospholipids and proteins and in a full-term (38-week) pregnancy appears in the lungs about five weeks before birth. Artificial surfactants are now used routinely to treat early preterm infants. For her contributions, Avery received the National Medal of Science in 1991.

CONCEPT CHECK 34.5

1. Why is the position of lung tissues *within* the body an advantage for terrestrial animals?
2. After a heavy rain, earthworms come to the surface. How would you explain this behavior in terms of an earthworm's requirements for gas exchange?
3. **MAKE CONNECTIONS** Describe the role of countercurrent exchange in facilitating both thermoregulation (see Concept 32.1) and respiration.

For suggested answers, see Appendix A.

Breathing ventilates the lungs

Having surveyed the route that air follows when we breathe, we turn now to the process of breathing itself. Like fishes, terrestrial vertebrates rely on ventilation to maintain high O_2 and low CO_2 concentrations at the gas exchange surface. The process that ventilates lungs is **breathing**, the alternating inhalation and exhalation of air. A variety of mechanisms for moving air in and out of lungs have evolved, as we will see by considering amphibians, birds, and mammals.

An amphibian such as a frog ventilates its lungs by **positive pressure breathing**, filling the lungs with forced airflow. During each cycle of ventilation, fresh air is first drawn through the nostrils into a specialized oral cavity. Next, this air-filled cavity is closed off while elastic recoil of the lungs directs stale air out through the mouth and nostrils. Finally, with the nostrils and mouth closed and the oral cavity open to the trachea, the floor of the oral cavity rises, forcing air into the lungs.

To bring fresh air to their lungs, birds use eight or nine air sacs situated on either side of the lungs. The air sacs do not function directly in gas exchange but act as bellows that keep air flowing through the lungs. Instead of alveoli, the sites of gas exchange in bird lungs are tiny channels called *parabronchi*. Passage of air through the entire system—lungs and air sacs—requires two cycles of inhalation and exhalation.

Two features of ventilation in birds make it highly efficient. First, when birds breathe, air passes over the gas exchange surface in only one direction. Second, incoming fresh air does not mix with air that has already carried out gas exchange.

How a Mammal Breathes

Mammals employ **negative pressure breathing**—pulling, rather than pushing, air into their lungs **(Figure 34.22)**. Using muscle contraction to actively expand their thoracic cavity, mammals lower air pressure in their lungs below that of the outside air. Because gas flows from a region of higher pressure to a region of lower pressure, air rushes through the nostrils and mouth and down the breathing tubes to the alveoli. During exhalation, the muscles controlling the thoracic cavity relax, and the cavity's volume is reduced. The increased air pressure in the alveoli forces air up the breathing tubes and out of the body. Thus, inhalation is always active and requires work, whereas exhalation is usually passive.

Expanding the thoracic cavity during inhalation involves rib muscles and the **diaphragm**, a sheet of skeletal muscle that forms the bottom wall of the cavity. Contracting one set of rib muscles expands the rib cage, the front wall of the thoracic cavity, by pulling the ribs upward and the sternum outward. At the same time, the diaphragm contracts, expanding the thoracic cavity downward. The effect of the descending diaphragm is similar to that of a plunger being drawn out of a syringe.

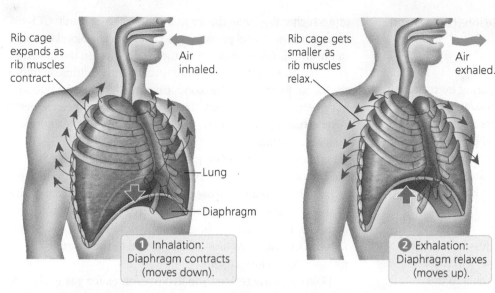

Rib cage expands as rib muscles contract.

Air inhaled.

Lung

Diaphragm

1 Inhalation: Diaphragm contracts (moves down).

Rib cage gets smaller as rib muscles relax.

Air exhaled.

2 Exhalation: Diaphragm relaxes (moves up).

▲ **Figure 34.22 Negative pressure breathing.** A mammal breathes by changing the air pressure within its lungs relative to the pressure of the outside atmosphere.

Within the thoracic cavity, a double membrane surrounds the lungs. The inner layer of this membrane adheres to the outside of the lungs, and the outer layer adheres to the wall of the thoracic cavity. A thin space filled with fluid separates the two layers. Surface tension in the fluid causes the two layers to stick together like two plates of glass separated by a film of water: The layers can slide smoothly past each other, but they cannot be pulled apart easily. Consequently, the volume of the thoracic cavity and the volume of the lungs change in unison.

The volume of air inhaled and exhaled with each breath is called **tidal volume**. It averages about 500 mL in resting adult humans. The tidal volume during maximal inhalation and exhalation is the **vital capacity**, which is about 3.4 L and 4.8 L for college-age women and men, respectively. The air that remains after a forced exhalation is called the **residual volume**. As humans age, our lungs lose their resilience, and residual volume increases at the expense of vital capacity.

Because the lungs in mammals do not completely empty with each breath, and because inhalation occurs through the same airways as exhalation, each inhalation mixes fresh air with oxygen-depleted residual air. As a result, the maximum P_{O_2} in alveoli is always considerably less than in the atmosphere. This is one reason mammals don't function as well as birds at high altitude. For example, humans have great difficulty obtaining enough O_2 when climbing Earth's highest peaks, such as Mount Everest (8,850 m) in the Himalayas. However, bar-headed geese and several other bird species easily fly through high Himalayan passes during their migrations.

Control of Breathing in Humans

Although you can voluntarily hold your breath or breathe faster and deeper, most of the time your breathing is regulated by involuntary mechanisms. These control mechanisms ensure

that gas exchange is coordinated with blood circulation and with metabolic demand.

The neurons mainly responsible for regulating breathing are in the medulla oblongata, near the base of the brain **(Figure 34.23)**. Neural circuits in the medulla form a breathing control center that establishes the breathing rhythm. When you breathe deeply, a negative-feedback mechanism prevents the lungs from overexpanding: During inhalation, sensors that detect stretching of the lung tissue send nerve impulses to the control circuits in the medulla, inhibiting further inhalation.

In regulating breathing, the medulla uses the pH of the surrounding tissue fluid as an indicator of blood CO_2 concentration. The reason pH can be used in this way is that blood CO_2 is the main determinant of the pH of cerebrospinal fluid, the fluid surrounding the brain and spinal cord. Carbon dioxide diffuses from the blood to the cerebrospinal fluid, where it reacts with water and forms carbonic acid (H_2CO_3). The

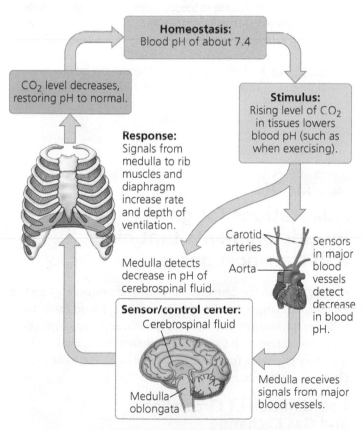

Homeostasis:
Blood pH of about 7.4

CO_2 level decreases, restoring pH to normal.

Stimulus:
Rising level of CO_2 in tissues lowers blood pH (such as when exercising).

Response:
Signals from medulla to rib muscles and diaphragm increase rate and depth of ventilation.

Carotid arteries

Aorta

Sensors in major blood vessels detect decrease in blood pH.

Medulla detects decrease in pH of cerebrospinal fluid.

Sensor/control center:
Cerebrospinal fluid

Medulla oblongata

Medulla receives signals from major blood vessels.

▲ **Figure 34.23 Homeostatic control of breathing.**

WHAT IF? *Suppose a person began breathing very rapidly while resting. Describe the effect on blood CO_2 levels and the steps by which the negative-feedback circuit would restore homeostasis.*

H_2CO_3 can then dissociate into a bicarbonate ion (HCO_3^-) and a hydrogen ion (H^+):

$$CO_2 + H_2O \rightleftharpoons H_2CO_3 \rightleftharpoons HCO_3^- + H^+$$

Increased metabolic activity, such as occurs during exercise, increases the concentration of CO_2 in the blood, and lowers pH through the reaction shown above. Sensors in the medulla as well as in major blood vessels detect this pH change. In response, the medulla's control circuits increase the depth and rate of breathing (see Figure 34.23). Both remain high until the excess CO_2 is eliminated in exhaled air and pH returns to a normal value.

The blood O_2 level usually has little effect on the breathing control centers. However, when the O_2 level drops very low (at high altitudes, for instance), O_2 sensors in the aorta and the carotid arteries in the neck send signals to the breathing control centers, which respond by increasing the breathing rate.

Breathing control is effective only if ventilation is matched to blood flow through the capillaries in the alveoli. During exercise, for instance, such coordination couples an increased breathing rate, which enhances O_2 uptake and CO_2 removal, with an increase in cardiac output.

CONCEPT CHECK 34.6

1. How does an increase in the CO_2 concentration in the blood affect the pH of cerebrospinal fluid?

2. A drop in blood pH causes an increase in heart rate. What is the function of this control mechanism?

3. **WHAT IF?** If an injury tore a small hole in the membranes surrounding your lungs, what effect on lung function would you expect?

For suggested answers, see Appendix A.

CONCEPT 34.7

Adaptations for gas exchange include pigments that bind and transport gases

The high metabolic demands of many animals necessitate the exchange of large quantities of O_2 and CO_2. Blood molecules called respiratory pigments facilitate this exchange through their interaction with O_2 and CO_2. Before exploring how respiratory pigments function, let's summarize the basic gas exchange circuit in humans.

Coordination of Circulation and Gas Exchange

The partial pressures of O_2 and CO_2 in the blood vary at different points in the circulatory system, as shown in **Figure 34.24**. Blood flowing through the alveolar capillaries has a lower P_{O_2}

and a higher P_{CO_2} than the air in the alveoli. As a result, CO_2 diffuses down its partial pressure gradient from the blood to the air in the alveoli. Meanwhile, O_2 in the air dissolves in the fluid that coats the alveolar epithelium and undergoes net diffusion into the blood. By the time the blood leaves the lungs in the pulmonary veins, its P_{O_2} has been raised and its P_{CO_2} has been lowered. After returning to the heart, this blood is pumped through the systemic circuit.

In the tissue capillaries, gradients of partial pressure favor the net diffusion of O_2 out of the blood and CO_2 into the blood. These gradients exist because cellular respiration in the mitochondria of cells near each capillary removes O_2 from and adds CO_2 to the surrounding interstitial fluid. After the blood unloads O_2 and loads CO_2, it is returned to the heart and pumped to the lungs again.

Having characterized the driving forces for gas exchange in different tissues, we will now introduce the critical role of the specialized carrier proteins—the respiratory pigments.

Respiratory Pigments

The low solubility of O_2 in water (and thus in blood) poses a problem for animals that rely on the circulatory system to deliver O_2. For example, a person requires almost 2 L of O_2 per minute during intense exercise, and all of it must be carried in the blood from the lungs to the active tissues. At normal body temperature and air pressure, however, only 4.5 mL of O_2 can dissolve into a

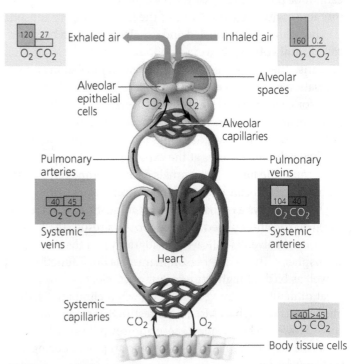

© 1998 Pearson Education, Inc.

▲ **Figure 34.24 Loading and unloading of respiratory gases.** The colored bars indicate the partial pressures (in mm Hg) of O_2 (P_{O_2}) and CO_2 (P_{CO_2}) in different locations.

WHAT IF? *If you consciously forced more air out of your lungs each time you exhaled, how would that affect the values shown above?*

liter of blood in the lungs. Even if 80% of the dissolved O_2 were delivered to the tissues (an unrealistically high percentage), the heart would still need to pump 555 L of blood per minute!

In fact, animals transport most of their O_2 bound to proteins called **respiratory pigments**. Respiratory pigments circulate with the blood or hemolymph and are often contained within specialized cells. The pigments greatly increase the amount of O_2 that can be carried in the circulatory fluid (to about 200 mL of O_2 per liter in mammalian blood). In our example of an exercising human with an O_2 delivery rate of 80%, the presence of a respiratory pigment reduces the cardiac output necessary for O_2 transport to a manageable 12.5 L of blood per minute.

A variety of respiratory pigments have evolved among the animal taxa. With a few exceptions, these molecules have a distinctive color (hence the term *pigment*) and consist of a metal bound to a protein. One example is the blue pigment hemocyanin, which has copper as its oxygen-binding component and is found in arthropods and many molluscs. The respiratory pigment of many invertebrates and almost all vertebrates is hemoglobin.

In vertebrates, hemoglobin is contained in the erythrocytes and consists of four subunits (polypeptide chains), each with a cofactor called a heme group that has an iron atom at its center. Each iron atom binds one molecule of O_2; hence, a single hemoglobin molecule can carry four molecules of O_2. Like all respiratory pigments, hemoglobin binds O_2 reversibly, loading O_2 in the lungs or gills and unloading it in other parts of the body. This process depends on cooperativity between the hemoglobin subunits (see Concept 6.5). When O_2 binds to one subunit, the others change shape slightly, increasing their affinity for O_2. Similarly, when four O_2 molecules are bound and one subunit unloads its O_2, the other three subunits more readily unload O_2, as an associated shape change lowers their affinity for O_2.

The cooperativity in both O_2 binding and release is evident in the dissociation curve for hemoglobin **(Figure 34.25a)**. Over the range of P_{O_2} where the dissociation curve has a steep slope, even a slight change in P_{O_2} causes hemoglobin to load or unload a substantial amount of O_2. Notice that the steep part of the curve corresponds to the range of P_{O_2} found in body tissues. When cells in a particular location begin working harder—during exercise, for instance—P_{O_2} dips in their vicinity as the O_2 is consumed in cellular respiration. Because of the effect of subunit cooperativity, a slight drop in P_{O_2} causes a relatively large increase in the amount of O_2 the blood unloads.

The production of CO_2 during cellular respiration promotes the unloading of O_2 by hemoglobin in active tissues. As we have seen, CO_2 reacts with water, forming carbonic acid, which lowers the pH of its surroundings. Low pH, in turn,

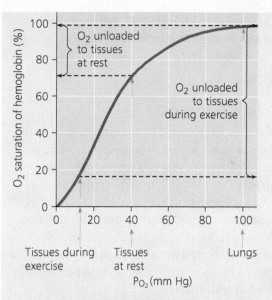

(a) **P_{O_2} and hemoglobin dissociation at pH 7.4.** The curve shows the relative amounts of O_2 bound to hemoglobin exposed to solutions with different P_{O_2}. At a P_{O_2} of 100 mm Hg, typical in the lungs, hemoglobin is about 98% saturated with O_2. At a P_{O_2} of 40 mm Hg, common in the vicinity of tissues at rest, hemoglobin is about 70% saturated, having unloaded nearly a third of its O_2. As shown in the above graph, hemoglobin can release much more O_2 to metabolically very active tissues, such as muscle tissue during exercise.

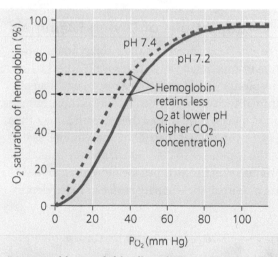

(b) **pH and hemoglobin dissociation.** Because hydrogen ions affect the shape of hemoglobin, a drop in pH shifts the O_2 dissociation curve toward the right (the Bohr shift). At a given P_{O_2}, say 40 mm Hg, hemoglobin gives up more O_2 at pH 7.2 than at pH 7.4, the normal pH of human blood. The pH is lower in very active tissues because the CO_2 produced by cellular respiration reacts with water, forming carbonic acid. There, hemoglobin releases more O_2, which supports the increased cellular respiration in the active tissues.

▲ Figure 34.25 **Dissociation curves for hemoglobin at 37°C.**

decreases the affinity of hemoglobin for O_2, an effect called the **Bohr shift (Figure 34.25b)**. Thus, where CO_2 production is greater, hemoglobin releases more O_2, which can then be used to support more cellular respiration.

In addition to its role in O_2 transport, hemoglobin assists in buffering the blood—that is, preventing harmful changes in pH. It also has a minor role in CO_2 transport, the topic we will explore next.

Carbon Dioxide Transport

Only about 7% of the CO_2 released by respiring cells is transported in solution in blood plasma. The remainder diffuses from the plasma into erythrocytes and reacts with water (assisted by the enzyme carbonic anhydrase), forming H_2CO_3. The H_2CO_3 readily dissociates into H^+ and HCO_3^-. Most of the H^+ binds to hemoglobin and other proteins, minimizing the change in blood pH. Most of the HCO_3^- diffuses out of the erythrocyte and is transported to the lungs in the plasma. The remainder, about 5% of the overall CO_2, binds to hemoglobin and is transported in erythrocytes.

When blood flows through the lungs, the relative partial pressures of CO_2 favor the diffusion of CO_2 out of the blood. As CO_2 diffuses from blood into alveoli, the amount of CO_2 in the blood decreases. This decrease shifts the chemical equilibrium in favor of the conversion of HCO_3^- to CO_2, enabling further net diffusion of CO_2 into alveoli. Overall, the P_{CO_2} gradient is sufficient to reduce P_{CO_2} by about 15% during passage of blood through the lungs.

Respiratory Adaptations of Diving Mammals

EVOLUTION Animals vary greatly in their ability to temporarily inhabit environments in which there is no access to their normal respiratory medium—for example, when an air-breathing mammal swims underwater. Whereas most humans, even well-trained divers, cannot hold their breath longer than 2 or 3 minutes or swim deeper than 20 m, the Weddell seal of Antarctica routinely plunges to 200–500 m and remains there for about 20 minutes (and sometimes for more than an hour). Another diving mammal, the elephant seal, can reach depths of 1,500 m—almost a mile—and stay submerged for as long as 2 hours! What evolutionary adaptations enable these animals to perform such amazing feats?

One adaptation of diving mammals to prolonged stays underwater is an ability to store large amounts of O_2. The Weddell seal has a greater volume of blood per kilogram of body mass than a human and has a high concentration of an oxygen-storing protein called **myoglobin** in its muscles. As a result, the Weddell seal can store about twice as much O_2 per kilogram of body mass as can a human.

Diving mammals not only have a relatively large O_2 stockpile but also have adaptations that conserve O_2. They swim with little muscular effort and glide passively upward or downward by changing their buoyancy. During a dive, most blood is routed to the brain, spinal cord, eyes, adrenal glands, and, in pregnant seals, the placenta. Blood supply to the muscles is restricted or, during the longest dives, shut off altogether. During dives of more than about 20 minutes, a Weddell seal's muscles deplete the O_2 stored in myoglobin and then derive their ATP from fermentation instead of respiration.

The unusual abilities of the Weddell seal and other air-breathing divers to power their bodies during long dives showcase two related themes in our study of organisms—the response to environmental challenges over the short term by physiological adjustments and over the long term as a result of natural selection.

CONCEPT CHECK 34.7

1. What determines whether the net diffusion of O_2 and CO_2 is into or out of the capillaries? Explain.
2. How does the Bohr shift help deliver O_2 to very active tissues?
3. **WHAT IF?** A doctor might give bicarbonate (HCO_3^-) to a patient who is breathing very rapidly. What assumption is the doctor making about the blood chemistry of the patient?

For suggested answers, see Appendix A.

34 Chapter Review

SUMMARY OF KEY CONCEPTS

CONCEPT 34.1

Circulatory systems link exchange surfaces with cells throughout the body (pp. 685–688)

- In animals with simple body plans, a **gastrovascular cavity** mediates exchange between the environment and body cells. Because net diffusion is slow over long distances, most complex animals have a circulatory system that moves fluid between cells and the organs that carry out exchange with the environment. Arthropods and most molluscs have an **open circulatory system**, in which **hemolymph** bathes organs directly.

- Vertebrates have a **closed circulatory system**, in which **blood** circulates in a closed network of pumps and vessels.
- The closed circulatory system of vertebrates consists of blood, blood vessels, and a two- to four-chambered **heart**. Blood pumped by a heart **ventricle** passes to **arteries** and then to **capillaries**, the sites of chemical exchange between blood and interstitial fluid. **Veins** return blood from capillaries to an **atrium**, which passes blood to a ventricle.
- Fishes, rays, and sharks have a single pump in their circulatory systems, whereas air-breathing vertebrates have two pumps combined in a single heart. Variations between different species in ventricle number and divisions reflect adaptations to different environments and metabolic needs.

CONCEPT 34.2

Coordinated cycles of heart contraction drive double circulation in mammals (pp. 689–690)

- The right ventricle pumps blood to the lungs, where it loads O_2 and unloads CO_2. Oxygen-rich blood from the lungs enters the heart at the left atrium and is pumped to the body tissues by the left ventricle. Blood returns to the heart through the right atrium.

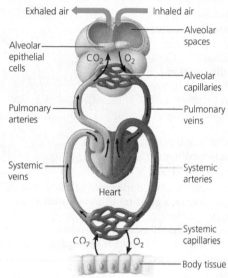

© 1998 Pearson Education, Inc.

- The **cardiac cycle**, one complete sequence of the heart's pumping and filling, consists of a period of contraction, called **systole**, and a period of relaxation, called **diastole**. The heartbeat originates with impulses at the **sinoatrial (SA) node** (pacemaker) of the right atrium. The impulses trigger atrial contraction, are delayed at the **atrioventricular (AV) node**, and then cause ventricular contraction. The nervous system, hormones, and body temperature influence pacemaker activity.

? What changes in cardiac function might you expect after surgical replacement of a defective heart valve?

CONCEPT 34.3

Patterns of blood pressure and flow reflect the structure and arrangement of blood vessels (pp. 690–694)

- Blood vessels have structures well adapted to function. Capillaries have narrow diameters and thin walls that facilitate exchange. Arteries contain thick elastic walls that maintain blood pressure. Veins contain one-way valves that contribute to the return of blood to the heart. Blood pressure is altered by changes in cardiac output and by variable constriction of arterioles.
- The velocity of blood flow is lowest in the capillary beds as a result of their large total cross-section area. Fluid leaks out of capillaries and is returned to blood by the **lymphatic system**, which also plays a vital role in defense against infection.

? If you rest your forearm on the top of your head rather than by your side, how, if at all, is blood pressure in that arm affected? Explain.

CONCEPT 34.4

Blood components function in exchange, transport, and defense (pp. 695–698)

- Whole blood consists of cells and cell fragments (**platelets**) suspended in a liquid matrix called **plasma**. Plasma proteins influence blood pH, osmotic pressure, and viscosity, and they function in lipid transport, immunity (antibodies), and blood clotting (fibrinogen). Red blood cells, or **erythrocytes**, transport O_2. White blood cells, or **leukocytes**, function in defense against microbes and foreign substances. Platelets function in blood clotting.
- A variety of diseases impair function of the circulatory system. In **sickle-cell disease**, an aberrant form of **hemoglobin** disrupts erythrocyte shape and function. In cardiovascular disease, inflammation of the arterial lining enhances deposition of lipids and cells, resulting in the potential for life-threatening damage to the heart or brain.

? In the absence of infection, what percentage of cells in human blood are leukocytes?

CONCEPT 34.5

Gas exchange occurs across specialized respiratory surfaces (pp. 698–704)

- At all sites of **gas exchange**, a gas undergoes net diffusion from where its **partial pressure** is higher to where it is lower. Air is more conducive to gas exchange than water because air has a higher O_2 content, lower density, and lower viscosity.
- Gills are outfoldings of the body specialized for gas exchange in water. The effectiveness of gas exchange in some gills, including those of fishes, is increased by **ventilation** and **countercurrent exchange** between blood and water. Gas exchange in insects relies on a **tracheal system**, a branched network of tubes that brings O_2 directly to cells. Spiders, land snails, and most terrestrial vertebrates have **lungs**. In mammals, inhaled air passes through the pharynx into the **trachea, bronchi,** bronchioles, and dead-end **alveoli**, where gas exchange occurs.

? Why does altitude have almost no effect on an animal's ability to rid itself of CO_2 through gas exchange?

CONCEPT 34.6

Breathing ventilates the lungs (pp. 704–706)

- Breathing mechanisms vary substantially among vertebrates. An amphibian ventilates its lungs by **positive pressure breathing**, which forces air down the trachea. Birds use a system of air sacs as bellows to keep air flowing through the lungs in one direction only. Mammals ventilate their lungs by **negative pressure breathing**, which pulls air into the lungs. Lung volume increases as the rib muscles and **diaphragm** contract. Incoming and outgoing air mix, decreasing the efficiency of ventilation.
- Control centers in the brain regulate the rate and depth of breathing. Sensors detect the pH of cerebrospinal fluid (reflecting CO_2 concentration in the blood), and the medulla adjusts breathing rate and depth to match metabolic demands. Secondary control is exerted by sensors in the aorta and carotid arteries that monitor blood levels of O_2 as well as CO_2 (via blood pH).

? How does air in the lungs differ from the fresh air that enters the body during inspiration?

Adaptations for gas exchange include pigments that bind and transport gases (pp. 706–708)

- In the lungs, gradients of partial pressure favor the net diffusion of O_2 into the blood and CO_2 out of the blood. The opposite situation exists in the rest of the body. **Respiratory pigments**, such as hemoglobin, bind O_2, greatly increasing the amount of O_2 transported by the circulatory system.
- Evolutionary adaptations enable some animals to satisfy extraordinary O_2 demands. Deep-diving air-breathers stockpile O_2 in blood and other tissues and deplete it slowly.

? *In what way is the role of a respiratory pigment like that of an enzyme?*

TEST YOUR UNDERSTANDING

Level 1: Knowledge/Comprehension

1. Which of the following respiratory systems is not closely associated with a blood supply?
   a. the lungs of a vertebrate
   b. the gills of a fish
   c. the tracheal system of an insect
   d. the skin of an earthworm
   e. the parapodia of a polychaete worm

2. Blood returning to the mammalian heart in a pulmonary vein drains first into the
   a. vena cava.
   b. left atrium.
   c. right atrium.
   d. left ventricle.
   e. right ventricle.

3. Pulse is a direct measure of
   a. blood pressure.
   b. stroke volume.
   c. cardiac output.
   d. heart rate.
   e. breathing rate.

4. When you hold your breath, which of the following blood gas changes first leads to the urge to breathe?
   a. rising O_2
   b. falling O_2
   c. rising CO_2
   d. falling CO_2
   e. rising CO_2 and falling O_2

Level 2: Application/Analysis

5. If a molecule of CO_2 released into the blood in your left toe is exhaled from your nose, it must pass through all of the following *except*
   a. the pulmonary vein.
   b. an alveolus.
   c. the trachea.
   d. the right atrium.
   e. the right ventricle.

6. Compared with the interstitial fluid that bathes active muscle cells, blood reaching these cells in arteries has a
   a. higher P_{O_2}.
   b. higher P_{CO_2}.
   c. greater bicarbonate concentration.
   d. lower pH.
   e. lower osmotic pressure.

7. Which of the following would *increase* the amount of oxygen undergoing net diffusion from the lungs into the blood?
   a. increasing the binding of oxygen to hemoglobin
   b. increasing the water vapor content of air in the lungs
   c. increasing the partial pressure of oxygen in the blood
   d. decreasing the red blood cell count of the blood
   e. decreasing the partial pressure of oxygen in the lung

Level 3: Synthesis/Evaluation

8. **DRAW IT** Plot blood pressure against time for one cardiac cycle in humans, drawing separate lines for the pressure in the aorta, the left ventricle, and the right ventricle. Below the time axis, add a vertical arrow pointing to the time when you expect a peak in atrial blood pressure.

9. **SCIENTIFIC INQUIRY**
   The hemoglobin of a human fetus differs from adult hemoglobin. Compare the dissociation curves of the two hemoglobins in the graph at right. Propose a hypothesis to explain the benefit of this difference between these two hemoglobins.

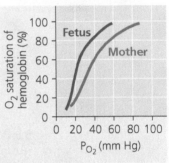

10. **FOCUS ON EVOLUTION**
    One of the many mutant opponents of the movie monster Godzilla is Mothra, a giant mothlike creature with a wingspan of several dozen meters. However, the largest known insects were Paleozoic dragonflies with half-meter wingspans. Focusing on respiration and gas exchange, explain why truly giant insects are improbable.

11. **FOCUS ON INTERACTIONS**
    Some athletes prepare for competition at sea level by sleeping in a tent in which P_{O_2} is kept artificially low. When climbing very high peaks, some mountaineers breathe from bottles of pure O_2. Relate these behaviors to humans' physiological interactions with our gaseous environment and control mechanisms governing oxygen delivery within the body.

For selected answers, see Appendix A.

MasteringBiology®

Students Go to **MasteringBiology** for assignments, the eText, and the Study Area with practice tests, animations, and activities.

Instructors Go to **MasteringBiology** for automatically graded tutorials and questions that you can assign to your students, plus Instructor Resources.

35

The Immune System

KEY CONCEPTS

35.1 In innate immunity, recognition and response rely on traits common to groups of pathogens

35.2 In adaptive immunity, receptors provide pathogen-specific recognition

35.3 Adaptive immunity defends against infection of body fluids and body cells

OVERVIEW

Recognition and Response

For a **pathogen**—a bacterium, fungus, virus, or other disease-causing agent—the internal environment of an animal is a nearly ideal habitat. The animal body offers a ready source of nutrients, a protected setting for growth and reproduction, and a means of transport to new environments. From the perspective of a cold or flu virus, we are wonderful hosts. From our vantage point, the situation is not so ideal. Fortunately, adaptations have arisen over the course of evolution that protect animals against many invaders.

Dedicated immune cells in the body fluids and tissues of most animals specifically interact with and destroy pathogens. For example, **Figure 35.1** shows an immune cell called a macrophage (brown) surrounding and engulfing a clump of bacteria (green). Immune cells also release defense molecules into body fluids, including proteins that punch holes in bacterial membranes or block viruses from entering body cells. Together, the body's defenses make up the **immune system**, which enables an animal to avoid or limit many infections. A foreign molecule or cell doesn't have to be pathogenic (disease-causing) to elicit an immune response, but we'll focus in this chapter on the immune system's role in defending against pathogens.

The first lines of defense offered by immune systems help prevent pathogens from gaining entrance to the body. For example, an outer covering, such as a shell or skin, blocks entry by many microbes. Sealing off the entire body surface is impossible, however, because gas exchange, nutrition, and reproduction require openings to the environment. Secretions that trap or kill microbes guard the body's entrances and exits, while the linings of the digestive tract, airway, and other exchange surfaces provide additional barriers to infection.

If a pathogen breaches barrier defenses and enters the body, the problem of how to fend off attack changes substantially. Housed within body fluids and tissues, the invader is no longer an outsider. To fight infections, an animal's immune system must detect foreign particles and cells within the body. In other words, a properly functioning immune system distinguishes nonself from self. How is this accomplished? Immune cells produce receptor molecules that bind specifically to molecules from foreign cells or viruses and activate defense responses. The specific binding of immune receptors to foreign molecules is a type of *molecular recognition* and is the central event in identifying nonself particles and cells.

▼ **Figure 35.1** What triggered this attack by an immune cell on a clump of rod-shaped bacteria?

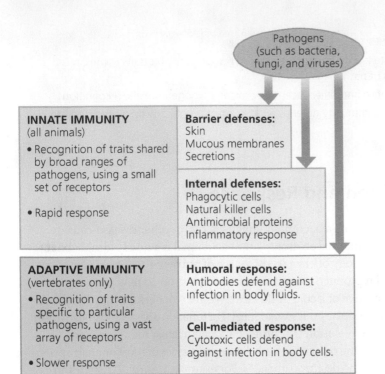

INNATE IMMUNITY (all animals) • Recognition of traits shared by broad ranges of pathogens, using a small set of receptors • Rapid response	**Barrier defenses:** Skin Mucous membranes Secretions
	Internal defenses: Phagocytic cells Natural killer cells Antimicrobial proteins Inflammatory response
ADAPTIVE IMMUNITY (vertebrates only) • Recognition of traits specific to particular pathogens, using a vast array of receptors • Slower response	**Humoral response:** Antibodies defend against infection in body fluids.
	Cell-mediated response: Cytotoxic cells defend against infection in body cells.

▲ **Figure 35.2 Overview of animal immunity.** Immune responses can be divided into innate and adaptive immunity. Some components of innate immunity help activate adaptive immune defenses.

Animal immune systems rely on either one or two major components for molecular recognition and defense. All animals have the component called **innate immunity**, which includes barrier defenses. Besides innate immunity, an additional component, called **adaptive immunity**, is found only in vertebrates. **Figure 35.2** provides an overview of the basic components of both innate and adaptive immunity.

Molecular recognition in innate immunity relies on a small set of receptors that bind to molecules or structures that are absent from animal bodies but common to a group of viruses, bacteria, or other microbes. Binding of an innate immune receptor to a foreign molecule activates internal defenses, enabling responses to a very broad range of pathogens.

In adaptive immunity, molecular recognition relies on a vast arsenal of receptors, each of which recognizes a feature typically found only on a particular part of a particular molecule in a particular pathogen. As a result, recognition and response in adaptive immunity occur with tremendous specificity.

The adaptive immune response, also known as the acquired immune response, is activated after the innate immune response and develops more slowly. The names *adaptive* and *acquired* reflect the fact that this immune response is enhanced by previous exposure to the infecting pathogen. Examples of adaptive responses include the synthesis of proteins that inactivate a bacterial toxin and the targeted killing of a virus-infected body cell.

In this chapter, we'll examine how each type of immunity protects animals from disease. You'll also learn how pathogens can avoid or overwhelm the immune system and how defects in the immune system can imperil an animal's health.

In innate immunity, recognition and response rely on traits common to groups of pathogens

Innate immunity is found in all animals (as well as in plants). In exploring innate immunity, we'll begin with invertebrates, which repel and fight infection with only this type of immunity. We'll then turn to vertebrates, in which innate immunity serves both as an immediate defense against infection and as the foundation for adaptive immune defenses.

Innate Immunity of Invertebrates

The great success of insects in terrestrial and freshwater habitats teeming with diverse microbes highlights the effectiveness of invertebrate innate immunity. In each of these environments, insects rely on their exoskeleton as a first line of defense against infection. Within the digestive system, **lysozyme**, an enzyme that breaks down bacterial cell walls, acts as a chemical barrier against pathogens ingested with food.

Any pathogen that breaches an insect's barrier defenses encounters a number of internal immune defenses. Immune cells called *hemocytes* travel throughout the body in the hemolymph, the insect circulatory fluid. Some hemocytes ingest and break down bacteria and other foreign substances, a process known as **phagocytosis (Figure 35.3)**. Other hemocytes

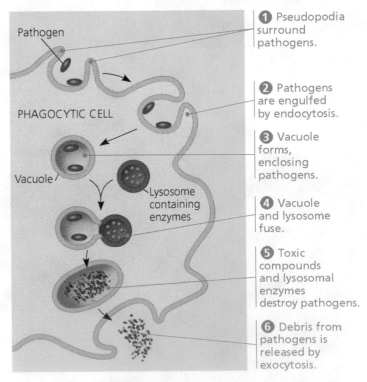

❶ Pseudopodia surround pathogens.

❷ Pathogens are engulfed by endocytosis.

❸ Vacuole forms, enclosing pathogens.

❹ Vacuole and lysosome fuse.

❺ Toxic compounds and lysosomal enzymes destroy pathogens.

❻ Debris from pathogens is released by exocytosis.

▲ **Figure 35.3 Phagocytosis.** This schematic depicts events in the ingestion and destruction of a microbe by a typical phagocytic cell.

release chemicals that kill pathogens and help entrap large invaders, such as *Plasmodium*, the parasite of mosquitoes that causes malaria in humans. One major class of defense molecules consists of antimicrobial peptides, which circulate throughout the body and inactivate or kill fungi and bacteria by disrupting their plasma membranes.

Immune cells of insects bind to molecules found only in the outer layers of fungi or bacteria. Fungal cell walls contain certain unique polysaccharides, whereas bacterial cell walls have polymers containing combinations of sugars and amino acids not found in animal cells. Such macromolecules serve as "identity tags" in the process of pathogen recognition. Insect immune cells secrete recognition proteins, each of which binds to a macromolecule characteristic of a broad class of bacteria or fungi. Once bound to a macromolecule, the recognition protein triggers an innate immune response specific for that class.

Innate Immunity of Vertebrates

Among jawed vertebrates, innate immune defenses coexist with the more recently evolved system of adaptive immunity. Because most of the recent discoveries regarding vertebrate innate immunity have come from studies of mice and humans, we'll focus here on mammals. We'll consider first the innate defenses that are similar to those found among invertebrates: barrier defenses, phagocytosis, and antimicrobial peptides. We'll then examine some unique aspects of vertebrate innate immunity, such as natural killer cells, interferons, and the inflammatory response.

Barrier Defenses

In mammals, barrier defenses block the entry of many pathogens. These defenses include the skin and the mucous membranes lining the digestive, respiratory, urinary, and reproductive tracts. The mucous membranes produce *mucus*, a viscous fluid that traps microbes and other particles. In the airway, ciliated epithelial cells sweep mucus and any entrapped microbes upward, helping prevent infection of the lungs.

Beyond their physical role in inhibiting microbial entry, body secretions create an environment that is hostile to many microbes. Lysozyme in tears, saliva, and mucous secretions destroys the cell walls of susceptible bacteria as they enter the openings around the eyes or the upper respiratory tract. Microbes in food or water and those in swallowed mucus must also contend with the acidic environment of the stomach, which kills most of them before they can enter the intestines. Similarly, secretions from oil and sweat glands give human skin a pH ranging from 3 to 5, acidic enough to prevent the growth of many bacteria.

Cellular Innate Defenses

Pathogens entering the mammalian body are engulfed by phagocytic cells that detect fungal or bacterial components using several types of receptors. Some mammalian receptors are very similar to Toll, a key activator of innate immunity in insects. Each mammalian **Toll-like receptor (TLR)** binds to fragments of molecules characteristic of a set of pathogens **(Figure 35.4)**. For example, TLR3 binds to double-stranded RNA, a form of nucleic acid characteristic of certain viruses. Similarly, TLR4 recognizes lipopolysaccharide, a molecule found on the surface of many bacteria, and TLR5 recognizes flagellin, the main protein of bacterial flagella. In each case, the recognized macromolecule is normally absent from the vertebrate body and is an essential component of certain groups of pathogens.

As in invertebrates, detection of invading pathogens in mammals triggers phagocytosis and destruction. The two main types of phagocytic cells in the mammalian body are neutrophils and macrophages. **Neutrophils**, which circulate in the blood, are attracted by signals from infected tissues. **Macrophages** ("big eaters") are larger phagocytic cells. Some migrate throughout the body, whereas others reside in organs and tissues where they are likely to encounter pathogens. For example, some macrophages are located in the spleen, where pathogens in the blood become trapped.

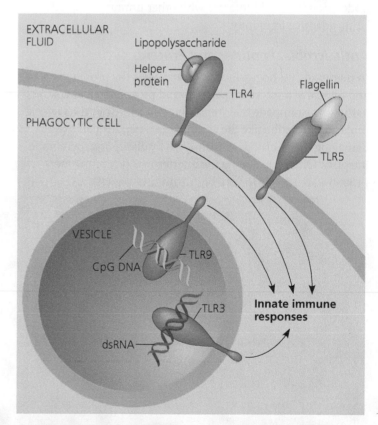

▲ **Figure 35.4 TLR signaling.** Each mammalian Toll-like receptor (TLR) recognizes a molecular pattern characteristic of a group of pathogens. Lipopolysaccharide, flagellin, CpG DNA (DNA containing unmethylated CG sequences), and double-stranded RNA (dsRNA) are all found in bacteria, fungi, or viruses, but not in animal cells. Together with other recognition and response factors, TLR proteins trigger internal innate immune defenses.

? *Some TLR proteins are on the cell surface, whereas others are inside vesicles. Suggest a possible benefit of this distribution.*

Two other types of cells—dendritic cells and eosinophils—provide additional functions in innate defense. *Dendritic cells* mainly populate tissues, such as skin, that contact the environment. They stimulate adaptive immunity against pathogens they encounter and engulf, as we'll explore shortly. *Eosinophils*, often found beneath mucous membranes, are important in defending against multicellular invaders, such as parasitic worms. Upon encountering such parasites, eosinophils discharge destructive enzymes.

Cellular innate defenses in vertebrates also involve **natural killer cells**. These cells circulate through the body and detect the abnormal array of surface proteins characteristic of some virus-infected and cancerous cells. Natural killer cells do not engulf stricken cells. Instead, they release chemicals that lead to cell death, inhibiting further spread of the virus or cancer.

Many cellular innate defenses in vertebrates involve the lymphatic system (see Figure 34.12). Some macrophages reside in lymph nodes, where they engulf pathogens that have entered the lymph from the interstitial fluid. Dendritic cells reside outside the lymphatic system but migrate to the lymph nodes after interacting with pathogens. Within the lymph nodes, dendritic cells interact with other immune cells, stimulating adaptive immunity.

Antimicrobial Peptides and Proteins

In mammals, pathogen recognition triggers the production and release of a variety of peptides and proteins that attack pathogens or impede their reproduction. Some of these defense molecules function like the antimicrobial peptides of insects, damaging broad groups of pathogens by disrupting membrane integrity. Others, including the interferons and complement proteins, are unique to vertebrate immune systems.

Interferons are proteins that provide innate defense by interfering with viral infections. Virus-infected body cells secrete interferons, which induce nearby uninfected cells to produce substances that inhibit viral reproduction. In this way, interferons limit the cell-to-cell spread of viruses in the body, helping control viral infections such as colds and influenza. Some white blood cells secrete a different type of interferon that helps activate macrophages, enhancing their phagocytic ability. Pharmaceutical companies now use recombinant DNA technology to mass-produce interferons to help treat certain viral infections, such as hepatitis C.

The infection-fighting **complement system** consists of roughly 30 proteins in blood plasma. These proteins circulate in an inactive state and are activated by substances on the surface of many microbes. Activation results in a cascade of biochemical reactions that can lead to lysis (bursting) of invading cells. The complement system also functions in the inflammatory response, our next topic, as well as in the adaptive defenses discussed later in the chapter.

Inflammatory Response

The pain and swelling that alert you to a splinter under your skin are the result of a local **inflammatory response**, the changes brought about by signaling molecules released upon injury or infection **(Figure 35.5)**. One important inflammatory signaling molecule is **histamine**, which is stored in the granules (vesicles) of **mast cells**, found in connective tissue. Histamine released at sites of damage triggers nearby blood vessels to dilate and become more permeable. Activated macrophages and neutrophils discharge **cytokines**, signaling molecules that in an immune response promote blood flow to the site of injury or infection. The increase in local blood supply causes the

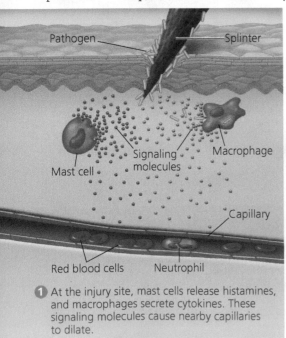

Mast cell · **Signaling molecules** · **Macrophage** · **Pathogen** · **Splinter** · **Capillary** · **Red blood cells** · **Neutrophil**

① At the injury site, mast cells release histamines, and macrophages secrete cytokines. These signaling molecules cause nearby capillaries to dilate.

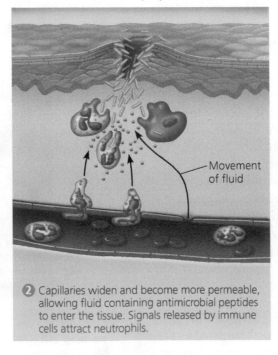

Movement of fluid

② Capillaries widen and become more permeable, allowing fluid containing antimicrobial peptides to enter the tissue. Signals released by immune cells attract neutrophils.

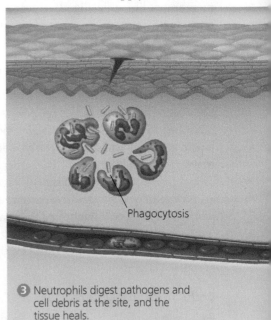

Phagocytosis

③ Neutrophils digest pathogens and cell debris at the site, and the tissue heals.

▲ **Figure 35.5 Major events in a local inflammatory response.**

redness and increased skin temperature typical of the inflammatory response (from the Latin *inflammare*, to set on fire). Blood-engorged capillaries leak fluid into neighboring tissues, causing swelling.

During inflammation, cycles of signaling and response transform the site. Activated complement proteins promote further release of histamine, attracting more phagocytic cells that enter injured tissues (see Figure 35.5) and carry out additional phagocytosis. At the same time, enhanced blood flow to the site helps deliver antimicrobial peptides. The result is an accumulation of pus, a fluid rich in white blood cells, dead pathogens, and cell debris from damaged tissue.

A minor injury or infection causes a local inflammatory response, but severe tissue damage or infection may lead to a response that is systemic (throughout the body). Cells in injured or infected tissue often secrete molecules that stimulate the release of additional neutrophils from the bone marrow. In a severe infection, such as meningitis or appendicitis, the number of white blood cells in the blood may increase several-fold within a few hours.

Another systemic inflammatory response is fever. In response to certain pathogens, substances released by activated macrophages cause the body's thermostat to reset to a higher temperature (see Concept 32.1). The benefits of the resulting fever are still a subject of debate. One hypothesis is that an elevated body temperature may enhance phagocytosis and, by speeding up chemical reactions, accelerate tissue repair.

Certain bacterial infections can induce an overwhelming systemic inflammatory response, leading to a life-threatening condition called *septic shock*. Characterized by very high fever, low blood pressure, and poor blood flow through capillaries, septic shock occurs most often in the very old and the very young. It is fatal in more than one-third of cases and kills more than 90,000 people each year in the United States alone.

Chronic (ongoing) inflammation can also threaten human health. For example, millions of individuals worldwide suffer from Crohn's disease and ulcerative colitis, often debilitating disorders in which an unregulated inflammatory response disrupts intestinal function.

Evasion of Innate Immunity by Pathogens

Adaptations have evolved in some pathogens that enable them to avoid destruction by phagocytic cells. For example, the outer capsule that surrounds certain bacteria interferes with molecular recognition and phagocytosis. One such bacterium, *Streptococcus pneumoniae*, played a critical role in the discovery that DNA can convey genetic information (see Figure 13.2). Other bacteria, after being engulfed by a host cell, resist breakdown within lysosomes. An example is the bacterium that causes tuberculosis (TB). Rather than being destroyed within host cells, this bacterium grows and reproduces, effectively hidden from the body's innate immune defenses. These and other mechanisms that prevent destruction by the innate immune system make certain fungi and

bacteria substantial pathogenic threats. Indeed, TB kills more than a million people a year worldwide.

CONCEPT 35.2

In adaptive immunity, receptors provide pathogen-specific recognition

Vertebrates are unique in having adaptive immunity in addition to innate immunity. The adaptive response relies on T cells and B cells, which are types of white blood cells called **lymphocytes**. Like all blood cells, lymphocytes originate from stem cells in the bone marrow. Some lymphocytes migrate from the bone marrow to the **thymus**, an organ in the thoracic cavity above the heart (see Figure 34.12). These lymphocytes mature into **T cells**. Lymphocytes that remain and mature in the bone marrow develop as **B cells**.

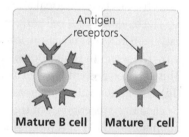

Mature B cell   Mature T cell

Any substance that elicits a response from a B cell or T cell is called an **antigen**. In adaptive immunity, recognition occurs when a B cell or T cell binds to an antigen, such as a bacterial or viral protein, via a protein called an **antigen receptor**. An antigen receptor is specific enough to bind to just one part of one molecule from a particular pathogen, such as a species of bacteria or strain of virus. Although the cells of the immune system produce millions of different antigen receptors, all of the antigen receptors made by a single B or T cell are identical. Infection by a virus, bacterium, or other pathogen triggers activation of B and T cells with antigen receptors specific for parts of that pathogen. B and T cells are shown in this text with only a few antigen receptors, but there are actually about 100,000 antigen receptors on the surface of a single B or T cell.

Antigens are usually foreign and are typically large molecules, either proteins or polysaccharides. The small, accessible portion of an antigen that binds to an antigen receptor is called an **epitope**, or *antigenic determinant*. An example is a group of amino acids in a particular protein. A single antigen usually has several different epitopes, each of which binds to a receptor with a different specificity. Because all antigen receptors produced by a single B cell or T cell are identical, they bind to the same epitope. Each B or T cell thus displays *specificity* for a particular epitope, enabling it to respond to any pathogen that produces molecules containing that same epitope.

The antigen receptors of B cells and T cells have similar components, but they encounter antigens in different ways. We'll consider the two processes in turn.

Antigen Recognition by B Cells and Antibodies

Each B cell antigen receptor is a Y-shaped molecule consisting of four polypeptide chains: two identical **heavy chains** and two identical **light chains**, with disulfide bridges linking the chains together **(Figure 35.6)**. A transmembrane region near one end of each heavy chain anchors the receptor in the cell's plasma membrane. A short tail region at the end of the heavy chain extends into the cytoplasm.

The light and heavy chains each have a *constant (C) region*, where amino acid sequences vary little among the receptors on different B cells. Within the two tips of the Y shape (see Figure 35.6), each chain has a *variable (V) region*, so named because its amino acid sequence varies extensively from one B cell to another. Together, parts of a heavy-chain V region and a light-chain V region form an asymmetric binding site for an antigen. As shown in Figure 35.6, each B cell antigen receptor has two identical antigen-binding sites.

The binding of a B cell antigen receptor to an antigen is an early step in B cell activation, leading eventually to formation of cells that secrete a soluble form of the receptor **(Figure 35.7a)**. This secreted protein is called an **antibody**, or **immunoglobulin (Ig)**. Antibodies have the same Y-shaped organization as B cell antigen receptors, but they are secreted rather than membrane-bound. It is the antibodies, rather than the B cells themselves, that actually help defend against pathogens.

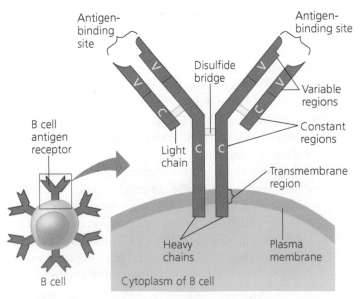

▲ **Figure 35.6 The structure of a B cell antigen receptor.**

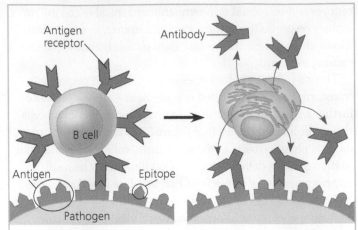

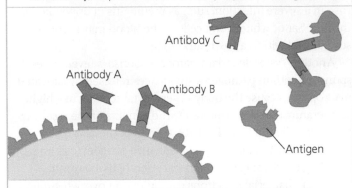

(a) B cell antigen receptors and antibodies. An antigen receptor of a B cell binds to an epitope, a particular part of an antigen. Following binding, the B cell gives rise to cells that secrete a soluble form of the antigen receptor. This soluble receptor, called an antibody, is specific for the same epitope as the original B cell.

(b) Antigen receptor specificity. Different antibodies can recognize distinct epitopes on the same antigen. Furthermore, antibodies can recognize free antigens as well as antigens on a pathogen's surface.

▲ **Figure 35.7 Antigen recognition by B cells and antibodies.**

MAKE CONNECTIONS *The interactions depicted here involve a highly specific binding between antigen and receptor. How is antigen-antibody binding similar to an enzyme-substrate interaction (see Figures 3.20 and 6.14)?*

The antigen-binding site of a membrane-bound receptor or antibody has a unique shape that provides a lock-and-key fit for a particular epitope. Many noncovalent bonds between an epitope and the binding surface provide a stable and specific interaction. Differences in the amino acid sequences of variable regions provide the variation in binding surfaces that enables this highly specific binding.

B cell antigen receptors and antibodies bind to intact antigens in the blood and lymph. As illustrated in **Figure 35.7b** for antibodies, they can bind to antigens on the surface of pathogens or free in body fluids.

Antigen Recognition by T Cells

For a T cell, the antigen receptor consists of two different polypeptide chains, an *α chain* and a *β chain*, linked by a disulfide bridge **(Figure 35.8)**. Near the base of the T cell antigen receptor

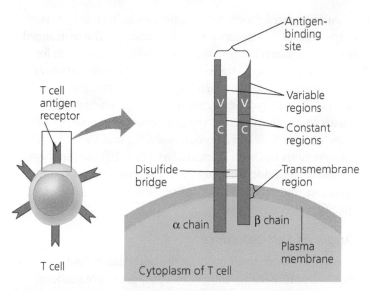

▲ **Figure 35.8 The structure of a T cell antigen receptor.**

(often called simply a T cell receptor) is a transmembrane region that anchors the molecule in the cell's plasma membrane. At the outer tip of the molecule, the variable (V) regions of α and β chains together form a single antigen-binding site. The remainder of the molecule is made up of the constant (C) regions.

Whereas the antigen receptors of B cells bind to epitopes of intact antigens on pathogens or circulating in body fluids, those of T cells bind only to fragments of antigens that are displayed, or presented, on the surface of host cells. The host protein that displays the antigen fragment on the cell surface is called a **major histocompatibility complex (MHC) molecule**.

Recognition of protein antigens by T cells begins when a pathogen or part of a pathogen either infects or is taken in by a host cell (**Figure 35.9**). Inside the host cell, enzymes in the cell

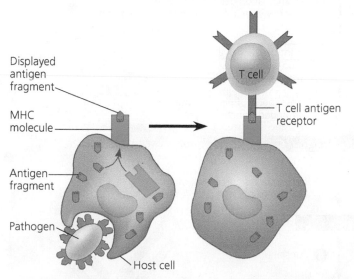

▲ **Figure 35.9 Antigen recognition by T cells.** Inside the host cell, an antigen fragment from a pathogen binds to an MHC molecule and is brought up to the cell surface, where it is displayed. The combination of MHC molecule and antigen fragment is recognized by a T cell.

cleave the antigen into smaller peptides. Each peptide, called an *antigen fragment*, then binds to an MHC molecule inside the cell. Movement of the MHC molecule and bound antigen fragment to the cell surface results in **antigen presentation**, the display of the antigen fragment in an exposed groove of the MHC protein.

In effect, antigen presentation advertises the fact that a host cell contains a foreign substance. If the cell displaying an antigen fragment encounters a T cell with the right specificity, the antigen receptor on the T cell can bind to both the antigen fragment and the MHC molecule. This interaction of an MHC molecule, an antigen fragment, and an antigen receptor is necessary for a T cell to participate in an adaptive immune response, as you'll see later.

B Cell and T Cell Development

Now that you know how B cells and T cells recognize antigens, let's consider four major characteristics of adaptive immunity. First, there is an immense diversity of lymphocytes and receptors, enabling the immune system to detect pathogens never before encountered. Second, adaptive immunity normally has self-tolerance, the lack of reactivity against an animal's own molecules and cells. Third, cell proliferation triggered by activation greatly increases the number of B and T cells specific for an antigen. Fourth, there is a stronger and more rapid response to an antigen encountered previously, due to a feature known as immunological memory.

Receptor diversity and self-tolerance arise as a lymphocyte matures. Cell proliferation and the formation of immunological memory occur later, after a mature lymphocyte encounters and binds to a specific antigen. We'll consider these four characteristics in the order in which they develop.

Generation of B Cell and T Cell Diversity

Each person makes more than 1 million different B cell antigen receptors and 10 million different T cell antigen receptors. Yet there are only about 20,000 protein-coding genes in the human genome. How, then, do we generate such remarkable diversity in antigen receptors? The answer lies in combinations. Think of selecting a car with a choice of three interior colors and six exterior colors. There are 18 (3 × 6) color combinations to consider. Similarly, by combining variable elements, the immune system assembles many different receptors from a much smaller collection of parts.

To understand the origin of receptor diversity, let's consider an immunoglobulin (Ig) gene that encodes the light chain of both secreted antibodies (immunoglobulins) and membrane-bound B cell antigen receptors. Although we'll analyze only a single Ig light-chain gene, all B and T cell antigen receptor genes undergo very similar transformations.

The capacity to generate diversity is built into the structure of Ig genes. A receptor light chain is encoded by three gene segments: a variable (*V*) segment, a joining (*J*) segment, and a constant (*C*) segment. The *V* and *J* segments together encode the variable region of the receptor chain, while the *C* segment encodes the constant region. The light-chain gene contains a single *C* segment, 40 different *V* segments, and 5 different *J* segments. These alternative copies of the *V* and *J* segments are arranged within the gene in a series **(Figure 35.10)**. Because a functional gene is built from one copy of each type of segment, the pieces can be combined in 200 different ways (40 *V* × 5 *J* × 1 *C*). The number of different heavy-chain combinations is even greater, resulting in even more diversity.

Assembling a functional Ig gene requires rearranging the DNA. Early in B cell development, an enzyme complex called *recombinase* links one light-chain *V* gene segment to one *J* gene segment. This recombination event eliminates the long stretch of DNA between the segments, forming a single exon that is part *V* and part *J*. Because there is only an intron between the *J* and *C* DNA segments, no further DNA rearrangement is required. Instead, the *J* and *C* segments of the RNA transcript will be joined when splicing removes the intervening RNA (see Figure 14.12 to review RNA splicing).

Recombinase acts randomly, linking any one of the 40 *V* gene segments to any one of the 5 *J* gene segments. Heavy-chain genes undergo a similar rearrangement. In any given cell, however, only one allele of a light-chain gene and one allele of a heavy-chain gene are rearranged. Furthermore, the rearrangements are permanent and are passed on to the daughter cells when the lymphocyte divides.

After both a light-chain and a heavy-chain gene have rearranged, antigen receptors can be synthesized. The rearranged genes are transcribed, and the transcripts are processed for translation. Following translation, the light chain and heavy chain assemble together, forming an antigen receptor (see Figure 35.10). Each pair of randomly rearranged heavy and light chains results in a different antigen-binding site. For the total population of B cells in a human body, the number of such combinations has been calculated as 3.5×10^6. Furthermore, mutations introduced during *VJ* recombination add additional variation, making the number of possible antigen-binding specificities even greater.

Origin of Self-Tolerance

In adaptive immunity, how does the body distinguish self from nonself? Because antigen receptor genes are randomly rearranged, some immature lymphocytes produce receptors specific for epitopes on the organism's own molecules. If these self-reactive lymphocytes were not eliminated or inactivated, the immune system could not distinguish self from nonself and would attack body proteins, cells, and tissues. Instead, as lymphocytes mature in the bone marrow or thymus, their antigen receptors are tested for self-reactivity. Some B and T cells with receptors specific for the body's own molecules are destroyed by programmed cell death. The remaining self-reactive lymphocytes are typically rendered nonfunctional, leaving only those lymphocytes that react to foreign molecules. Since the

▶ **Figure 35.10 Immunoglobulin (antibody) gene rearrangement.** The joining of randomly selected *V* and *J* gene segments (V_{39} and J_5 in this example) results in a functional gene that encodes the light-chain polypeptide of a B cell antigen receptor. Transcription, splicing, and translation result in a light chain that combines with a polypeptide produced from an independently rearranged heavy-chain gene to form a functional receptor. Mature B cells (and T cells) are exceptions to the generalization that all nucleated cells in the body have exactly the same DNA.

MAKE CONNECTIONS

Both alternative splicing and joining of V and J segments by recombination generate diverse gene products from a limited set of gene segments. How do these processes differ (see Figure 15.12)?

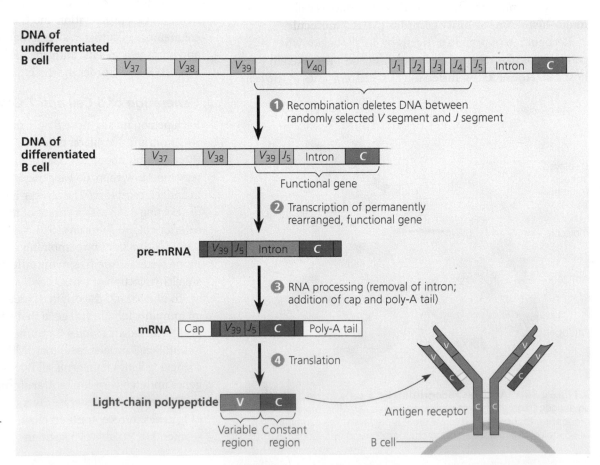

body normally lacks mature lymphocytes that can react against its own components, the immune system is said to exhibit self-tolerance.

Proliferation of B Cells and T Cells

Despite the enormous variety of antigen receptors, only a tiny fraction are specific for a given epitope. How, then, does an effective adaptive response develop? To begin with, an antigen is presented to a steady stream of lymphocytes in the lymph nodes (see Figure 34.12) until a match is made. A successful match then triggers changes in cell number and activity for the lymphocyte to which an antigen has bound.

The binding of an antigen receptor to an epitope initiates events that activate the lymphocyte. Once activated, a B cell or T cell undergoes multiple cell divisions. For each activated cell, the result of this proliferation is a clone, a population of cells that are identical to the original cell. Some cells from this clone become **effector cells**, short-lived cells that take effect immediately against the antigen and any pathogens producing that antigen. The effector forms of B cells are **plasma cells**, which secrete antibodies. The effector forms of T cells are helper T cells and cytotoxic T cells, whose roles we'll explore in Concept 35.3. The remaining cells in the clone become **memory cells**, long-lived cells that can give rise to effector cells if the same antigen is encountered later in the animal's life.

Figure 35.11 summarizes the proliferation of a lymphocyte into a clone of cells in response to binding to an antigen, using B cells as an example. This process is called **clonal selection** because an encounter with an antigen *selects* which lymphocyte will divide to produce a *clonal* population of thousands of cells specific for a particular epitope.

Immunological Memory

Immunological memory is responsible for the long-term protection that a prior infection provides against many diseases, such as chickenpox. This type of protection was noted almost 2,400 years ago by the Greek historian Thucydides. He observed that individuals who had recovered from the plague could safely care for those who were sick or dying, "for the same man was never attacked twice—never at least fatally."

Prior exposure to an antigen alters the speed, strength, and duration of the immune response. The production of effector cells from a clone of lymphocytes during the first exposure to an antigen is the basis for the **primary immune response**. The primary response peaks about 10–17 days after the initial exposure. During this time, selected B cells and T cells give rise to their effector forms. If an individual is exposed again to the same antigen, the response is faster (typically peaking only 2–7 days after exposure), of greater magnitude, and more prolonged. This is the **secondary immune response**, a hallmark of adaptive, or acquired, immunity. Because selected

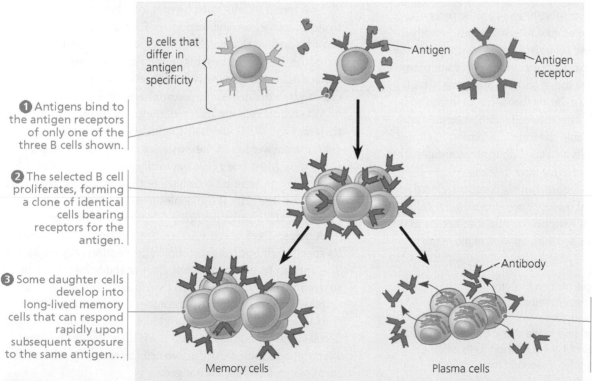

1 Antigens bind to the antigen receptors of only one of the three B cells shown.

2 The selected B cell proliferates, forming a clone of identical cells bearing receptors for the antigen.

3 Some daughter cells develop into long-lived memory cells that can respond rapidly upon subsequent exposure to the same antigen...

B cells that differ in antigen specificity

Antigen

Antigen receptor

Antibody

3 (continued) ...and other daughter cells develop into short-lived plasma cells that secrete antibodies specific for the antigen.

Memory cells

Plasma cells

▲ **Figure 35.11 Clonal selection.** This figure illustrates clonal selection, using B cells as an example. In response to a specific antigen and to immune cell signals (not shown), one B cell divides and forms a clone of cells. The remaining B cells, which have antigen receptors specific for other antigens, do not respond. The clone of cells formed by the selected B cell gives rise to memory B cells and antibody-secreting plasma cells. T cells also undergo clonal selection, generating memory T cells and effector T cells (cytotoxic T cells and helper T cells).

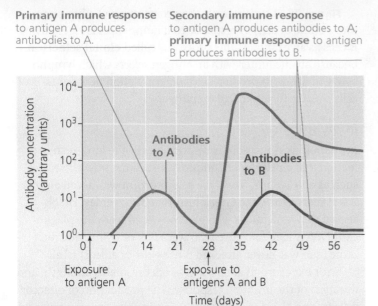

Primary immune response to antigen A produces antibodies to A.

Secondary immune response to antigen A produces antibodies to A; primary immune response to antigen B produces antibodies to B.

Antibodies to A

Antibodies to B

Exposure to antigen A

Exposure to antigens A and B

Time (days)

▲ Figure 35.12 The specificity of immunological memory. Long-lived memory cells generated in the primary response to antigen A give rise to a heightened secondary response to the same antigen, but do not affect the response to a different antigen (B).

B cells give rise to antibody-secreting effector cells, measuring the concentrations of specific antibodies in blood over time distinguishes the primary and secondary immune responses (Figure 35.12).

The secondary immune response relies on the reservoir of T and B memory cells generated following initial exposure to an antigen. Because these cells are long-lived, they provide the basis for immunological memory, which can span many decades. (Effector cells have much shorter life spans, which is why the immune response diminishes after an infection is overcome.) If an antigen is encountered again, memory cells specific for that antigen enable the rapid formation of clones of thousands of effector cells also specific for that antigen, thus generating a greatly enhanced immune defense.

Although the processes for antigen recognition, clonal selection, and immunological memory are similar for B cells and T cells, these two classes of lymphocytes fight infection in different ways and in different settings, as we'll explore next.

CONCEPT CHECK 35.2

1. **DRAW IT** Sketch a B cell antigen receptor. Label the V and C regions of the light and heavy chains. Label the antigen-binding sites, disulfide bridges, and transmembrane region. Where are these features located relative to the V and C regions?

2. Explain two advantages of having memory cells when a pathogen is encountered for a second time.

3. **WHAT IF?** If both copies of a light-chain gene and a heavy-chain gene recombined in each (diploid) B cell, how would this affect B cell development and function?

For suggested answers, see Appendix A.

Adaptive immunity defends against infection of body fluids and body cells

Having considered how clones of lymphocytes arise, we now explore how these cells help fight infections and minimize damage by pathogens. The defenses provided by B and T lymphocytes can be divided into a humoral immune response and a cell-mediated immune response. The **humoral immune response** occurs in the blood and lymph (once called body humors, or fluids). In the humoral response, antibodies help neutralize or eliminate toxins and pathogens in the blood and lymph. In the **cell-mediated immune response**, specialized T cells destroy infected host cells. Both responses include a primary immune response and a secondary immune response, with memory cells enabling the secondary response.

Helper T Cells: A Response to Nearly All Antigens

A type of T cell called a **helper T cell** triggers both the humoral and cell-mediated immune responses. Helper T cells themselves do not carry out those responses. Instead, signals from helper T cells initiate production of antibodies that neutralize pathogens and activate T cells that kill infected cells.

Two requirements must be met for a helper T cell to activate adaptive immune responses. First, a foreign molecule must be present that can bind specifically to the antigen receptor of the T cell. Second, this antigen must be displayed on the surface of an **antigen-presenting cell**. The antigen-presenting cell can be a dendritic cell, macrophage, or B cell.

When host cells are infected, they, too, display antigens on their surface. What, then, distinguishes an antigen-presenting cell? The answer lies in the existence of two classes of MHC molecules. Most body cells have only class I MHC molecules, but antigen-presenting cells have both class I and class II MHC molecules. The class II molecules provide a molecular signature by which an antigen-presenting cell is recognized.

A helper T cell and the antigen-presenting cell displaying its specific epitope have a complex interaction (**Figure 35.13**). The antigen receptors on the surface of the helper T cell bind to the antigen fragment and to the class II MHC molecule displaying that fragment on the antigen-presenting cell. At the same time, an accessory protein called CD4 on the helper T cell surface binds to the class II MHC molecule, helping keep the cells joined. As the two cells interact, signals in the form of cytokines are exchanged.

Antigen-presenting cells interact with helper T cells in several different contexts. Antigen presentation by a dendritic cell or macrophage activates a helper T cell, which then proliferates, forming a clone of activated cells. B cells present antigens

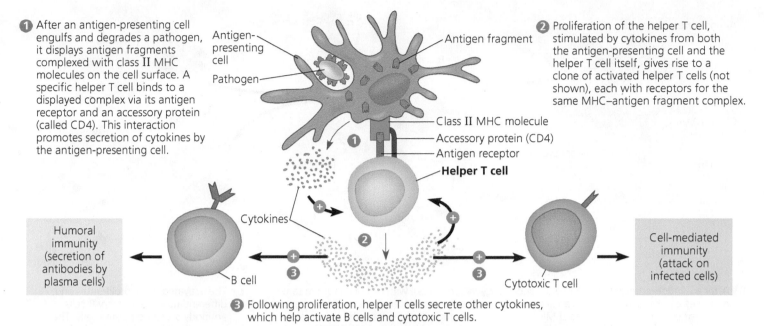

① After an antigen-presenting cell engulfs and degrades a pathogen, it displays antigen fragments complexed with class II MHC molecules on the cell surface. A specific helper T cell binds to a displayed complex via its antigen receptor and an accessory protein (called CD4). This interaction promotes secretion of cytokines by the antigen-presenting cell.

Antigen-presenting cell

Pathogen

Antigen fragment

② Proliferation of the helper T cell, stimulated by cytokines from both the antigen-presenting cell and the helper T cell itself, gives rise to a clone of activated helper T cells (not shown), each with receptors for the same MHC–antigen fragment complex.

Class II MHC molecule
Accessory protein (CD4)
Antigen receptor
Helper T cell

Humoral immunity (secretion of antibodies by plasma cells)

Cytokines

B cell

Cell-mediated immunity (attack on infected cells)

Cytotoxic T cell

③ Following proliferation, helper T cells secrete other cytokines, which help activate B cells and cytotoxic T cells.

▲ **Figure 35.13 The central role of helper T cells in humoral and cell-mediated immune responses.** In this example, a helper T cell responds to a dendritic cell displaying a microbial antigen.

to *already* activated helper T cells, which in turn activate the B cells themselves. Activated helper T cells also help stimulate cytotoxic T cells, as we'll discuss next.

Cytotoxic T Cells: A Response to Infected Cells

In the cell-mediated immune response, **cytotoxic T cells** use toxic proteins to kill cells infected by viruses or other intracellular pathogens. To become active, cytotoxic T cells require signals from helper T cells and interaction with an antigen-presenting cell. Fragments of foreign proteins produced in infected host cells associate with class I MHC

molecules and are displayed on the cell surface, where they can be recognized by activated cytotoxic T cells **(Figure 35.14)**. As with helper T cells, cytotoxic T cells have an accessory protein that binds to the MHC molecule. This accessory protein, called CD8, helps keep the two cells in contact while the cytotoxic T cell is activated.

The targeted destruction of an infected host cell by a cytotoxic T cell involves the secretion of proteins that disrupt membrane integrity and trigger cell death (see Figure 35.14). The death of the infected cell not only deprives the pathogen of a place to reproduce, but also exposes cell contents to circulating antibodies, which mark released antigens for disposal.

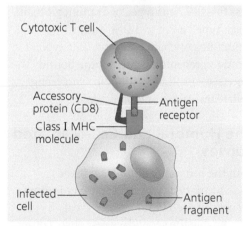

Cytotoxic T cell
Accessory protein (CD8)
Antigen receptor
Class I MHC molecule
Infected cell
Antigen fragment

① An activated cytotoxic T cell binds to a class I MHC–antigen fragment complex on an infected cell via its antigen receptor and an accessory protein (called CD8).

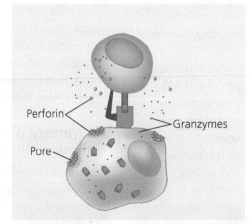

Perforin
Granzymes
Pore

② The T cell releases perforin molecules, which form pores in the infected cell membrane, and granzymes, enzymes that break down proteins. Granzymes enter the infected cell by endocytosis.

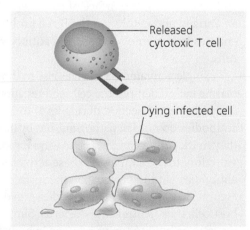

Released cytotoxic T cell
Dying infected cell

③ The granzymes initiate apoptosis within the infected cell, leading to fragmentation of the nucleus and cytoplasm and eventual cell death. The released cytotoxic T cell can attack other infected cells.

▲ **Figure 35.14 The killing action of cytotoxic T cells on an infected host cell.** An activated cytotoxic T cell releases molecules that make pores in an infected cell's membrane and enzymes that break down proteins, promoting the cell's death.

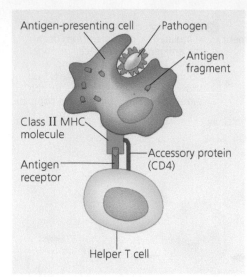
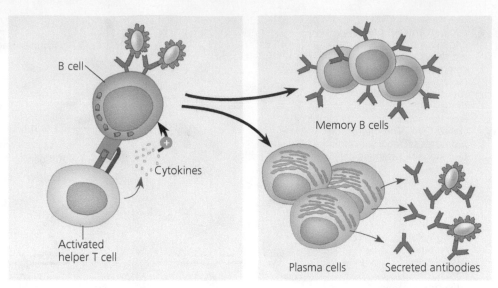

Antigen-presenting cell — Pathogen

Antigen fragment

Class II MHC molecule

Antigen receptor

Accessory protein (CD4)

Helper T cell

B cell

Cytokines

Activated helper T cell

Memory B cells

Plasma cells   Secreted antibodies

① After an antigen-presenting cell engulfs and degrades a pathogen, it displays an antigen fragment complexed with a class II MHC molecule. A helper T cell that recognizes the complex is activated with the aid of cytokines secreted from the antigen-presenting cell.

② When a B cell with receptors for the same epitope internalizes the antigen, it displays an antigen fragment on the cell surface in a complex with a class II MHC molecule. An activated helper T cell bearing receptors specific for the displayed fragment binds to and activates the B cell.

③ The activated B cell proliferates and differentiates into memory B cells and antibody-secreting plasma cells. The secreted antibodies are specific for the same antigen that initiated the response.

▲ **Figure 35.15 Activation of a B cell in the humoral immune response.** Most protein antigens require activated helper T cells to trigger a humoral response. A macrophage (shown here) or a dendritic cell can activate a helper T cell, which in turn can activate a B cell to give rise to antibody-secreting plasma cells.

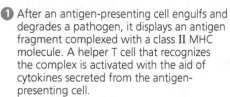 *What function do cell-surface antigen receptors play for memory B cells?*

B Cells and Antibodies: A Response to Extracellular Pathogens

The secretion of antibodies by clonally selected B cells is the hallmark of the humoral immune response. As illustrated in **Figure 35.15**, activation of B cells involves both helper T cells and proteins on the surface of pathogens. Stimulated by both an antigen and cytokines, the B cell proliferates and differentiates into memory B cells and antibody-secreting plasma cells.

A single activated B cell gives rise to thousands of identical plasma cells. Each plasma cell secretes approximately 2,000 antibodies every second during its 4- to 5-day life span. The antibodies do not kill pathogens, but by binding to antigens, they mark pathogens in various ways for inactivation or destruction. In the simplest of these activities, *neutralization*, antibodies bind to proteins on the surface of a virus (see Figure 35.7b). The bound antibodies prevent infection of a host cell, thus neutralizing the virus. Similarly, antibodies sometimes bind to toxins released in body fluids, preventing the toxins from entering body cells. Because each antibody has two antigen-binding sites, antibodies can also facilitate phagocytosis by linking bacterial cells, viruses, or other foreign substances into aggregates.

Antibodies sometimes work together with the proteins of the complement system. (The name *complement* reflects the fact that these proteins increase the effectiveness of antibody-directed attacks on bacteria.) Binding of a complement protein to an antigen-antibody complex on a foreign cell triggers events leading to formation of a pore in the membrane of the cell. Ions and water rush into the cell, causing it to swell and lyse.

B cells can express five different types of immunoglobulin. For a given B cell, each type has an identical antigen-binding specificity but a distinct heavy-chain C region. One type of B cell Ig, the B cell antigen receptor, is membrane bound. The other four Ig types consist of soluble antibodies, including those found in blood, tears, saliva, and breast milk.

Summary of the Humoral and Cell-Mediated Immune Responses

As noted earlier, both the humoral and cell-mediated responses can include primary as well as secondary immune responses. Memory cells of each type—helper T cell, B cell, and cytotoxic T cell—enable the secondary response. For example, when body fluids are reinfected by a pathogen encountered previously, memory B cells and memory helper T cells initiate a secondary humoral response. **Figure 35.16** reviews the events that initiate humoral and cell-mediated immune responses, highlights the central role of the helper T cell, and serves as a helpful summary of adaptive immunity.

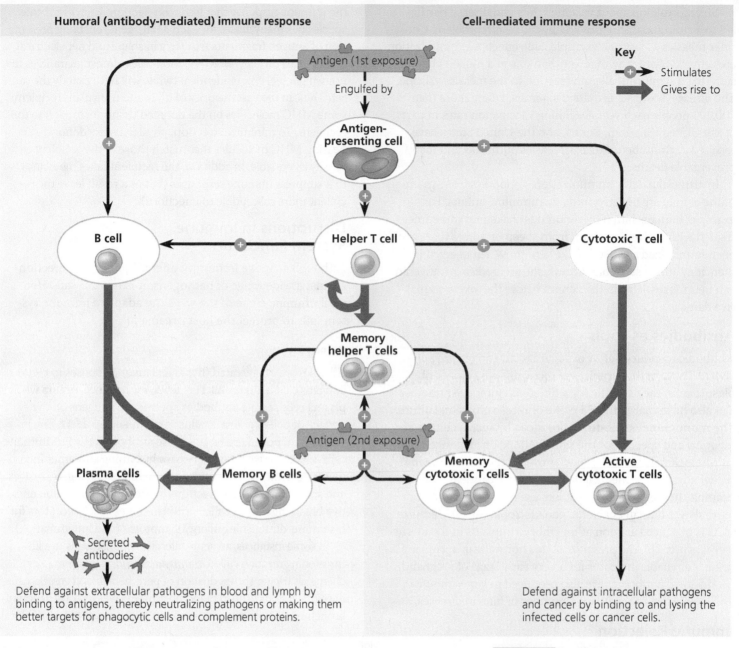

Key

⊕→ Stimulates

⟹ Gives rise to

Antigen (1st exposure)

Engulfed by

Antigen-presenting cell

B cell

Helper T cell

Cytotoxic T cell

Memory helper T cells

Antigen (2nd exposure)

Plasma cells

Memory B cells

Memory cytotoxic T cells

Active cytotoxic T cells

Secreted antibodies

Defend against extracellular pathogens in blood and lymph by binding to antigens, thereby neutralizing pathogens or making them better targets for phagocytic cells and complement proteins.

Defend against intracellular pathogens and cancer by binding to and lysing the infected cells or cancer cells.

▲ **Figure 35.16 An overview of the adaptive immune response.**

[?] *Identify each black or brown arrow as representing part of the primary or secondary response.*

ANIMATION

BioFlix Visit the Study Area in **MasteringBiology** for the BioFlix® 3-D Animation on Immunology.

Active and Passive Immunization

Our discussion of adaptive immunity has focused to this point on **active immunity**, the defenses that arise when a pathogen infects the body. A different type of immunity results when, for example, antibodies in the blood of a pregnant female cross the placenta to her fetus. This protection is called **passive immunity** because the antibodies in the recipient (in this case, the fetus) are produced by another individual (the mother). Antibodies present in breast milk provide additional passive immunity to the infant's digestive tract while the infant's immune system develops. Because passive immunity does not involve the recipient's B and T

cells, it persists only as long as the transferred antibodies last (a few weeks to a few months).

Both active immunity and passive immunity can be induced artificially. Active immunity is induced when antigens are introduced into the body in *vaccines*, which may be made from inactivated bacterial toxins, killed or weakened pathogens, or even genes encoding microbial proteins. This process, called **immunization** (or vaccination), induces a primary immune response and immunological memory. As a result, any subsequent encounter with the pathogen from which the vaccine was derived triggers a rapid and strong secondary immune response (see Figure 35.12).

Misinformation about vaccine safety and disease risk has led to a substantial and growing public health problem. Consider measles as just one example. Side effects of immunization are remarkably rare, with fewer than one in a million children suffering a significant allergic reaction to the measles vaccine. The disease, however, is quite dangerous, killing more than 200,000 people each year. Declining vaccination rates in parts of the United Kingdom, Russia, and the United States have resulted in a number of recent measles outbreaks and many preventable deaths.

In artificial passive immunization, antibodies from an immune animal are injected into a nonimmune animal. For example, humans bitten by venomous snakes are sometimes treated with antivenin, serum from sheep or horses that have been immunized against a snake venom. When injected immediately after a snakebite occurs, the antibodies in antivenin can neutralize toxins in the venom before the toxins do massive damage.

Antibodies as Tools

Antibodies produced after exposure to an antigen are *polyclonal*: They are the products of many different clones of plasma cells, each specific for a different epitope. Antibodies can also be prepared from a clone of B cells grown in culture. The **monoclonal antibodies** produced by such a culture are identical and specific for the same epitope on an antigen.

Monoclonal antibodies have provided the basis for many recent advances in medical diagnosis and treatment. For example, home pregnancy test kits use monoclonal antibodies to detect human chorionic gonadotropin (hCG). Because hCG is produced as soon as an embryo implants in the uterus (see Chapter 36), the presence of this hormone in a woman's urine is a reliable indicator for a very early stage of pregnancy. Monoclonal antibodies are also produced in large amounts and injected as a therapy for a number of human diseases.

Immune Rejection

Like pathogens, cells from another person can be recognized as foreign and attacked by immune defenses. For example, skin transplanted from one person to a genetically nonidentical person will look healthy for a week or so but will then be destroyed (rejected) by the recipient's immune response. Carbohydrates on the surface of transfused blood cells can also be recognized as foreign by the recipient's immune system, triggering an immediate and devastating reaction. To avoid this danger, the so-called ABO blood groups of the donor and recipient must be taken into account.

In the case of tissue and organ transplants, or grafts, MHC molecules stimulate the immune response that leads to rejection. Each vertebrate species has many alleles for each MHC gene, enabling presentation of antigen fragments that vary in shape and net electrical charge. This diversity of MHC molecules almost guarantees that no two people, except identical twins, will have exactly the same set. Thus, in the vast majority of graft and transplant recipients, some MHC molecules on the donated tissue are foreign to the recipient. To minimize rejection, physicians use donor tissue bearing MHC molecules that match those of the recipient as closely as possible. In addition, the recipient takes medicines that suppress immune responses (but as a result leave the recipient more susceptible to infections).

Disruptions in Immune System Function

Although adaptive immunity offers significant protection against a wide range of pathogens, it is not fail-safe. Here we'll examine some of the ways the adaptive immune system fails to protect the host organism.

Allergies

Allergies are exaggerated (hypersensitive) responses to certain antigens called **allergens**. Hay fever, for instance, occurs when plasma cells secrete antibodies specific for antigens on the surface of pollen grains, as illustrated in **Figure 35.17**. The interaction of pollen grains and these antibodies triggers immune cells in connective tissue to release histamine and other inflammatory chemicals. The results can include sneezing, teary eyes, and smooth muscle contractions in the lungs that inhibit effective breathing. Drugs called antihistamines block receptors for histamine, diminishing allergy symptoms (and inflammation).

In some instances, an acute allergic response leads to a life-threatening reaction called *anaphylactic shock*. Inflammatory chemicals trigger abrupt dilation of peripheral blood vessels, causing a precipitous drop in blood pressure, as well as constriction of bronchioles. Death may occur within minutes due to lack of blood

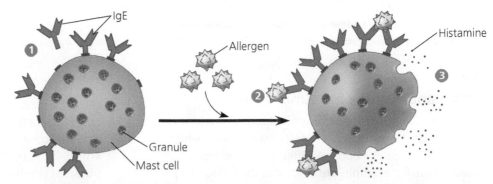

1 IgE antibodies produced in response to initial exposure to an allergen bind to receptors on mast cells.

2 On subsequent exposure to the same allergen, IgE molecules attached to a mast cell recognize and bind the allergen.

3 Cross-linking of adjacent IgE molecules triggers release of histamine and other chemicals, leading to allergy symptoms.

▲ **Figure 35.17 Mast cells and the allergic response.** In this example, pollen grains act as the allergen, and the immunoglobulins that mediate the response are of a type called IgE.

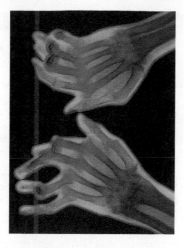

► **Figure 35.18 Colored X-ray of hands deformed by rheumatoid arthritis.**

flow and the inability to breathe. Substances that can cause anaphylactic shock in allergic individuals include bee venom, penicillin, peanuts, and shellfish. People with severe hypersensitivities often carry syringes containing the hormone epinephrine, which counteracts this allergic response.

Autoimmune Diseases

In some people, the immune system is active against particular molecules of the body, causing an **autoimmune disease**. In systemic lupus erythematosus, commonly called lupus, the immune system generates antibodies against histones and DNA. Other targets of autoimmunity are the insulin-producing beta cells of the pancreas (in type 1 diabetes) and the myelin sheaths that encase many neurons (in multiple sclerosis).

Gender, genetics, and environment all influence susceptibility to autoimmune disorders. For example, many autoimmune diseases afflict females more often than males. Women are nine times as likely as men to suffer from lupus and two to three times as likely to develop rheumatoid arthritis, a damaging and painful inflammation of the cartilage and bone in joints **(Figure 35.18)**. The cause of this sex bias, as well as the rise in autoimmune disease frequency in industrialized countries, are areas of active research and debate.

Immune System Avoidance

EVOLUTION Just as immune systems that ward off pathogens have evolved in animals, mechanisms that thwart immune responses have evolved in pathogens. In one such mechanism, a pathogen alters how it appears to the immune system. If a pathogen changes the epitopes it expresses to ones that a host has not previously encountered, it can reinfect or remain in the host without triggering the rapid and robust response mediated by memory cells. Such changes in epitope expression are called *antigenic variation*. The parasite that causes sleeping sickness provides an extreme example, periodically switching at random among 1,000 different versions of the protein found over its entire surface. In the **Scientific Skills Exercise**, you will interpret data related to this example of antigenic variation and the body's response.

Antigenic variation is the major reason the influenza, or "flu," virus remains a major public health problem. As it replicates in one human host after another, the human flu virus undergoes frequent mutations. Because any change that lessens recognition by the immune system provides a selective advantage, the virus steadily accumulates such alterations. These changes are the reason that a new flu vaccine must be distributed each year. In addition, the human flu virus occasionally exchanges genes with influenza viruses that infect domesticated animals, such as pigs or chickens. If the new strain expresses surface epitopes of the animal rather than the human virus, it may not be recognized by any of the memory cells in the human population. The resulting outbreak can be deadly: The 1918–1919 influenza outbreak killed more than 20 million people.

Some viruses avoid an immune response by infecting cells and then entering a largely inactive state called *latency*. The viral genome integrates into the chromosome of the host cell, which ceases making most viral proteins and typically releases no free viruses. Latency typically persists until conditions arise that are favorable for viral transmission or unfavorable for host survival.

Herpes simplex viruses provide a good example of latency. The type 1 virus causes most oral herpes infections, whereas the sexually transmitted type 2 virus is responsible for most cases of genital herpes. These viruses remain latent in sensory neurons until a stimulus such as fever, emotional stress, or menstruation reactivates the viruses. Activation of the type 1 virus can result in blisters around the mouth that are inaccurately called "cold" sores. Infections of the type 2 virus pose a serious threat to the babies of infected mothers and can increase transmission of HIV.

The **human immunodeficiency virus (HIV)**, the pathogen that causes AIDS (acquired immune deficiency syndrome), both escapes and attacks the adaptive immune response. Once introduced into the body, HIV infects helper T cells with high efficiency. Although the body responds to HIV with an immune response sufficient to eliminate most viral infections, some HIV invariably escapes. One reason HIV persists is that it has a very high mutation rate. Altered proteins on the surface of some mutated viruses reduce interaction with antibodies and cytotoxic T cells. Such viruses survive, proliferate, and mutate further. The virus thus evolves within the body. The continued presence of HIV is also helped by latency.

Over time, an untreated HIV infection not only avoids the adaptive immune response but also abolishes it. Viral reproduction and cell death triggered by the virus lead to loss of helper T cells, impairing both humoral and cell-mediated immune responses. The result is a progression to AIDS, characterized by susceptibility to infections and cancers that a healthy immune system would usually defeat. For example, *Pneumocystis carinii*, a common fungus that does not cause disease in healthy individuals, can result in severe pneumonia in people with AIDS. Such opportunistic diseases, as well as nerve damage and wasting, are the primary causes of death from AIDS, not HIV itself.

Comparing Two Variables on a Common x-Axis

How Does the Immune System Respond to a Changing Pathogen? Natural selection favors parasites that are able to maintain a low-level infection in a host for a long time. *Trypanosoma*, the unicellular parasite that causes sleeping sickness, is one example (see Figure 25.12). The glycoproteins covering a trypanosome's surface are encoded by a gene that is duplicated more than a thousand times in the organism's genome. Each copy is slightly different. By periodically switching among these genes, the trypanosome can change the molecular structure of its surface glycoproteins. In this exercise, you will interpret two data sets to explore hypotheses about the benefits of the trypanosome's ever-shifting surface glycoproteins and the host's immune response.

Part A: Data from a Study of Parasite Levels This study measured the abundance of parasites in the blood of one human patient during the first few weeks of a chronic infection.

Day	Number of Parasites (in millions) per mL of Blood
4	0.1
6	0.3
8	1.2
10	0.2
12	0.2
14	0.9
16	0.6
18	0.1
20	0.7
22	1.2
24	0.2

Part A: Interpret the Data

1. Plot the data in the above table as a line graph. Which column is the independent variable, and which is the dependent variable? Put the independent variable on the x-axis. (For additional information about graphs, see the Scientific Skills Review in Appendix F and in the Study Area in MasteringBiology.)

2. Visually displaying data in a graph can help make patterns in the data more noticeable. Describe any patterns revealed by your graph.

3. Assume that a drop in parasite abundance reflects an effective immune response by the host. Formulate a hypothesis to explain the pattern you described in question 2.

Part B: Data from a Study of Antibody Levels Many decades after scientists first observed the pattern of *Trypanosoma* abundance over the course of infection, researchers identified antibodies specific to different forms of the parasite's surface glycoprotein. The table below lists the relative abundance of two such antibodies during the early period of chronic infection, using an index ranging from 0 to 1.

Day	Antibody Specific to Glycoprotein Variant A	Antibody Specific to Glycoprotein Variant B
4	0	0
6	0	0
8	0.2	0
10	0.5	0
12	1	0
14	1	0.1
16	1	0.3
18	1	0.9
20	1	1
22	1	1
24	1	1

Part B: Interpret the Data

4. Note that these data were collected over the same period of infection (days 4–24) as the parasite abundance data you graphed in Part A. Therefore, you can incorporate these new data into your first graph, using the same x-axis. However, since the antibody level data are measured in a different way than the parasite abundance data, add a second set of y-axis labels on the right side of your graph. Then, using different colors or sets of symbols, add the data for the two antibody types. Labeling the y-axis two different ways enables you to compare how two dependent variables change relative to a shared independent variable.

5. Describe any patterns you observe by comparing the two data sets over the same period. Do these patterns support your hypothesis from Part A? Do they prove that hypothesis? Explain.

6. Scientists can now also distinguish the abundance of trypanosomes recognized specifically by antibodies type A and type B. How would incorporating such information change your graph?

(MB) A version of this Scientific Skills Exercise can be assigned in MasteringBiology.

HIV transmission requires the transfer of virus particles or infected cells via body fluids such as semen, blood, or breast milk. Unprotected sex (that is, without a condom) and transmission via HIV-contaminated needles (often among intravenous drug users) account for the vast majority of HIV infections. People infected with HIV can transmit the disease in the first few weeks of infection, *before* they express HIV-specific antibodies that can be detected in a blood test. Although HIV infection cannot be cured, drugs have been developed that can significantly slow HIV replication and the progression to AIDS. New drugs continue to be needed as HIV's high mutation rate results in the frequent appearance of drug-resistant strains.

Cancer and Immunity

When adaptive immunity is inactivated, the frequency of certain cancers increases dramatically. For example, the risk of developing Kaposi's sarcoma is 20,000 times greater for untreated AIDS patients than for healthy people. This observation was unanticipated. If the immune system recognizes only nonself, it should fail to recognize the uncontrolled growth of self cells that is the hallmark of cancer. It turns out, however, that viruses are involved in about 15–20% of all human cancers. Because the immune system can recognize viral proteins as foreign, it can act as a defense against viruses that can cause cancer and against cancer cells that harbor viruses.

Scientists have identified six viruses that can cause cancer in humans. The Kaposi's sarcoma herpesvirus is one such virus. Hepatitis B virus, which can trigger liver cancer, is another. A vaccine introduced in 1986 for hepatitis B virus was the first vaccine shown to help prevent a specific human cancer. Rapid progress on developing vaccines for virus-induced cancers continues. In 2006, the release of a vaccine against cervical cancer, specifically human papillomavirus (HPV), marked a major victory against a disease that afflicts more than half a million women worldwide every year.

35 Chapter Review

SUMMARY OF KEY CONCEPTS

CONCEPT 35.1

In innate immunity, recognition and response rely on traits common to groups of pathogens (pp. 712–715)

- In both invertebrates and vertebrates, **innate immunity** is mediated by physical and chemical barriers as well as cell-based defenses. Activation of innate immune responses relies on recognition proteins specific for broad classes of pathogens. Microbes that penetrate barrier defenses are ingested by phagocytic cells, which in vertebrates include **macrophages** and **dendritic cells**. In the **inflammatory response**, **histamine** and other chemicals released at the injury site promote changes in blood vessels that enhance immune cell access and action.

> **?** *In what ways does innate immunity protect the mammalian digestive tract?*

CONCEPT 35.2

In adaptive immunity, receptors provide pathogen-specific recognition (pp. 715–720)

- **Adaptive immunity** relies on two types of **lymphocytes** that arise from stem cells in the bone marrow: **T cells** and **B cells**. Lymphocytes have cell-surface **antigen receptors** for foreign molecules (**antigens**). Some T cells help other lymphocytes; others kill infected host cells. B cells called **plasma cells** produce soluble receptor proteins called **antibodies**, which bind to foreign molecules and cells. Activated T and B lymphocytes called **memory cells** defend against future infections by the same pathogen.
- Recognition of foreign molecules involves the binding of variable regions of receptors to an **epitope**, a small region of an antigen. B cells and antibodies recognize epitopes on the surface of antigens. T cells recognize protein epitopes in small antigen fragments (peptides) that are presented on the surface of host cells by proteins called **major histocompatibility complex (MHC) molecules**.
- The four major characteristics of B and T cell development are the generation of cell diversity, self-tolerance, proliferation, and immunological memory. Proliferation and memory are both based on **clonal selection**, illustrated here for B cells.

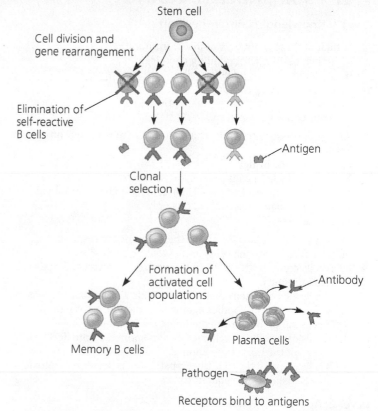

> **?** *Why is the adaptive immune response to an initial infection slower than the innate response?*

CONCEPT 35.3

Adaptive immunity defends against infection of body fluids and body cells (pp. 720–727)

- **Helper T cells** interact with antigen fragments displayed by class II MHC molecules on the surface of **antigen-presenting cells**: dendritic cells, macrophages, and B cells. Activated helper T cells secrete **cytokines** that stimulate other lymphocytes. In the **cell-mediated immune response**, activated **cytotoxic T cells** trigger destruction of infected cells. In the **humoral immune response**, antibodies help eliminate antigens by phagocytosis and complement-mediated lysis.

- **Active immunity** develops in response to infection or to **immunization**. The transfer of antibodies in **passive immunity** provides immediate, short-term protection.
- In tissue grafts and organ transplants, MHC molecules stimulate immune rejection.
- In allergies, such as hay fever, the interaction of antibodies and **allergens** triggers immune cells to release histamine and other mediators that cause vascular changes and allergic symptoms. Loss of self-tolerance can lead to **autoimmune diseases**, such as multiple sclerosis.
- Antigenic variation, latency, and direct assault on the immune system allow some pathogens to thwart immune responses. **HIV** infection destroys helper T cells, leaving the patient prone to disease. Immune defense against cancer appears to primarily involve action against viruses that can cause cancer and cancer cells that harbor viruses.

? *Do natural infection and immunization result in different types of immunological memory? Explain.*

TEST YOUR UNDERSTANDING

Level 1: Knowledge/Comprehension

1. Which of these is not part of insect immunity?
   a. activation of microbe-killing chemicals
   b. activation of natural killer cells
   c. phagocytosis by hemocytes
   d. production of antimicrobial peptides
   e. a protective exoskeleton

2. An epitope associates with which part of an antigen receptor or antibody?
   a. the disulfide bridge
   b. the heavy-chain constant regions only
   c. variable regions of a heavy chain and light chain combined
   d. the light-chain constant regions only
   e. the tail

3. Which statement best describes the difference in responses of effector B cells (plasma cells) and cytotoxic T cells?
   a. B cells confer active immunity; cytotoxic T cells confer passive immunity.
   b. B cells kill pathogens directly; cytotoxic T cells kill host cells.
   c. B cells secrete antibodies against a pathogen; cytotoxic T cells kill pathogen-infected host cells.
   d. B cells carry out the cell-mediated response; cytotoxic T cells carry out the humoral response.
   e. B cells respond the first time a pathogen is present; cytotoxic T cells respond subsequent times.

Level 2: Application/Analysis

4. Which of the following statements is *not* true?
   a. An antibody has more than one antigen-binding site.
   b. An antigen can have different epitopes.
   c. A pathogen makes more than one antigen.
   d. A lymphocyte has receptors for multiple different antigens.
   e. A liver cell makes one class of MHC molecule.

5. Which of the following should be the same in identical twins?
   a. the set of antibodies produced
   b. the set of MHC molecules produced
   c. the set of T cell antigen receptors produced
   d. the susceptibility to a particular virus
   e. the set of immune cells eliminated as self-reactive

Level 3: Synthesis/Evaluation

6. Vaccination increases the number of
   a. different receptors that recognize a pathogen.
   b. lymphocytes with receptors that can bind to the pathogen.
   c. epitopes that the immune system can recognize.
   d. macrophages specific for a pathogen.
   e. MHC molecules that can present an antigen.

7. Which of the following would *not* help a virus avoid triggering an adaptive immune response?
   a. having frequent mutations in genes for surface proteins
   b. infecting cells that produce very few MHC molecules
   c. producing proteins very similar to those of other viruses
   d. infecting and killing helper T cells
   e. building the viral shell from host proteins

8. **DRAW IT** Consider a pencil-shaped protein with two epitopes, Y (the "eraser" end) and Z (the "point" end). They are recognized by antibodies A1 and A2, respectively. Draw and label a picture showing the antibodies linking proteins into a complex that could trigger endocytosis by a macrophage.

9. **MAKE CONNECTIONS** Contrast the clonal selection of lymphocytes with Lamarck's idea for the inheritance of acquired characteristics (see Concept 19.1).

10. **SCIENTIFIC INQUIRY**
    A diagnostic test for tuberculosis (TB) involves injecting antigen (from the bacterium that causes TB) under the skin and then waiting a few days for a reaction to appear. This test is *not* useful for diagnosing TB in AIDS patients. Why?

11. **FOCUS ON EVOLUTION**
    Describe one invertebrate defense mechanism and discuss how it is an evolutionary adaptation retained in vertebrates.

12. **FOCUS ON INFORMATION**
    Among all nucleated body cells, only B and T cells lose DNA during their development and maturation. In a short essay (100–150 words), discuss the relationship between this loss and the theme of DNA as heritable biological information, focusing on similarities between cellular and organismal generations.

For selected answers, see Appendix A.

MasteringBiology®

Students Go to **MasteringBiology** for assignments, the eText, and the Study Area with practice tests, animations, and activities.

Instructors Go to **MasteringBiology** for automatically graded tutorials and questions that you can assign to your students, plus Instructor Resources.

36

Reproduction and Development

KEY CONCEPTS

36.1 Both asexual and sexual reproduction occur in the animal kingdom

36.2 Reproductive organs produce and transport gametes

36.3 The interplay of tropic and sex hormones regulates reproduction in mammals

36.4 Fertilization, cleavage, and gastrulation initiate embryonic development

OVERVIEW

Pairing Up for Sexual Reproduction

The sea slugs, or nudibranchs (*Nembrotha chamberlaini*), in **Figure 36.1** are mating. If not disturbed, these marine molluscs may remain joined for hours as sperm are transferred and eggs are fertilized. A few weeks later, new individuals will hatch, and sexual reproduction will be complete—but which parent is the mother of these offspring? The answer is simple yet probably unexpected: both. In fact, each sea slug produces eggs *and* sperm, and so is both a mother and a father to the next generation.

As humans, we tend to think of reproduction in terms of the mating of males and females and the fusion of sperm and eggs. Across the animal kingdom, however, reproduction takes many forms. In some species, individuals change their sex during their lifetime; in other species, such as sea slugs, an individual is both male and female. There are animals that can fertilize their own eggs, as well as others that can reproduce without any form of sex. In certain species, such as honeybees, only a few members of a large population reproduce.

A population outlives its members only by reproduction, the generation of new individuals from existing ones. In this chapter, we'll compare the diverse reproductive mechanisms that have evolved in the animal kingdom. Then we'll examine details of reproduction in mammals, with particular emphasis on the intensively studied example of humans. Lastly, we'll explore fundamental events in the earliest stages of an animal's development.

▼ **Figure 36.1** How can each of these sea slugs be both male and female?

CONCEPT 36.1

Both asexual and sexual reproduction occur in the animal kingdom

There are two modes of animal reproduction—sexual and asexual. In **sexual reproduction**, the fusion of haploid gametes forms a diploid cell, the **zygote**. The animal that develops from a zygote can in turn give rise to gametes by meiosis (see Figure 10.8). The female gamete, the **egg**, is large and nonmotile, whereas the male gamete, the **sperm**, is generally much smaller and motile. In contrast, in **asexual reproduction**, new individuals are generated without the

▲ **Figure 36.2 Asexual reproduction of a sea anemone (*Anthopleura elegantissima*).** The large individual in the center of this photograph is undergoing fission, a type of asexual reproduction. Two smaller individuals will form as the parent divides approximately in half. Each offspring will be a genetic copy of the parent.

fusion of egg and sperm. For most asexual animals, reproduction relies entirely on mitotic cell division.

Mechanisms of Asexual Reproduction

Several simple forms of asexual reproduction are found only among invertebrates. One of these is *budding*, in which new individuals arise from outgrowths of existing ones (see Figure 10.2). In stony corals, for example, buds form and remain attached to the parent. The eventual result is a colony more than 1 m across, consisting of thousands of connected individuals. Also common among invertebrates is *fission*, the separation of a parent organism into two individuals of approximately equal size **(Figure 36.2)**.

Asexual reproduction can also be a two-step process: *fragmentation*, the breaking of the body into several pieces, followed by *regeneration*, the regrowth of lost body parts. If more than one piece grows and develops into a complete animal, the net effect is reproduction. For example, certain annelid worms can split their body into several fragments, each regenerating a complete worm in less than a week. Many sponges, cnidarians, bristle worms, and sea squirts also reproduce by fragmentation and regeneration.

A particularly intriguing form of asexual reproduction is **parthenogenesis**, in which an egg develops without being fertilized. Among invertebrates, parthenogenesis occurs in certain species of bees, wasps, and ants. The progeny can be either haploid or diploid. If haploid, the offspring develop into adults that produce eggs or sperm without meiosis. Among vertebrates,

parthenogenesis has been observed in about one in every thousand species. Recently, zookeepers discovered evidence of parthenogenesis in Komodo dragons and in a species of hammerhead shark: In both cases, females had been kept completely isolated from males of their species but nevertheless produced offspring.

The rest of this chapter focuses on sexual reproduction, the existence of which is in fact puzzling, at least from an evolutionary perspective.

Sexual Reproduction: An Evolutionary Enigma

EVOLUTION   Sex must enhance reproductive success or survival because it would otherwise rapidly disappear. To see why, consider an animal population in which half the females reproduce sexually and half reproduce asexually **(Figure 36.3)**. We'll assume that the number of offspring per female is a constant, two in this case. The two offspring of an asexual female will both be daughters that will each give birth to two more daughters that can reproduce. In contrast, on average, half of a sexual female's offspring will be male. The number of sexual offspring will remain the same at each generation, because both a male and a female are required to reproduce. Thus, the asexual condition will increase in frequency at each generation. Yet despite this "twofold cost," sex is maintained even in animal species that can also reproduce asexually.

What advantage does sex provide? The answer remains elusive. Most hypotheses focus on the unique combinations of parental genes formed during meiotic recombination and fertilization. By producing offspring of varied genotypes, sexual reproduction may enhance the reproductive success of parents when environmental factors, such as pathogens, change relatively rapidly. In contrast, we would expect asexual reproduction to be most advantageous in stable, favorable environments because it perpetuates successful genotypes precisely.

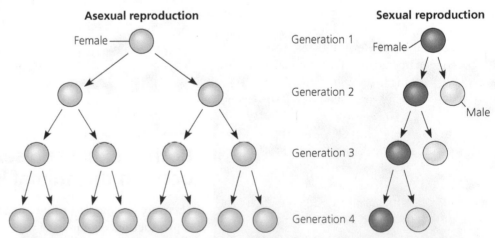

▲ **Figure 36.3 The "reproductive handicap" of sex.** These diagrams contrast the reproductive output of females (blue spheres) over four generations for asexual versus sexual reproduction, assuming two surviving offspring per female. The asexual portion of the population rapidly outgrows the sexual one.

▲ **Figure 36.4 Caribou (*Rangifer tarandus*) mother and calf.** As a result of warming due to global climate change, the number of caribou offspring in a West Greenland study site has fallen fourfold.

Reproductive Cycles

Most animals exhibit cycles in reproductive activity, often related to changing seasons. These cycles are controlled by hormones, whose secretion in turn is regulated by environmental cues. In this way, animals conserve resources, reproducing only when sufficient energy sources or stores are available and when environmental conditions favor the survival of offspring.

Because seasonal temperature is often an important cue for reproduction, climate change can decrease reproductive success. Researchers have demonstrated just such an effect on caribou (wild reindeer) in Greenland. In spring, caribou migrate to calving grounds to eat sprouting green plants, give birth, and care for their calves **(Figure 36.4)**. Prior to 1993, the arrival of caribou at the calving grounds coincided with the brief period during which the plants were nutritious and digestible. From 1993 to 2006, however, average spring temperatures in the calving grounds increased by more than 4°C, and the plants now sprout two weeks earlier. Because caribou migration is triggered by day length, not temperature, there is a mismatch between the timing of new plant growth and caribou birthing. Without adequate nutrition for the nursing females, production of caribou offspring has declined by 75% since 1993.

A different kind of reproductive cycle occurs in some species of whiptail lizards in the genus *Aspidoscelis*, in which members of breeding pairs alternate roles. In these species, reproduction is exclusively asexual, and there are no males. Nevertheless, these lizards have courtship and mating behaviors very similar to those of sexual species of *Aspidoscelis*. During the breeding season, one female of each mating pair mimics a male **(Figure 36.5a)**. Each member of the pair alternates roles two or three times during the season. An individual adopts female behavior prior to ovulation, when the level of the female sex hormone estradiol is high, and then switches to male-like behavior after ovulation, when the level of progesterone is

(a) Both lizards in this photograph are *A. uniparens* females. The one on top is playing the role of a male. Every two or three weeks during the breeding season, individuals switch sex roles.

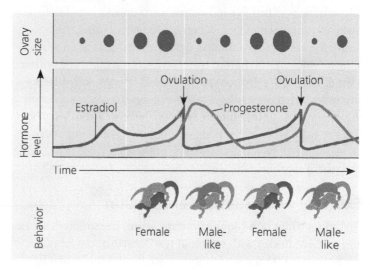

(b) The sexual behavior of *A. uniparens* is correlated with the cycle of ovulation mediated by sex hormones. As the blood level of estradiol rises, the ovaries grow, and the lizard behaves as a female. After ovulation, the estradiol level drops abruptly, and the progesterone level rises; these hormone levels correlate with male-like behavior.

▲ **Figure 36.5 Sexual behavior in parthenogenetic lizards.** The desert-grassland whiptail lizard (*Aspidoscelis uniparens*) is an all-female species. These reptiles reproduce by parthenogenesis, the development of an unfertilized egg. Nevertheless, ovulation is stimulated by mating behavior.

high **(Figure 36.5b)**. Ovulation is more likely to occur if the individual is mounted during the critical time of the hormone cycle; isolated lizards lay fewer eggs than those that go through the motions of sex. These observations support the hypothesis that these parthenogenetic lizards evolved from species having two sexes and still require certain sexual stimuli for maximum reproductive success.

Variation in Patterns of Sexual Reproduction

For many animals, finding a partner for sexual reproduction can be challenging. Adaptations that arose during the evolution of some species meet this challenge in a novel way—by blurring the strict distinction between male and female. One such adaptation arose among sessile (stationary) animals, such as barnacles; burrowing animals, such as clams; and some parasites, including tapeworms. Largely lacking mobility, these animals have little opportunity to find a mate. The evolutionary solution in this case is **hermaphroditism**, in which each individual has both male and female reproductive systems (the term *hermaphrodite* merges the names Hermes and Aphrodite, a Greek god and goddess). Because each hermaphrodite reproduces as both a male and a female, *any* two individuals can mate. Each animal donates and receives sperm during mating, as the sea slugs in Figure 36.1 are doing. In some species, hermaphrodites are also capable of self-fertilization, allowing a form of sexual reproduction that doesn't require any partner.

The bluehead wrasse (*Thalassoma bifasciatum*) provides an example of a quite different variation in sexual reproduction. These coral reef fish live in harems, each consisting of a single male and several females. When the lone male dies, the opportunity for sexual reproduction would appear lost. Instead, the largest female in the harem transforms into a male and within a week begins to produce sperm instead of eggs. What selective pressure in the evolution of the bluehead wrasse resulted in sex reversal for that female with the largest body? Because it is the male wrasse that defends a harem against intruders, a larger size may be particularly important for a male in ensuring successful reproduction.

External and Internal Fertilization

Sexual reproduction requires **fertilization**, the union of sperm and egg. In species with **external fertilization**, the female releases eggs into the environment, where the male then fertilizes them. Other species have **internal fertilization**: Sperm are deposited in or near the female reproductive tract, and fertilization occurs within the tract.

A moist habitat is almost always required for external fertilization, both to prevent the gametes from drying out and to allow the sperm to swim to the eggs. Many aquatic invertebrates simply shed their eggs and sperm into the surroundings, and fertilization occurs without the parents making physical contact. However, timing is crucial to ensure that mature sperm and eggs encounter one another.

Among some species with external fertilization, individuals clustered in the same area release their gametes into the water at the same time, a process known as *spawning*. In some cases, chemical signals generated by one individual as it releases gametes trigger other individuals to release gametes. In other cases, environmental cues, such as temperature or day length, cause a whole population to release gametes at one time. For example, the palolo worm, native to coral reefs of the South Pacific, coordinates its spawning to both the season and the lunar cycle. Sometime in spring when the moon is in its last quarter, palolo worms break in half, releasing tail segments engorged with sperm or eggs. These packets rise to the ocean surface and burst in such vast numbers that the sea appears milky with gametes.

▲ **Figure 36.6 External fertilization.** Many species of amphibians reproduce by external fertilization. In most of these species, behavioral adaptations ensure that a male is present when the female releases eggs. Here, a female frog (on bottom) has released a mass of eggs in response to being clasped by a male. The male released sperm (not visible) at the same time, and external fertilization has already occurred in the water.

The sperm quickly fertilize the floating eggs, and within hours, the palolo's once-a-year reproductive frenzy is complete.

When external fertilization is not synchronous across a population, individuals may exhibit specific "courtship" behaviors leading to the fertilization of the eggs of one female by one male **(Figure 36.6)**. By triggering the release of both sperm and eggs, these behaviors increase the probability of successful fertilization.

Internal fertilization is an adaptation that enables sperm to reach an egg efficiently, even when the environment is dry. It typically requires cooperative behavior that leads to copulation, as well as sophisticated and compatible reproductive systems. The male copulatory organ delivers sperm, and the female reproductive tract often has receptacles for storage and delivery of sperm to mature eggs.

No matter how fertilization occurs, the mating animals may make use of *pheromones*, chemicals released by one organism that can influence the physiology and behavior of other individuals of the same species. Pheromones are small, volatile or water-soluble molecules that disperse into the environment and, like hormones, are active in tiny amounts. Many pheromones function as mate attractants, enabling some female insects to be detected by males more than a kilometer away.

Ensuring the Survival of Offspring

Internal fertilization typically is associated with the production of fewer gametes than external fertilization but results in the survival of a higher fraction of zygotes. Better zygote survival is due in part to the fact that eggs fertilized internally are sheltered from potential predators. However, internal fertilization is also more often associated with mechanisms that provide

▲ **Figure 36.7 Parental care in an invertebrate.** Compared with many other insects, giant water bugs of the genera *Abedus* and *Belostoma* produce relatively few offspring, but offer much greater parental protection. Following internal fertilization, the female glues her fertilized eggs to the back of the male (shown here). The male carries them for days, frequently fanning water over them to keep the eggs moist, aerated, and free of parasites.

greater protection of the embryos and parental care of the young. For example, the eggs of birds and other reptiles have calcium- and protein-containing shells and internal membranes that protect against water loss and physical damage (see Figure 27.25). In contrast, the eggs of fishes and amphibians have only a gelatinous coat and lack internal membranes.

Rather than secreting a protective eggshell, some animals retain the embryo for a portion of its development within the female's reproductive tract. Embryos of marsupial mammals, such as kangaroos and opossums, spend only a short period in the uterus; the embryos then crawl out and complete fetal development attached to a mammary gland in the mother's pouch. Embryos of eutherian (placental) mammals, such as humans, remain in the uterus throughout fetal development. There they are nourished by the mother's blood supply through a temporary organ, the placenta. The embryos of some fishes and sharks also complete development internally.

When an eagle hatches out of an egg or when a human is born, the newborn is not yet capable of independent existence. Instead, adult birds feed their young and adult mammals nurse their offspring. Parental care is in fact widespread among animals, including invertebrate species **(Figure 36.7)**.

CONCEPT CHECK 36.1

1. How does internal fertilization facilitate life on land?
2. **WHAT IF?** If a hermaphrodite self-fertilizes, will the offspring be identical to the parent? Explain.
3. **MAKE CONNECTIONS** What examples of plant reproduction are most similar to asexual reproduction in animals? (See Concept 30.2.)

For suggested answers, see Appendix A.

Reproductive organs produce and transport gametes

Sexual reproduction in animals relies on sets of cells that are precursors for eggs and sperm. A group of cells dedicated to this function is often established early in the formation of the embryo and remains inactive while the body takes shape. Cycles of growth and mitosis then increase, or *amplify*, the number of cells available for making eggs or sperm.

Variation in Reproductive Systems

In producing gametes and making them available for fertilization, animals employ a variety of reproductive systems. **Gonads**, organs that produce gametes, are found in many but not all animals. Exceptions include the palolo worm, discussed earlier. The palolo and most other polychaete worms (phylum Annelida; see Figure 27.11) have separate sexes but lack distinct gonads; rather, the eggs and sperm develop from undifferentiated cells lining the coelom (body cavity). As the gametes mature, they are released from the body wall and fill the coelom. Depending on the species, mature gametes in these worms may be shed through the excretory opening, or the swelling mass of eggs may split a portion of the body open, spilling the eggs into the environment.

More elaborate reproductive systems include sets of accessory tubes and glands that carry, nourish, and protect the gametes and sometimes the developing embryos. Most insects, for example, have separate sexes with complex reproductive systems. In many insect species, the female reproductive system includes one or more *spermathecae*, sacs in which sperm may be stored for extended periods, a year or more in some species. Because the female releases male gametes from the spermathecae only in response to the appropriate stimuli, fertilization occurs under conditions likely to be well suited to embryonic development.

Vertebrate reproductive systems display limited but significant variations. In many nonmammalian vertebrates, the digestive, excretory, and reproductive systems have a common opening to the outside, the **cloaca**, a structure probably present in the ancestors of all vertebrates. Males of these species lack a well-developed penis and release sperm by turning the cloaca inside out.

In contrast, mammals generally lack a cloaca and have a separate opening for the digestive tract. Most female mammals also have separate openings for the excretory and reproductive systems. In some vertebrates, the uterus is divided into two chambers; in others, including humans and birds, it is a single structure.

Having surveyed some general features of animal reproduction, we turn now to human reproduction, beginning with the reproductive anatomy of males.

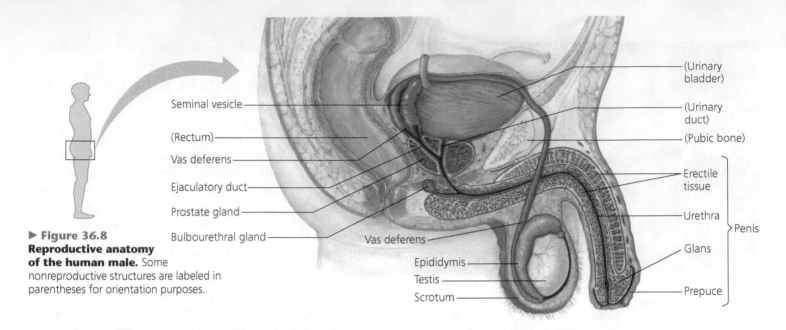

▶ **Figure 36.8**
Reproductive anatomy of the human male. Some nonreproductive structures are labeled in parentheses for orientation purposes.

Labels (from figure): Seminal vesicle, (Rectum), Vas deferens, Ejaculatory duct, Prostate gland, Bulbourethral gland, Vas deferens, Epididymis, Testis, Scrotum, (Urinary bladder), (Urinary duct), (Pubic bone), Erectile tissue, Urethra, Glans, Prepuce, Penis

Human Male Reproductive Anatomy

The human male's external reproductive organs are the scrotum and penis. The internal reproductive organs consist of gonads that produce both sperm and reproductive hormones, accessory glands that secrete products essential to sperm movement, and ducts that carry the sperm and glandular secretions **(Figure 36.8)**.

Testes

The male gonads, or **testes** (singular, *testis*), produce sperm in highly coiled tubes called **seminiferous tubules**. Most mammals produce sperm properly only when the testes are cooler than the rest of the body. In humans and many other mammals, the **scrotum**, a fold of the body wall, maintains testis temperature about 2°C below the core body temperature.

The testes develop in the abdominal cavity and descend into the scrotum just before birth (a testis within a scrotum is a *testicle*). In many rodents, the testes are drawn back into the cavity between breeding seasons, interrupting sperm maturation. Some mammals whose body temperature is low enough to allow sperm maturation—such as whales and elephants—retain the testes in the abdominal cavity at all times.

Ducts

From the seminiferous tubules of a testis, the sperm pass into the coiled duct of an **epididymis**. In humans, it takes 3 weeks for sperm to travel the 6 m length of this duct, during which time the sperm complete maturation and become motile. During **ejaculation**, the sperm are propelled from each epididymis through a muscular duct, the **vas deferens**. Each vas deferens (one from each epididymis) extends around and behind the urinary bladder, where it joins a duct from the seminal vesicle, forming a short **ejaculatory duct**. The ejaculatory ducts open into the **urethra**, the outlet tube for both the excretory system and the reproductive system. The urethra runs through the penis and opens to the outside at the tip of the penis.

Accessory Glands

Three sets of accessory glands—the seminal vesicles, the prostate gland, and the bulbourethral glands—produce secretions that combine with sperm to form **semen**, the fluid that is ejaculated. Two **seminal vesicles** contribute about 60% of the volume of semen. The fluid from the seminal vesicles is thick, yellowish, and alkaline. It contains mucus, the sugar fructose (which provides most of the sperm's energy), a coagulating enzyme, ascorbic acid, and local regulators called prostaglandins.

The **prostate gland** secretes its products into the urethra through small ducts. This fluid is thin and milky; it contains anticoagulant enzymes and citrate (a sperm nutrient). This gland undergoes benign (noncancerous) enlargement in more than half of all men over age 40 and in almost all men over 70. In addition, prostate cancer, which most often afflicts men 65 and older, is one of the most common human cancers.

The *bulbourethral glands* are a pair of small glands along the urethra below the prostate. Before ejaculation, they secrete clear mucus that neutralizes any acidic urine remaining in the urethra. Bulbourethral fluid also carries some sperm released before ejaculation, which is one reason for the high failure rate of the withdrawal method of birth control (coitus interruptus).

Penis

The human **penis** contains the urethra as well as three cylinders of spongy erectile tissue. During sexual arousal, the erectile tissue, which is derived from modified veins and capillaries, fills with blood from the arteries. As this tissue fills, the increasing pressure seals off the veins that drain the penis, causing it to engorge with blood. The resulting erection enables the penis to be inserted into the vagina. Alcohol consumption, certain drugs, emotional issues, and aging all can cause an inability to achieve an erection (erectile dysfunction). For individuals with long-term erectile dysfunction, drugs such as Viagra promote the vasodilating action of nitric oxide; the resulting relaxation of smooth muscles in the blood vessels of the penis enhances blood flow into the erectile tissues. Although all mammals rely on penile erection for mating, the penis of rodents, raccoons, walruses, whales, and several other mammals also contains a bone, the baculum, which is thought to further stiffen the penis for mating.

The main shaft of the penis is covered by relatively thick skin. The head, or **glans**, of the penis has a much thinner covering and is consequently more sensitive to stimulation. The human glans is covered by a fold of skin called the **prepuce**, or foreskin, which is removed when a male is circumcised.

Human Female Reproductive Anatomy

The human female's external reproductive structures are the clitoris and two sets of labia, which surround the clitoris and vaginal opening. The internal organs are the gonads, which produce both eggs and reproductive hormones, and a system of ducts and chambers, which receive and carry gametes and house the embryo and fetus **(Figure 36.9)**.

Ovaries

The female gonads are a pair of **ovaries** that flank the uterus and are held in place in the abdominal cavity by ligaments. The outer layer of each ovary is packed with **follicles**, each consisting of an **oocyte**, a partially developed egg, surrounded by support cells. The surrounding cells nourish and protect the oocyte during much of its formation and development.

Oviducts and Uterus

An **oviduct**, or fallopian tube, extends from the uterus toward a funnel-like opening at each ovary. The dimensions of this tube vary along its length, with the inside diameter near the uterus being as narrow as a human hair. Upon **ovulation**, the release of a mature egg, cilia on the epithelial lining of the oviduct help collect the egg by drawing fluid from the body cavity into the oviduct. Together with wavelike contractions of the oviduct, the cilia convey the egg down the duct to the **uterus**, also known as the womb. The uterus is a thick, muscular organ that can expand during pregnancy to accommodate a 4-kg fetus. The inner lining of the uterus, the **endometrium**, is richly supplied with blood vessels. The neck of the uterus, called the **cervix**, opens into the vagina.

Vagina and Vulva

The **vagina** is a muscular but elastic chamber that is the site for insertion of the penis and deposition of sperm during copulation. The vagina, which also serves as the birth canal through which a baby is born, opens to the outside at the **vulva**, the collective term for the external female genitalia.

A pair of thick, fatty ridges, the **labia majora**, encloses and protects the rest of the vulva. The vaginal opening and the separate opening of the urethra are located within a cavity bordered by a pair of slender skin folds, the **labia minora**. A thin piece of tissue called the **hymen** partly covers the vaginal opening in humans at birth and usually until sexual intercourse or vigorous physical activity ruptures it. Located at the top of the labia minora, the **clitoris** consists of erectile tissue supporting a rounded glans, or head, covered by a small hood of skin, the prepuce. During sexual arousal, the clitoris, vagina, and labia minora all engorge with blood and enlarge. Richly supplied with nerve endings, the clitoris is one of the most sensitive points of sexual stimulation. Sexual arousal also induces the vestibular glands near the vaginal opening to secrete lubricating mucus, thereby facilitating intercourse.

Mammary Glands

The **mammary glands** are present in both sexes, but they normally produce milk only in females. Though not part of the reproductive system, the female mammary glands are important to reproduction. Within the glands, small sacs of epithelial tissue secrete milk, which drains into a series of ducts that open at the nipple. The breasts contain connective and fatty (adipose) tissue in addition to the mammary glands. Because the low level of the hormone estradiol in males limits the development of the fat deposits, male breasts usually remain small.

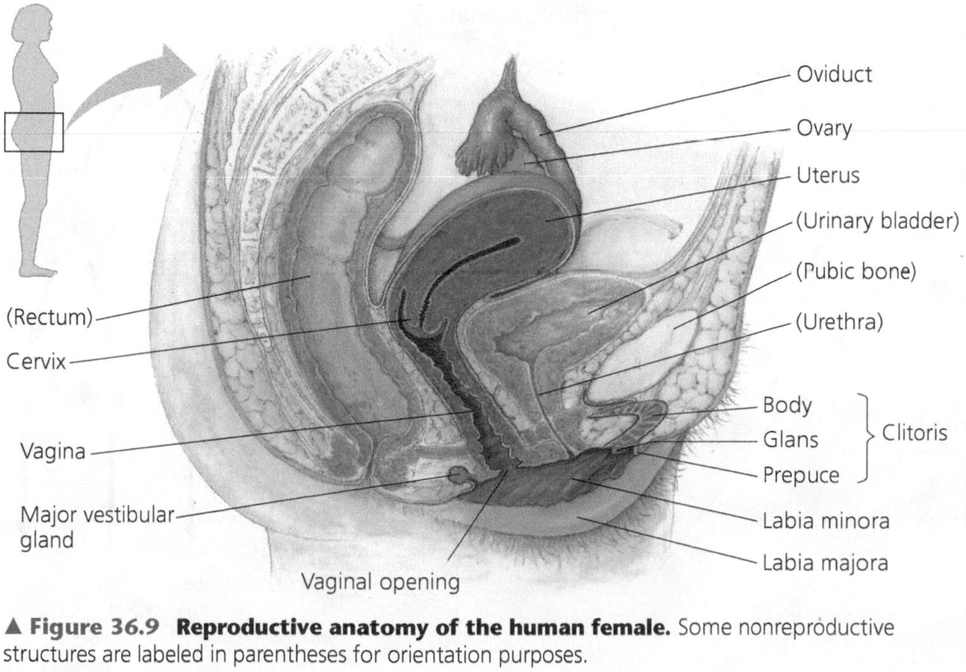

Gametogenesis

With this overview of reproductive anatomy in mind, we turn now to **gametogenesis**, the production of gametes. **Figure 36.10** compares this process in human males and females, highlighting the close relationship between the gonads' structure and their function.

Spermatogenesis, the formation and development of sperm, is continuous and

▲ **Figure 36.9 Reproductive anatomy of the human female.** Some nonreproductive structures are labeled in parentheses for orientation purposes.

Labels: Oviduct, Ovary, Uterus, (Urinary bladder), (Pubic bone), (Urethra), Body, Glans, Prepuce, Clitoris, Labia minora, Labia majora, (Rectum), Cervix, Vagina, Major vestibular gland, Vaginal opening

Spermatogenesis

These drawings correlate the mitotic and meiotic divisions in sperm development with the microscopic structure of seminiferous tubules.

The initial, or *primordial*, germ cells of the embryonic testes divide and differentiate into stem cells that divide by mitosis to form **spermatogonia**, which in turn generate spermatocytes, also by mitosis. Each spermatocyte gives rise to four spermatids through meiosis, reducing the chromosome number from diploid ($2n = 46$ in humans) to haploid ($n = 23$). Spermatids undergo extensive changes in cell shape and organization in differentiating into sperm.

Within the seminiferous tubules, there is a concentric organization of the steps of spermatogenesis. Stem cells are situated near the outer edge of the tubules. As spermatogenesis proceeds, cells move steadily inward as they pass through the spermatocyte stage and the spermatid stage. In the last step, mature sperm are released into the lumen (fluid-filled cavity) of the tubule. The sperm travel along the tubule into the epididymis, where they become motile.

The structure of a sperm cell fits its function. In humans, as in most species, a head containing the haploid nucleus is tipped with a specialized vesicle, the **acrosome**, which contains enzymes that help the sperm penetrate an egg. Behind the head, many mitochondria (or one large mitochondrion in some species) provide ATP for movement of the flagellar tail.

Oogenesis

Oogenesis begins in the female embryo with the production of **oogonia** from primordial germ cells. The oogonia divide by mitosis to form cells that begin meiosis, but stop the process at prophase I before birth. These developmentally arrested cells, called **primary oocytes**, each reside within a small follicle, a cavity lined with protective cells. Beginning at puberty, follicle-stimulating hormone (FSH) periodically stimulates a small group of follicles to resume growth and development. Cells of the follicle produce the primary female sex hormone, estradiol (a type of estrogen). Typically, only one follicle fully matures each month, with its primary oocyte completing meiosis I. The second meiotic division begins, but stops at metaphase. Thus arrested in meiosis II,

the **secondary oocyte** is released at ovulation, when its follicle breaks open. Only if a sperm penetrates the oocyte does meiosis II resume. (In other animal species, the sperm may enter the oocyte at the same stage, earlier, or later.) Each of the two meiotic divisions involves unequal cytokinesis, with the smaller cells becoming polar bodies that eventually degenerate (the first polar body may or may not divide again). Thus, the functional product of complete oogenesis is a single mature egg already containing a sperm head; fertilization is defined strictly as the fusion of the haploid nuclei of the sperm and secondary oocyte, although we often use it loosely to mean the entry of the sperm head into the egg.

The ruptured follicle left behind after ovulation develops into a mass called the **corpus luteum** ("yellow body"). The corpus luteum secretes additional estradiol as well as progesterone, a hormone that helps maintain the uterine lining during pregnancy. If the egg is not fertilized, the corpus luteum degenerates, and a new follicle matures during the next cycle.

At birth, the ovaries together contain 1–2 million primary oocytes, of which about 500 fully mature between puberty and menopause. To the best of our current knowledge, women are born with all the primary oocytes they will ever have. It is worth noting, however, that a similar conclusion regarding most other mammals was overturned in 2004 when researchers discovered that the ovaries of adult mice contain multiplying oogonia that develop into oocytes. If the same is true of humans, then the marked decline in fertility that occurs as women age might result from both a depletion of oogonia and the degeneration of aging oocytes.

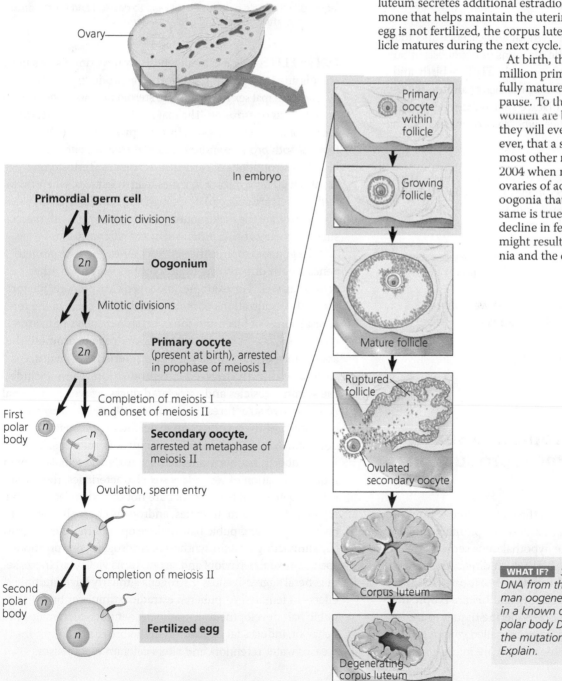

WHAT IF? *Suppose you are analyzing the DNA from the polar bodies formed during human oogenesis. If the mother has a mutation in a known disease gene, would analyzing the polar body DNA allow you to infer whether the mutation is present in the mature oocyte? Explain.*

prolific in adult males. To produce hundreds of millions of sperm each day, cell division and maturation occur throughout the seminiferous tubules coiled within the two testes. For a single sperm, the process takes about 7 weeks from start to finish.

Oogenesis, the development of mature oocytes (eggs), is a prolonged process in the human female. Immature eggs form in the ovary of the female embryo but do not complete their development until years, and often decades, later.

Spermatogenesis differs from oogenesis in three significant ways:

* Only in spermatogenesis do all four products of meiosis develop into mature gametes. In oogenesis, cytokinesis during meiosis is unequal, with almost all the cytoplasm segregated to a single daughter cell. This large cell is destined to become the egg; the other products of meiosis, smaller cells called polar bodies, degenerate.
* Spermatogenesis occurs throughout adolescence and adulthood. In contrast, the mitotic divisions of oogenesis in human females are thought to be complete before birth, and the production of mature gametes ceases at about age 50.
* Spermatogenesis produces mature sperm from precursor cells in a continuous sequence, whereas oogenesis has long interruptions.

CONCEPT CHECK 36.2

1. In what ways are a second polar body and an early spermatid similar? In what ways are they dissimilar?
2. Why might using a hot tub frequently make it harder for a couple to conceive a child?
3.  How are the uterus of an insect and the ovary of a flowering plant similar in function? How are they different? (See Figure 30.7.)
4. **WHAT IF?** If each vas deferens in a male was surgically sealed off, what changes would you expect in sexual response and ejaculate composition?

For suggested answers, see Appendix A.

36.3

The interplay of tropic and sex hormones regulates reproduction in mammals

In both male and female humans, the coordinated actions of hormones from the hypothalamus, anterior pituitary, and gonads govern reproduction. The hypothalamus secretes *gonadotropin-releasing hormone (GnRH)*, which directs the anterior pituitary to secrete **follicle-stimulating hormone (FSH)** and **luteinizing hormone (LH)**. FSH and LH are **tropic hormones**, meaning that they act on endocrine tissues to trigger the release of other hormones. They are called *gonadotropins* because the endocrine tissues they act on are in the gonads.

▲ **Figure 36.11 Androgen-dependent male anatomy and behavior in a lizard.** A male anole (*Norops ortoni*) extends his dewlap, a brightly colored skinflap beneath the throat. Testosterone is required in the male both for the dewlap to develop and for the anole to display it to attract mates and guard his territory.

FSH and LH regulate gametogenesis by targeting tissues in the gonads and by regulating sex hormone production.

The principal sex hormones are steroid hormones classified as *androgens* or *estrogens*. The major androgen is **testosterone**; the major estrogens are **estradiol** and **progesterone**. Males and females both produce androgens and estrogens, but differ in their blood concentrations of particular hormones. Testosterone levels are about 10 times higher in males than in females, whereas estradiol levels are about 10 times higher in females than in males. The gonads are the major source of sex hormones, with much smaller amounts being produced by the adrenal gland.

Like gonadotropins, the sex hormones regulate gametogenesis both directly and indirectly, but they have other actions as well. For example, androgens are responsible for the male vocalizations of many vertebrates, such as the territorial songs of birds and the courtship displays of lizards **(Figure 36.11)**. In human embryos, androgens promote the appearance of the primary sex characteristics of males, the structures directly involved in reproduction. These include the seminal vesicles and associated ducts, as well as external reproductive structures. In the **Scientific Skills Exercise**, you can interpret the results of an experiment investigating the development of reproductive structures in mammals.

At puberty, sex hormones in both male and female humans induce formation of secondary sex characteristics, the physical and behavioral features that are not directly related to the reproductive system. In males, androgens cause the voice to deepen, facial and pubic hair to develop, and muscles to grow (by stimulating protein synthesis). Androgens also promote specific sexual behaviors and sex drive, as well as an increase in general aggressiveness. Estrogens similarly have multiple effects in females. At puberty, estradiol stimulates breast and pubic hair development. Estradiol also influences female sexual behavior, induces fat deposition in the breasts and hips, increases water retention, and alters calcium metabolism.

Making Inferences and Designing an Experiment

What Role Do Hormones Play in Making a Mammal Male or Female? In non-egg-laying mammals, females have two X chromosomes, whereas males have one X chromosome and one Y chromosome. In the 1940s, French physiologist Alfred Jost wondered whether development of mammalian embryos as female or male in accord with their chromosome set requires instructions in the form of hormones produced by the gonads. In this exercise, you will interpret the results of an experiment that Jost performed to answer this question.

How the Experiment Was Done Working with rabbit embryos still in the mother's uterus at a stage before sex differences are observable, Jost surgically removed the portion of each embryo that would form the ovaries or testes. When the baby rabbits were born, he made note of their chromosomal sex and whether their genital structures were male or female.

Data from the Experiment

Chromosome Set	Appearance of Genitalia	
	No Surgery	Embryonic Gonad Removed
XY (male)	Male	Female
XX (female)	Female	Female

Interpret the Data

1. This experiment is an example of a research approach in which scientists infer how something works normally based on what happens when the normal process is blocked. What normal process was blocked in Jost's experiment? From the results, what inference can you make about the role of the gonads in controlling the development of mammalian genitalia?

2. The data in Jost's experiment could be explained if some aspect of the surgery other than gonad removal caused female genitalia to develop. If you were to repeat Jost's experiment, how might you test the validity of such an explanation?

3. What result would Jost have obtained if female development also required a signal from the gonad?

4. Design another experiment to determine whether the signal that controls male development is a hormone. Make sure to identify your hypothesis, prediction, data collection plan, and controls.

Data from A. Jost, Recherches sur la differenciation sexuelle de l'embryon de lapin (Studies on the sexual differentiation of the rabbit embryo), *Archives d'Anatomie Microscopique et de Morphologie Experimentale* 36:271–316 (1947).

(MB) A version of this Scientific Skills Exercise can be assigned in MasteringBiology.

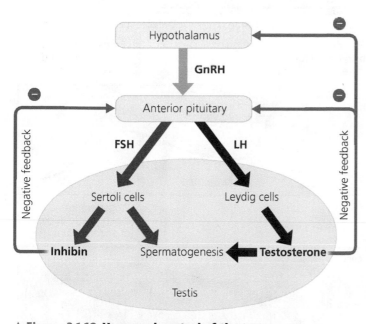

▲ **Figure 36.12 Hormonal control of the testes.**

We turn now to the role of gonadotropins and sex hormones in gametogenesis, beginning with males.

Hormonal Control of the Male Reproductive System

FSH and LH, released by the anterior pituitary in response to GnRH secretion by the hypothalamus, act on different types of cells in the testes to direct spermatogenesis **(Figure 36.12)**. **Sertoli cells**, located within the seminiferous tubules, respond to FSH by nourishing developing sperm (see Figure 36.10). **Leydig cells**, scattered in connective tissue between the tubules, respond to LH by producing testosterone and other androgens, which promote spermatogenesis in the tubules.

Two negative-feedback mechanisms control sex hormone production in males (see Figure 36.12). Testosterone regulates blood levels of GnRH, FSH, and LH through inhibitory effects on the hypothalamus and anterior pituitary. In addition, *inhibin*, a hormone that in males is produced by Sertoli cells, acts on the anterior pituitary gland to reduce FSH secretion. Together, these negative-feedback circuits maintain androgen production at optimal levels.

Hormonal Control of Female Reproductive Cycles

Whereas human males produce sperm continuously, human females produce eggs in cycles. Ovulation occurs only after the endometrium (lining of the uterus) has started to thicken and develop a rich blood supply, preparing the uterus for the possible implantation of an embryo. If pregnancy does not occur, the uterine lining is sloughed off, and another cycle begins. The cyclic shedding of the blood-rich endometrium from the uterus, a process that occurs in a flow through the cervix and vagina, is called **menstruation**.

There are two closely linked reproductive cycles in human females. Changes in the uterus define the **menstrual cycle**, also called the **uterine cycle**. Menstrual cycles average 28 days (although cycles vary, ranging from about 20 to 40 days). The cyclic events in the ovaries define the **ovarian cycle**. Hormone activity links the two cycles to each other, synchronizing ovarian follicle growth and ovulation with the establishment of a uterine lining that can support embryonic development.

Figure 36.13 outlines the major events of the female reproductive cycles, illustrating the close coordination across different tissues in the body.

The Ovarian Cycle

The ovarian cycle begins **1** with the release from the hypothalamus of GnRH, which stimulates the anterior pituitary to **2** secrete small amounts of FSH and LH. **3** Follicle-stimulating hormone (as its name implies) stimulates follicle growth, aided by LH, and **4** the cells of the growing follicles start to make estradiol. There is a slow rise in estradiol secreted during most of the **follicular phase**, the part of the ovarian cycle during which follicles grow and oocytes mature. (Several follicles begin to grow with each cycle, but usually only one matures; the others disintegrate.) The low levels of estradiol inhibit secretion of the pituitary hormones, keeping the levels of FSH and LH relatively low. During this portion of the cycle, regulation of the hormones controlling reproduction closely parallels the regulation observed in males.

5 When estradiol secretion by the growing follicle begins to rise steeply, **6** the FSH and LH levels increase markedly. Whereas a low level of estradiol inhibits the secretion of pituitary gonadotropins, a high concentration has the opposite effect: It stimulates gonadotropin secretion by acting on the hypothalamus to increase its output of GnRH. The effect is greater for LH because the high concentration of estradiol increases the GnRH sensitivity of LH-releasing cells in the pituitary. In addition, follicles respond more

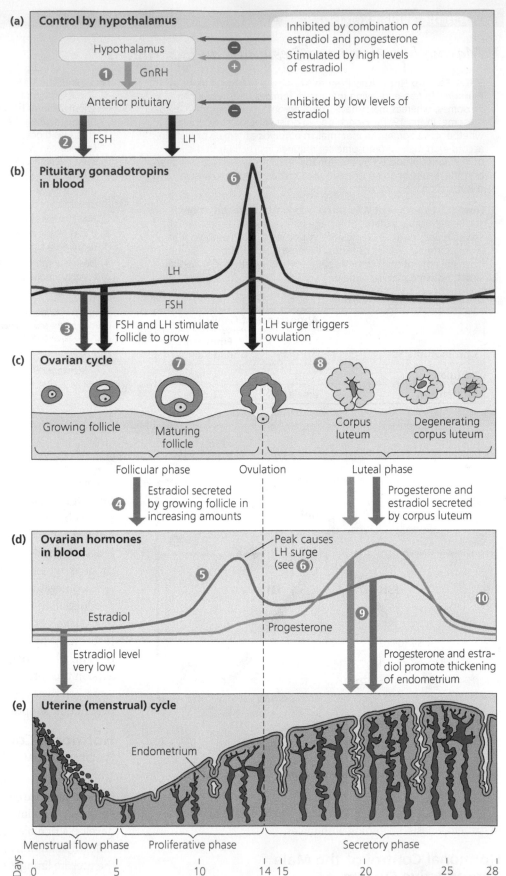

▲ **Figure 36.13 The reproductive cycles of the human female.** This figure shows how **(c)** the ovarian cycle and **(e)** the uterine (menstrual) cycle are regulated by changing hormone levels in the blood, depicted in parts **(a)**, **(b)**, and **(d)**. The time scale at the bottom of the figure applies to parts **(b)–(e)**.

strongly to LH at this stage because more of their cells have receptors for this hormone.

The increase in LH concentration caused by increased estradiol secretion from the growing follicle is an example of positive feedback. The result is final maturation of the follicle. ❼ The maturing follicle, containing a fluid-filled cavity, enlarges, forming a bulge near the surface of the ovary. The follicular phase ends at ovulation, about a day after the LH surge. In response to the peak in LH levels, the follicle and adjacent wall of the ovary rupture, releasing the secondary oocyte. There is sometimes a distinctive pain in the lower abdomen at or near the time of ovulation; this pain is felt on the left or right side, corresponding to whichever ovary has matured a follicle during that cycle.

The *luteal phase* of the ovarian cycle follows ovulation. ❽ LH stimulates the follicular tissue left behind in the ovary to transform into a corpus luteum, a glandular structure. Under continued stimulation by LH, the corpus luteum secretes progesterone and estradiol, which in combination exert negative feedback on the hypothalamus and pituitary. This feedback reduces the secretion of LH and FSH to very low levels, preventing another egg from maturing when a pregnancy may already be under way.

Near the end of the luteal phase, low gonadotropin levels cause the corpus luteum to disintegrate, triggering a sharp decline in estradiol and progesterone concentrations. The decreasing levels of ovarian steroid hormones liberate the hypothalamus and pituitary from the negative-feedback effect of these hormones. The pituitary can then begin to secrete enough FSH to stimulate the growth of new follicles in the ovary, initiating the next ovarian cycle.

The Uterine (Menstrual) Cycle

Prior to ovulation, ovarian steroid hormones stimulate the uterus to prepare for support of an embryo. Estradiol secreted in increasing amounts by growing follicles signals the endometrium to thicken. In this way, the follicular phase of the ovarian cycle is coordinated with the *proliferative phase* of the uterine cycle. After ovulation, ❾ the estradiol and progesterone secreted by the corpus luteum stimulate maintenance of the uterine lining, as well as further development, including enlargement of arteries and growth of endometrial glands. These glands secrete a nutrient fluid that can sustain an early embryo even before it implants in the uterine lining. Thus, the luteal phase of the ovarian cycle is coordinated with what is called the *secretory phase* of the uterine cycle.

Once the corpus luteum has disintegrated, ❿ the rapid drop in ovarian hormone levels causes arteries in the endometrium to constrict. Deprived of its circulation, the uterine lining largely disintegrates, and the uterus, in response to prostaglandin secretion, contracts. Small endometrial blood vessels constrict, releasing blood that is shed along with endometrial tissue and fluid. The result is menstruation—the *menstrual flow phase* of the uterine cycle. During this phase, which

usually lasts a few days, a new group of ovarian follicles begin to grow. By convention, the first day of flow is designated day 1 of the new uterine (and ovarian) cycle.

Overall, the hormonal cycles in females coordinate egg maturation and release with changes in the uterus, the organ that must accommodate an embryo if the egg cell is fertilized. If an embryo has not implanted in the endometrium by the end of the secretory phase, a new menstrual flow commences, marking the start of the next cycle. Later in the chapter, you'll learn about override mechanisms that prevent disintegration of the endometrium in pregnancy.

Menopause

After about 500 cycles, a woman undergoes **menopause**, the cessation of ovulation and menstruation. Menopause usually occurs between the ages of 46 and 54. During this interval, the ovaries lose their responsiveness to FSH and LH, resulting in a decline in estradiol production.

Menopause is an unusual phenomenon. In most other species, females and males retain their reproductive capacity throughout life. Is there an evolutionary explanation for menopause? One intriguing hypothesis proposes that during early human evolution, undergoing menopause after bearing several children allowed a mother to provide better care for her children and grandchildren, thereby increasing the survival of individuals who share much of her genetic makeup.

Menstrual Versus Estrous Cycles

In all female mammals, the endometrium thickens before ovulation, but only humans and some other primates have menstrual cycles. Other mammals have **estrous cycles**, in which in the absence of a pregnancy, the uterus reabsorbs the endometrium and no extensive fluid flow occurs. Whereas human females may engage in sexual activity throughout the menstrual cycle, mammals with estrous cycles usually copulate only during the period surrounding ovulation. This period, called estrus (from the Latin *oestrus*, frenzy, passion), is the only time the female is receptive to mating. It is often called "heat," and the female's temperature does increase slightly.

The length and frequency of estrous cycles vary widely among mammals. Bears and wolves have one estrous cycle per year; elephants typically have multiple cycles lasting 14–16 weeks each. Rats have estrous cycles throughout the year, each lasting only 5 days.

Human Sexual Response

In humans, the arousal of sexual interest is complex, involving a variety of psychological as well as physical factors. Although many reproductive structures in the male and female are quite different in appearance, they often serve similar functions in arousal, reflecting their shared developmental origin. For example, the same embryonic tissues give rise to the glans of the penis and the clitoris, to the scrotum and the labia majora, and to the skin on the penis and the labia minora. Furthermore, the general pattern of human sexual response is similar in males and females. Two types of physiological reactions predominate in both sexes: *vasocongestion*, the filling of a tissue with blood, and *myotonia*, increased muscle tension.

The sexual response cycle can be divided into four phases: excitement, plateau, orgasm, and resolution. An important function of the excitement phase is to prepare the vagina and penis for *coitus* (sexual intercourse). During this phase, vasocongestion is particularly evident in erection of the penis and clitoris; and in enlargement of the testicles, labia, and breasts. The vagina becomes lubricated and myotonia may occur, resulting in nipple erection or tension of the arms and legs.

In the plateau phase, these responses continue as a result of direct stimulation of the genitalia. In females, the outer third of the vagina becomes vasocongested, while the inner two-thirds slightly expands. This change, coupled with the elevation of the uterus, forms a depression for receiving sperm at the back of the vagina. Breathing increases and heart rate rises, sometimes to 150 beats per minute—not only in response to the physical effort of sexual activity, but also as an involuntary response to stimulation of the autonomic nervous system (see Chapter 38).

Orgasm is characterized by rhythmic, involuntary contractions of the reproductive structures in both sexes. Male orgasm has two stages. The first, emission, occurs when the glands and ducts of the reproductive tract contract, forcing semen into the urethra. Expulsion, or ejaculation, occurs when the urethra contracts and the semen is expelled. During female orgasm, the uterus and outer vagina contract, but the inner two-thirds of the vagina does not. Orgasm is the shortest phase of the sexual response cycle, usually lasting only a few seconds. In both sexes, contractions occur at about 0.8-second intervals and may also involve the anal sphincter and several abdominal muscles.

The resolution phase completes the cycle and reverses the responses of the earlier stages. Vasocongested organs return to their normal size and color, and muscles relax. Most of the changes of resolution are completed within 5 minutes, but some may take as long as an hour. Following orgasm, the male typically enters a refractory period, lasting anywhere from a few minutes to hours, during which erection and orgasm cannot be achieved. Females do not have a refractory period, making possible multiple orgasms within a short period of time.

CONCEPT CHECK 36.3

1. FSH and LH get their names from events of the female reproductive cycle, but they also function in males. How are their functions in females and males similar?

2. How does an estrous cycle differ from a menstrual cycle, and in what animals are the two types of cycles found?

3. **WHAT IF?** If a human female were to take estradiol and progesterone immediately after the start of a new menstrual cycle, how would ovulation be affected? Explain.

For suggested answers, see Appendix A.

Fertilization, cleavage, and gastrulation initiate embryonic development

Having explored gamete production and mating, we turn our attention now to development. Across a range of animal species, embryonic development involves common stages that occur in a set order. As shown in **Figure 36.14**, the first is fertilization, which forms a zygote. Development proceeds with the cleavage stage, during which a series of mitoses divide, or cleave, the zygote into a many-celled embryo. These cleavage divisions, which typically are rapid and lack accompanying cell growth, convert the embryo to a hollow ball of cells called a blastula. Next, the blastula folds in on itself, rearranging into a three-layered embryo, the gastrula, in a process called gastrulation. During organogenesis, the last major stage of embryonic development, local changes in cell shape and large-scale changes in cell location generate the rudimentary organs from which adult structures grow.

With this overview of embryonic development in mind, let's take a brief look at the early events of development—fertilization, cleavage, and gastrulation—in sea urchins (phylum Echinodermata; see Figure 27.11). Why sea urchins? Their gametes are easy to collect, and they have external fertilization; as a result, researchers can observe fertilization and subsequent events simply by combining eggs and sperm in seawater. Sea urchins are one example of a model organism, a species chosen for in-depth research in part because it is easy to study in the laboratory.

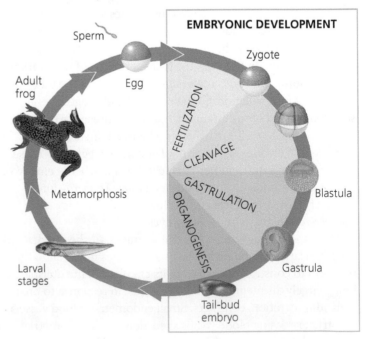

▲ **Figure 36.14 Developmental events in the life cycle of a frog.**

Fertilization

Molecules and events at the egg surface play a crucial role in each step of fertilization. First, sperm dissolve or penetrate any protective layer surrounding the egg to reach the plasma membrane. Next, molecules on the sperm surface bind to receptors on the egg, helping ensure that a sperm of the same species fertilizes the egg. Finally, changes at the surface of the egg prevent **polyspermy**, the entry of multiple sperm nuclei into the egg. If polyspermy were to occur, the resulting abnormal number of chromosomes in the embryo would be lethal. **Figure 36.15** illustrates the events that provide a fast and slow block to polyspermy in sea urchins, ensuring that only one sperm nucleus crosses the egg plasma membrane.

A major function of fertilization is combining haploid sets of chromosomes from two individuals into a single diploid cell, the zygote. However, the events of fertilization also initiate metabolic reactions that trigger the onset of embryonic development, thus "activating" the egg. As shown in **Figure 36.16**, activation leads to a number of events, such as an increase in protein synthesis, that precede the formation of a diploid nucleus.

What triggers egg activation? Studies show that sperm entry triggers release of internal Ca^{2+} stores into the egg cytoplasm and that injecting Ca^{2+} into an unfertilized egg activates

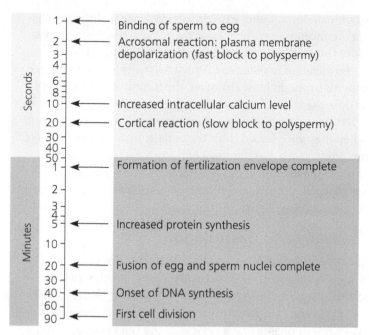

▲ **Figure 36.16 Timeline for the fertilization of sea urchin eggs.** The process begins when a sperm cell binds the egg (top of chart). Notice that the scale is logarithmic.

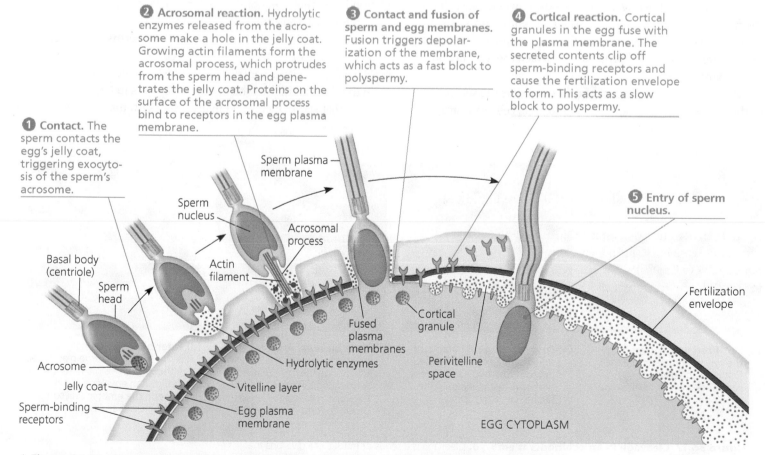

▲ **Figure 36.15 The acrosomal and cortical reactions during sea urchin fertilization.** The events following contact of a single sperm and egg ensure that the nucleus of only one sperm enters the egg cytoplasm.

egg metabolism. Other experiments indicate that the rise in Ca^{2+} concentration also causes the cortical reaction that provides the slow block to polyspermy (see Figure 36.15).

Fertilization in other species shares many features with the process in sea urchins. However, the timing of events differs, as does the stage of meiosis the egg has reached by the time it is fertilized. Sea urchin eggs have already completed meiosis when they are released from the female. In other species, eggs are arrested at a specific stage of meiosis and do not complete the meiotic divisions until fertilization occurs. Human eggs, for example, are arrested at metaphase of meiosis II prior to fertilization (see Figure 36.10).

Cleavage and Gastrulation

Once fertilization is complete, many animal species undergo a succession of rapid cell divisions that characterize the **cleavage** stage of early development. During cleavage, the cell cycle consists primarily of the S (DNA synthesis) and M (mitosis) phases (for a review of the cell cycle, see Figure 9.6). Cells essentially skip the G_1 and G_2 (gap) phases, and little or no protein synthesis occurs. As a result, cleavage partitions the cytoplasm of the large fertilized egg into many smaller cells. The first five to seven cleavage divisions produce a hollow ball of cells, the **blastula**, surrounding a fluid-filled cavity called the **blastocoel (Figure 36.17)**.

After cleavage, the rate of cell division slows considerably as the normal cell cycle is restored. The remaining stages of embryonic development are responsible for **morphogenesis**, the cellular and tissue-based processes by which the animal body takes shape.

During **gastrulation**, a set of cells at or near the surface of the blastula moves to an interior location, cell layers are established, and a primitive digestive tube is formed. Gastrulation reorganizes the hollow blastula into a two-layered or three-layered embryo called a **gastrula**. The cell layers produced by gastrulation are collectively called the embryonic *germ layers* (from the Latin *germen*, to sprout or germinate). In the late gastrula, **ectoderm** forms the outer layer and **endoderm** lines the embryonic digestive compartment or tract. In vertebrates and other animals with bilateral symmetry, a third germ layer, the **mesoderm**, forms between the ectoderm and the endoderm.

Gastrulation in the sea urchin begins at the vegetal pole of the blastula **(Figure 36.18)**. There, *mesenchyme cells* individually detach from the blastocoel wall and enter the blastocoel. The remaining cells near the vegetal pole flatten slightly and cause that end of the embryo to buckle inward. This process—the infolding of a sheet of cells into the embryo—is called *invagination*. Extensive rearrangement of cells transforms the shallow depression into a deeper, narrower, blind-ended tube called the *archenteron*. The open end of the archenteron, which will become the anus, is called the *blastopore*. A second opening, which will become the mouth, forms when the opposite end of the archenteron touches the inside of the ectoderm and the two layers fuse, producing a rudimentary digestive tube.

The cell movements and interactions that form the germ layers vary considerably among species. One basic distinction is whether the mouth develops from the first opening that forms in the embryo (protostomes) or the second (deuterostomes). Sea urchins and other echinoderms are deuterostomes, as are chordates like ourselves and other vertebrates.

Each germ layer contributes to a distinct set of structures in the adult animal, as shown for vertebrates in **Figure 36.19**. Note that some organs and many organ systems of the adult

(a) Fertilized egg. Shown here is the zygote shortly before the first cleavage division, surrounded by the fertilization envelope.

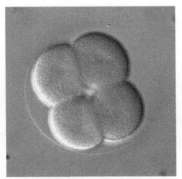

(b) Four-cell stage. Remnants of the mitotic spindle can be seen between the two pairs of cells that have just completed the second cleavage division.

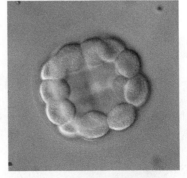

(c) Early blastula. After further cleavage divisions, the embryo is a multicellular ball that is still surrounded by the fertilization envelope. The blastocoel has begun to form in the center.

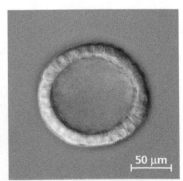

50 µm

(d) Later blastula. A single layer of cells surrounds a large blastocoel. Although not visible here, the fertilization envelope is still present; the embryo will soon hatch from it and begin swimming.

▲ **Figure 36.17 Cleavage in an echinoderm embryo.** Cleavage is a series of mitotic cell divisions that transform the zygote into a blastula, a hollow ball of cells called blastomeres. These light micrographs show the embryonic stages of a sand dollar, which are virtually identical to those of a sea urchin.

► **Figure 36.18 Gastrulation in a sea urchin embryo.** The movement of cells during gastrulation forms an embryo with a primitive digestive tube and three germ layers. Some of the mesenchyme cells that migrate inward (step 1) will eventually secrete calcium carbonate and form a simple internal skeleton. Embryos in steps 1–3 are viewed from the front, those in steps 4 and 5 from the side.

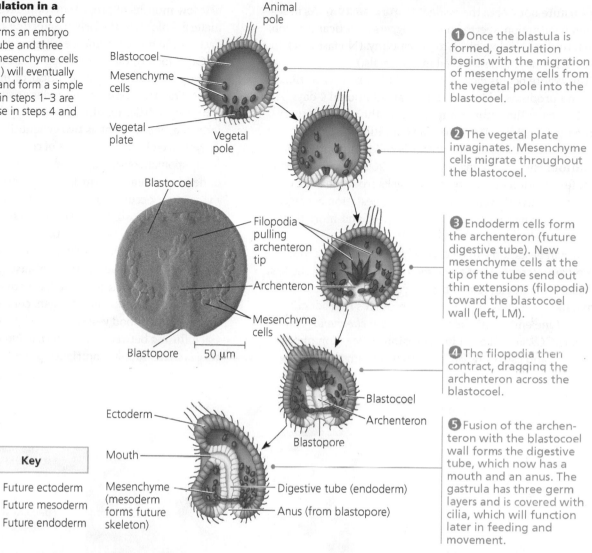

Animal pole

Blastocoel

Mesenchyme cells

Vegetal plate

Vegetal pole

❶ Once the blastula is formed, gastrulation begins with the migration of mesenchyme cells from the vegetal pole into the blastocoel.

❷ The vegetal plate invaginates. Mesenchyme cells migrate throughout the blastocoel.

Blastocoel

Filopodia pulling archenteron tip

Archenteron

Mesenchyme cells

Blastopore

50 µm

❸ Endoderm cells form the archenteron (future digestive tube). New mesenchyme cells at the tip of the tube send out thin extensions (filopodia) toward the blastocoel wall (left, LM).

❹ The filopodia then contract, dragging the archenteron across the blastocoel.

Blastocoel

Archenteron

Ectoderm

Mouth

Mesenchyme (mesoderm forms future skeleton)

Blastopore

Digestive tube (endoderm)

Anus (from blastopore)

❺ Fusion of the archenteron with the blastocoel wall forms the digestive tube, which now has a mouth and an anus. The gastrula has three germ layers and is covered with cilia, which will function later in feeding and movement.

Key

■ Future ectoderm
■ Future mesoderm
□ Future endoderm

ECTODERM (outer layer of embryo)

- Epidermis of skin and its derivatives (including sweat glands, hair follicles)
- Nervous and sensory systems
- Pituitary gland, adrenal medulla
- Jaws and teeth
- Germ cells

MESODERM (middle layer of embryo)

- Skeletal and muscular systems
- Circulatory and lymphatic systems
- Excretory and reproductive systems (except germ cells)
- Dermis of skin
- Adrenal cortex

ENDODERM (inner layer of embryo)

- Epithelial lining of digestive tract and associated organs (liver, pancreas)
- Epithelial lining of respiratory, excretory, and reproductive tracts and ducts
- Thymus, thyroid, and parathyroid glands

▲ **Figure 36.19 Major derivatives of the three embryonic germ layers in vertebrates.**

derive from more than one germ layer. For example, the adrenal glands have both ectodermal and mesoderm tissue, and many other endocrine glands contain endodermal tissue.

Having introduced the developmental stages of fertilization, cleavage, and gastrulation, using the sea urchin as the primary example, we now return to our consideration of human reproduction.

Human Conception, Embryonic Development, and Birth

During human copulation, the male delivers 2–5 mL of semen containing hundreds of millions of sperm. When first ejaculated, the semen coagulates, which may serve to keep the ejaculate in place until sperm reach the cervix. Soon after, anticoagulants liquefy the semen, and the sperm begin swimming through the uterus and oviducts. Fertilization—also called **conception** in humans—occurs when a sperm fuses with an egg (mature oocyte) in the oviduct **(Figure 36.20a)**. As in sea urchin fertilization, sperm binding triggers a cortical reaction, which results in a slow block to polyspermy. (No fast block to polyspermy has been identified in mammals.)

The zygote begins cleavage about 24 hours after fertilization and produces a blastocyst after an additional 4 days. A few days later, the embryo implants into the endometrium of the uterus **(Figure 36.20b)**. The condition of carrying one or more embryos in the uterus is called *pregnancy*, or **gestation**. Human pregnancy averages 266 days (38 weeks) from fertilization of the egg, or 40 weeks from the start of the last menstrual cycle. In comparison, gestation averages 21 days in many rodents, 270 days in cows, and more than 600 days in elephants.

Human gestation can be divided for convenience into three *trimesters* of about three months each. During the first trimester, the implanted embryo secretes hormones that signal its presence and regulate the mother's reproductive system. One embryonic hormone, *human chorionic gonadotropin (hCG)*, acts like LH in maintaining secretion of progesterone and estrogens by the corpus luteum through the first few months of pregnancy. Some hCG passes from the maternal blood to the urine; detecting hCG in the urine is the basis of the most common early pregnancy test.

Occasionally, the embryo splits during the first month of development, resulting in identical, or *monozygotic* (one-egg), twins. Fraternal, or *dizygotic*, twins arise in a very different way: Two follicles mature in a single cycle, are independently fertilized, and implant as two genetically distinct embryos.

Not all embryos are capable of completing development. Many spontaneously stop developing as a result of chromosomal or developmental abnormalities. Such spontaneous abortion, or *miscarriage*, occurs in as many as one-third of all pregnancies, often before the woman is even aware she is pregnant.

During its first 2–4 weeks of development, the embryo obtains nutrients directly from the endometrium. Meanwhile, the outer layer, called the **trophoblast**, grows outward and mingles with the endometrium, eventually helping form the **placenta**. This disk-shaped organ, containing both embryonic and maternal blood vessels, can weigh close to 1 kg. Material diffusing between the maternal and embryonic circulatory systems supplies nutrients, provides immune protection,

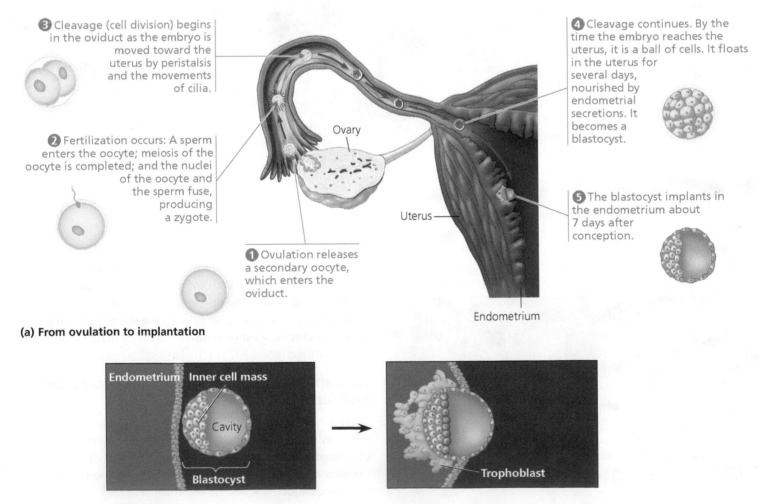

3 Cleavage (cell division) begins in the oviduct as the embryo is moved toward the uterus by peristalsis and the movements of cilia.

2 Fertilization occurs: A sperm enters the oocyte; meiosis of the oocyte is completed; and the nuclei of the oocyte and the sperm fuse, producing a zygote.

1 Ovulation releases a secondary oocyte, which enters the oviduct.

Ovary

4 Cleavage continues. By the time the embryo reaches the uterus, it is a ball of cells. It floats in the uterus for several days, nourished by endometrial secretions. It becomes a blastocyst.

5 The blastocyst implants in the endometrium about 7 days after conception.

Uterus

Endometrium

(a) From ovulation to implantation

Endometrium Inner cell mass

Cavity

Blastocyst

Trophoblast

(b) Implantation of blastocyst

▲ **Figure 36.20 Formation of a human zygote and early postfertilization events.**

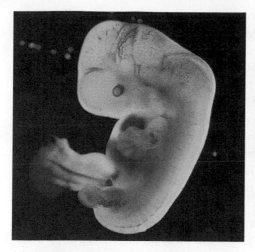 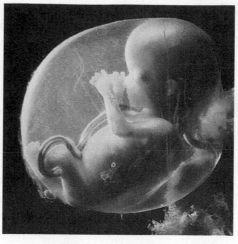 

(a) 5 weeks. Limb buds, eyes, the heart, the liver, and rudiments of all other organs have started to develop in the embryo, which is only about 1 cm long.

(b) 14 weeks. Growth and development of the offspring, now called a fetus, continue during the second trimester. This fetus is about 6 cm long.

(c) 20 weeks. Growth to nearly 20 cm in length requires adoption of the fetal position (head at knees) due to the limited space available.

▲ **Figure 36.21 Human fetal development.**

exchanges respiratory gases, and disposes of metabolic wastes for the embryo.

The first trimester is the main period of **organogenesis**, the development of the body organs. During organogenesis, the embryo is particularly susceptible to damage. For example, alcohol that passes through the placenta and reaches the developing central nervous system of the embryo can cause fetal alcohol syndrome, a disorder that can result in mental retardation and other serious birth defects. At 8 weeks, all the major structures of the adult are present in rudimentary form, and the embryo is called a **fetus**. The heart begins beating by the fourth week; a heartbeat can be detected at 8–10 weeks. At the end of the first trimester, the fetus, although well differentiated, is only 5 cm long **(Figure 36.21)**.

During the second trimester, the fetus grows to about 30 cm in length and is very active. The mother may feel fetal movements as early as one month into the second trimester. During the third trimester, the fetus grows to about 3–4 kg in weight and 50 cm in length. Fetal activity may decrease as the fetus fills the available space.

Childbirth begins with *labor*, a series of strong, rhythmic uterine contractions that push the fetus and placenta out of the body. Once labor begins, local regulators (prostaglandins) and hormones (chiefly estradiol and oxytocin) induce and regulate further contractions of the uterus. A positive-feedback loop (see Chapter 32) is central to this regulation: Uterine contractions stimulate secretion of oxytocin, which in turn stimulates further contractions.

One aspect of postnatal care unique to mammals is *lactation*, the production of mother's milk. In response to suckling by the newborn, as well as changes in estradiol levels after birth, the hypothalamus signals the anterior pituitary to secrete prolactin, which stimulates the mammary glands to produce milk. Suckling also stimulates the secretion of oxytocin from the posterior pituitary, which triggers milk release from the mammary glands (see Figure 32.12).

Contraception

Contraception, the deliberate prevention of pregnancy, can be achieved in a number of ways. Some contraceptive methods prevent gamete development or release from female or male gonads; others prevent fertilization by keeping sperm and egg apart; and still others prevent implantation of an embryo. For complete information on contraceptive methods, you should consult a health-care provider. The following brief introduction to the biology of the most common methods and the corresponding diagram in **Figure 36.22** make no pretense of being a contraception manual.

Fertilization can be prevented by abstinence from sexual intercourse or by any of several barriers that keep live sperm from contacting the egg. Temporary abstinence, often called the *rhythm method* of birth control or *natural family planning*, depends on refraining from intercourse when conception is most likely. Because the egg can survive in the oviduct for 24–48 hours and sperm for up to 5 days, a couple practicing temporary abstinence should not engage in intercourse for a significant number of days before and after ovulation. Contraceptive methods based on fertility awareness require that the couple be knowledgeable about physiological indicators associated with ovulation, such as changes in cervical mucus. Note that a pregnancy rate of 10–20% is typically reported for couples practicing natural family planning. (Pregnancy rate is the average number of women who become pregnant during a year for every 100 women using a particular pregnancy prevention method, expressed as a percentage.)

As a method of preventing fertilization, *coitus interruptus*, or withdrawal (removal of the penis from the vagina before

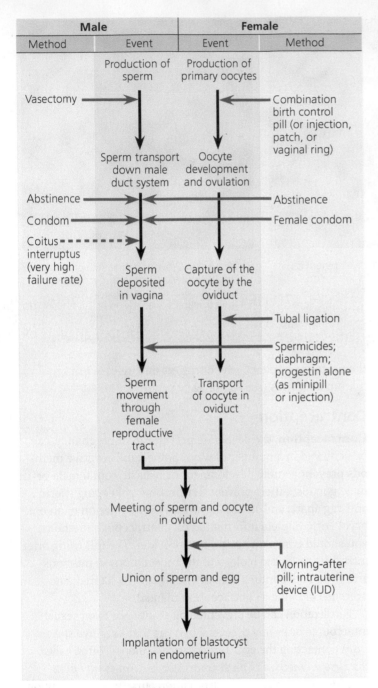

Male		Female	
Method	Event	Event	Method

Production of sperm — Production of primary oocytes

Vasectomy → Sperm transport down male duct system

Combination birth control pill (or injection, patch, or vaginal ring) → Oocyte development and ovulation

Abstinence → ← Abstinence

Condom → ← Female condom

Coitus interruptus (very high failure rate) ⇢ Sperm deposited in vagina

Capture of the oocyte by the oviduct

Tubal ligation →

Spermicides; diaphragm; progestin alone (as minipill or injection) →

Sperm movement through female reproductive tract — Transport of oocyte in oviduct

Meeting of sperm and oocyte in oviduct

Union of sperm and egg ← Morning-after pill; intrauterine device (IUD)

Implantation of blastocyst in endometrium

▲ **Figure 36.22 Mechanisms of several contraceptive methods.** Red arrows indicate where these methods, devices, or products interfere with events from the production of sperm and primary oocytes to an implanted, developing embryo.

ejaculation), is unreliable. Sperm from a previous ejaculate may be transferred in secretions that precede ejaculation. Furthermore, a split-second lapse in timing or willpower can result in tens of millions of sperm being transferred before withdrawal.

Used properly, several barrier methods of contraception that block the sperm from meeting the egg have pregnancy rates of less than 10%. The *condom* is a thin, latex rubber or natural membrane sheath that fits over the penis to collect the semen. For sexually active individuals, latex condoms are the only contraceptives that are highly effective in preventing the spread of sexually transmitted diseases, including AIDS. (This protection is, however, not absolute.) Another common barrier device is the *diaphragm*, a dome-shaped rubber cap inserted into the upper portion of the vagina before intercourse. Both devices have lower pregnancy rates when used in conjunction with a spermicidal (sperm-killing) foam or jelly. Other barrier devices include the vaginal pouch, or "female condom."

Except for complete abstinence from sexual intercourse, the most effective means of birth control are sterilization, intrauterine devices (IUDs), and hormonal contraceptives. Sterilization (vasectomy in males or tubal ligation in females) is almost 100% effective. The IUD has a pregnancy rate of 1% or less and is the most commonly used reversible method of birth control outside the United States. Placed in the uterus by a doctor, the IUD interferes with fertilization and implantation. Hormonal contraceptives, most often in the form of *birth control pills*, also have pregnancy rates of 1% or less.

The most commonly prescribed hormonal contraceptives are a combination of a synthetic estrogen and a synthetic progestin (progesterone-like hormone). This combination mimics negative feedback in the ovarian cycle, stopping the release of GnRH by the hypothalamus and thus of FSH and LH by the pituitary. The prevention of LH release blocks ovulation. In addition, the inhibition of FSH secretion by the low dose of estrogens in the pills prevents follicles from developing. Such combination birth control pills can also act as "morning-after" pills. Taken within 3 days after unprotected intercourse, they prevent fertilization or implantation with an effectiveness of about 75%.

A different type of hormonal contraceptive contains only progestin. Progestin causes a woman's cervical mucus to thicken so that it blocks sperm from entering the uterus. Progestin also decreases the frequency of ovulation and causes changes in the endometrium that may interfere with implantation if fertilization occurs. Progestin can be administered as injections that last for three months or as a tablet ("minipill") taken daily. Pregnancy rates for progestin treatment are very low.

Hormonal contraceptives have both beneficial and harmful side effects. Women who regularly smoke cigarettes face a three to ten times greater risk of dying from cardiovascular disease if they also use oral contraceptives. Among nonsmokers, birth control pills slightly raise a woman's risk of abnormal blood clotting, high blood pressure, heart attack, and stroke. Although oral contraceptives increase the risk for these cardiovascular disorders, they eliminate the dangers of pregnancy; women on birth control pills have mortality rates about one-half those of pregnant women. Also, the pill decreases the risk of ovarian and endometrial cancers.

Infertility and *In Vitro* Fertilization

Infertility—an inability to conceive offspring—is quite common, affecting about one in ten couples both in the United States and worldwide. The causes of infertility are varied, and the likelihood of a reproductive defect is nearly the same for men and women. Among preventable causes of infertility, the most significant is *sexually transmitted disease* (*STD*). In women 15–24 years old, approximately 700,000 cases of chlamydia and gonorrhea are reported annually in the United States. The actual number infected is considerably higher because most women with these STDs have no symptoms and are therefore unaware of their infection. Up to 40% of women who remain untreated for chlamydia or gonorrhea develop an inflammatory disorder that can lead to infertility or to potentially fatal complications during pregnancy.

Some forms of infertility are treatable. Hormone therapy can sometimes increase sperm or egg production, and surgery can often correct ducts that have failed to form properly or have become blocked. In some cases, doctors recommend *in vitro* **fertilization** (**IVF**), which involves mixing oocytes and sperm in culture dishes. Fertilized eggs are incubated until they have formed at least eight cells and are then transferred to the woman's uterus for implantation. If mature sperm are defective or low in number, a sperm nucleus is sometimes injected directly into an oocyte. Though costly, IVF procedures have enabled more than a million couples to conceive children.

CONCEPT CHECK 36.4

1. Where does fertilization occur in the human female?
2. Which of the three germ layers contributes least to tissues lining the interior or exterior of the body?
3. **WHAT IF?** If an STD led to complete blockage of both oviducts, what effect would you expect on the menstrual cycle and on fertility?

For suggested answers, see Appendix A.

36 Chapter Review

SUMMARY OF KEY CONCEPTS

CONCEPT 36.1

Both asexual and sexual reproduction occur in the animal kingdom (pp. 729–733)

- **Sexual reproduction** requires the fusion of male and female gametes, forming a diploid **zygote**. **Asexual reproduction** is the production of offspring without gamete fusion. Mechanisms of asexual reproduction include budding, fission, and fragmentation with regeneration. Variations on the mode of reproduction are achieved through **parthenogenesis**, **hermaphroditism**, and sex reversal. Hormones and environmental cues control reproductive cycles.
- **Fertilization**, whether external or internal, requires coordinated timing, which may be mediated by environmental cues, pheromones, or courtship behavior. **Internal fertilization** is typically often associated both with fewer offspring and with greater protection of offspring by the parents.

> **?** *Would a pair of haploid offspring produced by parthenogenesis be genetically identical? Explain.*

CONCEPT 36.2

Reproductive organs produce and transport gametes (pp. 733–738)

- Systems for gamete production and delivery range from undifferentiated cells in the body cavity to complex **gonads** with accessory tubes and glands that carry and protect gametes and embryos. In human males, **sperm** are produced in **testes**, which are suspended outside the body in the **scrotum**. Ducts connect the testes to internal accessory glands and to the **penis**. The reproductive system of the human female consists principally of the **ovaries**, **oviducts**, **uterus**, and **vagina** internally and the **labia** and the **glans** of the **clitoris** externally. **Eggs** are produced in the ovaries and upon fertilization develop in the uterus.
- **Gametogenesis**, or gamete production, consists of **spermatogenesis** in males and **oogenesis** in females. Meiosis generates one large egg in oogenesis, but four sperm in spermatogenesis. In humans, sperm develop continuously, whereas oocyte maturation is discontinuous and cyclic.

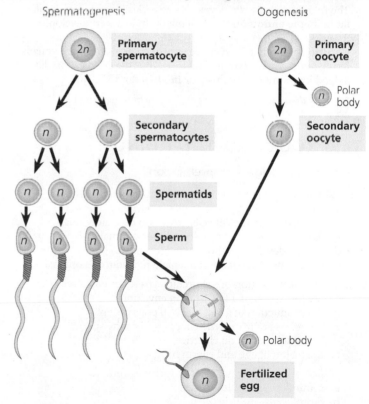

Human gametogenesis

> **?** *How does the difference in size and cellular contents between sperm and eggs relate to their specific functions?*

CONCEPT 36.3

The interplay of tropic and sex hormones regulates reproduction in mammals (pp. 738–742)

- In mammals, GnRH from the hypothalamus regulates the release of **FSH** and **LH** from the anterior pituitary, which in turn orchestrates gametogenesis. In males, secretion of androgens (chiefly testosterone) and sperm production are both controlled by FSH and LH. In the female **menstrual cycle**, cyclic secretion of FSH and LH brings about changes in the ovary and uterus via estrogens, primarily **estradiol**, and **progesterone**. The follicle and corpus luteum also secrete hormones, with positive and negative feedback coordinating the uterine and ovarian cycles.

- In **estrous cycles**, the endometrial lining is reabsorbed, and sexual receptivity is limited to a heat period.

? *Why do anabolic steroids lead to reduced sperm count?*

CONCEPT 36.4

Fertilization, cleavage, and gastrulation initiate embryonic development (pp. 742–749)

- Fertilization brings together the nuclei of sperm and egg, forming a diploid zygote, and activates the egg, initiating embryonic development. Changes at the egg surface triggered by sperm entry help block **polyspermy** in many animals.
- Fertilization is followed by **cleavage**, a period of rapid cell division without growth, which results in the production of a large number of cells. In many species, cleavage creates a multicellular ball called the **blastula**, which contains a fluid-filled cavity, the **blastocoel**.
- **Gastrulation** converts the blastula to a **gastrula**, which has a primitive digestive cavity and three germ layers: **ectoderm**, **mesoderm**, and **endoderm**.
- The mammalian zygote develops into a blastocyst before implantation in the **endometrium**. All major organs start developing by 8 weeks.
- **Contraception** may prevent release of gametes from the gonads, fertilization, or embryo implantation. Infertile couples may be helped by hormonal methods or *in vitro* **fertilization**.

? *How does the fertilization envelope form in sea urchins? What is its function?*

TEST YOUR UNDERSTANDING

Level 1: Knowledge/Comprehension

1. Which of the following characterizes parthenogenesis?
   a. An individual may change its sex during its lifetime.
   b. Specialized groups of cells grow into new individuals.
   c. An organism is first a male and then a female.
   d. An egg develops without being fertilized.
   e. Both mates have male and female reproductive organs.

2. The cortical reaction of sea urchin eggs functions directly in
   a. the formation of a fertilization envelope.
   b. the production of a fast block to polyspermy.
   c. the release of hydrolytic enzymes from the sperm.
   d. the generation of an electrical impulse by the egg.
   e. the fusion of egg and sperm nuclei.

3. Which of the following is *not* properly paired?
   a. seminiferous tubule—cervix
   b. Sertoli cells—follicle cells
   c. testosterone—estradiol
   d. scrotum—labia majora
   e. vas deferens—oviduct

4. Peaks of LH and FSH production occur during
   a. the menstrual flow phase of the uterine cycle.
   b. the beginning of the follicular phase of the ovarian cycle.
   c. the period just before ovulation.
   d. the end of the luteal phase of the ovarian cycle.
   e. the secretory phase of the menstrual cycle.

5. During human gestation, rudiments of all organs develop
   a. in the first trimester.
   b. in the second trimester.
   c. in the third trimester.
   d. while the embryo is in the oviduct.
   e. during the blastocyst stage.

Level 2: Application/Analysis

6. Which of the following is a true statement?
   a. All mammals have menstrual cycles.
   b. The endometrial lining is shed in menstrual cycles but reabsorbed in estrous cycles.
   c. Estrous cycles are more frequent than menstrual cycles.
   d. Estrous cycles are not controlled by hormones.
   e. Ovulation occurs before the endometrium thickens in estrous cycles.

7. For which is the number the same in males and females?
   a. interruptions in meiotic divisions
   b. functional gametes produced by meiosis
   c. meiotic divisions required to produce each gamete
   d. gametes produced in a given time period
   e. different cell types produced by meiosis

8. Which statement about human reproduction is false?
   a. Fertilization occurs in the oviduct.
   b. Effective hormonal contraceptives are currently available only for females.
   c. An oocyte completes meiosis after a sperm penetrates it.
   d. The earliest stages of spermatogenesis occur closest to the lumen of the seminiferous tubules.
   e. Spermatogenesis and oogenesis require different temperatures.

Level 3: Synthesis/Evaluation

9. **DRAW IT** In human spermatogenesis, mitosis of a stem cell gives rise to one cell that remains a stem cell and one cell that becomes a spermatogonium. (a) Draw four rounds of mitosis for a stem cell, and label the daughter cells. (b) For one spermatogonium, draw the cells it would produce from one round of mitosis followed by meiosis. Label the cells, and label mitosis and meiosis. (c) What would happen if stem cells divided like spermatogonia?

10. **SCIENTIFIC INQUIRY**
    Suppose that you discover a new egg-laying worm species. You dissect four adults and find both oocytes and sperm in each. Cells outside the gonad contain five chromosome pairs. Lacking genetic variants, how would you determine whether the worms can self-fertilize?

11. **FOCUS ON EVOLUTION**
    Hermaphroditism is often found in animals that are fixed to a surface. Motile species are less often hermaphroditic. Why?

12. **FOCUS ON ENERGY AND MATTER**
    In a short essay (100–150 words), discuss how investments of energy by females contribute to the reproductive success of a frog (see Figure 36.6) and of a human.

For selected answers, see Appendix A.

MasteringBiology®

Students Go to **MasteringBiology** for assignments, the eText, and the Study Area with practice tests, animations, and activities.

Instructors Go to **MasteringBiology** for automatically graded tutorials and questions that you can assign to your students, plus Instructor Resources.

Unit 7 Ecology

40 Population Ecology and the Distribution of Organisms

Ecologists study the interactions of organisms and the environment to understand the **distribution of species**. **Populations** of a species may be relatively stable in size or fluctuate greatly, driven by ecological and evolutionary factors.

41 Species Interactions

Populations of different species interact in ecological **communities** through processes such as competition, predation, and symbiosis.

42 Ecosystems and Energy

Energy flow and chemical cycling occur in an **ecosystem**, the community of organisms living in an area and the physical factors with which they interact. Within an ecosystem, organisms transfer **energy** through trophic levels and are characterized by their main source of nutrition and energy.

43 Global Ecology and Conservation Biology

Throughout the biosphere, human activities are altering trophic structures, energy flow, and natural disturbance—ecosystem processes on which all species depend. Efforts to sustain ecosystem processes and to preserve biodiversity from habitat loss and other threats comprise the fields of **global ecology** and **conservation biology**.

Ecology

40 Population Ecology and the Distribution of Organisms
41 Species Interactions
42 Ecosystems and Energy
43 Global Ecology and Conservation Biology

Animals
Plants
History of Life
Evolution
Genetics
Chemistry and Cells

Population Ecology and the Distribution of Organisms

40

KEY CONCEPTS

40.1 Earth's climate influences the structure and distribution of terrestrial biomes

40.2 Aquatic biomes are diverse and dynamic systems that cover most of Earth

40.3 Interactions between organisms and the environment limit the distribution of species

40.4 Dynamic biological processes influence population density, dispersion, and demographics

40.5 The exponential and logistic models describe the growth of populations

40.6 Population dynamics are influenced strongly by life history traits and population density

OVERVIEW

Discovering Ecology

When University of Delaware undergraduate Justin Yeager spent his summer abroad in Costa Rica, all he wanted was to see the tropical rain forest and to practice his Spanish. Instead, he discovered a population of the variable harlequin toad (*Atelopus varius*), a species thought to be extinct in the mountain slopes of Costa Rica and Panama, where it once lived **(Figure 40.1)**. During the 1980s and 1990s, roughly two-thirds of the 82 known species of harlequin toads vanished. Scientists think that a disease-causing chytrid fungus, *Batrachochytrium dendrobatidis*, contributed to many of these extinctions. Why was the fungus suddenly thriving in the rain forest? Cloudier days and warmer nights associated with global warming appear to have created an environment ideal for its success. As of early 2012, the toad species that Yeager found was surviving as a single known population of fewer than 100 individuals.

What environmental factors limit the geographic distribution of harlequin toads? How do variations in their food supply or interactions with other species, such as pathogens, affect the size of their population? Questions like these are the subject of **ecology** (from the Greek *oikos*, home, and *logos*, study), the scientific study of the interactions between organisms and the environment. As shown in **Figure 40.2**, ecological interactions occur at a hierarchy of scales from single organisms to the globe.

Ecology is a rigorous experimental science that requires a breadth of biological knowledge. Ecologists observe nature, generate hypotheses, manipulate environmental variables, and observe outcomes. Figure 40.2 provides a conceptual framework for the field of ecology as well as an organizational framework for this unit. In this chapter, we'll first consider how Earth's climate and other factors determine the location of major life zones on land and in the oceans. We'll then examine how ecologists investigate what controls the distribution of species and the density and size of populations. The next three chapters focus on community, ecosystem, and global ecology, as we explore how ecologists apply biological knowledge to predict the global consequences of human activities and to conserve Earth's biodiversity.

▼ **Figure 40.1** What threatens this amphibian's survival?

Ecologists work at different levels of the biological hierarchy, from individual organisms to the planet. Here we present a sample research question for each level of the hierarchy.

Global Ecology

The **biosphere** is the global ecosystem—the sum of all the planet's ecosystems and landscapes. **Global ecology** examines how the regional exchange of energy and materials influences the functioning and distribution of organisms across the biosphere.

◀ How does ocean circulation affect the global distribution of crustaceans?

Landscape Ecology

A **landscape** (or seascape) is a mosaic of connected ecosystems. Research in **landscape ecology** focuses on the factors controlling exchanges of energy, materials, and organisms across multiple ecosystems.

◀ To what extent do the trees lining a river serve as corridors of dispersal for animals?

Ecosystem Ecology

An **ecosystem** is the community of organisms in an area and the physical factors with which those organisms interact. **Ecosystem ecology** emphasizes energy flow and chemical cycling between organisms and the environment.

◀ What factors control photosynthetic productivity in a temperate grassland ecosystem?

Community Ecology

A **community** is a group of populations of different species in an area. **Community ecology** examines how interactions between species, such as predation and competition, affect community structure and organization.

◀ What factors influence the diversity of species that make up a forest?

Population Ecology

A **population** is a group of individuals of the same species living in an area. **Population ecology** analyzes factors that affect population size and how and why it changes through time.

◀ What environmental factors affect the reproductive rate of flamingos?

Organismal Ecology

Organismal ecology, which includes the subdisciplines of physiological, evolutionary, and behavioral ecology, is concerned with how an organism's structure, physiology, and behavior meet the challenges posed by its environment.

◀ How do hammerhead sharks select a mate?

Latitudinal Variation in Sunlight Intensity

Earth's curved shape causes latitudinal variation in the intensity of sunlight. Because sunlight strikes the **tropics** (those regions that lie between 23.5° north latitude and 23.5° south latitude) most directly, more heat and light per unit of surface area are delivered there. At higher latitudes, sunlight strikes Earth at an oblique angle, and thus the light energy is more diffuse on Earth's surface.

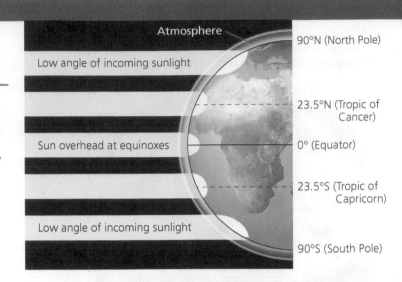

Global Air Circulation and Precipitation Patterns

Intense solar radiation near the equator initiates a global pattern of air circulation and precipitation. High temperatures in the tropics evaporate water from Earth's surface and cause warm, wet air masses to rise (blue arrows) and flow toward the poles. As the rising air masses cool, they release much of their water content, creating abundant precipitation in tropical regions. The high-altitude air masses, now dry, descend (tan arrows) toward Earth around 30° north and south, absorbing moisture from the land and creating an arid climate conducive to the development of the deserts that are common at those latitudes. Some of the descending air then flows toward the poles. At latitudes around 60° north and south, the air masses again rise and release abundant precipitation (though less than in the tropics). Some of the cold, dry rising air then flows to the poles, where it descends and flows back toward the equator, absorbing moisture and creating the comparatively rainless and bitterly cold climates of the polar regions.

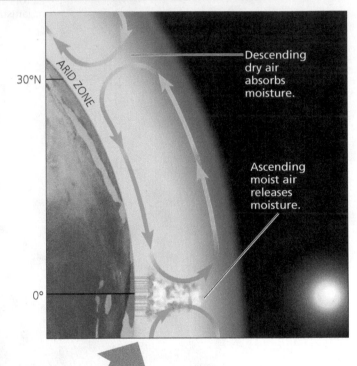

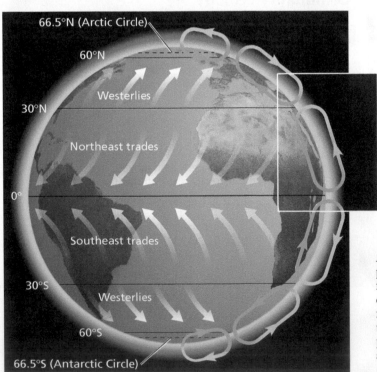

Air flowing close to Earth's surface creates predictable global wind patterns. As Earth rotates on its axis, land near the equator moves faster than that at the poles, deflecting the winds from the vertical paths shown above and creating the more easterly and westerly flows shown at left. Cooling trade winds blow from east to west in the tropics; prevailing westerlies blow from west to east in the temperate zones, defined as the regions between the Tropic of Cancer and the Arctic Circle and between the Tropic of Capricorn and the Antarctic Circle.

Earth's climate influences the structure and distribution of terrestrial biomes

The most significant influence on the distribution of organisms on land and in the oceans is **climate**, the long-term, prevailing weather conditions in a given area. Four physical factors—temperature, precipitation, sunlight, and wind—are particularly important components of climate. Such **abiotic**, or nonliving, factors are the chemical and physical attributes of the environment that influence the distribution and abundance of organisms. **Biotic**, or living, factors—the other organisms that are part of an individual's environment—similarly influence the distribution and abundance of life.

We begin by describing patterns in **macroclimate**, which is climate at the global, regional, and landscape levels.

Global Climate Patterns

Global climate patterns are determined largely by the input of solar energy and Earth's movement in space. The sun warms the atmosphere, land, and water. This warming establishes the temperature variations, cycles of air and water movement, and evaporation of water that cause dramatic latitudinal variations in climate. **Figure 40.3** summarizes Earth's climate patterns and how they are formed.

Regional Effects on Climate

Climate patterns include seasonal variation and can be modified by other factors, such as large bodies of water and mountain ranges.

Seasonality

As described in **Figure 40.4**, Earth's tilted axis of rotation and its annual passage around the sun cause strong seasonal cycles in middle to high latitudes. In addition to global changes in day length, solar radiation, and temperature, the changing angle of the sun over the course of the year affects local environments. For example, the belts of wet and dry air on either side of the equator move slightly northward and southward with the changing angle of the sun, producing marked wet and dry seasons around 20° north and 20° south latitude, where many tropical deciduous forests grow. In addition, seasonal changes in wind patterns alter ocean currents, sometimes causing the upwelling of cold water from deep ocean layers. This nutrient-rich water stimulates the growth of surface-dwelling phytoplankton and the organisms that feed on them.

Bodies of Water

Ocean currents influence climate along the coasts of continents by heating or cooling overlying air masses that pass across the land. Coastal regions are also generally wetter than inland areas at the same latitude. The cool, misty climate produced by the cold California Current that flows southward along western North America supports a coniferous rain forest

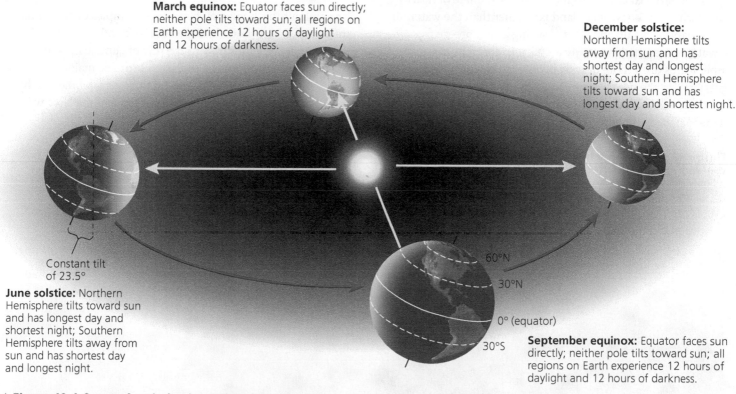

March equinox: Equator faces sun directly; neither pole tilts toward sun; all regions on Earth experience 12 hours of daylight and 12 hours of darkness.

December solstice: Northern Hemisphere tilts away from sun and has shortest day and longest night; Southern Hemisphere tilts toward sun and has longest day and shortest night.

Constant tilt of 23.5°

June solstice: Northern Hemisphere tilts toward sun and has longest day and shortest night; Southern Hemisphere tilts away from sun and has shortest day and longest night.

60°N
30°N
0° (equator)
30°S

September equinox: Equator faces sun directly; neither pole tilts toward sun; all regions on Earth experience 12 hours of daylight and 12 hours of darkness.

▲ **Figure 40.4 Seasonal variation in sunlight intensity.** Because Earth is tilted on its axis relative to its plane of orbit around the sun, the intensity of solar radiation varies seasonally. This variation is smallest in the tropics and increases toward the poles.

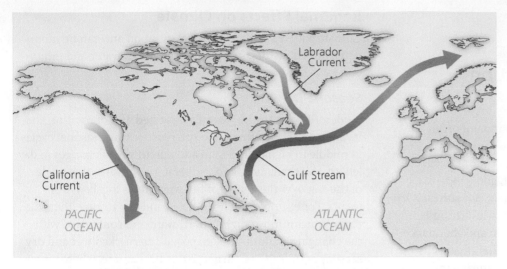

▲ **Figure 40.5 Circulation of surface water in the oceans around North America.** The California Current carries cold water southward along the western coast of North America. Along the eastern coast, the warm water of the Gulf Stream flows northward to northern Europe.

ecosystem along much of the continent's Pacific coast and large redwood groves farther south **(Figure 40.5)**. Conversely, the west coast of northern Europe has a mild climate because the Gulf Stream carries warm water from the equator to the North Atlantic. As a result, northwestern Europe is warmer during winter than southeastern Canada, which is farther south but is cooled by the Labrador Current flowing south from the coast of Greenland.

Because of the high specific heat of water (see Chapter 2), oceans and large lakes tend to moderate the climate of nearby land. During a hot day, when land is warmer than the water, air over the land heats up and rises, drawing a cool breeze from the water across the land **(Figure 40.6)**. In contrast, because temperatures drop more quickly over land than over water at night, air over the now warmer water rises, drawing cooler air from the land back out over the water and replacing it with warmer air from offshore.

Mountains

Like large bodies of water, mountains influence air flow over land. When warm, moist air approaches a mountain, the air rises and cools, releasing moisture on the windward side of the peak (see Figure 40.6). On the leeward side, cooler, dry air descends, absorbing moisture and producing a "rain shadow" that determines where many deserts are found, including the Mojave Desert of western North America and the Gobi Desert of Asia.

Mountains also affect the amount of sunlight reaching an area and the local temperature. South-facing slopes in the Northern Hemisphere receive more sunlight than north-facing slopes and are therefore warmer and drier. In many mountains of western North America, spruce and other conifers grow on the cooler north-facing slopes, but shrubby, drought-resistant plants inhabit the south-facing slopes. In addition, every 1,000-m increase in elevation produces an average temperature drop of approximately 6°C, equivalent to that produced by an 880-km increase in latitude. This is one reason that high-elevation communities at a given latitude can be similar to communities at lower elevations much farther from the equator.

Climate and Terrestrial Biomes

Throughout this book, you have seen examples of how climate influences where individual species are found (see Figure 8.18, for instance). We turn now to the role of climate in determining the nature and location of Earth's **biomes**, major life zones characterized by vegetation type (in terrestrial biomes) or by the physical environment (in aquatic biomes, which we will survey in Concept 40.2).

▶ **Figure 40.6 How large bodies of water and mountains affect climate.** This figure illustrates what can happen on a hot summer day.

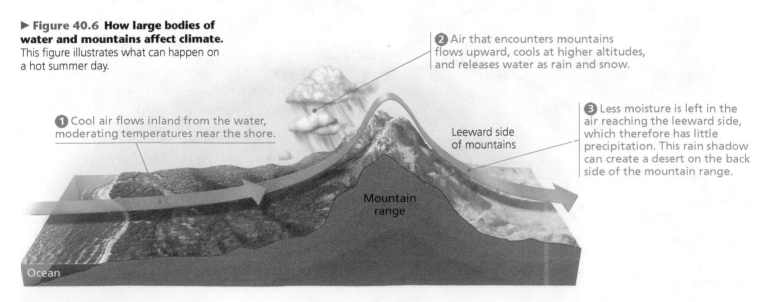

❶ Cool air flows inland from the water, moderating temperatures near the shore.

❷ Air that encounters mountains flows upward, cools at higher altitudes, and releases water as rain and snow.

❸ Less moisture is left in the air reaching the leeward side, which therefore has little precipitation. This rain shadow can create a desert on the back side of the mountain range.

Leeward side of mountains

Mountain range

Ocean

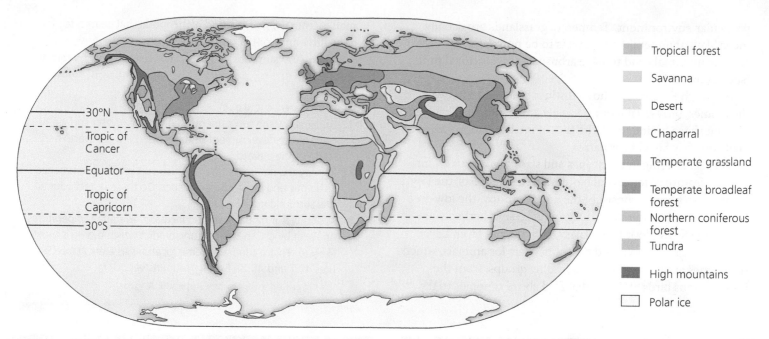

▲ **Figure 40.7 The distribution of major terrestrial biomes.** Although biomes are mapped here with sharp boundaries, biomes actually grade into one another, sometimes over large areas.

Because of the latitudinal patterns of climate described in Figure 40.3, the locations of terrestrial biomes also show strong latitudinal patterns **(Figure 40.7)**. One way to highlight the importance of climate on the distribution of biomes is to construct a **climograph**, a plot of the annual mean temperature and precipitation in a particular region **(Figure 40.8)**. Notice, for instance, that grasslands in North America are typically drier than forests and that deserts are drier still.

Factors other than mean temperature and precipitation also play a role in determining where biomes exist. For example, some areas in North America with a particular combination of temperature and precipitation support a temperate broadleaf forest, but other areas with similar values for these variables support a coniferous forest (see the overlap in Figure 40.8). One reason for this variation is that climographs are based on annual *averages*, but the *pattern* of climatic variation is often as important as the average climate. Some areas may receive regular precipitation throughout the year, whereas other areas may have distinct wet and dry seasons.

Natural and human-caused disturbances also alter the distribution of biomes. A **disturbance** is an event such as a storm, fire, or human activity that changes a community, removing organisms from it and altering resource availability. For instance, frequent fires can kill woody plants and keep a savanna from becoming the woodland that climate alone would support. Hurricanes and other storms create openings for new species in many tropical and temperate forests. Human-caused disturbances have altered much of Earth's surface, replacing natural communities with urban and agricultural ones.

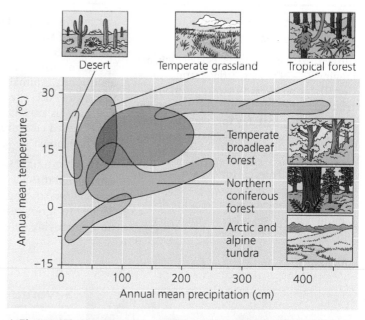

▲ **Figure 40.8 A climograph for some major types of biomes in North America.** The areas plotted here encompass the ranges of annual mean temperature and precipitation in the biomes.

General Features of Terrestrial Biomes

Most terrestrial biomes are named for major physical or climatic features and for their predominant vegetation. Temperate grasslands, for instance, are generally found in middle latitudes, where the climate is more moderate than in the tropics or polar regions, and are dominated by various grass species (see Figure 40.7). Each biome is also characterized by microorganisms, fungi, and animals adapted to that particular environment. Temperate grasslands are usually more likely than temperate forests to be populated by large grazing mammals and to have arbuscular mycorrhizal fungi (see Figure 29.13).

Although Figure 40.7 shows distinct boundaries between the biomes, terrestrial biomes usually grade into each other without sharp boundaries. The area of intergradation, called an **ecotone**, may be wide or narrow.

Vertical layering in the shapes and sizes of plants is an important feature of terrestrial biomes. In many forests, the layers from top to bottom consist of the upper **canopy**, the low-tree layer, the shrub understory, the ground layer of herbaceous plants, the forest floor (litter layer), and the root layer. Layering of vegetation provides different habitats for animals, which sometimes exist in well-defined feeding groups, from the insectivorous birds and bats that feed above canopies to the small mammals, worms, and arthropods that search for food in the litter and root layers below.

Figure 40.9 summarizes the major features of terrestrial biomes.

CONCEPT CHECK 40.1

1. Explain how the sun's unequal heating of Earth's surface leads to the development of deserts around 30° north and south of the equator.
2. Identify the natural biome in which you live, and summarize its abiotic and biotic characteristics. Do these reflect your actual surroundings? Explain.
3. **WHAT IF?** If global warming increases average temperatures on Earth by 4°C in this century, predict which biome is most likely to replace tundra in some locations as a result (see Figures 40.7 and 40.8). Explain your answer.

For suggested answers, see Appendix A.

▼ **Figure 40.9** | **Exploring Terrestrial Biomes**

Tropical Forest

Distribution Equatorial and sub-equatorial regions

Climate Temperature is usually high, averaging 25–29°C with little seasonal variation. In **tropical rain forests**, rainfall is relatively constant, about 200–400 cm annually. In **tropical dry forests**, precipitation averages about 150–200 cm annually, with a six- to seven-month dry season.

Organisms Tropical forests are vertically layered, and plants compete strongly for light. Broadleaf evergreen trees are dominant in rain forests, whereas many dry forest trees drop their leaves during the dry season. Tropical forests are home to millions of animal species, including an estimated 5–30 million still undescribed species of insects, spiders, and other arthropods. Animal diversity is higher than in any other terrestrial biome. The animals are adapted to the vertically layered environment and are often inconspicuous.

Human Impact Humans long ago established thriving communities in tropical forests. Rapid population growth leading to agriculture and development is now destroying many tropical forests.

A tropical rain forest in Costa Rica

Savanna

Distribution Equatorial and sub-equatorial regions

Climate Rainfall averages 30–50 cm per year in **savannas** and is seasonal, with a dry season that can last up to nine months. Temperature averages 24–29°C but varies seasonally more than in tropical forests.

Organisms Scattered trees often are thorny and have small leaves, an apparent adaptation to the relatively dry conditions. Fires are common in the dry season, and the dominant plant species are fire-adapted and tolerant of seasonal drought. Grasses and small nonwoody plants called forbs make up most of the ground cover. Large plant-eating mammals, such as wildebeests and zebras, and predators, including lions and hyenas, are common inhabitants. However, the dominant herbivores are insects, especially termites.

Human Impact The earliest humans likely lived in savannas. Overly frequent fires set by humans reduce tree regeneration by killing the seedlings and saplings. Cattle ranching and overhunting have led to declines in large-mammal populations.

A savanna in Kenya

Organ Pipe Cactus National Monument, Arizona

Desert

Distribution **Deserts** occur in bands near 30° north and south latitude or at other latitudes in the interior of continents (for instance, the Gobi Desert of north-central Asia).

Climate Precipitation is low and highly variable, generally less than 30 cm per year. Temperature varies seasonally and daily. It may exceed 50°C in hot deserts and fall below −30°C in cold deserts.

Organisms Desert landscapes are dominated by low, widely scattered vegetation. Common plants include succulents such as cacti or euphorbs, deeply rooted shrubs, and herbs that grow during the infrequent moist periods. Desert plant adaptations include tolerance to heat and desiccation, water storage, reduced leaf surface area, and physical defenses such as spines and toxins in leaves. Many desert plants carry out C_4 or CAM photosynthesis. Common desert animals include scorpions, ants, beetles, snakes, lizards, migratory and resident birds, and seed-eating rodents. Many species in hot deserts are active at night, when the air is cooler. Water conservation is a common adaptation, and some animals can obtain all their water by breaking down carbohydrates in seeds.

Human Impact Long-distance transport of water and deep groundwater wells have allowed humans to maintain substantial populations in deserts. Urbanization and conversion to irrigated agriculture have reduced the natural biodiversity of some deserts.

Chaparral

Distribution Midlatitude coastal regions on several continents

Climate Annual precipitation is typically 30–50 cm and is highly seasonal, with rainy winters and dry summers. Fall, winter, and spring are cool, with average temperatures of 10–12°C. Average summer temperature can reach 30°C.

Organisms **Chaparral** is dominated by shrubs and small trees adapted to frequent fires. Some fire-adapted shrubs produce seeds that will germinate only after a hot fire; food reserves stored in their roots enable them to resprout quickly and use nutrients released by the fire. Adaptations to drought include tough, evergreen leaves, which reduce water loss. Animals include browsers, such as deer and goats, that feed on twigs and buds of woody vegetation; there are also many species of insects, amphibians, small mammals, and birds.

Human Impact Chaparral areas have been heavily settled and reduced through conversion to agriculture and urbanization. Humans contribute to the fires that sweep across the chaparral.

An area of chaparral in California

Temperate Grassland

Distribution Typically at midlatitudes, often in the interior of continents

Climate Annual precipitation in **temperate grasslands** generally averages 30 to 100 cm and can be highly seasonal, with relatively dry winters and wet summers. Average temperatures frequently are below −10°C in winter and reach 30°C in summer.

Organisms Dominant plants are grasses and forbs, which vary in height from a few centimeters to 2 m in tallgrass prairie. Many grassland plants have adaptations that help them survive periodic, protracted droughts and fire. Grazing by large mammals such as bison and wild horses helps prevent establishment of woody shrubs and trees. Burrowing mammals, such as prairie dogs in North America, are also common.

Human Impact Because of their deep, fertile soils, temperate grasslands in North America and Eurasia have frequently been converted to farmland. In some drier grasslands, cattle and other grazers have turned parts of the biome into desert.

A grassland in Mongolia

Northern Coniferous Forest

Distribution In a broad band across northern North America and Eurasia to the edge of the arctic tundra, the **northern coniferous forest**, or *taiga*, is the largest terrestrial biome.

Climate Annual precipitation generally ranges from 30 to 70 cm. Winters are cold. Some areas of coniferous forest in Siberia typically range in temperature from −50°C in winter to over 20°C in summer.

Organisms Cone-bearing trees (conifers), such as pine, spruce, fir, and hemlock, are common, and some species depend on fire to regenerate. The conical shape of many conifers prevents snow from accumulating and breaking their branches, and their needle-like or scalelike leaves reduce water loss. Plant diversity in the shrub and herb layers is lower than in temperate broadleaf forests. Many migratory birds nest in northern coniferous forests. Mammals include moose, brown bears, and Siberian tigers. Periodic outbreaks of insects can kill vast tracts of trees.

Human Impact Although they have not been heavily settled by human populations, northern coniferous forests are being logged at a fast rate, and old-growth stands may soon disappear.

A coniferous forest in Norway

A temperate broadleaf forest in New Jersey

Temperate Broadleaf Forest

Distribution Midlatitudes in the Northern Hemisphere, with smaller areas in Chile, South Africa, Australia, and New Zealand

Climate Precipitation averages about 70 to 200 cm annually. Significant amounts fall during all seasons, with winter snow in some forests. Winter temperatures average around 0°C. Summers are humid, with maximum temperatures near 35°C.

Organisms The dominant plants of **temperate broadleaf forests** in the Northern Hemisphere are deciduous trees, which drop their leaves before winter, when low temperatures would reduce photosynthesis. In Australia, evergreen eucalyptus trees are common. In the Northern Hemisphere, many mammals hibernate in winter, while many bird species migrate to areas with warmer climates.

Human Impact Temperate broadleaf forests have been heavily settled globally. Logging and land clearing for agriculture and urban development have destroyed virtually all the original deciduous forests in North America, but these forests are returning over much of their former range.

Tundra

Distribution **Tundra** covers expansive areas of the Arctic, amounting to 20% of Earth's land surface. High winds and low temperatures produce alpine tundra on very high mountaintops at all latitudes, including the tropics.

Climate Precipitation averages 20 to 60 cm annually in arctic tundra but may exceed 100 cm in alpine tundra. Winters are cold, with average temperatures in some areas below −30°C. Summer temperatures generally average less than 10°C.

Organisms The vegetation of tundra is mostly herbaceous, typically a mixture of mosses, grasses, and forbs, with some dwarf shrubs, trees, and lichens. A permanently frozen soil layer called permafrost restricts the growth of plant roots. Large grazing musk oxen are resident, while caribou and reindeer are migratory. Predators include bears, wolves, foxes, and snowy owls. Many bird species migrate to the tundra for summer nesting.

Human Impact Tundra is sparsely settled but has become the focus of significant mineral and oil extraction in recent years.

Dovrefjell National Park, Norway

Aquatic biomes are diverse and dynamic systems that cover most of Earth

All types of aquatic biomes are found across the globe and show far less latitudinal variation than terrestrial biomes (**Figure 40.10**). Ecologists distinguish between freshwater and marine biomes on the basis of physical and chemical differences. Salt concentrations generally average 3% in marine biomes but are less than 0.1% in freshwater biomes.

The oceans make up the largest marine biome, covering about 75% of Earth's surface. Water evaporated from the oceans provides most of the planet's rainfall, and ocean temperatures have a major effect on global climate and wind patterns (see Figure 40.3). Marine algae and photosynthetic bacteria also supply a substantial portion of the world's oxygen and consume large amounts of atmospheric carbon dioxide.

Freshwater biomes are closely linked to the soils and biotic components of the surrounding terrestrial biome. Freshwater biomes are also influenced by the patterns and speed of water flow and the climate to which the biome is exposed.

Zonation in Aquatic Biomes

Many aquatic biomes are divided into vertical and horizontal zones, as illustrated for a lake in **Figure 40.11**. Light is

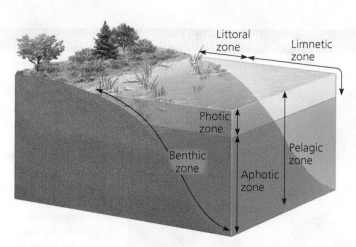

▲ **Figure 40.11 Zonation in a lake.** A lake environment is generally classified on the basis of three physical criteria: light penetration (photic and aphotic zones), distance from shore and water depth (littoral and limnetic zones), and whether the environment is open water (pelagic zone) or bottom (benthic zone).

absorbed by water and by photosynthetic organisms, so its intensity decreases rapidly with depth. The upper **photic zone** is where there is sufficient light for photosynthesis, and the lower **aphotic zone** is where little light penetrates. These two zones together make up the **pelagic zone**. At the bottom of these zones, deep or shallow, is the **benthic zone**, which consists of organic and inorganic sediments and is occupied by communities of organisms called the **benthos**.

Thermal energy from sunlight warms surface waters, but the deeper waters remain cold. In the ocean and in most

---

▼ **Figure 40.10** | **Exploring Aquatic Biomes**

Wetlands and Estuaries

Physical and Chemical Environment
Wetlands are inundated by water at least sometimes and support plants adapted to water-saturated soil. In an **estuary**, the transition zone between a river and the sea, seawater flows up and down the estuary channel during the changing tides. Nutrients from upstream make wetlands and estuaries among the most productive habitats on Earth. Because of high organic production by plants and decomposition by microorganisms, the water and soils are often low in dissolved oxygen. Both habitats filter dissolved nutrients and chemical pollutants.

Geologic Features Wetlands develop in diverse habitats, including shallow basins, the flooded banks of rivers and streams, and lake coasts. Along seacoasts, sediments from rivers and tidal waters create channels, islands, and mudflats in estuaries.

Organisms Water-saturated soils favor the growth of plants, such as cattails and sedges in wetlands and saltmarsh grasses in estuaries, that can grow in water or in soil that is anaerobic at times. In freshwater wetlands, herbivores may include crustaceans, aquatic insect larvae, and muskrats, and carnivores may include dragonflies, frogs, alligators, and herons. Estuaries support an abundance of oysters, crabs, and fish species that humans eat. Many marine invertebrates and fishes use estuaries as breeding grounds.

A basin wetland in the United Kingdom

Human Impact Draining and filling have destroyed up to 90% of wetlands. Filling, dredging, and upstream pollution have disrupted estuaries worldwide.

An oligotrophic lake in Alberta, Canada

Lakes

thousands of square kilometers. Light decreases with depth, creating photic and aphotic zones. Temperate lakes may have a seasonal thermocline; tropical lowland lakes have a thermocline year-round. **Oligotrophic lakes** are nutrient-poor and generally oxygen-rich; **eutrophic lakes** are nutrient-rich and often depleted of oxygen in the deepest zone in summer and if covered with ice in winter. High rates of decomposition in deeper layers of eutrophic lakes cause periodic oxygen depletion.

Geologic Features Oligotrophic lakes may become more eutrophic over time as runoff adds sediments and nutrients. They tend to have less surface area relative to their depth than eutrophic lakes.

Organisms Rooted and floating aquatic plants live in the **littoral zone**, the shallow, well-lit waters close to shore. The **limnetic zone**, where water is too deep to support rooted aquatic plants, is inhabited by a variety of phytoplankton, including cyanobacteria, and small drifting heterotrophs, or zooplankton, that graze on the phytoplankton. The benthic zone is inhabited by assorted invertebrates whose species composition depends partly on oxygen levels. Fishes live in all zones with sufficient oxygen.

Human Impact Runoff from fertilized land and dumping of wastes lead to nutrient enrichment, which can produce algal blooms, oxygen depletion, and fish kills.

Physical and Chemical Environment Standing bodies of water range from ponds a few square meters in area to lakes covering

Streams and Rivers

Physical and Chemical Environment Headwater streams are generally cold, clear, turbulent, and swift. Downstream in larger rivers, the water is generally warmer and more turbid because of suspended sediment. The salt and nutrient content of streams and rivers increases from the headwaters to the mouth, but oxygen content typically decreases.

Geologic Features Headwater stream channels are often narrow, have a rocky bottom, and alternate between shallow sections and deeper pools. Rivers are generally wide and meandering. River

bottoms are often silty from sediments deposited through time.

Organisms Headwater streams that flow through grasslands or deserts may be rich in phytoplankton or rooted aquatic plants. Diverse fishes and invertebrates inhabit unpolluted rivers and streams. In streams flowing through forests, organic matter from terrestrial vegetation is the primary source of food for aquatic consumers.

Human Impact Municipal, agricultural, and industrial pollution degrade water quality and can kill aquatic organisms. Dams impair the natural flow of streams

A headwater stream in Washington

and rivers and threaten migratory species such as salmon.

Intertidal Zones

A rocky intertidal zone on the Oregon coast

Physical and Chemical Environment An **intertidal zone** is periodically submerged and exposed by the tides, twice daily on

most marine shores. Upper strata experience longer exposures to air and greater variations in temperature and salinity, conditions that limit the distributions of many organisms to particular strata. Oxygen and nutrient levels are generally high and are renewed with each turn of the tides.

Geologic Features The rocky or sandy substrates of intertidal zones select for particular behavior and anatomy among intertidal organisms. The configuration of bays or coastlines influences the magnitude of tides and the exposure of intertidal zones to waves.

Organisms Diverse and plentiful marine algae grow on rocks in intertidal zones. Sandy intertidal zones exposed to waves

generally lack attached plants or algae, while those in protected bays or lagoons often support rich beds of seagrass and algae. Some animals have structural adaptations that enable them to attach to rocks. Many animals in sandy or muddy intertidal zones, such as worms, clams, and predatory crustaceans, bury themselves and feed as the tides bring food. Other common animals are sponges, sea anemones, and small fishes.

Human Impact Oil pollution has disrupted many intertidal areas. Rock walls and barriers built to reduce erosion from waves and storm surges disrupt some areas.

Coral Reefs

A coral reef in the Red Sea

Physical and Chemical Environment
Coral reefs are formed largely from the calcium carbonate skeletons of corals. Shallow reef-building corals live in the clear photic zone of tropical oceans, primarily near islands and along the edge of some continents. They are sensitive to temperatures below about 18–20°C and above 30°C. Deep-sea coral reefs are found at a depth of 200–1,500 m. Corals require high oxygen levels and are excluded by high inputs of fresh water and nutrients.

Geologic Features Corals require a solid substrate for attachment. A typical coral reef begins as a fringing reef on a young, high island, forms an offshore barrier reef later, and becomes a coral atoll as the older island submerges.

Organisms Unicellular algae live within the tissues of the corals in a mutualism that provides the corals with organic molecules. Diverse multicellular red and green algae also contribute substantial amounts of photosynthesis. Corals are the predominant animals on coral reefs, but fish and invertebrate diversity is also exceptionally high. Animal diversity on coral reefs rivals that of tropical forests.

Human Impact Collecting of coral skeletons and overfishing have reduced populations of corals and reef fishes. Global warming and pollution may be contributing to large-scale coral death. Development of coastal mangroves for aquaculture has also reduced spawning grounds for many species of reef fishes.

Oceanic Pelagic Zone

Physical and Chemical Environment The **oceanic pelagic zone** is a vast realm of open blue water, whose surface is constantly mixed by wind-driven currents. Because of higher water clarity, the photic zone extends to greater depths than in coastal marine waters. Oxygen content is generally high. Nutrient levels are generally lower than in coastal waters. Mixing of surface and deeper waters in fall and spring renews nutrients in the photic zones of temperate and high-latitude ocean areas.

Geologic Features This biome covers approximately 70% of Earth's surface. It has an average depth of nearly 4,000 m and a maximum depth of more than 10,000 m.

Organisms The dominant photosynthetic organisms are bacteria and other phytoplankton, which drift with the currents and account for half of global productivity. Zooplankton, including protists, worms, krill, jellies, and small larvae of invertebrates and fishes, eat the phytoplankton. Free-swimming animals include large squids, fishes, sea turtles, and marine mammals.

Human Impact Overfishing has depleted fish stocks in all oceans, which have also been polluted by waste dumping.

Open ocean near Iceland

Marine Benthic Zone

A deep-sea hydrothermal vent community

Physical and Chemical Environment
The **marine benthic zone** consists of the seafloor. Except for shallow, near-coastal areas, the marine benthic zone is dark. Water temperature declines with depth, while pressure increases. Organisms in the very deep benthic, or abyssal, zone are adapted to continuous cold (about 3°C) and high water pressure. Oxygen concentrations are generally sufficient to support diverse animal life.

Geologic Features Soft sediments cover most of the benthic zone, but there are areas of rocky substrate on reefs, submarine mountains, and new oceanic crust.

Organisms Photosynthetic organisms, mainly seaweeds and filamentous algae, live in shallow benthic areas with sufficient light. In the dark, hot environments near **deep-sea hydrothermal vents**, the food producers are chemoautotrophic prokaryotes. Coastal benthic communities include numerous invertebrates and fishes. Below the photic zone, most consumers depend entirely on organic matter raining down from above. Among the animals of the deep-sea hydrothermal vent communities are giant tube worms (pictured at left), some more than 1 m long. They are nourished by chemoautotrophic prokaryotes that live as symbionts within their bodies. Many other invertebrates, including arthropods and echinoderms, are also abundant around the hydrothermal vents.

Human Impact Overfishing has decimated important benthic fish populations, such as the cod of the Grand Banks off Newfoundland. Dumping of organic wastes has created oxygen-deprived benthic areas.

lakes, a narrow layer of abrupt temperature change called a **thermocline** separates the more uniformly warm upper layer from more uniformly cold deeper waters.

In both freshwater and marine environments, communities are distributed according to water depth, degree of light penetration, distance from shore, and whether they are found in open water or near the bottom. Plankton and many fish species live in the relatively shallow photic zone (see Figure 40.11). Most of the deep ocean is virtually devoid of light (the aphotic zone) and harbors relatively little life.

CONCEPT CHECK 40.2

The first two questions refer to Figure 40.10.

1. Why are phytoplankton, and not benthic algae or rooted aquatic plants, the dominant photosynthetic organisms of the oceanic pelagic zone?

2. **MAKE CONNECTIONS** Many organisms living in estuaries experience both freshwater and saltwater conditions each day with the rising and falling of tides. Explain how these changing conditions challenge the survival of these organisms (see Concept 32.3).

3. **WHAT IF?** Water leaving a reservoir behind a dam is often taken from deep layers of the reservoir. Would you expect fish found in a river below a dam in summer to be species that prefer colder or warmer water than fish found in an undammed river? Explain.

For suggested answers, see Appendix A.

CONCEPT 40.3

Interactions between organisms and the environment limit the distribution of species

So far in this chapter, we've examined Earth's climate and the characteristics of terrestrial and aquatic biomes. We've also introduced the range of biological levels at which ecologists work (see Figure 40.2). In this section, we'll examine how ecologists determine what factors control the distribution of species, such as the harlequin toad shown in Figure 40.1.

Species distributions are a consequence of both ecological and evolutionary interactions through time. The differential survival and reproduction of individuals that lead to evolution occur in *ecological time*, the minute-to-minute time frame of interactions between organisms and the environment. Through natural selection, organisms adapt to their environment over the time frame of many generations, in *evolutionary time*. One example of how events in ecological time have led to evolution is the selection for beak depth in Galápagos finches (see Figures 21.1 and 21.2). On the island of Daphne Major, finches with larger, deeper beaks were better able to survive during a drought because they could eat the large, hard seeds that were available. Finches with shallower beaks, which required smaller, softer seeds that were in short supply, were less likely to survive and reproduce. Because beak depth is hereditary in this species, the generation of finches born after the drought had beaks that were deeper than those of previous generations.

Biologists have long recognized global and regional patterns in the distribution of organisms (see the discussion of biogeography in Chapter 19). Kangaroos, for instance, are found in Australia but nowhere else on Earth. Ecologists ask not only *where* species occur, but also *why* species occur where they do: What factors determine their distribution? Ecologists generally need to consider multiple factors and alternative hypotheses when attempting to explain the distribution of species. To see how ecologists might arrive at such an explanation, let's work our way through the series of questions in the flowchart in **Figure 40.12**.

Dispersal and Distribution

One factor that contributes greatly to the global distribution of organisms is **dispersal**, the movement of individuals or gametes away from their area of origin or from centers of high population density. A biogeographer who studies the distributions of species in the context of evolutionary theory might consider dispersal in hypothesizing why there are no kangaroos in North America: A barrier may have kept them from reaching the continent. While land-bound kangaroos have not reached North America under their own power, other organisms that disperse more readily, such as some birds, have. The dispersal of organisms is critical to understanding the role of geographic isolation in evolution (see Chapter 22) as well as the broad patterns of species distribution that we see around the world today.

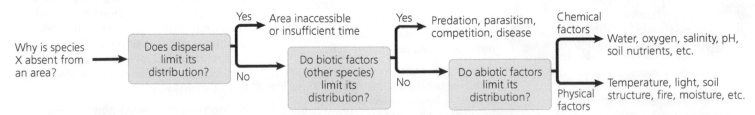

▲ **Figure 40.12 Flowchart of factors limiting geographic distribution.** As ecologists study the factors limiting a species' distribution, they often consider a series of questions like the ones shown here.

? *How might the importance of various abiotic factors differ for aquatic and terrestrial ecosystems?*

To determine if dispersal is a key factor limiting the distribution of a species, ecologists observe the results of intentional or accidental transplants of the species to areas where it was previously absent. For a transplant to be considered successful, some of the organisms must not only survive in the new area but also reproduce there sustainably. If a transplant is successful, then we can conclude that the *potential* range of the species is larger than its *actual* range; in other words, the species *could* live in certain areas where it currently does not.

Species introduced to new geographic locations often disrupt the communities and ecosystems to which they have been introduced and spread far beyond the area of introduction (see Chapter 43). Consequently, ecologists rarely move species to new geographic regions. Instead, they document the outcome when a species has been transplanted for other purposes, such as to introduce game animals or predators of pest species, or when a species has been accidentally transplanted.

Biotic Factors

If dispersal does not limit the distribution of a species, our next question is whether biotic factors—other species—are responsible. Often, negative interactions with predators (organisms that kill their prey) or herbivores (organisms that eat plants or algae) restrict the ability of a species to survive and reproduce. **Figure 40.13** describes a specific case of an herbivore, a sea urchin, limiting the distribution of a food species, a seaweed.

In addition to predation and herbivory, the presence or absence of pollinators, food resources, parasites, pathogens, or competing organisms can act as a biotic limitation on species distribution. Some of the most striking cases of limitation occur when humans accidentally or intentionally introduce exotic predators or pathogens into new areas and wipe out native species. You will encounter examples of these impacts in Chapter 43, where we discuss conservation biology.

Abiotic Factors

The last question in the flowchart in Figure 40.12 considers whether abiotic factors, such as temperature, water, oxygen, salinity, sunlight, or soil, might be limiting a species'

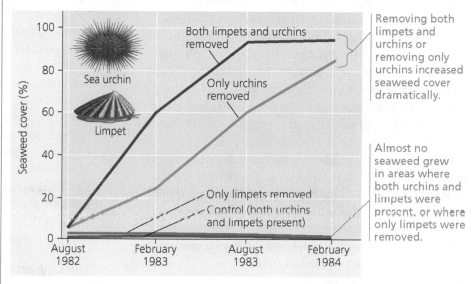

▼ **Figure 40.13** **Inquiry**

Does feeding by sea urchins limit seaweed distribution?

Experiment W. J. Fletcher, of the University of Sydney, Australia, reasoned that if sea urchins are a limiting biotic factor for seaweed growth in a particular ecosystem, then more seaweeds should invade an area from which sea urchins have been removed. To isolate the effect of sea urchins from that of a seaweed-eating mollusc, the limpet, he removed only urchins, only limpets, or both from study areas adjacent to a control site.

Results Fletcher observed a large difference in seaweed growth between areas with and without sea urchins.

Conclusion Removing both limpets and urchins resulted in the greatest increase in seaweed cover, indicating that both species have some influence on seaweed distribution. But since removing only urchins greatly increased seaweed growth while removing only limpets had little effect, Fletcher concluded that sea urchins have a much greater effect than limpets in limiting seaweed distribution.

Source W. J. Fletcher, Interactions among subtidal Australian sea urchins, gastropods, and algae: Effects of experimental removals, *Ecological Monographs* 57:89–109 (1987).

WHAT IF? If removing only limpets did not result in an increase in seaweed growth compared to the control, suggest a reason why removing both urchins and limpets resulted in greater seaweed growth than removing only urchins.

distribution. If the physical conditions at a site do not allow a species to survive and reproduce, then the species will not be found there.

- **Temperature** Environmental temperature is an important factor in the distribution of organisms because of its effect on biological processes. Cells may rupture if the water they contain freezes (at temperatures below 0°C), and the proteins of most organisms denature at temperatures above 45°C. Most organisms function best within a specific range of environmental temperature.

- **Water and Oxygen** Variation in water availability among habitats is another important factor in species distribution. Species living at the seashore or in tidal wetlands can dry out as the tide recedes, and terrestrial organisms face a nearly constant threat of drying. Many amphibians, such as the harlequin toad in Figure 40.1, are particularly vulnerable to drying because they use their moist, delicate skin for gas exchange.

Water affects oxygen availability in aquatic environments and in flooded soils. Because oxygen diffuses slowly in water, its concentration can be low in certain aquatic systems and soils, limiting cellular respiration and other physiological processes. Oxygen concentrations can be particularly low in both deep ocean and deep lake waters and sediments where organic matter is abundant.

- **Salinity** The salt concentration of water in the environment affects the water balance of organisms through osmosis. Most aquatic organisms are restricted to either freshwater or saltwater habitats by their limited ability to osmoregulate (see Chapter 32). Although most terrestrial organisms can excrete excess salts from specialized glands or in feces or urine, salt flats and other high-salinity habitats typically have few species of plants or animals.

- **Sunlight** Sunlight absorbed by photosynthetic organisms provides the energy that drives most ecosystems, and too little sunlight can limit the distribution of photosynthetic species. In forests, shading by leaves in the treetops makes competition for light especially intense, particularly for seedlings growing on the forest floor. In aquatic environments, most photosynthesis occurs near the surface, where sunlight is more available.

- **Rocks and Soil** On land, the pH, mineral composition, and physical structure of rocks and soil limit the distribution of plants and therefore of the animals that feed on them. The pH of soil can limit the distribution of organisms directly, through extreme acidic or basic conditions, or indirectly, by affecting the solubility of nutrients and toxins.

In a river, the composition of the rocks and soil that make up the substrate (riverbed) can affect water chemistry, which in turn influences the resident organisms. In freshwater and marine environments, the structure of the substrate determines the organisms that can attach to it or burrow into it.

So far in this chapter, you have seen how the distributions of biomes and organisms depend on abiotic and biotic factors. In the rest of the chapter, we'll continue to work our way through the hierarchy outlined in Figure 40.2, focusing on how abiotic and biotic factors influence the ecology of populations.

CONCEPT CHECK 40.3

1. Give examples of human actions that could expand a species' distribution by changing (a) its dispersal or (b) its biotic interactions.

2. **WHAT IF?** You suspect that deer are restricting the distribution of a tree species by preferentially eating the seedlings of the tree. How might you test this hypothesis?

For suggested answers, see Appendix A.

Dynamic biological processes influence population density, dispersion, and demographics

Population ecology explores how biotic and abiotic factors influence the density, distribution, and size of populations. A population is a group of individuals of a single species living in the same general area. Members of a population rely on the same resources, are influenced by similar environmental factors, and are likely to interact and breed with one another. Populations evolve as natural selection acts on heritable variations among individuals, changing the frequencies of alleles and traits over time. Evolution remains a central theme as we view populations in the context of ecology.

Populations are often described by their boundaries and size (number of individuals). Ecologists investigating a population define boundaries appropriate to the organism and to the questions being asked. A population's boundaries may be natural ones, as in the case of kangaroos in Australia, or they may be arbitrarily defined by an investigator—for example, a specific county in Minnesota for a study of oak trees.

Density and Dispersion

The **density** of a population is the number of individuals per unit area or volume: the number of oak trees per square kilometer in the Minnesota county or the number of bacteria per milliliter in a culture. **Dispersion** is the pattern of spacing among individuals within the boundaries of the population.

Density: A Dynamic Perspective

In rare cases, population size and density can be determined by counting all individuals within the boundaries of the population.

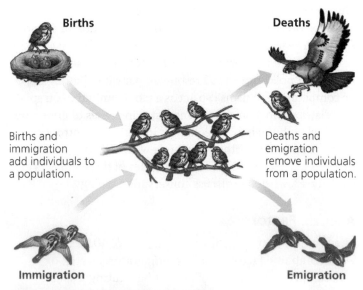

Births

Deaths

Births and immigration add individuals to a population.

Deaths and emigration remove individuals from a population.

Immigration

Emigration

▲ **Figure 40.14 Population dynamics.**

We could count all the sea stars in a tide pool, for instance. Large mammals that live in herds, such as elephants, can sometimes be counted accurately from airplanes. In most cases, however, it is impractical or impossible to count all individuals in a population. Instead, ecologists use a variety of sampling techniques to estimate densities and total population sizes. For example, they might count the number of oak trees in several randomly located 100×100 m plots, calculate the average density in the plots, and then extend the estimate to the population size in the entire area. Such estimates are most accurate when there are many sample plots and when the habitat is fairly homogeneous. In other cases, instead of counting single organisms, population ecologists estimate density from an indicator of population size, such as the number of nests, burrows, tracks, or fecal droppings.

Density is not a static property but changes as individuals are added to or removed from a population **(Figure 40.14)**. Additions occur through birth (which we define here to include all forms of reproduction) and **immigration**, the influx of new individuals from other areas. The factors that remove individuals from a population are death (mortality) and **emigration**, the movement of individuals out of a population and into other locations.

While birth and death rates influence the density of all populations, immigration and emigration also alter the density of many populations. Long-term studies of Belding's ground squirrels (*Spermophilus beldingi*) in the vicinity of Tioga Pass, in the Sierra Nevada of California, showed that some of the squirrels moved nearly 2 km from where they were born. This long-distance movement made them immigrants to other populations. In fact, immigrants made up 1–8% of the males and 0.7–6% of the females in the study population. Such immigration is a meaningful biological exchange between populations over time.

Patterns of Dispersion

Within a population's geographic range, local densities may differ substantially, creating contrasting patterns of dispersion. Differences in local density are among the most important characteristics for a population ecologist to study, since they provide insight into the environmental associations and social interactions of individuals in the population.

The most common pattern of dispersion is *clumped*, in which individuals are aggregated in patches. Plants and fungi are often clumped where soil conditions and other environmental factors favor germination and growth. Insects and salamanders may be clumped under a rotting log because of the higher humidity there. Clumping of animals may also be associated with mating behavior. Sea stars group together in tide pools, where food is readily available and where they can breed successfully **(Figure 40.15a)**. Forming groups may also increase the effectiveness of predation or defense; for example, a wolf pack is more likely than a single wolf to subdue a moose, and a flock of birds is more likely than a single bird to warn of a potential attack.

(a) Clumped. Sea stars group together where food is abundant.

(b) Uniform. Nesting king penguins exhibit uniform spacing, maintained by aggressive interactions between neighbors.

(c) Random. Dandelions grow from windblown seeds that land at random and later germinate.

▲ **Figure 40.15 Patterns of dispersion within a population's geographic range.**

A *uniform*, or evenly spaced, pattern of dispersion may result from direct interactions between individuals in the population. Some plants secrete chemicals that inhibit the germination and growth of nearby individuals that could compete for resources. Animals often exhibit uniform dispersion as a result of antagonistic social interactions, such as **territoriality**—the defense of a bounded physical space against encroachment by other individuals **(Figure 40.15b)**. Uniform patterns are rarer than clumped patterns.

In *random* dispersion (unpredictable spacing), the position of each individual in a population is independent of other individuals. This pattern occurs in the absence of strong attractions or repulsions among individuals or where key physical or chemical factors are relatively constant across the study area. Plants established by windblown seeds, such as dandelions, may be randomly distributed in a fairly uniform habitat (Figure 40.15c).

Demographics

The factors that influence population density and dispersion patterns—ecological needs of a species, structure of the environment, and interactions among individuals within the population—also influence other characteristics of populations. **Demography** is the study of the vital statistics of populations and how they change over time. Of particular interest to demographers are birth rates and death rates. A useful way to summarize some of the vital statistics of a population is to make a life table.

Life Tables

About a century ago, when life insurance first became available, insurance companies began to estimate how long, on average, people of a given age could be expected to live. To do this, demographers developed **life tables**, age-specific summaries of the survival pattern of a population. Population ecologists adapted this approach to the study of populations in general.

The best way to construct a life table is to follow the fate of a **cohort**, a group of individuals of the same age, from birth until all of the individuals are dead. To build the life table, we need to determine the number of individuals that die in each age-group and to calculate the proportion of the cohort surviving from one age class to the next.

Survivorship Curves

A graphic method of representing some of the data in a life table is a **survivorship curve**, a plot of the proportion or numbers in a cohort still alive at each age. Generally, a survivorship curve begins with a cohort of a convenient size—say, 1,000 individuals. Though diverse, survivorship curves can be classified into three general types (**Figure 40.16**). A Type I curve is flat at the start, reflecting low death rates during early and middle life, and then drops steeply as death rates increase among older age-groups. Many large mammals, including humans, that produce few offspring but provide them with good care exhibit this kind of curve. In contrast, a Type III curve drops sharply at the start, reflecting very high death rates for the young, but flattens out as death rates decline for those few individuals that survive the early period of die-off. This type of curve is usually associated with organisms that produce very large numbers of offspring but provide little or no care, such as long-lived plants, many fishes, and most marine invertebrates. An oyster, for example, may release millions of eggs, but most larvae hatched from fertilized

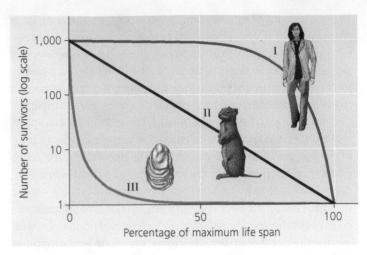

▲ **Figure 40.16 Idealized survivorship curves: Types I, II, and III.** The *y*-axis is logarithmic and the *x*-axis is on a relative scale so that species with widely varying life spans can be presented together on the same graph.

eggs die from predation or other causes. Those few offspring that survive long enough to attach to a suitable substrate and begin growing a hard shell tend to survive for a relatively long time. Type II curves are intermediate, with a constant death rate over the organism's life span. This kind of survivorship occurs in some rodents, invertebrates, lizards, and annual plants.

Many species fall somewhere between these basic types of survivorship or show more complex patterns. In birds, mortality is often high among the youngest individuals (as in a Type III curve) but fairly constant among adults (as in a Type II curve). Some invertebrates, such as crabs, may show a "stair-stepped" curve, with brief periods of increased mortality during molts, followed by periods of lower mortality when their protective exoskeleton is hard.

In populations not experiencing immigration or emigration, survivorship is one of the two key factors determining changes in population size. The other key factor determining population trends is reproductive rate.

Reproductive Rates

Demographers who study sexually reproducing species generally ignore the males and concentrate on the females in a population because only females produce offspring. Therefore, demographers view populations in terms of females giving rise to new females. The simplest way to describe the reproductive pattern of a population is to ask how reproductive output varies with the ages of females.

A **reproductive table**, or fertility schedule, is an age-specific summary of the reproductive rates in a population. It is constructed by measuring the reproductive output of a cohort from birth until death. For a sexual species, the reproductive table tallies the number of female offspring produced by each age-group. **Table 40.1** illustrates a reproductive table for Belding's ground squirrels. Reproductive output for sexual organisms such as birds and mammals is the product of the

Table 40.1 Reproductive Table for Belding's Ground Squirrels at Tioga Pass

Age (years)	Proportion of Females Weaning a Litter	Mean Size of Litters (Males + Females)	Mean Number of Females in a Litter	Average Number of Female Offspring*
0–1	0.00	0.00	0.00	0.00
1–2	0.65	3.30	1.65	1.07
2–3	0.92	4.05	2.03	1.87
3–4	0.90	4.90	2.45	2.21
4–5	0.95	5.45	2.73	2.59
5–6	1.00	4.15	2.08	2.08
6–7	1.00	3.40	1.70	1.70
7–8	1.00	3.85	1.93	1.93
8–9	1.00	3.85	1.93	1.93
9–10	1.00	3.15	1.58	1.58

Source P. W. Sherman and M. L. Morton, Demography of Belding's ground squirrel, *Ecology* 65:1617–1628 (1984).

*The average number of female offspring is the proportion weaning a litter multiplied by the mean number of females in a litter.

proportion of females of a given age that are breeding and the number of female offspring of those breeding females. Multiplying these numbers gives the average number of female offspring for each female in a given age-group (the last column in Table 40.1). For Belding's ground squirrels, which begin to reproduce at age 1 year, reproductive output rises to a peak at 4 years of age and then falls off in older females.

Reproductive tables vary considerably by species. Squirrels, for example, have a litter of two to six young once a year for less than a decade, whereas oak trees may drop thousands of acorns a year for hundreds of years. Mussels and other invertebrates may release millions of eggs and sperm in a spawning cycle. However, a high reproductive rate will not lead to rapid population growth unless conditions are near ideal for the growth and survival of offspring, as you'll learn in the next section.

CONCEPT CHECK 40.4

1. **DRAW IT** Each female of a particular fish species produces millions of eggs per year. Draw and label the most likely survivorship curve for this species, and explain your choice.

2. Imagine that you are constructing a life table for a different population of Belding's ground squirrels than the one shown in Table 40.1. If the proportion of females aged 5–6 years weaning a litter is 0.74 and the mean number of females in a litter is 3.01, what is the average number of female offspring for this cohort in a year?

3. **MAKE CONNECTIONS** A male stickleback fish attacks other males that invade its nesting territory (see Figure 39.15a). Predict the likely pattern of dispersion for male sticklebacks, and explain your reasoning.

For suggested answers, see Appendix A.

The exponential and logistic models describe the growth of populations

Populations of all species have the potential to expand greatly when resources are abundant. To appreciate the potential for population increase, consider a bacterium that can reproduce by fission every 20 minutes under ideal laboratory conditions. There would be two bacteria after 20 minutes, four after 40 minutes, and eight after 60 minutes. If reproduction continued at this rate for a day and a half without mortality, there would be enough bacteria to form a layer 30 cm deep over the entire globe. Unlimited growth cannot occur for long in nature, however. As population density increases, each individual has access to fewer resources. Ecologists study population growth in idealized conditions and in the more realistic conditions where different factors limit growth. We'll examine both scenarios in this section.

Per Capita Rate of Increase

Imagine a population consisting of a few individuals living in an ideal, unlimited environment. Under these conditions, there are no external limits on the abilities of individuals to harvest energy, grow, and reproduce. The population will increase in size with every birth and with the immigration of individuals from other populations, and it will decrease in size with every death and with the emigration of individuals out of the population. We can thus define a change in population size during a fixed time interval with the following verbal equation:

$$\begin{matrix} \text{Change in} \\ \text{population} \\ \text{size} \end{matrix} = \text{Births} + \begin{matrix} \text{Immigrants} \\ \text{entering} \\ \text{population} \end{matrix} - \text{Deaths} - \begin{matrix} \text{Emigrants} \\ \text{leaving} \\ \text{population} \end{matrix}$$

For now, we will simplify the equation by ignoring the effects of immigration and emigration.

We can use mathematical notation to express our simplified equation more concisely. If N represents population size and t represents time, then ΔN is the change in population size and Δt is the time interval (appropriate to the life span or generation time of the species) over which we are evaluating population growth. (The Greek letter delta, Δ, indicates change, such as change in time.) Using B for the number of births in the population during the time interval and D for the number of deaths, we can rewrite the verbal equation:

$$\frac{\Delta N}{\Delta t} = B - D$$

Next, we can convert this simple model to one in which births and deaths are expressed as the average number of births and deaths per individual (per capita) during the specified time interval. The *per capita birth rate* is the number of offspring produced per unit time by an average member of the population. If, for example, there are 34 births per year in a population of 1,000 individuals, the annual per capita birth rate is 34/1,000, or 0.034. If we know the annual per capita birth rate (symbolized by b), we can use the formula $B = bN$ to calculate the expected number of births per year in a population of any size. For example, if the annual per capita birth rate is 0.034 and the population size is 500,

$$B = bN = 0.034 \times 500 = 17 \text{ per year}$$

Similarly, the *per capita death rate* (symbolized by m, for mortality) allows us to calculate the expected number of deaths per unit time in a population, using the formula $D = mN$. If $m = 0.016$ per year, we would expect 16 deaths per year in a population of 1,000 individuals. The per capita birth and death rates can be calculated from estimates of population size and data in life tables and reproductive tables (for example, Table 40.1).

Now we can revise the population growth equation again, using per capita birth and death rates rather than the numbers of births and deaths:

$$\frac{\Delta N}{\Delta t} = bN - mN$$

One final simplification is in order. Population ecologists are most interested in the *difference* between the per capita birth rate and the per capita death rate. This difference is the *per capita rate of increase*, or r:

$$r = b - m$$

The value of r indicates whether a given population is growing ($r > 0$) or declining ($r < 0$). **Zero population growth (ZPG)** occurs when the per capita birth and death rates are equal ($r = 0$). Births and deaths still occur in such a population, of course, but they balance each other exactly.

Using the per capita rate of increase, we can now rewrite the equation for change in population size as

$$\frac{\Delta N}{\Delta t} = rN$$

Remember that this equation is for a specific time interval (often one year) and does not include immigration or emigration. Most ecologists prefer to use differential calculus to express population growth *instantaneously*:

$$\frac{dN}{dt} = r_{\text{inst}}N$$

Here, r_{inst} is the instantaneous per capita rate of increase.

Exponential Growth

Earlier we described a population whose members all have access to abundant food and are free to reproduce at their physiological capacity. Population increase under these ideal conditions is called **exponential population growth**. Under these conditions, the per capita rate of increase may assume the maximum rate for the species, denoted as r_{max}. The equation for exponential population growth is

$$\frac{dN}{dt} = r_{\text{max}}N$$

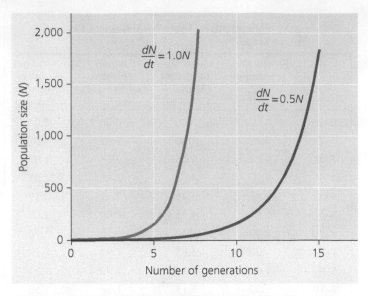

▲ **Figure 40.17 Population growth predicted by the exponential model.** This graph compares growth in two populations with different values of r_{max}. Increasing the value of r_{max} from 0.5 to 1.0 increases the rate of rise in population size over time, as reflected by the relative slopes of the curves at any given population size.

The size of a population that is growing exponentially increases at a constant rate, resulting eventually in a J-shaped growth curve when population size is plotted over time **(Figure 40.17)**. Although the maximum *rate* of increase is constant, the population grows more quickly when it is large than when it is small; thus, the curves in Figure 40.17 get progressively steeper over time. This occurs because population growth depends on N as well as r_{max}, and larger populations experience more births (and deaths) than small ones growing at the same per capita rate. It is also clear from Figure 40.17 that a population with a higher maximum rate of increase ($dN/dt = 1.0N$) will grow faster than one with a lower rate of increase ($dN/dt = 0.5N$).

The J-shaped curve of exponential growth is characteristic of some populations that are introduced into a new environment or whose numbers were drastically reduced and are rebounding. For example, the population of elephants in Kruger National Park, South Africa, grew exponentially for approximately 60 years after they were first protected from hunting **(Figure 40.18)**. The increasing number of elephants eventually caused enough damage to vegetation in the park that a collapse in their food supply was likely. To protect other species and the ecosystem before that happened, park managers began limiting the elephant population by using birth control and exporting elephants to other countries.

Carrying Capacity

The exponential growth model assumes that resources are unlimited, which is rarely the case in the real world. Ultimately, there is a limit to the number of individuals that can occupy a habitat. Ecologists define **carrying capacity**, symbolized by K,

▲ **Figure 40.18 Exponential growth in the African elephant population of Kruger National Park, South Africa.**

Table 40.2 Logistic Growth of a Hypothetical Population ($K = 1,500$)

Population Size (N)	Maximum Rate of Increase (r_{max})	$\dfrac{K - N}{K}$	Per Capita Rate of Increase $r_{max}\dfrac{(K - N)}{K}$	Population Growth Rate* $r_{max}N\dfrac{(K - N)}{K}$
25	1.0	0.98	0.98	+25
100	1.0	0.93	0.93	+93
250	1.0	0.83	0.83	+208
500	1.0	0.67	0.67	+333
750	1.0	0.50	0.50	+375
1,000	1.0	0.33	0.33	+333
1,500	1.0	0.00	0.00	0

*Rounded to the nearest whole number.

as the maximum population size that a particular environment can sustain. Carrying capacity varies over space and time with the abundance of limiting resources. Energy, shelter, refuge from predators, nutrient availability, water, and suitable nesting sites can all be limiting factors. For example, the carrying capacity for bats may be high in a habitat with abundant flying insects and roosting sites, but lower where there is abundant food but fewer suitable shelters.

Crowding and resource limitation can have a profound effect on population growth rate. If individuals cannot obtain sufficient resources to reproduce, the per capita birth rate (b) will decline. If they cannot consume enough energy to maintain themselves or if disease increases with density, the per capita death rate (m) may increase. A decrease in b or an increase in m lowers the per capita rate of increase (r).

The Logistic Growth Model

We can modify our mathematical model to include changes in growth rate as N increases. In the **logistic population growth** model, the per capita rate of increase approaches zero as the population size nears its carrying capacity.

To construct the logistic model, we start with the exponential population growth model and add an expression that reduces the per capita rate of increase as N increases. If the maximum sustainable population size (carrying capacity) is K, then $K - N$ is the number of additional individuals the environment can support, and $(K - N)/K$ is the fraction of K that is still available for population growth. By multiplying the exponential rate of increase $r_{max}N$ by $(K - N)/K$, we modify the change in population size as N increases:

$$\frac{dN}{dt} = r_{max}N\frac{(K - N)}{K}$$

When N is small compared with K, the term $(K - N)/K$ is close to 1, and the per capita rate of increase, $r_{max}(K - N)/K$, approaches the maximum rate of increase. But when N is large and resources are limiting, then $(K - N)/K$ is close to 0, and

the per capita rate of increase is small. When N equals K, the population stops growing. **Table 40.2** shows calculations of population growth rate for a hypothetical population growing according to the logistic model, with $r_{max} = 1.0$ per individual per year. Notice that the overall population growth rate is highest, +375 individuals per year, when the population size is 750, or half the carrying capacity. At a population size of 750, the per capita rate of increase remains relatively high (one-half the maximum rate), but there are more reproducing individuals (N) in the population than at lower population sizes.

As shown in **Figure 40.19**, the logistic model of population growth produces a sigmoid (S-shaped) growth curve when N

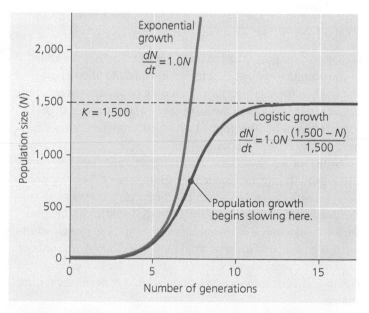

▲ **Figure 40.19 Population growth predicted by the logistic model.** The rate of population growth decreases as population size (N) approaches the carrying capacity (K) of the environment. The red line shows logistic growth in a population where $r_{max} = 1.0$ and $K = 1,500$ individuals. For comparison, the blue line illustrates a population continuing to grow exponentially with the same r_{max}.

Using the Logistic Equation to Model Population Growth

What Happens to the Size of a Population When It Overshoots Its Carrying Capacity? In the logistic population growth model, the per capita rate of population increase approaches zero as the population size (N) approaches the carrying capacity (K). Under some conditions, however, a population in the laboratory or the field can overshoot K, at least temporarily. For instance, if food becomes limiting to a population, there may be a delay before reproduction declines, and N may briefly exceed K. In this exercise, you will use the logistic equation to model the growth of the hypothetical population in Table 40.2 when $N > K$.

Interpret the Data

1. Assuming that $r_{max} = 1.0$ and $K = 1,500$, calculate the population growth rate for four cases where population size (N) is greater than carrying capacity (K): $N = 1,510$; $1,600$; $1,750$; and $2,000$ individuals. To do this, first write the equation for population growth rate given in Table 40.2. Plug in the values for each of the four cases, starting with $N = 1,510$, and solve the equation for each one. Which population size has the highest growth rate?

2. If r_{max} is doubled, predict how the population growth rates will change for the four population sizes given in question 1. Now calculate the population growth rate for the same four cases, this time assuming that $r_{max} = 2.0$ (and K still $= 1,500$).

3. Now let's see how the growth of a real-world population of *Daphnia* corresponds to this model. At what times in Figure 40.20b is the *Daphnia* population changing in ways that correspond to the values you calculated? Hypothesize why the population drops below the carrying capacity briefly late in the experiment.

(MB) A version of this Scientific Skills Exercise can be assigned in MasteringBiology.

is plotted over time (the red line). New individuals are added to the population most rapidly at intermediate population sizes, when there is not only a breeding population of substantial size, but also lots of available space and other resources in the environment. The population growth rate decreases dramatically as N approaches K.

Note that we haven't said anything yet about *why* the population growth rate decreases as N approaches K. For a population's growth rate to decrease, the birth rate b must decrease, the death rate m must increase, or both. Later in this chapter, we'll consider some of the factors affecting these rates, including the presence of disease, predation, and limited amounts of food and other resources. In the **Scientific Skills Exercise**, you can model what happens to a population if N becomes *greater* than K.

The Logistic Model and Real Populations

The growth of laboratory populations of some small animals, such as beetles and crustaceans, and of some microorganisms, such as bacteria, *Paramecium*, and yeasts, fits an

▶ **Figure 40.20**
How well do these populations fit the logistic growth model? In each graph, the black dots plot the measured growth of the population, and the red curve is the growth predicted by the logistic model.

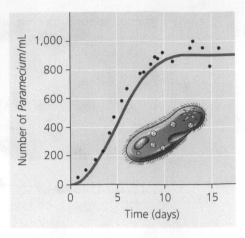

(a) A *Paramecium* population in the lab. Growth in a small culture closely approximates logistic growth if the researcher maintains a constant environment.

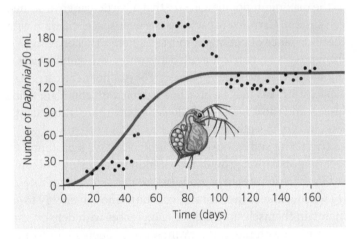

(b) A *Daphnia* (water flea) population in the lab. Growth in a small culture does not correspond well to the logistic model. This population overshoots the carrying capacity of its artificial environment before reaching an approximately stable size.

S-shaped curve fairly well under conditions of limited resources **(Figure 40.20a)**. These populations are grown in a constant environment lacking predators and competing species that may reduce growth of the populations, conditions that rarely occur in nature.

The assumptions built into the logistic model clearly do not apply to all populations. It assumes that populations adjust instantaneously to growth and approach carrying capacity smoothly. In reality, there is often a delay before the negative effects of an increasing population are realized. If food becomes limiting for a population, for instance, reproduction will decline eventually, but females may use their energy reserves to continue reproducing for a short time. This may cause the population to overshoot its carrying capacity temporarily, as shown for the water fleas in **Figure 40.20b**. Other populations fluctuate greatly, making it difficult even to define carrying capacity. We'll examine some possible reasons for such fluctuations later in this chapter.

The logistic model is a useful starting point for thinking about how populations grow and for constructing more complex models. The model is also important in conservation biology for predicting how rapidly a particular population might increase in numbers after it has been reduced to a small size and for estimating sustainable harvest rates for wildlife populations. Conservation biologists can use the model to estimate the critical size below which populations of certain organisms may become extinct.

CONCEPT CHECK 40.5

1. Why does a constant rate of increase (r_{max}) for a population produce a growth curve that is J-shaped?
2. Explain why a population that fits the logistic growth model increases more rapidly at intermediate size than at relatively small and large sizes.
3. **MAKE CONNECTIONS** Many viruses are pathogens of animals and plants (see Concept 17.3). How might the presence of pathogens alter the carrying capacity of a population? Explain.

For suggested answers, see Appendix A.

CONCEPT 40.6

Population dynamics are influenced strongly by life history traits and population density

EVOLUTION What environmental factors keep populations from growing indefinitely? Why are some populations fairly stable in size, while others are not? The answers to these questions depend in part on the traits of individuals, influenced through time by natural selection, and also on factors in the environment that vary with population density.

In every species, there are trade-offs between survival and reproductive traits such as frequency of reproduction, number of offspring (number of seeds produced by plants; litter or clutch size for animals), and investment in parental care. The traits that affect an organism's schedule of reproduction and survival make up its **life history**. A life history entails three main variables: when reproduction begins (the age at first reproduction or age at maturity), how often the organism reproduces, and how many offspring are produced per reproductive episode.

"Trade-offs" and Life Histories

No organism could produce unlimited numbers of offspring *and* provision them well. There is a trade-off between reproduction and survival. For instance, researchers in Scotland found that female red deer that reproduced in a given summer were more likely to die the next winter than were females that did not reproduce.

Selective pressures influence the trade-off between the number and size of offspring. Plants and animals whose young are likely to die often produce many small offspring. Plants that colonize disturbed environments, for example, usually produce many small seeds, only a few of which may reach a suitable habitat. Small size may also increase the chance of seedling establishment by enabling the seeds to be carried longer distances to a broader range of habitats **(Figure 40.21a)**. Animals that suffer high predation rates, such as quail, sardines, and mice, also tend to produce large numbers of offspring.

In other organisms, extra investment on the part of the parent greatly increases the offspring's chance of survival. Walnut and Brazil nut trees provision large seeds with nutrients that help the seedlings become established **(Figure 40.21b)**. Primates generally bear only one or two offspring at a time; parental care and an extended period of learning in the first several years of life are very important to offspring fitness. Such provisioning and extra care can be especially important in habitats with high population densities.

Ecologists have attempted to connect differences in favored traits at different population densities with the logistic growth model discussed in Concept 40.5. Selection for traits that are sensitive to population density and are favored at high densities

(a) Dandelions grow quickly and release a large number of tiny fruits, each containing a single seed. Producing numerous seeds ensures that at least some will grow into plants that eventually produce seeds themselves.

(b) Some plants, such as the Brazil nut tree (right), produce a moderate number of large seeds in pods (above). Each seed's large endosperm provides nutrients for the embryo, an adaptation that helps a relatively large fraction of offspring survive.

▲ **Figure 40.21 Variation in the size of seed crops in plants.**

is known as **K-selection**, or density-dependent selection. In contrast, selection for traits that maximize reproductive success in uncrowded environments (low densities) is called **r-selection**, or density-independent selection. These names follow from the variables of the logistic equation. *K*-selection is said to operate in populations living at a density near the limit imposed by their resources (the carrying capacity, *K*), where competition among individuals is stronger. Mature trees growing in an old-growth forest are an example of *K*-selected organisms. In contrast, *r*-selection is said to maximize *r*, the per capita rate of increase, and occurs in environments in which population densities are well below carrying capacity or individuals face little competition. Weeds growing in an abandoned agricultural field are an example of *r*-selected organisms.

Population Change and Population Density

Similar to the case of *r-selection*, a birth rate or death rate that does *not* change with population density is said to be **density independent**. In a classic study of population regulation, Andrew Watkinson and John Harper, of the University of Wales, found that the mortality of dune fescue grass (*Vulpia membranacea*) is mainly due to physical factors that kill similar proportions of a local population, regardless of its density. For example, drought stress that arises when the roots of the grass are uncovered by shifting sands is a density-independent factor. In contrast, and similar to *K-selection*, a death rate that rises as population density rises is said to be **density dependent**, as is a birth rate that falls with rising density. Watkinson and Harper found that reproduction by dune fescue declines as population density increases, in part because water or nutrients become more scarce. Thus, the key factors regulating birth rate in this population are density dependent, while death rate is largely regulated by density-independent factors. **Figure 40.22** shows how the combination of density-dependent reproduction and density-independent mortality can stop population growth, leading to an equilibrium population density in species such as dune fescue.

Mechanisms of Density-Dependent Population Regulation

Without some type of negative feedback between population density and the rates of birth and death, a population would never stop growing. Density-dependent regulation provides that feedback, halting population growth through mechanisms that reduce birth rates or increase death rates. Several mechanisms of density-dependent population regulation are described in **Figure 40.23**.

These various examples of population regulation by negative feedback show how increased densities cause population growth rates to decline by affecting reproduction, growth, and survival. But while negative feedback helps explain why populations stop growing, it does not address why some populations fluctuate dramatically while others remain relatively stable. That is the topic we address next.

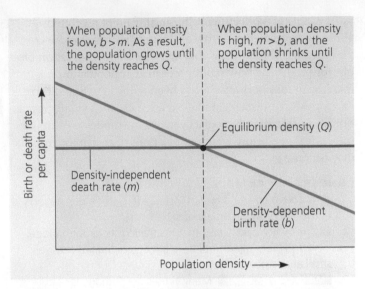

▲ **Figure 40.22 Determining equilibrium for population density.** This simple model considers only birth and death rates. (Immigration and emigration rates are assumed to be either zero or equal.) In this example, the birth rate changes with population density, while the death rate is constant. At the equilibrium density (*Q*), the birth and death rates are equal.

DRAW IT *Redraw this figure for the case where the birth and death rates are both density dependent, as occurs for many species.*

Population Dynamics

All populations for which we have long-term data show some fluctuation in size. Such population fluctuations from year to year or place to place, called **population dynamics**, are influenced by many factors and in turn affect other species, including our own. For example, fluctuations in fish populations influence seasonal harvests of commercially important species. The study of population dynamics focuses on the complex interactions between biotic and abiotic factors that cause variation in population sizes.

Stability and Fluctuation

Populations of large mammals were once thought to remain relatively stable over time, but long-term studies have challenged that idea. For instance, the moose population on Isle Royale in Lake Superior fluctuates substantially from year to year. What causes the size of this population to change so dramatically? Harsh weather, particularly cold winters, can weaken the moose and reduce food availability, decreasing the size of the population. When moose numbers are high, other factors, such as an increase in the density of ticks and other parasites, also cause the population to shrink.

Predation is an additional factor that regulates the population. Moose from the mainland colonized the island around 1900 by walking across the frozen lake. Wolves, which rely on moose for most of their food, followed around 1950. Because the lake has not frozen over in recent years, both populations have been isolated from immigration and emigration. Despite this isolation, the moose population experienced two major

As population density increases, many density-dependent mechanisms slow or stop population growth by decreasing birth rates or increasing death rates.

Competition for Resources

Increasing population density intensifies competition for nutrients and other resources, reducing reproductive rates. Farmers minimize the effect of resource competition on the growth of grains such as wheat (*Triticum aestivum*) and other crops by applying fertilizers to reduce nutrient limitations on crop yield.

Predation

Predation can be an important cause of density-dependent mortality if a predator captures more food as the population density of the prey increases. As a prey population builds up, predators may also feed preferentially on that species. Population increases in the collared lemming (*Dicrotonyx groenlandicus*) lead to density-dependent predation by several predators, including the snowy owl (*Bubo scandiacus*).

Disease

If the transmission rate of a disease increases as a population becomes more crowded, then the disease's impact is density dependent. In humans, the respiratory diseases influenza (flu) and tuberculosis are spread through the air when an infected person sneezes or coughs. Both diseases strike a greater percentage of people in densely populated cities than in rural areas.

Toxic Wastes

Yeasts, such as the brewer's yeast *Saccharomyces cerevisiae*, are used to convert carbohydrates to ethanol in winemaking. The ethanol that accumulates in the wine is toxic to yeasts and contributes to density-dependent regulation of yeast population size. The alcohol content of wine is usually less than 13% because that is the maximum concentration of ethanol that most wine-producing yeast cells can tolerate.

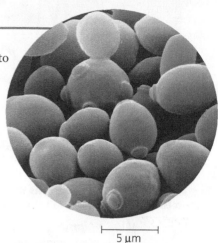

5 μm

Territoriality

Territoriality can limit population density when space becomes the resource for which individuals compete. Cheetahs (*Acinonyx jubatus*) use a chemical marker in urine to warn other cheetahs of their territorial boundaries. The presence of surplus, or nonbreeding, individuals is a good indication that territoriality is restricting population growth.

Intrinsic Factors

Intrinsic physiological factors sometimes regulate population size. Reproductive rates of white-footed mice (*Peromyscus leucopus*) in a field enclosure can drop even when food and shelter are abundant. This drop in reproduction at high population density is associated with aggressive interactions and hormonal changes that delay sexual maturation and depress the immune system.

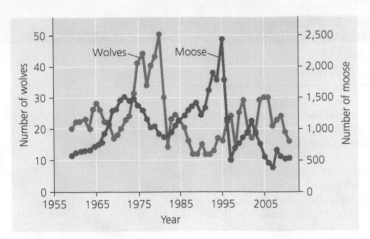

▲ **Figure 40.24 Fluctuations in moose and wolf populations on Isle Royale, 1959–2011.**

 ANIMATION

BioFlix Visit the Study Area in **MasteringBiology** for the BioFlix® 3-D Animation on Population Ecology.

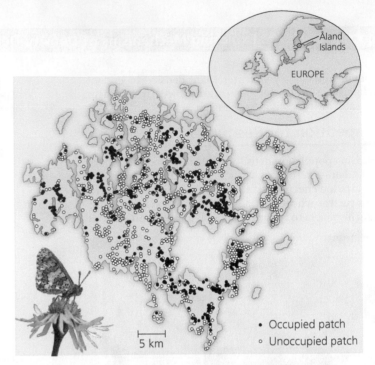

▲ **Figure 40.25 The Glanville fritillary: a metapopulation.** On the Åland Islands, local populations of this butterfly (filled circles) are found in only a fraction of the suitable habitat patches (open circles) at any given time. Individuals can move between local populations and colonize unoccupied patches.

increases and collapses during the last 50 years **(Figure 40.24)**. The first collapse coincided with a peak in the numbers of wolves from 1975 to 1980. The second collapse, around 1995, coincided with harsh winter weather, which increased the energy needs of the animals and made it harder for the moose to find food under the deep snow.

Immigration, Emigration, and Metapopulations

So far, our discussion of population dynamics has focused mainly on the contributions of births and deaths. However, immigration and emigration also influence populations. When a population becomes crowded and resource competition increases (see Figure 40.23), emigration often increases.

Immigration and emigration are particularly important when a number of local populations are linked, forming a **metapopulation**. Local populations in a metapopulation can be thought of as occupying discrete patches of suitable habitat in a sea of otherwise unsuitable habitat. Such patches vary in size, quality, and isolation from other patches, factors that influence how many individuals move among the populations. If one population becomes extinct, the patch it occupied can be recolonized by immigrants from another population.

The Glanville fritillary (*Melitaea cinxia*) illustrates the movement of individuals between populations. This butterfly is found in about 500 meadows across the Åland Islands of Finland, but its potential habitat in the islands is much larger, approximately 4,000 suitable patches. New populations of the butterfly regularly appear and existing populations become extinct, constantly shifting the locations of the 500 colonized patches **(Figure 40.25)**. The species persists in a balance of extinctions and recolonizations.

The metapopulation concept underscores the significance of immigration and emigration in the butterfly populations.

It also helps ecologists understand population dynamics and gene flow in patchy habitats, providing a framework for the conservation of species living in a network of habitat fragments and reserves. In fact, many aspects of population ecology that you have studied in this chapter have practical applications. Farmers may want to reduce the abundance of insect pests or stop the growth of an invasive weed that is spreading rapidly. Conservation ecologists need to know what environmental factors create favorable feeding or breeding habitats for endangered species, such as the white rhinoceros and the whooping crane. Management programs based on population-regulating factors have helped prevent the extinction of many endangered species.

CONCEPT CHECK 40.6

1. In the fish called the peacock wrasse (*Symphodus tinca*), females disperse some of their eggs widely and lay other eggs in a nest. Only the latter receive parental care. Explain the trade-offs in reproduction that this behavior illustrates.

2. **WHAT IF?** Mice that experience stress such as a food shortage will sometimes abandon their young. Explain how this behavior might have evolved in the context of reproductive trade-offs and life history.

3. **MAKE CONNECTIONS** Negative feedback is a process that regulates biological systems (see Concept 32.1). Explain how the density-dependent birth rate of dune fescue grass exemplifies negative feedback.

For suggested answers, see Appendix A.

40 Chapter Review

SUMMARY OF KEY CONCEPTS

CONCEPT 40.1

Earth's climate influences the structure and distribution of terrestrial biomes (pp. 820–826)

- Global **climate** patterns are largely determined by the input of solar energy and Earth's revolution around the sun.
- The changing angle of the sun over the year, bodies of water, and mountains exert seasonal, regional, and local effects on **macroclimate**.
- **Climographs** show that temperature and precipitation are correlated with **biomes**. Other factors also influence biome location.
- Terrestrial biomes are often named for major physical or climatic factors and for their predominant vegetation. Vertical layering is an important feature of terrestrial biomes.

> **?** *Some arctic tundra ecosystems receive as little rainfall as deserts but have much more dense vegetation. Based on Figure 40.8, what climatic factor might explain this difference?*

CONCEPT 40.2

Aquatic biomes are diverse and dynamic systems that cover most of Earth (pp. 827–830)

- Aquatic biomes are characterized primarily by their physical environment rather than by climate and are often layered with regard to light penetration, temperature, and community structure.
- In the ocean and in most lakes, an abrupt temperature change called a **thermocline** separates a more uniformly warm upper layer from more uniformly cold deeper waters.

> **?** *In which aquatic biomes might you find an aphotic zone?*

CONCEPT 40.3

Interactions between organisms and the environment limit the distribution of species (pp. 830–832)

- Ecologists want to know not only *where* species occur but also *why* those species occur where they do.
- The distribution of species may be limited by **dispersal**, **biotic** (living) factors, and **abiotic** (physical) factors, such as temperature extremes, salinity, and water availability.

> **?** *If you were an ecologist studying the chemical and physical limits to the distributions of species, how might you rearrange the flowchart in Figure 40.12?*

CONCEPT 40.4

Dynamic biological processes influence population density, dispersion, and demographics (pp. 832–835)

- Population **density**—the number of individuals per unit area or volume—reflects the interplay of births, deaths, immigration, and emigration. Environmental and social factors influence the **dispersion** of individuals.
- Populations increase from births and **immigration** and decrease from deaths and **emigration**. **Life tables**, **survivorship curves**, and **reproductive tables** summarize specific trends in **demography**.

> **?** *Gray whales* (Eschrichtius robustus) *gather each winter near Baja California to give birth. How might such behavior make it easier for ecologists to estimate birth and death rates for the species?*

CONCEPT 40.5

The exponential and logistic models describe the growth of populations (pp. 835–839)

- If immigration and emigration are ignored, a population's growth rate (the per capita rate of increase) equals its birth rate minus its death rate.
- The **exponential growth** equation $dN/dt = r_{max}N$ represents a population's potential growth in an unlimited environment, where r_{max} is the maximum per capita rate of increase and N is the number of individuals in the population.

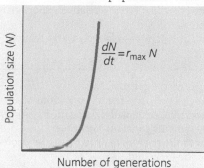

$$\frac{dN}{dt} = r_{max}N$$

Population size (N) / Number of generations

- Exponential growth cannot be sustained for long in any population. A more realistic population model limits growth by incorporating **carrying capacity** (K), the maximum population size the environment can support. According to the **logistic growth** equation $dN/dt = r_{max}N(K - N)/K$, growth levels off as population size approaches the carrying capacity.

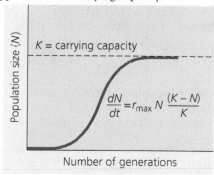

K = carrying capacity

$$\frac{dN}{dt} = r_{max}N\frac{(K - N)}{K}$$

Population size (N) / Number of generations

- The logistic model fits few real populations perfectly, but it is useful for estimating possible growth.

> **?** *As an ecologist who manages a wildlife preserve, you want to increase the preserve's carrying capacity for a particular endangered species. How might you go about accomplishing this?*

CONCEPT 40.6

Population dynamics are influenced strongly by life history traits and population density (pp. 839–842)

- **Life history** traits are evolutionary outcomes reflected in the development, physiology, and behavior of organisms.
- **Density-dependent** changes in birth and death rates curb population increase through negative feedback. Density-dependent limiting factors include intraspecific competition for limited food or space, increased predation, disease, intrinsic physiological factors, and buildup of toxic substances.

- All populations exhibit some size fluctuations, and many undergo regular boom-and-bust cycles influenced by complex interactions between biotic and abiotic factors. A **metapopulation** is a group of populations linked by immigration and emigration.

? *Name one biotic and one abiotic factor that contribute to yearly fluctuations in the size of the human population.*

TEST YOUR UNDERSTANDING

Level 1: Knowledge/Comprehension

1. Which of the following biomes is correctly paired with the description of its climate?
   a. savanna—low temperature, precipitation uniform during the year
   b. tundra—long summers, mild winters
   c. temperate broadleaf forest—relatively short growing season, mild winters
   d. temperate grasslands—relatively warm winters, most rainfall in summer
   e. tropical forests—nearly constant day length and temperature

2. A population's carrying capacity
   a. may change as environmental conditions change.
   b. can be accurately calculated using the logistic growth model.
   c. generally remains constant over time.
   d. increases as the per capita growth rate (r) decreases.
   e. can never be exceeded.

Level 2: Application/Analysis

3. When climbing a mountain, we can observe transitions in biological communities that are analogous to the changes
   a. in biomes at different latitudes.
   b. in different depths in the ocean.
   c. in a community through different seasons.
   d. in an ecosystem as it evolves over time.
   e. across the United States from east to west.

4. According to the logistic growth equation
   $$\frac{dN}{dt} = r_{max}N\frac{(K - N)}{K}$$
   a. the number of individuals added per unit time is greatest when N is close to zero.
   b. the per capita growth rate (r) increases as N approaches K.
   c. population growth is zero when N equals K.
   d. the population grows exponentially when K is small.
   e. the birth rate (b) approaches zero as N approaches K.

Level 3: Synthesis/Evaluation

5. **DRAW IT** After examining Figure 40.13, you decide to study feeding relationships among sea otters, sea urchins, and kelp. You know that sea otters prey on sea urchins and that urchins eat kelp. At four coastal sites, you measure kelp abundance. Then you spend one day at each site and mark whether otters are present or absent every 5 minutes during the day. Make a graph that shows how otter density depends on kelp abundance, using the data below. Then formulate a hypothesis to explain the pattern you observed.

Site	Kelp Abundance (% cover)	Otter Density (# sightings per day)
1	75	98
2	15	18
3	60	85
4	25	36

6. **WHAT IF** If the direction of Earth's rotation reversed, the most predictable effect would be
   a. no more night and day.
   b. a big change in the length of the year.
   c. winds blowing from west to east along the equator.
   d. a loss of seasonal variation at high latitudes.
   e. the elimination of ocean currents.

7. **SCIENTIFIC INQUIRY**
   Jens Clausen and colleagues, at the Carnegie Institution of Washington, studied how the size of yarrow plants (*Achillea lanulosa*) growing on the slopes of the Sierra Nevada varied with elevation. They found that plants from low elevations were generally taller than plants from high elevations, as shown below:

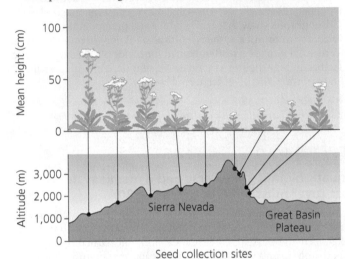

Source J. Clausen et al., Experimental studies on the nature of species. III. Environmental responses of climatic races of *Achillea*, Carnegie Institution of Washington Publication No. 581 (1948).

   Clausen and colleagues proposed two hypotheses to explain this variation within a species: (1) There are genetic differences between populations of plants found at different elevations. (2) The species has developmental flexibility and can assume tall or short growth forms, depending on local abiotic factors. If you had seeds from yarrow plants found at low and high elevations, what experiments would you perform to test these hypotheses?

8. **FOCUS ON EVOLUTION**
   Discuss how the concept of time applies to ecological situations and evolutionary changes. Do ecological time and evolutionary time ever overlap? If so, what are some examples?

9. **FOCUS ON INTERACTIONS**
   In a short essay (100–150 words), identify the factor or factors in Figure 40.23 that you think may ultimately be most important for density-dependent population regulation in humans, and explain your reasoning.

For selected answers, see Appendix A.

MasteringBiology®

Students Go to **MasteringBiology** for assignments, the eText, and the Study Area with practice tests, animations, and activities.

Instructors Go to **MasteringBiology** for automatically graded tutorials and questions that you can assign to your students, plus Instructor Resources.

41

Species Interactions

KEY CONCEPTS

41.1 Interactions within a community may help, harm, or have no effect on the species involved

41.2 Diversity and trophic structure characterize biological communities

41.3 Disturbance influences species diversity and composition

41.4 Biogeographic factors affect community diversity

41.5 Pathogens alter community structure locally and globally

OVERVIEW

Communities in Motion

Deep in the Lembeh Strait of Indonesia, a crab in the family Homolidae scuttles across the ocean floor holding a large sea urchin on its back (Figure 41.1). When a predatory fish arrives, the crab settles quickly into the sediments and puts its living shield to use. The fish darts in and tries to bite the crab. In response, the crab tilts the spiny sea urchin toward whichever side the fish attacks. The fish eventually gives up and swims away.

The "carrier crab" in Figure 41.1 clearly benefits from having the sea urchin on its back. But how does the sea urchin fare in this relationship? Its association with the crab might harm it, help it, or have no effect on its survival and reproduction. For example, the sea urchin may be harmed if the crab sets it down in an unsuitable habitat or in a place where it is vulnerable to predators. On the other hand, the crab may also protect the sea urchin from predators while carrying it. Additional observations or experiments would be needed before ecologists could answer this question.

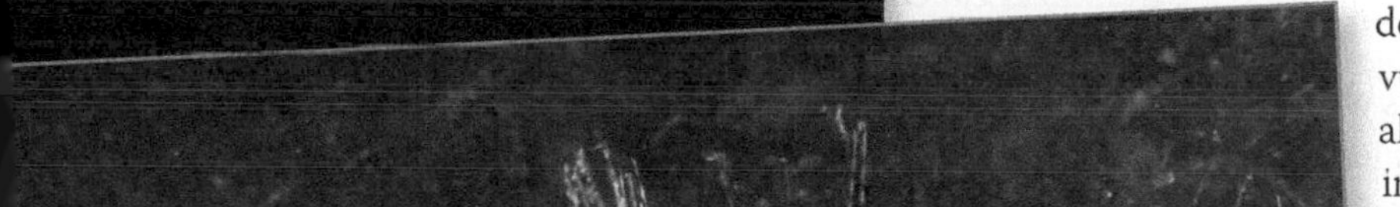

In Chapter 40, you learned how individuals within a population can affect other individuals of the same species. This chapter will examine ecological interactions between populations of different species. A group of populations of different species living close enough to interact is called a biological **community**. Ecologists define the boundaries of a particular community to fit their research questions: They might study the community of decomposers and other organisms living on a rotting log, the benthic community in Lake Superior, or the community of trees and shrubs in Great Smoky Mountains National Park in North Carolina and Tennessee.

We begin this chapter by exploring the kinds of interactions that occur between species in a community, such as the crab and sea urchin in Figure 41.1. We'll then consider several factors that are most significant in structuring a community—in determining how many species there are, which particular species are present, and the relative abundance of these species. Finally, we'll apply some of the principles of community ecology to the study of human disease.

▼ **Figure 41.1** Which species benefits from this interaction?

Interactions within a community may help, harm, or have no effect on the species involved

Some key relationships in the life of an organism are its interactions with individuals of other species in the community. These **interspecific interactions** include competition, predation, herbivory, symbiosis (including parasitism, mutualism, and commensalism), and facilitation. In this section, we will define and describe each of these interactions, recognizing that ecologists do not always agree on the precise boundaries of each type of interaction.

We will use the symbols + and − to indicate how each interspecific interaction affects the survival and reproduction of the two species engaged in the interaction. For example, predation is a +/− interaction, with a positive effect on the survival and reproduction of the predator population and a negative effect on that of the prey population. Mutualism is a +/+ interaction because the survival and reproduction of both species are increased in the presence of the other. We use a 0 to indicate that a population is not affected by the interaction in any known way.

Competition

Interspecific competition is a −/− interaction that occurs when individuals of different species compete for a resource that limits their growth and survival. Weeds growing in a garden compete with garden plants for nutrients and water. Lynx and foxes in the northern forests of Alaska and Canada compete for prey such as snowshoe hares. In contrast, some resources, such as oxygen, are rarely in short supply, at least on land; most terrestrial species use this resource, but they do not usually compete for it.

Competitive Exclusion

What happens in a community when two species compete for limited resources? In 1934, Russian ecologist G. F. Gause studied this question using laboratory experiments with two species of ciliated protists, *Paramecium aurelia* and *Paramecium caudatum*. He cultured the species under stable conditions, adding a constant amount of food each day. When Gause grew the two species separately, each population grew rapidly and then leveled off at the apparent carrying capacity of the culture (see Figure 40.20a for an illustration of the logistic growth of *P. aurelia*). But when Gause grew the two species together, *P. caudatum* became extinct in the culture. Gause inferred that *P. aurelia* had a competitive edge in obtaining food. He concluded that two species competing for the same limiting resources cannot coexist permanently in the same place. In the absence of disturbance, one species will use the resources more efficiently and reproduce more rapidly than the other. Even a slight reproductive advantage will eventually lead to

local elimination of the inferior competitor, an outcome called **competitive exclusion**.

Ecological Niches and Natural Selection

EVOLUTION The influence of evolution is evident in the concept of the **ecological niche**, the specific set of biotic and abiotic resources that an organism uses in its environment. American ecologist Eugene Odum used the following analogy to explain the niche concept: If an organism's habitat is its "address," the niche is the organism's "profession." The niche of a tropical tree lizard, for instance, includes the temperature range it tolerates, the size of branches on which it perches, the time of day when it is active, and the sizes and kinds of insects it eats. Such factors define the lizard's niche, or ecological role—how it fits into an ecosystem.

We can use the niche concept to restate the principle of competitive exclusion: Two species cannot coexist permanently in a community if their niches are identical. However, ecologically similar species *can* coexist in a community if one or more significant differences in their niches arise through time. Evolution by natural selection can result in one of the species using a different set of resources or similar resources at different times of the day or year. The differentiation of niches that enables similar species to coexist in a community is called **resource partitioning (Figure 41.2)**. You can think of resource partitioning in a community as "the

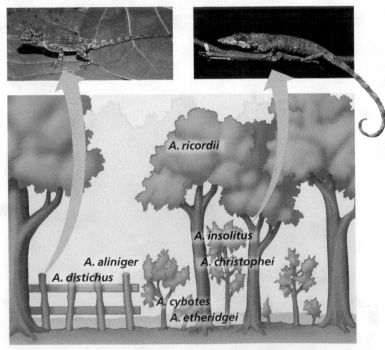

A. distichus perches on fence posts and other sunny surfaces.

A. insolitus usually perches on shady branches.

A. ricordii

A. insolitus

A. aliniger

A. distichus

A. christophei

A. cybotes

A. etheridgei

▲ **Figure 41.2 Resource partitioning among Dominican Republic lizards.** Seven species of *Anolis* lizards live in close proximity, and all feed on insects and other small arthropods. However, competition for food is reduced because each lizard species has a different preferred perch, thus occupying a distinct niche.

ghost of competition past"—the indirect evidence of earlier interspecific competition resolved by the evolution of niche differentiation.

As a result of competition, a species' *fundamental niche*, which is the niche potentially occupied by that species, is often different from its *realized niche*, the portion of its fundamental niche that it actually occupies in a particular environment. Ecologists can identify the fundamental niche of a species by testing the range of conditions in which it grows and reproduces in the absence of competitors. They can also test whether a potential competitor limits a species' realized niche by removing the competitor and seeing if the first species expands into the newly available space. The classic experiment depicted in **Figure 41.3** clearly showed that competition between two barnacle species kept one species from occupying part of its fundamental niche.

Character Displacement

Closely related species whose populations are sometimes allopatric (geographically separate; see Chapter 22) and sometimes sympatric (geographically overlapping) provide more evidence for the importance of competition in structuring communities. In some cases, the allopatric populations of such species are morphologically similar and use similar resources. By contrast, sympatric populations, which would potentially compete for resources, show differences in body structures and in the resources they use. This tendency for characteristics to diverge more in sympatric than in allopatric populations of two species is called **character displacement**. An example of character displacement in Galápagos finches is shown in **Figure 41.4**.

▼ Figure 41.3 Inquiry

Can a species' niche be influenced by interspecific competition?

Experiment Ecologist Joseph Connell studied two barnacle species—*Chthamalus stellatus* and *Balanus balanoides*—that have a stratified distribution on rocks along the coast of Scotland. *Chthamalus* is usually found higher on the rocks than *Balanus*. To determine whether the distribution of *Chthamalus* is the result of interspecific competition with *Balanus*, Connell removed *Balanus* from the rocks at several sites.

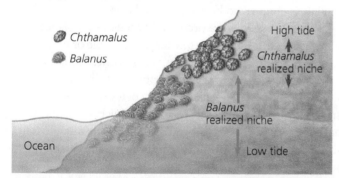

Results *Chthamalus* spread into the region formerly occupied by *Balanus*.

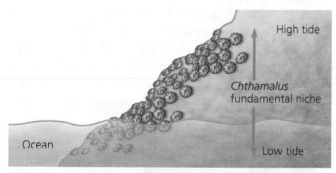

Conclusion Interspecific competition makes the realized niche of *Chthamalus* much smaller than its fundamental niche.

Source J. H. Connell, The influence of interspecific competition and other factors on the distribution of the barnacle *Chthamalus stellatus*, *Ecology* 42:710–723 (1961).

(MB) See the related Experimental Inquiry Tutorial in MasteringBiology.

WHAT IF? Other observations showed that *Balanus* cannot survive high on the rocks because it dries out during low tides. How would *Balanus*'s realized niche compare with its fundamental niche?

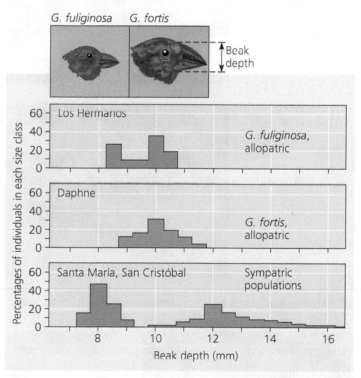

▲ **Figure 41.4 Character displacement: indirect evidence of past competition.** Allopatric populations of *Geospiza fuliginosa* and *Geospiza fortis* on Los Hermanos and Daphne Islands have similar beak morphologies (top two graphs) and presumably eat similarly sized seeds. However, where the two species are sympatric on Santa María and San Cristóbal, *G. fuliginosa* has a shallower, smaller beak and *G. fortis* a deeper, larger one (bottom graph), adaptations that favor eating different-sized seeds.

? *If the beak length of* G. fortis *is typically 12% longer than the beak depth, what is the predicted beak length of* G. fortis *individuals with the smallest beak depths observed on Santa María and San Cristóbal Islands?*

Predation

Predation refers to a +/− interaction between species in which one species, the predator, kills and eats the other, the prey. Though the term *predation* generally elicits such images as a lion attacking and eating an antelope, it applies to a wide range of interactions. An animal that kills a plant by eating the plant's tissues can also be considered a predator. Because eating and avoiding being eaten are prerequisite to reproductive success, the adaptations of both predators and prey tend to be refined through natural selection (see Concept 27.5). In the **Scientific Skills Exercise**, you can interpret data from an experiment investigating a specific predator-prey interaction.

Many important feeding adaptations of predators are obvious and familiar. Most predators have acute senses that enable them to find and identify potential prey. Rattlesnakes and other pit vipers, for example, find their prey with a pair of heat-sensing organs located between their eyes and nostrils (see Figure 38.17a). Many predators also have adaptations such as claws, teeth, stingers, or poison that help them catch and subdue their food. Predators that pursue their prey are generally fast and agile, whereas those that lie in ambush are often disguised in their environments.

Just as predators possess adaptations for capturing prey, potential prey animals have adaptations that help them avoid being eaten. Some common behavioral defenses are hiding, fleeing, and forming herds or schools. Active self-defense is less common, though some large grazing mammals vigorously defend their young from predators such as lions.

Animals also display a variety of morphological and physiological defensive adaptations. **Cryptic coloration**, or camouflage, makes prey difficult to see **(Figure 41.5a)**. Mechanical or chemical defenses protect species such as porcupines and skunks. Some animals, including the European fire salamander, can synthesize toxins, whereas others accumulate toxins passively from the plants they eat. Animals with effective chemical defenses often exhibit bright **aposematic coloration**, or warning coloration, such as that of the poison dart frog **(Figure 41.5b)**. Such coloration seems to be adaptive because predators often avoid brightly colored prey.

Some prey species are protected by their resemblance to other species. In **Batesian mimicry**, a palatable or harmless species mimics an unpalatable or harmful one. The larva of the hawkmoth *Hemeroplanes ornatus* puffs up its head and thorax when disturbed, looking like the head of a small venomous snake **(Figure 41.5c)**. In this case, the mimicry even involves behavior; the larva weaves its head back and forth and hisses like a snake. In **Müllerian mimicry**, two or more unpalatable species, such as the cuckoo bee and yellow jacket, resemble each other **(Figure 41.5d)**. In an example of convergent evolution, unpalatable animals in several different taxa have similar patterns of coloration: Black and yellow or red stripes characterize unpalatable animals as diverse as yellow jackets and coral snakes.

Many predators also use mimicry. The alligator snapping turtle has a tongue that resembles a wriggling worm, which is used to lure small fish. Any fish that tries to eat the "bait" is itself quickly consumed as the turtle's strong jaws snap closed.

▼ **Figure 41.5 Examples of defensive coloration in animals.**

(a) Cryptic coloration

► Canyon tree frog

(b) Aposematic coloration

► Poison dart frog

(c) Batesian mimicry: A harmless species mimics a harmful one.

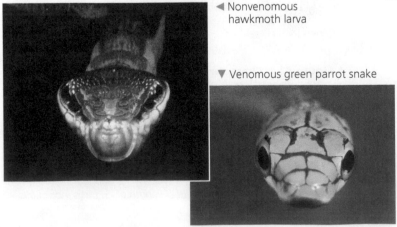

◄ Nonvenomous hawkmoth larva

▼ Venomous green parrot snake

(d) Müllerian mimicry: Two unpalatable species mimic each other.

◄ Cuckoo bee

▼ Yellow jacket

Herbivory

Ecologists use the term **herbivory** to refer to a +/− interaction in which an organism eats parts of a plant or alga. While large mammalian herbivores such as cattle, sheep, and water buffalo may be most familiar, most herbivores are actually invertebrates, such as grasshoppers, caterpillars, and beetles. In the ocean, herbivores include sea urchins, some tropical fishes, and certain mammals, including the manatee **(Figure 41.6)**.

Like predators, herbivores have many specialized adaptations. Many herbivorous insects have chemical sensors on their feet that enable them to distinguish between toxic and nontoxic plants or between more nutritious and less nutritious plants. Some mammalian herbivores, such as goats, use their sense of smell to examine plants. They may also eat just a specific part of a plant, such as the flowers. Many herbivores also have specialized teeth or digestive systems adapted to processing vegetation (see Chapter 33).

Unlike prey animals, plants cannot run away to avoid being eaten. Instead, a plant's arsenal against herbivores may feature chemical toxins or structures such as spines and thorns. Among the plant compounds that serve as chemical weapons are the poison strychnine, produced by the tropical

▲ **Figure 41.6 A marine herbivore.** This West Indies manatee (*Trichechus manatus*) in Florida is grazing on hydrilla, an introduced plant.

vine *Strychnos toxifera*, and nicotine, from the tobacco plant. Compounds that are not toxic to humans but may be distasteful to many herbivores are responsible for the familiar flavors of cinnamon, cloves, and peppermint.

Symbiosis

When individuals of two or more species live in direct and intimate contact with one another, their relationship is called **symbiosis**. In this text, we define symbiosis to include all such

---

Scientific Skills Exercise

Using Bar Graphs and Scatter Plots to Present and Interpret Data

Can a Native Predator Species Adapt Rapidly to an Introduced Prey Species? Cane toads (*Bufo marinus*) were introduced to Australia in 1935 in a failed attempt to control an insect pest. Since then, the toads have spread throughout northeastern Australia, reaching a population of over 200 million today. Cane toads have glands that produce a toxin that is poisonous to snakes and other potential predators of the toads. In this exercise, you will graph and interpret data from a two-part experiment conducted to determine whether native Australian predators have developed resistance to the cane toad toxin.

How the Experiment Was Done In part 1, researchers collected 12 black snakes (*Pseudechis porphyriacus*) from areas where cane toads had existed for 40–60 years and another 12 from areas free of cane toads. They offered the snakes either a freshly killed native frog (*Limnodynastes peronii*, a species the snakes commonly eat) or a freshly killed cane toad from which the toxin gland had been removed (making the toad nonpoisonous). In part 2, researchers collected snakes from areas where cane toads had been present for 5–60 years. To assess how cane toad toxin affected these snakes, they injected small amounts of the toxin into the snakes' stomachs and measured the snakes' swimming speed in a small pool.

Data from the Experiment, Part 1

Type of Prey Offered	Percentage of Snakes That Ate Prey Offered in Each Area	
	Cane Toads Present in Area for 40–60 Years	No Cane Toads in Area
Native frog	100	100
Cane toad	0	48

© 2006 The Royal Society

Data from the Experiment, Part 2

Time Since First Exposure to Cane Toads (years)	5	10	10	20	50	60	60	60	60	60
Percentage Reduction in Swimming Speed	52	19	30	30	5	5	9	11	12	22

© 2006 The Royal Society

Interpret the Data

1. Make a bar graph of the data in part 1. (For additional information about graphs, see the Scientific Skills Review in Appendix F and in the Study Area in MasteringBiology.)

2. What do the data represented in the graph suggest about the effects of cane toads on the predatory behavior of black snakes in areas where the toads are and are not currently found?

3. Suppose an enzyme that deactivates the cane toad toxin evolves in black snakes living in areas with cane toads. If the researchers repeated part 1 of this study, predict how the results would change.

4. Identify the dependent and independent variables in part 2. Make a scatter plot of the data.

5. Based on the scatter plot, what conclusion would you draw about whether exposure to cane toads is having a selective effect on black snakes in this study? Explain.

6. Explain why a bar graph is an appropriate type of graph for presenting the data in part 1 and a scatter plot is an appropriate type for presenting the data in part 2.

Data from B. L. Phillips and R. Shine, An invasive species induces rapid adaptive change in a native predator: cane toads and black snakes in Australia, *Proceedings of the Royal Society B* 273:1545–1550 (2006).

⓶ A version of this Scientific Skills Exercise can be assigned in MasteringBiology.

interactions, whether they are harmful, helpful, or neutral. Some biologists define symbiosis more narrowly as a synonym for mutualism, an interaction in which both species benefit.

Parasitism

Parasitism is a +/− symbiotic interaction in which one organism, the **parasite**, derives its nourishment from another organism, its **host**, which is harmed in the process. Parasites that live within the body of their host, such as tapeworms, are called **endoparasites**; parasites that feed on the external surface of a host, such as ticks and lice, are called **ectoparasites**. In one particular type of parasitism, parasitoid insects—usually small wasps—lay eggs on or in living hosts. The larvae then feed on the body of the host, eventually killing it. Some ecologists have estimated that at least one-third of all species on Earth today are parasites.

Many parasites have complex life cycles involving multiple hosts. The blood fluke, which infects approximately 200 million people around the world, requires two hosts for its development: humans and freshwater snails. Some parasites change the behavior of their current host in ways that increase the likelihood that the parasite will reach its next host. For instance, crustaceans that are parasitized by acanthocephalan (spiny-headed) worms leave protective cover and move into the open, where they are more likely to be eaten by the birds that are the second host in the worm's life cycle.

Parasites can significantly affect the survival, reproduction, and density of their host population, either directly or indirectly. For example, ticks that live as ectoparasites on moose weaken their hosts by withdrawing blood and causing hair breakage and loss. In their weakened condition, the moose have a greater chance of dying from cold stress or predation by wolves (see Figure 40.24).

Mutualism

Mutualistic symbiosis, or **mutualism**, is an interspecific interaction that benefits both species (+/+). You have seen examples of mutualism in previous chapters: Examples of mutualism include nitrogen fixation by bacteria in the root nodules of legumes; cellulose digestion by microorganisms in the digestive systems of termites and ruminant mammals; and photosynthesis by unicellular algae in corals. In the acacia-ant example shown in **Figure 41.7**, both species can survive alone. In some other cases, though, such as termites and microorganisms, both species have lost the ability to survive on their own.

Mutualism typically involves the coevolution of related adaptations in both species, with changes in either species likely to affect the survival and reproduction of the other. For example, most flowering plants have adaptations such as nectar or fruit that attract animals that pollinate flowers or disperse seeds (see Chapter 30). In turn, many animals have adaptations that help them find and consume nectar.

(a) Certain species of acacia trees in Central and South America have hollow thorns that house stinging ants of the genus *Pseudomyrmex*. The ants feed on nectar produced by the tree and on protein-rich swellings (orange in the photograph) at the tips of leaflets.

(b) The acacia benefits because the pugnacious ants, which attack anything that touches the tree, remove fungal spores, small herbivores, and debris. They also clip vegetation that grows close to the acacia.

▲ **Figure 41.7 Mutualism between acacia trees and ants.**

Commensalism

An interaction between species that benefits one of the species but neither harms nor helps the other (+/0) is called **commensalism**. Commensal interactions are difficult to document in nature because any close association between species likely affects both species, even if only slightly. For instance, "hitchhiking" species, such as algae that live on the shells of aquatic turtles or barnacles that attach to whales, are sometimes considered commensal. The hitchhikers gain a place to grow while having seemingly little effect on their ride. However, they may reduce the hosts' efficiency of movement in searching for food or escaping from predators. Conversely, the hitchhikers may help camouflage the hosts.

Some commensal associations involve one species obtaining food that is inadvertently exposed by another. Cattle egrets feed on insects flushed out of the grass by grazing bison, cattle,

▲ **Figure 41.8 A possible example of commensalism between cattle egrets and African buffalo.**

and other herbivores **(Figure 41.8)**. Because the birds typically find more prey when they follow herbivores, they clearly benefit from the association. Much of the time, the herbivores may be unaffected by the birds. However, they, too, may derive some benefit; the birds occasionally remove and eat ticks and other ectoparasites from the herbivores or may warn the herbivores of a predator's approach.

Facilitation

Species can have positive effects (+/+ or 0/+) on the survival and reproduction of other species without necessarily living in the direct and intimate contact of a symbiosis. This type of interaction, called **facilitation**, is particularly common in plant ecology. For instance, the black rush *Juncus gerardi* makes the soil more hospitable for other plant species in some zones of New England salt marshes **(Figure 41.9a)**. *Juncus* helps prevent salt buildup in the soil by shading the soil surface, which reduces evaporation. *Juncus* also prevents the salt marsh soils

(a) Salt marsh with *Juncus* (foreground)

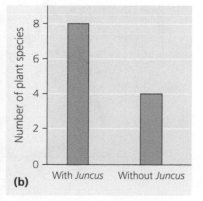

(b)

▲ **Figure 41.9 Facilitation by black rush (*Juncus gerardi*) in New England salt marshes.** Black rush increases the number of plant species that can live in the upper middle zone of the marsh.

from becoming oxygen depleted as it transports oxygen to its belowground tissues. In one study, when *Juncus* was removed from areas in the upper middle intertidal zone, those areas supported 50% fewer plant species **(Figure 41.9b)**.

All five types of interactions discussed so far—competition, predation, herbivory, symbiosis, and facilitation—strongly influence the structure of communities. You'll see other examples of these interactions throughout this chapter.

CONCEPT CHECK 41.1

1. Explain how interspecific competition, predation, and mutualism differ in their effects on the interacting populations of two species.
2. According to the principle of competitive exclusion, what outcome is expected when two species with identical niches compete for a resource? Why?
3. **MAKE CONNECTIONS** Figure 22.12 illustrates the formation of and possible outcomes for a hybrid zone over time. Imagine that two finch species colonize a new island and are capable of hybridizing (mating and producing viable offspring). The island contains two plant species, one with large seeds and one with small seeds, growing in isolated habitats. If the two finch species specialize in eating different plant species, would reproductive barriers be reinforced, weakened, or unchanged in this hybrid zone? Explain.

For suggested answers, see Appendix A.

CONCEPT 41.2

Diversity and trophic structure characterize biological communities

Along with the specific interactions described in the previous section, communities are also characterized by more general attributes, including how diverse they are and the feeding relationships of their species. In this section, you'll see why such ecological attributes are important. You'll also learn how a few species sometimes exert strong control on a community's structure, particularly on the composition, relative abundance, and diversity of its species.

Species Diversity

The **species diversity** of a community—the variety of different kinds of organisms that make up the community—has two components. One is **species richness**, the number of different species in the community. The other is the **relative abundance** of the different species, the proportion each species represents of all individuals in the community.

Imagine two small forest communities, each with 100 individuals distributed among four tree species (A, B, C, and D) as follows:

Community 1: 25A, 25B, 25C, 25D

Community 2: 80A, 5B, 5C, 10D

The species richness is the same for both communities because they both contain four species of trees, but the relative abundance is very different **(Figure 41.10)**. You would easily notice the four types of trees in community 1, but without looking carefully, you might see only the abundant species A in the second forest. Most observers would intuitively describe community 1 as the more diverse of the two communities.

Ecologists use many tools to compare the diversity of communities across time and space. They often calculate indexes of diversity based on species richness and relative abundance. One widely used index is the **Shannon diversity index** (H):

$$H = -(p_A \ln p_A + p_B \ln p_B + p_C \ln p_C + ...)$$

where A, B, C . . . are the species in the community, p is the relative abundance of each species, and ln is the natural logarithm. A higher value of H indicates a more diverse community. Let's use this equation to calculate the Shannon diversity index of the two communities in Figure 41.10. For community 1, $p = 0.25$ for each species, so

$$H = -4(0.25 \ln 0.25) = 1.39$$

For community 2,

$$H = -[0.8 \ln 0.8 + 2(0.05 \ln 0.05) + 0.1 \ln 0.1] = 0.71$$

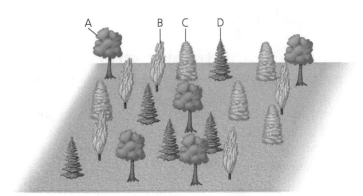

Community 1
A: 25% B: 25% C: 25% D: 25%

Community 2
A: 80% B: 5% C: 5% D: 10%

▲ **Figure 41.10 Which forest is more diverse?** Ecologists would say that community 1 has greater species diversity, a measure that includes both species richness and relative abundance.

These calculations confirm that community 1 is more diverse.

Determining the number and relative abundance of species in a community can be challenging. Because most species in a community are relatively rare, it may be hard to obtain a sample size large enough to be representative. It is also difficult to census highly mobile or less visible organisms, such as insects and nocturnal species. The small size of microorganisms makes them particularly difficult to sample, so ecologists now use molecular tools to help determine microbial diversity **(Figure 41.11)**.

▼ **Figure 41.11   Research Method**

Determining Microbial Diversity Using Molecular Tools

Application Ecologists are increasingly using molecular techniques, such as the analysis of restriction fragment length polymorphisms (RFLPs), to determine microbial diversity and richness in environmental samples. Noah Fierer and Rob Jackson, of Duke University, used RFLP analysis to compare the diversity of soil bacteria in 98 habitats across North and South America to help identify environmental variables associated with high bacterial diversity.

Technique Researchers first extract and purify DNA from the microbial community in each sample. They use the polymerase chain reaction (see Chapter 13) to amplify specific DNA (such as that encoding small ribosomal subunit RNA) and label the DNA with a fluorescent dye. Restriction enzymes then cut the amplified, labeled DNA into fragments of different lengths, which are separated by gel electrophoresis. The number and abundance of these fragments characterize the DNA profile of the sample. Based on their RFLP analysis, Fierer and Jackson calculated the Shannon diversity index (H) of each sample. They then looked for a correlation between H and several environmental variables.

Results The diversity of the sampled bacteria was related almost exclusively to soil pH, with the Shannon diversity index being highest in neutral soils and lowest in acidic soils.

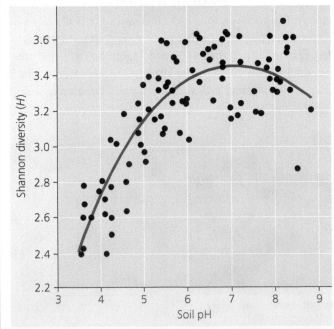

Source N. Fierer and R. B. Jackson, The diversity and biogeography of soil bacterial communities, *Proceedings of the National Academy of Sciences USA* 103:626–631 (2006).

▲ **Figure 41.12 Study plots at the Cedar Creek Natural History Area, site of long-term experiments in which researchers have manipulated plant diversity.**

Diversity and Community Stability

In addition to measuring species diversity, ecologists manipulate diversity in experimental communities in nature and in the laboratory. They do this to examine the potential benefits of diversity, including increased productivity and stability of biological communities.

Researchers at the Cedar Creek Natural History Area, in Minnesota, have been manipulating plant diversity in experimental communities for two decades **(Figure 41.12)**. Higher-diversity communities generally are more productive and are better able to withstand and recover from environmental stresses, such as droughts. More diverse communities are also more stable year to year in their productivity. In one decade-long experiment, for instance, researchers at Cedar Creek created 168 plots, each containing 1, 2, 4, 8, or 16 perennial grassland species. The most diverse plots produced **biomass** (the total mass of all individuals in a population) much more consistently than the single-species plots each year.

Higher-diversity communities are often more resistant to **invasive species**, which are organisms that become established outside their native range. Scientists working in Long Island Sound, off the coast of Connecticut, created communities with different levels of diversity consisting of sessile marine invertebrates, including tunicates (see Concept 27.3). They then examined how vulnerable these experimental communities were to invasion by an exotic tunicate. They found that the exotic tunicate was four times more likely to survive in lower-diversity communities than in higher-diversity ones. The researchers concluded that relatively diverse communities captured more of the resources available in the system, leaving fewer resources for the invader and decreasing its survival.

Trophic Structure

Experiments like the ones just described often examine the importance of diversity within one trophic level. The

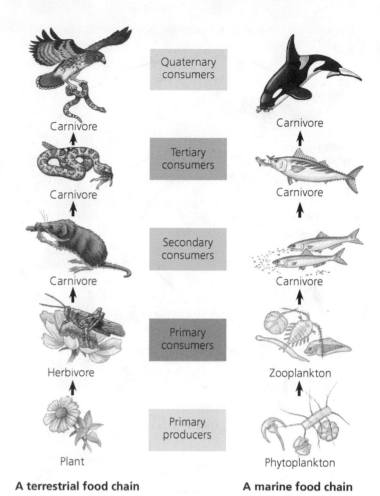

A terrestrial food chain   **A marine food chain**

▲ **Figure 41.13 Examples of terrestrial and marine food chains.** The arrows trace energy and nutrients that pass through the trophic levels of a community when organisms feed on one another. Decomposers, which "feed" on organisms from all trophic levels, are not shown here.

structure and dynamics of a community also depend on the feeding relationships between organisms in different trophic levels. These relationships together make up the **trophic structure** of the community. The transfer of food energy up the trophic levels from its source in plants and other autotrophs (primary producers) through herbivores (primary consumers) to carnivores (secondary, tertiary, and quaternary consumers) and eventually to decomposers is referred to as a **food chain (Figure 41.13)**.

In the 1920s, Oxford University biologist Charles Elton recognized that food chains are not isolated units but are linked together in **food webs**. Ecologists diagram the trophic relationships of a community using arrows linking species according to who eats whom. In an Antarctic pelagic community, for example, the primary producers are phytoplankton, which serve as food for the dominant grazing zooplankton, especially krill and copepods, both of which are crustaceans. These zooplankton species are in turn eaten by carnivores, including other plankton, penguins, seals, fishes, and baleen whales. Squids, which are carnivores that feed on fish and zooplankton, are another

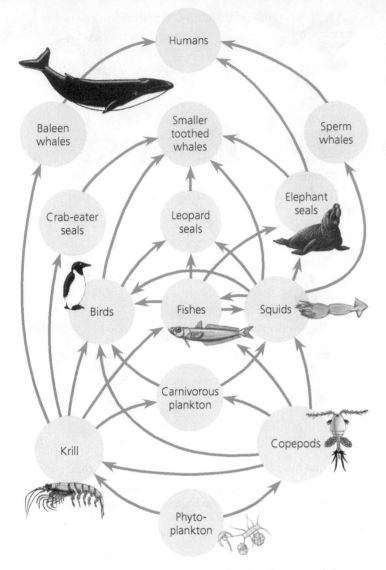

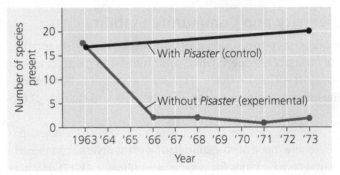

▲ **Figure 41.14 An Antarctic marine food web.** Arrows follow the transfer of food from the producers (phytoplankton) up through the trophic levels. For simplicity, this diagram omits decomposers.

▼ **Figure 41.15 Inquiry**

Is *Pisaster ochraceus* a keystone predator?

Experiment In rocky intertidal communities of western North America, the relatively uncommon sea star *Pisaster ochraceus* preys on mussels such as *Mytilus californianus*, a dominant species and strong competitor for space.

Robert Paine, of the University of Washington, removed *Pisaster* from an area in the intertidal zone and examined the effect on species richness.

Results In the absence of *Pisaster*, species richness declined as mussels monopolized the rock face and eliminated most other invertebrates and algae. In a control area where *Pisaster* was not removed, species richness changed very little.

Conclusion *Pisaster* acts as a keystone species, exerting an influence on the community that is not reflected in its abundance.

Source R. T. Paine, Food web complexity and species diversity, *American Naturalist* 100:65–75 (1966).

WHAT IF? Suppose that an invasive fungus killed most individuals of *Mytilus* at these sites. Predict how species richness would be affected if *Pisaster* were then removed.

---

important link in these food webs, as they are in turn eaten by seals and toothed whales **(Figure 41.14)**.

Note that a given species may weave into the web at more than one trophic level. In the food web shown in Figure 41.14, krill feed on phytoplankton as well as on other grazing zooplankton, such as copepods.

Species with a Large Impact

Certain species have an especially large impact on the structure of entire communities because they are highly abundant or play a pivotal role in community dynamics. The impact of these species occurs through trophic interactions and their influence on the physical environment.

Dominant species in a community are the species that are the most abundant or that collectively have the highest biomass. There is no single explanation for why a species becomes dominant in a community. One hypothesis suggests that dominant species are competitively superior in exploiting limited resources such as water or nutrients. Another hypothesis is that dominant species are most successful at avoiding predation or the impact of disease. This latter idea could explain the high biomass attained by some invasive species. Such species may not face the natural predators or parasites that would otherwise hold their populations in check.

In contrast to dominant species, **keystone species** are not usually abundant in a community. They exert strong control on community structure not by numerical might but by their pivotal ecological roles, or niches. **Figure 41.15** highlights the importance of a keystone species, a sea star, in maintaining the diversity of an intertidal community.

Other organisms exert their influence on a community not through trophic interactions but by changing their physical

▲ **Figure 41.16 Beavers as ecosystem engineers.** By felling trees, building dams, and creating ponds, beavers can transform large areas of forest into flooded wetlands.

environment. Species that dramatically alter their environment are called **ecosystem engineers** or, to avoid implying conscious intent, "foundation species." A familiar ecosystem engineer is the beaver **(Figure 41.16)**. The effects of ecosystem engineers on other species can be positive or negative, depending on the needs of the other species.

Bottom-Up and Top-Down Controls

Simplified models based on relationships between adjacent trophic levels are useful for describing community organization. For example, consider the three possible relationships between plants (*V* for vegetation) and herbivores (*H*):

$$V \rightarrow H \qquad V \leftarrow H \qquad V \leftrightarrow H$$

The arrows indicate that a change in the biomass of one trophic level causes a change in the other trophic level. $V \rightarrow H$ means that an increase in vegetation will increase the numbers or biomass of herbivores, but not vice versa. In this situation, herbivores are limited by vegetation, but vegetation is not limited by herbivory. In contrast, $V \leftarrow H$ means that an increase in herbivore biomass will decrease the abundance of vegetation, but not vice versa. A double-headed arrow indicates that each trophic level is sensitive to changes in the biomass of the other.

Two models of community organization are common: the bottom-up model and the top-down model. The $V \rightarrow H$ linkage suggests a **bottom-up model**, which postulates a unidirectional influence from lower to higher trophic levels. In this case, the presence or absence of mineral nutrients (*N*) controls plant (*V*) numbers, which control herbivore (*H*) numbers, which in turn control predator (*P*) numbers. The simplified bottom-up model is thus $N \rightarrow V \rightarrow H \rightarrow P$. To change the community structure of a bottom-up community, you need to alter biomass at the lower trophic levels, allowing those changes to propagate up through the food web. If you add mineral nutrients to stimulate plant growth, then the higher trophic levels should also increase in biomass. If you change predator abundance, however, the effect should not extend down to the lower trophic levels.

In contrast, the **top-down model** postulates the opposite: Predation mainly controls community organization because predators limit herbivores, herbivores limit plants, and plants limit nutrient levels through nutrient uptake. The simplified top-down model, $N \leftarrow V \leftarrow H \leftarrow P$, is also called the *trophic cascade model.* In a lake community with four trophic levels, the model predicts that removing the top carnivores will increase the abundance of primary carnivores, in turn decreasing the number of herbivores, increasing phytoplankton abundance, and decreasing concentrations of mineral nutrients. The effects thus move down the trophic structure as alternating +/− effects.

Ecologists have applied the top-down model to improve water quality in polluted lakes. This approach, called **biomanipulation**, attempts to prevent algal blooms and eutrophication by altering the density of higher-level consumers instead of using chemical treatments. In lakes with three trophic levels, removing fish should improve water quality by increasing zooplankton density, thereby decreasing algal populations. In lakes with four trophic levels, adding top predators should have the same effect. We can summarize the scenario of three trophic levels with the following diagram:

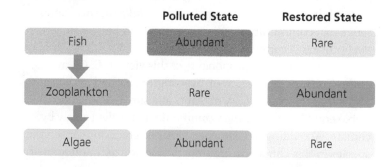

Ecologists in Finland used biomanipulation to help purify Lake Vesijärvi, a large lake that was polluted with city sewage and industrial wastewater until 1976. After pollution controls reduced these inputs, the water quality of the lake began to improve. By 1986, however, massive blooms of cyanobacteria started to occur in the lake. These blooms coincided with an increase in the population of roach, a fish species that eats zooplankton, which otherwise keep the cyanobacteria and algae in check. To reverse these changes, ecologists removed nearly a million kilograms of fish between 1989 and 1993, reducing roach abundance by about 80%. At the same time, they added a fourth trophic level by stocking the lake with pike perch, a predatory fish that eats roach. The water became clear, and the last cyanobacterial bloom was in 1989. The lake remains clear even though roach removal ended in 1993.

As these examples show, communities vary in their degree of bottom-up and top-down control. To manage agricultural landscapes, parks, reservoirs, and fisheries, we need to understand each particular community's dynamics.

CONCEPT CHECK 41.2

1. What two components contribute to species diversity? Explain how two communities with the same number of species can differ in species diversity.
2. Based on the food web in Figure 41.14, identify all of the organisms that eat and are eaten by elephant seals.
3. **WHAT IF?** Consider a grassland with five trophic levels: grasses, mice, snakes, raccoons, and bobcats. If you released additional bobcats into the grassland, how would plant biomass change if the bottom-up model applied? If the top-down model applied? Explain.

For suggested answers, see Appendix A.

CONCEPT 41.3

Disturbance influences species diversity and composition

Decades ago, most ecologists favored the traditional view that biological communities are at equilibrium, a more or less stable balance, unless seriously disturbed by human activities. The "balance of nature" view focused on interspecific competition as a key factor determining community composition and maintaining stability in communities. *Stability* in this context refers to a community's tendency to reach and maintain a relatively constant composition of species.

One of the earliest proponents of this view, F. E. Clements, of the Carnegie Institution of Washington, argued in the early 1900s that the community of plants at a site had only one stable equilibrium, a *climax community* controlled solely by climate. According to Clements, biotic interactions caused the species in this climax community to function as an integrated unit. His argument was based on the observation that certain species of plants are consistently found together, such as the oaks, maples, birches, and beeches in deciduous forests of the northeastern United States.

Other ecologists questioned whether most communities were at equilibrium or functioned as integrated units. A. G. Tansley, of Oxford University, challenged the concept of a climax community, arguing that differences in soils, topography, and other factors created many potential communities that were stable within a region. H. A. Gleason, of the University of Chicago, saw communities more as chance assemblages of species with similar abiotic requirements—for example, for temperature, rainfall, and soil type. Gleason and other ecologists also realized that disturbance keeps many communities from reaching a stable state in species diversity or composition. A **disturbance** is an event, such as a storm, fire, drought, or human activity, that changes a community by removing organisms from it or altering resource availability.

This recent emphasis on change has produced the **nonequilibrium model**, which describes most communities as constantly changing after disturbance. Even relatively stable communities can be rapidly transformed into nonequilibrium communities. Let's examine some of the ways disturbance influences community structure and composition.

Characterizing Disturbance

The types of disturbances and their frequency and severity vary among communities. Storms disturb most communities, even many in the oceans through the action of waves. Fire is a significant disturbance; in fact, chaparral and some grassland biomes require regular burning to maintain their structure and species composition. Many streams and ponds are disturbed by spring flooding and seasonal drying. A high level of disturbance is generally the result of frequent *and* intense disturbance, while low disturbance levels can result from either a low frequency or low intensity of disturbance.

The **intermediate disturbance hypothesis** states that moderate levels of disturbance foster greater species diversity than do low or high levels of disturbance. High levels of disturbance reduce diversity by creating environmental stresses that exceed the tolerances of many species or by disturbing the community so often that slow-growing or slow-colonizing species are excluded. At the other extreme, low levels of disturbance can reduce species diversity by allowing competitively dominant species to exclude less competitive ones. Meanwhile, intermediate levels of disturbance can foster greater species diversity by opening up habitats for occupation by less competitive species. Such intermediate disturbance levels rarely create conditions so severe that they exceed the environmental tolerances or recovery rates of potential community members.

The intermediate disturbance hypothesis is supported by many terrestrial and aquatic studies. In one study, ecologists in New Zealand compared the richness of invertebrate taxa living in the beds of streams exposed to different frequencies and intensities of flooding **(Figure 41.17)**. When floods occurred

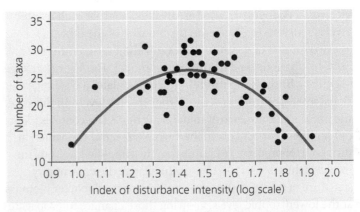

▲ Figure 41.17 Testing the intermediate disturbance hypothesis. Researchers identified the taxa (species or genera) of invertebrates at two locations in each of 27 New Zealand streams. They assessed the intensity of flooding at each location using an index of streambed disturbance. The number of invertebrate taxa peaked where the intensity of flooding was at intermediate levels.

either very frequently or rarely, invertebrate richness was low. Frequent floods made it difficult for some species to become established in the streambed, while rare floods resulted in species being displaced by superior competitors. Invertebrate richness peaked in streams that had an intermediate frequency or intensity of flooding, as predicted by the hypothesis.

Although moderate levels of disturbance appear to maximize species diversity, small and large disturbances often have important effects on community structure. Small-scale disturbances can create patches of different habitats across a landscape, which help maintain diversity in a community. Large-scale disturbances are also a natural part of many communities. Much of Yellowstone National Park, for example, is dominated by lodgepole pine, a tree species that requires the rejuvenating influence of periodic fires. Lodgepole pinecones remain closed until exposed to intense heat. When a forest fire burns the trees, the cones open and the seeds are released. The new generation of lodgepole pines can then thrive on nutrients released from the burned trees and in the sunlight that is no longer blocked by taller trees.

In the summer of 1988, extensive areas of Yellowstone burned during a severe drought. By 1989, burned areas in the park were largely covered with new vegetation, suggesting that the species in this community are adapted to rapid recovery after fire (Figure 41.18). In fact, large-scale fires have periodically swept through the lodgepole pine forests of Yellowstone and other northern areas for thousands of years.

Studies of the Yellowstone forest community and many others indicate that they are nonequilibrium communities, changing continually because of natural disturbances and the internal processes of growth and reproduction. Mounting evidence suggests that nonequilibrium conditions are in fact the norm for most communities.

Ecological Succession

Changes in the composition and structure of terrestrial communities are most apparent after some severe disturbance, such as a volcanic eruption or a glacier, strips away all the existing vegetation. The disturbed area may be colonized by a variety of species, which are gradually replaced by other species, which are in turn replaced by still other species—a process called **ecological succession**. When this process begins in a virtually lifeless area where soil has not yet formed, such as on a new volcanic island or on the rubble (moraine) left by a retreating glacier, it is called **primary succession**. In contrast, **secondary succession** occurs when an existing community has been cleared by some disturbance that leaves the soil intact, as in Yellowstone following the 1988 fires (see Figure 41.18).

During primary succession, the only life-forms initially present are often prokaryotes and protists. Lichens and mosses, which grow from windblown spores, are commonly the first macroscopic photosynthesizers to colonize such areas. Soil develops gradually as rocks weather and organic matter accumulates from the decomposed remains of the early colonizers. Once soil is present, the lichens and mosses are usually overgrown by grasses, shrubs, and trees that sprout from seeds blown in or carried in by animals. Eventually, an area is colonized by plants that become the community's dominant vegetation. Producing such a community through primary succession may take hundreds or thousands of years.

Early-arriving species and later-arriving ones are often linked by one of three processes. The early arrivals may *facilitate* the appearance of the later species by making the environment more favorable—for example, by increasing the fertility of the soil. Alternatively, the early species may *inhibit* establishment of the later species, so that successful colonization occurs in spite of, rather than because of, the activities of the early species.

(a) Soon after fire. The fire has left a patchy landscape. Note the unburned trees in the far distance.

(b) One year after fire. The community has begun to recover. Herbaceous plants, different from those in the former forest, cover the ground.

▲ **Figure 41.18 Recovery following a large-scale disturbance.** The 1988 Yellowstone National Park fires burned large areas of forests dominated by lodgepole pines.

Finally, the early species may be completely independent of the later species, which *tolerate* conditions created early in succession but are neither helped nor hindered by early species.

Ecologists have conducted some of the most extensive research on primary succession at Glacier Bay in southeastern Alaska, where glaciers have retreated more than 100 km since 1760 **(Figure 41.19)**. By studying the communities on moraines at different distances from the mouth of the bay, ecologists can examine different stages in succession. ❶ The exposed moraine is colonized first by pioneering species that include liverworts, mosses, fireweed, scattered *Dryas* (a mat-forming shrub), and willows. ❷ After about three decades, *Dryas* dominates the plant community. ❸ A few decades later, the area is invaded by alder, which forms dense thickets up to 9 m tall. ❹ In the next two centuries, these alder stands are overgrown first by Sitka spruce and later by a combination of western hemlock and mountain hemlock. In areas of poor drainage, the forest floor of this spruce-hemlock forest is invaded by sphagnum moss, which holds water and acidifies the soil, eventually killing the trees. Thus, by about 300 years after glacial retreat, the vegetation consists of sphagnum bogs on the poorly drained flat areas and spruce-hemlock forest on the well-drained slopes.

Succession on glacial moraines is related to environmental changes in soil nutrients and other environmental factors caused by transitions in the vegetation. Because the bare soil after glacial retreat is low in nitrogen, almost all the pioneer plants begin succession with poor growth and yellow leaves due to nitrogen deficiency. The exceptions are *Dryas* and alder, which have symbiotic bacteria that fix atmospheric nitrogen (see Chapter 29). Soil nitrogen increases quickly during the alder stage of succession and keeps increasing during the spruce stage. By altering soil properties, pioneer plant species can facilitate colonization by new plant species during succession.

Human Disturbance

Ecological succession is a response to disturbance of the environment, and the strongest agent of disturbance today is human activity. Agricultural development has disrupted what were once the vast grasslands of the North American prairie. Tropical rain forests are quickly disappearing as a result of clear-cutting for lumber, cattle grazing, and farmland. Centuries of overgrazing and agricultural disturbance have contributed to famine in parts of Africa by turning seasonal grasslands into vast barren areas.

Humans disturb marine ecosystems as well as terrestrial ones. The effects of ocean trawling, where boats drag weighted nets across the seafloor, are similar to those of clear-cutting a forest or plowing a field **(Figure 41.20)**. The trawls scrape and scour corals and other life on the seafloor. In a typical year, ships trawl an area about the size of South America, 150 times larger than the area of forests that are clear-cut annually.

❶ **Pioneer stage**

❷ *Dryas* **stage**

❹ **Spruce stage**

❸ **Alder stage**

1941
1907
1860
Glacier Bay
Alaska
1760

0 5 10 15
Kilometers

▲ **Figure 41.19 Glacial retreat and primary succession at Glacier Bay, Alaska.** The different shades of blue on the map show retreat of the glacier since 1760, based on historical descriptions.

▲ **Figure 41.20 Disturbance of the ocean floor by trawling.**
These photos show the seafloor off northwestern Australia before (top) and after (bottom) deep-sea trawlers have passed.

Because disturbance by human activities is often severe, it reduces species diversity in many communities. In Chapter 43, we'll take a closer look at how human-caused disturbance is affecting the diversity of life.

CONCEPT CHECK 41.3

1. Why do high and low levels of disturbance usually reduce species diversity? Why does an intermediate level of disturbance promote species diversity?

2. During succession, how might the early species facilitate the arrival of other species?

3. **WHAT IF?** Most prairies experience regular fires, typically every few years. These disturbances tend to be relatively modest. How would the species diversity of a prairie likely be affected if no burning occurred for 100 years? Explain your answer.

For suggested answers, see Appendix A.

CONCEPT 41.4

Biogeographic factors affect community diversity

So far, we have examined relatively small-scale or local factors that influence the diversity of communities, including the effects of species interactions, dominant species, and many types of disturbances. Ecologists also recognize that large-scale biogeographic factors contribute to the range of diversity observed in communities. The contributions of two biogeographic factors in particular—the latitude of a community and the area it occupies—have been investigated for more than a century.

Latitudinal Gradients

In the 1850s, both Charles Darwin and Alfred Wallace pointed out that plant and animal life was generally more abundant and diverse in the tropics than in other parts of the globe. Since

that time, many researchers have confirmed this observation. One study found that a 6.6-hectare plot (1 ha = 10,000 m^2) in tropical Malaysia contained 711 tree species, while a 2-ha plot of deciduous forest in Michigan typically contained just 10 to 15 tree species. Many groups of animals show similar latitudinal gradients.

The two key factors affecting latitudinal gradients of species richness are probably evolutionary history and climate. Over the course of evolutionary time, species richness may increase in a community as more speciation events occur (see Chapter 22). Tropical communities are generally older than temperate or polar communities, which have repeatedly "started over" after major disturbances from glaciations. Also, the growing season in tropical forests is about five times as long as in the tundra communities of high latitudes. In effect, biological time runs about five times as fast in the tropics as near the poles, so intervals between speciation events are shorter in the tropics.

Climate is a primary cause of the latitudinal gradient in richness and diversity. In terrestrial communities, the two main climatic factors correlated with diversity are sunlight and precipitation, both of which are relatively abundant in the tropics. These factors can be considered together by measuring a community's rate of **evapotranspiration**, the evaporation of water from soil and plants together. Evapotranspiration, a function of solar radiation, temperature, and water availability, is much higher in hot areas with abundant rainfall than in areas with low temperatures or low precipitation. *Potential evapotranspiration*, a measure of potential water loss that assumes that water is readily available, is determined by the amount of solar radiation and temperature and is highest in regions where both are plentiful. The species richness of animals as well as plants correlates well with both measures, as shown for vertebrates and potential evapotranspiration in **Figure 41.21**.

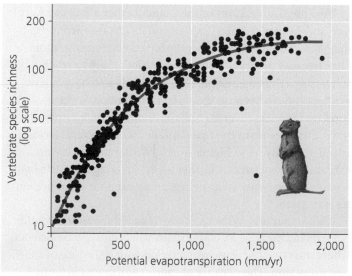

© 1991 University of Chicago Press

▲ **Figure 41.21 Energy, water, and species richness.** Vertebrate species richness in North America increases most predictably with potential evapotranspiration, expressed as rainfall equivalents (mm/yr).

Area Effects

In 1807, naturalist and explorer Alexander von Humboldt described one of the first patterns of species richness to be recognized, the **species-area curve**: All other factors being equal, the larger the geographic area of a community, the more species it has, in part because larger areas offer a greater diversity of habitats and microhabitats. The basic concept of diversity increasing with increasing area applies in many situations, from surveys of ant diversity in New Guinea to studies of plant species richness on islands of different sizes.

Because of their isolation and limited size, islands provide excellent opportunities for studying the biogeographic factors that affect the species diversity of communities. By "islands," we mean not only oceanic islands, but also habitat islands on land, such as lakes, mountain peaks, and habitat fragments—any patch surrounded by an environment not suitable for the "island" species. American ecologists Robert MacArthur and E. O. Wilson developed a general model of island biogeography, identifying the key determinants of species diversity on an island with a given set of physical characteristics.

Consider a newly formed oceanic island that receives colonizing species from a distant mainland. Two factors that determine the number of species on the island are the rate at which new species immigrate to the island and the rate at which species become extinct on the island. At any given time, an island's immigration and extinction rates are affected by the number of species already present. As the number of species on the island increases, the immigration rate of new species decreases, because any individual reaching the island is less likely to represent a species that is not already present. At the same time, as more species inhabit an island, extinction rates on the island increase because of the greater likelihood of competitive exclusion.

Two physical features of the island further affect immigration and extinction rates: its size and its distance from the mainland. Small islands generally have lower immigration rates because potential colonizers are less likely to reach a small island than a large one. Small islands also have higher extinction rates because they generally contain fewer resources, have less diverse habitats, and have smaller populations. Distance from the mainland is also important; a closer island generally has a higher immigration rate and lower extinction rate than one farther away. Arriving colonists help sustain the presence of a species on a near island and prevent its extinction.

MacArthur and Wilson's model is called the *island equilibrium model* because an equilibrium will eventually be reached where the rate of species immigration equals the rate of species extinction. The number of species at this equilibrium point is correlated with the island's size and distance from the mainland. Like any ecological equilibrium, this species equilibrium is dynamic; immigration and extinction continue, and the exact species composition may change over time.

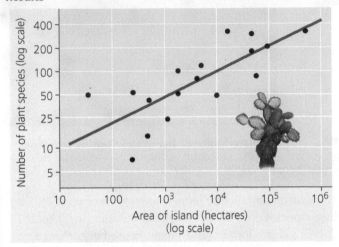

▼ **Figure 41.22** **Inquiry**

How does species richness relate to area?

Field Study Ecologists Robert MacArthur and E. O. Wilson studied the number of plant species on the Galápagos Islands in relation to the area of the different islands.

Results

Conclusion Plant species richness increases with island size, supporting the island equilibrium model.

Source R. H. MacArthur and E. O. Wilson, *The Theory of Island Biogeography*, Princeton University Press, Princeton, NJ (1967).

WHAT IF? Four islands in this study ranging in area from about 40 to 10,000 ha each contained about 50 plant species. What does such variation tell you about the simple assumptions of the island equilibrium model?

MacArthur and Wilson's studies of the diversity of plants and animals on island chains support the prediction that species richness increases with island size, in keeping with the island equilibrium model **(Figure 41.22)**. Species counts also fit the prediction that the number of species decreases with increasing remoteness of the island.

Over long periods, disturbances such as storms, adaptive evolutionary changes, and speciation generally alter the species composition and community structure on islands. Nonetheless, the island equilibrium model is widely applied in ecology. Conservation biologists in particular use it when designing habitat reserves or establishing a starting point for predicting the effects of habitat loss on species diversity.

CONCEPT CHECK 41.4

1. Describe two hypotheses that explain why species diversity is greater in tropical regions than in temperate and polar regions.
2. Describe how an island's size and distance from the mainland affect the island's species richness.
3. **WHAT IF?** Based on MacArthur and Wilson's island equilibrium model, how would you expect the richness of birds on islands to compare with the richness of snakes and lizards?

For suggested answers, see Appendix A.

Pathogens alter community structure locally and globally

Now that we have examined several important factors that structure biological communities, we will finish the chapter by examining community interactions involving **pathogens**—disease-causing organisms and viruses. Scientists have only recently come to appreciate how universal the effects of pathogens are in structuring ecological communities.

Effects on Community Structure

Pathogens produce especially clear effects on community structure when they are introduced into new habitats. Coral reef communities, for example, are increasingly susceptible to the influence of newly discovered pathogens. White-band disease, caused by an unknown pathogen, has resulted in dramatic changes in the structure and composition of Caribbean reefs. The disease kills corals by causing their tissue to slough off in a band from the base to the tip of the branches. Because of the disease, staghorn coral (*Acropora cervicornis*) has virtually disappeared from the Caribbean since the 1980s. Populations of elkhorn coral (*Acropora palmata*) have also been decimated. Such corals provide key habitat for lobsters as well as snappers and other fish species. When the corals die, they are quickly overgrown by algae. Surgeonfish and other herbivores that feed on algae come to dominate the fish community. Eventually, the corals topple because of damage from storms and other disturbances. The complex, three-dimensional structure of the reef disappears, and diversity plummets.

Pathogens also influence community structure in terrestrial ecosystems. In the forests and savannas of California, trees of several species are dying from sudden oak death (SOD). This recently discovered disease is caused by the fungus-like protist *Phytophthora ramorum* (see Chapter 25). SOD was first described in California in 1995, when hikers noticed trees dying around San Francisco Bay. By 2011, it had spread more than 1,000 km. During that time, it killed more than a million oaks and other trees from the central California coast to southern Oregon. The loss of these oaks has led to the decreased abundance of at least five bird species, including the acorn woodpecker and the oak titmouse, that rely on the oaks for food and habitat. Although there is currently no cure for SOD, scientists recently sequenced the genome of *P. ramorum* in hopes of finding a way to fight the pathogen.

Community Ecology and Zoonotic Diseases

Three-quarters of emerging human diseases and many of the most devastating diseases are caused by **zoonotic pathogens**—those that are transferred to humans from other animals, either through direct contact with an infected animal or by means of an intermediate species, called a **vector**. The vectors that spread zoonotic diseases are often parasites, including ticks, lice, and mosquitoes.

Identifying the community of hosts and vectors for a pathogen can help prevent diseases such as Lyme disease, which is spread by ticks. For years, scientists thought that the primary host for the Lyme pathogen was the white-footed mouse because mice are heavily parasitized by young ticks **(Figure 41.23)**. When researchers vaccinated mice against Lyme disease and released them into the wild, however, the number of infected ticks hardly changed. Further investigation in New York revealed that two inconspicuous shrew species were the hosts of more than half the ticks collected in the field. Identifying the dominant hosts for a pathogen provides information that may be used to control the hosts most responsible for spreading diseases.

Ecologists also use their knowledge of community interactions to track the spread of zoonotic diseases. For example, avian flu is caused by highly contagious viruses transmitted through the saliva and feces of birds (see Chapter 17). Most of these viruses affect wild birds mildly, but they often cause stronger symptoms in domesticated birds, the most common source of human infections. Since 2003, one particular viral strain, called H5N1, has killed hundreds of millions of poultry and more than 300 people. Millions more people are at risk of infection.

Control programs that quarantine domestic birds or monitor their transport may be ineffective if avian flu spreads naturally through the movements of wild birds. From 2003 to 2006, the H5N1 strain spread rapidly from southeast Asia into Europe and Africa, but by early 2012, it had not appeared in Australia or the Americas. The most likely place for infected wild birds to enter the Americas is Alaska, the entry point for ducks, geese, and shorebirds that migrate every year across the Bering Sea from Asia. Ecologists are studying the spread of the

▲ **Figure 41.23 Identifying Lyme disease host species.** A student researcher collects ticks from a white-footed mouse. Genetic analysis of the ticks from a variety of hosts enabled scientists to identify the former hosts of other ticks collected in the field that were no longer attached to a host.

▲ **Figure 41.24 Tracking avian flu.** Graduate student Travis Booms, of Boise State University, bands a young gyrfalcon as part of a project to monitor the spread of the disease.

virus by trapping and testing migrating and resident birds in Alaska **(Figure 41.24)**.

Human activities are transporting pathogens around the world at unprecedented rates. Genetic analyses suggest that *P. ramorum* likely came to North America from Europe in

nursery plants. Similarly, the pathogens that cause human diseases are spread by our global economy. H1N1, the virus that causes "swine flu" in humans, was first detected in Veracruz, Mexico, in early 2009. It quickly spread around the world when infected individuals flew on airplanes to other countries. By the time the outbreak ended in 2010, the first flu pandemic in 40 years had killed more than 17,000 people.

Community ecology provides the foundation for understanding the life cycles of pathogens and their interactions with hosts. Pathogen interactions are also greatly influenced by changes in the physical environment. To control pathogens and the diseases they cause, scientists need an ecosystem perspective—an intimate knowledge of how the pathogens interact with other species and with all aspects of their environment. Ecosystems are the subject of Chapter 42.

CONCEPT CHECK 41.5

1. What are pathogens?
2. **WHAT IF?** Rabies, a viral disease in mammals, is not currently found in the British Isles. If you were in charge of disease control there, what practical approaches might you employ to keep the rabies virus from reaching these islands?

For suggested answers, see Appendix A.

---

41 Chapter Review

SUMMARY OF KEY CONCEPTS

CONCEPT 41.1
Interactions within a community may help, harm, or have no effect on the species involved (pp. 846–851)

Interaction	Description
Competition (–/–)	Two or more species compete for a resource that is in short supply.
Predation (+/–)	One species, the predator, kills and eats the other, the prey.
Herbivory (+/–)	An herbivore eats part of a plant or alga.
Symbiosis	Individuals of two or more species live in close contact with one another. Symbiosis includes:
Parasitism (+/–)	The **parasite** derives its nourishment from a second organism, its **host**, which is harmed.
Mutualism (+/+)	Both species benefit from the interaction.
Commensalism (+/0)	One species benefits from the interaction, while the other is unaffected by it.
Facilitation (+/+ or 0/+)	A species has positive effects on other species without intimate contact.

- **Competitive exclusion** states that two species competing for the same resources cannot coexist permanently in the same place. **Resource partitioning** is the differentiation of **ecological niches** that enables species to coexist in a community.

? *Give an example of a pair of species that exhibit each interaction listed in the table above.*

CONCEPT 41.2
Diversity and trophic structure characterize biological communities (pp. 851–856)

- **Species diversity** measures the number of species in a community—its **species richness**—and their **relative abundance**. A community with similar abundances of species is more diverse than one in which one or two species are abundant and the remainder are rare.
- **Trophic structure** is a key factor in community dynamics. **Food chains** link the trophic levels in a community. Branching food chains and complex trophic interactions form **food webs**.
- **Dominant species** are the most abundant species in a community and possess high competitive abilities. **Keystone species** are usually less abundant species that exert a disproportionate influence on community structure because of their ecological niche. **Ecosystem engineers** influence community structure through their effects on the physical environment.
- The **bottom-up model** proposes a unidirectional influence from lower to higher trophic levels. **The top-down model** proposes that control of each trophic level comes from the trophic level above.

? *Based on indexes such as Shannon diversity, is a community of higher species richness always more diverse than a community of lower species richness? Explain.*

CONCEPT 41.3
Disturbance influences species diversity and composition (pp. 856–859)

- Increasing evidence suggests that **disturbance** and lack of equilibrium, rather than stability and equilibrium, are the norm for most communities. According to the **intermediate disturbance hypothesis**, moderate levels of disturbance can foster higher species diversity than can low or high levels of disturbance.

- **Ecological succession** is the sequence of community and ecosystem changes after a disturbance. **Primary succession** occurs where no soil exists when succession begins; **secondary succession** begins in an area where soil remains after a disturbance.
- Humans are the most widespread agents of disturbance, and their effects on communities often reduce species diversity.

? *Is the disturbance pictured in Figure 41.20 more likely to initiate primary or secondary succession? Explain.*

CONCEPT 41.4

Biogeographic factors affect community diversity (pp. 859–860)

- Species richness generally declines along a latitudinal gradient from the tropics to the poles. Climate influences the diversity gradient through energy (heat and light) and water.
- Species richness is directly related to a community's geographic size, a principle formalized in the **species-area curve**. The island equilibrium model maintains that species richness on an ecological island reaches an equilibrium where immigration is balanced by extinction.

? *How have periods of glaciation influenced latitudinal patterns of diversity?*

CONCEPT 41.5

Pathogens alter community structure locally and globally (pp. 861–862)

- Recent work has highlighted the role that **pathogens** play in structuring terrestrial and marine communities.
- **Zoonotic pathogens** are transferred from other animals to humans. Community ecology provides the framework for identifying key species interactions associated with such pathogens.

? *In what way can a vector of a zoonotic pathogen differ from a host of the pathogen?*

TEST YOUR UNDERSTANDING

Level 1: Knowledge/Comprehension

1. The feeding relationships among the species in a community determine the community's
   a. secondary succession.
   b. ecological niche.
   c. species richness.
   d. species-area curve.
   e. trophic structure.

2. Based on the intermediate disturbance hypothesis, a community's species diversity is increased by
   a. frequent massive disturbance.
   b. stable conditions with no disturbance.
   c. moderate levels of disturbance.
   d. human intervention to eliminate disturbance.
   e. intensive disturbance by humans.

Level 2: Application/Analysis

3. Which of the following could qualify as a top-down control on a grassland community?
   a. limitation of plant biomass by rainfall amount
   b. influence of temperature on competition among plants
   c. influence of soil nutrients on the abundance of grasses versus wildflowers
   d. effect of grazing intensity by bison on plant species diversity
   e. effect of humidity on plant growth rates

4. Community 1 contains 100 individuals distributed among four species (A, B, C, and D). Community 2 contains 100 individuals distributed among three species (A, B, and C).

   Community 1: 5A, 5B, 85C, 5D
   Community 2: 30A, 40B, 30C

   Calculate the Shannon diversity index (H) for each community. Which community is more diverse?

Level 3: Synthesis/Evaluation

5. **DRAW IT** An important species in the Chesapeake Bay estuary is the blue crab (*Callinectes sapidus*). It is an omnivore, eating eelgrass and other primary producers as well as clams. It is also a cannibal. In turn, the crabs are eaten by humans and by the endangered Kemp's Ridley sea turtle. Based on this information, draw a food web that includes the blue crab. Assuming that the top-down model holds for this system, what would happen to the abundance of eelgrass if humans stopped eating blue crabs?

6. **SCIENTIFIC INQUIRY**
   An ecologist studying plants in the desert performed the following experiment. She staked out two identical plots, each of which included a few sagebrush plants and numerous small annual wildflowers. She found the same five wildflower species in roughly equal numbers on both plots. She then enclosed one of the plots with a fence to keep out kangaroo rats, the most common grain-eaters of the area. After two years, four of the wildflower species were no longer present in the fenced plot, but one species had increased drastically. The control plot had not changed in species diversity. Using the principles of community ecology, propose a hypothesis to explain her results. What additional evidence would support your hypothesis?

7. **FOCUS ON EVOLUTION**
   Explain why adaptations of particular organisms to interspecific competition may not necessarily represent instances of character displacement. What would a researcher have to demonstrate about two competing species to make a convincing case for character displacement?

8. **FOCUS ON INFORMATION**
   In Batesian mimicry, a palatable species resembles an unpalatable one. Imagine that several individuals of a palatable, brightly colored fly species are carried by the wind to three remote islands. The first island has no predators of that species; the second has predators but no similarly colored, unpalatable species; and the third has both predators and a similarly colored, unpalatable species. In a short essay (100–150 words), predict what might happen to the coloration of the palatable species on each island over evolutionary time if coloration is a genetically controlled trait. Explain your predictions.

For selected answers, see Appendix A.

MasteringBiology®

Students Go to **MasteringBiology** for assignments, the eText, and the Study Area with practice tests, animations, and activities.

Instructors Go to **MasteringBiology** for automatically graded tutorials and questions that you can assign to your students, plus Instructor Resources.

42

Ecosystems and Energy

KEY CONCEPTS

42.1 Physical laws govern energy flow and chemical cycling in ecosystems

42.2 Energy and other limiting factors control primary production in ecosystems

42.3 Energy transfer between trophic levels is typically only 10% efficient

42.4 Biological and geochemical processes cycle nutrients and water in ecosystems

42.5 Restoration ecologists help return degraded ecosystems to a more natural state

OVERVIEW

Cool Ecosystem

Three hundred meters below Taylor Glacier, in Antarctica, an unusual community of bacteria lives on sulfur- and iron-containing ions. These organisms thrive in harsh conditions, without light or oxygen and at a temperature of −10°C, so low that the water would freeze if it weren't three times as salty as the ocean. How has this community survived, isolated from Earth's surface for at least 1.5 million years? The bacteria are chemoautotrophs, which obtain energy by oxidizing sulfur taken up from their sulfate-rich environment (see Chapter 24). They use iron as a final electron acceptor in their reactions. When the water flows from the base of the glacier and comes into contact with air, the reduced iron in the water is oxidized and turns red before the water freezes. The distinctive color gives this area of the glacier its name—Blood Falls (Figure 42.1).

Together, the bacterial community and surrounding environment make up an **ecosystem**, the sum of all the organisms living in a given area and the abiotic factors with which they interact. An ecosystem can encompass a vast area, such as a lake or forest, or a microcosm, such as the space under a fallen log or a desert spring (Figure 42.2). As with populations and communities, the boundaries of ecosystems are not always discrete. Many ecologists view the entire biosphere as a global ecosystem, a composite of all of the local ecosystems on Earth.

Regardless of an ecosystem's size, two key ecosystem processes cannot be fully described by the population or community phenomena you have studied so far: energy flow and chemical cycling. Energy enters most ecosystems as sunlight. It is converted to chemical energy by autotrophs, passed to heterotrophs in the organic compounds of food, and dissipated as heat. Chemical elements, such as carbon and nitrogen, are cycled among abiotic and biotic components of the ecosystem. Photosynthetic and chemosynthetic organisms take up these elements in inorganic form from the air, soil, and water and incorporate them into their biomass, some of which is consumed by animals. The elements are returned in inorganic form to the environment by the metabolism of plants and animals and by organisms such as bacteria and fungi that break down organic wastes and dead organisms.

▼ **Figure 42.1** Why is this Antarctic ice blood red?

▲ **Figure 42.2 A desert spring ecosystem.**

Both energy and matter are transformed in ecosystems through photosynthesis and feeding relationships. But unlike matter, energy cannot be recycled. An ecosystem must be powered by a continuous influx of energy from an external source—in most cases, the sun. Energy flows through ecosystems, whereas matter cycles within and through them.

Resources critical to human survival and welfare, ranging from the food we eat to the oxygen we breathe, are products of ecosystem processes. In this chapter, we'll explore the dynamics of energy flow and chemical cycling, emphasizing the results of ecosystem experiments. One way to study ecosystem processes is to alter environmental factors, such as temperature or the abundance of nutrients, and measure how ecosystems respond. We'll also consider some of the impacts of human activities on energy flow and chemical cycling. Finally, we'll explore the growing science of restoration ecology, which focuses on returning degraded ecosystems to a more natural state.

CONCEPT 42.1

Physical laws govern energy flow and chemical cycling in ecosystems

Cells transform energy and matter, subject to the laws of thermodynamics (see Concept 6.1). Like cell biologists, ecosystem ecologists study how energy and matter are transformed within a system and measure the amounts of both that cross the system's boundaries. By grouping the species in a community into trophic levels based on feeding relationships (see Concept 41.2), we can follow the transformations of energy in an ecosystem and map the movements of chemical elements.

Conservation of Energy

Because ecosystem ecologists study the interactions of organisms with the physical environment, many ecosystem approaches are based on laws of physics and chemistry. The first law of thermodynamics states that energy cannot be created or destroyed but only transferred or transformed. Plants and other photosynthetic organisms convert solar energy to chemical energy, but the total amount of energy does not change. The amount of energy stored in organic molecules must equal the total solar energy intercepted by the plant, minus the amounts reflected and dissipated as heat. Ecosystem ecologists often measure transfers within and across ecosystems, in part to understand how many organisms a habitat can support and how much food humans can harvest from a site.

One implication of the second law of thermodynamics, which states that every exchange of energy increases the entropy of the universe, is that energy conversions are inefficient. Some energy is always lost as heat. We can measure the efficiency of ecological energy conversions just as we measure the efficiency of light bulbs and car engines. Because the energy flowing through ecosystems is ultimately dissipated into space as heat, most ecosystems would vanish if the sun were not continuously providing energy to Earth.

Conservation of Mass

Matter, like energy, cannot be created or destroyed. This **law of conservation of mass** is as important for ecosystems as the laws of thermodynamics are. Because mass is conserved, we can determine how much of a chemical element cycles within an ecosystem or is gained or lost by that ecosystem over time.

Unlike energy, chemical elements are continually recycled within ecosystems. A carbon atom in CO_2 can be released from the soil by a decomposer, taken up by grass through photosynthesis, consumed by a bison or other grazer, and returned to the soil in the bison's waste. The measurement and analysis of chemical cycling are important aspects of ecosystem ecology.

Although most elements are not gained or lost on a global scale, they can be gained by or lost from a particular ecosystem. In a forest, mineral nutrients—the essential elements that plants obtain from soil—typically enter as dust or as solutes dissolved in rainwater or leached from rocks in the ground. Nitrogen is also supplied through the biological process of nitrogen fixation (see Figure 29.11). In terms of losses, some elements return to the atmosphere as gases, and others are carried out of the ecosystem by moving water or by wind. Like organisms, ecosystems are open systems, absorbing energy and mass and releasing heat and waste products.

In nature, most gains and losses to ecosystems are small compared to the amounts recycled within them. Still, the balance between inputs and outputs determines whether an ecosystem is a source or a sink for a given element. If a mineral nutrient's outputs exceed its inputs, it will eventually limit production in that system. Human activities often change the balance of inputs and outputs considerably, as we'll see later in this chapter.

Energy, Mass, and Trophic Levels

Ecologists group species into trophic levels based on their main source of nutrition and energy (see Concept 41.2). The trophic level that ultimately supports all others consists of autotrophs, also called the **primary producers** of the ecosystem. Most autotrophs are photosynthetic organisms that use light energy to synthesize sugars and other organic compounds, which they then use as fuel for cellular respiration and as building material for growth. Plants, algae, and photosynthetic prokaryotes are the most common autotrophs, although chemosynthetic prokaryotes are the primary producers in deep-sea hydrothermal vents (see Figure 40.10) and places deep underground or beneath ice (see Figure 42.1).

Organisms in trophic levels above the primary producers are heterotrophs, which depend directly or indirectly on the primary producers for their source of energy. Herbivores, which eat plants and other primary producers, are **primary consumers**. Carnivores that eat herbivores are **secondary consumers**, and carnivores that eat other carnivores are **tertiary consumers**.

Another group of heterotrophs is the **detritivores**, or **decomposers**, terms we use synonymously in this text to refer to consumers that get their energy from detritus. **Detritus** is nonliving organic material, such as the remains of dead organisms, feces, fallen leaves, and wood. Many detritivores are in turn eaten by secondary and tertiary consumers. Two important groups of detritivores are prokaryotes and fungi **(Figure 42.3)**. These organisms secrete enzymes that digest organic material; they then absorb the breakdown products, linking the consumers and primary producers in an ecosystem. In a forest, for instance, birds eat earthworms that have been feeding on leaf litter and its associated prokaryotes and fungi.

Detritivores also play a critical role in recycling chemical elements back to primary producers. Detritivores convert organic matter from all trophic levels to inorganic compounds usable by primary producers, closing the loop of an ecosystem's chemical cycling. Producers can then recycle these elements into organic compounds. If decomposition stopped, life would cease as detritus piled up and the supply of ingredients needed

▲ **Figure 42.3 Fungi decomposing a dead tree.**

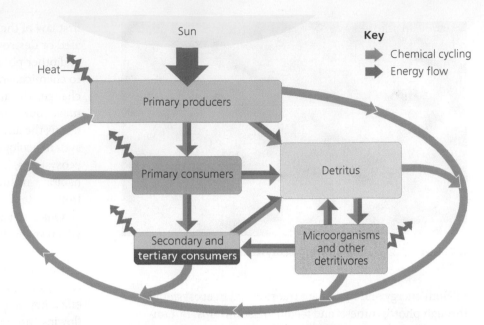

▲ **Figure 42.4 An overview of energy and nutrient dynamics in an ecosystem.** Energy enters, flows through, and exits an ecosystem, whereas chemical nutrients cycle primarily within it. In this generalized scheme, energy (dark orange arrows) enters from the sun as radiation, moves as chemical energy transfers through the food web, and exits as heat radiated into space. Most transfers of nutrients (blue arrows) through the trophic levels lead eventually to detritus; the nutrients then cycle back to the primary producers.

to synthesize new organic matter was exhausted. **Figure 42.4** summarizes the trophic relationships in an ecosystem.

CONCEPT CHECK 42.1

1. Why is the transfer of energy in an ecosystem referred to as energy flow, not energy cycling?
2. **WHAT IF?** You are studying nitrogen cycling on the Serengeti Plain in Africa. During your experiment, a herd of migrating wildebeests grazes through your study plot. What would you need to know to measure their effect on nitrogen balance in the plot?
3. **MAKE CONNECTIONS** How does the second law of thermodynamics explain why an ecosystem's energy supply must be continually replenished? (See Concept 6.1 to review the laws of thermodynamics.)

For suggested answers, see Appendix A.

CONCEPT 42.2

Energy and other limiting factors control primary production in ecosystems

The theme of energy transfer underlies all biological interactions (see Chapter 1). In most ecosystems, the amount of light energy converted to chemical energy—in the form of organic compounds—by autotrophs during a given time period is the ecosystem's **primary production**. These photosynthetic products are the starting point for most studies of ecosystem metabolism and energy flow. In ecosystems where the primary producers are chemoautotrophs, as described in the Overview, the initial energy input is chemical, and the initial products are the organic compounds synthesized by the microorganisms.

Ecosystem Energy Budgets

Since most primary producers use light energy to synthesize energy-rich organic molecules, consumers acquire their organic fuels secondhand (or even third- or fourthhand) through food webs (see Figure 41.14). Therefore, the total amount of photosynthetic production sets the spending limit for the entire ecosystem's energy budget.

The Global Energy Budget

Each day, Earth's atmosphere is bombarded by about 10^{22} joules of solar radiation (1 J = 0.239 cal). This is enough energy to supply the demands of the entire human population for approximately 20 years at 2010 energy consumption levels. The intensity of the solar energy striking Earth varies with latitude, with the tropics receiving the greatest input (see Figure 40.3). Most incoming solar radiation is absorbed, scattered, or reflected by clouds and dust in the atmosphere. The amount of solar radiation that ultimately reaches Earth's surface limits the possible photosynthetic output of ecosystems.

Only a small fraction of the sunlight that reaches Earth's surface is actually used in photosynthesis. Much of the radiation strikes materials that don't photosynthesize, such as ice and soil. Of the radiation that does reach photosynthetic organisms, only certain wavelengths are absorbed by photosynthetic pigments (see Figure 8.9); the rest is transmitted, reflected, or lost as heat. As a result, only about 1% of the visible light that strikes photosynthetic organisms is converted to chemical energy. Nevertheless, Earth's primary producers create about 150 billion metric tons (1.50×10^{14} kg) of organic material each year.

Gross and Net Production

Total primary production in an ecosystem is known as that ecosystem's **gross primary production (GPP)**—the amount of energy from light (or chemicals, in chemoautotrophic systems) converted to the chemical energy of organic molecules per unit time. Not all of this production is stored as organic material in the primary producers because they use some of the molecules as fuel in their own cellular respiration. **Net primary production (NPP)** is equal to gross primary production minus the energy used by the primary producers for their "autotrophic respiration" (R_a):

$$NPP = GPP - R_a$$

On average, NPP is about one-half of GPP. To ecologists, NPP is the key measurement because it represents the storage of chemical energy that will be available to consumers in the ecosystem.

Net primary production can be expressed as energy per unit area per unit time (J/m²·yr) or as biomass (mass of vegetation) added per unit area per unit time (g/m²·yr). (Note that biomass is usually expressed in terms of the dry mass of organic material.) An ecosystem's NPP should not be confused with the total biomass of photosynthetic autotrophs present, a measure called the *standing crop*. Net primary production is the amount of *new* biomass added in a given period of time. Although a forest has a large standing crop, its NPP may actually be less than that of some grasslands; grasslands do not accumulate as much biomass as forests because animals consume the plants rapidly and because grasses and herbs decompose more quickly than trees do.

Satellites provide a powerful tool for studying global patterns of primary production **(Figure 42.5)**. Images produced from satellite data show that different ecosystems vary considerably in their NPP. Tropical rain forests are among the most productive terrestrial ecosystems and contribute a large portion of the planet's NPP. Estuaries and coral reefs also have

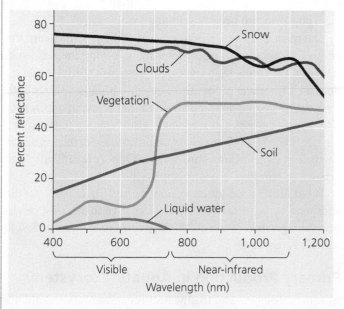

▼ Figure 42.5   Research Method

Determining Primary Production with Satellites

Application Because chlorophyll captures visible light (see Figure 8.9), photosynthetic organisms absorb more light at visible wavelengths (about 380–750 nm) than at near-infrared wavelengths (750–1,100 nm). Scientists use this difference in absorption to estimate the rate of photosynthesis in different regions of the globe using satellites.

Technique Most satellites determine what they "see" by comparing the ratios of wavelengths reflected back to them. Vegetation reflects much more near-infrared radiation than visible radiation, producing a reflectance pattern very different from that of snow, clouds, soil, and liquid water.

Results Scientists use the satellite data to help produce maps of primary production like the one in Figure 42.6.

► **Figure 42.6 Global net primary production.** This map is based on data collected by satellites, such as amount of sunlight absorbed by vegetation. Note that tropical land areas have the highest rates of production (yellow and red on the map).

Net primary production (kg carbon/m^2· yr)

? *Does this global map accurately illustrate the importance of some highly productive habitats, such as wetlands, coral reefs, and coastal zones? Explain.*

very high NPP, but their contribution to the global total is small because these ecosystems cover only about one-tenth the area covered by tropical rain forests. In contrast, while the open oceans are relatively unproductive **(Figure 42.6)**, their vast size means that together they contribute about as much global NPP as terrestrial systems do.

Whereas NPP can be stated as the amount of new biomass added in a given period of time, **net ecosystem production (NEP)** is a measure of the *total biomass accumulation* during that time. Net ecosystem production is defined as gross primary production minus the total respiration of all organisms in the system (R_T)—not just primary producers, as for the calculation of NPP, but decomposers and other heterotrophs as well:

$$NEP = GPP - R_T$$

NEP is useful to ecologists because its value determines whether an ecosystem is gaining or losing carbon over time. A forest may have a positive NPP but still lose carbon if heterotrophs release it as CO_2 more quickly than primary producers incorporate it into organic compounds.

The most common way to estimate NEP is to measure the net flux (flow) of CO_2 or O_2 entering or leaving the ecosystem. If more CO_2 enters than leaves, the system is storing carbon. Because O_2 release is directly coupled to photosynthesis and respiration (see Figure 7.2), a system that is giving off O_2 is also storing carbon. On land, ecologists typically measure only the net flux of CO_2 from ecosystems because detecting small changes in O_2 in a large atmospheric O_2 pool is difficult. In the oceans, researchers use both approaches.

What limits production in ecosystems? To ask this question another way, what factors could we change to increase production for a given ecosystem? We'll address this question first for aquatic ecosystems.

Primary Production in Aquatic Ecosystems

In aquatic (marine and freshwater) ecosystems, both light and nutrients are important in controlling primary production.

Light Limitation

Because solar radiation drives photosynthesis, you would expect light to be a key variable in controlling primary production in oceans. Indeed, the depth of light penetration affects primary production throughout the photic zone of an ocean or lake (see Figure 40.11). About half of the solar radiation is absorbed in the first 15 m of water. Even in "clear" water, only 5–10% of the radiation may reach a depth of 75 m.

If light were the main variable limiting primary production in the ocean, we would expect production to increase along a gradient from the poles toward the equator, which receives the greatest intensity of light. However, you can see in Figure 42.6 that there is no such gradient. Another factor must strongly influence primary production in the ocean.

Nutrient Limitation

More than light, nutrients limit primary production in most oceans and lakes. A **limiting nutrient** is the element that must be added for production to increase. The nutrient most often limiting marine production is either nitrogen or phosphorus. Concentrations of these nutrients are typically low in the photic zone because they are rapidly taken up by phytoplankton and because detritus tends to sink.

As detailed in **Figure 42.7**, nutrient enrichment experiments confirmed that nitrogen was limiting phytoplankton growth off the south shore of Long Island, New York. One practical application of this work is in preventing algal "blooms" caused by excess nitrogen runoff that fertilizes the phytoplankton. Prior to this research, phosphate contamination was thought to cause many such blooms in the ocean, but eliminating phosphates alone may not help unless nitrogen pollution is also controlled.

The macronutrients nitrogen and phosphorus are not the only nutrients that limit aquatic production. Several large areas of the ocean have low phytoplankton densities despite relatively high nitrogen concentrations. The Sargasso Sea,

Which nutrient limits phytoplankton production along the coast of Long Island?

Experiment Pollution from duck farms concentrated near Moriches Bay adds both nitrogen and phosphorus to the coastal water off Long Island, New York. To determine which nutrient limits phytoplankton growth in this area, John Ryther and William Dunstan, of the Woods Hole Oceanographic Institution, cultured the phytoplankton *Nannochloris atomus* with water collected from several sites, identified as A–G. They added either ammonium (NH_4^+) or phosphate (PO_4^{3-}) to some of the cultures.

Results The addition of ammonium caused heavy phytoplankton growth in the cultures, but the addition of phosphate did not.

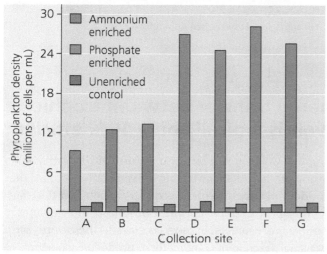

© 1971 AAAS

Conclusion Since adding phosphorus, which was already in rich supply, did not increase *Nannochloris* growth, whereas adding nitrogen increased phytoplankton density dramatically, the researchers concluded that nitrogen is the nutrient that limits phytoplankton growth in this ecosystem.

Source J. H. Ryther and W. M. Dunstan, Nitrogen, phosphorus, and eutrophication in the coastal marine environment, *Science* 171:1008–1013 (1971).

WHAT IF? How would you expect the results of this experiment to change if new duck farms substantially increased the amount of pollution in the water? Explain your reasoning.

Table 42.1 Nutrient Enrichment Experiment for Sargasso Sea Samples

Nutrients Added to Experimental Culture	Relative Uptake of ^{14}C by Cultures*
None (controls)	1.00
Nitrogen (N) + phosphorus (P) only	1.10
N + P + metals (excluding iron)	1.08
N + P + metals (including iron)	12.90
N + P + iron	12.00

*^{14}C uptake by cultures measures primary production.

Source D. W. Menzel and J. H. Ryther, Nutrients limiting the production of phytoplankton in the Sargasso Sea, with special reference to iron, *Deep Sea Research* 7:276–281 (1961).

upwelling stimulates growth of the phytoplankton that form the base of marine food webs, upwelling areas typically host highly productive, diverse ecosystems and are prime fishing locations. The largest areas of upwelling occur in the Southern Ocean (also called the Antarctic Ocean), along the equator, and in the coastal waters off Peru, California, and parts of western Africa.

In freshwater lakes, nutrient limitation is also common. During the 1970s, scientists showed that sewage and fertilizer runoff from farms and lawns adds large amounts of nutrients to lakes. Cyanobacteria and algae grow rapidly in response to these added nutrients. When the primary producers die, detritivores can reduce or even use up the available oxygen in the water through decomposition, also reducing the clarity of the water. The ecological impacts of this process, known as **eutrophication** (from the Greek *eutrophos*, well nourished), include the loss of many fish species from the lakes (see Figure 40.10).

Controlling eutrophication requires knowing which polluting nutrient is responsible. While nitrogen rarely limits primary production in lakes, a series of whole-lake experiments showed that phosphorus availability limits cyanobacterial growth. This and other ecological research led to the use of phosphate-free detergents and other important water quality reforms.

Primary Production in Terrestrial Ecosystems

At regional and global scales, temperature and moisture are the main factors controlling primary production in terrestrial ecosystems. Tropical rain forests, with their warm, wet conditions that promote plant growth, are the most productive of all terrestrial ecosystems (see Figure 42.6). In contrast, low-productivity systems are generally hot and dry, like many deserts, or cold and dry, like arctic tundra. Between these extremes lie the temperate forest and grassland ecosystems, which have moderate climates and intermediate productivity.

The climate variables of moisture and temperature are very useful for predicting NPP in terrestrial ecosystems. Primary production is greater in wetter ecosystems, as shown for the

a subtropical region of the Atlantic Ocean, has some of the clearest water in the world because of its low phytoplankton density. Nutrient enrichment experiments have revealed that the availability of the micronutrient iron limits primary production there **(Table 42.1)**. Windblown dust from land supplies most of the iron to the oceans but is relatively scarce in this and certain other regions compared to the oceans as a whole.

Areas of *upwelling*, where deep, nutrient-rich waters circulate to the ocean surface, have exceptionally high primary production. This fact supports the hypothesis that nutrient availability determines marine primary production. Because

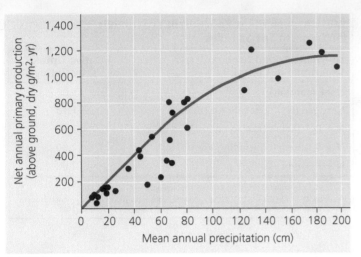

© 1970 Pearson Education, Inc.

▲ **Figure 42.8 A global relationship between net primary production and mean annual precipitation for terrestrial ecosystems.**

plot of NPP and annual precipitation in **Figure 42.8**. Along with mean annual precipitation, a second useful predictor is *actual evapotranspiration*, the total amount of water transpired by plants and evaporated from a landscape. Evapotranspiration increases with the temperature and amount of solar energy available to drive evaporation and transpiration.

Nutrient Limitations and Adaptations That Reduce Them

EVOLUTION Mineral nutrients in the soil also limit primary production in terrestrial ecosystems. As in aquatic systems, nitrogen and phosphorus are the nutrients that most commonly limit terrestrial production. Globally, nitrogen limits plant growth most. Phosphorus limitations are common in older soils where phosphate molecules have been leached away by water, such as in many tropical ecosystems. Phosphorus availability is also often low in the soils of deserts and other ecosystems with a basic pH, where some phosphorus precipitates and becomes unavailable to plants. Adding a nonlimiting nutrient, even one that is scarce, will not stimulate production. Conversely, adding more of the limiting nutrient will increase production until some other nutrient becomes limiting.

Various adaptations have evolved in plants that can increase their uptake of limiting nutrients. One important adaptation is the symbiosis between plant roots and nitrogen-fixing bacteria. Another is the mycorrhizal association between plant roots and fungi that supply phosphorus and other limiting elements to plants (see Concept 29.4). Plants also have root hairs and other anatomical features that increase their area of contact with the soil (see Chapter 28). Many plants release enzymes and other substances into the soil that increase the availability of limiting nutrients; such enzymes include phosphatases, which cleave a phosphate group from larger molecules and make it more soluble in the soil.

Studies relating nutrients to terrestrial primary production have practical applications in agriculture. Farmers maximize their crop yields by using fertilizers with the right balance of nutrients for the local soil and type of crop. This knowledge of limiting nutrients helps us feed billions of people today.

CONCEPT CHECK 42.2

1. Why is only a small portion of the solar energy that strikes Earth's atmosphere stored by primary producers?
2. How can ecologists experimentally determine the factor that limits primary production in an ecosystem?
3. **MAKE CONNECTIONS** Explain how nitrogen and phosphorus, the nutrients that most often limit primary production, are necessary for the Calvin cycle to function in photosynthesis (see Concept 8.3).

For suggested answers, see Appendix A.

CONCEPT 42.3

Energy transfer between trophic levels is typically only 10% efficient

The amount of chemical energy in consumers' food that is converted to new biomass during a given period is called the **secondary production** of the ecosystem. Consider the transfer of organic matter from primary producers to herbivores, the primary consumers. In most ecosystems, herbivores eat only a small fraction of plant material produced; globally, they consume only about one-sixth of total plant production. Moreover, they cannot digest all the plant material that they *do* eat, as anyone who has walked through a dairy farm will attest. Most of an ecosystem's production is eventually consumed by detritivores. Let's analyze the process of energy transfer and cycling more closely.

Production Efficiency

First we'll examine secondary production in one organism—a caterpillar. When a caterpillar feeds on a leaf, only about 33 J out of 200 J, or one-sixth of the potential energy in the leaf, is used for secondary production, or growth **(Figure 42.9)**. The caterpillar stores some of the remaining energy in organic compounds that will be used for cellular respiration and passes the rest in its feces. Most of the energy in feces is eventually lost as heat after the feces are consumed by detritivores. The energy used for the caterpillar's respiration is also lost from the ecosystem as heat. This is why energy is said to flow through, not cycle within, ecosystems. Only the chemical energy stored by herbivores as biomass, through growth or the production of offspring, is available as food to secondary consumers.

We can measure the efficiency of animals as energy transformers using the following equation:

$$\text{Production efficiency} = \frac{\text{Net secondary production} \times 100\%}{\text{Assimilation of primary production}}$$

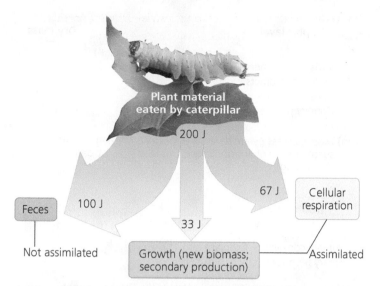

▲ Figure 42.9 Energy partitioning within a link of the food chain. Less than 17% of the caterpillar's food is actually used for secondary production (growth).

Net secondary production is the energy stored in biomass represented by growth and reproduction. Assimilation consists of the total energy taken in, not including losses in feces, used for growth, reproduction, and respiration. **Production efficiency**, therefore, is the percentage of energy stored in assimilated food that is *not* used for respiration. For the caterpillar in Figure 42.9, production efficiency is 33%; 67 J of the 100 J of assimilated energy is used for respiration. (The 100 J of energy lost as undigested material in feces does not count toward assimilation.) Birds and mammals typically have low production efficiencies, in the range of 1–3%, because they use so much energy in maintaining a constant, high body temperature. Insects and microorganisms are much more efficient, with production efficiencies averaging 40% or more.

Trophic Efficiency and Ecological Pyramids

Let's scale up now from the production efficiencies of individual consumers to the flow of energy through trophic levels.

Trophic efficiency is the percentage of production transferred from one trophic level to the next. Trophic efficiencies must always be less than production efficiencies because they take into account not only the energy lost through respiration and contained in feces, but also the energy in organic material in a lower trophic level that is not consumed by the next trophic level. Trophic efficiencies are generally only about 10%. In other words, 90% of the energy available at one trophic level typically is *not* transferred to the next. This loss is multiplied over the length of a food chain. For example, if 10% of available energy is transferred from primary producers to primary consumers, such as caterpillars, and 10% of that energy is transferred to secondary consumers, called carnivores, then only 1% of net primary production is available to secondary consumers (10% of 10%). In the **Scientific Skills Exercise**, you

Scientific Skills Exercise

Interpreting Quantitative Data in a Table

How Efficient Is Energy Transfer in a Salt Marsh Ecosystem? In a classic experiment, John Teal studied the flow of energy through the producers, consumers, and detritivores in a salt marsh. In this exercise, you will use the data from this study to calculate some measures of energy transfer between trophic levels in this ecosystem.

How the Study Was Done Teal measured the amount of solar radiation entering a salt marsh in Georgia over a year. He also measured the aboveground biomass of the dominant primary producers, which were grasses, as well as the biomass of the dominant consumers, including insects, spiders, and crabs, and of the detritus that flowed out of the marsh to the surrounding coastal waters. To determine the amount of energy in each unit of biomass, he dried the biomass, burned it in a calorimeter, and measured the amount of heat produced.

Data from the Study

Form of Energy	$kcal/m^2 \cdot yr$
Solar radiation	600,000
Gross grass production	34,580
Net grass production	6,585
Gross insect production	305
Net insect production	81
Detritus leaving marsh	3,671

Interpret the Data

1. What proportion of the solar energy that reaches the marsh is incorporated into gross primary production? Into net primary production? (A proportion is the same as a percentage divided by 100. Both measures are useful for comparing relative efficiencies across different ecosystems.)

2. How much energy is lost by primary producers as respiration in this ecosystem? How much is lost as respiration by the insect population?

3. If all of the detritus leaving the marsh is plant material, what proportion of all net primary production leaves the marsh as detritus each year?

Data from J. M. Teal, Energy flow in the salt marsh ecosystem of Georgia, *Ecology* 43:614–624 (1962).

(MB) A version of this Scientific Skills Exercise can be assigned in MasteringBiology.

can calculate trophic efficiency and other measures of energy flow in a salt marsh ecosystem.

The progressive loss of energy along a food chain severely limits the abundance of top-level carnivores that an ecosystem can support. Only about 0.1% of the chemical energy fixed by photosynthesis can flow all the way through a food web to a tertiary consumer, such as a snake or a shark. This explains why most food webs include only about four or five trophic levels (see Chapter 41).

The loss of energy with each transfer in a food chain can be represented by a *pyramid of net production*, in which the

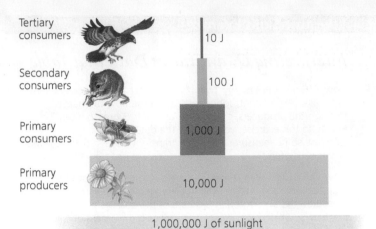

Tertiary consumers — 10 J

Secondary consumers — 100 J

Primary consumers — 1,000 J

Primary producers — 10,000 J

1,000,000 J of sunlight

▲ **Figure 42.10 An idealized pyramid of net production.** This example assumes a trophic efficiency of 10% for each link in the food chain. Notice that primary producers convert only about 1% of the energy available to them to net primary production.

trophic levels are arranged in tiers **(Figure 42.10)**. The width of each tier is proportional to the net production, expressed in joules, of each trophic level. The highest level, which represents top-level predators, contains relatively few individuals. The small population size typical of top predator species is one reason they tend to be vulnerable to extinction (as well as to the evolutionary consequences of small population size, discussed in Concept 21.3).

One important ecological consequence of low trophic efficiencies is represented in a *biomass pyramid*, in which each tier represents the standing crop (the total dry mass of all organisms) in one trophic level. Most biomass pyramids narrow sharply from primary producers at the base to top-level carnivores at the apex because energy transfers between trophic levels are so inefficient **(Figure 42.11a)**. Certain aquatic ecosystems, however, have inverted biomass pyramids: Primary consumers outweigh the producers **(Figure 42.11b)**. Such inverted biomass pyramids occur because the producers—phytoplankton—grow, reproduce, and are consumed so quickly by the zooplankton that they never develop a large population size, or standing crop. In other words, the phytoplankton have a short **turnover time**, which means they have a small standing crop compared to their production:

$$\text{Turnover time} = \frac{\text{Standing crop (g/m}^2)}{\text{Production (g/m}^2 \cdot \text{day)}}$$

Because the phytoplankton continually replace their biomass at such a rapid rate, they can support a biomass of zooplankton bigger than their own biomass. Nevertheless, because phytoplankton have much higher production than zooplankton, the pyramid of *production* for this ecosystem is still bottom-heavy, like the one in Figure 42.10.

The dynamics of energy flow through ecosystems have important implications for human consumers. Eating meat is a relatively inefficient way of tapping photosynthetic production. The same pound of soybeans that a person could eat for

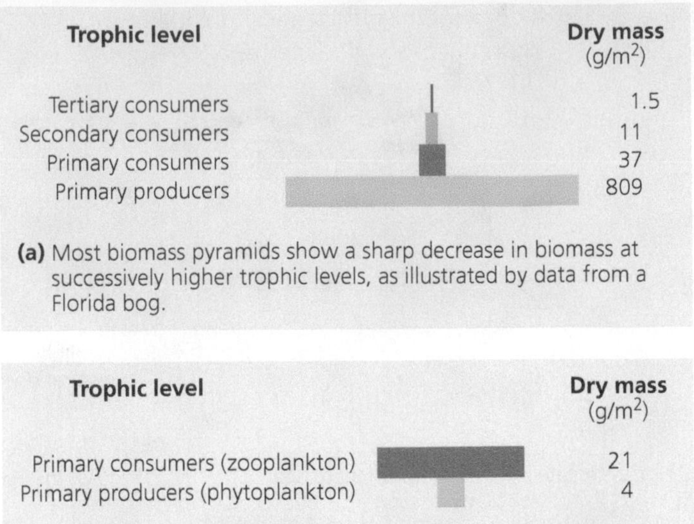

Trophic level	Dry mass (g/m^2)
Tertiary consumers	1.5
Secondary consumers	11
Primary consumers	37
Primary producers	809

(a) Most biomass pyramids show a sharp decrease in biomass at successively higher trophic levels, as illustrated by data from a Florida bog.

Trophic level	Dry mass (g/m^2)
Primary consumers (zooplankton)	21
Primary producers (phytoplankton)	4

(b) In some aquatic ecosystems, such as the English Channel, a small standing crop of primary producers (phytoplankton) supports a larger standing crop of primary consumers (zooplankton).

▲ **Figure 42.11 Pyramids of biomass (standing crop).** Numbers denote the dry mass of all organisms at each trophic level.

protein produces only a fifth of a pound of beef or less when fed to a cow. Worldwide agriculture could, in fact, feed many more people and require less land if we all fed more efficiently—as primary consumers, eating plant material.

In the next section, we'll look at how the transfer of nutrients and energy through food webs is part of a larger picture of chemical cycling in ecosystems.

CONCEPT CHECK 42.3

1. If an insect that eats plant seeds containing 100 J of energy uses 30 J of that energy for respiration and excretes 50 J in its feces, what is the insect's net secondary production? What is its production efficiency?

2. Tobacco leaves contain nicotine, a poisonous compound that is energetically expensive for the plant to make. What advantage might the plant gain by using some of its resources to produce nicotine?

3. **WHAT IF?** Detritivores are consumers that obtain their energy from detritus. How many joules of energy are potentially available to detritivores in the ecosystem represented in Figure 42.10?

For suggested answers, see Appendix A.

CONCEPT 42.4

Biological and geochemical processes cycle nutrients and water in ecosystems

Although most ecosystems receive abundant solar energy, chemical elements are available only in limited amounts. Life therefore depends on the recycling of essential chemical elements. Much of an organism's chemical stock is replaced continuously as nutrients are assimilated and waste products are released. When the organism dies, the atoms in its body are returned to the atmosphere, water, or soil by decomposers. Decomposition replenishes the pools of inorganic nutrients that plants and other autotrophs use to build new organic matter.

Decomposition and Nutrient Cycling Rates

Decomposition is controlled by the same factors that limit primary production in aquatic and terrestrial ecosystems (see Concept 42.2). These factors include temperature, moisture, and nutrient availability. Decomposers usually grow faster and decompose material more quickly in warmer ecosystems **(Figure 42.12)**. In tropical rain forests, most organic material decomposes in a few months to a few years, while in temperate forests, decomposition takes four to six years, on average. The difference is largely the result of the higher temperatures and more abundant precipitation in tropical rain forests.

Because decomposition in a tropical rain forest is rapid, relatively little organic material accumulates as leaf litter on the forest floor; about 75% of the nutrients in the ecosystem is present in the woody trunks of trees, and only about 10% is contained in the soil. Thus, the relatively low concentrations of some nutrients in the soil of tropical rain forests result from a short cycling time, not from a lack of these elements in the ecosystem. In temperate forests, where decomposition is much slower, the soil may contain as much as 50% of all the organic material in the ecosystem. The nutrients that are present in temperate forest detritus and soil may remain there for long periods before plants assimilate them.

Decomposition on land is also slower when conditions are either too dry for decomposers to thrive or too wet to supply them with enough oxygen. Ecosystems that are both cold and wet, such as peatlands, store large amounts of organic matter. Decomposers grow poorly there, and net primary production greatly exceeds decomposition.

In aquatic ecosystems, decomposition in anaerobic muds can take 50 years or longer. Bottom sediments are comparable to the detritus layer in terrestrial ecosystems; however, algae and aquatic plants usually assimilate nutrients directly from the water. Thus, the sediments often constitute a nutrient sink, and aquatic ecosystems are very productive only when there is exchange between the bottom layers of water and the water at the surface.

Biogeochemical Cycles

Because nutrient cycles involve both biotic and abiotic components, they are called **biogeochemical cycles**. For convenience, we can recognize two general categories of biogeochemical cycles: global and local. Gaseous forms of carbon, oxygen, sulfur, and nitrogen occur in the atmosphere, and

▼ **Figure 42.12** **Inquiry**

How does temperature affect litter decomposition in an ecosystem?

Experiment Researchers with the Canadian Forest Service placed identical samples of organic material—litter—on the ground in 21 sites across Canada (marked by letters on the map below). Three years later, they returned to see how much of each sample had decomposed.

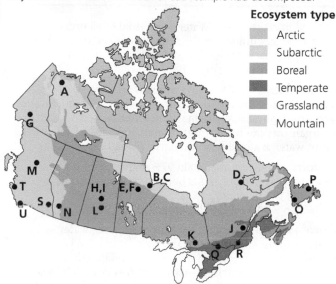

Results The mass of litter decreased four times faster in the warmest ecosystem than in the coldest ecosystem.

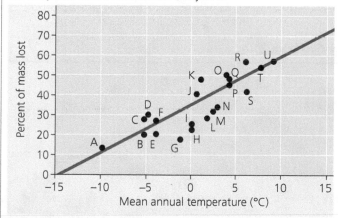

Conclusion Decomposition rate increases with temperature across much of Canada.

Source T. R. Moore et al., Litter decomposition rates in Canadian forests, *Global Change Biology* 5:75–82 (1999).

WHAT IF? What factors other than temperature might also have varied across these 21 sites? How might this variation have affected the interpretation of the results?

cycles of these elements are essentially global. Other elements, including phosphorus, potassium, and calcium, are too heavy to occur as gases at Earth's surface. They cycle locally in terrestrial ecosystems and more broadly in aquatic ecosystems.

Figure 42.13 provides a detailed look at the cycling of water, carbon, nitrogen, and phosphorus. When you study each

Examine each cycle closely, considering the major reservoirs of water, carbon, nitrogen, and phosphorus and the processes that drive each cycle. The widths of the arrows in the diagrams approximately reflect the relative contribution of each process to the movement of water or a nutrient in the biosphere.

The Water Cycle

Biological importance   Water is essential to all organisms, and its availability influences the rates of ecosystem processes, particularly primary production and decomposition in terrestrial ecosystems.

Forms available to life   All organisms are capable of exchanging water directly with their environment. Liquid water is the primary physical phase in which water is used, though some organisms can harvest water vapor. Freezing of soil water can limit water availability to terrestrial plants.

Reservoirs   The oceans contain 97% of the water in the biosphere. Approximately 2% is bound in glaciers and polar ice caps, and the remaining 1% is in lakes, rivers, and groundwater, with a negligible amount in the atmosphere.

Key processes   The main processes driving the water cycle are evaporation of liquid water by solar energy, condensation of water vapor into clouds, and precipitation. Transpiration by terrestrial plants also moves large volumes of water into the atmosphere. Surface and groundwater flow can return water to the oceans, completing the water cycle.

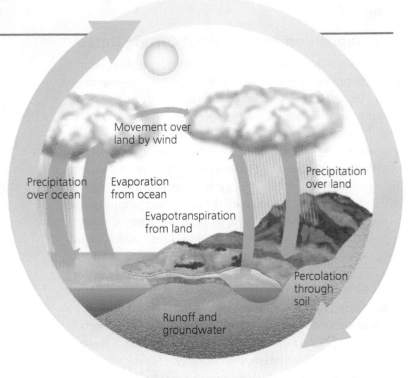

The Carbon Cycle

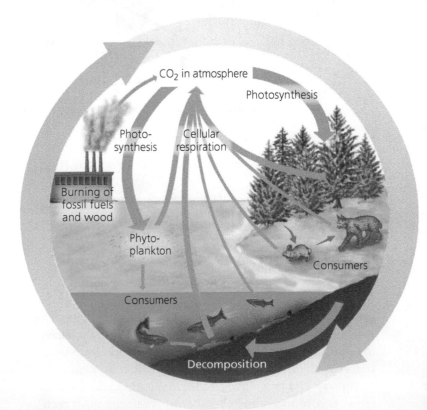

Biological importance   Carbon forms the framework of the organic molecules essential to all organisms.

Forms available to life   Photosynthetic organisms utilize CO_2 during photosynthesis and convert the carbon to organic forms that are used by consumers, including animals, fungi, and heterotrophic protists and prokaryotes.

Reservoirs   The major reservoirs of carbon include fossil fuels, soils, the sediments of aquatic ecosystems, the oceans (dissolved carbon compounds), plant and animal biomass, and the atmosphere (CO_2). The largest reservoir is sedimentary rocks such as limestone; however, this pool turns over very slowly. All organisms are capable of returning carbon directly to their environment in its original form (CO_2) through respiration.

Key processes   Photosynthesis by plants and phytoplankton removes substantial amounts of atmospheric CO_2 each year. This quantity is approximately equaled by CO_2 added to the atmosphere through cellular respiration by producers and consumers. The burning of fossil fuels and wood is adding significant amounts of additional CO_2 to the atmosphere. Over geologic time, volcanoes are also a substantial source of CO_2.

ANIMATION   ***BioFlix***   Visit the Study Area in **MasteringBiology** for the BioFlix® 3-D Animation on The Carbon Cycle.

The Nitrogen Cycle

Biological importance Nitrogen is part of amino acids, proteins, and nucleic acids and is often a limiting plant nutrient.

Forms available to life Plants can assimilate (use) two inorganic forms of nitrogen—ammonium (NH_4^+) and nitrate (NO_3^-)—and some organic forms, such as amino acids. Various bacteria can use all of these forms as well as nitrite (NO_2^-). Animals can use only organic forms of nitrogen.

Reservoirs The main reservoir of nitrogen is the atmosphere, which is 80% free nitrogen gas (N_2). The other reservoirs of inorganic and organic nitrogen compounds are soils and the sediments of lakes, rivers, and oceans; surface water and groundwater; and the biomass of living organisms.

Key processes The major pathway for nitrogen to enter an ecosystem is via *nitrogen fixation*, the conversion of N_2 to forms that can be used to synthesize organic nitrogen compounds. Certain bacteria, as well as lightning and volcanic activity, fix nitrogen naturally. Nitrogen inputs from human activities now outpace natural inputs on land. Two major contributors are industrially produced fertilizers and legume crops that fix nitrogen via bacteria in their root nodules. Other bacteria in soil convert nitrogen to different forms. Some bacteria carry out denitrification, the reduction of nitrate to nitrogen gases. Human activities also release large quantities of reactive nitrogen gases, such as nitrogen oxides, to the atmosphere.

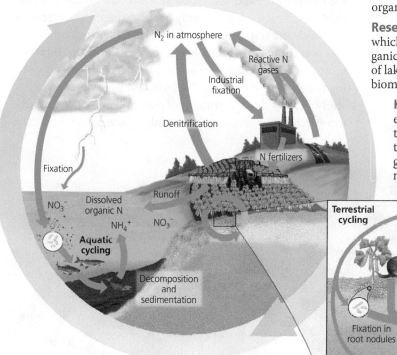

The Phosphorus Cycle

Biological importance Organisms require phosphorus as a major constituent of nucleic acids, phospholipids, and ATP and other energy-storing molecules and as a mineral constituent of bones and teeth.

Forms available to life The most biologically important inorganic form of phosphorus is phosphate (PO_4^{3-}), which plants absorb and use in the synthesis of organic compounds.

Reservoirs The largest accumulations of phosphorus are in sedimentary rocks of marine origin. There are also large quantities of phosphorus in soil, in the oceans (in dissolved form), and in organisms. Because soil particles bind PO_4^{3-}, the recycling of phosphorus tends to be quite localized in ecosystems.

Key processes Weathering of rocks gradually adds PO_4^{3-} to soil; some leaches into groundwater and surface water and may eventually reach the sea. Phosphate taken up by producers and incorporated into biological molecules may be eaten by consumers. Phosphate is returned to soil or water by either decomposition of biomass or excretion by consumers. Because there are no significant phosphorus-containing gases, only relatively small amounts of phosphorus move through the atmosphere, usually in the forms of dust and sea spray.

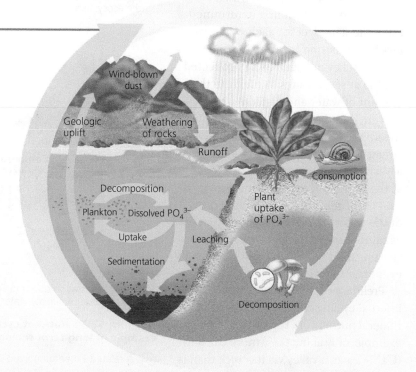

cycle, consider which steps are driven primarily by biological processes. For the carbon cycle, for instance, plants, animals, and other organisms control most of the key steps, including photosynthesis and decomposition. For the water cycle, however, purely physical processes control many key steps, such as evaporation from the oceans.

How have ecologists worked out the details of chemical cycling in various ecosystems? One common method is to follow the movement of naturally occurring, nonradioactive isotopes through the biotic and abiotic components of an ecosystem. Another method involves adding tiny amounts of radioactive isotopes of specific elements and tracing their progress. Scientists have also been able to make use of the radioactive carbon (^{14}C) released into the atmosphere during atom bomb testing in the 1950s and early 1960s. Scientists use this "spike" of ^{14}C to trace where and how quickly carbon flows into ecosystem components, including plants, soils, and ocean water.

Case Study: Nutrient Cycling in the Hubbard Brook Experimental Forest

Since 1963, ecologists Herbert Bormann, Gene Likens, and their colleagues have been studying nutrient cycling at the Hubbard Brook Experimental Forest in the White Mountains of New Hampshire. Their research site is a deciduous forest that grows in six small valleys, each drained by a single creek. Impenetrable bedrock underlies the soil of the forest.

The research team first determined the mineral budget for each of six valleys by measuring the input and outflow of several key nutrients. They collected rainfall at several sites to measure the amount of water and dissolved minerals added to the ecosystem. To monitor the loss of water and minerals, they constructed a small concrete dam with a V-shaped spillway across the creek at the bottom of each valley (**Figure 42.14a**). They found that about 60% of the water added to the ecosystem as rainfall and snow exits through the stream, and the remaining 40% is lost by evapotranspiration.

Preliminary studies confirmed that internal cycling conserved most of the mineral nutrients in the system. For example, only about 0.3% more calcium (Ca^{2+}) leaves a valley via its creek than is added by rainwater, and this small net loss is probably replaced by chemical decomposition of the bedrock. During most years, the forest even registers small net gains of a few mineral nutrients, including nitrogen.

Experimental deforestation of a watershed dramatically increased the flow of water and minerals leaving the watershed (**Figure 42.14b**). Over three years, water runoff from the newly deforested watershed was 30–40% greater than in a control watershed, apparently because there were no plants to absorb and transpire water from the soil. Most remarkable was the loss of nitrate, whose concentration in the creek increased 60-fold, reaching levels considered unsafe for drinking water (**Figure 42.14c**). The Hubbard Brook deforestation study showed that the amount of nutrients leaving an intact forest ecosystem is controlled mainly by the plants. Retaining

(a) Concrete dams and weirs built across streams at the bottom of watersheds enabled researchers to monitor the outflow of water and nutrients from the ecosystem.

(b) One watershed was clear-cut to study the effects of the loss of vegetation on drainage and nutrient cycling. All of the original plant material was left in place to decompose.

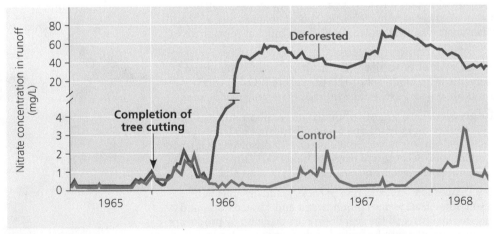

(c) The concentration of nitrate in runoff from the deforested watershed was 60 times greater than in a control (unlogged) watershed.

▲ **Figure 42.14 Nutrient cycling in the Hubbard Brook Experimental Forest: an example of long-term ecological research.**

(MB) A related Experimental Inquiry Tutorial can be assigned in MasteringBiology.

nutrients in ecosystems helps to maintain the productivity of the systems and, in some cases, to avoid problems caused by excess nutrient runoff (see Figure 42.7).

CONCEPT CHECK 42.4

1. **DRAW IT** For each of the four biogeochemical cycles detailed in Figure 42.12, draw a simple diagram that shows one possible path for an atom of that chemical from abiotic to biotic reservoirs and back.
2. Why does deforestation of a watershed increase the concentration of nitrates in streams draining the watershed?
3. **WHAT IF?** Why is nutrient availability in a tropical rain forest particularly vulnerable to logging?

For suggested answers, see Appendix A.

CONCEPT 42.5

Restoration ecologists help return degraded ecosystems to a more natural state

Ecosystems can recover naturally from most disturbances (including the experimental deforestation at Hubbard Brook) through the stages of ecological succession (see Concept 41.3). Sometimes that recovery takes centuries, though, particularly when human activities have degraded the environment. Tropical areas that are cleared for farming may quickly become unproductive because of nutrient losses. Mining activities may last for several decades, and the lands are often abandoned in a degraded state. Ecosystems can also be damaged by salts that build up in soils from irrigation and by toxic chemicals or oil spills. Biologists increasingly are called on to help restore and repair damaged ecosystems.

One of the basic assumptions of restoration ecology is that environmental damage is at least partly reversible. This optimistic view must be balanced by a second assumption—that ecosystems are not infinitely resilient. Restoration ecologists therefore work to identify and manipulate the processes that most limit recovery of ecosystems from disturbances. Where disturbance is so severe that restoring all of a habitat is impractical, ecologists try to reclaim as much of a habitat or ecological process as possible, within the limits of the time and money available to them.

In extreme cases, the physical structure of an ecosystem may need to be restored before biological restoration can occur. If a stream was straightened to channel water quickly through a suburb, restoration ecologists may reconstruct a meandering channel to slow down the flow of water eroding the stream bank. To restore an open-pit mine, engineers may first grade the site with heavy equipment to reestablish a gentle slope, spreading topsoil when the slope is in place **(Figure 42.15)**.

Once physical reconstruction of the ecosystem is complete—or when it is not needed—biological restoration is the next step. Two key strategies in biological restoration are bioremediation and biological augmentation.

Bioremediation

Using organisms—usually prokaryotes, fungi, or plants—to detoxify polluted ecosystems is known as **bioremediation**. Some plants and lichens adapted to soils containing heavy metals can accumulate high concentrations of toxic metals such as zinc, lead, and cadmium in their tissues. Restoration ecologists can introduce such species to sites polluted by mining and other human activities and then harvest these organisms to remove the metals from the ecosystem. For instance, researchers in the United Kingdom have discovered a lichen species that grows on soil polluted with uranium dust left over from mining. The lichen concentrates uranium in a dark pigment, making it useful as a biological monitor and potentially as a remediator.

(a) In 1991, before restoration

(b) In 2000, near the completion of restoration

▲ **Figure 42.15 A gravel and clay mine site in New Jersey before and after restoration.**

Ecologists already use the abilities of many prokaryotes to carry out bioremediation of soils and water. Scientists have sequenced the genomes of at least ten prokaryotic species specifically for their bioremediation potential. One of the species, the bacterium *Shewanella oneidensis*, appears particularly promising. It can metabolize a dozen or more elements under aerobic and anaerobic conditions. In doing so, it converts soluble forms of uranium, chromium, and nitrogen to insoluble forms that are less likely to leach into streams or groundwater. Researchers at Oak Ridge National Laboratory, in Tennessee, stimulated the growth of *Shewanella* and other uranium-reducing bacteria by adding ethanol to groundwater contaminated with uranium; the bacteria can use ethanol as an energy source. In just five months, the concentration of soluble uranium in the ecosystem dropped by 80% **(Figure 42.16)**.

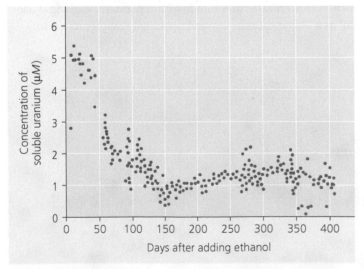

(a) Wastes containing uranium were dumped in these four unlined pits for more than 30 years, contaminating soils and groundwater.

(b) After ethanol was added, microbial activity decreased the concentration of soluble uranium in groundwater near the pits.

▲ **Figure 42.16 Bioremediation of groundwater contaminated with uranium at Oak Ridge National Laboratory, Tennessee.**

Biological Augmentation

In contrast to bioremediation, which is a strategy for removing harmful substances from an ecosystem, **biological augmentation** uses organisms to *add* essential materials to a degraded ecosystem. To augment ecosystem processes, restoration ecologists need to determine which factors, such as chemical nutrients, have been lost from a system and are limiting its recovery.

Encouraging the growth of plants that thrive in nutrient-poor soils often speeds up succession and ecosystem recovery. In alpine ecosystems of the western United States, nitrogen-fixing plants such as lupines are often planted to raise nitrogen concentrations in soils disturbed by mining and other activities. Once these nitrogen-fixing plants become established, other native species are better able to obtain enough soil nitrogen to survive. In other systems where the soil has been severely disturbed or where topsoil is missing entirely, plant roots may lack the mycorrhizal symbionts that help them meet their nutritional needs (see Chapter 26). Ecologists restoring a tallgrass prairie in Minnesota recognized this limitation and enhanced the recovery of native species by adding mycorrhizal symbionts to the soil they seeded.

Restoring the physical structure and plant community of an ecosystem does not necessarily ensure that animal species will recolonize a site and persist there. Because animals provide critical ecosystem services, including pollination and seed dispersal, restoration ecologists sometimes help wildlife reach and use restored ecosystems. They might release animals at a site or establish habitat corridors that connect a restored site to other places where the animals are found. They sometimes establish artificial perches for birds or dig burrows for other animals to use. Such efforts can improve the biodiversity of restored ecosystems and help the community persist.

Restoration Projects Worldwide

Because restoration ecology is a relatively new discipline and because ecosystems are complex, many restoration ecologists advocate adaptive management: experimenting with several promising types of management to learn what works best.

The long-term objective of restoration is to return an ecosystem as much as possible to its predisturbance state. **Figure 42.17** explores four ambitious and successful restoration projects. The great number of such projects around the world and the dedication of the people engaged in them suggest that restoration ecology will continue to grow as a discipline for many years.

CONCEPT CHECK 42.5

1. Identify the main goal of restoration ecology.
2. How do bioremediation and biological augmentation differ?
3. **WHAT IF?** In what way is the Kissimmee River project a more complete ecological restoration than the Maungatautari project (see Figure 42.17)?

For suggested answers, see Appendix A.

The examples highlighted on this page are just a few of the many restoration ecology projects taking place around the world.

▶ Kissimmee River, Florida

The Kissimmee River was converted from a meandering river to a 90-km canal, threatening many fish and wetland bird populations. Kissimmee River restoration has filled 12 km of drainage canal and reestablished 24 km of the original 167 km of natural river channel. Pictured here is a section of the Kissimmee canal that has been plugged (wide, light strip on the right side of the photo), diverting flow into remnant river channels (center of the photo). The project will also restore natural flow patterns, which will foster self-sustaining populations of wetland birds and fishes.

◀ Succulent Karoo, South Africa

In this desert region of southern Africa, as in many arid regions, overgrazing by livestock has damaged vast areas. Private landowners and government agencies in South Africa are restoring large areas of this unique region, revegetating the land and employing more sustainable resource management. The photo shows a small sample of the exceptional plant diversity of the Succulent Karoo; its 5,000 plant species include the highest diversity of succulent plants in the world.

▶ Maungatautari, New Zealand

Weasels, rats, pigs, and other introduced species pose a serious threat to New Zealand's native plants and animals, including kiwis, a group of flightless, ground-dwelling bird species. The goal of the Maungatautari restoration project is to exclude all exotic mammals from a 3,400-ha reserve located on a forested volcanic cone. A specialized fence around the reserve eliminates the need to continue setting traps and using poisons that can harm native wildlife. In 2006, a pair of critically endangered takahe (a species of flightless rail) were released into the reserve in hopes of reestablishing a breeding population of this colorful bird on New Zealand's North Island.

◀ Coastal Japan

Seaweed and seagrass beds are important nursery grounds for a wide variety of fishes and shellfish. Once extensive but now reduced by development, these beds are being restored in the coastal areas of Japan. Techniques include constructing suitable seafloor habitat, transplanting from natural beds using artificial substrates, and hand seeding (shown in this photograph).

SUMMARY OF KEY CONCEPTS

CONCEPT 42.1

Physical laws govern energy flow and chemical cycling in ecosystems (pp. 865–866)

- An **ecosystem** consists of all the organisms in a community and the abiotic factors with which they interact. The laws of physics and chemistry apply to ecosystems, particularly for the conservation of energy. Energy is conserved but degraded to heat during ecosystem processes.
- Based on the **law of conservation of mass**, ecologists study how much of a chemical element enters and leaves an ecosystem and cycles within it. Inputs and outputs are generally small compared to recycled amounts, but their balance determines whether the ecosystem gains or loses an element over time.

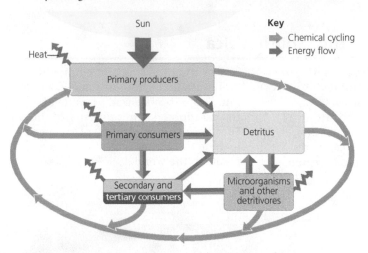

? *Based on the second law of thermodynamics, would you expect the typical biomass of primary producers in an ecosystem to be greater than or less than the biomass of secondary producers in the same ecosystem? Explain your reasoning.*

CONCEPT 42.2

Energy and other limiting factors control primary production in ecosystems (pp. 866–870)

- **Primary production** sets the spending limit for the global energy budget. **Gross primary production** is the total energy assimilated by an ecosystem in a given period. **Net primary production**, the energy accumulated in autotroph biomass, equals gross primary production minus the energy used by the primary producers for respiration. **Net ecosystem production** is the total biomass accumulation of an ecosystem, defined as the difference between gross primary production and total ecosystem respiration.
- In aquatic ecosystems, light and nutrients limit primary production. In terrestrial ecosystems, climatic factors such as temperature and moisture affect primary production on a large geographic scale, but a soil nutrient is often the limiting factor in primary production locally.

? *What additional variable do you need to know the value of in order to estimate NEP from NPP? Why might measuring this variable be difficult, for instance, in a sample of ocean water?*

CONCEPT 42.3

Energy transfer between trophic levels is typically only 10% efficient (pp. 870–872)

- The amount of energy available to each trophic level is determined by the net primary production and the **production efficiency**, the efficiency with which food energy is converted to biomass at each link in the food chain.
- The percentage of energy transferred from one trophic level to the next, called **trophic efficiency**, is typically 10%. Pyramids of net production and biomass reflect low trophic efficiency.

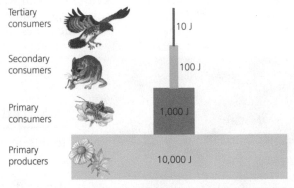

? *Why would runners have a lower production efficiency when running a long-distance race than when they are sedentary?*

CONCEPT 42.4

Biological and geochemical processes cycle nutrients and water in ecosystems (pp. 872–877)

- Water moves in a global cycle driven by solar energy. The carbon cycle primarily reflects the reciprocal processes of photosynthesis and cellular respiration.
- The proportion of a nutrient in a particular form and its cycling in that form vary among ecosystems, largely because of differences in the rate of decomposition.
- Nutrient cycling is strongly regulated by vegetation. The Hubbard Brook case study showed that logging increases water runoff and can cause large losses of minerals.

? *If decomposers usually grow faster and decompose material more quickly in warmer ecosystems, why is decomposition in hot deserts so slow?*

CONCEPT 42.5

Restoration ecologists help return degraded ecosystems to a more natural state (pp. 877–879)

- Restoration ecologists harness organisms to detoxify polluted ecosystems through the process of **bioremediation**.
- In **biological augmentation**, ecologists use organisms to add essential materials to ecosystems.

? *In preparing a site for surface mining and later restoration, what would be the advantage of removing the shallow topsoil first and setting it aside separately from the deeper soil, rather than removing all soil at once and mixing it in a single pile?*

TEST YOUR UNDERSTANDING

Level 1: Knowledge/Comprehension

1. Which of the following organisms is *incorrectly* paired with its trophic level?
   a. cyanobacterium—primary producer
   b. grasshopper—primary consumer
   c. zooplankton—primary producer
   d. eagle—tertiary consumer
   e. fungus—detritivore

2. Which of these ecosystems has the *lowest* net primary production per square meter?
   a. a salt marsh
   b. an open ocean
   c. a coral reef
   d. a grassland
   e. a tropical rain forest

3. The discipline that applies ecological principles to returning degraded ecosystems to a more natural state is known as
   a. population viability analysis.
   b. landscape ecology.
   c. conservation ecology.
   d. restoration ecology.
   e. resource conservation.

Level 2: Application/Analysis

4. Nitrifying bacteria participate in the nitrogen cycle mainly by
   a. converting nitrogen gas to ammonia.
   b. releasing ammonium from organic compounds, thus returning it to the soil.
   c. converting ammonia to nitrogen gas, which returns to the atmosphere.
   d. converting ammonium to nitrate, which plants absorb.
   e. incorporating nitrogen into amino acids and organic compounds.

5. Which of the following has the greatest effect on the rate of chemical cycling in an ecosystem?
   a. the ecosystem's rate of primary production
   b. the production efficiency of the ecosystem's consumers
   c. the rate of decomposition in the ecosystem
   d. the trophic efficiency of the ecosystem
   e. the location of the nutrient reservoirs in the ecosystem

6. The Hubbard Brook watershed deforestation experiment yielded all of the following results *except*:
   a. Most minerals were recycled within a forest ecosystem.
   b. The flow of minerals out of a natural watershed was offset by minerals flowing in.
   c. Deforestation increased water runoff.
   d. The nitrate concentration in waters draining the deforested area became dangerously high.
   e. Calcium levels remained high in the soil of deforested areas.

7. Which of the following would be considered an example of bioremediation?
   a. adding nitrogen-fixing microorganisms to a degraded ecosystem to increase nitrogen availability
   b. using a bulldozer to regrade a strip mine
   c. dredging a river bottom to remove contaminated sediments
   d. reconfiguring the channel of a river
   e. adding seeds of a chromium-accumulating plant to soil contaminated by chromium

8. If you applied a fungicide to a cornfield, what would you expect to happen to the rate of decomposition and net ecosystem production (NEP)?
   a. Both decomposition rate and NEP would decrease.
   b. Both decomposition rate and NEP would increase.
   c. Neither would change.
   d. Decomposition rate would increase and NEP would decrease.
   e. Decomposition rate would decrease and NEP would increase.

Level 3: Synthesis/Evaluation

9. **DRAW IT** Draw a simplified global water cycle showing ocean, land, atmosphere, and runoff from the land to the ocean. Add these annual water fluxes to your drawing: ocean evaporation, 425 km^3; ocean evaporation that returns to the ocean as precipitation, 385 km^3; ocean evaporation that falls as precipitation on land, 40 km^3; evapotranspiration from plants and soil that falls as precipitation on land, 70 km^3; runoff to the oceans, 40 km^3. Based on these global numbers, how much precipitation falls on land in a typical year?

10. **SCIENTIFIC INQUIRY**
   Using two neighboring ponds in a forest as your study site, design a controlled experiment to measure the effect of falling leaves on net primary production in a pond.

11. **FOCUS ON EVOLUTION**
   Some biologists have suggested that ecosystems are emergent, "living" systems capable of evolving. One manifestation of this idea is environmentalist James Lovelock's Gaia hypothesis, which views Earth itself as a living, homeostatic entity—a kind of superorganism. If ecosystems are capable of evolving, would this be a form of Darwinian evolution? Why or why not?

12. **FOCUS ON ENERGY AND MATTER**
   As described in Concept 42.4, decomposition typically occurs quickly in moist tropical forests. However, waterlogging in the soil of some moist tropical forests results over time in a buildup of organic matter called "peat." In a short essay (100–150 words), discuss the relationship of net primary production, net ecosystem production, and decomposition for such an ecosystem. Are NPP and NEP likely to be positive? What do you think would happen to NEP if a landowner drained the water from a tropical peatland, exposing the organic matter to air?

For selected answers, see Appendix A.

MasteringBiology®

Students Go to **MasteringBiology** for assignments, the eText, and the Study Area with practice tests, animations, and activities.

Instructors Go to **MasteringBiology** for automatically graded tutorials and questions that you can assign to your students, plus Instructor Resources.

43
Global Ecology and Conservation Biology

KEY CONCEPTS

43.1 Human activities threaten Earth's biodiversity

43.2 Population conservation focuses on population size, genetic diversity, and critical habitat

43.3 Landscape and regional conservation help sustain biodiversity

43.4 Earth is changing rapidly as a result of human actions

43.5 The human population is no longer growing exponentially but is still increasing rapidly

43.6 Sustainable development can improve human lives while conserving biodiversity

OVERVIEW

Psychedelic Treasure

Scurrying across a rocky outcrop, a lizard stops abruptly in a patch of sunlight. A conservation biologist senses the motion and turns to find a gecko splashed with rainbow colors, its bright orange legs and tail blending into a striking blue body, its head splotched with yellow and green. The psychedelic rock gecko (*Cnemaspis psychedelica*) was discovered in 2010 during an expedition to the Greater Mekong region of southeast Asia **(Figure 43.1)**. Its known habitat is restricted to Hon Khoai, an island occupying just 8 km² (3 square miles) in southern Vietnam. Other new species found during the same series of expeditions include the Elvis monkey, which sports a hairdo like that of a certain legendary musician. Between 2000 and 2010, biologists identified more than a thousand new species in the Greater Mekong region alone.

To date, scientists have described and named about 1.8 million species of organisms. Some biologists think that about 10 million more species currently exist; others estimate the number to be as high as 100 million. The greatest concentrations of species are found in the tropics. Unfortunately, tropical forests are being cleared at an alarming rate to support a burgeoning human population. In Vietnam, rates of deforestation are among the very highest in the world **(Figure 43.2)**. What will become of the psychedelic rock gecko and other newly discovered species if such activities continue unchecked?

Throughout the biosphere, human activities are altering trophic structures, energy flow, chemical cycling, and natural disturbance—ecosystem processes on which we and all other species depend (see Chapter 42). We have physically altered nearly half of Earth's land surface, and we use over half of all accessible surface fresh water. In the oceans, stocks of most major fisheries are shrinking because of overharvesting. By some estimates, we may be pushing more species toward extinction than the large asteroid that triggered the mass extinctions at the close of the Cretaceous period 65.5 million years ago (see Figure 23.10).

In this chapter, we apply a global perspective to the changes happening across Earth, focusing on a discipline that seeks to preserve life: **Conservation biology** integrates ecology, evolutionary biology, molecular biology, genetics,

▼ **Figure 43.1** What will be the fate of this newly described lizard species?

▲ **Figure 43.2 Tropical deforestation in Vietnam.**

and physiology to conserve biological diversity at all levels. Efforts to sustain ecosystem processes and stem the loss of biodiversity also connect the life sciences with the social sciences, economics, and humanities.

We'll begin by taking a closer look at the biodiversity crisis and examining some of the conservation strategies being adopted to slow the rate of species loss. We'll also examine how human activities are altering the environment through climate change and other global processes, and we'll investigate the link between these alterations and the growing human population. Finally, we'll consider how decisions about long-term conservation priorities could affect life on Earth.

CONCEPT 43.1

Human activities threaten Earth's biodiversity

Extinction is a natural phenomenon that has been occurring since life first evolved; it is the high *rate* of extinction that is responsible for today's biodiversity crisis (see Chapter 23). Because we can only estimate the number of species currently existing, we cannot determine the exact rate of species loss. However, we do know that human activities threaten Earth's biodiversity at all levels.

Three Levels of Biodiversity

Biodiversity—short for biological diversity—can be considered at three main levels: genetic diversity, species diversity, and ecosystem diversity **(Figure 43.3)**.

Genetic Diversity

Genetic diversity comprises not only the individual genetic variation *within* a population, but also the genetic variation *between* populations that is often associated with adaptations to local conditions (see Chapter 21). If one population becomes extinct, then a species may have lost some of the

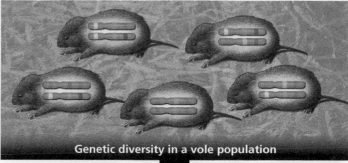

Genetic diversity in a vole population

Species diversity in a coastal redwood ecosystem

Community and ecosystem diversity across the landscape of an entire region

▲ **Figure 43.3 Three levels of biodiversity.** The oversized chromosomes in the top diagram symbolize the genetic variation within the population.

genetic diversity that makes microevolution possible. This erosion of genetic diversity in turn reduces the adaptive potential of the species.

Species Diversity

Public awareness of the biodiversity crisis centers on species diversity—the variety of species in an ecosystem or across the biosphere (see Chapter 41). As more species are lost to extinction, species diversity decreases. The U.S. Endangered Species Act defines an **endangered species** as one that is "in danger of extinction throughout all or a significant portion of its range." **Threatened species** are those considered likely to become endangered in the near future. The following are just a few statistics that illustrate the problem of species loss:

• According to the International Union for Conservation of Nature and Natural Resources (IUCN), 12% of the 10,000 known species of birds and 21% of the 5,500 known species of mammals are threatened.

Philippine eagle

Yangtze River dolphin

▲ **Figure 43.4 A hundred heartbeats from extinction.** These are two members of what E. O. Wilson calls the Hundred Heartbeat Club, species with fewer than 100 individuals remaining on Earth. The Yangtze River dolphin was even thought to be extinct, but a few individuals were reportedly sighted in 2007.

? *To document that a species has actually become extinct, what factors would you need to consider?*

- A survey by the Center for Plant Conservation showed that of the nearly 20,000 known plant species in the United States, 200 have become extinct since such records have been kept, and 730 are endangered or threatened.
- In North America, at least 123 freshwater animal species have become extinct since 1900, and hundreds more species are threatened. The extinction rate for North American freshwater fauna is about five times as high as that for terrestrial animals.

Extinction of species may also be local; for example, a species may be lost in one river system but survive in an adjacent one. Global extinction of a species means that it is lost from *all* the ecosystems in which it lived, leaving them permanently impoverished **(Figure 43.4)**.

Ecosystem Diversity

The variety of the biosphere's ecosystems is a third level of biological diversity. Because of the many interactions between populations of different species in an ecosystem, the local extinction of one species can have a negative impact on other species in the ecosystem (see Figure 41.15). For instance, bats called "flying foxes" are important pollinators and seed dispersers in the Pacific Islands, where they are increasingly hunted as a luxury food **(Figure 43.5)**. Conservation biologists fear that the extinction of flying foxes would also harm the native plants of the Samoan Islands, where four-fifths of the tree species depend on flying foxes for pollination or seed dispersal.

▲ **Figure 43.5 The endangered Marianas "flying fox" bat (*Pteropus mariannus*), an important pollinator.**

Some ecosystems have already been heavily affected by humans, and others are being altered at a rapid pace. Since European colonization, more than half of the wetlands in the contiguous United States have been drained and converted to agricultural and other uses. In California, Arizona, and New Mexico, roughly 90% of native riparian (streamside) communities have been affected by overgrazing, flood control, water diversions, lowering of water tables, and invasion by non-native plants.

Biodiversity and Human Welfare

Why should we care about the loss of biodiversity? One reason is what Harvard biologist E. O. Wilson calls *biophilia*, our sense of connection to nature and all life. The belief that other species are entitled to life is a pervasive theme of many religions and the basis of a moral argument that we should protect biodiversity. There is also a concern for future human generations. Paraphrasing an old proverb, G. H. Brundtland, a former prime minister of Norway, said: "We must consider our planet to be on loan from our children, rather than being a gift from our ancestors." In addition to such philosophical and moral justifications, species and genetic diversity bring us many practical benefits.

Benefits of Species and Genetic Diversity

Many species that are threatened could potentially provide medicines, food, and fibers for human use, making biodiversity a crucial natural resource. Products from aspirin to antibiotics were originally derived from natural sources. In food production, if we lose wild populations of plants closely related to agricultural species, we lose genetic resources that could be used to improve crop qualities, such as disease resistance. For instance, plant breeders responded to devastating outbreaks of the grassy stunt virus in rice (*Oryza sativa*) by screening 7,000 populations of this species and its close relatives for resistance to the virus. One population of a single relative, Indian rice (*Oryza nivara*), was found to be resistant to the virus, and scientists succeeded in breeding the resistance trait into commercial rice varieties. Today, the original disease-resistant population has apparently become extinct in the wild.

In the United States, about 25% of the prescriptions dispensed from pharmacies contain substances originally derived from plants. In the 1970s, researchers discovered that the rosy periwinkle (*Catharanthus roseus*), which grows on the island of Madagascar, off the coast of Africa, contains alkaloids that inhibit

Rosy periwinkle

cancer cell growth. This discovery led to treatments for two deadly forms of cancer, Hodgkin's lymphoma and childhood leukemia, resulting in remission in most cases.

Each loss of a species means the loss of unique genes, some of which may code for enormously useful proteins. The enzyme Taq polymerase was first extracted from a bacterium, *Thermus aquaticus*, found in hot springs at Yellowstone National Park. This enzyme is essential for the polymerase chain reaction (PCR) because it is stable at the high temperatures required for automated PCR (see Figure 13.25). However, because millions of species may become extinct before we discover them, we stand to lose the valuable genetic potential held in their unique libraries of genes.

Ecosystem Services

The benefits that individual species provide to humans are substantial, but saving individual species is only part of the reason for preserving ecosystems. We humans evolved in Earth's ecosystems, and we rely on these systems and their inhabitants for our survival. **Ecosystem services** encompass all the processes through which natural ecosystems help sustain human life. Ecosystems purify our air and water. They detoxify and decompose our wastes and reduce the impacts of extreme weather and flooding. The organisms in ecosystems pollinate our crops, control pests, and create and preserve our soils. Moreover, these diverse services are provided for free.

Perhaps because we don't attach a monetary value to the services of natural ecosystems, we generally undervalue them. In 1997, ecologist Robert Costanza and his colleagues estimated the value of Earth's ecosystem services at $33 trillion per year, nearly twice the gross national product of all the countries on Earth at the time ($18 trillion). It may be more realistic to do the accounting on a smaller scale. In 1996, New York City invested more than $1 billion to buy land and restore habitat in the Catskill Mountains, the source of much of the city's fresh water. This investment was spurred by increasing pollution of the water by sewage, pesticides, and fertilizers. By harnessing

ecosystem services to purify its water naturally, the city saved $8 billion it would have otherwise spent to build a new water treatment plant and $300 million a year to run the plant.

There is growing evidence that the functioning of ecosystems, and hence their capacity to perform services, is linked to biodiversity. As human activities reduce biodiversity, we are reducing the capacity of the planet's ecosystems to perform processes critical to our own survival.

Threats to Biodiversity

Many different human activities threaten biodiversity on local, regional, and global scales. The threats posed by these activities are of four major types: habitat loss, introduced species, overharvesting, and global change.

Habitat Loss

Human alteration of habitat is the single greatest threat to biodiversity throughout the biosphere. Habitat loss has been brought about by agriculture, urban development, forestry, mining, and pollution. As discussed later in this chapter, global climate change is already altering habitats today and will have an even larger effect later this century. When no alternative habitat is available or a species is unable to move, habitat loss may mean extinction. The IUCN implicates destruction of physical habitat for 73% of the species that have become extinct, endangered, vulnerable, or rare in the last few hundred years.

Habitat loss and fragmentation may occur over large regions. Approximately 98% of the tropical dry forests of Central America and Mexico have been cut down. The clearing of tropical rain forest in the state of Veracruz, Mexico, mostly for cattle ranching, has resulted in the loss of more than 90% of the original forest, leaving relatively small, isolated patches of forest. Other natural habitats have also been fragmented by human activities **(Figure 43.6)**.

▲ **Figure 43.6 Habitat fragmentation in the foothills of Los Angeles.** Development in the valleys may confine the organisms that inhabit the narrow strips of hillside.

In almost all cases, habitat fragmentation leads to species loss because the smaller populations in habitat fragments have a higher probability of local extinction. Prairie covered about 800,000 hectares (ha) of southern Wisconsin when Europeans first arrived in North America but occupies less than 800 ha today; most of the original prairie in this area is now used to grow crops. Plant diversity surveys of 54 Wisconsin prairie remnants conducted in 1948–1954 and repeated in 1987–1988 showed that the remnants lost between 8% and 60% of their plant species in the time between the two surveys.

Habitat loss is also a major threat to aquatic biodiversity. About 70% of coral reefs, among Earth's most species-rich aquatic communities, have been damaged by human activities. At the current rate of destruction, 40–50% of the reefs, home to one-third of marine fish species, could disappear in the next 30 to 40 years. Freshwater habitats are also being lost, often as a result of the dams, reservoirs, channel modification, and flow regulation now affecting most of the world's rivers. For example, the more than 30 dams and locks built along the Mobile River basin in the southeastern United States changed river depth and flow. While providing the benefits of hydroelectric power and increased ship traffic, these dams and locks also helped drive more than 40 species of mussels and snails to extinction.

Introduced Species

Introduced species, also called exotic species, are those that humans move intentionally or accidentally from the species' native locations to new geographic regions. Human travel by ship and airplane has accelerated the transplant of species. Free from the predators, parasites, and pathogens that limit their populations in their native habitats, such transplanted species may spread rapidly through a new region.

Some introduced species disrupt their new community, often by preying on native organisms or outcompeting them for resources. The brown tree snake was accidentally introduced to the island of Guam from other parts of the South Pacific after World War II: It was a "stowaway" in military cargo. Since then, 12 species of birds and 6 species of lizards that the snakes ate have become extinct on Guam, which had no native snakes. The devastating zebra mussel, a filter-feeding mollusc, was introduced into the Great Lakes of North America in 1988, most likely in the ballast water of ships arriving from Europe. Zebra mussels form dense colonies and have disrupted freshwater ecosystems, threatening native aquatic species. They have also clogged water intake structures, causing billions of dollars in damage to domestic and industrial water supplies.

Humans have deliberately introduced many species with good intentions but disastrous effects. An Asian plant called kudzu, which the U.S. Department of Agriculture once introduced in the southern United States to help control erosion, has taken over large areas of the landscape there **(Figure 43.7)**.

▲ **Figure 43.7 Kudzu, an introduced species, thriving in South Carolina.**

Introduced species are a worldwide problem, contributing to approximately 40% of the extinctions recorded since 1750 and costing billions of dollars each year in damage and control efforts. There are more than 50,000 introduced species in the United States alone.

Overharvesting

The term *overharvesting* refers generally to the harvesting of wild organisms at rates exceeding the ability of their populations to rebound. Species with restricted habitats, such as small islands, are particularly vulnerable to overharvesting. One such species was the great auk, a large, flightless seabird found on islands in the North Atlantic Ocean. By the 1840s, the great auk had been hunted to extinction to satisfy the human demand for its feathers, eggs, and meat.

Also susceptible to overharvesting are large organisms with low reproductive rates, such as elephants, whales, and rhinoceroses. The decline of Earth's largest terrestrial animals, the African elephants, is a classic example of the impact of overhunting. Largely because of the trade in ivory, elephant populations have been declining in most of Africa for the last 50 years. An international ban on the sale of new ivory resulted in increased poaching (illegal hunting), so the ban had little effect in much of central and eastern Africa. Only in South Africa, where once-decimated herds have been well protected for nearly a century, have elephant populations been stable or increasing (see Figure 40.18).

Conservation biologists increasingly use the tools of molecular genetics to track the origins of tissues harvested from endangered species. Researchers at the University of Washington have constructed a DNA reference map for the African elephant using DNA isolated from elephant dung. By comparing this reference map with DNA isolated from samples of ivory harvested either legally or by poachers, they can determine to within a few hundred kilometers where the elephants were killed **(Figure 43.8)**. Such work in Zambia suggested that poaching rates were 30 times higher than previously estimated,

▲ Figure 43.8 Ecological forensics and elephant poaching. These severed tusks were part of an illegal shipment of ivory intercepted on its way from Africa to Singapore in 2002. DNA-based evidence showed that the thousands of elephants killed for the tusks came from a relatively narrow east-west band centered in Zambia rather than from across Africa.

leading to improved antipoaching efforts by the Zambian government. Similarly, biologists using phylogenetic analyses of mitochondrial DNA (mtDNA) showed that some whale meat sold in Japanese fish markets came from illegally harvested endangered species (see Figure 20.6).

Many commercially important fish populations, once thought to be inexhaustible, have been decimated by overfishing. Demands for protein-rich food from an increasing human population, coupled with new harvesting technologies, such as long-line fishing and modern trawlers, have reduced these fish populations to levels that cannot sustain further exploitation. Until the past few decades, the North Atlantic bluefin tuna had little commercial value—just a few cents per pound for use in cat food. In the 1980s, however, wholesalers began airfreighting fresh, iced bluefin to Japan for sushi and sashimi. In that market, the fish now brings up to $100 per pound **(Figure 43.9)**. With increased harvesting spurred by such high prices, it took just ten years to reduce the western North Atlantic bluefin population to less than 20% of its 1980 size.

▲ Figure 43.9 Overharvesting. North Atlantic bluefin tuna are auctioned in a Japanese fish market.

Global Change

The fourth threat to biodiversity, global change, alters the fabric of Earth's ecosystems at regional to global scales. Global change includes alterations in climate, atmospheric chemistry, and broad ecological systems that reduce the capacity of Earth to sustain life.

One of the first types of global change to cause concern was *acid precipitation*, which is rain, snow, sleet, or fog with a pH less than 5.2. The burning of wood and fossil fuels releases oxides of sulfur and nitrogen that react with water in air, forming sulfuric and nitric acids. The acids eventually fall to Earth's surface, harming some aquatic and terrestrial organisms.

In the 1960s, ecologists determined that lake-dwelling organisms in eastern Canada were dying because of air pollution from factories in the midwestern United States. Newly hatched lake trout, for instance, die when the pH drops below 5.4. Lakes and streams in southern Norway and Sweden were losing fish because of pollution generated in Great Britain and central Europe. By 1980, the pH of precipitation in large areas of North America and Europe averaged 4.0–4.5 and sometimes dropped as low as 3.0. (To review pH, see Concept 2.5.)

Environmental regulations and new technologies have enabled many countries to reduce sulfur dioxide emissions in recent decades. In the United States, sulfur dioxide emissions decreased more than 40% between 1993 and 2009, gradually reducing the acidity of precipitation **(Figure 43.10)**. However, ecologists estimate that it will take decades for aquatic ecosystems to recover. Meanwhile, emissions of nitrogen oxides are increasing in the United States, and emissions of sulfur dioxide and acid precipitation continue to damage forests in Europe.

We will explore the importance of global change for Earth's biodiversity in more detail in Concept 43.4, where we examine such factors as climate change.

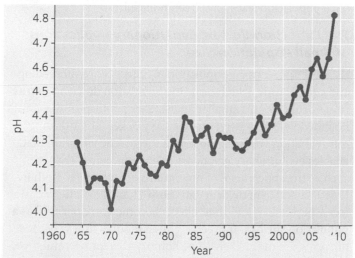

© 2001 Pearson Education, Inc.

▲ Figure 43.10 Changes in the pH of precipitation at Hubbard Brook, New Hampshire. Although still acidic, the precipitation in this northeastern U.S. forest has been increasing in pH for more than three decades.

CONCEPT CHECK 43.1

1. Explain why it is too narrow to define the biodiversity crisis as simply a loss of species.

2. Identify the four main threats to biodiversity and explain how each damages diversity.

3. **WHAT IF?** Imagine two populations of a fish species, one in the Mediterranean Sea and one in the Caribbean Sea. Now imagine two scenarios: (1) The populations breed separately, and (2) adults of both populations migrate yearly to the North Atlantic to interbreed. Which scenario would result in a greater loss of genetic diversity if the Mediterranean population were harvested to extinction? Explain your answer.

For suggested answers, see Appendix A.

CONCEPT 43.2

Population conservation focuses on population size, genetic diversity, and critical habitat

Biologists who work on conservation at the population and species levels use two main approaches. One approach focuses on populations that are small and hence often vulnerable. The other emphasizes populations that are declining rapidly, even if they are not yet small.

Small-Population Approach

Small populations are particularly vulnerable to overharvesting, habitat loss, and the other threats to biodiversity that you read about in Concept 43.1. After such factors have reduced a population's size, the small size itself can push the population to extinction. Conservation biologists who adopt the small-population approach study the processes that cause extinctions once population sizes have been reduced.

The Extinction Vortex: Evolutionary Implications of Small Population Size

EVOLUTION A small population is vulnerable to inbreeding and genetic drift, which draw the population down an **extinction vortex** toward smaller and smaller population size until no individuals survive **(Figure 43.11)**. A key factor driving the extinction vortex is the loss of the genetic variation that enables evolutionary responses to environmental change, such as the appearance of new strains of pathogens. Both inbreeding and genetic drift can cause a loss of genetic variation (see Chapter 21), and their effects become more harmful as a population shrinks. Inbreeding often reduces fitness because offspring are more likely to be homozygous for harmful recessive traits.

Not all small populations are doomed by low genetic diversity, and low genetic variability does not automatically lead to permanently small populations. For instance, overhunting of northern elephant seals in the 1890s reduced the species to only 20 individuals—clearly a bottleneck with reduced genetic

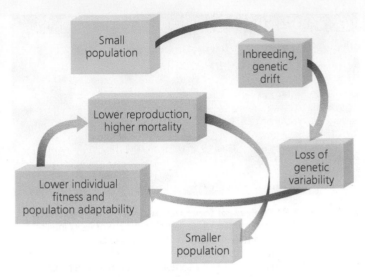

▲ **Figure 43.11 Processes driving an extinction vortex.**

variation. Since that time, however, the northern elephant seal populations have rebounded to about 150,000 individuals today, though their genetic variation remains relatively low. Thus, low genetic diversity does not always impede population growth.

Case Study: The Greater Prairie Chicken and the Extinction Vortex

When Europeans arrived in North America, the greater prairie chicken (*Tympanuchus cupido*) was common from New England to Virginia and across the western prairies of the continent. Land cultivation for agriculture fragmented the populations of this species, and its abundance decreased rapidly (see Chapter 21). Illinois had millions of greater prairie chickens in the 19th century but fewer than 50 by 1993. Researchers found that the decline in the Illinois population was associated with a decrease in fertility. As a test of the extinction vortex hypothesis, scientists increased the genetic variation of the Illinois population by importing 271 birds from larger populations elsewhere **(Figure 43.12)**. The Illinois population rebounded, confirming that it had been on its way to extinction until rescued by the transfusion of genetic variation.

Minimum Viable Population Size

How small does a population have to be before it starts down an extinction vortex? The answer depends on the type of organism and other factors. Large predators that feed high on the food chain usually require extensive individual ranges, resulting in low population densities. Therefore, not all rare species concern conservation biologists. All populations, however, require some minimum size to remain viable.

The minimal population size at which a species is able to sustain its numbers is known as the **minimum viable population (MVP)**. MVP is usually estimated for a given species using computer models that integrate many factors. The calculation may include, for instance, an estimate of how many individuals in a small population are likely to be killed by a natural

What caused the drastic decline of the Illinois greater prairie chicken population?

Experiment Researchers had observed that the population collapse of the greater prairie chicken was mirrored in a reduction in fertility, as measured by the hatching rate of eggs. Comparison of DNA samples from the Jasper County, Illinois, population with DNA from feathers in museum specimens showed that genetic variation had declined in the study population (see Figure 21.11). In 1992, Ronald Westemeier, Jeffrey Brawn, and colleagues began translocating prairie chickens from Minnesota, Kansas, and Nebraska in an attempt to increase genetic variation.

Results After translocation (blue arrow), the viability of eggs rapidly increased, and the population rebounded.

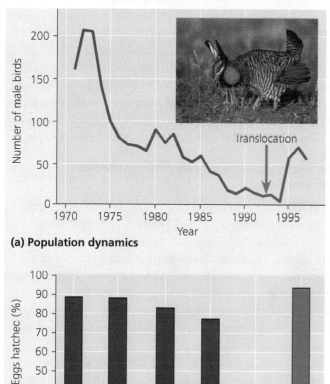

(a) Population dynamics

(b) Hatching rate

© 1998 AAAS

Conclusion Reduced genetic variation had started the Jasper County population of prairie chickens down the extinction vortex.

Source R. L. Westemeier et al., Tracking the long-term decline and recovery of an isolated population, *Science* 282:1695–1698 (1998).

Inquiry in Action Read and analyze the original paper in *Inquiry in Action: Interpreting Scientific Papers.*

WHAT IF? Given the success of using transplanted birds as a tool for increasing the percentage of hatched eggs in Illinois, why wouldn't you transplant additional birds immediately to Illinois?

catastrophe such as a storm. Once in the extinction vortex, two or three consecutive years of bad weather could finish off a population that is already below its MVP.

Effective Population Size

Genetic variation is the key issue in the small-population approach. The *total* size of a population may be misleading because only certain members of the population breed successfully and pass their alleles on to offspring. Therefore, a meaningful estimate of MVP requires the researcher to determine the **effective population size**, which is based on the breeding potential of the population.

The following formula incorporates the sex ratio of breeding individuals into the estimate of effective population size, abbreviated N_e:

$$N_e = \frac{4N_f N_m}{N_f + N_m}$$

where N_f and N_m are, respectively, the number of females and the number of males that successfully breed. If we apply this formula to an idealized population whose total size is 1,000 individuals, N_e will also be 1,000 if every individual breeds and the sex ratio is 500 females to 500 males. In this case, $N_e = (4 \times 500 \times 500)/(500 + 500) = 1,000$. Any deviation from these conditions (not all individuals breed or there is not a 1:1 sex ratio) reduces N_e. For instance, if the total population size is 1,000 but only 400 females and 400 males breed, then $N_e = (4 \times 400 \times 400)/(400 + 400) = 800$, or 80% of the total population. Numerous life history traits can influence N_e, and alternative formulas for estimating N_e take into account factors such as family size, age at maturation, genetic relatedness among population members, the effects of gene flow between geographically separated populations, and population fluctuations.

In actual study populations, N_e is always some fraction of the total population. Thus, simply determining the total number of individuals in a small population does not provide a good measure of whether the population is large enough to avoid extinction. Whenever possible, conservation programs attempt to sustain total population sizes that include at least the minimum viable number of *reproductively active* individuals. The conservation goal of sustaining effective population size (N_e) above MVP stems from the concern that populations retain enough genetic diversity to adapt as their environment changes.

Case Study: *Analysis of Grizzly Bear Populations*

One of the first population viability analyses was conducted in 1978 by Mark Shaffer, of Duke University, as part of a long-term study of grizzly bears in Yellowstone National Park and its surrounding areas (**Figure 43.13**). A threatened species in the United States, the grizzly bear (*Ursus arctos horribilis*) is currently found in only 4 of the 48 contiguous states. Its populations in those states have been drastically reduced and fragmented. In 1800, an estimated 100,000 grizzlies ranged over about 500 million ha of habitat, while today only about 1,000 individuals in six relatively isolated populations range over less than 5 million ha.

▲ **Figure 43.13 Long-term monitoring of a grizzly bear population.** The ecologist is fitting this tranquilized bear with a radio collar so that the bear's movements can be compared with those of other grizzlies in the Yellowstone National Park population.

Shaffer attempted to determine viable sizes for the Yellowstone grizzly population. Using life history data obtained for individual Yellowstone bears over a 12-year period, he simulated the effects of environmental factors on survival and reproduction. His models predicted that, given a suitable habitat, a Yellowstone grizzly bear population of 70–90 individuals would have about a 95% chance of surviving for 100 years. A slightly larger population of only 100 bears would have a 95% chance of surviving for twice as long, about 200 years.

How does the actual size of the Yellowstone grizzly population compare with Shaffer's predicted MVP? A current estimate puts the total grizzly bear population in the greater Yellowstone ecosystem at about 500 individuals. The relationship of this estimate to the effective population size, N_e, depends on several factors. Usually, only a few dominant males breed, and it may be difficult for them to locate females, since individuals inhabit such large areas. Moreover, females may reproduce only when there is abundant food. As a result, N_e is only about 25% of the total population size, or about 125 bears.

Because small populations tend to lose genetic variation over time, researchers have analyzed proteins, mtDNA, and short tandem repeats (see Chapter 18) to assess genetic variability in the Yellowstone grizzly bear population. All results to date indicate that the Yellowstone population has less genetic variability than other grizzly bear populations in North America.

How might conservation biologists increase the effective size and genetic variation of the Yellowstone grizzly bear population? Migration between isolated populations of grizzlies could increase both effective and total population sizes. Computer models predict that introducing only two unrelated bears each decade into a population of 100 individuals would reduce the loss of genetic variation by about half. For the grizzly bear, and probably for many other species with small populations, finding ways to promote dispersal among populations may be one of the most urgent conservation needs.

This case study and that of the greater prairie chicken bridge small-population models and practical applications in conservation. Next, we look at an alternative approach to understanding the biology of extinction.

Declining-Population Approach

The declining-population approach focuses on threatened and endangered populations that show a downward trend, even if the population is far above its minimum viable population. The distinction between a declining population, which may not be small, and a small population, which may not be declining, is less important than the different priorities of the two approaches. The small-population approach emphasizes smallness itself as an ultimate cause of a population's extinction, especially through the loss of genetic diversity. In contrast, the declining-population approach emphasizes the environmental factors that caused a population decline in the first place. If, for instance, an area is deforested, then species that depend on trees will decline in abundance and become locally extinct, whether or not they retain genetic variation. The following case study is one example of how the declining-population approach has been applied to the conservation of an endangered species.

Case Study: *Decline of the Red-Cockaded Woodpecker*

The red-cockaded woodpecker (*Picoides borealis*) is found only in the southeastern United States. It requires mature pine forests, preferably ones dominated by the longleaf pine, for its habitat. Most woodpeckers nest in dead trees, but the red-cockaded woodpecker drills its nest holes in mature, living pine trees. It also drills small holes around the entrance to its nest cavity, which causes resin from the tree to ooze down the trunk. The resin seems to repel predators, such as corn snakes, that eat bird eggs and nestlings.

Another critical habitat factor for the red-cockaded woodpecker is that the undergrowth of plants around the pine trunks must be low **(Figure 43.14a)**. Breeding birds tend to abandon nests when vegetation among the pines is thick and higher than about 4.5 m **(Figure 43.14b)**. Apparently, the birds need a clear flight path between their home trees and the neighboring feeding grounds. Periodic fires have historically swept through longleaf pine forests, keeping the undergrowth low.

One factor leading to the woodpecker's decline has been the destruction or fragmentation of suitable habitats by logging and agriculture. By recognizing key habitat factors, protecting some longleaf pine forests, and using controlled fires to reduce forest undergrowth, conservation managers have helped restore habitat that can support viable populations.

Sometimes conservation managers also help species colonize restored habitats. Because red-cockaded woodpeckers take months to excavate nesting cavities, researchers performed an experiment to see whether providing cavities for the birds would make them more likely to use a site. The

Red-cockaded
woodpecker

(a) Forests that can sustain red-cockaded woodpeckers have low undergrowth.

(b) Forests that cannot sustain red-cockaded woodpeckers have high, dense undergrowth that interferes with the woodpeckers' access to feeding grounds.

▲ **Figure 43.14 A habitat requirement of the red-cockaded woodpecker.**

? *How is habitat disturbance necessary for the long-term survival of the woodpecker?*

researchers constructed cavities in pine trees at 20 restored sites and compared nesting rates there with rates in sites without constructed cavities. The results were dramatic. Cavities in 18 of the 20 sites with constructed cavities were colonized by red-cockaded woodpeckers, and new breeding groups formed only in those sites. Based on this experiment, conservationists initiated a habitat maintenance program that included controlled burning and excavation of new nesting cavities, enabling this endangered species to begin to recover.

Weighing Conflicting Demands

Determining population numbers and habitat needs is only part of a strategy to save species. Scientists also need to weigh a species' needs against other conflicting demands.

Conservation biology often highlights the relationship between science, technology, and society. For example, an ongoing, sometimes bitter debate in the western United States pits habitat preservation for wolf, grizzly bear, and bull trout populations against job opportunities in the grazing and resource extraction industries. Programs that restocked wolves in Yellowstone National Park remain controversial for people concerned about human safety and for many ranchers concerned with potential loss of livestock outside the park.

Large, high-profile vertebrates are not always the focal point in such conflicts, but habitat use is almost always the issue. Should work proceed on a new highway bridge if it destroys the only remaining habitat of a species of freshwater mussel? If you owned a coffee plantation growing varieties that thrive in bright sunlight, would you be willing to change to shade-tolerant varieties that produce less coffee per hectare but can grow beneath trees that support large numbers of songbirds?

Another important consideration is the ecological role of a species. Because we cannot save every endangered species, we must determine which species are most important for conserving biodiversity as a whole. Identifying keystone species and finding ways to sustain their populations can be central to maintaining communities and ecosystems. In most situations, we must look beyond a species and consider the whole community and ecosystem as an important unit of biodiversity.

CONCEPT CHECK 43.2

1. How does the reduced genetic diversity of small populations make them more vulnerable to extinction?
2. If there was a total of 50 individuals in the two Illinois populations of greater prairie chickens in 1993, what was the effective population size if 15 females and 5 males bred?
3. **WHAT IF?** In 2011, at least ten grizzly bears in the greater Yellowstone ecosystem were killed through contact with people. Three things caused many of these deaths: collisions with automobiles, hunters (of other animals) shooting when charged by a female grizzly bear with cubs nearby, and conservation managers killing bears that attacked livestock repeatedly. If you were a conservation manager, what steps might you take to minimize such encounters in Yellowstone?

For suggested answers, see Appendix A.

CONCEPT 43.3

Landscape and regional conservation help sustain biodiversity

Although conservation efforts historically focused on saving individual species, efforts today often seek to sustain the biodiversity of entire communities, ecosystems, and landscapes. Such a broad view requires applying not just the principles of community, ecosystem, and landscape ecology but aspects of human population dynamics and economics as well. The goals of landscape ecology (see Chapter 40) include projecting future patterns of landscape use and making biodiversity conservation part of land-use planning.

Landscape Structure and Biodiversity

The biodiversity of a given landscape is in large part a function of the structure of the landscape. Understanding landscape structure is critically important in conservation because many species use more than one kind of ecosystem, and many live on the borders between ecosystems.

Fragmentation and Edges

The boundaries, or *edges*, between ecosystems—such as between a lake and the surrounding forest or between cropland and suburban housing tracts—are defining features of landscapes **(Figure 43.15)**. An edge has its own set of physical conditions, which differ from those on either side of it. The soil surface of an edge between a forest patch and a burned area receives more sunlight and is usually hotter and drier than the forest interior, but it is cooler and wetter than the soil surface in the burned area.

Some organisms thrive in edge communities because they gain resources from both adjacent areas. The ruffed grouse (*Bonasa umbellus*) is a bird that needs forest habitat for nesting, winter food, and shelter, but it also needs forest openings with dense shrubs and herbs for summer food.

Ecosystems in which edges arise from human alterations often have reduced biodiversity and a preponderance of edge-adapted species. For example, white-tailed deer thrive in edge habitats, where they can browse on woody shrubs; deer populations often expand when forests are logged and more edges are generated. The brown-headed cowbird (*Molothrus ater*) is

an edge-adapted species that lays its eggs in the nests of other birds, often migratory songbirds. Cowbirds need forests, where they can parasitize the nests of other birds, and open fields, where they forage on seeds and insects. Consequently, their populations are growing where forests are being cut and fragmented, creating more edge habitat and open land. Increasing cowbird parasitism and habitat loss are correlated with declining populations of several of the cowbird's host species.

The influence of fragmentation on the structure of communities has been explored since 1979 in the long-term Biological Dynamics of Forest Fragments Project. Located in the heart of the Amazon River basin, the study area consists of isolated fragments of tropical rain forest separated from surrounding continuous forest by distances of 80–1,000 m **(Figure 43.16)**. Numerous researchers working on this project have clearly documented the effects of this fragmentation on organisms ranging from bryophytes to beetles to birds. They have consistently found that species adapted to forest interiors show the greatest declines when patches are the smallest, suggesting that landscapes dominated by small fragments will support fewer species.

Corridors That Connect Habitat Fragments

In fragmented habitats, the presence of a **movement corridor**, a narrow strip or series of small clumps of habitat connecting otherwise isolated patches, can be extremely important for conserving biodiversity. Riparian habitats often serve as corridors, and in some nations, government policy prohibits altering these habitats. In areas of heavy human use, artificial corridors are sometimes constructed. Bridges or tunnels, for instance, can reduce the number of animals killed trying to cross highways **(Figure 43.17)**.

Movement corridors can also promote dispersal and reduce inbreeding in declining populations. Corridors have been

▲ **Figure 43.15 Edges between ecosystems.** Grasslands give way to forest ecosystems in Yellowstone National Park.

▲ **Figure 43.16 Amazon rain forest fragments created as part of the Biological Dynamics of Forest Fragments Project.**

▲ **Figure 43.17 An artificial corridor.** This bridge in Banff National Park, Canada, helps animals cross a human-created barrier.

shown to increase the exchange of individuals among populations of many organisms, including butterflies, voles, and aquatic plants. Corridors are especially important to species that migrate between different habitats seasonally. However, a corridor can also be harmful—for example, by allowing the spread of disease. In a 2003 study, a scientist at the University of Zaragoza, Spain, showed that habitat corridors facilitate the movement of disease-carrying ticks among forest patches in northern Spain. All the effects of corridors are not yet understood, and their impact is an area of active research in conservation biology.

Establishing Protected Areas

Conservation biologists are applying their understanding of landscape dynamics in establishing protected areas to slow biodiversity loss. Currently, governments have set aside about 7% of the world's land in various forms of reserves. Choosing where to place nature reserves and how to design them poses many challenges. Should the reserve be managed to minimize the risks of fire and predation to a threatened species? Or should the reserve be left as natural as possible, with such processes as fires ignited by lightning allowed to play out on their own? This is just one of the debates that arise among people who share an interest in the health of national parks and other protected areas.

Preserving Biodiversity Hot Spots

In deciding which areas are of highest conservation priority, biologists often focus on hot spots of biodiversity. A **biodiversity hot spot** is a relatively small area with numerous endemic species (species found nowhere else in the world) and a large number of endangered and threatened species (**Figure 43.18**). Nearly 30% of all bird species can be found in hot spots that make up only about 2% of Earth's land area. Together,

the "hottest" of the terrestrial biodiversity hot spots total less than 1.5% of Earth's land but are home to more than a third of all species of plants, amphibians, reptiles (including birds), and mammals. Aquatic ecosystems also have hot spots, such as coral reefs and certain river systems.

Biodiversity hot spots are good choices for nature reserves, but identifying them is not always simple. One problem is that a hot spot for one taxonomic group, such as butterflies, may not be a hot spot for some other taxonomic group, such as birds. Designating an area as a biodiversity hot spot is often biased toward saving vertebrates and plants, with less attention paid to invertebrates and microorganisms. Some biologists are also concerned that the hot-spot strategy places too much emphasis on such a small fraction of Earth's surface.

Global change makes the task of preserving hot spots even more challenging because the conditions that favor a particular community may not be found in the same location in the future. The biodiversity hot spot in the southwest corner of Australia (see Figure 43.18) holds thousands of species of endemic plants and numerous endemic vertebrates. Researchers recently concluded that between 5% and 25% of the plant species they examined may become extinct by 2080 because the plants will be unable to tolerate the increased dryness predicted for this region.

Philosophy of Nature Reserves

Nature reserves are biodiversity islands in a sea of habitat degraded by human activity. Protected "islands" are not isolated from their surroundings, however, and the nonequilibrium model (see Chapter 41) applies to nature reserves as well as to the larger landscapes around them.

An earlier policy—that protected areas should be set aside to remain unchanged forever—was based on the concept that ecosystems are balanced, self-regulating units. However, disturbance is common in all ecosystems (see Chapter 41), and management policies that ignore natural disturbances or attempt to prevent them have generally failed. For instance, setting aside an area of a fire-dependent community, such as a

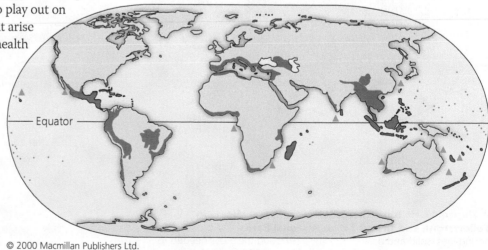

© 2000 Macmillan Publishers Ltd.

▲ **Figure 43.18 Earth's terrestrial (■) and marine (▲) biodiversity hot spots.**

portion of a tallgrass prairie, chaparral, or dry pine forest, with the intention of saving it is unrealistic if periodic burning is excluded. Without the dominant disturbance, the fire-adapted species are usually outcompeted and biodiversity is reduced.

An important conservation question is whether to create numerous small reserves or fewer large reserves. Small, unconnected reserves may slow the spread of disease between populations. One argument for large reserves is that large, far-ranging animals with low-density populations, such as the grizzly bear, require extensive habitats. Large reserves also have proportionately smaller perimeters than small reserves and are therefore less affected by edges.

As conservation biologists have learned more about the requirements for achieving minimum viable populations for endangered species, they have realized that most national parks and other reserves are far too small. The area needed for the long-term survival of the Yellowstone grizzly bear population, for instance, is more than ten times the combined area of Yellowstone and Grand Teton National Parks **(Figure 43.19)**. Areas of private and public land surrounding reserves will likely have to contribute to biodiversity conservation.

Zoned Reserves

Several nations have adopted a zoned reserve approach to landscape management. A **zoned reserve** is an extensive region that includes relatively undisturbed areas surrounded by areas that have been changed by human activity and are used for economic gain. The key challenge of the zoned reserve approach is to develop a social and economic climate in the surrounding lands that is compatible with the long-term viability of the protected core. These surrounding areas continue to support human activities, but regulations prevent the types of extensive alterations likely to harm the protected area. As a result, the surrounding habitats serve as buffer zones against further intrusion into the undisturbed area.

The small Central American nation of Costa Rica has become a world leader in establishing zoned reserves **(Figure 43.20)**. An agreement initiated in 1987 reduced Costa Rica's international debt in return for land preservation there. The agreement resulted in eight zoned reserves, called "conservation areas," that contain designated national park land. Costa Rica is making progress toward managing its zoned reserves, and the buffer zones provide a steady, lasting supply of forest products, water, and hydroelectric power while also supporting sustainable agriculture and tourism, both of which employ local people.

Although marine ecosystems have also been heavily affected by human exploitation, reserves in the ocean are far less common than reserves on land. Many fish populations around the world have collapsed as increasingly sophisticated equipment puts nearly all potential fishing grounds within human reach. In response, scientists at the University of York, England, have proposed establishing marine reserves around the world that would be off limits to fishing. They present strong evidence that a patchwork of marine reserves can serve as a means of both increasing fish populations within the reserves and improving fishing success in nearby areas. Their proposed system is a modern application of a centuries-old practice in the Fiji Islands in which some areas have historically remained closed to fishing—a traditional example of the zoned reserve concept.

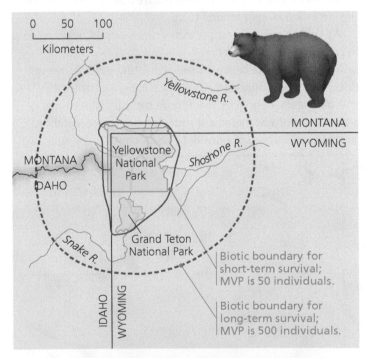

▲ **Figure 43.19 Biotic boundaries for grizzly bears in Yellowstone and Grand Teton National Parks.** The biotic boundaries (solid and dashed red lines) surround the areas needed to support minimum viable populations of 50 and 500 bears. Even the smaller of these areas is larger than the two parks.

▲ **Figure 43.20 Zoned reserves in Costa Rica.** Boundaries of the zoned reserves are indicated by black outlines.

▲ **Figure 43.21 A diver measuring coral in the Florida Keys National Marine Sanctuary.**

The United States adopted such a system in creating a set of 13 national marine sanctuaries, including the Florida Keys National Marine Sanctuary, which was established in 1990 **(Figure 43.21)**. Populations of marine organisms, including fishes and lobsters, recovered quickly after harvests were banned in the 9,500-km² reserve. Larger and more abundant fish now produce larvae that help repopulate reefs and improve fishing outside the sanctuary. The increased marine life within the sanctuary also makes it a favorite for recreational divers, increasing the economic value of this zoned reserve.

CONCEPT CHECK 43.3

1. What is a biodiversity hot spot?
2. How do zoned reserves provide economic incentives for long-term conservation of protected areas?
3. **WHAT IF?** Suppose a developer proposes to clear-cut a forest that serves as a corridor between two parks. To compensate, the developer also proposes to add the same area of forest to one of the parks. As a professional ecologist, how might you argue for retaining the corridor?

For suggested answers, see Appendix A.

CONCEPT 43.4

Earth is changing rapidly as a result of human actions

As we've discussed, landscape and regional conservation help protect habitats and preserve species. However, environmental changes that result from human activities are creating new challenges. As a consequence of human-caused climate change, for example, the place where a vulnerable species is found today may not be the same as the one needed for preservation in the future. What would happen if *many* habitats on Earth changed so quickly that the locations of preserves today were unsuitable for their species in 10, 50, or 100 years? Such a scenario is increasingly likely.

The rest of this section describes three types of environmental change that threaten biodiversity: nutrient enrichment, toxin accumulation, and climate change. The impacts of these and other changes are evident not just in human-dominated ecosystems, such as cities and farms, but also in the most remote ecosystems on Earth.

Nutrient Enrichment

Human activity often removes nutrients from one part of the biosphere and adds them to another. Someone eating strawberries in Washington, DC, consumes nutrients that only days before were in the soil in California; a short time later, some of these nutrients will be in the Potomac River, having passed through the person's digestive system and a local sewage treatment facility.

Farming is an example of how human activities are altering the environment through the enrichment of nutrients. After vegetation is cleared from an area, the existing reserve of nutrients in the soil is sufficient to grow crops for only a brief period because a substantial fraction of these nutrients is exported from the area in crop biomass. For this reason, farmers typically add fertilizers to increase crop yields.

Nitrogen is the main nutrient element lost through agriculture (see Figure 42.13). Plowing mixes the soil and speeds up decomposition of organic matter, releasing nitrogen that is then removed when crops are harvested. Applied fertilizers make up for the loss of usable nitrogen from agricultural ecosystems. However, without plants to take up nitrates from the soil, the nitrates are likely to be leached from the ecosystem (see Figure 42.14). Recent studies indicate that human activities have more than doubled Earth's supply of fixed nitrogen available to primary producers.

A problem arises when the nutrient level in an ecosystem exceeds the **critical load**, the amount of added nutrient, usually nitrogen or phosphorus, that can be absorbed by plants without damaging ecosystem integrity. For example, nitrogenous minerals in the soil that exceed the critical load eventually leach into groundwater or run off into freshwater and marine ecosystems, sometimes contaminating water supplies and killing fish. Nitrate concentrations in groundwater are increasing in most agricultural regions, sometimes reaching levels that are unsafe for drinking.

Many rivers contaminated with nitrates and ammonium from agricultural runoff and sewage drain into the Atlantic Ocean, with the highest inputs coming from northern Europe and the central United States. The Mississippi River carries nitrogen pollution to the Gulf of Mexico, fueling a phytoplankton bloom each summer. When the phytoplankton die, their decomposition by oxygen-using organisms creates an extensive "dead zone" of low oxygen concentrations along the Gulf coast

▲ **Figure 43.22 A phytoplankton bloom arising from nitrogen pollution in the Mississippi basin that leads to a dead zone.** In this satellite image from 2004, red and orange represent high concentrations of phytoplankton in the Gulf of Mexico.

(Figure 43.22). Fish and other marine animals disappear from some of the most economically important waters in the United States. To reduce the size of the dead zone, farmers have begun using fertilizers more efficiently, and managers are restoring wetlands in the Mississippi watershed, two changes stimulated by the results of ecosystem experiments.

Nutrient runoff can also lead to the eutrophication of lakes (see Concept 42.2). The bloom and subsequent die-off of algae and cyanobacteria and the ensuing depletion of oxygen are similar to what occurs in a marine dead zone. Such conditions threaten the survival of organisms. For example, eutrophication of Lake Erie coupled with overfishing wiped out commercially important fishes such as blue pike, whitefish, and lake trout by the 1960s. Since then, tighter regulations on the dumping of sewage and other wastes into the lake have enabled some fish populations to rebound, but many native species of fish and invertebrates have not recovered.

Toxins in the Environment

Human activities release an immense variety of toxic chemicals, including thousands of synthetic compounds previously unknown in nature, with little regard for the ecological consequences. Organisms acquire toxic substances from the environment along with nutrients and water. Some of the poisons are metabolized or excreted, but others accumulate in specific tissues, often fat. One of the reasons accumulated toxins are particularly harmful is that they become more concentrated in successive trophic levels of a food web. This phenomenon, called **biological magnification**, occurs because the biomass at any given trophic level is produced from a much larger biomass ingested from the level below (see Concept 42.3). Thus, top-level carnivores tend to be most severely affected by toxic compounds in the environment.

One class of industrially synthesized compounds that have demonstrated biological magnification are the chlorinated hydrocarbons, which include the industrial chemicals called PCBs (polychlorinated biphenyls) and many pesticides, such as DDT. Current research implicates many of these compounds in endocrine system disruption in a large number of animal

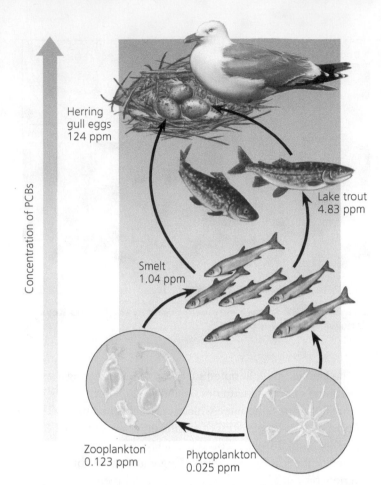

▲ **Figure 43.23 Biological magnification of PCBs in a Great Lakes food web.**

species, including humans. Biological magnification of PCBs has been found in the food web of the Great Lakes, where the concentration of PCBs in herring gull eggs, at the top of the food web, is nearly 5,000 times that in phytoplankton, at the base of the food web **(Figure 43.23).**

An infamous case of biological magnification that harmed top-level carnivores involved DDT, a chemical used to control insects such as mosquitoes and agricultural pests. In the decade after World War II, the use of DDT grew rapidly; its ecological consequences were not yet fully understood. By the 1950s, scientists were learning that DDT persists in the environment and is transported by water to areas far from where it is applied. One of the first signs that DDT was a serious environmental problem was a decline in the populations of pelicans, ospreys, and eagles, birds that feed at the top of food webs. The accumulation of DDT (and DDE, a product of its breakdown) in the tissues of these birds interfered with the deposition of calcium in their eggshells. When the birds tried to incubate their eggs, the weight of the parents broke the shells of affected eggs, resulting in catastrophic declines in the birds' reproduction rates. Rachel Carson's book *Silent Spring* helped bring the problem to public attention in the 1960s **(Figure 43.24)**, and DDT was banned in the United States in 1971. A dramatic recovery in populations of the affected bird species followed.

► Figure 43.24
Rachel Carson.
Through her writing and her testimony before the U.S. Congress, biologist and author Rachel Carson helped promote a new environmental ethic. Her efforts led to a ban on DDT use in the United States and stronger controls on the use of other chemicals.

In countries throughout much of the tropics, DDT is still used to control the mosquitoes that spread malaria and other diseases. Societies there face a trade-off between saving human lives and protecting other species. The best approach seems to be to apply DDT sparingly and to couple its use with mosquito netting and other low-technology solutions. The complicated history of DDT illustrates the importance of understanding the ecological connections between diseases and communities (see Concept 41.5).

Pharmaceuticals make up another group of toxins in the environment, one that is a growing concern among ecologists. The use of over-the-counter and prescription drugs has risen in recent years, particularly in industrialized nations. People who consume such products excrete residual chemicals in their waste and may also dispose of unused drugs improperly, such as in their toilets or sinks. Drugs that are not broken down in sewage treatment plants may then enter rivers and lakes with the material discharged from these plants. Growth-promoting drugs given to farm animals can also enter rivers and lakes with agricultural runoff. As a consequence, many pharmaceuticals are spreading in low concentrations across the world's freshwater ecosystems **(Figure 43.25)**.

Among the pharmaceuticals that ecologists are studying are the sex steroids, including forms of estrogen used for birth control. Some fish species are so sensitive to certain estrogens that concentrations of a few parts per trillion in their water can alter sexual differentiation and shift the female-to-male sex ratio toward females. Researchers in Ontario, Canada, conducted a seven-year experiment in which they applied the synthetic estrogen used in contraceptives to a lake in very low concentrations (5–6 ng/L). They found that chronic exposure of the fathead minnow (*Pimephales promelas*) to the estrogen led to feminization of males and a near extinction of the species from the lake.

Many toxins cannot be degraded by microorganisms and persist in the environment for years or even decades. In other cases, chemicals released into the environment may be relatively harmless but are converted to more toxic products by reaction with other substances, by exposure to light, or by the metabolism of microorganisms. Mercury, a by-product of plastic production and coal-fired power generation, has been routinely expelled into rivers and the sea in an insoluble form. Bacteria in the bottom mud convert the waste to methylmercury (CH_3Hg^+), an extremely toxic water-soluble compound that accumulates in the tissues of organisms, including humans who consume fish from the contaminated waters.

Greenhouse Gases and Climate Change

Human activities release a variety of gaseous waste products. People once thought that the vast atmosphere could absorb these materials indefinitely, but we now know that such additions can cause fundamental changes to the atmosphere and to its interactions with the rest of the biosphere. In this section, we'll examine how increasing concentrations of carbon dioxide and other greenhouse gases may affect species and ecosystems.

Since the Industrial Revolution, the concentration of CO_2 in the atmosphere has been increasing as a result of the burning of fossil fuels and deforestation. Scientists estimate that the average CO_2 concentration in the atmosphere before 1850 was about 274 ppm. In 1958, a monitoring station began taking very accurate measurements on Hawaii's Mauna Loa peak, a location far from cities and high enough for the atmosphere to be well mixed. At that time, the CO_2 concentration was 316 ppm **(Figure 43.26)**. Today, it exceeds 390 ppm, an increase of more than 40% since the mid-19th century. In the **Scientific Skills Exercise**, you can graph and interpret changes in CO_2 concentration that occur during the course of a year and over longer periods.

The marked increase in the concentration of atmospheric CO_2 over the last 150 years concerns scientists because

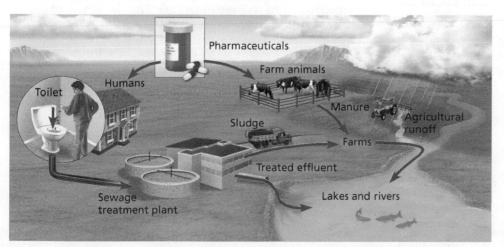

▲ **Figure 43.25 Sources and movements of pharmaceuticals in the environment.**

► **Figure 43.26 Increase in atmospheric carbon dioxide concentration at Mauna Loa, Hawaii, and average global temperatures.** Aside from normal seasonal fluctuations, the CO_2 concentration (blue curve) has increased steadily from 1958 to 2011. Though average global temperatures (red curve) fluctuated a great deal over the same period, there is a clear warming trend.

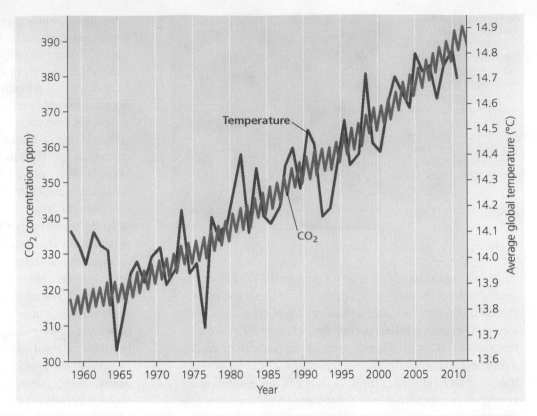

Scientific Skills Exercise

Graphing Cyclic Data

How Does the Atmospheric CO_2 Concentration Change During a Year and from Decade to Decade? The blue curve in Figure 43.26 shows how the concentration of CO_2 in Earth's atmosphere has changed over a span of more than 50 years. For each year in that span, two data points are plotted, one in May and one in November. A more detailed picture of the change in CO_2 concentration can be obtained by looking at measurements made at more frequent intervals. In this exercise, you'll graph monthly CO_2 concentrations for three years over three decades.

Data from the Study The data in the table below are average CO_2 concentrations (in parts per million) at the Mauna Loa monitoring station for each month in 1990, 2000, and 2010.

Month	1990	2000	2010
January	353.79	369.25	388.45
February	354.88	369.50	389.82
March	355.65	370.56	391.08
April	356.27	371.82	392.46
May	359.29	371.51	392.95
June	356.32	371.71	392.06
July	354.88	369.85	390.13
August	352.89	368.20	388.15
September	351.28	366.91	386.80
October	351.59	366.91	387,18
November	353.05	366.99	388.59
December	354.27	369.67	389.68

Interpret the Data

1. Plot the data for all three years on one graph. Select a type of graph that is appropriate for these data, and choose a vertical-axis scale that allows you to clearly see the patterns of CO_2 concentration changes, both during each year and from decade to decade. (For additional information about graphs, see the Scientific Skills Review in Appendix F and in the Study Area in MasteringBiology.)

2. Within each year, what is the pattern of change in CO_2 concentration? Why does this pattern occur?

3. The measurements taken at Mauna Loa represent average atmospheric CO_2 concentrations for the Northern Hemisphere. Suppose you could measure CO_2 concentrations under similar conditions in the Southern Hemisphere. What pattern would you expect to see in those measurements over the course of a year? Explain.

4. In addition to the changes within each year, what changes in CO_2 concentration occurred between 1990 and 2010? Calculate the average CO_2 concentration for the 12 months of each year. By what percentage did this average change from 1990 to 2000 and from 1990 to 2010?

Data from National Oceanic & Atmospheric Administration, Earth System Research Laboratory, Global Monitoring Division

(MB) A version of this Scientific Skills Exercise can be assigned in MasteringBiology.

of its link to increased global temperature. Much of the solar radiation that strikes the planet is reflected back into space. Although CO_2, methane, water vapor, and other gases in the atmosphere are transparent to visible light, they intercept and absorb much of the infrared radiation Earth emits, re-reflecting some of it back toward Earth. This process retains some of the solar heat. If it were not for this **greenhouse effect**, the average air temperature at Earth's surface would be a frigid $-18°C$ $(-0.4°F)$, and most life as we know it could not exist.

For more than a century, scientists have studied how greenhouse gases warm Earth and how fossil fuel burning could contribute to the warming. Most scientists are convinced that such warming is already occurring and will increase rapidly this century (see Figure 43.26). Global models predict that by the end of the 21st century, the atmospheric CO_2 concentration will have more than doubled, increasing average global temperature by about $3°C$ ($5°F$).

Supporting these models is a correlation between CO_2 levels and temperatures in prehistoric times. One way climatologists estimate past CO_2 concentrations is to measure CO_2 levels in bubbles trapped in glacial ice, some of which are 700,000 years old. Prehistoric temperatures are inferred by several methods, including analysis of past vegetation based on fossils and the chemical isotopes in sediments and corals. An increase of only $1.3°C$ would make the world warmer than at any time in the past 100,000 years. A warming trend would also alter the geographic distribution of precipitation, likely making agricultural areas of the central United States much drier, for example.

The ecosystems where the largest warming has *already* occurred are those in the far north, particularly northern coniferous forests and tundra. As snow and ice melt and uncover darker, more absorptive surfaces, these systems reflect less radiation back to the atmosphere and warm further. Arctic sea ice in the summer of 2007 covered the smallest area on record. Climate models suggest that there may be no summer ice there within a few decades, decreasing habitat for polar bears, seals, and seabirds. Higher temperatures also increase the likelihood of fires. In boreal forests of western North America and Russia, fires have burned twice the usual area in recent decades.

Range Shifts and Climate Change

Many organisms, especially plants that cannot disperse rapidly over long distances, may not be able to survive the rapid climate change projected to result from global warming. Furthermore, many habitats today are more fragmented than ever (see Concept 43.3), further limiting the ability of many organisms to migrate.

One way to predict the possible effects of future climate change on geographic ranges is to look back at the changes that have occurred in temperate regions since the last ice age ended. Until about 16,000 years ago, continental glaciers covered much of North America and Eurasia. As the climate warmed and the glaciers retreated, tree distributions expanded northward. A detailed record of these changes is captured in fossil pollen deposited in lakes and ponds. (Recall from Chapter 26 that wind and animals sometimes disperse pollen and seeds over great distances.) If researchers can determine the climatic limits of current distributions of organisms, they can make predictions about how those distributions may change with continued climatic warming.

A fundamental question when applying this approach to plants is whether seeds can disperse quickly enough to sustain the range shift of each species as climate changes. Fossil pollen shows that species with winged seeds that disperse relatively far from a parent tree, such as the sugar maple (*Acer saccharum*), expanded rapidly into the northeastern United States and Canada after the last ice age ended. In contrast, the northward range expansion of the eastern hemlock (*Tsuga canadensis*), whose seeds lack wings, was delayed nearly 2,500 years compared with the shift in suitable habitat.

Will plants and other species be able to keep up with the much more rapid warming projected for this century? Ecologists have attempted to answer this question for the American beech (*Fagus grandifolia*). Their models predict that the northern limit of the beech's range may move 700–900 km northward in the next century, and its southern range limit will shift even more. The current and predicted geographic ranges of this species under two different climate-change scenarios are illustrated in **Figure 43.27**. If these predictions are even approximately correct, the beech's range must shift 7–9 km northward per year to keep pace with the warming climate. However, since the end of the last ice age, the beech has moved at a rate of only 0.2 km per year. Without human help in moving to new habitats, species such as the American beech may have much smaller ranges or even become extinct.

Changes in the distributions of species are already evident in many well-studied groups of terrestrial, marine, and freshwater organisms, consistent with the signature of a warmer world. In Europe, for instance, the northern range limits of

(a) Current range
© 1989 AAAS

(b) 4.5°C warming over next century

(c) 6.5°C warming over next century

▲ **Figure 43.27 Current range and predicted range for the American beech under two climate-change scenarios.**

? *The predicted range in each scenario is based on climate factors alone. What other factors might alter the distribution of this species?*

22 of 35 butterfly species studied had shifted farther north by 35–240 km over the time periods for which records exist, in some cases beginning in 1900. Other research shows that a Pacific diatom, *Neodenticula seminae*, recently has colonized the Atlantic Ocean for the first time in 800,000 years. As Arctic sea ice has receded in the past decade, the increased flow of water from the Pacific has swept these diatoms around Canada and into the Atlantic, where they quickly became established.

Climate Change Solutions

We will need a variety of approaches to slow global warming and climate change in general. Quick progress can be made by using energy more efficiently and by replacing fossil fuels with renewable solar and wind power and, more controversially, with nuclear power. Today, coal, gasoline, wood, and other organic fuels remain central to industrialized societies and cannot be burned without releasing CO_2. Stabilizing CO_2 emissions will require concerted international effort and changes in both personal lifestyles and industrial processes.

Another important approach to slowing global warming is to reduce deforestation around the world, particularly in the tropics. Deforestation currently accounts for about 12% of greenhouse gas emissions. Recent research shows that paying countries *not* to cut forests could decrease the rate of deforestation by half within 10 to 20 years. Reduced deforestation would not only slow the buildup of greenhouse gases in our atmosphere but sustain native forests and preserve biodiversity, a positive outcome for all.

CONCEPT CHECK 43.4

1. How can the addition of excess mineral nutrients to a lake threaten its fish population?
2. **MAKE CONNECTIONS** There are vast stores of organic matter in the soils of northern coniferous forests and tundra around the world. Suggest an explanation for why scientists who study global warming are closely monitoring these stores (see Figure 42.12).

For suggested answers, see Appendix A.

CONCEPT 43.5

The human population is no longer growing exponentially but is still increasing rapidly

Global environmental problems, such as climate change, arise from the intersection of two factors. One is the growing amount of goods and resources that each of us consumes. The other is the increasing size of the human population, which has grown at an unprecedented rate in the last few centuries. No population can grow indefinitely, however. In this section,

we'll apply ecological concepts to the specific case of the human population.

The Global Human Population

The exponential growth model (see Figure 40.19) approximates the human population explosion over the last four centuries **(Figure 43.28)**. Ours is a singular case; no other population of large animals has likely ever sustained so much growth for so long. The human population increased relatively slowly until about 1650, at which time approximately 500 million people inhabited Earth. Our population doubled to 1 billion within the next two centuries, doubled again to 2 billion by 1930, and doubled still again by 1975 to more than 4 billion. The global population is now more than 7 billion and is increasing by about 78 million each year. At this rate, it takes only about four years to add the equivalent of another United States to the world population. Ecologists predict a population of 8.1–10.6 billion people on Earth by the year 2050.

Though the global population is still growing, the *rate* of growth did begin to slow during the 1960s **(Figure 43.29)**. The annual rate of increase in the global population peaked at 2.2% in 1962 but had declined to 1.1% by 2011. Current models project a continued decline in the annual growth rate to roughly 0.5% by 2050, a rate that would still add 45 million more people per year if the population climbs to a projected 9 billion. The reduction in growth rate over the past four decades shows that the human population has departed from true exponential growth, which assumes a constant rate. This departure is the result of fundamental changes in population dynamics due to diseases, including AIDS, and to voluntary population control.

The growth rates of individual nations vary with their degree of industrialization. In industrialized nations, populations are

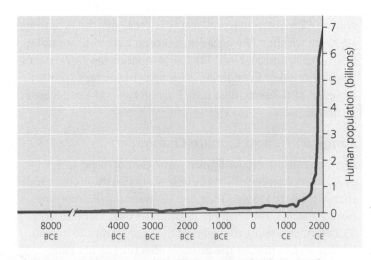

▲ **Figure 43.28 Human population growth (data as of 2011).** The global human population has grown almost continuously throughout history, but it skyrocketed after the Industrial Revolution. The rate of population growth has slowed in recent decades, mainly as a result of decreased birth rates throughout the world.

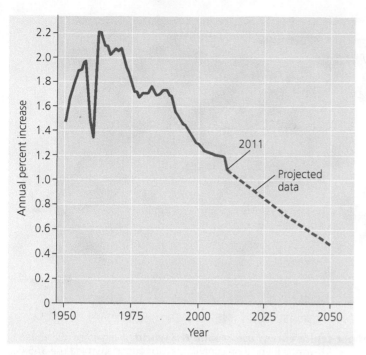

▲ **Figure 43.29 Annual percent increase in the global human population (data as of 2011).** The sharp dip in the 1960s is due mainly to a famine in China in which about 60 million people died.

near equilibrium, with growth rates of about 0.1% per year and reproductive rates near the replacement level (total fertility rate = 2.1 children per female). In countries such as Canada, Germany, Japan, and the United Kingdom, total reproductive rates are in fact *below* replacement. These populations will eventually decline if there is no immigration and if the birth rate does not change. In fact, the population is already declining in many eastern and central European countries.

In contrast, most of the current global population growth of 1.1% per year is concentrated in less industrialized nations, where about 80% of the world's people now live. Countries such as Afghanistan, Uganda, and Jordan had populations that grew by more than 3% *per year* between 2005 and 2010. Although death rates have declined rapidly since 1950 in many less industrialized countries, birth rates have declined substantially in only some of them. The fall in birth rate has been most dramatic in China. Largely because of the Chinese government's strict one-child policy, the expected total fertility rate (children per woman per lifetime) decreased from 5.9 in 1970 to 1.6 in 2011. The transition to lower birth rates has also been rapid in some African countries, though birth rates remain high in most of sub-Saharan Africa. In India, birth rates have fallen more slowly.

A unique feature of human population growth is our ability to control family sizes using family planning and voluntary contraception. Social change and the rising educational and career aspirations of women in many cultures encourage women to delay marriage and postpone reproduction. Delayed reproduction helps to decrease population growth rates and to

move a society toward zero population growth under conditions of low birth rates and low death rates. However, there is a great deal of disagreement as to how much support should be provided for global family planning efforts.

Global Carrying Capacity

No ecological question is more important than the future size of the human population. The projected worldwide population size depends on assumptions about future changes in birth and death rates. As we noted earlier, population ecologists project a global population of approximately 8.1–10.6 billion people in 2050. In other words, without some catastrophe, an estimated 1–4 billion people will be added to the population in the next four decades because of the momentum of population growth. But just how many humans can the biosphere support? Will the world be overpopulated in 2050? Is it *already* overpopulated?

Estimates of Carrying Capacity

Estimates of the human carrying capacity of Earth have varied from less than 1 billion to more than 1,000 billion (1 trillion), with an average of 10–15 billion. Carrying capacity is difficult to estimate, and scientists use different methods to produce their estimates. Some current researchers use curves like that produced by the logistic equation (see Figure 40.19) to predict the future maximum of the human population. Others generalize from existing "maximum" population density and multiply this number by the area of habitable land. Still others base their estimates on a single limiting factor, such as food, and consider variables such as the amount of available farmland, the average yield of crops, the prevalent diet—vegetarian or meat based—and the number of calories needed per person per day.

Limits on Human Population Size

A more comprehensive approach to estimating the carrying capacity of Earth is to recognize that humans have multiple constraints: We need food, water, fuel, building materials, and other resources, such as clothing and transportation. The **ecological footprint** concept summarizes the aggregate land and water area required by each person, city, or nation to produce all the resources it consumes and to absorb all the waste it generates. One way to estimate the ecological footprint of the entire human population is to add up all the ecologically productive land on the planet and divide by the population. This calculation yields approximately 2 ha per person (1 ha = 2.47 acres). Reserving some land for parks and conservation means reducing this allotment to 1.7 ha per person—the benchmark for comparing actual ecological footprints. Anyone who consumes resources that require more than 1.7 ha to produce is said to be using an unsustainable share of Earth's resources. A typical ecological footprint for a person in the United States is about 10 ha.

Ecologists sometimes calculate ecological footprints using other currencies besides land area, such as energy use. Average energy use differs greatly in developed and developing nations (**Figure 43.30**). A typical person in the United States, Canada, or Norway consumes roughly 30 times the energy that a person in central Africa does. Moreover, fossil fuels, such as oil, coal, and natural gas, are the source of 80% or more of the energy used in most developed nations. This unsustainable reliance on fossil fuels is changing Earth's climate and increasing the amount of waste that each of us produces. Ultimately, the combination of resource use per person and population density determines our global ecological footprint.

How many people our planet can sustain depends on the quality of life each of us enjoys and the distribution of wealth across people and nations, topics of great concern and political debate. Unlike other organisms, we can decide whether zero population growth will be attained through social changes based on human choices or, instead, through increased mortality due to resource limitation, plagues, war, and environmental degradation.

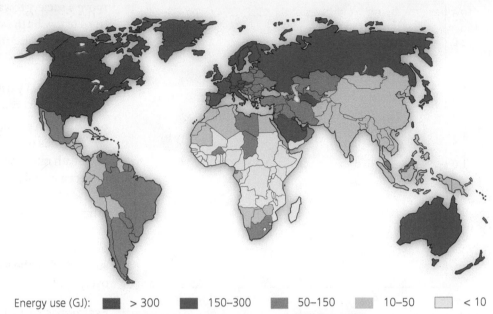

Energy use (GJ): ■ > 300   ■ 150–300   ■ 50–150   ■ 10–50   □ < 10

▲ **Figure 43.30 Annual per capita energy use around the world.** A gigajoule (GJ) equals 10⁹ J. For comparison, leaving a 100-watt light bulb on continuously for one year would use 3.15 GJ.

CONCEPT CHECK 43.5

1. How has the growth of Earth's human population changed in recent decades? Answer in terms of growth rate and the number of people added each year.
2. **WHAT IF?** What choices can you make to influence your own ecological footprint?

For suggested answers, see Appendix A.

CONCEPT 43.6

Sustainable development can improve human lives while conserving biodiversity

With the loss and fragmentation of habitats, changes in Earth's physical environment and climate, and increasing human population, we face difficult trade-offs in managing the world's resources. Preserving all habitat patches isn't feasible, so biologists must help societies set conservation priorities by identifying which habitat patches are most crucial. Ideally, implementing these priorities should also improve the quality of life for local people. Ecologists use the concept of *sustainability* as a tool to establish long-term conservation priorities.

Sustainable Development

We need to understand the interconnections of the biosphere if we are to protect species from extinction and improve the quality of human life. To this end, many nations, scientific societies, and other groups have embraced the concept of **sustainable development**, economic development that meets the needs of people today without limiting the ability of future generations to meet their needs.

Achieving sustainable development is an ambitious goal. To sustain ecosystem processes and stem the loss of biodiversity, we must connect life science with the social sciences, economics, and the humanities. We must also reassess our personal values. As you learned in Concept 43.5, those of us living in developed nations have a larger ecological footprint than do people living in developing nations. By including the long-term costs of consumption in profit-and-loss calculations, we can learn to value the natural processes that sustain us. The following case study illustrates how the combination of scientific and personal efforts can make a significant difference in creating a truly sustainable world.

Case Study: *Sustainable Development in Costa Rica*

The success of conservation in Costa Rica that we discussed in Concept 43.3 has required a partnership between the national government, nongovernment organizations (NGOs), and private citizens. Many nature reserves established by individuals have been recognized by the government as national wildlife reserves and given significant tax benefits. However, conservation and restoration of biodiversity make up only one facet of sustainable development; the other key facet is improving the human condition.

How have the living conditions of the Costa Rican people changed as the country has pursued its conservation goals? Two of the most fundamental indicators of living conditions are infant mortality rate and life expectancy. From 1930 to 2010, the infant mortality rate in Costa Rica declined from 170 to 9 per 1,000 live births; over the same period, life expectancy increased from about 43 years to 79 years. Another indicator of living conditions is the literacy rate, which was 96% in 2011, compared to 97% in the United States. Such statistics show that living conditions in Costa Rica have improved greatly over the period in which the country has dedicated itself to conservation and restoration. While this result does not prove that conservation *causes* an improvement in human welfare, we can say with certainty that development in Costa Rica has attended to both nature *and* people.

(a) Detail of animals in a 17,000-year-old cave painting, Lascaux, France

(b) A 30,000-year-old ivory carving of a water bird, found in Germany

(c) Nature lovers on a wildlife-watching expedition

▲ **Figure 43.31 Biophilia, past and present.**

The Future of the Biosphere

Our modern lives are very different from those of early humans, who hunted and gathered to survive. Their reverence for the natural world is evident in the early murals of wildlife they painted on cave walls **(Figure 43.31a)** and in the stylized visions of life they sculpted from bone and ivory **(Figure 43.31b)**.

Our lives reflect remnants of our ancestral attachment to nature and the diversity of life—the concept of *biophilia* that was introduced early in this chapter. We evolved in natural environments rich in biodiversity, and we still have an affinity for such settings **(Figure 43.31c, d)**. E. O. Wilson makes the case that our biophilia is innate, an evolutionary product of natural selection acting on a brainy species whose survival depended on a close connection to the environment and a practical appreciation of plants and animals.

Our appreciation of life guides the field of biology today. We celebrate life by deciphering the genetic code that makes each species unique. We embrace life by using fossils and DNA to chronicle evolution through time. We preserve life through our efforts to classify and protect the millions of species on Earth. We respect life by using nature responsibly and reverently to improve human welfare.

Biology is the scientific expression of our desire to know nature. We are most likely to protect what we appreciate, and we are most likely to appreciate what we understand. By learning about the processes and diversity of life, we also become more aware of ourselves and our place in the biosphere. We hope this book has served you well in this lifelong adventure.

CONCEPT CHECK 43.6

1. What is meant by the term *sustainable development*?
2. How might biophilia inspire us to conserve species and restore ecosystems?
3. **WHAT IF?** Suppose a new fishery is discovered, and you are put in charge of developing it sustainably. What ecological data might you want on the fish population? What criteria would you apply for the fishery's development?

For suggested answers, see Appendix A.

(d) A young biologist holding a songbird

SUMMARY OF KEY CONCEPTS

43.1

Human activities threaten Earth's biodiversity (pp. 883–888)

- Biodiversity can be considered at three main levels:

Genetic diversity: source of variations that enable populations to adapt to environmental changes

Species diversity: important in maintaining structure of communities and food webs

Ecosystem diversity: provides life-sustaining services such as nutrient cycling and waste decomposition

- Our biophilia enables us to recognize the value of biodiversity for its own sake. Other species also provide humans with food, fiber, medicines, and **ecosystem services**.
- Four major threats to biodiversity are habitat loss, **introduced species**, overharvesting, and global change.

> **?** *Give at least three examples of key ecosystem services that nature provides for people.*

43.2

Population conservation focuses on population size, genetic diversity, and critical habitat (pp. 888–891)

- When a population drops below a **minimum viable population (MVP)** size, its loss of genetic variation due to nonrandom mating and genetic drift can trap it in an **extinction vortex**.
- The declining-population approach focuses on the environmental factors that cause decline, regardless of absolute population size. It follows a step-by-step conservation strategy.
- Conserving species often requires resolving conflicts between the habitat needs of **endangered species** and human demands.

> **?** *Why is the minimum viable population size smaller for a population that is more genetically diverse than it is for a less genetically diverse population?*

43.3

Landscape and regional conservation help sustain biodiversity (pp. 891–895)

- The structure of a landscape can strongly influence biodiversity. As habitat fragmentation increases and edges become more extensive, biodiversity tends to decrease. **Movement corridors** can promote dispersal and help sustain populations.
- **Biodiversity hot spots** are also hot spots of extinction and thus prime candidates for protection. Sustaining biodiversity in parks and reserves requires management to ensure that human activities in the surrounding landscape do not harm the protected habitats. The **zoned reserve** model recognizes that conservation efforts often involve working in landscapes that are greatly affected by human activity.

> **?** *Give two examples that show how habitat fragmentation can harm species in the long term.*

43.4

Earth is changing rapidly as a result of human actions (pp. 895–900)

- Agriculture removes plant nutrients from ecosystems, so large supplements are usually required. The nutrients in fertilizer can pollute groundwater and surface water, where they can stimulate excess algal growth (eutrophication).
- The release of toxic wastes has polluted the environment with harmful substances that often persist for long periods and become increasingly concentrated in successively higher trophic levels of food webs (**biological magnification**).
- Because of the burning of wood and fossil fuels and other human activities, the atmospheric concentration of CO_2 and other greenhouse gases has been steadily increasing. The ultimate effects include significant global warming and other changes in climate.

> **?** *In the face of biological magnification of toxins, is it healthier to feed at a lower or higher trophic level? Explain.*

43.5

The human population is no longer growing exponentially but is still increasing rapidly (pp. 900–902)

- Since about 1650, the global human population has grown exponentially, but within the last 50 years, the rate of growth has fallen by half. While some nations' populations are growing rapidly, those of others are stable or declining in size.
- The carrying capacity of Earth for humans is uncertain. **Ecological footprint** is the aggregate land and water area needed to produce all the resources a person or group of people consume and to absorb all of their wastes. It is one measure of how close we are to the carrying capacity of Earth. With a world population of more than 7 billion people, we are already using many resources in an unsustainable manner.

> **?** *How are we humans different from other species in being able to "choose" a carrying capacity?*

Sustainable development can improve human lives while conserving biodiversity (pp. 902–903)

- The goal of the Sustainable Biosphere Initiative is to acquire the ecological information needed for the development, management, and conservation of Earth's resources.
- Costa Rica's success in conserving tropical biodiversity has involved a partnership between the government, other organizations, and private citizens. Human living conditions in Costa Rica have improved along with ecological conservation.
- By learning about biological processes and the diversity of life, we become more aware of our close connection to the environment and the value of other organisms that share it.

? *Why is sustainability such an important goal for conservation biologists?*

TEST YOUR UNDERSTANDING

Level 1: Knowledge/Comprehension

1. One characteristic that distinguishes a population in an extinction vortex from most other populations is that
   a. its habitat is fragmented.
   b. it is a rare, top-level predator.
   c. its effective population size is much lower than its total population size.
   d. its genetic diversity is very low.
   e. it is not well adapted to edge conditions.

2. The main cause of the increase in the amount of CO_2 in Earth's atmosphere over the past 150 years is
   a. increased worldwide primary production.
   b. increased worldwide standing crop.
   c. an increase in the amount of infrared radiation absorbed by the atmosphere.
   d. the burning of larger amounts of wood and fossil fuels.
   e. additional respiration by the rapidly growing human population.

3. What is the single greatest threat to biodiversity?
   a. overharvesting of commercially important species
   b. introduced species that compete with native species
   c. pollution of Earth's air, water, and soil
   d. disruption of trophic relationships as more and more prey species become extinct
   e. habitat alteration, fragmentation, and destruction

Level 2: Application/Analysis

4. Which of the following is a consequence of biological magnification?
   a. Toxic chemicals in the environment pose greater risk to top-level predators than to primary consumers.
   b. Populations of top-level predators are generally smaller than populations of primary consumers.
   c. The biomass of producers in an ecosystem is generally higher than the biomass of primary consumers.
   d. Only a small portion of the energy captured by producers is transferred to consumers.
   e. The amount of biomass in the producer level of an ecosystem decreases if the producer turnover time increases.

5. Which of the following strategies would most rapidly increase the genetic diversity of a population in an extinction vortex?
   a. Capture all remaining individuals in the population for captive breeding followed by reintroduction to the wild.
   b. Establish a reserve that protects the population's habitat.
   c. Introduce new individuals transported from other populations of the same species.
   d. Sterilize the least fit individuals in the population.
   e. Control populations of the endangered population's predators and competitors.

6. Of the following statements about protected areas that have been established to preserve biodiversity, which one is *not* correct?
   a. About 25% of Earth's land area is now protected.
   b. National parks are one of many types of protected areas.
   c. Most protected areas are too small to protect species.
   d. Management of a protected area should be coordinated with management of the land surrounding the area.
   e. It is especially important to protect biodiversity hot spots.

Level 3: Synthesis/Evaluation

7. **DRAW IT** Using Figure 43.26 as a starting point, extend the *x*-axis to the year 2100. Then extend the CO_2 curve, assuming that the CO_2 concentration continues to rise as fast as it did from 1974 to 2011. What will be the approximate CO_2 concentration in 2100? What ecological factors and human decisions will influence the actual rise in CO_2 concentration? How might additional scientific data help societies predict this value?

8. **SCIENTIFIC INQUIRY**
   DRAW IT Suppose that you are managing a forest reserve, and one of your goals is to protect local populations of woodland birds from parasitism by the brown-headed cowbird. You know that female cowbirds usually do not venture more than about 100 m into a forest and that nest parasitism is reduced when woodland birds nest away from forest edges. The reserve you manage extends about 6,000 m from east to west and 1,000 m from north to south. It is surrounded by a deforested pasture on the west, an agricultural field for 500 m in the southwest corner, and intact forest everywhere else. You must build a road, 10 m by 1,000 m, from the north to the south side of the reserve and construct a maintenance building that will take up 100 m² in the reserve. Draw a map of the reserve, showing where you would put the road and the building to minimize cowbird intrusion along edges. Explain your reasoning.

9. **FOCUS ON EVOLUTION**
   One factor favoring rapid population growth by an introduced species is the absence of the predators, parasites, and pathogens that controlled its population in the region where it evolved. In a short essay (100–150 words), explain how evolution by natural selection would influence the rate at which native predators, parasites, and pathogens in a region of introduction attack an introduced species.

10. **FOCUS ON INTERACTIONS**
    In a short essay (100–150 words), identify the factor or factors that you think may ultimately be most important in regulating the human population, and explain your reasoning.

For selected answers, see Appendix A.

MasteringBiology®

Students Go to **MasteringBiology** for assignments, the eText, and the Study Area with practice tests, animations, and activities.

Instructors Go to **MasteringBiology** for automatically graded tutorials and questions that you can assign to your students, plus Instructor Resources.

Chapter 1

Figure Questions

Figure 1.4 Possible answers include the match between the shape of the humming-bird's beak and the flower from which it obtains nectar; the streamlined shape of the bird's body and ability to fold up its legs for efficient flight; the color and appearance of the flower, which attract the hummingbird. **Figure 1.19** The resident mice might have a more reddish coat color. Both beach and mainland mice would probably suffer higher predation rates in the new habitat than in their native environments. However, the dark mainland mice might be expected to stand out less against the background, so might have a lower predation rate than the beach mice.

Concept Check 1.1

1. Examples: A molecule consists of *atoms* bonded together. Each organelle has an orderly arrangement of *molecules*. Photosynthetic plant cells contain *organelles* called chloroplasts. A tissue consists of a group of similar *cells*. Organs such as the heart are constructed from several *tissues*. A complex multicellular organism, such as a plant, has several types of *organs*, such as leaves and roots. A population is a set of *organisms* of the same species. A community consists of *populations* of the various species inhabiting a specific area. An ecosystem consists of a biological *community* along with the nonliving factors important to life, such as air, soil, and water. The biosphere is made up of all of Earth's *ecosystems*. **2.** (a) Organization: Structure and function are correlated. (b) Organization: The cell is an organism's basic unit of structure and function, *and* Information: Life's processes involve the expression and transmission of genetic information. (c) Energy and Matter: Life requires transfer and transformation of energy, *and* Interactions: Organisms interact with other organisms and with the physical environment. **3.** Some possible answers: *Organization*: The ability of a human heart to pump blood requires an intact heart; it is not a capability of any of the heart's tissues or cells working alone. *Information*: Human eye color is determined by the combination of genes inherited from the two parents. *Energy and Matter*: A plant, such as a grass, absorbs energy from the sun and transforms it into molecules that act as stored fuel. Animals can eat parts of the plant and use the food for energy to carry out their activities. *Interactions*: A mouse eats food, such as nuts or grasses, and deposits some of the food material as feces and urine. Construction of a nest rearranges the physical environment and may hasten degradation of some of its components. The mouse may also act as food for a predator. *Evolution*: All plants have chloroplasts, indicating their descent from a common ancestor.

Concept Check 1.2

1. An address pinpoints a location by tracking from broader to narrower categories—a state, city, zip, street, and building number. This is analogous to the groups-subordinate-to-groups structure of biological taxonomy. **2.** The naturally occurring heritable variation in a population is "edited" by natural selection because individuals with heritable traits better suited to the environment survive and reproduce more successfully than others. Over time, better-suited individuals persist and their percentage in the population increases, while less suited individuals become less prevalent—a type of population editing.
3.

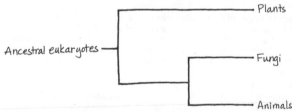

Concept Check 1.3

1. Inductive reasoning derives generalizations from specific cases; deductive reasoning predicts specific outcomes from general premises. **2.** The coloration pattern of the mice **3.** A scientific theory is usually more general than a hypothesis and substantiated by a much greater amount of evidence. Natural selection is an explanatory idea that applies to all kinds of organisms and is supported by vast amounts of evidence of various kinds. **4.** Science aims to understand natural phenomena and how they work, while technology involves application of scientific discoveries for a particular purpose or to solve a specific problem.

Summary of Key Concepts Questions

1.1 Finger movements rely on the coordination of the many structural components of the hand (muscles, nerves, bones, etc.), each of which is composed of elements from lower levels of biological *organization* (cells, molecules). The development of the hand relies on the genetic *information* encoded in chromosomes found in cells throughout the body. To power the finger movements that result in a text message, muscle and nerve cells require chemical *energy* that they transform in powering muscle contraction or in propagating nerve impulses. Finally, all of the anatomical and physiological features that allow the activity of texting are the outcome of a process of natural selection that resulted in the *evolution* of hands and of the mental facilities for use of language. **1.2** Ancestors of the beach mouse may have exhibited variations in their coat color. Because of the prevalence of visual predators, the better camouflaged (lighter) mice may have survived longer and been able to produce more offspring. Over time,

a higher and higher proportion of individuals in the population would have had the adaptation of lighter fur that acted to camouflage the mouse. **1.3** Inductive reasoning is used in forming hypotheses, while deductive reasoning leads to predictions that are used to test hypotheses.

Test Your Understanding

1. b **2.** c **3.** b **4.** c **5.** c **6.** d
7. Your figure should show the following: (1) For the biosphere, Earth with an arrow coming out of a tropical ocean; (2) for the ecosystem, a distant view of a coral reef; (3) for the community, a collection of reef animals and algae, with corals, fishes, some seaweed, and any other organisms you can think of; (4) for the population, a group of fish of the same species; (5) for the organism, one fish from your population; (6) for the organ, the fish's stomach, and for the organ system, the whole digestive tract (see Chapter 33 for help); (7) for a tissue, a group of similar cells from the stomach; (8) for a cell, one cell from the tissue, showing its nucleus and a few other organelles; (9) for an organelle, the nucleus, where most of the cell's DNA is located; and (10) for a molecule, a DNA double helix. Your sketches can be very rough!

Chapter 2

Figure Questions

Figure 2.6 Atomic number = 12; 12 protons, 12 electrons; 3 electron shells; 2 valence electrons **Figure 2.15** The plant is submerged in water (H_2O), in which the CO_2 is dissolved. The sun's energy is used to make sugar, which is found in the plant and can act as food for the plant itself, as well as for animals that eat the plant. The oxygen (O_2) is present in the bubbles. **Figure 2.16** One possible answer:

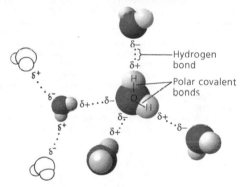

Figure 2.20 Without hydrogen bonds, water would behave like other small molecules, and the solid phase (ice) would be denser than liquid water. The ice would sink to the bottom and would no longer insulate the body of water. All the water would eventually freeze because the average annual temperature at the South Pole is −50°C. The shrimp could not survive. **Figure 2.21** Heating the solution would cause the water to evaporate faster than it is evaporating at room temperature. At a certain point, there wouldn't be enough water molecules to dissolve the salt ions. The salt would start coming out of solution and re-forming crystals. Eventually, all the water would evaporate, leaving behind a pile of salt like the original pile. **Figure 2.24** By causing the loss of coral reefs, a decrease in the ocean's carbonate concentration would have a ripple effect on noncalcifying organisms. Some of these organisms depend on the reef structure for protection, while others feed on species associated with reefs.

Concept Check 2.1

1. Yes, because an organism requires trace elements, even though only in small amounts **2.** A person with an iron deficiency will probably show fatigue and other effects of a low oxygen level in the blood. (The condition is called anemia and can also result from too few red blood cells or abnormal hemoglobin.)

Concept Check 2.2

1. $^{15}_{7}N$ **2.** 9 electrons; two electron shells; 1 electron is needed to fill the valence shell. **3.** The elements in a row all have the same number of electron shells. In a column, all the elements have the same number of electrons in their valence shells.

Concept Check 2.3

1. Each carbon atom has only three covalent bonds instead of the required four. **2.** The attraction between oppositely charged ions, forming ionic bonds **3.** If you could synthesize molecules that mimic these shapes, you might be able to treat diseases or conditions caused by the inability of affected individuals to synthesize such molecules.

Concept Check 2.4

1. At equilibrium, the forward and reverse reactions occur at the same rate. **2.** $C_6H_{12}O_6 + 6 O_2 \rightarrow 6 CO_2 + 6 H_2O + Energy$. Glucose and oxygen react to form carbon dioxide and water, releasing energy. We breathe in oxygen because we need it for this reaction to occur, and we breathe out carbon dioxide because it is a product of this reaction. (This reaction is called cellular respiration, and you will learn more about it in Chapter 7.)

Concept Check 2.5

1. Hydrogen bonds hold neighboring water molecules together. This cohesion helps the chain of water molecules move upward against gravity in water-conducting cells as water evaporates from the leaves. Adhesion between water molecules and the walls of the water-conducting cells also helps counter gravity. **2.** As water freezes, it expands because water molecules move farther apart in forming ice crystals. When there is water in a crevice of a boulder, expansion due to freezing may crack the boulder. **3.** A liter of blood would contain 7.8×10^{13} molecules of ghrelin (1.3×10^{-10} moles per liter $\times 6.02 \times 10^{23}$ molecules per mole). **4.** 10^5, or 100,000 **5.** The covalent bonds of water molecules would not be polar, and water molecules would not form hydrogen bonds with each other.

Summary of Key Concepts Questions

2.1 Iodine (part of a thyroid hormone) and iron (part of hemoglobin in blood) are both trace elements, required in minute quantities. Calcium and phosphorus (components of bones and teeth) are needed by the body in much greater quantities.

2.2

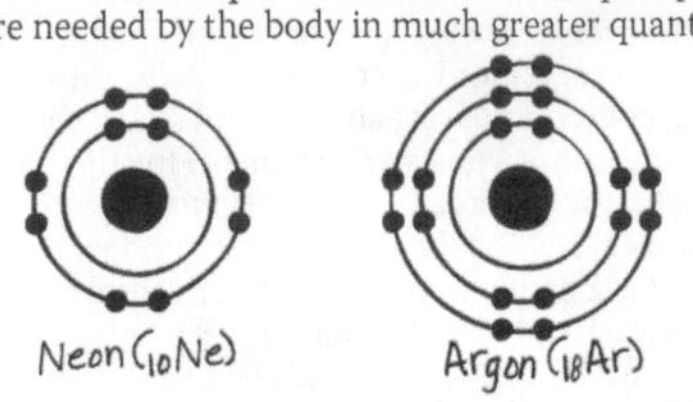

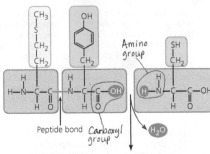

Both neon and argon are unreactive because they have completed valence shells. They do not have unpaired electrons that could participate in chemical bonds. **2.3** Electrons are shared equally between the two atoms in a nonpolar covalent bond. In a polar covalent bond, the electrons are drawn closer to the more electronegative atom. In the formation of ions, one or more electrons are completely transferred from one atom to a much more electronegative atom. **2.4** The concentration of products would increase as the added reactants were converted to products. Eventually, an equilibrium would again be reached in which the forward and reverse reactions were proceeding at the same rate and the relative concentrations of reactants and products returned to where they were before the addition of more reactants. **2.5** The polar covalent bonds of a water molecule allow it to form hydrogen bonds with other water molecules and other polar molecules as well. The sticking together of water molecules, called cohesion, and the sticking of water to other molecules, called adhesion, help water rise from the roots of plants to their leaves, among other biological benefits. Hydrogen bonding between water molecules is responsible for water's high specific heat (resistance to temperature change), which helps moderate temperature on Earth. Hydrogen bonding is also responsible for water's high heat of vaporization, which makes water useful for evaporative cooling. A lattice of stable hydrogen bonds in ice makes it less dense than liquid water, so that it floats, creating an insulating surface on bodies of water that allows organisms to live underneath. Finally, the polarity of water molecules resulting from their polar covalent bonds makes water an excellent solvent; polar and ionic atoms and molecules that are needed for life can exist in a dissolved state and participate in chemical reactions.

Test Your Understanding

1. b **2.** d **3.** d **4.** e **5.** c **6.** c **7.** e **8.** e

9.

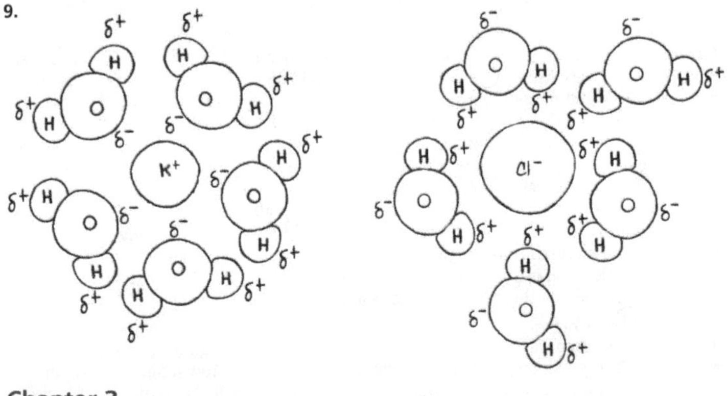

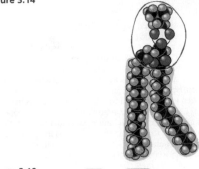

Chapter 3

Figure Questions
Figure 3.8

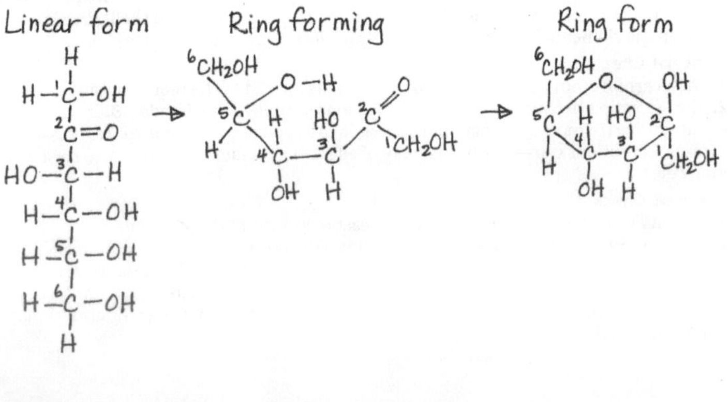

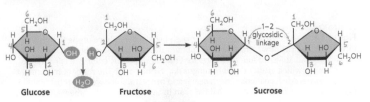

Note that the oxygen on carbon 5 lost its proton and that the oxygen on carbon 2, which used to be the carbonyl oxygen, gained a proton. Four carbons are in the fructose ring, and two are not. (The latter two carbons are attached to carbons 2 and 5, which are in the ring.) The fructose ring differs from the glucose ring, which has five carbons in the ring and one that is not. (Note that the orientation of this fructose molecule is flipped relative to that of the one in Figure 3.9.)

Figure 3.9

Figure 3.14

Figure 3.18

Figure 3.22 The R group of glutamic acid is acidic and hydrophilic, whereas that of valine is nonpolar and hydrophobic. Therefore, it is unlikely that valine can participate in the same intramolecular interactions that glutamic acid can. A change in these interactions would be expected to cause a disruption of molecular structure. **Figure 3.24** The spirals are α helices.

Concept Check 3.1

1. Both consist largely of hydrocarbon chains. **2.** It has both an amino group ($-NH_2$), which makes it an amine, and a carboxyl group ($-COOH$), which makes it a carboxylic acid. **3.** A chemical group that can act as a base (by picking up H^+) has been replaced with a group that can act as an acid, increasing the acidic properties of the molecule. The shape of the molecule would also change, likely changing the molecules with which it can interact.

Concept Check 3.2

1. Nine, with one water molecule required to hydrolyze each connected pair of monomers **2.** The amino acids in the fish protein must be released in hydrolysis reactions and incorporated into other proteins in dehydration reactions.

Concept Check 3.3

1. $C_3H_6O_3$ **2.** $C_{12}H_{22}O_{11}$ **3.** The antibiotic treatment is likely to have killed the cellulose-digesting prokaryotes in the cow's stomach. The absence of these prokaryotes would hamper the cow's ability to obtain energy from food and could lead to weight loss and possibly death. Thus, prokaryotic species are reintroduced, in appropriate combinations, in the gut culture given to treated cows.

Concept Check 3.4

1. Both have a glycerol molecule attached to fatty acids. The glycerol of a fat has three fatty acids attached, whereas the glycerol of a phospholipid is attached to two fatty acids and one phosphate group. **2.** Human sex hormones are steroids, a type of hydrophobic compound. **3.** The oil droplet membrane could consist of a single layer of phospholipids rather than a bilayer, because an arrangement in which the hydrophobic tails of the membrane phospholipids were in contact with the hydrocarbon regions of the oil molecules would be more stable.

Concept Check 3.5

1. The function of a protein is a consequence of its specific shape, which is lost when a protein becomes denatured. **2.** Secondary structure involves hydrogen bonds between atoms of the polypeptide backbone. Tertiary structure involves interactions between atoms of the side chains of the amino acid monomers. **3.** These are all nonpolar amino acids, so you would expect this region to be located in the interior of the folded polypeptide, where it would not contact the aqueous environment inside the cell.

Concept Check 3.6

1.

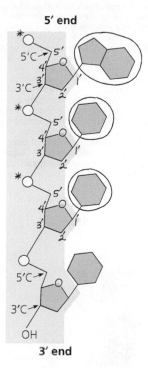

5′ end

3′ end

2.

5′–T A G G C C T–3′
3′–A T C C G G A–5′

3. a.

Mismatch

5′–T A A G C C T–3′
3′–A T C C G G A–5′

b.

3′–A T T C G G A–5′

Summary of Key Concepts Questions

3.1 The methyl group is nonpolar and not reactive. The other six groups are called functional groups and can participate in chemical reactions. Except for the sulfhydryl group, these functional groups are hydrophilic; they increase the solubility of organic compounds in water. **3.2** The polymers of carbohydrates, proteins, and nucleic acids are built from three different types of monomers: monosaccharides, amino acids, and nucleotides, respectively. **3.3** Both starch and cellulose are polymers of glucose, but the glucose monomers are in the α configuration in starch and the β configuration in cellulose. The glycosidic linkages thus have different geometries, giving the polymers different shapes and thus different properties. Starch is an energy-storage compound in plants; cellulose is a structural component of plant cell walls. Humans can hydrolyze starch to provide energy but cannot hydrolyze cellulose. Cellulose aids in the passage of food through the digestive tract. **3.4** Lipids are not polymers because they do not exist as a chain of linked monomers. They are not considered macromolecules because they do not reach the giant size of many polysaccharides, proteins, and nucleic acids. **3.5** A polypeptide, which may consist of hundreds of amino acids in a specific sequence (primary structure), has regions of coils and pleats (secondary structure), which are then folded into irregular contortions (tertiary structure) and may be non-covalently associated with other polypeptides (quaternary structure). The linear order of amino acids, with the varying properties of their side chains (R groups), determines what secondary and tertiary structures will form to produce a protein. The resulting unique three-dimensional shapes of proteins are key to their specific and diverse functions. **3.6** The complementary base pairing of the two strands of DNA makes possible the precise replication of DNA every time a cell divides, ensuring that genetic information is faithfully transmitted. In some types of RNA, complementary base pairing enables RNA molecules to assume specific three-dimensional shapes that facilitate diverse functions.

Test Your Understanding

1. b **2.** d **3.** d **4.** b **5.** a **6.** d **7.** c

8.

	Monomers or Components	Polymer or larger molecule	Type of linkage
Carbohydrates	Monosaccharides	Polysaccharides	Glycosidic linkages
Lipids	Fatty acids	Triacylglycerols	Ester linkages
Proteins	Amino acids	Polypeptides	Peptide bonds
Nucleic acids	Nucleotides	Polynucleotides	Phosphodiester linkages

Chapter 4

Figure Questions

Figure 4.5 A phospholipid is a lipid, consisting of a glycerol molecule joined to two fatty acids and one phosphate group. Together, the glycerol and phosphate end of the phospholipid form the "head," which is hydrophilic, while the hydrocarbon chains on the fatty acids form hydrophobic "tails." The presence in a single molecule of both a hydrophilic and a hydrophobic region makes the molecule ideal as the main building block of a membrane. **Figure 4.8** The DNA in a chromosome dictates synthesis of a messenger RNA (mRNA) molecule, which then moves out to the cytoplasm. There, the information is used for the production, on ribosomes, of proteins that carry out cellular functions. **Figure 4.9** Any of the bound ribosomes (attached to the endoplasmic reticulum) could be circled, because any could be making a protein that will be secreted.

Figure 4.22

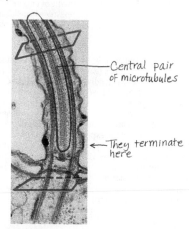

Each centriole has 9 sets of 3 microtubules, so the entire centrosome (two centrioles) has 54 microtubules. Each microtubule consists of a helical array of tubulin dimers (as shown in Table 4.1).

Figure 4.23

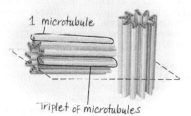

The two central microtubules terminate above the basal body, so they aren't present at the level of the cross section through the basal body, indicated by the lower red rectangle in (a).

Concept Check 4.1

1. Stains used for light microscopy are colored molecules that bind to cell components, affecting the light passing through, while stains used for electron microscopy involve heavy metals that affect the beams of electrons. **2.** (a) Light microscope, (b) scanning electron microscope

Concept Check 4.2

1. See Figure 4.7.

2.

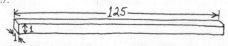

This cell would have the same volume as the large cell in column 2 and the collection of small cells in column 3 but proportionally more surface area than that in column 2 and less than that in column 3. Thus, the surface-to-volume ratio should be greater

than 1.2 but less than 6. To obtain the surface area, add the area of the six sides (the top, bottom, sides, and ends): $125 + 125 + 125 + 125 + 1 + 1 = 502$. The surface-to-volume ratio equals 502 divided by a volume of 125, or 4.0.

Concept Check 4.3

1. Ribosomes in the cytoplasm translate the genetic message, carried from the DNA in the nucleus by mRNA, into a polypeptide chain. **2.** Nucleoli consist of DNA and the ribosomal RNA (rRNA) made according to its instructions, as well as proteins imported from the cytoplasm. Together, the rRNA and proteins are assembled into large and small ribosomal subunits. (These are exported through nuclear pores to the cytoplasm, where they will participate in polypeptide synthesis.) **3.** Each chromosome consists of one long DNA molecule attached to numerous protein molecules, a combination called chromatin. The chromosomes are "condensing" as their thin strands of chromatin coil up to form shorter, thicker structures.

Concept Check 4.4

1. The primary distinction between rough and smooth ER is the presence of bound ribosomes on the rough ER. Both types of ER make phospholipids, but membrane proteins and secretory proteins are all produced on the ribosomes of the rough ER. The smooth ER also functions in detoxification, carbohydrate metabolism, and storage of calcium ions. **2.** Transport vesicles move membranes and substances they enclose between other components of the endomembrane system. **3.** The mRNA is synthesized in the nucleus and then passes out through a nuclear pore to be translated on a bound ribosome, attached to the rough ER. The protein is synthesized into the lumen of the ER and perhaps modified there. A transport vesicle carries the protein to the Golgi apparatus. After further modification in the Golgi, another transport vesicle carries it back to the ER, where it will perform its cellular function.

Concept Check 4.5

1. Both organelles are involved in energy transformation, mitochondria in cellular respiration and chloroplasts in photosynthesis. They both have multiple membranes that separate their interiors into compartments. In both organelles, the innermost membranes—cristae, or infoldings of the inner membrane, in mitochondria, and the thylakoid membranes in chloroplasts—have large surface areas with embedded enzymes that carry out their main functions. **2.** Yes. Plant cells are able to make their own sugar by photosynthesis, but mitochondria in these eukaryotic cells are the organelles that are able to generate energy from sugars, a function required in all cells. **3.** Mitochondria and chloroplasts are not derived from the ER, nor are they connected physically or via transport vesicles to organelles of the endomembrane system. Mitochondria and chloroplasts are structurally quite different from vesicles derived from the ER, which are bounded by a single membrane.

Concept Check 4.6

1. Dynein arms, powered by ATP, move neighboring doublets of microtubules relative to each other. Because they are anchored within the organelle and with respect to one another, the doublets bend instead of sliding past each other. Synchronized bending of the nine microtubule doublets brings about bending of both structures. **2.** Such individuals have defects in the microtubule-based movement of cilia and flagella. Thus, the sperm can't move because of malfunctioning or nonexistent flagella, and the airways are compromised because cilia that line the trachea malfunction or don't exist, and so mucus can't be cleared from the lungs.

Concept Check 4.7

1. One obvious difference is the presence of direct cytoplasmic connections between cells of plants (plasmodesmata) and animals (gap junctions). These connections result in the cytoplasm being continuous between adjacent cells. **2.** The cell would not be able to function properly and would probably soon die, as the cell wall or ECM must be permeable to allow the exchange of matter between the cell and its external environment. Molecules involved in energy production and use must be allowed entry, as well as those that provide information about the cell's environment. Other molecules, such as products synthesized by the cell for export and the by-products of cellular respiration, must be allowed to exit. **3.** The parts of the protein that face aqueous regions would be expected to have polar or charged (hydrophilic) amino acids, while the parts that go through the membrane would be expected to have nonpolar (hydrophobic) amino acids. You would predict polar or charged amino acids at each end (tail), in the region of the cytoplasmic loop, and in the regions of the two extracellular loops. You would predict nonpolar amino acids in the four regions inside the membrane between the tails and loops.

Summary of Key Concepts Questions

4.1 Both light and electron microscopy allow cells to be studied visually, thus helping us understand internal cellular structure and the arrangement of cell components. Cell fractionation techniques separate out different groups of cell components, which can then be analyzed biochemically to determine their function. Performing microscopy on cell fractions helps to correlate the biochemical function of the cell with the cell component responsible. **4.2** The separation of different functions in different organelles has several advantages. Reactants and enzymes can be concentrated in one area instead of spread throughout the cell. Reactions that require specific conditions, such as a lower pH, can be compartmentalized. And enzymes for specific reactions may be embedded in the membranes that enclose or partition an organelle. **4.3** The nucleus contains the genetic material of the cell in the form of DNA, which codes for messenger RNA, which in turn provides instructions for the synthesis of proteins (including the proteins that make up part of the ribosomes). DNA also codes for ribosomal RNA, which is combined with proteins in the nucleolus into the subunits of ribosomes. Within the cytoplasm, ribosomes join with mRNA to build polypeptides, using the genetic information in the mRNA. **4.4** Transport vesicles move proteins and membrane synthesized by the rough ER to the Golgi for further processing and then to the plasma membrane, lysosomes, or other locations in the cell, including back to the ER. **4.5** According to the endosymbiont theory, mitochondria originated from an oxygen-using prokaryotic cell that was engulfed by an ancestral eukaryotic

cell. Over time, the host and endosymbiont evolved into a single unicellular organism. Chloroplasts originated when at least one of these mitochondria-containing eukaryotic cells engulfed and then retained a photosynthetic prokaryote. **4.6** Inside the cell, motor proteins interact with components of the cytoskeleton to move cellular parts. Motor proteins may "walk" vesicles along microtubules. The movement of cytoplasm within a cell involves interactions of the motor protein myosin and microfilaments (actin filaments). Whole cells can be moved by the rapid bending of flagella or cilia, which is caused by the motor-protein-powered sliding of microtubules within these structures. Some cells move by amoeboid movement, which involves interactions of microfilaments with myosin. Interactions of motor proteins and microfilaments in muscle cells can propel multicellular organisms. **4.7** A plant cell wall is primarily composed of microfibrils of cellulose embedded in other polysaccharides and proteins. The ECM of animal cells is primarily composed of the glycoproteins collagen and fibronectin, as well as other protein fibers. These fibers are embedded in a network of carbohydrate-rich proteoglycans. A plant cell wall provides structural support for the cell and, collectively, for the plant body. In addition to giving support, the ECM of an animal cell allows for communication of environmental changes into the cell.

Test Your Understanding

1. b **2.** d **3.** b **4.** e **5.** a **6.** d **7.** c **8.** See Figure 4.7.

Chapter 5

Figure Questions

Figure 5.3

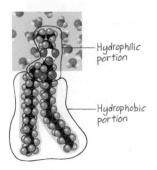

The hydrophilic portion is in contact with an aqueous environment (cytosol or extracellular fluid), and the hydrophobic portion is in contact with the hydrophobic portions of other phospholipids in the interior of the bilayer. **Figure 5.4** You couldn't rule out movement of proteins within membranes of the same species. You might propose that the membrane lipids and proteins from one species weren't able to mingle with those from the other species because of some incompatibility. **Figure 5.7** A transmembrane protein like the dimer in (c) might change its shape upon binding to a particular ECM molecule. The new shape might enable the interior portion of the protein to bind to a second, cytoplasmic protein that would relay the message to the inside of the cell, as shown in (f).

Figure 5.8

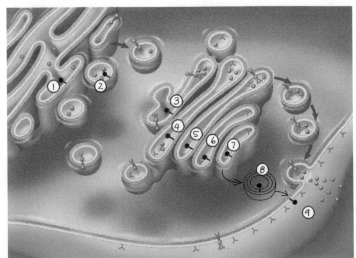

The protein would contact the extracellular fluid. **Figure 5.10** The orange dye would be evenly distributed throughout the solution on both sides of the membrane. The solution levels would not be affected because the orange dye can diffuse through the membrane and equalize its concentration. Thus, no additional osmosis would take place in either direction. **Figure 5.15** The diamond solutes are moving into the cell (down), and the round solutes are moving out of the cell (up); both are moving against their concentration gradient. **Figure 5.23** The testosterone molecule is hydrophobic and can therefore pass directly through the lipid bilayer of the plasma membrane into the cell. (Hydrophilic molecules cannot do this.) **Figure 5.24** The active form of protein kinase 1

Concept Check 5.1

1. They are on the inner side of the transport vesicle membrane. **2.** The grasses living in the cooler region would be expected to have more unsaturated fatty acids in their membranes because those fatty acids remain fluid at lower temperatures. The grasses living immediately adjacent to the hot springs would be expected to have more saturated fatty acids, which would allow the fatty acids to "stack" more closely, making the membranes less fluid and therefore helping them to stay intact at higher temperatures. (Cholesterol could not moderate the effects of temperature on membrane fluidity in this case because it is not found in appreciable quantities in plant cell membranes.)

Concept Check 5.2

1. O_2 and CO_2 are both small, nonpolar molecules that can easily pass through the hydrophobic interior of a membrane. **2.** Water is a polar molecule, so it cannot pass very rapidly through the hydrophobic region in the middle of a phospholipid bilayer. **3.** The hydronium ion is charged, while glycerol is not. Charge is probably more significant than size as a basis for exclusion by the aquaporin channel.

Concept Check 5.3

1. CO_2 is a nonpolar molecule that can diffuse through the plasma membrane. As long as it diffuses away so that the concentration remains low outside the cell, other CO_2 molecules will continue to exit the cell in this way. (This is the opposite of the case for O_2, described in this section.) **2.** The water is hypotonic to the plant cells, so the plant cells take up water. Thus, the cells of the vegetable remain turgid, and the vegetable (for example, lettuce or spinach) remains crisp and does not wilt. **3.** The *Paramecium caudatum*'s contractile vacuole becomes less active. The vacuole pumps out excess water that accumulates in the cell; this accumulation occurs only in a hypotonic environment.

Concept Check 5.4

1. The pump uses ATP. To establish a voltage, ions have to be pumped against their gradients, which requires energy. **2.** Each ion is being transported against its electrochemical gradient. If either ion were transported down its electrochemical gradient, this process *would* be considered cotransport. **3.** The internal environment of a lysosome is acidic, so it has a higher concentration of H^+ than does the cytosol. Therefore, you might expect the membrane of the lysosome to have a proton pump such as that shown in Figure 5.16 to pump H^+ into the lysosome.

Concept Check 5.5

1. Exocytosis. When a transport vesicle fuses with the plasma membrane, the vesicle membrane becomes part of the plasma membrane.
2.

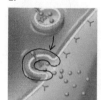

3. The glycoprotein would be synthesized in the ER lumen, move through the Golgi apparatus, and then travel in a vesicle to the plasma membrane, where it would undergo exocytosis and become part of the ECM.

Concept Check 5.6

1. The secretion of neurotransmitter molecules at a synapse is an example of local signaling. The electrical signal that travels along a very long nerve cell and is passed to the next nerve cell can be considered an example of long-distance signaling. (Note, however, that local signaling at the synapse between two cells is necessary for the signal to pass from one cell to the next.) **2.** Protein phosphatases reverse the effects of the kinases. **3.** At each step in a cascade of sequential activations, one molecule or ion may activate numerous molecules functioning in the next step.

Summary of Key Concepts Questions

5.1 Plasma membranes define the cell by separating the cellular components from the external environment. This allows conditions inside cells to be controlled by membrane proteins, which regulate entry and exit of molecules and even cell function (see Figure 5.7). The processes of life can be carried out inside the controlled environment of the cell, so membranes are crucial. In eukaryotes, membranes also function to subdivide the cytoplasm into different compartments where distinct processes can occur, even under differing conditions such as pH. **5.2** Aquaporins are channel proteins that greatly increase the permeability of a membrane to water molecules, which are polar and therefore do not readily diffuse through the hydrophobic interior of the membrane. **5.3** There will be a net diffusion of water out of a cell into a hypertonic solution. The free water concentration is higher inside the cell than in the solution (where many water molecules are not free, but clustered around the higher concentration of solute particles). **5.4** One of the solutes moved by the cotransporter is actively transported against its concentration gradient. The energy for this transport comes from the concentration gradient of the other solute, which was established by an electrogenic pump that used energy (usually provided by ATP) to transport the other solute across the membrane. **5.5** Receptor-mediated endocytosis. In this process, specific molecules bind to receptors on the plasma membrane in a region where a coated pit develops. The cell can acquire bulk quantities of those specific molecules when the coated pit forms a vesicle and carries the bound molecules into the cell. **5.6** A cell is able to respond to a hormone only if it has a receptor protein on the cell surface or inside the cell that can bind to the hormone. The response to a hormone depends on the specific cellular activity that a signal transduction pathway triggers within the cell. The response can vary for different types of cells.

Test Your Understanding

1. b **2.** a **3.** c **4.** b **5.** d **6.** b

Chapter 6

Figure Questions

Figure 6.9 The R group of glutamine (Gln, or Q) is like that of glutamic acid (Glu, or E), except it has an amino group ($-NH_2$) in place of a hydroxyl group ($-OH$), so Gln is drawn as a Glu with an attached $-NH_2$.

Figure 6.12

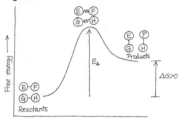

Figure 6.16

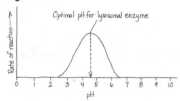

Concept Check 6.1

1. The second law is the trend toward randomization, or increasing entropy. When the concentrations of a substance on both sides of a membrane are equal, the distribution is more random than when they are unequal. Diffusion of a substance to a region where it is initially less concentrated increases entropy, making it an energetically favorable (spontaneous) process as described by the second law. (This explains the process seen in Figure 5.9.) **2.** The apple has potential energy in its position hanging on the tree, and the sugars and other nutrients it contains have chemical energy. The apple has kinetic energy as it falls from the tree to the ground. Finally, when the apple is digested and its molecules broken down, some of the chemical energy is used to do work, and the rest is lost as thermal energy.

Concept Check 6.2

1. Cellular respiration is a spontaneous and exergonic process. The energy released from glucose is used to do work in the cell or is lost as heat. **2.** The reaction is exergonic because it releases energy—in this case, in the form of light. (This is a nonbiological version of the bioluminescence seen in Figure 6.1.)

Concept Check 6.3

1. ATP usually transfers energy to endergonic processes by phosphorylating (adding phosphate groups to) other molecules. (Exergonic processes phosphorylate ADP to regenerate ATP.) **2.** A set of coupled reactions can transform the first combination into the second. Since this is an exergonic process overall, ΔG is negative and the first combination must have more free energy (see Figure 6.9). **3.** Active transport. The solute is being transported against its concentration gradient, which requires energy, provided by ATP hydrolysis.

Concept Check 6.4

1. A spontaneous reaction is a reaction that is exergonic. However, if it has a high activation energy that is rarely attained, the rate of the reaction may be low. **2.** Only the specific substrate(s) will fit properly into the active site of an enzyme, the part of the enzyme that carries out catalysis. **3.** In the presence of malonate, increase the concentration of the normal substrate (succinate) and see whether the rate of reaction increases. If it does, malonate is a competitive inhibitor.

Concept Check 6.5

1. The activator binds in such a way that it stabilizes the active form of an enzyme, whereas the inhibitor stabilizes the inactive form.

Summary of Key Concepts Questions

6.1 The process of "ordering" a cell's structure is accompanied by an increase in the entropy, or disorder, of the universe. For example, an animal cell takes in highly ordered organic molecules as the source of matter and energy used to build and maintain its structures. In the same process, however, the cell releases heat and the simple molecules of carbon dioxide and water to the surroundings. The increase in entropy of the latter process offsets the entropy decrease in the former. **6.2** Spontaneous reactions supply the energy to perform cellular work. **6.3** The free energy released from the hydrolysis of ATP may drive endergonic reactions through the transfer of a phosphate group to a reactant molecule, forming a more reactive phosphorylated intermediate. ATP hydrolysis also powers the mechanical and transport work of a cell, often by powering shape changes in the relevant motor proteins. Cellular respiration, the catabolic breakdown of glucose, provides the energy for the endergonic regeneration of ATP from ADP and P_i. **6.4** Activation energy barriers prevent the complex molecules of the cell, which are rich in free energy, from spontaneously breaking down to less ordered, more stable molecules. Enzymes permit a regulated metabolism by binding to specific substrates and forming enzyme-substrate complexes that selectively lower the E_A for the chemical reactions in a cell. **6.5** A cell tightly regulates its metabolic pathways in response to fluctuating needs for energy and materials. The binding of activators or inhibitors to regulatory sites on allosteric enzymes stabilizes either the active or inactive form of the subunits. For example, the binding of ATP to a catabolic enzyme in a cell with excess ATP would inhibit that pathway. Such types of feedback inhibition preserve chemical resources within a cell. If ATP supplies are depleted, binding of ADP to the regulatory site of catabolic enzymes will activate that pathway.

Test Your Understanding
1. b **2.** c **3.** b **4.** a **5.** c **6.** e **7.** c

8.

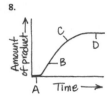

A. The substrate molecules are entering the cells, so no product is made yet.
B. There is sufficient substrate, so the reaction is proceeding at a maximum rate.
C. As the substrate is used up, the rate decreases (the slope is less steep).
D. The line is flat because no new substrate remains and thus no new product appears.

Chapter 7

Figure Questions

Figure 7.7 Because there is no external source of energy for the reaction, it must be exergonic, and the reactants must be at a higher energy level than the products.
Figure 7.14 At first, some ATP could be made, since electron transport could proceed as far as complex III, and a small H^+ gradient could be built up. Soon, however, no more electrons could be passed to complex III because it could not be reoxidized by passing its electrons to complex IV.   **Figure 7.15** First, there are 2 NADH from the oxidation of pyruvate plus 6 NADH from the citric acid cycle (CAC); 8 NADH × 2.5 ATP/ NADH = 20 ATP. Second, there are 2 FADH$_2$ from the CAC; 2 FADH$_2$ × 1.5 ATP/ FADH$_2$ = 3 ATP. Third, the 2 NADH from glycolysis enter the mitochondrion through one of two types of shuttle. They pass their electrons either to 2 FAD, which become FADH$_2$ and result in 3 ATP, or to 2 NAD$^+$, which become NADH and result in 5 ATP. Thus, 20 + 3 + 3 = 26 ATP or 20 + 3 + 5 = 28 ATP from all NADH and FADH$_2$.

Concept Check 7.1

1. Both processes include glycolysis, the citric acid cycle, and oxidative phosphorylation. In aerobic respiration, the final electron acceptor is molecular oxygen (O_2); in anaerobic respiration, the final electron acceptor is a different substance.
2. Substrate-level phosphorylation, which occurs during glycolysis and the citric acid cycle, involves the direct transfer of a phosphate group from an organic substrate to ADP by an enzyme. Oxidative phosphorylation occurs during the third stage of cellular respiration, which is called oxidative phosphorylation. In this process, the synthesis of ATP from ADP and inorganic phosphate (P$_i$) is powered by the redox reactions of the electron transport chain.   **3.** C$_4$H$_6$O$_5$ would be oxidized, and NAD$^+$ would be reduced.

Concept Check 7.2

1. NAD$^+$ acts as the oxidizing agent in step 6, accepting electrons from glyceraldehyde 3-phosphate, which thus acts as the reducing agent.

Concept Check 7.3

1. NADH and FADH$_2$; they will donate electrons to the electron transport chain.
2. CO$_2$ is released from the pyruvate that is the end product of glycolysis, and CO$_2$ is also released during the citric acid cycle.

Concept Check 7.4

1. Oxidative phosphorylation would eventually stop entirely, resulting in no ATP production by this process. Without oxygen to "pull" electrons down the electron transport chain, H$^+$ would not be pumped into the mitochondrion's intermembrane space and chemiosmosis would not occur.   **2.** Decreasing the pH means the addition of H$^+$. This would establish a proton gradient even without the function of the electron transport chain, and we would expect ATP synthase to function and synthesize ATP. (In fact, it was experiments like this that provided support for chemiosmosis as an energy-coupling mechanism.)   **3.** One of the components of the electron transport chain, ubiquinone (Q), must be able to diffuse within the membrane. It could not do so if the membrane were locked rigidly into place.

Concept Check 7.5

1. A derivative of pyruvate, such as acetaldehyde during alcohol fermentation, or pyruvate itself during lactic acid fermentation; oxygen during aerobic respiration   **2.** The cell would need to consume glucose at a rate about 16 times the consumption rate in the aerobic environment (2 ATP are generated by fermentation versus up to 32 ATP by cellular respiration).

Concept Check 7.6

1. The fat is much more reduced; it has many —CH$_2$— units, and in all these bonds the electrons are equally shared. The electrons present in a carbohydrate molecule are already somewhat oxidized (shared unequally in bonds), as quite a few of them are bound to oxygen.   **2.** When you consume more food than necessary for metabolic processes, your body synthesizes fat as a way of storing energy for later use.
3. When oxygen is present, the fatty acid chains containing most of the energy of a fat

are oxidized and fed into the citric acid cycle and the electron transport chain. During intense exercise, however, oxygen is scarce in muscle cells, so ATP must be generated by glycolysis alone. A very small part of the fat molecule, the glycerol backbone, can be oxidized via glycolysis, but the amount of energy released by this portion is insignificant compared with that released by the fatty acid chains. (This is why moderate exercise, staying below 70% maximum heart rate, is better for burning fat—because enough oxygen remains available to the muscles.)

Summary of Key Concepts Questions

7.1 Most of the ATP produced in cellular respiration comes from oxidative phosphorylation, in which the energy released from redox reactions in an electron transport chain is used to produce ATP. In substrate-level phosphorylation, an enzyme directly transfers a phosphate group to ADP from an intermediate substrate. All ATP production in glycolysis occurs by substrate-level phosphorylation; this form of ATP production also occurs at one step in the citric acid cycle.   **7.2** The oxidation of the three-carbon sugar glyceraldehyde 3-phosphate yields energy. In this oxidation, electrons and H$^+$ are transferred to NAD$^+$, forming NADH, and a phosphate group is attached to the oxidized substrate. ATP is then formed by substrate-level phosphorylation when this phosphate group is transferred to ADP.   **7.3** The release of six molecules of CO$_2$ represents the complete oxidation of glucose. During the processing of two pyruvates to acetyl CoA, the fully oxidized carboxyl group (—COO$^-$) is given off as CO$_2$. The remaining four carbons are released as CO$_2$ in the citric acid cycle as citrate is oxidized back to oxaloacetate.   **7.4** The flow of H$^+$ through the ATP synthase complex causes the rotor and attached rod to rotate, exposing catalytic sites in the knob portion that produce ATP from ADP and P$_i$. ATP synthases are found in the inner mitochondrial membrane, the plasma membrane of prokaryotes, and membranes within chloroplasts.   **7.5** Anaerobic respiration yields more ATP. The 2 ATP produced by substrate-level phosphorylation in glycolysis represent the total energy yield of fermentation. NADH passes its "high-energy" electrons to pyruvate or a derivative of pyruvate, recycling NAD$^+$ and allowing glycolysis to continue. Anaerobic respiration uses an electron transport chain to capture the energy of the electrons in NADH via a series of redox reactions; ultimately, the electrons are transferred to an electronegative molecule other than oxygen. Also, additional molecules of NADH are produced in anaerobic respiration as pyruvate is oxidized.   **7.6** The ATP produced by catabolic pathways is used to drive anabolic pathways. Also, many of the intermediates of glycolysis and the citric acid cycle are used in the biosynthesis of a cell's molecules.

Test Your Understanding

1. d **2.** c **3.** c **4.** a **5.** e **6.** a **7.** b
8.

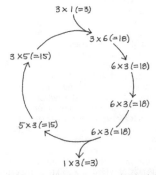

Chapter 8

Figure Questions

Figure 8.9 Red, but not violet-blue, wavelengths would pass through the filter, so the bacteria would not congregate where the violet-blue light normally comes through. Therefore, the left "peak" of bacteria would not be present, but the right peak would be observed because the red wavelengths passing through the filter would be used for photosynthesis.
Figure 8.17

Three carbon atoms enter the cycle, one by one, as individual CO$_2$ molecules and leave the cycle in one three-carbon molecule (G3P) per three turns of the cycle.

Concept Check 8.1

1. CO$_2$ enters the leaves via stomata, and water enters the plant via roots and is carried to the leaves through veins.   **2.** Using ^{18}O, a heavy isotope of oxygen, as a label, researchers were able to confirm van Niel's hypothesis that the oxygen produced during photosynthesis originates in water, not in carbon dioxide.   **3.** The light reactions could *not* keep producing NADPH and ATP without the NADP$^+$, ADP, and P$_i$ that the Calvin cycle generates. The two cycles are interdependent.

Concept Check 8.2

1. Green, because green light is mostly transmitted and reflected—not absorbed—by photosynthetic pigments **2.** Water (H_2O) is the initial electron donor; $NADP^+$ accepts electrons at the end of the electron transport chain, becoming reduced to NADPH. **3.** The rate of ATP synthesis would slow and eventually stop. Because the added compound would not allow a proton gradient to build up across the membrane, ATP synthase could not catalyze ATP production.

Concept Check 8.3

1. The amount of energy and reducing power required to form a molecule determines the amount of potential energy that molecule stores. Glucose is a valuable energy source because it is highly reduced, storing lots of potential energy in its electrons. To reduce CO_2 to glucose, a large amount of energy and reducing power are required in the form of large numbers of ATP and NADPH molecules, respectively. **2.** The light reactions require ADP and $NADP^+$, which would not be formed in sufficient quantities from ATP and NADPH if the Calvin cycle stopped. **3.** Photorespiration decreases photosynthetic output by adding oxygen, instead of carbon dioxide, to the Calvin cycle.

Summary of Key Concepts Questions

8.1 CO_2 and H_2O are the products of respiration; they are the reactants in photosynthesis. In respiration, glucose is oxidized to CO_2 as electrons are passed through an electron transfer chain from glucose to O_2, producing H_2O. In photosynthesis, H_2O is the source of electrons, which are energized by light, temporarily stored in NADPH, and used to reduce CO_2 to carbohydrate. **8.2** The action spectrum of photosynthesis shows that some wavelengths of light that are not absorbed by chlorophyll *a* are still effective at promoting photosynthesis. The light-harvesting complexes of photosystems contain accessory pigments, such as chlorophyll *b* and carotenoids, which absorb different wavelengths and pass the energy to chlorophyll *a*, broadening the spectrum of light useful for photosynthesis. **8.3**

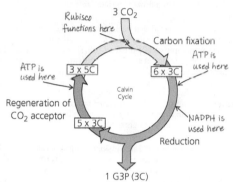

In the reduction phase of the Calvin cycle, ATP phosphorylates a three-carbon compound, and NADPH then reduces this compound to G3P. ATP is also used in the regeneration phase, when five molecules of G3P are converted to three molecules of the five-carbon compound RuBP. Rubisco catalyzes the first step of carbon fixation—the addition of CO_2 to RuBP.

Test Your Understanding

1. d **2.** b **3.** c **4.** d **5.** c **6.** b **7.** d **8.** 6; 18; 12
9.

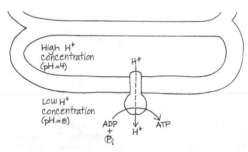

The ATP would end up outside the thylakoid. The thylakoids were able to make ATP in the dark because the researchers set up an artificial proton concentration gradient across the thylakoid membrane; thus, the light reactions were not necessary to establish the H^+ gradient required for ATP synthesis by ATP synthase.

Chapter 9

Figure Questions
Figure 9.4

One sister chromatid

Circling the other chromatid instead would also be correct. **Figure 9.5** The chromosome has four chromatid arms. **Figure 9.7** 12; 2; 2; 1
Figure 9.8

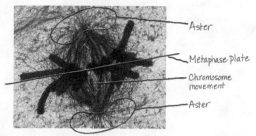

Figure 9.9 The mark would have moved toward the nearer pole. The lengths of fluorescent microtubules between that pole and the mark would have decreased, while the lengths between the chromosomes and the mark would have remained the same.
Figure 9.14 In both cases, the G_1 nucleus would have remained in G_1 until the time it normally would have entered the S phase. Chromosome condensation and spindle formation would not have occurred until the S and G_2 phases had been completed.
Figure 9.16 The cell would divide under conditions where it was inappropriate to do so. If the daughter cells and their descendants also ignored the checkpoint and divided, there would soon be an abnormal mass of cells. (This type of inappropriate cell division can contribute to the development of cancer.) **Figure 9.17** The cells in the vessel with PDGF would not be able to respond to the growth factor signal and thus would not divide. The culture would resemble the culture without the added PDGF.

Concept Check 9.1
1. 2 **2.** 39; 39; 78

Concept Check 9.2
1. 6 chromosomes, duplicated; 12 chromatids **2.** Following mitosis, cytokinesis results in two genetically identical daughter cells in both plant cells and animal cells. However, the mechanism of dividing the cytoplasm is different in animals and plants. In an animal cell, cytokinesis occurs by cleavage, which divides the parent cell in two with a contractile ring of actin filaments. In a plant cell, a cell plate forms in the middle of the cell and grows until its membrane fuses with the plasma membrane of the parent cell. A new cell wall grows inside the cell plate. **3.** They elongate the cell during anaphase. **4.** From the end of S phase in interphase through the end of metaphase in mitosis

Concept Check 9.3
1. The nucleus on the right was originally in the G_1 phase; therefore, it had not yet duplicated its chromosome. The nucleus on the left was in the M phase, so it had already duplicated its chromosome. **2.** Most body cells are in a nondividing state called G_0. **3.** Both types of tumors consist of abnormal cells, but their characteristics are different. A benign tumor stays at the original site and can usually be surgically removed; the cells have some genetic and cellular changes from normal, non-tumor cells. Cancer cells from a malignant tumor have more significant genetic and cellular changes, can spread from the original site by metastasis, and may impair the functions of one or more organs. **4.** The cells might divide even in the absence of PDGF. In addition, they would not stop when the surface of the culture vessel was covered; they would continue to divide, piling on top of one another.

Summary of Key Concepts Questions
9.1 The DNA of a eukaryotic cell is packaged into structures called *chromosomes*. Each chromosome is a long molecule of DNA, which carries hundreds to thousands of genes, with associated proteins that maintain chromosome structure and help control gene activity. This DNA-protein complex is called *chromatin*. The chromatin of each chromosome is long and thin when the cell is not dividing. Prior to cell division, each chromosome is duplicated, and the resulting sister *chromatids* are attached to each other by proteins at the centromeres and, for many species, all along their lengths (sister chromatid cohesion). **9.2** Chromosomes exist as single DNA molecules in G_1 of interphase and in anaphase and telophase of mitosis. During S phase, DNA replication produces sister chromatids, which persist during G_2 of interphase and through prophase, prometaphase, and metaphase of mitosis. **9.3** Checkpoints allow cellular surveillance mechanisms to determine whether the cell is prepared to go to the next stage. Internal and external signals move a cell past these checkpoints. The G_1 checkpoint, called the "restriction point" in mammalian cells, determines whether a cell will complete the cell cycle and divide or switch into the G_0 phase. The signals to pass this checkpoint often are external—such as growth factors. Regulation of the cell cycle is carried out by a molecular system, including kinases and proteins called cyclins. The signal to pass the M phase checkpoint is not activated until all chromosomes are attached to kinetochore fibers and are aligned at the metaphase plate. Only then will sister chromatid separation occur.

Test Your Understanding

1. b **2.** a **3.** b **4.** a **5.** e **6.** See Figure 9.7 for a description of major events.

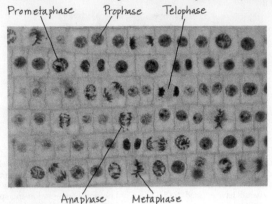

Prometaphase   Prophase   Telophase

Anaphase   Metaphase

Only one cell is indicated for each stage, but other correct answers are also present in this micrograph.

7.

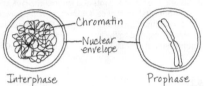

Chromatin

Nuclear envelope

Interphase   Prophase

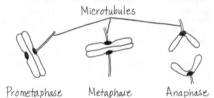

Microtubules

Prometaphase   Metaphase   Anaphase

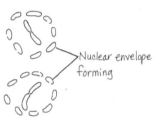

Nuclear envelope forming

Telophase and cytokinesis

Chapter 10

Figure Questions

Figure 10.4 The haploid number, *n*, is 3; a set is always haploid; two sets; two sets.

Figure 10.7

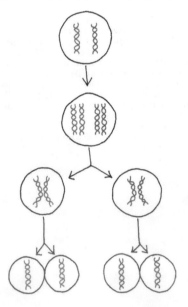

(A short strand of DNA is shown here for simplicity, but each chromosome or chromatid contains a very long coiled and folded DNA molecule.)   **Figure 10.8** If the two cells in Figure 9.7 underwent another round of mitosis, each of the four resulting cells would have six chromosomes, while the four cells resulting from meiosis in Figure 10.8 each have three chromosomes. In mitosis, DNA replication (and thus chromosome duplication) precedes each prophase, ensuring that daughter cells have the same number of chromosomes as the parent cell. In meiosis, in contrast, DNA replication occurs only before prophase I (not before prophase II). Thus, in two rounds of mitosis, the chromosomes duplicate twice and divide twice, while in meiosis, the chromosomes duplicate once and divide twice.   **Figure 10.9** Yes. Each of the six chromosomes (three per cell) shown in telophase I has one nonrecombinant chromatid and one recombinant chromatid. Therefore, eight possible sets of chromosomes can be generated for the cell on the left and eight for the cell on the right. See metaphase II in Figure 10.8; note that the chromosomes can line up in different arrangements.

Concept Check 10.1

1. Parents pass genes to their offspring; the genes program cells to make specific enzymes and other proteins, whose cumulative action produces an individual's inherited traits.   **2.** Such organisms reproduce by mitosis, which generates offspring whose genomes are exact copies of the parent's genome (without taking mutation into account).   **3.** She should clone it. Cross-breeding it with another plant would generate offspring that have additional variation, which she no longer desires, now that she has obtained her ideal orchid.

Concept Check 10.2

1. Each of the six chromosomes is duplicated, so each contains two DNA double helices. Therefore, there are 12 DNA molecules in the cell.   **2.** In meiosis, the chromosome count is reduced from diploid to haploid; the union of two haploid gametes in fertilization restores the diploid chromosome count.   **3.** The haploid number (*n*) is 7; the diploid number (2*n*) is 14.   **4.** This organism has the life cycle shown in Figure 10.6c. Therefore, it must be a fungus or a protist, perhaps an alga.

Concept Check 10.3

1. The chromosomes are similar in that each is composed of two sister chromatids, and the individual chromosomes are positioned similarly at the metaphase plate. The chromosomes differ in that in a mitotically dividing cell, sister chromatids of each chromosome are genetically identical, but in a meiotically dividing cell, sister chromatids are genetically distinct because of crossing over in meiosis I. Moreover, the chromosomes in metaphase of mitosis can be a diploid set or a haploid set, but the chromosomes in metaphase of meiosis II always consist of a haploid set.   **2.** If crossing over did not occur, the two homologs would not be associated in any way. This might result in incorrect arrangement of homologs during metaphase I and ultimately in formation of gametes with an abnormal number of chromosomes.

Concept Check 10.4

1. Mutations in a gene lead to the different versions (alleles) of that gene.   **2.** If the segments of the maternal and paternal chromatids that undergo crossing over are genetically identical and thus have the same two alleles for every gene, then the recombinant chromosomes will be genetically equivalent to the parental chromosomes. Crossing over contributes to genetic variation only when it involves the rearrangement of different alleles.

Summary of Key Concepts Questions

10.1 Genes program specific traits, and offspring inherit genes from each parent, accounting for similarities in their appearance to one or the other parent. Humans reproduce sexually, which ensures new combinations of genes (and thus traits) in the offspring. Consequently, the offspring are not clones of their parents (which would be the case if humans reproduced asexually).   **10.2** Animals and plants both reproduce sexually, alternating meiosis with fertilization. Both have haploid gametes that unite to form a diploid zygote, which then goes on to divide mitotically, forming a diploid multicellular organism. In animals, haploid cells become gametes and don't undergo mitosis, while in plants, the haploid cells resulting from meiosis undergo mitosis to form a haploid multicellular organism, the gametophyte. This organism then goes on to generate haploid gametes. (In plants such as trees, the gametophyte is quite reduced in size and not obvious to the casual observer.)   **10.3** At the end of meiosis I, the two members of a homologous pair end up in different cells, so they cannot pair up and undergo crossing over.   **10.4** First, during independent assortment in metaphase I, each pair of homologous chromosomes lines up independent of every other pair at the metaphase plate, so a daughter cell of meiosis I randomly inherits either a maternal or paternal chromosome. Second, due to crossing over, each chromosome is not exclusively maternal or paternal, but includes regions at the ends of the chromatid from a nonsister chromatid (a chromatid of the other homolog). (The nonsister segment can also be in an internal region of the chromatid if a second crossover occurs beyond the first one before the end of the chromatid.) This provides much additional diversity in the form of new combinations of alleles. Third, random fertilization ensures even more variation, since any sperm of a large number containing many possible genetic combinations can fertilize any egg of a similarly large number of possible combinations.

Test Your Understanding

1. a **2.** b **3.** d **4.** c **5.** d

6. (a)

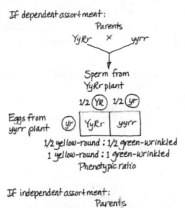

Labels:
Centromere
Kinetochore
Nonsister chromatids (different colors)
Gene loci
Chromosome (duplicated)
Sister chromatids (same color)
Sister chromatid cohesion
Chiasma
Homologous pair OR Pair of homologs

(b) Metaphase I (c) A haploid set is made up of one long, one medium, and one short chromosome. In this diagram, the haploid set going to each side of the cell is either blue or red, except for small segments of the other color due to crossing over. A diploid set is made up of all red and blue chromosomes together.
7. This cell must be undergoing meiosis because homologous chromosomes are associated with each other at the metaphase plate; this does not occur in mitosis.

Chapter 11

Figure Questions
Figure 11.3 All offspring would have purple flowers. (The ratio would be one purple to zero white.) The P generation plants are true-breeding, so mating two purple-flowered plants produces the same result as self-pollination: All the offspring have the same trait.
Figure 11.8

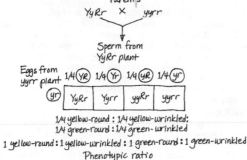

If dependent assortment:
Parents
YyRr × yyrr
Sperm from YyRr plant
1/2 YR   1/2 yr
Eggs from yyrr plant   yr
YyRr   yyrr
1/2 yellow-round : 1/2 green-wrinkled
1 yellow-round : 1 green-wrinkled
Phenotypic ratio

If independent assortment:
Parents
YyRr × yyrr
Sperm from YyRr plant
1/4 YR   1/4 Yr   1/4 yR   1/4 yr
Eggs from yyrr plant   yr
YyRr   Yyrr   yyRr   yyrr
1/4 yellow-round : 1/4 yellow-wrinkled : 1/4 green-round : 1/4 green-wrinkled
1 yellow-round : 1 yellow-wrinkled : 1 green-round : 1 green-wrinkled
Phenotypic ratio

Yes, this cross would also have allowed Mendel to make different predictions for the two hypotheses, thereby allowing him to distinguish the correct one. **Figure 11.10** Your classmate would probably point out that the F_1 generation hybrids show an intermediate phenotype between those of the homozygous parents, which supports the blending hypothesis. You could respond that crossing the F_1 hybrids results in the reappearance of the white phenotype, rather than identical pink offspring, which fails to support the idea of traits blending during inheritance. **Figure 11.11** Both the I^A and I^B alleles are dominant to the i allele, which is recessive and results in no attached carbohydrate. The I^A and I^B alleles are codominant; both are expressed in the phenotype of $I^A I^B$ heterozygotes, who have type AB blood. **Figure 11.15** In the Punnett square, two of the three individuals with normal coloration are carriers, so the probability is $^2/_3$. (Note that you must take into account everything you know when you calculate probability: You know she is not aa, so there are only three possible genotypes to consider.)

Concept Check 11.1
1. There are 423 round peas and 133 wrinkled peas, a ratio of 3.18:1, or roughly 3:1.
2. According to the law of independent assortment, 25 plants ($^1/_{16}$ of the offspring) are predicted to be $aatt$, or recessive for both characters. The actual result is likely to differ slightly from this value.

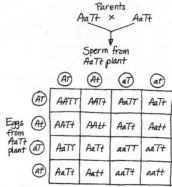

Parents
AaTt × AaTt
Sperm from AaTt plant
AT   At   aT   at
Eggs from AaTt plant:
AT: AATT   AATt   AaTT   AaTt
At: AATt   AAtt   AaTt   Aatt
aT: AaTT   AaTt   aaTT   aaTt
at: AaTt   Aatt   aaTt   aatt

3. The plant could make eight different gametes (*YRI, YRi, Yrl, Yri, yRI, yRi, yrI,* and *yri*). To fit all the possible gametes in a self-pollination, a Punnett square would need 8 rows and 8 columns. It would have spaces for the 64 possible unions of gametes in the offspring. **4.** Self-pollination is sexual reproduction because meiosis is involved in forming gametes, which unite during fertilization. As a result, the offspring in self-pollination are genetically different from the parent.

Concept Check 11.2
1. $^1/_2$ homozygous dominant (AA), 0 homozygous recessive (aa), and $^1/_2$ heterozygous (Aa) **2.** $^1/_4 BBDD$; $^1/_4 BbDD$; $^1/_4 BBDd$; $^1/_4 BbDd$ **3.** The genotypes that fulfill this condition are $ppyyIi$, $ppYyii$, $Ppyyii$, $ppYYii$, and $ppyyii$. Use the multiplication rule to find the probability of getting each genotype, and then use the addition rule to find the overall probability of meeting the conditions of this problem:

ppyyIi	1/2 (probability of pp) × 1/4 (yy) × 1/2 (Ii)	= 1/16
ppYyii	1/2 (pp) × 1/2 (Yy) × 1/2 (ii)	= 2/16
Ppyyii	1/2 (Pp) × 1/4 (yy) × 1/2 (ii)	= 1/16
ppYYii	1/2 (pp) × 1/4 (YY) × 1/2 (ii)	= 1/16
ppyyii	1/2 (pp) × 1/4 (yy) × 1/2 (ii)	= 1/16
Fraction predicted to have at least two recessive traits		= 6/16 or 3/8

Concept Check 11.3
1. Incomplete dominance describes the relationship between two alleles of a single gene, whereas epistasis relates to the genetic relationship between two genes (and the respective alleles of each). **2.** Half of the children would be expected to have type A blood and half type B blood. **3.** The black and white alleles are incompletely dominant, with heterozygotes being gray in color. A cross between a gray rooster and a black hen should yield approximately equal numbers of gray and black offspring.

Concept Check 11.4
1. $^1/_9$ (Since cystic fibrosis is caused by a recessive allele, Beth and Tom's siblings who have CF must be homozygous recessive. Therefore, each parent must be a carrier of the recessive allele. Since neither Beth nor Tom has CF, this means they each have a $^2/_3$ chance of being a carrier. If they are both carriers, there is a $^1/_4$ chance that they will have a child with CF. $^2/_3 \times ^2/_3 \times ^1/_4 = ^1/_9$.) 0 (Both Beth and Tom would have to be carriers to produce a child with the disease.) **2.** In the monohybrid cross involving flower color, the ratio is 3.15 purple : 1 white, while in the human family in the pedigree, the ratio in the third generation is 1 free : 1 attached earlobe. The difference is due to the small sample size (two offspring) in the human family. If the second-generation couple in this pedigree were able to have 929 offspring as in the pea plant cross, the ratio would likely be closer to 3:1. (Note that none of the pea plant crosses in Table 11.1 yielded *exactly* a 3:1 ratio.)

Summary of Key Concepts Questions
11.1 Alternative versions of genes, called alleles, are passed from parent to offspring during sexual reproduction. In a cross between purple- and white-flowered homozygous parents, the F_1 offspring are all heterozygous, each inheriting a purple allele from one parent and a white allele from the other. Because the purple allele is dominant, it determines the phenotype of all F_1 offspring to be purple, while the expression of the recessive white allele is masked. Only in the F_2 generation is it possible for a white allele to exist in a homozygous state, which causes the white trait to be expressed.
11.2

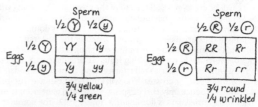

Sperm
1/2 Y   1/2 y
Eggs
1/2 Y: YY   Yy
1/2 y: Yy   yy
3/4 yellow   1/4 green

Sperm
1/2 R   1/2 r
Eggs
1/2 R: RR   Rr
1/2 r: Rr   rr
3/4 round   1/4 wrinkled

3/4 yellow × 3/4 round = 9/16 yellow-round
3/4 yellow × 1/4 wrinkled = 3/16 yellow-wrinkled
1/4 green × 3/4 round = 3/16 green-round
1/4 green × 1/4 wrinkled = 1/16 green-wrinkled

= 9 yellow-round : 3 yellow-wrinkled : 3 green-round : 1 green-wrinkled

11.3 The ABO blood group is an example of multiple alleles because this single gene has more than two alleles (I^A, I^B, and i). Two of the alleles, I^A and I^B, exhibit codominance, since both carbohydrates (A and B) are present when these two alleles exist together in a genotype. I^A and I^B each exhibit complete dominance over the i allele. This situation is not an example of incomplete dominance because each allele affects the phenotype in a distinguishable way, so the result is not intermediate between the two phenotypes. Because this situation involves a single gene, it is not an example of epistasis or polygenic inheritance. **11.4** The chance of the fourth child having cystic fibrosis is $^1/_4$, as it was for each of the other children, because each birth is an independent event. We already know that both parents are carriers, so whether their first three children are carriers or not has no bearing on the probability that their next child will have the disease. The parents' genotypes provide the only relevant information.

Test Your Understanding

1. Gene, l; Allele, e; Character, g; Trait, b; Dominant allele, j; Recessive allele, a; Genotype, k; Phenotype, h; Homozygous, c; Heterozygous, f; Testcross, i; Monohybrid cross, d
2.

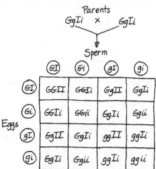

3. Man I^Ai; woman I^Bi; child ii. Genotypes for future children are predicted to be $^1/_4$ I^AI^B, $^1/_4$ I^Ai, $^1/_4$ I^Bi, $^1/_4$ ii. **4.** $^1/_2$ **5.** A cross of $Ii \times ii$ would yield offspring with a genotypic ratio of 1 Ii : 1 ii (2:2 is an equivalent answer) and a phenotypic ratio of 1 inflated : 1 constricted (2:2 is equivalent).

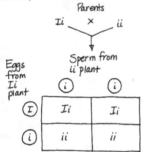

6. (a) $^1/_{64}$; (b) $^1/_{64}$; (c) $^1/_8$; (d) $^1/_{32}$ **7.** Albino (b) is a recessive trait; black (B) is dominant. First cross: parents $BB \times bb$; gametes B and b; offspring all Bb (black coat). The black guinea pig in the second cross is a heterozygote. Second cross: parents $Bb \times bb$; gametes $^1/_2 B$ and $^1/_2 b$ (heterozygous parent) and b; offspring $^1/_2 Bb$ (black) and $^1/_2 bb$ (white). **8.** Parental cross is $AAC^RC^R \times aaC^WC^W$. F_1 genotype is AaC^RC^W, phenotype is all axial-pink. F_2 genotypes are 1 AAC^RC^R : 2 AAC^RC^W : 1 AAC^WC^W : 2 AaC^RC^R : 4 AaC^RC^W : 2 AaC^WC^W : 1 aaC^RC^R : 2 aaC^RC^W : 1 aaC^WC^W. F_2 phenotypes are 3 axial-red : 6 axial-pink : 3 axial-white : 1 terminal-red : 2 terminal-pink : 1 terminal-white. **9.** (a) $PPLl \times PPLl$, $PPLl \times PpLl$, $PpLl \times PpLl$; (b) $ppLl \times ppLl$; (c) $PPLL \times$ any of the 9 possible genotypes or $PPll \times ppLL$; (d) $PpLl \times Ppll$; (e) $PpLl \times PPLl$ **10.** (a) $^3/_4 \times ^3/_4 \times ^3/_4 = ^{27}/_{64}$; (b) $1 - ^{27}/_{64} = ^{37}/_{64}$; (c) $^1/_4 \times ^1/_4 \times ^1/_4 = ^1/_{64}$; (d) $1 - ^1/_{64} = ^{63}/_{64}$ **11.** (a) $^1/_{256}$; (b) $^1/_{16}$; (c) $^1/_{256}$; (d) $^1/_{64}$; (e) $^1/_{128}$ **12.** (a) 1; (b) $^1/_{32}$; (c) $^1/_8$; (d) $^1/_2$ **13.** $^1/_9$ **14.** 25%, will be cross-eyed; all (100%) of the cross-eyed offspring will also be white. **15.** Matings of the original mutant cat with true-breeding noncurl cats will produce both curl and noncurl F_1 offspring if the curl allele is dominant, but only noncurl offspring if the curl allele is recessive. Whether the curl trait is dominant or recessive, you would obtain true-breeding offspring homozygous for the curl allele from matings between the F_1 cats resulting from the original curl × noncurl crosses. If dominant, you wouldn't be able to tell truebreeding, homozygous offspring from heterozygotes without further crosses. You will know that cats are true-breeding when curl × curl matings produce only curl offspring for several generations. As it turns out, the allele that causes curled ears is dominant. **16.** $^1/_{16}$ **17.** The dominant allele I is epistatic to the P/p locus, and thus the genotypic ratio for the F_1 generation will be 9 $I–P–$ (colorless) : 3 $I–pp$ (colorless) : 3 $iiP–$ (purple) : 1 $iipp$ (red). Overall, the phenotypic ratio is 12 colorless : 3 purple : 1 red. **18.** Recessive. All affected individuals (Arlene, Tom, Wilma, and Carla) are homozygous recessive aa. George is Aa, since

some of his children with Arlene are affected. Sam, Ann, Daniel, and Alan are each Aa, since they are all unaffected children with one affected parent. Michael also is Aa, since he has an affected child (Carla) with his heterozygous wife Ann. Sandra, Tina, and Christopher can each have either the AA or Aa genotype. **19.** $^1/_6$ **20.** 9 $B–A–$ (agouti) : 3 $B–aa$ (black) : 3 $bbA–$ (white) : 1 $bbaa$ (white). Overall, 9 agouti : 3 black : 4 white.

Chapter 12

Figure Questions

Figure 12.2 The ratio would be 1 yellow-round : 1 green-round : 1 yellow-wrinkled : 1 green-wrinkled. **Figure 12.4** About $^3/_4$ of the F_2 offspring would have red eyes and about $^1/_4$ would have white eyes. About half of the white-eyed flies would be female and half would be male; similarly, about half of the red-eyed flies would be female and half would be male. **Figure 12.7** All the males would be color-blind, and all the females would be carriers. **Figure 12.9** The two largest classes would still be the parental-type offspring (offspring with the phenotypes of the true-breeding P generation flies), but now they would be gray-vestigial and black-normal because those were the specific allele combinations in the P generation. **Figure 12.10** The two chromosomes below, left, are like the two chromosomes inherited by the F_1 female, one from each P generation fly. They are passed by the F_1 female intact to the offspring and thus could be called "parental" chromosomes. The other two chromosomes result from crossing over during meiosis in the F_1 female. Because they have combinations of alleles not seen in either of the F_1 female's chromosomes, they can be called "recombinant" chromosomes. (Note that in this example, the alleles on the recombinant chromosomes, $b^+ vg^+$ and $b vg$, are the allele combinations that were on the parental chromosomes in the cross shown in Figures 12.9 and 12.10. The basis for calling them parental chromosomes is the combination of alleles that was present on the P generation chromosomes.)

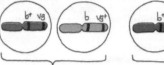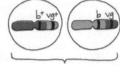

Parental chromosomes   Recombinant chromosomes

Concept Check 12.1

1. The law of segregation relates to the inheritance of alleles for a single character. The law of independent assortment of alleles relates to the inheritance of alleles for two characters. **2.** The physical basis for the law of segregation is the separation of homologs in anaphase I. The physical basis for the law of independent assortment is the alternative arrangements of all the homologous chromosome pairs in metaphase I. **3.** To show the mutant phenotype, a male needs to possess only one mutant allele. If this gene had been on a pair of autosomes, *both* alleles would have had to be mutant for an individual to show the recessive mutant phenotype, a much less probable situation.

Concept Check 12.2

1. Because the gene for this eye-color character is located on the X chromosome, all female offspring will be red-eyed and heterozygous ($X^{w+} X^w$); all male offspring will inherit a Y chromosome from the father and be white-eyed (X^wY). **2.** $^1/_4$ ($^1/_2$ chance that the child will inherit a Y chromosome from the father and be male × $^1/_2$ chance that he will inherit the X carrying the disease allele from his mother); if the child is a boy, there is a $^1/_2$ chance he will have the disease; a female will have zero chance (but $^1/_2$ chance of being a carrier). **3.** With a disorder caused by a dominant allele, there is no such thing as a "carrier," since those with the allele have the disorder. Because the allele is dominant, the females lose any "advantage" in having two X chromosomes, since one disorder-associated allele is sufficient to result in the disorder. All fathers who have the dominant allele will pass it along to *all* their daughters, who will also have the disorder. A mother who has the allele (and thus the disorder) will pass it to half of her sons and half of her daughters.

Concept Check 12.3

1. Crossing over during meiosis I in the heterozygous parent produces some gametes with recombinant genotypes for the two genes. Offspring with a recombinant phenotype arise from fertilization of the recombinant gametes by homozygous recessive gametes from the double-mutant parent. **2.** In each case, the alleles contributed by the female parent (in the egg) determine the phenotype of the offspring because the male in this cross contributes only recessive alleles. **3.** No. The order could be A-C-B or C-A-B. To determine which possibility is correct, you need to know the recombination frequency between B and C.

Concept Check 12.4

1. In meiosis, a combined 14-21 chromosome behaves as one chromosome. If a gamete receives the combined 14-21 chromosome and a normal copy of chromosome 21, trisomy 21 will result when this gamete combines with a normal gamete during fertilization. **2.** No. The child can be either I^AI^Ai or I^Aii. A sperm of genotype I^AI^A could result from nondisjunction in the father during meiosis II, while an egg with the genotype ii could result from nondisjunction in the mother during either meiosis I or meiosis II. **3.** Activation of this gene could lead to the production of too much of this kinase. If the kinase is involved in a signaling pathway that triggers cell division, too much of it could trigger unrestricted cell division, which in turn could contribute to the development of a cancer (in this case, a cancer of one type of white blood cell). **4.** The inactivation of two X chromosomes in XXX women would leave them with one genetically active X, as in women with the normal number of chromosomes. Microscopy should reveal two Barr bodies in XXX women.

Summary of Key Concepts Questions

12.1 Because the sex chromosomes are different from each other and because they determine the sex of the offspring, Morgan could use the sex of the offspring as a phenotypic characteristic to follow the parental chromosomes. (He could also have followed them under a microscope, as the X and Y chromosomes look different.) At the same time, he could record eye color to follow the eye-color alleles. **12.2** Males have only one X chromosome, along with a Y chromosome, while females have two X chromosomes. The Y chromosome has very few genes on it, while the X has about 1,000. When a recessive X-linked allele that causes a disorder is inherited by a male on the X from his mother, there isn't a second allele present on the Y (males are hemizygous), so the male has the disorder. Because females have two X chromosomes, they must inherit two recessive alleles in order to have the disorder, a rarer occurrence. **12.3** Crossing over results in new combinations of alleles. Crossing over is a random occurrence, and the more distance there is between two genes, the more chances there are for crossing over to occur, leading to a new combination of alleles. **12.4** In inversions and reciprocal translocations, the same genetic material is present in the same relative amount but just organized differently. In aneuploidy, duplications, and deletions, the balance of genetic material is upset, as large segments are either missing or present in more than one copy. Apparently, this type of imbalance is very damaging to the organism. (Although it isn't lethal in the developing embryo, the reciprocal translocation that produces the Philadelphia chromosome can lead to a form of cancer by altering the expression of important genes.)

Test Your Understanding

1. 0; $\frac{1}{2}$; $\frac{1}{16}$ **2.** Recessive; if the disorder were dominant, it would affect at least one parent of a child born with the disorder. The disorder's inheritance is sex-linked because it is seen only in boys. For a girl to have the disorder, she would have to inherit recessive alleles from *both* parents. This would be very rare, since males with the recessive allele on their X chromosome die in their early teens. **3.** Between T and A, 12%; between A and S, 5% **4.** Between T and S, 18%; sequence of genes is T–A–S. **5.** $\frac{1}{4}$ for each daughter ($\frac{1}{2}$ chance that the child will be female × $\frac{1}{2}$ chance of a homozygous recessive genotype); $\frac{1}{2}$ for first son **6.** About one-third of the distance from the vestigial-wing locus to the brown-eye locus **7.** 6%; wild-type heterozygous for normal wings and red eyes × recessive homozygous for vestigial wings and purple eyes **8.** Fifty percent of the offspring will show phenotypes resulting from crossovers. These results would be the same as those from a cross where A and B were *not* on the same chromosome. Further crosses involving other genes on the same chromosome would reveal the genetic linkage and map distances. **9.** 450 each of blue-oval and white-round (parentals) and 50 each of blue-round and white-oval (recombinants)

Chapter 13

Figure Questions

Figure 13.2 The living S cells found in the blood sample were able to reproduce to yield more S cells, indicating that the S trait is a permanent, heritable change, rather than just a one-time use of the dead S cells' capsules. **Figure 13.4** The radioactivity would have been found in the pellet when proteins were labeled (batch 1) because proteins would have had to enter the bacterial cells to program them with genetic instructions. It's hard for us to imagine now, but the DNA might have played a structural role that allowed some of the proteins to be injected while it remained outside the bacterial cell (resulting in no radioactivity in the pellet in batch 2). **Figure 13.11** The tube from the first replication would look the same, with a middle band of hybrid ^{15}N-^{14}N DNA, but the second tube would not have the upper band of two light blue strands. Instead, it would have a bottom band of two dark blue strands, like the bottom band in the result predicted after one replication in the conservative model. **Figure 13.13** In the bubble at the top of the micrograph in (b), arrows should be drawn pointing left and right to indicate the two replication forks. **Figure 13.14** Looking at any of the DNA strands, we see that one end is called the 5′ end and the other the 3′ end. If we proceed from the 5′ end to the 3′ end on the left-most strand, for example, we list the components in this order: phosphate group → 5′ C of the sugar → 3′ C → phosphate → 5′ C → 3′ C. Going in the opposite direction on the same strand, the components proceed in the reverse order: 3′ C → 5′ C → phosphate. Thus, the two directions are distinguishable, which is what we mean when we say that the strands have directionality. (Review Figure 13.5 if necessary.)

Figure 13.17

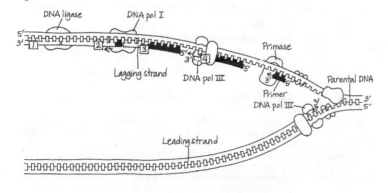

Figure 13.23

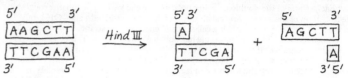

Concept Check 13.1

1. You can't tell which end is the 5′ end. You need to know which end has a phosphate group on the 5′ carbon (the 5′ end) or which end has an —OH group on the 3′ carbon (the 3′ end). **2.** He was expecting that the mouse injected with the mixture of heat-killed S cells and living R cells would survive, since neither type of cell alone would kill the mouse.

Concept Check 13.2

1. Complementary base pairing ensures that the two daughter molecules are exact copies of the parental molecule. When the two strands of the parental molecule separate, each serves as a template on which nucleotides are arranged, by the base-pairing rules, into new complementary strands.

2.

Protein	Function
Helicase	Unwinds parental double helix at replication forks
Single-strand binding protein	Binds to and stabilizes single-stranded DNA until it can be used as a template
Topoisomerase	Relieves "overwinding" strain ahead of replication forks by breaking, swiveling, and rejoining DNA strands
Primase	Synthesizes an RNA primer at 5′ end of leading strand and at 5′ end of each Okazaki fragment of lagging strand
DNA pol III	Using parental DNA as a template, synthesizes new DNA strand by covalently adding nucleotides to 3′ end of a preexisting DNA strand or RNA primer
DNA pol I	Removes RNA nucleotides of primer from 5′ end and replaces them with DNA nucleotides
DNA ligase	Joins 3′ end of DNA that replaces primer to rest of leading strand and joins Okazaki fragments of lagging strand

3. In the cell cycle, DNA synthesis occurs during the S phase, between the G_1 and G_2 phases of interphase. DNA replication is therefore complete before the mitotic phase begins.

Concept Check 13.3

1. A nucleosome is made up of eight histone proteins, two each of four different types, around which DNA is wound. Linker DNA runs from one nucleosome to the next. **2.** Euchromatin is chromatin that becomes less compacted during interphase and is accessible to the cellular machinery responsible for gene activity. Heterochromatin, on the other hand, remains quite condensed during interphase and contains genes that are largely inaccessible to this machinery.

Concept Check 13.4

1. The covalent sugar-phosphate bonds of the DNA strands **2.** Yes, *Pvu*I will cut the molecule (at the position indicated by the dashed red line).

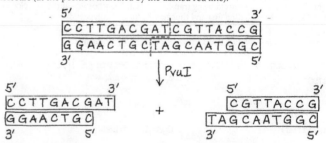

3. Cloning requires joining two pieces of DNA—a cloning vector, such as a bacterial plasmid, and a gene or DNA fragment from another source. Both pieces must be cut with the same restriction enzyme, creating sticky ends that will base-pair with complementary ends on other fragments. (The sugar-phosphate backbones will then be ligated together.) In DNA sequencing, primers base-pair to the template, allowing DNA synthesis to start, and then nucleotides are added to the growing strand based on complementarity of base pairing. In PCR, the primers must base-pair with their target sequences in the DNA mixture, locating one specific region among many, and complementary base pairing is the basis for the building of the new strand during the extension step.

Summary of Key Concepts Questions

13.1 Each strand in the double helix has polarity, the end with a phosphate group on the 5′ carbon of the sugar being called the 5′ end, and the end with an —OH group on the 3′ carbon of the sugar being called the 3′ end. The two strands run in opposite directions, so each end of the molecule has both a 5′ and a 3′ end. This arrangement is called "antiparallel." If the strands were parallel, they would both run 5′ → 3′ in the same direction, so an end of the molecule would have either two 5′ ends or two 3′ ends. **13.2** On both the leading and lagging strands, DNA polymerase adds onto the 3′ end of an RNA primer synthesized by primase, synthesizing DNA in the 5′ → 3′ direction. Because the parental strands are antiparallel, however, only on the leading strand does synthesis proceed continuously into the replication fork. The lagging strand is synthesized bit by bit in the direction away from the fork as a series of shorter Okazaki fragments, which are later joined together by DNA ligase. Each fragment is initiated by synthesis of an RNA primer by primase as soon as a given stretch of single-stranded template strand is opened up. Although both strands are synthesized at the same rate, synthesis of the lagging strand is delayed because initiation of each fragment begins only when sufficient template strand is available. **13.3** Most of the chromatin in an interphase nucleus is condensed. Much is present as the 30-nm fiber, with some in the form of the 10-nm fiber and some as looped domains of the 30-nm fiber. (These different levels of chromatin packing may reflect differences in gene expression occurring in these regions.) Also, a small percentage of the chromatin, such as that at the centromeres and telomeres, is highly condensed heterochromatin. **13.4** A plasmid vector and a source of foreign DNA to be cloned are both cut with the same restriction enzyme, generating restriction fragments with sticky ends. These fragments are mixed together, ligated, and reintroduced into bacterial cells, which can then make many copies of the foreign DNA or its product.

Test Your Understanding

1. c **2.** c **3.** b **4.** d **5.** c **6.** c **7.** d **8.** b **9.** a

10. Like histones, the *E. coli* proteins would be expected to contain many basic (positively charged) amino acids, such as lysine and arginine, which can form weak bonds with the negatively charged phosphate groups on the sugar-phosphate backbone of the DNA molecule.

11.

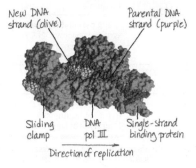

New DNA strand (olive)   Parental DNA strand (purple)

Sliding clamp   DNA pol III   Single-strand binding protein

Direction of replication

Chapter 14

Figure Questions

Figure 14.2 The mutant would not grow in the absence of arginine. **Figure 14.5** The mRNA sequence (5′-UGGUUUGGCUCA-3′) is the same as the nontemplate DNA strand sequence (5′-TGGTTTGGCTCA-3′), except there is U in the mRNA and T in the DNA. **Figure 14.8** The processes are similar in that polymerases form polynucleotides complementary to an antiparallel DNA template strand. In replication, however, both strands act as templates, whereas in transcription, only one DNA strand acts as a template. **Figure 14.9** The RNA polymerase would bind directly to the promoter, rather than depending on the previous binding of other factors. **Figure 14.23** The mRNA on the right (the longest one) started being transcribed first. The ribosome at the top, closest to the DNA, started translating first and thus has the longest polypeptide.

Concept Check 14.1

1. Recessive **2.** A polypeptide made up of 10 Gly (glycine) amino acids
3.

"Template sequence" (from nontemplate sequence in problem, written 3′ → 5′):   3′-ACGACTGAA-5′

mRNA sequence: 5′-UGCUGACUU-3′

Translated: Cys-STOP-Leu

(Remember that the mRNA is antiparallel to the DNA strand.) A protein translated from the nontemplate sequence would have a completely different amino acid sequence and would most likely be nonfunctional. (It would also be shorter because of the stop signal shown in the mRNA sequence above—and possibly others earlier in the mRNA sequence.)

Concept Check 14.2

1. The promoter is the region of DNA to which RNA polymerase binds to begin transcription, and it is at the upstream end of the gene (transcription unit). **2.** In a bacterial cell, RNA polymerase recognizes the gene's promoter and binds to it. In a eukaryotic cell, transcription factors mediate the binding of RNA polymerase to the promoter. In both cases, sequences in the promoter bind precisely to the RNA polymerase, so the enzyme is in the right location and orientation. **3.** The transcription

factor that recognizes the TATA sequence would be unable to bind, so RNA polymerase could not bind, and transcription of that gene probably would not occur.

Concept Check 14.3

1. Due to alternative splicing of exons, each gene can result in multiple different mRNAs and can thus direct synthesis of multiple different proteins. **2.** In editing a video, segments are cut out and discarded (like introns), and the remaining segments are joined together (like exons) so that the regions of joining ("splicing") are not noticeable. **3.** Once the mRNA has exited the nucleus, the cap prevents it from being degraded by hydrolytic enzymes and facilitates its attachment to ribosomes. If the cap were removed from all mRNAs, the cell would no longer be able to synthesize any proteins and would probably die.

Concept Check 14.4

1. First, each aminoacyl-tRNA synthetase specifically recognizes a single amino acid and attaches it only to an appropriate tRNA. Second, a tRNA charged with its specific amino acid binds only to an mRNA codon for that amino acid. **2.** The structure and function of the ribosome seem to depend more on the rRNAs than on the ribosomal proteins. Because it is single-stranded, an RNA molecule can hydrogen-bond with itself and with other RNA molecules. RNA molecules make up the interface between the two ribosomal subunits, so presumably RNA-RNA binding helps hold the ribosome together. The binding site for mRNA in the ribosome includes rRNA that can bind the mRNA. Also, complementary bonding within an RNA molecule allows it to assume a particular three-dimensional shape and, along with the RNA's functional groups, presumably enables rRNA to catalyze peptide bond formation during translation. **3.** A signal peptide on the leading end of the polypeptide being synthesized is recognized by a signal-recognition particle that brings the ribosome to the ER membrane. There the ribosome attaches and continues to synthesize the polypeptide, depositing it in the ER lumen. **4.** Because of wobble, the tRNA could bind to either 5′-GCA-3′ or 5′-GCG-3′, both of which code for alanine (Ala). Alanine would be attached to the tRNA.

tRNA   Ala

3′ CGU 5′

GCA   GCG
5′ ⊔⊔⊔ 3′   5′ ⊔⊔⊔ 3′

Concept Check 14.5

1. In the mRNA, the reading frame downstream from the deletion is shifted, leading to a long string of incorrect amino acids in the polypeptide, and in most cases, a stop codon will arise, leading to premature termination. The polypeptide will most likely be nonfunctional. **2.** Heterozygous individuals, said to have sickle-cell trait, have a copy each of the wild-type allele and the sickle-cell allele. Both alleles will be expressed, so these individuals will have both normal and sickle-cell hemoglobin molecules. Apparently, having a mix of the two forms of β-globin has no effect under most conditions, but during prolonged periods of low blood oxygen (such as at higher altitudes), these individuals can show some signs of sickle-cell disease.

3.

Normal DNA sequence
(template strand is on top):   3′-TACTTGTCCGATATC-5′
                                 5′-ATGAACAGGCTATAG-3′

mRNA sequence: 5′-AUGAACAGGCUAUAG-3′

Amino acid sequence: Met-Asn-Arg-Leu-STOP

Mutated DNA sequence
(template strand is on top):   3′-TACTTGTCCAATATC-5′
                                 5′-ATGAACAGGTTATAG-3′

mRNA sequence: 5′-AUGAACAGGUUAUAG-3′

Amino acid sequence: Met-Asn-Arg-Leu-STOP

No effect: The amino acid sequence is Met-Asn-Arg-Leu both before and after the mutation because the mRNA codons 5′-CUA-3′ and 5′-UUA-3′ both code for Leu. (The fifth codon is a stop codon.)

Summary of Key Concepts Questions

14.1 A gene contains genetic information in the form of a nucleotide sequence. The gene is first transcribed into an RNA molecule, and a messenger RNA molecule is ultimately translated into a polypeptide. The polypeptide makes up part or all of a protein, which performs a function in the cell and contributes to the phenotype of the organism. **14.2** Both bacterial and eukaryotic genes have promoters, regions where RNA polymerase ultimately binds and begins transcription. In bacteria, RNA polymerase binds directly to the promoter; in eukaryotes, transcription factors bind first to the promoter, and then RNA polymerase binds to the transcription factors and promoter together.   **14.3** Both the 5′ cap and the poly-A tail help the mRNA exit from the nucleus and then, in the cytoplasm, help ensure mRNA stability and allow it to bind to ribosomes.   **14.4** tRNAs function as translators between the nucleotide-based language of mRNA and the amino-acid-based language of polypeptides. A tRNA carries a specific amino acid, and the anticodon on the tRNA is complementary to the codon on the mRNA that codes for that amino acid. In the ribosome, the tRNA binds to the A site, where the polypeptide being synthesized is joined to the new amino acid, which becomes the new (C-terminal) end of the polypeptide. Next, the tRNA moves to the P site. When the next amino acid is added via transfer of the polypeptide to the new tRNA, the now empty tRNA moves to the E site, where it exits the ribosome.   **14.5** When a nucleotide base is altered chemically, its base-pairing characteristics may be changed. When that happens, an incorrect nucleotide is likely to be incorporated into the complementary strand during the next replication of the DNA, and successive rounds of replication will perpetuate the mutation. Once the gene is transcribed, the mutated codon may code for a different amino acid that inhibits or changes the function of a protein. If the chemical change in the base is detected and repaired by the DNA repair system before the next replication, no mutation will result.

Test Your Understanding
1. b   **2.** d   **3.** a   **4.** a   **5.** b   **6.** d   **7.** e
8.

Type of RNA	Functions
Messenger RNA (mRNA)	Carries information specifying amino acid sequences of proteins from DNA to ribosomes
Transfer RNA (tRNA)	Serves as translator molecule in protein synthesis; translates mRNA codons into amino acids
Ribosomal RNA (rRNA)	Plays catalytic (ribozyme) roles and structural roles in ribosomes
Primary transcript	Is a precursor to mRNA, rRNA, or tRNA, before being processed; some intron RNA acts as a ribozyme, catalyzing its own splicing
Small RNAs in spliceosome	Play structural and catalytic roles in spliceosomes, the complexes of protein and RNA that splice pre-mRNA

Chapter 15

Figure Questions
Figure 15.3 As the concentration of tryptophan in the cell falls, eventually there will be none bound to repressor molecules, which will then take on their inactive shapes and dissociate from the operator, allowing transcription of the operon to resume. The enzymes for tryptophan synthesis will be made, and they will begin to synthesize tryptophan again in the cell.   **Figure 15.11** The albumin gene enhancer has the three control elements colored yellow, gray, and red. The sequences in the liver and lens cells would be identical, since the cells are in the same organism.

Concept Check 15.1
1. Binding by the *trp* corepressor (tryptophan) activates the *trp* repressor, shutting off transcription of the *trp* operon; binding by the *lac* inducer (allolactose) inactivates the *lac* repressor, leading to transcription of the *lac* operon.   **2.** When glucose is scarce, cAMP is bound to CAP and CAP is bound to the promoter, favoring the binding of RNA polymerase. However, in the absence of lactose, the repressor is bound to the operator, blocking RNA polymerase binding to the promoter. Therefore, the operon genes are not transcribed.   **3.** The cell would continuously produce β-galactosidase and the two other enzymes for lactose utilization, even in the absence of lactose, thus wasting cell resources.

Concept Check 15.2
1. Histone acetylation is generally associated with gene expression, while DNA methylation is generally associated with lack of expression.   **2.** General transcription factors function in assembling the transcription initiation complex at the promoters for all genes. Specific transcription factors bind to control elements associated with a particular gene and, once bound, either increase (activators) or decrease (repressors) transcription of that gene.   **3.** The three genes should have some similar or identical sequences in the control elements of their enhancers. Because of this similarity, the same specific transcription factors in muscle cells could bind to the enhancers of all three genes and stimulate their expression coordinately.

Concept Check 15.3
1. The mRNA would persist and be translated into the cell division–promoting protein, and the cell would probably divide. If the intact miRNA is necessary for inhibition of cell division, then division of this cell might be inappropriate. Uncontrolled cell division could lead to formation of a mass of cells (tumor) that prevents proper functioning of the organism and could contribute to the development of cancer.   **2.** The *XIST* RNA is transcribed from the *XIST* gene on the X chromosome that will be inactivated. It then binds to that chromosome and induces heterochromatin formation. A likely model is that the *XIST* RNA somehow recruits chromatin modification enzymes that lead to formation of heterochromatin.

Concept Check 15.4
1. In RT-PCR, the primers must base-pair with their target sequences in the DNA mixture, locating one specific region among many. In microarray analysis, the labeled probe binds only to the specific target sequence owing to complementary nucleic acid hybridization (DNA-DNA hybridization).   **2.** As a researcher interested in cancer development, you would want to study genes represented by spots that are green or red because these are genes for which the expression level differs between the two types of tissues. Some of these genes may be expressed differently as a result of cancer, but others might play a role in causing cancer.

Summary of Key Concepts Questions
15.1 A corepressor and an inducer are both small molecules that bind to the repressor protein in an operon, causing the repressor to change shape. In the case of a corepressor (like tryptophan), this shape change allows the repressor to bind to the operator, blocking transcription. In contrast, an inducer causes the repressor to dissociate from the operator, allowing transcription to begin.   **15.2** The chromatin must not be tightly condensed because it must be accessible to transcription factors. The appropriate specific transcription factors (activators) must bind to the control elements in the enhancer of the gene, while repressors must not be bound. The DNA must be bent by a bending protein so the activators can contact the mediator proteins and form a complex with general transcription factors at the promoter. Then RNA polymerase must bind and begin transcription.   **15.3** miRNAs do not "code" for the amino acids of a protein—they are never translated. Each miRNA associates with a group of proteins to form a complex. Binding of the complex to an mRNA with a complementary sequence causes that mRNA to be degraded or blocks its translation. This is considered gene regulation because it controls the amount of a particular mRNA that can be translated into a functional protein.   **15.4** The genes that are expressed in a given tissue or cell type determine the proteins (and ncRNAs) that are the basis of the structure and functions of that tissue or cell type. Understanding which groups of interacting genes establish particular structures and allow certain functions will help us learn how the parts of an organism form and are maintained and help us treat diseases that occur when faulty gene expression leads to malfunctioning tissues.

Test Your Understanding
1. d   **2.** a   **3.** c   **4.** d   **5.** e   **6.** b
7. (a)

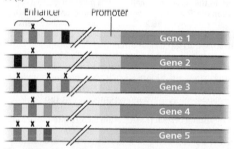

The purple, blue, and red activator proteins would be present.
(b)

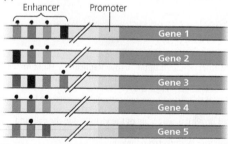

Only gene 4 would be transcribed.
(c) In nerve cells, the orange, blue, green, and black activators would have to be present, thus activating transcription of genes 1, 2, and 4. In skin cells, the red, black, purple, and blue activators would have to be present, thus activating genes 3 and 5.

Chapter 16

Figure Questions
Figure 16.4 Even if the mutant MyoD protein couldn't activate the *myoD* gene, it could still turn on genes for the other proteins in the pathway (other transcription factors, which would turn on the genes for muscle-specific proteins, for example). Therefore, some differentiation would occur. But unless there were other activators that could compensate for the loss of the MyoD protein's activation of the *myoD* gene, the cell would not be able to maintain its differentiated state.   **Figure 16.10** Normal Bicoid protein would be made in the anterior end and compensate for the presence of mutant *bicoid* mRNA put into the egg by the mother. Development should be normal, with a head present. (This is what was observed.)   **Figure 16.11** None of the eggs with the transplanted nuclei from the four-cell embryo at the upper left would have developed into a tadpole. Also, the resulting samples might include only some of the tissues of a tadpole. The tissues that develop might differ from treatment to treatment, depending on which of the four nuclei was transplanted. (This assumes that there was some way to tell the four cells apart, as one can in some frog species.)

Concept Check 16.1

1. Cells undergo differentiation during embryonic development, becoming different from each other. Therefore, in the adult organism, there are many highly specialized cell types.   **2.** By binding to a receptor on the receiving cell's surface and triggering a signal transduction pathway involving intracellular molecules such as second messengers and transcription factors that affect gene expression   **3.** Because their products, made and deposited into the egg by the mother, determine the head and tail ends, as well as the back and belly, of the embryo (and eventually the adult fly)

Concept Check 16.2

1. The state of chromatin modification in the nucleus from the intestinal cell was undoubtedly less similar to that of a nucleus from a fertilized egg, explaining why many fewer of these nuclei were able to be reprogrammed. In contrast, the chromatin in a nucleus from a cell at the four-cell stage would have been much more like that of a nucleus in a fertilized egg and therefore much more easily programmed to direct development.   **2.** No, primarily because of subtle (and perhaps not so subtle) differences in their environments. A technique would have to be worked out for turning a human iPS cell into a pancreatic cell (probably by inducing expression of pancreas-specific regulatory genes in the cell).

Concept Check 16.3

1. Apoptosis is signaled by p53 protein when a cell has extensive DNA damage, so apoptosis plays a protective role in eliminating a cell that might contribute to cancer. If mutations in the genes in the apoptotic pathway blocked apoptosis, a cell with such damage could continue to divide and might lead to tumor formation.   **2.** When an individual has inherited an oncogene or a mutant allele of a tumor-suppressor gene   **3.** A cancer-causing mutation in a proto-oncogene usually makes the gene product overactive, whereas a cancer-causing mutation in a tumor-suppressor gene usually makes the gene product nonfunctional.

Summary of Key Concepts Questions

16.1 The first process involves cytoplasmic determinants, including mRNAs and proteins, placed into specific locations in the egg by the mother. The cells that are formed from different regions in the egg during early cell divisions will have different proteins in them, which will direct different programs of gene expression. The second process involves how the cells respond to signaling molecules secreted by neighboring cells. The signaling pathways in the responding cells lead to different patterns of gene expression. The coordination of these two processes results in each cell following a unique pathway in the developing embryo.   **16.2** Cloning a mouse involves transplanting a nucleus from a differentiated mouse cell into a mouse egg cell that has had its own nucleus removed. Activating the egg cell and promoting its development into an embryo in a surrogate mother results in a mouse that is genetically identical to the mouse that donated the nucleus. In this case, the differentiated nucleus has been reprogrammed by factors in the egg cytoplasm. Mouse ES cells are generated from inner cells in mouse blastocysts, so in this case the cells are "naturally" reprogrammed by the process of reproduction and development. (Cloned mouse embryos can also be used as a source of ES cells.) iPS cells can be generated without the use of embryos from a differentiated adult mouse cell by adding certain transcription factors into the cell. In this case, the transcription factors are reprogramming the cells to become pluripotent.   **16.3** The protein product of a proto-oncogene is usually involved in a pathway that stimulates cell division. The protein product of a tumor-suppressor gene is usually involved in a pathway that inhibits cell division.

Test Your Understanding

1. a **2.** a **3.** d **4.** c **5.** b

Chapter 17

Figure Questions

Figure 17.3 Top vertical arrow: Infection. Left upper arrow: Replication. Right upper arrow: Transcription. Right middle arrow: Translation. Lower left and right arrows: Self-assembly. Bottom middle arrow: Exit.   **Figure 17.7** There are many steps that could be interfered with: binding of the virus to the cell, reverse transcriptase function, integration into the host cell chromosome, genome synthesis (in this case, transcription of RNA from the integrated provirus, assembly of the virus inside the cell, and budding of the virus. (Many of these, if not all, are targets of actual medical strategies to block progress of the infection in HIV-infected people.)

Concept Check 17.1

1. TMV consists of one molecule of RNA surrounded by a helical array of proteins. The influenza virus has eight molecules of RNA, each surrounded by a helical array of proteins, similar to the arrangement of the single RNA molecule in TMV. Another difference between the viruses is that the influenza virus has an outer envelope and TMV does not.   **2.** The T2 phages were an excellent choice for use in the Hershey-Chase experiment because they consist of only DNA surrounded by a protein coat, and DNA and protein were the two candidates for macromolecules that carried genetic information. Hershey and Chase were able to radioactively label each type of molecule alone and follow it during separate infections of *E. coli* cells with T2. Only the DNA entered the bacterial cell during infection, and only labeled DNA showed up in some of the progeny phage. Hershey and Chase concluded that the DNA must carry the genetic information necessary for the phage to reprogram the cell and produce progeny phages.

Concept Check 17.2

1. Lytic phages can only carry out lysis of the host cell, whereas lysogenic phages may either lyse the host cell or integrate into the host chromosome. In the latter case, the viral DNA (prophage) is simply replicated along with the host chromosome. Under certain conditions, a prophage may exit the host chromosome and initiate a lytic cycle.   **2.** Both the viral RNA polymerase and the cellular RNA polymerase in Figure 14.10 synthesize an RNA molecule complementary to a template strand. However, the cellular RNA polymerase in Figure 14.10 uses one of the strands of the DNA double helix as a template, whereas the viral RNA polymerase uses the RNA of the viral genome as a template.

3. Because it synthesizes DNA from its RNA genome. This is the reverse ("retro") of the usual DNA → RNA information flow.

Concept Check 17.3

1. Mutations can lead to a new strain of a virus that can no longer be effectively fought by the immune system, even if an animal had been exposed to the original strain; a virus can jump from one species to a new host; and a rare virus can spread if a host population becomes less isolated.   **2.** In horizontal transmission, a plant is infected from an external source of the virus, which can enter through a break in the plant's epidermis due to damage by herbivores. In vertical transmission, a plant inherits viruses from its parent either via infected seeds (sexual reproduction) or via an infected cutting (asexual reproduction).   **3.** Humans are not within the host range of TMV, so they can't be infected by the virus.

Summary of Key Concepts Questions

17.1 Viruses are generally considered nonliving because they are not capable of replicating outside of a host cell. To replicate, they depend completely on host enzymes and resources.   **17.2** Single-stranded RNA viruses require an RNA polymerase that can make RNA using an RNA template. (Cellular RNA polymerases make RNA using a DNA template.) Retroviruses require reverse transcriptases to make DNA using an RNA template. (Once the first DNA strand has been made, the same enzyme can promote synthesis of the second DNA strand.)   **17.3** The mutation rate of RNA viruses is higher than that of DNA viruses because RNA polymerase has no proofreading function, so errors in replication are not corrected. Their higher mutation rate means that RNA viruses change faster than DNA viruses, allowing them to have an altered host range and to evade immune defenses in possible hosts.

Test Your Understanding

1. c **2.** d **3.** c **4.** d **5.** b

6. As shown in the sketch, the viral genome would be translated into capsid proteins and envelope glycoproteins directly, rather than after a complementary RNA copy was made. A complementary RNA strand would still be made, however, that could be used as a template for many new copies of the viral genome.

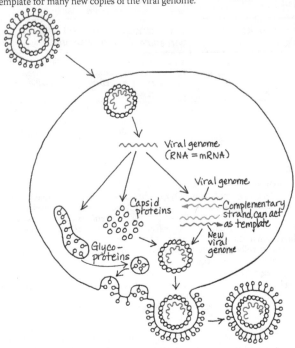

Chapter 18

Figure Questions

Figure 18.2 In stage 2 of this figure, the order of the fragments relative to each other is not known and will be determined later by computer. The unordered nature of the fragments is reflected by their scattered arrangement in the diagram.   **Figure 18.7** The transposon would be cut out of the DNA at the original site rather than copied, so the figure would show the original stretch of DNA without the transposon after the mobile transposon had been cut out.   **Figure 18.9** The RNA transcripts extending from the DNA in each transcription unit are shorter on the left and longer on the right. This means that RNA polymerase must be starting on the left end of the unit and moving toward the right.

Figure 18.12

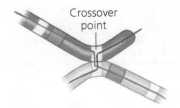

Crossover point

Figure 18.13 Pseudogenes are nonfunctional. They could have arisen by any mutations in the second copy that made the gene product unable to function. Examples would be base changes that introduce stop codons in the sequence, alter amino acids, or change a region of the gene promoter so that the gene can no longer be expressed. **Figure 18.14** Let's say a transposable element (TE) existed in the intron to the left of the indicated EGF exon in the EGF gene, and the same TE was present in the intron to the right of the indicated F exon in the fibronectin gene. During meiotic recombination, these TEs could cause nonsister chromatids on homologous chromosomes to pair up incorrectly, as seen in Figure 18.12. One gene might end up with an F exon next to an EGF exon. Further mistakes in pairing over many generations might result in these two exons being separated from the rest of the gene and placed next to a single or duplicated K exon. In general, the presence of repeated sequences in introns and between genes facilitates these processes because it allows incorrect pairing of nonsister chromatids, leading to novel exon combinations.

Concept Check 18.1

1. In the whole-genome shotgun approach, short fragments generated by multiple restriction enzymes are cloned and sequenced and then ordered by computer programs that identify overlapping regions. In this way, a composite sequence is obtained.

Concept Check 18.2

1. The Internet allows centralization of databases such as GenBank and software resources such as BLAST, making them freely accessible. Having all the data in a central database, easily accessible on the Internet, minimizes the possibility of errors and of researchers working with different data. It streamlines the process of science, since all researchers are able to use the same software programs, rather than each having to obtain their own, possibly different, software. It speeds up dissemination of data and ensures as much as possible that errors are corrected in a timely fashion. These are just a few answers; you can probably think of more. **2.** Cancer is a disease caused by multiple factors. To focus on a single gene or a single defect would ignore other factors that may influence the cancer and even the behavior of the single gene being studied. The systems approach, because it takes into account many factors at the same time, is more likely to lead to an understanding of the causes and most useful treatments for cancer. **3.** Some of the transcribed region is accounted for by introns. The rest is transcribed into noncoding RNAs, including small RNAs, such as microRNAs (miRNAs). These RNAs help regulate gene expression by blocking translation, causing degradation of mRNA, binding to the promoter and repressing transcription, or causing remodeling of chromatin structure. The functions of the remainder are not yet known.

Concept Check 18.3

1. Alternative splicing of RNA transcripts from a gene and post-translational processing of polypeptides **2.** The total number of completed genomes is found by clicking on "Complete Projects" under "Isolate Genomes"; the numbers of completed genomes for each domain are at the top of this page. The number of genomes "in progress" is visible if you click on "Incomplete Projects" under "Isolate Genomes" on the home page; the number is broken down by domains and also by the status of the project. (*Note*: Back at the home page, you can click on "Phylogenetic" under "Genome Distribution" to see how the numbers of sequenced genomes are distributed among phylogenetic groups at the phylum level. Note the number of Chordate genomes near the bottom of the table.) **3.** Prokaryotes are generally smaller cells than eukaryotic cells, and they reproduce by binary fission. The evolutionary process involved is natural selection for more quickly reproducing cells: The faster they can replicate their DNA and divide, the more likely they will be able to dominate a population of prokaryotes. The less DNA they have to replicate, then, the faster they will reproduce.

Concept Check 18.4

1. The number of genes is higher in mammals, and the amount of noncoding DNA is greater. Also, the presence of introns in mammalian genes makes them larger, on average, than prokaryotic genes. **2.** The copy-and-paste transposon mechanism and retrotransposition **3.** In the rRNA gene family, identical transcription units for the three different RNA products are present in long, tandemly repeated arrays. The large number of copies of the rRNA genes enables organisms to produce the rRNA for enough ribosomes to carry out active protein synthesis, and the single transcription unit ensures that the relative amounts of the different rRNA molecules produced are correct. Each globin gene family consists of a relatively small number of nonidentical genes. The differences in the globin proteins encoded by these genes result in production of hemoglobin molecules adapted to particular developmental stages of the organism.

Concept Check 18.5

1. If meiosis is faulty, two copies of the entire genome can end up in a single cell. Errors in crossing over during meiosis can lead to one segment being duplicated while another is deleted. During DNA replication, slippage backward along the template strand can result in segment duplication. **2.** For either gene, a mistake in crossing over during meiosis could have occurred between the two copies of that gene, such that one ended up with a duplicated exon. This could have happened several times, resulting in the multiple copies of a particular exon in each gene. **3.** Homologous transposable elements scattered throughout the genome provide sites where recombination can occur between different chromosomes. Movement of these elements into coding or regulatory sequences may change expression of genes. Transposable elements also can carry genes with them, leading to dispersion of genes and in some cases different patterns of expression. Transport of an exon during transposition and its insertion into a gene may add a new functional domain to the originally encoded protein, a type of exon shuffling. (For any of these changes to be heritable, they must happen in germ cells, cells that will give rise to gametes.)

Concept Check 18.6

1. Because both humans and macaques are primates, their genomes are expected to be more similar than the macaque and mouse genomes are. The mouse lineage diverged from the primate lineage before the human and macaque lineages diverged. **2.** Homeotic genes differ in their *non*homeobox sequences, which determine the interactions of homeotic gene products with other transcription factors and hence which genes are regulated by the homeotic genes. These nonhomeobox sequences differ in the two organisms, as do the expression patterns of the homeobox genes.

Summary of Key Concepts Questions

18.1 One focus of the Human Genome Project was to improve sequencing technology in order to speed up the process. During the project, many advances in sequencing technology allowed faster reactions, which were therefore less expensive. **18.2** The most significant finding was that more than 90% of the human genomic region studied was transcribed, which suggested that the transcribed RNA (and thus the DNA from which it was produced) was performing some unknown functions. The project has been expanded to include other species because to determine the functions of these transcribed DNA elements, it is necessary to carry out this type of analysis on the genomes of species that can be used in laboratory experiments. **18.3** (a) In general, bacteria and archaea have smaller genomes, lower numbers of genes, and higher gene density than eukaryotes. (b) Among eukaryotes, there is no apparent systematic relationship between genome size and phenotype. The number of genes is often lower than would be expected from the size of the genome—in other words, the gene density is often lower in larger genomes. (Humans are an example.) **18.4** Transposable elements can move from place to place in the genome, and some of these sequences make a new copy of themselves when they do so. Thus, it is not surprising that they make up a significant percentage of the genome, and this percentage might be expected to increase over evolutionary time. **18.5** Chromosomal rearrangements within a species lead to some individuals having different chromosomal arrangements. Each of these individuals could still undergo meiosis and produce gametes, and fertilization involving gametes with different chromosomal arrangements could result in viable offspring. However, during meiosis in the offspring, the maternal and paternal chromosomes might not be able to pair up, causing gametes with incomplete sets of chromosomes to form. Most often, when zygotes are produced from such gametes, they do not survive. Ultimately, a new species could form if two different chromosomal arrangements became prevalent within a population and individuals could mate successfully only with other individuals having the same arrangement. **18.6** Comparing the genomes of closely related species can reveal information about more recent evolutionary events, perhaps events that resulted in the distinguishing characteristics of the species. Comparing the genomes of very distantly related species can tell us about evolutionary events that occurred a very long time ago. For example, genes that are shared between distantly related species must have arisen before those species diverged.

Test Your Understanding

1. c **2.** a **3.** a **4.** c **5.** b

6.
1. ATFTI...PKSSD...TSSTT...NARRD
2. ATETI...PKSS[E]...TSS[T]T...[N]ARRD
3. ATETI...PKSSD...TSSTT...NARRD
4. ATETI...PKSSD...TSSNT...SARRD
5. ATETI...PKSSD...TSSTT...NARRD
6. VTETI...PKSSD...TSSTT...NARRD

(a) Lines 1, 3, and 5 are the C, G, R species. (b) See the underlined amino acids in Line 4. (Line 4 is the human sequence.) (c) Line 6 is the orangutan sequence. (d) There is one amino acid difference between the mouse (the E on line 2) and the C, G, R species (which have a D in that position). There are three amino acid differences between the mouse and the human. (The E, T, and N in the mouse sequence are instead D, N, and S, respectively, in the human sequence.) (e) Because only one amino acid difference arose during the 60–100 million years since the mouse and C, G, R species diverged, it is somewhat surprising that two additional amino acid differences resulted during the 6 million years since chimpanzees and humans diverged. This indicates that the *FOXP2* gene has been evolving faster in the human lineage than in the lineages of other primates.

Chapter 19

Figure Questions

Figure 19.6 The cactus-eater is more closely related to the seed-eater; Figure 1.16 shows that they share a more recent common ancestor (a seed-eater) than the cactus-eater shares with the insect-eater. **Figure 19.9** The common ancestor lived more than 5.5 million years ago. **Figure 19.13** The colors and body forms of these mantids allow them to blend into their surroundings, providing an example of how organisms are well matched to life in their environments. The mantids also share features with one another (and with all other mantids), such as six legs, grasping forelimbs, and large eyes. These shared features illustrate another key observation about life: the unity of life that results from descent from a common ancestor. Over time, as these mantids diverged from a common ancestor, they accumulated different adaptations that made them well suited for life in their different environments. Eventually, these differences became large enough that new species were formed, thus contributing to the great diversity of life. **Figure 19.14** These results show that being reared from the egg stage on one plant species or the other did not result in the adult having a beak length appropriate for that host; instead, adult beak lengths were determined primarily by the population from which the eggs were obtained. Because an egg from a balloon vine population likely had long-beaked parents, while an egg from a goldenrain tree population likely had short-beaked parents, these results indicate that beak length is an inherited trait. **Figure 19.20** Hind limb structure changed first. *Rodhocetus* lacked flukes, but its pelvic bones and hind limbs had changed substantially from how those bones were shaped and arranged in *Pakicetus*. For example, in *Rodhocetus*, the pelvis and hind limbs appear to be oriented for paddling, whereas they were oriented for walking in *Pakicetus*.

Concept Check 19.1

1. Hutton and Lyell proposed that geologic events in the past were caused by the same processes operating today, at the same gradual rate. This principle suggested that Earth must be much older than a few thousand years, the age that was widely accepted at that time. Hutton and Lyell's ideas also stimulated Darwin to reason that the slow accumulation of small changes could ultimately produce the profound changes documented in the fossil record. In this context, the age of Earth was important to Darwin, because unless Earth was very old, he could not envision how there would have been enough time for evolution to occur.   **2.** By these criteria, Cuvier's explanation of the fossil record and Lamarck's hypothesis of evolution are both scientific. Cuvier thought that species did not evolve over time. He also suggested that sudden, catastrophic events caused extinctions in particular areas. These assertions can be tested against the fossil record, and his assertion that species do not evolve has been falsified. With respect to Lamarck, his principle of use and disuse can be used to make testable predictions for fossils of groups such as whale ancestors as they adapted to a new habitat. Lamarck's principle of use and disuse and his associated principle of the inheritance of acquired characteristics can also be tested directly in living organisms; these principles have been falsified.

Concept Check 19.2

1. Organisms share characteristics (the unity of life) because they share common ancestors; the great diversity of life occurs because new species have repeatedly formed when descendant organisms gradually adapted to different environments, becoming different from their ancestors.   **2.** The fossil mammal species (or its ancestors) would most likely have colonized the Andes from within South America, whereas ancestors of mammals currently found in African mountains would most likely have colonized those mountains from other parts of Africa. As a result, the Andes fossil species would share a more recent common ancestor with South American mammals than with mammals in Africa. Thus, for many of its traits, the fossil mammal species would probably more closely resemble mammals that live in South American jungles than mammals that live on African mountains. It is also possible, however, that the fossil mammal species could resemble the African mountain mammals because similar environments selected for similar adaptations (even though they were only distantly related to one another).   **3.** As long as the white phenotype (encoded by the genotype *pp*) continues to be favored by natural selection, the frequency of the *p* allele will likely increase over time in the population. If the proportion of white individuals increases relative to purple individuals, the frequency of the recessive *p* allele will also increase relative to that of the *P* allele, which only appears in purple individuals (some of whom also carry a *p* allele).

Concept Check 19.3

1. An environmental factor such as a drug does not create new traits, such as drug resistance, but rather selects for traits among those that are already present in the population.   **2.** (a) Despite their different functions, the forelimbs of different mammals are structurally similar because they all represent modifications of a structure found in the common ancestor. (b) This is a case of convergent evolution. The similarities between the sugar glider and flying squirrel indicate that similar environments selected for similar adaptations despite different ancestry.   **3.** At the time that dinosaurs originated, Earth's landmasses formed a single large continent, Pangaea. Because many dinosaurs were large and mobile, it is likely that early members of these groups lived on many different parts of Pangaea. When Pangaea broke apart, fossils of these organisms would have moved with the rocks in which they were deposited. As a result, we would predict that fossils of early dinosaurs would have a broad geographic distribution (this prediction has been upheld).

Summary of Key Concepts Questions

19.1 Darwin thought that descent with modification occurred as a gradual, steplike process. The age of Earth was important to him because if Earth were only a few thousand years old (as conventional wisdom suggested), there wouldn't have been sufficient time for major evolutionary change.   **19.2** All species have the potential to overreproduce—that is, to produce more offspring than can be supported by the environment. This ensures that there will be what Darwin called a "struggle for existence" in which many of the offspring are eaten, starved, diseased, or unable to reproduce for a variety of other reasons. Members of a population exhibit a range of heritable variations, some of which make it likely that their bearers will leave more offspring than other individuals (for example, the bearer may escape predators more effectively or be more tolerant of the physical conditions of the environment). Over time, natural selection resulting from factors such as predators, lack of food, or the physical conditions of the environment can increase the proportion of individuals with favorable traits in a population (evolutionary adaptation).   **19.3** The hypothesis that cetaceans originated from a terrestrial mammal and are closely related to even-toed ungulates is supported by several lines of evidence. For example, fossils document that early cetaceans had hind limbs, as expected for organisms that descended from a land mammal; these fossils also show that cetacean hind limbs became reduced over time. Other fossils show that early cetaceans had a type of ankle bone that is otherwise found only in even-toed ungulates, providing strong evidence that even-toed ungulates are the land mammals to which cetaceans are most closely related. DNA sequence data also indicate that even-toed ungulates are the land mammals to which cetaceans are most closely related.

Test Your Understanding

1. b   **2.** d   **3.** d   **4.** c   **5.** a
6. (a)

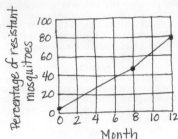

(b) The rapid rise in the percentage of mosquitoes resistant to DDT was most likely caused by natural selection in which mosquitoes resistant to DDT could survive and reproduce while other mosquitoes could not. (c) In India—where DDT resistance first appeared—natural selection would have caused the frequency of resistant mosquitoes to increase over time. If resistant mosquitoes then migrated from India (for example, transported by wind or in planes, trains, or ships) to other parts of the world, the frequency of DDT resistance would increase there as well.

Chapter 21

Figure Questions

Figure 21.4 The genetic code is redundant, meaning that more than one codon can specify the same amino acid. As a result, a substitution at a particular site in a coding region of the *Adh* gene might change the codon but not the translated amino acid and thus not the resulting protein encoded by the gene. One way an insertion in an exon would not affect the gene produced is if it occurs in an untranslated region of the exon. This is the case for the insertion at location 1,703.   **Figure 21.8** The predicted frequencies are 36% $C^R C^R$, 48% $C^R C^W$, and 16% $C^W C^W$.   **Figure 21.13** Directional selection. Goldenrain tree has smaller fruit than does the native host, balloon vine. Thus, in soapberry bug populations feeding on goldenrain tree, bugs with shorter beaks had an advantage, resulting in directional selection for shorter beak length.   **Figure 21.16** Crossing a single female's eggs with both an SC and an LC male's sperm allowed the researchers to directly compare the effects of the males' contribution to the next generation, since both batches of offspring had the same maternal contribution. This isolation of the male's impact enabled researchers to draw conclusions about differences in genetic "quality" between the SC and LC males.   **Figure 21.18** The researchers measured the percentage of successfully reproducing adults in the breeding population that had each phenotype. This approach of determining which phenotype was favored by selection assumes that reproduction was a sufficient indicator of relative fitness (as opposed to counting the number of eggs laid or offspring hatched, for example) and that mouth phenotype was the driving factor determining the fish's ability to reproduce.

Concept Check 21.1

1. Within a population, genetic differences among individuals provide the raw material on which natural selection and other mechanisms can act. Without such differences, allele frequencies could not change over time—and hence the population could not evolve.   **2.** Many mutations occur in somatic cells, which do not produce gametes and so are lost when the organism dies. Of mutations that do occur in cell lines that produce gametes, many do not have a phenotypic effect on which natural selection can act. Others have a harmful effect and are thus unlikely to increase in frequency because they decrease the reproductive success of their bearers.   **3.** Its genetic variation (whether measured at the level of the gene or at the level of nucleotide sequences) would probably drop over time. During meiosis, crossing over and the independent assortment of chromosomes produce many new combinations of alleles. In addition, a population contains a vast number of possible mating combinations, and fertilization brings together the gametes of individuals with different genetic backgrounds. Thus, via crossing over, independent assortment of chromosomes, and fertilization, sexual reproduction reshuffles alleles into fresh combinations each generation. Without sexual reproduction, the rate of forming new combinations of alleles would be vastly reduced, causing the overall amount of genetic variation to drop.

Concept Check 21.2

1. Each individual has two alleles, so the total number of alleles is 1,400. To calculate the frequency of allele A, note that each of the 85 individuals of genotype AA has two A alleles, each of the 320 individuals of genotype Aa has one A allele, and each of the 295 individuals of genotype aa has zero A alleles. Thus, the frequency (p) of allele A is

$$p = \frac{(2 \times 85) + (1 \times 320) + (0 \times 295)}{1,400} = 0.35$$

There are only two alleles (A and a) in our population, so the frequency of allele a must be $q = 1 - p = 0.65$.   **2.** Because the frequency of allele a is 0.45, the frequency of allele A must be 0.55. Thus, the expected genotype frequencies are $p^2 = 0.3025$ for genotype AA, $2pq = 0.495$ for genotype Aa, and $q^2 = 0.2025$ for genotype aa.
3. There are 120 individuals in the population, so there are 240 alleles. Of these, there are 124 V alleles—32 from the 16 VV individuals and 92 from the 92 Vv individuals. Thus, the frequency of the V allele is $p = 124/240 = 0.52$; hence, the frequency of the v allele is $q = 0.48$. Based on the Hardy-Weinberg equation, if the population were not evolving, the frequency of genotype VV should be $p^2 = 0.52 \times 0.52 = 0.27$; the frequency of genotype Vv should be $2pq = 2 \times 0.52 \times 0.48 = 0.5$; and the frequency of genotype vv should be $q^2 = 0.48 \times 0.48 = 0.23$. In a population of 120 individuals, these expected genotype frequencies lead us to predict that there would be 32 VV individuals (0.27×120), 60 Vv individuals (0.5×120), and 28 vv individuals (0.23×120). The actual numbers for the population (16 VV, 92 Vv, 12 vv) deviate from these expectations (fewer homozygotes and more heterozygotes than expected). This indicates that the population is not in Hardy-Weinberg equilibrium and hence may be evolving at this locus.

Concept Check 21.3

1. Natural selection is more "predictable" in that it alters allele frequencies in a nonrandom way: It tends to increase the frequency of alleles that increase the organism's reproductive success in its environment and decrease the frequency of alleles that decrease the organism's reproductive success. Alleles subject to genetic drift increase or decrease in frequency by chance alone, whether or not they are advantageous.   **2.** Genetic drift results from chance events that cause allele frequencies to fluctuate at random from generation to generation; within a population, this process tends to decrease genetic variation over time. Gene flow is the transfer of alleles between populations, a process that can introduce new alleles to a population and hence may increase its genetic variation (albeit slightly, since rates of gene flow are often low).   **3.** Selection is not important at this locus; furthermore, the populations are not small, and hence the effects of genetic drift should not be pronounced. Gene flow is occurring via the movement of pollen and seeds. Thus, allele and genotype frequencies in these populations should become more similar over time as a result of gene flow.

Concept Check 21.4

1. Zero, because fitness includes reproductive contribution to the next generation, and a sterile mule cannot produce offspring   **2.** Although both gene flow and genetic drift can increase the frequency of advantageous alleles in a population, they can also decrease the frequency of advantageous alleles or increase the frequency of harmful alleles. Only natural selection *consistently* results in an increase in the frequency of alleles that enhance survival or reproduction. Thus, natural selection is the only mechanism that consistently leads to adaptive evolution.   **3.** The three modes of natural selection (directional, stabilizing, and disruptive) are defined in terms of the selective advantage of different *phenotypes*, not different genotypes. Thus, the type of selection represented by heterozygote advantage depends on the phenotype of the heterozygotes. In this question, because heterozygous individuals have a more extreme phenotype than either homozygote, heterozygous advantage represents directional selection.   **4.** Under prolonged low-oxygen conditions, some of the red blood cells of a heterozygote may sickle, leading to harmful effects (see Figure 3.22). This does not occur in individuals with two normal hemoglobin alleles, suggesting that there may be selection against heterozygotes in malaria-free regions (where heterozygote advantage does not occur). However, since heterozygotes are healthy under most conditions, selection against them is unlikely to be strong.

Summary of Key Concepts Questions

21.1 Much of the nucleotide variability at a genetic locus occurs within introns. Nucleotide variation at these sites typically does not affect the phenotype because introns do not code for the protein product of the gene. (Note to students: In certain circumstances, it is possible that a change in an intron could affect RNA splicing and ultimately have some phenotypic effect on the organism, but such mechanisms are not covered in this introductory text.) There are also many variable nucleotide sites within exons. However, most of the variable sites within exons reflect changes to the DNA sequence that do not change the sequence of amino acids encoded by the gene (and hence may not affect the phenotype).   **21.2** No, this is not an example of circular reasoning. Calculating p and q from observed genotype frequencies does not imply

that those genotype frequencies must be in Hardy-Weinberg equilibrium. Consider a population that has 195 individuals of genotype AA, 10 of genotype Aa, and 195 of genotype aa. Calculating p and q from these values yields $p = q = 0.5$. Using the Hardy-Weinberg equation, the predicted equilibrium frequencies are $p^2 = 0.25$ for genotype AA, $2pq = 0.5$ for genotype Aa, and $q^2 = 0.25$ for genotype aa. Since there are 400 individuals in the population, these predicted genotype frequencies indicate that there should be 100 AA individuals, 200 Aa individuals, and 100 aa individuals—numbers that differ greatly from the values that we used to calculate p and q.   **21.3** It is unlikely that two such populations would evolve in similar ways. Since their environments are very different, the alleles favored by natural selection would probably differ between the two populations. Although genetic drift may have important effects in each of these small populations, drift causes unpredictable changes in allele frequencies, so it is unlikely that drift would cause the populations to evolve in similar ways. Both populations are geographically isolated, suggesting that little gene flow would occur between them (again making it less likely that they would evolve in similar ways).   **21.4** Compared to males, it is likely that the females of such species would be larger, more colorful, endowed with more elaborate ornamentation (for example, a large morphological feature such as the peacock's tail), and more apt to engage in behaviors intended to attract mates or prevent other members of their sex from obtaining mates.

Test Your Understanding

1. e   **2.** c   **3.** e   **4.** b   **5.** a   **6.** d
7. The frequency of the lap^{94} allele increases as one moves from southwest to northeast across Long Island Sound.

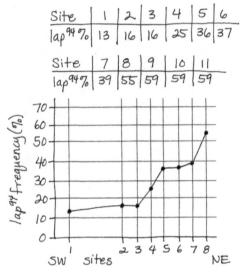

A hypothesis that explains the shape of the graph and accounts for the observations stated in the question is that the frequency of the lap^{94} allele at different sites results from an interaction between selection and gene flow. Under this hypothesis, in the southwest portion of the Sound, salinity is relatively low, and selection against the lap^{94} allele is strong. Moving toward the northeast and into the open ocean, where salinity is relatively high, selection favors a high frequency of the lap^{94} allele. However, because mussel larvae disperse long distances, gene flow prevents the lap^{94} allele from becoming fixed in the open ocean or from declining to zero in the southwestern portion of Long Island Sound.

Chapter 22

Figure Questions

Figure 22.8 If this had not been done, the strong preference of "starch flies" and "maltose flies" to mate with like-adapted flies could have occurred simply because the flies could detect (for example, by sense of smell) what their potential mates had eaten as larvae—and they preferred to mate with flies that had a similar smell to their own.   **Figure 22.10** In murky waters where females distinguish colors poorly, females of each species might mate often with males of the other species. Hence, since hybrids between these species are viable and fertile, the gene pools of the two species could become more similar over time.   **Figure 22.11** The graph suggests there has been gene flow of some fire-bellied toad alleles into the range of the yellow-bellied toad. Otherwise, all individuals located to the left of the hybrid zone portion of the graph would have allele frequencies very close to 1.   **Figure 22.12** Because the populations had only just begun to diverge from one another at this point in the process, it is likely that any existing barriers to reproduction would weaken over time.   **Figure 22.16** Over time, the chromosomes of the experimental hybrids came to resemble those of *H. anomalus*. This occurred even though conditions in the laboratory differed greatly from conditions in the field, where *H. anomalus* is found, suggesting that selection for laboratory conditions was not strong. Thus, it is unlikely that the observed rise in the fertility of the experimental hybrids was due to selection for life under laboratory conditions.   **Figure 22.17** The presence of *M. cardinalis* plants that carry the *M. lewisii yup* allele would make it more likely that bumblebees would transfer pollen between the two monkey flower species. As a result, we would expect the number of hybrid offspring to increase.

Concept Check 22.1

1. (a) All except the biological species concept can be applied to both asexual and sexual species because they define species on the basis of characteristics other than the ability to reproduce. In contrast, the biological species concept can be applied only to sexual species. (b) The easiest species concept to apply in the field would be the morphological species concept because it is based only on the appearance of the organism. Additional information about its ecological habits, evolutionary history, and reproduction are not required. **2.** Because these birds live in fairly similar environments and can breed successfully in captivity, the reproductive barrier in nature is probably prezygotic; given the species' differences in habitat preference, this barrier could result from habitat isolation.

Concept Check 22.2

1. In allopatric speciation, a new species forms while in geographic isolation from its parent species; in sympatric speciation, a new species forms in the absence of geographic isolation. Geographic isolation greatly reduces gene flow between populations, whereas ongoing gene flow is more likely in sympatric populations. As a result, sympatric speciation is less common than allopatric speciation. **2.** Gene flow between subsets of a population that live in the same area can be reduced in a variety of ways. In some species—especially plants—changes in chromosome number can block gene flow and establish reproductive isolation in a single generation. Gene flow can also be reduced in sympatric populations by habitat differentiation (as seen in the apple maggot fly, *Rhagoletis*) and sexual selection (as seen in Lake Victoria cichlids). **3.** Allopatric speciation would be less likely to occur on a nearby island than on an isolated island of the same size. The reason we expect this result is that continued gene flow between mainland populations and those on a nearby island reduces the chance that enough genetic divergence will take place for allopatric speciation to occur. **4.** If all of the homologs failed to separate during anaphase I of meiosis, some gametes would end up with an extra set of chromosomes (and others would end up with no chromosomes). If a gamete with an extra set of chromosomes fused with a normal gamete, a triploid would result; if two gametes with an extra set of chromosomes fused with each other, a tetraploid would result.

Concept Check 22.3

1. Hybrid zones are regions in which members of different species meet and mate, producing some offspring of mixed ancestry. Such regions can be viewed as "natural laboratories" in which to study speciation because scientists can directly observe factors that cause (or fail to cause) reproductive isolation. **2.** (a) If hybrids consistently survived and reproduced poorly compared with the offspring of intraspecific matings, reinforcement could occur. If it did, natural selection could cause prezygotic barriers to reproduction between the parent species to strengthen over time, decreasing the production of unfit hybrids and leading to a completion of the speciation process. (b) If hybrid offspring survived and reproduced as well as the offspring of intraspecific matings, indiscriminate mating between the parent species would lead to the production of large numbers of hybrid offspring. As these hybrids mated with each other and with members of both parent species, the gene pools of the parent species could fuse over time, reversing the speciation process.

Concept Check 22.4

1. The time between speciation events includes (1) the length of time that it takes for populations of a newly formed species to begin diverging reproductively from one another and (2) the time it takes for speciation to be complete once this divergence begins. Although speciation can occur rapidly once populations have begun to diverge from one another, it may take millions of years for that divergence to begin. **2.** Investigators transferred alleles at the *yup* locus (which influences flower color) from each parent species to the other. *M. lewisii* plants with an *M. cardinalis yup* allele received many more visits from hummingbirds than usual; hummingbirds usually pollinate *M. cardinalis* but avoid *M. lewisii*. Similarly, *M. cardinalis* plants with an *M. lewisii yup* allele received many more visits from bumblebees than usual; bumblebees usually pollinate *M. lewisii* and avoid *M. cardinalis*. Thus, alleles at the *yup* locus can influence pollinator choice, which in these species provides the primary barrier to interspecific mating. Nevertheless, the experiment does not prove that the *yup* locus alone controls barriers to reproduction between *M. lewisii* and *M. cardinalis*; other genes might enhance the effect of the *yup* locus (by modifying flower color) or cause entirely different barriers to reproduction (for example, gametic isolation or a postzygotic barrier). **3.** Crossing over. If crossing over did not occur, each chromosome in an experimental hybrid would remain as in the F_1 generation: composed entirely of DNA from one parent species or the other.

Summary of Key Concepts Questions

22.1 According to the biological species concept, a species is a group of populations whose members interbreed and produce viable, fertile offspring; thus, gene flow occurs between populations of a species. In contrast, members of different species do not interbreed, and hence no gene flow occurs between their populations. Overall, then, in the biological species concept, species can be viewed as designated by the *absence* of gene flow—making gene flow of central importance to the biological species concept. **22.2** Sympatric speciation can be promoted by factors such as polyploidy, habitat shifts, and sexual selection, all of which can reduce gene flow between the subpopulations of a larger population. But such factors can also occur in allopatric populations and hence can also promote allopatric speciation. **22.3** If the hybrids are selected against, the hybrid zone could persist if individuals from the parent species regularly travel into the zone, where they mate to produce hybrid offspring. If hybrids are not selected against, there is no cost to the continued production of hybrids, and large numbers of hybrid offspring may be produced. Natural selection for life in different environments may keep the gene pools of the two parent species distinct, thus preventing the loss (by fusion) of the parent species and once again causing the hybrid zone to be stable over time. **22.4** As the goatsbeard plant, Bahamas mosquitofish, and apple maggot fly illustrate, speciation continues to happen today. A new species

can begin to form whenever gene flow is reduced between populations of the parent species. Such reductions in gene flow can occur in many ways: A new, geographically isolated population may be founded by a few colonists; some members of the parent species may begin to utilize a new habitat; and sexual selection may isolate formerly connected populations or subpopulations. These and many other such events are happening today.

Test Your Understanding

1. b **2.** c **3.** c **4.** a **5.** e **6.** d
7. Here is one possibility:

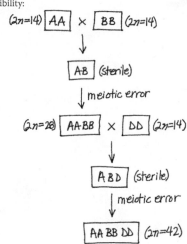

Concept Check 34.1

1. In both an open circulatory system and a fountain, fluid is pumped through a tube and then returns to the pump after collecting in a pool. **2.** The ability to shut off blood supply to the lungs when the animal is submerged **3.** The O_2 content would be abnormally low because some oxygen-depleted blood returned to the right atrium from the systemic circuit would mix with the oxygen-rich blood in the left atrium.

Concept Check 34.2

1. The pulmonary veins carry blood that has just passed through capillary beds in the lungs, where it accumulated O_2. The venae cavae carry blood that has just passed through capillary beds in the rest of the body, where it lost O_2 to the tissues. **2.** The delay allows the atria to empty completely, filling ventricles fully before they contract. **3.** The heart, like any other muscle, becomes stronger through regular exercise. You would expect a stronger heart to have a greater stroke volume, which would allow for the decrease in heart rate.

Concept Check 34.3

1. The large total cross-sectional area of the capillaries **2.** An increase in blood pressure and cardiac output combined with the diversion of more blood to the skeletal muscles would increase the capacity for action by increasing the rate of blood circulation and delivering more O_2 and nutrients to the skeletal muscles. **3.** Additional hearts could be used to improve blood return from the legs. However, it might be difficult to coordinate the activity of multiple hearts and to maintain adequate blood flow to hearts far from the gas exchange organs.

Concept Check 34.4

1. An increase in the number of white blood cells (leukocytes) may indicate that the person is combating an infection. **2.** Clotting factors do not initiate clotting but are essential steps in the clotting process. Also, the clots that form a thrombus typically result from an inflammatory response to an atherosclerotic plaque, not from clotting at a wound site. **3.** The chest pain results from inadequate blood flow in coronary arteries. Vasodilation promoted by nitric oxide from nitroglycerin increases blood flow, providing the heart muscle with additional oxygen and thus relieving the pain.

Concept Check 34.5

1. Their interior position helps them stay moist. If the respiratory surfaces of lungs extended out into the terrestrial environment, they would quickly dry out, and diffusion of O_2 and CO_2 across these surfaces would stop. **2.** Earthworms need to keep their skin moist for gas exchange, but they need air outside this moist layer. If they stay in their waterlogged tunnels after a heavy rain, they will suffocate because they cannot get as much O_2 from water as from air. **3.** In the extremities of some vertebrates, blood flows in opposite directions in neighboring veins and arteries; this countercurrent arrangement maximizes the recapture of heat from blood leaving the body core in arteries, which is important for thermoregulation in cold environments. Similarly, in the gills of fish, water passes over the gills in the direction opposite to that of blood flowing through the gill capillaries, maximizing the extraction of oxygen from the water along the length of the exchange surface.

Concept Check 34.6

1. An increase in blood CO_2 concentration causes an increase in the rate of CO_2 diffusion into the cerebrospinal fluid, where the CO_2 combines with water to form carbonic acid. Dissociation of carbonic acid releases hydrogen ions, decreasing the pH of the cerebrospinal fluid. **2.** Increased heart rate increases the rate at which CO_2-rich blood is delivered to the lungs, where CO_2 is removed. **3.** A hole would allow air to enter the space between the inner and outer layers of the double membrane, resulting in a condition called a pneumothorax. The two layers would no longer stick together, and the lung on the side with the hole would collapse and cease functioning.

Concept Check 34.7

1. The direction of net diffusion is determined by the difference in partial pressure. Net diffusion of gases occurs from a region of higher partial pressure to a region of lower partial pressure. **2.** The Bohr shift causes hemoglobin to release more O_2 at a lower pH, such as found in the vicinity of tissues with high rates of cellular respiration and CO_2 release. **3.** The doctor is assuming that the rapid breathing is the body's response to low blood pH. Metabolic acidosis, the lowering of blood pH, can have many causes, including complications of certain types of diabetes, shock (extremely low blood pressure), and poisoning.

Summary of Key Concepts Questions

34.1 In a closed circulatory system, an ATP-driven muscular pump generally moves fluids in one direction on a scale of millimeters to meters. Exchange between cells and their environment relies on diffusion, which involves random movements of molecules. Concentration gradients of molecules across exchange surfaces can drive rapid net diffusion on a scale of 1 mm or less. **34.2** Replacement of a defective valve should increase stroke volume. A lower heart rate would therefore be sufficient to maintain the same cardiac output. **34.3** Blood pressure in the arm would fall by 25 to 30 mm Hg, the same difference as is normally seen between your heart and your brain. **34.4** One microliter of blood contains about 5 million erythrocytes and 5,000 leukocytes, so leukocytes make up only about 0.1% of the cells in the absence of infection. **34.5** Because CO_2 is such a small fraction of atmospheric gas (0.29 mm Hg/760 mm Hg, or less than 0.04%), the partial pressure gradient of CO_2 between the respiratory surface and the environment always strongly favors the release of CO_2 to the atmosphere. **34.6** Because the lungs do not empty completely with each breath, incoming and outgoing air mix, resulting in a lower P_{O_2} in the lungs than in the air that enters the body during inspiration. **34.7** An enzyme speeds up a reaction without changing the equilibrium and without being consumed. Similarly, a respiratory pigment speeds up the movement of gases in the body without changing the equilibrium and without being consumed.

Test Your Understanding

1. c **2.** b **3.** d **4.** c **5.** a **6.** a **7.** a

Chapter 34

Figure Questions

Figure 34.8 Each feature of the ECG recording, such as the sharp upward spike, occurs once per cardiac cycle. Using the x-axis to measure the time in seconds between successive spikes and dividing that number into 60 would yield the heart rate as the number of cycles per minute. **Figure 34.21** The reduction in surface tension results from the presence of surfactant. Therefore, for all the infants that had died of RDS, you would expect the amount of surfactant to be near zero. For infants that had died of other causes, you would expect the amount of surfactant to be near zero for weights less than 1,200 g but much greater than zero for weights above 1,200 g.
Figure 34.23 Breathing at a rate greater than that needed to meet metabolic demand (hyperventilation) would lower blood CO_2 levels. Sensors in major blood vessels and the medulla would signal the breathing control centers to decrease the rate of contraction of the diaphragm and rib muscles, decreasing the breathing rate and restoring normal CO_2 levels in the blood and other tissues. **Figure 34.24** The resulting increase in tidal volume would enhance ventilation within the lungs, increasing P_{O_2} and decreasing P_{CO_2} in the alveoli.

8.

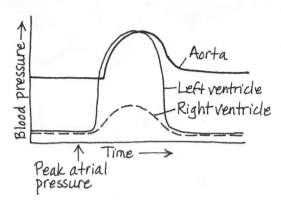

Chapter 35

Figure Questions

Figure 35.4 Cell-surface TLRs recognize pathogens identifiable by surface molecules, whereas TLRs in vesicles recognize pathogens identifiable by internal molecules after the pathogens are broken down.   **Figure 35.7** Part of the enzyme or antigen receptor provides a structural "backbone" that maintains overall shape, while interaction occurs at a surface with a close fit to the substrate or antigen. The combined effect of multiple noncovalent interactions at the active site or binding site is a high-affinity interaction of tremendous specificity.   **Figure 35.10** After gene rearrangement, a lymphocyte and its daughter cells make a single version of the antigen receptor. In contrast, alternative splicing is not heritable and can give rise to diverse gene products in a single cell.   **Figure 35.15** These receptors enable memory cells to present antigen on their cell surface to a helper T cell. This presentation of antigen is required to activate memory cells in a secondary immune response.   **Figure 35.16** Primary response: arrows extending from Antigen (1st exposure), Antigen-presenting cell, Helper T cell, B cell, Plasma cells, Cytotoxic T cell, and Active cytotoxic T cells; secondary response: arrows extending from Antigen (2nd exposure), Memory helper T cells, Memory B cells, and Memory cytotoxic T cells

Concept Check 35.1

1. Because pus contains white blood cells, fluid, and cell debris, it indicates an active and at least partially successful inflammatory response against invading microbes.
2. Whereas the ligand for the TLR receptor is a foreign molecule, the ligand for many signal transduction pathways is a molecule produced by the animal itself.   **3.** Bacteria with a human host would likely grow optimally at normal human body temperature or, if fever were often induced, at a temperature a few degrees higher.

Concept Check 35.2

1. See Figure 35.6. The transmembrane regions lie within the C regions, which also form the disulfide bridges. In contrast, the antigen-binding sites are in the V regions.
2. Generating memory cells ensures both that a receptor specific for a particular epitope will be present and that there will be more lymphocytes with this specificity than in a host that had never encountered the antigen.   **3.** If each B cell produced two different light and heavy chains for its antigen receptor, different combinations would make four different receptors. If any one was self-reactive, the lymphocyte would be eliminated in the generation of self-tolerance. For this reason, many more B cells would be eliminated, and those that could respond to a foreign antigen would be less effective at doing so due to the variety of receptors (and antibodies) they express.

Concept Check 35.3

1. Myasthenia gravis is considered an autoimmune disease because the immune system produces antibodies against self molecules (certain receptors on muscle cells).
2. A child lacking a thymus would have no functional T cells. Without helper T cells to help activate B cells, the child would be unable to produce antibodies against extracellular bacteria. Furthermore, without cytotoxic T cells or helper T cells, the child's immune system would be unable to kill virus-infected cells.   **3.** If the handler developed immunity to proteins in the antivenin, another injection could provoke a severe immune response. The handler's immune system might also now produce antibodies that could neutralize the venom in the absence of antivenin.

Summary of Key Concepts Questions

35.1 Lysozyme in saliva destroys bacterial cell walls; the viscosity of mucus helps trap bacteria; acidic pH in the stomach kills many bacteria; and the tight packing of cells lining the gut provides a physical barrier to infection.   **35.2** Sufficient numbers of cells to mediate an innate immune response are always present, whereas an adaptive response requires selection and proliferation of an initially very small cell population specific for the infecting pathogen.   **35.3** No. Immunological memory after a natural infection and immunological memory after immunization are very similar. There may be minor differences in the particular antigens that can be recognized in a subsequent infection.

Test Your Understanding

1. b   **2.** c   **3.** c   **4.** d   **5.** b   **6.** b   **7.** c

---

8. One possible answer:

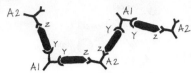

Chapter 36

Figure Questions

Figure 36.10 The analysis would be informative because the polar bodies contain all of the maternal chromosomes that don't end up in the mature egg. For example, finding two copies of the disease gene in the polar bodies would indicate its absence in the egg. This method of genetic testing is sometimes carried out when oocytes collected from a female are fertilized with sperm in a laboratory dish.

Concept Check 36.1

1. Internal fertilization allows the sperm to reach the egg without either gamete drying out.   **2.** No. Owing to random assortment of chromosomes during meiosis, the offspring may receive the same copy or different copies of a particular parental chromosome from the sperm and the egg. Furthermore, genetic recombination during meiosis will result in reassortment of genes between pairs of parental chromosomes.   **3.** Both fragmentation and budding in animals have direct counterparts in the asexual reproduction of plants.

Concept Check 36.2

1. Both have a haploid DNA content and very little cytoplasm. However, the early spermatid will develop into a functional gamete, whereas a polar body is a by-product of oocyte production.   **2.** Spermatogenesis occurs normally only when the testicles are cooler than the rest the body. Extensive use of a hot tub (or of very tight-fitting underwear) can cause a decrease in sperm quality and number.   **3.** Like the uterus of an insect, the ovary of a plant is the site of fertilization. Unlike the plant ovary, the uterus is not the site of egg production, which occurs in the insect ovary. In addition, the fertilized insect egg is expelled from the uterus, whereas the plant embryo develops within a seed in the ovary.   **4.** The only effect of sealing off each vas deferens is an absence of sperm in the ejaculate. Sexual response and ejaculate volume are unchanged. The cutting and sealing off of these ducts, a *vasectomy*, is a common surgical procedure for men who do not wish to produce any (more) offspring.

Concept Check 36.3

1. In both females and males, FSH encourages the growth of cells that support and nourish developing gametes (follicle cells in females and Sertoli cells in males), and LH stimulates the production of sex hormones that promote gametogenesis (estrogens, primarily estradiol, in females and androgens, especially testosterone, in males).   **2.** In estrous cycles, which occur in most female mammals, the endometrium is reabsorbed (rather than shed) if fertilization does not occur. Estrous cycles often occur just one or a few times a year, and the female is usually receptive to copulation only during the period around ovulation. Menstrual cycles are about four weeks in length, do not restrict receptivity to copulation to a particular interval, and are found only in humans and some other primates.   **3.** The combination of estradiol and progesterone would have a negative-feedback effect on the hypothalamus, blocking release of GnRH. This would interfere with LH secretion by the pituitary, thus preventing ovulation. This is in fact one basis of action of the most common hormonal contraceptives.

Concept Check 36.4

1. Fertilization occurs in one of the two oviducts.   **2.** Mesoderm. Skin is derived from ectoderm, and the lining of many internal organs is endodermal in origin.   **3.** The menstrual cycle would be unaffected because it is controlled by hormones, which circulate in the bloodstream. However, the woman could not become pregnant naturally because the oviduct blockage would prevent sperm from reaching her eggs.

Summary of Key Concepts Questions

36.1 Not necessarily. Because parthenogenesis involves meiosis, the mother would pass on to each offspring a random and therefore typically distinct combination of the chromosomes she inherited from her mother and father.   **36.2** The small size and lack of cytoplasm characteristic of a sperm are adaptations well suited to its function as a delivery vehicle for DNA. The large size and rich cytoplasmic contents of eggs support the growth and development of the embryo.   **36.3** Circulating anabolic steroids mimic the feedback regulation of testosterone, turning off pituitary signaling to the testes and thereby blocking the release of signals required for spermatogenesis.
36.4 The fertilization envelope forms after cortical granules release their contents outside the egg, causing the vitelline membrane to rise and harden. The fertilization envelope serves as a barrier to fertilization by more than one sperm.

Test Your Understanding

1. d   **2.** a   **3.** a   **4.** c   **5.** a   **6.** b   **7.** c   **8.** d

9.

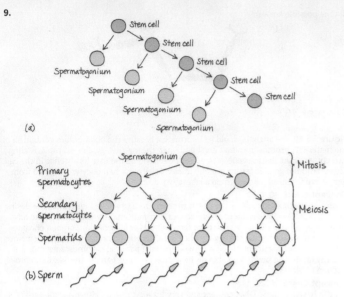

(a)

Spermatogonium → Primary spermatocytes → Secondary spermatocytes → Spermatids

Mitosis

Meiosis

(b) Sperm

(c) The supply of stem cells would be used up and spermatogenesis would not be able to continue.

Chapter 40

Figure Questions

Figure 40.12 Some factors, such as fire, are relevant only for terrestrial systems. At first glance, water availability is primarily a terrestrial factor, too. However, species living along the intertidal zone of oceans or along the edge of lakes also suffer desiccation. Salinity stress is important for species in some aquatic and terrestrial systems. Oxygen availability is an important factor primarily for species in some aquatic systems and in soils and sediments.   **Figure 40.13** When only urchins were removed, limpets may have increased in abundance and reduced seaweed cover somewhat (the difference between the purple and blue lines on the graph).

Figure 40.22

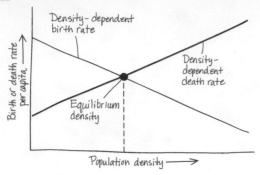

Concept Check 40.1

1. In the tropics, high temperatures evaporate water and cause warm, moist air to rise. The rising air cools and releases much of its water as rain over the tropics. The remaining dry air descends at approximately 30° north and south, causing deserts to occur in those regions. **2.** Answers will vary by location but should be based on the information in Figure 40.9. How much your local area has been altered from its natural state will influence how much it reflects the expected characteristics of your biome, particularly the expected plants and animals. **3.** Northern coniferous forest is likely to replace tundra along the boundary between these biomes. To see why, note that northern coniferous forest is adjacent to tundra throughout North America, northern Europe, and Asia (see Figure 40.7) and that the temperature range for northern coniferous forest is just above that for tundra (see Figure 40.8).

Concept Check 40.2

1. In the oceanic pelagic zone, the ocean bottom lies below the photic zone, so there is too little light to support benthic algae or rooted plants. **2.** As explained in Concept 32.3, aquatic organisms either gain or lose water by osmosis if the osmolarity of their environment differs from their internal osmolarity. Water gain can cause cells to swell, and water loss can cause them to shrink. To avoid excessive changes in cell volume, organisms that live in estuaries must be able to compensate for both water gain (under freshwater conditions) and water loss (under saltwater conditions). **3.** In a river below a dam, the fish are more likely to be species that prefer colder water. In summer, the deep layers of a reservoir are colder than the surface layers, so a river below a dam will be colder than an undammed river.

Concept Check 40.3

1. (a) Humans might transplant a species to a new area that it could not previously reach because of a geographic barrier. (b) Humans might eliminate a predator or herbivore species, such as sea urchins, from an area. **2.** One test would be to build a fence around a plot of land in an area that has trees of that species, excluding all deer from the plot. You could then compare the abundance of tree seedlings inside and outside the fenced plot over time.

Concept Check 40.4

1.

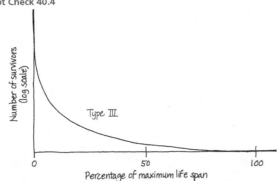

A Type III survivorship curve is most likely because very few of the young probably survive. **2.** The average number of female offspring is $0.74 \times 3.01 = 2.23$. **3.** Male sticklebacks would likely have a uniform pattern of dispersion, with antagonistic interactions maintaining a relatively constant spacing between them.

Concept Check 40.5

1. Though r_{max} is constant, N, the population size, is increasing. As r_{max} is applied to an increasingly large N, population growth ($r_{max}N$) accelerates, producing the J-shaped curve. **2.** When N (population size) is small, there are relatively few individuals producing offspring. When N is large, near the carrying capacity, the per capita growth rate is relatively small because it is limited by available resources. The steepest part of the logistic growth curve corresponds to a population with a number of reproducing individuals that is substantial but not yet near carrying capacity. **3.** If a population becomes too crowded, the likelihood of disease and mortality may increase because of the effects of pathogens. Thus, pathogens can reduce the long-term carrying capacity of a population.

Concept Check 40.6

1. By preferentially investing in the eggs it lays in the nest, the peacock wrasse increases their probability of survival. The eggs it disperses widely and does not provide care for are less likely to survive but require a lower investment by the adults. (In a sense, the adults avoid the risk of placing all their eggs in one basket.) **2.** If a parent's survival is compromised greatly by caring for young during times of stress, the animal's fitness may increase if it abandons its current offspring and survives to produce healthier offspring at a later time. **3.** In negative feedback, the output, or product, of a process slows that process. In populations that have a density-dependent birth rate, such as dune fescue grass, an accumulation of product (more individuals, resulting in a higher population density) slows the process of population growth by decreasing the birth rate.

Summary of Key Concepts Questions

40.1 Because tundra is much cooler than deserts (see Figure 40.8), less water evaporates during the growing season and the tundra stays more moist. **40.2** An aphotic zone is most likely to be found in the deep waters of a lake, the oceanic pelagic zone, or the marine benthic zone. **40.3** You could make a flowchart that begins with abiotic limitations—first determining the physical and chemical conditions under which a species could survive—and then moves through the other factors listed in the flowchart. **40.4** Ecologists can potentially estimate birth rates by counting the number of young whales born each year, and they can estimate death rates by seeing how the number of adults changes each year. **40.5** There are many things you could do to increase the carrying capacity of the species, including increasing its food supply, protecting it from predators, and providing more sites for nesting or reproduction. **40.6** An example of a biotic factor would be disease caused by a pathogen; natural disasters, such as floods and storms, are examples of abiotic factors.

Test Your Understanding

1. e **2.** a **3.** a **4.** c
5.

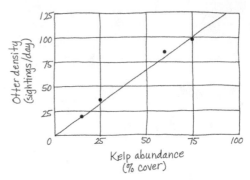

Based on what you learned from Figure 40.13 and on the positive relationship you observed in the field between kelp abundance and otter density, you could hypothesize that otters lower sea urchin density, reducing feeding of the urchins on kelp.
6. c

Chapter 41

Figure Questions

Figure 41.3 Its realized and fundamental niches would be similar, unlike those of *Chthamalus*. **Figure 41.4** The smallest beak depth observed for *G. fortis* on Santa María and San Cristóbal Islands is 10 mm. Therefore, the predicted beak length is $10 \text{ mm} \times 1.12 = 11 \text{ mm}$. **Figure 41.15** The death of individuals of *Mytilus*, a dominant species, should open up space for other species and increase species richness even in the absence of *Pisaster*. **Figure 41.22** Other factors not included in the model must contribute to the number of species.

Concept Check 41.1

1. Interspecific competition has negative effects on both species (−/−). In predation, the predator population benefits at the expense of the prey population (+/−). Mutualism is a symbiosis in which both species benefit (+/+). **2.** One of the competing species will become locally extinct because of the greater reproductive success of the more efficient competitor. **3.** By specializing in eating seeds of a single plant species, individuals of the two finch species may be less likely to come into contact in the separate habitats, reinforcing a reproductive barrier to hybridization.

Concept Check 41.2

1. Species richness, the number of species in the community, and relative abundance, the proportions of the community represented by the various species, both contribute to species diversity. Compared to a community with a very high proportion of one species, a community with a more even proportion of species is considered more diverse. **2.** The food web in Figure 41.14 indicates that elephant seals eat fish and squid and are in turn eaten by smaller toothed whales and humans. **3.** According to the bottom-up model, adding extra predators would have little effect on lower trophic levels, particularly grasses. If the top-, increased bobcat numbers would decrease raccoon numbers, increase snake numbers, decrease mouse numbers, and increase grass biomass.

Concept Check 41.3

1. High levels of disturbance are generally so disruptive that they eliminate many species from communities, leaving the community dominated by a few tolerant species.

Low levels of disturbance permit competitively dominant species to exclude other species from the community. Moderate levels of disturbance, however, can facilitate coexistence of a greater number of species in a community by preventing competitively dominant species from becoming abundant enough to eliminate other species from the community. **2.** Early successional species can facilitate the arrival of other species in many ways, including increasing the fertility or water-holding capacity of soils or providing shelter to seedlings from wind and intense sunlight. **3.** The absence of fire for 100 years would represent a change to a low level of disturbance. According to the intermediate disturbance hypothesis, this change should cause diversity to decline as competitively dominant species gained sufficient time to exclude less competitive species.

Concept Check 41.4
1. Ecologists propose that the greater species richness of tropical regions is the result of their longer evolutionary history and the greater solar energy input and water availability in tropical regions. **2.** Immigration of species to islands declines with distance from the mainland and increases with island area. Extinction of species is lower on larger islands and on less isolated islands. Since the number of species on islands is largely determined by the difference between rates of immigration and extinction, the number of species will be highest on large islands near the mainland and lowest on small islands far from the mainland. **3.** Because of their greater mobility, birds disperse to islands more often than snakes and lizards, so birds should have greater richness.

Concept Check 41.5
1. Pathogens are microorganisms or viruses that cause disease. **2.** To keep the rabies virus out, you could ban imports of all mammals, including pets. Potentially, you could also attempt to vaccinate all dogs in the British Isles against the virus. A more practical approach might be to quarantine all pets brought into the country that are potential carriers of the disease, the approach the British government actually takes.

Summary of Key Concepts Questions
41.1 Competition: a fox and a bobcat competing for prey. Predation: an orca eating a sea otter. Herbivory: a bison grazing in a prairie. Parasitism: a parasitoid wasp that lays its eggs on a caterpillar. Mutualism: a fungus and an alga that make up a lichen. Commensalism: a remora attached to a whale. Facilitation: a flowering plant and its pollinator. **41.2** Not necessarily if the more species-rich community is dominated by only one or a few species **41.3** Because of the presence of species initially, the disturbance would initiate secondary succession in spite of its severe appearance. **41.4** Glaciations have severely reduced diversity in northern temperate, boreal, and Arctic ecosystems, compared to tropical ecosystems. **41.5** A host is required to complete the pathogen's life cycle, but a vector is not. A vector is an intermediate species that merely transports a pathogen to its host.

Test Your Understanding
1. e **2.** c **3.** d **4.** Community 1: $H = -(0.05 \ln 0.05 + 0.05 \ln 0.05 + 0.85 \ln 0.85 + 0.05 \ln 0.05) = 0.59$. Community 2: $H = -(0.30 \ln 0.30 + 0.40 \ln 0.40 + 0.30 \ln 0.30) = 1.1$. Community 2 is more diverse. **5.** Crab numbers should increase, reducing the abundance of eelgrass.

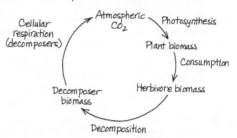

Chapter 42

Figure Questions
Figure 42.6 Wetlands, coral reefs, and coastal zones cover areas too small to be seen clearly on global maps. **Figure 42.7** If the new duck farms made nitrogen available in rich supply, as phosphorus already is, then adding extra nitrogen in the experiment would not increase phytoplankton density. **Figure 42.12** Water availability is probably another factor that varied across the sites. Such factors not included in the experimental design could make the results more difficult to interpret. Multiple factors can also change together in nature, so ecologists must be careful that the factor they are studying is actually causing the observed response and is not just correlated with it.

Concept Check 42.1
1. Energy passes through an ecosystem, entering as sunlight and leaving as heat. It is not recycled within the ecosystem. **2.** You would need to know how much biomass the wildebeests ate from your plot and how much nitrogen was contained in that biomass. You would also need to know how much nitrogen they deposited in urine or feces. **3.** The second law states that in any energy transfer or transformation, some of the energy is dissipated to the surroundings as heat. This "escape" of energy from an ecosystem is offset by the continuous influx of solar radiation.

Concept Check 42.2
1. Only a fraction of solar radiation strikes plants or algae, only a portion of that fraction is of wavelengths suitable for photosynthesis, and much energy is lost as a result of reflection or heating of plant tissue. **2.** By manipulating the level of the factors of interest, such as phosphorus availability or soil moisture, and measuring responses by primary producers **3.** The enzyme rubisco, which catalyzes the first step in the Calvin cycle, is the most abundant protein on Earth. Photosynthetic organisms require considerable nitrogen to make rubisco. Phosphorus is also needed as a component of

several metabolites in the Calvin cycle and as a component of both ATP and NADPH (see Figure 8.17).

Concept Check 42.3
1. 20 J; 40% **2.** Nicotine protects the plant from herbivores. **3.** Total net production $= 10,000 + 1,000 + 100 + 10$ J $= 11,110$ J; at steady state, this is the amount of energy theoretically available to detritivores.

Concept Check 42.4
1. For example, for the carbon cycle:

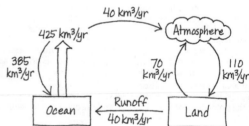

Cycling of a carbon atom

2. Removal of the trees stops nitrogen uptake from the soil, allowing nitrate to accumulate there. The nitrate is washed away by precipitation and enters the streams. **3.** Most of the nutrients in a tropical rain forest are contained in the trees, so removing the trees by logging rapidly depletes nutrients from the ecosystem. The nutrients that remain in the soil are quickly carried away into streams and groundwater by the abundant precipitation.

Concept Check 42.5
1. The main goal is to restore degraded ecosystems to a more natural state. **2.** Bioremediation uses organisms, generally prokaryotes, fungi, or plants, to detoxify or remove pollutants from ecosystems. Biological augmentation uses organisms, such as nitrogen-fixing plants, to add essential materials to degraded ecosystems. **3.** The Kissimmee River project returns the flow of water to the original channel and restores natural flow, a self-sustaining outcome. Ecologists at the Maungatautari reserve will need to maintain the integrity of the fence indefinitely, an outcome that is not self-sustaining in the long term.

Summary of Key Concepts Questions
42.1 Because energy conversions are inefficient, with some energy inevitably lost as heat, you would expect that a given mass of primary producers would support a smaller biomass of secondary producers. **42.2** For estimates of NEP, you need to measure the respiration of all organisms in an ecosystem, not just the respiration of primary producers. In a sample of ocean water, primary producers and other organisms are usually mixed together, making their respective respirations hard to separate. **42.3** Runners use much more energy in respiration when they are running than when they are sedentary, reducing their production efficiency. **42.4** Factors other than temperature, including a shortage of water and nutrients, slow decomposition in hot deserts. **42.5** If the topsoil and deeper soil are kept separate, you could return the deeper soil to the site first and then apply the more fertile topsoil to improve the success of revegetation and other restoration efforts.

Test Your Understanding
1. c **2.** b **3.** d **4.** d **5.** c **6.** e **7.** e **8.** e
9.

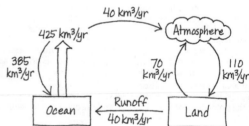

Based on these global numbers, approximately 110 km³ of precipitation falls over land each year.

Chapter 43

Figure Questions
Figure 43.4 You would need to know the complete range of the species and that it is missing across all of that range. You would also need to be certain that the species isn't hidden, as might be the case for an animal that is hibernating underground or a plant that is present in the form of seeds or spores. **Figure 43.12** Because the population of Illinois birds has a different genetic makeup than birds in other regions, you would want to maintain to the greatest extent possible the frequency of beneficial genes or alleles found only in that population. In restoration, preserving genetic diversity in a species is as important as increasing organism numbers. **Figure 43.14** The natural disturbance regime in this habitat includes frequent fires that clear undergrowth but do not kill mature pine trees. Without these fires, the undergrowth quickly fills in and the habitat becomes unsuitable for red-cockaded woodpeckers. **Figure 43.27** Dispersal limitations, the activities of people (such as a broad-scale conversion of forests to agriculture or selective harvesting), or many other factors (including those discussed in Concept 40.3)

Concept Check 43.1

1. In addition to species loss, the biodiversity crisis includes the loss of genetic diversity within populations and species and the degradation of entire ecosystems. **2.** Habitat destruction, such as deforestation, channelizing of rivers, or conversion of natural ecosystems to agriculture or cities, deprives species of places to live. Introduced species, which are transported by humans to regions outside their native range, where they are not controlled by their natural pathogens or predators, often reduce the population sizes of native species through competition or predation. Overharvesting has reduced populations of plants and animals or driven them to extinction. Finally, global change is altering the environment to the extent that it reduces the capacity of Earth to sustain life. **3.** If both populations breed separately, then gene flow between the populations would not occur and genetic differences between them would be greater. As a result, the loss of genetic diversity would be greater than if the populations interbreed.

Concept Check 43.2

1. Reduced genetic variation decreases the capacity of a population to evolve in the face of change. **2.** The effective population size, N_e, was $4(15 \times 5)/(15 + 5) = 15$ birds. **3.** Because millions of people use the greater Yellowstone ecosystem each year, it would be impossible to eliminate all contact between people and bears. Instead, you might try to reduce the kinds of encounters where bears are killed. You might recommend lower speed limits on roads in the park, adjust the timing or location of hunting seasons (where hunting is allowed outside the park) to minimize contact with mother bears and cubs, and provide financial incentives for livestock owners to try alternative means of protecting livestock, such as using guard dogs.

Concept Check 43.3

1. A small area supporting numerous endemic species as well as a large number of endangered and threatened species **2.** Zoned reserves may provide sustained supplies of forest products, water, hydroelectric power, educational opportunities, and income from tourism. **3.** Habitat corridors can increase the rate of movement or dispersal of organisms between habitat patches and thus the rate of gene flow between subpopulations. They thus help prevent a decrease in fitness attributable to inbreeding. They can also minimize interactions between organisms and humans as the organisms disperse; in cases involving potential predators, such as bears or large cats, minimizing such interactions is desirable.

Concept Check 43.4

1. Adding nutrients causes population explosions of algae and the organisms that feed on them. Increased respiration by algae and consumers, including detritivores, depletes the lake's oxygen, which the fish require. **2.** Because higher temperatures lead to faster decomposition, organic matter in these soils could be quickly decomposed to CO_2, speeding up global warming.

Concept Check 43.5

1. The growth rate of Earth's human population has dropped by half since the 1960s, from 2.2% in 1962 to 1.1% today. Nonetheless, growth has not slowed much because the smaller growth rate is counterbalanced by increased population size; the number of additional people on Earth each year remains enormous—approximately 78 million. **2.** Each of us influences our ecological footprint by how we live—what we eat, how much energy we use, and the amount of waste we generate—as well as by how many children we have. Making choices that reduce our demand for resources makes our ecological footprint smaller.

Concept Check 43.6

1. Sustainable development is an approach to development that works toward the long-term prosperity of human societies and the ecosystems that support them, which requires linking the biological sciences with the social sciences, economics, and humanities. **2.** Biophilia, our sense of connection to nature and all forms of life, may act as a significant motivation for the development of an environmental ethic that resolves not to allow species to become extinct or ecosystems to be destroyed. Such an ethic is necessary if we are to become more attentive and effective custodians of the environment. **3.** At a minimum, you would want to know the size of the population and the average reproductive rate of its individuals. To develop the fishery sustainably, you would seek a harvest rate that maintains the population near its original size and maximizes its harvest in the long term rather than the short term.

Summary of Key Concepts Questions

43.1 Nature provides us with many beneficial services, including a supply of reliable, clean water, the production of food and fiber, and the dilution and detoxification of

our pollutants. **43.2** A more genetically diverse population is better able to withstand pressures from disease or environmental change, making it less likely to become extinct over a given period of time. **43.3** Habitat fragmentation can isolate populations, leading to inbreeding and genetic drift, and it can make populations more susceptible to local extinctions resulting from the effects of pathogens, parasites, or predators. **43.4** It's healthier to feed at a lower trophic level because biological magnification increases the concentration of toxins at higher levels. **43.5** We are unique in our potential ability to reduce global population through contraception and family planning. We also are capable of consciously choosing our diet and personal lifestyle, and these choices influence the number of people Earth can support. **43.6** One goal of conservation biology is to preserve as many species as possible. Sustainable approaches that maintain the quality of habitats are required for the long-term survival of organisms.

Test Your Understanding

1. d **2.** d **3.** e **4.** a **5.** c **6.** a

7.

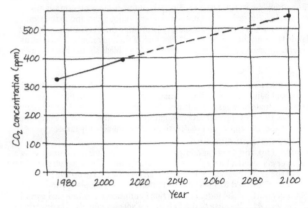

Between 1974 and 2011, Earth's atmospheric CO_2 concentration increased from approximately 330 ppm to 394 ppm. If this rate of increase of approximately 1.7 ppm/yr continues, the concentration in 2100 will be about 545 ppm. The actual rise in CO_2 concentration could be larger or smaller, depending on Earth's human population, our per capita energy use, and the extent to which societies take steps to reduce CO_2 emissions, including replacing fossil fuels with renewable or nuclear fuels. Additional scientific data will be important for many reasons, including determining how quickly greenhouse gases such as CO_2 are removed from the atmosphere by the biosphere.

8.

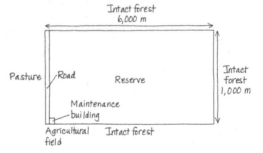

To minimize the area of forest into which the cowbirds penetrate, you should locate the road along one edge of the reserve. Any other location would increase the area of affected habitat. Similarly, the maintenance building should be in a corner of the reserve to minimize the area susceptible to cowbirds.

Appendix B Periodic Table of the Elements

Atomic number (number of protons) → 6
Element symbol → C
Atomic mass (number of protons plus number of neutrons averaged over all isotopes) → 12.01

Metals   Metalloids   Nonmetals

Representative elements

Groups: Elements in a vertical column have the same number of electrons in their valence (outer) shell and thus have similar chemical properties.

Periods: Each horizontal row contains elements with the same total number of electron shells. Across each period, elements are ordered by increasing atomic number.

Alkali metals   Alkaline earth metals

Halogens   Noble gases

Transition elements

Period number	1 Group 1A	2 Group 2A	3 3B	4 4B	5 5B	6 6B	7 7B	8 8B	9 8B	10 8B	11 1B	12 2B	13 Group 3A	14 Group 4A	15 Group 5A	16 Group 6A	17 Group 7A	18 Group 8A
1	1 H 1.008																	2 He 4.003
2	3 Li 6.941	4 Be 9.012											5 B 10.81	6 C 12.01	7 N 14.01	8 O 16.00	9 F 19.00	10 Ne 20.18
3	11 Na 22.99	12 Mg 24.31											13 Al 26.98	14 Si 28.09	15 P 30.97	16 S 32.07	17 Cl 35.45	18 Ar 39.95
4	19 K 39.10	20 Ca 40.08	21 Sc 44.96	22 Ti 47.87	23 V 50.94	24 Cr 52.00	25 Mn 54.94	26 Fe 55.85	27 Co 58.93	28 Ni 58.69	29 Cu 63.55	30 Zn 65.41	31 Ga 69.72	32 Ge 72.64	33 As 74.92	34 Se 78.96	35 Br 79.90	36 Kr 83.80
5	37 Rb 85.47	38 Sr 87.62	39 Y 88.91	40 Zr 91.22	41 Nb 92.91	42 Mo 95.94	43 Tc (98)	44 Ru 101.1	45 Rh 102.9	46 Pd 106.4	47 Ag 107.9	48 Cd 112.4	49 In 114.8	50 Sn 118.7	51 Sb 121.8	52 Te 127.6	53 I 126.9	54 Xe 131.3
6	55 Cs 132.9	56 Ba 137.3	57* La 138.9	72 Hf 178.5	73 Ta 180.9	74 W 183.8	75 Re 186.2	76 Os 190.2	77 Ir 192.2	78 Pt 195.1	79 Au 197.0	80 Hg 200.6	81 Tl 204.4	82 Pb 207.2	83 Bi 209.0	84 Po (209)	85 At (210)	86 Rn (222)
7	87 Fr (223)	88 Ra (226)	89† Ac (227)	104 Rf (261)	105 Db (262)	106 Sg (266)	107 Bh (264)	108 Hs (269)	109 Mt (268)	110 Ds (271)	111 Rg (272)	112 Cn (285)	113 — (284)	114 — (289)	115 — (288)	116 — (293)	117 — (294?)	118 — (294)

*Lanthanides	58 Ce 140.1	59 Pr 140.9	60 Nd 144.2	61 Pm (145)	62 Sm 150.4	63 Eu 152.0	64 Gd 157.3	65 Tb 158.9	66 Dy 162.5	67 Ho 164.9	68 Er 167.3	69 Tm 168.9	70 Yb 173.0	71 Lu 175.0
†Actinides	90 Th 232.0	91 Pa 231.0	92 U 238.0	93 Np (237)	94 Pu (244)	95 Am (243)	96 Cm (247)	97 Bk (247)	98 Cf (251)	99 Es 252	100 Fm 257	101 Md 258	102 No 259	103 Lr 260

Name (Symbol)	Atomic Number	Name (Symbol)	Atomic Number	Name (Symbol)	Atomic Number	Name (Symbol)	Atomic Number	Name (Symbol)	Atomic Number
Actinium (Ac)	89	Copernicium (Cn)	112	Iridium (Ir)	77	Palladium (Pd)	46	Sodium (Na)	11
Aluminum (Al)	13	Copper (Cu)	29	Iron (Fe)	26	Phosphorus (P)	15	Strontium (Sr)	38
Americium (Am)	95	Curium (Cm)	96	Krypton (Kr)	36	Platinum (Pt)	78	Sulfur (S)	16
Antimony (Sb)	51	Darmstadtium (Ds)	110	Lanthanum (La)	57	Plutonium (Pu)	94	Tantalum (Ta)	73
Argon (Ar)	18	Dubnium (Db)	105	Lawrencium (Lr)	103	Polonium (Po)	84	Technetium (Tc)	43
Arsenic (As)	33	Dysprosium (Dy)	66	Lead (Pb)	82	Potassium (K)	19	Tellurium (Te)	52
Astatine (At)	85	Einsteinium (Es)	99	Lithium (Li)	3	Praseodymium (Pr)	59	Terbium (Tb)	65
Barium (Ba)	56	Erbium (Er)	68	Lutetium (Lu)	71	Promethium (Pm)	61	Thallium (Tl)	81
Berkelium (Bk)	97	Europium (Eu)	63	Magnesium (Mg)	12	Protactinium (Pa)	91	Thorium (Th)	90
Beryllium (Be)	4	Fermium (Fm)	100	Manganese (Mn)	25	Radium (Ra)	88	Thulium (Tm)	69
Bismuth (Bi)	83	Fluorine (F)	9	Meitnerium (Mt)	109	Radon (Rn)	86	Tin (Sn)	50
Bohrium (Bh)	107	Francium (Fr)	87	Mendelevium (Md)	101	Rhenium (Re)	75	Titanium (Ti)	22
Boron (B)	5	Gadolinium (Gd)	64	Mercury (Hg)	80	Rhodium (Rh)	45	Tungsten (W)	74
Bromine (Br)	35	Gallium (Ga)	31	Molybdenum (Mo)	42	Roentgenium (Rg)	111	Uranium (U)	92
Cadmium (Cd)	48	Germanium (Ge)	32	Neodymium (Nd)	60	Rubidium (Rb)	37	Vanadium (V)	23
Calcium (Ca)	20	Gold (Au)	79	Neon (Ne)	10	Ruthenium (Ru)	44	Xenon (Xe)	54
Californium (Cf)	98	Hafnium (Hf)	72	Neptunium (Np)	93	Rutherfordium (Rf)	104	Ytterbium (Yb)	70
Carbon (C)	6	Hassium (Hs)	108	Nickel (Ni)	28	Samarium (Sm)	62	Yttrium (Y)	39
Cerium (Ce)	58	Helium (He)	2	Niobium (Nb)	41	Scandium (Sc)	21	Zinc (Zn)	30
Cesium (Cs)	55	Holmium (Ho)	67	Nitrogen (N)	7	Seaborgium (Sg)	106	Zirconium (Zr)	40
Chlorine (Cl)	17	Hydrogen (H)	1	Nobelium (No)	102	Selenium (Se)	34		
Chromium (Cr)	24	Indium (In)	49	Osmium (Os)	76	Silicon (Si)	14		
Cobalt (Co)	27	Iodine (I)	53	Oxygen (O)	8	Silver (Ag)	47		

Metric Prefixes:
10^9 = giga (G) 10^{-2} = centi (c) 10^{-9} = nano (n)
10^6 = mega (M) 10^{-3} = milli (m) 10^{-12} = pico (p)
10^3 = kilo (k) 10^{-6} = micro (µ) 10^{-15} = femto (f)

Measurement	Unit and Abbreviation	Metric Equivalent	Metric-to-English Conversion Factor	English-to-Metric Conversion Factor
Length	1 kilometer (km)	= 1,000 (10^3) meters	1 km = 0.62 mile	1 mile = 1.61 km
	1 meter (m)	= 100 (10^2) centimeters = 1,000 millimeters	1 m = 1.09 yards 1 m = 3.28 feet 1 m = 39.37 inches	1 yard = 0.914 m 1 foot = 0.305 m
	1 centimeter (cm)	= 0.01 (10^{-2}) meter	1 cm = 0.394 inch	1 foot = 30.5 cm 1 inch = 2.54 cm
	1 millimeter (mm)	= 0.001 (10^{-3}) meter	1 mm = 0.039 inch	
	1 micrometer (µm) (formerly micron, µ)	= 10^{-6} meter (10^{-3} mm)		
	1 nanometer (nm) (formerly millimicron, mµ)	= 10^{-9} meter (10^{-3} µm)		
	1 angstrom (Å)	= 10^{-10} meter (10^{-4} µm)		
Area	1 hectare (ha)	= 10,000 square meters	1 ha = 2.47 acres	1 acre = 0.405 ha
	1 square meter (m^2)	= 10,000 square centimeters	1 m^2 = 1.196 square yards 1 m^2 = 10.764 square feet	1 square yard = 0.8361 m^2 1 square foot = 0.0929 m^2
	1 square centimeter (cm^2)	= 100 square millimeters	1 cm^2 = 0.155 square inch	1 square inch = 6.4516 cm^2
Mass	1 metric ton (t)	= 1,000 kilograms	1 t = 1.103 tons	1 ton = 0.907 t
	1 kilogram (kg)	= 1,000 grams	1 kg = 2.205 pounds	1 pound = 0.4536 kg
	1 gram (g)	= 1,000 milligrams	1 g = 0.0353 ounce 1 g = 15.432 grains	1 ounce = 28.35 g
	1 milligram (mg)	= 10^{-3} gram	1 mg = approx. 0.015 grain	
	1 microgram (µg)	= 10^{-6} gram		
Volume (solids)	1 cubic meter (m^3)	= 1,000,000 cubic centimeters	1 m^3 = 1.308 cubic yards 1 m^3 = 35.315 cubic feet	1 cubic yard = 0.7646 m^3 1 cubic foot = 0.0283 m^3
	1 cubic centimeter (cm^3 or cc)	= 10^{-6} cubic meter	1 cm^3 = 0.061 cubic inch	1 cubic inch = 16.387 cm^3
	1 cubic millimeter (mm^3)	= 10^{-9} cubic meter = 10^{-3} cubic centimeter		
Volume (liquids and gases)	1 kiloliter (kL or kl)	= 1,000 liters	1 kL = 264.17 gallons	
	1 liter (L or l)	= 1,000 milliliters	1 L = 0.264 gallon 1 L = 1.057 quarts	1 gallon = 3.785 L 1 quart = 0.946 L
	1 milliliter (mL or ml)	= 10^{-3} liter = 1 cubic centimeter	1 mL = 0.034 fluid ounce 1 mL = approx. ¼ teaspoon 1 mL = approx. 15–16 drops (gtt.)	1 quart = 946 mL 1 pint = 473 mL 1 fluid ounce = 29.57 mL 1 teaspoon = approx. 5 mL
	1 microliter (µL or µl)	= 10^{-6} liter (10^{-3} milliliter)		
Pressure	1 megapascal (MPa)	= 1,000 kilopascals	1 MPa = 10 bars	1 bar = 0.1 MPa
	1 kilopascal (kPa)	= 1,000 pascals	1 kPa = 0.01 bar	1 bar = 100 kPa
	1 pascal (Pa)	= 1 newton/m^2 (N/m^2)	1 Pa = 1.0×10^{-5} bar	1 bar = 1.0×10^5 Pa
Time	1 second (s or sec)	= ⅟₆₀ minute		
	1 millisecond (ms or msec)	= 10^{-3} second		
Temperature	Degrees Celsius (°C) (0 K [Kelvin] = −273.15°C)		°F = ⁹⁄₅°C + 32	°C = ⁵⁄₉ (°F − 32)

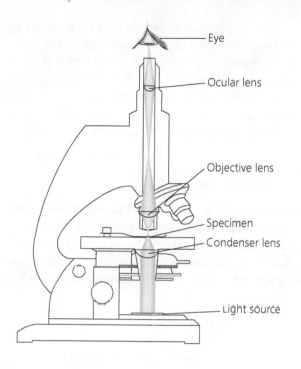

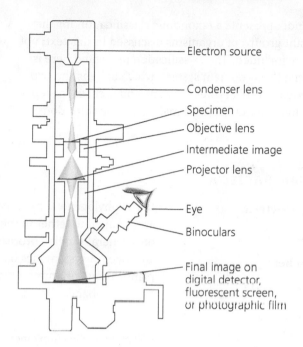

Light Microscope

In light microscopy, light is focused on a specimen by a glass condenser lens; the image is then magnified by an objective lens and an ocular lens, for projection on the eye, digital camera, digital video camera, or photographic film.

Electron Microscope

In electron microscopy, a beam of electrons (top of the microscope) is used instead of light, and electromagnets are used instead of glass lenses. The electron beam is focused on the specimen by a condenser lens; the image is magnified by an objective lens and a projector lens for projection on a digital detector, fluorescent screen, or photographic film.

Appendix E Classification of Life

This appendix presents a taxonomic classification for the major extant groups of organisms discussed in this text; not all phyla are included. The classification presented here is based on the three-domain system, which assigns the two major groups of prokaryotes, bacteria and archaea, to separate domains (with eukaryotes making up the third domain).

Various alternative classification schemes are discussed in Unit Four of the text. The taxonomic turmoil includes debates about the number and boundaries of kingdoms and about the alignment of the Linnaean classification hierarchy with the findings of modern cladistic analysis. In this review, asterisks (*) indicate currently recognized phyla thought by some systematists to be paraphyletic.

DOMAIN BACTERIA

- **Proteobacteria**
- **Chlamydia**
- **Spirochetes**
- **Cyanobacteria**
- **Gram-positive bacteria**

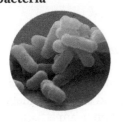

DOMAIN ARCHAEA

- **Korarchaeota**
- **Euryarchaeota**
- **Crenarchaeota**
- **Nanoarchaeota**

DOMAIN EUKARYA

In the phylogenetic hypothesis we present in Chapter 25, major clades of eukaryotes are grouped together in the four "supergroups" listed in blue type. Formerly, all the eukaryotes generally called protists were assigned to a single kingdom, Protista. However, advances in systematics have made it clear that some protists are more closely related to plants, fungi, or animals than they are to other protists. As a result, the kingdom Protista has been abandoned.

Excavata
- Diplomonadida (diplomonads)
- Parabasala (parabasalids)
- Euglenozoa (euglenozoans)
    Kinetoplastida (kinetoplastids)
    Euglenophyta (euglenids)

The "SAR" clade
- Stramenopila (stramenopiles)
    Bacillariophyta (diatoms)
    Phaeophyta (brown algae)
- Alveolata (alveolates)
    Dinoflagellata (dinoflagellates)
    Apicomplexa (apicomplexans)
    Ciliophora (ciliates)

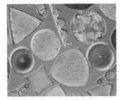

- Rhizaria
    Foraminifera (forams)
    Cercozoa (cercozoans)

Archaeplastida
- Rhodophyta (red algae)
- Chlorophyta (green algae: chlorophytes)
- Charophyta (green algae: charophytes)
- Plantae
    Phylum Hepatophyta (liverworts) ⎫
    Phylum Bryophyta (mosses)   ⎬ Nonvascular plants (bryophytes)
    Phylum Anthocerophyta (hornworts) ⎭
    Phylum Lycophyta (lycophytes)   ⎫
    Phylum Monilophyta (ferns, horsetails, ⎬ Seedless vascular plants
      whisk ferns)   ⎭
    Phylum Ginkgophyta (ginkgo)   ⎫
    Phylum Cycadophyta (cycads)   ⎬ Gymnosperms ⎫
    Phylum Gnetophyta (gnetophytes) ⎪   ⎬ Seed plants
    Phylum Coniferophyta (conifers) ⎭   ⎪
    Phylum Anthophyta (flowering   ⎫ Angiosperms ⎭
      plants)   ⎭

Unikonta
- Amoebozoa (amoebozoans)
  - Dictyostelida (cellular slime molds)
  - Entamoeba (entamoebas)
- Nucleariida (nucleariids)
- Fungi
  - *Phylum Chytridiomycota (chytrids)
  - *Phylum Zygomycota (zygomycetes)
  - Phylum Glomeromycota (glomeromycetes)
  - Phylum Ascomycota (sac fungi)
  - Phylum Basidiomycota (club fungi)

- Choanoflagellata (choanoflagellates)
- Animalia
  - Phylum Porifera (sponges)
  - Phylum Ctenophora (comb jellies)
  - Phylum Cnidaria (cnidarians)
  - Lophotrochozoa (lophotrochozoans)
    - Phylum Platyhelminthes (flatworms)
    - Phylum Ectoprocta (ectoprocts)
    - Phylum Brachiopoda (brachiopods)
    - Phylum Rotifera (rotifers)
    - Phylum Mollusca (molluscs)
    - Phylum Annelida (segmented worms)

Ecdysozoa (ecdysozoans)
- Phylum Nematoda (roundworms)
- Phylum Arthropoda (This survey groups arthropods into a single phylum, but some zoologists now split the arthropods into multiple phyla.)
  - Subphylum Chelicerata (horseshoe crabs, arachnids)
  - Subphylum Myriapoda (millipedes, centipedes)
  - Subphylum Hexapoda (insects, springtails)
  - Subphylum Crustacea (crustaceans)
- Phylum Onychophora (velvet worms)

Deuterostomia (deuterostomes)
- Phylum Hemichordata (hemichordates)
- Phylum Echinodermata (echinoderms)
- Phylum Chordata (chordates)
  - Subphylum Cephalochordata (lancelets)
  - Subphylum Urochordata (tunicates)
  - Subphylum Craniata (craniates)
    - Myxini (hagfishes)
    - Petromyzontida (lampreys)
    - Chondrichthyes (sharks, rays, chimaeras)
    - Actinopterygii (ray-finned fishes)
    - Actinistia (coelacanths)
    - Dipnoi (lungfishes)
    - Amphibia (amphibians)
    - Reptilia (tuataras, lizards, snakes, turtles, crocodilians, birds)
    - Mammalia (mammals)

} Vertebrates

Graphs

Graphs provide a visual representation of numerical data. They may reveal patterns or trends in the data that are not easy to recognize in a table. A graph is a diagram that shows how one variable in a data set is related (or perhaps not related) to another variable. If one variable is dependent on the other, the dependent variable is typically plotted on the *y*-axis and the independent variable on the *x*-axis. Types of graphs that are frequently used in biology include scatter plots, line graphs, bar graphs, and histograms.

A **scatter plot** is used when the data for all variables are numerical and continuous. Each piece of data is represented by a point. In a **line graph**, each data point is connected to the next point in the data set with a straight line, as in the graph below.

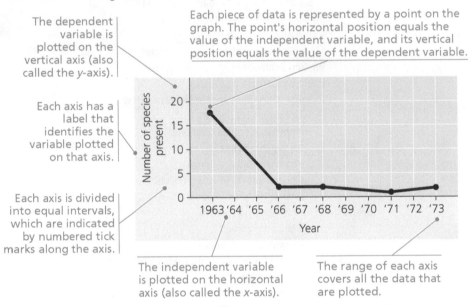

The dependent variable is plotted on the vertical axis (also called the *y*-axis).

Each axis has a label that identifies the variable plotted on that axis.

Each axis is divided into equal intervals, which are indicated by numbered tick marks along the axis.

Each piece of data is represented by a point on the graph. The point's horizontal position equals the value of the independent variable, and its vertical position equals the value of the dependent variable.

The independent variable is plotted on the horizontal axis (also called the *x*-axis).

The range of each axis covers all the data that are plotted.

Two or more data sets can be plotted on the same line graph to show how two dependent variables are related to the same independent variable.

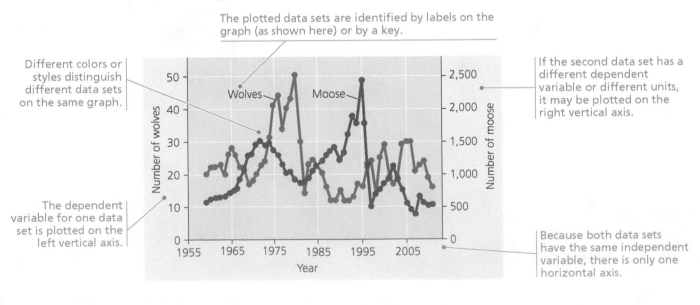

The plotted data sets are identified by labels on the graph (as shown here) or by a key.

Different colors or styles distinguish different data sets on the same graph.

The dependent variable for one data set is plotted on the left vertical axis.

If the second data set has a different dependent variable or different units, it may be plotted on the right vertical axis.

Because both data sets have the same independent variable, there is only one horizontal axis.

In some scatter-plot graphs, a straight or curved line is drawn through the entire data set to show the general trend in the data. A straight line that mathematically fits the data best is called a *regression line*. Alternatively, a mathematical function that best fits the data may describe a curved line, often termed a *best-fit curve*.

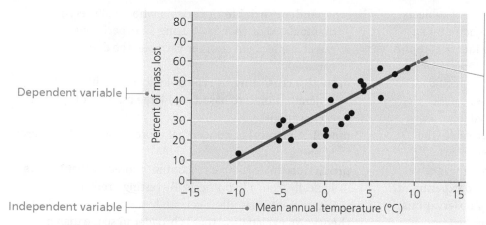

Dependent variable

Independent variable

The regression line can be expressed as a mathematical equation. It allows you to predict the value of the dependent variable for any value of the independent variable within the range of the data set and, less commonly, beyond the range of the data.

A **bar graph** is a kind of graph in which the independent variable represents groups or nonnumerical categories and the values of the dependent variable(s) are shown by bars.

Each piece of data is represented by a bar on the graph. The top of the bar aligns with the value of the dependent variable.

As in a line graph or scatter plot, the vertical axis is usually used for the dependent variable.

The axis for the dependent variable is labeled and divided into equal intervals indicated by numbered tick marks.

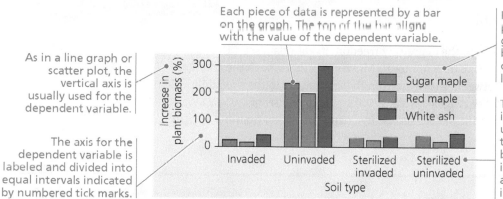

If multiple data sets are plotted on the same bar graph, they are distinguished by bars of different colors or styles and identified by labels or a key.

The groups or categories of the independent variable are usually spaced equally along the horizontal axis. (In some bar graphs, the horizontal axis is used for the dependent variable and the vertical axis for the independent variable.)

A variant of a bar graph called a **histogram** can be made for numeric data by first grouping, or "binning," the variable plotted on the *x*-axis into intervals of equal width. The "bins" may be integers or ranges of numbers. In the histogram below, the intervals are 25 mg/dL wide. The height of each bar shows the percent (or alternatively, the number) of experimental subjects whose characteristics can be described by one of the intervals plotted on the *x*-axis.

The height of this bar shows the percent of individuals (about 4%) whose plasma LDL cholesterol levels are in the range indicated on the *x*-axis.

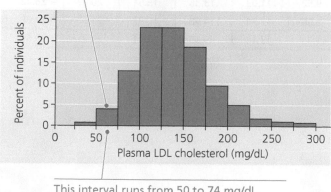

This interval runs from 50 to 74 mg/dL.

Glossary of Scientific Inquiry Terms

See Concept 1.3 for more discussion of the process of scientific inquiry.

control group In a controlled experiment, a set of subjects that lacks (or does not receive) the specific factor being tested. Ideally, the control group should be identical to the experimental group in other respects.

deductive reasoning A type of logic in which specific results are predicted from a general premise.

dependent variable In an experiment, a variable whose value is influenced by changes in another variable (the independent variable).

experimental group A set of subjects that has (or receives) the specific factor being tested in a controlled experiment.

hypothesis A testable explanation for a set of observations based on the available data and guided by inductive reasoning. A hypothesis is narrower in scope than a theory.

independent variable A variable whose value is manipulated or changed during an experiment to reveal possible effects on another variable (the dependent variable).

inductive reasoning A type of logic in which generalizations are based on a large number of specific observations.

model A physical or conceptual representation of a natural phenomenon.

prediction In deductive reasoning, a forecast that follows logically from a hypothesis. By testing predictions, experiments may allow certain hypotheses to be rejected.

theory An explanation that is broader in scope than a hypothesis, generates new hypotheses, and is supported by a large body of evidence.

Chi-Square (χ^2) Distribution Table

To use the table, find the row that corresponds to the degrees of freedom in your data set. (The degrees of freedom is the number of categories of data minus 1.) Move along that row to the pair of values that your calculated χ^2 value lies between. Move up from those numbers to the probabilities at the top of the columns to find the probability range for your χ^2 value. A probability of 0.05 or less is generally considered significant. See the Scientific Skills Exercise in Chapter 12 for an example of how to use the chi-square test.

Degrees of Freedom (df)	Probability										
	0.95	0.90	0.80	0.70	0.50	0.30	0.20	0.10	0.05	0.01	0.001
1	0.004	0.02	0.06	0.15	0.45	1.07	1.64	2.71	3.84	6.64	10.83
2	0.10	0.21	0.45	0.71	1.39	2.41	3.22	4.61	5.99	9.21	13.82
3	0.35	0.58	1.01	1.42	2.37	3.66	4.64	6.25	7.82	11.34	16.27
4	0.71	1.06	1.65	2.19	3.36	4.88	5.99	7.78	9.49	13.28	18.47
5	1.15	1.61	2.34	3.00	4.35	6.06	7.29	9.24	11.07	15.09	20.52
6	1.64	2.20	3.07	3.83	5.35	7.23	8.56	10.64	12.59	16.81	22.46
7	2.17	2.83	3.82	4.67	6.35	8.38	9.80	12.02	14.07	18.48	24.32
8	2.73	3.49	4.59	5.53	7.34	9.52	11.03	13.36	15.51	20.09	26.12
9	3.33	4.17	5.38	6.39	8.34	10.66	12.24	14.68	16.92	21.67	27.88
10	3.94	4.87	6.18	7.27	9.34	11.78	13.44	15.99	18.31	23.21	29.59

Glossary

Pronunciation Key

Pronounce

ā	as in	ace
a/ah		ash
ch		chose
ē		meet
e/eh		bet
g		game
ī		ice
i		hit
ks		box
kw		quick
ng		song
ō		robe
o		ox
oy		boy
s		say
sh		shell
th		thin
ū		boot
u/uh		up
z		zoo

′ = primary accent
′ = secondary accent

5′ cap A modified form of guanine nucleotide added onto the end of a pre-mRNA molecule.

A site One of a ribosome's three binding sites for tRNA during translation. The A site holds the tRNA carrying the next amino acid to be added to the polypeptide chain. (A stands for aminoacyl tRNA.)

ABC hypothesis A model of flower formation identifying three classes of organ identity genes that direct formation of the four types of floral organs.

abiotic (ā′-bī-ot′-ik) Nonliving; referring to the physical and chemical properties of an environment.

abscisic acid (ABA) (ab-sis′-ik) A plant hormone that slows growth, often antagonizing the actions of growth hormones. Two of its many effects are to promote seed dormancy and facilitate drought tolerance.

absorption The third stage of food processing in animals: the uptake of small nutrient molecules by an organism's body.

absorption spectrum The range of a pigment's ability to absorb various wavelengths of light; also a graph of such a range.

accessory fruit A fruit, or assemblage of fruits, in which the fleshy parts are derived largely or entirely from tissues other than the ovary.

acetyl CoA Acetyl coenzyme A; the entry compound for the citric acid cycle in cellular respiration, formed from a two-carbon fragment of pyruvate attached to a coenzyme.

acetylcholine (as′-uh-til-kō′-lēn) One of the most common neurotransmitters; functions by binding to receptors and altering the permeability of the postsynaptic membrane to specific ions, either depolarizing or hyperpolarizing the membrane.

acid A substance that increases the hydrogen ion concentration of a solution.

acrosome (ak′-ruh-sōm) A vesicle in the tip of a sperm containing hydrolytic enzymes and other proteins that help the sperm reach the egg.

actin (ak′-tin) A globular protein that links into chains, two of which twist helically about each other, forming microfilaments (actin filaments) in muscle and other kinds of cells.

action potential An electrical signal that propagates (travels) along the membrane of a neuron or other excitable cell as a nongraded (all-or-none) depolarization.

action spectrum A graph that profiles the relative effectiveness of different wavelengths of radiation in driving a particular process.

activation energy The amount of energy that reactants must absorb before a chemical reaction will start; also called free energy of activation.

activator A protein that binds to DNA and stimulates gene transcription. In prokaryotes, activators bind in or near the promoter; in eukaryotes, activators generally bind to control elements in enhancers.

active immunity Long-lasting immunity conferred by the action of B cells and T cells and the resulting B and T memory cells specific for a pathogen. Active immunity can develop as a result of natural infection or immunization.

active site The specific region of an enzyme that binds the substrate and that forms the pocket in which catalysis occurs.

active transport The movement of a substance across a cell membrane against its concentration or electrochemical gradient, mediated by specific transport proteins and requiring an expenditure of energy.

adaptation Inherited characteristic of an organism that enhances its survival and reproduction in a specific environment.

adaptive evolution A process in which traits that enhance survival or reproduction tend to increase in frequency in a population over time.

adaptive immunity A vertebrate-specific defense that is mediated by B lymphocytes (B cells) and T lymphocytes (T cells) and that exhibits specificity, memory, and self-nonself recognition; also called acquired immunity.

adaptive radiation Period of evolutionary change in which groups of organisms form many new species whose adaptations allow them to fill different ecological roles in their communities.

addition rule A rule of probability stating that the probability of any one of two or more mutually exclusive events occurring can be determined by adding their individual probabilities.

adenosine triphosphate *See* ATP (adenosine triphosphate).

adhesion The clinging of one substance to another, such as water to plant cell walls, by means of hydrogen bonds.

aerobic respiration A catabolic pathway for organic molecules, using oxygen (O_2) as the final electron acceptor in an electron transport chain and ultimately producing ATP. This is the most efficient catabolic pathway and is carried out in most eukaryotic cells and many prokaryotic organisms.

aggregate fruit A fruit derived from a single flower that has more than one carpel.

AIDS (acquired immunodeficiency syndrome) The symptoms and signs present during the late stages of HIV infection, defined by a specified reduction in the number of T cells and the appearance of characteristic secondary infections.

alcohol fermentation Glycolysis followed by the reduction of pyruvate to ethyl alcohol, regenerating NAD^+ and releasing carbon dioxide.

aldosterone (al-dos′-tuh-rōn) A steroid hormone that acts on tubules of the kidney to regulate the transport of sodium ions (Na^+) and potassium ions (K^+).

alga (plural, **algae**) Member of a diverse collection of photosynthetic protists that includes unicellular and multicellular forms. Algal species are included in three eukaryote supergroups (Excavata, "SAR" clade, and Archaeplastida).

alimentary canal (al′-uh-men′-tuh-rē) A complete digestive tract, consisting of a tube running between a mouth and an anus.

allele (uh-lē′-ul) Any of the alternative versions of a gene that may produce distinguishable phenotypic effects.

allergen An antigen that triggers an exaggerated immune response.

allopatric speciation (al'-uh-pat'-rik) The formation of new species in populations that are geographically isolated from one another.

allopolyploid (al'-ō-pol'-ē-ployd) A fertile individual that has more than two chromosome sets as a result of two different species interbreeding and combining their chromosomes.

allosteric regulation The binding of a regulatory molecule to a protein at one site that affects the function of the protein at a different site.

alpha (α) helix (al'-fuh hē'-liks) A coiled region constituting one form of the secondary structure of proteins, arising from a specific pattern of hydrogen bonding between atoms of the polypeptide backbone (not the side chains).

alternation of generations A life cycle in which there is both a multicellular diploid form, the sporophyte, and a multicellular haploid form, the gametophyte; characteristic of plants and some algae.

alternative RNA splicing A type of eukaryotic gene regulation at the RNA-processing level in which different mRNA molecules are produced from the same primary transcript, depending on which RNA segments are treated as exons and which as introns.

altruism (al'-trū-iz-um) Selflessness; behavior that reduces an individual's fitness while increasing the fitness of another individual.

alveolates (al-vē'-uh-lets) One of the three major subgroups for which the "SAR" eukaryotic supergroup is named. Alveolate protists have membrane-enclosed sacs (alveoli) located just under the plasma membrane.

alveolus (al-vē'-uh-lus) (plural, **alveoli**) One of the dead-end air sacs where gas exchange occurs in a mammalian lung.

amino acid (uh-mēn'-ō) An organic molecule possessing both a carboxyl and an amino group. Amino acids serve as the monomers of polypeptides.

amino group A chemical group consisting of a nitrogen atom bonded to two hydrogen atoms; can act as a base in solution, accepting a hydrogen ion and acquiring a charge of 1+.

aminoacyl-tRNA synthetase An enzyme that joins each amino acid to the appropriate tRNA.

ammonia A small, toxic molecule (NH_3) produced by nitrogen fixation or as a metabolic waste product of protein and nucleic acid metabolism.

amniote (am'-nē-ōt) Member of a clade of tetrapods named for a key derived character, the amniotic egg, which contains specialized membranes, including the fluid-filled amnion, that protect the embryo. Amniotes include mammals as well as birds and other reptiles.

amniotic egg An egg that contains specialized membranes that function in protection, nourishment, and gas exchange. The amniotic egg was a major evolutionary innovation, allowing embryos to develop on land in a fluid-filled sac, thus reducing the dependence of tetrapods on water for reproduction.

amoeba (uh-mē'-buh) A member of one of several groups of unicellular eukaryotes that have pseudopodia.

amoebocyte (uh-mē'-buh-sīt') An amoeba-like cell that moves by pseudopodia and is found in most animals. Depending on the species, it may digest and distribute food, dispose of wastes, form skeletal fibers, fight infections, or change into other cell types.

amoebozoan (uh-mē'-buh-zō'-an) A protist in a clade that includes many species with lobe- or tube-shaped pseudopodia.

amphibian Member of the tetrapod class Amphibia, including salamanders, frogs, and caecilians.

amphipathic (am'-fē-path'-ik) Having both a hydrophilic region and a hydrophobic region.

amplification The strengthening of stimulus energy during transduction.

amylase (am'-uh-lās') An enzyme that hydrolyzes starch (a glucose polymer from plants) and glycogen (a glucose polymer from animals) into smaller polysaccharides and the disaccharide maltose.

anabolic pathway (an'-uh-bol'-ik) A metabolic pathway that consumes energy to synthesize a complex molecule from simpler molecules.

anaerobic respiration (an-er-ō'-bik) A catabolic pathway in which inorganic molecules other than oxygen accept electrons at the "downhill" end of electron transport chains.

analogous Having characteristics that are similar because of convergent evolution, not homology.

analogy (an-al'-uh-jē) Similarity between two species that is due to convergent evolution rather than to descent from a common ancestor with the same trait.

anaphase The fourth stage of mitosis, in which the chromatids of each chromosome have separated and the daughter chromosomes are moving to the poles of the cell.

anatomy The structure of an organism.

anchorage dependence The requirement that a cell must be attached to a substratum in order to initiate cell division.

aneuploidy (an'-yū-ploy'-dē) A chromosomal aberration in which one or more chromosomes are present in extra copies or are deficient in number.

angiosperm (an'-jē-ō-sperm) A flowering plant, which forms seeds inside a protective chamber called an ovary.

angiotensin II A peptide hormone that stimulates constriction of precapillary arterioles and increases reabsorption of NaCl and water by the proximal tubules of the kidney, increasing blood pressure and volume.

anion (an'-ī-on) A negatively charged ion.

anterior Pertaining to the front, or head, of a bilaterally symmetric animal.

anterior pituitary A portion of the pituitary that develops from non-neural tissue; consists of endocrine cells that synthesize and secrete several tropic and nontropic hormones.

anther In an angiosperm, the terminal pollen sac of a stamen, where pollen grains containing sperm-producing male gametophytes form.

antibody A protein secreted by plasma cells (differentiated B cells) that binds to a particular antigen; also called immunoglobulin. All antibodies have the same Y-shaped structure and in their monomer form consist of two identical heavy chains and two identical light chains.

anticodon (an'-tī-kō'-don) A nucleotide triplet at one end of a tRNA molecule that base-pairs with a particular complementary codon on an mRNA molecule.

antidiuretic hormone (ADH) (an'-tī-dī-yū-ret'-ik) A peptide hormone, also known as vasopressin, that promotes water retention by the kidneys. Produced in the hypothalamus and released from the posterior pituitary, ADH also functions in the brain.

antigen (an'-ti-jen) A substance that elicits an immune response by binding to receptors of B cells, antibodies, or T cells.

antigen presentation The process by which an MHC molecule binds to a fragment of an intracellular protein antigen and carries it to the cell surface, where it is displayed and can be recognized by a T cell.

antigen receptor The general term for a surface protein, located on B cells and T cells, that binds to antigens, initiating adaptive immune responses. The antigen receptors on B cells are called B cell receptors, and the antigen receptors on T cells are called T cell receptors.

antigen-presenting cell A cell that upon ingesting pathogens or internalizing pathogen proteins generates peptide fragments that are bound by class II MHC molecules and subsequently displayed on the cell surface to T cells. Macrophages, dendritic cells, and B cells are the primary antigen-presenting cells.

antiparallel Referring to the arrangement of the sugar-phosphate backbones in a DNA double helix (they run in opposite 5' → 3' directions).

aphotic zone (ā'-fō'-tik) The part of an ocean or lake beneath the photic zone, where light does not penetrate sufficiently for photosynthesis to occur.

apical bud (ā'-pik-ul) A bud at the tip of a plant stem; also called a terminal bud.

apical dominance (ā'-pik-ul) Tendency for growth to be concentrated at the tip of a plant shoot, because the apical bud partially inhibits axillary bud growth.

apical meristem (ā'-pik-ul mār'-uh-stem) Embryonic plant tissue in the tips of roots and buds of shoots. The dividing cells of an apical meristem enable the plant to grow in length.

apicomplexan (ap'-ē-kom-pleks'-un) A protist in a clade that includes many species that parasitize animals. Some apicomplexans cause human disease.

apomixis (ap'-uh-mik'-sis) The ability of some plant species to reproduce asexually through seeds without fertilization by a male gamete.

apoplast (ap'-ō-plast) Everything external to the plasma membrane of a plant cell, including cell walls, intercellular spaces, and the space within dead structures such as xylem vessels and tracheids.

apoptosis (ā-puh-tō'-sus) A type of programmed cell death that is brought about by activation of enzymes that break down many chemical components in the cell.

aposematic coloration (ap'-ō-si-mat'-ik) The bright warning coloration of many animals with effective physical or chemical defenses.

appendix A small, finger-like extension of the vertebrate cecum; contains a mass of white blood cells that contribute to immunity.

aquaporin A channel protein in the plasma membrane of a plant, animal, or microorganism cell that specifically facilitates osmosis, the diffusion of free water across the membrane.

aqueous solution (ā'-kwē-us) A solution in which water is the solvent.

arbuscular mycorrhizae (ar-bus'-kyū-lur mī'-kō-rī'-zē) Associations of a fungus with a plant root system in which the fungus causes the invagination of the host (plant) cells' plasma membranes.

arbuscular mycorrhizal fungus A symbiotic fungus whose hyphae grow through the cell wall of plant roots and extend into the root cell (enclosed in tubes formed by invagination of the root cell plasma membrane).

Archaea (ar'-kē'-uh) One of two prokaryotic domains, the other being Bacteria.

Archaeplastida (ar'-kē-plas'-tid-uh) One of four supergroups of eukaryotes proposed in a current hypothesis of the evolutionary history of eukaryotes. This monophyletic group, which includes red algae, green algae, and land plants, descended from an ancient protist ancestor that engulfed a cyanobacterium. *See also* Excavata, "SAR" clade, and Unikonta.

artery A vessel that carries blood away from the heart to organs throughout the body.

arthropod A segmented, molting bilaterian animal with a hard exoskeleton and jointed appendages. Familiar examples include insects, spiders, millipedes, and crabs.

artificial selection The selective breeding of domesticated plants and animals to encourage the occurrence of desirable traits.

asexual reproduction The generation of offspring from a single parent that occurs without the fusion of gametes (by budding, division of a single cell, or division of the entire organism into two or more parts). In most cases, the offspring are genetically identical to the parent.

associative learning The acquired ability to associate one environmental feature (such as a color) with another (such as danger).

aster A radial array of short microtubules that extends from each centrosome toward the plasma membrane in an animal cell undergoing mitosis.

astrocyte A glial cell with diverse functions, including providing structural support for neurons, regulating the interstitial environment, facilitating synaptic transmission, and assisting in regulating the blood supply to the brain.

atherosclerosis A cardiovascular disease in which fatty deposits called plaques develop in the inner walls of the arteries, obstructing the arteries and causing them to harden.

atom The smallest unit of matter that retains the properties of an element.

atomic mass The total mass of an atom, which is the mass in grams of 1 mole of the atom.

atomic nucleus An atom's dense central core, containing protons and neutrons.

atomic number The number of protons in the nucleus of an atom, unique for each element and designated by a subscript.

ATP (adenosine triphosphate) (a-den'-ō-sēn trī-fos'-fāt) An adenine-containing nucleoside triphosphate that releases free energy when its phosphate bonds are hydrolyzed. This energy is used to drive endergonic reactions in cells.

ATP synthase A complex of several membrane proteins that functions in chemiosmosis with adjacent electron transport chains, using the energy of a hydrogen ion (proton) concentration gradient to make ATP. ATP synthases are found in the inner mitochondrial membranes of eukaryotic cells and in the plasma membranes of prokaryotes.

atrioventricular (AV) node A region of specialized heart muscle tissue between the left and right atria where electrical impulses are delayed for about 0.1 second before spreading to both ventricles and causing them to contract.

atrioventricular (AV) valve A heart valve located between each atrium and ventricle that prevents a backflow of blood when the ventricle contracts.

atrium (ā'-trē-um) (plural, **atria**) A chamber of the vertebrate heart that receives blood from the veins and transfers blood to a ventricle.

autoimmune disease An immunological disorder in which the immune system turns against self.

autonomic nervous system (ot'-ō-nom'-ik) An efferent branch of the vertebrate peripheral nervous system that regulates the internal environment; consists of the sympathetic, parasympathetic, and enteric divisions.

autopolyploid (ot'-ō-pol'-ē-ployd) An individual that has more than two chromosome sets that are all derived from a single species.

autosome (ot'-ō-sōm) A chromosome that is not directly involved in determining sex; not a sex chromosome.

autotroph (ot'-ō-trōf) An organism that obtains organic food molecules without eating other organisms or substances derived from other organisms. Autotrophs use energy from the sun or from oxidation of inorganic substances to make organic molecules from inorganic ones.

auxin (ôk'-sin) A term that primarily refers to indoleacetic acid (IAA), a natural plant hormone that has a variety of effects, including cell elongation, root formation, secondary growth, and fruit growth.

avirulent Describing a pathogen that can mildly harm, but not kill, the host.

axillary bud (ak'-sil-ar-ē) A structure that has the potential to form a lateral shoot, or branch. The bud appears in the angle formed between a leaf and a stem.

axon (ak'-son) A typically long extension, or process, of a neuron that carries nerve impulses away from the cell body toward target cells.

B cells The lymphocytes that complete their development in the bone marrow and become effector cells for the humoral immune response.

Bacteria One of two prokaryotic domains, the other being Archaea.

bacteriophage (bak-tēr'-ē-ō-fāj) A virus that infects bacteria; also called a phage.

bacteroid A form of the bacterium *Rhizobium* contained within the vesicles formed by the root cells of a root nodule.

balancing selection Natural selection that maintains two or more phenotypic forms in a population.

bar graph A graph in which the independent variable represents groups or nonnumerical categories. Each piece of data is represented by a bar, whose height (or length) represents the value of the independent variable for the group or category indicated.

bark All tissues external to the vascular cambium, consisting mainly of the secondary phloem and layers of periderm.

Barr body A dense object lying along the inside of the nuclear envelope in cells of female mammals, representing a highly condensed, inactivated X chromosome.

basal body (bā′-sul) A eukaryotic cell structure consisting of a "9 + 0" arrangement of microtubule triplets. The basal body may organize the microtubule assembly of a cilium or flagellum and is structurally very similar to a centriole.

basal taxon In a specified group of organisms, a taxon whose evolutionary lineage diverged early in the history of the group.

base A substance that reduces the hydrogen ion concentration of a solution.

Batesian mimicry (bāt′-zē-un mim′-uh-krē) A type of mimicry in which a harmless species looks like a species that is poisonous or otherwise harmful to predators.

behavior Individually, an action carried out by muscles or glands under control of the nervous system in response to a stimulus; collectively, the sum of an animal's responses to external and internal stimuli.

behavioral ecology The study of the evolution of and ecological basis for animal behavior.

benign tumor A mass of abnormal cells with specific genetic and cellular changes such that the cells are not capable of surviving at a new site and generally remain at the site of the tumor's origin.

benthic zone The bottom surface of an aquatic environment.

benthos (ben′-thōz) The communities of organisms living in the benthic zone of an aquatic biome.

beta (β) pleated sheet One form of the secondary structure of proteins in which the polypeptide chain folds back and forth. Two regions of the chain lie parallel to each other and are held together by hydrogen bonds between atoms of the polypeptide backbone (not the side chains).

beta oxidation A metabolic sequence that breaks fatty acids down to two-carbon fragments that enter the citric acid cycle as acetyl CoA.

bicoid A maternal effect gene that codes for a protein responsible for specifying the anterior end in *Drosophila melanogaster*.

bilateral symmetry Body symmetry in which a central longitudinal plane divides the body into two equal but opposite halves.

bilaterian (bī′-luh-ter′-ē-uhn) Member of a clade of animals with bilateral symmetry and three germ layers.

bile A mixture of substances that is produced in the liver and stored in the gallbladder; enables formation of fat droplets in water as an aid in the digestion and absorption of fats.

binary fission A method of asexual reproduction by "division in half." In prokaryotes, binary fission does not involve mitosis, but in single-celled eukaryotes that undergo binary fission, mitosis is part of the process.

binomial A common term for the two-part, latinized format for naming a species, consisting of the genus and specific epithet; also called a binomen.

biodiversity hot spot A relatively small area with numerous endemic species and a large number of endangered and threatened species.

bioenergetics (1) The overall flow and transformation of energy in an organism. (2) The study of how energy flows through organisms.

biofilm A surface-coating colony of one or more species of prokaryotes that engage in metabolic cooperation.

biofuel A fuel produced from biomass.

biogenic amine A neurotransmitter derived from an amino acid.

biogeochemical cycle Any of the various chemical cycles that involve both biotic and abiotic components of ecosystems.

biogeography The scientific study of the past and present geographic distributions of species.

bioinformatics The use of computers, software, and mathematical models to process and integrate biological information from large data sets.

biological augmentation An approach to restoration ecology that uses organisms to add essential materials to a degraded ecosystem.

biological clock An internal timekeeper that controls an organism's biological rhythms. The biological clock marks time with or without environmental cues but often requires signals from the environment to remain tuned to an appropriate period. *See also* circadian rhythm.

biological magnification A process in which retained substances become more concentrated at each higher trophic level in a food chain.

biological species concept Definition of a species as a group of populations whose members have the potential to interbreed in nature and produce viable, fertile offspring, but do not produce viable, fertile offspring with members of other such groups.

biology The scientific study of life.

biomanipulation An approach that applies the top-down model of community organization to alter ecosystem characteristics. For example, ecologists can prevent algal blooms and eutrophication by altering the density of higher-level consumers in lakes instead of by using chemical treatments.

biomass The total mass of organic matter comprising a group of organisms in a particular habitat.

biome (bī′-ōm) Any of the world's major ecosystem types, often classified according to the predominant vegetation for terrestrial biomes and the physical environment for aquatic biomes and characterized by

adaptations of organisms to that particular environment.

bioremediation The use of organisms to detoxify and restore polluted and degraded ecosystems.

biosphere The entire portion of Earth inhabited by life; the sum of all the planet's ecosystems.

biotic (bī-ot′-ik) Pertaining to the living factors—the organisms—in an environment.

blade (1) A leaflike structure of a seaweed that provides most of the surface area for photosynthesis. (2) The flattened portion of a typical leaf.

blastocoel (blas′-tuh-sēl) The fluid-filled cavity that forms in the center of a blastula.

blastula (blas′-tyū-luh) A hollow ball of cells that marks the end of the cleavage stage during early embryonic development in animals.

blood A connective tissue with a fluid matrix called plasma in which red blood cells, white blood cells, and cell fragments called platelets are suspended.

blue-light photoreceptor A type of light receptor in plants that initiates a variety of responses, including phototropism and slowing of hypocotyl elongation.

body plan In multicellular eukaryotes, a set of morphological and developmental traits that are integrated into a functional whole—the living organism.

Bohr shift A lowering of the affinity of hemoglobin for oxygen, caused by a drop in pH. It facilitates the release of oxygen from hemoglobin in the vicinity of active tissues.

bolus A lubricated ball of chewed food.

bottleneck effect Genetic drift that occurs when the size of a population is reduced, as by a natural disaster or human actions. Typically, the surviving population is no longer genetically representative of the original population.

bottom-up model A model of community organization in which mineral nutrients influence community organization by controlling plant or phytoplankton numbers, which in turn control herbivore numbers, which in turn control predator numbers.

Bowman's capsule (bō′-munz) A cup-shaped receptacle in the vertebrate kidney that is the initial, expanded segment of the nephron where filtrate enters from the blood.

brain Organ of the central nervous system where information is processed and integrated.

brainstem A collection of structures in the vertebrate brain, including the midbrain, the pons, and the medulla oblongata; functions in homeostasis, coordination of movement, and conduction of information to higher brain centers.

branch point The representation on a phylogenetic tree of the divergence of two or more taxa from a common ancestor. A branch point is usually shown as a dichotomy in which a branch representing the ancestral lineage splits (at the branch point) into two branches, one for each of the two descendant lineages.

brassinosteroid A steroid hormone in plants that has a variety of effects, including inducing cell elongation, retarding leaf abscission, and promoting xylem differentiation.

breathing Ventilation of the lungs through alternating inhalation and exhalation.

bronchus (brong′-kus) (plural, **bronchi**) One of a pair of breathing tubes that branch from the trachea into the lungs.

brown alga A multicellular, photosynthetic protist with a characteristic brown or olive color that results from carotenoids in its plastids. Most brown algae are marine, and some have a plantlike body.

bryophyte (brī′-uh-fīt) An informal name for a moss, liverwort, or hornwort; a nonvascular plant that lives on land but lacks some of the terrestrial adaptations of vascular plants.

buffer A solution that contains a weak acid and its corresponding base. A buffer minimizes changes in pH when acids or bases are added to the solution.

bulk feeder An animal that eats relatively large pieces of food.

bulk flow The movement of a fluid due to a difference in pressure between two locations.

C₃ plant A plant that uses the Calvin cycle for the initial steps that incorporate CO_2 into organic material, forming a three-carbon compound as the first stable intermediate.

C₄ plant A plant in which the Calvin cycle is preceded by reactions that incorporate CO_2 into a four-carbon compound, the end product of which supplies CO_2 for the Calvin cycle.

callus A mass of dividing, undifferentiated cells growing in culture.

calorie (cal) The amount of heat energy required to raise the temperature of 1 g of water by 1°C; also the amount of heat energy that 1 g of water releases when it cools by 1°C. The Calorie (with a capital C), usually used to indicate the energy content of food, is a kilocalorie.

Calvin cycle The second of two major stages in photosynthesis (following the light reactions), involving fixation of atmospheric CO_2 and reduction of the fixed carbon into carbohydrate.

CAM plant A plant that uses crassulacean acid metabolism, an adaptation for photosynthesis in arid conditions. In this process, carbon dioxide entering open stomata during the night is converted to organic acids, which release CO_2 for the Calvin cycle during the day, when stomata are closed.

Cambrian explosion A relatively brief time in geologic history when many present-day phyla of animals first appeared in the fossil record. This burst of evolutionary change occurred about 535–525 million years ago and saw the emergence of the first large, hard-bodied animals.

canopy The uppermost layer of vegetation in a terrestrial biome.

capillary (kap′-il-ār′-ē) A microscopic blood vessel that penetrates the tissues and consists of a single layer of endothelial cells that allows exchange between the blood and interstitial fluid.

capillary bed A network of capillaries in a tissue or organ.

capsid The protein shell that encloses a viral genome. It may be rod-shaped, polyhedral, or more complex in shape.

capsule (1) In many prokaryotes, a dense and well-defined layer of polysaccharide or protein that surrounds the cell wall and is sticky, protecting the cell and enabling it to adhere to substrates or other cells. (2) The sporangium of a bryophyte (moss, liverwort, or hornwort).

carbohydrate (kar′-bō-hī′-drāt) A sugar (monosaccharide) or one of its dimers (disaccharides) or polymers (polysaccharides).

carbon fixation The initial incorporation of carbon from CO_2 into an organic compound by an autotrophic organism (a plant, another photosynthetic organism, or a chemoautotrophic prokaryote).

carbonyl group (kar-buh-nēl′) A chemical group present in aldehydes and ketones and consisting of a carbon atom double-bonded to an oxygen atom.

carboxyl group (kar-bok′-sil) A chemical group present in organic acids and consisting of a single carbon atom double-bonded to an oxygen atom and also bonded to a hydroxyl group.

cardiac cycle (kar′-dē-ak) The alternating contractions and relaxations of the heart.

cardiac muscle A type of striated muscle that forms the contractile wall of the heart. Its cells are joined by intercalated disks that relay the electrical signals underlying each heartbeat.

cardiovascular system A closed circulatory system with a heart and branching network of arteries, capillaries, and veins. The system is characteristic of vertebrates.

carnivore An animal that mainly eats other animals.

carotenoid (kuh-rot′-uh-noyd′) An accessory pigment, either yellow or orange, in the chloroplasts of plants and in some prokaryotes. By absorbing wavelengths of light that chlorophyll cannot, carotenoids broaden the spectrum of colors that can drive photosynthesis.

carpel (kar′-pul) The ovule-producing reproductive organ of a flower, consisting of the stigma, style, and ovary.

carrier In genetics, an individual who is heterozygous at a given genetic locus for a recessively inherited disorder. The heterozygote is generally phenotypically normal for the disorder but can pass on the recessive allele to offspring.

carrying capacity The maximum population size that can be supported by the available resources, symbolized as K.

Casparian strip (ka-spār′-ē-un) A water-impermeable ring of wax in the endodermal cells of plants that blocks the passive flow of water and solutes into the stele by way of cell walls.

catabolic pathway (kat′-uh-bol′-ik) A metabolic pathway that releases energy by breaking down complex molecules to simpler molecules.

catalyst (kat′-uh-list) A chemical agent that selectively increases the rate of a reaction without being consumed by the reaction.

cation (cat′-ī′-on) A positively charged ion.

cation exchange A process in which positively charged minerals are made available to a plant when hydrogen ions in the soil displace mineral ions from the clay particles.

cecum (sē′-kum) (plural, **ceca**) The blind pouch forming one branch of the large intestine.

cell body The part of a neuron that houses the nucleus and most other organelles.

cell cycle An ordered sequence of events in the life of a cell, from its origin in the division of a parent cell until its own division into two. The eukaryotic cell cycle is composed of interphase (including G_1, S, and G_2 subphases) and M phase (including mitosis and cytokinesis).

cell cycle control system A cyclically operating set of molecules in the eukaryotic cell that both triggers and coordinates key events in the cell cycle.

cell division The reproduction of cells.

cell fractionation The disruption of a cell and separation of its parts by centrifugation at successively higher speeds.

cell plate A membrane-bounded, flattened sac located at the midline of a dividing plant cell, inside which the new cell wall forms during cytokinesis.

cell wall A protective layer external to the plasma membrane in the cells of plants, prokaryotes, fungi, and some protists. Polysaccharides such as cellulose (in plants and some protists), chitin (in fungi), and peptidoglycan (in bacteria) are important structural components of cell walls.

cell-mediated immune response The branch of adaptive immunity that involves the activation of cytotoxic T cells, which defend against infected cells.

cellular respiration The catabolic pathways of aerobic and anaerobic respiration, which break down organic molecules and use an electron transport chain for the production of ATP.

cellulose (sel´-yū-lōs) A structural polysaccharide of plant cell walls, consisting of glucose monomers joined by β glycosidic linkages.

central nervous system (CNS) The portion of the nervous system where signal integration occurs; in vertebrate animals, the brain and spinal cord.

central vacuole In a mature plant cell, a large membranous sac with diverse roles in growth, storage, and sequestration of toxic substances.

centriole (sen´-trē-ōl) A structure in the centrosome of an animal cell composed of a cylinder of microtubule triplets arranged in a 9 + 0 pattern. A centrosome has a pair of centrioles.

centromere (sen´-trō-mēr) In a duplicated chromosome, the region on each sister chromatid where they are most closely attached to each other by proteins that bind to specific DNA sequences; this close attachment causes a constriction in the condensed chromosome. (An uncondensed, unduplicated chromosome has a single centromere, identified by its DNA sequence.)

centrosome (sen´-trō-sōm) A structure present in the cytoplasm of animal cells that functions as a microtubule-organizing center and is important during cell division. A centrosome has two centrioles.

cercozoan An amoeboid or flagellated protist that feeds with threadlike pseudopodia.

cerebellum (sār´-ruh-bel´-um) Part of the vertebrate hindbrain located dorsally; functions in unconscious coordination of movement and balance.

cerebral cortex (suh-rē´-brul) The surface of the cerebrum; the largest and most complex part of the mammalian brain, containing nerve cell bodies of the cerebrum; the part of the vertebrate brain most changed through evolution.

cerebral hemisphere The right or left side of the cerebrum.

cerebrum (suh-rē´-brum) The dorsal portion of the vertebrate forebrain, composed of right and left hemispheres; the integrating center for memory, learning, emotions, and other highly complex functions of the central nervous system.

cervix (ser´-viks) The neck of the uterus, which opens into the vagina.

chaparral A scrubland biome of dense, spiny evergreen shrubs found at midlatitudes along coasts where cold ocean currents circulate offshore; characterized by mild, rainy winters and long, hot, dry summers.

character An observable heritable feature that may vary among individuals.

character displacement The tendency for characteristics to be more divergent in sympatric populations of two species than in allopatric populations of the same two species.

checkpoint A control point in the cell cycle where stop and go-ahead signals can regulate the cycle.

chemical bond An attraction between two atoms, resulting from a sharing of outer-shell electrons or the presence of opposite charges on the atoms. The bonded atoms gain complete outer electron shells.

chemical energy Energy available in molecules for release in a chemical reaction; a form of potential energy.

chemical equilibrium In a chemical reaction, the state in which the rate of the forward reaction equals the rate of the reverse reaction, so that the relative concentrations of the reactants and products do not change with time.

chemical reaction The making and breaking of chemical bonds, leading to changes in the composition of matter.

chemiosmosis (kem´-ē-oz-mō´-sis) An energy-coupling mechanism that uses energy stored in the form of a hydrogen ion gradient across a membrane to drive cellular work, such as the synthesis of ATP. Under aerobic conditions, most ATP synthesis in cells occurs by chemiosmosis.

chemoreceptor A sensory receptor that responds to a chemical stimulus, such as a solute or an odorant.

chiasma (plural, **chiasmata**) (kī-az´-muh, kī-az´-muh-tuh) The X-shaped, microscopically visible region where crossing over has occurred earlier in prophase I between homologous nonsister chromatids. Chiasmata become visible after synapsis ends, with the two homologs remaining associated due to sister chromatid cohesion.

chitin (kī´-tin) A structural polysaccharide, consisting of amino sugar monomers, found in many fungal cell walls and in the exoskeletons of all arthropods.

chlorophyll (klōr´-ō-fil) A green pigment located in membranes within the chloroplasts of plants and algae and in the membranes of certain prokaryotes. Chlorophyll *a* participates directly in the light reactions, which convert solar energy to chemical energy.

chlorophyll *a* A photosynthetic pigment that participates directly in the light reactions, which convert solar energy to chemical energy.

chlorophyll *b* An accessory photosynthetic pigment that transfers energy to chlorophyll *a*.

chloroplast (klōr´-ō-plast) An organelle found in plants and photosynthetic protists that absorbs sunlight and uses it to drive the synthesis of organic compounds from carbon dioxide and water.

choanocyte (kō-an´-uh-sīt) A flagellated feeding cell found in sponges. Also called a collar cell, it has a collar-like ring that traps food particles around the base of its flagellum.

cholesterol (kō-les´-tuh-rol) A steroid that forms an essential component of animal cell membranes and acts as a precursor molecule for the synthesis of other biologically important steroids, such as many hormones.

chondrichthyan (kon-drik´-thē-an) Member of the class Chondrichthyes, vertebrates with skeletons made mostly of cartilage, such as sharks and rays.

chordate Member of the phylum Chordata, animals that at some point during their development have a notochord; a dorsal, hollow nerve cord; pharyngeal slits or clefts; and a muscular, post-anal tail.

chromatin (krō´-muh-tin) The complex of DNA and proteins that makes up eukaryotic chromosomes. When the cell is not dividing, chromatin exists in its dispersed form, as a mass of very long, thin fibers that are not visible with a light microscope.

chromosome (krō´-muh-sōm) A cellular structure consisting of one DNA molecule and associated protein molecules. (In some contexts, such as genome sequencing, the term may refer to the DNA alone.) A eukaryotic cell typically has multiple, linear chromosomes, which are located in the nucleus. A prokaryotic cell often has a single, circular chromosome, which is found in the nucleoid, a region that is not enclosed by a membrane. *See also* chromatin.

chromosome theory of inheritance A basic principle in biology stating that genes are located at specific positions (loci) on chromosomes and that the behavior of chromosomes during meiosis accounts for inheritance patterns.

chylomicron (kī´-lō-mī´-kron) A lipid transport globule composed of fats mixed with cholesterol and coated with proteins.

chyme (kīm) The mixture of partially digested food and digestive juices formed in the stomach.

ciliate (sil´-ē-it) A type of protist that moves by means of cilia.

cilium (sil´-ē-um) (plural, **cilia**) A short appendage containing microtubules in eukaryotic cells. A motile cilium is specialized for locomotion or moving fluid past the cell; it is formed from a core of nine outer doublet microtubules and two inner single microtubules (the "9 + 2" arrangement) ensheathed in an extension of the plasma membrane. A primary cilium is usually nonmotile and plays a sensory and signaling role; it lacks the two inner microtubules (the "9 + 0" arrangement).

circadian rhythm (ser-kā´-dē-un) A physiological cycle of about 24 hours that persists even in the absence of external cues.

citric acid cycle A chemical cycle involving eight steps that completes the metabolic breakdown of glucose molecules begun in glycolysis by oxidizing acetyl CoA (derived from pyruvate) to carbon dioxide; occurs within the mitochondrion in eukaryotic cells and in the cytosol of prokaryotes; together with pyruvate oxidation, the second major stage in cellular respiration.

clade (klayd) A group of species that includes an ancestral species and all of its descendants.

cladistics (kluh-dis′-tiks) An approach to systematics in which organisms are placed into groups called clades based primarily on common descent.

class In Linnaean classification, the taxonomic category above the level of order.

cleavage (1) The process of cytokinesis in animal cells, characterized by pinching of the plasma membrane. (2) The succession of rapid cell divisions without significant growth during early embryonic development that converts the zygote to a ball of cells.

cleavage furrow The first sign of cleavage in an animal cell; a shallow groove around the cell in the cell surface near the old metaphase plate.

climate The long-term prevailing weather conditions at a given place.

climograph A plot of the temperature and precipitation in a particular region.

clitoris (klit′-uh-ris) An organ at the upper intersection of the labia minora that engorges with blood and becomes erect during sexual arousal.

cloaca (klō-ā′-kuh) A common opening for the digestive, urinary, and reproductive tracts found in many nonmammalian vertebrates but in few mammals.

clonal selection The process by which an antigen selectively binds to and activates only those lymphocytes bearing receptors specific for the antigen. The selected lymphocytes proliferate and differentiate into a clone of effector cells and a clone of memory cells specific for the stimulating antigen.

clone (1) A lineage of genetically identical individuals or cells. (2) In popular usage, an individual that is genetically identical to another individual. (3) As a verb, to make one or more genetic replicas of an individual or cell. *See also* gene cloning.

cloning vector In genetic engineering, a DNA molecule that can carry foreign DNA into a host cell and replicate there. Cloning vectors include plasmids.

closed circulatory system A circulatory system in which blood is confined to vessels and is kept separate from the interstitial fluid.

cochlea (kok′-lē-uh) The complex, coiled organ of hearing that contains the organ of Corti.

codominance The situation in which the phenotypes of both alleles are exhibited in the heterozygote because both alleles affect the phenotype in separate, distinguishable ways.

codon (kō′-don) A three-nucleotide sequence of DNA or mRNA that specifies a particular amino acid or termination signal; the basic unit of the genetic code.

coefficient of relatedness The fraction of genes that, on average, are shared by two individuals.

coenzyme (kō-en′-zīm) An organic molecule serving as a cofactor. Most vitamins function as coenzymes in metabolic reactions.

cofactor Any nonprotein molecule or ion that is required for the proper functioning of an enzyme. Cofactors can be permanently bound to the active site or may bind loosely and reversibly, along with the substrate, during catalysis.

cognition The process of knowing that may include awareness, reasoning, recollection, and judgment.

cognitive map A neural representation of the abstract spatial relationships between objects in an animal's surroundings.

cohesion The linking together of like molecules, often by hydrogen bonds.

cohesion-tension hypothesis The leading explanation for the ascent of xylem sap. It states that transpiration exerts pull on xylem sap, putting the sap under negative pressure or tension, and that the cohesion of water molecules transmits this pull along the entire length of the xylem from shoots to roots.

cohort A group of individuals of the same age in a population.

coleoptile (kō′-lē-op′-tul) The covering of the young shoot of the embryo of a grass seed.

coleorhiza (kō′-lē-uh-rī′-zuh) The covering of the young root of the embryo of a grass seed.

collagen A glycoprotein in the extracellular matrix of animal cells that forms strong fibers, found extensively in connective tissue and bone; the most abundant protein in the animal kingdom.

collecting duct The location in the kidney where processed filtrate, called urine, is collected from the renal tubules.

collenchyma cell (kō-len′-kim-uh) A flexible plant cell type that occurs in strands or cylinders that support young parts of the plant without restraining growth.

colon (kō′-len) The largest section of the vertebrate large intestine; functions in water absorption and formation of feces.

commensalism (kuh-men′-suh-lizm) A symbiotic relationship in which one organism benefits but the other is neither helped nor harmed.

communication In animal behavior, a process involving transmission of, reception of, and response to signals. The term is also used in connection with other organisms, as well as individual cells of multicellular organisms.

community All the organisms that inhabit a particular area; an assemblage of populations of different species living close enough together for potential interaction.

community ecology The study of how interactions between species affect community structure and organization.

companion cell A type of plant cell that is connected to a sieve-tube element by many plasmodesmata and whose nucleus and ribosomes may serve one or more adjacent sieve-tube elements.

competitive exclusion The concept that when populations of two similar species compete for the same limited resources, one population will use the resources more efficiently and have a reproductive advantage that will eventually lead to the elimination of the other population.

competitive inhibitor A substance that reduces the activity of an enzyme by entering the active site in place of the substrate, whose structure it mimics.

complement system A group of about 30 blood proteins that may amplify the inflammatory response, enhance phagocytosis, or directly lyse extracellular pathogens.

complementary DNA (cDNA) A double-stranded DNA molecule made *in vitro* using mRNA as a template and the enzymes reverse transcriptase and DNA polymerase. A cDNA molecule corresponds to the exons of a gene.

complete dominance The situation in which the phenotypes of the heterozygote and dominant homozygote are indistinguishable.

complete flower A flower that has all four basic floral organs: sepals, petals, stamens, and carpels.

compound A substance consisting of two or more different elements combined in a fixed ratio.

compound eye A type of multifaceted eye in insects and crustaceans consisting of up to several thousand light-detecting, focusing ommatidia.

concentration gradient A region along which the density of a chemical substance increases or decreases.

conception The fertilization of an egg by a sperm in humans.

cone A cone-shaped cell in the retina of the vertebrate eye, sensitive to color.

conformer An animal for which an internal condition conforms to (changes in accordance with) changes in an environmental variable.

conifer Member of the largest gymnosperm phylum. Most conifers are cone-bearing trees, such as pines and firs.

conjugation (kon′-jū-gā′-shun) (1) In prokaryotes, the direct transfer of DNA between two cells that are temporarily joined. When the two cells are members of different species, conjugation results in horizontal gene transfer. (2) In ciliates, a sexual process in which two cells exchange haploid micronuclei but do not reproduce.

connective tissue Animal tissue that functions mainly to bind and support other tissues, having a sparse population of cells scattered through an extracellular matrix.

conservation biology The integrated study of ecology, evolutionary biology, physiology, molecular biology, and genetics to sustain biological diversity at all levels.

contraception The deliberate prevention of pregnancy.

contractile vacuole A membranous sac that helps move excess water out of certain freshwater protists.

control element A segment of noncoding DNA that helps regulate transcription of a gene by serving as a binding site for a transcription factor. Multiple control elements are present in a eukaryotic gene's enhancer.

control group In a controlled experiment, a set of subjects that lacks (or does not receive) the specific factor being tested. Ideally, the control group should be identical to the experimental group in other respects.

controlled experiment An experiment in which an experimental group is compared with a control group that varies only in the factor being tested.

convergent evolution The evolution of similar features in independent evolutionary lineages.

cooperativity A kind of allosteric regulation whereby a shape change in one subunit of a protein caused by substrate binding is transmitted to all the other subunits, facilitating binding of additional substrate molecules to those subunits.

coral reef Typically a warm-water, tropical ecosystem dominated by the hard skeletal structures secreted primarily by corals. Some coral reefs also exist in cold, deep waters.

corepressor A small molecule that binds to a bacterial repressor protein and changes the protein's shape, allowing it to bind to the operator and switch an operon off.

cork cambium (kam'-bē-um) A cylinder of meristematic tissue in woody plants that replaces the epidermis with thicker, tougher cork cells.

corpus callosum (kor'-pus kuh-lō'-sum) The thick band of nerve fibers that connects the right and left cerebral hemispheres in mammals, enabling the hemispheres to process information together.

corpus luteum (kor'-pus lū'-tē-um) A secreting tissue in the ovary that forms from the collapsed follicle after ovulation and produces progesterone.

cortex (1) The outer region of cytoplasm in a eukaryotic cell, lying just under the plasma membrane, that has a more gel-like consistency than the inner regions due to the presence of multiple microfilaments. (2) In plants, ground tissue that is between the vascular tissue and dermal tissue in a root or eudicot stem.

cortical nephron In mammals and birds, a nephron with a loop of Henle located almost entirely in the renal cortex.

cotransport The coupling of the "downhill" diffusion of one substance to the "uphill" transport of another against its own concentration gradient.

countercurrent exchange The exchange of a substance or heat between two fluids flowing in opposite directions. For example, blood in a fish gill flows in the opposite direction of water passing over the gill, maximizing diffusion of oxygen into and carbon dioxide out of the blood.

countercurrent multiplier system A countercurrent system in which energy is expended in active transport to facilitate exchange of materials and generate concentration gradients.

covalent bond (kō-vā'-lent) A type of strong chemical bond in which two atoms share one or more pairs of valence electrons.

crassulacean acid metabolism (CAM) An adaptation for photosynthesis in arid conditions, first discovered in the family Crassulaceae. In this process, a plant takes up CO_2 at night when stomata are open and incorporates it into a variety of organic acids; during the day, when stomata are closed, CO_2 is released from the organic acids for use in the Calvin cycle.

crista (plural, **cristae**) (kris'-tuh, kris'-tē) An infolding of the inner membrane of a mitochondrion. The inner membrane houses electron transport chains and molecules of the enzyme catalyzing the synthesis of ATP (ATP synthase).

critical load The amount of added nutrient, usually nitrogen or phosphorus, that can be absorbed by plants without damaging ecosystem integrity.

cross-fostering study A behavioral study in which the young of one species are placed in the care of adults from another species.

crossing over The reciprocal exchange of genetic material between nonsister chromatids during prophase I of meiosis.

cryptic coloration Camouflage that makes a potential prey difficult to spot against its background.

culture A system of information transfer through social learning or teaching that influences the behavior of individuals in a population.

cuticle (kyū'-tuh-kul) (1) A waxy covering on the surface of stems and leaves that prevents desiccation in terrestrial plants. (2) The exoskeleton of an arthropod, consisting of layers of protein and chitin that are variously modified for different functions. (3) A tough coat that covers the body of a nematode.

cyclic AMP (cAMP) Cyclic adenosine monophosphate, a ring-shaped molecule made from ATP that is a common intracellular signaling molecule (second messenger) in eukaryotic cells. It is also a regulator of some bacterial operons.

cystic fibrosis (sis'-tik fī-brō'-sis) A human genetic disorder caused by a recessive allele for a chloride channel protein; characterized by an excessive secretion of mucus and consequent vulnerability to infection; fatal if untreated.

cytochrome (sī'-tō-krōm) An iron-containing protein that is a component of electron transport chains in the mitochondria and chloroplasts of eukaryotic cells and the plasma membranes of prokaryotic cells.

cytogenetic map A map of a chromosome that locates genes with respect to chromosomal features distinguishable in a microscope.

cytokine (sī'-tō-kīn') Any of a group of small proteins secreted by a number of cell types, including macrophages and helper T cells, that regulate the function of other cells.

cytokinesis (sī'-tō-kuh-nē'-sis) The division of the cytoplasm to form two separate daughter cells immediately after mitosis, meiosis I, or meiosis II.

cytokinin (sī'-tō-kī'-nin) Any of a class of related plant hormones that retard aging and act in concert with auxin to stimulate cell division, influence the pathway of differentiation, and control apical dominance.

cytoplasm (sī'-tō-plaz'-um) The contents of the cell enclosed by the plasma membrane; in eukaryotes, the portion exclusive of the nucleus.

cytoplasmic determinant A maternal substance, such as a protein or RNA, that when placed into an egg influences the course of early development by regulating the expression of genes that affect the developmental fate of cells.

cytoskeleton A network of microtubules, microfilaments, and intermediate filaments that extends throughout the cytoplasm and serves a variety of mechanical, transport, and signaling functions.

cytosol (sī'-tō-sol) The semifluid portion of the cytoplasm.

cytotoxic T cell A type of lymphocyte that, when activated, kills infected cells as well as certain cancer cells and transplanted cells.

dalton A measure of mass for atoms and subatomic particles; the same as the atomic mass unit, or amu.

data Recorded observations.

day-neutral plant A plant in which flower formation is not controlled by photoperiod or day length.

decomposer An organism that absorbs nutrients from nonliving organic material such as corpses, fallen plant material, and the wastes of living organisms and converts them to inorganic forms; a detritivore.

deductive reasoning A type of logic in which specific results are predicted from a general premise.

deep-sea hydrothermal vent A dark, hot, oxygen-deficient environment associated with volcanic activity on or near the sea-floor. The producers in a vent community are chemoautotrophic prokaryotes.

de-etiolation The changes a plant shoot undergoes in response to sunlight; also known informally as greening.

dehydration reaction A chemical reaction in which two molecules become covalently bonded to each other with the removal of a water molecule.

deletion (1) A deficiency in a chromosome resulting from the loss of a fragment through breakage. (2) A mutational loss of one or more nucleotide pairs from a gene.

demography The study of changes over time in the vital statistics of populations, especially birth rates and death rates.

denaturation (dē-nā′-chur-ā′-shun) In proteins, a process in which a protein loses its native shape due to the disruption of weak chemical bonds and interactions, thereby becoming biologically inactive; in DNA, the separation of the two strands of the double helix. Denaturation occurs under extreme (noncellular) conditions of pH, salt concentration, or temperature.

dendrite (den′-drīt) One of usually numerous, short, highly branched extensions of a neuron that receive signals from other neurons.

density The number of individuals per unit area or volume.

density dependent Referring to any characteristic that varies with population density.

density independent Referring to any characteristic that is not affected by population density.

density-dependent inhibition The phenomenon observed in normal animal cells that causes them to stop dividing when they come into contact with one another.

deoxyribonucleic acid (DNA) (dē-ok′-sē-rī′-bō-nū-klā′-ik) A nucleic acid molecule, usually a double-stranded helix, in which each polynucleotide strand consists of nucleotide monomers with a deoxyribose sugar and the nitrogenous bases adenine (A), cytosine (C), guanine (G), and thymine (T); capable of being replicated and determining the inherited structure of a cell's proteins.

deoxyribose (dē-ok′-si-rī′-bōs) The sugar component of DNA nucleotides, having one fewer hydroxyl group than ribose, the sugar component of RNA nucleotides.

dependent variable In an experiment, a variable whose value is influenced by changes in another variable (the independent variable).

depolarization A change in a cell's membrane potential such that the inside of the membrane is made less negative relative to the outside. For example, a neuron membrane is depolarized if a stimulus decreases its voltage from the resting potential of −70 mV in the direction of zero voltage.

dermal tissue system The outer protective covering of plants.

desert A terrestrial biome characterized by very low precipitation.

desmosome A type of intercellular junction in animal cells that functions as a rivet, fastening cells together.

determinate growth A type of growth characteristic of most animals and some plant organs, in which growth stops after a certain size is reached.

determination The progressive restriction of developmental potential whereby the possible fate of each cell becomes more limited as an embryo develops. At the end of determination, a cell is committed to its fate.

detritivore (deh-trī′-tuh-vōr) A consumer that derives its energy and nutrients from nonliving organic material such as corpses, fallen plant material, and the wastes of living organisms; a decomposer.

detritus (di-trī′-tus) Dead organic matter.

diabetes mellitus (dī′-uh-bē′-tis mel′-uh-tus) An endocrine disorder marked by an inability to maintain glucose homeostasis. The type 1 form results from autoimmune destruction of insulin-secreting cells; treatment usually requires daily insulin injections. The type 2 form most commonly results from reduced responsiveness of target cells to insulin; obesity and lack of exercise are risk factors.

diaphragm (dī′-uh-fram′) (1) A sheet of muscle that forms the bottom wall of the thoracic cavity in mammals. Contraction of the diaphragm pulls air into the lungs. (2) A dome-shaped rubber cup fitted into the upper portion of the vagina before sexual intercourse. It serves as a physical barrier to the passage of sperm into the uterus.

diastole (dī-as′-tō-lē) The stage of the cardiac cycle in which a heart chamber is relaxed and fills with blood.

diatom A photosynthetic protist in the stramenopile clade; diatoms have a unique glass-like wall made of silicon dioxide embedded in an organic matrix.

differential gene expression The expression of different sets of genes by cells with the same genome.

differentiation The process by which a cell or group of cells becomes specialized in structure and function.

diffusion The random thermal motion of particles of liquids, gases, or solids. In the presence of a concentration or electrochemical gradient, diffusion results in the net movement of a substance from a region where it

is more concentrated to a region where it is less concentrated.

digestion The second stage of food processing in animals: the breaking down of food into molecules small enough for the body to absorb.

dihybrid (dī′-hī′-brid) An organism that is heterozygous with respect to two genes of interest. All the offspring from a cross between parents doubly homozygous for different alleles are dihybrids. For example, parents of genotypes *AABB* and *aabb* produce a dihybrid of genotype *AaBb*.

dihybrid cross A cross between two organisms that are each heterozygous for both of the characters being followed (or the self-pollination of a plant that is heterozygous for both characters).

dinoflagellate (dī-nō-flaj′-uh-let) Member of a group of mostly unicellular photosynthetic algae with two flagella situated in perpendicular grooves in cellulose plates covering the cell.

dioecious (dī-ē′-shus) In plant biology, having the male and female reproductive parts on different individuals of the same species.

diploid cell (dip′-loyd) A cell containing two sets of chromosomes ($2n$), one set inherited from each parent.

diplomonad A protist that has modified mitochondria and multiple flagella.

directional selection Natural selection in which individuals at one end of the phenotypic range survive or reproduce more successfully than do other individuals.

disaccharide (dī-sak′-uh-rīd) A double sugar, consisting of two monosaccharides joined by a glycosidic linkage formed by a dehydration reaction.

dispersal The movement of individuals or gametes away from their parent location. This movement sometimes expands the geographic range of a population or species.

dispersion The pattern of spacing among individuals within the boundaries of a population.

disruptive selection Natural selection in which individuals on both extremes of a phenotypic range survive or reproduce more successfully than do individuals with intermediate phenotypes.

distal tubule In the vertebrate kidney, the portion of a nephron that helps refine filtrate and empties it into a collecting duct.

disturbance A natural or human-caused event that changes a biological community and usually removes organisms from it. Disturbances, such as fires and storms, play a pivotal role in structuring many communities.

disulfide bridge A strong covalent bond formed when the sulfur of one cysteine monomer bonds to the sulfur of another cysteine monomer.

DNA (deoxyribonucleic acid) (dē-ok′-sē-rī′-bō-nū-klā′-ik) A nucleic acid molecule, usually a double-stranded helix, in which each polynucleotide strand consists of nucleotide monomers with a deoxyribose sugar and the nitrogenous bases adenine (A), cytosine (C), guanine (G), and thymine (T); capable of being replicated and determining the inherited structure of a cell's proteins.

DNA ligase (lī′-gās) A linking enzyme essential for DNA replication; catalyzes the covalent bonding of the 3′ end of one DNA fragment (such as an Okazaki fragment) to the 5′ end of another DNA fragment (such as a growing DNA chain).

DNA methylation The presence of methyl groups on the DNA bases (usually cytosine) of plants, animals, and fungi. (The term also refers to the process of adding methyl groups to DNA bases.)

DNA microarray assay A method to detect and measure the expression of thousands of genes at one time. Tiny amounts of a large number of single-stranded DNA fragments representing different genes are fixed to a glass slide and tested for hybridization with samples of labeled cDNA.

DNA polymerase (puh-lim′-er-ās) An enzyme that catalyzes the elongation of new DNA (for example, at a replication fork) by the addition of nucleotides to the 3′ end of an existing chain. There are several different DNA polymerases; DNA polymerase III and DNA polymerase I play major roles in DNA replication in *E. coli.*

DNA replication The process by which a DNA molecule is copied; also called DNA synthesis.

DNA sequencing Determining the order of nucleotide bases in a gene or DNA fragment.

domain (1) A taxonomic category above the kingdom level. The three domains are Archaea, Bacteria, and Eukarya. (2) A discrete structural and functional region of a protein.

dominant allele An allele that is fully expressed in the phenotype of a heterozygote.

dominant species A species with substantially higher abundance or biomass than other species in a community. Dominant species exert a powerful control over the occurrence of other species.

dopamine A neurotransmitter that is a catecholamine, like epinephrine and norepinephrine.

dormancy A condition typified by extremely low metabolic rate and a suspension of growth and development.

dorsal Pertaining to the top of an animal with radial or bilateral symmetry.

double bond A double covalent bond; the sharing of two pairs of valence electrons by two atoms.

double circulation A circulatory system consisting of separate pulmonary and systemic circuits, in which blood passes through the heart after completing each circuit.

double fertilization A mechanism of fertilization in angiosperms in which two sperm cells unite with two cells in the female gametophyte (embryo sac) to form the zygote and endosperm.

double helix The form of native DNA, referring to its two adjacent antiparallel polynucleotide strands wound around an imaginary axis into a spiral shape.

Down syndrome A human genetic disease usually caused by the presence of an extra chromosome 21; characterized by developmental delays and heart and other defects that are generally treatable or non-life-threatening.

Duchenne muscular dystrophy (duh-shen′) A human genetic disease caused by a sex-linked recessive allele; characterized by progressive weakening and a loss of muscle tissue.

duodenum (dū′-uh-dēn′-um) The first section of the small intestine, where chyme from the stomach mixes with digestive juices from the pancreas, liver, and gallbladder as well as from gland cells of the intestinal wall.

duplication An aberration in chromosome structure due to fusion with a fragment from a homologous chromosome, such that a portion of a chromosome is duplicated.

dynein (dī′-nē-un) In cilia and flagella, a large motor protein extending from one microtubule doublet to the adjacent doublet. ATP hydrolysis drives changes in dynein shape that lead to bending of cilia and flagella.

E site One of a ribosome's three binding sites for tRNA during translation. The E site is the place where discharged tRNAs leave the ribosome. (E stands for exit.)

ecological footprint The aggregate land and water area required by a person, city, or nation to produce all of the resources it consumes and to absorb all of the wastes it generates.

ecological niche (nich) The sum of a species' use of the biotic and abiotic resources in its environment.

ecological species concept Definition of a species in terms of ecological niche, the sum of how members of the species interact with the nonliving and living parts of their environment.

ecological succession Transition in the species composition of a community following a disturbance; establishment of a community in an area virtually barren of life.

ecology The study of how organisms interact with each other and their environment.

ecosystem All the organisms in a given area as well as the abiotic factors with which they interact; one or more communities and the physical environment around them.

ecosystem ecology The study of energy flow and the cycling of chemicals among the various biotic and abiotic components in an ecosystem.

ecosystem engineer An organism that influences community structure by causing physical changes in the environment.

ecosystem service A function performed by an ecosystem that directly or indirectly benefits humans.

ecotone The transition from one type of habitat or ecosystem to another, such as the transition from a forest to a grassland.

ectoderm (ek′-tō-durm) The outermost of the three primary germ layers in animal embryos; gives rise to the outer covering and, in some phyla, the nervous system, inner ear, and lens of the eye.

ectomycorrhizae (ek′-tō-mī′-kō-rī′-zē) Associations of a fungus with a plant root system in which the fungus surrounds the roots but does not cause invagination of the host (plant) cell's plasma membrane.

ectomycorrhizal fungus A symbiotic fungus that forms sheaths of hyphae over the surface of plant roots and also grows into extracellular spaces of the root cortex.

ectoparasite A parasite that feeds on the external surface of a host.

ectothermic Referring to organisms for which external sources provide most of the heat for temperature regulation.

Ediacaran biota (ē′-dē-uh-keh′-run bī-ō′-tuh) An early group of macroscopic, soft-bodied, multicellular eukaryotes known from fossils that range in age from 635 million to 535 million years old.

effective population size An estimate of the size of a population based on the numbers of females and males that successfully breed; generally smaller than the total population.

effector cell (1) A muscle cell or gland cell that performs the body's response to stimuli as directed by signals from the brain or other processing center of the nervous system. (2) A lymphocyte that has undergone clonal selection and is capable of mediating an adaptive immune response.

egg The female gamete.

egg-polarity gene A gene that helps control the orientation (polarity) of the egg; also called a maternal effect gene.

ejaculation The propulsion of sperm from the epididymis through the muscular vas deferens, ejaculatory duct, and urethra.

ejaculatory duct In mammals, the short section of the ejaculatory route formed by the convergence of the vas deferens and a duct from the seminal vesicle. The ejaculatory duct transports sperm from the vas deferens to the urethra.

electrocardiogram (ECG or EKG) A record of the electrical impulses that travel through heart muscle during the cardiac cycle.

electrochemical gradient The diffusion gradient of an ion, which is affected by both the concentration difference of an ion across a membrane (a chemical force) and the ion's tendency to move relative to the membrane potential (an electrical force).

electrogenic pump An active transport protein that generates voltage across a membrane while pumping ions.

electromagnetic receptor A receptor of electromagnetic energy, such as visible light, electricity, or magnetism.

electromagnetic spectrum The entire spectrum of electromagnetic radiation, ranging in wavelength from less than a nanometer to more than a kilometer.

electron A subatomic particle with a single negative electrical charge and a mass about 1/2,000 that of a neutron or proton. One or more electrons move around the nucleus of an atom.

electron microscope (EM) A microscope that uses magnets to focus an electron beam on or through a specimen, resulting in a practical resolution a hundredfold greater than that of a light microscope using standard techniques. A transmission electron microscope (TEM) is used to study the internal structure of thin sections of cells. A scanning electron microscope (SEM) is used to study the fine details of cell surfaces.

electron shell An energy level of electrons at a characteristic average distance from the nucleus of an atom.

electron transport chain A sequence of electron carrier molecules (membrane proteins) that shuttle electrons down a series of redox reactions that release energy used to make ATP.

electronegativity The attraction of a given atom for the electrons of a covalent bond.

element Any substance that cannot be broken down to any other substance by chemical reactions.

elimination The fourth and final stage of food processing in animals: the passing of undigested material out of the body.

embryo sac (em'-brē-ō) The female gametophyte of angiosperms, formed from the growth and division of the megaspore into a multicellular structure that typically has eight haploid nuclei.

embryonic lethal A mutation with a phenotype leading to death of an embryo or larva.

embryophyte Alternate name for land plants that refers to their shared derived trait of multicellular, dependent embryos.

emergent properties New properties that arise with each step upward in the hierarchy of life, owing to the arrangement and interactions of parts as complexity increases.

emigration The movement of individuals out of a population.

endangered species A species that is in danger of extinction throughout all or a significant portion of its range.

endemic (en-dem'-ik) Referring to a species that is confined to a specific geographic area.

endergonic reaction (en'-der-gon'-ik) A nonspontaneous chemical reaction, in which free energy is absorbed from the surroundings.

endocrine gland (en'-dō-krin) A gland that secretes hormones directly into the interstitial fluid, from which they diffuse into the bloodstream.

endocrine system The internal system of communication involving hormones, the ductless glands that secrete hormones, and the molecular receptors on or in target cells that respond to hormones; functions in concert with the nervous system to effect internal regulation and maintain homeostasis.

endocytosis (en'-dō-sī-tō'-sis) Cellular uptake of biological molecules and particulate matter via formation of vesicles from the plasma membrane.

endoderm (en'-dō-durm) The innermost of the three primary germ layers in animal embryos; lines the archenteron and gives rise to the liver, pancreas, lungs, and the lining of the digestive tract in species that have these structures.

endodermis In plant roots, the innermost layer of the cortex that surrounds the vascular cylinder.

endomembrane system The collection of membranes inside and surrounding a eukaryotic cell, related either through direct physical contact or by the transfer of membranous vesicles; includes the plasma membrane, the nuclear envelope, the smooth and rough endoplasmic reticulum, the Golgi apparatus, lysosomes, vesicles, and vacuoles.

endometrium (en'-dō-mē'-trē-um) The inner lining of the uterus, which is richly supplied with blood vessels.

endoparasite A parasite that lives within a host.

endophyte A fungus that lives inside a leaf or other plant part without causing harm to the plant.

endoplasmic reticulum (ER) (en'-dō-plaz'-mik ruh-tik'-yū-lum) An extensive membranous network in eukaryotic cells, continuous with the outer nuclear membrane and composed of ribosome-studded (rough) and ribosome-free (smooth) regions.

endorphin (en-dōr'-fin) Any of several hormones produced in the brain and anterior pituitary that inhibit pain perception.

endoskeleton A hard skeleton buried within the soft tissues of an animal.

endosperm In angiosperms, a nutrient-rich tissue formed by the union of a sperm with two polar nuclei during double fertilization. The endosperm provides nourishment to the developing embryo in angiosperm seeds.

endospore A thick-coated, resistant cell produced by some bacterial cells when they are exposed to harsh conditions.

endosymbiont theory The theory that mitochondria and plastids, including chloroplasts, originated as prokaryotic cells engulfed by host cells. The engulfed cell and its host cell then evolved into a single organism. *See also* endosymbiosis.

endosymbiosis A mutually beneficial relationship between two species in which one organism lives inside the cell or cells of another organism.

endothelium (en'-dō-thē'-lē-um) The simple squamous layer of cells lining the lumen of blood vessels.

endothermic Referring to organisms that are warmed by heat generated by their own metabolism. This heat usually maintains a relatively stable body temperature higher than that of the external environment.

endotoxin A toxic component of the outer membrane of certain gram-negative bacteria that is released only when the bacteria die.

energy The capacity to cause change, especially to do work (to move matter against an opposing force).

energy coupling In cellular metabolism, the use of energy released from an exergonic reaction to drive an endergonic reaction.

enhancer A segment of eukaryotic DNA containing multiple control elements, usually located far from the gene whose transcription it regulates.

enteric division One of three divisions of the autonomic nervous system; consists of networks of neurons in the digestive tract, pancreas, and gallbladder; normally regulated by the sympathetic and parasympathetic divisions of the autonomic nervous system.

entropy A measure of disorder, or randomness.

enzyme (en'-zīm) A macromolecule serving as a catalyst, a chemical agent that increases the rate of a reaction without being consumed by the reaction. Most enzymes are proteins.

enzyme-substrate complex A temporary complex formed when an enzyme binds to its substrate molecule(s).

epicotyl (ep'-uh-kot'-ul) In an angiosperm embryo, the embryonic axis above the point of attachment of the cotyledon(s) and below the first pair of miniature leaves.

epidemic A general outbreak of a disease.

epidermis (1) The dermal tissue system of nonwoody plants, usually consisting of a single layer of tightly packed cells. (2) The outermost layer of cells in an animal.

epididymis (ep'-uh-did'-uh-mus) A coiled tubule located adjacent to the mammalian testis where sperm are stored.

epigenetic inheritance Inheritance of traits transmitted by mechanisms not directly involving the nucleotide sequence of a genome.

epinephrine (ep′-i-nef′-rin) A catecholamine that, when secreted as a hormone by the adrenal medulla, mediates "fight-or-flight" responses to short-term stresses; also released by some neurons as a neurotransmitter; also known as adrenaline.

epiphyte (ep′-uh-fīt) A plant that nourishes itself but grows on the surface of another plant for support, usually on the branches or trunks of trees.

epistasis (ep′-i-stā′-sis) A type of gene interaction in which the phenotypic expression of one gene alters that of another independently inherited gene.

epithelial tissue (ep′-uh-thē′-lē-ul) Sheets of tightly packed cells that line organs and body cavities as well as external surfaces; also called epithelium.

epithelium An epithelial tissue.

epitope A small, accessible region of an antigen to which an antigen receptor or antibody binds; also called an antigenic determinant.

equilibrium potential (E_{ion}) The magnitude of a cell's membrane voltage at equilibrium, calculated using the Nernst equation.

erythrocyte (eh-rith′-ruh-sīt) A blood cell that contains hemoglobin, which transports oxygen; also called a red blood cell.

esophagus (eh-sof′-uh-gus) A muscular tube that conducts food, by peristalsis, from the pharynx to the stomach.

essential amino acid An amino acid that an animal cannot synthesize itself and must be obtained from food in prefabricated form.

essential element A chemical element required for an organism to survive, grow, and reproduce.

essential fatty acid An unsaturated fatty acid that an animal needs but cannot make.

essential nutrient A substance that an organism cannot synthesize from any other material and therefore must absorb in preassembled form.

estradiol (es′-truh-dī′-ol) A steroid hormone that stimulates the development and maintenance of the female reproductive system and secondary sex characteristics; the major estrogen in mammals.

estrous cycle (es′-trus) A reproductive cycle characteristic of female mammals except humans and certain other primates, in which the nonpregnant endometrium is reabsorbed rather than shed, and sexual response occurs only during mid-cycle at estrus.

estuary The area where a freshwater stream or river merges with the ocean.

ethylene (eth′-uh-lēn) A gaseous plant hormone involved in responses to mechanical stress, programmed cell death, leaf abscission, and fruit ripening.

etiolation Plant morphological adaptations for growing in darkness.

euchromatin (yū-krō′-muh-tin) The less condensed form of eukaryotic chromatin that is available for transcription.

euglenozoan Member of a diverse clade of flagellated protists that includes predatory heterotrophs, photosynthetic autotrophs, and pathogenic parasites.

Eukarya (yū-kar′-ē-uh) The domain that includes all eukaryotic organisms.

eukaryotic cell (yū′-ker-ē-ot′-ik) A type of cell with a membrane-enclosed nucleus and membrane-enclosed organelles. Organisms with eukaryotic cells (protists, plants, fungi, and animals) are called eukaryotes.

eumetazoan (yū′-met-uh-zō′-un) Member of a clade of animals with true tissues. All animals except sponges and a few other groups are eumetazoans.

Eustachian tube (yū-stā′-shun) The tube that connects the middle ear to the pharynx.

eutherian (yū-thēr′-ē-un) Placental mammal; mammal whose young complete their embryonic development within the uterus, joined to the mother by the placenta.

eutrophic lake (yū-trōf′-ik) A lake that has a high rate of biological productivity supported by a high rate of nutrient cycling.

eutrophication A process by which nutrients, particularly phosphorus and nitrogen, become highly concentrated in a body of water, leading to increased growth of organisms such as algae or cyanobacteria.

evaporative cooling The process in which the surface of an object becomes cooler during evaporation, a result of the molecules with the greatest kinetic energy changing from the liquid to the gaseous state.

evapotranspiration The total evaporation of water from an ecosystem, including water transpired by plants and evaporated from a landscape, usually measured in millimeters and estimated for a year.

evo-devo Evolutionary developmental biology; a field of biology that compares developmental processes of different multicellular organisms to understand how these processes have evolved and how changes can modify existing organismal features or lead to new ones.

evolution Descent with modification; the idea that living species are descendants of ancestral species that were different from the present-day ones; also defined more narrowly as the change in the genetic composition of a population from generation to generation.

Excavata One of four supergroups of eukaryotes proposed in a current hypothesis of the evolutionary history of eukaryotes. Excavates have unique cytoskeletal features, and some species have an "excavated" feeding groove on one side of the cell body. *See also* "SAR" clade, Archaeplastida, and Unikonta.

excitatory postsynaptic potential (EPSP) An electrical change (depolarization) in the membrane of a postsynaptic cell caused by the binding of an excitatory neurotransmitter from a presynaptic cell to a postsynaptic receptor; makes it more likely for a postsynaptic cell to generate an action potential.

excretion The disposal of nitrogen-containing metabolites and other waste products.

exergonic reaction (ek′-ser-gon′-ik) A spontaneous chemical reaction, in which there is a net release of free energy.

exocrine gland (ek′-sō-krin) A gland that secretes substances through a duct onto a body surface or into a body cavity.

exocytosis (ek′-sō-sī-tō′-sis) The cellular secretion of biological molecules by the fusion of vesicles containing them with the plasma membrane.

exon A sequence within a primary transcript that remains in the RNA after RNA processing; also refers to the region of DNA from which this sequence was transcribed.

exoskeleton A hard encasement on the surface of an animal, such as the shell of a mollusc or the cuticle of an arthropod, that provides protection and points of attachment for muscles.

exotoxin (ek′-sō-tok′-sin) A toxic protein that is secreted by a prokaryote or other pathogen and that produces specific symptoms, even if the pathogen is no longer present.

expansin Plant enzyme that breaks the cross-links (hydrogen bonds) between cellulose microfibrils and other cell wall constituents, loosening the wall's fabric.

experimental group A set of subjects that has (or receives) the specific factor being tested in a controlled experiment.

exponential population growth Growth of a population in an ideal, unlimited environment, represented by a J-shaped curve when population size is plotted over time.

external fertilization The fusion of gametes that parents have discharged into the environment.

extinction vortex A downward population spiral in which inbreeding and genetic drift combine to cause a small population to shrink and, unless the spiral is reversed, become extinct.

extracellular matrix (ECM) The meshwork surrounding animal cells, consisting of glycoproteins, polysaccharides, and proteoglycans synthesized and secreted by the cells.

extreme halophile An organism that lives in a highly saline environment, such as the Great Salt Lake or the Dead Sea.

extreme thermophile An organism that thrives in hot environments (often 60–80°C or hotter).

extremophile An organism that lives in environmental conditions so extreme that few other species can survive there. Extremophiles include extreme halophiles ("salt lovers") and extreme thermophiles ("heat lovers").

F factor In bacteria, the DNA segment that confers the ability to form pili for conjugation and associated functions required for the transfer of DNA from donor to recipient. The F factor may exist as a plasmid or be integrated into the bacterial chromosome.

F plasmid The plasmid form of the F factor.

F$_1$ generation The first filial, hybrid (heterozygous) offspring arising from a parental (P generation) cross.

F$_2$ generation The offspring resulting from interbreeding (or self-pollination) of the hybrid F$_1$ generation.

facilitated diffusion The passage of molecules or ions down their electrochemical gradient across a biological membrane with the assistance of specific transmembrane transport proteins, requiring no energy expenditure.

facilitation An interaction in which one species has a positive effect on the survival and reproduction of another species without the intimate association of a symbiosis.

facultative anaerobe (fak′-ul-tā′-tiv an′-uh-rōb) An organism that makes ATP by aerobic respiration if oxygen is present but that switches to anaerobic respiration or fermentation if oxygen is not present.

family In Linnaean classification, the taxonomic category above genus.

fast-twitch fiber A muscle fiber used for rapid, powerful contractions.

fat A lipid consisting of three fatty acids linked to one glycerol molecule; also called a triacylglycerol or triglyceride.

fatty acid A carboxylic acid with a long carbon chain. Fatty acids vary in length and in the number and location of double bonds; three fatty acids linked to a glycerol molecule form a fat molecule, also known as a triacylglycerol or triglyceride.

feces (fē′-sēz) The wastes of the digestive tract.

feedback inhibition A method of metabolic control in which the end product of a metabolic pathway acts as an inhibitor of an enzyme within that pathway.

fermentation A catabolic process that makes a limited amount of ATP from glucose (or other organic molecules) without an electron transport chain and that produces a characteristic end product, such as ethyl alcohol or lactic acid.

fertilization (1) The union of haploid gametes to produce a diploid zygote. (2) The addition of mineral nutrients to the soil.

fetus (fē′-tus) A developing mammal that has all the major structures of an adult. In humans, the fetal stage lasts from the 9th week of gestation until birth.

fiber A lignified cell type that reinforces the xylem of angiosperms and functions in mechanical support; a slender, tapered sclerenchyma cell that usually occurs in bundles.

fibronectin An extracellular glycoprotein secreted by animal cells that helps them attach to the extracellular matrix.

filtrate Cell-free fluid extracted from the body fluid by the excretory system.

filtration In excretory systems, the extraction of water and small solutes, including metabolic wastes, from the body fluid.

fimbria (plural, **fimbriae**) A short, hairlike appendage of a prokaryotic cell that helps it adhere to the substrate or to other cells.

first law of thermodynamics The principle of conservation of energy: Energy can be transferred and transformed, but it cannot be created or destroyed.

fixed action pattern In animal behavior, a sequence of unlearned acts that is essentially unchangeable and, once initiated, usually carried to completion.

flaccid (flas′-id) Limp. Lacking turgor (stiffness or firmness), as in a plant cell in surroundings where there is a tendency for water to leave the cell. (A walled cell becomes flaccid if it has a higher water potential than its surroundings, resulting in the loss of water.)

flagellum (fluh-jel′-um) (plural, **flagella**) A long cellular appendage specialized for locomotion. Like motile cilia, eukaryotic flagella have a core with nine outer doublet microtubules and two inner single microtubules (the "9 + 2" arrangement) ensheathed in an extension of the plasma membrane. Prokaryotic flagella have a different structure.

florigen A flowering signal, probably a protein, that is made in leaves under certain conditions and that travels to the shoot apical meristems, inducing them to switch from vegetative to reproductive growth.

flower In an angiosperm, a specialized shoot with up to four sets of modified leaves, bearing structures that function in sexual reproduction.

fluid feeder An animal that lives by sucking nutrient-rich fluids from another living organism.

fluid mosaic model The currently accepted model of cell membrane structure, which envisions the membrane as a mosaic of protein molecules drifting laterally in a fluid bilayer of phospholipids.

follicle (fol′-uh-kul) A microscopic structure in the ovary that contains the developing oocyte and secretes estrogens.

follicle-stimulating hormone (FSH) A tropic hormone that is produced and secreted by the anterior pituitary and that stimulates the production of eggs by the ovaries and sperm by the testes.

follicular phase That part of the ovarian cycle during which follicles are growing and oocytes maturing.

food chain The pathway along which food energy is transferred from trophic level to trophic level, beginning with producers.

food vacuole A membranous sac formed by phagocytosis of microorganisms or particles to be used as food by the cell.

food web The interconnected feeding relationships in an ecosystem.

foraging The seeking and obtaining of food.

foram (foraminiferan) An aquatic protist that secretes a hardened shell containing calcium carbonate and extends pseudopodia through pores in the shell.

forebrain One of three ancestral and embryonic regions of the vertebrate brain; develops into the thalamus, hypothalamus, and cerebrum.

fossil A preserved remnant or impression of an organism that lived in the past.

founder effect Genetic drift that occurs when a few individuals become isolated from a larger population and form a new population whose gene pool composition is not reflective of that of the original population.

fovea (fō′-vē-uh) The place on the retina at the eye's center of focus, where cones are highly concentrated.

fragmentation A means of asexual reproduction whereby a single parent breaks into parts that regenerate into whole new individuals.

frameshift mutation A mutation occurring when nucleotides are inserted in or deleted from a gene and the number inserted or deleted is not a multiple of three, resulting in the improper grouping of the subsequent nucleotides into codons.

free energy The portion of a biological system's energy that can perform work when temperature and pressure are uniform throughout the system. The change in free energy of a system (ΔG) is $G_{\text{final state}} - G_{\text{initial state}}$. It can be calculated by the equation $\Delta G = \Delta H - T\Delta S$, where ΔH is the change in enthalpy (in biological systems, equivalent to total energy), T is the absolute temperature, and ΔS is the change in entropy.

frequency-dependent selection Selection in which the fitness of a phenotype depends on how common the phenotype is in a population.

fruit A mature ovary of a flower. The fruit protects dormant seeds and often aids in their dispersal.

functional group A specific configuration of atoms commonly attached to the carbon skeletons of organic molecules and involved in chemical reactions.

G protein A GTP-binding protein that relays signals from a plasma membrane signal receptor, known as a G protein-coupled receptor, to other signal transduction proteins inside the cell.

G protein-coupled receptor (GPCR) A signal receptor protein in the plasma membrane that responds to the binding of a signaling molecule by activating a G protein. Also called a G protein-linked receptor.

G_0 phase A nondividing state occupied by cells that have left the cell cycle, sometimes reversibly.

G_1 phase The first gap, or growth phase, of the cell cycle, consisting of the portion of interphase before DNA synthesis begins.

G_2 phase The second gap, or growth phase, of the cell cycle, consisting of the portion of interphase after DNA synthesis occurs.

gallbladder An organ that stores bile and releases it as needed into the small intestine.

gamete (gam′-ēt) A haploid reproductive cell, such as an egg or sperm. Gametes unite during sexual reproduction to produce a diploid zygote.

gametogenesis The process by which gametes are produced.

gametophyte (guh-mē′-tō-fīt) In organisms (plants and some algae) that have alternation of generations, the multicellular haploid form that produces haploid gametes by mitosis. The haploid gametes unite and develop into sporophytes.

gamma-aminobutyric acid (GABA) An amino acid that functions as a neurotransmitter in the central nervous system of vertebrates.

ganglion (gang′-glē-uhn) (plural, **ganglia**) A cluster (functional group) of nerve cell bodies in a centralized nervous system.

gap junction A type of intercellular junction in animal cells, consisting of proteins surrounding a pore that allows the passage of materials between cells.

gas exchange The uptake of molecular oxygen from the environment and the discharge of carbon dioxide to the environment.

gas exchange circuit The branch of the circulatory system that supplies the organs where gases are exchanged with the environment; in many amphibians, it supplies the lungs and skin and is called a *pulmocutaneous circuit*, whereas in birds and mammals, it supplies only the lungs and is called a *pulmonary circuit.*

gastric juice A digestive fluid secreted by the stomach.

gastrovascular cavity A central cavity with a single opening in the body of certain animals, including cnidarians and flatworms, that functions in both the digestion and distribution of nutrients.

gastrula (gas′-trū-luh) An embryonic stage in animal development encompassing the formation of three layers: ectoderm, mesoderm, and endoderm.

gastrulation (gas′-trū-lā′-shun) In animal development, a series of cell and tissue movements in which the blastula-stage embryo folds inward, producing a three-layered embryo, the gastrula.

gated channel A transmembrane protein channel that opens or closes in response to a particular stimulus.

gated ion channel A gated channel for a specific ion. The opening or closing of such channels may alter a cell's membrane potential.

gel electrophoresis (ē-lek′-trō-fōr-ē′-sis) A technique for separating nucleic acids or proteins on the basis of their size and electrical charge, both of which affect their rate of movement through an electric field in a gel made of agarose or another polymer.

gene A discrete unit of hereditary information consisting of a specific nucleotide sequence in DNA (or RNA, in some viruses).

gene cloning The production of multiple copies of a gene.

gene expression The process by which information encoded in DNA directs the synthesis of proteins or, in some cases, RNAs that are not translated into proteins and instead function as RNAs.

gene flow The transfer of alleles from one population to another, resulting from the movement of fertile individuals or their gametes.

gene pool The aggregate of all copies of every type of allele at all loci in every individual in a population. The term is also used in a more restricted sense as the aggregate of alleles for just one or a few loci in a population.

gene-for-gene recognition A widespread form of plant disease resistance involving recognition of pathogen-derived molecules by the protein products of specific plant disease resistance genes.

genetic drift A process in which chance events cause unpredictable fluctuations in allele frequencies from one generation to the next. Effects of genetic drift are most pronounced in small populations.

genetic engineering The direct manipulation of genes for practical purposes.

genetic map An ordered list of genetic loci (genes or other genetic markers) along a chromosome.

genetic profile An individual's unique set of genetic markers, detected most often today by PCR.

genetic recombination General term for the production of offspring with combinations of traits that differ from those found in either parent.

genetic variation Differences among individuals in the composition of their genes or other DNA segments.

genetics The scientific study of heredity and hereditary variation.

genome (jē′-nōm) The genetic material of an organism or virus; the complete complement of an organism's or virus's genes along with its noncoding nucleic acid sequences.

genomics (juh-nō′-miks) The study of whole sets of genes and their interactions within a species, as well as genome comparisons between species.

genotype (jē′-nō-tīp) The genetic makeup, or set of alleles, of an organism.

genus (jē′-nus) (plural, **genera**) A taxonomic category above the species level, designated by the first word of a species' two-part scientific name.

geologic record A standard time scale dividing Earth's history into time periods grouped into four eons—Hadean, Archaean, Proterozoic, and Phanerozoic—and further subdivided into eras, periods, and epochs.

gestation (jes-tā′-shun) Pregnancy; the state of carrying developing young within the female reproductive tract.

gibberellin (jib′-uh-rel′-in) Any of a class of related plant hormones that stimulate growth in the stem and leaves, trigger the germination of seeds and breaking of bud dormancy, and (with auxin) stimulate fruit development.

glans The rounded structure at the tip of the clitoris or penis that is involved in sexual arousal.

glia (glial cells) Cells of the nervous system that support, regulate, and augment the functions of neurons.

global ecology The study of the functioning and distribution of organisms across the biosphere and how the regional exchange of energy and materials affects them.

glomerulus (glō-mār′-yū-lus) A ball of capillaries surrounded by Bowman's capsule in the nephron and serving as the site of filtration in the vertebrate kidney.

glutamate An amino acid that functions as a neurotransmitter in the central nervous system.

glyceraldehyde 3-phosphate (G3P) (glis′-er-al′-de-hīd) A three-carbon carbohydrate that is the direct product of the Calvin cycle; it is also an intermediate in glycolysis.

glycogen (glī′-kō-jen) An extensively branched glucose storage polysaccharide found in the liver and muscle of animals; the animal equivalent of starch.

glycolipid A lipid with one or more covalently attached carbohydrates.

glycolysis (glī-kol′-uh-sis) A series of reactions that ultimately splits glucose into pyruvate. Glycolysis occurs in almost all living cells, serving as the starting point for fermentation or cellular respiration.

glycoprotein A protein with one or more covalently attached carbohydrates.

glycosidic linkage A covalent bond formed between two monosaccharides by a dehydration reaction.

gnathostome (na′-thu-stōm) Member of the vertebrate subgroup possessing jaws.

Golgi apparatus (gol′-jē) An organelle in eukaryotic cells consisting of stacks of flat membranous sacs that modify, store, and route products of the endoplasmic reticulum and synthesize some products, notably noncellulose carbohydrates.

gonad (gō′-nad) A male or female gamete-producing organ.

graded potential In a neuron, a shift in the membrane potential that has an amplitude proportional to signal strength and that decays as it spreads.

gram-negative Describing the group of bacteria that have a cell wall that is structurally more complex and contains less peptidoglycan than the cell wall of gram-positive bacteria. Gram-negative bacteria are often more toxic than gram-positive bacteria.

gram-positive Describing the group of bacteria that have a cell wall that is structurally less complex and contains more peptidoglycan than the cell wall of gram-negative bacteria. Gram-positive bacteria are usually less toxic than gram-negative bacteria.

granum (gran′-um) (plural, **grana**) A stack of membrane-bounded thylakoids in the chloroplast. Grana function in the light reactions of photosynthesis.

gravitropism (grav′-uh-trō′-pizm) A response of a plant or animal to gravity.

gray matter Regions of dendrites and clustered neuron cell bodies within the CNS.

green alga A photosynthetic protist, named for green chloroplasts that are similar in structure and pigment composition to the chloroplasts of land plants. Green algae are a paraphyletic group; some members are more closely related to land plants than they are to other green algae.

greenhouse effect The warming of Earth due to the atmospheric accumulation of carbon dioxide and certain other gases, which absorb reflected infrared radiation and reradiate some of it back toward Earth.

gross primary production (GPP) The total primary production of an ecosystem.

ground tissue system Plant tissues that are neither vascular nor dermal, fulfilling a variety of functions, such as storage, photosynthesis, and support.

growth factor (1) A protein that must be present in the extracellular environment (culture medium or animal body) for the growth and normal development of certain types of cells. (2) A local regulator that acts on nearby cells to stimulate cell proliferation and differentiation.

guard cells The two cells that flank the stomatal pore and regulate the opening and closing of the pore.

gustation The sense of taste.

gymnosperm (jim′-nō-sperm) A vascular plant that bears naked seeds—seeds not enclosed in protective chambers.

hair cell A mechanosensory cell that alters output to the nervous system when hairlike projections on the cell surface are displaced.

half-life The amount of time it takes for 50% of a sample of a radioactive isotope to decay.

Hamilton's rule The principle that for natural selection to favor an altruistic act, the benefit to the recipient, devalued by the coefficient of relatedness, must exceed the cost to the altruist.

haploid cell (hap′-loyd) A cell containing only one set of chromosomes (n).

Hardy-Weinberg principle The principle that frequencies of alleles and genotypes in a population remain constant from generation to generation, provided that only Mendelian segregation and recombination of alleles are at work.

haustorium (plural, **haustoria**) (ho-stōr′-ē-um, ho-stōr′-ē-uh) In certain symbiotic fungi, a specialized hypha that can penetrate the tissues of host organisms.

heart A muscular pump that uses metabolic energy to elevate the hydrostatic pressure of the circulatory fluid (blood or hemolymph). The fluid then flows down a pressure gradient through the body and eventually returns to the heart.

heart attack The damage or death of cardiac muscle tissue resulting from prolonged blockage of one or more coronary arteries.

heart murmur A hissing sound that most often results from blood squirting backward through a leaky valve in the heart.

heat Thermal energy in transfer from one body of matter to another.

heat of vaporization The quantity of heat a liquid must absorb for 1 g of it to be converted from the liquid to the gaseous state.

heat-shock protein A protein that helps protect other proteins during heat stress. Heat-shock proteins are found in plants, animals, and microorganisms.

heavy chain One of the two types of polypeptide chains that make up an antibody molecule and B cell receptor; consists of a variable region, which contributes to the antigen-binding site, and a constant region.

helicase An enzyme that untwists the double helix of DNA at replication forks, separating the two strands and making them available as template strands.

helper T cell A type of T cell that, when activated, secretes cytokines that promote the response of B cells (humoral response) and cytotoxic T cells (cell-mediated response) to antigens.

hemoglobin (hē′-mō-glō′-bin) An iron-containing protein in red blood cells that reversibly binds oxygen.

hemolymph (hē′-mō-limf′) In invertebrates with an open circulatory system, the body fluid that bathes tissues.

hemophilia (hē′-muh-fil′-ē-uh) A human genetic disease caused by a sex-linked recessive allele, resulting in the absence of one or more blood-clotting proteins; characterized by excessive bleeding following injury.

hepatic portal vein A large vessel that conveys nutrient-laden blood from the small intestine to the liver, which regulates the blood's nutrient content.

herbivore (hur′-bi-vōr′) An animal that mainly eats plants or algae.

herbivory An interaction in which an organism eats parts of a plant or alga.

heredity The transmission of traits from one generation to the next.

hermaphroditism (hur-maf′-rō-dī-tizm) A condition in which an individual has both female and male gonads and functions as both a male and female in sexual reproduction by producing both sperm and eggs.

heterochromatin (het′-er-ō-krō′-muh-tin) Eukaryotic chromatin that remains highly compacted during interphase and is generally not transcribed.

heterochrony (het′-uh-rok′-ruh-nē) Evolutionary change in the timing or rate of an organism's development.

heterocyst (het′-er-ō-sist) A specialized cell that engages in nitrogen fixation in some filamentous cyanobacteria; also called a heterocyte.

heterotroph (het′-er-ō-trōf) An organism that obtains organic food molecules by eating other organisms or substances derived from them.

heterozygote advantage Greater reproductive success of heterozygous individuals compared with homozygotes; tends to preserve variation in a gene pool.

heterozygous (het′-er-ō-zī′-gus) Having two different alleles for a given gene.

high-density lipoprotein (HDL) A particle in the blood made up of thousands of cholesterol molecules and other lipids bound to a protein. HDL scavenges excess cholesterol.

hindbrain One of three ancestral and embryonic regions of the vertebrate brain; develops into the medulla oblongata, pons, and cerebellum.

histamine (his′-tuh-mēn) A substance released by mast cells that causes blood vessels to dilate and become more permeable in inflammatory and allergic responses.

histogram A variant of a bar graph in which a numerical independent variable is divided into equal intervals (or groups called "bins"). The height (or length) of each bar represents the value of the dependent variable for a particular interval.

histone (his′-tōn) A small protein with a high proportion of positively charged amino acids that binds to the negatively charged DNA and plays a key role in chromatin structure.

histone acetylation The attachment of acetyl groups to certain amino acids of histone proteins.

HIV (human immunodeficiency virus) The infectious agent that causes AIDS. HIV is a retrovirus.

holdfast A rootlike structure that anchors a seaweed.

homeobox (hō′-mē-ō-boks′) A 180-nucleotide sequence within homeotic genes and some other developmental genes that is widely conserved in animals. Related sequences occur in plants and yeasts.

homeostasis (hō′-mē-ō-stā′-sis) The steady-state physiological condition of the body.

homeotic gene (hō-mē-o′-tik) Any of the master regulatory genes that control placement and spatial organization of body parts in animals, plants, and fungi by controlling the developmental fate of groups of cells.

homologous chromosomes (hō-mol′-uh-gus) A pair of chromosomes of the same length, centromere position, and staining pattern that possess genes for the same characters at corresponding loci. One homologous chromosome is inherited from the organism's father, the other from the mother. Also called homologs, or a homologous pair.

homologous structures Structures in different species that are similar because of common ancestry.

homology (hō-mol′-ō-jē) Similarity in characteristics resulting from a shared ancestry.

homoplasy (hō′-muh-play′-zē) A similar (analogous) structure or molecular sequence that has evolved independently in two species.

homozygous (hō′-mō-zī′-gus) Having two identical alleles for a given gene.

horizontal gene transfer The transfer of genes from one genome to another through mechanisms such as transposable elements, plasmid exchange, viral activity, and perhaps fusions of different organisms.

hormone In multicellular organisms, one of many types of secreted chemicals that are formed in specialized cells, travel in body fluids, and act on specific target cells in other parts of the body, changing the target cells' functioning. Hormones are thus important in long-distance signaling.

host The larger participant in a symbiotic relationship, often providing a home and food source for the smaller symbiont.

host range The limited number of species whose cells can be infected by a particular virus.

Human Genome Project An international collaborative effort to map and sequence the DNA of the entire human genome.

human immunodeficiency virus (HIV) The pathogen that causes AIDS (acquired immune deficiency syndrome).

humoral immune response (hyū′-mer-ul) The branch of adaptive immunity that involves the activation of B cells and that leads to the production of antibodies, which defend against bacteria and viruses in body fluids.

humus (hyū′-mus) Decomposing organic material that is a component of topsoil.

Huntington's disease A human genetic disease caused by a dominant allele; characterized by uncontrollable body movements and degeneration of the nervous system; usually fatal 10 to 20 years after the onset of symptoms.

hybrid Offspring that results from the mating of individuals from two different species or from two true-breeding varieties of the same species.

hybrid zone A geographic region in which members of different species meet and mate, producing at least some offspring of mixed ancestry.

hybridization In genetics, the mating, or crossing, of two true-breeding varieties.

hydration shell The sphere of water molecules around a dissolved ion.

hydrocarbon An organic molecule consisting of only carbon and hydrogen.

hydrogen bond A type of weak chemical bond that is formed when the slightly positive hydrogen atom of a polar covalent bond in one molecule is attracted to the slightly negative atom of a polar covalent bond in another molecule or in another region of the same molecule.

hydrogen ion A single proton with a charge of $1+$. The dissociation of a water molecule (H_2O) leads to the generation of a hydroxide ion (OH^-) and a hydrogen ion (H^+); in water, H^+ is not found alone but associates with a water molecule to form a hydronium ion.

hydrolysis (hī-drol′-uh-sis) A chemical reaction that breaks bonds between two molecules by the addition of water; functions in disassembly of polymers to monomers.

hydronium ion A water molecule that has an extra proton bound to it; H_3O^+, commonly represented as H^+.

hydrophilic (hī′-drō-fil′-ik) Having an affinity for water.

hydrophobic (hī′-drō-fō′-bik) Having no affinity for water; tending to coalesce and form droplets in water.

hydrophobic interaction A type of weak chemical interaction caused when molecules that do not mix with water coalesce to exclude water.

hydroponic culture A method in which plants are grown in mineral solutions rather than in soil.

hydrostatic skeleton A skeletal system composed of fluid held under pressure in a closed body compartment; the main skeleton of most cnidarians, flatworms, nematodes, and annelids.

hydroxide ion A water molecule that has lost a proton; OH^-.

hydroxyl group (hī-drok′-sil) A chemical group consisting of an oxygen atom joined to a hydrogen atom. Molecules possessing this group are soluble in water and are called alcohols.

hymen A thin membrane that partly covers the vaginal opening in the human female. The hymen is ruptured by sexual intercourse or other vigorous activity.

hyperpolarization A change in a cell's membrane potential such that the inside of the membrane becomes more negative relative to the outside. Hyperpolarization reduces the chance that a neuron will transmit a nerve impulse.

hypersensitive response A plant's localized defense response to a pathogen, involving the death of cells around the site of infection.

hypertension A disorder in which blood pressure remains abnormally high.

hypertonic Referring to a solution that, when surrounding a cell, will cause the cell to lose water.

hypha (plural, **hyphae**) (hī′-fuh, hī′-fē) One of many connected filaments that collectively make up the mycelium of a fungus.

hypocotyl (hī′-puh-cot′-ul) In an angiosperm embryo, the embryonic axis below the point of attachment of the cotyledon(s) and above the radicle.

hypothalamus (hī′-pō-thal′-uh-mus) The ventral part of the vertebrate forebrain; functions in maintaining homeostasis, especially in coordinating the endocrine and nervous systems; secretes hormones of the posterior pituitary and releasing factors that regulate the anterior pituitary.

hypothesis (hī-poth′-uh-sis) A testable explanation for a set of observations based on the available data and guided by inductive reasoning. A hypothesis is narrower in scope than a theory.

hypotonic Referring to a solution that, when surrounding a cell, will cause the cell to take up water.

imbibition The physical adsorption of water onto the internal surfaces of structures.

immigration The influx of new individuals into a population from other areas.

immune system An animal body's system of defenses against agents that cause disease.

immunization The process of generating a state of immunity by artificial means. In active immunization, also called vaccination, an inactive or weakened form of a pathogen is administered, inducing B and T cell responses and immunological memory. In passive immunization, antibodies specific for a particular microbe are administered, conferring immediate but temporary protection.

immunoglobulin (Ig) (im′-yū-nō-glob′-yū-lin) *See* antibody.

imprinting In animal behavior, the formation at a specific stage in life of a long-lasting behavioral response to a specific individual or object.

in situ **hybridization** A technique using nucleic acid hybridization with a labeled probe to detect the location of a specific mRNA in an intact organism.

in vitro **fertilization (IVF)** (vē′-trō) Fertilization of oocytes in laboratory containers followed by artificial implantation of the early embryo in the mother's uterus.

inclusive fitness The total effect an individual has on proliferating its genes by producing its own offspring and by providing aid that enables other close relatives to increase production of their offspring.

incomplete dominance The situation in which the phenotype of heterozygotes is intermediate between the phenotypes of individuals homozygous for either allele.

incomplete flower A flower in which one or more of the four basic floral organs (sepals, petals, stamens, or carpels) are either absent or nonfunctional.

independent variable A variable whose value is manipulated or changed during an experiment to reveal possible effects on another variable (the dependent variable).

indeterminate growth A type of growth characteristic of plants, in which the organism continues to grow as long as it lives.

induced fit Caused by entry of the substrate, the change in shape of the active site of an enzyme so that it binds more snugly to the substrate.

inducer A specific small molecule that binds to a bacterial repressor protein and changes the repressor's shape so that it cannot bind to an operator, thus switching an operon on.

induction The process in which one group of embryonic cells influences the development of another, usually by causing changes in gene expression.

inductive reasoning A type of logic in which generalizations are based on a large number of specific observations.

inflammatory response An innate immune defense triggered by physical injury or infection of tissue involving the release of substances that promote swelling, enhance the infiltration of white blood cells, and aid in tissue repair and destruction of invading pathogens.

inflorescence A group of flowers tightly clustered together.

ingestion The first stage of food processing in animals: the act of eating.

ingroup A species or group of species whose evolutionary relationships are being examined in a given analysis.

inhibitory postsynaptic potential (IPSP) An electrical change (usually hyperpolarization) in the membrane of a postsynaptic neuron caused by the binding of an inhibitory neurotransmitter from a presynaptic cell to a postsynaptic receptor; makes it more difficult for a postsynaptic neuron to generate an action potential.

innate behavior Animal behavior that is developmentally fixed and under strong genetic control. Innate behavior is exhibited in virtually the same form by all individuals in a population despite internal and external environmental differences during development and throughout their lifetimes.

innate immunity A form of defense common to all animals that is active immediately upon exposure to pathogens and that is the same whether or not the pathogen has been encountered previously.

inner ear One of three main regions of the vertebrate ear; includes the cochlea (which in turn contains the organ of Corti) and the semicircular canals.

inquiry The search for information and explanation, often focusing on specific questions.

insertion A mutation involving the addition of one or more nucleotide pairs to a gene.

integral protein A transmembrane protein with hydrophobic regions that extend into and often completely span the hydrophobic interior of the membrane and with hydrophilic regions in contact with the aqueous solution on one or both sides of the membrane (or lining the channel in the case of a channel protein).

integrin In animal cells, a transmembrane receptor protein with two subunits that interconnects the extracellular matrix and the cytoskeleton.

integument (in-teg′-yū-ment) Layer of sporophyte tissue that contributes to the structure of an ovule of a seed plant.

intercalated disk (in-ter′-kuh-lā′-ted) A specialized junction between cardiac muscle cells that provides direct electrical coupling between the cells.

interferon (in′-ter-fēr′-on) A protein that has antiviral or immune regulatory functions. Interferon-α and interferon-β, secreted by virus-infected cells, help nearby cells resist viral infection; interferon-γ, secreted by T cells, helps activate macrophages.

intermediate disturbance hypothesis The concept that moderate levels of disturbance can foster greater species diversity than low or high levels of disturbance.

intermediate filament A component of the cytoskeleton that includes filaments intermediate in size between microtubules and microfilaments.

internal fertilization The fusion of eggs and sperm within the female reproductive tract. The sperm are typically deposited in or near the tract.

interneuron An association neuron; a nerve cell within the central nervous system that forms synapses with sensory and/or motor neurons and integrates sensory input and motor output.

internode A segment of a plant stem between the points where leaves are attached.

interphase The period in the cell cycle when the cell is not dividing. During interphase, cellular metabolic activity is high, chromosomes and organelles are duplicated, and cell size may increase. Interphase often accounts for about 90% of the cell cycle.

interspecific competition Competition for resources between individuals of two or more species when resources are in short supply.

interspecific interaction A relationship between individuals of two or more species in a community.

interstitial fluid The fluid filling the spaces between cells in most animals.

intertidal zone The shallow zone of the ocean adjacent to land and between the high- and low-tide lines.

introduced species A species moved by humans, either intentionally or accidentally, from its native location to a new geographic region; also called a non-native or exotic species.

intron (in′-tron) A noncoding, intervening sequence within a primary transcript that is removed from the transcript during RNA processing; also refers to the region of DNA from which this sequence was transcribed.

invasive species A species, often introduced by humans, that takes hold outside its native range.

inversion An aberration in chromosome structure resulting from reattachment of a chromosomal fragment in a reverse orientation to the chromosome from which it originated.

invertebrate An animal without a backbone. Invertebrates make up 95% of animal species.

ion (ī′-on) An atom or group of atoms that has gained or lost one or more electrons, thus acquiring a charge.

ion channel A transmembrane protein channel that allows a specific ion to diffuse across the membrane down its concentration or electrochemical gradient.

ionic bond (ī-on′-ik) A chemical bond resulting from the attraction between oppositely charged ions.

ionic compound A compound resulting from the formation of an ionic bond; also called a salt.

iris The colored part of the vertebrate eye, formed by the anterior portion of the choroid.

isotonic (ī′-sō-ton′-ik) Referring to a solution that, when surrounding a cell, causes no net movement of water into or out of the cell.

isotope (ī′-sō-tōp′) One of several atomic forms of an element, each with the same number of protons but a different number of neutrons, thus differing in atomic mass.

iteroparity Reproduction in which adults produce offspring over many years; also known as repeated reproduction.

joule (J) A unit of energy: 1 J = 0.239 cal; 1 cal = 4.184 J.

juxtaglomerular apparatus (JGA) (juks'-tuh-gluh-mār'-yū-ler) A specialized tissue in nephrons that releases the enzyme renin in response to a drop in blood pressure or volume.

juxtamedullary nephron In mammals and birds, a nephron with a loop of Henle that extends far into the renal medulla.

karyogamy (kār'-ē-og'-uh-mē) In fungi, the fusion of haploid nuclei contributed by the two parents; occurs as one stage of sexual reproduction, preceded by plasmogamy.

karyotype (kār'-ē-ō-tīp) A display of the chromosome pairs of a cell arranged by size and shape.

keystone species A species that is not necessarily abundant in a community yet exerts strong control on community structure by the nature of its ecological role or niche.

kidney In vertebrates, one of a pair of excretory organs where blood filtrate is formed and processed into urine.

kilocalorie (kcal) A thousand calories; the amount of heat energy required to raise the temperature of 1 kg of water by 1°C.

kin selection Natural selection that favors altruistic behavior by enhancing the reproductive success of relatives.

kinetic energy (kuh-net'-ik) The energy associated with the relative motion of objects. Moving matter can perform work by imparting motion to other matter.

kinetochore (kuh-net'-uh-kōr) A structure of proteins attached to the centromere that links each sister chromatid to the mitotic spindle.

kingdom A taxonomic category, the second broadest after domain.

K-selection Selection for life history traits that are sensitive to population density; also called density-dependent selection.

labia majora A pair of thick, fatty ridges that encloses and protects the rest of the vulva.

labia minora A pair of slender skin folds that surrounds the openings of the vagina and urethra.

lacteal (lak'-tē-ul) A tiny lymph vessel extending into the core of an intestinal villus and serving as the destination for absorbed chylomicrons.

lactic acid fermentation Glycolysis followed by the reduction of pyruvate to lactate, regenerating NAD^+ with no release of carbon dioxide.

lagging strand A discontinuously synthesized DNA strand that elongates by means of Okazaki fragments, each synthesized in a $5' \rightarrow 3'$ direction away from the replication fork.

landscape An area containing several different ecosystems linked by exchanges of energy, materials, and organisms.

landscape ecology The study of how the spatial arrangement of habitat types affects the distribution and abundance of organisms and ecosystem processes.

large intestine The portion of the vertebrate alimentary canal between the small intestine and the anus; functions mainly in water absorption and the formation of feces.

larynx (lār'-inks) The portion of the respiratory tract containing the vocal cords; also called the voice box.

lateral meristem (mār'-uh-stem) A meristem that thickens the roots and shoots of woody plants. The vascular cambium and cork cambium are lateral meristems.

lateral root A root that arises from the pericycle of an established root.

lateralization Segregation of functions in the cortex of the left and right cerebral hemispheres.

law of conservation of mass A physical law stating that matter can change form but cannot be created or destroyed. In a closed system, the mass of the system is constant.

law of independent assortment Mendel's second law, stating that each pair of alleles segregates, or assorts, independently of each other pair during gamete formation; applies when genes for two characters are located on different pairs of homologous chromosomes or when they are far enough apart on the same chromosome to behave as though they are on different chromosomes.

law of segregation Mendel's first law, stating that the two alleles in a pair segregate (separate from each other) into different gametes during gamete formation.

leading strand The new complementary DNA strand synthesized continuously along the template strand toward the replication fork in the mandatory $5' \rightarrow 3'$ direction.

leaf The main photosynthetic organ of vascular plants.

leaf primordium A finger-like projection along the flank of a shoot apical meristem, from which a leaf arises.

learning The modification of behavior based on specific experiences.

lens The structure in an eye that focuses light rays onto the photoreceptors.

lenticel (len'-ti-sel) A small raised area in the bark of stems and roots that enables gas exchange between living cells and the outside air.

leukocyte (lū'-kō-sīt') A blood cell that functions in fighting infections; also called a white blood cell.

Leydig cell (lī'-dig) A cell that produces testosterone and other androgens and is located between the seminiferous tubules of the testes.

lichen A mutualistic association between a fungus and a photosynthetic alga or cyanobacterium.

life cycle The generation-to-generation sequence of stages in the reproductive history of an organism.

life history The traits that affect an organism's schedule of reproduction and survival.

life table An age-specific summary of the survival pattern of a population.

ligand (lig'-und) A molecule that binds specifically to another molecule, usually a larger one.

ligand-gated ion channel A transmembrane protein containing a pore that opens or closes as it changes shape in response to a signaling molecule (ligand), allowing or blocking the flow of specific ions; also called an ionotropic receptor.

light chain One of the two types of polypeptide chains that make up an antibody molecule and B cell receptor; consists of a variable region, which contributes to the antigen-binding site, and a constant region.

light microscope (LM) An optical instrument with lenses that refract (bend) visible light to magnify images of specimens.

light reactions The first of two major stages in photosynthesis (preceding the Calvin cycle). These reactions, which occur on the thylakoid membranes of the chloroplast or on membranes of certain prokaryotes, convert solar energy to the chemical energy of ATP and NADPH, releasing oxygen in the process.

light-harvesting complex A complex of proteins associated with pigment molecules (including chlorophyll *a*, chlorophyll *b*, and carotenoids) that captures light energy and transfers it to reaction-center pigments in a photosystem.

lignin (lig'-nin) A strong polymer embedded in the cellulose matrix of the secondary cell walls of vascular plants that provides structural support in terrestrial species.

limiting nutrient An element that must be added for production to increase in a particular area.

limnetic zone In a lake, the well-lit, open surface waters far from shore.

linear electron flow A route of electron flow during the light reactions of photosynthesis that involves both photosystems (I and II) and produces ATP, NADPH, and O_2. The net electron flow is from H_2O to $NADP^+$.

line graph A two-dimensional graph in which each data point is connected to the next point in the data set with a straight line.

linkage map A genetic map based on the frequencies of recombination between markers during crossing over of homologous chromosomes.

linked genes Genes located close enough together on a chromosome that they tend to be inherited together.

lipid (lip'-id) Any of a group of large biological molecules, including fats, phospholipids, and steroids, that mix poorly, if at all, with water.

littoral zone In a lake, the shallow, well-lit waters close to shore.

liver A large internal organ in vertebrates that performs diverse functions, such as producing bile, maintaining blood glucose level, and detoxifying poisonous chemicals in the blood.

loam The most fertile soil type, made up of roughly equal amounts of sand, silt, and clay.

lobe-fin Member of a clade of osteichthyans having rod-shaped muscular fins. The group includes coelacanths, lungfishes, and tetrapods.

local regulator A secreted molecule that influences cells near where it is secreted.

locomotion Active motion from place to place.

locus (plural, **loci**) (lō′-kus, lō′-sī) A specific place along the length of a chromosome where a given gene is located.

logistic population growth Population growth that levels off as population size approaches carrying capacity.

long-day plant A plant that flowers (usually in late spring or early summer) only when the light period is longer than a critical length.

long-term memory The ability to hold, associate, and recall information over one's lifetime.

loop of Henle The hairpin turn, with a descending and ascending limb, between the proximal and distal tubules of the vertebrate kidney; functions in water and salt reabsorption.

low-density lipoprotein (LDL) A particle in the blood made up of thousands of cholesterol molecules and other lipids bound to a protein. LDL transports cholesterol from the liver for incorporation into cell membranes.

lung An infolded respiratory surface of a terrestrial vertebrate, land snail, or spider that connects to the atmosphere by narrow tubes.

luteinizing hormone (LH) (lū′-tē-uh-nī′-zing) A tropic hormone that is produced and secreted by the anterior pituitary and that stimulates ovulation in females and androgen production in males.

lycophyte (lī′-kuh-fīt) An informal name for a member of the phylum Lycophyta, a group of seedless vascular plants that includes club mosses and their relatives.

lymph The colorless fluid, derived from interstitial fluid, in the lymphatic system of vertebrates

lymph node An organ located along a lymph vessel. Lymph nodes filter lymph and contain cells that attack viruses and bacteria.

lymphatic system A system of vessels and nodes, separate from the circulatory system, that returns fluid, proteins, and cells to the blood.

lymphocyte A type of white blood cell that mediates immune responses. The two main classes are B cells and T cells.

lysogenic cycle (lī′-sō-jen′-ik) A type of phage replicative cycle in which the viral genome becomes incorporated into the bacterial host chromosome as a prophage, is replicated along with the chromosome, and does not kill the host.

lysosome (lī′-suh-sōm) A membrane-enclosed sac of hydrolytic enzymes found in the cytoplasm of animal cells and some protists.

lysozyme (lī′-sō-zīm) An enzyme that destroys bacterial cell walls; in mammals, found in sweat, tears, and saliva.

lytic cycle (lit′-ik) A type of phage replicative cycle resulting in the release of new phages by lysis (and death) of the host cell.

macroclimate Large-scale patterns in climate; the climate of an entire region.

macroevolution Evolutionary change above the species level. Examples of macroevolutionary change include the origin of a new group of organisms through a series of speciation events and the impact of mass extinctions on the diversity of life and its subsequent recovery.

macromolecule A giant molecule formed by the joining of smaller molecules, usually by a dehydration reaction. Polysaccharides, proteins, and nucleic acids are macromolecules.

macronutrient An essential element that an organism must obtain in relatively large amounts. *See also* micronutrient.

macrophage (mak′-rō-fāj) A phagocytic cell present in many tissues that functions in innate immunity by destroying microbes and in acquired immunity as an antigen-presenting cell.

major histocompatibility complex (MHC) molecule A host protein that functions in antigen presentation. Foreign MHC molecules on transplanted tissue can trigger T cell responses that may lead to rejection of the transplant.

malignant tumor A cancerous tumor containing cells that have significant genetic and cellular changes and are capable of invading and surviving in new sites. Malignant tumors can impair the functions of one or more organs.

mammal Member of the class Mammalia, amniotes that have hair and mammary glands (glands that produce milk).

mammary gland An exocrine gland that secretes milk to nourish the young. Mammary glands are characteristic of mammals.

map unit A unit of measurement of the distance between genes. One map unit is equivalent to a 1% recombination frequency.

marine benthic zone The ocean floor.

marsupial (mar-sū′-pē-ul) A mammal, such as a koala, kangaroo, or opossum, whose young complete their embryonic development inside a maternal pouch.

mass extinction The elimination of a large number of species throughout Earth, the result of global environmental changes.

mass number The sum of the number of protons and neutrons in an atom's nucleus.

mast cell A vertebrate body cell that produces histamine and other molecules that trigger inflammation in response to infection and in allergic reactions.

maternal effect gene A gene that, when mutant in the mother, results in a mutant phenotype in the offspring, regardless of the offspring's genotype. Maternal effect genes, also called egg-polarity genes, were first identified in *Drosophila melanogaster*.

matter Anything that takes up space and has mass.

maximum parsimony A principle that states that when considering multiple explanations for an observation, one should first investigate the simplest explanation that is consistent with the facts.

mechanoreceptor A sensory receptor that detects physical deformation in the body's environment associated with pressure, touch, stretch, motion, or sound.

medulla oblongata (meh-dul′-uh ōb′-long-go′-tuh) The lowest part of the vertebrate brain, commonly called the medulla; a swelling of the hindbrain anterior to the spinal cord that controls autonomic, homeostatic functions, including breathing, heart and blood vessel activity, swallowing, digestion, and vomiting.

megapascal (MPa) (meg′-uh-pas-kal′) A unit of pressure equivalent to about 10 atmospheres of pressure.

megaspore A spore from a heterosporous plant species that develops into a female gametophyte.

meiosis (mī-ō′-sis) A modified type of cell division in sexually reproducing organisms consisting of two rounds of cell division but only one round of DNA replication. It results in cells with half the number of chromosome sets as the original cell.

meiosis I The first division of a two-stage process of cell division in sexually reproducing organisms that results in cells with half the number of chromosome sets as the original cell.

meiosis II The second division of a two-stage process of cell division in sexually reproducing organisms that results in cells with half the number of chromosome sets as the original cell.

membrane potential The difference in electrical charge (voltage) across a cell's plasma membrane due to the differential distribution of ions. Membrane potential affects the activity of excitable cells and the transmembrane movement of all charged substances.

memory cell One of a clone of long-lived lymphocytes, formed during the primary immune response, that remains in a lymphoid organ until activated by exposure to the same antigen that triggered its formation. Activated memory cells mount the secondary immune response.

menopause The cessation of ovulation and menstruation, marking the end of a human female's reproductive years.

menstrual cycle (men'-stru̅-ul) In humans and certain other primates, a type of reproductive cycle in which the nonpregnant endometrium is shed through the cervix into the vagina; also called the uterine cycle.

menstruation The shedding of portions of the endometrium during a uterine (menstrual) cycle.

meristem (mãr'-uh-stem) Plant tissue that remains embryonic as long as the plant lives, allowing for indeterminate growth.

mesoderm (mez'-o̅-derm) The middle primary germ layer in a triploblastic animal embryo; develops into the notochord, the lining of the coelom, muscles, skeleton, gonads, kidneys, and most of the circulatory system in species that have these structures.

mesophyll (mez'-o̅-fil) Leaf cells specialized for photosynthesis. In C_3 and CAM plants, mesophyll cells are located between the upper and lower epidermis; in C_4 plants, they are located between the bundle-sheath cells and the epidermis.

messenger RNA (mRNA) A type of RNA, synthesized using a DNA template, that attaches to ribosomes in the cytoplasm and specifies the primary structure of a protein. (In eukaryotes, the primary RNA transcript must undergo RNA processing to become mRNA.)

metabolic pathway A series of chemical reactions that either builds a complex molecule (anabolic pathway) or breaks down a complex molecule to simpler molecules (catabolic pathway).

metabolic rate The total amount of energy an animal uses in a unit of time.

metabolism (muh-tab'-uh-lizm) The totality of an organism's chemical reactions, consisting of catabolic and anabolic pathways, which manage the material and energy resources of the organism.

metagenomics The collection and sequencing of DNA from a group of species, usually an environmental sample of microorganisms. Computer software sorts partial sequences and assembles them into genome sequences of individual species making up the sample.

metaphase The third stage of mitosis, in which the spindle is complete and the chromosomes, attached to microtubules at their kinetochores, are all aligned at the metaphase plate.

metaphase plate An imaginary structure located at a plane midway between the two poles of a cell in metaphase on which the centromeres of all the duplicated chromosomes are located.

metapopulation A group of spatially separated populations of one species that interact through immigration and emigration.

metastasis (muh-tas'-tuh-sis) The spread of cancer cells to locations distant from their original site.

methanogen (meth-an'-o̅-jen) An organism that produces methane as a waste product of the way it obtains energy. All known methanogens are in domain Archaea.

methyl group A chemical group consisting of a carbon bonded to three hydrogen atoms. The methyl group may be attached to a carbon or to a different atom.

microevolution Evolutionary change below the species level; change in the allele frequencies in a population over generations.

microfilament A cable composed of actin proteins in the cytoplasm of almost every eukaryotic cell, making up part of the cytoskeleton and acting alone or with myosin to cause cell contraction; also known as an actin filament.

micronutrient An essential element that an organism needs in very small amounts. *See also* macronutrient.

microRNA (miRNA) A small, single-stranded RNA molecule, generated from a hairpin structure on a precursor RNA transcribed from a particular gene. The miRNA associates with one or more proteins in a complex that can degrade or prevent translation of an mRNA with a complementary sequence.

microspore A spore from a heterosporous plant species that develops into a male gametophyte.

microtubule A hollow rod composed of tubulin proteins that makes up part of the cytoskeleton in all eukaryotic cells and is found in cilia and flagella.

microvillus (plural, **microvilli**) One of many fine, finger-like projections of the epithelial cells in the lumen of the small intestine that increase its surface area.

midbrain One of three ancestral and embryonic regions of the vertebrate brain; develops into sensory integrating and relay centers that send sensory information to the cerebrum.

middle ear One of three main regions of the vertebrate ear; in mammals, a chamber containing three small bones (the malleus, incus, and stapes) that convey vibrations from the eardrum to the oval window.

middle lamella (luh-mel'-uh) In plants, a thin layer of adhesive extracellular material, primarily pectins, found between the primary walls of adjacent young cells.

migration A regular, long-distance change in location.

mineral In nutrition, a simple nutrient that is inorganic and therefore cannot be synthesized in the body.

minimum viable population (MVP) The smallest population size at which a species is able to sustain its numbers and survive.

mismatch repair The cellular process that uses specific enzymes to remove and replace incorrectly paired nucleotides.

missense mutation A nucleotide-pair substitution that results in a codon that codes for a different amino acid.

mitochondrial matrix The compartment of the mitochondrion enclosed by the inner membrane and containing enzymes and substrates for the citric acid cycle, as well as ribosomes and DNA.

mitochondrion (mi̅'-to̅-kon'-dre̅-un) (plural, **mitochondria**) An organelle in eukaryotic cells that serves as the site of cellular respiration; uses oxygen to break down organic molecules and synthesize ATP.

mitosis (mi̅-to̅'-sis) A process of nuclear division in eukaryotic cells conventionally divided into five stages: prophase, prometaphase, metaphase, anaphase, and telophase. Mitosis conserves chromosome number by allocating replicated chromosomes equally to each of the daughter nuclei.

mitotic (M) phase The phase of the cell cycle that includes mitosis and cytokinesis.

mitotic spindle An assemblage of microtubules and associated proteins that is involved in the movement of chromosomes during mitosis.

mixotroph An organism that is capable of both photosynthesis and heterotrophy.

model A physical or conceptual representation of a natural phenomenon.

model organism A particular species chosen for research into broad biological principles because it is representative of a larger group and usually easy to grow in a lab.

molarity A common measure of solute concentration, referring to the number of moles of solute per liter of solution.

mole (mol) The number of grams of a substance that equals its molecular weight in daltons and contains Avogadro's number of molecules.

molecular clock A method for estimating the time required for a given amount of evolutionary change, based on the observation that some regions of genomes evolve at constant rates.

molecular mass The sum of the masses of all the atoms in a molecule; sometimes called molecular weight.

molecule Two or more atoms held together by covalent bonds.

monilophyte An informal name for a member of the phylum Monilophyta, a group of seedless vascular plants that includes ferns and their relatives.

monoclonal antibody (mon'-ō-klŏn'-ul) Any of a preparation of antibodies that have been produced by a single clone of cultured cells and thus are all specific for the same epitope.

monogamous (muh-nog'-uh-mus) Referring to a type of relationship in which one male mates with just one female.

monohybrid An organism that is heterozygous with respect to a single gene of interest. All the offspring from a cross between parents homozygous for different alleles are monohybrids. For example, parents of genotypes *AA* and *aa* produce a monohybrid of genotype *Aa*.

monohybrid cross A cross between two organisms that are heterozygous for the character being followed (or the self-pollination of a heterozygous plant).

monomer (mon'-uh-mer) The subunit that serves as the building block of a polymer.

monophyletic (mon'-ō-fī-let'-ik) Pertaining to a group of taxa that consists of a common ancestor and all of its descendants. A monophyletic taxon is equivalent to a clade.

monosaccharide (mon'-ō-sak'-uh-rīd) The simplest carbohydrate, active alone or serving as a monomer for disaccharides and polysaccharides. Also known as simple sugars, monosaccharides have molecular formulas that are generally some multiple of CH_2O.

monosomic Referring to a diploid cell that has only one copy of a particular chromosome instead of the normal two.

monotreme An egg-laying mammal, such as a platypus or echidna. Like all mammals, monotremes have hair and produce milk, but they lack nipples.

morphogen A substance, such as Bicoid protein in *Drosophila*, that provides positional information in the form of a concentration gradient along an embryonic axis.

morphogenesis (mōr'-fō-jen'-uh-sis) The development of the form of an organism and its structures.

morphological species concept Definition of a species in terms of measurable anatomical criteria.

motor neuron A nerve cell that transmits signals from the brain or spinal cord to muscles or glands.

motor protein A protein that interacts with cytoskeletal elements and other cell components, producing movement of the whole cell or parts of the cell.

motor system An efferent branch of the vertebrate peripheral nervous system composed of motor neurons that carry signals to skeletal muscles in response to external stimuli.

motor unit A single motor neuron and all the muscle fibers it controls.

movement corridor A series of small clumps or a narrow strip of quality habitat (usable by organisms) that connects otherwise isolated patches of quality habitat.

mucus A viscous and slippery mixture of glycoproteins, cells, salts, and water that moistens and protects the membranes lining body cavities that open to the exterior.

Müllerian mimicry (myū-lār'-ē-un) Reciprocal mimicry by two unpalatable species.

multifactorial Referring to a phenotypic character that is influenced by multiple genes and environmental factors.

multigene family A collection of genes with similar or identical sequences, presumably of common origin.

multiple fruit A fruit derived from an entire inflorescence.

multiplication rule A rule of probability stating that the probability of two or more independent events occurring together can be determined by multiplying their individual probabilities.

muscle tissue Tissue consisting of long muscle cells that can contract, either on its own or when stimulated by nerve impulses.

mutagen (myū'-tuh-jen) A chemical or physical agent that interacts with DNA and can cause a mutation.

mutation (myū-tā'-shun) A change in the nucleotide sequence of an organism's DNA or in the DNA or RNA of a virus.

mutualism (myū'-chū-ul-izm) A symbiotic relationship in which both participants benefit.

mycelium (mī-sē'-lē-um) The densely branched network of hyphae in a fungus.

mycorrhiza (plural, **mycorrhizae**) (mī'-kō-rī'-zuh, mī'-kō-rī'-zē) A mutualistic association of plant roots and fungus.

myelin sheath (mī'-uh-lin) Wrapped around the axon of a neuron, an insulating coat of cell membranes from Schwann cells or oligodendrocytes. It is interrupted by nodes of Ranvier, where action potentials are generated.

myofibril (mī'-ō-fī'-bril) A longitudinal bundle in a muscle cell (fiber) that contains thin filaments of actin and regulatory proteins and thick filaments of myosin.

myoglobin (mī'-uh-glō'-bin) An oxygen-storing, pigmented protein in muscle cells.

myosin (mī'-uh-sin) A type of motor protein that associates into filaments that interact with actin filaments, causing cell contraction.

NAD$^+$ Nicotinamide adenine dinucleotide, a coenzyme that cycles easily between oxidized (NAD$^+$) and reduced (NADH) states, thus acting as an electron carrier.

NADP$^+$ Nicotinamide adenine dinucleotide phosphate, an electron acceptor that, as NADPH, temporarily stores energized electrons produced during the light reactions.

natural killer cell A type of white blood cell that can kill tumor cells and virus-infected cells as part of innate immunity.

natural selection A process in which individuals that have certain inherited traits tend to survive and reproduce at higher rates than other individuals *because of* those traits.

negative feedback A form of regulation in which accumulation of an end product of a process slows the process; in physiology, a primary mechanism of homeostasis, whereby a change in a variable triggers a response that counteracts the initial change.

negative pressure breathing A breathing system in which air is pulled into the lungs.

nephron (nef'-ron) The tubular excretory unit of the vertebrate kidney.

nerve A fiber composed primarily of the bundled axons of neurons.

nerve net A weblike system of neurons, characteristic of radially symmetric animals, such as hydras.

nervous system The fast-acting internal system of communication involving sensory receptors, networks of nerve cells, and connections to muscles and glands that respond to nerve signals; functions in concert with the endocrine system to effect internal regulation and maintain homeostasis.

nervous tissue Tissue made up of neurons and supportive cells.

net ecosystem production (NEP) The gross primary production of an ecosystem minus the energy used by all autotrophs and heterotrophs for respiration.

net primary production (NPP) The gross primary production of an ecosystem minus the energy used by the producers for respiration.

neural plasticity The capacity of a nervous system to change with experience.

neuron (nyūr'-on) A nerve cell; the fundamental unit of the nervous system, having structure and properties that allow it to conduct signals by taking advantage of the electrical charge across its plasma membrane.

neuropeptide A relatively short chain of amino acids that serves as a neurotransmitter.

neurotransmitter A molecule that is released from the synaptic terminal of a neuron at a chemical synapse, diffuses across the synaptic cleft, and binds to the postsynaptic cell, triggering a response.

neutral variation Genetic variation that does not provide a selective advantage or disadvantage.

neutron A subatomic particle having no electrical charge (electrically neutral), with a mass of about 1.7×10^{-24} g, found in the nucleus of an atom.

neutrophil The most abundant type of white blood cell. Neutrophils are phagocytic and tend to self-destruct as they destroy foreign invaders, limiting their life span to a few days.

nitrogen cycle The natural process by which nitrogen, either from the atmosphere or from decomposed organic material, is converted by soil bacteria to compounds assimilated by plants. This incorporated nitrogen is then taken in by other organisms and subsequently released, acted on by bacteria, and made available again to the nonliving environment.

nitrogen fixation The conversion of atmospheric nitrogen (N_2) to ammonia (NH_3). Biological nitrogen fixation is carried out by certain prokaryotes, some of which have mutualistic relationships with plants.

nociceptor (nō′-si-sep′-tur) A sensory receptor that responds to noxious or painful stimuli; also called a pain receptor.

node A point along the stem of a plant at which leaves are attached.

node of Ranvier (ron′-vē-ā′) Gap in the myelin sheath of certain axons where an action potential may be generated. In saltatory conduction, an action potential is regenerated at each node, appearing to "jump" along the axon from node to node.

nodule A swelling on the root of a legume. Nodules are composed of plant cells that contain nitrogen-fixing bacteria of the genus *Rhizobium*.

noncompetitive inhibitor A substance that reduces the activity of an enzyme by binding to a location remote from the active site, changing the enzyme's shape so that the active site no longer effectively catalyzes the conversion of substrate to product.

nondisjunction An error in meiosis or mitosis in which members of a pair of homologous chromosomes or sister chromatids fail to separate properly from each other.

nonequilibrium model A model that maintains that communities change constantly after being buffeted by disturbances.

nonpolar covalent bond A type of covalent bond in which electrons are shared equally between two atoms of similar electronegativity.

nonsense mutation A mutation that changes an amino acid codon to one of the three stop codons, resulting in a shorter and usually nonfunctional protein.

norepinephrine A catecholamine that is chemically and functionally similar to epinephrine and that acts as a hormone or neurotransmitter; also known as noradrenaline.

northern coniferous forest A terrestrial biome characterized by long, cold winters and dominated by cone-bearing trees.

notochord (nō′-tuh-kord′) A longitudinal, flexible rod that runs along the anterior-posterior axis of a chordate in the dorsal part of the body.

nuclear envelope In a eukaryotic cell, the double membrane that surrounds the

nucleus, perforated with pores that regulate traffic with the cytoplasm. The outer membrane is continuous with the endoplasmic reticulum.

nuclear lamina A netlike array of protein filaments that lines the inner surface of the nuclear envelope and helps maintain the shape of the nucleus.

nucleariid Member of a group of unicellular, amoeboid protists that are more closely related to fungi than they are to other protists.

nuclease An enzyme that cuts DNA or RNA, either removing one or a few bases or hydrolyzing the DNA or RNA completely into its component nucleotides.

nucleic acid (nū-klā′-ik) A polymer (polynucleotide) consisting of many nucleotide monomers; serves as a blueprint for proteins and, through the actions of proteins, for all cellular activities. The two types are DNA and RNA.

nucleic acid hybridization The process of base pairing between a gene and a complementary sequence on another nucleic acid molecule.

nucleic acid probe In DNA technology, a labeled single-stranded nucleic acid molecule used to locate a specific nucleotide sequence in a nucleic acid sample. Molecules of the probe hydrogen-bond to the complementary sequence wherever it occurs; radioactive, fluorescent, or other labeling of the probe allows its location to be detected.

nucleoid (nū′-klē-oyd) A non-membrane-enclosed region in a prokaryotic cell where its chromosome is located.

nucleolus (nū-klē′-ō-lus) (plural, **nucleoli**) A specialized structure in the nucleus consisting of chromosomal regions containing ribosomal RNA (rRNA) genes along with ribosomal proteins imported from the cytoplasm; site of rRNA synthesis and ribosomal subunit assembly. *See also* ribosome.

nucleosome (nū′-klē-ō-sōm′) The basic, bead-like unit of DNA packing in eukaryotes, consisting of a segment of DNA wound around a protein core composed of two copies of each of four types of histone.

nucleotide (nū′-klē-ō-tīd′) The building block of a nucleic acid, consisting of a five-carbon sugar covalently bonded to a nitrogenous base and one or more phosphate groups.

nucleotide excision repair A repair system that removes and then correctly replaces a damaged segment of DNA using the undamaged strand as a guide.

nucleotide-pair substitution A type of point mutation in which one nucleotide in a DNA strand and its partner in the complementary strand are replaced by another pair of nucleotides.

nucleus (1) An atom's central core, containing protons and neutrons. (2) The organelle of a eukaryotic cell that contains the genetic ma-

terial in the form of chromosomes, made up of chromatin. (3) A cluster of neurons.

nutrition The process by which an organism takes in and makes use of food substances.

obligate anaerobe (ob′-lig-et an′-uh-rōb) An organism that only carries out fermentation or anaerobic respiration. Such organisms cannot use oxygen and in fact may be poisoned by it.

oceanic pelagic zone Most of the ocean's waters far from shore, constantly mixed by ocean currents.

odorant A molecule that can be detected by sensory receptors of the olfactory system.

Okazaki fragment (ō′-kah-zah′-kē) A short segment of DNA synthesized away from the replication fork on a template strand during DNA replication. Many such segments are joined together to make up the lagging strand of newly synthesized DNA.

olfaction The sense of smell.

oligodendrocyte A type of glial cell that forms insulating myelin sheaths around the axons of neurons in the central nervous system.

oligotrophic lake A nutrient-poor, clear lake with few phytoplankton.

ommatidium (ōm′-uh-tid′-ē-um) (plural, **ommatidia**) One of the facets of the compound eye of arthropods and some polychaete worms.

omnivore An animal that regularly eats animals as well as plants or algae.

oncogene (on′-kō-jēn) A gene found in viral or cellular genomes that is involved in triggering molecular events that can lead to cancer.

oocyte A cell in the female reproductive system that differentiates to form an egg.

oogenesis (ō′-uh-jen′-uh-sis) The process in the ovary that results in the production of female gametes.

oogonium (ō′-uh-gō′-nē-em) (plural, **oogonia**) A cell that divides mitotically to form oocytes.

open circulatory system A circulatory system in which fluid called hemolymph bathes the tissues and organs directly and there is no distinction between the circulating fluid and the interstitial fluid.

operator In bacterial and phage DNA, a sequence of nucleotides near the start of an operon to which an active repressor can attach. The binding of the repressor prevents RNA polymerase from attaching to the promoter and transcribing the genes of the operon.

operon (op′-er-on) A unit of genetic function found in bacteria and phages, consisting of a promoter, an operator, and a coordinately regulated cluster of genes whose products function in a common pathway.

opisthokont (uh-pis′-thuh-kont′) Member of an extremely diverse clade of eukaryotes that includes fungi, animals, and several closely-related groups of protists.

opsin A membrane protein bound to a light-absorbing pigment molecule.

oral cavity The mouth of an animal.

order In Linnaean classification, the taxonomic category above the level of family.

organ A specialized center of body function composed of several different types of tissues.

organ of Corti The actual hearing organ of the vertebrate ear, located in the floor of the cochlear duct in the inner ear; contains the receptor cells (hair cells) of the ear.

organ system A group of organs that work together in performing vital body functions.

organelle (ōr-guh-nel′) Any of several kinds of membrane-enclosed structures with specialized functions, suspended in the cytosol of eukaryotic cells.

organic compound A chemical compound containing carbon.

organismal ecology The branch of ecology concerned with the morphological, physiological, and behavioral ways in which individual organisms meet the challenges posed by their biotic and abiotic environments.

organogenesis (ōr-gan′-ō-jen′-uh-sis) The process in which organ rudiments develop from the three germ layers after gastrulation.

origin of replication Site where the replication of a DNA molecule begins, consisting of a specific sequence of nucleotides.

osmoconformer An animal that is isoosmotic with its environment.

osmolarity (oz′-mō-lār′-uh-tē) Solute concentration expressed as molarity.

osmoregulation Regulation of solute concentrations and water balance by a cell or organism.

osmoregulator An animal that controls its internal osmolarity independent of the external environment.

osmosis (oz-mō′-sis) The diffusion of free water molecules across a selectively permeable membrane.

osteichthyan (os′-tē-ik′-thē-an) Member of a vertebrate clade with jaws and mostly bony skeletons.

outer ear One of three main regions of the ear in reptiles (including birds) and mammals; made up of the auditory canal and, in many birds and mammals, the pinna.

outgroup A species or group of species from an evolutionary lineage that is known to have diverged before the lineage that contains the group of species being studied. An outgroup is selected so that its members are closely related to the group of species being studied, but not as closely related as any study-group members are to each other.

oval window In the vertebrate ear, a membrane-covered gap in the skull bone, through which sound waves pass from the middle ear to the inner ear.

ovarian cycle (ō-vār′-ē-un) The cyclic recurrence of the follicular phase, ovulation, and the luteal phase in the mammalian ovary, regulated by hormones.

ovary (ō′-vuh-rē) (1) In flowers, the portion of a carpel in which the egg-containing ovules develop. (2) In animals, the structure that produces female gametes and reproductive hormones.

oviduct (ō′-vuh-duct) A tube passing from the ovary to the vagina in invertebrates or to the uterus in vertebrates, where it is also known as a fallopian tube.

ovulation The release of an egg from an ovary. In humans, an ovarian follicle releases an egg during each uterine (menstrual) cycle.

ovule (o′-vyūl) A structure that develops within the ovary of a seed plant and contains the female gametophyte.

oxidation The complete or partial loss of electrons from a substance involved in a redox reaction.

oxidative phosphorylation (fos′-fōr-uh-lā′-shun) The production of ATP using energy derived from the redox reactions of an electron transport chain; the third major stage of cellular respiration.

oxidizing agent The electron acceptor in a redox reaction.

oxytocin (ok′-si-tō′-sen) A hormone produced by the hypothalamus and released from the posterior pituitary. It induces contractions of the uterine muscles during labor and causes the mammary glands to eject milk during nursing.

P generation The true-breeding (homozygous) parent individuals from which F_1 hybrid offspring are derived in studies of inheritance; P stands for "parental."

P site One of a ribosome's three binding sites for tRNA during translation. The P site holds the tRNA carrying the growing polypeptide chain. (P stands for peptidyl tRNA.)

p53 gene A tumor-suppressor gene that codes for a specific transcription factor that promotes the synthesis of proteins that inhibit the cell cycle.

paedomorphosis (pē′-duh-mōr′-fuh-sis) The retention in an adult organism of the juvenile features of its evolutionary ancestors.

pain receptor A sensory receptor that responds to noxious or painful stimuli; also called a nociceptor.

paleontology (pā′-lē-un-tol′-ō-jē) The scientific study of fossils.

pancreas (pan′-krē-us) A gland with exocrine and endocrine tissues. The exocrine portion functions in digestion, secreting enzymes and an alkaline solution into the small intestine via a duct; the ductless endocrine portion functions in homeostasis, secreting the hormones insulin and glucagon into the blood.

pandemic A global epidemic.

Pangaea (pan-jē′-uh) The supercontinent that formed near the end of the Paleozoic era, when plate movements brought all the landmasses of Earth together.

parabasalid A protist, such as a trichomonad, with modified mitochondria.

paraphyletic (pār′-uh-fī-let′-ik) Pertaining to a group of taxa that consists of a common ancestor and some, but not all, of its descendants.

parasite (pār′-uh-sīt) An organism that feeds on the cell contents, tissues, or body fluids of another species (the host) while in or on the host organism. Parasites harm but usually do not kill their host.

parasitism (pār′-uh-sit-izm) A symbiotic relationship in which one organism, the parasite, benefits at the expense of another, the host, by living either within or on the host.

parasympathetic division One of three divisions of the autonomic nervous system; generally enhances body activities that gain and conserve energy, such as digestion and reduced heart rate.

parenchyma cell (puh-ren′-ki-muh) A relatively unspecialized plant cell type that carries out most of the metabolism, synthesizes and stores organic products, and develops into a more differentiated cell type.

parental type An offspring with a phenotype that matches one of the true-breeding parental (P generation) phenotypes; also refers to the phenotype itself.

parthenogenesis (par′-thuh-nō′-jen′-uh-sis) A form of asexual reproduction in which females produce offspring from unfertilized eggs.

partial pressure The pressure exerted by a particular gas in a mixture of gases (for instance, the pressure exerted by oxygen in air).

passive immunity Short-term immunity conferred by the transfer of antibodies, as occurs in the transfer of maternal antibodies to a fetus or nursing infant.

passive transport The diffusion of a substance across a biological membrane with no expenditure of energy.

pathogen An organism or virus that causes disease.

pattern formation The development of a multicellular organism's spatial organization, the arrangement of organs and tissues in their characteristic places in three-dimensional space.

PCR *See* polymerase chain reaction.

pedigree A diagram of a family tree with conventional symbols, showing the occurrence of heritable characters in parents and offspring over multiple generations.

pelagic zone The open-water component of aquatic biomes.

penis The copulatory structure of male mammals.

pepsin An enzyme present in gastric juice that begins the hydrolysis of proteins. Pepsin is synthesized as an inactive precursor form, pepsinogen.

peptide bond The covalent bond between the carboxyl group on one amino acid and the amino group on another, formed by a dehydration reaction.

peptidoglycan (pep′-tid-ō-glī′-kan) A type of polymer in bacterial cell walls consisting of modified sugars cross-linked by short polypeptides.

perception The interpretation of sensory system input by the brain.

pericycle The outermost layer in the vascular cylinder, from which lateral roots arise.

periderm (pār′-uh-derm′) The protective coat that replaces the epidermis in woody plants during secondary growth, formed of the cork and cork cambium.

peripheral nervous system (PNS) The sensory and motor neurons that connect to the central nervous system.

peripheral protein A protein loosely bound to the surface of a membrane or to part of an integral protein and not embedded in the lipid bilayer.

peristalsis (pār′-uh-stal′-sis) (1) Alternating waves of contraction and relaxation in the smooth muscles lining the alimentary canal that push food along the canal. (2) A type of movement on land produced by rhythmic waves of muscle contractions passing from front to back, as in many annelids.

peritubular capillary One of the tiny blood vessels that form a network surrounding the proximal and distal tubules in the kidney.

peroxisome (puh-rok′-suh-sōm′) An organelle containing enzymes that transfer hydrogen atoms from various substrates to oxygen (O_2), producing and then degrading hydrogen peroxide (H_2O_2).

petal A modified leaf of a flowering plant. Petals are the often colorful parts of a flower that advertise it to insects and other pollinators.

petiole (pet′-ē-ōl) The stalk of a leaf, which joins the leaf to a node of the stem.

pH A measure of hydrogen ion concentration equal to $-\log [H^+]$ and ranging in value from 0 to 14.

phage (fāj) A virus that infects bacteria; also called a bacteriophage.

phagocytosis (fag′-ō-sī-tō′-sis) A type of endocytosis in which large particulate substances or small organisms are taken up by a cell. It is carried out by some protists and by certain immune cells of animals (in mammals, mainly macrophages, neutrophils, and dendritic cells).

pharyngeal cleft (fuh-rin′-jē-ul) In chordate embryos, one of the grooves that separate a series of pouches along the sides of the pharynx and may develop into a pharyngeal slit.

pharyngeal slit (fuh-rin′-jē-ul) In chordate embryos, one of the slits that form from the pharyngeal clefts and communicate to the outside, later developing into gill slits in many vertebrates.

pharynx (fār′-inks) (1) An area in the vertebrate throat where air and food passages cross. (2) In flatworms, the muscular tube that protrudes from the ventral side of the worm and ends in the mouth.

phenotype (fē′-nō-tīp) The observable physical and physiological traits of an organism, which are determined by its genetic makeup.

pheromone (fār′-uh-mōn) In animals and fungi, a small molecule released into the environment that functions in communication between members of the same species. In animals, it acts much like a hormone in influencing physiology and behavior.

phloem (flō′-em) Vascular plant tissue consisting of living cells arranged into elongated tubes that transport sugar and other organic nutrients throughout the plant.

phloem sap The sugar-rich solution carried through a plant's sieve tubes.

phosphate group A chemical group consisting of a phosphorus atom bonded to four oxygen atoms; important in energy transfer.

phospholipid (fos′-fō-lip′-id) A lipid made up of glycerol joined to two fatty acids and a phosphate group. The hydrocarbon chains of the fatty acids act as nonpolar, hydrophobic tails, while the rest of the molecule acts as a polar, hydrophilic head. Phospholipids form bilayers that function as biological membranes.

phosphorylated intermediate A molecule (often a reactant) with a phosphate group covalently bound to it, making it more reactive (less stable) than the unphosphorylated molecule.

photic zone (fō′-tic) The narrow top layer of an ocean or lake, where light penetrates sufficiently for photosynthesis to occur.

photomorphogenesis Effects of light on plant morphology.

photon (fō′-ton) A quantum, or discrete quantity, of light energy that behaves as if it were a particle.

photoperiodism (fō′-tō-pēr′-ē-ō-dizm) A physiological response to photoperiod, the relative lengths of night and day. An example of photoperiodism is flowering.

photophosphorylation (fō′-tō-fos′-fōr-uh-lā′-shun) The process of generating ATP from ADP and phosphate by means of chemiosmosis, using a proton-motive force generated across the thylakoid membrane of the chloroplast or the membrane of certain prokaryotes during the light reactions of photosynthesis.

photoreceptor An electromagnetic receptor that detects the radiation known as visible light.

photorespiration A metabolic pathway that consumes oxygen and ATP, releases carbon dioxide, and decreases photosynthetic output. Photorespiration generally occurs on hot, dry, bright days, when stomata close and the O_2/CO_2 ratio in the leaf increases, favoring the binding of O_2 rather than CO_2 by rubisco.

photosynthesis (fō′-tō-sin′-thi-sis) The conversion of light energy to chemical energy that is stored in sugars or other organic compounds; occurs in plants, algae, and certain prokaryotes.

photosystem A light-capturing unit located in the thylakoid membrane of the chloroplast or in the membrane of some prokaryotes, consisting of a reaction-center complex surrounded by numerous light-harvesting complexes. There are two types of photosystems, I and II; they absorb light best at different wavelengths.

photosystem I (PS I) One of two light-capturing units in a chloroplast's thylakoid membrane or in the membrane of some prokaryotes; it has two molecules of P700 chlorophyll *a* at its reaction center.

photosystem II (PS II) One of two light-capturing units in a chloroplast's thylakoid membrane or in the membrane of some prokaryotes; it has two molecules of P680 chlorophyll *a* at its reaction center.

phototropism (fō′-tō-trō′-pizm) Growth of a plant shoot toward or away from light.

phyllotaxy (fil′-uh-tak′-sē) The pattern of leaf attachment to the stem of a plant.

phylogenetic species concept Definition of a species as the smallest group of individuals that share a common ancestor, forming one branch on the tree of life.

phylogenetic tree A branching diagram that represents a hypothesis about the evolutionary history of a group of organisms.

phylogeny (fī-loj′-uh-nē) The evolutionary history of a species or group of related species.

phylum (fī′-lum) (plural, **phyla**) In Linnaean classification, the taxonomic category above class.

physiology The processes and functions of an organism.

phytochrome (fī′-tuh-krōm) A type of light receptor in plants that mostly absorbs red light and regulates many plant responses, such as seed germination and shade avoidance.

pilus (plural, **pili**) (pī′-lus, pī′-lī) In bacteria, a structure that links one cell to another at the start of conjugation; also known as a sex pilus or conjugation pilus.

pinocytosis (pī′-nō-sī-tō′-sis) A type of endocytosis in which the cell ingests extracellular fluid and its dissolved solutes.

pistil A single carpel or a group of fused carpels.

pith Ground tissue that is internal to the vascular tissue in a stem; in many monocot roots, parenchyma cells that form the central core of the vascular cylinder.

pituitary gland (puh-tū′-uh-tār′-ē) An endocrine gland at the base of the hypothalamus; consists of a posterior lobe, which stores and releases two hormones produced by the hypothalamus, and an anterior lobe, which produces and secretes many hormones that regulate diverse body functions.

placenta (pluh-sen′-tuh) A structure in the pregnant uterus for nourishing a viviparous fetus with the mother's blood supply; formed from the uterine lining and embryonic membranes.

plasma (plaz′-muh) The liquid matrix of blood in which the blood cells are suspended.

plasma cell The antibody-secreting effector cell of humoral immunity. Plasma cells arise from antigen-stimulated B cells.

plasma membrane The membrane at the boundary of every cell that acts as a selective barrier, regulating the cell's chemical composition.

plasmid (plaz′-mid) A small, circular, double-stranded DNA molecule that carries accessory genes separate from those of a bacterial chromosome; in DNA cloning, can be used as a vector carrying up to about 10,000 base pairs (10 kb) of DNA.

plasmodesma (plaz′-mō-dez′-muh) (plural, **plasmodesmata**) An open channel through the cell wall that connects the cytoplasm of adjacent plant cells, allowing water, small solutes, and some larger molecules to pass between the cells.

plasmogamy (plaz-moh′-guh-mē) In fungi, the fusion of the cytoplasm of cells from two individuals; occurs as one stage of sexual reproduction, followed later by karyogamy.

plasmolysis (plaz-mol′-uh-sis) A phenomenon in walled cells in which the cytoplasm shrivels and the plasma membrane pulls away from the cell wall; occurs when the cell loses water to a hypertonic environment.

plastid One of a family of closely related organelles that includes chloroplasts, chromoplasts, and amyloplasts. Plastids are found in the cells of photosynthetic eukaryotes.

plate tectonics The theory that the continents are part of great plates of Earth's crust that float on the hot, underlying portion of the mantle. Movements in the mantle cause the continents to move slowly over time.

platelet A pinched-off cytoplasmic fragment of a specialized bone marrow cell. Platelets circulate in the blood and are important in blood clotting.

pleiotropy (plī′-o-truh-pē) The ability of a single gene to have multiple effects.

pluripotent Describing a cell that can give rise to many, but not all, parts of an organism.

point mutation A change in a single nucleotide pair of a gene.

polar covalent bond A covalent bond between atoms that differ in electronegativity. The shared electrons are pulled closer to the more electronegative atom, making it slightly negative and the other atom slightly positive.

polar molecule A molecule (such as water) with an uneven distribution of charges in different regions of the molecule.

pollen grain In seed plants, a structure consisting of the male gametophyte enclosed within a pollen wall.

pollen tube A tube that forms after germination of the pollen grain and that functions in the delivery of sperm to the ovule.

pollination (pol′-uh-nā′-shun) The transfer of pollen to the part of a seed plant containing the ovules, a process required for fertilization.

poly-A tail A sequence of 50–250 adenine nucleotides added onto the 3′ end of a pre-mRNA molecule.

polygamous Referring to a type of relationship in which an individual of one sex mates with several of the other.

polygenic inheritance (pol′-ē-jen′-ik) An additive effect of two or more genes on a single phenotypic character.

polymer (pol′-uh-mer) A long molecule consisting of many similar or identical monomers linked together by covalent bonds.

polymerase chain reaction (PCR) (puh-lim′-uh-rās) A technique for amplifying DNA *in vitro* by incubating it with specific primers, a heat-resistant DNA polymerase, and nucleotides.

polynucleotide (pol′-ē-nū′-klē-ō-tīd) A polymer consisting of many nucleotide monomers in a chain. The nucleotides can be those of DNA or RNA.

polypeptide (pol′-ē-pep′-tīd) A polymer of many amino acids linked together by peptide bonds.

polyphyletic (pol′-ē-fī-let′-ik) Pertaining to a group of taxa derived from two or more different ancestors.

polyploidy (pol′-ē-ploy′-dē) A chromosomal alteration in which the organism possesses more than two complete chromosome sets. It is the result of an accident of cell division.

polyribosome (pol′-ē-rī′-buh-sōm′) A group of several ribosomes attached to, and translating, the same messenger RNA molecule; also called a polysome.

polysaccharide (pol′-ē-sak′-uh-rīd) A polymer of many monosaccharides, formed by dehydration reactions.

polyspermy Fusion of the egg with more than one sperm.

polytomy (puh-lit′-uh-mē) In a phylogenetic tree, a branch point from which more than two descendant taxa emerge. A polytomy indicates that the evolutionary relationships between the descendant taxa are not yet clear.

pons A portion of the brain that participates in certain automatic, homeostatic functions, such as regulating the breathing centers in the medulla.

population A group of individuals of the same species that live in the same area and interbreed, producing fertile offspring.

population dynamics The study of how complex interactions between biotic and abiotic factors influence variations in population size.

population ecology The study of populations in relation to their environment, including environmental influences on population density and distribution, age structure, and variations in population size.

positional information Molecular cues that control pattern formation in an animal or plant embryonic structure by indicating a cell's location relative to the organism's body axes. These cues elicit a response by genes that regulate development.

positive feedback A form of regulation in which an end product of a process speeds up that process; in physiology, a control mechanism in which a change in a variable triggers a response that reinforces or amplifies the change.

positive pressure breathing A breathing system in which air is forced into the lungs.

posterior Pertaining to the rear, or tail end, of a bilaterally symmetric animal.

posterior pituitary An extension of the hypothalamus composed of nervous tissue that secretes oxytocin and antidiuretic hormone made in the hypothalamus; a temporary storage site for these hormones.

postzygotic barrier (pōst′-zī-got′-ik) A reproductive barrier that prevents hybrid zygotes produced by two different species from developing into viable, fertile adults.

potential energy The energy that matter possesses as a result of its location or spatial arrangement (structure).

predation An interaction between species in which one species, the predator, eats the other, the prey.

prediction In deductive reasoning, a forecast that follows logically from a hypothesis. By testing predictions, experiments may allow certain hypotheses to be rejected.

prepuce (prē′-pyūs) A fold of skin covering the head of the clitoris or penis.

pressure potential (Ψ_P) A component of water potential that consists of the physical pressure on a solution, which can be positive, zero, or negative.

prezygotic barrier (prē′-zī-got′-ik) A reproductive barrier that impedes mating between species or hinders fertilization if interspecific mating is attempted.

primary cell wall In plants, a relatively thin and flexible layer that surrounds the plasma membrane of a young cell.

primary consumer An herbivore; an organism that eats plants or other autotrophs.

primary electron acceptor In the thylakoid membrane of a chloroplast or in the membrane of some prokaryotes, a specialized molecule that shares the reaction-center complex with a pair of chlorophyll *a* molecules and that accepts an electron from them.

primary growth Growth produced by apical meristems, lengthening stems and roots.

primary immune response The initial adaptive immune response to an antigen, which appears after a lag of about 10–17 days.

primary oocyte (ō′-uh-sīt) An oocyte prior to completion of meiosis I.

primary producer An autotroph, usually a photosynthetic organism. Collectively, autotrophs make up the trophic level of an ecosystem that ultimately supports all other levels.

primary production The amount of light energy converted to chemical energy (organic compounds) by the autotrophs in an ecosystem during a given time period.

primary structure The level of protein structure referring to the specific linear sequence of amino acids.

primary succession A type of ecological succession that occurs in an area where there were originally no organisms present and where soil has not yet formed.

primary transcript An initial RNA transcript from any gene; also called pre-mRNA when transcribed from a protein-coding gene.

primase An enzyme that joins RNA nucleotides to make a primer during DNA replication, using the parental DNA strand as a template.

primer A short stretch of RNA with a free 3′ end, bound by complementary base pairing to the template strand and elongated with DNA nucleotides during DNA replication.

problem solving The cognitive activity of devising a method to proceed from one state to another in the face of real or apparent obstacles.

producer An organism that produces organic compounds from CO_2 by harnessing light energy (in photosynthesis) or by oxidizing inorganic chemicals (in chemosynthetic reactions carried out by some prokaryotes).

product A material resulting from a chemical reaction.

production efficiency The percentage of energy stored in assimilated food that is not used for respiration or eliminated as waste.

progesterone A steroid hormone that prepares the uterus for pregnancy; the major progestin in mammals.

prokaryote An organism that has a prokaryotic cell; an informal term for an organism in either domain Bacteria or domain Archaea.

prokaryotic cell (prō′-kār′-ē-ot′-ik) A type of cell lacking a membrane-enclosed nucleus and membrane-enclosed organelles. Organisms with prokaryotic cells (bacteria and archaea) are called prokaryotes.

prometaphase The second stage of mitosis, in which the nuclear envelope fragments and the spindle microtubules attach to the kinetochores of the chromosomes.

promoter A specific nucleotide sequence in the DNA of a gene that binds RNA polymerase, positioning it to start transcribing RNA at the appropriate place.

prophage (prō′-fāj) A phage genome that has been inserted into a specific site on a bacterial chromosome.

prophase The first stage of mitosis, in which the chromatin condenses into discrete chromosomes visible with a light microscope, the mitotic spindle begins to form, and the nucleolus disappears but the nucleus remains intact.

prostate gland (pros′-tāt) A gland in human males that secretes an acid-neutralizing component of semen.

protease An enzyme that digests proteins by hydrolysis.

protein (prō′-tēn) A biologically functional molecule consisting of one or more polypeptides folded and coiled into a specific three-dimensional structure.

protein kinase An enzyme that transfers phosphate groups from ATP to a protein, thus phosphorylating the protein.

protein phosphatase An enzyme that removes phosphate groups from (dephosphorylates) proteins, often functioning to reverse the effect of a protein kinase.

proteoglycan (prō′-tē-ō-glī′-kan) A large molecule consisting of a small core protein with many carbohydrate chains attached, found in the extracellular matrix of animal cells. A proteoglycan may consist of up to 95% carbohydrate.

proteomics (prō′-tē-ō′-miks) The systematic study of the full protein sets (proteomes) encoded by genomes.

protist An informal term applied to any eukaryote that is not a plant, animal, or fungus. Most protists are unicellular, though some are colonial or multicellular.

protocell An abiotic precursor of a living cell that had a membrane-like structure and that maintained an internal chemistry different from that of its surroundings.

proton (prō′-ton) A subatomic particle with a single positive electrical charge, with a mass of about 1.7×10^{-24} g, found in the nucleus of an atom.

proton pump An active transport protein in a cell membrane that uses ATP to transport hydrogen ions out of a cell against their concentration gradient, generating a membrane potential in the process.

proton-motive force The potential energy stored in the form of a proton electrochemical gradient, generated by the pumping of hydrogen ions (H^+) across a biological membrane during chemiosmosis.

proto-oncogene (prō′-tō-on′-kō-jēn) A normal cellular gene that has the potential to become an oncogene.

protoplast The living part of a plant cell, which also includes the plasma membrane.

provirus A viral genome that is permanently inserted into a host genome.

proximal tubule In the vertebrate kidney, the portion of a nephron immediately downstream from Bowman's capsule that conveys and helps refine filtrate.

pseudogene (sū′-dō-jēn) A DNA segment that is very similar to a real gene but does not yield a functional product; a DNA segment that formerly functioned as a gene but has become inactivated in a particular species because of mutation.

pseudopodium (sū′-dō-pō′-dē-um) (plural, **pseudopodia**) A cellular extension of amoeboid cells used in moving and feeding.

pulse The rhythmic bulging of the artery walls with each heartbeat.

punctuated equilibria In the fossil record, long periods of apparent stasis, in which a species undergoes little or no morphological change, interrupted by relatively brief periods of sudden change.

Punnett square A diagram used in the study of inheritance to show the predicted genotypic results of random fertilization in genetic crosses between individuals of known genotype.

pupil The opening in the iris, which admits light into the interior of the vertebrate eye. Muscles in the iris regulate its size.

purine (pyū′-rēn) One of two types of nitrogenous bases found in nucleotides, characterized by a six-membered ring fused to a five-membered ring. Adenine (A) and guanine (G) are purines.

pyrimidine (puh-rim′-uh-dēn) One of two types of nitrogenous bases found in nucleotides, characterized by a six-membered ring. Cytosine (C), thymine (T), and uracil (U) are pyrimidines.

quantitative character A heritable feature that varies continuously over a range rather than in an either-or fashion.

quaternary structure (kwot′-er-nār′-ē) The particular shape of a complex, aggregate protein, defined by the characteristic three-dimensional arrangement of its constituent subunits, each a polypeptide.

R plasmid A bacterial plasmid carrying genes that confer resistance to certain antibiotics.

radial symmetry Symmetry in which the body is shaped like a pie or barrel (lacking a left side and a right side) and can be divided into mirror-imaged halves by any plane through its central axis.

radicle An embryonic root of a plant.

radioactive isotope An isotope (an atomic form of a chemical element) that is unstable;

the nucleus decays spontaneously, giving off detectable particles and energy.

radiometric dating A method for determining the absolute age of rocks and fossils, based on the half-life of radioactive isotopes.

ras **gene** A gene that codes for Ras, a G protein that relays a growth signal from a growth factor receptor on the plasma membrane to a cascade of protein kinases, ultimately resulting in stimulation of the cell cycle.

ray-finned fish Member of the class Actinopterygii, aquatic osteichthyans with fins supported by long, flexible rays, including tuna, bass, and herring.

reabsorption In excretory systems, the recovery of solutes and water from filtrate.

reactant A starting material in a chemical reaction.

reaction-center complex A complex of proteins associated with a special pair of chlorophyll *a* molecules and a primary electron acceptor. Located centrally in a photosystem, this complex triggers the light reactions of photosynthesis. Excited by light energy, the pair of chlorophylls donates an electron to the primary electron acceptor, which passes an electron to an electron transport chain.

reading frame On an mRNA, the triplet grouping of ribonucleotides used by the translation machinery during polypeptide synthesis.

receptacle The base of a flower; the part of the stem that is the site of attachment of the floral organs.

reception The binding of a signaling molecule to a receptor protein, activating the receptor by causing it to change shape. *See also* sensory reception.

receptor potential An initial response of a receptor cell to a stimulus, consisting of a change in voltage across the receptor membrane proportional to the stimulus strength.

receptor-mediated endocytosis (en′-dō-sī-tō′-sis) The movement of specific molecules into a cell by the inward budding of vesicles containing proteins with receptor sites specific to the molecules being taken in; enables a cell to acquire bulk quantities of specific substances.

recessive allele An allele whose phenotypic effect is not observed in a heterozygote.

recombinant chromosome A chromosome created when crossing over combines DNA from two parents into a single chromosome.

recombinant DNA A DNA molecule made *in vitro* with segments from different sources.

recombinant type (recombinant) An offspring whose phenotype differs from that of the true-breeding P generation parents; also refers to the phenotype itself.

rectum The terminal portion of the large intestine, where the feces are stored prior to elimination.

red alga A photosynthetic protist, named for its color, which results from a red pigment that masks the green of chlorophyll. Most red algae are multicellular and marine.

redox reaction (rē′-doks) A chemical reaction involving the complete or partial transfer of one or more electrons from one reactant to another; short for **red**uction-**ox**idation reaction.

reducing agent The electron donor in a redox reaction.

reduction The complete or partial addition of electrons to a substance involved in a redox reaction.

reflex An automatic reaction to a stimulus, mediated by the spinal cord or lower brain.

refractory period (rē-frakt′-ōr-ē) The short time immediately after an action potential in which the neuron cannot respond to another stimulus, owing to the inactivation of voltage-gated sodium channels.

regression line A line drawn through a scatter plot that shows the general trend of the data. It represents an equation that is calculated mathematically to best fit the data and can be used to predict the value of the dependent variable for any value of the independent variable.

regulator An animal for which mechanisms of homeostasis moderate internal changes in a particular variable in the face of external fluctuation of that variable.

regulatory gene A gene that codes for a protein, such as a repressor, that controls the transcription of another gene or group of genes.

reinforcement In evolutionary biology, a process in which natural selection strengthens prezygotic barriers to reproduction, thus reducing the chances of hybrid formation. Such a process is likely to occur only if hybrid offspring are less fit than members of the parent species.

relative abundance The proportional abundance of different species in a community.

relative fitness The contribution an individual makes to the gene pool of the next generation, relative to the contributions of other individuals in the population.

renal cortex The outer portion of the vertebrate kidney.

renal medulla The inner portion of the vertebrate kidney, beneath the renal cortex.

renal pelvis The funnel-shaped chamber that receives processed filtrate from the vertebrate kidney's collecting ducts and is drained by the ureter.

renin-angiotensin-aldosterone system (RAAS) A hormone cascade pathway that helps regulate blood pressure and blood volume.

repetitive DNA Nucleotide sequences, usually noncoding, that are present in many copies in a eukaryotic genome. The repeated units may be short and arranged tandemly (in series) or long and dispersed in the genome.

replication fork A Y-shaped region on a replicating DNA molecule where the parental strands are being unwound and new strands are being synthesized.

repressor A protein that inhibits gene transcription. In prokaryotes, repressors bind to the DNA in or near the promoter. In eukaryotes, repressors may bind to control elements within enhancers, to activators, or to other proteins in a way that blocks activators from binding to DNA.

reproductive isolation The existence of biological factors (barriers) that impede members of two species from producing viable, fertile offspring.

reproductive table An age-specific summary of the reproductive rates in a population.

reptile Member of the clade of amniotes that includes tuataras, lizards, snakes, turtles, crocodilians, and birds.

residual volume The amount of air that remains in the lungs after forceful exhalation.

resource partitioning The division of environmental resources by coexisting species such that the niche of each species differs by one or more significant factors from the niches of all coexisting species.

respiratory pigment A protein that transports oxygen in blood or hemolymph.

response (1) In cellular communication, the change in a specific cellular activity brought about by a transduced signal from outside the cell. (2) In feedback regulation, a physiological activity triggered by a change in a variable.

resting potential The membrane potential characteristic of a nonconducting excitable cell, with the inside of the cell more negative than the outside.

restriction enzyme An endonuclease (type of enzyme) that recognizes and cuts DNA molecules foreign to a bacterium (such as phage genomes). The enzyme cuts at specific nucleotide sequences (restriction sites).

restriction fragment A DNA segment that results from the cutting of DNA by a restriction enzyme.

restriction site A specific sequence on a DNA strand that is recognized and cut by a restriction enzyme.

retina (ret′-i-nuh) The innermost layer of the vertebrate eye, containing photoreceptor cells (rods and cones) and neurons; transmits images formed by the lens to the brain via the optic nerve.

retinal The light-absorbing pigment in rods and cones of the vertebrate eye.

retrotransposon (re′-trō-trans-pō′-zon) A transposable element that moves within a genome by means of an RNA intermediate, a transcript of the retrotransposon DNA.

retrovirus (re′-trō-vī′-rus) An RNA virus that replicates by transcribing its RNA into

DNA and then inserting the DNA into a cellular chromosome; an important class of cancer-causing viruses.

reverse transcriptase (tran-skrip′-tās) An enzyme encoded by certain viruses (retroviruses) that uses RNA as a template for DNA synthesis.

reverse transcriptase–polymerase chain reaction (RT-PCR) A technique for determining expression of a particular gene. It uses reverse transcriptase and DNA polymerase to synthesize cDNA from all the mRNA in a sample and then subjects the cDNA to PCR amplification using primers specific for the gene of interest.

rhizarians (rī-za′-rē-uhns) One of the three major subgroups for which the "SAR" eukaryotic supergroup is named. Many species in this clade are amoebas characterized by threadlike pseudopodia.

rhizobacterium A soil bacterium whose population size is much enhanced in the rhizosphere, the soil region close to a plant's roots.

rhizoid (rī′-zoyd) A long, tubular single cell or filament of cells that anchors bryophytes to the ground. Unlike roots, rhizoids are not composed of tissues, lack specialized conducting cells, and do not play a primary role in water and mineral absorption.

rhizosphere The soil region close to plant roots and characterized by a high level of microbiological activity.

rhodopsin (rō-dop′-sin) A visual pigment consisting of retinal and opsin. Upon absorbing light, the retinal changes shape and dissociates from the opsin.

ribonucleic acid (RNA) (rī′-bō-nū-klā′-ik) A type of nucleic acid consisting of a polynucleotide made up of nucleotide monomers with a ribose sugar and the nitrogenous bases adenine (A), cytosine (C), guanine (G), and uracil (U); usually single-stranded; functions in protein synthesis, gene regulation, and as the genome of some viruses.

ribose The sugar component of RNA nucleotides.

ribosomal RNA (rRNA) (rī′-buh-sō′-mul) RNA molecules that, together with proteins, make up ribosomes; the most abundant type of RNA.

ribosome (rī′-buh-sōm′) A complex of rRNA and protein molecules that functions as a site of protein synthesis in the cytoplasm; consists of a large subunit and a small subunit. In eukaryotic cells, each subunit is assembled in the nucleolus. *See also* nucleolus.

ribozyme (rī′-buh-zīm) An RNA molecule that functions as an enzyme, such as an intron that catalyzes its own removal during RNA splicing.

RNA interference (RNAi) A technique used to silence the expression of selected genes. RNAi uses synthetic double-stranded RNA molecules that match the sequence of a particular gene to trigger the breakdown of the gene's messenger RNA.

RNA polymerase An enzyme that links ribonucleotides into a growing RNA chain during transcription, based on complementary binding to nucleotides on a DNA template strand.

RNA processing Modification of RNA primary transcripts, including splicing out of introns, joining together of exons, and alteration of the 5′ and 3′ ends.

RNA splicing After synthesis of a eukaryotic primary RNA transcript, the removal of portions of the transcript (introns) that will not be included in the mRNA and the joining together of the remaining portions (exons).

rod A rodlike cell in the retina of the vertebrate eye, sensitive to low light intensity.

root An organ in vascular plants that anchors the plant and enables it to absorb water and minerals from the soil.

root cap A cone of cells at the tip of a plant root that protects the apical meristem.

root hair A tiny extension of a root epidermal cell, growing just behind the root tip and increasing surface area for absorption of water and minerals.

root system All of a plant's roots, which anchor it in the soil, absorb and transport minerals and water, and store food.

rooted Describing a phylogenetic tree that contains a branch point (often, the one farthest to the left) representing the most recent common ancestor of all taxa in the tree.

rough ER That portion of the endoplasmic reticulum with ribosomes attached.

round window In the mammalian ear, the point of contact where vibrations of the stapes create a traveling series of pressure waves in the fluid of the cochlea.

r-selection Selection for life history traits that maximize reproductive success in uncrowded environments; also called density-independent selection.

rubisco (rū-bis′-kō) Ribulose bisphosphate (RuBP) carboxylase, the enzyme that catalyzes the first step of the Calvin cycle (the addition of CO_2 to RuBP).

ruminant (rū′-muh-nent) An animal, such as a cow or a sheep, with multiple stomach compartments specialized for an herbivorous diet.

S phase The synthesis phase of the cell cycle; the portion of interphase during which DNA is replicated.

saccule In the vertebrate ear, a chamber in the vestibule behind the oval window that participates in the sense of balance.

salicylic acid (sal′-i-sil′-ik) A signaling molecule in plants that may be partially responsible for activating systemic acquired resistance to pathogens.

salivary gland A gland associated with the oral cavity that secretes substances that lubricate food and begin the process of chemical digestion.

salt A compound resulting from the formation of an ionic bond; also called an ionic compound.

saltatory conduction (sol′-tuh-tōr′-ē) Rapid transmission of a nerve impulse along an axon, resulting from the action potential jumping from one node of Ranvier to another, skipping the myelin-sheathed regions of membrane.

"SAR" clade One of four supergroups of eukaryotes proposed in a current hypothesis of the evolutionary history of eukaryotes. This supergroup contains a large, extremely diverse collection of protists from three major subgroups: stramenopiles, alveolates, and rhizarians. *See also* Excavata, Archaeplastida, and Unikonta.

sarcomere (sar′-kō-mēr) The fundamental, repeating unit of striated muscle, delimited by the Z lines.

sarcoplasmic reticulum (SR) (sar′-kō-plaz′-mik ruh-tik′-yū-lum) A specialized endoplasmic reticulum that regulates the calcium concentration in the cytosol of muscle cells.

saturated fatty acid A fatty acid in which all carbons in the hydrocarbon tail are connected by single bonds, thus maximizing the number of hydrogen atoms attached to the carbon skeleton.

savanna A tropical grassland biome with scattered trees and large herbivores and maintained by occasional fires and drought.

scanning electron microscope (SEM) A microscope that uses an electron beam to scan the surface of a sample, coated with metal atoms, to study details of its topography.

scatter plot A graph in which each piece of data is represented by a point, but individual points are not connected by lines.

Schwann cell A type of glial cell that forms insulating myelin sheaths around the axons of neurons in the peripheral nervous system.

science An approach to understanding the natural world.

scion (sī′-un) The twig grafted onto the stock when making a graft.

sclereid (sklār′-ē-id) A short, irregular sclerenchyma cell in nutshells and seed coats. Sclereids are scattered throughout the parenchyma of some plants.

sclerenchyma cell (skluh-ren′-kim-uh) A rigid, supportive plant cell type usually lacking a protoplast and possessing thick secondary walls strengthened by lignin at maturity.

scrotum A pouch of skin outside the abdomen that houses the testes; functions in maintaining the testes at the lower temperature required for spermatogenesis.

second law of thermodynamics The principle stating that every energy transfer or transformation increases the entropy of the universe. Usable forms of energy are at least partly converted to heat.

second messenger A small, nonprotein, water-soluble molecule or ion, such as a calcium ion (Ca^{2+}) or cyclic AMP, that relays a signal to a cell's interior in response to a signaling molecule bound by a signal receptor protein.

secondary cell wall In plant cells, a strong and durable matrix that is often deposited in several laminated layers around the plasma membrane and that provides protection and support.

secondary consumer A carnivore that eats herbivores.

secondary endosymbiosis A process in eukaryotic evolution in which a heterotrophic eukaryotic cell engulfed a photosynthetic eukaryotic cell, which survived in a symbiotic relationship inside the heterotrophic cell.

secondary growth Growth produced by lateral meristems, thickening the roots and shoots of woody plants.

secondary immune response The adaptive immune response elicited on second or subsequent exposures to a particular antigen. The secondary immune response is more rapid, of greater magnitude, and of longer duration than the primary immune response.

secondary oocyte (ō′-uh-sīt) An oocyte that has completed the first of the two meiotic divisions.

secondary production The amount of chemical energy in consumers' food that is converted to their own new biomass during a given time period.

secondary structure Regions of repetitive coiling or folding of the polypeptide backbone of a protein due to hydrogen bonding between constituents of the backbone (not the side chains).

secondary succession A type of succession that occurs where an existing community has been cleared by some disturbance that leaves the soil or substrate intact.

secretion (1) The discharge of molecules synthesized by a cell. (2) The discharge of wastes from the body fluid into the filtrate.

seed An adaptation of some terrestrial plants consisting of an embryo packaged along with a store of food within a protective coat.

seed coat A tough outer covering of a seed, formed from the outer coat of an ovule. In a flowering plant, the seed coat encloses and protects the embryo and endosperm.

seedless vascular plant An informal name for a plant that has vascular tissue but lacks seeds. Seedless vascular plants form a paraphyletic group that includes the phyla Lycophyta (club mosses and their relatives) and Monilophyta (ferns and their relatives).

selective permeability A property of biological membranes that allows them to regulate the passage of substances across them.

self-incompatibility The ability of a seed plant to reject its own pollen and sometimes the pollen of closely related individuals.

semelparity Reproduction in which an organism produces all of its offspring in a single event; also known as big-bang reproduction.

semen (sē′-mun) The fluid that is ejaculated by the male during orgasm; contains sperm and secretions from several glands of the male reproductive tract.

semicircular canals A three-part chamber of the inner ear that functions in maintaining equilibrium.

semiconservative model Type of DNA replication in which the replicated double helix consists of one old strand, derived from the parental molecule, and one newly made strand.

semilunar valve A valve located at each exit of the heart, where the aorta leaves the left ventricle and the pulmonary artery leaves the right ventricle.

seminal vesicle (sem′-i-nul ves′-i-kul) A gland in males that secretes a fluid component of semen that lubricates and nourishes sperm.

seminiferous tubule (sem′-i-nif′-er-us) A highly coiled tube in the testis in which sperm are produced.

senescence (se-nes′-ens) The growth phase in a plant or plant part (as a leaf) from full maturity to death.

sensitive period A limited phase in an animal's development when learning of particular behaviors can take place; also called a critical period.

sensor In homeostasis, a receptor that detects a stimulus.

sensory adaptation The tendency of sensory neurons to become less sensitive when they are stimulated repeatedly.

sensory neuron A nerve cell that receives information from the internal or external environment and transmits signals to the central nervous system.

sensory reception The detection of a stimulus by sensory cells.

sensory receptor An organ, cell, or structure within a cell that responds to specific stimuli from an organism's external or internal environment.

sensory transduction The conversion of stimulus energy to a change in the membrane potential of a sensory receptor cell.

sepal (sē′-pul) A modified leaf in angiosperms that helps enclose and protect a flower bud before it opens.

serial endosymbiosis A hypothesis for the origin of eukaryotes consisting of a sequence of endosymbiotic events in which mitochondria, chloroplasts, and perhaps other cellular structures were derived from small prokaryotes that had been engulfed by larger cells.

serotonin (ser′-uh-tō′-nin) A neurotransmitter, synthesized from the amino acid tryptophan, that functions in the central nervous system.

Sertoli cell A support cell of the seminiferous tubule that surrounds and nourishes developing sperm.

set point In homeostasis in animals, a value maintained for a particular variable, such as body temperature or solute concentration.

sex chromosome A chromosome responsible for determining the sex of an individual.

sex-linked gene A gene located on either sex chromosome. Most sex-linked genes are on the X chromosome and show distinctive patterns of inheritance; there are very few genes on the Y chromosome.

sexual dimorphism (dī-mōr′-fizm) Differences between the secondary sex characteristics of males and females of the same species.

sexual reproduction A type of reproduction in which two parents give rise to offspring that have unique combinations of genes inherited from both parents via the gametes.

sexual selection A form of natural selection in which individuals with certain inherited characteristics are more likely than other individuals to obtain mates.

Shannon diversity index An index of community diversity symbolized by H and represented by the equation $H = -(p_A \ln p_A + p_B \ln p_B + p_C \ln p_C + \ldots)$, where A, B, C . . . are species, p is the relative abundance of each species, and ln is the natural logarithm.

shared ancestral character A character that is shared by members of a particular clade but that originated in an ancestor that is not a member of that clade.

shared derived character An evolutionary novelty that is unique to a particular clade.

shoot system The aerial portion of a plant body, consisting of stems, leaves, and (in angiosperms) flowers.

short tandem repeat (STR) Simple sequence DNA containing multiple tandemly repeated units of two to five nucleotides. Variations in STRs act as genetic markers in STR analysis, used to prepare genetic profiles.

short-day plant A plant that flowers (usually in late summer, fall, or winter) only when the light period is shorter than a critical length.

short-term memory The ability to hold information, anticipations, or goals for a time and then release them if they become irrelevant.

sickle-cell disease A recessively inherited human blood disorder in which a single nucleotide change in the β-globin gene causes hemoglobin to aggregate, changing red blood cell shape and causing multiple symptoms in afflicted individuals.

sieve plate An end wall in a sieve-tube element, which facilitates the flow of phloem sap in angiosperm sieve tubes.

sieve-tube element A living cell that conducts sugars and other organic nutrients in the phloem of angiosperms; also called a sieve-tube member. Connected end to end, they form sieve tubes.

sign stimulus An external sensory cue that triggers a fixed action pattern by an animal.

signal In animal behavior, transmission of a stimulus from one animal to another. The term is also used in the context of communication in other kinds of organisms and in cell-to-cell communication in all multicellular organisms.

signal peptide A sequence of about 20 amino acids at or near the leading (amino) end of a polypeptide that targets it to the endoplasmic reticulum or other organelles in a eukaryotic cell.

signal transduction pathway A series of steps linking a mechanical, chemical, or electrical stimulus to a specific cellular response.

signal-recognition particle (SRP) A protein-RNA complex that recognizes a signal peptide as it emerges from a ribosome and helps direct the ribosome to the endoplasmic reticulum (ER) by binding to a receptor protein on the ER.

silent mutation A nucleotide-pair substitution that has no observable effect on the phenotype; for example, within a gene, a mutation that results in a codon that codes for the same amino acid.

simple fruit A fruit derived from a single carpel or several fused carpels.

simple sequence DNA A DNA sequence that contains many copies of tandemly repeated short sequences.

single bond A single covalent bond; the sharing of a pair of valence electrons by two atoms.

single circulation A circulatory system consisting of a single pump and circuit, in which blood passes from the sites of gas exchange to the rest of the body before returning to the heart.

single nucleotide polymorphism (SNP) A single base-pair site in a genome where nucleotide variation is found in at least 1% of the population.

single-lens eye The camera-like eye found in some jellies, polychaete worms, spiders, and many molluscs.

single-strand binding protein A protein that binds to the unpaired DNA strands during DNA replication, stabilizing them and holding them apart while they serve as templates for the synthesis of complementary strands of DNA.

sinoatrial (SA) node A region in the right atrium of the heart that sets the rate and timing at which all cardiac muscle cells contract; the pacemaker.

sister chromatids Two copies of a duplicated chromosome attached to each other by proteins at the centromere and, sometimes, along the arms. While joined, two sister chromatids make up one chromosome. Chromatids are eventually separated during mitosis or meiosis II.

sister taxa Groups of organisms that share an immediate common ancestor and hence are each other's closest relatives.

skeletal muscle A type of striated muscle that is generally responsible for the voluntary movements of the body.

sliding-filament model The idea that muscle contraction is based on the movement of thin (actin) filaments along thick (myosin) filaments, shortening the sarcomere, the basic unit of muscle organization.

slow-twitch fiber A muscle fiber that can sustain long contractions.

small interfering RNA (siRNA) One of multiple small, single-stranded RNA molecules generated by cellular machinery from a long, linear, double-stranded RNA molecule. The siRNA associates with one or more proteins in a complex that can degrade or prevent translation of an mRNA with a complementary sequence. In some cases, siRNA can also block transcription by promoting chromatin modification.

small intestine The longest section of the alimentary canal, so named because of its small diameter compared with that of the large intestine; the principal site of the enzymatic hydrolysis of food macromolecules and the absorption of nutrients.

smooth ER That portion of the endoplasmic reticulum that is free of ribosomes.

smooth muscle A type of muscle lacking the striations of skeletal and cardiac muscle because of the uniform distribution of myosin filaments in the cells; responsible for involuntary body activities.

social learning Modification of behavior through the observation of other individuals.

sodium-potassium pump A transport protein in the plasma membrane of animal cells that actively transports sodium out of the cell and potassium into the cell.

solute (sol′-yūt) A substance that is dissolved in a solution.

solute potential (Ψ_S) A component of water potential that is proportional to the molarity of a solution and that measures the effect of solutes on the direction of water movement; also called osmotic potential, it can be either zero or negative.

solution A liquid that is a homogeneous mixture of two or more substances.

solvent The dissolving agent of a solution. Water is the most versatile solvent known.

somatic cell (sō-mat′-ik) Any cell in a multicellular organism except a sperm or egg or their precursors.

spatial learning The establishment of a memory that reflects the environment's spatial structure.

spatial summation A phenomenon of neural integration in which the membrane potential of the postsynaptic cell is determined by the combined effect of EPSPs or IPSPs produced nearly simultaneously by different synapses.

speciation (spē′-sē-ā′-shun) An evolutionary process in which one species splits into two or more species.

species (spē′-sēz) A population or group of populations whose members have the potential to interbreed in nature and produce viable, fertile offspring, but do not produce viable, fertile offspring with members of other such groups.

species diversity The number and relative abundance of species in a biological community.

species richness The number of species in a biological community.

species-area curve The biodiversity pattern that shows that the larger the geographic area of a community is, the more species it has.

specific heat The amount of heat that must be absorbed or lost for 1 g of a substance to change its temperature by 1°C.

spectrophotometer An instrument that measures the proportions of light of different wavelengths absorbed and transmitted by a pigment solution.

sperm The male gamete.

spermatogenesis The continuous and prolific production of mature sperm cells in the testis.

spermatogonium (plural, **spermatogonia**) A cell that divides mitotically to form spermatocytes.

sphincter (sfink′-ter) A ringlike band of muscle fibers that controls the size of an opening in the body, such as the passage between the esophagus and the stomach.

spliceosome (spli′-sō-sōm) A large complex made up of proteins and RNA molecules that splices RNA by interacting with the ends of an RNA intron, releasing the intron and joining the two adjacent exons.

spontaneous process A process that occurs without an overall input of energy; a process that is energetically favorable.

sporangium (spōr-an′-jē-um) (plural, **sporangia**) A multicellular organ in fungi and plants in which meiosis occurs and haploid cells develop.

spore (1) In the life cycle of a plant or alga undergoing alternation of generations, a haploid cell produced in the sporophyte by meiosis. A spore can divide by mitosis to develop into a multicellular haploid individual, the gametophyte, without fusing with another cell. (2) In fungi, a haploid cell, produced either sexually or asexually, that produces a mycelium after germination.

sporophyte (spō-ruh-fīt′) In organisms (plants and some algae) that have alternation of generations, the multicellular diploid form that results from the union of gametes. The sporophyte produces haploid spores by meiosis that develop into gametophytes.

sporopollenin (spōr-uh-pol′-eh-nin) A durable polymer that covers exposed zygotes of charophyte algae and forms the walls of plant spores, preventing them from drying out.

stabilizing selection Natural selection in which intermediate phenotypes survive or reproduce more successfully than do extreme phenotypes.

stamen (stā′-men) The pollen-producing reproductive organ of a flower, consisting of an anther and a filament.

starch A storage polysaccharide in plants, consisting entirely of glucose monomers joined by α glycosidic linkages.

start point In transcription, the nucleotide position on the promoter where RNA polymerase begins synthesis of RNA.

statocyst (stat′-uh-sist′) A type of mechanoreceptor that functions in equilibrium in invertebrates by use of statoliths, which stimulate hair cells in relation to gravity.

statolith (stat′-uh-lith′) (1) In plants, a specialized plastid that contains dense starch grains and may play a role in detecting gravity. (2) In invertebrates, a dense particle that settles in response to gravity and is found in sensory organs that function in equilibrium.

stele (stēl) The vascular tissue of a stem or root.

stem A vascular plant organ consisting of an alternating system of nodes and internodes that support the leaves and reproductive structures.

stem cell Any relatively unspecialized cell that can produce, during a single division, one identical daughter cell and one more specialized daughter cell that can undergo further differentiation.

steroid A type of lipid characterized by a carbon skeleton consisting of four fused rings with various chemical groups attached.

sticky end A single-stranded end of a double-stranded restriction fragment.

stigma (plural, **stigmata**) The sticky part of a flower's carpel, which receives pollen grains.

stimulus In feedback regulation, a fluctuation in a variable that triggers a response.

stipe A stemlike structure of a seaweed.

stock The plant that provides the root system when making a graft.

stoma (stō′-muh) (plural, **stomata**) A microscopic pore surrounded by guard cells in the epidermis of leaves and stems that allows gas exchange between the environment and the interior of the plant.

stomach An organ of the digestive system that stores food and performs preliminary steps of digestion.

stramenopiles One of the three major subgroups for which the "SAR" eukaryotic supergroup is named. This clade arose by secondary endosymbiosis and includes diatoms and brown algae.

stratum (strah′-tum) (plural, **strata**) A rock layer formed when new layers of sediment cover older ones and compress them.

stroke The death of nervous tissue in the brain, usually resulting from rupture or blockage of arteries in the head.

stroma (strō′-muh) The dense fluid within the chloroplast surrounding the thylakoid membrane and containing ribosomes and DNA; involved in the synthesis of organic molecules from carbon dioxide and water.

stromatolite Layered rock that results from the activities of prokaryotes that bind thin films of sediment together.

style The stalk of a flower's carpel, with the ovary at the base and the stigma at the top.

substrate The reactant on which an enzyme works.

substrate feeder An animal that lives in or on its food source, eating its way through the food.

substrate-level phosphorylation The enzyme-catalyzed formation of ATP by direct transfer of a phosphate group to ADP from an intermediate substrate in catabolism.

sugar sink A plant organ that is a net consumer or storer of sugar. Growing roots, shoot tips, stems, and fruits are examples of sugar sinks supplied by phloem.

sugar source A plant organ in which sugar is being produced by either photosynthesis or the breakdown of starch. Mature leaves are the primary sugar sources of plants.

sulfhydryl group A chemical group consisting of a sulfur atom bonded to a hydrogen atom.

suprachiasmatic nucleus (SCN) A group of neurons in the hypothalamus of mammals that functions as a biological clock.

surface tension A measure of how difficult it is to stretch or break the surface of a liquid.

surfactant A substance secreted by alveoli that decreases surface tension in the fluid that coats the alveoli.

survivorship curve A plot of the number of members of a cohort that are still alive at each age; one way to represent age-specific mortality.

suspension feeder An animal that feeds by capturing small organisms or food particles suspended in the surrounding medium.

sustainable development Development that meets the needs of people today without limiting the ability of future generations to meet their needs.

symbiont (sim′-bē-ont) The smaller participant in a symbiotic relationship, living in or on the host.

symbiosis An ecological relationship between organisms of two different species that live together in direct and intimate contact.

sympathetic division One of three divisions of the autonomic nervous system; generally increases energy expenditure and prepares the body for action.

sympatric speciation (sim-pat′-rik) The formation of new species in populations that live in the same geographic area.

symplast In plants, the continuum of cytoplasm connected by plasmodesmata between cells.

synapse (sin′-aps) The junction where a neuron communicates with another cell across a narrow gap via a neurotransmitter or an electrical coupling.

synapsid Member of an amniote clade distinguished by a single hole on each side of the skull. Synapsids include the mammals.

synapsis (si-nap′-sis) The pairing and physical connection of duplicated homologous chromosomes during prophase I of meiosis.

systematics A scientific discipline focused on classifying organisms and determining their evolutionary relationships.

systemic acquired resistance A defensive response in infected plants that helps protect healthy tissue from pathogenic invasion.

systemic circuit The branch of the circulatory system that supplies oxygenated blood to and carries deoxygenated blood away from organs and tissues throughout the body.

systems biology An approach to studying biology that aims to model the dynamic behavior of whole biological systems based on a study of the interactions among the system's parts.

systole (sis′-tō-lē) The stage of the cardiac cycle in which a heart chamber contracts and pumps blood.

T cells The class of lymphocytes that mature in the thymus; they include both effector cells for the cell-mediated immune response and helper cells required for both branches of adaptive immunity.

taproot A main vertical root that develops from an embryonic root and gives rise to lateral (branch) roots.

tastant Any chemical that stimulates the sensory receptors in a taste bud.

taste bud A collection of modified epithelial cells on the tongue or in the mouth that are receptors for taste in mammals.

TATA box A DNA sequence in eukaryotic promoters crucial in forming the transcription initiation complex.

taxis (tak′-sis) An oriented movement toward or away from a stimulus.

taxon (plural, **taxa**) A named taxonomic unit at any given level of classification.

taxonomy (tak-son′-uh-mē) A scientific discipline concerned with naming and classifying the diverse forms of life.

Tay-Sachs disease A human genetic disease caused by a recessive allele for a dysfunctional enzyme, leading to accumulation of certain lipids in the brain. Seizures, blindness, and degeneration of motor and mental performance usually become manifest a few months after birth, followed by death within a few years.

technology The application of scientific knowledge for a specific purpose, often involving industry or commerce but also including uses in basic research.

telophase The fifth and final stage of mitosis, in which daughter nuclei are forming and cytokinesis has typically begun.

temperate broadleaf forest A biome located throughout midlatitude regions where there is sufficient moisture to support the growth of large, broadleaf deciduous trees.

temperate grassland A terrestrial biome that exists at midlatitude regions and is dominated by grasses and forbs.

temperate phage A phage that is capable of replicating by either a lytic or lysogenic cycle.

temperature A measure in degrees of the average kinetic energy (thermal energy) of the atoms and molecules in a body of matter.

template strand The DNA strand that provides the pattern, or template, for ordering, by complementary base pairing, the sequence of nucleotides in an RNA transcript.

temporal summation A phenomenon of neural integration in which the membrane potential of the postsynaptic cell in a chemical synapse is determined by the combined effect of EPSPs or IPSPs produced in rapid succession.

terminator In bacteria, a sequence of nucleotides in DNA that marks the end of a gene and signals RNA polymerase to release the newly made RNA molecule and detach from the DNA.

territoriality A behavior in which an animal defends a bounded physical space against encroachment by other individuals, usually of its own species.

tertiary consumer (ter'-shē-ār'-ē) A carnivore that eats other carnivores.

tertiary structure The overall shape of a protein molecule due to interactions of amino acid side chains, including hydrophobic interactions, ionic bonds, hydrogen bonds, and disulfide bridges.

test In foram protists, a porous shell that consists of a single piece of organic material hardened with calcium carbonate.

testcross Breeding an organism of unknown genotype with a homozygous recessive individual to determine the unknown genotype. The ratio of phenotypes in the offspring reveals the unknown genotype.

testis (plural, **testes**) The male reproductive organ, or gonad, in which sperm and reproductive hormones are produced.

testosterone A steroid hormone required for development of the male reproductive system, spermatogenesis, and male secondary sex characteristics; the major androgen in mammals.

tetanus (tet'-uh-nus) The maximal, sustained contraction of a skeletal muscle, caused by a very high frequency of action potentials elicited by continual stimulation.

tetrapod Member of a vertebrate clade characterized by limbs with digits. Tetrapods include mammals, amphibians, and birds and other reptiles.

thalamus (thal'-uh-mus) An integrating center of the vertebrate forebrain. Neurons with cell bodies in the thalamus relay neural input to specific areas in the cerebral cortex and regulate what information goes to the cerebral cortex.

theory An explanation that is broader in scope than a hypothesis, generates new hypotheses, and is supported by a large body of evidence.

thermal energy Kinetic energy due to the random motion of atoms and molecules; energy in its most random form. *See also* heat.

thermocline A narrow stratum of abrupt temperature change in the ocean and in many temperate-zone lakes.

thermodynamics (ther'-mō-dī-nam'-iks) The study of energy transformations that occur in a collection of matter. *See* first law of thermodynamics; second law of thermodynamics.

thermoreceptor A receptor stimulated by either heat or cold.

thermoregulation The maintenance of internal body temperature within a tolerable range.

thick filament A filament composed of staggered arrays of myosin molecules; a component of myofibrils in muscle fibers.

thigmomorphogenesis A response in plants to chronic mechanical stimulation, resulting from increased ethylene production. An example is thickening stems in response to strong winds.

thigmotropism (thig-mo'-truh-pizm) A directional growth of a plant in response to touch.

thin filament A filament consisting of two strands of actin and two strands of regulatory protein coiled around one another; a component of myofibrils in muscle fibers.

threatened species A species that is considered likely to become endangered in the foreseeable future.

threshold The potential that an excitable cell membrane must reach for an action potential to be initiated.

thrombus A fibrin-containing clot that forms in a blood vessel and blocks the flow of blood.

thylakoid (thī'-luh-koyd) A flattened, membranous sac inside a chloroplast. Thylakoids often exist in stacks called grana that are interconnected; their membranes contain molecular "machinery" used to convert light energy to chemical energy.

thymus (thī'-mus) A small organ in the thoracic cavity of vertebrates where maturation of T cells is completed.

tidal volume The volume of air a mammal inhales and exhales with each breath.

tight junction A type of intercellular junction between animal cells that prevents the leakage of material through the space between cells.

tissue An integrated group of cells with a common structure, function, or both.

Toll-like receptor (TLR) A membrane receptor on a phagocytic white blood cell that recognizes fragments of molecules common to a set of pathogens.

tonicity The ability of a solution surrounding a cell to cause that cell to gain or lose water.

top-down model A model of community organization in which predation influences community organization by controlling herbivore numbers, which in turn control plant or phytoplankton numbers, which in turn control nutrient levels; also called the trophic cascade model.

topoisomerase A protein that breaks, swivels, and rejoins DNA strands. During DNA replication, topoisomerase helps to relieve strain in the double helix ahead of the replication fork.

totipotent (tō'-tuh-pōt'-ent) Describing a cell that can give rise to all parts of the embryo and adult, as well as extraembryonic membranes in species that have them.

trace element An element indispensable for life but required in extremely minute amounts.

trachea (trā'-kē-uh) The portion of the respiratory tract that passes from the larynx to the bronchi; also called the windpipe.

tracheal system In insects, a system of branched, air-filled tubes that extends throughout the body and carries oxygen directly to cells.

tracheid (trā'-kē-id) A long, tapered water-conducting cell found in the xylem of nearly all vascular plants. Functioning tracheids are no longer living.

trait One of two or more detectable variants in a genetic character.

transcription The synthesis of RNA using a DNA template.

transcription factor A regulatory protein that binds to DNA and affects transcription of specific genes.

transcription initiation complex The completed assembly of transcription factors and RNA polymerase bound to a promoter.

transcription unit A region of DNA that is transcribed into an RNA molecule.

transduction (1) A process in which phages (viruses) carry bacterial DNA from one bacterial cell to another. When these two cells are members of different species, transduction results in horizontal gene transfer. (2) In cellular communication, the conversion of a signal from outside the cell to a form that can bring about a specific cellular response; also called signal transduction.

transfer RNA (tRNA) An RNA molecule that functions as a translator between nucleic acid and protein languages by carrying specific amino acids to the ribosome, where they recognize the appropriate codons in the mRNA.

transformation (1) The conversion of a normal animal cell to a cancerous cell. (2) A change in genotype and phenotype due to the assimilation of external DNA by a cell. When the external DNA is from a member of a different species, transformation results in horizontal gene transfer.

transgenic Pertaining to an organism whose genome contains a gene introduced from another organism of the same or a different species.

translation The synthesis of a polypeptide using the genetic information encoded in an mRNA molecule. There is a change of "language" from nucleotides to amino acids.

translocation (1) An aberration in chromosome structure resulting from attachment of a chromosomal fragment to a nonhomologous chromosome. (2) During protein synthesis, the third stage in the elongation cycle, when the RNA carrying the growing polypeptide moves from the A site to the P site on the ribosome. (3) The transport of organic nutrients in the phloem of vascular plants.

transmission electron microscope (TEM) A microscope that passes an electron beam through very thin sections stained with metal atoms and is primarily used to study the internal ultrastructure of cells.

transpiration The evaporative loss of water from a plant.

transport epithelium One or more layers of specialized epithelial cells that carry out and regulate solute movement.

transport protein A transmembrane protein that helps a certain substance or class of closely related substances to cross the membrane.

transport vesicle A small membranous sac in a eukaryotic cell's cytoplasm carrying molecules produced by the cell.

transposable element A segment of DNA that can move within the genome of a cell by means of a DNA or RNA intermediate; also called a transposable genetic element.

transposon A transposable element that moves within a genome by means of a DNA intermediate.

transverse (T) tubule An infolding of the plasma membrane of skeletal muscle cells.

triacylglycerol (trī-as′-ul-glis′-uh-rol) A lipid consisting of three fatty acids linked to one glycerol molecule; also called a fat or triglyceride.

triple response A plant growth maneuver in response to mechanical stress, involving slowing of stem elongation, thickening of the stem, and a curvature that causes the stem to start growing horizontally.

triplet code A genetic information system in which a set of three-nucleotide-long words specifies the amino acids for polypeptide chains.

trisomic Referring to a diploid cell that has three copies of a particular chromosome instead of the normal two.

trophic efficiency The percentage of production transferred from one trophic level to the next.

trophic structure The different feeding relationships in an ecosystem, which determine the route of energy flow and the pattern of chemical cycling.

trophoblast The outer epithelium of a mammalian blastocyst. It forms the fetal part of the placenta, supporting embryonic development but not forming part of the embryo proper.

tropic hormone A hormone that has an endocrine gland or endocrine cells as a target.

tropical dry forest A terrestrial biome characterized by relatively high temperatures and precipitation overall but with a pronounced dry season.

tropical rain forest A terrestrial biome characterized by relatively high precipitation and temperatures year-round.

tropics Latitudes between 23.5° north and south.

tropism A growth response that results in the curvature of whole plant organs toward or away from stimuli due to differential rates of cell elongation.

tropomyosin The regulatory protein that blocks the myosin-binding sites on actin molecules.

troponin complex The regulatory proteins that control the position of tropomyosin on the thin filament.

true-breeding Referring to organisms that produce offspring of the same variety over many generations of self-pollination.

tumor-suppressor gene A gene whose protein product inhibits cell division, thereby preventing the uncontrolled cell growth that contributes to cancer.

tundra A terrestrial biome at the extreme limits of plant growth. At the northernmost limits, it is called arctic tundra, and at high altitudes, where plant forms are limited to low shrubby or matlike vegetation, it is called alpine tundra.

turgid (ter′-jid) Swollen or distended, as in plant cells. (A walled cell becomes turgid if

it has a lower water potential than its surroundings, resulting in entry of water.)

turgor pressure The force directed against a plant cell wall after the influx of water and swelling of the cell due to osmosis.

turnover time The time required to replace the standing crop of a population or group of populations (for example, of phytoplankton), calculated as the ratio of standing crop to production.

twin study A behavioral study in which researchers compare the behavior of identical twins raised apart with that of identical twins raised in the same household.

tympanic membrane Another name for the eardrum, the membrane between the outer and middle ear.

Unikonta (yū′-ni-kon′-tuh) One of four supergroups of eukaryotes proposed in a current hypothesis of the evolutionary history of eukaryotes. This clade, which is supported by studies of myosin proteins and DNA, consists of amoebozoans and opisthokonts. *See also* Excavata, "SAR" clade, and Archaeplastida.

unsaturated fatty acid A fatty acid that has one or more double bonds between carbons in the hydrocarbon tail. Such bonding reduces the number of hydrogen atoms attached to the carbon skeleton.

urea A soluble nitrogenous waste produced in the liver by a metabolic cycle that combines ammonia with carbon dioxide.

ureter (yū-rē′-ter) A duct leading from the kidney to the urinary bladder.

urethra (yū-rē′-thruh) A tube that releases urine from the mammalian body near the vagina in females and through the penis in males; also serves in males as the exit tube for the reproductive system.

uric acid A product of protein and purine metabolism and the major nitrogenous waste product of insects, land snails, and many reptiles. Uric acid is relatively nontoxic and largely insoluble.

urinary bladder The pouch where urine is stored prior to elimination.

uterine cycle The changes that occur in the uterus during the reproductive cycle of the human female; also called the menstrual cycle.

uterus A female organ where eggs are fertilized and/or development of the young occurs.

utricle In the vertebrate ear, a chamber in the vestibule behind the oval window that opens into the three semicircular canals.

vaccine A harmless variant or derivative of a pathogen that stimulates a host's immune system to mount defenses against the pathogen.

vacuole (vak′-yū-ōl′) A membrane-bounded vesicle whose specialized function varies in different kinds of cells.

vagina Part of the female reproductive system between the uterus and the outside opening;

the birth canal in mammals. During copulation, the vagina accommodates the male's penis and receives sperm.

valence The bonding capacity of a given atom; the number of covalent bonds an atom can form usually equals the number of unpaired electrons in its outermost (valence) shell.

valence electron An electron in the outermost electron shell.

valence shell The outermost energy shell of an atom, containing the valence electrons involved in the chemical reactions of that atom.

van der Waals interactions Weak attractions between molecules or parts of molecules that results from transient local partial charges.

variation Differences between members of the same species.

vas deferens In mammals, the tube in the male reproductive system in which sperm travel from the epididymis to the urethra.

vasa recta The capillary system in the kidney that serves the loop of Henle.

vascular cambium A cylinder of meristematic tissue in woody plants that adds layers of secondary vascular tissue called secondary xylem (wood) and secondary phloem.

vascular plant A plant with vascular tissue. Vascular plants include all living plant species except liverworts, mosses, and hornworts.

vascular tissue Plant tissue consisting of cells joined into tubes that transport water and nutrients throughout the plant body.

vascular tissue system A transport system formed by xylem and phloem throughout a vascular plant. Xylem transports water and minerals; phloem transports sugars, the products of photosynthesis.

vasoconstriction A decrease in the diameter of blood vessels caused by contraction of smooth muscles in the vessel walls.

vasodilation An increase in the diameter of blood vessels caused by relaxation of smooth muscles in the vessel walls.

vector An organism that transmits pathogens from one host to another.

vegetative propagation Cloning of plants by humans.

vegetative reproduction Cloning of plants in nature.

vein (1) In animals, a vessel that carries blood toward the heart. (2) In plants, a vascular bundle in a leaf.

ventilation The flow of air or water over a respiratory surface.

ventral Pertaining to the underside, or bottom, of an animal with radial or bilateral symmetry.

ventricle (ven'-tri-kul) (1) A heart chamber that pumps blood out of the heart. (2) A

space in the vertebrate brain, filled with cerebrospinal fluid.

vernalization The use of cold treatment to induce a plant to flower.

vertebrate A chordate animal with a backbone. Vertebrates include sharks and rays, ray-finned fishes, coelacanths, lungfishes, amphibians, reptiles, and mammals.

vesicle (ves'-i-kul) A membranous sac in the cytoplasm of a eukaryotic cell.

vessel A nonliving, water-conducting tube found in most angiosperms and a few nonflowering vascular plants that is formed by the end-to-end connection of vessel elements.

vessel element A short, wide, water-conducting cell found in the xylem of most angiosperms and a few nonflowering vascular plants. Dead at maturity, vessel elements are aligned end to end to form vessels.

vestigial structure A feature of an organism that is a historical remnant of a structure that served a function in the organism's ancestors.

villus (plural, **villi**) (1) A finger-like projection of the inner surface of the small intestine. (2) A finger-like projection of the chorion of the mammalian placenta. Large numbers of villi increase the surface areas of these organs.

viral envelope A membrane, derived from membranes of the host cell, that cloaks the capsid, which in turn encloses a viral genome.

virulent Describing a pathogen against which an organism has little specific defense.

virulent phage A phage that replicates only by a lytic cycle.

virus An infectious particle incapable of replicating outside of a cell, consisting of an RNA or DNA genome surrounded by a protein coat (capsid) and, for some viruses, a membranous envelope.

visible light That portion of the electromagnetic spectrum that can be detected as various colors by the human eye, ranging in wavelength from about 380 nm to about 750 nm.

vital capacity The maximum volume of air that a mammal can inhale and exhale with each breath.

vitamin An organic molecule required in the diet in very small amounts. Many vitamins serve as coenzymes or parts of coenzymes.

voltage-gated ion channel A specialized ion channel that opens or closes in response to changes in membrane potential.

vulva Collective term for the female external genitalia.

water potential (Ψ) The physical property predicting the direction in which water will flow, governed by solute concentration and applied pressure.

wavelength The distance between crests of waves, such as those of the electromagnetic spectrum.

wetland A habitat that is inundated by water at least some of the time and that supports plants adapted to water-saturated soil.

white matter Tracts of axons within the CNS.

whole-genome shotgun approach Procedure for genome sequencing in which the genome is randomly cut into many overlapping short segments that are sequenced; computer software then assembles the complete sequence.

wild type The phenotype most commonly observed in natural populations; also refers to the individual with that phenotype.

wilting The drooping of leaves and stems that occurs when plant cells become flaccid.

wobble Flexibility in the base-pairing rules in which the nucleotide at the 5' end of a tRNA anticodon can form hydrogen bonds with more than one kind of base in the third position (3' end) of a codon.

xerophyte A plant adapted to an arid climate.

X-linked gene A gene located on the X chromosome; such genes show a distinctive pattern of inheritance.

X-ray crystallography A technique used to study the three-dimensional structure of molecules. It depends on the diffraction of an X-ray beam by the individual atoms of a crystallized molecule.

xylem (zi'-lum) Vascular plant tissue consisting mainly of tubular dead cells that conduct most of the water and minerals upward from the roots to the rest of the plant.

xylem sap The dilute solution of water and dissolved minerals carried through vessels and tracheids.

yeast Single-celled fungus. Yeasts reproduce asexually by binary fission or by the pinching of small buds off a parent cell. Many fungal species can grow both as yeasts and as a network of filaments; relatively few species grow only as yeasts.

zero population growth (ZPG) A period of stability in population size, when additions to the population through births and immigration are balanced by subtractions through deaths and emigration.

zoned reserve An extensive region that includes areas relatively undisturbed by humans surrounded by areas that have been changed by human activity and are used for economic gain.

zoonotic pathogen A disease-causing agent that is transmitted to humans from other animals.

zygote (zi'-gōt) The diploid cell produced by the union of haploid gametes during fertilization; a fertilized egg.

2,4-D (2,4-dichlorophenoxyacetic acid), 622
3-D shape, protein, 40*f*
3-Phosphoglycerate, 168, 171*f*
3' end, 59*f*, 61, 276–277
5-Methyl cytosine, 43*f*
5' cap, **276**–277
5' end, polynucleotide, 61
10-nm fibers, 259–260*f*
30-nm fibers, 259, 261*f*
300-nm fibers, 261*f*

A

ABC hypothesis, **599**
abdominal-A (abd-A) gene, 536*f*
Abedus, 733*f*
Abiotic factors, **821**
 in community equilibrium, 856
 in pollination, 602*f*
 in species distributions, 821, 831–832
Abiotic stresses, plant responses to, **633**–635
Abiotic synthesis, organic compound, 459–460
Abnormal chromosome numbers, 240–242
ABO blood groups, 98, 216, 724
Abomasum, 677*f*
Abortions, spontaneous, 240, 746
Abscisic acid (ABA), **591**–592, 620*t*, **623**–624, 633
Abscission, leaf, 623, 625–626
Absolute dating, 438
Absorption, **668**
 as food processing stage, 668
 fungal feeding by, 508–509
 in large intestine, 675–676
 in small intestine, 675
 soil as source of essential elements for, by
  roots, 578–582
 of water and minerals by root cells, 587
Absorption spectrum, **160**, 161*f*
Abstinence, contraception by, 747–748*f*
Acacia trees, 850*f*
Acceleration, mechanoreceptors and, 784–785*f*
Accessory fruits, **606**
Accessory glands, male reproductive, 734
Acclimatization, thermoregulation and, 647
Acer rubrum, 560
Acetic acid, 43*f*
Acetone, 43*f*
Acetylation, histone, 299
Acetylcholine, **763**–764, 795–797
Acetylcholinesterase, 764, 797
Acetyl CoA (acetyl coenzyme A), **142**–143*f*
Achondroplasia, 222
Acid growth hypothesis, 620–621
Acidic amino acids, 54
Acidic side chains, 53*f*
Acidification, ocean, 36–37

Acid reflux, 674
Acids, **34**–37
Acinonyx jubatus, 841*f*
Acorn worm, 535*f*
Acquired immunity. *See* Adaptive immunity
Acquired immunodeficiency syndrome (AIDS).
  See AIDS (acquired immunodeficiency
  syndrome)
Acquired traits, inheritance of, 367
Acropora cervicornis and *Acropora palmata*, 861
Acrosomal reaction, animal fertilization, 743*f*
Acrosomes, **736***f*
Actin, **87**–88, 180. *See also* Thick filaments (actin)
Actinistia, 538*f*
Actinomycetes, 473*f*
Actinopterygii, 538*f*
Action potentials
 conduction of, in neurons, 759
 evolution of axon structure and, 760
 generation of, in neurons, 757–759
 graded potentials and, 756–757
 of heart muscles, 798
 hyperpolarization and depolarization of, 756,
  757*f*
 in plants, **633**
 in sensory transmission, 780
 of skeletal muscles, 796*f*, 797
 of smooth muscles, 798–799
Action spectrum, **160**, 161*f*, **626**–627
Activation, allosteric, 131–132
Activation energy, **125**–126
Activators, 131, **297**, 301–302
Active immunity, **723**–724
Active sites, enzyme, **126**–128
Active transport, **103**–106, 574–575. *See also*
  Passive transport; Sodium-potassium
  pump
Actual evapotranspiration, 870
Actual range, species, 831
Acyclovir, 338
Adaptation, sensory, 780
Adaptations, **369**. *See also* Evolution; Natural
  selection
 adaptive evolution and, 410–415
 for animal gas exchange, 706–708
 artificial selection, natural selection, and,
  371–372
 of axon structure, 760
 circulatory, for thermoregulation, 646
 of digestive compartments, 670–671
 evolution and, 7
 floral, that prevent self-fertilization, 609
 fungal, for feeding by absorption, 508–509
 in herbivory, 849
 mouse coat coloration, 1, 12–15
 of pathogens, 725–726
 in predation, 848–849
 prokaryotic, 458–459*f*, 462–467
 that reduce terrestrial nutrient limitations, 870
 research by Charles Darwin on, 369
 respiratory, of diving mammals, 708
 sexual reproduction patterns as, 731–732
 terrestrial, of fungi and plants, 505–508,
  511–513, 516–517
 terrestrial, of land plants, 169–171
 vascular plant evaporative water loss, 592–593
 vascular plant nutritional, 582–586*f*
 vascular plant resource acquisition, 571–574

 of vertebrate digestive systems, 676–678
 of vertebrate kidneys to diverse environments,
  660–661
Adaptive evolution, **410**–415. *See also*
  Adaptations
 biodiversity from, 450
 directional, disruptive, and stabilizing selection
  in, 411–412
 natural selection in, 407, 412, 414–415
 preservation of genetic variation in, 413–414
 relative fitness and, 411
 sexual selection in, 412–413
Adaptive immunity, **712**, 715–727
 active and passive immunization in, 723–724
 allergies and, 724–725
 antibodies produced by, as tools in medicine,
  724
 antigen recognition by B cells and T cells in,
  715–717
 autoimmune diseases and, 725
 B cell and T cell development in, 717–720
 B cells and antibodies as responses to
  extracellular pathogens in, 722
 cancer and, 726–727
 cytotoxic T cells as responses to infected cells
  in, 721
 evolution of immune system avoidance and
  immunodeficiency, 725–726
 helper T cells as responses to antigens in,
  720–721
 human disorders from disruptions in, 724–727
 humoral and cell-mediated immune responses
  in, 720–723*f*
 immune rejection in, 724
 immunological memory in, 719–720
 innate immunity vs., 712
 origin of self-tolerance in, 718–719
 overview of, 723*f*
Adaptive management, 878–879*f*
Adaptive radiations, **447**–449. *See also* Radiations
Addiction, brain reward system and, 775
Addition rule, **213**
Adenine, 60–62, 248, 250–251, 622
Adenoid, 694*f*
Adenomatous polyposis coli (APC) gene, 327
Adenosine diphosphate. *See* ADP (adenosine
  diphosphate)
Adenosine triphosphate. *See* ATP (adenosine
  triphosphate)
Adenoviruses, 331
Adenylyl cyclase, 112–113*f*
Adhesion, **30**, 590
Adipose tissue, animal, 680
ADP (adenosine diphosphate)
 as enzyme activator, 131
 hydrolysis of ATP to, 122–124
 in muscle contraction, 794*f*–795
 as organic phosphate compound, 44
 synthesis of ATP from, 124, 139–140, 145–147
Adrenal cortex, **650***f*
Adrenal glands, **650***f*, 652
Adrenal medulla, **650***f*
Adult stem cells, 323–324
Adventitious roots, 555, 610
Aerial roots, 555*f*
Aerobic prokaryotes, 464*f*
Aerobic respiration, **136**, 150–151, 798. *See also*
  Cellular respiration

Afferent neurons, 770–771
African-Americans, sickle-cell disease in, 222
African buffalo, 851*f*
African elephants, 370–371, 836, 837*f*
African gray parrots, 778
Africans
    genomes of, 360
    malaria and sickle-cell alleles in, 414*f*
    sickle-cell disease in, 222
Agar, 270
Aggregate fruits, **606**
Aggression, animal, 804, 806
Aging
    cytokinins in plant, 622
    telomeric DNA and, 259
Aglaophyton major, 508*f*, 511*f*
Agonistic behavior, 811
Agriculture. *See also* Crop plants
    artificial selection and breeding in, 597,
        611–612
    C₃ plants in, 169
    C₄ plants in, 170
    community disturbances by, 858
    effects of atmospheric carbon dioxide on
        productivity of, 170
    fertilizers in, 870, 875*f*
    importance of mycorrhizae to, 585
    plant biotechnology and genetic engineering
        in, 612–614
    polyploidy in, 426
    prokaryotes in disease-suppressive soil and,
        477
    soil management in, 580–581
    viral diseases and, 341
Agrobacterium, 472*f*, 477, 485
AIDS (acquired immunodeficiency syndrome),
    336, 337*f*, 402, 725–726. *See also* HIV
    (human immunodeficiency virus)
Ailuropoda melanoleuca, 348*t*
Ain, Michael C., 222*f*
Air circulation patterns, global, 820*f*
Airfoils, wings as, 802
Air roots, 555*f*
Air sacs, 704
Åland Islands, 842
Alanine, 53*f*
Alarm signals, 806
Alberta, Canada, lake biome in, 828*f*
Albinism, 221*f*, 268
Albumin, 302*f*
Alcohol, 43*f*, 841*f*
Alcohol fermentation, **150**–151
Aldehyde compounds, 43*f*
Alder, 858
Aldosterone, **662**
Aleuria aurantia, 512*f*
Algae, **482***f*
    biomanipulation of, 855
    blooms of, 868
    cells of, 73*f* (*see also* Eukaryotic cells)
    chloroplasts of, 83*f*
    in eukaryotic phylogeny, 491*f*
    evolution of land plants from green, 505,
        507–508
    in fossil record, 482*f*–483*f*
    land animals vs., 540*f*
    lichens as symbioses of fungi and, 522*f*
    in marine ecosystems, 547
    oceans and photosynthetic, 827
    origins of photosynthetic, 486–487
    as photoautotrophs, 156*f*
    sexual life cycles in, 196–197
    wavelengths of light driving photosynthesis
        in, 161*f*
Alimentary canals, **670**–671, 677–678
Alkaptonuria, 269

Allantois, **544***f*
Alleles, **209**
    alteration in frequencies of, in populations,
        400, 406–410
    behavior of recessive, 221
    correlating behavior of chromosome pairs
        with, 230–231
    dominant, 215–216
    dominant vs. recessive, 209–210, 220
    evolution and genetic variation from, 204
    frequencies of, in populations, 402–406
    genetic variation and, 201, 413–414
    homozygous vs. heterozygous organisms and,
        210
    in meiosis, 197
    microevolution as alteration in frequencies of,
        in populations, 400
    multiple, and ABO blood groups, 216
    mutations as sources of new, 240, 401–402
Allergens, 613, **724**–725
Allergies, 724–725
Alligators, 545*f*
Alligator snapping turtles, 848
Allolactose, 296
Allopatric speciation, **423**
    character displacement and, 847
    continental drift and, 444
    evidence of, 424–425
    identifying dependent and independent
        variables, making scatter plots, and
        interpreting data on, 427
    process of, 423
    sympatric speciation vs., 423*f*
Allopolyploids, **425**–426
All-or-none responses, 757
Allosteric regulation, 130, **131**, 132
Alpha (α) carbon, 52
Alpha cells, pancreatic, 680
α chain, antigen Alpha, 716–717
α-globin genes, 352–353
α glucose ring structure, polysaccharide, 48*f*
α-helix, **56***f*, 97*f*
α-lactalbumin, 355
Alpha proteobacteria, 472*f*, 485–486
Alpheus genus, 424–425
Alpine woodsorrel, 609*f*
Alternate phyllotaxy, 573
Alternation of generations, **196**, **506***f*–507
Alternative RNA splicing, **278**, **304**, 348
Altruism, 813–814
Alu elements, 351, 359
Aluminum, bioremediation of, 877
Alveolates, 491*f*, **494**, 501–502
Alveoli, 494, **703**, 704
Alzheimer's disease, 59
Amacrine cells, 786*f*, 788
Amanita muscaria, 512*f*
Amborella trichopoda, 520*f*–521
American black bears, 8
Amine compounds, 43*f*
Amino acids, **52**
    abiotic synthesis of, 459–460
    activation of, in translation, 280–281, 283*f*,
        287*f*
    analyzing sequence data of, 63
    in catabolic and anabolic pathways, 152
    essential, for animal nutrition, 666
    evolution of human globin gene sequences of,
        356
    genetic code for, 272–274
    neurotransmitters as, 764
    polypeptides as polymers of, 52–54
    in protein structure, 56*f*
    in sickle-cell disease, 58
    side chains (R groups) of, 53*f*
Amino acid sequence identity tables, 356

Aminoacyl-tRNA synthetases, **280**–281, 283*f*
Amino end, polypeptide, 54
Amino group, 43*f*, 52
Amitochondriate protists, 489
Ammonia, 27*f*, 34–35, 653, **654**–655
Ammonifying bacteria, 582
Ammonium, 875*f*
Ammonium chloride, 26
Amnion, 543–**544***f*
Amniotes, **543**–547
Amniotic eggs, **543**–544
Amoebas, 88, 175*f*, 491*f*, **494**, 497
Amoebocytes, 529*f*–**530**
Amoeboid movement, 88
Amoebozoans, **497**–498
Amphibians, **543**
    circulatory systems of, 687
    external fertilization in, 732*f*
    gills of, 684
    parental care in, 811
    skin as respiratory organ for, 701
    as terrestrial vertebrates, 543
Amphipathic molecules, **95**
Amplification, cancer gene, 324–325
Amplification, DNA, 264–265
Amplification, PCR. *See* Polymerase chain
    reaction (PCR)
Amplification, sensory, **780**
Amygdala, 775
Amylase, **671**, 673*f*
Amylopectin, 47
Amyloplasts, 83
Amylose, 47*f*
Amyotrophic lateral sclerosis (ALS), 795
Anabaena, 466
Anableps anableps, 293
Anabolic pathways, **117**, 152
Anabrus simplex, 348
Anacystis nidulans, 485
Anaerobic respiration, 136, 148–151, **465**–466
Analgesics, 764
Analogies, **385**–386
Analogous structures, **376**
Anaphase, **177**, 179*f*, 181*f*, 182*f*, 200*f*
Anaphase I, 198*f*, 200*f*
Anaphase II, 199*f*
Anaphylactic shock, 724–725
Anatomical homologies, 375–376
Anatomy, **641**
Ancestry, common, 375–376, 384, 387–388
Anchorage dependence, **186**–187
Anchoring cell junctions, 90*f*
Androgens, **650***f*, 738–739
Anemia, 696
Aneuploidies, **240**–242
Angina pectoris, 698
Angiosperms, **516**, 597–616. *See also* Crop plants;
    Plant(s)
    artificial selection, breeding, and genetic
        engineering of, 597, 611–614
    bulk flow translocation in, 594
    development of (*see* Plant development)
    evolution of, 519–521
    evolution of organs of, 554–556
    flowers, seeds, and fruits of, 518–519, 597–
        607*f* (*see also* Flowers; Fruits; Seeds)
    gametophyte-sporophyte relationship in, 515*f*
    insect radiations and radiation of, 541–542
    life cycles of, 600*f*
    meristematic control of flowering of, 562
    monocot vs. eudicot, 553–554*f*
    overview of structure of, 555*f*
    phylogeny of, 513*f*, 520–521
    as seed plants, 516
    sexual and asexual reproduction in, 608–611
Angiotensin converting enzyme (ACE), 662

Angiotensin II, **662**
Angraecum sesquipedale, 548*f*
Angular motion, mechanoreceptors and, 784–785*f*
Animal(s), 528–551. *See also* Birds; Eukaryotes; Fishes; Human(s); Insects; Invertebrates; Mammals; Vertebrates
    anatomy-physiology correlation in, 641
    aquatic vs. terrestrial, 540*f* (*see also* Aquatic animals; Land animals)
    behaviors of (*see* Animal behaviors)
    body plans of, 532–533
    brains of (*see* Brains)
    Cambrian explosion and bilaterian radiation of, 530–532
    catabolism and diets of, 135–136, 151–152
    cells of (*see* Animal cells)
    chitin as structural polysaccharide of, 49
    circulation and gas exchange in (*see* Cardiovascular systems; Circulatory systems; Gas exchange)
    cloning of, 320–322
    colonization of land by, 539–547
    comparing genomes of, 358–360 (*see also* Genome(s))
    development processes of, 360–361 (*see also* Embryonic development)
    diseases and disorders of, 331–332, 338–340, 841*f*
    domain Eukarya and, 8*f*
    ecological and evolutionary effects of, 547–550
    embryonic development of (*see* Embryonic development)
    in energy flow and chemical cycling, 6–7
    evolution of, 488–489
    flower pollination by, 602*f*
    in fossil record, 436–440, 441*f*, 531*f*
    fruit and seed dispersal by, 607*f*
    herbivore adaptations in, 373–374
    hierarchical organization of tissues, organs, and organ systems in, 642–643*f*
    homeostasis in (*see* Homeostasis)
    immune systems of (*see* Immune systems)
    land plant interactions with, 524
    nutrition in (*see* Animal nutrition; Digestive systems)
    origination of, in sponges and cnidarians, 528–530
    phylogeny of, 533–534
    plant defenses against herbivory by, 636
    in predation (*see* Predation)
    radiations of aquatic, 532–539 (*see also* Aquatic animals)
    reproduction of (*see* Animal reproduction)
    saturated and unsaturated fats of, 49–50
    storage polysaccharides of, 47*f*–48
    Unikonta supergroup and, 496–497
    viruses affecting, 335–340
Animal behaviors, 792–816
    altruism and inclusive fitness in, 813–814
    animal brains and, 778
    behavioral ecology and, 803
    behavioral rhythms of, 804–805
    experience, learning, and, 806–809 (*see also* Learning)
    fixed action patterns, 804
    foraging, 809–810
    genetics and evolution of, 809–811, 812–814
    innate, 806
    mating and mate choice, 810–811
    migration, 804
    muscle function in, 793–799 (*see also* Muscle contraction; Muscles)
    nervous systems, motor systems, and, 792–793 (*see also* Motor systems; Nervous systems)
    sensory inputs stimulating, 803–806

    signals and communication, 805–806
    skeletal systems and locomotion in, 799–803
    species dispersal, 830–831
Animal cells. *See also* Eukaryotic cells
    active transport in, 105
    apoptosis of, 315
    blood, 695*f*, 696–697
    cell cycle of, 174*f*, 175, 180–182 (*see also* Cell cycle)
    cell junctions in, 90–91
    cellular respiration by mitochondria of, 81, 82–83
    circulatory systems, gas exchange surfaces, and, 684–685
    cotransport in, 105–106
    endocytosis in, 106–107*f*
    extracellular matrix of, 88–90
    local and long-distance cell signaling of, 108*f*
    meiosis in, 198*f*–199*f*
    microtubules of, 85
    nuclear transplantation of differentiated, in cloning, 320–322
    organelles of, 72*f*
    plasma membranes of, 72*f*, 95*f* (*see also* Plasma membranes)
    reproductive cloning of mammalian, 321–322
    of sponges, 529*f*–530
    stem cells, 322–324
    thyroid hormone level and oxygen consumption of, 149
    water balance and tonicity of, 101*f*
Animalia, kingdom, 8*f*, 395
Animal nutrition, **665**–683
    dietary requirements for, 666–668
    evolutionary adaptations of vertebrate digestive systems in, 676–678
    feeding mechanisms in, 669*f*
    food processing stages in, 668–671
    herbivore, carnivore, and omnivore diets for, 665
    mammalian/human digestive system organs and, 671–676
    obesity genes and appetite regulation in, 681
    regulation of digestion, energy allocation, and appetite in, 678–682
Animal reproduction, 729–750. *See also* Human reproduction
    asexual, 729–730
    asexual vs. sexual, 193
    embryonic development in, 742–749
    fertilization mechanisms in, 732
    hormonal regulation of mammalian/human, 738–742
    reproductive cycles in, 731
    reproductive organs in, 733–738
    reproductive tables and rates of, 834–835
    sexual, as evolutionary enigma, 730
    sexual life cycles in, 196–197 (*see also* Sexual life cycles)
    variations in patterns of sexual, 731–732
Anions, **26**, 104–105
Ankle bones, 376*f*
Annelids, 535*f*, 685*f*, 799–800
Annuals, plant, 562
Anolis lizards, 846*f*
Anomalocaris, 531*f*
Anopheles mosquitoes, 501–502
Anser anser, 807
Antagonistic functions, autonomic nervous system, 771
Antagonistic interactions, 833
Antagonistic muscle pairs, 799
Antarctica, 436, 853–854, 864
Antennae, mechanoreceptors and, 782
Anterior ends, **533**
Anterior pituitary gland, **649**, 650*f*, 738–742, 747

Anthers, 519, **598**
Anthoceros, 514*f*
Anthopleura elegantissima, 730*f*
Anthozoa, 530*f*
Anthrax, 473*f*
Anti-aging, plant cytokinins in, 622
Antibiotic drugs
    bacterial infections and, 281
    bacterial resistance to, 374–375, 469–470, 476, 548
    for cystic fibrosis, 222
    as enzyme inhibitors, 130
    gram-positive bacteria and, 473*f*
    peptidoglycan and, 463
    prokaryotic ribosomes and, 465
    viruses and, 338
Antibodies, **716**
    antigen recognition by, 716
    in B cell and T cell diversity, 717–718
    binding of, to proteins, 55*f*
    in humoral immune response, 722–723*f*
    in medical diagnosis and treatment, 724
    role of, in immunity, 723–724
Anticodons, **279**–280
Antidiuretic hormone (ADH), **650***f*, **652**, 661–663
Antifreeze proteins, 635
Antigen fragments, 717
Antigenic determinant, 715
Antigenic variation, 725–726
Antigen presentation, **717**
Antigen-presenting cells, **720**–721
Antigen receptors, **715**–717
Antigens, **715**
    helper T cells as responses to, 720–721
    recognition of, by B cells and T cells, 715–717
    variations in, and immune system evasion, 725–726
Antihistamines, 724
Antimicrobial peptides, 714
Antiparallel DNA backbones, 61–**62**, 248, **250**–251
Antivenin, 724
Antiviral drugs, 338
Ants, 541*f*, 607*f*, 850*f*
Anus, 670*f*, 671*f*, 676, 744, 745*f*
Anvil (incus), 783*f*
Apes, 546
Aphelocoma californica, 778
Aphids, 669*f*
Aphotic zone, **827**
Apical buds, 555*f*, **556**
Apical dominance, **564**, 621–622
Apical meristems, **507**, **560**–561*f*
Apical surface, epithelial, 643*f*
Apicomplexans, **501**–502
Apicoplast, 501–502
Apomixis, **608**, 614
Apoplast, **574**
Apoplastic transport route, 574, 587*f*
Apoptosis, **315**
    cancer and, 187
    p53 gene and, 326
    as plant response to flooding, 633–634, 635*f*
    plant senescence and, 625
Aposematic coloration, **848**
Appendages, arthropod jointed, 536
Appendix, **676**, 694*f*
Appetite, regulation of, 681–682
Apple fruit, 606*f*
Apple maggot flies, 426
Aquaporins, **99**, **577**, **658**
    in facilitated diffusion, 102
    in kidney function, 659*f*
    in passive transport, 100
    in plasma membrane selective permeability, 94

Aquatic animals, 532–539. *See also* Animal(s)
  body plans of, 532–533
  diversification of, 533–534
  gills for gas exchange in, 701
  radiation of invertebrate, 534–536
  radiation of vertebrate, 536–539
  terrestrial animals vs., 540f
Aquatic biomes
  decomposition and nutrient cycling rates in, 873
  inverted biomass pyramids of, 872
  locomotion in, 802
  photosynthetic protists in, 499–500
  primary production in, 868–869
  zonation in, 827–830
Aquatic lobe-fin, 540f
Aqueous humor, 786f
Aqueous solutions, **33**–36, 100
Arabidopsis thaliana (mustard plant), 561–562
  altering gene expression by touch in, 632f
  genetic engineering of herbivore defenses in, 636
  genome size of, 348t
  photoreceptors of, 627–628
  stem elongation of, 623f
  triple response in, 625
Arachnids, 535f
Arbuscular mycorrhizae, **509**–510, 512f, **584**–585
Archaea, 8f. *See also* Archaea, domain;
    Prokaryotes
  cells of (*see* Prokaryotic cells)
  eukaryotic features derived from, 484t
  genome size and number of genes in species of, 348
  membrane lipid composition in, 96
Archaea, domain, **8**. *See also* Archaea
  compared to Bacteria and Eukarya, 471t
  evolutionary relationships of, 358f
  horizontal gene transfer and, 395–396
  phylogeny of, 471, 474
Archaean eon, 439t
Archaefructus sinensis, 519–520
Archaeoglobus fulgidus, 348t
Archaeplastida, 490f–491f, **495**–496
Archenteron, 744, 745f
Arctic, 826f
Ardipithecus ramidus, 546
Area effects, 860
Arginine, 53f, 270
Arid conditions, plants and, 169–171
Ariolimax californicus, 812–813
Aristotle, 366
Arizona, desert biome in, 825f
Arms, chromatid, 176
Arms race, evolutionary, 548
Arousal, brain functions and, 771
Artemia, 451
Arteries, **686**–687f, 691–693, 698
Arterioles, 686, 691f, 692–693
Arthropods, **536**
  chitin as structural polysaccharide of, 49
  compound eyes of, 786
  exoskeletons of, 800
  general characteristics of, 540–541
  insects as, 541–542
  nervous systems of, 769f
  origins of, 534–536
  skeletal muscles of, 799
Artificial selection, **371**–372, 597, 611–612
Ascocarps, 512f
Ascomycetes, **512**f
Asexual reproduction, **193**, **608**, **729**
  allocation of energy in, 610
  angiosperm, 608–611
  fungal, 510, 511f
  mechanisms of, 730

reproductive cycles in, 731
  sexual reproduction vs., 193, 204, 729–730 (*see also* Sexual reproduction)
  of single-cell eukaryotes, 182
Asian elephants, 370–371
Asian ladybird beetles, 371f
A site (aminoacyl-tRNA binding site), **281**, 283f
Asparagine, 53f
Aspartic acid, 53f
Aspen trees, 608
Aspidoscelis uniparens, 731
Aspirin, 781
Assembly stage, phage lytic cycle, 333f
Assessment, nutritional, 668
Associative learning, **808**, 809f
Asteroid collision, mass extinction from, 445–446
Asters, **177**, 178f
Asthma, 694
Astragalus bones, 376f
Astrocytes, **769**f, **770**
Atelopus varius, 818
Atherosclerosis, 51, 106, **697**–698
Athletes, blood doping by, 696
Athyrium filix-femina, 515f
Atmosphere
  Cambrian explosion and changes to Earth's, 531
  Earth's early, 459–462
Atomic mass, **21**
Atomic nucleus, **21**
Atomic number, **21**
Atoms, **20**–28
  atomic number and atomic mass of, 21
  electron distribution of, and chemical properties of, 23–24
  energy levels of electrons of, 22–23
  formation and function of molecules by chemical bonding of, 24–28
  isotopes of, 21–22
  subatomic particles of, 20–21
  tracking, through photosynthesis, 157–158
ATP (adenosine triphosphate), **44**, **122**
  aminoacyl-tRNA synthetases and, 280
  conversion of, to cyclic AMP, 112–113f
  dietary requirements for synthesis of, 666, 679
  in DNA replication, 255
  as energy for active transport, 103–104
  mitochondria and, 81, 83, 91
  in muscle contraction, 794f–795
  as organic phosphate compound, 44
  regeneration of, 124
  regulation of regeneration of, 131
  structure and hydrolysis of, 122–123
  synthesis of, by cellular respiration, 135, 139–148 (*see also* Cellular respiration)
  synthesis of, by fermentation and anaerobic respiration, 148–151
  synthesis of, in catabolic pathways, 136–140
  synthesis of, in chemiosomosis, 145–147
  synthesis of, in light reactions of photosynthesis, 158, 164–167f, 171
  types of work and energy coupling by, 122
  yield of, at each stage of cellular respiration, 147–148
  yield of, by fermentation, 150–151
ATP cycle, 124f
ATP synthase, **145**–147, 167f
Atria, heart, **686**, 687f, 688–689
Atrioventricular (AV) node, **690**
Atrioventricular (AV) valve, **689**
Attached earlobes, pedigree analysis case, 219–220
Attachment function, membrane protein, 97f
Attachment stage, phage lytic cycle, 333f
Auditory canal, 783f
Auditory communication, 806

Auditory cortex, 776f
Auditory nerve, 783f, 784
Australia, 376, 830, 849
Australian moles, 385–386
Autism, 778
Autoimmune diseases, **725**
Autonomic nervous system, vertebrate, **771**
Autophagy, 80
Autopolyploids, **425**–426
Autosomal aneuploidies, 242
Autosomes, **195**
Autotrophs, **155**, 465t, 474, 866. *See also* Primary producers
Auxin, **620**
  in cell differentiation, 622
  in cell elongation, 620–621
  discovery of, 618–619
  in leaf abscission, 626
  overview of, 620t
  in plant development, 621–622
  in plant gravitropism, 632
  polar transport of, 621f
  practical uses of, 622
Avery, Mary Ellen, 703–704
Avery, Oswald, 246
Avian brains, 778
Avian flu, 339, 861–862
Avirulent pathogens, **636**–637
Avogadro's number, 34
Avr (avirulence) genes, 636–637
Axel, Richard, 781–782
Axillary buds, 555f, **556**
Axis establishment, 318–320
Axolotl salamanders, 450, 684
Axon hillocks, 752f, 759, 762
Axons, **752**. *See also* Action potentials, of neurons
  in central nervous systems, 770
  evolution of structure of, 760
  motor proteins and cytoskeleton of squid giant, 85f
  in nervous system signaling, 648
  nervous tissue and, 643f
  structure and function of, 752
Azidothymidine (AZT), 338
Azure vase sponge, 529f

B
Bacilli, 462f
Bacillus anthracis, 473f
Bacillus coagulans, 70f
Bacillus thuringiensis, 612
Backbones, nucleic acid, 61–62, 248, 250–251
Backbones, polypeptide, 54, 56f–57f
Bacteria, 8f. *See also* Bacteria, domain;
    Prokaryotes
  alcohol fermentation and, 150–151
  anaerobic respiration in, 148–149
  antibiotic drugs and infections by, 281
  antibiotic resistance in, 374–375, 469–470, 476, 548
  bioremediation using, 878
  cells of (*see* Prokaryotic cells)
  chemoautotrophic, in Antarctica, 864
  chromosomes and binary fission in, 182–183
  diversity of, in soil, 852f
  in DNA cloning, 262
  DNA packing in chromosomes of, 259
  DNA replication in, 252–257f
  in energy flow and chemical cycling, 6f–7
  eukaryotic features derived from, 484t
  evidence that DNA can transform, 246
  gene expression in, 271f (*see also* Bacterial gene regulation)
  genome size and number of genes in species of, 348

glycolysis in ancient, 151
Gram staining of, 462–463
horizontal gene transfer and, 395–396, 468, 476
inhabiting human bodies, 458–459f
innate immunity and, 712–715
macrophages and, 91f, 711
membrane lipid composition in, 96
mutations in, 290
mutualistic, in vertebrate digestion, 677
mutualistic and pathogenic, 475–476
origin of mitochondria and plastids in, 484–487
origins of photosynthesis in, 156
in Permian mass extinction, 445
as photoautotrophs, 156f
photosynthetic (see Cyanobacteria)
phylogeny of, 471, 472f–473f
plant nutrition and soil, 582–584
synthesis of multiple polypeptides during translation in, 286
transcription in, 274–276
viral infections of (see Phages (bacteriophages))
Bacteria, domain, 8. See also Bacteria
compared to Archaea and Eukarya, 471t
evolutionary relationships of, 358f
genome size and number of genes for, 348t
horizontal gene transfer and, 395–396
phylogeny of, 471, 472f–473f
Bacterial gene regulation. See also Gene regulation
negative, 295–297
operon model of, 294–295
positive, 297–298
regulation of metabolic pathways in, 293–294
Bacteriophages (phages). See Phages (bacteriophages)
Bacteriorhodopsin, 97f
Bacteroides thetaiotaomicron, 476
Bacteroids, 583–584
Baculum, 735
Baker, C. S., 384f
Balance
locomotion and, 802
mechanoreceptors for hearing and, 782–785f
Balance of nature view, 856
Balancing selection, 414, 415f
Balanus balanoides, 847f
Baleen, 669f
Ball-and-socket joints, 801f
Ball-and-stick models, 27f, 41f
Banana slugs, 812–813
Banglomorpha, 482f–483f
Barbiturates, 77
Bar graphs, 15, 149, 477, 560, 634, 849, F-2
Bark, 569
Barley, 617, 623f
Barnacles, 847f
Barr, Murray, 233
Barr body, 233–234
Barrier defenses, immune system, 711–712, 713
Barrier methods, contraceptive, 748
Basal animals, 530, 534
Basal body, 86, 87f
Basal lamina, epithelial, 643f
Basal metabolic rate (BMR), 679
Basal nuclei, 773f
Basal taxon, 383–384
Base pairing, nucleic acid, 62, 250–252, 253f
Bases, 34–36, 61–62
Basic amino acids, 54
Basic side chains, 53f
Basidiomycetes, 512f
Basilar membrane, 783f, 784
Basophils, 695f, 696f
Batesian mimicry, 848
Batrachochytrium dendrobatidis, 818

Bats, 385–386, 450, 602f, 801–802
B cells, 715
antigen recognition by, 715–716
development of, 717–720
in humoral immune response, 722–723f
T cells and, 696f
Bdelloid rotifers, 204
Bdellovibrios, 472f
Beach mouse, 1, 12–14
Beadle, George, 269–270
Beagle, Charles Darwin's voyage on HMS, 368–369
Beaks
finch, 10, 14, 369f, 399
soapberry bug, 373–374
Beans, 604f, 605f, 629, 666
Bears, 8, 118f, 394, 422, 665
Beavers, 855
Bed bugs, 541f
Bees, 541f, 597, 602f, 805–806
Beetles, 19f, 371f
Behavior, 792. See also Animal behaviors
Behavioral ecology, 803, 806
Behavioral isolation, 420f
Belding's ground squirrels, 813, 833, 834–835
Belostoma, 733f
Beluga whales, 781f
Benign tumors, 187–189
Bennettitales, 520
Benthic zone, 827, 829f
Benthos, 827
Best-fit curve, F-2
β chain, antigen, 716–717
β-galactosidase, 296
β-globin, 63, 352–353
β glucose ring structure, polysaccharide, 48f
β pleated sheet, protein, 56f
Beta-carotene, 612, 613f, 667
Beta cells, pancreatic, 680
Beta oxidation, 152
Beta proteobacteria, 472f
BGI (formerly Beijing Genome Institute), 345
Bicarbonate ions, 36–37, 649
Biceps, 799f
bicoid gene, 319–320, 361
Biennials, plant, 562
Bilateral symmetry, 317, 524, 532–533
Bilaterians, 531, 534. See also Invertebrates; Vertebrates
in animal phylogeny, 534
Cambrian explosion and origins of, 531–532
in predation, 548–549
tissue layers in, 533f
Bilayers, phospholipid. See Phospholipid bilayers
Bile, 674
Binary fission, 182–183, 466–468
Binding sites, ribosome, 281, 283f, 284
Binomial names (taxonomy), 366, 382
Biochemistry, 69. See also Chemistry
Biodiversity. See also Species diversity
from adaptive evolution by natural selection, 450
branching phylogeny and, 550
effects of adaptive radiations on, 447–449
effects of mass extinctions on, 445f, 446–447
evolution of, 365–366, 370
Bioenergetics, 117, 679–681, 803. See also Energy; Energy flow
Biofilms, 466
Biofuels, 613
Biogenic amines, 764
Biogeochemical cycles, 873–876
Biogeographic factors, community
area effects, 860
latitudinal gradients, 859
Biogeography, 377–378, 830–832, 859–860

Bioinformatics, 6, 344
centralized resources for, 345, 346f
genomics and, 6, 343–344 (see also Genomics)
protein structure and function and, 59
proteomics, systems biology, medicine, and, 347
understanding functions of protein-coding genes in, 345–346
understanding genes and gene expression in, 346–347
Biological augmentation, 878
Biological clocks, 628–629, 774. See also Circadian rhythms
Biological molecules
carbohydrates, 45–49 (see also Carbohydrates)
lipids, 49–51 (see also Lipids)
nucleic acids, 60–63 (see also Nucleic acids)
as organic compounds and macromolecules, 40, 44–45 (see also Organic compounds)
proteins, 51–59 (see also Proteins)
shape and function of, 27–28
Biological species concept, 419–422
Biology, 1
behavioral ecology in, 803 (see also Animal behaviors)
biogeography in, 377–378, 830–832 (see also Species distributions)
cells in (see Cell(s))
connection of chemistry and, 19 (see also Chemistry)
conservation biology in, 839, 860
cytology and biochemistry in, 69
demography in, 834–835
evolution as core theme of, 1, 7–11 (see also Evolution)
genetics in, 5–6, 192 (see also Bioinformatics; Genetics; Genomics)
island biogeography in, 860
metagenomics in, 470
molecular genealogy and molecular, 62–63
radioactive isotopes in, 21–22
science and inquiry in, 11–16 (see also Research methods; Science; Scientific skills)
systematics and taxonomy in, 382–385, 390f (see also Systematics; Taxonomy)
themes of, 2–7 (see also Life)
Bioluminescence, 116, 475f
Biomanipulation, 855
Biomass, 613, 853
pyramid of, and standing crop, 872
in secondary production, 870–871
standing crop measure of, 867
total accumulation of, 868
Biomass pyramid, 872
Biomes, 822–823. See also Aquatic biomes; Biosphere; Ecosystems; Terrestrial biomes
Bioremediation, 20, 477–478, 868–869, 877–878
Biosphere, 2f, 474–475, 819f. See also Biomes; Ecosystems
Biosphere 2, 37
Biosynthesis, 117, 152, 666, 679
Biotechnology. See also Genetic engineering
organismal cloning, 320–324
plant, 612–614
prokaryotes in, 477
Biotic factors, 821
in community equilibrium, 856
in pollination, 602f
in species distributions, 821, 831
Biotic interactions. See Interactions, ecological
Biotic stresses, plant responses to, 633, 636–638
Bipedal animals, 802
Bipolar cells, 786f, 788

Birds, 544, 545f
  alimentary canals in, 670f
  applying parsimony in molecular systematics
      of, 390f
  avian flu and, 339, 861–862
  bats vs., 385–386
  breathing by, 704
  as descended from dinosaurs, 391–392
  double circulation in, 687–688
  evolution of brains and cognition in, 778
  field research on, by Charles Darwin, 368–369
  flower pollination by, 433, 602f
  gene flow in great tits, 409–410
  genetic drift in greater prairie chickens,
      408–409
  kidney adaptations of, 661
  learning by, 807–809
  locomotion by, 801–802
  natural selection in, 10, 14, 399
  problem solving of, 809
  production efficiency of, 871f
  structure and function correlation in, 4
  unity and diversity among, 9f
Birth control. See Contraception, human
Birth control pills, 748
Birth defects, human, 668
Birth rates
  population change and, 840–842
  population dynamics and, 832f–835
  population growth and, 835–839
Bitter tastants, 782
Bivalves, 800
Black bears, 8
Black rush plants, 851
Black snakes, 849
Black-tipped reef shark, 538f
Blades, 493
Blades, leaf, 555f, 556
Blastocoel, 744–745
Blastocysts, 323, 746
Blastopore, 744
BLAST program, 345
Blastula, 742, 744–745
Blebbing, 315
Blending hypothesis in heredity, 206
Blindness, 408, 473f, 667
Blindness, color, 789
Blind spot, 786f
Blood, 685. See also Blood pressure; Blood vessels
  ABO blood groups for human, 98, 216, 724
  apoptosis of human white blood cells of, 315f
  cell division of bone marrow cells and,
      174–175f
  in closed circulatory systems, 685–688
  clotting of, 233, 696–697
  components of, 695f
  composition and function of, 695–698
  connective tissue of, 643f
  countercurrent exchange and fish, 701
  flow velocity of, 691–692
  gas exchange adaptations of, 706–708
  hemophilia and clotting of, 233
  immune rejection of transfusions of, 724
  osmolarity of, 661–662
  pH of human, 35f, 36
  processing of filtrate from, by kidneys,
      658–659
  volume and pressure of, in kidney regulation,
      662–663
Blood-brain barrier, 770
Blood doping, 696
Blood Falls, 864f
Blood flukes, 850
Blood poisoning, 472f
Blood pressure

  in cardiovascular systems, 692–693
  in circulatory systems, 686–687
  hypertension and, 698
  kidney homeostasis and, 662
Blood proteins, 56f–57f
Blood types, human, 98, 216, 724
Blood vessels
  adaptations of, for thermoregulation, 646
  blood flow velocity in, 691–692
  blood pressure in, 692–693
  capillary function, 693
  in circulatory systems, 685–686
  diseases of, 697–698
  structure and function of, 691
Blooms, 494, 499
Blooms, algal, 868
Blue-footed boobies, 420f
Bluehead wrasse, 732
Blue jays, 808, 809f
Blue-light photoreceptors, 627, 629
Blue whales, 384f
Bodies, animal, 642–643f, 800–801
Body cavities, animal, 533
Body fat, 681–682
Body hairs, insect, 782
Body plans, 532
  animal, 532
  apoptosis and, 315
  arthropod, 536
  fungal, 509
  homeotic genes and, 360–361
  macroevolution of, from changes in
      developmental genes, 449–452f
  pattern formation and, 317–320
  symmetry, tissues, and body cavities in animal,
      532–533
  unity and diversity of bird, 9f
Bohr shift, 707
Boletus edulis, 509f
Bolting, 622–623f
Bolus, 672
Bombardier beetle, 19f
Bombina, 428–429, 430
Bonding, parental, 807
Bone marrow, 174–175f, 715
Bones. See also Skeletal systems
  endoskeletons of, 800
  human middle ear, 783f, 784
  of human skeleton, 800, 801f
Bonneia, 483f
Bonnemaisonia hamifera, 496f
Booms, Travis, 862f
Borisy, Gary, 181f
Bormann, Herbert, 876
Borrelia burgdorferi, 473f, 476f
Botox, 764
Bottleneck effect, 408–409
Bottlenose dolphins, 771
Bottom-up model, trophic control, 855
Botulism, 110, 335, 473f, 476, 764
Boundaries
  community, 845
  ecosystem, 864
  population, 832
Bound ribosomes, 76, 285
Boveri, Theodor, 228
Bowman's capsule, 657f, 658
Boysen-Jensen, Peter, 618f–619
Bradybaena, 420f
Brain cancer, 188, 347
Brain cells, human, 66f
Brains, 751
  arousal and sleep functions of, 771
  biological clock regulation by, 774
  breathing control centers in human, 705–706

  cancer of, 188, 347
  in central nervous systems, 769, 770
  cerebral cortex functions in, 776–779
  drug addiction and reward system of, 775
  evolution of cognition in, 777–778
  frontal lobe function in, 777
  functional brain imaging of, 775–776
  information processing by, 777
  language and speech functions of, 776–777
  lateralization of cortical function in, 777
  limbic system of, and emotions, 775
  mammalian, 752f
  memory and learning in, 778–779
  neural plasticity of, 778, 779f
  neurons in, 751–753 (see also Neurons)
  opiate receptors in mammalian, 765
  organization of human, 772f–773f
  sensory systems and, 779–789 (see also
      Sensory systems)
  songbird vs. human, 778f
  strokes in, 698
  structure of human, 772f–773f, 778f
  vertebrate, 771–779
  visual information processing in, 788–789
Brainstem, 772f, 773f
Brain waves, 771
Branching, carbon skeleton, 42f
Branching, plant stem, 508, 564, 573
Branching evolution, 455
Branch length, phylogenetic tree, 388–391
Branch points, phylogenetic tree, 383–384
Brassinosteroids, 620t, 623
Brazil nut trees, 839f
BRCA1 and BRCA2 genes, 327
Bread mold (Neurospora crassa), 269–270
Breakdown pathways, 117
Breast cancer, 188f, 189, 327
Breasts, human, 735
Breathing, 704–706
Breathing control centers, 705–706
Breeding, 371–372, 597, 611–612
Brewer's yeast. See Saccharomyces cerevisiae
Briggs, Robert, 320
Brightfield (unstained specimen) microscopy, 68f
Brightfield (stained specimen) microscopy, 68f
Brine shrimp, 361f, 451
Broca, Pierre, 776–777
Broca's area, 776f, 777
Bronchi, 703
Bronchioles, 703
Brooding, 391–392
Brown algae, 493–494
Brown bears, 8, 118f
Brown fat, 148
Brush border, 674f, 675
Bryophytes, 513f, 514, 515f
Bt toxin, 612
Bubo scandiacus, 841f
Buchloe dactyloides, 573–574
Buck, Linda, 781–782
Budding, 72f, 193f, 202, 730
Buffalo grass, 573–574
Buffers, 36
Bufo marinus, 849
Bugs, 541f
Bulbourethral glands, 734
Bulk feeders, 669f
Bulk flow, vascular plant, 577–578, 588–590, 594
Bulk transport, 106–107f
Bumblebees, 433
Bundle-sheath cells, 169–170, 565
Burgess Shale fossil bed, 437f
Burkholderia glathei, 475f
Burkitt's lymphoma, 328
Butterflies, 541f, 602f, 613–614, 808, 809f, 842

C

C₃ plants, **169**
C₄ plants, **169**–170
Cactus, 592*f*, 602*f*, 825*f*
Cactus-eater finches, 369*f*
Cadherins, 488–489
Caecilians, 543
Caenorhabditis elegans (soil worm), 315, 346, 348
Calcification, 37
Calcitonin, **650***f*
Calcium, 20
Calcium carbonate, 36–37
Calcium ions, 77–78, 112, 744–745, 761, 795–797, 798, 799
California, chaparral biome in, 825*f*
California Current, 821–822
California mouse, 806
Callus, **610**
Callyspongia plicifera, 529*f*
Calmodulin, 799
Calorie (cal), **31**
Calorie (C), 679
Calvin, Melvin, 159
Calvin cycle, **158**. *See also* Photosynthesis
    evolution of alternative mechanisms of, 169–171
    overview of, 159*f*, 168*f*, 171*f*
    phases of, 167–169
    as stage of photosynthesis, 158–159
Cambrian explosion, 450, 483*f*, **530**–539
Camouflage, 1, 12–14, 15, 372*f*, 848
cAMP (cyclic adenosine monophosphate), 112–113*f*, 297, 763
CAM (crassulacean acid metabolism) plants, **170**–171, **592**–593
Campylobacter, 472*f*
Canada goose, 646*f*
Canadian Forest Service, 873*f*
Cancer
    brain cancer, 118, 347
    breast cancer, 188*f*, 189, 327
    carcinogen screening and, 290
    chromosomal translocations and, 242–243*f*
    development of, from abnormal cell cycle control, 324–328
    development of, from interference with normal cell-signaling pathways, 325–326
    DNA mismatch repair and colon, 257
    HIV and, 726
    immunity and, 726–727
    inherited predisposition and other factors contributing to, 327–328
    interpreting histograms on inhibition of cell cycle of, 188
    loss of cell cycle controls in, 186*f*–189
    multistep model of development of, 326–327
    obesity and, 681
    protein kinases in, 112
    radioactive isotopes in PET scans for, 22
    skin, 258
    systems biology approach to, 347
    telomeres and treatment of, 259
    types of genes associated with, 324–325
Cancer Genome Atlas, 347
Cane toads, 849
Canopy, **573**, **824**
Canyon tree frog, 848*f*
Capillaries, 650*f*, **686**–687*f*, 691–693, 706
Capillary beds, **686**, 691–693
Capsaicin, 781
Capsids, **331**–332
Capsomeres, 331
Capsule, 70*f*, **463**
Carbohydrates, **45**
    catabolism of, 136, 151–152

digestion of, 673*f*, 674–675
    membrane, 95*f*, 98
    monosaccharide and disaccharide sugars, 45–47*f*
    as organic compounds and macromolecules, 40
    as polymers of monomers, 44–45
    polysaccharides, 47–49
    as product of photosynthesis, 171
    types of, 45
Carbon
    as essential element, 20
    isotopes of, 21, 438
    in organic compounds, 40–42 (*see also* Organic compounds)
Carbon-12, 438
Carbon-14, 438
Carbonate ions, 37
Carbon cycle, 522, 874*f*
Carbon dioxide
    in alternative carbon fixation mechanisms, 169–171
    in capillaries, 693
    carbon bonds in, 42
    in carbon cycle, 874*f*
    catabolic pathways and, 136
    in circulation and gas exchange, 684–685 (*see also* Circulatory systems; Gas exchange)
    effects of atmospheric, on crop productivity, 170
    effects of removal of, by land plants, 522
    gas exchange adaptations for transport of, 708
    in global climate change, 499–500
    inhibition of fruit ripening with, 624
    in interspecific interactions, 7
    in mammalian cardiovascular systems, 688
    net ecosystem production and, 868
    in ocean acidification, 36–37
    in photosynthesis, 28–29*f*, 159
    in regulation of human breathing, 705–706
    rubisco as acceptor for, in Calvin cycle, 168–169
    as stimulus for stomatal opening and closing, 591
Carbon fixation, **159**, 169–171
Carbonic acid, 35–37
Carboniferous period, 516, 518, 522
Carbon monoxide, 766
Carbon skeletons
    biosynthesis of, in anabolic pathways, 152
    of fatty acids, 49
    of organic compounds, 42
    of steroid lipids, 50
    of sugars, 46
Carbonyl group, 43*f*, 45–46
Carboxyl end, polypeptide, 54
Carboxyl group, 43*f*, 49, 52
Carboxylic acid, 43*f*
Carcinogens, 290. *See also* Cancer
Carcinus maenas, 548
Cardiac cycle, **689**, 692
Cardiac muscle, 643*f*, **798**
Cardiovascular diseases, 681, 697–699, 748
Cardiovascular systems, **686**. *See also* Circulatory systems
    blood composition and function in, 695–698
    blood vessels, blood flow, and blood pressure in, 690–694
    coordination of gas exchange and, 706 (*see also* Gas exchange)
    gas exchange adaptations in, 706–708
    hearts in mammalian, 688–690
    human diseases of, 681, 697–699, 748
    lymphatic systems and, 693–694
    single and double circulation in vertebrate, 686–688

Caribou, 403*f*, 731
Carnivores, 383*f*, **665**, 676*f*, 677–678, 866
Carnivorous plant, 586*f*
Carotenoids, **162**
Carpellate flowers, 609
Carpels, 207, **519**, **598**
Carrier crabs, 845
Carrier proteins
    in cotransport, 105*f*
    in facilitated diffusion, 102
    as transport proteins, 99
Carriers, **221**
Carroll, Scott, 373*f*
Carrying capacity, **836**–837
Cartilage, 800
Cartilage fish, 539
Casparian strip, **588**
Cassava, 612, 613*f*
Castor bean seeds, 604*f*
Catabolic pathways, **117**. *See also* Cellular respiration
    ATP production by, 136
    cellular respiration as, 117
    redox reactions in, 136–139
    versatility of, 151–152
Catabolite activator protein (CAP), 297
Catalysts, **51**, **125**. *See also* Enzymatic catalysis
Catalytic cycle, 127–128
Caterpillars, 365, 401*f*, 636, 669*f*, 870–871
Cation exchange, **581**
Cations, **26**, 104–105
Cats, 233–234, 322, 781
Cattle, 850–851
Cattle egrets, 850–851
Caulerpa, 496*f*
Cavalier-Smith, Thomas, 497*f*
CC (Carbon Copy, cloned cat), 322
CCD protein domain, 488–489
Cecum, **675**–676
Cedar Creek Natural History Area, 853
Cell(s), 66–93. *See also* Animal cells; Plant cells
    apoptosis (programmed death) of, 315
    auxin in differentiation of, 622
    auxin in elongation of, 620–621
    cell cycle of (*see* Cell cycle)
    cellular integration of, 91
    cellular respiration and fermentation by (*see* Cellular respiration; Fermentation)
    cytokinins in division and differentiation of, 622
    differentiation of (*see* Differentiation, cellular)
    eukaryotic vs. prokaryotic, 69–71 (*see also* Eukaryotic cells; Prokaryotic cells)
    as fundamental units of life, 4, 66
    in hierarchy of biological organization, 3*f*
    locations of enzymes in, 132
    metabolism of (*see* Metabolism)
    microscopy and biochemistry in study of, 67–69
    pH of, 36
    photosynthesis by (*see* Photosynthesis)
    plasma membranes of (*see* Plasma membranes)
    prokaryotic, as Earth's first, 458
    protein synthesis in, 60
    protocells as first, 459–460
    size range of, 67*f*
    surface-to-volume ratios of, 71*f*
    using scale bars to calculate volume and surface area of, 74
    viral infections of, 330 (*see also* Viruses)
    water balance of, 100–102
Cell body, neuron, **752**, 762
Cell-cell recognition, 97*f*, 98

Cell cycle, **174**–190
    bacterial binary fission in, 182–183
    cancer development from abnormal regulation
        of, 324–328
    cytokinesis in, 180–182
    depolymerization of kinetochore microtubules
        during, 181*f*
    determining phase of, arrested by inhibitor,
        188
    evolution of mitosis in, 183
    genetic material and cell division process in,
        174–176
    phases of, 177–183
    phases of, in animal cells, 178*f*–179*f*
    phases of, in plant cells, 182*f*
    regulation of, by cell cycle control system,
        183–189
Cell cycle control system, **184**
    cancer development from abnormal, 324–328
    checkpoints of, 184–187
    evidence for cytoplasmic signals in, 184
    loss of, in cancer cells, 187–189
    regulation of cell division by, 183–184
Cell cycle–inhibiting pathway, 326*f*
Cell cycle–stimulating pathway, 325*f*
Cell differentiation. *See* Differentiation, cellular
Cell division, **174**. *See also* Cell cycle
    bacterial fission, 182–183
    cancer development and, 324–328
    cytokinins and, 622
    DNA replication in, 5
    effects of platelet-derived growth factor
        (PDGF) on, 186
    embryonic development and, 312
    evolution of mitosis in, 183
    fluorescence micrographs of, 174*f*
    genetic material and, 174–176
    in meiosis, 197–201
    in mitosis, 200*f*–201
    time required for human, 177
Cell fractionation, **69**
Cell junctions, 90–91, 97*f*
Cell-mediated immune response, 712*f*, **720**–723*f*
Cell plate, 181*f*, **182**
Cell sap, 80
Cell signaling, 108–113
    cancer development from interference with,
        325–326
    in cell cycle control system, 184–189
    cilia and, 86
    in endocrine signaling, 648
    evolution of, 113
    extracellular matrix in, 89–90
    induction in, 313
    local and long-distance, 108–109
    membrane proteins in, 97*f*
    plant responses to (*see* Plant responses)
    receptor proteins and reception stage of,
        109–111
    response stage of, 113
    three stages of, 109
    transduction stage of, 111–113*f*
Cell-type specific transcription, 302*f*
Cellular innate immune defenses, 713–714
Cellular membranes. *See also* Plasma membranes
    of chloroplasts, 83
    internal eukaryotic, 71
    of mitochondria, 82–83
    in plant response to cold stress, 635
    specialized prokaryotic, 464–465
    synthesis of, by rough ER, 78
    vesicles and, 76–77
Cellular respiration, **136**. *See also* Metabolism
    ATP yield at each stage of, 147–148
    biosynthesis in anabolic pathways and, 152
    as catabolic pathway, 117

diffusion in, 100
effect of thyroid hormone on cellular oxygen
    consumption and, 149
energy flow, chemical cycling, photosynthesis,
    and, 135
evolutionary significance of glycolysis in, 151
fermentation and anaerobic respiration vs.,
    148–151
glucose in, 46
glycolysis in, 139–141
mitochondria in, 71, 81, 82–83
overall reaction for, 120
overview of, 135*f*, 139*f*
oxidative phosphorylation in, 143–148
photosynthesis vs., 157, 158
pyruvate oxidation and citric acid cycle in,
    142–143*f*
redox reactions in catabolic pathways and ATP
    production of, 136–140
in secondary production, 870–871
stages of, 136, 139–140
versatility of catabolism in, 151–152
Cellular slime molds, 497–498
Cellulose, **48**
    in cell walls, 88
    digestion of, 677
    as hydrophilic substance, 33–34
    as product of photosynthesis, 171
    proteins synthesizing, 505
    as storage polysaccharide, 47*f*
    as structural polysaccharide, 48*f*
Cellulose synthase, 88
Cell walls, **88**
    cellulose-synthesizing proteins and, 505
    functions of, 88, 89*f*, 90
    plant, 73*f*
    prokaryotic, 70*f*, 462–464
    protistan, 73*f*
    water balance and, 101–102
Cenozoic era, 439*t*, 444*f*
Central canal, central nervous system, 770
Central dogma, DNA, 272
Centralized resources, genomic, 345, 346*f*
Central nervous system (CNS), **753**, **769**
    neural plasticity of, 778
    neurotransmitters and, 763–765
    peripheral nervous systems and, 769, 770*f*
    sensory systems and, 779 (*see also* Sensory
        systems)
    vertebrate, 769–770
Central vacuoles, 73*f*, **80**–81
Centrifuges, 69
Centrioles, **85**, 177
Centromeres, **176**
Centromeric DNA, 352
Centrosomes, 72*f*, **85**, 177–181*f*, 198*f*
Cephalization, 769
Cercozoans, **495**
Cerebellum, **772***f*–773*f*, 776*f*
Cerebral cortex, **773***f*, 776–779, 789
Cerebral hemispheres, **773***f*
Cerebral palsy, 773*f*
Cerebrospinal fluid, 705–706, 770
Cerebrum, **772***f*–773*f*
Certainty of paternity, 810–811
Cervix, **735**
Cetaceans, 376–377
Chamois, 667*f*
Chance, natural selection and, 415
Channel proteins, 99, 102. *See also* Aquaporins
Chaparral, **825***f*
Chara, 505*f*
Character displacement, **847**
Characters, **207**
    construction of phylogenetic trees from
        shared, 387–392

genetic variation of phenotypic, 400–401
    multifactorial, 218
    taxonomy and, 383
    traits and, 207
Character tables, 388*f*
Chargaff, Edwin, 248, 249
Chargaff's rules, 248, 250–251
Charophytes, 496, 505
Chase, Martha, 246–248
Checkpoints, cell cycle control system, **184**–187
Cheetahs, 841*f*
Chemical bonds, **24**–29
    covalent bonds, 24–25
    hydrogen bonds, 27
    ionic bonds, 25–26
    making and breaking of, by chemical reactions,
        28–29
    molecular shape and function from, 27–28
    van der Waals interactions, 27
    weak, 26–27
Chemical cycling
    biogeochemical cycles in, 873–876 (*see also*
        Biogeochemical cycles)
    as biological theme, 6–7
    cellular respiration, photosynthesis, energy
        flow, and, 135
    conservation of mass and, 865
    decomposition and rates of, 873
    effects of land plants and fungi on, 521–522
    energy flow and, in ecosystems, 864–866 (*see
        also* Energy flow)
    Hubbard Brook Experimental Forest case
        study in, 876–877
    overview of, 866*f*
    primary production and, 866–870
    prokaryotic, 474–475
    secondary production efficiency of, between
        trophic levels, 870–872
    trophic levels and, 864–866
Chemical digestion, 668, 671–675. *See also*
        Digestion; Digestive systems
Chemical energy, **117**
    conversion of, by mitochondria, 82–83 (*see
        also* Cellular respiration)
    conversion of light energy to, by chloroplasts,
        83, 156–159 (*see also* Photosynthesis)
    diet and, 666
    in energy flow and chemical cycling, 6*f*–7 (*see
        also* Energy flow)
Chemical equilibrium, **29**, 119–122
Chemical groups, 42–44
Chemical mutagens, 290
Chemical reactions, **28**
    activation energy barrier of, 125–126
    in aqueous solutions, 34
    chemical energy in, 117
    free energy and, 120*f*
    functional groups in, 44
    making and breaking of chemical bonds by,
        28–29
    metabolism and, 116
    in photosynthesis, 157–158
Chemical recycling. *See* Chemical cycling
Chemical signaling, extracellular matrix in, 89–90
Chemical signaling, neurons and, 751–753. *See
        also* Chemical synapses; Neurons
Chemical signals, plant, 636
Chemical structure, DNA, 250*f*
Chemical synapses, 761–765
    generation of postsynaptic potentials at, 762
    modulated signaling at, 762–763
    neurotransmitters and, 761–762, 763–765
    overview of, 761*f*
    summation of postsynaptic potentials at, 762,
        763*f*
Chemical work, 122, 123–124

Chemiosmosis, **145**
 ATP yield from, 147–148
 in cellular respiration, 139–140, 145–147
 in chloroplasts vs. in mitochondria, 165–167f
 in light reactions of photosynthesis, 158
Chemistry, 19–39
 atomic structure and properties of elements
  in, 20–24
 connection of biology and, 19 (*see also*
  Biology)
 formation and function of molecules through
  chemical bonding of atoms in, 24–28 (*see*
  also Molecules)
 hydrogen bonding and properties of water in,
  29–37 (*see also* Water)
 making and breaking of chemical bonds
  through chemical reactions in, 28–29
 matter as pure elements and compounds in,
  19–20
 organic compounds in (*see* Organic
  compounds)
Chemoautotrophs, 465t, 829f, 864–866
Chemoheterotrophs, 465t
Chemoreceptors, **781**–782
Chemosynthetic organisms, 864–866
Chemotaxis, 464
Chemotherapy, 188
Chemotrophs, 465t
Chen caerulescens, 547
Chestnut blight, 524, 638
Chiasmata, **198**f, 201
Chicxulub crater, 446
Chief cells, 672f, 673
Childbirth, human, 747
Chimpanzees
 comparison of chromosome sequences of
  humans and, 353
 comparison of human genome with genome
  of, 358–359
 complete genome sequence for, 343
 heterochrony and growth rates in skulls of, 449
 HIV in, 393
 observations of, by Jane Goodall, 11
 problem solving of, 809
 skulls of humans vs., 386
Chips, human gene microarray, 347
Chi-square (χ^2) distribution table, F-3
Chi-square (χ^2) test, 238, F-3
Chitin, **49, 508, 800**
Chlamydia, 473f, 749
Chlamydomonas, 73f, 487f–488, 496
Chlorarachniophytes, 486, 495
Chloride cells, 654
Chloride ions, 754t
Chloride transport channels, cystic fibrosis and,
 221
Chlorine, 20f, 26f
Chlorophyll, 83, **157**, 160–164, 867f
Chlorophyll *a*, **160**–164
Chlorophyll *b*, **160**–162
Chlorophytes, 496
Chloroplasts, **81**
 chemiosmosis in, 146–147
 chemiosmosis in, vs. in mitochondria,
  165–167f
 endosymbiont theory on evolutionary origins
  of, 82
 light reactions in thylakoids of, 159
 in photosynthesis, 3–4, 81, 83, 156–157, 171
 pigments of, 160–162
 plant cell, 73f
 protistan cell, 73f
 transgenic crops and DNA in, 614
Chlorosis, 579
Choanocytes, 529f–**530**
Choanoflagellates, 488–489, 498

Cholera, 110, 472f, 476
Cholesterol, **50**
 in cardiovascular diseases, 697–698, 699
 effects of, on membrane fluidity, 96
 in plasma membranes, 95f
 receptor-mediated endocytosis of, 106–107f
 as steroid lipid, 50–51
Chondrichthyans, 538f–**539**
Chordates, 534, **536**–537, 800
Chorion, 543–**544**f
Choroid, 786f
Chromatin, **75, 175, 259**
 animal cell, 72f
 cell division and, 174–176
 in cell nucleus, 75–76
 packing of, in chromosomes, 259–261f
 plant cell, 73f
 regulation of structure of, 299
 remodeling of, by noncoding RNAs, 306
Chromium, bioremediation of, 878
Chromoplasts, 83
Chromosomal alterations, 240–243f
 abnormal chromosome numbers, 240–242
 of chromosome structure, 241
 human disorders due to, 241–243
Chromosomal basis of inheritance, 228–244
 behavior of chromosomes as physical basis of
  Mendelian inheritance in, 228–231
 chromosomal alterations as cause of genetic
  disorders in, 240–243f
 constructing linkage maps in, 239f
 determining gene linkage using chi-square (χ^2)
  test in, 238
 evolution of gene concept from, 290
 genes on chromosomes as Gregor Mendel's
  hereditary factors, 228
 linked genes and linkage in, 234–240
 sex-linked genes in, 231–234
Chromosomes, **75, 175**
 alterations of, and genetic disorders, 240–243f
 bacterial, 70f
 bacterial binary fission and, 182–183
 behavior of, as physical basis of Mendelian
  inheritance, 228–231 (*see also*
  Chromosomal basis of inheritance)
 behavior of, in human life cycle, 195–196
 in cancer cells, 187–189
 cell division and distribution of, 174–176
 in cell nucleus, 75–76
 in cells, 69
 correlating behavior of alleles with pairs of,
  230–231
 DNA molecules in, 60
 as DNA molecules packed with proteins,
  259–261f
 duplication and alteration of, in genome
  evolution, 353–355
 fluorescence micrographs of, 174f, 228f
 genetic variation from mutations in, 402
 genetic variation from sexual reproduction and
  homologous, 402
 independent assortment of, 201–203f
 inheritance of genes in, 193
 locating genes along, 228
 mapping distance between genes on, 237–240
 Gregor Mendel's model and, 209
 movement of, on kinetochore microtubules,
  180, 181f
 number of, in human cells, 175–176
 in plant mitosis, 182f
 preparing karyotypes of, 194f
 in prokaryotic cells, 464–465
 prokaryotic genetic recombination and,
  468–470
 recombinant, 203
 reduction of number of, by meiosis, 197f

 sets of human, 194–195
 sex (*see* Sex chromosomes)
Chromosome theory of inheritance, **228**–231. *See*
 also Chromosomal basis of inheritance
Chronic inflammation, 715
Chronic myelogenous leukemia (CML), 242–243f
Chthamalus stellatus, 847f
Chylomicrons, **675**
Chyme, **672**–673, 674
Chymotrypsin, 673f, 674
Chytridium, 512f
Chytrids, **512**f
Cichlid fish, 426–427, 430
Cigarette smoke, 327, 748
Cilia, **85, 495**f
 bronchial, 703
 mechanoreceptors and, 780–781, 782
 as microtubule-containing cellular extensions,
  85–87
 structure of, 87f
Ciliates, **494**, 495f
Circadian rhythms, **591**, 591, 628–**629**, 774,
 804–805
Circannual rhythms, 804–805
Circulatory systems, 684–698
 adaptations of, for thermoregulation, 646
 blood composition and function in, 695–698
 blood vessels, blood flow, and blood pressure
  in, 690–694
 cardiovascular diseases of, 697–698, 699
 cells and exchange surfaces of, 685–688
 coordination of gas exchange and, 706
 evolutionary variation in, 687–688
 gas exchange adaptations in, 706–708
 gas exchange and, 684, 686–688, 698–708 (*see*
  also Gas exchange)
 genetic factors in cardiovascular disease of,
  699 (*see also* Cardiovascular diseases)
 hearts in mammalian double circulation,
  688–690
 mammalian, 642t
cis face, Golgi apparatus, 79, 81f
cis-retinal, 787–788
Cisternae, Golgi apparatus, 78–79
Cisternal maturation model, 79
Citric acid cycle, **139**
 ATP yield from, 147–148
 in catabolic and anabolic pathways, 151–152
 in cellular respiration, 139–140, 142–143f
Citrulline, 270f
Clades, **387**
Cladistics, **387**–388
Clams, 799, 800
Clark's nutcracker, 808
Classes (taxonomy), **382**
Classification. *See* Taxonomy
Claw-waving behavior, 792, 804–805
Clear cutting, 524
Cleavage, **180**–182, 742, **744**–746
Cleavage furrows, 179f, **180**–182
Clements, F. E., 856
Climate, **821**. *See also* Global climate change
 community equilibrium and, 856
 continental drift and changes in, 444
 effects of large bodies of water on, 31f
 global patterns of, 820f–821 (*see also*
  Macroclimate)
 latitudinal gradients of species diversity and,
  859
 mass extinctions and changes in, 446
 using dendrochronolgy to study changes in,
  568
Climax communities, 856
Climographs, **823**
Clitoris, **735**, 741–742
Cloaca, **733**

Clocks, biological. *See* Biological clocks
Clocks, molecular. *See* Molecular clocks
Clonal selection, **719**
Clones. *See also* DNA cloning; Gene cloning;
    Organismal cloning
    asexual reproduction and, 193
    meaning of term, 320
    plant cuttings as, 610
    plant fragmentation and, 608
    plant test-tube or *in vitro*, 610–611
Cloning vectors, DNA, **264**
Closed circulatory systems, **685**–688. *See also*
    Cardiovascular systems
Clostridium botulinum, 473f, 476
Club fungi, 512f
Club mosses, 514, 515f
Clumped dispersion, 833
Clutch size, 839
Cnidarians, 529–530, 533, 799–800, 802
    nerve nets of, 768–769
Coastal Japan restoration project, 879f
Coat coloration, mouse, 1, 12–14, 15
Coated pits, endocytosis and, 107f
Coat proteins, 107f
Cocci, 462f
Coccosteus cuspidatus, 437f
Cochlea, 783f, 784
Cochlear duct, 783f
Cocklebur, 630–631
Cocktails, drug, 338, 402
Cod, 548
Coding strands, DNA, 272
Codominance, **215**–216
Codon recognition stage, translation elongation
    cycle, 283f
Codons, **272**
    anticodons and, 279–280
    genetic code and, 272–274
Coefficient of relatedness (r), **813**–814
Coelacanths, 534, 538f–539
Coelom, 533
Coenzyme Q (CoQ), 144
Coenzymes, **129**
Coevolution, plant host-pathogen, 636–637
Cofactors, **129**
Cognition, 776–778, **808**–809. *See also* Cerebral
    cortex
Cognitive maps, **808**
Cohesins, 176, 179f, 201
Cohesion, 30, 590
Cohesion-tension hypothesis, **588**–590
Cohorts, **834**
Coitus, 741–742. *See also* Sexual intercourse,
    human
Coitus interruptus, 734, 747–748
Cold
    plant response to, 635
    thermoreceptors and, 781
Cold viruses, 332
Coleochaete, 505f
Coleoptile, **604**, 618–619
Coleorhiza, **604**
Collagen, 57f, **88**–90, 355
Collapses, population, 840–842
Collar cells, sponge, 530
Collecting ducts, **657**f, **658**, 659f
Collenchyma cells, **558**f, 565
Colon, **675**–676
Colon cancer, 257
Colonial organisms, 530f
Colonies, multicellular, 487
Colonization of land. *See* Land animals; Land
    plants
Coloration
    mate choice by, 427
    predation, natural selection, and, in guppies, 378

    predation and, 848
    predation and mouse coat, 1, 12–14, 15
Color blindness, 232–233f, 789
Colorectal cancer, 326–327
Color vision, 789
Comamonas testosteroni, 485
Combinatorial gene activation control, 302
Comet collision, mass extinction from, 445–446
Commensalism, **475**, 512f, **850**–851
Common arrowhead, 610f
Communicating cell junctions, 90f
Communication
    animal nervous system, 648
    between drought-stressed plants, 634
    between herbivore-stressed plants, 636
    by neurons, 751–753 (*see also* Neurons)
Communication, animal, **805**–806
Communities, **819**f, **845**–863
    biogeographic factors affecting diversity in,
        859–860
    community ecology and, 861–862
    comparing genomes of human, 360
    disturbances in, affecting species diversity and
        composition, 856–859
    ecological interactions between species in, 845
        (*see also* Interactions, ecological)
    in hierarchy of biological organization, 2f
    interspecific interactions in, 845–851
    metagenomics and genome sequencing of, 345
    pathogens in, 860–861
    scientific, 14–16
    species diversity and trophic structure in,
        851–856
Community ecology, **819**f, 861–862
Companion cells, **559**f
Compartments, digestive, 670–671
Competition
    as density-dependent population regulation
        mechanism, 841f
    in population dynamics, 840–842
    sexual, 413, 811
    in theory of evolution by natural selection,
        9–10
    uniform dispersion and, 833
Competitive exclusion, **846**
Competitive inhibitors, **129**–130f
Complementary base pairing, nucleic acid, 62. *See*
    also Base pairing, nucleic acid
Complementary DNA (cDNA), **308**
    DNA microarray assays and, 308–309
    reverse transcriptase-polymerase chain
        reaction (RT-PCR) and, 307–308
Complement system, **714**, 722
Complete digestive tracts, 670–671
Complete dominance, **215**–216
Complete flowers, **598**. *See also* Flowers
Complete growth medium, 269, 270
Complete proteins, 666
Complex camera lens-type eye, 453f
Complex eyes, 453
Compound eyes, **786**
Compounds, **20**. *See also* Organic compounds
    in aqueous solutions, 33
    dissolving of, in aqueous solutions, 33–34
    emergent properties and elements in, 20
    pure elements vs., 25
Compromises, evolutionary, 415
Computational tools, 343–347. *See also*
    Bioinformatics
Computer model, ribosome, 281f
Concentration, chemical reactant, 29
Concentration gradients, **100**
    cotransport down, 105–106
    diffusion down, 99–100
    electrochemical gradients as, 104–105
Conception, human, **745**–748

Condoms, 748
Conduction, **646**f, 759–760
Cones, **787**f, 788, 789
Cone snails, 751
Confocal light microscopy, 68f, 69
Conformers, **644**
Congenital disorders, 194f
Conifers, **518**
Conjugation, prokaryotic, **468**–470
Conjunctiva, 786f
Connective tissue, 57f, **643**f
Connell, Joseph, 847f
Conodonts, 537–539
Consanguineous mating, human, 221
Conservation biology, 839, 842, 860
Conservation of energy, 118, 865
Conservation of mass, 865
Conservative model, DNA replication, 252, 253f
Conserved Domain Database (CDD), 346f
Constant (C) region, antigen, 716–717
Constipation, 676
Consumers
    in ecosystem trophic structure, 853f, 866
    in energy flow and chemical cycling, 6–7
    producers and, 155 (*see also* Producers)
Consumption, regulation of, 681–682
Contact, animal fertilization, 743f
Continental drift, 377–378, 442–444
Contraception, human, 734, **747**–748
Contractile proteins, 52f
Contractile vacuoles, **80**, 101
Contraction, muscle. *See* Muscle contraction
Contrast, microscope, 67
Control center, homeostatic, 645
Control elements, **300**–304
Control groups, 14
Controlled experiments, **14**
Conus geographus, 751
Convection, **646**f
Convergent evolution, **376**, 381f, 385–386
Cooksonia, 508f
Cooling, evaporative, 32
Cooper, Vaughn, 467f
Cooperation, metabolic, 466
Cooperativity, **131**–132
Coordinately controlled genes
    bacterial, 294–295
    eukaryotic, 302–304
Coprophagy, 677
Copulation, human, 741–742, 745–748, 766
Copy-number variants (CNVs), 359–360
CoQ (coenzyme Q), 144
Coral reefs, 37, **829**f, 861
Corepressors, **295**–297
Cork cambium, **560**–561f, 566, 569
Cormorant, flightless, 418
Corn
    action spectrum for, 627f
    artificial selection of, 611f
    complete genome sequence for, 343, 348t
    cytokinin in, 622
    health of transgenic *Bt*, 613
    mineral deficiency in, 580f
    precocious germination in, 624f
    proteins in, 666
    response of, to flooding and oxygen
        deprivation, 635f
    seed germination of, 605f
    seed structure of, 604f
    transposable elements and, 350
Cornea, 786f
Corn smut fungus, 523f
Corpus callosum, **773**f, 777
Corpus luteum, **737**f, 746
Correlation, form-function, 4, 7–8, 641
Correlations, positive and negative, 610

Cortex, plant, **557**, 563f–564, 566
Cortical nephrons, **657**f
Cortical reaction, animal fertilization, 743f
Cortical reaction, human fertilization, 746
Costa Rica
  sustainable development in, 902–903
  tropical rain forest biome in, 824f
  zoned reserves in, 894f
Cost-benefit behavior analysis, 813–814
Cotransport, **105**–106
Cotransport proteins, 105–106
Cotton, 33–34
Cotyledons, 553–554f, 603–604
Counseling, genetic, 223
Countercurrent exchange, **646**, **701**
Countercurrent multiplier system, **660**
Courtship behaviors. See also Mating
  behavioral isolation and, 420f
  external fertilization and, 732
  forms of animal communication in, 805
  sexual selection and, 413
  sexual selection and female mate choice in, 811
Covalent bonds, **24**–25, 41–42, 57f
Cows, 48, 677
Coyotes, 677f–678
Crabs, 792, 804–805, 845
Cranes, 807
Crassulacean acid metabolism (CAM) plants, **170**–171, **592**–593
Crawling, 799–800, 802
Crayfish, 782
C-reactive protein (CRP), 698
Creatine phosphate, 794f–795
Creeping juniper, 518f
Crenarchaeota clade, 474–475
Cretaceous mass extinction, 445–446, 447
Crick, Francis
  central dogma of, 272
  discovery of DNA molecular structure by, 245, 249–251
  model of DNA replication of, 252, 253f
  reductionism of, 3
Crickets, 348
Cri du chat, 242
Cristae, mitochondrial, 82–**83**
Critical night length, plants and, 630–631
Crocodiles, 391–392, 544, 545f
Crop, bird alimentary canal, 670f
Crop plants. See also Agriculture; Angiosperms
  artificial selection and breeding of, 597, 611–612
  biotechnology and genetic engineering of, 612–614
  determining effects of atmospheric carbon dioxide on productivity of, 170
  eudicots as, 521f
  polyploidy in, 426
  prokaryotes in disease-suppressive soil and, 477
  soil fertilization for, 580, 841f, 870, 875f
  viral diseases and, 341
Cross-fostering studies, **806**
Crossing over, **198**f, **236**
  gene duplication due to unequal, 354f
  in meiosis, 198f, 201
  recombinant chromosomes from, in sexual life cycles, 203
  in recombination of linked genes, 236, 237f
Cross-pollination (crossing), 207–208, 611–612
Crustaceans, 361f, 451, 540f
Crustose lichens, 522f
Cryolophosaurus, 436
Cryphonectria parasitica, 524
Cryptic coloration, **848**
Cryptochromes, 627

Crypts, 592f
Crystallin, 302f
Crystalline ice, 32–33
C-terminus, 54, 97f, 282
ctr mutants, ethylene and, 625
Cuckoo bee, 848f
Cucurbita pepo, 521f
Culex pipiens, 410
Culture, **809**
Cupula, 784–785f
Currents, ocean, 821–822
Cuticle, arthropod, **540**, 800
Cuticle, plant, **507**, **556**
Cuttings, plant, 610, 622
Cuvier, Georges, 367
Cyanide, 613f
Cyanobacteria
  as bacterial group, 473f
  chemical cycling by, 474
  in eutrophication, 869
  fossils of, 461–462
  marine biomes and, 827
  in marine ecosystems, 547
  metabolic cooperation in, 466
  origin of photosynthetic plastids in, 486–487
  as photoautotrophs, 156f
Cycas revolute, 518f
Cyclic AMP (cyclic adenosine monophosphate, cAMP), **112**–113f, **297**, 763
Cyclic GMP, 788
Cycling, chemical. See Chemical cycling
Cyclins, 185
Cyclins, 185
Cynodonts, 441f, 447f
Cysteine, 43f, 53f
Cystic fibrosis, 217, **221**–222
Cytochromes, **144**–145, 146f, 165–167f
Cytogenetic maps, **240**
Cytokines, **714**–715
Cytokinesis, **176**, 179f, 198f–199f
Cytokinins, 620t, **622**
Cytology, 69, 228
Cytoplasm, **70**
  cell cycle control signals in, 184
  in cells, 70
  cytokinesis and division of, 176
  intracellular receptor proteins in, 110–111
Cytoplasmic determinants, **312**–313f
Cytosine, 60–61, 62, 248, 250–251
Cytoskeletons, **84**
  animal cell, 72f
  components of, 85–88
  in eukaryotic cells, 482
  membrane proteins and attachment to, 97f
  plant cell, 73f
  structure and function of components of, 86t
  structure of, 84f
  support and motility roles of, 84–85
Cytosol, **69**, 76, 80–81, 285–286
Cytotoxic T cells, **721**, 723f

D
Dalton, **21**
Dance language, honeybee, 805
Dandelions, 602f, 607f, 833f, 834, 839f
Daphnia, 838f
Darkness, plant flowering and, 630–631
Darwin, Charles
  Beagle voyage and field research of, on adaptations, 368–370
  on genetic variation and evolution, 204
  on grandeur of evolutionary process, 379
  historical context of life and ideas of, 366–367
  on island species, 378
  on Madagascar orchid pollinator, 548f
  on mystery of speciation, 418

  on natural selection, inheritance, and evolution, 400
  on origin of angiosperms, 519
  publication of The Origin of Species by, 365, 369–370
  scientific evidence supporting theory of, 373–379
  on species diversity of tropics, 859
  study by, of phototropism in grass coleoptiles, 618–619
  theoretical aspects of theory of, 379
  theory of, on descent with modification by natural selection, 370–372
  theory of, on evolution by natural selection, 9–10
Darwin, Francis, 618–619
Data, scientific, **11**. See also Scientific skills exercises
Databases, genomic, 345, 346f
Dating, fossil record, 438
Daughter cells, 174–176
Day-neutral plants, **630**
db gene, 681
DDT pesticide, 130, 407
Deamination, amino acid, 152
Death rates
  population change and, 840–842
  population dynamics and, 832f–835
  population growth and, 835–839
December solstice, 821f
Deciduous forest, nutrient cycling in, 876–877
Decision making, 777
Decomposers (detritivores), **474**, **866**
  effects of, on ecosystems, 547
  in energy flow and chemical cycling, 6f–7, 873
  fungi as, 512f, 522
  as heterotrophs, 155
  nutrient cycling rates and, 873
  prokaryotic, 474–475
Decomposition
  effects of temperature on, 873f
  nutrient cycling rates and, 873
Deductive reasoning, **12**
Deep-sea hydrothermal vents, 459, 471, 474–475, **829**f, 866
Deer, 268, 677, 839
DEET insect repellant, 781
De-etiolation (greening), **626**
Defecation, 676
Defensive adaptations, predation and, 848
Defensive proteins, 52f
Deficiencies, plant mineral, 579–580
Deforestation
  as community disturbance, 858
  experimental, and nutrient cycling, 876–877
Degradation
  mRNA, 304
  protein, 305
Dehydration, animal, 654
Dehydration, plant, 169–171, 505, 507
Dehydration reactions, **44**
  in disaccharide formation, 46–47f
  in polymer synthesis, 44–45f
  in polypeptide formation, 54f
  in triacylglycerol (fat) synthesis, 49f
Dehydrogenases, 137–138, 139
Deinococcus radiodurans, 458
Deletions, chromosome, **241**, 242
Deletions, nucleotide-pair, **289**–290
Delta proteobacteria, 472f
Demographics, 834–835
Demography, **834**–835
Denaturation, protein, **58**–59
Dendrites, 643f, **752**, 762, 781
Dendritic cells, 714
Dendrochronology, 568

Density, population, **832**–834, 839–840, 841*f*
Density-dependent inhibition, **186**–187
Density-dependent population regulation, 839–**840**, 841*f*
Density-independent population regulation, **840**
Dentition
    diet and adaptations of, 676–677
    mammalian, 440, 441*f*
Deoxyribose, **61**, 248
Dependent variables, identifying, 427
Dephosphorylation, protein, 111–112
Depolarization, **756**–759, 788, 798
Depolymerization, 177
Depression, 764
Derivatives, plant cell, 561
Derived characters, shared, 387–388
Derived traits, land plant, 507
Dermal tissue system, plant, **556**–557
Descent with modification theory, 9–10, 365–366, 370–372. *See also* Evolution; Natural selection
Desert ant, 641
Desert iguana, 647
Desert mouse, 655
Deserts, 592–593, 822, **825***f*
Desert spring ecosystem, 865*f*
Desmodus rotundas, 661
Desmognathus ochrophaeus, 427
Desmosomes, **90***f*
Determinate growth, **560**, 562
Determination, **313**–315
Detoxification
    by peroxisomes, 84
    smooth ER and, 77
    by sunflowers, 20
Detritivores. *See* Decomposers (detritivores)
Detritus, **866**, 873
Deuterostomia, 534–535*f*, 536–537
Development, 311–329
    cancer development from abnormal cell cycle control in, 324–328
    as cell division function, 174–175*f*
    comparing genomes and processes of, 360–361
    DNA in, 5
    embryonic (*see* Embryonic development)
    in human life cycle, 195*f*
    macroevolution of, from changes in developmental genes, 449–452*f*
    model organisms in study of, 311
    organismal cloning and stem cells in, 320–324
    plant (*see* Plant development)
    postzygotic barriers and, 419, 421*f*
    vascular plant, 516
Developmental genes, 449–452*f*
Devonian period, 518
DHFR (dihydrofolate reductase) enzyme, 497*f*
Diabetes mellitus, 323–324, **680**–681, 725
Diacodexis, 377
Diagnosis, antibodies as tools in, 724
Diaphragm, respiratory system, **704**
Diaphragm, contraceptive, 748
Diarrhea, 105–106, 469, 476, 491*f*, 676
Diastole, **689**
Diastolic pressure, 692, 698
Diatoms, 183*f*, 491*f*, **493**
Diazepam, 764
Dichanthelium lanuginosum, 510
Dickinsonia costata, 437*f*
Dicrotonyx groenlandicus, 841*f*
Dictyostelium, 497–498
Didinium, 481
Diencephalon, 772*f*, 773*f*
Dietary fiber, 48
Diets. *See also* Animal nutrition
    allopatric speciation and divergence in, 424*f*

cellular respiration and animal, 135–136, 151–152
    evolutionary adaptations of vertebrate digestive systems for, 676–678
    genetic variation in prey selection and, 812–813
    herbivore, carnivore, and omnivore, 665
    nonheritable variation and, 401*f*
    nutritional requirements for, 666–668
    phenylketonuria and, 405–406
    trophic efficiency and human, 872
Differential gene expression, 293, **298**–299, 312. *See also* Embryonic development; Gene expression; Gene regulation
Differential-interference contrast (Nomarski) microscopy, 68*f*
Differential reproductive success, 204
Differential speciation success, 455
Differentiation, cellular, **312**
    auxin and cytokinins in, 622
    cytoplasmic determinants and inductive signals in, 312–313
    in embryonic development, 312
    in plants, 561–562
    sequential gene regulation during, 313–315
Diffusion, 99, **685**
    across respiratory surfaces, 700–701
    in circulation and gas exchange, 685
    free energy and, 120*f*
    as passive transport down concentration gradients, 99–100, 104*f*
    transport proteins in facilitated, 102
    of water across plant plasma membranes, 575–577
    of water and minerals into root cells, 587–588
Digestion, **668**. *See also* Digestive systems; Food processing
    absorption in large intestine after, 675–676
    digestive compartments in, 670–671
    endocrine signaling in, 649
    as food processing stage, 668
    fungal, 508–509
    human, as hydrolysis, 45
    by lysosomes, 79–80
    regulation of, 678–679
    in small intestine, 673, 674–675
    in stomach, 672–674
    vertebrate adaptations for, 676–677
Digestive compartments, 670–671
Digestive systems, 670–678. *See also* Animal nutrition
    dental adaptations in, 676–677
    digestive compartments in, 670–671
    evolutionary adaptations of vertebrate, 676–678
    large intestine in, 675–676
    mammalian, 642*t*
    mutualistic adaptations in, 677
    oral cavity, pharynx, and esophagus in, 671–672
    organs of mammalian/human, 671–676
    small intestine in, 674–675
    stomach and intestinal adaptations in, 677–678
    stomach in, 672–674
Digger wasps, 808
Dihybrid crosses, **211**–212, 214
Dihybrids, **211**–212
Dihydrofolate reductase (DHFR) enzyme, 497*f*
Dihydroxyacetone phosphate (DHAP), 152
Dimers, 85
Dimetrodon, 437*f*
Dimorphism, sexual, 810
Dinoflagellates, 183*f*, **494**
Dinosaurs
    adaptive radiation of mammals after extinction of, 448

birds as descended from, 391–392
flying, 801
in fossil record, 9*f*, 436, 437*f*
in geologic record, 440
mass extinction of, 445–446
as reptiles, 544, 545*f*
Dioecious species, **609**
Diphasiastrum tristachyum, 515*f*
Diphtheria, 335
Diploid cells, **195**
    genetic variation preserved in recessive alleles of, 413–414
    mitosis vs. meiosis in, 200*f*–201
    in sexual life cycles, 195–197
Diploidy, 413–414
Diplomonads, 491*f*, **492**
Dipnoi, 538*f*
Dipsosaurus dorsalis, 647
Directionality, DNA replication, 255–256
Directional selection, **411**
Disaccharides, **46**–47*f*
Diseases and disorders, animal
    as density-dependent population regulation mechanisms, 847*f*
    viral, 331–332, 335–336, 338–340
Diseases and disorders, human
    alkaptonuria, 269
    allergies, 724–725
    aneuploidy of sex chromosomes and, 242
    atherosclerosis and familial hypercholesterolemia, 51, 106
    autism, 778
    autoimmune, 725
    bacterial, 281, 335, 464, 468–470, 472*f*–473*f*, 476
    cardiovascular diseases, 697–698, 699
    chromosomal alterations and genetic, 240–243*f*
    color blindness, 232–233*f*, 789
    community ecology and zoonotic, 861–862
    cri du chat and chronic myelogenous leukemia (CML), 242–243*f*
    cystic fibrosis, 217, 221–222
    as density-dependent population regulation mechanisms, 841*f*
    depression, 764
    diabetes, 323–324, 680–681
    diarrhea and constipation, 105–106, 491*f*, 676
    dietary deficiencies, 667–668
    dominantly inherited, 222–223
    Down syndrome, 194*f*, 240, 242
    drug addiction, 775
    Duchenne muscular dystrophy, 232–233
    dysentery, 469
    edema, asthma, and lymphatic system, 694
    epilepsy, 759, 777
    erectile dysfunction, 765
    fetal alcohol syndrome, 747
    food poisoning (botulism), 764
    gastric ulcers and acid reflux, 673–674
    glaucoma, 786*f*
    gonorrhea, 464
    G protein-coupled receptors in, 110
    heart murmurs, 689
    hemophilia, 233, 697
    HIV/AIDS (*see* AIDS (acquired immunodeficiency syndrome); HIV (human immunodeficiency virus))
    hormonal contraceptives and, 748
    Huntington's disease, 222–223, 323
    hypertension, 662, 698
    immune system disruptions and, 724–727
    immunization against, 723–724
    immunodeficiency, 725–726
    infertility, 748–749
    influenza (*see* Influenza viruses)

iodine deficiencies, 20
jaundice, 674
Klinefelter syndrome and Turner syndrome, 242
Lou Gehrig's disease (amyotrophic lateral sclerosis, ALS), 795
lysosomal storage diseases, 80
malaria, 222, 414, 491f, 501–502, 713
measles, 724
methicillin-resistant *S. aureus* (MRSA) and flesh-eating disease, 374–375
from misfolding of proteins, 59
mosaicism, 233–234
multifactorial, 223
myasthenia gravis, 795–796
myotonia, 759
neurotransmitters and, 763–765
Parkinson's disease, 323–324, 764
phenylketonuria, 405–406
pleiotropy and, 217
pneumonia, 468, 725
polydactyly, 216
protists and, 500–502
recessively inherited, 220–222
respiratory, 841f
respiratory distress syndrome (RDS), 703–704
retinitis pigmentosa, 408
sexually transmitted diseases, 464, 473f, 725–726, 748, 749
sickle-cell disease (*see* Sickle-cell disease)
sleeping sickness, 492–493f, 500–501
stem cells in treatments for, 323–324
Tay-Sachs disease, 216
thyroid disease, 651f
tuberculosis, 715, 841f
viral, 331–332, 335–336, 337f, 338–340
xeroderma pigmentosum, 258
X-linked disorders, 232–233
Diseases and disorders, plant
epidemics of, 638
fungal, 523–524
plant defenses against, 636–638
plant pathogens and, 861
viral, 331, 341
Disease-suppressive soil, 477
Disorder, entropy and, 118–119
Disorders. *See* Diseases and disorders, animal; Diseases and disorders, human; Diseases and disorders, plant
Dispersal, fruit and seed, 607f
Dispersal, species, **830–831**
Dispersal modes, evolutionary rates and, 443
Dispersion, population, **832–834**
Dispersive model, DNA replication, 252, 253f
Disruptive selection, 411f–**412**
Distal control elements, 300–302
Distal tubule, 657f, **658**, 659f
Distribution of species. *See* Species distributions
Distribution patterns, making histograms and analyzing, 219
Disturbances, **823**, **856**
in biomes, 823
characterizing, 856–857
ecological succession and, 857–858
human, 858–859
restoration ecology and ecosystem restoration after, 877–879f
Disulfide bridges, **57f**
Diurnal animals, 806
Divergence
of closely related species, 358–359
of gene-sized regions of DNA, 354–355
morphological, 385
Divergent evolution
allopatric speciation and, 423–425
speciation rates and, 432

Diversity. *See also* Biodiversity; Species diversity
B cell and T cell, 717–718
eukaryotic, 482f
evolution and unity in, 7, 8–9
evolution of, 365–366
within species, 419f
three domains of life in classification of, 7–9
Diversity, species. *See* Species diversity
Diving mammals, respiratory adaptations of, 708
Dizygotic twins, 746
Dizziness, 784
DNA (deoxyribonucleic acid), **5, 60**
analyzing viral evolution using phylogenetic tree based on, 340
in animal cells, 72f
in bacteria, 182–183
cell division and distribution of, 174–176
in cell nucleus, 74–76
changes in, during meiosis of budding yeast cells, 202
in chloroplasts, 83f
components of, 60–61
constructing phylogenetic trees using, 388–391
discovery of structure of, 245, 248–251
evaluating molecular homologies in, 386–387
evidence for, as genetic material, 245–248
evidence of, for origination of animals, 529
evolutionary significance of mutations of, 258
evolution of genomes from duplication, rearrangement, and mutation of, 353–357
gene density and noncoding, in genomes, 349
genetic engineering and (*see* Genetic engineering)
genetic variability as nucleotide variability in, 400–401f
genetic variation due to mutations in, 401–402
genomes as complete sequences of, 6 (*see also* Genome(s))
homeoboxes in, 360–361
homologies and, 385–387
human gene microarray chips containing, 347
inheritance of, in genes and chromosomes, 193
interpreting data from, in phylogenetic trees, 394
interpreting sequence logos for, 284
introns and exons, 277
as measure of evolution, 62–63
methylation of, in eukaryotic gene regulation, 299
in mitochondria and chloroplasts, 82
as molecular homology, 376
monitoring gene expression and, 307–309
p53 gene and repair of, 326
packing of proteins and, into chromosomes, 259–261f
phylogenies based on, 381f
in plant cells, 73f
prokaryotic, 464–465
in prokaryotic and eukaryotic cells, 69–70
prokaryotic genetic recombination of, 468–470
recombinant (*see* Recombinant DNA)
repetitive and noncoding, in genomes, 349–352
replication of, 245, 251–259 (*see also* DNA replication)
role of, in protein synthesis, 60
species identity in mitochondrial, 384f
structure and function of, 5–6
structure of molecules of, 62
template strands of, 272–273
transcription by, 274–276
viral (*see* DNA viruses)
DNA chips, 308–309
DNA cloning, 262–265
DNA Data Bank of Japan, 345

DNA deletion experiments, analyzing, 303
DNA ligase, **256**, 258, 263f–264
DNA methylation, **299**
DNA microarray assays, **308**–309
DNA polymerases, **254**–257f, 265, 307–308
DNA replication, **245**, 251–259
antiparallel elongation of DNA strands in, 254–256
base pairing to template strands in, 251–252, 253f
cell division and, 5
DNA replication complex of, 256–257f
errors in, and genome evolution, 354–355
evolutionary significance of mutations during, 258
models of, 251f, 252f, 253f, 257f
proofreading and repairing of DNA during, 257–258
start of, at origins of replication, 253–254
steps of, 252–257f
synthesizing new DNA strands in, 254–255
of telomeres at ends of molecules, 258–259
DNA replication complex, 256–257f
DNA sequences
amino acid sequences of polypeptides and, 63
changes in, of developmental genes, 450–451
genes as, 290
genomes as, 6 (*see also* Genome(s))
interpreting data from, in phylogenetic trees, 394
interpreting sequence logos for, 284
noncoding, 349–352
in taxonomy, 8
types of, in human genome, 349f
DNA sequencing, **265**
analyzing viral evolution using phylogenetic tree and, 340
in cancer treatment, 188–189
DNA microarray assays and, 308
genome sequencing and, 6, 344–345 (*see also* Genome sequencing)
of ribosomal RNA of prokaryotes related to mitochondria, 485
technology of, 6
DNA strands, 249–250
DNA technology
animal stem cells in, 322–324
organismal cloning in, 320–322
in study of bacterial binary fission, 182–183
DNA viruses
evolution of, 336–337
as pathogens, 338–341
replicative cycles of, 332–335, 337f
structure of, 330–332
Dodder, 586f
Dolly (cloned lamb), 321–322
Dolphins, 376–377, 771
Domains, protein, **355**–357, 488–489
Domains, taxonomy, 7–9, 348t, 358f, **382**, 395–396, 470f, 471t. *See also* Archaea domain; Bacteria domain; Eukarya domain
Domestication, plant, 597, 611–612
Dominance, degrees of, 215–216
Dominant alleles, **209**–210, 216, 220, 222–223
Dominantly inherited disorders, human, 222–223
Dominant species, **854**
Dominant traits, 208, 220, 222–223
Dominican Republic lizards, 846f
Donkeys, 421f
Dopamine, **764**, 775
Dormancy, 463, **604**–605, 624
Dorsal, hollow nerve cords, 537
Dorsal sides, **532**–533
Double bonds, **24**
in carbon skeletons, 42f
covalent bonds as, 24–25f

in organic compounds, 41–42
of unsaturated fatty acids, 49–50
Double circulation, **686**–690. *See also*
Cardiovascular systems
Double fertilization, angiosperm, **601**, 603*f*
Double helix, DNA, 5, **62**, 245, **249**–251, 260*f*
Double membrane, nuclear envelope as, 74–75
Douglas fir tree, 518*f*
Dovrefjell National Park, 826*f*
Down syndrome, 194*f*, 240, **242**
Drift, genetic, 407–409
Drosophila melanogaster (fruit fly)
changes in developmental genes of, 451
complete genome sequence for, 343, 346
correlation of allele behavior and
chromosomes in, 230–231
courtship behaviors of, 805
crossing over in, 236, 237*f*
diploid and haploid numbers of, 195
eye color of, 269
foraging genes of, 809–810
genetic variability of, 400–401*f*
genome size of, 348
homeotic genes in, 360–361
linkage map of, 239*f*
linked genes and, 234–235
as model organism, 230
as model organism for study of development,
311
natural selection and insecticide resistance in,
407
pattern formation and body plan of, 317–320
phylogenetic tree of, 388, 389*f*
studying expression of single genes in, 307–308
Drosophila pseudoobscura (fruit fly), 424*f*, 433
Drought
abscisic acid in plant tolerance of, 624
plant responses to, 633, 634
Drugs
addiction to, 775
antibiotic (*see* Antibiotic drugs)
antiviral, 338
cancer chemotherapy, 188
cocktails of, in AIDS treatment, 402
evolution of resistance to, 374–375
opiates (*see* Opiates)
peroxisomes and, 84
smooth ER and, 77
Dryas, 858
Dry fruits, 606
Duchenne muscular dystrophy, **232**–233
Duckweed, 73*f*
Ducts, male reproductive, 734
Dune fescue grass, 840
Dunstan, William, 869*f*
Duodenum, 649, 671*f*, **674**
Duoshantuophyton, 483*f*
Duplications, chromosome, **241**, 353–355
Duplications, gene, 402
Dusky salamanders, 427
Dwarfism, 222
Dyes, microscopy and, 68*f*
Dyneins, **86**–87*f*
Dysentery, 469

E

Eagles, 733
Eardrums
human, 783*f*, 784
invertebrate, 782
Ears
bones of mammalian, 440, 441*f*
human, 783*f*
insect, 782
Ear stones (otoliths), 784–785*f*

Earth
climate of (*see* Climate; Macroclimate)
conditions on early, and development of life,
459–462
mass extinctions of life on, 444–447
plate tectonics of, 442–444
prokaryotic cells as first cells of life on, 458
Earthworms, 582, 670*f*, 685*f*, 701, 800. *See also*
Caenorhabditis elegans
Eastern glass lizard, 381–382
Ebola virus, 338
Ecdysozoa, 534–535*f*
Echinoderms, 535*f*, 800
Ecological interactions. *See* Interactions,
ecological
Ecological niches, 422, **846**–847
Ecological pyramids, 871–872
Ecological species concept, **422**
Ecological succession, **857**–858
Ecological time, 830
Ecology, **818**–844
aquatic biomes in, 827–830 (*see also* Aquatic
biomes)
climate, macroclimate, and, 820*f*–823 (*see also*
Climate; Macroclimate)
ecological effects of animals, 547
ecological interactions and species
distributions in, 818, 830–832 (*see
also* Interactions, ecological; Species
distributions)
importance of mycorrhizae in, 585
mass extinctions and, 447
populations in, 819*f*, 832–842 (*see also*
Population(s))
prokaryotic roles in, 474–475
scope and fields of, 819*f* (*see also* Community
ecology; Ecosystem ecology; Global
ecology; Landscape ecology; Organismal
ecology; Population ecology)
terrestrial biomes in, 822–826*f* (*see also*
Terrestrial biomes)
Ecosystem ecology, 819*f*, 876
Ecosystem engineers, **855**
Ecosystems, 819*f*, **864**–881
effects of animals on, 547
effects of mass extinctions on, 445*f*, 446–447
energy flow and chemical cycling in trophic
structure of, 6–7, 135*f*, 864–866 (*see also*
Chemical cycling; Energy flow; Trophic
structure)
genome sequencing of metagenomes in,
345–346
in hierarchy of biological organization, 2*f*
importance of mycorrhizae to, 585
prokaryotic roles in, 474–475
regulation of primary production in, 866–870
restoration ecology and restoration of
degraded, 877–879*f*
secondary production efficiency in, between
trophic levels, 870–872
soil as, 581–582
water and nutrient cycling in, 872–877 (*see
also* Biogeochemical cycles)
Ecotones, **824**
Ectoderm, **533**, 744–745
Ectomycorrhizae, **509**, 512*f*, **584**
Ectoparasites, **850**
Ectoprocts, 535*f*
Ectothermic organisms, **544**, **645**, 647, 679
Edema, 694
Ediacaran biota, 437*f*, 450, **483***f*, **529**, 531–532
Edidin, Michael, 96*f*
Effector cells, **719**
Effectors, 636–637
Efferent neurons, 771
Egg-polarity genes, **318**–320

Eggs, **729**
in animal fertilization, 743–744
of birds and dinosaurs, 391–392
chromosomes in human, 175
embryo survival and, 733
as female gametes, 729
in human fertilization, 745–748
in human oogenesis, 737*f*–738
ein mutants, ethylene and, 625
Ejaculation, **734**, 742, 745
Ejaculatory duct, **734**
Electrically charged side chains, 53*f*
Electrical membrane potential, 104–105
Electrical signaling, neurons and, 751–753. *See
also* Neurons
Electrical synapses, 761
Electrocardiogram (ECG or EKG), **690**
Electrochemical gradients, **104**–105
Electroencephalogram (EEG), 771
Electrogenic pumps, **105**
Electrolytes, blood, 695
Electromagnetic energy or radiation, 160
Electromagnetic receptors, **781**
Electromagnetic spectrum, **160**
Electron distribution diagrams, 23*f*, 25*f*
Electronegativity, **25**
Electron microscope (EM), **67**, 68*f*, D-1*f*
Electrons, **20**
configuration of carbon, in organic
compounds, 41–42
distribution of, and chemical properties of
atoms, 23–24
electron shells and energy of, 22–23
in excitation of chlorophyll by light, 162–163
ionic bonding and transfer of, 26*f*
in light reactions of photosynthesis, 163–167*f*
redox reactions and, 136–137
as subatomic particles, 20–21
Electron shells, **22**–23
Electron transport chains, **139**
in aerobic and anaerobic respiration vs. in
fermentation, 150–151
ATP yield from, 147–148
in catabolic pathways and cellular respiration,
137–140
chemiosmosis and, 145–147
in light reactions of photosynthesis, 163–167*f*
in oxidative phosphorylation, 144–145
Electrophysiologists, 756*f*
Electroreceptors, 781
Elements, **20**–24
Elephantiasis, 694
Elephants, 370–371, 836, 837*f*
Elevation, climate and, 822
Elimination, **668**, 676
Elk, 810*f*
Elkhorn coral, 861
Elodea, 29*f*
Elongation, antiparallel DNA, 254–256
Elongation factors, 282
Elongation stage
transcription, 275
translation, 282, 283*f*
Elton, Charles, 853
Embryo(s)
anatomical similarities in vertebrate, 375
development of plant, 603–604
ensuring survival of, 732–733
land plant, 506*f*
monocot vs. eudicot, 554*f*
mortality rates for hybrid, 241
Embryonic development, 311–320
analyzing quantitative and spatial data on *Hox*
genes in, 316
cytoplasmic determinants and inductive
signals in, 312–313

genetic program for, 312
model organisms in study of development and, 311
pattern formation and body plans in, 317–320
sequential gene regulation in, 313–315
Embryonic development, animal, 742–749
cleavage and gastrulation in, 744–745
fertilization in, 743–744
human, 745–749
sex determination in, 739
stages of, 742
Embryonic germ layers, 744–745
Embryonic lethals, **318**
Embryonic stem (ES) cells, 323–324
Embryophytes, **506**f
Embryo sacs, plant, **600**–601
Emergent properties, **3**–4, 20, 218–219
Emerging diseases, 861–862
Emerging viruses, 338–340
Emigration, 832f–**833**, 842
Emission, human, 742
Emotions
limbic system and, 775
prefrontal cortex and, 777
Encephalitis, 338
ENCODE (Encyclopedia of DNA Elements), 346
Endangered species
imprinting for, 807
molluscs as, 549
population dynamics and, 842
Endemic species, **378**
Endergonic reactions, **121**
energy coupling of, 122–124
metabolism and, 120–121
Endocarp, 607f
Endocrine cells, 648
Endocrine glands, human, 650f–651f
Endocrine pathways, 649
Endocrine signaling, 648–653
evolution of hormone function in, 653
feedback regulation in, 652
functions of endocrine and nervous systems in, 648–649
human endocrine system in, 650f–651f
in local cell signaling, 108
multiple effects of hormones in, 652
neuroendocrine pathways in, 649–652
pathways of water-soluble and lipid-soluble hormones in, 652
simple endocrine pathways in, 649
Endocrine systems, **648**
coordination and control functions of, 648
hormonal regulation of digestion by, 678–679
human, 650f–651f
mammalian, 642t
Endocytosis, **106**–107f
Endoderm, **533**, **744**–745
Endodermis, **564**, **587**–588
Endomembrane system, **76**–81
bound ribosomes and, 285
components and functions of, 76–77, 81, 91
endoplasmic reticulum of, 77–78
Golgi apparatus of, 78–79
lysosomes of, 79–80
organelles and functions of, 81f
targeting polypeptides to, 285–286
vacuoles of, 80–81
vesicles of, 76–77
Endometrium, **735**
Endoparasites, **850**
Endophytes, **523**–524
Endoplasmic reticulum (ER), **77**
animal cell, 72f
plant cell, 73f
ribosomes and, 76
rough ER functions, 78

smooth ER functions, 77–78
synthesis of membrane proteins and lipids in, 98f
targeting polypeptides to, 285–286
Endorphins, 28, 55, **764**
Endoskeletons, 799f, **800**, 801f
Endosperm, **601**
Endospores, **463**
Endosymbionts, 82, 484
Endosymbiont theory, **82**, 156, **484**–486
Endosymbiosis, **484**–487
Endothelin, 693
Endothelium, **691**
Endothermic organisms, **544**, **645**, 647, 679, 687–688
Endotoxins, **476**
Energy, **22**, **117**. *See also* Energy flow
allocation of, in angiosperm reproduction, 610
biofuel technology to reduce dependence on fossil fuels for, 612–613
chemiosmosis as energy-coupling mechanism, 145–147
conservation of, 865
of diffusion, 99–100
electron shells and levels of, 22–23
forms of, 117–118
of hydrocarbons and fats, 42, 50
input and output of, in glycolysis, 140f–141f
locomotion and, 802, 803
metabolism and cellular, 116 (*see also* Metabolism)
regulation of allocation of, in animal nutrition, 679–681
thermodynamics and laws of transformation of, 118–119, 865
transfer and transformation of matter and, as biological theme, 6–7
transformation of, by mitochondria, chloroplasts, and peroxisomes, 81–84
Energy coupling, **122**–124, 145–147
Energy flow
as biological theme, 6–7
cellular respiration, photosynthesis, chemical cycling, and, 135
chemical cycling and, in ecosystems, 864–866 (*see also* Chemical cycling)
conservation of energy and, 865
ecosystem energy budgets and, 867–868
overview of, 866f
primary production and, 866–870
secondary production efficiency of, between trophic levels, 870–872
Engelmann, Theodor W., 161f
Engineering, genetic. *See* Genetic engineering
Enhancers, **300**–302
Ensatina genus, 421f
Entamoebas, 497
Enteric division, peripheral nervous system, 678, **771**
Entropy, **118**–119
Entry stage, phage lytic cycle, 333f
Enveloped viruses, 335–336
Environment
adaptations of vertebrate kidneys to diversity in, 660–661
adaptive evolution as fitness to, 407, 411 (*see also* Homeostasis)
animal maintenance of internal, 644–645 (*see also* Homeostasis)
animal regulating and conforming responses to, 644
aquatic physical and chemical (*see* Aquatic biomes)
behavior and stimuli from, 804
bottleneck effect and changes in, 408–409
Cambrian explosion and changes in, 531
cancer development and, 327–328

chemical cycling and, 6f–7
differential gene expression and, 293 (*see also* Gene regulation)
Earth's early, and origin of life, 459–462
effects of, on protein structure, 58–59
effects on cell division of, 185
enzymatic catalysis and factors of, 129–130
ethylene in plant responses to stresses from, 624–625
genetics vs., in animal behaviors, 807
genome sequencing of metagenomes in, 345–346
impact of, on phenotypes, 218
impacts of evolution of land plants and fungi on, 521–524
induction from, in cellular differentiation, 313
interaction of chance, natural selection, and, 415
interaction of organisms with other organisms and, as biological theme, 7 (*see also* Interactions, ecological)
ionic bond strength and, 26
multifactorial disorders and, 223
plant responses to, 631–638
reproductive cycles and cues from, 731
spatial learning and, 807–808
species distributions and, 818
as surroundings and organisms, 365
Environmental issues
ecology and, 818 (*see also* Ecology)
extinctions (*see* Extinctions)
honeybee population decline, 602f
restoration of degraded ecosystems, 877–879f
Enzymatic catalysis, 125–132
activation energy barrier and, 125–126
calculating rate of, 128
cofactors, coenzymes, and, 129
effects of environmental factors on, 129–130
effects of temperature and pH on, 129
in enzyme active sites, 127–128
enzyme inhibitors and, 129–130
evolution of enzymes and, 130
lowering of activation energy barriers by, 125–126
regulating, 130–132
substrate specificity of enzymes in, 126–127
Enzymatic hydrolysis, 668. *See also* Chemical digestion
Enzyme complexes, 132
Enzymes, **44**, **125**
3-D structure of, 59f
allosteric regulation of, 130–132
autophagy by lysosomal, 80
in cardiovascular diseases, 699
as catalysts, 125 (*see also* Enzymatic catalysis)
in chemical digestion, 671–674
in chemiosmosis, 145–147
evolution of, 130
fungal, 508
gene relationship with, in protein synthesis, 269–270
inducible and repressible, 296–297
locations of, in cells, 132
membrane functions of, 97f
in mitochondria, 71, 83
nonenzyme proteins and, 270
in peroxisomes, 84
phagocytosis by lysosomal, 79–80
as protein catalysts, 51, 52f
in protein phosphorylation and dephosphorylation, 111–112
regulation of, in bacterial gene regulation, 294–297
restriction, 262–264, 334
in saliva, 671–672
of smooth ER, 77–78

specialized proteins as, in synthesis and breakdown of organic compounds, 44–45
structure of, 55*f*
in substrate-level phosphorylation, 140
Enzyme-substrate complexes, **126**–127
Eons, geologic, 439*t*
Eosinophils, 695*f*, 696*f*, 714
Ependymal cells, 769*f*
Ephrussi, Boris, 269
Epicotyl, **604**
Epidemics, **338**–340, 638
Epidemiology, 668
Epidermis, plant, **556**, 561–562, 636
Epididymis, **734**
Epigenetic inheritance, **299**
Epilepsy, 759, 777
Epinephrine (adrenaline), **650***f*, **652**
    as biogenic amine, 764
    in cell signaling, 112–113*f*
    in fight-or-flight responses, 690
    in glycogen breakdown, 109
    in nervous systems, 771
Epiphytes, **586***f*
Epistasis, **217**
Epithalamus, 773*f*
Epithelial tissue, 90–91, **643***f*, 674
Epithelium, **643***f*
Epitopes, **715**–716, 724
Epochs, geologic, 439*t*
Epsilon proteobacteria, 472*f*
Epstein-Barr virus, 328
Equational division, 201
Equilibrium
    chemical (*see* Chemical equilibrium)
    community, 856
    Hardy-Weinberg, 403–405
    mechanoreceptors for hearing and, 782–785*f*
    population, 840*f*
Equilibrium potential, **755**–756
Equinoxes, 821*f*
Equus, 454–455
Eras, geologic, 439*t*
Erectile dysfunction, 734–735, 765
Erectile tissue, 734*f*
Erection, penile, 734–735, 765
Ergot fungus, 523*f*
Errors, DNA replication, 257–258
Erythrocytes, 695*f*, **696**, 708. *See also* Red blood cells
Erythropoietin (EPO), 696
Escherichia coli (E. coli) bacteria
    binary fission in, 182–183*f*
    complete genome sequence for, 343
    in DNA cloning, 262
    DNA replication using, 252–257*f*
    gene regulation in, 294–297
    genetic recombination and conjugation in, 468–470
    genome size of, 348
    in human digestive system, 676
    pathogenic strains of, 476
    phages and, 246–248
    as proteobacteria, 472*f*
    rapid reproduction and mutation of, 467–468
    in research on origin of mitochondria, 485
    viral infection of, 332, 334–335
E site (exit site), **281**, 283*f*
Esophagus, 670*f*, 671*f*, **672**, 673*f*, 674, 677*f*, 702
Essential amino acids, **666**
Essential elements, **20**, **578**–582
Essential fatty acids, **666**
Essential nutrients, **666**–668
Ester linkages, 49
Estradiol, 652, 737*f*, **738**, 740–741, 747
Estrogens, 50, **650***f*, 738–742, 746, 748
Estrous cycles, **741**

Estuaries, **827***f*
Ethane, 41*f*
Ethanol (ethyl alcohol), 43*f*, 150, 477, 841*f*, 878
Ethical issues, plant biotechnology, 613–614
Ethylene (ethene), 41*f*, 108, 620*t*, **624**–626, 633–635*f*
Etiolation, **626**
eto mutants, ethylene and, 625
Eucera longicornis, 597
Euchromatin, **259**
Eudicots
    in angiosperm phylogenies, 520*f*–521*f*
    embryo development in, 603*f*
    monocots vs., 553–554*f*
    overview of structure of, 555*f*
    primary growth of roots of, 562*f*
    roots of, 563
    seed structure of, 604*f*
Euglenids, 492
Euglenozoans, **492**–493*f*
Euhadra, 432–433
Eukarya, domain, **8**. *See also* Eukaryotes
    compared with Bacteria and Archaea, 471*t*
    evolutionary relationships of, 358*f*
    genome size and number of genes for, 348*t*
    horizontal gene transfer and, 395–396
Eukaryotes, 481–503. *See also* Animal(s); Eukarya domain; Plant(s)
    Cambrian explosion and evolution of, 530–532
    cell structure of, 481–482 (*see also* Eukaryotic cells)
    early evolution of, 482*f*–483*f*
    embryonic development of (*see* Embryonic development)
    endosymbiosis in evolution of, 484–487
    fossil record of, 482–484
    four supergroups in phylogeny of, 489–498
    genomes of (*see* Eukaryotic genomes)
    in geologic record, 440
    origination of animals in, 528–530
    origination of multicellularity in, 483*f*, 487–489
    origins of key features of, 484*t*
    phylogenetic tree of, 490*f*–491*f*
    protists as unicellular, 481, 499–502 (*see also* Protists)
    taxonomy of, 395–397
    Unikonta as root of phylogenetic tree of, 497
Eukaryotic cells, **4**, **69**. *See also* Cell(s)
    cell cycle of (*see* Cell cycle)
    cellular integration of, 91
    characteristics of, 71, 481–482
    chromatin packing in chromosomes of, 259–261*f*
    cytoskeletons of, 84–88
    DNA replication in, 253, 254*f* (*see also* DNA replication)
    electron transport chains in, 139, 144
    endomembrane systems of, 76–81
    extracellular components of, and connections between, 88–91
    gene expression in, 271*f*
    genetic instructions for, in nucleus and ribosomes of, 74–76
    internal membranes and functions of, 71
    microscopy and biochemistry in study of, 67–69
    mitochondria, chloroplasts, and peroxisomes of, 81–84
    mutations in, 290
    organelles of animal and plant, 72*f*–73*f* (*see also* Animal cells; Plant cells)
    prokaryotic cells vs., 4, 69–71 (*see also* Prokaryotic cells)
    protein synthesis in, 60
    regulation of gene expression in (*see* Eukaryotic gene regulation)

replication of telomeres at ends of DNA molecules of, 258–259
    RNA processing after transcription in, 276–278
    synthesis of multiple polypeptides in translation of, 286–287*f*
    transcription in, 274–276
    translation in, 278–287*f*
    using scale bars to calculate volume and surface area of, 74
Eukaryotic gene regulation, 298–305. *See also* Gene regulation
    analyzing DNA deletion experiments on, 303
    differential gene expression and, 298–299
    mechanisms of post-transcriptional, 304–305
    regulation of chromatin structure in, 299
    regulation of transcription initiation in, 299–304
    stages of gene expression and, 298*f*
Eukaryotic genomes, 348–357. *See also* Genome(s)
    evolution of, from DNA duplication, rearrangement, and mutation, 353–357
    genes and multigene families in, 352–353
    pseudogenes and repetitive DNA in, 349–350
    simple sequence DNA and short tandem repeats in, 351–352
    size, number of genes, and gene density of, 348–349
    transposable elements and related sequences in, 350–351
Eumetazoans, 530, **534**
European green crab, 548
European honeybees, 805
European Molecular Biology Laboratory, 345
Euryarchaeota clade, 474
Eustachian tube, **783***f*
Eutherians, 447*f*, **546**
Eutrophication, 855, **869**
Eutrophic lakes, **828***f*
Evaporation, 32, **646***f*, 874*f*
Evaporative cooling, **32**
Evapotranspiration, **859**
Even-toed ungulates, 376–377
Evo-devo (evolutionary developmental biology), **360**–361, 449
Evolution, **1**, **365**–380. *See also* Adaptations; Natural selection
    adaptive (*see* Adaptive evolution)
    of alternative plant carbon-fixation mechanisms, 169–171
    of animal hormone function, 653
    of animals from sponges and cnidarians, 528–530
    of asexual and sexual reproduction, 608–609
    associative learning and, 808
    of axon structure, 760
    of biological order, 119
    of cells, 66
    of cell signaling, 113
    classification of three domains of life in, 7–9
    of cognition and cerebral cortex, 777–778
    comparing genome sequences to study, 357–361
    convergent, 376, 381*f*, 385–386
    as core theme of biology, 1, 7–11
    as descent with modification by natural selection, 365–366, 370–372
    of differences in membrane lipid composition, 96
    of digestive compartments, 670–671
    divergent, 423–425, 432
    DNA and proteins as measures of, 62–63
    early, of eukaryotes, 481–487
    of ecological niches, 846–847
    effects of animals on, 548, 549

effects of humans on, 548–550
endosymbiont theory on origins of mitochondria and chloroplasts in, 82
of enzymes, 130
evidence for, in biogeography and geographical distribution of species, 377–378
evidence for, in direct observations of evolutionary change, 373–375
evidence for, in fossil record, 376–377
evidence for, in homologies, 375–376
field research on, by Charles Darwin, 368–370
of foraging behaviors, 809–810
of fungi, 510–513
of genetic code, 273–274
genetic variation and, of animal behavior, 812–814
of genetic variation from genetic recombination and natural selection, 236
genetic variation within populations and, 204
of genomes from DNA duplication, rearrangement, and mutation, 353–357
of glycolysis, 151
of gymnosperms, 518
historical context of Darwinian revolution in, 366–367
of human globin gene amino acid sequences, 356
of immune system avoidance mechanisms, 725–726
J.-B. de Lamarck's theory of, 367
of land plants, 513f
of land plants and fungi, 505–508
of leaves, 557f
life history diversity and, 839–840
macroevolution (see Macroevolution)
of mitosis, 183
molecular clocks and rates of, 392–394
of pathogens that evade immune systems, 725–726
of patterns of sexual reproduction, 731–732
phylogenies as evolutionary histories, 381–382 (see also Phylogenies)
of plant antifreeze proteins, 635
of plant defense systems, 636–638
of populations (see Microevolution)
predation and natural selection in, 378
of roots, 555f
of roots and leaves, 516
scientific evidence supporting theory of, 373–379
of seeds, 517
sexual reproduction as enigma of, 730
significance of altered DNA nucleotides and mutations in, 258
speciation as bridge between microevolution and macroevolution, 418 (see also Speciation)
species distributions and, 830
of stems, 556f
taxonomy and relationships in, 382–385
theoretical aspects of Charles Darwin's theory on, 379
theory of natural selection and, 9–10
of tolerance to toxic elements, 20
tree of life and, 10–11
of variations in double circulation circulatory systems, 687–688
of vascular plant organs, 554–556
of vertebrate digestive systems, 676–678
of viruses, 336–337, 340
of visual perception, 785–787
Evolutionary developmental biology (evo-devo), 360–361, 449
Evolutionary time, 830
Evolutionary trees, 370–371
Exaptations, 453, 464

Excavata, 490f, **492–493**f
Excitatory postsynaptic potential (EPSP), **762**
Excited state, pigment molecule, 162
Excitement phase, human sexual response, 742
Excretion, **653**, **655**f
Excretory systems
    four functions of, 655–656
    invertebrate, 656
    kidney function in, 658–663
    mammalian, 642t, 657f
    vertebrate, 656
Executive functions, brain, 777
Exergonic reactions, **120**
    catabolic pathways and, 136
    energy coupling of, 122–124
    energy profile of, 125f–126
    metabolism and, 120–121
Exhalation, 704–705
Exit tunnel, ribosome, 281
Exocrine cells, 648
Exocytosis, 98f, **106**
Exon duplication, 355–357
Exons, **277–278**, 348, 355–357, 400–401f
Exon shuffling, 355–357
Exoskeletons, **800**
    arthropod, 536, 540–541
    chitin as structural polysaccharide of, 49
    in skeletal systems, 799f
Exotoxins, **476**
Expansins, **620–621**
Experimental groups, 14
Experiments
    controlled, 14
    designing, 548, 739, 774
Exponential population growth, **836**
Extant lineages, 513
Extension, muscle, 799f
Extensor muscles, 799f
External fertilization, **732**, 742, 811. See also Fertilization, reproductive
External skeletons, 799f
Extinctions
    ecology and, 818 (see also Ecology)
    in fossil record, 376–377, 440
    island equilibrium model and, 860
    mass extinctions vs., 444–445 (see also Mass extinctions)
    population dynamics and, 842
Extracellular digestion, 670–671
Extracellular matrix (ECM), **88–90**, 97f
Extreme halophiles, **471**, 474
Extreme thermophiles, **471**, 474
Extremophiles, **471**, 474
Eyecups, 453f
Eyes
    color of fruit fly, 230–231, 269
    compound, 786
    differential gene expression in fish, 293
    evolution of, 453
    ocelli (eyespots) as, 785
    single-lens, 786–787
    visual information processing in vertebrate, 787–788
Eyespots, 785

F
F_1 (first filial) generations, **207**–208
F_2 (second filial) generations, **207**–208
Facilitated diffusion, **102**
    as passive transport, 104f
    transport proteins in, 102
    of water across plant plasma membranes, 577
Facilitation, **851**
Facultative anaerobes, **151**, 466
FAD (flavin adenine dinucleotide), 142–143f

FADH$_2$, 139f, 142–143f, 147–148
Falling phase, action potential, 758–759
Familial hypercholesterolemia, 106
Families (taxonomy), **382**
Family histories, 219–220
Family resemblence, heredity and, 192f
Fangs, 677
Fast-twitch fibers, **798**
Fats, **49**
    in cardiovascular diseases, 698
    catabolism of, 136, 152
    digestion of, 673f, 674–675
    energy storage in, 680
    fatty acids and, 49–50
    as hydrophobic, 42
Fat-soluble vitamins, 667
Fatty acids, **49–50**, 152, 666
Feces, **676**
    production efficiency and, 870–871
    seed dispersal in, 607f
Feedback inhibition, **132**, 294
Feedback regulation. See also Regulation
    of animal digestion, energy storage, and appetite, 678–682
    in endocrine signaling pathways, 652
Feeding mechanisms, 508–509, 669f. See also Food processing
Feeding relationships, 547–548
Female condoms, 748
Female gametophytes, angiosperm, 600–601
Females
    autoimmune diseases in human, 725
    hormonal control of reproductive systems of human, 739–741
    inactivation of X-linked genes in mammalian, 233–234
    mammalian sex determination of, 739
    mate choice by, 413, 811
    oogenesis in human, 737f–738
    parental care and, 810–811
    reproductive anatomy of human, 735
    reproductive rates and, 834–835
    sex determination of, 231–232
Fermentation, **136**
    anaerobic and aerobic respiration vs., 148–149, 150–151
    as catabolic, 136
    types of, 150
Ferns, 513f, 514, 515f
Ferrets, 339
Fertility, human, 747–749
Fertility, hybrid, 421f
Fertility schedules, 834–835
Fertilization, reproductive, **195**, **598**, **732**
    angiosperm, 598
    angiosperm double fertilization, 601, 603f
    in animal embryonic development, 742–744
    cell division and, 176
    ensuring offspring survival following, 732–733
    external versus internal, 732
    human, 745–746
    human contraception and, 747–748
    human in vitro fertilization, 749
    in human life cycle, 195–196
    mechanisms preventing angiosperm self-fertilization, 609
    G. Mendel's techniques of, 207–208
    parental care and internal vs. external, 811
    parthenogenic self-fertilization, 730, 732
    prezygotic barriers and, 419–421f
    random, 203
    in sexual life cycles, 196–197
Fertilization, soil, 580, 841f, 870, 875f
Fescue grass, 840
Fetal alcohol syndrome, 747
Fetus, human, **747**

Fever, 647, 715
F factor, **469**
Fiber, dietary, 48
Fiber cells, **558f**
Fibers, muscle, 793–795, 797–798
Fibrin, 696–697
Fibrinogen, 696–697
Fibroblasts, 186, 643f
Fibronectin, **89f**
Fibrous proteins, 55, 56f–57f
Fibrous root systems, 554–555
Fiddler crabs, 792, 804–805
Fields, visual, 788, 789
Fierer, Noah, 852f
Fight-or-flight responses, 652, 690, 771
Filaments
    flagellum, 464f
    flower, 519, 598
    fungi as, 509
    muscle, 793–795
Filopodia, 91, 745f
Filter feeders, **530**, 547, 669f
Filtrate, **656**, 658–659
Filtration, **655f, 656**
Fimbriae, 70f, **463–464**
Finches, 10, 14, 369, 399, 830, 847f
Finland, 842, 855
Fin whales, 384f
Fire, 856–857
Fire-bellied toad, 428–429
Firefly gene transplantation, 274f
Firefly squid, 116
Fireworm, 535f
First law of thermodynamics, **118**, 865
Fishapod (*Tiktaalik*), 542
Fishes
    allopatric speciation in, 423
    animal diets and, 665
    changes in developmental gene regulation in,
        451–452f
    circulatory systems of, 687
    convergent evolution in, 376
    differential gene expression in, 293
    effects of human overfishing on sexual
        maturation of, 548
    electromagnetic receptors of, 781
    fixed action patterns in, 804
    frequency-dependent selection and,
        414, 415f
    gills for gas exchange in, 701
    membrane lipid composition in, 96
    osmoregulation by, 653–654
    parental care in, 811
    pheromones as alarm signals for, 806
    population fluctuations in, 840
    sex reversal in, 732
Fission, 730
Fitness, relative, 411
FitzRoy, Robert, 368
Fixed action patterns, **804**
Fixed alleles, 403, 409
Flaccid cells, **102, 577**
Flagella, **85**
    animal cell, 72f
    euglenozan, 492
    as microtubules, 85–87
    prokaryotic, 70f, 464
    protistan cell, 73f
    structure of, 87f
Flame bulbs, 656
Flashlight fish, 475f
Flatworms, 656, 685f, 769f, 799–800
Flavin mononucleotide (FMN), 144
Flemming, Walther, 177
Flesh-eating disease, 374

Fleshy fruits, 606
Fletcher, W. J., 831f
Flexion, muscle, 799f
Flexor muscles, 799f
Flies, 602f, 786f, 811. *See also Drosophila
        melanogaster*
Flight, 541, 801–802
Flightless cormorant, 418
Flooding
    as ecosystem disturbance, 856–857
    plant responses to, 633–634, 635f
Florida, restoration project in, 879f
Florida panther, 408f
Florigen, **631**
Fluorescence microscopy, 174f
FLOWERING LOCUS T (FT) gene, 631
Flowering plants. *See* Angiosperms
Flower mantids, 372f
Flowers, **518**, 597–607f
    adaptations that prevent self-fertilization of,
        609
    double fertilization of, 601, 603f
    flowering of, 562, 629–631
    fruit development from, 605–607f
    hypothetical florigen hormone in flowering
        of, 631
    impact of pollinators on, 524
    monocot vs. eudicot, 554f
    photoperiodism and flowering of, 630–631
    pollination of, 597, 601, 602f
    preventing transgene escape with genetically
        engineered, 614
    seed development from, 601–605, 607f
    structure and function of, 598–601
    structure of, 518–519
Fluctuation, population, 840–842
Fluid-based skeletons, 799–800
Fluid feeders, 669f
Fluidity, membrane, 95–96
Fluid mosaic model, 94–**95**, 99
Fluorescence, 163
Fluorescence microscopy, 68f, 69, 182
Flu viruses. *See* Influenza viruses
Fly agaric, 512f
Flying. *See* Flight
Flying squirrels, 376
Focusing, visual, 789
Folding, protein, 58–59, 282
Folic acid, 668
Foliose lichens, 522f
Follicles, **735**
Follicle-stimulating hormone (FSH), 737f,
    738–742
Follicular phase, ovarian cycle, **740**–741
Food
    in cellular respiration, 135–136, 151–152
    genetically modified organisms (GMOs) as,
        613
    natural selection by source of, 399
Food chains, **853**
    production efficiency in, 870–871
    trophic efficiency in, 871–872
Food poisoning, 335, 472f, 476, 764
Food processing. *See also* Animal nutrition;
        Digestive systems
    evolutionary adaptations of vertebrate
        digestive systems for, 676–678
    feeding mechanisms in, 669f
    mammalian/human digestive system organs
        for, 671–676
    stages of, 668–671
Food vacuoles, 79–**80**, 107f, **495f**, 670
Food webs, 499–500, **853**–854
Foolish seedling disease, 622
Foraging, **809**–810

Foraminiferans, **494**–495
Forams, **494**
Forebrain, **772f**
Foregut, 670f
Forelimbs, mammalian, 375
Forest fires, 856–857
Forests
    case study on nutrient cycling in, 876–877
    clear-cutting of, 524
    northern coniferous, 826f
    temperate broadleaf, 826f
    tropical, 824f
Form-function correlation, 4, 7–8, 641
Fossil fuels
    biofuel technology to reduce dependence on,
        612–613
    carbon skeletons of, 42
    ocean acidification and, 36–37
Fossil record
    adaptive radiations in, 447–449
    angiosperms in, 519–520
    biogeography and, 377–378
    bryophytes in, 514
    Cambrian explosion and bilaterian radiation
        in, 530–532
    dating of rocks and fossils in, 438
    dinosaurs in, 436
    as documentation of history of life, 436–438
    early eukaryotes in, 482–484
    early land animals in, 539–540
    evidence for evolution in, 370–371,
        376–377
    evidence in, of dinosaurs as ancestors of birds,
        391–392
    evolutionary trends in, 454–455
    fungi in, 511
    geologic record and, 438–440
    gymnosperms in, 518
    homologies vs. analogies in, 386
    human evolution in, 546–547
    insects in, 541–542
    land plant origin and diversification in,
        507–508
    mass extinctions in, 445–447
    origin of animals in, 528–529
    origin of mammals in, 441f
    origins of new groups of organisms in, 440
    phylogenetic trees and, 388–391
    prokaryotes in, as evidence of early life,
        461–462
    representative organisms in, 437f
    seedless vascular plants in, 514–516
    speciation patterns in, 430–431
    strata in, 438
    tetrapods in, 542–543f
Fossils, 9f, **366–368**, 458. *See also* Fossil record
Founder effect, **408**
Four-chambered hearts, 688
Fovea, 786f, **789**
FOXP2 gene, 359
F plasmids, **469**
Fractals, 553
Fragmentation, plant, **608**
Fragmentation, reproductive, 730
Frameshift mutations, **290**
Franklin, Rosalind, 249–251
Fraternal twins, human, 746
Free energy, **119**–122
    ATP hydrolysis, energy coupling, and, 123f
    in electron transport chains, 144f
    free-energy change, stability, equilibrium, and,
        119–120
    metabolism and, 120–122
Free energy of activation, 125–126
Free ribosomes, 76, 285

Freezing
    as ecosystem disturbance, 857–858
    plant responses to, 635
Frequency, sound, 784
Frequency-dependent selection, **414**, 415*f*
Freshwater biomes, 827–830, 868–869. *See also* Aquatic biomes
Freshwater fishes, 653–654
Friction, locomotion and, 801, 802
Fritillaria assyriaca, 348
Fritillaries, 842
Frogs
    as amphibians, 543
    continental drift and speciation of, 444
    external fertilization of, 732*f*
    intersexual selection and mate choice among tree, 413*f*
    life cycle of, 742*f*
    nuclear transplantation in, 320–321
    polyploidy in, 425
    in predation, 848*f*
Fronds, 515*f*
Frontal lobe, 776*f*, 777
Frontal lobotomy, 777
Frost-tolerant plants, 635
Fructose
    hydrolysis of sucrose to, 125–126
    as monosaccharide, 46*f*
    synthesis of sucrose from, 47*f*
Fruit fly. *See Drosophila melanogaster*
Fruiting bodies, fungal, 512*f*
Fruitlets, 606
Fruit ripening, 108
Fruits, **519**, **605**
    auxin in growth of, 622
    dispersal of, 607*f*
    ethylene in ripening of, 626
    form and function of, 605–606
    gibberellins in growth of, 622–623*f*
Frye, Larry, 96*f*
Fumonisin, 613
Functional groups, 43*f*, **44**
Functional magnetic resonance imaging (fMRI), 775–776
Function-form correlation, 4, 7–8, 641
Fundamental niches, 847
Fungi
    alcohol fermentation and, 150–151
    calculate volume and surface area of, 74
    cells of, 72*f* (*see also* Eukaryotic cells)
    chemical cycling and biotic interactions of, 521–524
    colonization of land by plants and, 504, 508–513
    in ecosystem trophic structure, 866
    in energy flow and chemical cycling, 6*f*–7
    Eukarya domain and, 8*f*
    evolution of, 510–513
    innate immunity and, 712–715
    life cycles of, 511*f*
    mycorrhizal, 574, 584–585
    mycorrhizal, and land plants, 504, 508, 509–510
    nutritional adaptations of, 508–509
    origin of, in protists, 510–511
    overproduction of, 371*f*
    phylogeny of, 512*f*, 513
    plant toxin as, 613
    sexual and asexual reproduction of, 510, 511*f*
    sexual life cycles of, 196–197
    structural polysaccharides and, 48–49
    in toad extinctions, 818
    Unikonta supergroup and, 496–497
    vacuoles in, 80
Fungi, kingdom, **8***f*, 395

Fusarium, 613
Fusiform shape, 802
Fusion, hybrid zone, 429*f*–**430**

G
G_0 phase, **185**
G_1 phase, **177**
G_1 phase checkpoint, 184*f*, 185–186
G_2 phase, **177**
G_2 phase checkpoint, 184*f*, 185
Gage, Phineas, 777
Galápagos Islands, 10, 368–369, 418, 830, 860*f*. *See also* Finches
Gallbladder, 671*f*, **674**
Gälweiler, Leo, 621*f*
Gambusia hubbsi, 423
Gametes, **175**, **193**
    chromosomes in human, 175
    gametogenesis and, 735–738
    as haploid cells, 195
    inheritance of genes in, 193
    law of segregation of, 209–210
    meiotic nondisjunction of, 240–241
Gametic isolation, 421*f*
Gametogenesis, **735–738**
Gametophytes, **506***f*
    of bryophytes, 514
    development of male and female, in angiosperms, 600–601
    of land plants, 506*f*
    reduced, in seed plants, 517
    of seedless vascular plants, 514–516
    sporophytes and, 196
Gamma-aminobutyric acid (GABA), **764**
Gamma proteobacteria, 472*f*
Ganglia, **751**, 769
Ganglion cells, 786*f*, 788
Gap junctions, **90***f*, 108
Garden peas, G. Mendel's, 207–212
Garlic mustard, 585*f*
Garrod, Archibald, 269
Garter snakes, 420*f*, 812–813
Gas chromatography, 624
Gases, as neurotransmitters, 765–766
Gas exchange, **698**–708
    arthropod, 541
    breathing, lung ventilation, and, 704–706
    circulatory systems and, 684, 686–688 (*see also* Circulatory systems)
    coordination of circulation and, 706
    gills for, in aquatic animals, 701
    lungs for, in mammalian respiratory systems, 702–704 (*see also* Respiratory systems)
    partial pressure gradients in, 699–700
    respiratory adaptations for, in diving mammals, 708
    respiratory media in, 700
    respiratory pigments and adaptations for, 706–708
    respiratory surfaces in, 700–701
    tracheal systems for, in insects, 702
Gas exchange circuits, **686**–687
Gasterosteus aculeatus, 804
Gastric cecae, 670*f*
Gastric glands, 672, 672*f*
Gastric juice, **672**–673
Gastric ulcers, 673
Gastrovascular cavities, **530**, **670**–671, **685**, 800
Gastrula, **744**–745
Gastrulation, 742, **744**–745
Gated channels, **102**
Gated ion channels, **756**–760
Gause, G. F., 846
Gecko lizards, 27

Geese, 547, 807
Gel electrophoresis, **263**, 308*f*
Genbank, 345
Gene(s), **5**, **60**, **193**. *See also* Chromosomes; DNA (deoxyribonucleic acid)
    alleles as alternative versions of, 209–210 (*see also* Alleles)
    analyzing distribution patterns in human skin pigmentation as polygenic trait, 219
    animal behavior and, 812–814
    appetite regulation by *ob* and *db*, in mice, 681
    bacterial resistance, 469–470
    B cell and T cell diversity and, 717–718
    calibrating molecular clocks of, 392–393
    in cell nucleus, 74–76
    cloning, 262, 265
    color vision and, 789
    concept of, 290
    coordinately controlled bacterial, 294–295
    coordinately controlled eukaryotic, 302–304
    dating origin of HIV using, 393–394
    developmental (*see* Developmental genes)
    DNA, nucleic acids, and, 60
    in DNA structure and function, 5–6
    enzyme relationship with, in protein synthesis, 269–270
    epistasis of, 217
    evolution of, with related and with novel functions, 354–355
    evolution of FOXP2, 359
    extending Mendelian inheritance for multiple, 217–218
    extending Mendelian inheritance for single, 215–217
    fluorescent dye and, 228*f*
    foraging, 809–810
    gene expression and protein synthesis (*see* Gene expression)
    genetic variation due to alterations of number or position of, 402
    homeotic, 318, 360–361, 450, 451
    homologous, 376
    horizontal gene transfer of, 395–396
    inheritance of, 193
    locating, along chromosomes, 228
    mapping distance between, on chromosomes, 237–240
    master regulatory, 314–315
    maternal effect (egg-polarity), as axis establishment, 318–320
    as measures of evolution, 62–63
    in G. Mendel's particulate hypothesis of inheritance, 206
    multigene families and, in genomes, 352–353
    mutations, faulty proteins, and faulty, 268
    mutations of (*see* Mutations)
    mycorrhizal, 511–513
    number and density of, in genomes, 348–349
    organization of typical eukaryotic, 300
    pleiotropy of, 217
    in prokaryotic and eukaryotic cells, 69–70
    pseudogenes, 349–350
    ras and *p53* and *p21*, in cancer development, 325–326
    rearrangement of parts of, through exon duplication and shuffling, 355–357
    regulation of (*see* Gene regulation)
    regulatory, 295, 297
    RNA splicing and split, 277–278
    sex-linked, 232–234
    speciation and, 432–433
    systems biology in study of, 346–347
    transcription factors and, 111
    transplanting of, into different species, 274*f*
    transposable elements, 350–351

types of cancer, 324–325
understanding functions for protein-coding, 345–346
using chi-square (χ^2) test to determine linkage of, 238
Genealogy, molecular, 62–63
Gene cloning, **262**, 265. *See also* DNA cloning
Gene expression, **6**, **268**–292
analyzing quantitative and spatial data on, 316
basic principles of transcription and translation in, 270–272
discovery of principles of, 269–270
gene concept and, 290
genetic code in, 272–274
genetic engineering and transgenic, 273–274
genetic information flow, protein synthesis, and, 268
interpreting DNA sequence logos for translation initiation in, 284
monitoring, for single genes and for groups of genes, 307–309
mutations of nucleotides affecting protein structure and function in, 268, 288–290
overview of, 271*f*
in plant development, 561–562
protein synthesis in, 6
regulation of (*see* Gene regulation)
RNA processing after transcription in, 276–278
summary of eukaryotic transcription and translation in, 287*f*
systems biology in study of, 346–347
transcription as, 299
transcription as DNA-directed RNA synthesis in, 274–276
translation as RNA-directed polypeptide synthesis in, 278–287*f*
Gene flow, **409**
biological species concept and, 419, 422
as cause of microevolution, 409–410
geographic separation and, 427–428
Hardy-Weinberg equilibrium and, 405
speciation and, 432
Gene-for-gene recognition, **636**–637
Gene pools, **403**–406
General transcription factors, 300
Generative cells, 601
Gene regulation
analyzing DNA deletion experiments on, 303
analyzing quantitative and spatial gene data on, 316
by auxin, 620
bacterial, 293–298
changes in, of developmental genes, 451–452*f*
development, embryonic development, and, 311 (*see also* Development; Embryonic development)
differential gene expression and, 293, 298–299
eukaryotic, 298–305
faulty, in cloned animals, 322
flowering and, 631
monitoring of gene expression and, 307–309
noncoding RNAs in, 305–306
in plant development, 561–562
Genetically modified (GM) organisms, 612–614
Genetic code
amino acid dictionary of, 273*f*
codons and triplet code of, 272–273
deciphering, 273
DNA structure and function and, 6
evolution of, 273–274
as molecular homology, 376
mutations and, 288–290
Genetic counseling, 223

Genetic disorders
from abnormal chromosome numbers, 240–241
alkaptonuria, 269
from alterations of chromosome structure, 241–243
counseling for, 223
dominantly inherited, 222–223
multifactorial, 223
mutations and, 288–290
preparing karyotypes for, 194*f*
recessively inherited, 220–222
Genetic diversity, prokaryotic, 467–470
Genetic drift, **407**–409
Genetic engineering, **261**
amplifying DNA for cloning using polymerase chain reaction in, 264–265
DNA cloning in, 262
DNA sequencing in, 265
gene transplantation in, 273–274
nucleic acid hybridization in, 261
organismal cloning in, 320–324
plant biotechnology and, 612–614
of plant defenses against herbivores, 636
prokaryotes in, 477
using restriction enzymes to make recombinant DNA in, 262–264
Genetic maps, **237**–240
Genetic mutants, 681, 774
Genetic profiles, **351**
Genetic prospecting, 470
Genetic recombination, **235**
evolution of genetic variation from natural selection and, 236
linked genes and, 235–237*f*
of linked genes through crossing over, 236, 237*f*
in prokaryotes, 468–470
transposable elements and, 357
of unlinked genes through independent assortment of chromosomes, 236
Genetics, **192**
B cell and T cell diversity and, 717–718
chromosomal basis of inheritance in (*see* Chromosomal basis of inheritance)
color vision and, 789
cytology and, 228
DNA structure and function in, 5–6 (*see also* DNA (deoxyribonucleic acid))
environment vs., in animal behaviors, 807
experiments on obesity genes in mice, 681
expression and transmission of genetic information as biological theme, 5–6
foraging behavior and, 809–810
gene regulation in (*see* Gene regulation)
genetic basis of animal behavior, 812–814
genetic engineering and (*see* Genetic engineering)
genomics and bioinformatics in, 6 (*see also* Bioinformatics; Genomics)
heredity, inheritance, variation, and, 192 (*see also* Genetic variation; Inheritance)
Hox genes in Cambrian explosion, 531
macroevolution of development from changes in developmental genes in, 449–452*f*
of mammalian sex determination, 739
Mendelian inheritance in (*see* Mendelian inheritance)
molecular basis of inheritance in (*see* Molecular basis of inheritance)
molecular evidence for origination of animals, 529–530
of mycorrhizae, 511–513
in phylogeny of animals, 533–534
prokaryotic, 464–465, 467–470

Punnett square and, 210
research on prokaryotes related to mitochondria, 485
sexual life cycles in (*see* Sexual life cycles)
solving complex problems in, with rules of probability, 214
of speciation, 432–433
twin studies in, 806
Genetic variation, **400**
crossing over, recombinant chromosomes, and, 203
evolution and, within populations, 204
evolution of, from genetic recombination and natural selection, 236
genetic drift and loss of, 408–409
independent assortment of chromosomes and, 201–203*f*
microevolution and sources of, 401–402
phylogenetic tree branch lengths and, 388–391
preservation of, 413–414, 415*f*
in prey selection, 812–813
random fertilization and, 203
types of, 400–401
Gene trees, 384*f*
Gene variability, 400
Genome(s), **6**, **175**, 343–363
bioinformatics, proteomics, and systems biology in analysis of, 345–347
cell division and, 175–176
comparing, to study evolution and development of species, 357–361
differential gene expression for identical, 293, 298–299 (*see also* Gene regulation)
evolution of, from DNA duplication, rearrangement, and mutation, 353–357
genomics and bioinformatics in study of, 6, 343–344 (*see also* Bioinformatics; Genomics; Metagenomics)
horizontal gene transfer between, 395–396
Human Genome Project and genome sequencing techniques, 344–345
monitoring gene expression of groups of genes in, 308–309
noncoding DNA and multigene families in eukaryotic, 349–353 (*see also* Eukaryotic genomes)
noncoding RNAs in, 305–306
p53 gene as guardian angel of, 326
predicting percentages of nucleotides in, using data in tables, 249
prokaryotic, 464–465, 470
reading amino acid sequence identity tables for, 356
size and estimated number of genes in, for organisms in three domains, 348*t*
species with complete sequences available, 343, 346, 348, 476
variations in size, number of genes, and gene density of, 347–349
viral, 331, 333–334, 336–337
Genome sequencing, 6, 340, 344–345. *See also* DNA sequencing
Genomic equivalence, 320
Genomics, **6**, **343**, 470. *See also* Bioinformatics; Genetics; Metagenomics; Systems biology
Genotypes, **210**
gene expression as link between phenotypes and, 268
in Hardy-Weinberg equilibrium, 404–406
heterozygote advantage and, 414
phenotypes vs., 210, 218 (*see also* Phenotypes)
relative fitness and, 411
transformation and, 246
Genus/genera (taxonomy), **382**

Geographical separation, speciation with and without, 423–428
Geographic distribution of species. *See* Species distributions
Geologic record, **438**–440
Geospiza fuliginosa and *Geospiza fortis*, 847*f*
Germ cells, 195, 259, 736*f*–737*f*
Germination, seed, 605, 623, 627–628
Germ layers, 533, 534, 744–745
Gestation, human, **746**–747
Ghrelin, 681
Giant panda, 348*t*
Giant squid, 85*f*
Giant water bugs, 733*f*
Giardia intestinalis, 491*f*, 492
Gibberellins, 620*t*, **622**–623
Gibbons, 63
Gibbs, J. Willard, 119
Gibbs free energy, 119, G-12. *See also* Free energy
Gills
    for gas exchange in aquatic animals, 684, 701
    single circulation and, 686, 687*f*
Giraffes, 135*f*, 538*f*, 693
Gizzard, 670*f*
GLABRA-2 gene, 561–562
Glaciation, ecological succession after, 858
Glacier Bay, Alaska, glacial retreat and primary succession in, 858*f*
Glans, clitoris, 735
Glans, penis, 734*f*, **735**
Glanville fritillaries, 842
Glaucoma, 786*f*
Gleason, H. A., 856
Glia (glial cells), **643***f*, **752**, 769–771
Glioblastoma (brain cancer), 188, 347
Global climate change. *See also* Climate
    land plants and, 522
    photosynthetic protists and, 499–500
    volcanism and, 445
Global climate patterns, 820*f*–821. *See also* Macroclimate
Global ecology, **819***f*
Global energy budget, 867
Global net primary production, 867–868
Global warming. *See* Global climate change
Globigerina, 491*f*
Globin genes, 352–356
Globular proteins, 55, 56*f*–57*f*
Glomeromycetes, **512***f*
Glomerulus, **657***f*, 662
Glomus mosseae, 512*f*
Glucagon, **650***f*, 680
Glucocorticoids, **650***f*
Glucose
    in anaerobic respiration and fermentation, 148–151
    calculating rate of enzymatic catalysis of, 128
    in cellular respiration, 136, 139–141 (*see also* Cellular respiration)
    in diabetes mellitus, 680–681
    effects of age on transport of, 103
    homeostasis of, 680
    hydrolysis of sucrose to, 125–126
    linear and ring forms of, 46*f*
    as monosaccharide, 45–46
    in photosynthesis, 28–29, 157, 171
    in positive gene regulation, 297
    in storage polysaccharides, 47–48
    in structural polysaccharides, 48*f*
    synthesis of sucrose from, 47*f*
    transport proteins for, 99, 102
Glucose-6-phosphatase, 128
Glucose 6-phosphate, 128
Glutamate, **764**, 788

Glutamic acid, 53*f*, 58, 123*f*, 273
Glutamine, 53*f*, 123*f*
Glyceraldehyde, 46*f*
Glyceraldehyde 3-phosphate (G3P), **167**–169, 171*f*
Glycerol, 49–50
Glycerol phosphate, 43*f*
Glycine, 43*f*, 53*f*, 764
Glycogen, 47*f*, **48**, 109, 151–152, 680, 795
Glycolipids, 95*f*, **98**
Glycolysis, **139**
    ATP yield from, 147–148
    in catabolic and anabolic pathways, 151–152
    in cellular respiration, 139–141
    evolutionary significance of, 151
    fermentation and, 149–151
    glycolytic muscle fibers and, 798
    overview of, 140*f*–141*f*
Glycolytic muscle fibers, 798
Glycoproteins, **78**, **98**
    in cell-cell recognition, 97*f*
    in extracellular matrix, 88–89
    in Golgi apparatus, 79
    membrane carbohydrates in, 98
    in plasma membranes, 95*f*
    rough ER and, 78
    synthesis of, in endoplasmic reticulum, 98*f*
    viruses and, 331*f*–332, 335–336
Glycosidic linkages, **46**–49
Gnathostomes, **537**–539
Goatsbeard plants, 426
Gobi Desert, 822
Golden rice, 612
Golgi apparatus, **78**
    animal cell, 72*f*
    in endomembrane system, 81*f*
    functions of, 78–79
    plant cell, 73*f*
    in synthesis of membrane components, 98*f*
Gonadotropin-releasing hormone (GnRH), 738–742
Gonadotropins, 738–742
Gonads, 176, 195–196, **733**, 739. *See also* Reproductive organs, human
Gonium, 487*f*
Gonorrhea, 464, 749
Goodall, Jane, 11
G protein-coupled receptors (GPCRs), **110**, 112–113
G proteins, **110**, 112–113, 788
Graded muscle contraction, 797
Graded potentials, neuron, **756**–757, 762
Grafting, plant, 610
Grafts, tissue, 724
Gram, Hans Christian, 462
Gram-negative bacteria, **462**–463, 472*f*
Gram-positive bacteria, **462**–463, 473*f*
Gram staining technique, 462–463
Grant, Peter and Rosemary, 14, 399
Granum (grana), **83**, 157
Grapes, 623*f*
Graphs
    bar graphs, 15, 149, 477, 560, 634, 849
    comparing two variables on common x-axis of, 726
    estimating quantitative data from, and developing hypotheses, 443
    histograms, 188, 219, 699
    interpreting, with two sets of data, 103
    line graphs, 128, 202
    scatter plots, 37, 170, 427, 849
Grass, 618–619, 840
Grasshoppers, 361*f*, 670*f*, 685*f*, 702*f*, 799*f*
Grasslands, temperate, 825*f*
Gravitational motion, free energy and, 120*f*
Gravitropism, **632**

Gravity
    blood pressure and, 693
    locomotion and, 801–802
    mechanoreceptors for sensing, in humans, 784–785*f*
    mechanoreceptors for sensing, in invertebrates, 782
    plant responses to, 632
Graylag geese, 807
Gray matter, **770**
Gray tree frogs, 413*f*, 425
Greater prairie chickens, 408–409
Great Salt Lake, 471
Great tits (birds), 409–410
Green algae, **496**
    cells of, 73*f*
    chloroplasts of, 83*f*
    in eukaryotic phylogeny, 491*f*
    evolution of land plants from, 486–488, 505, 507–508
    land animals vs., 540*f*
Greening, plant, 626
Green parrot snake, 848*f*
Griffith, Frederick, 246
Grizzly bears, 422
Grolar bears, 422
Gross primary production (GPP), **867**–868
Ground squirrels, 813, 833, 834–835
Ground tissue system, plant, 556–**557**
Groups, control and experimental, 14
Growth factors, **185**
    in cancer, 186–187
    in cell cycle control system, 185–186
    in cell signaling, 108, 113
    coordinate control by, 303–304
Growth inhibitors, plant, 619, 623–624
Growth rates, heterochrony and, 449–450
Growth regulators, plant. *See* Hormones, plant
Growth rings, tree, 568
Grus americana, 807
Grus canadensis, 807
GTP (guanosine triphosphate), 142–143, 281–282, 283*f*
Guanine, 60–61, 62, 248, 250–251
Guanosine triphosphate (GTP). *See* GTP (guanosine triphosphate)
Guard cells, **565**, 590–592
Guinea pigs, 103
Gulf Stream, 822
Gulls, 807, 810*f*
Guppies, 378
Gurdon, John, 320–321
Gustation, **781**
Gutenberg, Johannes, 15–16
Gymnosperms, **516**
    evolution of, 518
    gametophyte-sporophyte relationship in, 515*f*
    ovule to seed in, 517*f*
    phylogeny of, 513*f*
    as seed plants, 516

H
H1N1 virus, 338–340, 862
H5N1 virus, 339, 861–862
Habitat
    carrying capacity of, 836–837
    islands as, 860
    metapopulations and, 842
    sympatric speciation and differentiation of, 426
Habitat corridors, 878
Habitat isolation, 420*f*
Hadean eon, 439*t*
Haemophilus influenzae, 348*t*
Hagfishes, 537–538*f*

Hair cells, 783f, **784–785f**
Hairs
    invertebrate, 782
    mechanoreceptors and, 780–781
Haldane, J. B. S., 459, 814
Half-life, **438**
Hallucigenia, 437f, 531f, 536
Halobacterium, 471
Halophiles, 471, 474
Halophytes, 634
Hamilton, William, 813–814
Hamilton's rule, **813–814**
Hammer (malleus), 783f
Hammerhead sharks, 730
Hamsters, 774
Haploid cells, **195**–197
Hardy-Weinberg equilibrium, 403–405
Hardy-Weinberg principle, **403–406**
Harlequin toads, 818
Harper, John, 840
Haustoria, **509**
Hawaiian Islands
    allopatric speciation in, 424
    adaptive radiation in, 448–449
Hawaiian silversword plants, 385
Hawkmoth, 848f
Hazel, 602f
Head, clitoris, 735
Head, penis, 735
Head structure morphogen, 319–320
Headwater streams, 828f
Hearing, mechanoreceptors for equilibrium and,
    782–785f
Heart attack, **698**
Heartbeat rhythm, 690
Heartburn, 674
Heart rate, 690
Hearts, **685**
    blood pressure and, 692–693
    cardiac muscle of, 798
    in circulatory systems, 685–688
    control of rhythmic beating of, 690
    evolutionary variation in, 687–688
    fetal, 747
    mammalian, 688–690
Heartwood, 568
Heat, **31**, 117
    as byproduct of cellular respiration, 148
    in denaturation of proteins, 59
    in energy flow and chemical cycling, 6f–7, 135
    extreme thermophiles and, 471, 474
    plant response to, 635
    as thermal energy, 31
    thermoreceptors and, 781
Heat of vaporization, **32**
Heat-shock proteins, **635**
Heavy chains, **716**, 718
Heimlich maneuver, 672
HeLa cells, 187
Helianthus genus, 431–432
Helical viruses, 331
Helicases, **253**
Helicobacter pylori, 472f, 673
Helium, 21f
Helper T cells, **720**–721, 723f
Heme group, 144
Heme oxygenase, 766
Hemeroplanes ornatus, 848
Hemichordates, 535f
Hemipterans, 541f
Hemispheres, brain, 773f, 777, 789
Hemizygous organisms, 232
Hemocyanin, 707
Hemocytes, 712–713
Hemoglobin, **696**
    analyzing polypeptide sequence data of, 63

cooperativity as allosteric regulation in,
    131–132
    in erythrocytes, 696
    in gas exchange, 707–708
    globin gene families and, 352–353, 356
    sickle-cell disease and, 58f, 222
    structure of, 57f
Hemolymph, 656, **685**–686, 712–713
Hemophilia, **233**, 697
Hemorrhagic fever, 338
Henslow, John, 368
Hepatic portal vein, **675**, 680
Hepatitis B virus, 727
Herbicides
    auxin in, 622
    transgenic, 612
Herbivores, **665**
    as biotic factors limiting species distribution,
        831
    dentition and diet in, 676f
    digestive adaptations of, 677–678
    in ecosystem trophic structure, 866
    effects of, on ecosystems, 547
    evolution by natural selection in, due to food
        source changes, 373–374
    land plant interactions with, 524
    plant defenses against, 636
    production efficiency of, 870–871
Herbivory, **849**
Hereditary factors, genes as, 228
Heredity, **192**. *See also* Genetics; Genetic
        variation; Inheritance
Hermaphroditism, **732**
Heroin, 28, 55
Herpes simplex virus, 725
Herpesviruses, 336, 338, 725
Hershey, Alfred, 246–248
Heterocephalus glaber, 813
Heterochromatin, **259**, 306
Heterochrony, **449**–450
Heterocysts (heterocytes), **466**
Heterotrophs, **155**, 465t, 499, 508, 866
Heterozygote advantage, **414**
Heterozygous organisms, **210**
Hexoses, 46f
Hibernation, 148, 804
High-density lipoprotein (HDL), **697**
Highly conserved genes, 358, 360–361
High-throughput sequencing technology, 6, 344
Hindbrain, **772f**
Hindgut, 670f
Hinge joints, 801f
Hinges, mammalian jaw, 441f
Hippocampus, 775, 778–779
Histamine, **714**–715, 724
Histidine, 53f, 666
Histograms, 188, 219, 699, F-2
Histone acetylation, **299**
Histones, **260f**
Hitchhiking commensalism, 850–851
HIV (human immunodeficiency virus), **336**
    AIDS and, 336, 337f
    antiviral drugs and, 338
    attacks on immune system by, **725**–726
    cell infection by, 330
    dating origin of, using molecular clock,
        393–394
    as emerging virus, 338–339
    rapid reproduction of, 402
    replicative cycle of, 337f
    specificity of, 332
HIV-1 M strain, 393–394
Hoekstra, Hopi, 13, 15
Holdfasts, 483f, **493**
Homeoboxes, **360**–361
Homeodomains, 360–361

Homeostasis, **644**
    anatomy, physiology, and, 641
    of blood glucose levels, 680
    endocrine signaling in, 648–653
    hierarchical organization of tissues, organs,
        and organ systems in, 642–643f
    of human breathing, 705–706
    kidney function in, 656–663
    osmoregulation and excretion in, 653–658 (*see
        also* Excretory systems; Osmoregulation)
    regulating and conforming mechanisms of,
        644–645
    thermoregulation in, 645–647
Homeotic genes, **318**, 360–361, **450**, 561–562
Homing pigeons, 804
Homo genus, 546
Homo habilis, 546
Homolidae family, 845
Homologies, **375**
    analogies vs., 385–386
    anatomical and molecular, 375–376
    convergent evolution and analogies vs., 376
    evaluating molecular, 386–387
    as evidence for evolution, 375–376
    morphological and molecular, 385
Homologous chromosomes (homologs), **194**
    behavior of, as basis of law of segregation,
        228–229f
    human, 194–195
    independent assortment of, 201–203f
    in meiosis, 197, 198f, 201
Homologous structures, **375**–376
Homo neanderthalensis, 343
Homoplasies, **386**
Homo sapiens (human), 382, 546–547. *See also*
        Human(s)
Homozygous organisms, **210**
Honeybees, 602f, 805–806, 808–809
Hook, flagellum, 464f
Hooke, Robert, 67
Hopping, 802
Horizontal cells, 786f, 788
Horizontal gene transfer, 395–**396**, 468, 476
Horizontal transmission, viral, 341
Hormonal contraceptives, 748
Hormonal signaling, 108
Hormone cascade pathways, 651f
Hormone-receptor complexes, 111
Hormones, animal, **108**, **648**. *See also* Hormones,
        human
    antidiuretic hormone (ADH), 661–662
    cellular oxygen consumption and thyroid, 149
    coordinate control by, 303–304
    endocrine system, 648
    evolution of functions of, 653
    in long-distance cell signaling, 108–109
    in mammalian sex determination, 739
    membrane proteins and, 97f
    pathways of water-soluble and lipid-soluble,
        652
    proteins as, 52f
    regulation of appetite by, 681–682
    regulation of digestion by, 678–679
    regulation of mammalian/human reproduction
        by, 738–742
    sex (*see* Sex hormones)
Hormones, human. *See also* Hormones, animal
    birth control, 748
    in embryonic development and childbirth,
        746–747
    human endocrine system and, 650f–651f
    reproductive, 737f, 738–742
Hormones, plant, **617**–626
    abscisic acid, 620t, 623–624
    auxin, 618–619, 620–622
    brassinosteroids, 620t, 623

cytokinins, 620t, 622
discovery of, 618–619
ethylene, 620t, 624–626
florigen, 631
gibberellins, 620t, 622–623
overview of, 620t
as plant growth regulators, 617–618
Hornworts, 513f, 514
Horowitz, Norman, 269
Horses, 400f, 421f, 454–455, 677
Host cells, endosymbiont, 484
Host-pathogen coevolution, plant, 636–637
Host ranges, viral, **332**
Hosts, **475**, 548, **850**, 861
Hoxd gene, 316
Hox genes
  analyzing quantitative and spatial data on, 316
  as animal development genes, 361
  in arthropod body plans, 536
  in macroevolution of development, 450, 451
  origins of, 531
  in phylogeny of animals, 533
HTLV-1 virus, 328
Hubbard Brook Experimental Forest, 876–877
Human(s). *See also* Animal(s); Mammals
  apoptosis of white blood cells of, 315f
  behavior of chromosome sets in life cycle of, 195–196
  biological species concept and diversity of, 419f
  biosynthesis of amino acids in anabolic pathways of, 152
  blood types of, 98
  body (*see* Human body)
  brain and nerve cells of, 66f (*see also* Brains; Neurons)
  cardiovascular systems of (*see* Cardiovascular systems)
  cells of, 72f (*see also* Animal cells)
  cellular respiration and diets of, 136, 151–152
  chromosome sets in cells of, 194–195
  chromosomes in somatic cells and gametes of, 175–176
  communication forms of, 806
  detecting pregnancy in, 724
  digestion as hydrolysis in, 45
  digestive system organs of, 671–676 (*see also* Digestive systems)
  disease as density-dependent population regulation mechanism for, 841f
  disturbances by, 858–859
  ecological impacts of, 499–500
  embryonic development, 745–749 (*see also* Embryonic development)
  endocrine system of, 650f–651f
  environmental impacts of (*see* Human environmental impacts)
  evaporative cooling of, by sweat, 32
  evolutionary effects of, 548–550
  evolutionary tree of, 546f
  evolution of, as primate, 546–547
  excretory system of, 657f
  fermentation in muscle cells of, 150
  gene flow and evolution of, 410
  genetics of (*see* Human genetics)
  genome of (*see* Human genome)
  homeostasis in (*see* Homeostasis)
  hormones (*see* Hormones, human)
  immune systems of (*see* Immune systems)
  impacts of, on land plants, 524
  inactivated olfactory receptor genes of, 402
  karyotypes of chromosomes of, 194f
  kidneys of (*see* Kidneys)
  locomotion of, 802
  lymphatic systems of, 693–694
  nervous systems of (*see* Nervous systems; Neurons)

noncoding RNAs in genomes of, 305–306
  number of chromosomes of, 75–76
  nutrition (*see* Animal nutrition)
  phagocytosis in, 80
  pH of blood of, 35f, 36
  phylogenetic trees of, 389f
  receptor-mediated endocytosis in, 106–107f
  reducing hunger and malnutrition in, with transgenic crops, 612
  regulation of breathing in, 705–706
  relatedness of rhesus monkeys, gibbons, and, 63
  reproduction of (*see* Human reproduction)
  sex-linked genes of, and inheritance, 231–234
  skulls of chimpanzees vs., 386
  species name of, 382
  spread of pathogens by, 861–862
  stem cells of, 322
  transgenic crops and health of, 613
  trophic efficiency and meat eating by, 872
  twin studies of behavior of, 806
Human body
  bacterial prokaryotes and, 458–459f, 472f–473f, 475–478
  bones and joints of skeleton of, 801f
  evolution of human eye, 453
  heterochrony and growth rates in skulls of, 449
  interaction of muscles and skeletons in locomotion in, 799
  skin of, 781
  structure of ears of, 783f
  structure of eyes of, 786f–787f
Human chorionic gonadotropin (hCG), 724, 746
Human environmental impacts
  biome disturbances, 823
  community disturbances, 858–859
  in nitrogen cycle, 875f
  ocean acidification, 36–37
  restoration ecology and, 877–879f
Human genetics, 219–223
  counseling in, based on Mendelian genetics, 223
  dominant alleles and disorders in, 216
  dominantly inherited disorders in, 222–223
  multifactorial disorders in, 223
  multiple alleles and ABO blood groups in, 216
  pedigree analysis in, 220
  pleiotropy and disorders in, 217
  recessively inherited disorders in, 220–222
  skin pigmentation and polygenic inheritance in, 217–218, 219
Human genome. *See also* Genome(s)
  comparing, of different communities, 360
  comparing genomes of other species to, 346, 350, 353–354, 358–359
  function of FOXP2 gene in, 359
  globin gene families in, 352–356
  Human Genome Project and complete sequence for, 343–345
  microarray chips containing, 347
  size, number of genes, and gene density of, 348–349
  types of DNA sequences in, 349f
Human Genome Project, **344**–345
Human immunodeficiency virus (HIV). *See* HIV (human immunodeficiency virus)
Human papillomavirus (HPV), 727
Human reproduction. *See also* Animal reproduction
  birth rates (*see* Birth rates)
  conception, embryonic development, and birth in, 745–747
  contraception in, 734, 747–748
  female reproductive anatomy in, 735
  female reproductive cycles in, 739–742
  gametogenesis in, 735–738

hormonal regulation of, 738–742
  infertility and *in vitro* fertilization in, 748–749
  male reproductive anatomy in, 734
  male reproductive system in, 739
  reproductive organs in, 734–735
  sexual response in, 741–742
Hummingbirds, 4, 433, 602f
Humoral immune response, 712f, **720**–723f
Humpback whales, 384f, 669f
Humus, **580**, 582
Hunger, transgenic crops and reducing human, 612
Huntington's disease, **222**–223, 323
Hutton, James, 367
Hybrid breakdown, 421f
Hybridization, **207**–208, 307–308, 611–612
Hybrids, **419**
  bears, 422
  gene flow and, 419
  hybrid zones and, 428–430
  postzygotic reproductive barriers and, 421f
  sterility of, 421f, 433
  sympatric speciation and, 425–426
Hybrid zones, **428**–430
Hydras, 193f, 670, 769f, 800
Hydration shells, **33**
Hydrocarbons, **42**
Hydrocarbon tails, 96f, 162f
Hydrochloric acid, 34, 672–673
Hydrogen
  covalent bonding of molecules of, 24f, 25f
  as essential element, 20
  in hydrocarbons, 42
  oxidation of organic molecules containing, 137
  peroxisomes and, 84
  as pure element, 25
  in saturated and unsaturated fatty acids, 49–50
  valence of, and organic compounds, 41f
Hydrogenated vegetable oils, 50
Hydrogen bonds, **27**
  in DNA and tRNA structure, 62f
  of ice, 32
  properties of water from, 29–30
  in protein structure, 56f–57f
  as weak chemical bonds, 27
Hydrogen ions, **34**–36, 575
Hydrogenosomes, 492
Hydrogen peroxide, 84
Hydrogen sulfide gas, 445
Hydrolysis, **44**
  of ATP, 122–124
  breakdown of polymers by, 44–45f
  of cellulose, 48
  enzymatic, 668
  by lysosomes, 79–80
  of storage polysaccharides, 47–48
  in translation, 283f
Hydrolytic enzymes
  fungal, 508
  lysosomal, 79–80
Hydronium ions, **34**–36
Hydrophilic substances, **33**
  phospholipid heads, 50
  plasma membranes and, 71f
  side chains, 53f, 54
  water and, 33–34
Hydrophobic interactions, 57f
Hydrophobic substances, **34**
  hydrocarbons and fats as, 42
  phospholipid tails, 50
  plasma membranes and, 71f
  side chains, 53f, 54
  water and, 33–34
Hydroponic culture, **578**
Hydrostatic skeletons, **799**–800
Hydrothermal vents, 459, 471, 474–475, **829**f, 866

Hydroxide ions, **34**–36
Hydroxyl group, 43*f*, 44–45, 46, 49
Hydrozoa, **530***f*
Hyla versicolor, 413*f*, 425
Hymen, **735**
Hymenopterans, 541*f*
Hypercholesterolemia, 106
Hypermastigote, 500*f*
Hyperosmotic solutions, 653, 661
Hyperpolarization, **756**–759, 788
Hypersensitive response, plant, **637**
Hypertension, 662, **698**
Hypertonic solutions, **101**
Hyphae, fungal, **509**–511, 512*f*
Hypocotyl, **604**
Hypoosmotic solutions, 653
Hypothalamus, **647**, **650***f*, **773***f*
    drug addiction and reward system of, 775
    kidney regulation by, 661–662
    regulation of mammalian/human reproduction
        by, 738–742, 747
    suparchiasmatic nucleus (SCN) in, 774
    thermoregulation by, 647
Hypotheses
    estimating quantitative data from graphs and
        developing, 443
    phylogenetic trees as, 391–392
    science and, **12**, 14
    theory as, in science, 379
Hypotonic solutions, **101**
Hyracotherium, 454–455

I

Ibuprofen, 781
Ice, floating of, 32–33
Iceland, ocean pelagic biome near, 829*f*
Icosahedral viruses, 331
Identical DNA sequences, 352
Identical twins, human, 746
Ileum, 674
Illicium, 521*f*
Imaging, functional brain, 775–776
Imbibition, **605**
Immigration, 832*f*–**833**, 842, 860
Immune response, primary and secondary,
    719–720
Immune systems, **711**–728
    adaptive immunity in (*see* Adaptive immunity)
    blood plasma in, 695
    cardiovascular diseases and, 697
    human disorders from disruptions in, 724–727
    immunization and, 723–724
    innate immunity in, 712–715
    leukocytes in, 696
    lymphatic systems and, 694
    mammalian, 642*t*
    molecular recognition and response to
        pathogens by, 711–712
    overview of innate and adaptive immunity in,
        712
    responses of, to changing pathogens, 726
Immunization, **723**–724
Immunodeficiency, 725–726
Immunoglobulin (Ig), **716**, 717–718
Immunological memory, 719–720
Implantation, human, 746*f*
Imprinting, **807**
Inclusive fitness, **813**–814
Incomplete dominance, **215**–216
Incomplete flowers, **598**. *See also* Flowers
Incomplete metamorphosis, 541*f*
Incomplete proteins, 666
Incus (anvil), 783*f*
Independent assortment, law of, 211–**212**,
    228–229*f*, 236

Independent assortment of chromosomes,
    201–203*f*, 236
Independent variables, identifying, 427
Indeterminate growth, **560**, 562
Indian corn, 350
Indian pipe, 586*f*
Indoleacetic acid (IAA). *See* Auxin
Indolebutyric acid (IBA), 622
Induced fit, **127**
Induced pluripotent stem cells, 323–324
Inducers, **296**–297
Inducible enzymes, 296–297
Inducible innate immune response, 712–713
Inducible operons, 295–297
Induction, **313**
Inductive reasoning, **11**–12
Inert elements, 24
Inertia, 802
Infection
    bacterial, 281, 335, 464, 468–470, 472*f*–473*f*,
        476
    cellular innate defenses and, 713–714
    cytotoxic T cell response to, 721
    inflammatory response and, 714–715
    plant response to, 636–638
Infertility, human, 748–749
Inflammation, 472*f*, 697–698, 781
Inflammatory response, **714**–715
Inflorescences, **598**
Influenza viruses, 55*f*, 331*f*–332, 338–340, 725,
    841*f*, 861–862
Information, genetic, 5–6. *See also* DNA
    (deoxyribonucleic acid); Genetics
Information processing
    in birds, 778
    cerebral cortex and, 777
    neurons and, 753
    problem solving and, 809
    vertebrate visual, 787–788
Infrared receptors, 781*f*
Ingestion, 481, **668**–669*f*
Ingroups, **388**
Inhalation, 704–705
Inheritance. *See also* Genetics; Genetic variation
    blending and particulate hypotheses on, 206
    of cancer predisposition, 327–328
    chromosome theory of, 228–231 (*see also*
        Chromosomal basis of inheritance)
    C. Darwin on, 371–372
    DNA in, 5
    epigenetic, 299
    of genes and chromosomes, 193 (*see also*
        Sexual life cycles)
    genetic variation and, 400–402
    Mendelian (*see* Mendelian inheritance)
    polygenic, 217–218, 219
    of X-linked genes, 232–233
Inheritance of acquired characteristics principle,
    367
Inhibin, 739
Inhibiting hormones, **650***f*
Inhibition
    allosteric, 131
    cell division, 186–188
Inhibitors, enzyme, 129–130, 131
Inhibitors, plant growth, 619, 623–624
Inhibitory postsynaptic potential (IPSP), **762**
Initials, plant cell, 561
Initiation
    transcription, 274–276, 299–304
    translation, 281–282, 283*f*, 284, 304–305
Initiation factors, 282
Innate behavior, **806**
Innate immunity, **712**–715
    adaptive immunity vs., 712
    antimicrobial peptides and proteins, 714

barrier defenses, 711–712, 713
    cellular innate defenses, 713–714
    evasion of, by pathogens, 715
    inflammatory response, 714–715
    invertebrate, 712–713
    vertebrate, 713–715
Inner ear, **783***f*, 784–785*f*
Inner membrane, nuclear, 75*f*
Innocence Project, The, 351
Inorganic topsoil components, 581
Inquiry, scientific, **11**–16. *See also* Science
Insect-eater finches, 369*f*
Insecticide resistance, 407, 410, 501–502
Insects
    as arthropods, 541–542
    body plans of, 451
    camouflage in, 371*f*
    characteristics of, 540*f*
    circulatory systems of, 685*f*
    compound eyes of, 786
    evolution by natural selection in, due to food
        source changes, 373–374
    evolution of, 365
    excretory systems of, 656
    exoskeletons of, 800
    flying, 801–802
    Hox genes in, 361*f*
    innate immunity in, 712–713
    insecticide resistance of, 410
    nervous systems of, 769*f*
    nonheritable variation in, 401*f*
    parasitism in, 850
    plant response to herbivores and, 636
    as pollinators, 433, 524, 602*f*
    problem solving of, 808–809
    production efficiency of, 871
    reproductive organs of, 733
    skeletal muscles of, 799
    tracheal systems for gas exchange in, 702
    transgenic crops and, 612
Insertions, nucleotide-pair, **289**–290
In situ hybridization, **307**, 311
Insoluble fiber, 48
Instability, free energy and, 119–120
Instantaneous per capita rate of increase, 836
Insulin, 78, 106, 270, **650***f*, 680
Insulin-dependent diabetes, 681
Integral proteins, 95*f*, **97**
Integration
    cellular, 91
    nervous system, 753
Integrins, **89***f*
Integument, **517**, 601, 642*t*
Interactions, ecological. *See also* Ecology
    as biological theme, 7
    dispersion patterns and, 833–834
    in effects of animals on ecosystems, 547
    in effects of animals on evolution, 548
    effects of land plants and fungi on, 523–524
    interspecific, 846–851
    prokaryotic, 475
    in species distributions, 818
    in vascular plant nutrition, 582–586*f*
Intercalary meristems, 564
Intercalated disks, 798
Intercellular joining. *See* Cell junctions
Interdisciplinary genomics research teams, 6
Interferons, **714**
Intergradation, terrestrial biome, 824
Intermediate disturbance hypothesis, **856**–857
Intermediate filaments, **88**
    animal cell, 72*f*
    of desmosomes, 90*f*
    functions of, 85, 88
    plant cell, 73*f*
    structure and function of, 86*t*

Intermembrane space, chloroplast, 83f
Intermembrane space, mitochondrial, 82f–83
Internal cell membranes. *See* Cellular membranes
Internal defenses, 712f
Internal fertilization, **732**, 811. *See also*
    Fertilization, reproductive
Internal skeletons, 799f
Internet resources, genomic, 345, 346f
Interneurons, **753**
Internodes, 555f, **556**
Interphase, **177**, 178f, 197f
Intersexual selection, 413, 811
Interspecific competition, **846**–847
Interspecific interactions, **846**–851. *See also*
    Interactions, ecological
    competition, 846–847
    facilitation, 851
    herbivory, 849
    predation, 848–849
    symbiosis, 849–851
    symbols for, 846
Interspecific mating, 419
Interstitial fluid, **644**, 685, 693, 694f
Intertidal zone, **828f**
Intestinal bacteria, 475–476
Intestines, 670f, 677f. *See also* Large intestine;
    Small intestine
Intracellular digestion, 670
Intracellular receptor proteins, 110–111
Intracellular recording, 756f
Intrasexual selection, 413, 811
Intrauterine devices (IUDs), 748
Intrinsic physiological factors, density-dependent
    population regulation and, 841f
Introns, **277**–278, 350, 400–401f
Invagination, 744, 745f
Invasive species, **853**
Inversions, chromosome, **241**
Invertebrates, **534**
    action potential conduction speed in, 760
    in animal phylogeny, 534
    excretory systems of, 656
    hydrostatic skeletons of, 799–800
    innate immunity in, 712–713
    mechanoreceptors for sensing gravity and
        sound in, 782
    nervous systems of, 768–769
    parental care in, 733f, 811
    reproductive organs of, 733
    skeletal muscles of, 799
In vitro culturing, angiosperm, 610–611, 612
In vitro DNA amplification, 264–265
In vitro fertilization (IVF), **749**
In vitro recombinant DNA, 262
Involuntary nervous system, 771
Iodine, 20, 651f, 667
Iodine deficiencies, 20
Ion channel proteins, 759
Ion channels, **102**, **754**
    in facilitated diffusion, 102
    ligand-gated, as transmembrane receptors, 110
    mechanoreceptors and, 780–781
    neuron potentials and, 754–760
    neuron resting potential and, 754–756
    in vascular plant solute transport, 575
Ionic bonds, 25–**26**, 57f
Ionic compounds (salts), 25–**26**, 33–34
Ionotropic receptors, 762
Ion pumps
    in active transport, 104–105
    neuron resting potential and, 754–756
Ions, **26**
    blood electrolytes, 695
    concentrations of, inside and outside of
        mammalian neurons, 754t
    as second messengers in cell signaling, 112–113f

Iridium, 445
Iris, **786**–787
Irish potato famine, 638
Iron
    as essential element, 20
    human requirements for, 667
    ocean fertilization using, 869
    as plant deficiency, 579–580
Island biogeography and island equilibrium
    model, 860
Island species, 378
Isle Royale, fluctuations in moose and wolf
    populations on, 840
Isolated systems, 118, 121f
Isoleucine, 53f, 132f
Isomers, retinal, 787–788
Isoosmotic solutions, 653
Isotonic solutions, **101**
Isotopes, **21**–22, 158, 876

J

Jackson, Rob, 852f
Jacob, François, 294, 295, 453
Japan, restoration project in, 879f
Japanese snails, 432–433
Jaundice, 674
Jawed vertebrates, 537–539
Jawfish, 811f
Jawless vertebrates, 537–538f
Jaws
    mammalian, 440, 441f, 544
    snake, 412
Jejunum, 674
Jellies (jellyfish), 274f, 530f
Jet-propulsion locomotion, 802
Jointed appendages, arthropod, 536
Joints, human, 801f
Jost, Alfred, 739
Joule (J), **31**, 679
J-shaped exponential growth curve, 836
Jumping genes, 350
Juncus gerardi, 851
June solstice, 821f
Juniperus horizontalis, 518f
Juxtaglomerular apparatus (JGA), **662**
Juxtamedullary nephrons, **657f**

K

Kalahari Desert, stone plants (*Lithops*) in, 571
Kangaroos, 546f, 802, 830
Kaposi's sarcoma herpesvirus, 726–727
Karyogamy, **510**, 511f
Karyotypes, **194**
Kaufman, D. W., 15
Kenya, savanna biome in, 825f
Keratin, 88, 270
Ketone compounds, 43f
Keystone species, **854**
Kidneys, **656**, 657f
    adaptations of vertebrate, to diverse
        environments, 660–661
    concentration of urine in mammalian,
        659–660
    homeostatic regulation of, 661–663
    mammalian/human, 657f
    processing by, of blood filtrate to urine,
        658–659
    in vertebrate excretory systems, 656
Killifish, 378
Kilocalorie (kcal), **31**, 679
Kimberella, 532
Kinases. *See* Protein kinases
Kinetic energy, **30**–31, **117**–119. *See also* Energy;
    Potential energy

Kinetochore microtubules, 178f, 180, 181f,
    198f–199f
Kinetochores, **177**, 178f
Kinetoplastids, 492–493f
King, Thomas, 320
Kingdoms, taxonomic, 8f, **382**, 395, 505f
King penguins, 833f
Kin selection, 813–**814**
Kissimmee River restoration project, 879f
Kiwi bird, 879f
Klinefelter syndrome, 242
Knee-jerk reflex, 770
Knob, ATP synthase, **145f**
Knowledge, evolution of cognition and, 777–778
Koalas, 546f, 677f–678
Kodiak bears, 665
Komodo dragons, 730
Korarchaeota clade, 474
Kornberg, Roger, 59f
Krebs cycle. *See* Citric acid cycle
Kruger National Park, exponential growth in the
    African elephant population of, 836, 837f
K-selection, 839–**840**

L

Labeling, GMO food, 613
Labia majora, **735**
Labia minora, **735**
Labor, human childbirth, 747
Labrador Current, 822
Lacks, Henrietta, 187
Lac operon, 295–297
Lactate, 150
Lactation, human, 747
Lacteals, 674f, **675**
Lactic acid fermentation, **150**–151
Lactose, 46, 295–297
Lagging strand, DNA, **256**
Lakes
    biomanipulation of trophic levels in, 855
    as freshwater biomes, 828f
    nutrient limitation in, 869
Lake Vesijärvi, biomanipulation of, 855
Lake Victoria
    sympatric speciation and sexual selection in
        cichlids in, 426
    fusion of hybridized cichlid species in, 430
Lamarck, Jean-Baptiste de, 367
λ (lambda) phage, 334–335
Lampreys, 537–538f
Lancelets, 537
Land, locomotion on, 802
Land animals, 539–547. *See also* Animal(s)
    aquatic animals vs., 540f
    arthropods as, 540–542 (*see also* Arthropods)
    characteristics of, 540f
    evolution of, from aquatic animals, 539–540
    vertebrates as, 542–547 (*see also* Vertebrates)
Land plants, 504–527. *See also* Angiosperms;
    Plant(s)
    alternation of generations in, 506f–507
    characteristics of, 540f
    chemical cycling and biotic interactions of,
        521–524
    colonization of land by fungi and, 504,
        508–513 (*see also* Fungi)
    coping with high-temperature soils by
        mycorrhizal fungi and, 510
    development of (*see* Plant development)
    early evolution and terrestrial adaptations of,
        505–508
    evolution of alternative carbon fixation
        mechanisms in, 169–171
    nonvascular bryophytes, 514
    phylogeny of, 513–516

seedless vascular, 514–516
seed plants, 516–521 (*see also* Seed plants)
vascular (*see* Vascular plants)
Landscape ecology, **819**f
Landscapes, **819**f
Language
    cerebral cortex and, 776–777
    FOXP2 gene and, 359
Large intestine, 671f, **675–676**
Large-scale disturbances, 856–857
Large-scale mutations, 288
Larval stage, insect, 541f
Larynx, 672, **702–703**f
Latency, viral, 725
Lateralization, **777**
Lateral meristems, **560**–561f
Lateral roots, **554**–555f, 564
Latitude, sunlight intensity and, 820f, 821f
Latitudinal gradients, 859
Law of conservation of mass, **865**
Law of independent assortment, 211–**212**,
    228f–229, 236
Law of segregation, 207–208, **209**–211, 228–229f
Laws of thermodynamics, 118–119, 865
Leading strand, DNA, **255–256**
Leaf (leaves), **516**, **556**. *See also* Shoots
    anatomy of, 565f
    architecture of, for light capture, 573
    auxin in pattern formation of, 621
    brassinosteroids in abscission of, 623
    effects of transpiration on wilting and
        temperature of, 592
    ethylene and auxin in abscission of, 625–626
    evolution of, 516
    green color of, 160
    monocot vs. eudicot, 554f
    morphology of, 560
    photosynthesis in, 3f, 156–159
    primary growth of, 564–566
    in shoot systems, 555f
    structure of, 556
    tissue organization of, 565
Leaf mantids, 372f
Leaf primordia, **564**
Learning, **806–809**
    associative, 808, 809f
    cerebral cortex and, 778–779
    cognition, problem solving, and social,
        808–809
    cognitive maps and spatial, 807–808
    imprinting in, 807
Leefructus, 520
Left atrium, 688f, 689f
Left ventricle, 688f, 689f
Leggadina hermannsburgensis, 655
Legionnaires' disease, 472f
Legumes, 583–584, 629
Lemaitre, Bruno, 713
Lemming, 841f
Length, carbon skeleton, 42f
Lens, eye, 786f, 789
Lens cells, differential gene expression in, 302f
Lenses, microscope, 67
Lenski, Richard, 467f
Lenticels, **569**
Leopards, 382–383
Lepidopterans, 365, 541f
Leprosy, 473f
Leptin, 681–682
Lesser snow goose, 547
Lettuce seed germination, 628f
Leucine, 53f
Leukemia, 242–243f, 328
Leukocytes, 695f, **696**, 697
Lewis, Edward B., 318
Leydig cells, **739**

Lichen, **521–522**, 877
Life
    adaptive radiations of, 447–449
    biology as the study of, 1 (*see also* Biology)
    carbon in organic compounds of, 40
    cells as fundamental units of, 66 (*see also*
        Cell(s))
    cellular respiration and work of, 135
    chemistry and, 19
    classification of diversity of, 7–9 (*see also*
        Systematics; Taxonomy)
    conditions on early Earth for origin of, 459–462
    correlation of structure and function in, 4
    diversity of (*see* Biodiversity)
    emergent properties of, 3–4
    essential elements for, 20
    evolution of, 1, 7–11, 366, 370 (*see also*
        Evolution)
    expression and transmission of genetic
        information in, 5–6 (*see also* Genetics)
    extinctions and mass extinctions of, 444–447
    fossil record and geologic record as
        documentation of history of, 436–440,
        441f (*see also* Fossil record; Geologic
        record)
    hierarchy of biological organization of, 2f–3f
    importance of cell division and cell cycle to,
        174
    importance of water for, 29–30
    interaction of organisms with environment
        and other organisms in, 7 (*see also*
        Interactions, ecological; Organisms)
    limits of natural selection in history of, 415
    metabolism and energy for, 116 (*see also*
        Metabolism)
    order as characteristic of, 119
    phylogenies as evolutionary history of,
        381–382 (*see also* Phylogenies)
    prokaryotic cells as first cells of, 458
    themes of, 2–7
    three domains of, 395–396
    transfer and transformation of energy and
        matter in, 6–7 (*see also* Chemical cycling;
        Energy; Energy flow)
    tree of, 10–11
    viruses and characteristics of, 330
Life cycles, **194**
    angiosperm, 600f
    cellular slime mold, 498f
    Drosophila melanogaster, 317–318
    frog, 742f
    fungal, 511f
    human, 195–196
    land plant, 506f–507, 514–516
    of pathogens, 862
    Plasmodium, 501f
    sexual (*see* Sexual life cycles)
Life histories, **839–840**
Life spans, plant, 562
Life tables, **834**
Ligand-gated ion channels, **110**, 761f, **762**
Ligands, **109**
Light chains, **716**, 717–718
Light-detecting organs, 785. *See also* Visual
    systems
Light energy
    in energy flow and chemical cycling, 6f–7, 135
    in photosynthesis (*see* Photosynthesis)
    primary production in aquatic ecosystems and
        limitations of, 868
    properties of, 160
    reception of, by plants (*see* Light reception,
        plant)
    as stimulus for stomatal opening and closing,
        591
    sunlight as, 117, 122, 135 (*see also* Sunlight)

Light-harvesting complexes, **163–164**
Light microscope (LM), **67**, 68f, D-1f
Light reactions, **158–167**f. *See also* Photosynthesis
    chemiosmosis in chloroplasts vs. in
        mitochondria, 165–167f
    determining absorption spectrum for, 161f
    determining wavelengths of light driving, 161f
    excitation of chlorophyll by light energy in,
        162–163
    linear electron flow in, 164–165
    nature of sunlight and, 160
    overview of, 159f, 164f, 167f, 171f
    photosynthetic pigments as light receptors in,
        160–162
    photosystems of, 163–164
    as stage of photosynthesis, 158–159
Light reception, plant, **626–631**
    biological clocks and circadian rhythms in,
        628–629
    hypothetical flowering hormone and, 631
    photomorphogenesis and photoreceptors in,
        626–628
    photoperiodism and seasonal responses in,
        629–631
    phototropism and, 618–619
    signal transduction pathways and, 617
Lignin, **516**, **558**f
Likens, Eugene, 876
Lilies, 348
Limbic system, 775
Limbs
    evolution of tetrapod, 542–543f
    homologous structures in, 375–376
    mammalian, 10
Limiting nutrients, primary production and,
    868–869
Limnetic zone, **828**f
Limnodynastes peronii, 849
Limpets, 831f
LINE-1 (*L1*) retrotransposons, 351
Linear electron flow, **164–167**f
Linear regression lines, scatter plots with, 37, 170
Linear structure, glucose, 46f
Line graphs, 128, 202, 726, F-1
Linkage. *See* Linked genes
Linkage groups, 239
Linkage maps, **239–240**
Linked genes, **234–240**
    constructing linkage maps of, 239f
    genetic recombination and, 235–237f
    inheritance of, 234–235
    mapping of, 237–240
    sex-linked genes vs., 234
    using chi-square (χ^2) test to determine linkage
        of, 238
Linker DNA, 260f
Linnaean classification system, 382–383
Linnaeus, Carolus, 366, 382
Linoleic acid, 666
Lionfish, 538f
Lipids, **49**
    bilayers of (*see* Phospholipid bilayers)
    digestion of, 674–675
    fats, 49–50
    membrane (*see* Membrane lipids)
    in nuclear envelopes, 74–75
    as organic compounds, 40
    phospholipids, 50
    in protocells, 460
    steroids, 50–51
    structure of, 51f
    synthesis of, by smooth ER, 77
    Tay-Sachs disease and, 216
Lipid-soluble hormones, 652
Lipopolysaccharides, 463
Lithops, 571

Litter decomposition, 873f
Litter size, 839
Littoral zone, **828f**
Littorina obtusata, 548
Liver, **674**
    bile production by, 674
    differential gene expression in, 302f
    in glucose homeostasis, 680
    in human digestive system, 671f
Liverworts, 506f, 513f, 514
Living topsoil components, 582
Lizards, 27, 381–382, 544, 545f, 645f, 731, 846f
Loams, **581**. *See also* Soil
Lobe-fins, **539**, 540f
Lobes, brain, 776, 777
Lobopods, 536
Lobotomy, 777
Lobsters, 541f
Local biogeochemical cycles, 873
Local cell signaling, 108–109
Local inflammatory response, 714–715
Local regulators, **108**, 747
Lock-and-key specificity, viral, 332
Locomotion, **801**–803
Locus, gene, **193**, 209, 217
Lodgepole pines, 857
Logging, 858
Logistic population growth, **837**–839
Long-day plants, **630**
Long-distance cell signaling, 108–109
Long-distance signaling, neuron, 751–753. *See also* Neurons
Long-distance transport, plant, 577–578
Long-horned bees, 597
Long-term memory, **778**–779
Looped domains, DNA, 261f
Loop of Henle, **657f**, **658**, 659f
Lophotrochozoa, 534–535f
Lorenz, Konrad, 803, 807
Loudness, 784
Lou Gehrig's disease, 795
Low-density lipoprotein (LDL), 106, 109, **697**–699
LSD, 764
Lung cancer, 347
Lung cells, newt, 5f
Lungfishes, 538f–539
Lungs, **702**
    breathing and ventilation of, 704–706
    in mammalian respiratory systems, 702–704
Lupines, 878
Lupus, 725
Luteal phase, ovarian cycle, 741
Luteinizing hormone (LH), **738**–742
Lycophytes, 513f, **514**, 515f, 516
Lyell, Charles, 367, 368–369
Lyme disease, 473f, 476, 861
Lymph, **694**
Lymphatic systems, **693**
    cardiovascular systems and, 693–694
    cellular innate immune defenses and, 714
    lacteals in, 675
    mammalian, 642t
Lymph nodes, **694**
Lymphocytes, 695f, 696f, **715**. *See also* B cells; T cells
Lymphoid stem cells, 696f
Lymph vessels, 694
Lyon, Mary, 233–234
Lysine, 53f
Lysogenic cycle, **334**–335
Lysosomal storage diseases, 80
Lysosomes, **79**
    animal cell, 72f
    in endomembrane system, 81f
    functions of, 79–80, 91

    in phagocytosis, 107f
    viruses and, 338
Lysozymes, 33f, 55f, 355, **712**–713
Lytic cycle, **333**–334

M

MacArthur, Robert, 860
McClintock, Barbara, 350
Macroclimate, **821**–823. *See also* Climate; Global climate change
Macroevolution, **418**, **436**–456. *See also* Evolution
    adaptive radiations in, 447–449
    of development from changes in developmental genes, 449–452f
    effects on rates of, by differing modes of dispersal, 443
    fossil record and geologic record as documentation of, 436–440, 441f (*see also* Fossil record; Geologic record)
    mass extinctions in, 444–447
    novelties and trends in, 452–455
    plate tectonics and, 442–444
    speciation and, 418, 433 (*see also* Speciation)
    speciation and extinction rates in, 440–449
Macromolecules, **40**
    3-D structure of, 40f, 59f
    abiotic synthesis of, 460
    carbohydrates, 45–59 (*see also* Carbohydrates)
    diversity of polymers of, 45
    nucleic acids, 60–63 (*see also* Nucleic acids)
    as organic compounds and biological molecules, 40 (*see also* Organic compounds)
    as polymers of monomers, 44
    proteins, 51–59 (*see also* Proteins)
    synthesis and breakdown of polymers of, 44–45
Macronutrients, plant, **578**
Macrophages, 80, 91f, 643f, 697, 711, **713**–714
Madagascar orchid, 548f
Mad cow disease, 59
Mads-box genes, 450
Maggot flies, 426
Maggots, 669f
Magnesium chloride, 26
Magnesium deficiency, plant, 579–580
Magnetic field, Earth's, 781, 804
Magnetite, 781
Magnification, microscope, 67
Magnolia grandiflora, 521f
Magnoliids, 520f–521f
Mainland mouse, 1, 12–14
Maize. *See* Corn
Major histocompatibility complex (MHC) molecule, **717**, 724
Malaria, 222, 414, 491f, 501–502, 713
Male gametophytes, angiosperm, 601
Males
    female mate choice and competition between, 811
    hormonal control of reproductive systems of human, 739
    parental care by, 810–811
    reproductive anatomy of human, 734
    sex determination of, 231–232, 739
    sexual competition between, 413
    spermatogenesis in human, 735–738
    territorial responses of, 804
Malignant tumors, **187**–189
Malleus (hammer), 783f
Malnutrition, 612, 667–668
Malpighian tubules, 656
Malthus, Thomas, 372
Maltose, 46

Mammals, **544**. *See also* Animal(s)
    adaptive radiations of, 447–448
    bats as flying, 801–802
    breathing in, 704–705
    cardiovascular systems of (*see* Cardiovascular systems)
    cellular respiration in hibernating, 148
    characteristics and lineages of, 544–546
    digestive system organs of, 671–676
    ensuring survival of offspring of, 732
    excretory systems of, 657f
    glia in brains of, 752f (*see also* Neurons)
    homologous structures in, 375
    hormonal regulation of reproduction in, 738–742
    hormones in sex determination of, 739
    humans as, 546–547 (*see also* Human(s))
    inactivation of X-linked genes in female, 233–234
    innate immunity in, 713–715
    ion concentrations inside and outside of neurons of, 754t
    kidney adaptations of, 661
    kidney function in, 659–660
    lungs in respiratory systems of, 702–704
    mechanoreceptors for hearing and equilibrium in, 782–785f
    modeling neurons of, 755f
    molecular clock for, 393f
    natural selection of limbs of, 10
    organ systems of, 642t
    origination of cetaceans as land animals, 376–377
    origin of, 440, 441f
    production efficiency of, 871
    reproductive cloning of, 321–322
    reproductive organs of, 733 (*see also* Reproductive organs, human)
    respiratory adaptations of diving, 708
    sex determination in, 232
    specific opiate receptors in brains of, 765
    tissues of, 642–643f
Mammary glands, 544, **735**
Manatee, 849f
Mantellinae frogs, 444
Mantids, 372f
Mapping, brain activity, 775–776
Maps, genetic. *See* Genetic maps
Map units, **239**
Marchantia, 506f
March equinox, 821f
Marella, 531f
Marine benthic zone, **829f**
Marine biomes. *See also* Aquatic biomes
    characteristics and types of, 827–830
    effects of animals on, 547
    fishes in, 653–654 (*see also* Fishes)
    food chains in, 853f
    food webs in, 854f
    photosynthetic protists as producers in, 499–500
    primary production in, 868–869
Marine crustaceans, 540f
Marine worm, 700f
Marshall, Barry, 673
Marsh gas, 474
Marsupials, 447f, **546**, 733
Mass
    conservation of, 865
    ecosystems and, 865–866 (*see also* Chemical cycling)
Mass extinctions, **445**
    consequences of, 445f, 446–447
    extinctions vs., 444–445 (*see also* Extinctions)
    first five, 445–446
    human activities and, 548–550

human impacts and, 524
possibility of current sixth, 446
Mass number, **21**
Mast cells, **714**–715
Master regulatory genes, 314–315. *See also*
    Homeotic genes; *Hox* genes
Mate choice, 413, 426–427, 811
Mate recognition, 420*f*
Maternal age, Down syndrome and, 242
Maternal chromosomes, 201–202
Maternal effect genes, **318**–320
Mating. *See also* Courtship behaviors;
    Reproduction
    animal behaviors in, 792
    animal communication and, 805–806
    animal reproduction and, 729 (*see also* Animal
        reproduction)
    bioluminescence in, 116
    clumped dispersion and, 833
    fertilization and, 732
    genetic basis of behaviors in, 812
    human, 221, 741–742, 765 (*see also*
        Copulation, human)
    hybrids from interspecific, 419
    hybrid zones and, 428–430
    mating systems and parental care in, 810–811
    mating systems and sexual dimorphism in, 810
    G. Mendel's techniques of cross-pollination as,
        207–208
    pheromones and, 806
    prokaryotic mating bridges, 468–469
    reproductive barriers to, 420*f*–421*f*
    sexual selection and mate choice in, 811
Mating systems, 810–811
Matter, **19**
    in ecosystems, 865–866
    as pure elements and compounds, 19–20
    transfer and transformation of energy and, as
        biological theme, 6–7 (*see also* Chemical
        cycling)
Maungatautari restoration project, 879*f*
Maximum metabolic rates, animal, 679
Maximum parsimony, 390*f*, **391**
Maze experiments, 809
Meadowlarks, 419*f*
Meadow voles, 812
Mean annual precipitation, 870
Measles virus, 332, 724
Meat eating, human, 872
Mechanical digestion, 668, 670*f*, 671–672. *See also*
        Digestion; Digestive systems
Mechanical isolation, 420*f*
Mechanical signaling, extracellular matrix in, 89
Mechanical stimuli, plant responses to, 632–633
Mechanical stress, plant responses to, 624–625
Mechanical work, 122, 124*f*
Mechanoreceptors, **780**–781, 782–785*f*
Mediator proteins, 301–302
Medicine. *See also* Diseases and disorders, human
    antibodies as tools in, 724
    application of systems biology to, 347
    drugs in (*see* Drugs)
    G protein-coupled receptors in, 110
    radioactive isotopes as tracers in, 21–22
    stem cells in, 322–324
Mediterranean orchids, 597
Medium, growth, 269–270
Medulla oblongata, 705–706, **773***f*
Megapascal (MPa), **576**
Megaphylls, **516**
Megaspores, 517, **601**
Megasporocytes, 601
Meiosis, **196**
    in animal cells, 198*f*–199*f*
    changes in DNA of budding yeast cells during,
        202

errors in, 240–243*f*
genetic variation from errors in, 402
genome evolution and errors in, 354–355
human gametogenesis and, 736*f*–738
in human life cycle, 195*f*–196
mitosis vs., 200*f*–201
overview of, 197*f*
production of gametes by, 176
in sexual life cycles, 196–197
stages of, 197–199*f*
Meiosis I, **197**, 198*f*, 201–203*f*
Meiosis II, **197**, 199*f*, 201
Melatonin, **650***f*
Melitaea cinxia, 842
Membrane carbohydrates, 95*f*, 98
Membrane lipids, 94–96, 98
Membrane potentials, **104**–105, **754**, 780. *See
        also* Action potentials, neuron; Resting
        potentials, neuron
Membrane proteins. *See also* Plasma membranes
    animal cell, 89*f*
    aquaporins, 94, 102 (*see also* Aquaporins)
    in cell junctions, 90*f*
    exocytosis of secretory, 98*f*
    fluid mosaic model and movement of, 95–96
    in glycoproteins, 98
    phosphorylation and dephosphorylation of,
        111–112
    receptor proteins, 97*f*, 107*f*, 109–113*f*
    rough ER and, 78
    synthesis of, in endoplasmic reticulum, 98*f*
    targeting of polypeptides to, 285–286
    transport proteins, 99, 102, 103–106 (*see also*
        Transport proteins)
    types and functions of, 97–98
Membranes, cellular. *See* Cellular membranes
Memory, cerebral cortex and, 778–779
Memory cells, **719**, 723*f*
Mendel, Gregor, 206*f*. *See also* Mendelian
        inheritance
    experimental, quantitative approach of, 207
    genes as hereditary factors of, 206, 228
    on genetic variation, 204
    law of independent assortment by, 211–212,
        236
    law of segregation of, 207–211
    model of inheritance by, 400
Mendelian inheritance, 206–227
    C. Darwin's theory and, 400
    environmental impacts on phenotypes and,
        218
    evolution of gene concept from, 290
    extending, for multiple genes, 217–218
    extending, for single gene, 215–217
    genes in particulate hypothesis of, 206
    human genetics and, 219–223
    integrating, with emergent properties,
        218–219
    law of independent assortment of, 211–212
    law of segregation of, 207–211
    laws of probability governing, 213–214
    G. Mendel's experimental quantitative
        approach, 207
    physical basis of, in behavior of chromosomes,
        228–231
Menopause, **741**
Menstrual cycle, **740**–741
Menstrual flow phase, uterine cycle, 741
Menstruation, **739**–741
Meristems, **560**–562
Meselson, Matthew, 252, 253*f*
Mesencephalon, 772*f*
Mesenchyme cells, 744, 745*f*
Mesocricetus auratus, 774
Mesoderm, **533**, **744**–745
Mesophyll, **156**, 157*f*, 169–170, **565**

Mesozoic era, 439*t*, 444*f*
Messenger molecules, local cell signaling and, 108
Messenger RNA (mRNA), **271**
    in blocking translation, 304–305
    degradation of, 304
    effects of microRNAs and small interfering
        RNAs on, 305–306
    genetic code and synthesis of, 272–273
    growth factors in synthesis of, 113
    interpreting DNA sequence logos for
        ribosome-binding sites for, 284
    in maternal effect (egg-polarity) genes,
        319–320
    monitoring gene expression and, 307–309
    in protein synthesis, 76
    RNA processing of, after transcription,
        276–278
    role of, in protein synthesis, 60
    role of, in transcription and translation,
        271–272
    in situ hybridization and, 311
    transcription and synthesis of, 271*f*, 274–276
    transcription factors and, 111
    in translation, 278–287*f*
    viruses and, 336
Metabolic defects, 269–271
Metabolic pathways, **116**–117, 293–298
Metabolic rate, **679**
Metabolism, **116**–134. *See also* Cellular
        respiration
    ATP energy coupling of exergonic and
        endergonic reactions in, 122–124
    biosynthesis in anabolic pathways of, 152
    catabolic pathways of, 136–140, 151–152
    chemical energy of life and, 116
    enzymatic catalysis of, by lowering energy
        barriers, 125–130
    enzymatic regulation of, 51
    exergonic and endergonic reactions in,
        120–121
    forms of energy for, 117–118
    free-energy change and chemical equilibrium
        in, 119–122
    laws of thermodynamics and, 118–119
    metabolic pathways of, 116–117
    prokaryotic, 465–466
    protocell, 460
    regulating, by regulating enzyme activity,
        130–132
    regulation of, in animal nutrition, 679–681
Metabotropic receptors, 762–763
Metagenomics, **346**, 470
Metamorphosis, frog, 653
Metamorphosis, insect, 541*f*
Metaphase, **177**, 179*f*, 182*f*, 200*f*
Metaphase chromosomes, 261*f*
Metaphase I, 198*f*, 200*f*, 201–203*f*
Metaphase II, 199*f*
Metaphase plate, 179*f*, **180**
Metapopulations, **842**
Metastasis, 187*f*, **188**
Metencephalon, 772*f*
Meteorites, 460
Methane, 25*f*, 27*f*, 41*f*, 137*f*
Methanogens, **474**
Methanopyrus kandleri, 474*f*
Methanosarcina barkeri, 348*t*
Methicillin, 374
Methicillin-resistant *S. aureus* (MRSA), 374–375
Methionine, 53*f*, 273
Methods, research. *See* Research methods
Methylated compounds, 43*f*
Methylation, DNA, 299
Methyl group, 43*f*, 44
Methylsalicylic acid, 637–638
Metric system, 67*f*, C-1

MHC (major histocompatibility complex) molecule, 717, 724
Mice. *See* Mouse (mice)
Microarray chips, human genome, 347
Microbial diversity, 852*f*
Microevolution, **400**, **418**. *See also* Evolution
  evolution of populations as, 399–400
  gene flow as cause of, 409–410
  genetic drift as cause of, 407–409
  genetic variation and, 400–402, 413–414
  limits of natural selection in, 414–415
  natural selection as cause of, 407
  natural selection as cause of adaptive evolution in, 410–415
  sexual selection in, 412–413
  speciation and, 418 (*see also* Speciation)
  using Hardy-Weinberg equation to test, 402–406
Microfibrils, 48, 88
Microfilaments (actin filaments), **87**
  animal cell, 72*f*
  in animal cytokinesis, 180–182
  in cytoskeletons, 84*f*
  functions of, 85, 87–88
  plant cell, 73*f*
  structural role of, 87*f*
  structure and function of, 86*t*
Microglia, 769*f*
Micronutrients, plant, **579**
Microphylls, **516**
Micropyles, 601
MicroRNAs (miRNAs), **305**–306, 326
Microscopy, 67–69, 182, D-1*f*
Microsporangia, 601
Microspores, 517, **601**
Microsporocytes, 601
Microtubule-organizing center, 177
Microtubules (tubulin polymers), **85**
  animal cell, 72*f*
  of cilia and flagella, 87*f*
  in cytoskeletons, 84*f*
  functions of, 85–87
  in mitotic spindle, 177–181*f*
  plant cell, 73*f*
  structure and function of, 86*t*
Microtus ochrogaster, 812
Microtus pennsylvanicus, 812
Microvilli, 71, 72*f*, 87*f*, 674*f*, **675**
Midbrain, 772*f*
Middle ear, **783***f*, 784
Middle lamella, **88**, 89*f*
Midgut, 670*f*
Migration, **804**
  electromagnetic receptors and, 781
  as fixed action pattern, 804
Miller, Stanley, 459–460
Mimicry
  endorphin, 764
  in predation, 848
Mimics, molecular, 28, 55
Minimal medium, 269
Mimosa pudica, 633
Mimulus, 433, 610
Mineralocorticoids, **650***f*
Minerals
  plant symptoms of deficiencies in, 579–580
  root architecture and acquisition of, 573–574
  transpiration of, from roots to shoots via xylem, 587–590
  vascular plant transport of, across plasma membranes, 574–575
Minerals, animal dietary, **667**
Minimal medium, 269–270
Minimum metabolic rates, animal, 679
Minipill contraceptives, 748
Minke whales, 384*f*

Miscarriage, human, 746
Misfolding, protein, 59
Mismatch repairs, DNA, **257**
Missense mutations, **288**–289
Mistletoe, 586*f*
Mitchell, Peter, 147
Mites, 636
Mitochondria, **81**
  animal cell, 72*f*
  in cellular respiration, 71, 81, 82–83
  chemiosmosis in, 145–147
  chemiosmosis in, vs. in chloroplasts, 165–167*f*
  endosymbiont theory on evolutionary origins of, 82
  enzymes in, 132
  eukaryotic electron transport chains in, 139, 144
  genetic research on prokaryotes related to, 485
  in human sperm, 736*f*
  in mammal hibernation, 148
  origin of, in endosymbiosis, 484–486
  plant cell, 73*f*
  pyruvate oxidation in, 142
Mitochondrial DNA (mtDNA)
  in evidence for endosymbiosis, 485–486
  interpreting data from, in phylogenetic trees, 394
  species identity in, 384*f*
Mitochondrial matrix, 82*f*–83
Mitosis, **176**
  in animal cells, 177–182
  in chromatin packing, 261*f*
  evolution of, 183
  in human gametogenesis, 736*f*–737*f*
  in human life cycle, 195*f*
  meiosis vs., 200*f*–201
  in plant cells, 182*f*
  in sexual life cycles, 196–197
Mitotic (M) phase, **177**
Mitotic (M) phase checkpoint, 184*f*, 185
Mitotic spindles, **177**–181*f*
Mixotrophs, **494**, 495, 499
M line, skeletal muscle, 793
Mobile genetic elements, evolution of, 336
Mockingbirds, 368–369
Model organisms, **311**. *See also Arabidopsis thaliana* (mustard plant); *Caenorhabditis elegans* (soil worm); *Drosophila melanogaster* (fruit fly); Mouse; *Saccharomyces cerevisiae* (yeast)
  for DNA research, 246
  for embryonic development research, 311
  Neurospora crassa (bread mold), 269–270
Modified stems, 556*f*
Mojave Desert, rain shadow and, 822
Molarity, **34**
Molds, 269–270, 510
Mole (mol), **34**
Molecular basis of inheritance, 245–267
  chromosomes as DNA molecules packed with proteins in, 259–261*f*
  DNA as genetic material in, 245–251
  DNA structure and DNA replication in, 245
  evolution of gene concept from, 290
  genetic engineering and, 261–265
  predicting percentages of nucleotides in genomes and, 249
  proteins in DNA replication and repair in, 251–259
Molecular biology, 62–63, 337
Molecular clocks, **392**
  calibration of, 392–393
  dating origin of HIV using, 393–394
  differences in clock speed of, 393
  in evidence for origination of animals, 529
  fungal lineages determined by, 511

  for mammals, 393*f*
  potential problems with, 393
Molecular formulas, 24, 25*f*, 41*f*
Molecular genealogy, 62–63
Molecular homologies, 376, 385–387
Molecular homoplasies, 386
Molecular mass, **34**
Molecular recognition, immune system, 711–712
Molecular systematics. *See also* Systematics
  applying parsimony to problems in, 390*f*
  evaluating molecular homologies in, 386–387
  prokaryotic phylogenies of, 470–474
Molecules, **24**
  formation of, by chemical bonding of atoms, 24–27
  in hierarchy of biological organization, 3*f*
  origin of self-replicating, 459–462
  polar, of water, 29–30
  as second messengers in cell signaling, 112–113*f*
  shape and function of, 27–28
Mole rats, 813
Moles, 385–386
Molluscs
  as arthropods, 535*f*
  as endangered species, 549
  exoskeletons of, 800
  eye complexity in, 453*f*
  predator-prey relationships of, 548
Monarch butterflies, 613–614, 808, 809*f*
Monera, kingdom, 395
Mongolia, grassland biome in, 825*f*
Monilophytes, 513*f*, **514**, 515*f*
Monkey flower, 433, 610
Monkeys, 63, 546. *See also* Chimpanzees
Monoclonal antibodies, **724**
Monocots
  in angiosperm phylogenies, 520*f*–521*f*
  eudicots vs., 553–554*f*
  roots of, 563
  seed structure of, 604*f*
Monocytes, 695*f*, 696*f*
Monod, Jacques, 294, 295
Monogamous mating, **810**
Monohybrid crosses, **211**, 213
Monohybrids, **211**
Monomers, **44**–45
Monophyletic clades, **387**
Monosaccharides, **45**–46
Monosiga brevicollis, 488–489
Monosodium glutamate (MSG), 782
Monosomic zygotes, **240**–241
Monosomy X, 242
Monotremes, 447*f*, **546**
Monozygotic twins, 746
Montmorillonite, 460
Moose, 840–841
Morgan, Thomas Hunt, 230–231, 234–237*f*, 246
"Morning-after" pills, 748
Morphine, 28, 55
Morphogenesis, **312**, **744**–745. *See also* Embryonic development
Morphogen gradient hypothesis, 319–320
Morphogens, **319**–320
Morphological homologies, 385
Morphological isolation, 420*f*
Morphological species concept, **422**
Morphology
  fungal, 509
  macroevolution of, from changes in developmental genes, 449–452*f*
  species concepts and, 418–419, 422
Mortality rates
  population change and, 840–842
  population dynamics and, 832*f*–835
  population growth and, 835–839

Mosaicism, 233–234
Mosquitofish, 423
Mosquitoes, 410, 501–502, 669*f*
Mosses, 507*f*, 513*f*, 514, 515*f*
Moths, 541*f*, 548*f*, 602*f*
Motility
  cell, 84–88
  prokaryotic, 464 (*see also* Movement)
Motor, flagellum, 464*f*
Motor cortex, 776*f*
Motor neurons, **753**, 770, 795–797
Motor output, nervous system, 753
Motor proteins, 52*f*, 84, **85**–87, 124*f*, 180–181*f*
Motor systems, **771**, 793–803
  animal behaviors and, 792–793 (*see also*
    Animal behaviors)
  energy costs of locomotion in, 803
  muscle function in, 793–799 (*see also* Muscles)
  skeletal systems and locomotion in, 799–803
    (*see also* Skeletal systems)
  in vertebrate peripheral nervous systems, 771
Motor unit, **797**
Mountains, climate and, 822
Mouse (mice)
  animal behavior studies of, 806
  appetite regulation genes in, 681
  comparing human genome to genome of, 350,
    359
  comparison of chromosome sequences of
    humans and, 353–354
  complete genome sequence for, 343
  density-dependent population regulation of,
    841*f*
  FOXP2 gene evolution in, 359
  genome size of, 348*t*
  homeotic genes in, 360*f*
  osmotic homeostasis in desert, 655
  paw development of, 315*f*–316
  predation and coat coloration adaptations of,
    1, 12–14, 15
  selection modes and, 411*f*
Mouth, 670*f*, 744, 745*f*
Movement. *See also* Motility
  mechanoreceptors for sensing, in humans,
    784–785*f*
  prokaryotic, 464
Movement, cell. *See* Motility, cell
mPGES-1 gene, 303
Mucor, 512*f*
Mucous cells, 672*f*
Mucous membranes, 713
Mucus, **671**–672, 713
Mucus escalator, 703
Mules, 421*f*
Muller, Hermann, 290
Müllerian mimicry, **848**
Multicellular asexual reproduction, 193*f*
Multicellular organisms, 66, 483*f*, 487–489
Multifactorial characters, **218**
Multifactorial disorders, human, 223
Multigene families, **352**–353
Multiple fruits, **606**
Multiple sclerosis, 725
Multiplication rule, **213**
Murchison meteorite, 460
Muscle cells, 314–315, 648
Muscle contraction, 793–799
  nervous system regulation of tension in, 797
  of nonskeletal muscles, 798–799
  regulation of, 795–797
  skeletal muscle fibers and, 797–798
  skeletal muscle structure and, 793
  skeletons and, 799 (*see also* Skeletal systems)
  sliding-filament model of, 794–795
Muscles
  contraction of (*see* Muscle contraction)

energy storage in, 680
fermentation in cells of human, 150
mammalian, 642*t*
nonskeletal types of, 798–799
skeletal (*see* Skeletal muscles)
skeletal systems, locomotion, and, 799–803
stomach, 673–674
Muscle tissue, **643***f*
Muscular dystrophy, 232–233
Mushrooms, fungal, 509*f*, 512*f*
Mus musculus, 343, 348*t*. *See also* Mouse (mice)
Mustard plant. *See Arabidopsis thaliana*
Mutagens, **290**
Mutant phenotypes, 230–231
Mutants
  designing experiments using genetic, 774
  experiments on obesity in, 681
  nutritional, 269–270
Mutations, **288**
  cancer development from, 324–327
  color vision and, 789
  embryonic lethals and abnormal pattern
    formation by, 318
  emerging viruses and, 338–339
  as errors in DNA proofreading, 258
  evolution and genetic variation from, 204
  evolution of enzymes by, 130
  as faulty proteins from faulty genes, 268,
    288–290
  genome evolution and, 353–357
  Hardy-Weinberg equilibrium and, 405
  of ion channel protein genes, 759
  mutagens and spontaneous, 290
  mutant phenotypes and, 230–231
  nucleotide-pair insertions and deletions and
    frameshift mutations, 289–290
  nutritional mutants and, 269–270
  point mutations, 288
  prokaryotic, 467–468
  random, as source of alleles, 240
  silent, missense, and nonsense mutations as
    nucleotide-pair substitutions, 288–289
  as sources of genetic variation, 401–402
  systems biology and cancer-causing, 347
Mutualism, **475**, **850**
  in flower pollination, 597, 602*f*
  fungal, 509–510, 521–524
  in interspecific interactions, 7
  lichen as, 521–522
  mycorrhizae as (*see* Mycorrhizae)
  in plant nutrition, 582–585
  predation and, 845
  as symbiosis, 850
  symbols for, 845–846
  in vertebrate digestive adaptations, 677
Mutualistic bacteria, 475–476
Myasthenia gravis, 795–796
Mycelium (mycelia), **509**
Mycobacterium tuberculosis, 476
Mycoplasma capricolum, 485
Mycorrhizae, **509**, **584**
  agricultural and ecological importance
    of, 585
  associations of, 555
  bioremediation using, 878
  coping of, with high-temperature soils, 510
  disruption of, by garlic mustard, 585*f*
  evolution of, 511–513
  as mutualism, 850
  nutrient limitations and, 870
  as plant-fungi mutualism, 574
  types of, 509, 584–585
Myelencephalon, 772*f*
Myelination, 769*f*
Myelin sheath, **760**
Myeloid stem cells, 696*f*

Myllokunmingia fengjiaoa, 536–537
Myoblasts, 314–315
Myocardial infarctions, 698
MyoD protein, 301*f*, 314–315
Myofibrils, **793**
Myoglobin, **708**, **798**
Myosin, **88**, 180. *See also* Thin filaments (myosin)
Myotonia, 742, 759
Myxini, 538*f*
Myxobacteria, 472*f*
Myxospores, 472*f*

N

NAD$^+$ (nicotinamide adenine dinucleotide),
  137–139, 142–143*f*, 149–150
NADH, 139*f*, 142–143*f*, 147–150
NADP$^+$ (nicotinamide adenine dinucleotide
  phosphate), **158**, 171
NADP$^+$ reductase, 167*f*
NADPH, 159, 164–167*f*, 171
Naked mole rats, 813
Naloxone, 765
Nannochloris atomus, 869*f*
Nanoarchaeota clade, 474
National Cancer Institute, 347
National Center for Biotechnology Information
  (NCBI), 345, 346*f*
National Institutes of Health (NIH), 345, 347
National Library of Medicine, 345
National Medal of Science, 704
Natural family planning, 747–748*f*
Natural killer cells, **714**
Natural plastics, 477
Natural selection, **10**, **369**. *See also* Evolution
  biodiversity from adaptive evolution by, 450
  as cause of microevolution, 407
  C. Darwin's research focus on adaptations and,
    369 (*see also* Adaptations)
  C. Darwin's theory of descent with
    modification by, 365–366, 370–372
  directional, disruptive, and stabilizing selection
    in, 411–412
  of ecological niches, 846–847
  in evolution of drug resistance, 374–375
  evolution of genetic variation from genetic
    recombination and, 236
  Hardy-Weinberg equilibrium and, 405
  insect evolution by, due to food source
    changes, 373–374
  key role of, in adaptive evolution, 410, 412
  life histories and, 839–840
  limitations of, in adaptive evolution, 414–415
  making and testing predictions about
    predation, coloration of guppies, and, 378
  molecular level, 461
  mutations from altered DNA nucleotides and,
    258
  relative fitness and, 411
  species selection as, 455
  theory by C. Darwin on evolution by, 9–10
  tree of life and, 10–11
Nature vs. nurture, 218, 807
Navigation, migration and, 804
Negative correlations, 610
Negative feedback, **645**
  in endocrine signaling, 652
  in homeostasis, 645
  in population regulation, 840
Negative gene regulation, bacterial, 295–297
Negative gravitropism, 632
Negative pressure breathing, **704**–705
Nematodes, 535*f*, 799–800. *See also*
  Caenorhabditis elegans (soil worm)
Nembrotha chamberlaini, 729
Nephrons, **657***f*, 658

Nerve cells
  exocytosis by, 106 (*see also* Neurons)
  gated channels in, 102
  human, 66*f*
Nerve cords, chordate, 537
Nerve gas, 764
Nerve impulses, 648
Nerve nets, **769–770**
Nerves, **753**, **769**
Nervous systems, **648**, 768–791
  animal behaviors and, 792 (*see also* Animal behaviors)
  central nervous systems and peripheral nervous systems in human, 770*f*
  control of heart rhythm by, 690
  control of skeletal muscle tension by, 797
  coordination and control functions of, 648
  diversity of, 769*f*
  glia in, 769–770
  Huntington's disease and, 222–223
  ligand-gated ion channels in, 110
  mammalian, 642*t*
  neurons, nerves, and types of, 768–769
  neurons in information processing by, 753 (*see also* Neurons)
  regulation of digestion by enteric division of, 678
  sensory reception in, 779–789 (*see also* Sensory reception)
  synaptic signaling in, 108–109
  vertebrate brains in, 771–779 (*see also* Brains, vertebrate)
  vertebrate central nervous systems, 769–770 (*see also* Central nervous system (CNS))
  vertebrate peripheral nervous systems, 769–771 (*see also* Peripheral nervous system (PNS))
Nervous tissue, **643***f*
Nests, 391–392
Net ecosystem production (NEP), **868**
Net flux (flow), carbon dioxide, 868
Net primary production (NPP), **867**–868
Neural plasticity, **778**, 779*f*
Neural tube birth defects, human, 668
Neuroendocrine pathways, 649–652
Neuromuscular junctions, 763–764
Neurons, **643***f*, **751**–767
  axons and action potentials of, 756–761 (*see also* Action potentials, neuron)
  electrical and chemical signaling by, 751
  in endocrine signaling, 648
  glia and, in vertebrate nervous systems, 770
  in human ears, 783*f*
  in human eyes, 786*f*
  ion concentrations inside and outside of mammalian, 754*t*
  ion pumps, ion channels, and resting potentials of, 754–756
  major neurotransmitters for, 764*t*
  measuring membrane potentials of, using intracellular recording, 756*f*
  nervous system information processing and, 753
  in sensory transmission, 780
  signaling of, at chemical synapses, 761–765
  specific opiate receptors and, 765
  structure and function of, 752
  in vertebrate peripheral nervous systems, 770–771
Neuropeptides, **764**
Neurospora crassa (bread mold), 269–270
Neurotransmitters, **752**
  acetylcholine, 763–764
  amino acids, 764
  biogenic amines, 764
  chemical synapses and, 761–762 (*see also* Chemical synapses)

  clearing of, from synaptic clefts, 762
  gases, 765–766
  in genetic basis of animal behavior, 812
  in local cell signaling, 108
  major, 764*t*
  modulated signaling by, 762–763
  neuropeptides, 764
  in regulation of muscle contraction, 795–797
  in synaptic signaling, 108–109
  vision and, 788
Neutralization, 722
Neutral variation, **413**
Neutrons, **20**–21
Neutrophils, 695*f*, 696*f*, **713**–714
New Jersey, temperate broadleaf forest biome in, 826*f*
Newt lung cells, 5*f*
Newton, Sir Isaac, 14
New Zealand, restoration project in, 879*f*
Niches, ecological, 422, 846–847
Nicotine, 763, 849
Night length, flowering and, 630–631
Nirenberg, Marshall, 273
Nitrate, 875*f*
Nitric oxide, 111, 693, 765–766
Nitrification, 582
Nitrifying bacteria, 582
Nitrite, 875*f*
Nitrogen
  as essential element, 20
  as limiting nutrient in aquatic biomes, 868–869
  as limiting nutrient in terrestrial biomes, 870
  valence of, and organic compounds, 41*f*
Nitrogen cycle, 472*f*, **582**–584, 875*f*
Nitrogen deficiency, plant, 580
Nitrogen fixation, **466**, **583**
  biological augmentation and, 878
  conservation of mass and, 865
  cyanobacteria and, 473*f*
  as mutualism, 850
  nitrogen cycle and, 875*f*
  prokaryotic, 466
Nitrogen-fixing bacteria, 582–584
Nitrogenous bases, nucleic acid, 60–61, 62, 248
Nitrogenous wastes, 653, 654–655
Nitrosomonas, 472*f*
Nobel Prizes
  R. Axel and L. Buck, 781–782
  F. Jacob, 453
  R. Kornberg, 59*f*
  B. Marshall and R. Warren, 673
  B. McClintock, 350
  P. Mitchell, 147
  C. Nüsslein-Volhard and E. Wieschaus, 318
  for research on noncoding RNAs, 305
  N. Tinbergen, K. von Frisch, and K. Lorenz, 803
  J. Watson, F. Crick, and M. Wilkins, 251
Nociceptors, **781**
Nocturnal animals, 806
Nodes, lymph, 694
Nodes, plant, 555*f*, **556**
Nodes of Ranvier, **760**
Nodules, **583**–584
Nomarski (differential-interference contrast) microscopy, 68*f*
Nonbreeding adults, territoriality and, 841*f*
Noncoding DNA, 349–352
Noncoding RNAs (ncRNAs), 305–306
Noncompetitive inhibitors, **129**–130*f*
Nondisjunction, **240**–242
Nonequilibrium model, **856**
Nongonococcal urethritis, 473*f*
Nonheritable genetic variation, 401
Nonhomologous chromosomes, 228*f*–229

Nonidentical DNA sequences, 352–353
Non-insulin-dependent diabetes, 681
Nonkinetochore microtubules, 178*f*, 180
Nonparental types, 236
Nonpolar covalent bonds, **25**
Nonpolar molecules, 99
Nonpolar side chains, 53*f*, 54
Nonself recognition, immune system, 711–712
Nonsense mutations, **289**
Nonsister chromatids, 195*f*, 203
Nonspontaneous processes, 119
Nonsteroid hormones, 303–304
Nontemplate strands, DNA, 272
Nonvascular plants, 513*f*, 514, 515*f*
Norepinephrine (noradrenaline), **650***f*, 763, **764**
Nori, 496*f*
Normal range, homeostatic, 645
North America, biomes in, 822
North American moles, 385
Northern coniferous forests, **826***f*
Northern Hemisphere
  mountains and vegetation, 822
  seasonal variation in, 821*f*
Norway, coniferous forest and tundra biomes in, 826*f*
Norway spruce, 573*f*
Notochords, **537**
Novel functions, evolution of genes with, 355
Novelties, evolutionary, 453
N-P-K fertilizers, 580
N-terminus, 54, 97*f*, 260*f*, 282, 299
Nucifraga columbiana, 808
Nuclear envelope, **74**
  animal cell, 72*f*
  bound ribosomes and, 76
  in endomembrane system, 81*f*
  functions of, 74–75
  nuclear contents and, 75*f*
  plant cell, 73*f*
  in transcription, 271
Nucleariids, 498, **511**
Nuclear lamina, **75**
Nuclear magnetic resonance (NMR) spectroscopy, 59
Nuclear pores, 75*f*
Nuclear transplantation, animal cloning and, 320–322
Nucleases, **258**
Nucleic acid hybridization, **261**, **307**
Nucleic acid probes, **307**
Nucleic acids, **60**
  components of, 44–45, 60–61
  digestion of, 673*f*
  as genetic material, 245 (*see also* DNA (deoxyribonucleic acid))
  as measures of evolution, 62–63
  as organic compounds and macromolecules, 40
  roles of, 60
  structure of DNA and RNA molecules, 62
  viruses as, with protein coats, 330–332
Nucleoids, **69**–70, **259**, **465**
Nucleolus, 72*f*, 73*f*, 75*f*, **76**
Nucleomorph, 486
Nucleosides, 60–61
Nucleosomes, **260***f*
Nucleotide excision repairs, **258**
Nucleotide-pair insertions and deletions, 289–290
Nucleotide-pair substitutions, **288**–289
Nucleotides, **60**
  coding and noncoding, 277–278
  components of, 248
  in DNA structure and function, 5
  evolutionary significance of altered, 258
  genetic triplet code of, 272–274

mutations as base-pair insertions and deletions of, 289–290
mutations as base-pair substitutions of, 288–289
predicting percentages of, in genomes, 249
Nucleotide variability, 400–401f
Nucleus, cell, **74**
    animal and fungal, 72f
    DNA in, 69, 74–76
    in endomembrane system, 81f
    in eukaryotic cells, 482
    in eukaryotic cells vs. in prokaryotic cells, 4
    intracellular receptor proteins in, 110–111
    mechanisms of cell division in, 183
    mitosis and genetic material in, 176
    nuclear envelope and contents of, 75f
    organismal cloning by transplantation of, 320–322
    plant and protist, 73f
Nucleus accumbens, 776
Nudibranchs, 729
Nurture vs. nature, 807
Nurture vs. nature, Mendelian inheritance and, 218
Nüsslein-Volhard, Christiane, 318, 319f
Nutrient cycling. *See* Chemical cycling
Nutrient enrichment experiments, 868–869
Nutrient limitations
    primary production in aquatic ecosystems and, 868–869
    primary production in terrestrial ecosystems and, 869–870
Nutrition
    animal (*see* Animal nutrition)
    fungal, 508–509
    plant (*see* Plant nutrition)
    prokaryotic, 465–466
    protist, 499
Nutritional modes
    photosynthesis and, 155
    prokaryotic, 465t
Nutritional mutants, 269–270
Nymphaea, 520f–521
Nymphs, insect, 541f

O

Oak Ridge National Laboratory, bioremediation of contaminated groundwater at, 878
Oak trees, 861
Obesity, 681–682
ob gene, 681
Obligate aerobes, 465
Obligate anaerobes, **151**, 465–466
Observations, scientific, 11–12, 373–375
Occam's razor, 391
Occipital lobe, 776f
Ocean currents, climate and, 821–822
Oceanic pelagic zone, **829**f
Oceans
    acidification of, 36–37
    anoxia in, 445
    climate and currents of, 821–822
    disturbances of, 858–859f
    effects of, on climate, 31f
    iron fertilization of, 869
    as marine biomes, 827–830 (*see also* Aquatic biomes)
    mass extinctions and ecology of, 447
    primary production in, 868–869
Ocelli, 785
Ocotillo, 592f
Octopus, 535f
Odor, pheromones and, 806
Odorants, **781**–782

Odum, Eugene, 846
Offspring
    ensuring survival of, 732–733
    reproduction/survival trade-offs of, 839–840
Oil spills, 477–478
Okazaki fragments, **256**
Oleander, 592f
Olfaction, **781**–782
Olfactory bulb, brain, 775f
Olfactory communication, 806
Olfactory receptor genes, human, 402
Oligodendrocytes, **760**, 769f
Oligotrophic lakes, **828**f
Omasum, 677f
Ommatidia, **786**
Omnivores, **665**, 676f
Oncogenes, **324**–327
One gene–one enzyme hypothesis, 270
One gene–one polypeptide hypothesis, 270
One gene–one protein hypothesis, 270
On the Origin of Species by Means of Natural Selection (book). *See Origin of Species, The* (book)
Onychophorans, 536f
Oocytes, **735**, 737f–738, 746
Oogenesis, 737f–**738**
Oogonia, **737**f
Oparin, A. I., 459
Open circulatory systems, **685**–686
Open-pit mine restoration, 877
Open systems, 118, 121f
Operators, **294**–295
Operon model, 294–297
Operons, **294**–297
Ophisaurus ventralis, 381
Ophrys scolopax, 597
Opiates, 28, 55, 764, 765
Opisthokonts, **498**
Opossums, 546f
Opposite phyllotaxy, 573
Opsin, **787**f
Optic chiasm, 788–789
Optic disk, 786f
Optic nerves, 786f, 788–789
Optimal conditions, for enzyme activity, 129
Oral cavity, **671**–672, 673f
Oral contraceptives, 748
Orange peel fungus, 512f
Orangutan, 343
Orchids, 597
Order, as property of life, 119
Orders (taxonomy), **382**
Ordovician period, 511f
Oregon coast, intertidal zone biome on, 828f
Organelles, **67**
    autophagy of, by lysosomes, 80
    endomembrane system, 81f
    in eukaryotic cells, 482
    in hierarchy of biological organization, 3f
    microscopy in study of, 67–69
    in prokaryotic and eukaryotic cells, 69–73f
Organic acid, 43f
Organic compounds, **40**–65
    abiotic synthesis of, 459–460
    ATP as, 44
    bonds with carbon atoms in, 41–42
    carbohydrates, 45–49 (*see also* Carbohydrates)
    carbon in, and biological molecules of life, 40
    carbon skeletons of, 42f
    catabolic pathways and, 136
    chemical groups and properties of, 42–44
    diversity of, 45
    lipids, 49–51 (*see also* Lipids)
    nucleic acids, 60–63 (*see also* Nucleic acids)
    proteins, 51–59 (*see also* Proteins)

shape of simple, 41f
    synthesis and breakdown of, 44–45
    valences of elements of, 41f
Organic fertilizers, 580
Organic phosphate, 43f, 44
Organic topsoil components, 582
Organismal cloning, 320–324
Organismal ecology, **819**f
Organisms
    in aquatic biomes (*see* Aquatic biomes)
    aquatic vs. terrestrial, 540f
    cell as basic unit of structure and function for, 4
    correlation of structure and function in, 4
    DNA in development of, 5
    embryonic development of (*see* Embryonic development)
    in energy flow and chemical cycling, 6–7
    fossil record and geologic record as documentation of history of, 436–440, 441f
    in hierarchy of biological organization, 2f
    homozygous vs. heterozygous, 210
    interactions of, with environment and other organisms as biological theme, 7 (*see also* Interactions, ecological)
    model (*see* Model organisms)
    as open systems, 118
    organismal ecology and, 819f
    photosynthetic and chemosynthetic, in ecosystems, 864–866
    plant nutrition and relationships with, 582–586f
    prokaryotes as Earth's first, 458, 461–462
    speciation and extinction rates of, 440–449
    species distributions of (*see* Species distributions)
    in terrestrial biomes (*see* Terrestrial biomes)
    in topsoil, 582
    transgenic, 273–274, 477, 612–614
Organization, levels of biological, 2–4
Organ of Corti, **783**f
Organogenesis, 742, 744–745, **747**
Organ Pipe Cactus National Monument, 825f
Organs, 3f
Organs, animal, **642**
    animal reproductive, 733
    excretory, 657f
    eyes and light-detecting, 785–787
    heterochrony and development of reproductive, 450
    human reproductive, 734–738, 741–742
    immune system rejection of transplanted, 724
    mammalian/human digestive system, 671–676 (*see also* Digestive systems)
    organogenesis of, in animal embryonic development, 744–745
    organogenesis of, in human embryonic development, 747
    smooth muscle and vertebrate, 798–799
Organs, plant, **554**–556, 598
Organ systems, 3f
Organ systems, animal, **642**
Orgasm, 742
Orgasm phase, human sexual response, 742
Orientation, leaf, 573
Origin of replication, **182**–183f (*see also* Origins of replication)
Origin of Species, The (book), 9, 365, 369–370, 378, 379, 400
Origins of replication, **253**–254
Ornithine, 270f
Oryza sativa (rice), 348t (*see also* Rice)
Osmoconformers, **653**–654

Osmolarity, **653**
Osmoreceptors, 781
Osmoregulation, **101**, **653**
 challenges and mechanisms of, 653–654
 desert mice study on, 655
 excretion and, 653
 excretory systems and, 655–658
 kidney function in, 658–663
 nitrogenous wastes and, 654–655
 osmosis, osmolarity, and, 653
 salinity and, 832
 water balance and, 101
Osmoregulators, **653**–654
Osmosis, **101**, **575**
 diffusion of water by, across plant plasma
  membranes, 575–577
 osmoregulation and, 653–654
 in thigmotropism, 633
 water balance and, 100–102
Osmotic potential, 576
Osmotic pressure, blood, 693
Ossicles, 800
Osteichthyans, 538f–**539**
Otoliths (ear stones), 784–785f
Outer ear, **783**f, 783f
Outer membrane, nuclear, 75f
Outgroups, **388**
Oval window, **783**f, 784
Ovarian cancer, 347
Ovarian cycle, **740**–741
Ovaries, human, 176, 195–196, **650**f, **735**, 737f,
 740–741
Ovaries, plant, **519**, **598**, 605
Overfishing, 548
Overgrazing, 858, 879f
Overharvesting, pearl mussel, 548
Overnourishment, 681–682
Overproduction, offspring, 371–372
Oviducts, **735**
Oviraptor dinosaurs, 392
Ovulation, 731, **735**, 737f, 740–741, 746f
Ovules, **517**, **598**
Owls, 15
Oxidation, **136**–137
Oxidative muscle fibers, 798
Oxidative phosphorylation, **140**
 ATP yield from, 147–148
 in cellular respiration, 139–140
 chemiosmosis in, 145–147
 electron transport chains in, 144–145
 vs. photophosphorylation, 165–167f
Oxidizing agents, **136**–137
Oxygen
 aerobic respiration and, 136
 anaerobic respiration, fermentation, and,
  148–151
 Cambrian explosion and increase in
  atmospheric, 531
 in capillaries, 693
 in circulation and gas exchange, 684–685
  (see also Circulatory systems; Gas
  exchange)
 covalent bonding of molecules of, 25f
 as essential element, 20
 in interspecific interactions, 7
 in mammalian cardiovascular systems, 688
 net ecosystem production and, 868
 ocean anoxia and low levels of, 445
 in photosynthesis, 28–29f
 plant deprivation of, 633–634, 635f
 as product of photosynthesis, 155, 171
 as pure element, 25
 role of, in prokaryotic metabolism, 465–466
 species distributions and availability of,
  831–832

 storage of, by diving mammals, 708
 thyroid hormone level and cellular
  consumption of, 149
 valence of, and organic compounds, 41f
Oxytocin, **649**, **650**f, 652, 747

P

p21 gene, 326
p53 gene, **326**
P680 chlorophyll *a*, 164
P700 chlorophyll *a*, 164
Pacemaker, heart, 690
Pacman mechanism, 180
Paedomorphosis, **450**
Pain, 764
Pain receptors, **781**
Pair bonding, 807, 812
Paleontology, 9f, **367**, 438
Paleozoic era, 439t, 444f
Palisade mesophyll, 565
Pallium, bird, 778
Palumbi, S. R., 384f
Pancreas, **649**, **650**f, **674**
 in chemical digestion, 671f, 673f, 674
 exocytosis by, 106
 in glucose homeostasis, 680
 ribosomes in cells of, 76f
Pancreatic islets, 680
Pandas, 687
Pandemics, 338–340, 861–862
Pandorina, 487f
Pangaea, 377–378, **444**
Panthera pardus, 382–383
Panthers, 408f
Pan troglodytes (chimpanzee). See Chimpanzees
Papaya, 612
Papillae, 782
Papillomaviruses, 328
Parabasalids, 491f, **492**, 500f
Parabronchi, 704
Parachutes, seed and fruit, 607f
Paracrine signaling, 108
Parahippus, 455
Paralysis, 795
Paramecium, 101, 481, 495f, 838f, 846
Paramyosin, 799
Paraphyletic clades, **387**
Parasites, **475**, **850**
 evolutionary radiations and, 548
 fungal, 512f, 523–524
 human disorders from, 473f
 protists as, 492, 497, 500–502
Parasitic plants, 586f
Parasitism, **475**, **850**
Parasympathetic division, peripheral nervous
 system, **771**
Parathyroid glands, **650**f
Parathyroid hormone (PTH), **650**f
Parenchyma cells, **558**f, 563f–564, 565
Parental care
 ensuring offspring survival with, 733
 experience and behavior in, 806
 genetic basis of, 812
 mating systems and, 810–811
 reproduction/survival trade-offs and, 839
Parental types, **236**
Parietal cells, 672f, 673
Parietal lobe, 776f
Parkinson's disease, 59, 323–324, 764
Parrots, 778
Parsimony, molecular systematics and, 390f, 391
Parthenogenesis, **730**, 731
Parthion, 130
Partial pressure, 699–700

Particles, subatomic, 20–21
Particulate hypothesis of inheritance, 206. *See also*
 Mendelian inheritance
Parus major, 409–410
Passive immunity, **723**–724
Passive transport, **100**
 active transport vs., 104f (*see also* Active
  transport)
 diffusion down concentration gradients as,
  99–100
 down electrochemical gradients, 104–105
 transport proteins and facilitated diffusion as,
  102–103
 of water across plant plasma membranes,
  575–577
 water balance and osmosis as, 100–102
Paternal chromosomes, 201–202
Paternity, certainty of, 810–811
Pathogenicity, 246
Pathogens, **475**, **711**, **860**
 bacterial, 472f–473f, 476
 B cells and antibodies as responses to
  extracellular, 722
 in communities, 860–861
 cytotoxic T cell response to cells infected by,
  721
 evasion of innate immunity by, 715
 evolutionary adaptations of, that evade
  immune systems, 725–726
 fungal, 523–524
 identifying hosts and vectors for, 861
 immune system molecular recognition and
  response to, 711–717
 plant defenses against, 636–638
 prokaryotic, 475–476
 viruses as, 338–341
 zoonotic, and human diseases, 861–862
Pattern, evolutionary, 365–366, 379
Pattern formation, **317**–320, 621
Pattle, Richard, 703
Paulinella chromatophora, 495
Pauling, Linus, 249
PCR (polymerase chain reaction). *See* Polymerase
 chain reaction (PCR)
PCSK9 enzyme, 699
Peacocks, 412f
Pea fruit, 606f
Pea plants, 207–212, 634
Pearl mussels, 549
Peatlands, 873
Pectins, 79, 88
Pediastrum, 487f
Pedigrees, **220**
Pelagic zone, **827**, 829f
Penguins, 833f
Penicillin, 130, 374
Penis, **734**–735, 741–742, 747–748, 765
Penny bun fungus, 509f
Pentoses, 46f, 60–61
Pepsin, 672f, **673**
Pepsinogen, 672f
Peptide bonds, **54**, 283f
Peptides, 714, 715
Peptidoglycan, **462**–463
Per capita birth rate, 835–836
Per capita death rate, 836
Per capita rate of increase, 835–836, 838
Perception, sensory, **780**
Perception, visual. *See* Visual systems
Perennials, plant, 562
Pericycle, **564**
Periderm, **556**, 569, 636
Perilymph, 784, 785f
Periodic table of elements, 23f, B-1
Periods, geologic, 439t

Peripheral nervous system (PNS), **753**, **769**
  central nervous system and, 769, 770*f*
  divisions of vertebrate, 770–771
  neurotransmitters and, 763–765
Peripheral proteins, 95*f*, **97**
Peripheral vision, 789
Perissodus microlepis, 414, 415*f*
Peristalsis, **671**, 674, **800**, 802
Peritubular capillaries, **657***f*
Periwinkle, 548
Permian mass extinction, 445, 447
Peromyscus californicus, 806
Peromyscus leucopus, 806, 841*f*
Peromyscus polionotus, 1, 12–14
Peroxisomes, **84**
  animal cell, 72*f*
  functions of, 84
  plant cell, 73*f*
  structure of, 84*f*
Pertussis, 110
Pesticides
  as enzyme inhibitors, 130
  transgenic, 612
Petals, **518–519***f*, **598**
Petioles, leaf, **555***f*, **556**
Petromyzontida, 538*f*
PET (positron emission tomography) scanners, 22
Peyer's patches, 694*f*
Pfisteria shumwayae, 494
P (parental) generations, **207**–208
pH, **35**
  adjusting soil, 580–581
  in duodenum, 649
  ecological succession and soil, 858
  enzymatic catalysis and, 129
  hemoglobin dissociation and, 707–708
  of human cerebrospinal fluid, 705–706
  ocean acidification and, 36–37
  pH scale and, 35–36
  species distributions and soil, 832
PHA (polyhydroxyalkanoate), 477
Phages (bacteriophages), **246**, **332**
  evidence for viral DNA in, 246–248
  lysogenic cyle of prophages and temperate, 334–335
  lytic cycle of virulent, 333–334
  replicative cycles of, 333–335
  structure of, 331*f*–332
  transduction of, 468
Phagocytosis, **79**, **107***f*, **712**
  endocytosis as, 106–107*f*
  food processing and, 670
  innate immunity and, 712–715
  by lysosomes, 79–80
  macrophages in, 91*f*
  by sponges, 529*f*–530
Phalacrocorax harrisi, 418
Phanerozoic eon, 439*t*, 444*f*
Pharyngeal arches, 375*f*
Pharyngeal slits (clefts), **537**
Pharynx, 670*f*, 671*f*, **672**, 673*f*, 783*f*
Phase-contrast microscopy, 68*f*
Phenotypes, **210**
  dominance and, 216
  evolution of mutant, 258
  gene expression as link between genotypes and, 268
  genes and, 290
  genetic variation and, 400–401
  genotypes vs., 210, 218 (*see also* Genotypes)
  impact of environment on, 218
  mutant, 230–231
  pleiotropy and, 217
  relative fitness and, 411
  transformation and, 246
  types of natural selection and, 411–412

Phenylalanine, 53*f*, 273, 405–406
Phenylketonuria (PKU), 405–406
Pheromones, 732, **805**–806
Philadelphia chromosome, 242–243*f*
Philanthus triangulum, 808
Phloem, **516**, **557**, **572**
  evolution of, 572
  primary growth and, 563
  secondary, 566–569
  sugar-conducting cells of, 559*f*
  sugar translocation from sources to sinks via, 593–594
  in vascular plant transport, 516
  in vascular tissue system, 557
Phloem sap, **593**–594
Phoenix roebelenii, 521*f*
Phosphate deficiency, plant, 580
Phosphate group, 43*f*, 44, 112
Phosphodiesterase, 788
Phosphodiester linkages, 61
Phospholipid bilayers. *See also* Cellular membranes; Plasma membranes
  cellular membranes as, 71*f*, 74–75
  plasma membranes as, 94–95, 99
  structure of, 50, 51*f*
  synthesis of, by rough ER, 78
Phospholipids, **50**
Phosphorus, 20, 248, 868–869, 870
Phosphorus cycle, 875*f*
Phosphorylated intermediates, **124**
Phosphorylation
  in cellular respiration, 139–140
  in light reactions of photosynthesis, 158–159
  oxidative (*see* Oxidative phosphorylation)
  oxidative, vs. photophosphorylation, 165–167*f*
  protein, 111–112
Photic zone, **827**, 868
Photoautotrophs, 155, 156*f*, 465*t*, 499
Photoblepharon palpebratus, 475*f*
Photoheterotrophs, 465*t*
Photomorphogenesis, **626**–628
Photons, **160**, 162–163
Photoperiodism, **630**–631
Photophosphorylation, **158**–159, 165–167*f*
Photoprotection, 162
Photopsins, 789
Photoreceptors, **785**–789
Photorespiration, **169**
Photosynthates, 554
Photosynthesis, **155**–173
  Calvin cycle of, 158–159, 167–171
  in carbon cycle, 874*f*
  cellular respiration, energy flow, chemical cycling, and, 135
  as chemical reaction, 28–29
  chemiosmosis in, 146–147
  chloroplasts as sites of, 3–4, 81, 83, 156–157 (*see also* Chloroplasts)
  conversion of light energy to chemical energy of food by, 156–159
  cyanobacteria and, 473*f* (*see also* Cyanobacteria)
  determining rate of, with satellites, 867*f*
  effects of atmospheric carbon dioxide concentration on productivity of, 170
  electrons, electron shells, and, 23
  endosymbiont theory on evolution of chloroplasts and, 82*f*
  in energy flow and chemical cycling, 6–7, 864–866
  evolution of alternative carbon fixation mechanisms in, 169
  fossil evidence of prokaryotic, 461–462
  importance of, 155–156, 171
  levels of organization of leaves and, 3*f*

light reactions of, 158–167*f* (*see also* Light reactions)
  oceans and, 827
  origin of, 486–487, 495
  overview of, 157*f*, 159*f*, 171*f*
  prokaryotic, 464*f*
  by protists, 499–500
  sunlight availability and, 832
  tracking atoms through, 157–158
  two stages of, 158–159
  vascular plant adaptations for, 573
Photosystem I (PS I), **164**–167*f*, 171*f*
Photosystem II (PS II), **164**–165, 171*f*
Photosystems, **163**–167*f*
Phototrophs, 465*t*
Phototropin, 627
Phototropism, **618**–619
pH scale, 35–36. *See also* pH
Phycoerythrin, 495
Phyllotaxy, **573**, 621
Phylogenetic species concept, **422**
Phylogenetic trees, **383**, 387–392. *See also* Phylogenies
  analyzing viral evolution using DNA sequences and, 340
  applications of, 384–385
  applying parsimony to, 390*f*
  cladistics and, 387–388
  of eukaryotes, 490*f*–491*f*
  as hypotheses, 391–392
  identifying species identity of food sold as whale meat using, 384*f*
  interpreting, 384, 394
  linking taxonomy and phylogeny using, 383–384
  maximum parsimony in, 391
  prokaryotic, 470*f*
  proportional branch lengths of, 388–391
Phylogenies, **381**–398
  angiosperm, 520*f*–521*f*
  of animals, 533–534
  applying parsimony to help evaluate the most likely, 390*f*
  biodiversity and branching, 550
  of eukaryotes, 489–498
  of fungi, 512*f*
  horizontal gene transfer and, 395–396
  human, 546–547
  inferring, from morphological and molecular data, 385–387
  invertebrate, 535*f*
  investigating evolutionary history of life with, 381–382 (*see also* Evolution)
  land plant, 513*f*
  phylogenetic species concept and, 422
  prokaryotic, 470–474
  revising, from new information, 395–396
  systematics, taxonomy, and evolutionary relationships in, 382–385 (*see also* Systematics; Taxonomy)
  three domains of life and, 395
  using molecular clocks to track evolutionary time for, 392–394
  using shared characters to construct phylogenetic trees of, 387–392 (*see also* Phylogenetic trees)
  vertebrate, 538*f*
Phylum/phyla (taxonomy), **382**
Physical ecosystem reconstruction, 877
Physiological factors, density-dependent population regulation and, 841*f*
Physiology, **641**
Phytochromes, **627**–628, 629
Phytophthora, 500, 523, 861
Phytoplankton, **473***f*, 853, 868–869, 872
Pigeons, 804, 808

Pigmentation, human skin, 217–219
Pigmented cells, 453f
Pigmented epithelium, 786f–787f
Pigments
    bile, 674
    as photosynthetic light receptors, 160–162
    in photosystems, 163–164
    respiratory, 706–708
    visual, 787f, 789
Pigs, 274f, 339
Pikaia, 531f
Pili, **464**, 468–469
Pineal gland, **650f**, 773f
Pineapple fruit, 606f
Pin flower, 609f
Pinhole camera-type eye, 453f
Pinna, 783f
Pinocytosis, 106–**107f**, 670
Pistils, **598**
Pisum sativum, 634
Pitch, 784
Pitcher plants, 586f
Pith, **557**, 566
Pituitary gland, 28, **649**, **650f**, 738–742, 747, 773f
Pit vipers, 781f
Pitx1 gene, 451–452f
Pivot joints, 801f
Piwi-associated RNAs (piRNAs), 306
Placenta, **546f**, **746**–747
Placental transfer cells, 506f
Placoderms, 437f
Plagiochila deltoidea, 514f
Planarians, 685f, 769f, 785
Plankton, 830
Plant(s)
    adaptations of, that reduce nutrient
      limitations, 870
    adaptive radiation of terrestrial, 448–449
    angiosperms (*see* Angiosperms)
    Archaeplastida supergroup and, 495–496
    area effects and, 860f
    bioremediation using, 877–878
    cells of (*see* Plant cells)
    cellulose as structural polysaccharide of, 48
    colonization of land by (*see* Land plants)
    communication between, 634
    community stability and diversity of, 853
    crop (*see* Crop plants)
    defensive adaptations of, 849
    determining effects of atmospheric carbon
      dioxide on productivity of, 170
    development of (*see* Plant development)
    in energy flow and chemical cycling, 6–7
    Eukarya domain and, 8f
    evolution by natural selection in response to
      introduced species of, 373–374
    facilitation in, 851
    genomes of (*see* Genome(s))
    hormones of (*see* Hormones, plant)
    innate immunity in, 712
    insect radiations and radiation of, 541–542
    in nitrogen cycle, 875f
    nutrition in (*see* Plant nutrition)
    organismal cloning of, 320
    parasites of, 500
    as photoautotrophs, 155, 156f
    photosynthesis by (*see* Photosynthesis)
    polyploidy in, 241
    reproduction of (*see* Plant reproduction)
    resource acquisition in (*see* Plant resource
      acquisition)
    responses of (*see* Plant responses)
    storage polysaccharides of, 47f
    structural polysaccharides of, 48
    sympatric speciation in, 425–426
    transgenic, 477

transport in (*see* Plant transport)
    tumors in, 472f
    unsaturated fats of, 50
    vascular (*see* Vascular plants)
    water transport in, 30
Plantae, kingdom, **8f**, 395, 505f
Plant cells. See also Eukaryotic cells
    active transport in, 105
    cell cycle of, 181f, 182 (*see also* Cell cycle)
    cell walls of, 88, 89f
    common types of, 557–559f
    cotransport in, 105
    gene expression, gene regulation, and
      differentiation of, 561–562
    meristem generation of, 560–562
    microtubules of, 85
    organelles of, 73f
    plasma membranes of (*see* Plasma membranes)
    plasmodesmata in, 90
    solute and water transport across plasma
      membranes of, 574–578
    vacuoles of, 80–81
    water balance and tonicity of, 101f, 102
Plant development, 553–570
    auxin in, 621–622
    cells in, 557–559f
    hierarchy of organs, tissues, and cells in,
      554–560
    leaf morphology in, 560
    meristems in, 560–562
    of monocot vs. eudicot angiosperms, 553–554f
    organs in, 554–556
    primary growth of roots and shoots in,
      562–566
    secondary growth of stems and roots in woody
      plants, 566–569
Plant-growth-promoting rhizobacteria, 582
Plant growth regulators. See Hormones, plant
Plant nutrition. See also Plant transport
    essential macronutrients and micronutrients
      for, 578–579
    hydroponic culture and, 578
    mineral deficiency symptoms and, 579–580
    relationships with organisms and adaptations
      for, 582–586f
    resource acquisition adaptations for, 571–574
    soil management for, 580–581
    soil texture and composition for, 581–582
Plant reproduction. See also Reproduction
    allocation of energy in, 610
    alternation of generations in, 506f–507
    angiosperm, 608–611 (*see also* Flowers)
    artificial selection and breeding in, 597,
      611–612
    asexual vs. sexual, 193
    biotechnology and genetic engineering in,
      612–614
    double fertilization in, 601, 603f
    meristematic control of reproductive growth
      in, 562
    sexual life cycles in, 196–197 (*see also* Sexual
      life cycles)
Plant resource acquisition. See also Plant transport
    evolution of xylem and phloem as vascular
      tissues for, 572–573
    overview of transport and, 572f
    root architecture and acquisition of water and
      minerals in, 573–574
    shoot architecture and light capture in, 573
    underground plants and, 571
Plant responses, 617–639
    to attacks by herbivores and pathogens,
      636–638
    communication between plants in, 634
    to environmental stimuli other than light,
      631–635

    evolution of, 635
    to light, 617, 626–631 (*see also* Light reception,
      plant)
    plant hormones and, 617–626 (*see also*
      Hormones, plant)
    signal transduction pathways and, 617
Plant transport, 571–596
    apoplastic, symplastic, and transmembrane
      routes in, 574
    essential elements in, 579t
    hydroponic culture in study of, 578
    nutritional relationships with organisms and,
      582–586f
    overview of resource acquisition and, 572f
    regulation of transpiration rate by stomata in,
      590–593
    resource acquisition adaptations and, 571–574
    short-distance and long-distance mechanisms
      of, 574–578
    soil as source of essential elements for root
      absorption in, 578–582
    sugar-conducting cells in phloem in, 559f
    of sugars from sources to sinks via phloem in,
      593–594
    temperature and water uptake by seeds in, 576
    transpiration of water and minerals from roots
      to shoots via xylem in, 587–590
    water-conducting cells in xylem in, 559f
Plaque, 698
Plasma, **695**, 708
Plasma cells, **719**
Plasma membranes, **70**. See also Cellular
    membranes
    active transport across, 103–106
    animal cell, 72f, 89f
    bulk transport across, by exocytosis and
      endocytosis, 106–107f
    cardiac muscle, 798
    in cell signaling, 108–113
    in endomembrane system, 81f
    evolution of differences in lipid composition
      of, 96
    fluidity of, 95–96
    fluid mosaic model of, 94–95
    functions of, 69–70
    membrane carbohydrates of, in cell-cell
      recognition, 98
    membrane potentials of neuron, 754–760
    membrane protein types and functions in,
      94, 95–96, 97–98 (*see also* Membrane
      proteins)
    microfilaments and, 87f
    passive transport across, 99–103
    plant cell, 73f
    prokaryotic, 70f
    prokaryotic electron transport chains in, 139,
      144
    selective permeability of, 94, 99
    synthesis and sidedness of, 98
    targeting of polypeptides to, 285–286
    transport proteins in, 99
    vascular plant transport of solutes and water
      across, 574–578
Plasmids, **262**, 336–337, **465**, 468–470
Plasmodesmata, **90**
    in local cell signaling, 108
    plant cell, 73f, 88, 89f, 90
Plasmodium, 491f, 501–502, 713
Plasmogamy, **510**, 511f
Plasmolysis, **102**, **577**
Plastics, natural, 477
Plastids, **83**, 484–487
Plastocyanin, 165
Plastoquinone, 165
Plateau phase, human sexual response, 742
Platelet-derived growth factor (PDGF), 186

Platelets, 186, 695f, **696**
Plate tectonics, **442**–444
Platypus, 546f, 781
Pleasure, brain activity and, 775, 776
Pleiotropy, **217**
Plesiosaur, 437f
Plumule, 604
Pluripotent cells, **323**–324
Pneumatophores, 555f
Pneumocystis carinii, 725
Pneumonia, 246, 468, 725
Poecilia reticulata, 378
Point mutations, **288**
    cancer genes and, 324–325
    mutagens as cause of, 290
    as sources of genetic variation, 401–402
Poison dart frog, 848f
Poisons, detoxification of, 77
Polar bears, 8, 422
Polar bodies, 737f–738
Polar covalent bonds, **25**
Polar microtubules, 180, 181f
Polar molecules, **29**–30, 99
Polar side chains, 53f, 54
Polar transport, auxin, 620, 621f
Poliovirus, 338
Pollen grains, 207, **517**, 554f, **601**
Pollen tubes, **601**
Pollination, **517**, **601**
    asexual reproduction vs., 608
    breeding plants by cross-pollination, 611–612
    genetic engineering of flowers to force self-
        pollination, 614
    mechanisms of, 601, 602f
    G. Mendel's techniques of, 207–208
    mutualistic relationships in, 597
    reciprocal selection and, 548f
    reproductive isolation and pollinator choice
        for, 433
Pollution
    biomanipulation and, 855
    bioremediation of, 877–879f
    bioremediation of aquatic, 868–869
    prokaryotes and bioremediation of, 477–478
Polyadenylation signal sequence, 276
Polyandry, 810
Poly-A tail, **276**–277, 308
Polyclonal antibodies, 724
Polydactyly, 216
Polygamous mating, **810**
Polygenic inheritance, **217**–218, 219
Polygyny, 810
Polymerase chain reaction (PCR), **264**–265, 308f,
    470, 471
Polymerization, 177
Polymers, **44**–45
Polynucleotides, **60**–61
Polypeptides, **52**
    analyzing sequence data of, 63
    completion and targeting of, in translation,
        282–286
    hormones as, 652
    in levels of protein structure, 56f–57f
    mutations affecting structure and function of,
        288–290
    one gene–one polypeptide hypothesis and, 270
    peptide bonds in formation of, 54
    as polymers of amino acids, 52
    RNA-directed synthesis of, in translation,
        271–272, 281–282, 283f
    synthesis of multiple, in translation, 286–287f
Polyphyletic clades, **387**
Polyploidy, **241**, **425**–426
Polyps, 326–327, 530f
Polyribosomes (polysomes), 286–287f

Polysaccharides, **47**–49
    storage, 47–48
    structural, 48–49
Polyspermy, **743**, 746
Polytomies, **384**
Polytrichum commune, 514f
Pongo pigmaeus, 343
Pons, 772f, **773**f
Poplar trees, 613
Population(s), **402**, 819f, 832–842
    character displacement in, 847
    decline of honeybee, 602f
    declining amphibian, 543
    dynamics of, 832–835, 839–842
    ecological interactions and species
        distributions of, 818, 830–832 (see
        also Interactions, ecological; Species
        distributions)
    evolution and genetic variation within, 204
    growth of, 835–839
    in hierarchy of biological organization, 2f
    microevolution as evolution of, 399–400 (see
        also Microevolution)
    natural selection and evolution of, 371–372
    population ecology and, 819f, 832 (see also
        Ecology; Population ecology)
    in theory of evolution by natural selection,
        9–10
    using Hardy-Weinberg equation to test
        evolution in, 402–406
Population dynamics, **840**. See also Population(s)
    birth rates, death rates, and demographics in,
        834–835
    density and dispersion in, 832–834
    density factors in population regulation in,
        839–840, 841f
    immigration, emigration, and metapopulations
        in, 842
    overview of, 832f
    population growth in, 835–839 (see also
        Population growth)
    population stability and fluctuation in,
        840–842
    reproduction/survival trade-offs in life
        histories, 839–840
Population ecology, **819**f, 832. See also
        Population(s)
Population growth, 835–839. See also
        Population(s)
    carrying capacity in, 836–837
    exponential model of, 836
    logistic model of, 837–839
    modeling, using logistic equation, 838
    per capita rate of increase in, 835–836
Population regulation, 839–840, 841f. See also
        Population(s)
    density and mechanisms of, 839–840, 841f
    population dynamics and, 840–842
    reproduction/survival trade-offs in life
        histories and, 839–840
Pore complexes, 74–75
Pores, sponge, 529f
Porifera phylum, 534
Porphyra, 496f
Porphyrin ring, 162f
Porpoises, 376–377
Position, mechanoreceptors and sense of, 784–785f
Positional information, **317**
Positive correlations, 610
Positive feedback, **652**
    in endocrine signaling, 652
    action potentials and, 757
Positive gene regulation, bacterial, 297
Positive gravitropism, 632
Positive pressure breathing, **704**

Positron-emission tomography (PET), 22, 775
Post-anal tails, 537
Postelsia, 493f
Posterior ends, **533**
Posterior pituitary gland, **649**, **650**f, 747
Postsynaptic cells, 752
Postsynaptic neurons, 762
Postsynaptic potentials, 762, 763f
Post-transcriptional gene regulation, 304–305
Post-translational modifications, protein, 285
Postzygotic barriers, **419**, 421f
Potassium, 20
Potassium ions, 754–760
Potatoes, 626f
Potato late blight, 500, 638
Potential energy, **22**–23, 100, **117**–119. See also
        Energy; Kinetic energy
Potential evapotranspiration, 859
Potential range, species, 831
Prairie chickens, 408–409,
Prairie restoration, 878
Prairie voles, 812
Precipitation
    climate and, 821–823 (see also Climate)
    climographs of, 823
    global patterns of, 820f–821
    oceans and, 827
    primary production in terrestrial biomes and,
        869–870
    in water cycle, 874f
Precocious germination, 624
Predation, **848**
    in Cambrian explosion, 531
    camouflage and, 365
    clumped dispersion and, 833
    as density-dependent population regulation
        mechanism, 841f
    effects of, on ecosystems, 547–548
    effects of, on evolution, 528, 548, 549
    genetic variation in, 812–813
    as interspecific interaction, 848–849
    in molluscs, 548
    mouse coat coloration and, 1, 12–14, 15
    mutualism and, 845
    natural selection, coloration in guppies, and,
        378
    numbers of offspring and, 839
    population fluctuations and, 840–842
    in top-down model of trophic control, 855
Predators
    as biotic factors limiting species distribution,
        831
    feeding adaptations of, 848–849
    plant recruitment of, as herbivore defense, 636
Predictions
    making and testing, 378
    of paleontology, 438
    using Hardy-Weinberg equation to interpret
        data and make, 406
Prefrontal cortex, 776f, 777
Pregnancy, human, 724, 746–747
Pre-mRNA, 271, 276–278
Prenatally and Postnatally Diagnosed Conditions
        Awareness Act, 242
Prepuce, clitoris, 735
Prepuce, penis, 734f, **735**
Pressure, mechanoreceptors for, 780–781
Pressure flow, 594
Pressure potential, **576**–577
Pressure waves, 784
Presynaptic cells, 752
Presynaptic neurons, 762
Prey. See also Predation
    defensive adaptations of, 848
    genetic variation in selection of, 812–813

Prezygotic barriers, **419**–420*f*
Primary cell walls, **88**, 89*f*
Primary cilia, 86
Primary consumers, 853*f*, **866**
Primary electron acceptors, **163**
Primary growth, plant, **560**
    meristem generation of cells for, 560–562
    overview of, 561*f*
    of roots, 562–564
    of woody stems, 567*f*
Primary immune response, **719**–720
Primary oocytes, **737***f*
Primary producers, 853*f*, **865**
Primary production, **866**–870
    in aquatic biomes, 868–869
    determining, with satellites, 867*f*
    ecosystem energy budgets and, 867–868
    in terrestrial biomes, 869–870
Primary structure, protein, **56***f*
Primary succession, **857**–858
Primary transcripts, **271**
Primase, **254**
Primates. *See also* Chimpanzees
    cloning of, 322
    HIV in, 393
    humans as, 546–547
    parental care in, 839
Primer, RNA, **254**
Primitive lenses, 453*f*
Primordial germ cells, human, 736*f*–737*f*
Principle of conservation of energy, 118
Principles of Geology (book), 368
Printing press, 15–16
Probability, laws of, 213–214
Problem solving, 808–**809**
Process, evolutionary, 365–366, 379
Producers, **499**
    consumers and, 155 (*see also* Consumers)
    in ecosystem trophic structure, 853*f*, 866
    effects of, on ecosystems, 547
    in energy flow and chemical cycling, 6–7
    photosynthetic protists as, 499
Production efficiency, 870–**871**
Products, chemical reaction, **28**–29
Progesterone, **738**, 746
Progestin, 748
Progestins, **650***f*
Programmed cell death. *See* Apoptosis
Prokaryotes, **458**–480. *See also* Archaea; Bacteria
    anaerobic respiration by, 148
    bioremediation using, 878
    biosphere roles of, in chemical cycling and
        ecological interactions, 474–475
    cells of (*see* Prokaryotic cells)
    in disease-suppressive soil, 477
    diverse evolutionary adaptations of, 458–459*f*,
        462–467
    as Earth's first living organisms, 458, 461–462
    fossils of, 437*f*
    genetic diversity of, 466–470
    genetic research on, related to mitochondria,
        485
    glycolysis in ancient, 151
    impacts of, on humans, 475–478
    motility of, 464
    nutritional and metabolic adaptations of,
        465–466
    origin of mitochondria and plastids in,
        484–486
    origins of life and, 459–462
    origins of photosynthesis in, 156
    phylogenies of, 470–474
    rapid reproduction of, by binary fission,
        466–468
    shapes of, 462

size of and number of genes in genomes of,
    348
summary of adaptations of, 466
taxonomy of, 395–397
in trophic structure, 866
Prokaryotic cells, **4**, **69**. *See also* Prokaryotes
    binary fission in, 182–183
    cell-surface structures of, 462–464
    DNA replication in, 252–257*f*
    electron transport chains in, 139, 144
    in endosymbiont theory on origins of
        mitochondria and chloroplasts, 82
    eukaryotic cells vs., 4, 69–71 (*see also*
        Eukaryotic cells)
    gene expression in, 271*f*
    internal organization and DNA of, 464–465
    mutations in, 290
    organelles of, 70*f*
    protein synthesis in, 60
    regulation of gene expression in (*see* Bacterial
        gene regulation)
    synthesis of multiple polypeptides in
        translation of, 286
    transcription in, 274–276
Prolactin, 653, 747
Proliferative phase, uterine cycle, 741
Proline, 53*f*, 273
Prometaphase, **177**, 178*f*, 182*f*
Promoters, transcription, **274**
Proofreading, DNA, 257–258
Propagation, vegetative, 622
Propanal, 43*f*
Properties, emergent, 3–4
Prophages, **334**–335
Prophase, **177**, 178*f*, 182*f*, 200*f*
Prophase I, 198*f*, 200*f*
Prophase II, 199*f*
Prostaglandins, 747, 781
Prostate glands, **734**
Prosthetic groups, 144
Protease, **673**
Protection, skeletal, 799
Protein Data Bank, 345
Protein kinase A, 112, 652
Protein kinases, **111**–113*f*
Protein phosphatases, **112**
Proteins, **52**. *See also* Polypeptides
    3-D structure of, 40*f*, 59*f*
    amino acids of, 52–54
    analyzing polypeptide sequence data of, 63
    antibiotics and prokaryotic synthesis of, 465
    antibody binding to, 55*f*
    antimicrobial, 714
    in aqueous solutions, 33
    auxin transport, 621*f*
    in bacterial binary fission, 182–183
    in blood plasma, 695
    catabolism of, 136, 152
    in cell nucleus, 74–76
    cellulose-synthesizing, 505
    completing and targeting functional, after
        translation, 282–286
    Conserved Domain Database (CDD) of
        structures of, 346*f*
    denaturation and renaturation of, 58–59
    digestion of, 673*f*
    diversity of, 51
    DNA and RNA in synthesis of, 5–6
    in DNA replication and repair, 251–259
    DNA vs., as genetic material, 246–248
    as enzymes, 44, 51, 125, 130 (*see also*
        Enzymatic catalysis; Enzymes)
    essential amino acids and, 666
    eukaryotic gene regulation in processing and
        degradation of, 305

exon duplication and shuffling and domains of,
    355–357
in extracellular matrix, 88–90
heat-shock and antifreeze, 635
human dietary deficiencies in, 667
initiation factors, elongation factors, and
    release factors in translation, 282
levels of structure of, 55–58
as measure of evolution, 62–63
membrane (*see* Membrane proteins; Receptor
    proteins; Transport proteins)
motor (*see* motor proteins)
mutations from faulty genes and faulty, 268
nonenzyme, 270
as organic compounds and macromolecules,
    40
in origin of multicellular animals, 488–489
in pattern formation, 319–320
photopsin, 789
in photosystems, 163–164
in plant pathogen defenses, 636–637
in plasma membranes, 71*f*
as polymers of monomers, 44–45
in prokaryotic flagella, 464
Protein Data Bank of structures of, 345
proteomics as study of full sets (proteomes)
    of, 347
regulatory, in skeletal muscle contraction,
    795–797
repressor, 295–297
respiratory pigments, 706–708
ribosomes in synthesis of, 76, 91
role of nucleic acids in synthesis of, 60
rough ER and secretory, 78
sickle-cell disease as change in primary
    structure of, 58
signal transduction pathways in synthesis of,
    113
structure and function of, 54–59
synthesis of (*see* Gene expression)
tissue-specific, 313–315
transcription activator and mediator, 301–302
transcription factors, 275–276
types of, and functions of, 52*f*
understanding functions of genes for coding,
    345–346
viral coats of, 331–332, 333*f*
water-soluble, 33*f*
Proteobacteria, 472*f*, 484–486
Proteoglycans, **89***f*
Proteomes, 347
Proteomics, **347**
Proterocladus, 483*f*
Proterozoic eon, 439*t*
Protista, kingdom, 395, 489
Protists, 8*f*, **481**
    cells of, 73*f* (*see also* Eukaryotic cells)
    contractile vacuoles of, 80
    ecological roles and impacts of, 499–502
    effects of, on human health, 500–502
    endosymbiosis in evolution of photosynthetic,
        486–487
    Eukarya domain and, 8*f*
    four supergroups in phylogeny of eukaryotes
        and, 489–498
    origin of fungi in, 511
    origin of multicellular animals in, 488–489
    photosynthetic, 499–500
    sexual life cycles of, 196–197
    structural and functional diversity of, 499
    symbiotic, 500
    as unicellular eukaryotes, 481
Protocells, **459**–460
Protonephridia, 656
Proton-motive force, **146**, 148

Proton pumps, **105**
  in acid growth hypothesis, 620–621
  in vascular plant solute transport, 574–575
Protons, **20**–21, 34, 166
Proto-oncogenes, **324**–327
Protoplast, **577**
Proviruses, **336**, 337*f*
Proximal control elements, 300
Proximal tubule, **657***f*, **658**, 659*f*
Proximate causation, 803
Prozac, 764
PR (pathogenesis-related) proteins, 637
Pseudechis porphyriacus, 849
Pseudogenes, **349**–350, 376
Pseudomyrmex, 850*f*
Pseudopodia, 91, 491*f*, **494**, 497
Pseudotsuga menziesii, 518*f*
P site (peptidyl-tRNA binding site), **281**, 283*f*
Psittacus erithacus, 778
Psychoactive drugs, 764
Pterosaurs, 801
Puberty, human, 738
Puffball fungus, 371*f*
Pulmocutaneous circuits, 686–687*f*
Pulmonary circuits, 686–687*f*, 689*f*
Pulp, fruit, 606
Pulse, **692**
Puma concolor coryi, 408*f*
Punctuated equilibria, **431**
Puncture vine, 607*f*
Pundamilia genus, 426–427, 430
Punnett squares, **210**, 217*f*, 218*f*, 236
Pupil, **786**–787
Purines, **60**–61, 250–251
Purple sulfur bacteria, 156*f*
Pus, 715
Pygmy date palm, 521*f*
Pyramid of net production, 871–872
Pyrimidines, **60**–61, 250–251
Pyrococcus furiosus, 471
Pyruvate
  ATP yield from oxidation of, 147–148
  in fermentation, 150–151
  oxidation of, in cellular respiration, 139, 142–143*f*
  oxidation of glucose to, in cellular respiration, 141

Q
Qualitative data, 11
Quantitative characters, **217**–218
Quantitative data, 11, 655
Quantitative experimental approach, G. Mendel's, 207–212
Quaternary consumers, 853*f*
Quaternary structure, protein, 57*f*
Questions, scientific, 12

R
Rabbits, 677, 739
Radial glia, 770
Radial symmetry, **532**–533
Radiation. *See also* Ultraviolet (UV) radiation
  alterations of chromosome structure by, 241
  as cancer treatment, 188
  damage from radioactive, 22
  mutagenic, 290
Radiation process, thermoregulation and, **646***f*
Radiations
  adaptive, 447–449
  amniote, 544–546
  Cambrian explosion and bilaterian, 530–532
  ecological interactions and, 548
  invertebrate, 534–536

natural selection and, 10–11
  prokaryotic, 458, 478
  vertebrate, 536–539
Radicle, **604**
Radioactive isotopes, **21**–22, 248, 438, 876
Radiometric dating, **438**
Raft spiders, 31*f*
Rainbows, 155*f*
Rain shadows, 822
Rana pipiens, 320
Random dispersion, 833*f*, 834
Random fertilization, 203
Random mating, 403–405
Random mutations, 240. *See also* Mutations
Randomness, entropy and, 118–119
Ranges, species, 831
Rangifer tarandus, 731
Rapetosaurus krausei, 9*f*
Rapid eye movements (REMs), 771
Rare evolutionary event, eukaryotic, 497*f*
ras gene and Ras protein, **325**–326
Raspberry fruit, 606*f*
Rats, 775, 808
Rattlesnakes, 677, 781*f*, 848
Ravens, 809
Ray-finned fishes, 538*f*–**539**
Rays, 539
Reabsorption, **655***f*, **656**, 658
Reactants, chemical reaction, **28**–29
Reaction-center complexes, **163**–164
Reading frames, **273**, 289–290
Realized niches, 847
Reasoning
  evolution of cognition and, 777–778
  inductive and deductive, 11–12
Receptacle, flower, **598**
Reception, light. *See* Light reception, plant
Reception, sensory, 779–780
Reception stage, cell-signaling, **109**–111, 113*f*
Receptive fields, 788
Receptor-mediated endocytosis, 106–**107***f*
Receptor potential, **780**, 788*f*
Receptor proteins, 52*f*, 97*f*, 107*f*, 109–113*f*
Receptors
  cellular innate immune defense, 713–714
  dendrites as, 752
  opiate, 765
  sensory (*see* Sensory receptors)
Recessive alleles, **209**–210, 220–222, 413–414
Recessively inherited human disorders, 220–222
Recessive traits, 208, 220–222, 232–234
Reciprocal selection, 548*f*
Recombinant bacteria, 262
Recombinant chromosomes, crossing over and, **203**
Recombinant DNA, **262**–264
Recombinants (recombinant types), **236**
Recombinase, 718
Recombination frequencies, 237–240
Reconstruction, ecosystem, 877
Recording, intracellular, 756*f*
Recruitment, motor neuron, 797
Recruitment of animal predators, plant, 636
Rectum, 670*f*, 671*f*, **676**
Recycling, chemical. *See* Chemical cycling
Red algae, 486–487, 491*f*, **495**–496
Red blood cells, 58*f*, 103, 695*f*, 696, 708
Red deer, 839
Red mangrove, 624*f*
Red maple trees, 560
Red-necked phalaropes, 810*f*
Redox (oxidation-reduction) reactions, **136**
  in citric acid cycle, 143*f*
  oxidation of organic fuel molecules during cellular respiration, 137

photosynthesis as, 158
  stepwise energy harvest via NAD$^+$ and electron transport chain in, 137–139
Red Sea, coral reef biome in, 829*f*
Red tide, 494
Reduced hybrid fertility, 421*f*
Reduced hybrid viability, 421*f*
Reducing agents, **136**–137
Reduction, **136**–137
Reductional division, 201
Reductionism, 3, 4
Reduction phase, Calvin cycle, 168
Redundancy, genetic code, 273, 288
Redwood trees, 193*f*
Reflexes, 671–672, 678, **770**
Refractory period, **759**, 798
Regeneration, 730
Regeneration phase, Calvin cycle, 168–169
Regenerative medicine, 324
Regional adaptive radiations, 448–449
Regression lines, scatter plots with, 37, 170, F-2
Regulation
  of animal digestion, energy storage, and appetite, 678–682
  of biological clocks, 774
  of blood pressure, 692–693
  of cell cycle by cell cycle control system, 183–189
  cell signaling and cellular, 113
  of enzymatic catalysis, 130–132
  of gene expression (*see* Gene regulation)
  homeostatic (*see* Homeostasis)
  hormonal, of human embryonic development and childbirth, 746–747
  hormonal, of mammalian/human sexual reproduction, 738–742
  of human breathing, 705–706
  of muscle contraction, 795–797
  population (*see* Population regulation)
  of primary production in ecosystems, 866–870
Regulators, **644**
Regulatory genes, **295**, 297, 314–315. *See also* Homeotic genes
Regulatory proteins, 795–797
Reinforcement, hybrid zone, **429**
Rejection, immune, 724
Relatedness, altruism and, 813–814
Relationships, plant nutrition and, 582–586*f*
Relative abundance, **851**–852
Relative fitness, **411**
Relay molecules, 109, 111–112
Release factors, 282, 283*f*
Release stage, phage lytic cycle, 333*f*
Releasing hormones, **650***f*
Renal cortex, **657***f*
Renal medulla, **657***f*
Renal pelvis, **657***f*
Renaturation, protein, 58*f*–59
Renin-angiotensin-aldosterone system (RAAS), **662**–663
Repair, DNA, 257–258
Repenomamus giganticus, 448
Repetitive DNA, **350**–352
Replication, DNA. *See* DNA replication
Replication forks, **253**–254
Replicative cycles, viral, 333–336, 337*f*
Repolarization, 759
Repressible enzymes, 296–297
Repressible operons, 295–297
Repressors, **295**–297
Reproduction. *See also* Animal reproduction; Human reproduction; Plant reproduction
  binary fission as rapid prokaryotic, 466–468
  as cell division function, 174–175*f*
  evolution and success in, 204
  fungal, 510, 511*f*

heterochrony and development of organs of, 450
overreproduction of offspring and natural selection, 371–372
protocell, 460
rapid, as source of genetic variation in viruses, 402
sexual, as source of genetic variation, 402
sexual life cycles in (see Sexual life cycles)
sexual vs. asexual, 193, 729–730 (see also Asexual reproduction; Sexual reproduction)
in theory of evolution by natural selection, 9–10
trade-offs between survival and, in life histories, 839–840
Reproductive barriers
hybrid zones and, 428–430
prezygotic and postzygotic, 420f–421f
Reproductive cells. See Gametes
Reproductive cloning, 321–322
Reproductive cycles, animal, 731
Reproductive cycles, human, 738–742
Reproductive growth, plant, 562
Reproductive isolation, 419
allopatric speciation and, 423–425
geographic separation and, 427–428
hybrid zones and, 428–430
identifying dependent and independent variables, making scatter plots, and interpreting data on, 427
pollinator choice and, 433
prezygotic and postzygotic reproductive barriers in, 420f–421f
sympatric speciation and, 425–427
Reproductive leaves, 557f
Reproductive organs
animal, 733
heterochrony and development of, 450
Reproductive organs, human, 734–738
female, 735
gametogenesis in, 735–738
male, 734–735
in sexual intercourse, 741–742
Reproductive rates, 834–835, 841f
Reproductive shoots, 555f
Reproductive systems, mammalian, 642t
Reproductive tables, 834–835
Reproductive technology, 749
Reptiles, 544
circulatory systems of, 687
diversity of, 545f
flying, 801–802
in geologic record, 440
Research methods. See also Scientific skills exercises
applying parsimony to problems in molecular systematics, 390f
constructing linkage maps, 239f
crossing pea plants, 207f
determination absorption spectrum using spectrophotometers, 161f
determining microbial diversity using molecular tools, 852f
determining primary production with satellites, 867f
hydroponic culture, 578f
intracellular recording, 756, 756f
preparing karyotypes, 194f
polymerase chain reaction (PCR) technique, 264f
reproductive cloning of mammals by nuclear transplantation, 321f
reverse transcriptase-polymerase chain reaction (RT-PCR), 308f
testcrosses, 211f

Reservoirs, nutrient, 874f–875f
Residual volume, 705
Resistance genes, 469–470
Resistance (R) proteins, 636–637
Resolution, microscope, 67
Resolution phase, human sexual response, 742
Resource acquisition. See Plant resource acquisition
Resource competition, density-dependent population regulation and, 841f
Resource partitioning, 846–847
Respiratory diseases, human, 841f
Respiratory distress syndrome (RDS), 703–704
Respiratory media, 700
Respiratory pigments, 707
Respiratory surfaces, 700–701
Respiratory systems. See also Gas exchange
breathing in, 704–706
gas exchange adaptations in, 706–708
lungs in mammalian, 702–704
mammalian, 642t
respiratory distress syndrome (RDS) in, 703–704
Response, homeostatic, 645
Responses, plant. See Plant responses
Response stage, cell-signaling, 109, 113
Rest and digest responses, 771
Resting potentials, neuron, 754–756
Resting potential state, action potential, 758–759
Restoration ecology, 877–879f
biological augmentation in, 878
biomanipulation of trophic levels in, 855
bioremediation in, 877–878
worldwide restoration projects, 878–879f
Restriction enzymes, 262–264, 334
Restriction fragment length polymorphism (RFLP), 852f
Restriction fragments, 263
Restriction sites, 263
Reticulum, 677f
Retina, 786f, 788, 789
Retinal, 787f–788
Retinitis pigmentosa, 408
Retrotransposons, 350–351
Retroviruses, 307, 336, 337f
Reverse transcripts, 307–308
Reverse transcriptase, 307–308, 336, 337f
Reverse transcriptase-polymerase chain reaction (RT-PCR), 307–308
Reversibility
of chemical reactions, 29
of weak chemical bonds, 26–27
Reward system, brain, 775, 776
Rhacophorinae frogs, 444
Rhagoletis pomonella, 426
Rhesus monkeys, 63
Rheumatoid arthritis, 725
Rhizarians, 491f, 494
Rhizobacteria, 582
Rhizobium, 472f, 583–584
Rhizoctonia solani, 477
Rhizoids, 514, 572
Rhizomes, 556f
Rhizosphere, 477, 582
Rhodopsin, 787f, 788
Rhomaleosaurus victor, 437f
Rhythm method, contraception by, 747–748f
Rhythms, behavioral, 804–805
Ribbon models, 55f
Ribose, 46f, 61, 122
Ribosomal RNA (rRNA), 280
comparing genetic sequences of, for prokaryotes related to mitochondria, 485
eukaryotic ribosomes and, 76
gene family of, 352

horizontal gene transfer and, 395–396
in phylogeny of animals, 533
ribozymes and, 278
in translation, 280–282, 283f
Ribosomes, 76, 271
animal cell, 72f
in cells, 69
in chloroplasts, 83f
eukaryotic, 75f
free and bound, 76
interpreting DNA sequence logos to identify binding sites for, 284
in mitochondria and chloroplasts, 82
plant cell, 73f
polyribosomes, 286–287f
prokaryotic, 70f, 465
in protein synthesis, 60, 76, 91
as sites of translation, 271, 280–283f
Ribozymes, 125, 278, 460–461
Rice, 348t, 421f, 622
Right atrium, 688f, 689f
Right ventricle, 688f, 689f
Ring structures, 42f, 46f, 48f
Rising phase, action potential, 758–759
Risk factors, cardiovascular disease, 698
Rivers, 828f
RNA (ribonucleic acid)
ATP in, 122
components of, 60–61
development of self-replicating, 460–461
in DNA replication, 254–255
in evidence for endosymbiosis, 485
gene density in genomes and, 349
messenger (see Messenger RNA (mRNA))
as molecular homology, 376
noncoding (see Noncoding RNAs (ncRNAs))
in protein synthesis, 5–6
ribosomal (see Ribosomal RNA (rRNA))
ribosomal RNA gene family, 352
ribozymes of, as enzymes, 125
role of, in protein synthesis, 60
sequencing of, 308–309
structure of molecules of, 62
viral (see RNA viruses)
RNA interference (RNAi), 306
RNA polymerase, 274–276, 294–297, 333, 336
RNA polymerase II, 300
RNA polymerase II enzyme, 59f
RNA processing, 276–278, 287f, 300f, 304
RNA sequencing, 308–309
RNA splicing, 277–278
RNA splicing, alternative, 304
RNA viruses
emerging viruses as, 338–340
replicative cycles of, 332–333, 335–336, 337f
structure of, 330–332
Rock python, 669f
Rocks
dating of, 438
species distributions and, 832
weathering of, in phosphorus cycle, 875f
Rod, ATP synthase, 145f
Rodents (Rodentia), 677
Rodents, 781
Rods, 786f, 787f, 788, 789
Rod-shaped bacteria, 70f
Rod-shaped prokaryotes, 462f
Rolling circle replication, 469f
Romanesco, 553
Root caps, 562–563
Rooted phylogenetic trees, 383–384
Root hairs, 555f, 561–562, 587
Root pressure, 590
Roots, 516, 554
adventitious, 610
apical meristems of, 507

architecture of, and acquisition of water and minerals, 573–574

evolution of, 516

flooding responses of, 633–634, 635*f*

gravitropism in, 632

monocot vs. eudicot, 554*f*

mycorrhizal fungi and, 509, 574 (*see also* Mycorrhizae)

primary growth of, 562–564

in root systems, 555*f*

secondary growth of, 566–569

soil as source of essential elements for absorption by, 578–582 (*see also* Soil)

transpiration of water and minerals from, to shoots via xylem, 587–590

Root systems, **554–555**

Rotor, ATP synthase, **145***f*

Rough ER, **77**

   animal cell, 72*f*

   in endomembrane system, 81*f*

   in eukaryotic nucleus, 75*f*

   functions of, 78

   plant cell, 73*f*

Round dance, honeybee, 805*f*

Round window, **784**

Roundworms, 535*f*, 799–800

Rous, Peyton, 328

R plasmids, **469–470**

r-selection, **840**

Rubisco (RuBP carboxylase), **168–169**

Rule of multiplication, 404

Rumen, 677*f*

Ruminants, **677**

Running, 802

Rupicara rupicapra, 667*f*

Ryther, John, 869*f*

S

Saccharomyces cerevisiae (yeast)

   calculating volume and surface area of cells of, 74

   changes in DNA content of budding cells of, during meiosis, 202

   density-dependent population regulation of, 841*f*

   genome size of, 348

Saccule, **784–785***f*

Sac fungi, 512*f*

Safety issues, transgenic crop, 613–614

Sago palm, 518*f*

Salamanders, 421*f*, 427, 450, 543, 684, 769*f*

Salicylic acid, 637–**638**

Salinity

   aquatic biome, 827

   extreme halophiles and, 471, 474

   osmosis and, 101

   species distributions and, 832

   species facilitation and soil, 851

Salinity, plant responses to, 634

Saliva, 671–672

Salivary glands, **671–672**

Salmon, 665

Salmonella, 472*f*, 476

Salt, table. *See* Sodium chloride (table salt)

Saltatory conduction, **760**

Salt concentration. *See* Salinity

Salt marshes, 851, 871*f*

Salts (ionic compounds), **26**

   in blood plasma, 695

   nitrogenous wastes (*see* Filtrate)

Salty tastants, 782

Sand dollars, 175*f*, 744*f*

Sandhill cranes, 807

Sandy inland mouse, 655

Sapwood, 568

"SAR" clade, 490*f*–491*f*, **493**–495

Sarcomeres, **793–795**

Sarcoplasmic reticulum (SR), **795–798**

Sargasso Sea nutrient enrichment experiment, 869*t*

Sarin, 130, 764

Satellites, determining primary production with, 867*f*

Satiety center, 681–682

Saturated enzymes, 128

Saturated fats, 49–50*f*

Saturated fatty acids, **49**–50*f*

Saturated hydrocarbon tails, 96*f*

Savannas, **824***f*

Scala naturae (scale of nature), Aristotle's, 366

Scale, skeletal, 800–801

Scale bars, 74

Scale-eating fish, 414, 415*f*

Scales, 544

Scanning electron microscope (SEM), 68*f*, **69**

Scarlet fever, 335

Scatter plots, 37, 170, 427, 849, F-2

Schematic model, ribosome, 281*f*

Schmidt-Nielsen, Knut, 803

Schwann cells, **760**, 769*f*

Science, nature of **11**–16

   community and diversity in, as social process, 14–16

   controlled experiments, **14**

   data, **11**

   deductive reasoning, **12**

   hypothesis, **12**

   inductive reasoning, **11**

   inquiry, **11**

   observations, 11

   questions, 12

   research methods of (*see* Research methods)

   skills for (*see* Scientific skills exercises)

   technology, **15**

   theory, **14**, 379

Scientific notation, 765

Scientific skills exercises

   amino acid sequences, 63, 356

   bar graphs, 15, 149, 477, 549, 560, 634, 849, F-2

   chi-square (χ^2) test, 238, F-2

   converting units, 202

   DNA deletion experiments, 303

   DNA sequence logos, 284

   experimental design, 549, 739, 774

   gene expression data, 316

   genetic mutants, 681, 774

   genetic sequences, 485

   graphs, estimating quantitative data from, 443

   graph with two sets of data, 103, F-1

   graph with two variables on common *x*-axis, 726

   Hardy-Weinberg equation, 406

   histograms, 188, 219, 699, F-2

   making and testing predictions, 378

   line graphs, 128, 202, 549, 726, F-1

   logistic equation to model population growth, 838

   phylogenetic trees, 340, 394

   polypeptide sequence data, 63

   positive and negative correlations, 610

   quantitative data, 655

   ratios, 655

   regression lines, 37, 170, F-2

   scale bars, 74

   scatter plots, 37, 170, 427, 849, F-2

   scientific notation, 765

   slope, 128

   surface area, 74

   synthesizing information from multiple data sets, 510

   temperature coefficients, 576

tables with data, 249, 871

volume, 74

Scion, **610**

Sclera, 786*f*

Sclereids, **558***f*

Sclerenchyma cells, **558***f*

Scr gene, 450

Scrotum, **734**

Scrub jays, 778

Scutellum, 604

Scyphozoa, **530***f*

Sea anemones, 530*f*, 730*f*

Seagrass restoration project, 879*f*

Sea lettuce, 496*f*

Seals, 708

Sea palm, 493*f*

Sea slugs, 729

Seasons

   climate and, 821

   plant photoperiodism and responses to, 629–631

Sea stars, 535*f*, 700*f*, 833*f*

Sea turtles, 7*f*

Sea urchins, 119*f*, 421*f*, 535*f*, 742–745, 831*f*, 845

Seaweed, 491*f*, 493–494, 496, 831*f*, 879*f*

Secondary cell walls, **88**, 89*f*

Secondary consumers, 853*f*, **866**

Secondary endosymbiosis, **486–487**

Secondary growth, plant, **560**

   cork cambium and periderm production in, 569

   meristem generation of cells for, 560–562

   overview of, 561*f*

   of stems and roots in woody plants, 566–569

   tissues of, 566

   vascular cambium and secondary vascular tissue in, 568–569

   of woody stems, 567*f*

Secondary immune response, **719–720**

Secondary oocytes, **737***f*

Secondary production, **870–872**

Secondary structure, protein, **56***f*

Secondary succession, **857–858**

Secondary vascular tissue, 568–569

Second law of thermodynamics, **118–119**, 865

Second messengers, **112–113***f*, 762–763, 780

Secretin, 649, 652

Secretion function, **655***f*, **656**

   of excretory system, 655*f*, 656

   of liver and pancreas, 674

   of small intestine, 674

   of stomach, 672–673

Secretory phase, uterine cycle, 741

Secretory proteins, 78, 98*f*

Sedimentary rock, fossils in, 438

Seed coat, 517, **604**

Seed-eater finches, 369*f*

Seedless vascular plants, 513*f*, **514**–516

Seedling development, seed germination and, 605

Seed plants, 516–521. *See also* Angiosperms; Gymnosperms; Land plants

   angiosperms and gymnosperms as, 516

   gametophyte-sporophyte relationship in, 515*f*

   origin and diversification of angiosperms, 518–521

   origin and diversification of gymnosperms, 518

   phylogeny of, 513*f*

   terrestrial adaptations of, 516–517

Seeds, **516**, 517

   abscisic acid in dormancy of, 624

   dispersal of, 607*f*

   dormancy of, 604–605

   embryo development and, 603–604

   endosperm development and, 601–603

germination of, and seedling development, 605
gibberellins in germination of, 623
phytochromes and germination of, 627–628
structure of mature, 604
temperature and water uptake by, 576
Segmented bodies, arthropod, 536
Segmented worms, 799–800
Segregation, law of, 207–208, **209**–211, 228f–229
Seizures, 759
Selective breeding, 371–372
Selective inhibition, enzyme, 130
Selective permeability, **94**, 95, 99, 754–756
Selective protein degradation, 305
Self-fertilization (selfing), 609
Self-incompatibility, plant, **609**
Self-pollination, 207–208
Self-pruning, 573
Self-replicating molecules, 459–462
Self-thinning, 594
Self-tolerance, 718–719
Semen, **734**, 742, 745–746
Semicircular canals, **783**f, 784–785f
Semiconservative model, DNA replication, **252**, 253f
Semilunar valves, **689**
Seminal vesicles, **734**
Seminiferous tubules, **734**, 736f, 739
Senescence, plant, **625**
Senile dementia, 59
Sensitive period, **807**
Sensitive plants, 633
Sensors, homeostatic, **645**
Sensory adaptation, **780**
Sensory amplification, **780**
Sensory association cortex, 776f
Sensory input, nervous system, 753
Sensory neurons, **753**, 770, 783f
Sensory reception, **779**
Sensory receptors, **779**
chemoreceptors, 781–782
electromagnetic receptors, 781
mechanoreceptors, 780–785f
nociceptors (pain receptors), 781
photoreceptors, 785–789
somatosensory, 777
thermoreceptors, 781
Sensory systems, 779–789
cerebral cortex and, 777
hearing, equilibrium, and mechanoreceptors in, 782–785f
photoreceptors and vision in, 785–789 (see also Visual systems)
sensory amplification and adaptation in, 780
sensory perception in, 780
sensory reception and transduction in, 779–780
sensory transmission in, 780
types of sensory receptors in, 780–782
Sensory transduction, **780**, 787–788
Sepals, **518**–519f, **598**
Separase, 180
September equinox, 821f
Septic shock, 715
Septum, 687
Sequence logos, interpreting, 284
Sequences, amino acid, 63, 356
Sequences, DNA. See DNA sequences
Sequencing, DNA and genome. See DNA sequencing; Genome sequencing
Sequencing by synthesis technique, 344
Serial endosymbiosis hypothesis, **484**–486
Serial transfer, 467f
Serine, 53f
Serotonin, **764**
Sertoli cells, **739**
Serum, 695

Set point, homeostatic, **645**
Sex chromosomes, **195**
aneuploidy of human, 242
as chromosomal basis of sex, 231–232
human, 194–195
inactivation of X-linked genes in female mammals, 233–234
inheritance of X-linked genes, 232–233
mammalian, 232f
in mammalian sex determination, 739
patterns of inheritance of, 231–234
Sex determination, 231–232, 739
Sex hormones
chemical groups of, 44
cholesterol as steroid lipid of, 50
intracellular receptor proteins for, 111
production of, by smooth ER, 77
regulation of mammalian/human reproduction by, 738–742
as steroid hormones, 652
Sex-linked genes, **232**. See also Sex chromosomes
Sex pili, 464, 468–469
Sex reversal, 732
Sexual dimorphism, **412**, 810
Sexual intercourse, human, 741–742, 745–748, 765
Sexual life cycles, 192–205
asexual vs. sexual reproduction and, 193–194
behavior of chromosomes in, as physical basis of Mendelian inheritance, 228–231 (see also Chromosomal basis of inheritance)
changes in DNA content of budding yeast cells during meiosis, 202
chromosome sets in human, 194–196
genetics, heredity, variation, and, 192
genetic variation produced by, 201–204
inheritance of genes in, 193
land plant, 506f–507
meiosis in, 197–201
preparing karyotypes of chromosomes and, 194f
types of, 196–197
Sexually transmitted diseases (STDs), 464, 473f, 725–726, 748, 749
Sexual reproduction, **193**, 729
allocation of energy in angiosperm, 610
angiosperm, 608–609 (see also Flowers)
asexual reproduction vs., 193, 204, 729–730 (see also Asexual reproduction)
effects of human overfishing on cod, 548
as evolutionary enigma, 730
fungal, 510, 511f
gametogenesis in, 735–738
hormonal regulation of mammalian/human, 738–742
human (see Human reproduction)
reproductive cycles in, 731
sexual selection and, 412–413
as source of genetic variation, 402
variations in patterns of, 731–732
Sexual response, human, 741–742
Sexual selection, **412**
adaptive evolution and, 412–413
female mate choice in, 811
sympatric speciation and, 426–427
S genes, 609
Shade avoidance, plant, 628
Shannon diversity index, **852**
Shape
of carrier proteins, 102
chemical groups and, of organic compounds, 44
cytoskeletons and cell, 84
of enzymes and proteins, 40f, 59f
intermediate filaments and cell, 88
molecular, 27–28

of proteins, 55–59
swimming and body, 802
Shapes
allosteric regulation and enzyme, 130–131
enzyme, 126–127
prokaryotic, 462
protist, 481
Shared ancestral characters, **387**–388
Shared derived characters, **387**–388
Sharks, 538f–539, 730
Sheep, 321–322, 677
Shell, amniotic egg, 544
Shewanella oneidensis, 878
Shigella, 469
Shoots. See also Leaf (leaves); Stems
apical meristems of, 507
auxin polar transport in, 621f
gravitropism in, 632
light capture and architecture of, 573
primary growth of, 564–566
transpiration of water and minerals from roots to, via xylem, 587–590
Shoot systems, **554**–556
Short-day plants, **630**
Short-distance signaling, neuron, 751–753. See also Neurons
Short-distance transport, plant, 574–577
Short tandem repeats (STRs), **351**–352
Short-term memory, **778**–779
Shrimp, 361f, 424–425, 451
Sickle-cell disease, **58**, **222**, **696**
hemoglobin protein structure and, 58f
malaria and evolutionary implications of, 222
malaria and heterozygote advantage in, 414
pleiotropy and, 217
point mutations and, 288, 401
Side chains, amino acid, 52–54, 57f
Sieve plates, **559**f, 593
Sieve-tube elements, **559**f, 593
Signaling, neurons and, 751–753. See also Neurons
Signal peptides, **285**–286
Signal-recognition particles (SRPs), **285**–286
Signals, animal, **805**–806
Signal transduction pathways, **109**
in cell signaling, 109, 112–113f
coordinate control of, 304
induction in, 313
neurotransmitters and, 762–763
in plant light reception, 617 (see also Light reception, plant; Plant responses)
in sensory systems, 780
in visual sensory transduction, 787–788
water-soluble hormones and, 652
Sign stimulus, **804**
Silencing, transcription, 302
Silent mutations, **288**–289
Silicosis, 703
Silk fibers, 56f
Silversword plants, 385, 448f
Similarity, species and, 419f
Simple fruits, **606**
Simple sequence DNA, **351**–352
Single-celled organisms, 66. See also Prokaryotic cells; Protists
Single bonds, **24**, 25f, 41–42
Single circulation, **686**, 687f
Single-lens eyes, **786**–787
Single nucleotide polymorphisms (SNPs), **359**–360
Single-strand binding proteins, **253**
Sinoatrial (SA) node, **690**
Sister cells, 174–176
Sister chromatid cohesion, 176, 197, 201
Sister chromatids, **176**, 195f, 197–201, 228f
Sister species, 424

Sister taxa, **383**
Size
    area effects and island, 860
    of cells, 67*f*
    eukaryotic cell vs. prokaryotic cell, 4, 70
    evolution of axon, 760
    genetic drift and small population, 409
    of genomes, 348
    Hardy-Weinberg equilibrium and large
        population, 405
    population (*see* Population dynamics;
        Population growth)
    prokaryote, 462
    of skeletons, 800–801
Skeletal muscles, **643***f*, **793–798**
    nervous system regulation of tension of, 797
    regulation of contraction of, 795–797
    sliding-filament model of contraction for,
        794–795
    structure of, 793
    types of fibers of, 797–798
Skeletal systems, 799–803
    bones and joints of human, 801*f*
    endoskeletons, 800, 801*f*
    energy costs of locomotion in, 803
    exoskeletons, 800
    hydrostatic skeletons, 799–800
    locomotion in, 801–803
    mammalian, 642*t*
    size and scale of skeletons in, 800–801
    skeletal muscles (*see* Skeletal muscles)
    types of, 799–801
Skeletons, carbon, 42*f*
Skills, scientific. *See* Scientific skills
Skin
    as barrier defense, 713
    cancer of, 258
    as gas exchange tissue, 687
    human, 781
    mammalian, 642
    pigmentation of human, 217–219
    as respiratory organ, 701
Skulls
    human vs. chimpanzee, 386, 449
    mammalian, 440, 441*f*
Skunks, 420*f*
Sleep
    brain functions and, 771, 774
    memory and, 779
Sleeping sickness, 492–493*f*, 500–501
Sleep movements, plant, 629
Sliding-filament model, **794–**795
Slime layer, 463
Slime molds, 497–498
Slope, line graph, 128
Slow-twitch fibers, **798**
Slugs, 812–813
Small interfering RNAs (siRNAs), 305–**306**
Small intestine, **674**
    in alimentary canals, 671*f*
    digestion in, 673, 674–675
    evolutionary adaptations of, 677–678
Smallpox, 338
Small-scale mutations, 288–290
Smell
    pheromones and communication by, 806
    sense of, 781–782
Smoking, 327, 703, 748, 763
Smooth ER, **77**
    animal cell, 72*f*
    in endomembrane system, 81*f*
    functions of, 77–78
    plant cell, 73*f*
Smooth muscle, **643***f*, **798–**799
Snails, 420*f*, 432–433, 751, 753
Snakebite, 724

Snakes, 381–382, 412, 420*f*, 544–545*f*, 677, 781,
    802, 812–813, 848–849
Snapdragons, 215
Snapping shrimp, 424–425
Snowy owl, 841*f*
Soapberry bugs, 373–374
Social learning, **809**
Social process, science as, 14–16
Society, plant biotechnology and, 613–614
Sodium, 20*f*, 26*f*
Sodium chloride (table salt). *See also* Salinity
    in aqueous solutions, 33*f*
    emergent properties of, 20*f*
    human diets and excess, 667
    ionic bonds of, 26*f*
    kidney processing of, 658–660
    plant responses to excessive, 634
Sodium ions, 754–760
Sodium-potassium pump, **104-**105, **754–**756
Software, systems biology and, 345, 346*f*
Soil
    bacteria in, 472*f*, 474–475
    bacteria in, and plant nutrition, 582–584
    bioremediation of, 877–878
    coping with high-temperature, by land plants
        and mycorrhizal fungi, 510
    determining diversity of bacteria in, 852*f*
    disease-suppressive, 477
    in ecological succession, 857
    essential elements for plants in, 578–580
    land plants and formation of, 522
    management of, for plant nutrition, 580–581
    plant response to excessive salinity of, 634
    resource competition and fertilization of, 841*f*
    root architecture and acquisition of water and
        minerals from, 573–574
    species distributions and, 832
    species facilitation and salinity of, 851
    texture and composition of, 581–582
Soil worm, 315, 346, 348
Solar energy. *See* Light energy; Sunlight
Solstices, 821*f*
Solute potential, **576**
Solute potential equation, 596
Solutes, **33**
    chemoreceptors and, 781–782
    concentration of, in aqueous solutions, 34
    diffusion of, across plasma membranes, 99–100
    effects of, on water potential, 576–577
    nitrogenous wastes (*see* Filtrate)
    transpiration of, from roots to shoots via
        xylem, 587–590
    vascular plant transport of, across plasma
        membranes, 574–575, 577–578
Solutions, **33**, 100
Solvents, **33–**34
Somatic cells, **175**, **193**
Somatosensory cortex, 776*f*, 777
Somatosensory receptors, 777
Songbird brains, 778*f*
Sound, mechanoreceptors for hearing, 782–785*f*
Sour tastants, 782
South Africa, restoration project in, 879*f*
Southern Hemisphere seasonal variation, 821*f*
Southern magnolia, 521*f*
Soybean population, microevolution of, 406
Space-filling models, 25*f*, 27*f*, 41*f*, 51*f*, 55*f*, 250*f*
Spanish flu, 339
Spatial learning, **807–808**
Spatial pattern, homeotic genes and, 450
Spatial summation, **762**
Spawning, 732
Speciation, **418–435**
    adaptive radiations and, 447–449
    allopatric, 423–425
    allopatric vs. sympatric, 423*f*, 427–428

as conceptual bridge between microevolution
    and macroevolution, 418
continental drift and, 444
C. Darwin on, 370–372
differential, and species selection, 455
genetics of, 432–433
geographic separation and, 423–428
hybrid zones, reproductive isolation, and,
    428–430
identifying dependent and independent
    variables, making scatter plots, and
    interpreting data on reproductive isolation
    in, 427
macroevolution from, 433
morphological, ecological, and phylogenetic
    species concepts in, 422
reproductive isolation and biological species
    concept in, 418–422
sympatric, 425–427
time course of, 430–432
unity and species concepts in, 422
Species, **419**
    of animals, 528–529
    bilaterian invertebrate, 535*f*
    biological species concept of, 418–422
    classification of, 382–385
    in communities, 2*f*, 845 (*see also*
        Communities)
    comparing developmental processes of,
        360–361
    comparing genomes of, 358–360
    with complete genome sequences available,
        343, 346, 348
    C. Darwin's interest in geographic distribution
        of, 368–369
    C. Darwin's theory of origin of, 370–372
    diversity of (*see* Species diversity)
    dominant and keystone, 854–855
    endemic, 378
    extinctions of (*see* Extinctions)
    of fungi, 513
    genomes of (*see* Genome(s))
    geographic distribution of (*see* Species
        distributions)
    identifying, of whale meat, 384*f*
    interactions between (*see* Interactions,
        ecological; Interspecific interactions)
    loss of land plant, 524
    morphological, ecological, and phylogenetic
        species concepts of, 422
    morphology and, 418–419
    origin of (*see* Speciation)
    phylogenies as evolutionary histories of (*see*
        Phylogenies)
    taxonomic classification of, 366–367
Species-area curves, **860**
Species distributions
    abiotic factors in, 821, 831–832
    in aquatic biomes, 827–830 (*see also* Aquatic
        biomes)
    biogeography and, 377–378
    biotic factors in, 821, 831
    climate and, 821–823 (*see also* Climate;
        Macroclimate)
    dispersal factors in, 830–831
    ecological time vs. evolutionary time in, 830
    ecology, populations, and, 818 (*see also*
        Ecology; Population(s))
    factors limiting, 830*f*
    in terrestrial biomes, 823–824 (*see also*
        Terrestrial biomes)
Species diversity, **851**. *See also* Biodiversity
    biogeographical factors affecting, 859–860
    community stability and, 853, 856
    disturbances, ecological succession, and,
        856–859

human impacts on, 858–859
species richness, relative abundance, and, 851–852
trophic structure and, 853–855
Species richness, **851**
area effects and species-area curves of, 860
latitudinal gradients of, 859
Species selection, 455
Specific heat, **31**–32
Specificity
enzyme substrate, 126–127
viral, 332
Specific transcription factors, 300–302
Specimen preparation, 69
Spectrophotometer, **160**, 161f
Speech, cerebral cortex and, 776–777
Sperm, **729**
in animal fertilization, 743–744
chromosomes in human, 175
flagella of, 86
in human fertilization, 745–748
human reproductive organs and, 734
human sexual intercourse and, 742
human spermatogenesis and, 735–738
as male gametes, 729
mammalian sex determination and, 232
Spermathecae, 733
Spermatids, 736f
Spermatocytes, 736f
Spermatogenesis, human, **735**–738
Spermatogonia, **736**f
Spermicidal foam or jelly, 748
Spermophilus beldingi, 833–835
Sphagnum moss, 507f, 858
S phase, **177**
Spherical prokaryotes, 462f
Sphincters, **671**, 674
Spiders, 31f, 56f, 535f
Spilogale, 420f
Spinal cords, 769–770
Spines, 557f
Spiny acritarch, 532f
Spiny anteaters, 546f
Spiral phyllotaxy, 573
Spiral prokaryotes, 462f
Spirilla, 462f
Spirochetes, 462f, 473f
Spirodela oligorrhiza, 73f
Spirogyra crassa, 83f
Spleen, 694f
Spliceosomes, **278**
Split-brain effect, 777
Sponges, 529–530, 533–534
Spongy mesophyll, 565
Spontaneous abortions, 240, 746
Spontaneous mutations, 290
Spontaneous processes, **119**–122
Sporangia, **507**, 508
Spores, **506**f
fungal, 509f, 510, 511f
in plant reproduction, 196
of seedless vascular plants, 515f
seeds vs., 517
walled, in land plants, 507
Sporocytes, 507
Sporophytes, **506**f
of bryophytes, 514
of land plants, 506f
in plant reproduction, 196
of seedless vascular plants, 514–516
seed plant, 517
Sporopollenin, **505**, 507
Spotted skunks, 420f
Spriggina floundersi, 483f
Squamates, 544, 545f
Squids, 85f, 116

Srb, Adrian, 269
S-shaped logistic growth curve, 837–839
Stability
community, 853, 856
free energy and, 119–120
hybrid zone, 429f–**430**
population, 840–842
Stabilizing selection, 411f–**412**
Stable isotopes, 21
Staghorn coral, 861
Staghorn fern, 586f
Stahl, Franklin, 252, 253f
Stained specimen brightfield microscopy, 68f
Stalk-eyed flies, 811
Stamens, 207, **519**, **598**
Staminate flowers, 609
Standard metabolic rate (SMR), 679
Standing crop, 867, 872
Stapes (stirrup), 783f
Staphylococcus, 374–375, 459f, 473f
Star anise, 520f–521f
Starches, **47**
catabolism of, 136, 151–152
as product of photosynthesis, 171
structure of, 48f
Starfish. *See* Sea stars
Start codons, 273
Start point, transcription, **274**–275
Statins, 698
Statocysts, **782**
Statoliths, **632**, 782
Stator, ATP synthase, **145**f
Stechmann, Alexandra, 497f
Stele, **557**, 563
Stem cells, **322**–324, 561, **696**, 770
Stems, **556**. *See also* Shoots
architecture of, for light capture, 573
ethylene in triple response of, to mechanical stress, 624–625
gibberellins in elongation of, 622–623f
monocot vs. eudicot, 554f
primary and secondary growth of, 561f
primary and secondary growth of woody, 567f
primary growth of, 564–566
secondary growth of woody, 566–569
in shoot systems, 555f
structure of, 556
tissue organization of, 565–566
Stents, 698
Sterility, 421f, 433, 614
Sterilization, 748
Steroids, **50**
coordinate control by, 303–304
in evidence for origin of animals, 529
intracellular receptors and, 111
as lipids, 50–51
production of, by smooth ER, 77
regulation of mammalian/human reproduction by, 738–742
sex hormones as, 652
Steward, F. C., 320
Stickleback fish, 376, 451–452f, 804
Sticky ends, DNA, **264**
Stigma, 519, **598**
Stimulus
homeostatic, **645**
imprinting, 807
plant responses to environmental, 631–638
sensory reception and transduction of, 779–780
sign, 804
Stimulus-response chains, 805
Stink bugs, 541f
Stinson, Kristina, 585f
Stipes, **493**
Stirrup (stapes), 783f

Stock, plant, **610**
Stolons, 556f
Stomach, **672**
in alimentary canals, 670f, 671f
chemical digestion in, 672–673
dynamics of, 673–674
evolutionary adaptations of, 677–678
Stomach acid, 649
Stomach ulcers, 472f
Stomata, **156**, **507**, **565**
abscisic acid and, 624
of CAM plants, 170–171
in land plants, 507
regulation of transpiration by, 590–593
transpiration and, 169
Stone plants, 571
Stop codons, 273, 282, 283f
Storage, regulation of energy, 680
Storage leaves, 557f
Storage polysaccharides, 47–48
Storage proteins, 52f
Storage roots, 555f
Storms, 856
Stramenopiles, 491f, **493**–494, 500
Strands, DNA, 5
Strangling aerial roots, 555f
Strata, 366f–**367**, 438
Streams, 828f, 856
Streptococcus, 246, 463f, 468, 473f, 715
Streptomyces, 473f
Stresses, plant responses to environmental, 633–638
Stretch receptors, 780–781
Striated muscles, 643f, 793. *See also* Skeletal muscles
Strigolactones, 621
Strobili, 515f
Stroke, **698**
Stroke volume, blood, 689
Stroma, **83**, **156**, 166–167f
Stromatolites, 437f, **440**, 461–462
Structural formulas, 24, 25f, 41f, 51f
Structural polysaccharides, 48–49
Structural proteins, 52f
Structure-function correlation, 4, 7–8
Strychnine, 764, 849
Strychnos toxifera, 849
Sturtevant, Alfred H., 237–240
Style, flower, 519, **598**
Subatomic particles, 20–21
Suberin, 569
Substance P, 764
Substrate feeders, 669f
Substrate-level phosphorylation, **140**
Substrates, **126**–127
Succulent Karoo restoration project, 879f
Succulent plants, 119f, 170–171, 879f
Sucrase, 125–126
Sucrose
cotransport of, 105f
as disaccharide, 46–47f
hydrolysis of, to glucose and fructose, 125–126
as product of photosynthesis, 171
translocation of, in vascular plants, 593–594
Sudden oak death (SOD), 500, 638, 861
Sugar beets, 477
Sugar gliders, 376
Sugar-phosphate backbone, DNA, 61–62, 248, 250–251
Sugars
in aqueous solutions, 33
conduction of, in plant cells, 559f
monosaccharide and disaccharide, as carbohydrates, 45–47f
in nucleic acids, 60–61

as products of Calvin cycle, 159, 167–171
translocation of, from sources to sinks via phloem, 593–594
Sugar sinks, **593**–594
Sugar sources, **593**–594
Suicide genes, 326
Sulfate-reducing bacteria, 148–149
Sulfhydryl group, 43*f*
Sulfolobus, 471
Sulfur, 20, 248
Sulfur bacteria, 472*f*
Summation
muscle tension, 797, 798
postsynaptic potential, 762, 763*f*
Sundew, 586*f*
Sunflowers, 20, 431–432
Sunlight. *See also* Light energy; Ultraviolet (UV) radiation
aquatic biomes and, 827
cancer and, 327
climate and, 821–822 (*see also* Climate)
DNA damage from, 258
in energy flow and chemical cycling, 6–7
as energy for life, 135
global energy budget and, 867
latitudinal variation in intensity of, 820*f*, 821*f*
as light energy, 117, 122 (*see also* Light energy)
in photosynthesis, 28–29
primary production in aquatic ecosystems and limitations of, 868
properties of, 160
rainbows and, 155*f*
species distributions and availability of, 832
Supercontinent, 442, 444
Supergroups, eukaryotic, 489–492
Supplements, dietary, 667–668
Support, cell, 84
Support, skeletal, 799
Suprachiasmatic nucleus (SCN), **774**
Surface area, cell, 70–71*f*, 74
Surface area-to-volume ratios, 70–71*f*
Surface tension, **30**, 31*f*
Surfactants, **703**–704
Surgeonfish, 7*f*
Surroundings, system, 118
Survival
adaptations, natural selection and, 371–372
trade-offs between reproduction and, in life histories, 839–840
Survivorship curves, **834**
Suspension feeders, 669*f*
Suspensor cells, 603
Suspensory ligament, 786*f*
Sustainable resource management, 879*f*
Sutherland, Earl W., 109, 112
Sutton, Walter S., 228
Swallowing reflex, 672
Sweat, human, 32, 645
Sweet tastants, 782
Swimming, 802
Swine flu, 339, 862
Switchgrass, 613
Symbionts, **475**, 512*f*
Symbiosis, **475**, **849**
of anthozoans, 530*f*
commensalism as, 850–851
in flower pollination, 597
fungal, 523
lichen as, 521–522
mutualism as, 850
nutrient limitations and, 870
parasitism as, 850
protistan, 500
in vertebrate digestive adaptations, 677
Sym genes, 513
Symmetry, animal body, 532–533

Sympathetic division, peripheral nervous system, **771**
Sympatric populations, character displacement in, 847
Sympatric speciation, **425**
allopatric speciation vs., 423*f*, 427–428
habitat differentiation in, 426
polyploidy in, 425–426
sexual selection in, 426–427
Symplast, **574**
Symplastic transport route, 574, 587*f*
Synapses, **752**. *See also* Chemical synapses
electrical and chemical, 761
neural plasticity, memory, learning and, 778–779
regulation of muscle contraction and, 795–797
structure and function of, 752
Synapsids, 441*f*, **544**–546
Synapsis, **198***f*, 201
Synaptic signaling, 108–109
Synaptic terminals, 752
Synaptonemal complex, 198*f*
Synchlora aerata, 365
Syndromes, 242
Synergids, 601
Synthesis stage, lytic cycle, 333*f*
Synthesizing data from multiple datasets, 510
Synthetic estrogen or progestin, 748
Syphilis, 473*f*
Systematics, **382**. *See also* Phylogenies; Taxonomy
applying parsimony to problems in molecular, 390*f*
constructing phylogenetic trees from shared characters in, 387–392
evaluating molecular homologies in molecular, 386–387
interpreting phylogenies using, 382
prokaryotic phylogenies of molecular, 470–474
Systemic acquired resistance, **637**–638
Systemic circuits, **686**–687, 689*f*
Systemic inflammatory response, 715
Systemic lupus erythematosus, 725
Systems
chemical equilibrium and work in, 121–122
thermodynamics and, 118
Systems biology, **4**, **347**
emergent properties and reductionism in, 4
in study of genomes, genes, and gene expression, 346–347
Systole, **689**
Systolic pressure, 692, 698
Szent-Györgyi, Albert, 666

T

T2 phage, 246–248
T4 phage, 331*f*, 333*f*
Tables, data in, 249, 871
Table salt. *See* Sodium chloride (table salt)
Tags, molecular identification, 79
Taiga, 826*f*
Tails, histone, 260*f*
Tails, post-anal, 537
Takahe bird, 879*f*
Tallgrass prairie restoration, 878
Tansley, A. G., 856
Tappania, 437*f*, 482*f*
Taproots, **554**–555*f*
Taq polymerase, 265
Target cells, endocrine signaling, 648–649, 652
Tar spot fungus, 523*f*
Tastants, **781**
Taste
pheromones and communication by, 806
sense of, 781–782
Taste buds, **782**

TATA boxes, 275–**276**
Tatum, Edward, 269–270
Taxis, **464**
Taxol, 188
Taxonomy, **382**–385. *See also* Systematics
binomial nomenclature in, 382
early schemes of, 366–367
hierarchical classification in, 382–383
phylogenies and, 383–384 (*see also* Phylogenies)
possible plant kingdoms, 505*f*
three-domain system of, 7–9, 395
Taxon/taxa (taxonomy), **383**–384, 387
Taylor Glacier, 864
Tay-Sachs disease, 80, **216**, 221
T cells, **715**
antigen recognition by, 715–717
B cells and, 696*f*
cytotoxic T cells, 721
development of, 717–720
helper T cells, 720–721
in humoral and cell-mediated immune response, 720–723*f*
Teal, John, 871*f*
Technology, **15**
genomics, 344–347
prokaryotes in research and, 476–478
science and, 15–16
Tectonic plates, 443*f*
Tectorial membrane, 783*f*
Teeth
diet and adaptations of, 676–677
mammalian, 440, 441*f*, 544
Telencephalon, 772*f*
Telomerase, **259**
Telomeres, 258–259
Telomeric DNA, 352
Telophase, **177**, 179*f*, 182*f*, 200*f*
Telophase I, 198*f*, 200*f*
Telophase II, 199*f*
Temperate broadleaf forests, 826*f*
Temperate grasslands, 825*f*
Temperate phages, **334**–335
Temperature, **31**
aquatic biomes and, 827–830
calculating and interpreting coefficients for, 576
climate and, 821–823 (*see also* Climate)
climographs of, 823
in denaturation of proteins, 58–59
effects of, on litter decomposition in ecosystems, 873*f*
effects of transpiration on leaf, 592
enzymatic catalysis and, 129
as kinetic energy, 31
mass extinctions and, 446
membrane lipid composition and, 96
moderation of, by water, 30–32
oceans and, 827
plant responses to high and low, 635
species distributions and, 831
thermoreceptors and, 781
thermoregulation and (*see* Thermoregulation)
water uptake by seeds and, 576
Temperature coefficient, 576
Templates, viral, 336
Template strands, DNA, 252–253*f*, 254–257*f*, **272**–273
Tempo, speciation, 430–432
Temporal fenestra, 441*f*
Temporal isolation, 420*f*
Temporal lobe, 776*f*
Temporal summation, **762**
Tendrils, 557*f*
Tension, muscle, 797
Termination codons, 273

Termination stage
  transcription, 275
  translation, 282, 283*f*
Terminators, transcription, **274**
Termites, 48, 500
Terrestrial adaptations. *See also* Land plants
  of fungi and land plants, 505–508, 511–513
  of seed plants, 516–517
Terrestrial animals. *See* Land animals
Terrestrial biomes, 822–826*f*
  adaptations to, 169–171
  chaparral, 825*f*
  climate and, 822–823
  climographs for, 823*f*
  deserts, 825*f*
  effects of animals on, 547
  food chains in, 853*f*
  general features of, 823–824
  global distribution of, 823*f*
  locomotion in, 802
  natural and human-caused disturbances of,
    823
  northern coniferous forests, 826*f*
  primary production in, 869–870
  savannas, 824*f*
  temperate broadleaf forests, 826*f*
  temperate grasslands, 825*f*
  tropical forests, 824*f*
  tundra, 826*f*
Terrestrial plants. *See* Land plants
Terrestrial vertebrates, 540*f*, 542–546. *See also*
    Vertebrates
  amniotes, 543–547
  amphibians, 543
  tetrapods, 542–543*f*
Territoriality, **833**, 841*f*
Tertiary consumers, 853*f*, **866**
Tertiary structure, protein, **57***f*
Testcrosses, 210–**211**, 234–237*f*
Testes, 176, 195–196, **650***f*, **734**, 739
Testicles, 734
Testosterone, 50, 111, 652, **738**, 739
Tests (shells), **494**
Test-tube cloning, plant, 610–611
Tetanus, **797**, 798
Tetraploids, 241, 425–426
Tetrapods, 440, 441*f*, 538*f*–**539**, 542–543*f*
Texture, soil, 581
Thalamus, **773***f*, 775
Thalassoma bifasciatum, 732
Thamnophis, 420*f*, 812–813
Themes, biological, 2–7
  emergent properties at levels of biological
    organization, 2–4
  evolution, 7
  expression and transmission of genetic
    information, 5–6
  interaction of organisms with environment
    and other organisms, 7
  transfer and transformation of energy and
    matter, 6–7
Theobroma cacao, 523*f*
Theories, scientific, **14**, 379
Therapeutic cloning, 323–324
Therapsids, 441*f*
Thermal energy, 30, **31**, 99–100, **117**. *See also*
    Heat
Thermocline, **830**
Thermodynamics, **118**–119, 865
Thermoreceptors, **781**
Thermoregulation, **645**
  acclimatization in, 647
  balancing heat loss and gain in, 645–646
  circulatory adaptations for, 646
  desert ant, 641
  in endothermic and ectothermic animals, 645

  physiological thermostats and fever in, 647
  regulators and conformers in, 644
Thermus aquaticus, 265
Thick filaments (actin), 783*f*, **793**, 794–795
Thigmomorphogenesis, **632**–633
Thigmotropism, 632–**633**
Thin filaments (myosin), **793**, 794–795
Thiol compounds, 43*f*
Thiomargarita, 462, 472*f*
Thirst, 781
Thompson seedless grapes, 623*f*
Thoracic cavities, 704–705
Three-chambered hearts, 687
Three-spined stickleback fish, 451–452*f*, 804
Threonine, 53*f*, 132
Threshold, **757**
Thrombin, 696–697
Thrombus, **697**
Thrum flower, 609*f*
Thylakoid membranes, 156–157, 166–167*f*
Thylakoids, **83**, **156**–157, 159
Thylakoid space, 156–157
Thymine, 60–62, 248, 250–251, 270, 272
Thymine dimers, 258
Thymus, 694*f*, **715**
Thyroid disease, 651*f*
Thyroid gland, **650***f*
  hormones of, as chemical messengers, 111
  iodine deficiencies and, 20
Thyroid hormone (T_3 and T_4), 149, **650***f*–651*f*, 653
Thyroid-stimulating hormone (TSH), 651*f*
Thyrotropin, 651*f*
Thyrotropin-releasing hormone (TRH), 651*f*
Ticks, 861
Tidal rhythms, 804
Tidal volume, **705**
Tight junctions, **90***f*
Tiktaalik, 437*f*, 542
Time
  ecological and evolutionary, in species
    distributions, 830
  hybrid zones over, 429–430
  phylogenetic tree branch lengths and, 388–391
  required for human cell division, 177
  speciation over, 430–432
Timing, developmental, 449–450
Tinbergen, Niko, 803–804, 808
Tissue culture methods, plant, 622
Tissue plasminogen activator (TPA), 357
Tissues, 3*f*
Tissues, animal, **530**, **642**
  bilaterian, 533
  in hierarchical organization of animal bodies,
    642–643*f*
  immune system rejection of transplanted, 724
  lack of, by sponges, 530
  proteins specific to, 313–315
  renewal of, as cell division function, 174–175*f*
Tissues, plant, **554**
Tissue-specific proteins, 313–315
Tissue systems, plant, 556–557, 562–566
Toadfish, 798
Toads, 428–430, 543, 818, 849
Tobacco mosaic virus (TMV), 331, 341
Tobacco plant, 274*f*, 849
Tobacco smoke, 703, 763
Toll-like receptor (TLR), **713**–714
Tomatoes, 622
Tongues, 671*f*, 672
Tonicity, **101**
Tonsils, 694*f*
Tools
  antibodies as, 724
  computational, 343–347 (*see also*
    Bioinformatics)
Top-down model, trophic control, **855**

Topoisomerase, **253**
Topsoil composition, 581–582. *See also* Soil
Torpedo shape, 802
Tortoiseshell cats, 233–234
Total biomass accumulation, 868
Totipotent cells, **320**–321
Totipotent organisms, **609**
Touch
  animal mechanoreceptors and sense of,
    780–781
  plant response to, 632–633
Toxic elements
  in denaturation of proteins, 58–59
  evolution of tolerance to, 20
Toxic wastes
  bioremediation of, 877–878
  as density-dependent population regulation
    mechanism, 841*f*
Toxins
  Bt toxin and plant, 612
  enzymatic catalysis and, 130
  neurotransmission and, 764
  in predation, 848–849
Trace elements, **20**
Tracers, radioactive, 21–22, 158
Trachea (windpipe), 672, **702**–703
Tracheal systems, 541, **702**
Tracheids, **516**, **559***f*
Trade-offs, life history, 839–840
Tragopogon, 426
Traits, **207**
  characters and, 207
  dominant vs. recessive, 208, 220, 221*f*
  inheritance of, 371–372
  inheritance of acquired, 367
  inheritance of X-linked genes and recessive,
    232–234
  land plant derived, 507
  in theory of evolution by natural selection,
    9–10
Transcription, **271**
  analyzing DNA deletion experiments on
    eukaryotic, 303
  basic principles of translation and, 270–272
  coupling of translation and, 286
  DNA template strands in, 272–273
  effects of noncoding RNAs on eukaryotic,
    305–306
  eukaryotic gene regulation after, 304–305
  eukaryotic regulation of initiation of, 299–304
  as gene expression, 299
  genetic code and, 272*f*
  molecular components of, 274
  overview of, in gene expression, 271*f*
  regulation of bacterial, 293–298
  RNA processing after, 276–278
  summary of eukaryotic translation and, 287*f*
  three stages of synthesis of RNA transcripts in,
    274–276
Transcription factors, 111, **275**–276, 300–302, 316
Transcription initiation complex, **275**–276, 300,
    301*f*
Transcription units, **274**
Transduction, prokaryotic, **468**
Transduction, sensory, 779–780, 787–788
Transduction stage, cell-signaling, **109**, 111–113*f*
trans face, Golgi apparatus, 79, 81*f*
Trans fats, 698
Transfer RNA (tRNA), 62, **278**–283*f*, 287*f*
Transformation, cancer, **187**
Transformation, DNA, **246**
Transformation, energy, 117–119. *See also* Energy
Transformation, prokaryotic, **468**
Transfusions, blood, 724
Transgene escape, 614
Transgenes, 612, 614

Transgenic organisms, 273–274, 477, **612**–614
Transitional ER, 78
Transition state, 125
Translation, **271**, 278–287*f*
  basic concept of, 279*f*
  basic principles of transcription and, 270–272
  completing and targeting functional proteins in, 282–286
  coupling of transcription and, 286
  eukaryotic gene regulation at initiation of, 304–305
  genetic code and, 272*f*
  interpreting DNA sequence logos to identify ribosome-binding sites in, 284
  molecular components of, 278–281
  overview of, in gene expression, 271*f*
  summary of eukaryotic transcription and, 287*f*
  synthesis of multiple polypeptides in, 286–287*f*
  three stages of polypeptide synthesis in, 281–282, 283*f*
Translation initiation complex, 282
Translation initiation factors, 305
Translocation, cancer gene, 324–325
Translocation, chromosome, **241**, 242–243*f*
Translocation, vascular plant, **593**–594
Translocation stage, translation elongation cycle, 283*f*
Transmembrane proteins, 97
Transmembrane receptor proteins, 109–110
Transmembrane transport route, 574, 587*f*
Transmission, sensory, 780
Transmission electron microscope (TEM), 68*f*, **69**
Transmission rate, disease, 841*f*
Transpiration, **588**, 874*f*
  regulation of, by stomata, 590–593
  transport of water and minerals from roots to shoots via xylem by, 587–590
Transpirational pull, 588–589
Transplants, immune system rejection of tissue, 724
Transplants, species, 831
Transport, plant. *See* Plant transport
Transport epithelia, **655**, 658–659
Transport proteins, **99**. *See also* Aquaporins; Carrier proteins; Channel proteins
  in active transport, 103–105
  in cotransport, 105–106
  in facilitated diffusion, 102
  functions of, 52*f*, 97*f*
  transport work and, 124*f*
  types and functions of, in plasma membranes, 99
  in vascular plant transport, 574–575, 577
Transport vesicles, **78**
  in endomembrane system, 81*f*
  in exocytosis, 106
  Golgi apparatus and, 78–79
  in lysosomal autophagy, 80
Transport work, 122, 124*f*
Transposable elements, **350**
  contribution of, to genome evolution, 357
  transposons and retroposons in genomes, 350–351
Transposition process, 350, 357
Transposons, 306, 336, **350**–351
trans-retinal, 787–788
Transthyretin, 56*f*–57*f*, 59
Transverse (T) tubules, **795**–797
Tree frogs, mate selection of, 413*f*
Tree of life. *See also* Phylogenetic trees; Phylogenies
  C. Darwin's, 10–11, 370–371
  phylogenies and three-domain taxonomy of, 395–397
Tree rings, 568

Trees, 193*f*
  disruption of mycorrhizae of, by garlic mustard, 585*f*
  leaf morphology of, 560
  trunk anatomy of, 569*f*
Trends, evolutionary, 454–455
Treponema pallidum, 473*f*
Triacylglycerols, **49**–50
Triceps, 799*f*
Triceratium morlandii, 493*f*
Trichechus manatus, 849*f*
Trichomes, 556
Trichomonas vaginalis, 492
Triglycerides, 49–50, 675
Trimesters, human pregnancy, 746–747
Trioses, 46*f*
Triple response, plant, **624**–625
Triplet code, **272**–274
Triploidy, 241
Trisomic zygotes, **240**–242
Trisomy X (XXX), 242
Tristan da Cunha, 408
Triticum aestivum, 426, 841*f*
Trophic cascade model, 855
Trophic efficiency, **871**–872
Trophic structure, **853**
  bottom-up and top-down controls of, 855
  in ecosystem energy flow and chemical cycling, 864–866
  ecosystem energy flow and chemical cycling in, 864–866
  food chains and food webs in, 853–854
  secondary production efficiency in, 870–871
  species with large impact on, 854–855
  trophic efficiency and ecological pyramids in, 871–872
Trophoblast, **746**
Tropical dry forests, 824*f*
Tropical rain forests, 524, **824**f, 873
Tropic hormones (tropins), 651*f*, **738**–742
Tropic of Cancer, 820*f*
Tropic of Capricorn, 820*f*
Tropics, 820*f*, 859
Tropisms, **618**
Tropomyosin, **795**–797
Troponin complex, **795**–797
Troponin T gene, 304
Trp operon, 294–297
True-breeding organisms, **207**
Trypanosoma, 492–493*f*, 500–501
Trypsin, 673*f*, 674
Tryptophan, 53*f*, 294–297, 764
Tuataras, 544, 545*f*
Tubal ligation, 748
Tube cells, 601
Tuberculosis, 473*f*, 476, 715, 841*f*
Tubers, 556*f*
Tubulin protein, 85
Tumbleweeds, 607*f*
Tumors, cancer, 187–189
Tumor-suppressor genes, **325**
Tumor viruses, 328
Tundra, **826**f
Tunicates, 537
Turgid cells, **102**, **577**
Turgor movements, plant, 633*f*
Turgor pressure, 102, **577**, 591
Turner syndrome, 242
Turnover time, **872**
Turtles, 544, 545*f*, 848
Tutu, Desmond, 360
Twins, human, 746
Twin studies, **806**
Tympanic canal, 783*f*
Tympanic membrane (eardrum), **783**f, 784
  invertebrate, 782

Tympanuchus cupido, 408–409
Type 1 and type 2 diabetes, 681, 725
Typhoid fever, 476
Tyrosine, 53*f*, 764

U

Ubiquinone (Q), 144, 146*f*
Ubx gene, 450, 451, 536*f*
Ulcers, gastric, 673
Ultimate causation, 803
Ultraviolet (UV) radiation. *See also* Radiation; Sunlight
  cancer and, 327
  DNA damage from, 258
  mutagenic, 290
Ulva, 496*f*
Umami tastant, 782
Underground plants, 571
Undernutrition, 667
Undershoot phase, action potential, 758–759
Ungulates, 376–377
Unicellular eukaryotes. *See* Protists
Unicellular photoautotrophs, 156*f*
Uniform dispersion, 833
Unikonta, 490*f*–491*f*, **496**–498
Unisexual flowers, 598
United Kingdom, wetland biome in, 827*f*
Unity
  evolution of diversity and, 7, 8–9, 365–366, 370
  universality of genetic code and, 274
Unlinked genes
  mapping, 237–240
  recombination of, through independent assortment of chromosomes, 236
Unpaired electrons, 24
Unsaturated fats, 49–50*f*
Unsaturated fatty acids, **49**–50*f*
Unsaturated hydrocarbon tails, 96*f*
Unselfish behavior, 813–814
Unstained specimen brightfield microscopy, 68*f*
Untranslated regions (UTRs), 277, 304–305
Upwellings, 869
Uracil, 60–61, 62, 270, 272–273
Uranium, bioremediation of, 878
Uranium-238, 438
Urea, **655**, 659
Ureter, **657**f
Urethra, **657**f, **734**
Urey, Harold, 459–460
Uric acid, **655**
Urinary bladder, **657**f
Urine
  concentration of, in mammalian kidney, 659–660
  excretory system production of, 655–656
  kidney production of, 658–659
  territoriality and, 841*f*
Ursidae family, 394
Ursus, 8, 422
USA300 bacteria, 374
Use and disuse principle, 367
Uterine cycle, **740**–741
Uterus, 72*f*, 733, **735**, 740–742, 746–747
Utricle, **784**–785*f*

V

Vaccines, **338**, 723–724
Vacuoles, 73*f*, **80**–81, 101
Vagina, **735**, 742, 747–748
Vaginal pouch, 748
Valence, **24**–25, **41**–42
Valence electrons, 23–24
Valence shells, **23**–24

Valeria, 532f
Valine, 53f, 58
Valium, 764
Valves, heart, 689
Vampire bats, 661
van der Waals interactions, **27**, 57f
van Leeuwenhoek, Antoni, 67
van Niel, C. B., 158
Vaporization, 32
Variable (V) region, antigen, 716–717
Variables, identifying dependent and independent, 427
Variation, **192**. *See also* Genetic variation
Vasa recta, **657**f
Vascular bundles, 557, 566
Vascular cambium, **560**–561f, 566, 568–569
Vascular cylinder, 557
Vascular plants, **513**. *See also* Angiosperms; Plant(s)
  evolution of organs of, 554–556
  gametophyte-sporophyte relationship in, 515f
  overview of resource acquisition and transport in, 572f
  phylogeny of, 513f
  resource acquisition adaptations of, 571–574
  seedless, 514–516
  soil as source of essential elements for, 578–582
  transport in (*see* Plant transport)
  underground, 571
  vascular tissue in, 513, 516
Vascular rays, 568
Vascular tissue, plant, **513**, 516, 572–573. *See also* Phloem; Xylem
Vascular tissue system, plant, **556**–557, 559f
Vas deferens, **734**
Vasectomy, 748
Vasocongestion, 742
Vasoconstriction, 646, **692**–693
Vasodilation, 646, **692**–693, 734–735
Vasopressin, **650**f, 661–662, 812
Vectors, **860**–861
Vegetable oil, hydrophobic, 33–34
Vegetal plate, 745f
Vegetarian diets, 666
Vegetation, terrestrial biomes and, 823–824. *See also* Terrestrial biomes
Vegetative growth, plant, 562
Vegetative propagation, **610**, 622
Vegetative reproduction, **608**–611
Vegetative shoots, 555f
Veins, blood, **686**, 687f, 691, 693
Veins, leaf, 156, 554f, **556**, 565
Venomous snails, 751
Venter, Craig, 344
Ventilation, **701**
Ventral sides, **532**–533
Ventral tegmental area (VTA), 775
Ventricles, central nervous system, 769f, 770
Ventricles, heart, **686**, 687f, 688–689
Venules, 686, 691f
Venus flytrap, 586f, 633
Vernalization, 630–**631**
Vertebrates, **534**. *See also* Animal(s)
  action potential conduction speed in, 760
  adaptive immunity in, 712, 715 (*see also* Adaptive immunity)
  amniotes as terrestrial, 543–547 (*see also* Amniotes)
  amphibians as terrestrial, 543 (*see also* Amphibians)
  anatomical similarities in embryos of, 375
  in animal phylogeny, 534
  cardiovascular systems of (*see* Cardiovascular systems)

  evolutionary adaptations of digestive systems of, 676–678 (*see also* Digestive systems)
  evolution of brains of, 777–778 (*see also* Brains)
  excretory systems of, 656
  innate immunity in, 714–716
  mechanoreceptors for hearing and equilibrium in, 782–785f
  nervous systems of, 769–771
  origins of tetrapods as, 542–543f
  reproductive organs of, 733
  skeletal muscles of (*see* Skeletal muscles)
  swimming, 802
  terrestrial, 540f, 542–546
  visual systems of, 787–789
Vertical layering, terrestrial biome, 824
Vertical transmission, viral, 341
Vesicles, **76**
  abiotically produced, as protocells, 460
  in bulk transport, 106–107f
  in endomembrane system, 76–77, 81f
  in lysosomal autophagy, 80f
  in plant cytokinesis, 181f, 182
  transport (*see* Transport vesicles)
Vessel elements, **559**f
Vessels, circulatory, 685–686. *See also* Blood vessels
Vessels, lymphatic, 694f
Vessels, xylem, **559**f
Vestibular canal, 783f, 784
Vestibular glands, 735
Vestigial structures, **375**–376
Viagra, 734–735, 765
Vibrio cholerae, 472f, 476
Villi, 674f, **675**
Viral envelopes, **332**, 335–336
Viral integration, 328
Virchow, Rudolf, 174
Virulent pathogens, **636**
Virulent phages, **333**–334
Viruses, **246**, **330**–342
  analyzing DNA sequence-based phylogenetic tree to understand evolution of influenza, 340
  in cancer development, 328, 726–727
  characteristics of life and, 330
  community ecology and zoonotic, 853
  evidence for viral DNA in bacteriophages, 246–248
  evolution of, 336–337
  importance of, for molecular biology, 337
  interferons and, 714
  latency of, 725
  as pathogens, 338–341
  rapid reproduction of, as source of genetic variation, 402
  replicative cycles of, 332–336, 337f
  RNAi pathways and, 306
  structure of, 330–332
Visible light, **160**
Vision. *See* Visual systems
Visual association cortex, 776f
Visual communication, 806
Visual cortex, 776f
Visual fields, 788, 789
Visual pigments, 787f, 789
Visual systems, 785–789
  color vision in, 789
  compound eyes in, 786
  evolution of, 785–787
  light-detecting organs in, 785
  sensory transduction in, 787–788
  single-lens eyes in, 786–787
  structure of human eyes in, 786f–787f
  vertebrate, 787–789
  visual fields in, 789

  visual information processing in brain in, 788–789
  visual information processing in retina in, 788
Vital capacity, **705**
Vitamin A, 612, 613f, 667
Vitamin B$_9$, 668
Vitamin C, 666–667
Vitamin D, 667
Vitamins, **666**–668
Vitreous humor, 786f
Vocal cords and vocal folds, 702
Vocalization, *FOXP2* gene and, 359
Volcanic springs, 471
Volcanoes, 445, 459
Voles, 812
Voltage, membrane potential, 104–105
Voltage-gated ion channels, 756f, **757**–760, 761f
Volume, cell, 70–71f, 74
Volume, sound, 784
Voluntary nervous system, 771
Volvox, 487–488, 491f, 496
von Frisch, Karl, 803, 805
von Humboldt, Alexander, 860
Vulva, **735**

W

Waggle dance, honeybee, 805f
Waists, chromatid, 176
Wakefulness, brain functions and, 774
Walking, 802
Wallace, Alfred Russel, 369, 859
Walnut trees, 839
Walrus, 645f
Warren, Robin, 673
Washington, stream biome in, 828f
Wasps, 541f, 636, 808
Watasenia scintillans, 116
Water, 29–37
  acids, bases, buffers, and pH of solutions of, 34–36
  aqueous solutions and, as solvent of life, 33–34
  biomanipulation and quality of, 855
  bioremediation of polluted, 868–869, 878–879f
  in blood plasma, 695
  catabolic pathways and, 136
  cell balance of (*see* Water balance)
  cohesion and adhesion of, 30
  as compound, 25
  conduction of, in plant cells, 559f
  covalent bonding of, 25f
  effect of large bodies of, on climate, 31f
  evapotranspiration of, 859
  evolution of alternative plant mechanisms to reduce loss of, 169–171
  floating of ice on liquid, 32–33
  fruit and seed dispersal by, 607f
  hydrogen bonds and properties of, 27, 29–30
  kidney conservation of, 659–660
  land plants and, 505, 507
  moderation of temperature by, 30–32
  molecular shape of, 27f
  ocean acidification and, 36–37
  pH of, 35f
  in photosynthesis, 28–29f
  plant responses to flooding, 633–634, 635f
  regulation of plant transpiration and loss of, 590–593
  root architecture and acquisition of, 573–574
  seed germination and imbibition of, 605
  species distributions and availability of, 831–832
  splitting of, in photosynthesis, 158
  temperature and uptake of, by seeds, 576
  in thigmotropism, 633

transpiration of, from roots to shoots via xylem, 587–590
transport of, in plants, 30
vascular plant transport of, across plasma membranes, 575–578
Water balance
  kidney processing and, 658–659
  osmoregulation and, 653–655
  osmosis and, 100–102
Water bugs, 733f
Water cycle, 874f
Water fleas, 838f
Water lilies, 520f–521
Water potential, **575**
  in plant transpiration, 588–589
  in vascular plant water transport, 575–577
Water-soluble hormones, 652
Water-soluble vitamins, 666–667
Watkinson, Andrew, 840
Watson, James
  discovery of DNA molecular structure by, 245, 249–251
  model of DNA replication of, 252, 253f
  reductionism of, 3
Wavelengths, light, **160**–162
Weak acids, 35
Weak chemical bonds, 26–27
Weather, population fluctuations and, 840–842. See also Climate
Weathering, phosphorus cycle, 875f
Websites, genomic, 345, 346f
Weddell seals, 708
Weeds, transgene escape and, 614
Welch, Allison, 413f
Went, Friz, 619
Wernicke, Karl, 777
Wernicke's area, 776f, 777
Western garter snakes, 812–813
Western gulls, 810f
Western scrub jays, 778
West Nile virus, 332, 338, 410
Wetlands, **827**f
Whales, 376–377, 384f, 450, 669f, 781f
Wheat, 426, 485, 841f
Whiptail lizards, 731
Whiskers, 781
White-band disease, 861
White blood cells, 315f, 695f, 696
White-footed mouse, 806, 841f
White matter, **770**

Whole-genome shotgun genome-sequencing approach, **344**
Whooping cough, 110
Whooping cranes, 807
Whorled phyllotaxy, 573
Widow's peak pedigree analysis case, 219–220
Wieschaus, Eric, 318
Wildflowers, 407–408
Wildlife management, 839, 878
Wild mustard species, artificial selection of, 371f
Wild types, **230**–231
Wilkins, Maurice, 249–251
William of Occam, 391
Wilson, E. O., 860
Wilting, 102, **577**, 592, 624, 633
Wind
  climate and, 821–823 (see also Climate)
  flower pollination by, 602f
  fruit and seed dispersal by, 607f
  global patterns of, 820f–821
Windpipe (trachea), 672, **702**–703
Winged fruits and seeds, 607f
Wings
  bat vs. bird, 385–386
  bird, 545f
  insect, 799
  locomotion by, 801–802
Withdrawal method, contraceptive, 734, 747–748
Wobble, **280**
Wolves, 840–842
Wood, 88
Work
  ATP hydrolysis and, 123–124
  cellular respiration and, 135
  types of cellular, 122
World Health Organization (WHO), 338
Worldwide adaptive radiations, 447–449
Worms, 315, 346, 348, 799–800. See also Earthworms

X
Xanthopan morganii, 548f
X chromosomes, 194–195, 231–234, 739
Xenopus laevis, 320–321
Xeroderma pigmentosum, 258
Xerophytes, **592**
X-linked genes, **232**–234
X-ray crystallography, 55f, **59**, 249–250

X-rays, mutations and, 290
Xylem, **516**, **557**, **572**
  evolution of, 572
  primary growth and, 563
  secondary, 566–569
  transpiration of water and minerals from roots to shoots via, 587–590
  in vascular plant transport, 516
  in vascular tissue system, 557
  water-conducting cells of, 559f
Xylem sap, **588**–590
X-Y sex determination system, 232

Y
Y chromosomes, 194–195, 231–234, 739
Yeager, Justin, 818
Yeasts, **509**. See also Saccharomyces cerevisiae (yeast)
  alcohol fermentation and, 150–151
  cell division in, 183f
  cells of, 72f (see also Eukaryotic cells)
  fungi as, 509
Yellow-bellied toad, 428–429
Yellow jacket, 848f
Yellowstone National Park, 471f, 474, 510, 857
Y-linked genes, 232
Yolk sac, 543–**544**f

Z
Zea mays, 343, 348t. See also Corn
Zeatin, 622
Zero population growth (ZPG), **836**
Zinc deficiency, plant, 580
Z line, skeletal muscle, 793
Zonation, aquatic biome, 827–830
Zone of cell division, 562f, 563
Zone of differentiation, 562f, 563
Zone of elongation, 562f, 563
Zoonotic pathogens, **860**–861
Zooplankton, 829f, 853
Zucchini, 521f
Zygomycetes, **512**f
Zygotes, **195**, **729**
  abnormal chromosome numbers in, 240–242
  ensuring survival of, 732–733
  fertilization and, 742–744
  human, 195–196, 746